W9-BIV-287

Color and Black & White Television Theory and Servicing

second edition

ALVIN A. LIFF

Chairman of the Consumer Electronics Department
Technical Career Institutes Inc., New York, N.Y.

Prentice-Hall, Inc., Englewood Cliffs, New Jersey 07632

Library of Congress Cataloging in Publication Data

Liff, Alvin A.,
Color and black & white television theory and servicing.
Includes index.
1. Television—Repairing. 2. Television—Receivers and reception. 3. Video tape recorders and recording. I. Title. II. Title: Color and black and white telvision theory and servicing.
TK6642.L494 1985 621.388'87 84-11441
ISBN 0-13-151192-0

Editorial/production supervision: Mary Carnis
Cover design: Judy Matz Coniglio
Manufacturing buyer: Gordon Osbourne

Printed in the United States of America

10 9 8 7 6

ISBN 0-13-151192-0 01

Prentice-Hall International, Inc., *London*
Prentice-Hall of Australia Pty. Limited, *Sydney*
Editora Prentice-Hall do Brasil, Ltda., *Rio de Janeiro*
Prentice-Hall Canada Inc., *Toronto*
Prentice-Hall of India Private Limited, *New Delhi*
Prentice-Hall of Japan, Inc., *Tokyo*
Prentice-Hall of Southeast Asia Pte. Ltd., *Singapore*
Whitehall Books Limited, *Wellington, New Zealand*

In memory of my mother, Celia

Contents

3 Principles of Color TV 85

4 The Front End 125

5 AFT and Remote Control Circuits 176

6 Video IF Amplifiers 198

12 Sync Separators 457

13 Deflection Oscillators 481

14 Vertical Deflection Circuits 524

15 Horizontal Output Amplifier Systems 551

16 The Bandpass Amplifier and Associated Circuits 600

17 Color Sync Circuits 632

18 Color Demodulators and Output Amplifier Circuits 661

19 Video Tape Recorders 697

Index 745

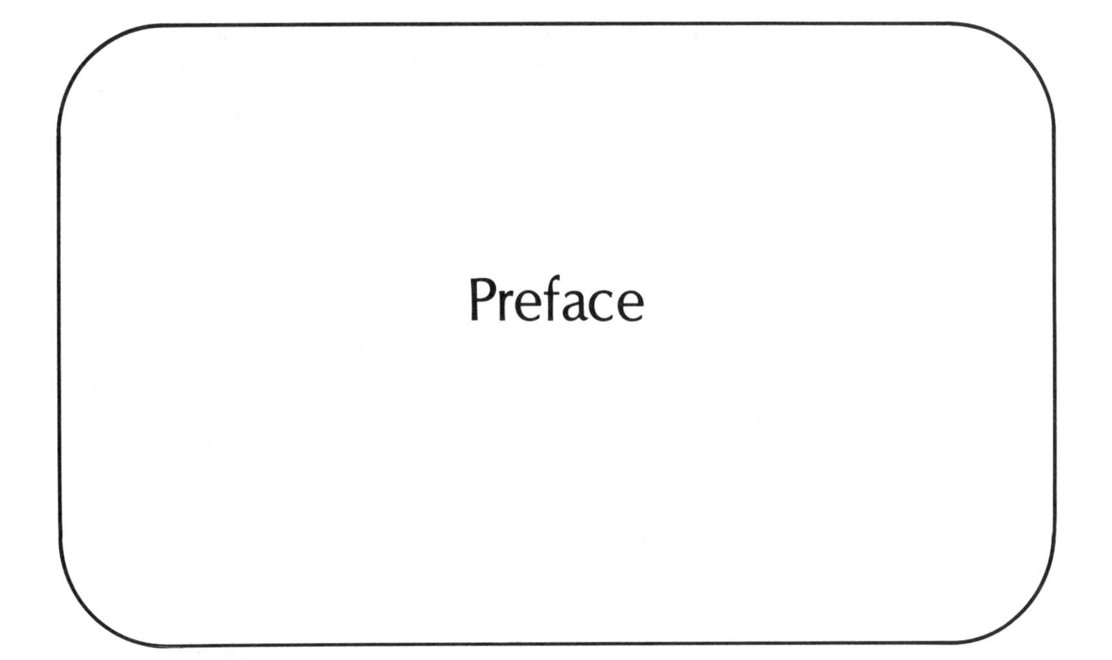

Preface

This book is directed to those who wish to become troubleshooting television electronic technicians. It is written with the assumption that the reader has already completed courses in basic electronics, and both AM and FM radio-receivers. The treatment is essentially nonmathematical and is presented in an easy-to-read format. It is suitable for courses given in high schools, vocational schools, and junior colleges.

In the five years since the first edition was published, the field of television has gone through revolutionary changes. The vacuum tube, once king of television, has all but disappeared. Vacuum tube receivers are still in use and represent a segment of the servicing industry. However, it will not be long before they are just a memory.

The integrated circuit, LSI technology, and the microprocessor have allowed rapid improvement in television quality, ease of operation, and reliability.

In the present edition I have endeavored to include those changes in the field which have had an important impact on the operation of the receiver, as well as the needs of the technicians who must service it. I have also removed that material which has become dated. However, at the end of each circuit chapter, I have included a discussion of one typical vacuum tube circuit.

The basic organization of the book remains the same as in the first edition. Important changes and additions have been introduced, the most extensive of which is the addition of Chapter 19, Video Tape Recorders. Other new material includes a new color block diagram preceding each circuit chapter, as well as the following partial list of important topics: enlarged discussion of triggered and dual trace oscilloscopes (Chapter 1); a discussion of cable, pay, and satellite TV systems (Chapter 2); modern channel selection methods (Chapter 4); modern microcomputer-controlled remote control systems (Chapter 5); SAW filters and synchronous detectors (Chapter 6); flat picture tubes and projection television (Chapter 9); the comb filter (Chapter 10); kick-start and pulse regulators (Chapter 11); vertical countdown circuits (Chapter 12); modern vertical output circuits (Chapter 14); gate-controlled switch horizontal output circuits, high-voltage regulator systems, and scan-derived low-voltage power supplies (Chapter 15); and a discussion of a VIR system (Chapter 18).

The nineteen chapters of this book are divided into four major areas: (1) the tools of the trade, (2) television standards, (3) circuit analysis and troubleshooting, and (4) video tape recorders.

Chapter 1 outlines the basic tools and skills needed by the service technician to perform his or her job. The chapter discusses the role of both the bench and the field television service personnel. In addition, the theory and application of the trigger and dual trace oscilloscopes and other types of test equipment are included. A generalized logical and systematic method of electronic troubleshooting is presented. Flowcharts are used to indicate the order of troubleshooting procedure as well as the test equipment to be used. Signal tracing and signal substitution testing procedures are reviewed.

RF and IF alignment procedures and the theory and use of the sweep generator are given in the appropriate chapters as required. Crosshatch and color bar generator theory and application are covered in detail in Chapters 9 and 16. Black & white and color picture tube characteristics and setup procedures are discussed in Chapter 9.

Included in Chapters 1 and 4 are two test equipment circuits that may be used by the instructor as class construction projects. The transistor tester circuit found in Chapter 1 is a highly effective in-circuit tester which can be constructed from scrap parts. The tuner subber described in Chapter 4 also can be constructed from a discarded solid-state receiver. The author has found both of the above to be successful projects based on his own teaching experience.

The second major area of the book concerns the basic television standards and systems for both black & white and color television. Chapter 2 is devoted to black & white television. In this chapter, raster development, video signal generation, and the basic television standards are developed. A block diagram of a typical black & white receiver is analyzed. The function of each stage is indicated, and the resultant effect on raster, picture, and sound of a defective stage is summarized in tabular form.

Chapter 3 covers the NTSC color television standards of transmission and reception. The composite video signal for a color bar pattern is developed. As in Chapter 2, a block diagram of a typical color receiver is analyzed. Here also, the function of each stage is indicated, and the resultant effects on raster, picture, color, and sound are again summarized in tabular form.

The third and major section of the book covers the circuits that comprise the television receiver. Beginning with Chapter 4, each chapter starts with a complete block diagram of a color receiver in which the area being discussed is shaded. This is accompanied by an introductory discussion which gives the student an overall view of the circuit under discussion with its relationship to the rest of the receiver. Circuits such as the video IF amplifier, the sound strip, the sync separators, and the deflection oscillators that do not differ substantially from one another in color or black & white receivers are discussed in common chapters. However, any differences that do exist between black & white and color usage are fully explained.

Chapters 16 to 18 are devoted to those circuits that are found only in color receivers. Chapter 18 also includes a brief discussion of other color television systems such as SECAM and PAL.

Video tape recorders are discussed in Chapter 19. In this chapter the basic characteristics of magnetic recording, as well as the methods of video recording, playback, and servo head and capstan control are presented.

In addition, each chapter provides a detailed discussion of troubleshooting problems common to the circuit being investigated, as well as the methods for locating these faults. Wherever possible actual circuits are used and analyzed. The circuits all have actual component values and typical voltages, thereby allowing the instructor to use meaningful practical circuits in classroom discussions. At the end of each chapter there are questions that may be used to test the students' knowledge of the material presented.

I would like to thank the following people for their help during the development of this second edition of my book: Mr. Greg Carey of Sencore for his valuable technical input and photos; Mr. James F. White of Zenith; Mr. Wally Bass of Quasar; and Mr. S.N. Zell of RCA for their rapid and generous response to my needs for information and materials.

Alvin A. Liff

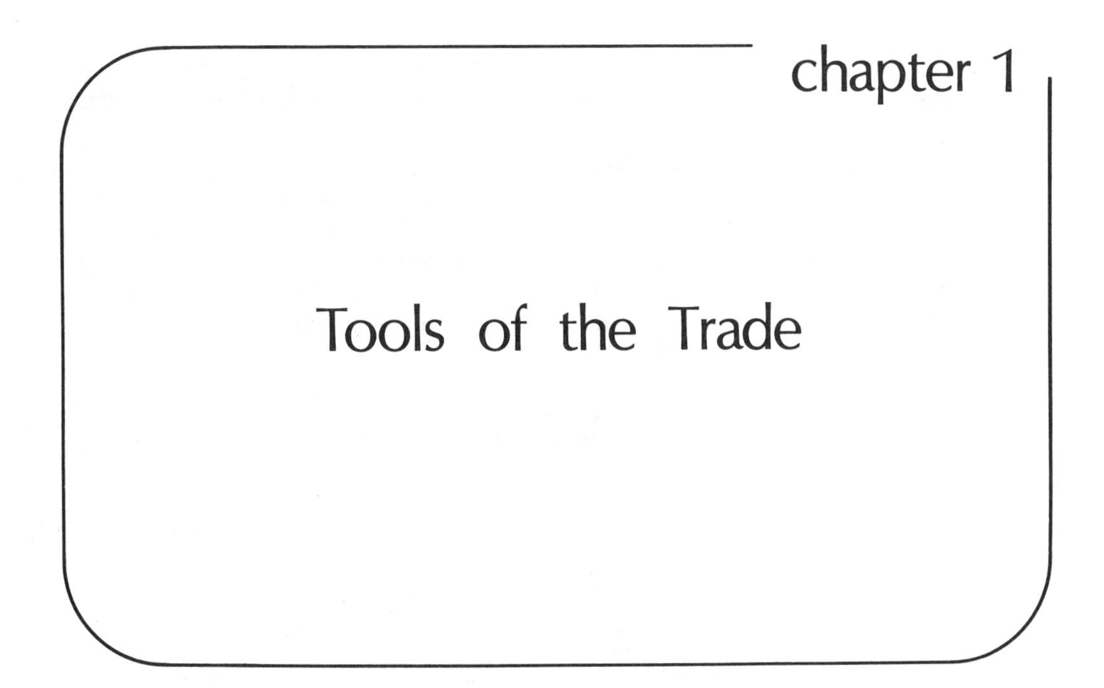

chapter 1

Tools of the Trade

1-1 introduction

The student of television receiver servicing must come to this field of work equipped with certain basic knowledge. He or she must have a basic understanding of dc and ac theory as well as a good understanding of vacuum tube and transistor circuit theory. To be successful, the service technician must also have some knowledge of test equipment and a practical knowledge of how to use it. It is also assumed that the service technician has a basic knowledge of AM and FM radio receivers.

This chapter is devoted to a discussion of the important aspects of electronic theory and practice that the television service technician should know.

Throughout his or her working career the television service technician will encounter the following three servicing diagrams:

1. The block diagram
2. The pictorial diagram
3. The schematic diagram

Each of these diagrams has useful information that the troubleshooting technician must be aware of. In general, servicing diagrams may be obtained from either the receiver manufacturer or from Sams Photofact.*

The block diagram. In a sense, the block diagram is the shorthand of the electronic technician for it enables the electronic technician to condense the overall operation of a system into

*Howard W. Sams and Co., Inc., Indianapolis, IN 46206.

a pictorial representation. It is usually designed to be read from left to right and to show the signal flow path functions of each stage as well as the function of each stage, the location of controls, waveforms, and frequencies. An example of the block diagram of a color television receiver is shown in Figure 3-17.

One of the first things that a troubleshooting television electronic technician must do to become an expert is to memorize the basic block diagram of the electronic system he or she is to specialize in. This is because the procedures used to isolate the defective stage of an electronic system and then the defective component are based upon a knowledge of how the system works. This knowledge can only be obtained through a thorough knowledge of the system's block diagram.

The pictorial diagram. The pictorial diagram is an actual photograph or line drawing of the physical appearance of the layout, position, size, and structure of the components and chassis that make up the electronic system. The photograph type of pictorial diagram, such as that of Figure 1-1, is mainly useful for indicating the location of parts, adjustments, and controls on the chassis of the electronic device.

Figure 1-2 illustrates a television receiver of modular construction. In this kind of construction, groups of circuits are placed on a plug-in type of printed circuit board. In modern television receivers, this type of construction is being replaced by a single unitized board with all the circuits mounted on it. This is an example of a line drawing pictorial diagram that serves the same functions as a photograph. However, as can be seen by comparing the two types, the line drawing shows the physical layout somewhat more clearly.

The schematic diagram. The schematic diagram is the electrical wiring diagram of the electronic industry. This diagram tells not only how electronic components are connected to one another, but it also indicates the electrical component values, voltages from component pins to ground, and pertinent wave shapes. Schematic diagrams may also include parts lists and alignment instructions as well as other miscellaneous adjustment procedures. An example of a black & white television schematic diagram is shown in Figure 1-3.

The wave shapes shown on the schematic usually indicate the peak-to-peak voltages that would be measured on a calibrated oscilloscope. Also associated with the waveform is the frequency that the sweep oscillator of the oscilloscope must be set to in order to view the waveform shown in the schematic diagram. Thus, if two cycles of one waveform are to be displayed on the scope screen, the frequency of the sweep oscillator must be made one-half of the input signal frequency.

Most schematic diagrams usually list the conditions used to obtain the voltages indicated on the schematic on one of the edges of the schematic diagram. These conditions must always be observed when making any measurements during troubleshooting procedures. Failure to do so may result in much wasted time and effort tracking down the cause of erroneous voltage measurements.

1-2 TV service job descriptions

In the television servicing industry there are basically two workers: the outside service technician and the bench troubleshooting technician.

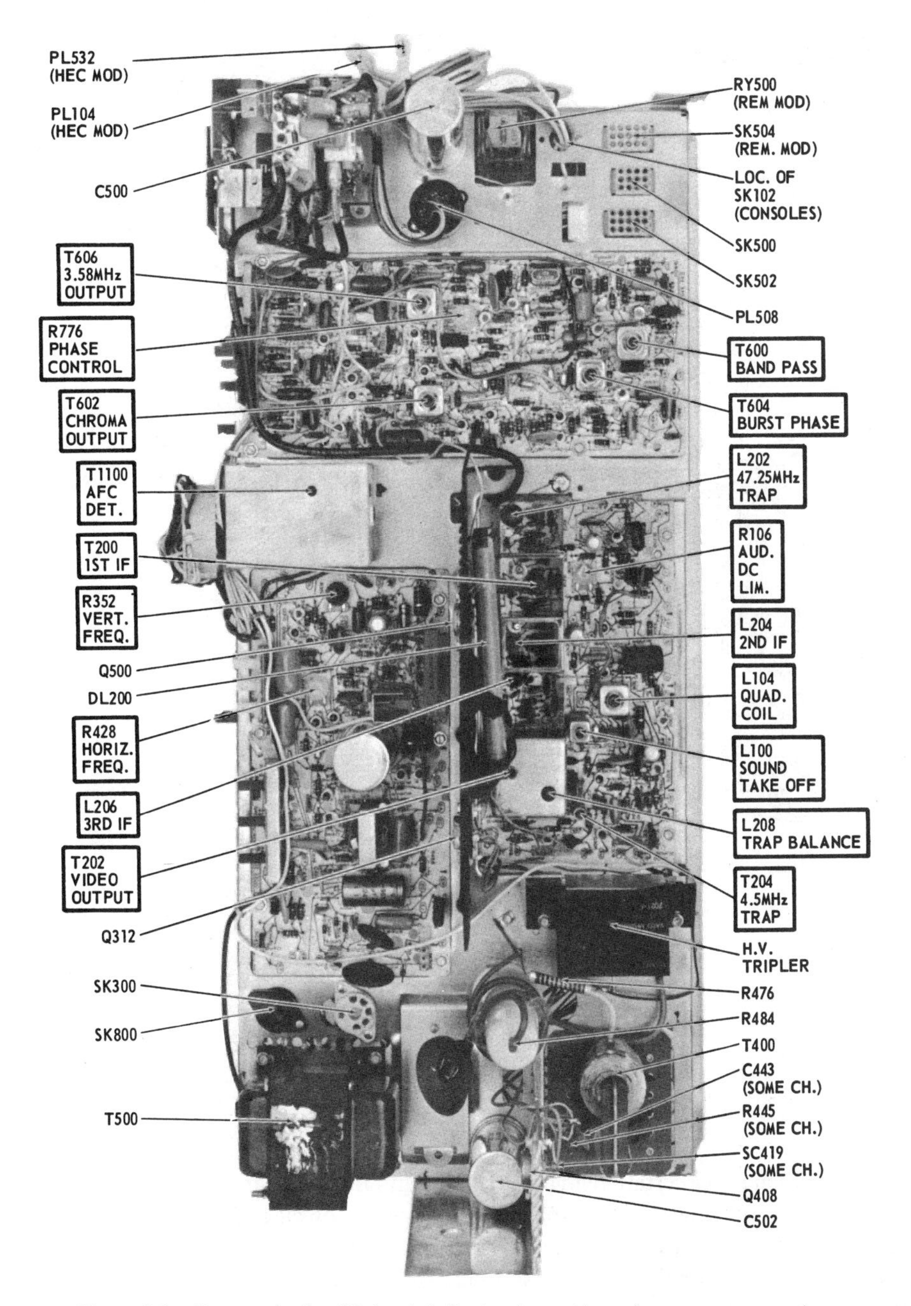

Figure 1-1 Photograph of a TV chassis indicating the positions of components, controls, and adjustments. (Courtesy of Electronic Components Group, GTE Sylvania.)

The outside service technician. Let us first consider the duties of the outside service technician. This person's job is to go to the customer's home to inspect the television system for faults or perhaps to install a new television system. The service technician is usually expected to

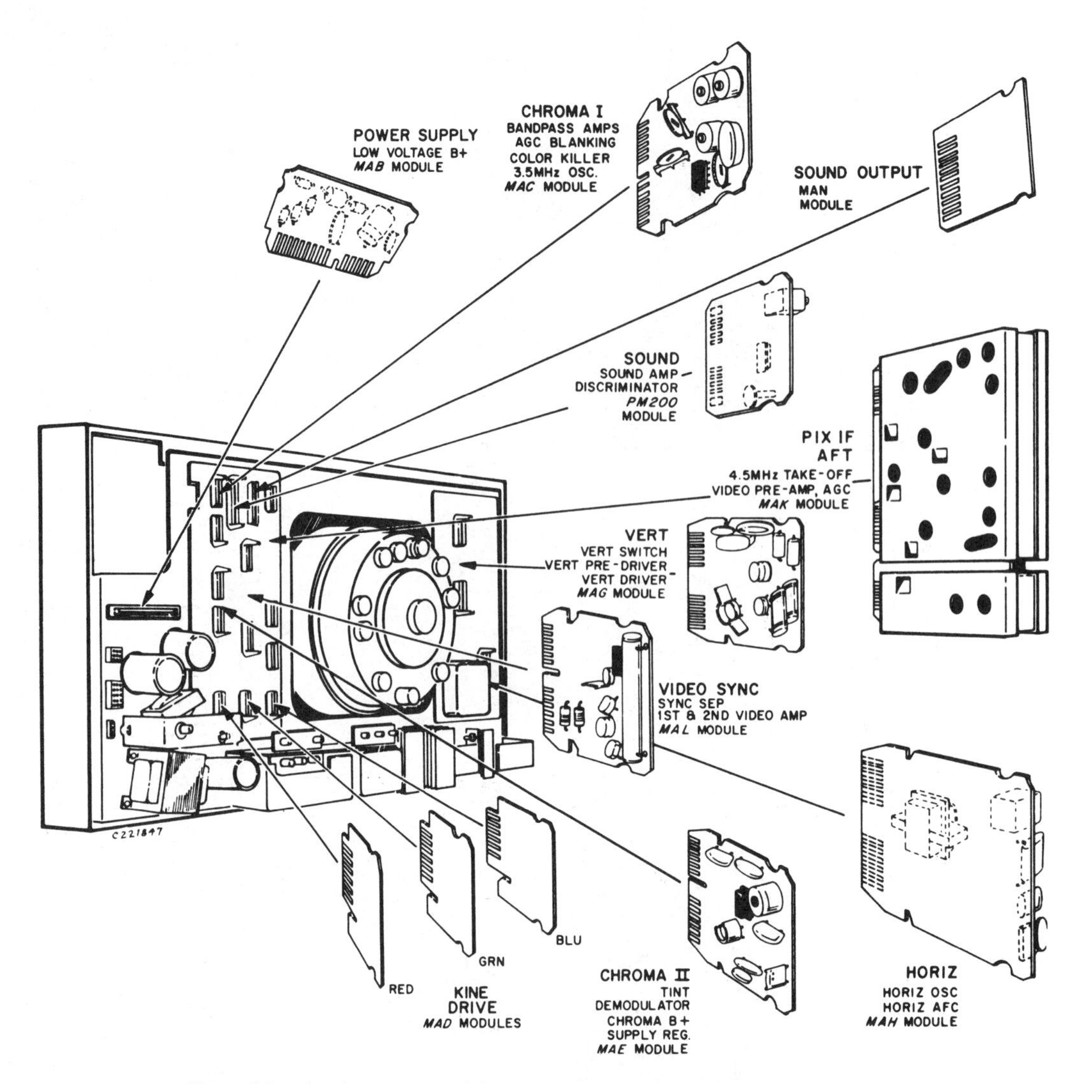

Figure 1-2 Line drawing pictorial diagram. (Courtesy of RCA, Consumer Electronics.)

do no more than make certain field adjustments and minor repairs. Such things as degaussing a color picture tube, adjusting dynamic convergence, or cleaning and adjusting the tuner, as well as replacing tubes, modules or transistors are the usual extent of the work. However, the outside service technician may also make certain minor repairs to both the receiver and the antenna systems. Such repairs might include replacing obviously defective components and repairing broken wires. Major repairs usually require that the receiver be pulled out and serviced at the shop where more extensive servicing facilities are available.

To perform these functions, the outside service technician's tool kit must contain an assortment of hand tools as well as electronic test equipment and parts. Table 1-1 lists some of the tools and supplies that might be found in a typical outside service technician's tool kit. This list is not necessarily the required minimum or maximum that must be taken

TABLE 1-1 Outside Service Technician's Tool Kit

Basic Tools to Be Carried for All Jobs

1. Pliers: long-nose and diagonals
2. Knife or razor blade
3. Screwdrivers: set of Phillips and set of blade types
4. Spin tights: long hollow-shaft type
5. Alignment screwdrivers: (a) hex, (b) plastic blade, (c) recessed metal blade
6. Small parts kit: (a) assorted resistors, (b) capacitors, (c) screws, (d) fuses and fusible resistors
7. Flashlight and/or trouble lamp
8. Tuner cleaner, anticorona dope, and lubricating grease
9. Picture tube brighteners: color and black & white
10. Set of jumper leads with alligator clips on both sides
11. Dropcloth and protective covering pads
12. Cheater cords: regular and polarized
13. Soldering gun (100 W); soldering iron (25 W)
14. Solder, 60/40 rosin core
15. 20,000 Ω/V VOM or FET VOM
16. Set carrier
17. Tube tester (picture tube rejuvenator, etc.)
18. Tube and module kit for black & white sets; tube and module substitution guides
19. Mirrors: hand, and on a stand

Color TV Servicing

1. Color bar generator
2. Degaussing coil
3. Tube and module kit for color receivers; substitution guide

Antenna Installation

1. Ladder
2. Hand power drill and set of drills, including carbide-tipped drills
3. 2-lb hammer
4. Channel lock pliers
5. Set of open-end wrenches
6. Set of box wrenches
7. Ratchet wrenches
8. Vise grip pliers
9. Heavy-duty screw drivers: Phillips and blade type
10. Citizen band walkie-talkies (2), or a portable television receiver

to any given job. Each job dictates the tools and supplies that will be needed. Black & white receiver servicing, for example, does not require the color bar generator, the degaussing coil, and the tube caddy needed for tubes, modules, and transistors found in color receivers. An outside service call devoted to black & white receiver servicing does not require an assortment of tools necessary for antenna installation. Hence, the ladder, power drill, vise grip, channel lock pliers, heavy-duty screwdrivers, and hammer are not needed for such work. Since it is not possible to tell in advance what additional work may be contracted once the service technician arrives at the home of the customer, many servicing establishments provide their outside service technician with a small panel truck to carry all the tools necessary to meet any situation that could arise. Since a small truck has considerable space, this arrangement has the additional advantage of allowing the outside serviceman to make a number of pickups and deliveries of TV sets before returning to the shop.

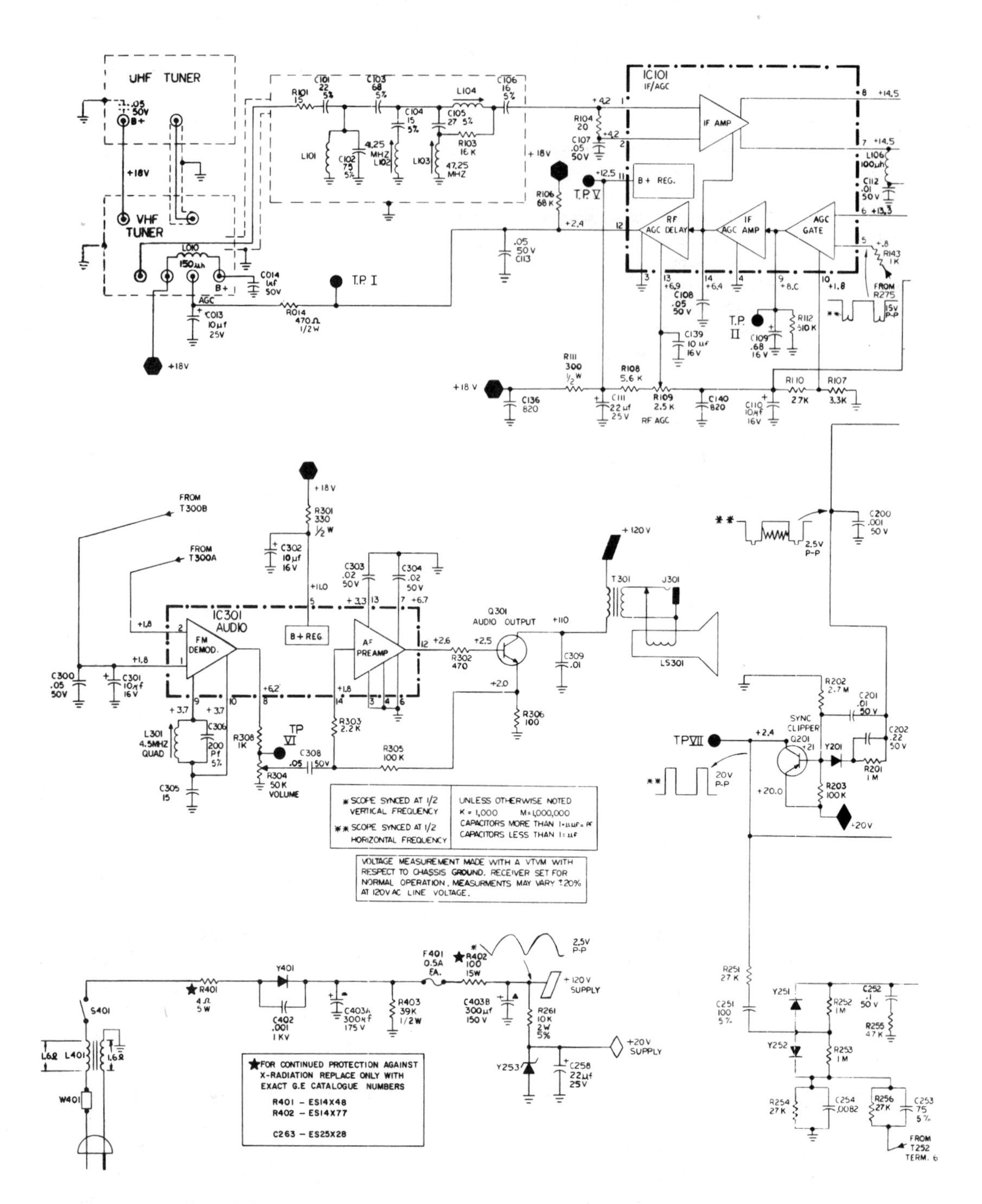

Figure 1-3 Schematic diagram of a black & white solid-state TV receiver (Courtesy of General Electric.)

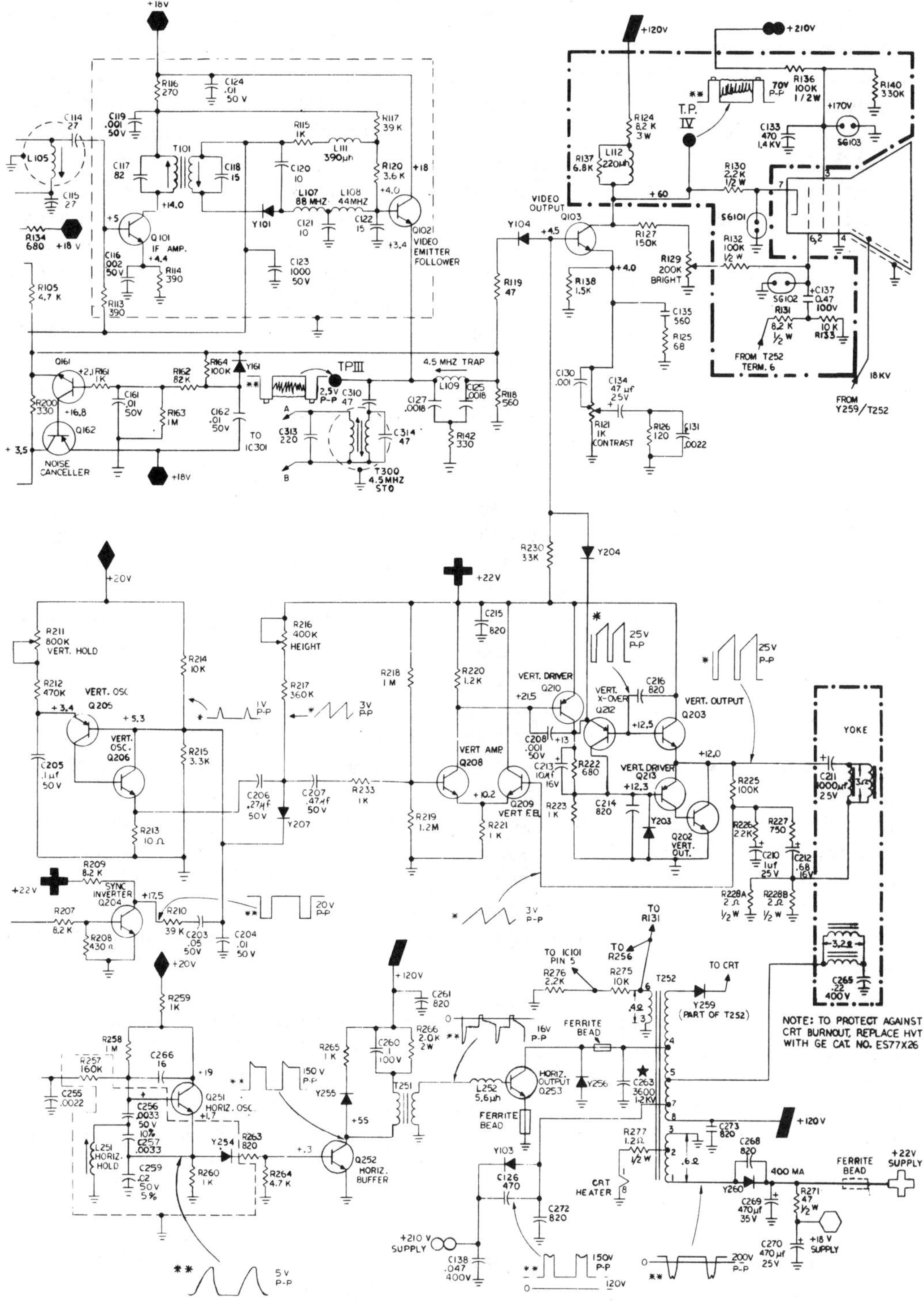

Figure 1-3 *(continued)*

Some of the tools and supplies listed in Table 1-1 need some additional comment. Today there are hundreds of different tubes, transistors and solid-state modules being used in all of the different TV sets that have been produced since 1946. Since it is clearly impossible to carry all of these different devices, most service technicians resort to tube, transistor, and module substitution guides as a means of minimizing the number of different types that must be carried to a given job. A number of trade magazines periodically publish lists of the most frequently used tubes and module interchange tables that may be helpful in determining which tubes or modules the outside service technician should carry on service calls.

The outside service technician is also the public relations representative of the TV servicing firm. Therefore, in order to establish a good public image, it is good practice for the outside service technician to use dropcloths and protective pads to prevent damage to the customer's rugs, furniture, floors, walls, and so on, and to use set carriers when removing heavy receivers. In addition, a courteous and knowledgeable attitude will leave the customer feeling confident that he or she is dealing with a trustworthy and professional servicing organization.

The bench troubleshooting technician. The bench troubleshooting technician is usually the most skilled of all TV service technicians. Since this person's main job is to find and repair faults in TV receivers and VCRs, the bench troubleshooting technician must have a thorough understanding of the theoretical aspects of the entire video system as well as a good theoretical grasp of the circuit operation of each circuit that makes up the system. He or she must also be expert in the use of a wide variety of test equipment. Last but not least, he or she usually has had years of practical experience.

Table 1-2 lists the tools and test equipment that the bench troubleshooting technician must have. By and large, the tools and test equipment used by the bench troubleshooting technician are similar to those used by the outside service technician. In addition, the bench troubleshooting technician must have such sophisticated devices as oscilloscopes, substitution signal generators, sweep generators, and other types of test equipment that allow him or her to quickly find the trouble and repair it. An absolute must for rapid TV servicing is a complete file of TV schematics. These schematics may be obtained from the manufacturer or through such dealers as Sams Photofact. If clear schematic diagrams are not available, servicing can become a difficult if not an impossible and time-consuming chore.

VCR servicing. Video cassette recorder servicing requires specialized test and adjustment equipment as well as the usual test equipment needed for successful TV servicing. Specialized VCR tool kits are available from several manufacturers. RCA provides a tool kit for use with VHS tape recorders, and Sony supplies a tool kit that may be used for Beta-type VCRs. Table 1-3 lists some of the specialized equipment needed for VCR adjustment and repair.

1-3 the oscilloscope

The oscilloscope is probably the most important test instrument at the TV service technician's disposal. A photograph of a commercial service oscilloscope is shown in Figure 1-4. The oscilloscope is important because it can measure the following quantities: (1) peak-

TABLE 1-2 Bench Repairman's Tool and Test Equipment List

1. ICs, transistors, modules, and tubes; substitution guides.
2. Sams Photofact schematics and manufacturers' schematics
3. Wide-band oscilloscope and probes (LCP, direct, demodulator)
4. VTVM and probes (direct, ac, RF), high-voltage probe
5. Audio generator
6. RF signal generator, tuner subber
7. Sweep generator with a built-in marker generator
8. In-circuit transistor checker
9. Deflection substitution generators; yoke/flyback checker
10. Degaussing coil
11. Color bar generator
12. Hand tools: (a) long-nose and diagonal pliers; (b) set of spin-tight, set of Phillips, set of blade type, and long screwdrivers; (c) low-wattage soldering iron; (d) set of small ratchet wrenches
13. Cheater cords: polarized and regular
14. Jumper leads with alligator clips on both ends of wire
15. Resistors, variable resistors, capacitors, and transistors, with alligator clips
16. Substitution boxes: R and C
17. Supply of small parts: R and C
18. Chemical aids, tuner cleaner, anticorona dope, chiller
19. Soldering gun, soldering iron, solder, soldering iron printed circuit tiplets, solder sucker
20. Bench light and flashlight
21. Tube and transistor tester
22. Bias supply for AGC work
23. Bench power supply (0 to ±300 V and 0 to ±25 V)
24. Color picture tube jig test assembly with long connecting cable and accessory adapters
25. Isolation transformer
26. PC board holding vise

to-peak ac voltages, (2) dc voltages, (3) wave shapes, (4) phase (vectorscope), (5) frequency, and (6) bandpass characteristic of amplifiers.

The oscilloscope is basically a peak-to-peak reading electrostatic voltmeter that also allows visual observation of wave shapes. The block diagrams of the three basic types of oscilloscopes are shown in Figure 1-5 shown on page 12.

TABLE 1-3 VCR Technician's Tool and Test Equipment List

1. Metric-size wrenches and feeler gauges
2. Metric alignment screwdrivers
3. Special screwdrivers and grip ring pliers
4. Test alignment tapes for VHS and Beta
5. Positioning fixtures for VHS and Beta
6. Torque gauges or cassette
7. Tension scales
8. Head-cleaning kit, long-stem swabs, and Freon TF solvent
9. Lubrication kits for VHS and Beta
10. Test equipment: (a) digital multimeter; (b) RF/IF alignment generators; (c) audio signal generator; (d) color bar generator of the NTSC type; (e) frequency counter, 30 MHz range, seven digits, 20-mV sensitivity; (f) good-quality dual trace oscilloscope having 15-MHz bandwidth or better, 5-mV sensitivity, and an XY feature for vector displays
11. Head demagnetizer

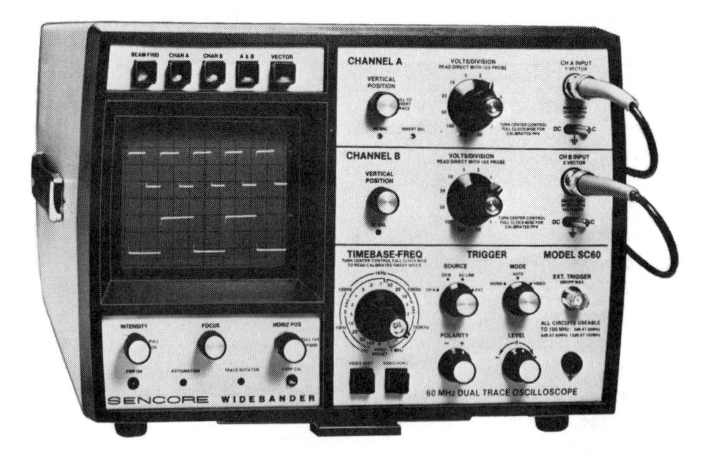

Figure 1-4 Dual channel service-type oscilloscope. (Courtesy of Sencore.)

1-4 the oscilloscope block diagram

The block diagram of the internal circuitry of the basic recurrent sweep type of servicing oscilloscope shown in Figure 1-5(a) reveals that it consists of three groups of basic circuits: (1) the vertical and horizontal amplifiers, (2) the sawtooth generator with its synchronizing circuits, and (3) the low- and high-voltage power supplies. Further inspection of the block diagram shows that these circuits are interconnected by means of two switches, the sync selector switch and the horizontal selector switch.

The vertical and horizontal amplifiers. The purpose of the vertical and horizontal amplifiers is to provide sufficient amplification so that weak signals may be measured and observed on the oscilloscope. The two amplifier systems are not usually identical. Almost all commercial servicing oscilloscopes provide higher voltage amplification for the vertical deflection amplifier than can be obtained from the horizontal amplifiers. Typical ratings are 10 mV peak to peak (p-p) per centimeter for the vertical deflection amplifiers, and 500 mV p-p per centimeter for the horizontal deflection amplifiers. The bandpass of each of these amplifiers also differs from one another. For an oscilloscope designed to service color television receivers, the bandpass of the vertical deflection amplifiers should be at least 4.5 MHz wide. The horizontal amplifiers may have a bandpass of only 500 kHz. These differences reflect the fact that the vertical amplifiers must be able to amplify very small complex wave shapes without introducing any distortion whereas the horizontal amplifiers are usually called upon to amplify only the output of the horizontal deflection sawtooth generator.

The vertical amplifier chain has three controls associated with it: (1) the step attenuator, (2) the vertical gain control, and (3) the vertical centering control. The step attenuator provides accurate voltage division in known steps; this is needed in order to allow the oscilloscope to be used as a voltmeter. Across each resistor of the voltage divider is a capacitor whose purpose is to provide compensation for frequency distortion. The vertical gain control is used in conjunction with the step attenuator to permit calibration.

The vertical centering control is able to shift the electron beam up or down by adjusting the dc balance of the push-pull deflection amplifier.

The horizontal amplifier chain also has three controls associated with it. It is similar to the vertical amplifiers in that the horizontal amplifiers include a horizontal gain control, a horizontal centering control, and in place of the step attenuator a horizontal selector switch. Since horizontal deflection is not used to measure voltage, accurate voltage dividers such as the step attenuator and calibration techniques used with it are not usually required. However, some amplitude control of horizontal deflection is desirable in order to allow expanded observation of portions of the input waveform. This is accomplished by the horizontal gain control.

The horizontal selector switch. The function of the horizontal selector switch is to allow the oscilloscope to be used in three different ways. When the switch is in the horizontal input position, any externally generated signal can be applied through the horizontal deflection amplifiers to the horizontal deflection plates. Thus, it is possible to use the oscilloscope for Lissajous and vectorscope pattern observation.

If the horizontal selector switch is in "*line*" position, the input to the horizontal amplifiers is a 60-Hz sine wave that is obtained from the filament winding of the oscilloscope power transformer. This ac voltage is amplified and is applied to the horizontal deflection plates, causing horizontal deflection of the electron beam at the line frequency rate. The primary application of the line position of the horizontal selector switch is for those applications in which the oscilloscope is used in conjunction with a sweep generator during visual alignment of the tuned amplifiers of a TV receiver. This alignment procedure will be discussed in a later chapter.

The *internal* position of the switch is often labeled in the abbreviated form INT, and it is used whenever it is desired to use the oscilloscope for waveform observations. In this position the output of the sawtooth generator is applied to the input of the horizontal deflection amplifiers, thereby providing horizontal deflection that is proportional to time.

The sawtooth generator. The sawtooth generator is usually a multivibrator oscillator arranged to produce a sawtooth-shaped output voltage. The frequency of the oscillator is made adjustable over a wide range by means of two controls: a *coarse frequency switch* and a *fine frequency potentiometer*. These controls are located on the front panel of the oscilloscope.

The sync selector switch. Also associated with the input to the sawtooth generator are the *sync selector switch* and the *sync level controls*. The purpose of the sync selector switch is to provide the necessary input signal to the sawtooth generator that will force its frequency to be a slave of the input signal. If this is accomplished, the observed oscilloscope wave shape will remain stationary, and the horizontal deflection sawtooth oscillator is said to be synchronized with the input signal.

The sync selector switch has three positions: line sync, external sync, and internal sync. In the line position the input to the sawtooth oscillator is a 60-Hz sine wave. This sync input position is useful when observing waveforms that are the same as or at some multiple of the ac line frequency.

The external sync position finds greatest application when dealing with low-level signals or signals that may be hum modulated and, therefore, may be difficult to synchronize. In this case, sync voltage is obtained from some other point in the circuit under investigation and is fed into the external sync input terminal.

The most useful position of the sync selector switch is the internal (INT) position.

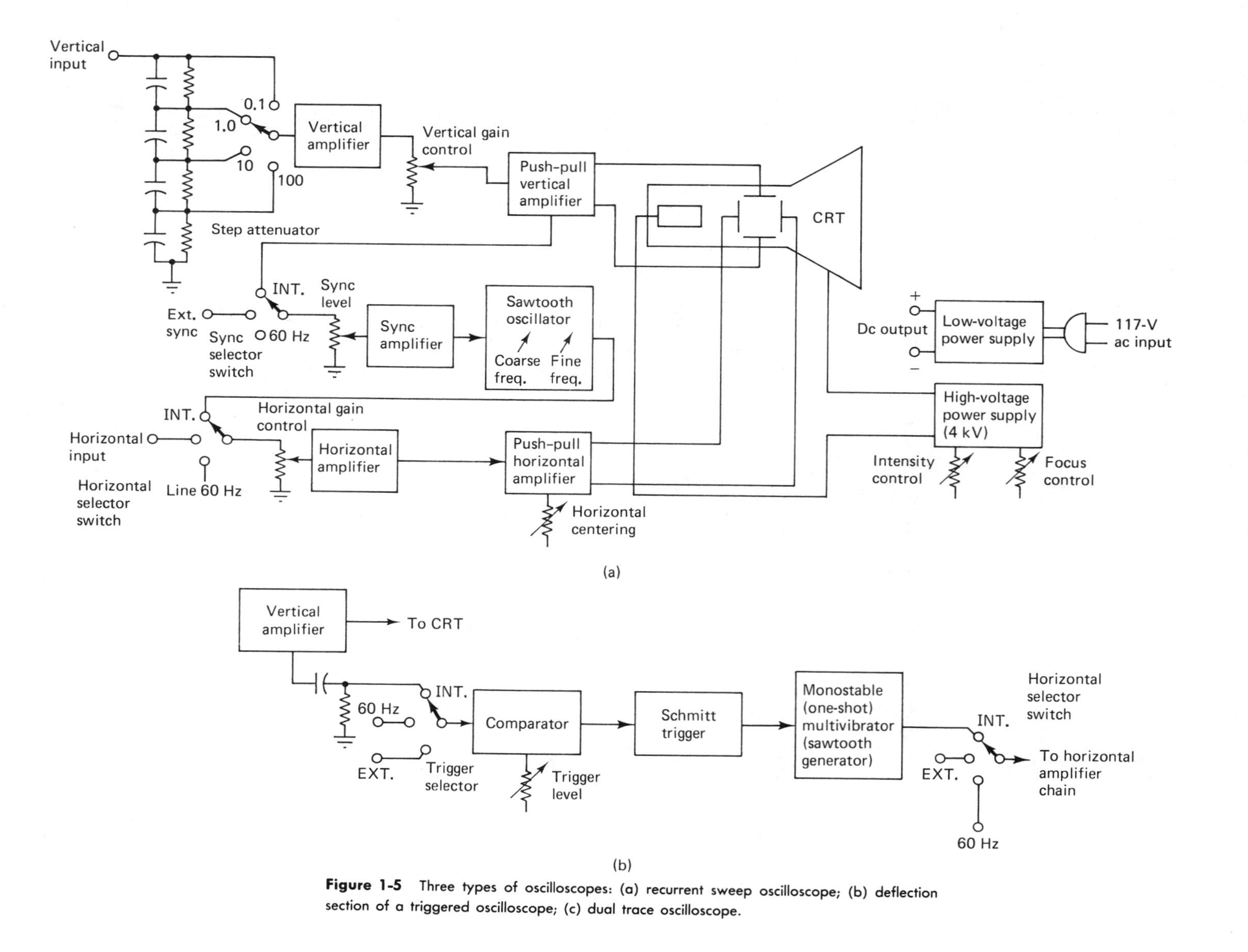

Figure 1-5 Three types of oscilloscopes: (a) recurrent sweep oscilloscope; (b) deflection section of a triggered oscilloscope; (c) dual trace oscilloscope.

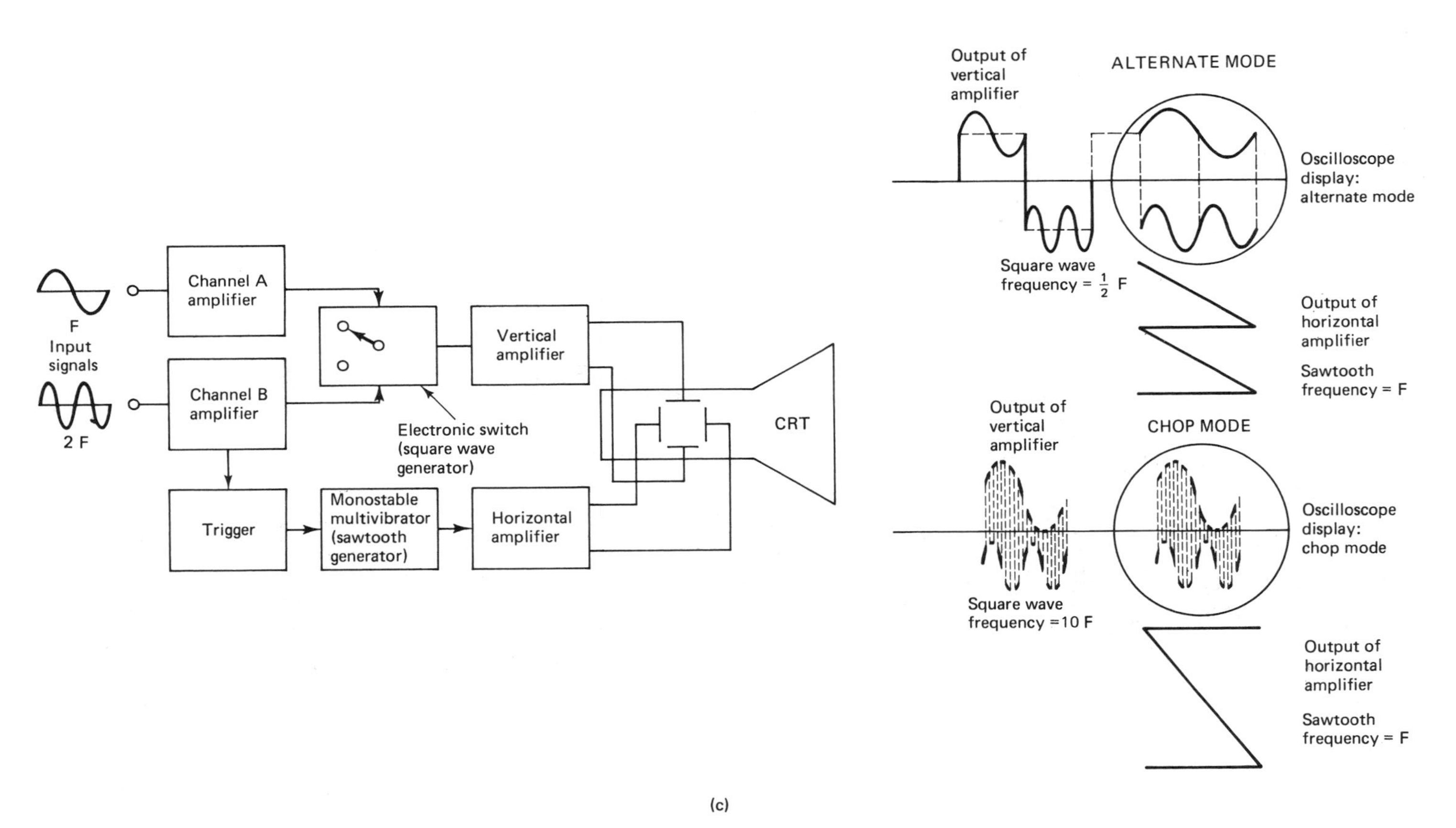

(c)

Figure 1-5 *(continued)*

When the switch is in the internal position, a portion of the output voltage of the vertical amplifier is fed to the sync input of the sawtooth generator. In this way the frequency of the sawtooth generator will be synchronized with the vertical input signal.

Synchronization. Synchronization of a sawtooth oscillator depends upon three factors. First, the frequency of the oscillator should be lower than the input signal frequency. This requirement is met in the oscilloscope by having two frequency controls that are able to control the frequency of the oscillator over a wide range. Some oscilloscopes also have front panel terminals available for further extending the frequency range of the oscilloscope. Second, synchronization is also sensitive to input signal polarity. Therefore, almost all oscilloscopes have a sync polarity reversing switch. Third, synchronization is very sensitive to the amplitude of the synchronizing voltage; therefore, a sync level control is included on all TV service oscilloscopes. Proper adjustment of this control occurs when the minimum sync level is used to produce synchronization.

The triggered oscilloscope. Figure 1-5(b) shows the block diagram of the deflection section of a typical triggered oscilloscope. In the recurrent sweep type of oscilloscope [Figure 1-5(a)], a free-running astable sawtooth oscillator is used to generate the deflection sawtooth. However, in the triggered oscilloscope a monostable sawtooth generator is used. This type of generator will function only when triggered by a pulse. When there are no trigger pulses the oscilloscope screen will be blank. A Schmitt trigger precedes the monostable to ensure that a consistently narrow rectangular pulse drives the monostable. The input to the Schmitt trigger is obtained from a comparator which produces an output only when its input signal level exceeds the dc level set by the trigger level control. The input signal to the comparator is selected by a three-position trigger selector switch. When this switch is in the INT (internal) position, a portion of the vertical deflection signal is used to drive the comparator. The EXT (external) and 60-Hz positions allow trigger signals to be obtained from sources outside the oscilloscope or from the power mains.

Two advantages of this system are that there are no overlapping displays and horizontal sweep expansion is easily provided. This allows small portions of the input signal to be examined. For example, the color burst in a color TV transmission can be more easily and accurately observed with a triggered sweep oscilloscope than with a conventional recurrent sweep oscilloscope. Another advantage of the triggered sweep oscilloscope is that sweep can be started at any point along the time axis of the waveform. This makes possible convenient observation of the rise or fall time or any other portion of an observed waveform.

Dual trace oscilloscopes. Many modern oscilloscopes are triggered dual trace oscilloscopes. These single-electron-beam CRT oscilloscopes are able to display two related waveforms one above the other. The basic block diagram of this type of oscilloscope is shown in Figure 1-5(c). Also shown in the diagram are the two types of displays that can be observed with this oscilloscope.

Aside from measuring voltages and waveforms the dual trace oscilloscope is useful for measuring phase or timing relationships between two signals. In addition, some of these oscilloscopes are arranged so that the sum of, or the difference between, the two input signals may also be displayed.

As indicated in the block diagram, the horizontal section of the oscilloscope is essentially the same as that of a triggered oscilloscope. To convert a single-vertical-channel device into a two-channel system requires that each of the two vertical channels be alternately gated into and out of conduction. When either channel is conducting, its output signal is conveyed to the vertical deflection plates. The gating is accomplished by a square-wave generator called an "electronic switch." On one half-cycle it biases one channel on and the other off. On the next half-cycle the amplifier states are reversed. With no input signal applied to either channel input, the outputs of the vertical amplifiers feeding the vertical deflection plates will be a square wave.

If the frequency of the horizontal deflection sawtooth is made equal to the frequency of the gating square wave, the oscilloscope will display a single square-wave pattern. When vertical input signals are applied to both channels, the positive peak of the square wave will vary at the channel A input signal rate. The negative peak will vary at the channel B input signal rate. See Figure 1-5(c).

To make the two waveforms appear one above the other requires that the square wave be broken into two segments. The positive-half-cycle segment of the square wave is deflected across the width of the face of the CRT. The negative-half-cycle segment will appear below the positive segment and will also cover the width of the CRT face. This display can be achieved by making the horizontal deflection sawtooth frequency twice the frequency of the gating square wave. See Figure 1-5(c). If the gating is done rapidly enough, the pattern will not flicker. The conditions producing this display are referred to as the "alternate mode" of operation.

A second display mode called "chop" is also possible. In this arrangement the frequency of the square-wave gating generator is made much higher in frequency than the horizontal deflection sawtooth or the vertical input signals. As a result, the input signals are chopped into small segments that are displayed on the oscilloscope face. See Figure 1-5(c). The chop mode is used when observing low-frequency signals, while the alternate mode is used for observing high frequencies.

Oscilloscope probes. The oscilloscope should be provided with three accessory probes that will expand the oscilloscope's usefulness.

The simplest probe used at the vertical input of the oscilloscope is the *direct probe.* This probe is used when making measurements in low-frequency, low-impedance circuits in which loading is no problem. The direct probe is simply a shielded wire that allows the vertical input to be connected into the electrical circuit under investigation. Unfortunately, this may allow the oscilloscope's input impedance (approximately a 1-MΩ resistor in shunt with a 30-pF capacitor) to load the circuit being measured.

To avoid this limitation, the *low-capacitance probe* (LCP) is used. This probe consists of a large resistor shunted by a variable capacitor. The effect of placing the parallel combination of a 9-MΩ resistor and a 0 to 30-pF capacitor in series with the oscilloscope's input impedance is to increase the input impedance of the oscilloscope's from 1 MΩ to 10 MΩ. At the same time it produces a divide-by-10 voltage divider that requires that all voltage being measured be multiplied by a factor of 10.

The variable capacitor that is in shunt with the 9-MΩ resistor is used to provide frequency compensation which makes the system's voltage division independent of frequency.

Proper adjustment of this capacitor is obtained by observing a known good-quality square wave, and adjusting the capacitor until the leading edge is free of any overshoot or rounding and is perfectly square.

The demodulator probe is simply a half-wave rectifier used as a detector. The prime function of the demodulator probe is to allow observation of *modulated signals* that may appear in the RF or IF amplifier stages of a receiver. Without this probe it would not be possible to signal trace in IF amplifiers operating at a frequency of about 44 MHz.

1-5 transistor testers

Transistor testers are extremely useful and important pieces of test equipment. They may be used either in the shop or in the field. Their prime function is to determine whether or not a transistor is defective and should be replaced. An alternative method of determining this is by direct substitution of the suspected defective device by a known good one. However, it is often necessary to know the condition of a transistor when a known good one is not available for substitution. Under these conditions a tester can provide the desired information.

Transistor testers may be classified into two basic types: the *in-circuit tester* and the *out-of-circuit tester.* The in-circuit tester is designed to indicate whether or not a transistor is able to amplify while it is still connected in its circuit. Therefore, it is able to provide either a go or a no-go indication without the bother of unsoldering the transistor from its circuit board. The go or no-go test is often the best method for isolating troubles, because from a servicing point of view it is only necessary to know if the transistor is able to operate; it is not necessary to know the value of the transistor's parameters.

"In-circuit" testers. There are two types of in-circuit transistor testers in current use today. One uses the transistor to be tested to complete an oscillator circuit. The other makes the circuit being tested into a bridge, balancing out all of the shunt resistance of the circuit and leaving only the transistor to be tested.

When the oscillating in-circuit tester is to be used for troubleshooting, the power to the defective equipment is first turned off and then the terminals of the suspected defective transistor are connected to the tester. The transistor under test completes the tester's circuit and converts it into an oscillator. If the transistor is good, the circuit will oscillate and its output ac voltage is indicated by the self-contained loudspeaker's producing a loud tone. If the transistor is no good or if the circuit elements shunting the transistor load the oscillator too heavily, the oscillator will not function. In general, in-circuit transistor testers are designed to be able to check transistors with as little as 100 Ω shunting the transistor. A schematic diagram of this tester is shown in Figure 1-6.

From the diagram it can be seen that the test circuit is essentially an Armstrong oscillator. In the emitter circuit there is a switch for selecting the transistor type that provides the proper dc voltages to the transistor.

If the tester does not produce a tone, this means that the transistor *may* be defective, not that it *is* defective. Such an indication will make it worthwhile to remove the transistor from the circuit and check it. It should be borne in mind, that this indication may also be the result of some other component failure, or of a low-value resistance shunting the transistor.

SEE THE TEXT FOR THE CAPTIONS FOR ALL FIGURES.

FIGURE 3-16A

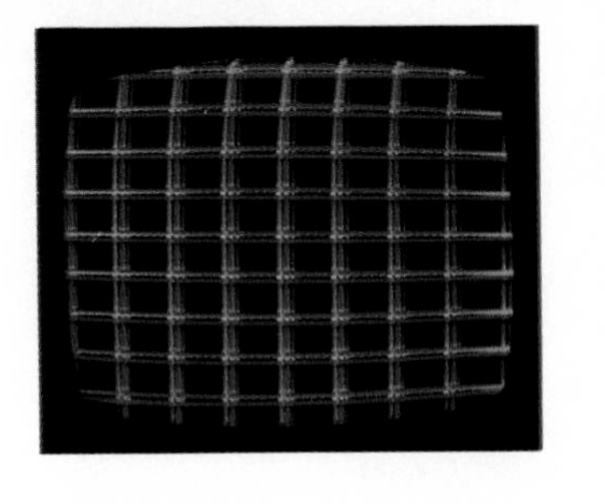

FIGURE 3-16B

FIGURE 5-2A

FIGURE 5-2B

FIGURE 5-2C

FIGURE 10-29C

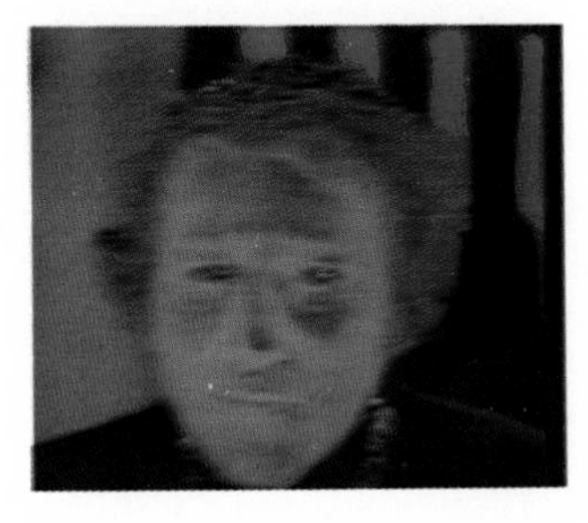

SEE THE TEXT FOR THE CAPTIONS FOR ALL FIGURES.

FIGURE 16-5

FIGURE 17-6

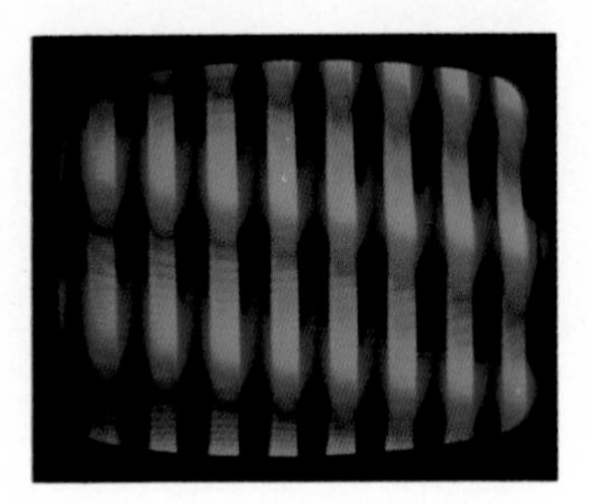

FIGURE 17-14

FIGURE 18-10

FIGURE 18-11

FIGURE 18-12

FIGURE 18-13

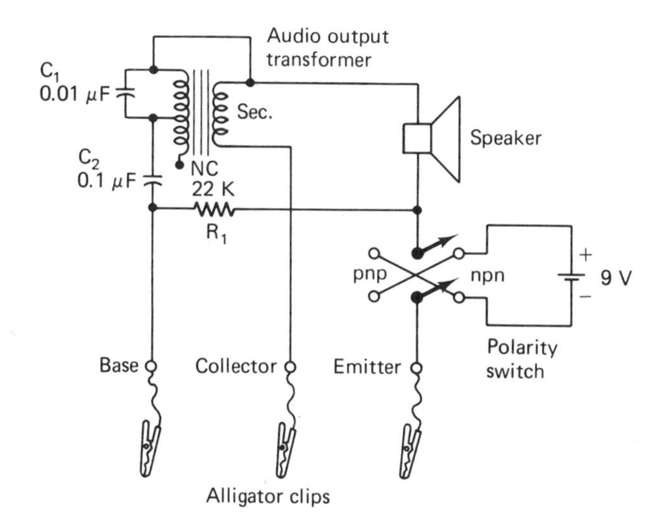

Figure 1-6 Schematic diagram for an in-circuit transistor tester.

1-6 basic troubleshooting procedures

The purpose of this section is to summarize most of the usual techniques and procedures used in troubleshooting defective television receivers. These techniques and procedures are general and may be used to service almost any kind of electronic equipment.

Assumptions. Troubleshooting always begins with the service technician making some assumptions that he or she may not even be aware of making. These assumptions must be understood and taken into account if successful and rapid servicing is to be achieved.

One assumption often made by the service technician is that the faulty equipment has not been worked on by others. An assumption like this can lead to wasted time and effort since one does not normally look for a miswired transistor, or other miswired connections on a printed circuit board.

Not only is it important for the service technician to know the extent of other people's work on a set before attempting to repair it, but the technician must also be aware that the customer does not always tell the complete story when he or she asks you to repair the receiver.

Another common assumption made by service technicians is that their test equipment is working properly. Test equipment is only as good as its reliability. A voltmeter that reads 100 V when the actual circuit voltage is 40 V is as useful as no meter at all.

The assumption that the schematic diagram of the defective receiver is correct or that the voltages associated with it are what will actually appear in the receiver can lead to a wrong diagnosis that in turn means wasted time. It does not matter whether the schematic diagrams are obtained from the manufacturer or from other sources. These diagrams can and do often contain errors of omission as well as errors in wiring, part values, and voltages.

To help minimize this problem, the service technician should always ask, "Is this voltmeter reading, part value, or circuit diagram reasonable in comparison to other similar circuits?"

Still one more assumption that may lead the service technician astray is that he or she may be working on the basis that there is only one fault in the receiver. The service technician must be aware of the possibility of multiple faults so that he or she does not use a troubleshooting procedure that might be misleading.

In short, a troubleshooting service technician must be a "doubter." That is, the technician must question all facts pertaining to the information provided by the customer, the measurements made by the test equipment, and the accuracy of the service literature. The service technician must never believe that there can be no more than one fault at a time. This attitude is only useful if it is coupled with a solid understanding of basic theory and a knowledge of the basic troubleshooting procedures.

The object of all troubleshooting procedures is to find the trouble and correct it. This is usually done in the following three steps:

1. Eliminating obvious defects
2. Isolating the defective stage
3. Isolating and replacing the defective component

The block diagram of Figure 1-7 summarizes the basic procedures used in troubleshooting. This diagram should be memorized as an aid to rapid servicing.

1-7 obvious defects

The elimination of obvious defects requires that *all senses* be used to find the fault. The *eyes* are used to see if there are any broken wires, broken or cracked printed circuit boards, charred resistors, loose tubes, defective line cords, and cold solder joints, etc.

The sense of *smell* is used to indicate the presence and location of burnt transformers, resistors, capacitors, and diodes.

The sense of *touch* is useful in locating improper circuit operation or defective coils and transformers. Improper circuit operation often results in very low or high circuit current, which in turn may cause underheating or overheating of the circuit components. The implication is that if the technician finds an unusually cool or an overheated component (coil, resistor, transistor), the defective stage has been isolated. However, for this method to be successfully used, it is necessary for the service technician to become familiar with the normal operating temperatures of components found in a television receiver. For example, a capacitor that is isolated from any heat source, such as a tube or a transistor, should feel cold to the touch. A coil or transformer should feel cool after 10 minutes of operation. Hot spots or undue overheating of inductors can be the result of shorted turns in the coil.

The sense of *hearing* is used to locate arcing, frying or hissing noises that may indicate high-voltage discharge or burning components. The quality of the sound output of the speaker may also be used to help isolate the defective stage. For example, a hum usually indicates power supply troubles. Distortion usually indicates a shift in operating conditions of the amplifier, implying that one or more components have changed value. Buzz is a complex fault that may be the result of improper alignment (of either the video or sound sections of the receiver), defective AGC, or clipping in the video strip.

Eliminating obvious defects also includes such things as testing all tubes and transistors in those stages of the receiver suspected, checking whether or not the ac line

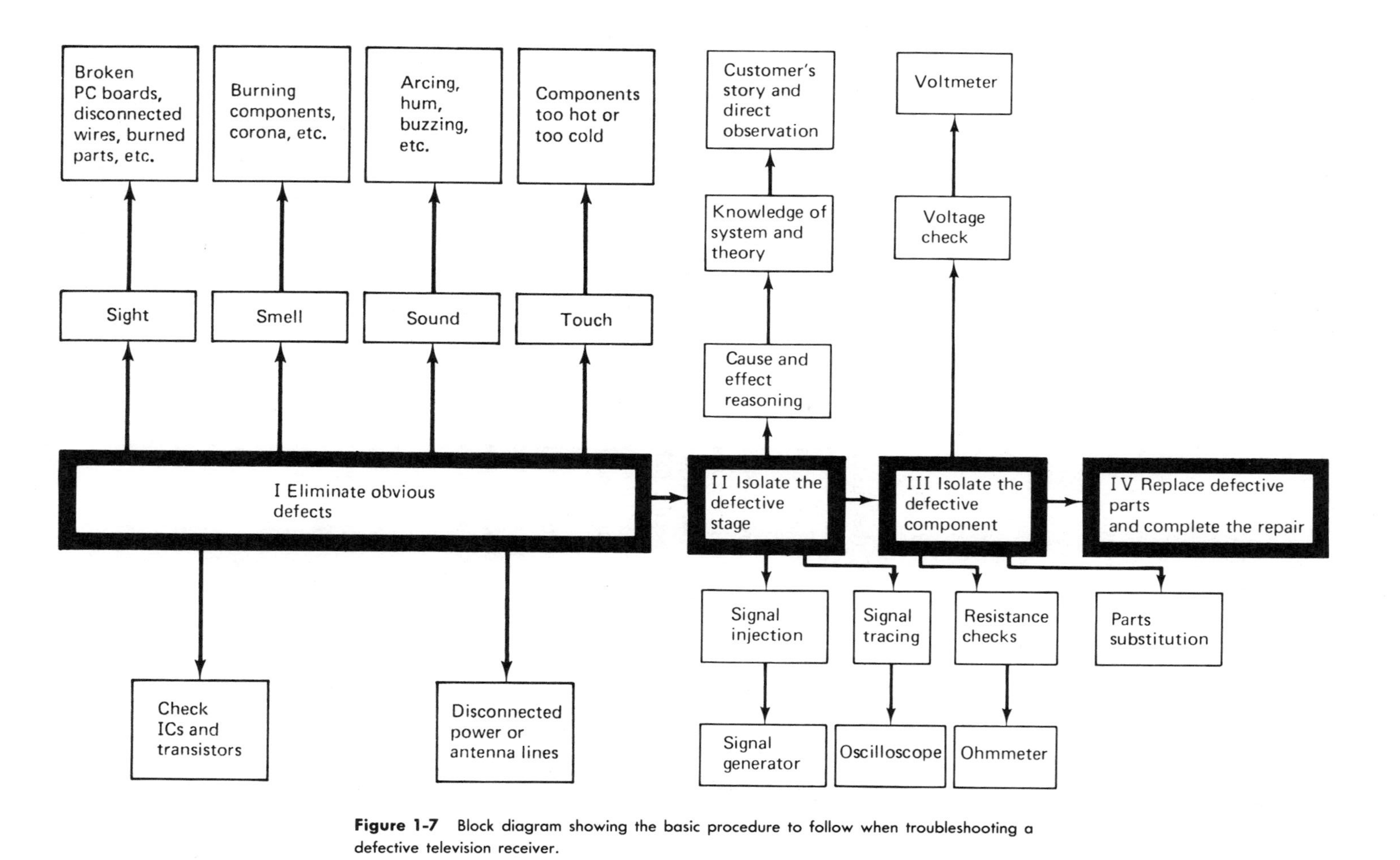

Figure 1-7 Block diagram showing the basic procedure to follow when troubleshooting a defective television receiver.

cord is plugged in, and checking whether or not the antenna is disconnected from the antenna terminals of the receiver.

1-8 isolating the defective stage

A television receiver may have as many as 40 or more individual stages of signal processing and generation. Because of this number of stages it becomes impractical to check each stage to find the defective one. A better approach would be to isolate the defective stage by one of the methods shown in the troubleshooting block diagram of Figure 1-7. These are:

1. Cause-and-effect reasoning
2. Signal tracing
3. Signal substitution

In TV servicing *cause-and-effect reasoning* is probably the most useful method for isolating the defective stage of the receiver. In this method it is assumed that the service technician has a knowledge of the circuit and its operation. In particular, the service technician must know the block diagram of the equipment being serviced. On the basis of this knowledge and the fault that exists in the receiver the service technician can isolate the defect.

For example, a television receiver may be shown to consist basically of three parts: a sound section, a picture section, and a scanning section. If a receiver has good sound but no picture and if a blank raster (light output) appears on the screen, the experienced service technician will conclude that a specific part of the picture section of the receiver must be at fault. Thus, by means of cause (bad picture section of the receiver) and effect (no picture) reasoning, the trouble has been narrowed from any one of 20 or 40 stages to one or two stages.

It is of the utmost importance that the television service technician be expert in this technique since it is the most rapid method of stage isolation there is. This technique is mastered by first learning the block diagram of the TV receiver as well as how the receiver works and then by learning how to interpret the symptoms that are heard or seen on the screen of the picture tube.

Signal tracing. *Signal tracing* is a fault-finding technique used to isolate a defective stage out of many that may exist in a complex TV system.

Signal tracing is based on the fact that throughout the *properly functioning* television receiver there are distinctive signals that when changed or missing may be used as clues for determining the location of faults in a defective receiver. The method requires that the service technician be equipped with an oscilloscope and the service literature for the receiver being worked on. The service literature may be similar to that shown in Figure 1-3.

The basic method used when signal tracing a basic amplifier is shown in Figure 1-8. First, it is necessary to apply a signal to the receiver (from an antenna or signal generator) and then using an oscilloscope and its proper probes follow the path of the signal as it passes through the amplifier. It should be kept in mind that not only is the amplitude of the observed signal important, but of equal importance is any change in wave shape from that indicated on the schematic diagram.

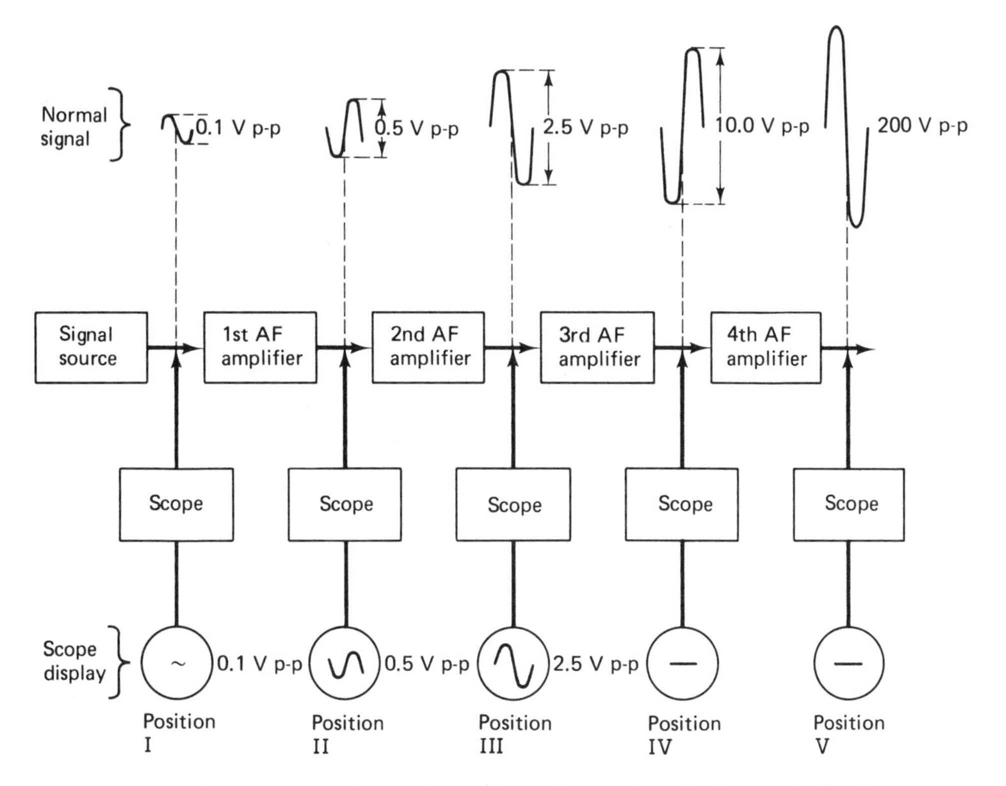

Figure 1-8 The top row of wave shapes indicates the normal signal voltages to be found at each point in a basic four-stage audio amplifier. Below the block diagram are shown the wave shapes as seen on an oscilloscope for the same amplifier when it is assumed that the third stage is defective.

Signal tracing usually begins at the signal source where it is known that a signal is present and then works back toward the output end of the amplifier. The faulty stage is located by determining the two points corresponding to the last observed normal signal and the first observed abnormal signal. The fault must lie in the circuit between these two points.

Television receivers use many circuits that are self-generating signal sources. These circuits are found in the deflection or scanning portion of the receiver. When signal tracing these stages it is not necessary to have any input signal source (generator or antenna) applied to the input of the receiver, but, it is necessary to know what wave shapes and amplitudes should be found at various points throughout the scanning portions of the receiver. This information is obtained from the service literature for the receiver.

Signal substitution. *Signal substitution,* or *signal injection* as it is sometimes called, is one of the more useful methods at the disposal of the service technician for isolating a defective stage in the television receiver. It may be considered to be the opposite of signal tracing. To use this technique, a signal whose frequency, wave shape, and amplitude are similar to the normal signal appearing at a given point in the receiver is injected into that point from a signal generator. See Figure 1-9. Replacing a lost signal with that generated by the signal

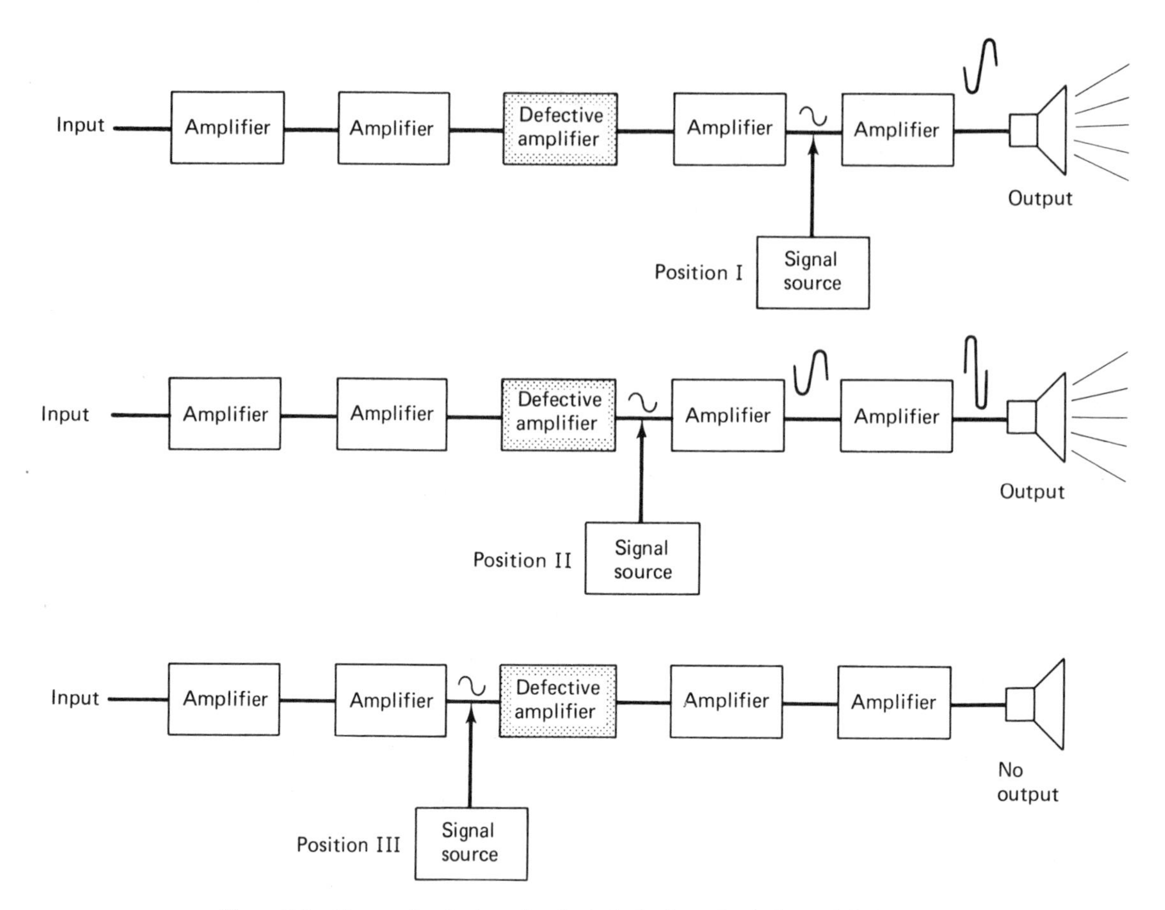

Figure 1-9 Diagram showing how signal substitution is used to isolate a defective stage.

generator should restore the receiver to its normal operation. If it does, this may indicate that the stage being replaced is defective.

A good source of substitution signals is from commercially available substitution generators. A photograph showing a commercially available generator is shown in Figure 1-10. This particular generator also supplies RF, IF, and audio outputs. The RF and IF outputs are actual video signals of a test pattern. This type of generator supplies two types of sweep outputs. One is a 60-Hz sawtooth for use in the vertical deflection circuits. The other is a 15,750-Hz sawtooth for use in the horizontal deflection circuits. Both output amplitudes are variable in order to allow the outputs to be adjusted to the levels required by the receiver under test.

Another method for obtaining substitution signals is to use sections of a known good receiver to substitute for those sections of the defective receiver that are suspected of being faulty. In this case, the good section will restore the defective receiver to normal.

For example, one of the most useful applications of this technique is for the restoration of high voltage in a defective television receiver. Television receivers require voltages ranging from 1.5 to 30 kV for proper picture tube operation. If this voltage is lost,

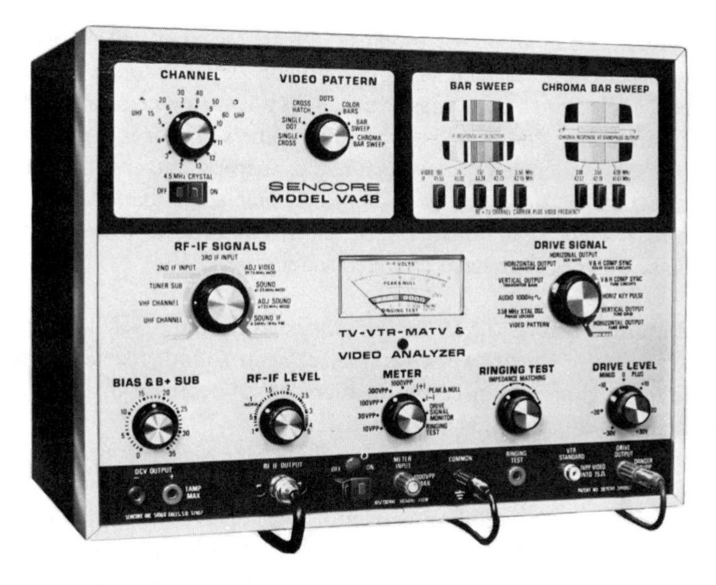

Figure 1-10 Signal substitution generator. (Courtesy of Sencore.)

the picture will also be lost, very often hiding the cause of the trouble. If the high voltage is restored by taking it from a good set, the service technician will be able to observe the picture and from it determine which circuit is at fault.

Other servicing techniques. The usual methods of stage isolation do *not always* provide the fastest route to defective stage isolation. Signal injection, signal tracing, and cause and effect reasoning may not work for such troubles as intermittents, oscillation, or misalignment. These methods may also be ineffective for stage isolation in circuits that use feedback over two or more stages.

Intermittents. *Intermittents* can be isolated by one of at least two techniques. In some instances, signal tracing can be used to isolate the intermittent stage. Here, the disturbance can be seen on the oscilloscope at the output stages. Then, by tracing the signal back toward the input, a point is reached where the disturbance is not present on the oscilloscope, but is observed at the output of the receiver. The stage immediately following the "good" indication must be where the intermittent is located.

Probably the most frequently used technique is to use a freezing aerosol spray (sometimes called *chiller*). The chiller is sprayed on each component while the receiver is on. The service technician observes the output of the receiver and looks for large output changes that may result when a component is sprayed. Such changes indicate a faulty component, which is probably the cause of the intermittent condition.

Oscillation. Amplifier stages that oscillate are generally forced into this state by one of three things:

1. Stage gain has increased
2. Undesired positive feedback has increased
3. Misalignment

Hence, troubleshooting *oscillation* (or its close cousin *regeneration*) revolves about finding those things that will cause an increase in stage gain or an increase in positive feedback. In vacuum tubes, such things as a gassy tube, leaky transformers or coupling capacitors, or any other circuit change that would decrease the tube's bias or increase its screen voltage, may be responsible for increasing stage gain and thereby causing the stage to oscillate. Transistor amplifiers may also tend to oscillate if their stage gain is increased excessively by a change in bias. Here again this may be the result of leaky transformers or coupling capacitors or bias resistors that have changed value.

Misalignment. Misalignment of the IF or RF stages can produce such trouble symptoms as no color, ringing, picture smear, weak picture, buzz in sound, regeneration or oscillation. This is because misalignment can cause an increase or decrease in stage gain, as well as adjusting the phase of any residual undesired feedback.

Oscillation or regeneration can be produced by increases in positive feedback. In vacuum tube RF and IF amplifiers open filament bypass capacitors, open screen bypass capacitors, open B supply decoupling capacitors, or open AGC filter capacitors may be responsible for increasing feedback to the point where oscillation sets in. Similarly, in solid-state receivers open power supply decoupling filter capacitors, AGC filters, or open neutralizing capacitors can be the cause of stage oscillation.

Other possible causes of oscillation or regeneration may be caused by poor lead dress, bad ground connection, and missing or disconnected tube or transistor shields.

1-9 isolating and replacing the defective component

In Figure 1-7 it can be seen that once the defective stage has been isolated, and assuming that the tube, transistors or ICs are good, it is then necessary to determine the defective component and replace it with a good one. Defective components fall into one of three categories: (1) they may be open, (2) they may be shorted, or (3) they may have changed value. Each one of these changes will have an effect on circuit operation and will usually leave its mark by making changes in the wave shape or voltage distribution throughout the circuit. If the service technician has a schematic diagram on which is listed the normal wave shapes and voltages found in the set being worked on, and if the technician has adjusted the receiver to the conditions that the manufacturer used when making its measurements, the service technician can then determine which wave shapes or voltages are in error. From this information it can usually be determined which component is at fault.

For instance, shown in Figure 1-11 is the schematic diagram of a video IF amplifier used in a television receiver. Included in the schematic diagram are the parts values and the normal voltages found at each pin in the circuit. If stage isolation techniques had traced the trouble to the video IF amplifier stage and if the base voltage measured is 5 V instead of the normal 1.8 V, this may be taken as an indication that the transistor is conducting heavily. The technician should then determine what might cause this condition. Possible causes might be (1) a defective transistor Q_2, (2) an open base resistor R_2, (3) a shorted or leaky coupling capacitor C_1, (4) a leaky feedback capacitor C_2, or (5) decreased bias resistor R_1. The service technician should then check these possibilities out by either an ohmmeter check or by direct parts substitution.

A similar line of reasoning may be used with IC circuits. Figure 1-12 shows a

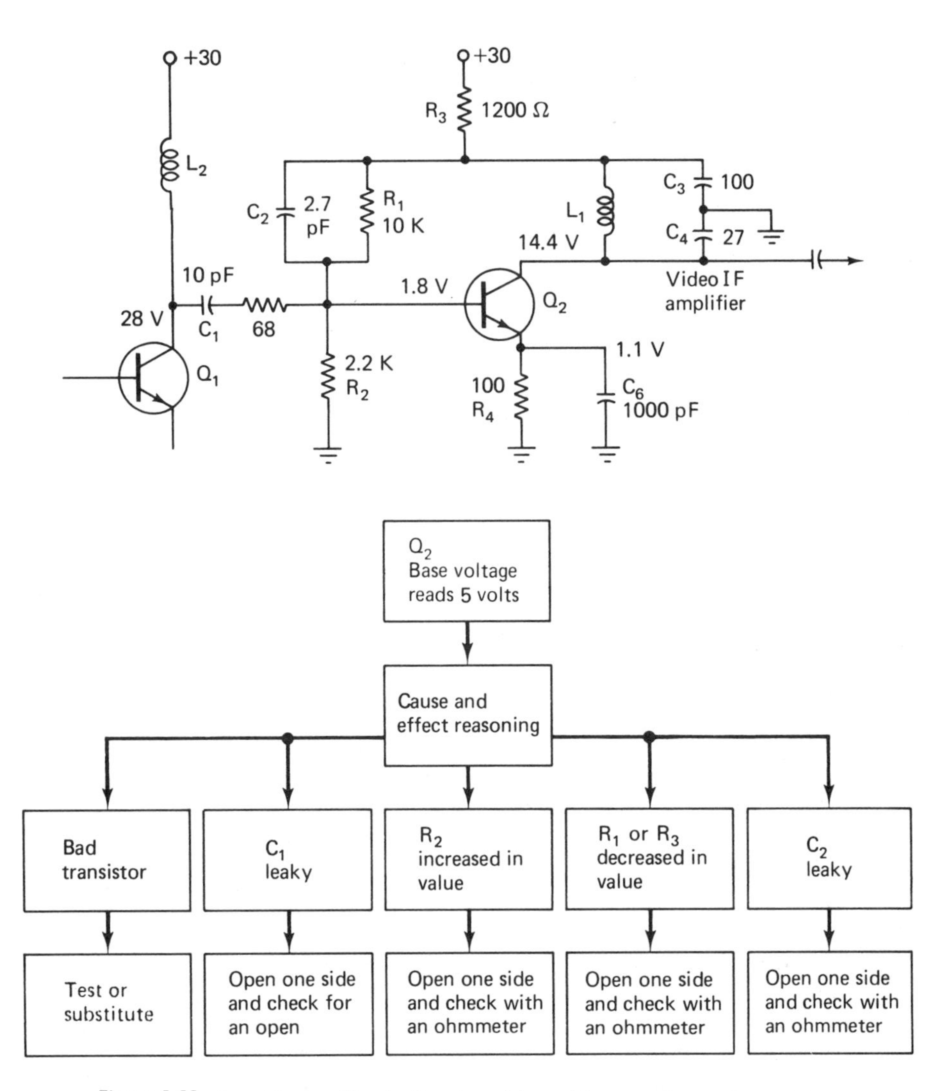

Figure 1-11 Transistor amplifier circuit and possible servicing procedures to be used for a given defect.

typical IC amplifier. Associated with the schematic diagram is a block diagram showing a possible defect and the servicing procedure used to isolate the defective component.

In the case of the transistor circuit, the first thing to be checked is the transistor since it has a high incidence of failure. If it is bad, it is a good idea to check all of those components that may have caused the transistor to break down. The usual procedure is to check each component with an ohmmeter to determine whether the resistors have changed value or the capacitors have become leaky. When using an ohmmeter in a circuit in which various shunt paths may exist, it is preferable to cut one side of the component out of the circuit before making a check. Since this may be difficult to do on some printed circuit boards, the printed circuit wiring may be cut with a razor blade or an in-circuit check may be used. In-circuit resistance checks may be acceptable in some cases but not in others.

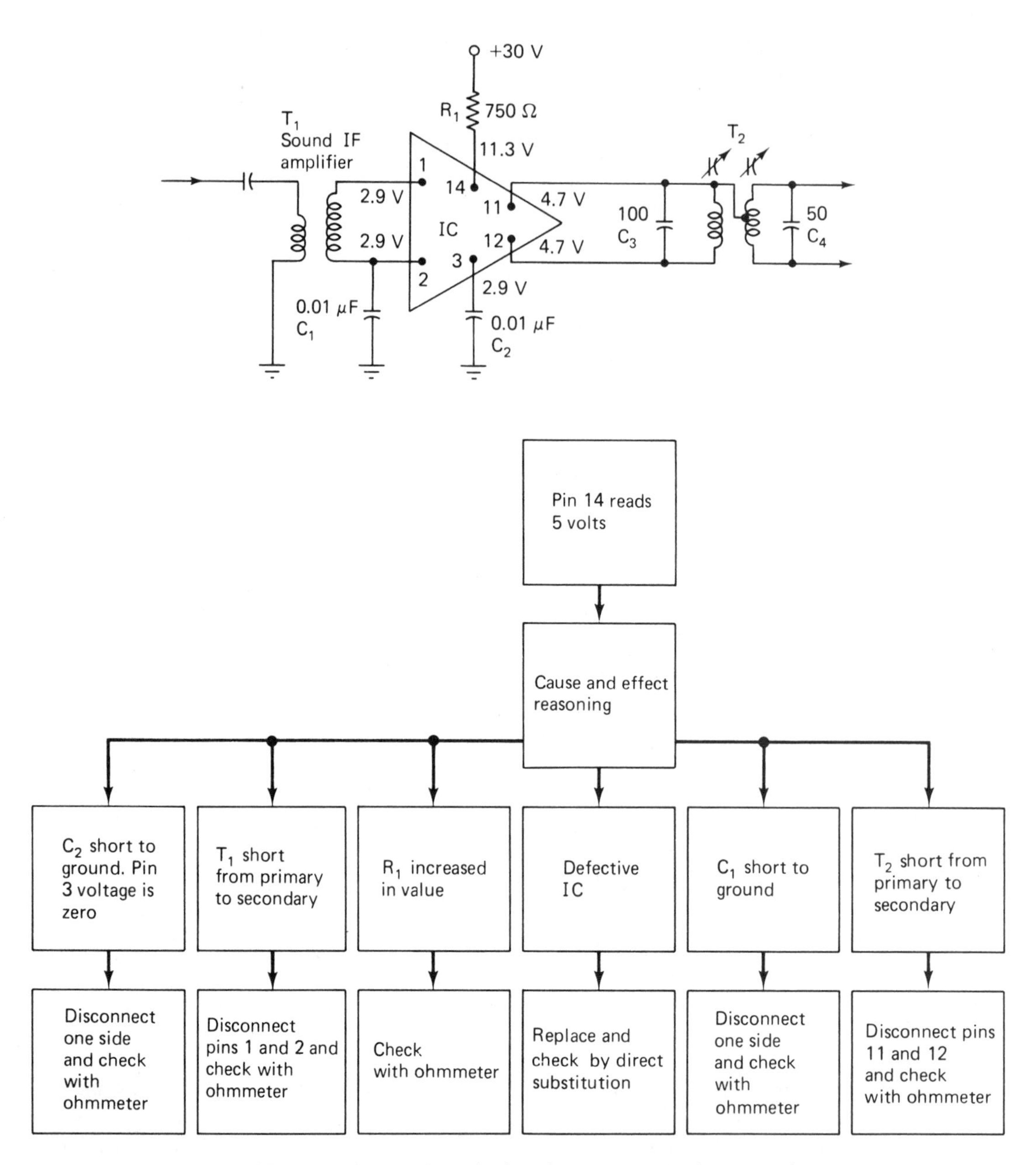

Figure 1-12 IC sound IF amplifier and a logical service procedure for a particular circuit defect.

For example, if an ohmmeter were placed across R_3 in Figure 1-11, and if it were assumed that B+ did not have a resistive voltage divider across it, the ohmmeter would read the proper value of 1200 Ω because the B+ point is open to ground. However, it is not possible to determine whether or not C_2 is leaky without removing one of its leads from the circuit, because R_1 is in parallel with it and would determine the resistance measured by the ohmmeter. Similarly, in-circuit resistance measurements of R_2 may cause the base to emitter junction to turn on, placing R_4 in parallel with R_2, thus drastically reducing the measured value.

In the case of ICs, similar procedures are followed (see Figure 1-12) except that greater care must be exercised in determining whether or not a component failure has caused the IC to fail before the IC is replaced. If this is not done, the service technician will have two or more ICs to replace.

1-10 TV receiver servicing

In some respects TV servicing is easier than servicing other types of electronic equipment. This is because the picture tube display can be used to isolate quickly the defective stage since it tells the story if only one learns how to interpret it. To do this, the service technician must know the block diagram of the television receiver. See Figure 2-25.

The picture and sound outputs of the TV receiver are generated by four basic sections of the receiver: (1) the power supply section, (2) the sound section, (3) the picture section, and (4) the raster (line pattern) generating deflection circuits. Obviously, if a single fault developed in any one section, it would affect the outputs in a distinctive way.

For example, if the sound section failed, the receiver would have no sound output, but the raster and picture would be present. See Figure 1-13(a).

If the video or picture section of the receiver failed, the sound and raster-forming portions of the receiver would still function and would produce the effect of a blank raster accompanied by sound, as shown in Figure 1-13(b).

The raster-forming section of the receiver is also responsible for generating some of the potentials needed to operate the picture tube; therefore, if the *horizontal* deflection section failed, both picture and raster would be lost, as shown in Figure 1-13(c). If the *vertical* deflection system failed, only horizontal deflection would exist, as shown in Figure 1-13(d).

The low-voltage power supply is common to all sections of the receiver. It is responsible for supplying the heater as well as the B supply requirements of each stage of the receiver. Therefore, if it failed, the receiver would go completely dead; there would be no sound, no picture, and no raster. See Figure 1-13(e).

The subject matter of the remaining portions of this book will to a large extent be an extension of this technique. Each chapter dealing with the 20 to 40 stages of a TV receiver will also include pictures showing how faults in these stages may affect the picture, raster, and sound outputs of the receiver.

The color receiver. Color receivers are essentially the same as black & white receivers except for two important areas. The color receiver includes a color circuit section and a convergence circuit section. Defects in the sound, picture, and deflection sections will result in symptoms similar to that of a black & white receiver.

Defects in the color circuits can cause such basic symptoms as poor convergence, no color, loss of color sync, color ok no black & white, and improper color. These symptoms may be seen in the color insert to be found in this book.

In actual practice, black & white receivers make use of approximately 20 stages, but color receivers make use of approximately 40 stages. It is vitally important for the

(a)

Pix OK.
Raster OK.
No sound.

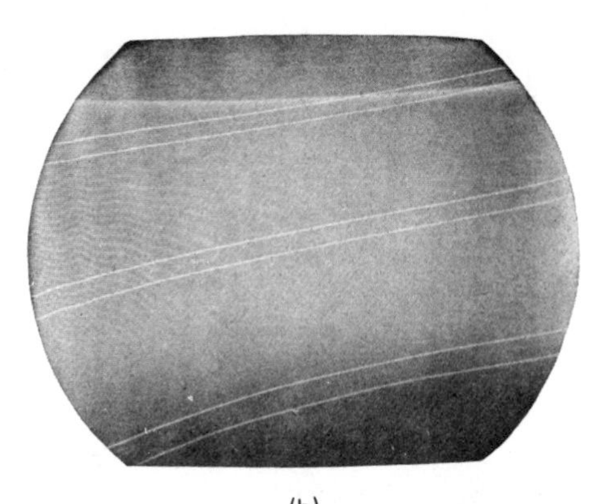

(b)

No pix.
Raster OK.
Sound OK.

(c) No horizontal deflection

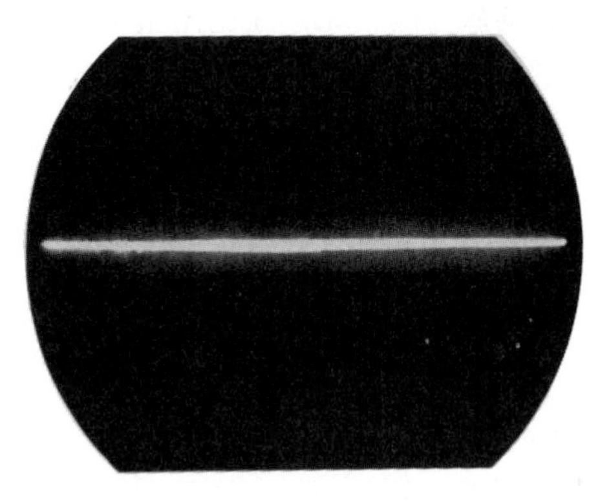

(d) No vertical deflection

(e)

No pix.
No raster.
No sound.

Figure 1-13 Defects that appear in a TV receiver if each of the five major sections on the TV receiver is disabled: (a) bad sound section; (b) bad picture section; (c) bad horizontal deflection section; (d) bad vertical deflection section; (e) bad power supply section. (Photos courtesy of General Electric.)

service technician to be able to recognize which portion of the receiver is defective by simply looking at the picture tube. Being able to do this will enable the service technician to narrow down the number of stages that may be potentially defective.

QUESTIONS

1. What are the three types of electrical diagrams used in television servicing? Explain how they are used.
2. Which diagram should be used if the service technician wants to locate the position of a component?
3. Which diagram should be used if the service technician wants to know the proper wave shape at the collector of a given transistor?
4. Explain how the block diagram is important in electronic servicing.
5. What are the duties of the outside TV technician?
6. What equipment must the outside service technician have when servicing a color receiver?
7. What test equipment and other facilities must a TV bench technician have to be able to do an efficient job?
8. What electrical quantities can an oscilloscope measure?
9. What are the differences between a conventional oscilloscope, a triggered oscilloscope, and a dual trace oscilloscope?
10. What are the three basic steps in troubleshooting?
11. Explain what is meant by "elimination of obvious defects."
12. What are the three techniques that may be used to isolate a defective stage? What knowledge and test equipment must the TV technician have in order to use these techniques?
13. Explain how signal tracing and signal subsitution are performed.
14. Describe two ways that signal substitution may be used to isolate a defective stage.
15. What stages might be the cause of the following symptoms in a black & white receiver? (a) Raster O.K., no pix, no sound; (b) no raster, no pix, no sound; (c) raster O.K., pix O.K., no sound.
16. What sections of a color TV receiver might cause the following symptoms? (a) Raster O.K., pix O.K., no color, sound O.K.; (b) poor convergence, pix O.K., color O.K., sound O.K.; (c) no raster, no pix, no color, no sound.

---

chapter 2

The Black & White Television System

2-1 introduction

Television is an electronic system for conveying pictures from one place to another. Television, just as any other system dealing with pictures, is dealing with the characteristics of the human eye. This is because it is the eye that determines whether or not the reproduced picture is a faithful rendition of the scene being displayed. Hence, any discussion of television must begin with a description of the eye and its characteristics.

The eye. From Figure 2-1(a) it can be seen that the eye is very similar in structure to a camera. They both have a *lens* which gathers the incoming light and focuses it on a light-sensitive surface. In a camera this surface is the photographic film; in the eye this surface is the *retina.* Both the camera and the eye have a structure called an *iris* that controls the amount of light that falls on the light-sensitive surface. In the eye the iris is located between the cornea and the lens. The iris will automatically open to a larger diameter when the scene is dark, admitting more light, and it will close to a smaller diameter and reduce the light reaching the retina when the scene is bright.

The retina of each eye consists of approximately 132 million microscopically small light-sensitive nerve endings that because of their shapes are called *rods* and *cones.* The rods are more or less uniformly distributed throughout the retina; the cones are concentrated in a rather small area called the *fovea centralis.* It has been determined that the rods are sensitive to light and dark and that the cones are sensitive to color.

Angle of resolution. The size of the rods and cones are largely responsible for determining the ability of the eye to separate two closely spaced objects. This is because any light falling between two rods, or two cones, will not be seen. The normal eye is able to resolve two

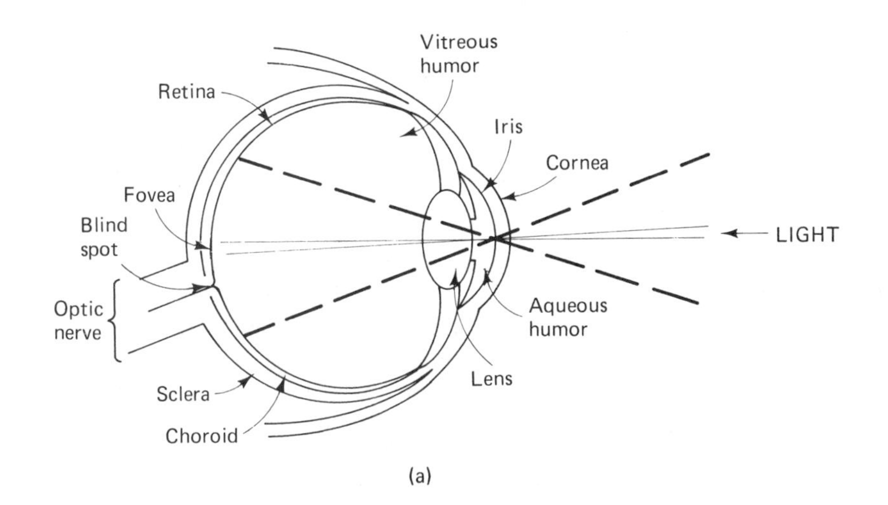

Figure 2-1 (a) Cross-sectional view of the eye; (b) Halftone photograph and enlargement of photograph showing dot structure.

objects that form an angle with the eye of approximately 1.5 minutes of arc, or approximately 1/40 of a degree. This is called *the angle of resolution.* If the objects form an angle smaller than this, the eye sees one object, not two objects. This fact is important in determining the number of lines that will make up the television picture. The angle of resolution and its relationship to the number of lines in a TV picture will be discussed in detail in the section of this chapter devoted to scanning.

Picture elements. Seeing is basically accomplished in three steps. First, the lens gathers the light from the scene and then focuses it as an image on the retina. Second, the image excites each rod and cone to generate an electrical impulse in proportion to the light falling on it. In effect, the image has been broken into millions of pieces called *picture elements.* In fact, all picture-reproducing systems use picture elements to build up the picture. Thus, for example, photographs use small crystals of varying density to construct a picture. Newspapers and photographs in magazines and books make use of small dots of various shades to construct a picture. An example of this is seen in Figure 2-1(b). Television, both black & white and color, also makes use of picture elements to construct the televised picture.

In the third step of seeing, the electrical impulses generated by the rods and cones are carried by the optic nerve to the brain where the information received by the eye is interpreted. An obvious illustration of this is the fact that all lenses, including the lens of the eye, invert an image. Therefore, we really see things upside down. It is the brain that interprets the picture information that it receives and allows us to see right side up.

Persistence of vision. A characteristic of the eye that is important to television picture reproduction is the ability of the eye to follow the activities of moving or changing things. For example, if a light is made to flash on and off at a relatively slow rate, say 5 flashes per second, the eye will be able to follow the flashing with relative ease. If, however, the flashing rate is increased to somewhat fewer than 25 flashes per second the eye does not see a distinct change of light but sees instead a *flicker.* If the flashing rate is then increased to 30 or more flashes per second, the eye sees a continuous flow of light. This characteristic of the eye is called *persistence of vision.* In television this characteristic of the eye allows the picture to be formed on the screen of a television receiver without the observer's seeing its construction. This quality of the eye and the high speed at which the TV picture is constructed allow motion pictures to be reproduced by television.

Angle of view. Another characteristic of the eye important to the television system is the *angle of view.* It has been determined that for most people the most pleasing vertical picture size subtends an angle of approximately 15°. This characteristic is evident when one goes to the movies. Few people like to sit very close or very far from the movie screen where the angle of view is either too large or too small. Most people prefer to sit midway between these two extremes. This midpoint position in the theater provides an angle of view of approximately 15° for the moviegoer.

Aspect ratio. The Federal Communications Commission has set up a standard that controls the relative vertical and horizontal size of television pictures. This standard is called the *aspect ratio* and it requires that the televised picture have a height to width relationship of 3 to 4. For example, this means that a television screen 9 in. high must be 12 in. wide.

Scene characteristics. Listed below are four characteristics of a scene that must be fully reproduced if complete reproduction and picture fidelity are to be achieved.

1. All shades of light and dark must be preserved by the reproducing system.
2. Motion in the scene must be recreated in the reproduction of the scene.
3. An essential ingredient to complete reproduction is the retention of natural color in the copy of the observed scene.
4. If a completely faithful reproduction of a scene is to be made, it is necessary that the scene's three-dimensional characteristics be preserved. Three-dimensional effects are the result of two-eyed vision because the vision of each eye does not superimpose exactly. The end result of this displacement is the visual impression of three dimensions.

The television system is a picture-reproducing scheme. As such, it must convey the information that the eye demands in order to be satisfactory. The simplest television system would be one in which only shades of light and dark are transmitted. Such a system would only be useful for transmitting black & white still pictures. A more reasonable system would be one in which information concerning black & white shading and motion is transmitted. In fact, this is what is done in commercial black & white TV transmission. If it is desired to achieve still greater realism, then color information is also transmitted. Of course, the receiver of color transmissions is somewhat more complex than that designed for black & white reception. Complete picture reproduction can be achieved if in addition to the above qualities three-dimensional information is also transmitted. A number of three-dimensional systems have been proposed but none have been commercially adopted.

2-2 picture transmission

The television picture is constructed in two steps. The first step consists of breaking up the picture into small pieces called *picture elements* and then transmitting these picture elements to the receiver. The second step consists of reconstructing the picture elements into a picture at the receiver. There are two ways by which picture elements may be transmitted: all at once or one at a time.

Parallel transmission. The process in which picture elements are all transmitted at once is called *parallel transmission.* In this process all picture elements are transmitted at the same time and are all reconstructed at the receiver at the same time. The eye can be considered to be an example of parallel transmission since all the rods and cones are excited simultaneously and break up the image into picture elements. This information is then transmitted simultaneously to the brain.

In a simple commercial parallel system, an image is focused onto a plate consisting of many light-sensitive receptors. Each of these receptors generates a voltage that represents a picture element and which is directly proportional to the light intensity falling on it. The output voltage of each receptor is fed into an amplifier chain that in turn feeds a display panel. The display panel is the receiver in this television system. For every light receptor there is one light source on the display panel. Therefore, any light pattern falling on the light sensor panel will produce a picture consisting of an equal number of light sources on the display panel. Of course, the greater the number of light sensors and sources, the greater will be the image resolution.

The disadvantage of this arrangement as a commercial system is primarily cost. In a commercial unit each light sensor must have an expensive amplifier system associated with it in order to drive a light source on the display panel.

With the advent of integrated circuits (IC), parallel systems of TV transmission are being developed. The IC allows many light-sensitive semiconductor devices to be packaged together with their necessary amplifiers in a very small space and at relatively low cost. At the receiving end light-emitting or liquid crystal semiconductor devices can form an integrated circuit display panel.

Sequential scanning. The process of sending picture elements from the transmitter to the receiver one at a time is called *sequential scanning,* or simply scanning. In this system the image is dissected into picture elements by means of an electron beam that converts the light intensity of an image into a varying electrical signal. The varying voltage obtained in this manner is called the *video signal.* In commercial television the dissecting pattern that is used is called a *raster.* The raster is used to break up the image into 525 horizontal lines. A close-up photograph of a raster and picture formed on the face of a TV picture tube showing the raster structure is shown in Figure 2-2. For this operation to be performed, the dissecting electron beam must move horizontally to form one line and then vertically to permit the formation of the next line. This procedure must be performed at both the transmitter, where the dissecting takes place, and at the receiver, where the image is reconstructed. If the picture at the receiver is to be faithfully reconstructed, the electron beams at both transmitter and receiver must be in step with one another. That is, the receiver scanning process must be synchronized with the scanning process at the transmitter. In practice, this is accomplished by using horizontal and vertical synchronizing pulses generated by the transmitter to notify the receiver at what point in the scanning process

Figure 2-2 (a) Photograph of a picture tube showing a normal picture and raster; (b) close-up picture of a raster showing its line structure. (Courtesy of General Electric.)

(a) (b)

the transmitter is at. These synchronizing pulses force the receiver to be a slave of the transmitter, thereby keeping the two scanning systems locked together.

Scanning lines. An important result of the angle of view and the angle of resolution is that they combine to determine the number of horizontal lines that make up the TV picture. The angle of resolution has been defined as being 1.5 minutes of arc. The angle of view is 15° or 900 minutes of arc (15° $\times$ 60 minutes per degree = 900 minutes of arc). It is, therefore, evident that if the line structure of the TV system is not to be seen when the picture is at a distance from the viewer such that the angle of view is produced, 600 horizontal lines (900 minutes, angle of view/1.5 minutes, angle of resolution) must make up the picture. In actual practice in the United States a complete picture called a *frame* consists of 525 lines. Other countries have TV systems that may have as many as 1000 horizontal lines. The greater the number of lines, the closer one may get to the picture before the line structure of the televised picture reveals itself. Another advantage of the increased numbers of horizontal lines is that finer detail may be seen.

In television, resolution is defined as the maximum number of lines that can be discerned on the screen in a distance equal to the tube height. For most modern receivers this ranges from 350 to 400 horizontal lines.

2-3 raster formation

Both the transmitter TV camera and the receiver's picture tube are essentially cathode-ray tubes. Both of these devices have an electron gun assembly and an electron beam deflection assembly. In the case of a TV camera tube, such as the vidicon, a light-sensitive surface replaces the light-emitting phosphor surface of the picture tube screen. A simplified cross-sectional view of each of these tubes is shown in Figure 2-3.

Television camera tubes convert light energy into electrical energy by means of photoelectric effects. There are three photoelectric effects:

1. *Photoemission*: the condition where electrons leave a surface when light falls on it.
2. *Photoconduction*: the condition where the resistance of a material decreases in the presence of light.
3. *Photovoltaic effect*: the condition where a voltage is generated by light falling on a surface. This phenomenon is not used in television camera tubes.

The older image orthicon television camera tube made use of photoemission, the vidicon television camera tube uses photoconduction. The vidicon is probably the most commonly used TV camera tube.

The vidicon. The vidicon is able to convert light into electrical voltage variations by means of a photoconductive material that coats the inner surface of its faceplate. This semiconductor material has the characteristic of decreasing its resistance as the light intensity falling on it increases, and vice versa. Therefore, in those regions of the faceplate where the light of the image is brightest the resistance of the photoconductive surface is lowest. In effect, this surface forms a light-sensitive variable resistor.

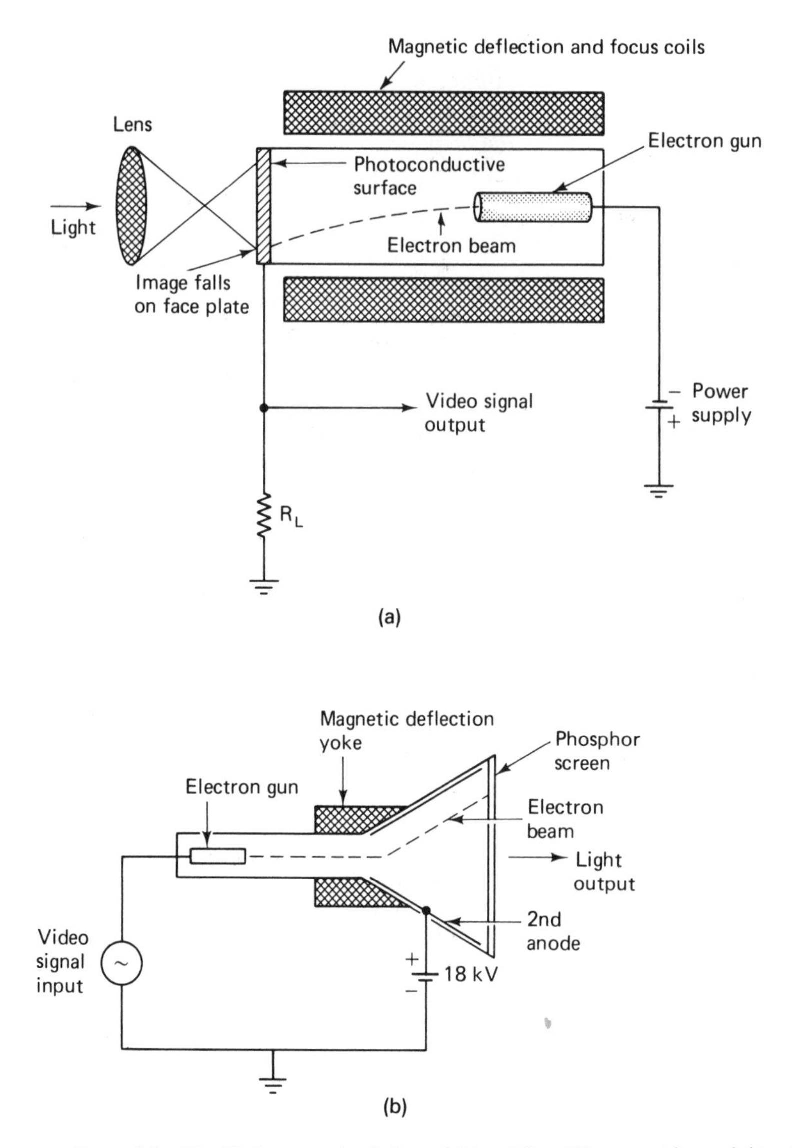

Figure 2-3 Simplified cross-sectional views of (a) a vidicon TV camera tube, and (b) a TV picture tube.

From the diagram of Figure 2-3(a) it can be seen that the vidicon electron beam constitutes part of the complete path for current flow. This comes about as follows: First, current leaves the battery and goes to the electron gun. There it becomes the electron beam which then flows through the photoconductive surface. The current continues on, passing through the load resistor R_L, and finally returns to the battery. In passing through the load resistor the current develops a voltage drop that becomes the video signal. Since the electron beam will be deflected to produce a raster, the current flow through R_L will vary depending upon the instantaneous resistance of the photoconductive surface. In short, the video signal will consist of a dc component and an ac component whose frequency and amplitude will vary depending upon picture detail.

CCD. In recent years solid-state charge-coupled-device (CCD) technology has made possible image sensors that are small, rugged, and extremely sensitive. These full-resolution devices may be used to replace the standard vidicon television camera. As such they may be used in high-quality studio broadcasting applications as well as surveillance and other similar work.

The CCD is organized as a matrix of image sensor elements (pixels) arranged in horizontal and vertical lines. See Figure 2-4(a). A typical CCD may have 488 image sensor elements arranged horizontally and 380 elements vertically, for a total of 185,440 photoelements. Other matrix sizes are available. Each pixel is typically 12 μm by 18μm in size. A typical array may be 9 mm by 11 mm. A CCD array mounted in an 18-pin DIP is shown in Figure 2-4(b).

The CCD employs a charge-transfer system in which charges are created by the light of the image being televised striking the pixels. These charges are contained in MOS capacitors fabricated on a single-crystal wafer. By varying the capacitor electrode voltages in the manner of a shift register, charge packets are moved from capacitor to capacitor to a signal output amplifier. If charge transfer is made while the array is illuminated by the scene, picture smearing will occur. To prevent this, line-by-line rapid charge transfer of the entire field is made during vertical retrace. The charges are transferred to an opaque storage array equal in size to the photosensitive array. This device is called a *frame transfer CCD* and is usually rectangular in shape to accommodate the equal-size storage area.

Another CCD arrangement used to avoid smearing is called *interline transfer.* Here the CCD is organized so that the photosensitive columns of pixels are interleaved with opaque column charge-transfer registers. This results in a smaller area imaging surface that suffers from some light loss.

CID. An alternative MOS structure used for solid-state imaging is the charge-injection device (CID). This device is similar to the CCD. It is a matrix array of pixel electrodes on a silicon

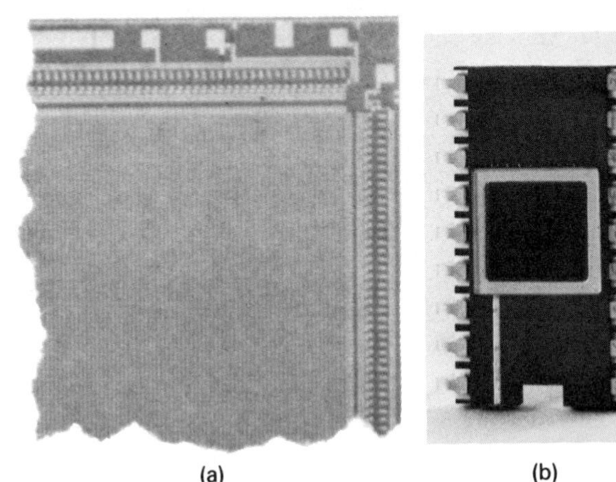
(a) (b)

Figure 2-4 (a) A section of the matrix of image sensors arranged in horizontal and vertical lines; (b) CCD array mounted on an 18-pin DIP IC. (Courtesy of RCA/Electro Optics and Devices.)

wafer. Light striking this surface generates minority carriers under each pair of electrodes. By removing the voltages on the electrodes, the photon-produced charges are injected into the wafer substrate. The substrate current output is proportional to the light intensity falling on the pixels and is the video signal.

Some of the advantages of the CID image sensor are low dark current, minimal opaque regions, and bloom resistance. Its chief limitations are large sensing capacitance and high fixed pattern noise.

Picture tube. The picture tube shown in Figure 2-3(b) is a device that is able to convert electrical energy into light energy. It is able to do this because an electron beam, generated by the electron gun, is first accelerated by the high voltage (1.5 to 30 kV) second anode to a considerable velocity, and then it is suddenly stopped by the phosphor coating on the face of the tube. The energy released by this impact forces the phosphor coating on the face of the tube to emit light. This light output is directly related to the amount of high voltage at the second anode and the number of electrons striking the phosphor screen. In a properly operating television receiver the second anode voltage is designed to remain constant. Therefore, the usual manner by which the light output of the picture tube is varied is by varying electron beam density. Since the light output of the picture tube must form a picture, the electron beam is *intensity modulated* by the video signal. That is, the video signal, obtained from the camera tube, is made to vary the grid to cathode bias of the picture tube, thereby varying its light output by controlling the number of electrons that reach the phosphor screen.

Magnetic deflection. To produce a raster, the electron beam must be simultaneously moved horizontally to form a horizontal line as it strikes the phosphor screen, and vertically to prepare for the next line. Electron beam deflection may be accomplished by electrostatic or magnetic means. All color and black & white TV receivers as well as most camera tubes use magnetic deflection as a means of generating a TV raster.

Magnetic deflection is used in both the vidicon and the picture tube of Figure 2-3 as a means of generating a raster in both devices. The deflection is produced by two sets of coils called a *yoke* that are spaced at right angles to the other around the neck of the tube. The horizontally positioned coil produces the vertical deflection and the vertically positioned coil produces the horizontal deflection. Figure 2-5 shows a photograph of a deflection yoke and also a cross-sectional view of its windings as well as its schematic representation. It can be seen that the deflection coil is an electromagnet whose air gap is occupied by the neck of the cathode-ray tube (CRT). If a current is passed through the *vertical* deflection yoke winding of Figure 2-5, a uniform magnetic field will be produced inside the neck of the tube. The direction of this field is determined by the direction of current flow through the yoke winding. If the direction of current flow is reversed, the magnetic field will also reverse direction.

Passing through the magnetic field of the yoke is the electron beam that is able to generate a magnetic field of its own since it is an electric current. The interaction of the yoke's magnetic field and the beam's magnetic field forces the beam to be deflected. The amount of deflection is directly proportional to the strength of the yoke's magnetic field. If the current through the yoke is reversed, then the interaction between the two magnetic fields will result in a reversed deflection. In actual practice, the current through the yoke

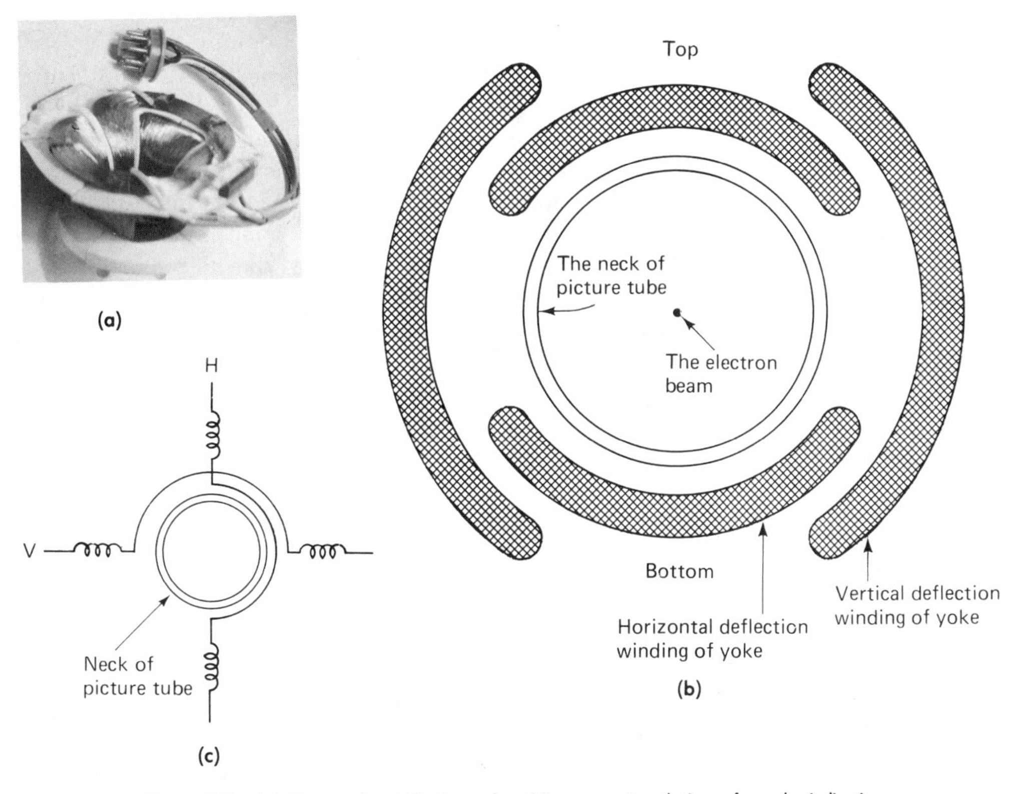

Figure 2-5 (a) Picture tube deflection yoke; (b) cross-sectional view of a yoke indicating the relative positions of the vertical and horizontal deflection windings; (c) schematic representation of the yoke winding around the neck of the CRT.

is ac. Therefore, the electron beam will periodically move up and down at a rate equal to the frequency of the ac current passing through the vertical deflection coils. In commercial television the vertical deflection rate is 60 Hz and the horizontal deflection rate is 15,750 Hz. The amount of deflection is determined by the amplitude of the ac current. A large current will produce a large deflection. A small current will produce a small deflection.

The sawtooth. One of the most important aspects of the deflecting current flowing through the yoke is its wave shape. The reasons for this are: (1) The writing speed of the beam is controlled by the deflection current wave shape as the beam moves across the face of the tube. This writing speed must be uniform in order to provide a constant light output as the beam generates the raster. If the beam is moving slowly, greater numbers of electrons hit the screen per unit time. Therefore, the light output will be high, and vice versa. A deflection current wave shape that will provide a uniform writing speed is a sawtooth current. (2) Since the electron beam must move from one end of the picture tube face to the other and then return, the deflection current wave shape should allow for a longer period of time for scanning and a shorter period of time for the return. This condition is again met by a sawtooth current.

2-4 horizontal deflection and interlace scanning

Horizontal deflection. Horizontal deflection is provided by passing a sawtooth current through the horizontal deflection windings of the yoke that are located above and below the neck of the tube. This current creates a vertically positioned magnetic field inside the neck of the tube. As the intensity of the field grows, the magnetic interaction between the electron beam's magnetic field and the magnetic field of the yoke forces the electron beam to move from the left side of the screen to the right side. At this point the sawtooth current through the horizontal winding of the yoke suddenly reverses direction and forces the electron beam to return rapidly (retrace) to the left side of the screen. In commercial television this procedure is repeated 15,750 times per second. Simultaneous vertical deflection forces the beam to move from the top to the bottom of the screen 60 times per second. It is the combined effects of vertical and horizontal deflection that results in a *raster.*

Two factors that help to determine the manner by which the TV raster is generated are as follows:

1. The formation of the raster must be fast enough so that persistence of vision does not result in flicker.
2. The system of raster generation should not result in a video frequency bandpass that is too large.

Interlace scanning. The system that meets these requirements uses *odd line interlaced scanning.* This system is referred to as being odd lined because the total number of lines forming the raster is 525, an odd number. In this system, all the odd lines from the top of the screen to the bottom are scanned first. Then the beam is returned to the top of the screen and all the even lines from the top to the bottom of the raster are scanned. The odd lines are called the *odd field* and the even lines are called the *even field.* The process of weaving the odd and even lines together to form a raster is called *interlacing.* The complete raster of both even and odd lines is referred to as a *frame.* Commercial television has standardized a complete picture (frame) as 525 lines. Therefore, each field consists of one-half this number of lines, or 262.5 lines. As a means of eliminating flicker, each field is standardized to take place in $^1/_{60}$ second. Therefore, both fields occupy $^2/_{60}$ second or $^1/_{30}$ second. The purpose of interlacing the odd and even fields is to provide a picture repetition rate of 30 frames per second that is flashed on the television screen 60 times (fields) per second. Since the flashing rate (60 Hz) exceeds the flicker rate ($\approx$30 flashes per second) of the average eye, no flicker is seen and the technique of television scanning is hidden from the observer.

Scanning frequencies. These standards determine the frequency of the sawtooth yoke currents. Thus, the vertical deflection current must be 60 Hz if 60 fields are to be scanned each second. Also, since there are 525 horizontal lines being scanned each frame, and since there are to be 30 frames per second, the total number of lines each second must be 525 lines $\times$ 30 = 15,750 horizontal lines per second. This means that the horizontal deflection sawtooth must have a frequency of 15,750 Hz.

Figures 2-6 and 2-7 summarize the scanning characteristics of commercial television. In Figure 2-6 the horizontal and vertical deflection sawtooth currents are shown with respect to the raster that they generate. Notice that one vertical deflection sawtooth is responsible for moving the electron beam from the top of the screen to the bottom and back again in

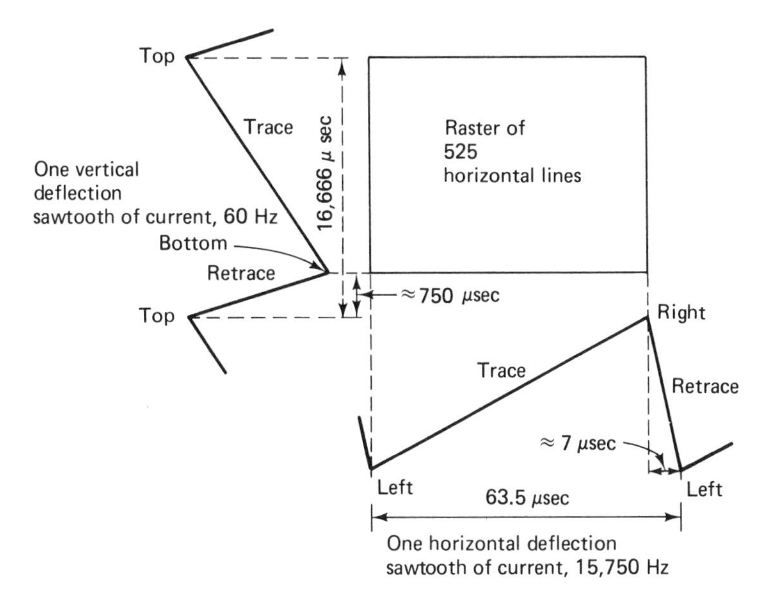

Figure 2-6 Horizontal and vertical deflection waveforms needed to form the raster.

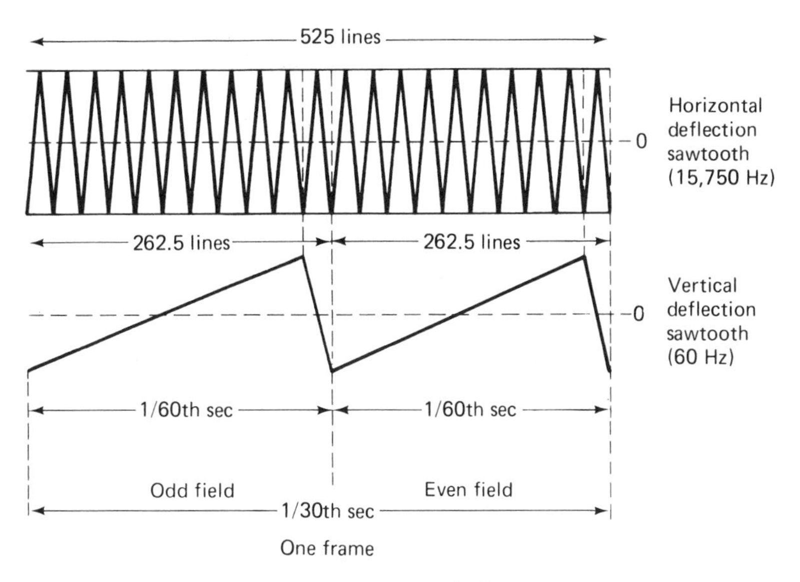

Figure 2-7 TV scanning standards.

1/60 second. During this time the horizontal deflection current causes the electron beam to move from the left side of the screen to the right and back again 262.5 times. Each vertical sawtooth occupies a time interval of 16,666 μs, of which approximately 800 μs are used for retrace. One horizontal sawtooth has a 63.5 μs duration, of which retrace is approximately 7 μs. In Figure 2-7 the horizontal and vertical deflection currents are positioned one over the other so that their time relationships become apparent.

2-5 the seven-line raster

The best way to understand how a commercial television raster of 525 lines is produced is to examine in detail how a simple seven-line raster is formed. Since the total number of lines forming this raster is to be seven, each field will contain 3.5 lines. In terms of deflection sawtooth currents, this means that one frame will consist of seven horizontal sawteeth of current and two vertical sawteeth of current. This is illustrated in Figure 2-8.

The horizontal deflection sawtooth is responsible for forcing the electron beam from the left side of the screen through the center to the right side of the screen. Therefore, the amplitude of the sawtooth is labeled left, center, and right. The vertical deflection sawtooth is also labeled to indicate the amplitude of the sawtooth that is responsible for

Figure 2-8 Formation of a seven-line raster.

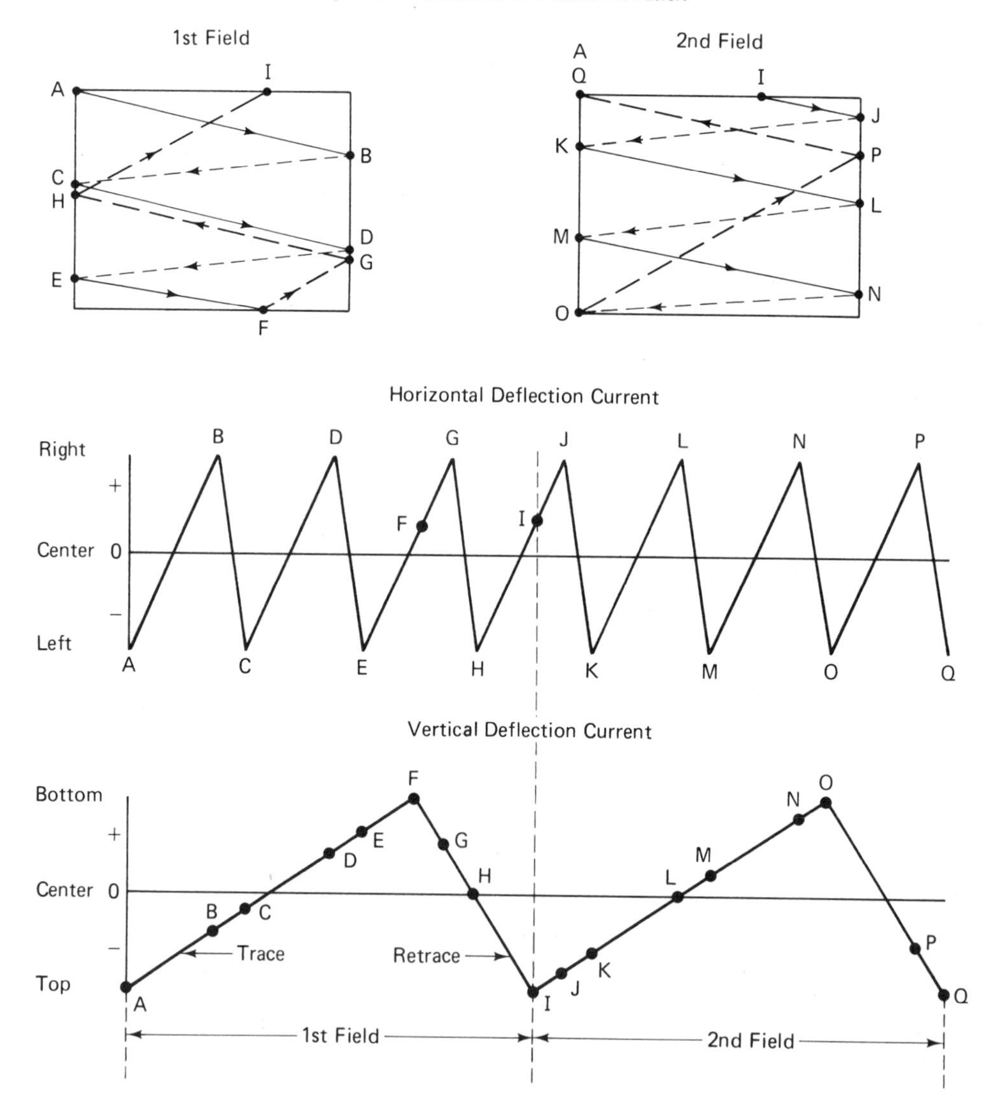

Figure 2-9 Blank raster showing vertical retrace lines. (Courtesy of General Electric.)

forcing the electron beam to the top, center, or bottom of the raster. Since both sawtooth currents are acting at the same time, the position of the electron beam depends upon the simultaneous action of both currents. If a vertical line is dropped between the two sawtooth currents shown in Figure 2-8, the intersection of the line on each sawtooth indicates the position of the beam on the face of the TV screen.

In this seven-line raster only one horizontal line (from *F* to *I* and *O* to *Q*) is used up during each vertical retrace. Thus, two lines are lost each frame making a total of five *active lines* seen in the raster each frame. In commercial television vertical retrace is responsible for 13 to 21 lines being lost each field or a total of 26 to 42 lines lost each frame. This leaves 483 to 499 active lines per frame.

Figure 2-9 is a TV raster showing vertical retrace lines. Notice that the vertical retrace lines are brighter than the background lines and that there are no horizontal retrace lines visible. Both of these effects are due to the writing speed of the electron beam (velocity modulation). The vertical retrace lines are brighter than the horizontal lines because vertical retrace is slower moving than horizontal scan. The horizontal retrace lines move faster than the horizontal scan; therefore, they are too dim in comparison to the horizontal scan lines to be seen.

Study of Figure 2-8 shows that at the end of the first field the electron beam is positioned at the top (*I*) of the raster. Thus, it is perfectly situated to begin the first half line of the second field. As a result of this positioning, the even lines will now fall between the odd lines formed by the first field, thereby producing an interlaced raster. This alignment of fields is the result of using an odd number of lines to produce the raster. If an even number of lines were used, interlacing would be very difficult to achieve since the lines of each field would overlap one another.

2-6 the video signal

Up to this point we have seen how the eye determines some of the standards of the TV system and how the TV raster is generated. We will now examine a simple television system and see how the video signal is generated and how the picture is reconstructed at the receiver. Figure 2-10(a) illustrates *the flying spot* scanning system for generating a video

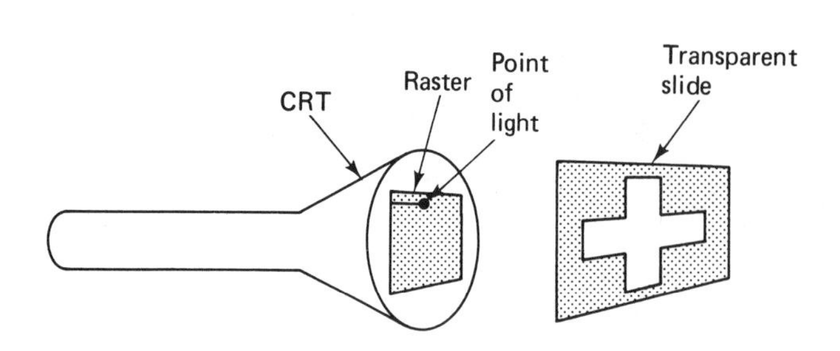

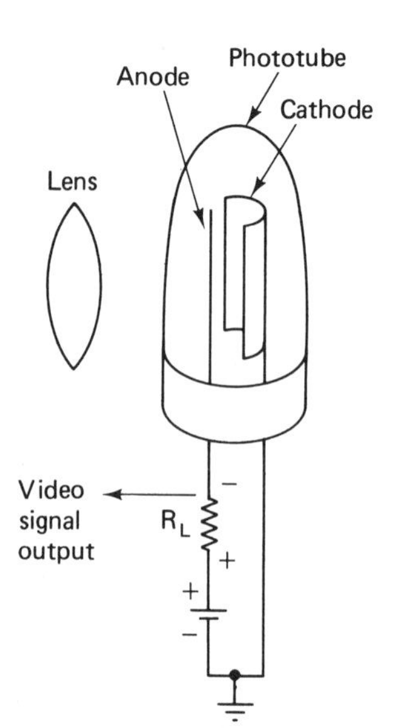

(a)

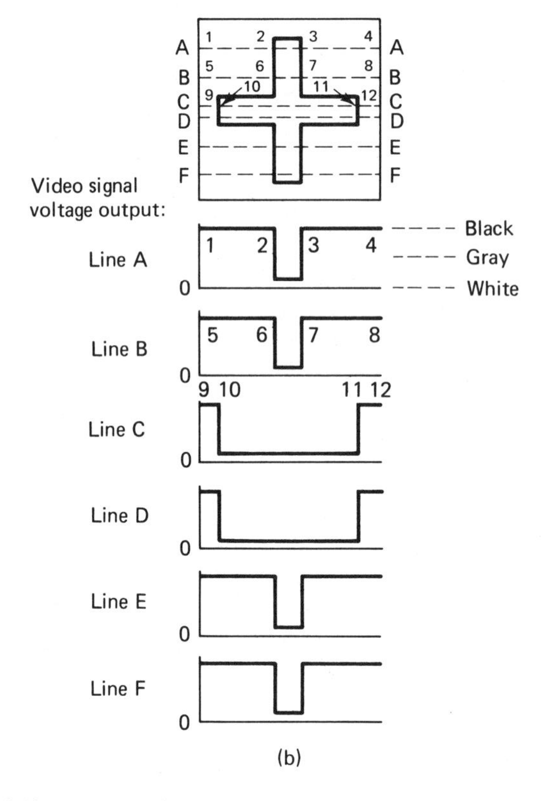

(b)

Figure 2-10 Formation of a video signal using a flying spot scanner: (a) the physical arrangement of the flying spot scanner system; (b) the video signals produced by the flying spot scanner for six horizontal lines.

signal. This method is used in many types of television servicing TV picture generators and is limited to televising of transparent slides or movie film. If live television is desired, television camera tubes such as the vidicon are used to generate the video signal.

In this system a raster is generated in the usual manner on the face of an oscilloscope type of CRT. A transparent slide is made to cover the face of the tube so that the light output of the tube passes through the slide. It is the pinpoint of light that forms the raster that is responsible for breaking up the transparent slide into picture elements. When the point of light is in front of a transparent section of the slide, all of the light passes through the film and makes its way via the focusing lens to the cathode of a phototube. The presence of light on the cathode of the phototube causes it to emit electrons. The amount of emission is directly proportioned to the light intensity. Since the anode of the phototube is returned to the positive side of a power supply, the photoelectrons are attracted to the anode. Current then flows from the cathode to the anode, through the video load resistor and the power supply, and then back to the cathode to complete the circuit. Hence, for maximum light (white) conditions, the video output voltage will be minimum.

If the raster point of light is opposite an opaque (black) portion of film, no light will reach the phototube. As a result, no photocurrent flows and the video output voltage will be maximum and equal to the supply voltage. Therefore, for zero light (black) conditions, the video output voltage is maximum. Of course, the video output voltage will fall between these two extremes for values of light intensity other than white or black (gray).

Figure 2-10(b) shows the video output voltage generated by the phototube for six arbitrarily selected horizontal lines as compared to the transparent film being scanned.

The six lines of video signal that have been generated by this scanning method are only a small part of the 262.5 lines that constitute the entire field. It should be understood that each active line of the field will generate a video signal whose amplitude is directly proportional to the light level. The duration of the video signal will, of course, depend upon the duration of a given level of brightness.

Video frequency range. It is possible to determine roughly the range of video signal frequencies generated by the flying spot scanner for a TV system using commercial standards. It is important to know this range of video frequencies because they play an important part in determining such things as the standards for scanning and the characteristics of IF and video amplifiers in the television receiver.

Assume that a video signal is to be generated of a picture that consists of a checkerboard pattern of white and black squares. Each square of this checkerboard is to be the size of one picture element. In practice, this means that each picture element must not be larger or smaller than the diameter of the scanning beam or approximately 0.05 in. in diameter. Since there are 525 horizontal lines in a complete raster, there must be 525 *vertical* picture elements. The aspect ratio is 4:3; therefore, one and one-third ($4/3$) times as many picture elements will occupy each horizontal line as compared to the vertical picture elements. Thus, there are $525 \times 4/3 = 700$ picture elements in each line. Since there are 525 lines for a complete picture, the entire frame will have $525 \times 700 = 367{,}500$ picture elements. Thirty complete frames occur each second; therefore, the 367,500 picture elements are scanned 30 times per second, or a total of 11,025,000 picture elements are scanned each second.

The video signal that will be produced by scanning a black & white checkerboard pattern is a square wave; the black region producing a positive-going output and the white region producing a negative-going voltage. Since two picture elements are required to generate one video square-wave cycle, the video frequency generated under ideal conditions will be ½ × 11,025,000 or 5,512,500 Hz. Since, no television system is perfect, such things

Figure 2-11 (a) Video signal produced by scanning a picture consisting of one black and one white horizontal bar; (b) the frequency response of an ideal TV camera.

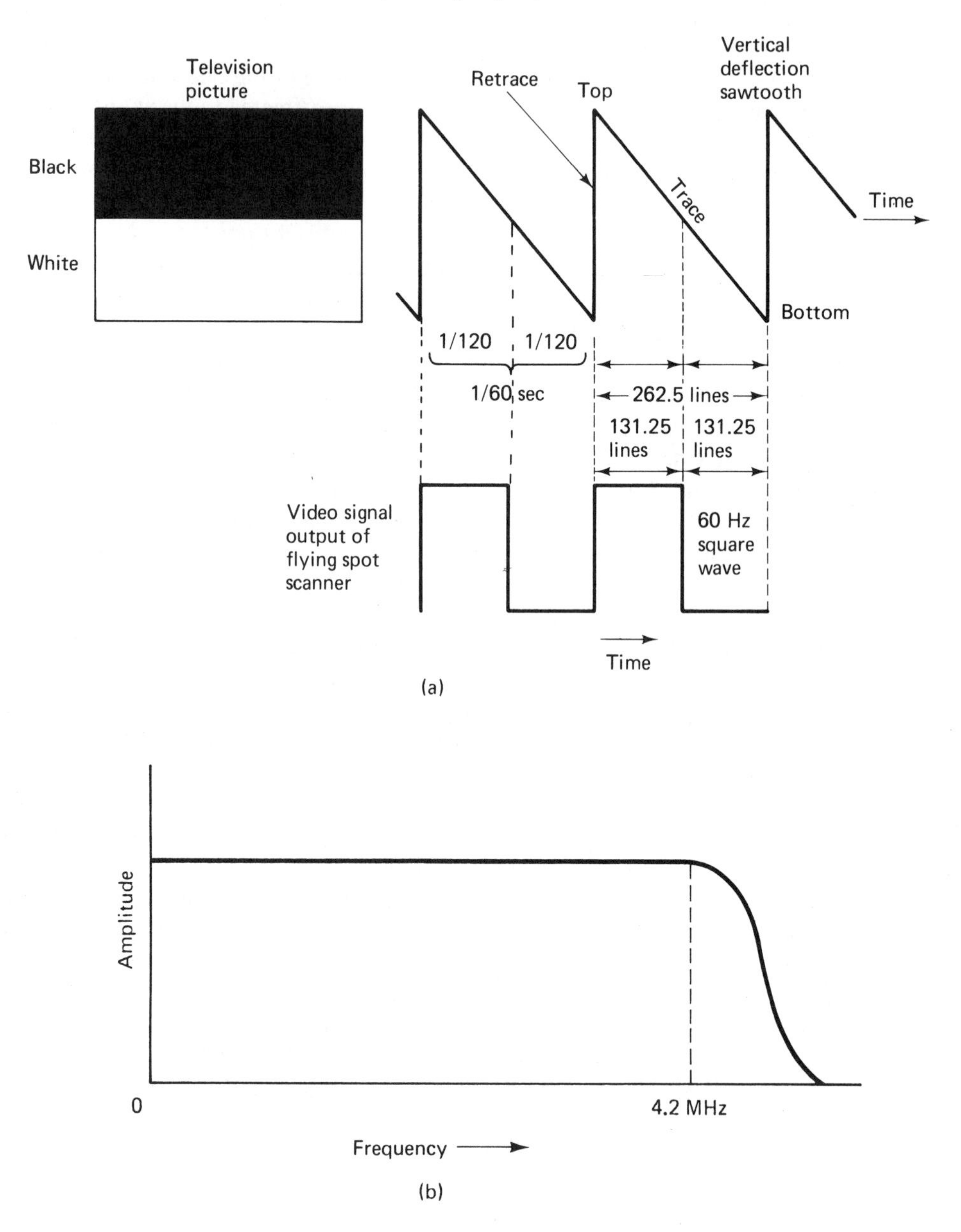

as imperfect electron beam shape and improper interlace can reduce the video frequency to as much as 75%. This percentage is called the *utilization ratio* and will reduce the video signal frequency generated in this system to 4,134,375 Hz. In practice, the FCC has set 4.2 MHz as the maximum video signal that the television system may transmit.

Since the 4.2-MHz video signal was generated by scanning a picture consisting of approximately 300,000 black & white patches, it can be said that the *high-frequency* components of the video signal represent the *fine detail* in the picture. In practice, the video signal is very complex and will contain low-frequency as well as high-frequency components. The *low-frequency* components correspond to *large areas* in the picture. For example, assume that a picture was being scanned that consisted of a black & white region such as that shown in Figure 2-11(a). It is known that the first field of 262.5 lines will scan the picture from the top to the bottom of the screen in 1/60 second. Therefore, the first 131.25 horizontal lines that occur during the black portion of the picture will generate a video signal voltage that will be maximum positive. The remaining 131.25 lines of the first field will scan that part of the picture that is white, and, therefore will generate little or no video output voltage. Since the next field will repeat this procedure, the video output voltage will be a square wave whose repetition rate will be 60 cycles per second. Hence, this square wave represents large areas of black & white in the picture. Thus, the video signal frequencies generated by the scanning process can extend from dc (a completely black, gray, or white picture) to 4.2 MHz. Figure 2-11(b) shows the frequency response of an ideal television camera.

2-7 picture reproduction

Intensity modulation. Up until now we have discussed how the video signal is generated. The question of how the picture is reproduced at the receiver from the video signal voltage generated at the transmitter still remains. The answer to this question is shown in Figure 2-12. Figure 2-12(a) illustrates the light output versus grid-to-cathode voltage characteristics of a typical television picture tube. Notice that the light output of the picture tube is maximum when the grid-to-cathode voltage is zero. This light level corresponds to white. When the grid of the picture tube is made approximately 100 V negative with respect to the cathode, the picture tube is cut off and no light output is produced. The absence of light corresponds to black. If a video signal corresponding to line *A* of Figure 2-10(b) is applied between grid and cathode of the picture tube, the video wave shape will fall on the light transfer curve as shown in Figure 2-12(a). In this case, the bias between grid and cathode and the amplitude of the video signal have been arranged so that regions 1 to 2 and 3 to 4 produce light cut off (black) and so that the region between 2 and 3 bias the picture tube in order to produce maximum light output (white). Thus, when a video signal is applied between grid and cathode of the picture tube, the light output of the CRT will vary from one point on the face of the tube to the next depending upon the instantaneous video signal amplitude. This process is called *intensity modulation.*

Picture tube display of video. An example of how the video signal that was generated in Figure 2-10 is reproduced into a television picture is shown in Figure 2-12(b). Here we see the video signal voltage of line *A* being applied between grid and cathode of the picture tube. This voltage intensity modulates the electron scanning beam so that from points 1 to 2 the light output is cut off. Between points 2 and 3 the electron beam is turned on and

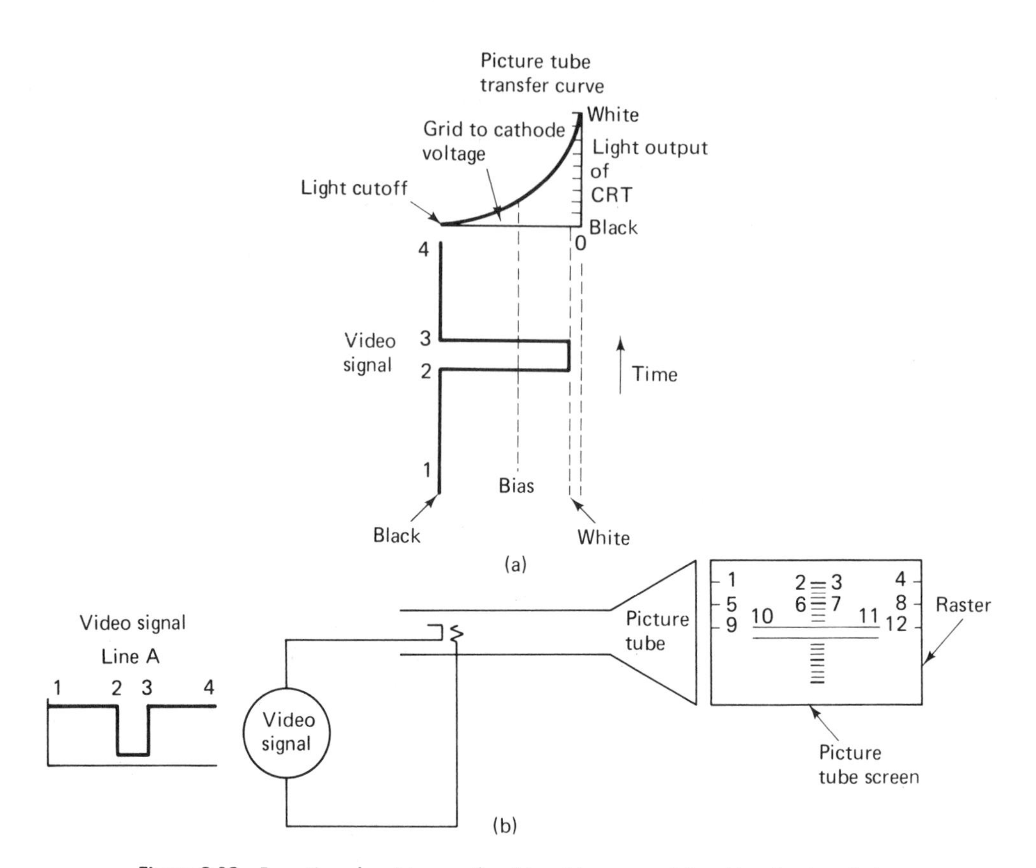

Figure 2-12 Formation of a picture on the picture tube screen: (a) a video signal applied to the light transfer curve of a picture tube; (b) the video signal applied between cathode and grid of the picture tube and the resultant picture produced.

a light output appears on the face of the tube. Between points 3 and 4 the tube is again cut off and no light appears on the screen. This process is repeated for each horizontal line. The composite result is the formation of a complete picture corresponding to the picture that was televised at the transmitter.

As was the case at the camera, the high-frequency video signal corresponds to fine detail and the low-frequency video components represent the large areas in the reproduced television picture that appears on the face of the picture tube. To illustrate, assume that a video signal square wave of 120 Hz is applied between grid and cathode of the picture tube as shown in Figure 2-13. In a television system there are three ac signals simultaneously applied to the picture tube at any one instant. These are the vertical and horizontal deflection sawtooth currents and the video signal. To be able to predict what will appear on the picture tube screen requires that the state of all three signals be known at any given time. In this case, the video signal frequency is twice that of the vertical scanning rate. Therefore, since each vertical sawtooth is responsible for 262.5 horizontal lines, each video cycle will occupy 131.25 horizontal lines, and each one-half cycle of video will represent approximately 65 complete horizontal lines. If the video signal is positive going and applied to the cathode of the picture tube, the picture tube will be black for these 65 lines. The next 65 lines will

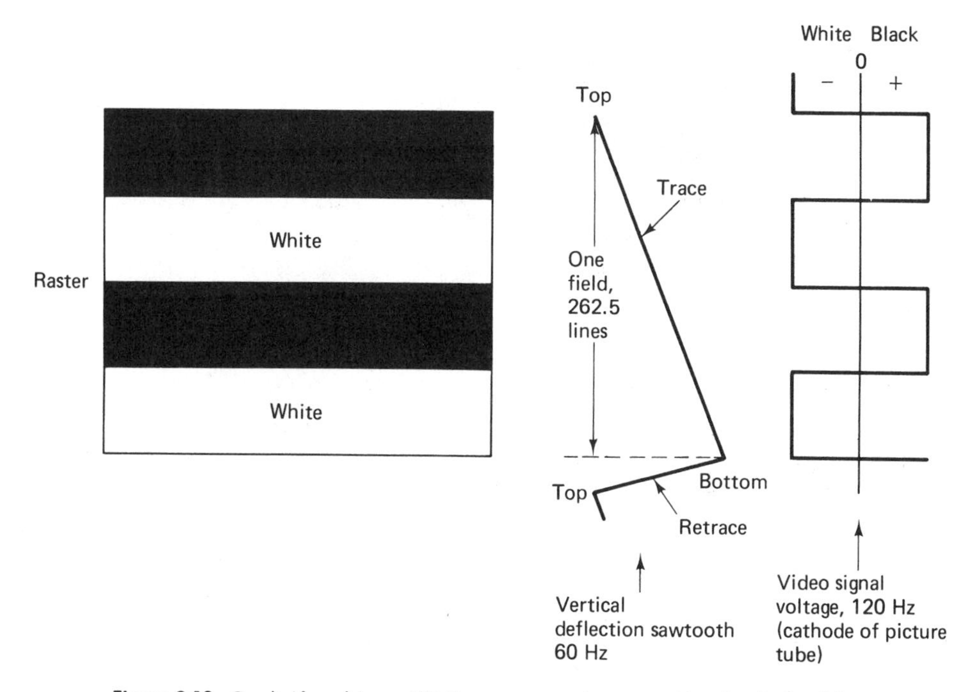

Figure 2-13 Result of applying a 120-Hz square wave between grid and cathode of the picture tube.

be white because the video signal will be negative going during this time. The remaining cycle of motor video signal voltage that occurs during this field will force the picture tube to repeat the pattern produced by the first video cycle. This pattern of two black and two white horizontal bars will repeat itself for each field. If the frequency relationship between the video signal and the vertical deflection sawtooth remains at exactly 2:1, the pattern will remain stationary. If, however, the frequency relationship is not exact, the pattern will appear to roll vertically. A stationary pattern will be obtained provided that the video signal is a whole-number multiple of the field rate. Thus, if the frequency of the video is 600 Hz, the pattern on the picture tube screen will be 10 black and 10 white horizontal bars.

If the frequency of the video signal is increased to 15,750 Hz, which is the same as the horizontal scanning rate, there will be a distinct change in the nature of the reproduced picture. See Figure 2-14. Notice that instead of horizontal bars, as was the previous case, there are two vertical bars, one black and the other white. This comes about because during the time occupied by *each horizontal line* one video square-wave cycle occurs that causes the picture tube to switch on or turn off. When the video signal is positive going, the picture tube will be cut off and no light will be emitted. However, when the video signal is negative going, the picture tube will turn on and light will be emitted from the face of the tube.

Of course, if the video frequency were to be increased to 157,500 Hz, the reproduced picture would consist of 10 black and 10 white vertical bars. The reasoning for this is the same as in the previous case. These bars will also remain stationary provided that the video signal is an exact multiple of the horizontal scanning rate.

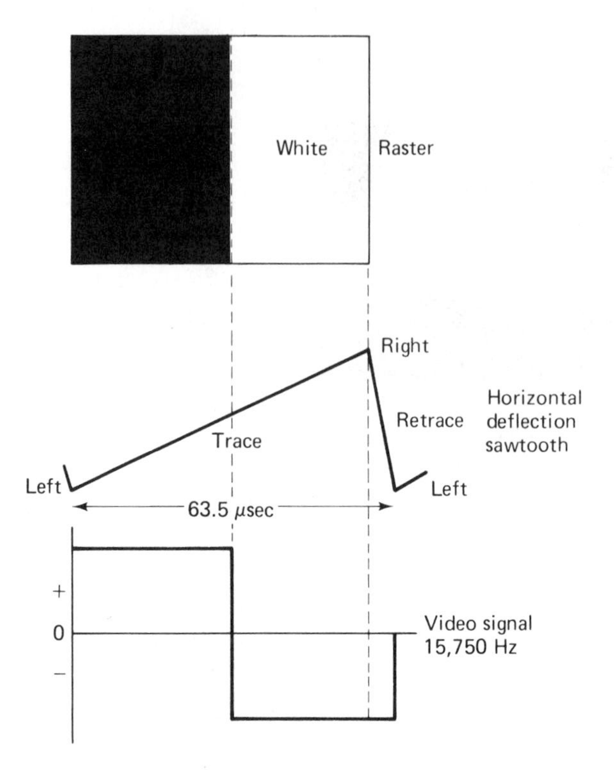

Figure 2-14 Result of a video frequency equal to the horizontal scanning rate being applied between grid and cathode of the picture tube.

2-8 the composite video signal

The video signal generated at the camera cannot be used by itself at the receiver to reproduce the televised picture because in the TV system so far discussed the scanning generators at the transmitter and receiver will not be at the same frequency at the same time. As a result, the picture at the receiver will roll vertically and tear horizontally. Such things as temperature variations, ac line fluctuations, and other circuit differences that exist between both transmitter and receiver circuits are responsible for this condition. Obviously, the way to remedy this situation is for the transmitter to transmit information that will force the receiver's scanning generators to synchronize with the transmitter's scanning generators. This is in fact what is actually done. Since there is a horizontal scanning generator and a vertical scanning generator to be synchronized, the transmitter will provide both a horizontal and a vertical synchronizing pulse. Each horizontal line must be synchronized and each field must begin at the proper time. To do this, the transmitter must contain a synchronizing generator that provides *a horizontal sync pulse* 15,750 times per second and *a vertical sync pulse* 60 times per second. The horizontal synchronizing pulses are timed so that the leading edge of the pulse ends trace and initiates retrace. The vertical sync pulse is also responsible for ending the trace portion of the vertical deflection sawtooth and for beginning its retrace. However, this action is somewhat more complex than that used for horizontal synchronization and will be discussed in detail in a later section of this chapter.

Blanking. Another problem associated with raster formation that is compensated for at the transmitter is the visual effects of retrace. Figure 2-9 shows a picture of a TV raster that has a number of bright vertical retrace lines.

The transmitter is able to compensate for these annoying effects by generating a pulse that is of sufficient amplitude and duration that it keeps the camera tube and the picture tube cut off during each horizontal and vertical retrace period. These pulses are called *blanking pulses* and are part of the composite video signal that the transmitter transmits to the receiver.

In summary, the black & white television transmitter must generate a *composite video signal* that contains the following components:

1. The video signal
2. Horizontal synchronizing pulses
3. Vertical synchronizing pulses
4. Horizontal blanking pulses
5. Vertical blanking pulses

An additional component that must also be transmitted is the sound information that is associated with the picture.

Horizontal sync and blanking. Figure 2-15 shows the composite video signal for three horizontal lines. Figure 2-16 is a photograph of an oscilloscope pattern showing how the composite video signal would actually appear. At the receiver this signal would normally

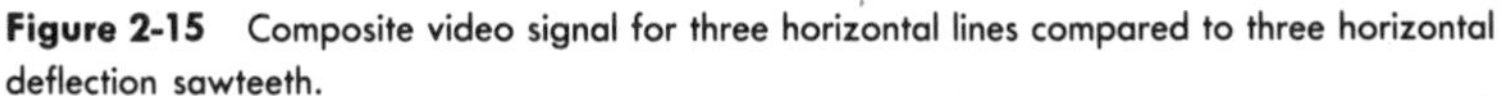

Figure 2-15 Composite video signal for three horizontal lines compared to three horizontal deflection sawteeth.

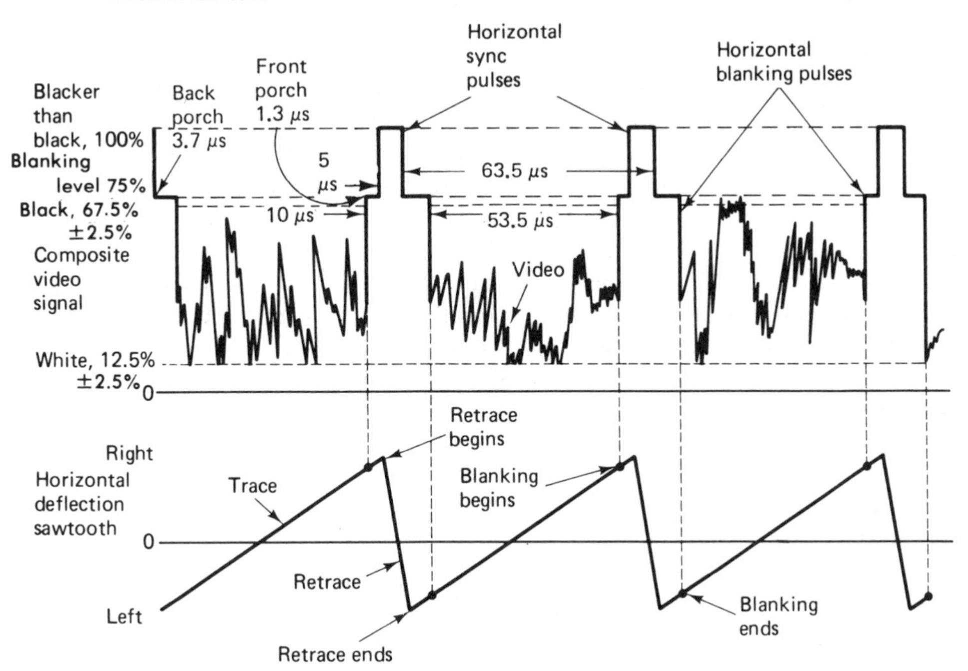

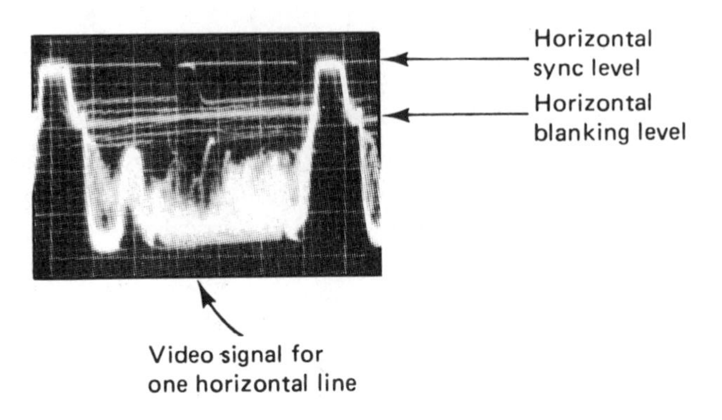

Figure 2-16 Oscilloscope pattern of the composite video signal for one horizontal line.

be found between grid and cathode of the CRT and thereby would intensity modulate the electron beam of the picture tube.

A close examination of the composite video signal of Figure 2-15 discloses a number of important facts. First, it can be seen that the amplitude of the signal defines brightness. Thus, the 12.5% level corresponds to white and the 67.5% level corresponds to black. In other words, when this signal is applied between grid and cathode of the picture tube, those video signal levels falling on or near the 12.5% level will produce maximum light output, whereas such video signals as horizontal blanking that fall on the 75% level will cut off the picture tube and no light output will occur. The horizontal sync pulse has an amplitude of 100% and is 25% of the total composite video signal. Since its level exceeds that of the black level, the sync pulse amplitude is often referred to as being *blacker than black.*

The white level is kept at approximately 12.5% in order to prevent a loss of transmitted picture carrier. Loss of picture carrier can occur if the video signal level were allowed to drop to zero during scanning of a bright white picture element. Since this loss of picture information will occur at a field rate, it is possible for this RF carrier interruption to be heard as a loud and unpleasant buzz in the sound at the receiver. The maintenance of the 12.5% level prevents this buzz from becoming a problem at the receiver.

The duration of one horizontal line is $1/15{,}750$ second or 63.5 μs. The blanking pulse interval occurs each line and lasts for approximately 10 μs; therefore, the active video portion of one horizontal line that is actually seen lasts for 53.5 μs.

Riding on the blanking pulse pedestal is the horizontal sync pulse. This pulse has a duration of approximately 5 μs. The horizontal sync pulse is not positioned exactly in the center of the blanking pulse pedestal. It is positioned more toward the leading edge of the blanking pulse than the trailing edge. The spaces on either side of the horizontal sync pulse are called the *front porch* and the *back porch.* The front porch has a time duration of 1.3 μs and the longer back porch has a time interval of 3.7 μs. The back porch interval is important during color TV transmission, because a burst of eight cycles of 3.58 MHz is placed here. This color burst acts to synchronize the receiver color circuits so that proper color reproduction is possible.

Directly under the composite video signal of Figure 2-15 are three horizontal deflection sawteeth. These waveforms are positioned in relation to the composite video signal as they would occur in a normally operating television receiver. Notice that the leading edge of the horizontal blanking pulse occurs before the deflection sawtooth has reached its maximum amplitude and retrace begins. As a result, the electron beam in the picture tube is turned off before it can reach the farthest point on the right side of the screen. Shortly after blanking has begun the horizontal sync pulse initiates horizontal retrace. The duration of retrace is determined by the characteristics of the horizontal deflection circuits of the receiver, and it may last longer than the sync pulse interval. However, this is of no importance, provided that retrace is not longer than the blanking interval, since the back porch of the blanking pulse will hide retrace. Proper television receiver design demands that horizontal retrace be approximately 7 μs in length; therefore, in the normally operating receiver horizontal blanking will extend beyond the start of trace. The reason for this is to allow room for design tolerances that might cause retrace to be longer than normal.

If horizontal retrace were too long, blanking would end before the beam would reach the left side of the screen. As a result, that part of the video signal immediately following blanking would appear as picture information moving from right to left, whereas the remaining video signal would produce picture information moving from left to right. Since this would occur for each horizontal line, the visual effect would be that the part of the picture on the left side of the screen would be folded over the remaining picture information.

In modern television receivers the horizontal sawtooth generator's frequency is controlled by a dc error voltage obtained by comparing the horizontal sync pulse frequency to the local generator's frequency. A change in this error voltage will cause a timing difference between the horizontal sync pulse interval and the retrace portion of the horizontal deflection sawtooth. Figure 2-17(a) shows the sync pulse and sawtooth relationship that occurs in a normally operating receiver whose raster has been made narrow to allow observation of the front and back porches of the blanking interval. Figure 2-17(b) shows the result of a change in error voltage such that the sync and blanking pulses occur in the center of the trace portion of the deflection sawtooth. This arrangement allows the horizontal sync and blanking pulses to become visible provided that the brightness control is advanced so that black becomes gray and the blacker than black sync pulses become black.

Vertical sync and blanking. The current FCC standards on television scanning require that one field of 262.5 horizontal lines be scanned each 1/60 second. A vertical retrace and blanking interval must be provided between each field to ensure that vertical retrace begins at the proper time and to hide the visual effects of retrace. To accomplish this, the transmitter generates a complex pulse such as that shown in Figure 2-18.

This pulse occurs at the end of each *field* and lasts until vertical retrace has ended and trace has started all over again. According to the FCC standards, each vertical blanking interval lasts for a minimum of 13 lines and a maximum of 21 horizontal lines. Since there are two such periods for each *frame*, a total of 26 to 42 horizontal lines are lost during both vertical blanking periods. The only difference between each of the vertical blanking periods is that the separation between the last horizontal blanking pulse and the first vertical blanking interval is a full horizontal line, whereas this separation is only one-half line for the second vertical blanking interval of the frame. This comes about because of the use of odd line interlace scanning as the means of constructing a raster.

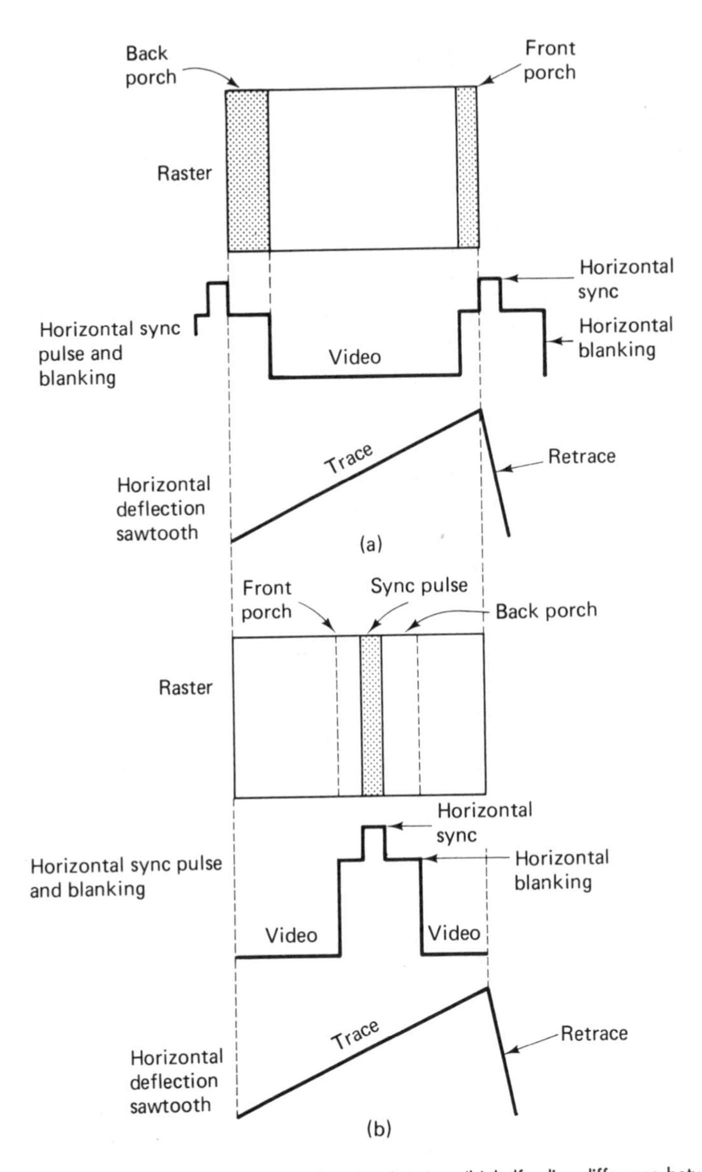

Figure 2-17 (a) Normal raster and video signal timing; (b) half a line difference between horizontal retrace and horizontal sync.

The vertical blanking interval has sufficient amplitude to keep the picture tube cut off and thereby hide vertical retrace. During this time, however, other pulses of greater amplitude are superimposed on the vertical blanking pulse and are also hidden since their amplitude places them in the region of blacker than black. Examination of Figure 2-18 shows that there are three types of pulses riding on the vertical blanking pulse during black & white transmission. These are equalizing pulses, vertical sync pulses, and horizontal sync pulses. During color transmission a color burst consisting of 8 to 10 cycles of 3.58 MHz is inserted following each horizontal sync pulse.

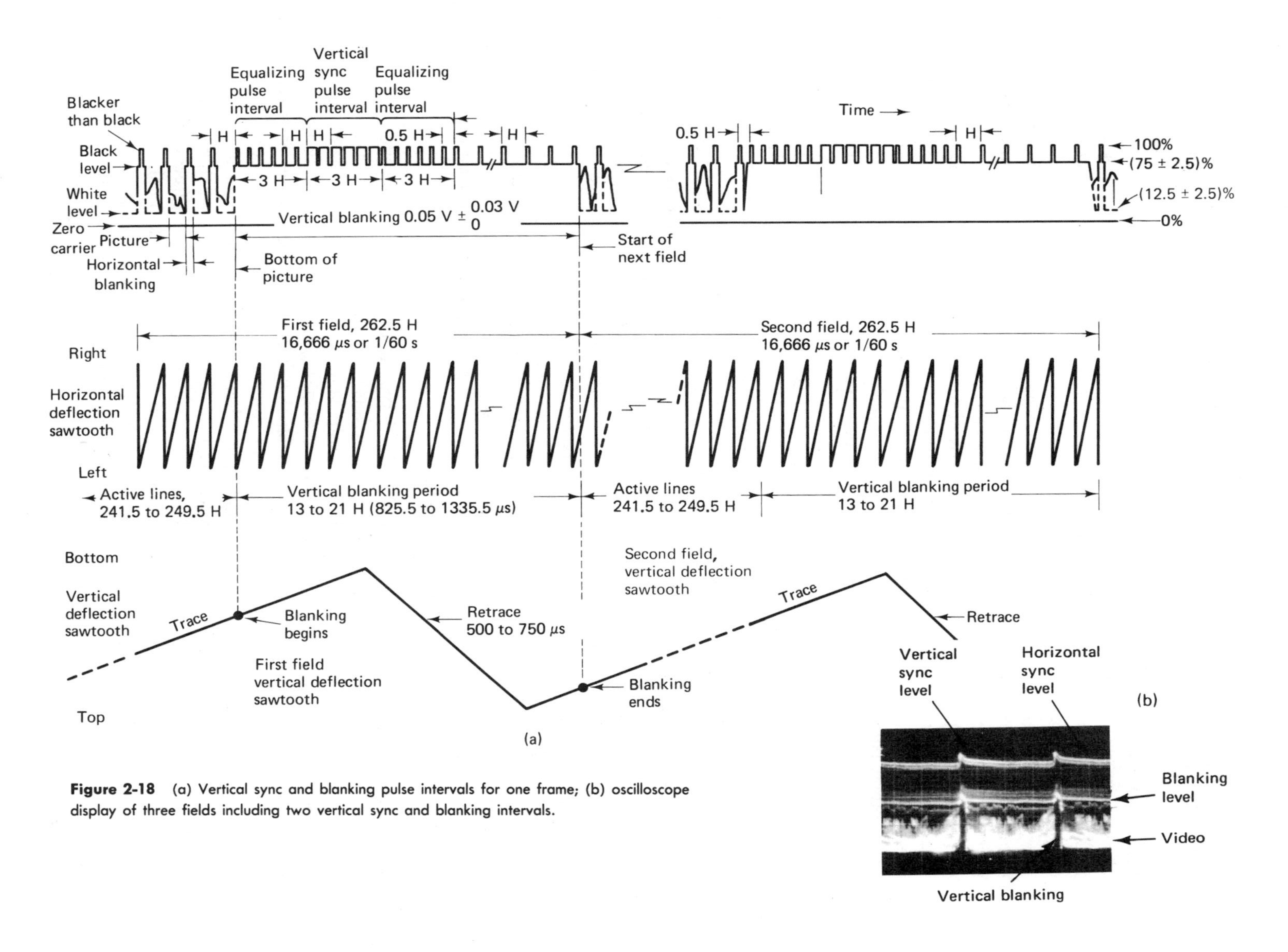

Figure 2-18 (a) Vertical sync and blanking pulse intervals for one frame; (b) oscilloscope display of three fields including two vertical sync and blanking intervals.

Starting at the leading edge of the vertical blanking pulse, six *equalizing pulses* occupying the time interval of 3 horizontal lines precede six vertical sync pulses that also occupy a time interval of 3 horizontal lines. This in turn is followed by six more equalizing pulses that also have a time interval of 3 horizontal lines. The remaining part of the vertical blanking interval has a time duration of 4 to 12 horizontal lines; this interval is devoted to the transmission of horizontal sync pulses whose purpose is to maintain horizontal sync during vertical blanking.

Equalizing pulses have a pulse width of 2.5 μs and occur at twice the horizontal line rate. Therefore, every other equalizing pulse can act like a horizontal sync pulse, since it will occur at the time that a horizontal pulse would have ordinarily appeared. The primary purpose of the equalizing pulse is to prevent loss of interlace scanning. This might occur because of the one-half line difference between odd line and even line vertical blanking intervals. Loss of interlace scanning would mean the the two fields would superimpose one on the other. This is sometimes referred to as *pairing of lines.* The visual effect of line pairing is the reduction of picture detail and an apparent increase in the spacing between horizontal lines.

The vertical sync pulses consist of six pulses of 27 μs duration each. This is almost the same length of time as one-half horizontal line. In between the vertical sync pulses there are six intervals of 4.4μs each, that are called *serrations.* Since two serrations occur each horizontal line, every second serration is used as a horizontal sync pulse.

As indicated in Figure 2-18, vertical blanking begins before the vertical deflection sawtooth has reached its maximum value and the field has reached the bottom of the raster. Approximately 5 or 6 horizontal lines after the beginning of vertical blanking, vertical retrace begins. Examination of Figure 2-18 shows that vertical retrace begins during the transmission of the vertical sync pulses. The precise time that retrace begins cannot be clearly defined and, therefore, the blanking interval must be long enough to hide it. In the average television receiver vertical retrace may take from 500 to 750 μs. Since vertical blanking lasts for 825 to 1335 μs, the vertical retrace period is effectively hidden. Immediately following vertical retrace the electron beam is at the top of the raster. The vertical blanking time duration, however, is long enough to keep the raster blanked out for at least 4 to 13 horizontal lines before active scanning begins.

The hammerhead pattern. The vertical blanking and sync interval of the composite video signal is not ordinarily visible on the TV screen because they occur during vertical retrace which is blanked out. If this portion of the composite video signal is to be seen, the relative timing between the vertical retrace and the vertical blanking interval will have to be shifted to that indicated in Figure 2-19 so that vertical blanking occurs near the center of trace. This can be done by adjusting the vertical hold control slightly. If the hold control is adjusted properly, the blanking bar will appear as a horizontal bar in the center of the TV screen as shown in Figure 2-19(b). Since the vertical blanking level will keep the picture tube beyond cut off and hide the equalizing and vertical sync pulses, the brightness control must be readjusted so that black becomes gray, and blacker than black becomes black. Under these conditions, a pattern that suggests a hammerhead becomes evident.

From the basic definition of video brightness levels it is known that the blackest shades in the pattern must correspond to the vertical sync and equalizing pulses. Since the vertical sync pulses have a time duration of approximately one-half line, the long black

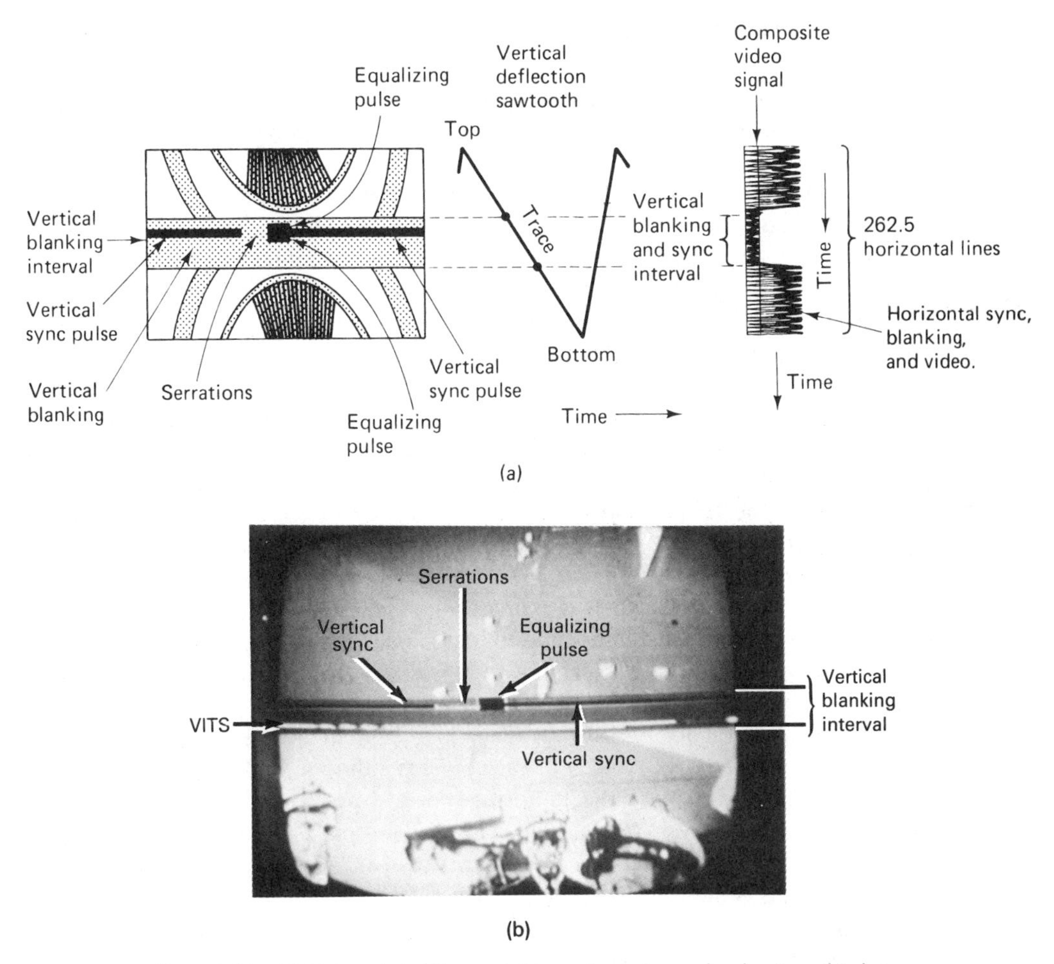

Figure 2-19 (a) Timing relationships needed to produce a hammerhead pattern; (b) photo of a hammerhead pattern and the VITS signal.

bars in the hammerhead pattern must correspond to the vertical sync pulses, and the short duration black bars must be due to the equalizing pulses. The space between the vertical sync pulses is gray; therefore, its level must correspond to blanking level. Since it appears between the two vertical sync pulses, it must correspond to the time interval of the serrations.

The vertical interval test signal (VITS). Notice in Figure 2-19(b) that during the vertical blanking interval below the hammerhead pattern there are a number of white horizontal bars of different lengths. The fact that these bars are of varying lengths means that they represent video pulses of progressively shorter duration. Hence, they can be used at the TV transmitter or receiver as a test of video frequency response or color capability. The fact that some of these test bars are of a constant white level enables broadcasters in network television transmissions to establish a reference white level. In this way all TV cameras in the network can be made to produce the same levels of video signal for white. Vertical interval test signals (VITS) may occur during lines 17 to 21 of each field and may differ from one field to the next.

The signal information impressed upon lines 17 and 18 of the vertical blanking interval changes its form on odd and even fields. Figure 2-20 shows the video content of lines 17 through 21 for both odd and even fields. These waveforms may be easily observed on a dual trace triggered oscilloscope. When this is done, it will be noticed that the horizontal sync pulses of the odd field are one-half line out of step with the horizontal sync pulses of the even field. This is the normal consequence of interlace scanning.

Line 17 of the first field of the VITS consists of a multiburst signal. The active time of this signal is divided into six bursts of video and two levels of luminance. Immediately following vertical blanking the video information rises to a white level (100 units IRE), remaining there for about 6 μs and then falls to a gray level of 50 IRE units. Following these luminance level changes are six equal-amplitude and equal-duration bursts of 0.5, 1.25, 2.0, 3.0, 3.58, and 4.2 MHz. At the transmitter this signal is used to determine the frequency response of the complete camera chain. This signal may also be used at the receiver for similar evaluation of the IF and video amplifier frequency response. The Sencore model VA48 signal generator is able to generate similar multiburst signals that are used for video IF and chroma alignment as well as video amplifier performance evaluation.

Line 17 of the second field is a stair-step luminance signal of eight levels corresponding to a gray scale ranging from white to black. Each step is modulated with a 3.58-MHz signal of varying phase corresponding to the colors yellow, cyan, green, magenta, red, and blue. This pattern is not too useful for TV service work.

Line 18 is identical for both fields. The video content of these lines consists of a staircase luminance signal of six levels. On each level is riding a 3.58-MHz color burst of zero phase. Following this signal the luminance level drops to black for about 4 μs. Superimposed on the black level are two signals, the $2T \sin^2$ pulse and the $12.5T$ modulated $\sin^2$ pulse. These have a half-amplitude time duration of 0.25 and 1.571 μs, respectively. Following these pulses is an 18μs white level "window."

Horizontal line 19 on both fields of the vertical blanking interval is the vertical reference signal (VIR). Line 20 is being developed as a network source identification signal using a 48-bit digital signal.

Line 21 is being used by some PBS (public broadcast service) stations as well as some other TV stations for digital programming of captions to aid the deaf in understanding TV program content. At the receiver a decoder is used to convert the digital signals into the proper form to display captions on the picture tube screen.

A number of TV stations are experimenting with systems of teltext. Lines 10 to 18 of the VITS are sometimes used for this type of transmission.

Summary. The composite video signal generated at the transmitter consists of nine signals combined into a single signal that is used to modulate the TV transmitter. Table 2-1 lists all of these components, their frequency range, and their purpose.

2-9 the modulated video signal

Up to this point in this chapter we have discussed the method of raster formation, the means for generating a video signal, and the nature of the composite video signal. The question of how the composite video signal and its associated sound reach the receiver still

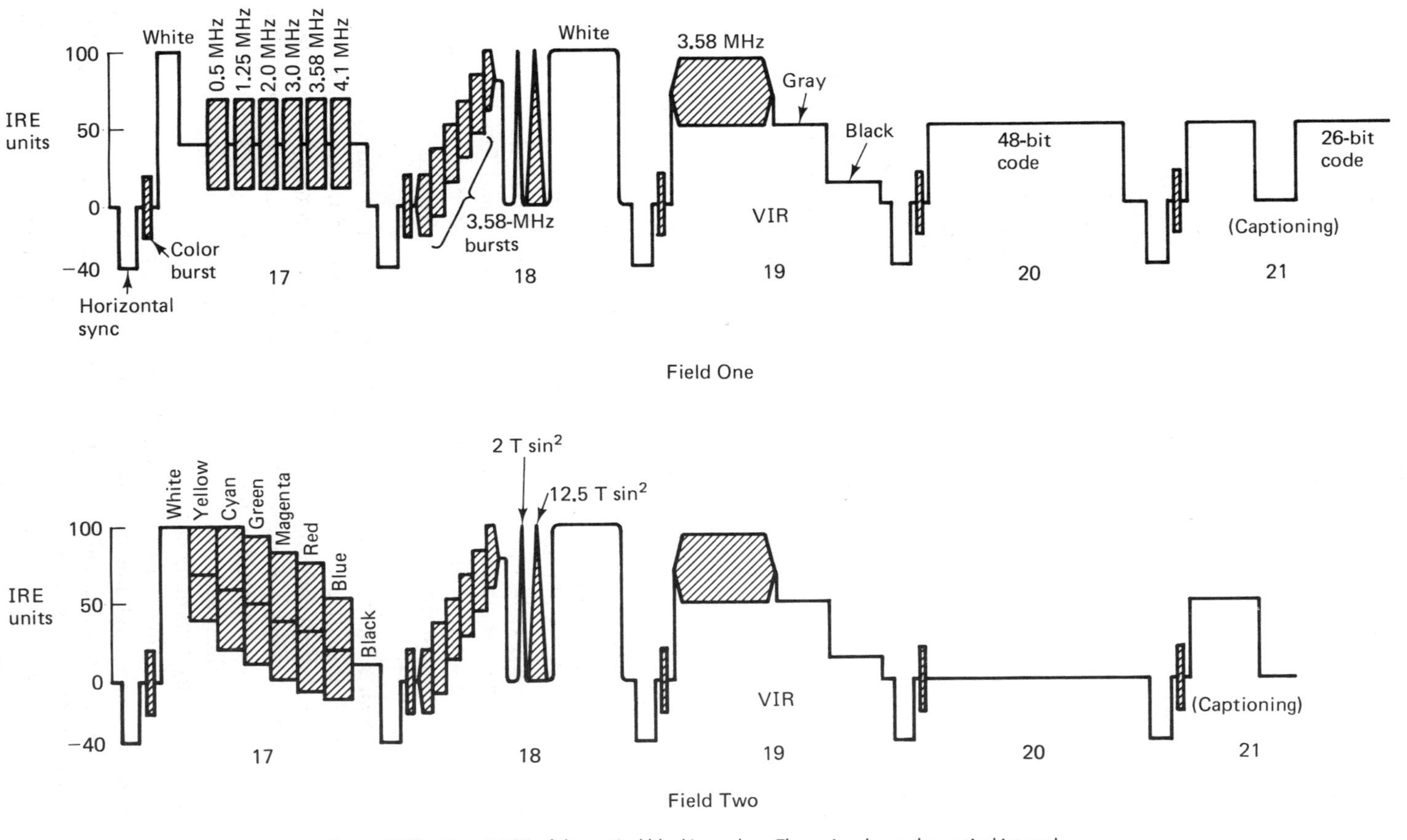

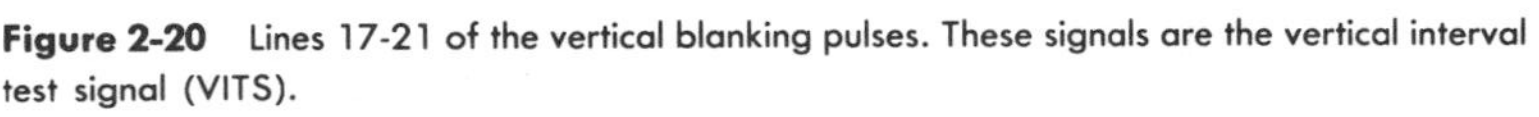

Figure 2-20 Lines 17-21 of the vertical blanking pulses. These signals are the vertical interval test signal (VITS).

TABLE 2-1 **Video Signal Components**

Signal	*Frequency*	*Purpose*
1. Black & white video signal	0–4.2 MHz	Represents the black & white picture information
2. Color signal	3.58 MHz + 0.5 MHz − 1.5 MHz	Represents the color picture information
3. Horizontal blanking pulses	15,750-Hz B&W, 15,734-Hz color	Turns off picture tube during horizontal retrace
4. Vertical blanking pulses	60-Hz B&W, 59.94-Hz color	Turns off picture tube during vertical retrace
5. Horizontal sync pulses	15,750-Hz B&W, 15,734-Hz color	Forces the receiver's horizontal oscillator to be in step with the transmitter's
6. Vertical sync pulses	60-Hz B&W, 59.94-Hz color	Forces the receiver's vertical oscillator to be in step with the transmitter's
7. Equalizing pulses	Twice the horizontal rate	Prevents the loss of interlace scanning
8. Color burst	3.58 MHz	Synchronizes the receiver color circuits with the transmitter's
9. VITS	Various frequencies	Studio test signals

remains. In this section we will discuss the commercial television transmission standards that are used in the United States.

All radio broadcasting techniques require that the information to be transmitted modulate a high-frequency carrier signal. Modulation is necessary because the information to be transmitted usually is not high enough in frequency to be efficiently transmitted or is at the wrong frequency to be transmitted. In the case of commercial television, the signal to be transmitted is the composite video signal and its associated sound. The FCC standards governing the transmission of these signals require that the composite video signal use amplitude modulation and that the associated sound signal use frequency modulation.

Amplitude modulation. Amplitude modulation (AM) is a system of conveying intelligence by means of radio transmission. In this system the amplitude of the intelligence controls the amplitude of a high-frequency signal called a *carrier*, and the frequency of the intelligence determines the rate at which the amplitude of the high-frequency carrier varies. Thus, the chief characteristic of this method of modulation is that the RF signal's amplitude varies with time. Figure 2-21 illustrates the appearance of an amplitude modulated TV signal that has been modulated by two horizontal lines of composite video.

Positive modulation is a process that results in a waveform where the sync pulses are at maximum power level and the white level approaches zero. This system is used in the United States. Negative modulation results in a waveform where the sync pulse level is zero and the white level is at maximum.

If an RF signal that has been modulated by a single modulating signal is analyzed into its component parts, it will be found that the amplitude modulated signal consists of

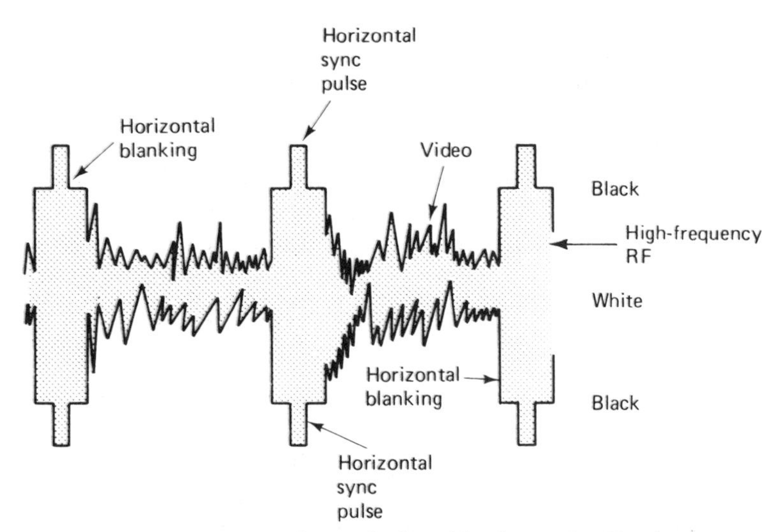

Figure 2-21 RF waveform of an amplitude-modulated composite video signal.

three sine-wave components. These are the carrier component, the upper sideband component, and the lower sideband component. The carrier component is an RF signal whose frequency is the same as the assigned frequency of the transmitter and whose amplitude remains constant during modulation. The amplitude of the carrier component is also larger than the amplitude of the sidebands.

The sidebands are sine-wave RF signals that differ in frequency from the carrier by an amount equal to the modulating signal frequency. For example, if a video signal of 100 kHz were modulating an RF signal operating at an assigned frequency of 61.25 MHz (channel 3), the carrier component would be at 61.25 MHz, the upper sideband would be above the carrier frequency by the modulating frequency or at 61.35 MHz, and the lower sideband would be at 61.15 MHz. If the modulating signal were increased to 4 MHz, the upper sideband would increase to 65.25 MHz and the lower sideband would now become 57.25 MHz. Thus, the process of modulation results in a band of frequencies that extends above and below the carrier frequency by an amount equal to the highest modulating frequency.

Since the highest video frequency is 4.2 MHz, the extreme lower and upper sidebands will each be separated from the carrier by 4.2 MHz. As a result, the total bandwidth occupied by the sidebands will be 8.4 MHz. The FCC has assigned a 72-MHz band of frequencies for television use; therefore, an 8.4-MHz sideband bandwidth will limit the total number of TV channels that can fit into this space to approximately nine channels. Obviously, if the bandwidth occupied by each channel were reduced, the total number of TV channels that could fit into the allotted 72 MHz would increase. The question then is how to reduce the total sideband bandwidth but at the same time not reduce the high frequency content of the video signal?

Vestigial sideband transmission. At first thought it would seem that if the high-frequency component of the video signal were reduced to say a maximum of 3 MHz, this would reduce the overall sideband bandwidth to 6 MHz. Thus, the number of TV channels that could fit into the available 72 MHz would increase from 9 to 12. Unfortunately, this cannot be done. This solution results in a degrading of picture quality since the fine detail in the picture is the result of the high-frequency video.

Another possible solution would be to take advantage of the fact that both the upper and the lower sidebands contain the same video information. Thus, if one of the sidebands were to be completely removed, the other would still be able to convey complete picture information to the receiver. In practice, complete removal of one of the sidebands is not practical or desirable; therefore, a form of partial sideband suppression is used. This form of radio transmission is called *vestigial sideband transmission.* The FCC standards require that the lower sideband be transmitted in vestigial sideband form and that the upper sideband be transmitted in its entirety.

There are at least two important reasons for retaining part of the suppressed lower sideband spectrum. First, the filters that are used at the transmitter cannot remove all of the lower sideband without adversely affecting the upper sideband structure and thereby introducing undesired distortion into the video signal. In this light, the vestigial sideband may be thought of as a guardband. Second, suppression of one of the sidebands in effect reduces the strength of the signal arriving at the receiver by one-half. In the case of strong signal areas, this reduction in signal strength will have no noticeable effect. In weak signal areas, however, the loss of one of the sidebands will result in a poor signal to noise ratio. It should be noted that in weak signal areas high-frequency video information is not needed, or for that matter seen, since it is obscured by electrical noise. Thus, if the lower-frequency sidebands that represent the low-frequency video signal are transmitted together with the upper sideband components, the strength of the low-frequency video signal will be doubled, resulting in an improved signal to noise ratio at the receiver.

In practice, the lower vestigial sideband that is transmitted has a range of 1.25 MHz, resulting in a total channel sideband spectrum of 6 MHz. The amplitude of the lower sideband is constant for 0.75 MHz and then drops toward zero for the remaining 0.5 MHz. See Figure 2-22. The importance of the vestigial sideband should not be underestimated since the vestigial sideband contains the video information for both horizontal and vertical sync and blanking pulses as well as the low-frequency video signal information related to large picture areas.

In addition to the amplitude modulated picture information that the TV transmitter must transmit, the transmitter also transmits the associated sound by means of frequency modulation. To do this, the transmitter requires the use of two separate carriers, one for the picture information and the other for the sound.

Frequency modulation. Frequency modulation (FM) differs from amplitude modulation in that modulating intelligence controls the *frequency* of the carrier instead of its amplitude. The *amplitude* of the audio modulating signal determines the amount of *frequency deviation* of the carrier, and the audio *frequency* determines the *rate of deviation.* During one cycle of audio modulating signal the RF carrier is forced to shift its frequency from rest or starting frequency to a maximum frequency, then back through rest frequency to a minimum frequency, and then finally to rest frequency again. The amount of frequency shift on one side of the rest frequency is termed the *deviation* of the carrier.

The standards governing commercial television allow a maximum deviation of 25 kHz. This represents 100% modulation. This means that for 100% modulation, the sound portion of the TV transmission occupies a total bandwidth of 50 kHz. The FCC standards allow the audio modulating signal to have a frequency range of 50 Hz to 15 kHz.

Stereo sound. In many countries TV sound is transmitted using a stereo format. In Japan, a multiplex system similar to that in use in American commercial FM broadcasting is used. In West Germany a nonmultiplex two-channel stereo system is in use. Stereo TV broadcasting may not lend itself to many TV programs. However, with two or more audio channels available, not only can stereo be broadcasted but the second channel may also be used for the transmission of a second monophonic program. Such programs may be another language dubbed onto the sound track of a foreign film, or a commentary concerning the broadcast material.

In the United States the stereo system adopted was developed by Zenith, and its noise reduction system was developed by dbx Corporation. This system is very similar to that currently used in commercial FM broadcasting. The main differences are: 1) the pilot frequency is at 15,734 Hz, 2) the L-R channel is amplitude modulated, and 3) a system of noise reduction is used. Figure 2-22 shows the frequency spectrum of the stereo system. Notice that in addition to the main L+R and L-R stereo difference channels, an amplitude modulated SAP (Second-Audio-Program) channel and a professional channel are also included. The distribution and number of SAP and professional channels are at the discretion of the TV operator. Professional channels may be used by the station for internal communication, or may be used in a manner similar to FM SCA services.

Television station frequency allocations. Figure 2-23 illustrates the sideband spectrum for a television station assigned to channel 4. Notice that the amplitude of the sidebands are constant throughout a range extending from 0.75 MHz below the picture carrier to 4.0 MHz above it. The amplitude of both the lower and upper sidebands then rapidly drops toward zero. In the case of the lower sideband, however, this drop requires 0.5 MHz, whereas for the upper sideband the drop off takes place in 0.2 MHz.

Located 4.5 MHz above the picture carrier is the frequency modulated sound carrier. This 4.5-MHz frequency separation between picture and sound carriers should be remembered because it plays an important role in TV receiver operation. Also important to color receiver operation is the 920-kHz separation between the color subcarrier and the sound carrier. The channel begins at the edge of the lower sideband and it ends 0.25 MHz above the sound carrier. Thus, the total channel bandwidth is 6 MHz.

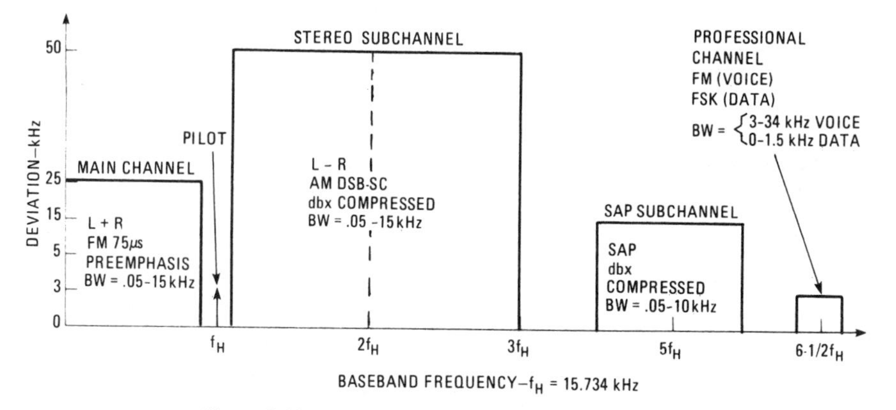

Figure 2-22 The frequency spectrum of the TV stereo system.

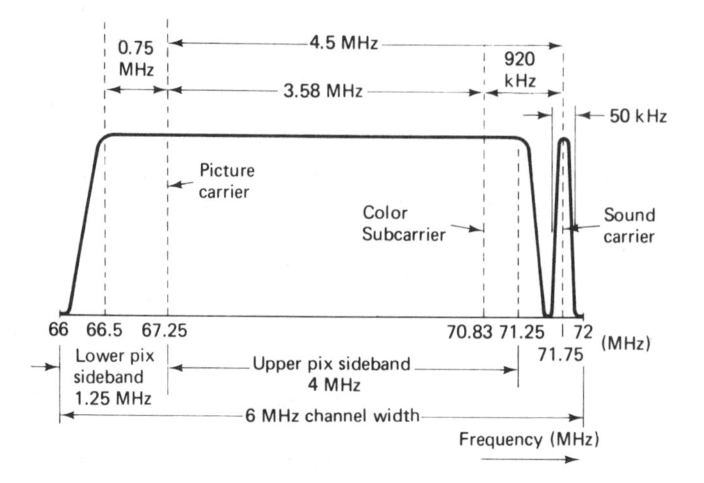

Figure 2-23 Sideband spectrum for TV channel 4.

Television VHF and UHF frequency allocations. In the United States there are a total of 82 channels allocated in VHF (very high frequency) and UHF (ultrahigh frequency) bands for the transmission of commercial television programs. Channels are numbered consecutively from 2 (there is no channel 1) through 83. Each of these channels occupies a bandwidth of 6 MHz. See Figure 2-24.

The first 12 channels (2 through 13) are assigned frequencies in the VHF band. Channel 2 begins at 54 MHz and channel 13 ends at 216 MHz. This band of frequencies is not continuous; there are two major gaps in the band that are assigned to other services. Thus, there is a 4-MHz gap between channel 4 and channel 5 and an 86-MHz gap between channel 6 and channel 7. Within the larger gap the FM band (88 to 108 MHz) and various aeronautical and navigational services may be found.

Channel 14 begins at 470 MHz and represents the beginning of the UHF allocations of TV channels. There are a total of 70 UHF channels, each having a 6-MHz bandwidth. Channel 83 is the highest-frequency channel and ends at 890 MHz. There are no gaps in the UHF band of the TV allocations. However, UHF channels 69 to 83 have not been assigned because these frequencies are used for mobile radio.

Table 2-2 lists all of the TV VHF, UHF, and cable channels and their frequency allocations for the picture carrier, the sound carrier, and the color carrier. Since cable television is in a rapid growth period, their frequency allocations are subject to change.

Figure 2-24 Complete VHF and UHF TV frequency spectrum. Each channel, VHF or UHF, is 6 MHz wide.

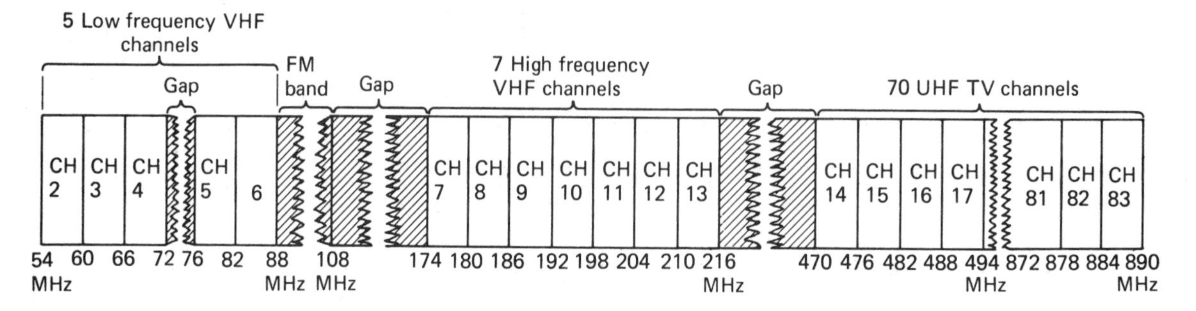

TABLE 2-2 U.S. TV Frequencies Including Midband, Superband and Hyperband

CHANNEL	BAND	CENTER FREQ.	VIDEO CARRIER	COLOR CARRIER	SOUND CARRIER	OSC FREQ.
			VHF LOW BAND			
2	54- 60	57	55.25	58.83	59.75	101
3	60- 66	63	61.25	64.83	65.75	107
4	66- 72	69	67.25	70.83	71.75	113
5	76- 82	79	77.25	80.83	81.75	123
6	82- 88	85	83.25	86.83	87.75	129
			MID BAND			
A-2(00)	108-114	111	109.25	112.83	113.75	155
A-1(01)	114-120	117	115.25	118.83	119.75	161
A- (14)	120-126	123	121.25	124.83	125.75	167
B- (15)	126-132	129	127.25	130.83	131.75	173
C- (16)	132-138	135	133.25	136.83	137.75	179
D- (17)	138-144	141	139.25	142.83	143.75	185
E- (18)	144-150	147	145.25	148.83	149.75	191
F- (19)	150-156	153	151.25	154.83	155.75	197
G- (20)	156-162	159	157.25	160.83	161.75	203
H- (21)	162-168	165	163.25	166.83	167.75	209
I- (22)	168-174	171	169.25	172.83	173.75	215
			VHF HIGH BAND			
7	174-180	177	175.25	178.83	179.75	221
8	180-186	183	181.25	184.83	185.75	227
9	186-192	189	187.25	190.83	191.75	233
10	192-198	195	193.25	196.83	197.75	239
11	198-204	201	199.25	202.83	203.75	245
12	204-210	207	205.25	208.83	209.75	251
13	210-216	213	211.25	214.83	215.75	257
			SUPER BAND			
J- (23)	216-222	219	217.25	220.83	221.75	263
K- (24)	222-228	225	223.25	226.83	227.75	269
L- (25)	228-234	231	229.25	232.83	233.75	275
M- (26)	234-240	237	235.25	238.83	239.75	281
N- (27)	240-246	243	241.25	244.83	245.75	287
O- (28)	246-252	249	247.25	250.83	251.75	293
P- (29)	252-258	255	253.25	256.83	257.75	299
Q- (30)	258-264	261	259.25	262.83	263.75	305
R- (31)	264-270	267	265.25	268.83	269.75	311
S- (32)	270-276	273	271.25	274.83	275.75	317
T- (33)	276-282	279	277.25	280.83	281.75	323
U- (34)	282-288	285	283.25	286.83	287.75	329
V- (35)	288-294	291	289.25	292.83	293.75	335
W- (36)	294-300	297	295.25	298.83	299.75	341
			HYPER BAND			
AA-(37)	300-306	303	301.25	304.83	305.75	347
BB-(38)	306-312	309	307.25	310.83	311.75	353
CC-(39)	312-318	315	313.25	316.83	317.75	359
DD-(40)	318-324	321	319.25	322.83	323.75	365
EE-(41)	324-330	327	325.25	328.83	329.75	371
FF-(42)	330-336	333	331.25	334.83	335.75	377
GG-(43)	336-342	339	337.25	340.83	341.75	383
HH-(44)	342-348	345	343.25	346.83	347.75	389
II-(45)	348-354	351	349.25	352.83	353.75	385
JJ-(46)	354-360	357	355.25	358.83	359.75	401
KK-(47)	360-366	363	361.25	364.83	365.75	407
LL-(48)	366-372	369	367.25	370.83	371.75	413
MM-(49)	372-378	375	373.25	376.83	377.75	419
NN-(50)	378-384	381	379.25	382.83	383.75	425
OO-(51)	384-390	387	385.25	388.83	389.75	431
PP-(52)	390-396	393	391.25	394.83	395.75	437
QQ-(53)	396-402	399	397.25	400.83	401.75	443
RR-(54)	402-408	405	403.25	406.83	407.75	449
SS-(55)	408-414	411	409.25	412.83	413.75	455
TT-(56)	414-420	417	415.25	418.83	419.75	461
UU-(57)	420-426	423	421.25	424.83	425.75	467
VV-(58)	426-432	429	427.25	430.83	431.75	473
WW-(59)	432-438	435	433.25	436.83	437.75	479
XX-(60)	438-444	441	439.25	442.83	443.75	485
YY-(61)	444-450	447	445.25	448.83	449.75	491
ZZ-(62)	450-456	453	451.25	454.83	455.75	497
AAA-(63)	456-462	459	457.25	460.83	461.75	503
BBB-(64)	462-468	465	463.25	466.83	467.75	509

TABLE 2-2 *(continued)*

CHANNEL	BAND	CENTER FREQ.	VIDEO CARRIER	COLOR CARRIER	SOUND CARRIER	OSC FREQ.
			UHF BAND			
14	470-476	473	471.25	474.83	475.75	517
15	476-482	479	477.25	480.83	481.75	523
16	482-488	485	483.25	486.83	487.75	529
17	488-494	491	489.25	492.83	493.75	535
18	494-500	497	495.25	498.83	499.75	541
19	500-506	503	501.25	504.83	505.75	547
20	506-512	509	507.25	510.83	511.75	553
21	512-518	515	513.25	516.83	517.75	559
22	518-524	521	519.25	522.83	523.75	565
23	524-530	527	525.25	528.83	529.75	571
24	530-536	533	531.25	534.83	535.75	577
25	536-542	539	537.25	540.83	541.75	583
26	542-548	545	543.25	546.83	547.75	589
27	548-554	551	549.25	552.83	553.75	595
28	554-560	557	555.25	558.83	559.75	601
29	560-566	563	561.25	564.83	565.75	607
30	566-572	569	567.25	570.83	571.75	613
31	572-578	575	573.25	576.83	577.75	619
32	578-584	581	579.25	582.83	583.75	625
33	584-590	587	585.25	588.83	589.75	631
34	590-596	593	591.25	594.83	595.75	637
35	596-602	599	597.25	600.83	601.75	643
36	602-608	605	603.25	606.83	607.75	649
37	608-614	611	609.25	612.83	613.75	655
38	614-620	617	615.25	618.83	619.75	661
39	620-626	623	621.25	624.83	625.75	667
40	626-632	629	627.25	630.83	631.75	673
41	632-638	635	633.25	636.83	637.75	679
42	638-644	641	639.25	642.83	643.75	685
43	644-650	647	645.25	648.83	649.75	691
44	650-656	653	651.25	654.83	655.75	697
45	656-662	659	657.25	660.83	661.75	703
46	662-668	665	663.25	666.83	667.75	709
47	668-674	671	669.25	672.83	673.75	715
48	674-680	677	675.25	678.83	679.75	721
49	680-686	683	681.25	684.83	685.75	727
50	686-692	689	687.25	690.83	691.75	733
51	692-698	695	693.25	696.83	697.75	739
52	698-704	701	699.25	702.83	703.75	745
53	704-710	707	705.25	708.83	709.75	751
54	710-716	713	711.25	714.83	715.75	757
55	716-722	719	717.25	720.83	721.75	763
56	722-728	725	723.25	726.83	727.75	769
57	728-734	731	729.25	732.83	733.75	775
58	734-740	737	735.25	738.83	739.75	781
59	740-746	743	741.25	744.83	745.75	787
60	746-752	749	747.25	750.83	751.75	793
61	752-758	755	753.25	756.83	757.75	799
62	758-764	761	759.25	762.83	763.75	805
63	764-770	767	765.25	768.83	769.75	811
64	770-776	773	771.25	774.83	775.75	817
65	776-782	779	777.25	780.83	781.75	823
66	782-788	785	783.25	786.83	787.75	829
67	788-794	791	789.25	792.83	793.75	835
68	794-800	797	795.25	798.83	799.75	841
69	800-806	803	801.25	804.83	805.75	847
70	806-812	809	807.25	810.83	811.75	853
71	812-818	815	813.25	816.83	817.75	859
72	818-824	821	819.25	822.83	823.75	865
73	824-830	827	825.25	828.83	829.75	871
74	830-836	833	831.25	834.83	835.75	877
75	836-842	839	837.25	840.83	841.75	883
76	842-848	845	843.25	846.83	847.75	889
77	848-854	851	849.25	852.83	853.75	895
78	854-860	857	855.25	858.83	859.75	901
79	860-866	863	861.25	864.83	865.75	907
80	866-872	869	867.25	870.83	871.75	913
81	872-878	875	873.25	876.83	877.75	919
82	878-884	881	879.25	882.83	883.75	925
83	884-890	887	885.25	888.83	889.75	931

2-10 block diagram of a black & white TV receiver

In earlier sections of this chapter we discussed raster formation, the video signal, and the modulated RF signal that leaves the transmitter. In this section we discuss the television receiver and the functions of the stages that make up the receiver.

The TV receiver consists of five basic sections. Each section performs a definite function in producing the televised picture and its associated sound. Thus, the television signal is picked up by the antenna and fed into a *signal processing* section of the receiver that is similar to an ordinary AM receiver and is able to change the high-frequency RF signal into its original composite video and its associated sound. The composite video signal is fed directly to the cathode of the picture tube.

The sound signal, however, requires additional processing provided by the *sound circuits* that select, detect, and amplify the frequency modulated sound signal.

A portion of the composite video is also fed into the *synchronizing circuits* that strip away the blanking and video signals leaving at its output only vertical and horizontal sync pulses. These pulses are then fed to the *deflection circuits.*

The deflection circuits operate with or without an input signal and provide to the deflection yoke both the horizontal and vertical deflection sawtooth currents that generate the raster. The sync pulses that are fed into the deflection circuits control the frequency of these sawtooth generators so that they remain in step with the sawtooth generators at the transmitter.

The *low- and high-voltage power supplies* form another section. The function of the low-voltage power supply is to energize not only the picture tube but also the other circuits in the receiver. The high-voltage power supply is a byproduct of horizontal deflection and is used solely to provide the accelerating potentials necessary for picture tube operation.

Of course, the actual black & white receiver is much more complex than the simplified block diagram just discussed and it requires approximately 20 different stages for its operation. In modern television receivers these twenty or more stages are combined into groups of circuits that are formed into several integrated circuits. Manufacturer's schematic diagrams usually draw the integrated circuit showing their internal structure in block diagram form.

A complete block diagram of a black & white television receiver is shown in Figure 2-25 on page 68. A brief description of each stage function, the effect of a defective stage on the raster, the picture, the sound, and the function of all controls are listed in Table 2-3 on page 69. The defective stage is assumed to be inoperative and is independent of other stages. In some cases, such as AGC, two possible faults can occur; both are listed. Whether or not the receiver uses vacuum tubes, transistors, or integrated circuits does not affect the block diagram.

2-11 systems of television transmission

Television signals may be transmitted from their point of origin to the television receiver by four methods; (1) broadcast TV, (2) cable TV, (3) subscriber TV, and (4) satellite TV.

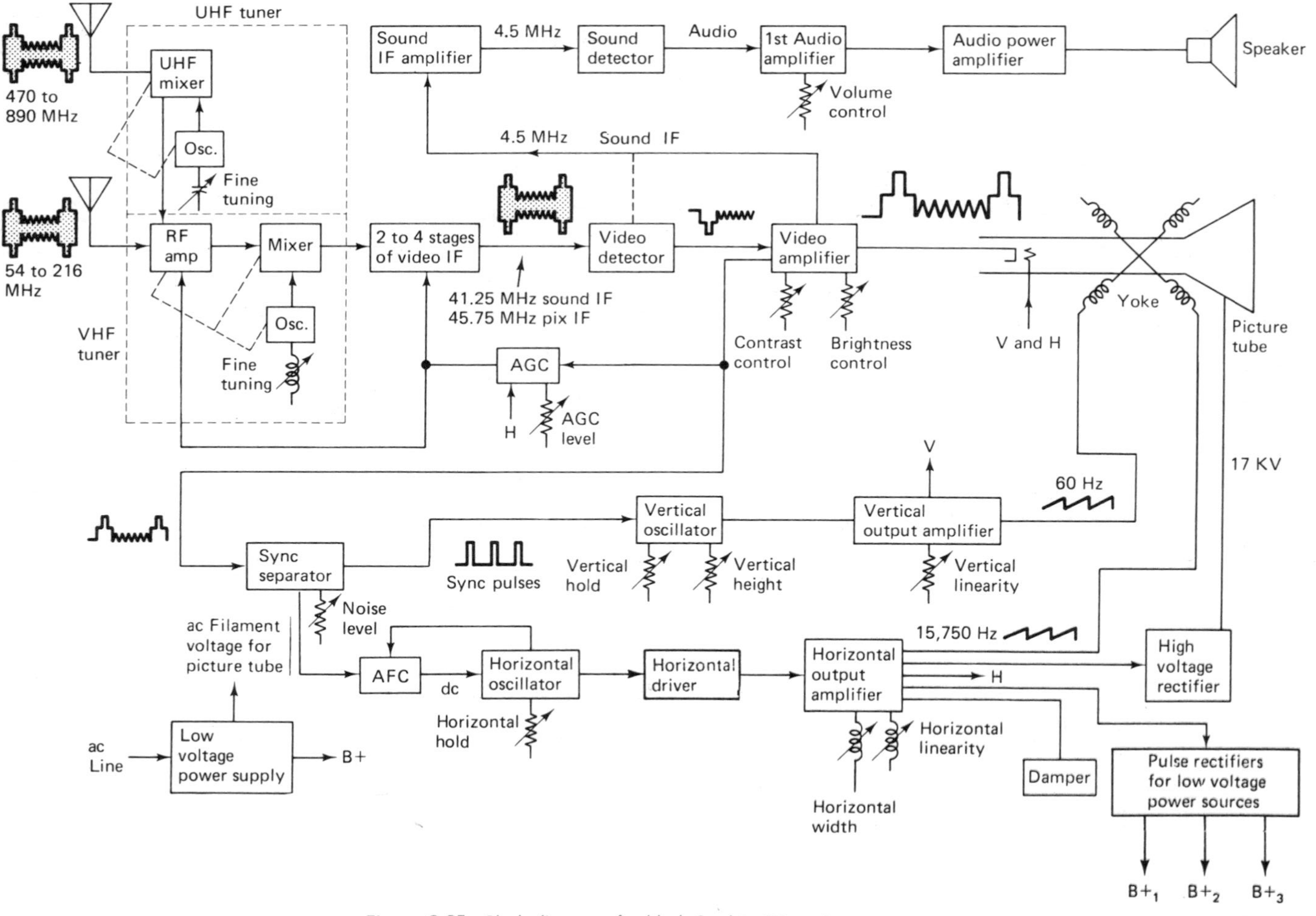

Figure 2-25 Block diagram of a black & white TV receiver.

TABLE 2-3

Stage	*Function*	*Raster*	*Picture*	*Sound*
RF amplifier	Minimizes local oscillator radiation, improves the signal-to-noise ratio of the receiver, and improves selectivity.	O.K.	Snow, weak, or no picture	Weak or no sound
Mixer/oscillator	Acts as a frequency changer that converts the VHF signal into a lower-frequency IF.	O.K.	No picture	No sound
Fine-tuning control	Adjusts local oscillator frequency, thereby controlling the video IF frequency.			
UHF mixer/oscillator	A frequency changer that converts the UHF RF signals into the IF.	O.K.	No picture on UHF; VHF O.K.	No sound on UHF; VHF O.K.
Fine-tuning control	Adjusts the local oscillator frequency, thereby controlling the IF output frequency of the mixer.			
Video IF	Two to four stages of amplification that provide most of the amplification and selectivity of the receiver.	O.K.	No picture	No sound
Video detector	A frequency changer that converts the video IF into a lower-frequency sound IF (4.5 MHz) and the composite video signal (0 to 4.2 MHz).	O.K.	No picture	No sound
Automatic gain control (AGC)	Tends to maintain a constant-amplitude signal at the *output* of the detector under conditions of *input* signal variation.	O.K.	Excessive AGC—weak or no picture	Weak or no sound
AGC level control	Provides an adjustment of AGC circuit action that compensates for local signal strength.		Insufficient AGC—overloaded or negative picture	Buzz in sound
Video amplifier	Provides wide-band amplification for the composite video signal so that the peak to peak swing of the video signal can drive the picture tube between beam current cutoff (black) and maximum beam current (white).	O.K.	No picture	No sound
Contrast control	Controls the gain of the video amplifier stage and thereby controls the peak-to-peak video signal output of the video amplifier.			

TABLE 2-3 *(contiuned)*

Stage	*Function*	*Raster*	*Picture*	*Sound*
Brightness control	Determines the dc bias between grid and cathode of the picture tube and thereby controls the average light output of the picture tube.			
Sound IF	The input to this amplifier may be taken from either the video detector or the video amplifier. This stage provides amplification and is a filter that separates the 4.5-MHz sound IF from the composite video signal.	O.K	O.K.	No sound
Sound detector	First converts the frequency modulated sound information into amplitude modulation, and then it detects it in the usual way, converting the sound IF into audio.	O.K.	O.K.	No sound
First audio amplifier	Provides sufficient amplification of the audio output of the sound detector so that the audio power amplifier will receive adequate drive.	O.K.	O.K.	No sound
Volume control	Controls the peak-to-peak amplitude of audio signal that feeds the audio amplifier.			
Audio power amplifier	Amplifies the audio signal to a level sufficient to drive the loudspeaker.	O.K.	O.K.	No sound
Sync separator	Separates the horizontal and vertical sync pulses from the composite video and feeds the sync pulses to the deflection circuits.	O.K.	Loss of both vertical and horizontal sync	O.K.
Vertical oscillator	Generates a 60-Hz sawtooth voltage.	No vertical deflection	Video signal reaches picture tube	O.K.
Vertical hold control	Controls the frequency of the vertical oscillator.			
Vertical height control	Determines the amplitude of the vertical deflection sawtooth.			
Vertical amplifier	Amplifies the vertical deflection sawtooth to a level sufficient to drive the deflection yoke.	No vertical deflection	Video signal reaches picture tube	O.K.
Vertical linearity control	Adjusts the operating point of the vertical output stage to			

TABLE 2-3 *(continued)*

Stage	*Function*	*Raster*	*Picture*	*Sound*
	compensate for a nonlinearity in the deflection sawtooth.			
Automatic frequency control (AFC)	Compares the horizontal sync pulse to the output of the horizontal oscillator and develops a dc error voltage that controls the horizontal oscillator frequency.	O.K.	Loss of horizontal sync	O.K.
Horizontal oscillator	Generates a horizontal deflection sawtooth.	No raster	No picture	No sound
Horizontal hold	Controls the frequency of the horizontal oscillator.			
Horizontal driver	Buffer amplifier that provides isolation, amplification, and power matching between the horizontal oscillator and the horizontal output amplifier.	No raster	No picture	No sound
Horizontal deflection amplifier	This stage has the greatest number of separate function of any stage in the receiver. Following is a list of these functions: (1) drives the horizontal deflection yoke; (2) provides the high-voltage pulses that are used as part of the high-voltage power supply; (3) develops pulses that when rectified supplement the low-voltage power supply; (4) develops pulses used in AGC, AFC, and in horizontal blanking circuits.	No raster	No picture	O.K.
Horizontal linearity control	Adjusts the linearity of the horizontal deflection sawtooth and thereby the horizontal linearity of the picture.			
Horizontal width control	Adjusts the amplitude of the horizontal deflection sawtooth, which in turn determines picture width.			
Damper	Works in conjunction with the horizontal output amplifier to provide horizontal deflection on the left side of the screen.	No raster	No picture	O.K.
Pulse rectifiers	Converts horizontal pulses generated by the horizontal flyback transformer into various low-voltage power sources.	Defects depend upon which circuits the pulse rectifier feeds.		

TABLE 2-3 *(continued)*

Stage	*Function*	*Raster*	*Picture*	*Sound*
High-voltage rectifier	Converts the high-voltage pulses generated by the horizontal output stage into high-voltage dc.	No raster	No picture	O.K.
Low-voltage power supply	Supplies the ac filament for the CRT and dc power supply requirements for some of the stages in the receiver.	No raster	No picture	No sound

Broadcast TV. In the case of broadcast or "free" TV, as it is sometimes called, the program material is used to modulate a high-power transmitter which then radiates this signal over a large broadcast area. Receivers pick up this program material and process it, producing color or black & white pictures with their associated sound. In this system the receiver owner pays no direct fee to the transmitter operators for the use of the transmitted materials. An ordinary television receiver is used to receive these signals.

Cable TV. In recent years master antenna (MATV) and community antenna television (CATV) systems have had widespread popularity. The purpose of a MATV system is to deliver clear, good-quality television pictures to every TV receiver connected to the system. In this application one or more antennas are used to pick up the local VHF and UHF signals. The signals are then amplified and distributed by 75-Ω coaxial cable to a large number of ordinary television receivers. Typical applications for MATV systems are motels, hotels, schools, apartment buildings, and so on.

CATV systems have certain similarities and differences when compared to the MATV system. The CATV system is a cable system that distributes good-quality television pictures to a very large number of receivers throughout an entire community. This system is under FCC regulation. In general, CATV systems distribute increased television programming to subscribers who pay a fee for this service. This programming may have as many as 31 active VHF and UHF channels and requires the use of a special receiver converter.

MATV. The block diagram of a basic MATV system is shown in Figure 2-26. The system may be divided into two basic parts: the head end and the distribution system. The head end refers to the antenna installation. This includes low-noise preamplifiers necessary to boost weak signals to a usable level, matching devices to avoid standing waves, traps to eliminate undesirable interference, attenuators to equalize channel signal levels, and antenna mixers for combining the outputs of all of the antennas and other signal sources used to feed the system.

The antenna. The antenna must be selected and installed with great care since reception cannot be better than the quality of the signal delivered by the antenna. The type and number of antennas used depend upon the distance and direction of each channel that is to be part of the system. In areas where all of the signals are coming from one direction,

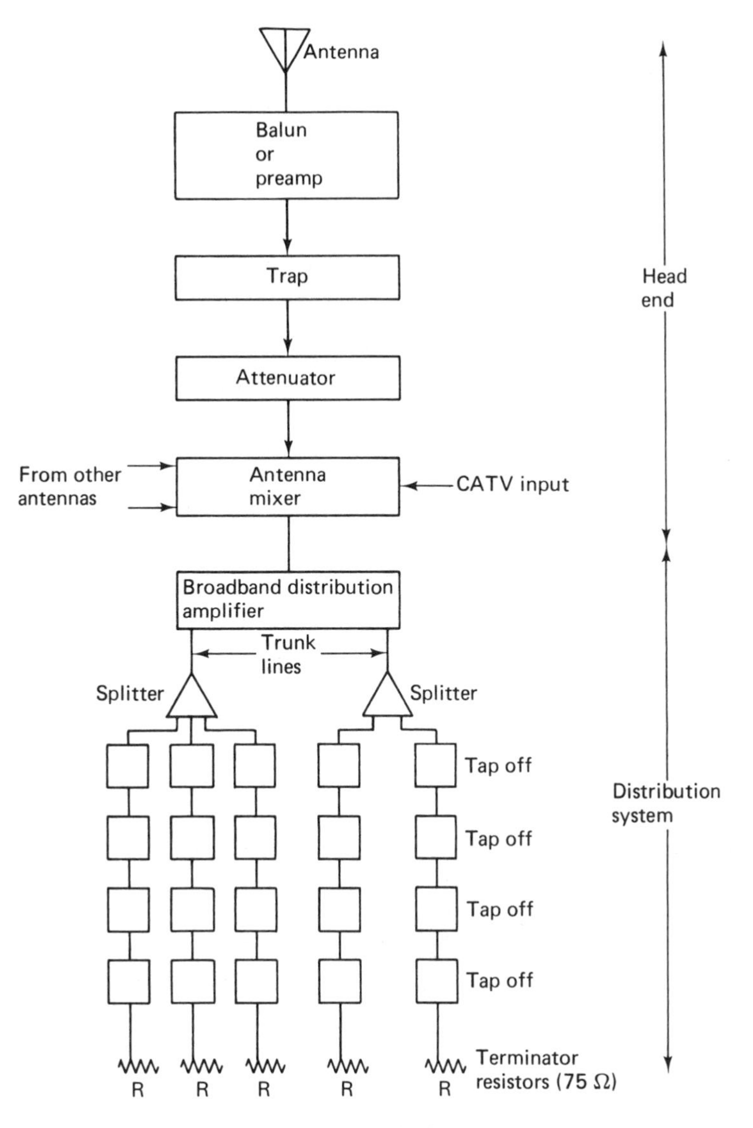

Figure 2-26 MATV distribution system.

a single broadband antenna (log-periodic) may be used. Where stations lie in different directions, single-channel antenna installations (yagi) may be required.

The balun. Most MATV equipment is designed to have 75-Ω impedance. This is to allow a convenient match between the coaxial transmission line and the components making up the system. Since most TV antennas have a 300-Ω impedance, a matching transformer called a *balun* must be used to convert the 300-Ω antenna impedance into 75-Ω. The balun is mounted as close to the antenna as possible. Outdoor baluns must be weatherproofed, but indoor baluns do not.

The preamplifier. In weak signal areas the signal must be amplified sufficiently to overcome the random noise generated in the MATV system and the TV receiver. To be effective, this amplification must be done right at the antenna, and it must be done by a preamplifier that generates less noise than the circuits it is feeding. The usual method for measuring noise performance is by the noise figure rating of the amplifier. This ranging is in db (decibels) and measures how much the internally generated noise of the amplifier adds to the signal during the amplification process. In general, the smaller the noise figure of an amplifier, the greater the sensitivity of the system. An excellent picture is defined as having a signal-to-noise ratio of 44 dB. The noise figure of VHF preamplifier is typically 3.6 dB and that of a UHF preamplifier is 6.0 dB.

Filters and traps. To provide interference-free reception, filters and traps are introduced at the output of the preamplifier. These traps may be either fixed or variable. Fixed traps are usually designed as band-stop filters that attenuate an undesired range of signals such as the FM band. Variable traps are adjustable single-frequency filters that are used to attenuate a signal that is causing interference. To be effective, these filters must have good skirt selectivity so that undesired signals close to the desired signal can be sharply attenuated. This type of filter requires high Q tuned circuits.

Attenuators. When the signal strength of the received channels has considerable variation, attenuators are used to equalize the channels. If the signals are not equalized, the stronger signals will override and interfere with the weaker ones. Attenuators reduce the signals by specific attenuation ratios and may be either fixed or variable.

Antenna mixers. In practice, the signal sources supplying the MATV system may come from a variety of sources. The simplest system is one in which a single broadband antenna is the prime signal source. A more complex system is one in which each channel has its own antenna and their outputs are then combined. Another possibility is to convert UHF channels into unused VHF channels and then to combine these signals without having them interfere with one another. In some cases, CATV signals containing as many as 31 VHF and UHF channels may be combined with locally generated UHF signals, but this requires that some of the original CATV channels must be filtered out before the desired channels can be added.

In those arrangements in which more than one antenna is used, the signals are combined in antenna mixing units before broadband amplification. In some cases, the outputs of the antenna mixing units may feed individual channel amplifiers before broadband amplification. Antenna mixers are bandpass filters that cover the low TV band (channels 2 through 6) or the high TV band (channels 6 through 13). Depending upon interference problems, filters and traps may be used with the antenna mixing units.

In many commercial MATV systems converters are used to change a UHF channel into a VHF channel that is not being used. Each UHF channel requires a separate UHF/VHF converter. In addition, UHF converters act as amplifier mixers that amplify the signal being converted and combine VHF signals from another converter or VHF antenna into a single output.

Distribution. The output of the antenna mixers feeds a distribution broadband amplifier. The function of this amplifier is to increase the signal level to a level sufficient to overcome the distribution system's losses while providing an acceptable signal to every set in the system. Distribution amplifiers are rated according to their gain (+46 dBMV is typical) and their output capability (typical rating 50 dBMV (0.4 V). If too large an input signal is applied to the amplifier or if the gain of the amplifier is excessive, the amplifier will overload and produce interference. Many broadband distribution amplifiers amplify both VHF and UHF signals simultaneously. The combined output feeds the main trunk line which in turn feeds all of the receivers in the system. The trunk line is usually fed into a splitter that divides the signal into two, three, or four separate lines. Splitters are impedance-matching resistive inductive devices that provide intertrunk isolation.

Taps. The output of the splitters feed distribution lines (75-Ω coaxial cable) that brings the television signal to a point of delivery. This point is called a *tapoff.* The tapoff provides isolation between sets on the same line, preventing interference. Some tapoffs provide variable isolation to balance the signal levels delivered to the TV receiver. Three types of tapoffs are commonly used: wall tap, line drop tap, and pressure tap.

Wall taps look like and are used in the same way as an ac outlet. They are mounted in a standard electrical outlet box in the wall. Wall taps may be obtained with 300-Ω output, 75-Ω output, and a dual output. The preferred method is to use a 75-Ω type with a matching transformer. The matching transformer is usually mounted at the antenna terminals of the receiver and will have a VHF output and a UHF output.

Line drop taps are used in attics and other such places. Line drop taps can provide as many as four outputs. Each output either runs directly to a TV receiver (via a matching transformer) or it can feed a wall tap. Most applications of line drop taps are where the main cable runs down a hallway and feeder lines are dropped down into each room.

Pressure taps are used outdoors when the distribution lines are strung between poles, under gables, or other external applications.

Terminators. Improperly terminated or unmatched 75-Ω transmission lines will develop standing waves. These waves are caused by reflections that result in ghosts appearing in an otherwise good TV picture. To prevent this from happening, the end of each 75-Ω distribution cable is terminated with a 75-Ω resistor called a *terminator.*

CATV. Community antenna television (CATV) systems may be used as one of the inputs to a MATV system or it may be a single input signal source for an individual TV receiver. Advances in technology, marketing considerations, and new FCC rules have prompted operators to increase the total number of channels made available to subscribers. Ordinary VHF/UHF television receivers are capable of receiving 12 standard TV channels. CATV systems may deliver a wide-band signal that has 31 or more channels ranging in frequency from 5 MHz to 408 MHz. To make UHF channels available, the CATV operators *down-convert* these channels to an available VHF channel in the wide-band signal spectrum. To receive all of the channels on the receiver, the CATV subscriber must have a special converter attached to the receiver, or a "cable ready" receiver that has the converter built in.

CATV frequency spectrum. Figure 2-27 shows the frequency allocations that have been used in the CATV system. Due to the rapid expansion of CATV, these frequency allocations are subject to change. The CATV channel allocations are divided into five basic ranges; the *low band* consisting of five VHF channels, 2 through 6 (54 to 88 MHz), and the FM

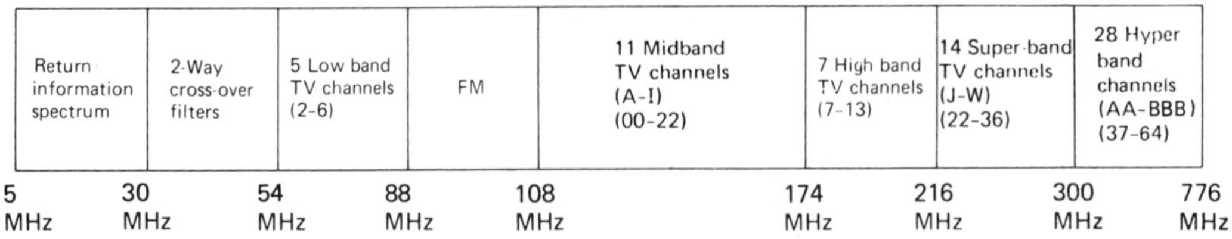

Return information spectrum	2-Way cross-over filters	5 Low band TV channels (2-6)	FM	11 Midband TV channels (A-I) (00-22)	7 High band TV channels (7-13)	14 Super-band TV channels (J-W) (22-36)	28 Hyper band channels (AA-BBB) (37-64)

5 MHz 30 MHz 54 MHz 88 MHz 108 MHz 174 MHz 216 MHz 300 MHz 776 MHz

Figure 2-27 Frequency system spectrum for CATV systems.

band (88 to 108 MHz), eleven *midband* channels designated A-2 through I (108 to 174 MHz), the seven *high-band* VHF channels 7 through 13 (174 to 216 MHz), fourteen *superband* channels designated J through W (216 to 300 MHz), and twenty-eight additional *hyperband* channels designated AA through BBB (300 to 776 MHz) are also available. In practice, many CATV systems limit the superband and hyperband to only five channels (J to N), making a total of 31 channels. Some CATV systems include *subband* allocations that range from 5.0 to 54 MHz. This portion of the CATV spectrum may be used for ordinary channel allocations as well as one or more nonvoice return communications channels. These channels allow the subscriber to send information (voting, purchase requests, burglar alarms, computer terminals, and so on) back to the CATV source.

The CATV converter. To convert the midband (A to I), the superband (J to W), and the hyperband (AA-BBB) channels to standard TV frequencies that may be processed by an ordinary TV receiver, the CATV subscriber must have a frequency converter unit. This unit may either be external to the receiver or it may be built in (cable ready).

A block diagram of a CATV converter is shown in Figure 2-28. The wide-band CATV signal is fed into the converter by a 75-Ω coaxial cable. This signal then passes through an FM trap that attenuates those signals that might cause interference problems. The CATV wide-band signal is then fed into an *up converter* that increases the frequency of the input signals into the UHF band. For example, one of the inputs to the up converter is the CATV band that ranges from 54 to 408 MHz. Mixing this with an oscillator signal that ranges from 530.5 to 182.50 MHz will produce an up converter UHF output *sum* frequency that is 585.75 MHz for the picture and 590.25 MHz for the sound. One of the advantages of this arrangement is that the VHF picture and sound carrier signals are not frequency inverted (picture carrier IF frequency is normally lower than sound IF carrier frequency), as is the case in an ordinary tuner mixer stage. However, the most important reason for increasing the frequency of the CATV signals is to permit ease of tuning. Notice that the CATV wide-band signal has a low-frequency to high-frequency ratio of more than 4 to 1. If channel selection is done in the VHF band, channel selection becomes very difficult. If channel frequencies are increased to the UHF band, the converter oscillator need only vary through an easily obtained frequency range of less than 2 to 1. The oscillator frequency is adjusted by varactor tuning of the up-converter oscillator.

After up converting, the signals are amplified and then fed into a *down converter.* It is the function of this stage to lower the UHF signal to that of a VHF channel, typically channels 2 or 3. In the block diagram the frequency of the fixed oscillator (524.5 MHz) is placed below the UHF signals so that the output of the mixer is not frequency inverted. As a result of the above, all CATV channels appear at the antenna terminals of the receiver as VHF channel 3.

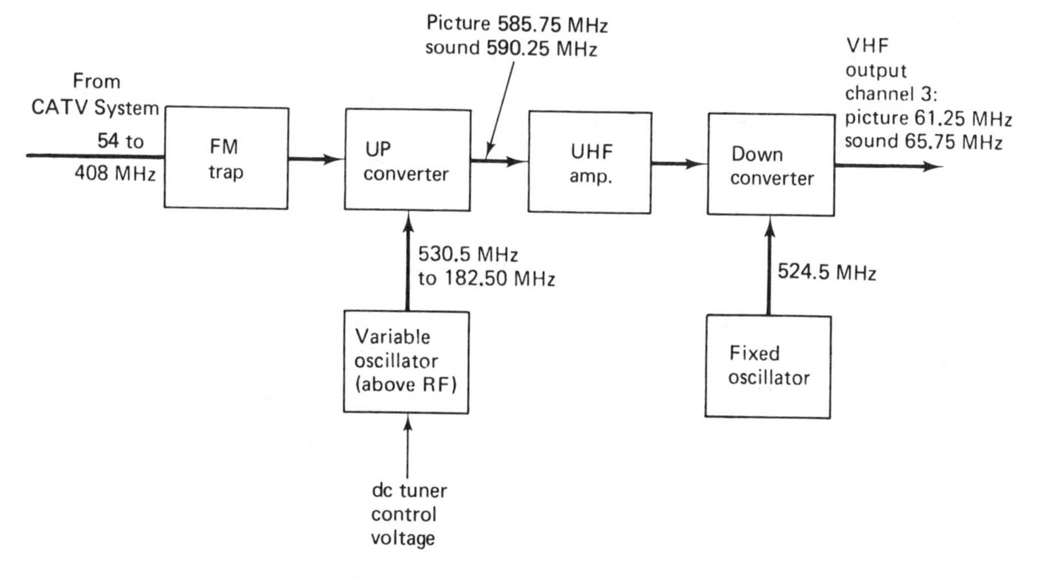

Figure 2-28 CATV wideband converter.

Other CATV converters make use of a frequency synthesis type of tuner similar to those discussed in Section 4-6.

Subscriber TV. Subscriber TV or pay TV uses a standard VHF or UHF channel to transmit an encoded or scrambled picture. Scrambling techniques may also be used on cable or satellite television. At the receiver a special decoder must be attached to the receiver to unscramble the transmitted signal. The decoder is connected to the receiver internally or to its antenna terminals.

The subscriber pays a fee to the originating company for the use of the decoder. The advantage of pay TV is its program material, which features full-length movies, special events, sporting events, and other kinds of entertainment free of commercials.

There are at least three types of subscriber TV systems of encoding. In one type the horizontal sync is suppressed or inverted. The associated audio is shifted upward in frequency and uses a 15,734-Hz subcarrier.

In a second system the video is mixed with a 15,734-Hz sine-wave and the program audio is on a 63,000-Hz subcarrier.

The third type is more complex. Here the sync is either suppressed or inverted and the video is further scrambled by randomly switching it from positive to negative. The audio uses a 39,335-Hz subcarrier.

In the arrangements above the audio signal has been shifted to a higher frequency by multiplex techniques. Thus the encoded signal is associated with a subcarrier and pilot signal similar to the L-R and pilot signals used in commercial FM multiplex transmission. This could result in a situation where nonsubscribers who tune to a scrambled transmission may hear a commercial extolling the advantages of pay TV, while a subscriber tuned to the same channel would hear the normal sound associated with the picture. An alternative method for encoding the sound is to modify the composite RF signal by adding a sound channel below the video carrier. In other schemes the audio is not altered at all.

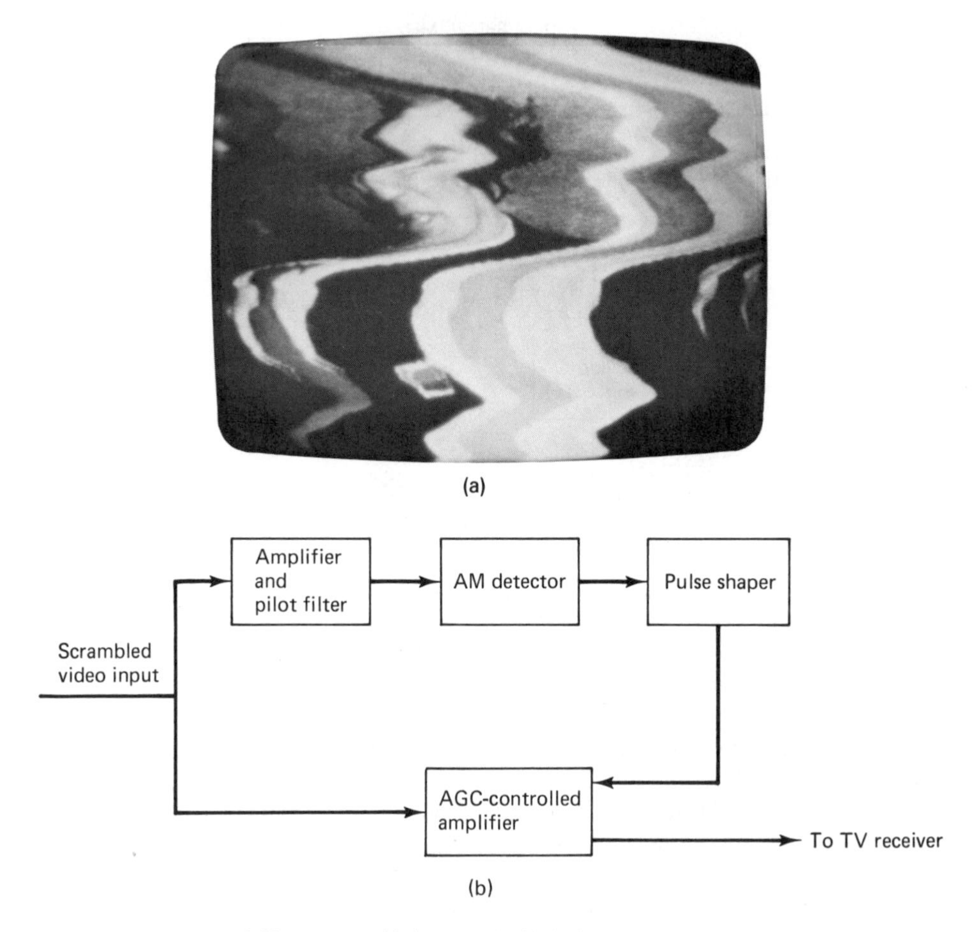

Figure 2-29 (a) Scrambled picture; (b) block diagram of a simple decoder.

The video signal is encoded in such a way that it disables the receiver sync separator, color burst gate, and AGC circuits. This is done by suppressing the composite video signal's sync and blanking pulses. The result of this on a normal receiver is to cause the picture to tear and pull horizontally, and to be overloaded to the point where the incoherent picture becomes negative (black & white reverse) as well as shifting picture color. See Figure 2-29(a).

At the receiver the decoder restores the suppressed blanking and sync pulses of the composite video signal to normal by increasing the IF amplifier gain during the horizontal blanking and sync interval. This is done by generating a pulse of the correct frequency (15,734.26 Hz) and proper duration (about 10 μs) and injecting it into the receiver AGC system. The AGC controlling pulse is generated from signals transmitted by the sound information. See Figure 2-29(b).

Depending upon the system, the audio is restored by using the subcarrier to generate a signal that is then used for two purposes: first, it is used to generate the horizontal rate AGC controlling pulse, and second, it is used to regenerate the sound subcarrier necessary for audio restoration.

Satellite TV. One of the major disadvantages of television transmission is its limited range. This is due to the "line of sight" VHF and UHF propagation characteristics of the transmitted TV signal. This limitation results in poor or no TV reception to millions of viewers who might otherwise wish to receive these programs.

A method for overcoming this limitation and for providing nationwide coverage for television transmission is the use of geosynchronous satellite relay stations. A geosynchronous satellite is a receiving-transmitting station that has been placed in an orbit directly above the earth's equator at a distance of 22,300 miles (35,888 km). At that altitude the satellite revolves around the earth once every 24 hours and appears to an earth observer to remain at a stationary point in the heavens. From this location the satellite can radiate 12 or 24 television channels per satellite of very high picture quality covering an area of 40% of the earth's surface. This radiation pattern is called a "Footprint." In addition, the microwave transmission of the satellite is relatively unaffected by such things as time of day, sunspot activity, or weather conditions. Satellites are solar powered and therefore are assured a long useful life.

In 1965 Intelsat was organized by 19 participating nations to fund a series of geosynchronous satellites. There are now more than 100 nations which have joined the system. Intelsat has satellites located in three groups: one is located over the Pacific Ocean, another over the Atlantic Ocean, and a third over the Indian Ocean.

The communication demands placed on Intelsat-type satellites by the United States, Canada, the USSR, and other nations have far outstripped the signal-handling capacity of these satellites. Since 1971 additional satellites have been added forming the Domestic Satellite System. These geosynchronous satellites have been activated for the United States, Canada, Indonesia, and the USSR. The North American orbit parking region is from about 61.5° west longitude to about 175° west longitude. Satellites are spaced at 4°, 2°, and 1 degree intervals. A 4° spacing corresponds to about 1800 miles between satellites. The older satellites operated in the C band (4 to 6 GHz*) and were spaced at 4° intervals. (See Figure 2-30). Recent FCC authorizations have decreased this spacing to 2°. Newer satellites and direct broadcast satellites (DBS) operate at higher frequencies (Ku band, 12 to 14 GHz and have spacings of 1°. As a result, 122 satellites of all classes are available. Currently 69 have been authorized. As of 1984, 33 have been launched into orbit. By 1987 all of the C-band authorizations will also be in orbit (35).

Uplink. Television program material or telephone voice signals are transmitted from earth to a satellite by high-power transmitters (1 to 3 kW) operating between 5.9 and 6.4 GHz. This leg of the communication system is called "uplink."

The uplink signal is a standard composite video signal that has had high-frequency preemphasis to improve signal-to-noise characteristics. This video signal and the FM sound subcarrier, frequency modulate the 6-GHz uplink carrier. After modulation the FM/FM signal is fed to a high-gain 10-m (32.5 ft in diameter) parabolic transmitting antenna. This type of antenna is similar in shape to an automobile headlight. In such an antenna all of the radiated signal is concentrated into an intense highly directive beam of microwave energy. Such antennas have gains of 30 to 80 dBW (decibels above 1 W), resulting in an effective isotropic radiated power (EIRP) of as much as 100,000,000 W or more, ensuring noise-free reception at the satellite.

*A gigahertz (GHz) is 1000 MHz.

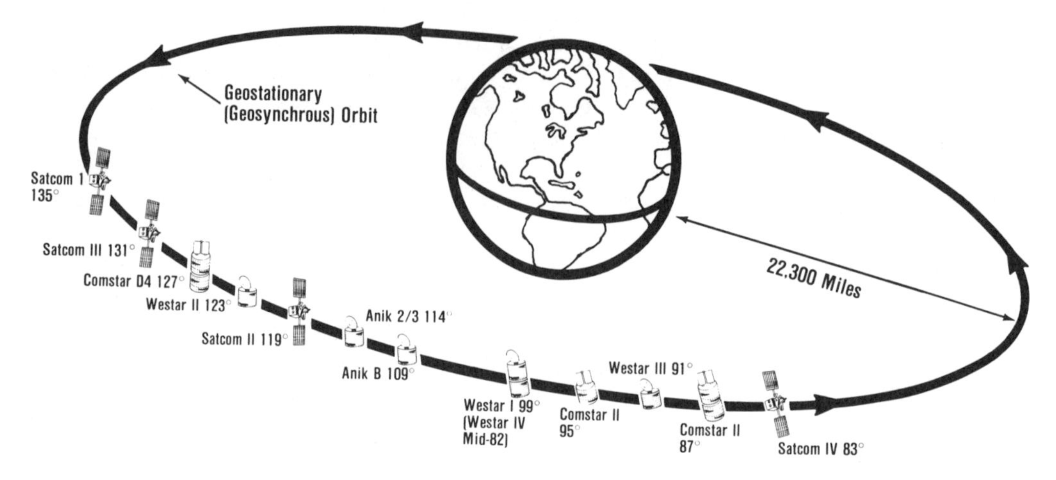

Figure 2-30 Domestic satellite "parking" spots.

Downlink. The signals received by the satellite are then lowered in frequency to a band between 3.7 and 4.3 GHz (a 500-MHz bandwidth) and reradiated back to earth. This part of the relay system is called "downlink." Satellites are designed to handle 12 or 24 TV channels simultaneously. The repeater circuit that receives, amplifies, and retransmits these channels is called a transponder. Each transponder has a power output of 5 W.

The 500-MHz downlink band is divided so that each channel is assigned a 40-MHz-wide band that also contains a 4-MHz guard band, leaving a 36-MHz active region for each video signal. Some spectrum space is also used for ground-to-satellite command signals and satellite-to-ground acknowledgment signals as well as beacons to permit satellite position-determination measurements.

Satellites that have 24 transponders must fit 24 channels of 36 MHz each (a potential bandwidth of 864 MHz) into a transmitted bandwidth of 500 MHz. To do this the channels are first spaced at 20-MHz steps. Thus, any one channel overlaps its two adjacent channels. To avoid interference, adjacent channels are broadcast using different polarizations. Odd-numbered channels are vertically polarized, and even-numbered channels are horizontally polarized. At the earth end of the downlink relay system a receiving antenna that is vertically polarized will only receive signals that are vertically polarized; horizontally polarized signals are ignored. This means that to change from odd to even channels, or vice versa, the receiving antenna must change its antenna polarization.

Receiving antenna. The signal transmitted by the satellite and received by an earth station is about 8000 times weaker than that received from a local TV station. In order to bring this very weak signal up to broadcast quality level a large (10 to 30 ft in diameter) aluminum or fiberglass dish-shaped antenna is used. See Figure 2-31. The antenna collects the signal and then focuses to a collecting point called a feed assembly. The shape of the antenna is critical. Any departure from a parabolic-shaped geometry will result in significant signal loss. A 0.1-in. deviation will result in a 1-dB loss of signal strength. Thus, a spherically shaped antenna must be made larger than a parabolic antenna for the same sensitivity. Because spherical antennas are less directional, additional feed assemblies can be used to pick up more than one satellite at a time.

Figure 2-31 Receiver-type parabolic antenna.

Mountings. To achieve maximum utility, the antenna can be built so that it may be pointed at different satellites. Two important types of antenna mountings are currently being used. These are elevation-over-azimuth (el-az), and the polar mounting.

An el-az mounting requires two adjustments to point the antenna at a satellite. These are elevation and azimuth. Elevation is expressed as an angle from the horizon (0°) to the zenith (90°). Thus, pointing the axis (feed assembly) of the antenna to 45° would place it at halfway between the horizon and the zenith. Azimuth is measured as an angle swept out on a horizontal plane in a clockwise direction. North is defined as zero degrees. The required el and az angles for a given satellite are computed based on antenna location and satellite position.

The polar mounting is identical to the type used for astronomical telescopes. It is a two-axis device; the polar axis must be accurately positioned so that it is parallel with the earth's axis (pointing at the north star). The antenna is affixed to the polar axis and tilted to a small declination angle so that it points to the satellite. Rotating the polar axis is all that is required to point the antenna to other satellites.

TVRO. Figure 2-32 shows a typical satellite communication system. The receiving part of this system is called the television-receive-only (TVRO) terminal. Notice that the TVRO consists of four basic parts: the antenna, a microwave receiver, a modulator, and a television receiver or monitor.

The antenna collects the signal and concentrates it at the focal point called the feed assembly. A typical 11-ft-diameter antenna can increase the signal strength delivered to the receiver 10,000 times (40 dBi) above that of a simple dipole antenna.

The signal delivered to the feed assembly then is fed to a low-noise amplifier (LNA). This stage is necessary to increase the very weak input signal level to a point where it can effectively override the electrical noise generated in the mixer stage of the receiver.

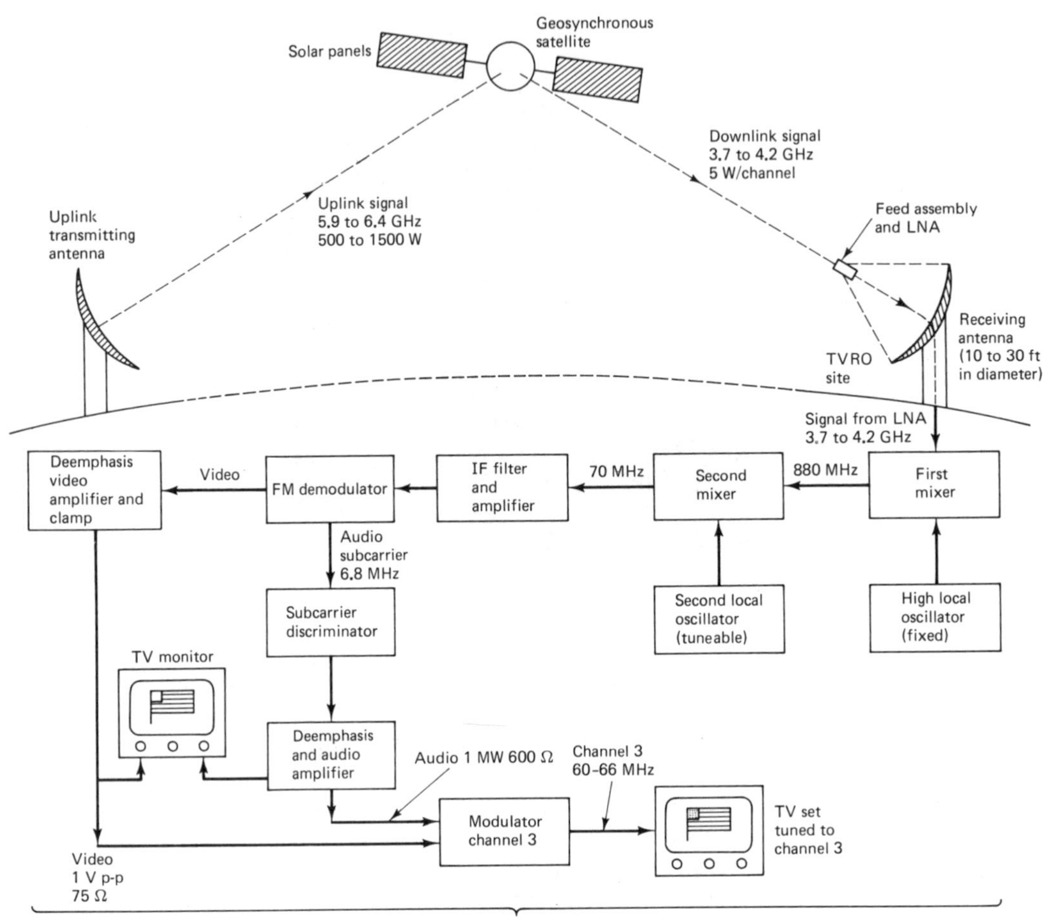

Figure 2-32 Satellite communication system using a double conversion TVRO.

The noise generated by the LNA may be expressed as a noise temperature in degrees Kelvin, or as a noise figure measured in decibels. Typical noise figures for a LNA used in a TVRO fall between 1.17 dB, corresponding to a noise temperature of 90°K, and 2.27 dB, or 200°K. This compares to a typical VHF TV receiver RF amplifier whose noise figure ranges from 3 to 6 dB. The low-noise figure of the LNA is due to the use of two expensive low-noise gallium arsenide field-effect transistor (GASFET) RF amplifiers connected in cascade.

The LNA is mounted on or near the feed assembly as a means of reducing transmission line losses and thereby improves the signal-to-noise ratio of the receiver. The LNA output feeds a mixer that lowers the input 3.7- to 4.2-GHz signal to a more manageable 880 MHz. The expensive transmission line feeding the mixer introduces some signal loss, which can be reduced if the mixer is made an integral part of the LNA. Such LNA-mixer combinations are called low-noise converters (LNC).

The output of the first converter (880 MHz) is then fed into a second tunable converter, which lowers the incoming signal to 70 MHz. The local oscillator of the second mixer is adjustable so that it can be used to select the desired channel. The advantage of a double-conversion system is a reduction of image response.

The output of the second converter feeds a wide-band 70-MHz IF amplifier. This stage must have sufficient bandpass if a good-quality picture is to be obtained. The signal is then fed to an FM demodulator, where it is converted into the original video information that modulated the uplink transmitter. Deemphasis is used to counteract the effects of preemphasis, and a clamp is used to eliminate the energy dispersal waveform introduced at the transmitter. The output of the clamp may be used directly to drive a TV monitor or a modulator. The output of the modulator is tuned to channel 3 or 4.

The FM demodulator produces a 6.8-MHz subcarrier output that carries the audio information associated with the picture. This signal is fed into an FM discriminator or PLL circuit and is converted into audio. Deemphasis brings the signal back to its original form, which is then used to drive the modulator.

Direct broadcast satellites (DBS). Direct broadcast satellites (DBSs) are geosynchronous high-powered (200 W per transponder) satellites operating at 12 GHz that can offer as many as 40 channels. As such, they allow the use of *small* (2 to 3 ft in diameter), relatively inexpensive parabolic antennas that can literally be mounted on the roof of a house or in the window of an apartment. DBS is a new technology which will bring to rural and noncable areas TV programming that would otherwise not be available. Programming will range from advertiser-supported to scrambled pay-TV services.

The basic arrangement of DBS equipment is the same as ordinary 4-GHz satellite equipment. A dish antenna, an amplifier/down converter, and a tuner are all the equipment needed.

QUESTIONS

1. What four characteristics of a scene must be satisfied if complete reproduction of a picture is to be achieved?
2. What is a picture element?
3. Of what importance to television is the angle of resolution and the angle of view?
4. What are the characteristics of the eye concerning flicker?
5. Why is persistence of vision important to the television system?
6. Which system of picture transmission is used in practical television today?
7. What is a video signal? How is it generated?
8. Why must the receiver scanning system be synchronized?
9. Describe how a flying spot scanner generates a video signal.
10. Describe how the video signal is used at the receiver to reproduce the original picture.
11. What is the frequency of the vertical and horizontal deflection currents used in black & white television?
12. What is a field and what is a frame? How often are they generated?
13. Draw an 11-line raster. Assume that two lines are lost during vertical retrace.
14. Why are the vertical retrace lines brighter than the background raster?
15. What factors determine (a) the output amplitude, (b) the highest frequency, and (c) the lowest frequency of the video signal?
16. Between what two electrodes of the CRT is the video signal delivered? Why?

17. What would be seen on the face of a CRT if a 600-Hz square wave was used to intensity modulate it? A 1,575,000-Hz square wave?
18. Why is the white level of the composite video signal maintained at 12.5%?
19. Explain how the hammerhead pattern may be made visible on a TV receiver. How can the horizontal blanking and sync pulse be made visible on a TV receiver picture?
20. What is the purpose of equalizing pulses and during what time do they appear?
21. What is the difference between the composite video signal and the modulated composite video signal?
22. What are the transmission standards for the sound portion of the television system?
23. Why is vestigial sideband transmission used for picture transmission?
24. Explain the differences between frequency modulation and amplitude modulation.
25. What are the picture carrier frequencies for channels 6, 10, 15, and 81?
26. Draw the complete block diagram of a black & white television receiver. Include all controls, wave shapes, and frequencies.
27. Explain the difference between free TV, pay TV, cable TV, and satellite TV.
28. What is the difference between CATV and MATV?
29. In an MATV system, what would be the effect on receiver operation of an open terminator resistor?
30. Why do cable converters "up convert" the VHF input signal to a UHF signal?
31. In pay TV, what is one method of altering the composite video signal so as to produce a scrambled picture? What is done to the associated sound?
32. What is done at the receiver to decode the scrambled picture and altered sound of a pay TV broadcast?
33. What is a geosynchronous satellite? What is its important advantage over ordinary methods of TV transmission?
34. What is the maximum number of geosynchronous satellites possible assuming 4° spacing? Assuming 2° spacing?
35. What is meant by "uplink," and at what frequency does it operate?
36. What is meant by "downlink," and at what frequency does it operate?
37. What kind of antenna is used to transmit to and receive from a satellite? What are its characteristics?
38. How are 24 channels fit into a bandwidth that would seem to be sufficient for only 12 channels?
39. What is an LNA? An LNC?
40. What is a TVRO?
41. Explain how a satellite receiver is tuned to receive different satellites as well as different channels on the same satellite.

chapter 3

Principles of Color TV

3-1 introduction

Television was defined in Chapter 2 as an electronic system for conveying a picture from one place to another. It was further pointed out that in the final analysis it is the characteristics of human eye that determine the characteristics of the black & white television system. This is also true in color television.

If the eye is to be satisfied that a picture has been *completely* reproduced, four characteristics of the scene must be present in the picture: (1) brightness information (black & white), (2) motion, (3) color, and (4) depth information.

In black & white television only brightness and motion information are transmitted to the receiver. In color television, however, color as well as brightness and motion information are transmitted. The addition of color information to the television system results in a picture of increased realism and beauty.

Depth information would allow a three-dimensional (3D) TV system. At present there are few regularly scheduled 3D systems broadcasting anywhere in the world. In Japan animated 3D cartoons have been regularly broadcasted, and in the United States several full-length movies have been transmitted in 3D.

These systems have produced 3D by means of colored glasses. The left eye views through a red or green filter, and the right eye views through a blue filter. A two-color picture is then transmitted in red and blue, and slightly displaced from one another. When viewed through the glasses, each eye sees slightly displaced pictures that are then interpreted by the brain as 3D.

The addition of color information to the existing black & white signal will of necessity result in a more complex composite video signal as well as a more complex receiver. Although many different systems of color TV have been proposed, the system that is currently used today combines the features of *compatibility* with the features of high quality. Compatibility means that color TV broadcasts are receivable on black & white receivers. Conversely, color receivers are able to receive black & white TV broadcasts. The following sections of this chapter will describe the basic principles of color TV transmission and reception.

3-2 color principles

Since it is the human eye that is the final judge of any reproduced color picture, the characteristics and limitations of the eye dictate the main features of the color TV system. Therefore, this chapter will begin, as did the chapter on black & white TV, with a short discussion of some of the characteristics of the eye, especially those concerning color.

Light is a form of electromagnetic energy identical in composition to that of radio waves. Figure 3-1 shows the electromagnetic spectrum extending from 300 Hz to the highest frequency detectable. Visible light represents a very narrow range of the entire spectrum. Light differs from radio waves in that visible light has a frequency much higher than that used in any commercial system of radio communication. For example, the UHF TV band extends from 470 to 890 MHz, whereas visible light has a frequency range of 430 terahertz (430×10^{12} Hz) to 750 terahertz (750×10^{12} Hz). In a real sense, the eye may be thought of as a kind of radio receiver whose bandpass falls between 430 and 750 terahertz.

The frequency or wavelength of light determines the color or hue that we see. Thus, 430×10^{12} Hz (a wavelength of about 700 nm) corresponds to the color red, and the higher frequency 750×10^{12} Hz (a wavelength of 400 nm) is seen as a deep blue. Of course, some colors do not correspond to any one given wavelength of light. Brown and purple are examples of such colors. These colors are the result of a mixture of different proportions of red, green, and blue. The interpretation of this mixture as brown or purple is done in the observer's mind.

Color characteristics. Color may be described by three characteristics: *hue, brightness,* and *saturation.* Hue is defined as the color sensation produced when light of one or more wavelengths is viewed. Brightness refers to the intensity of the light being observed. Thus, any given color can be viewed as being brighter or darker depending upon its intensity. Obviously, if the intensity of a color were reduced to the point where it was no longer visible, it would be placed on a brightness scale corresponding to black. Similarly, if its intensity were increased sufficiently, its brightness level would correspond to white.

Saturation describes the degree of purity of a color and its freedom from white light. For example, red is a fundamental hue that when diluted by the addition of white becomes pink. In short, pale or pastel shades of hue are less saturated than are the vivid shades of color. Saturation is related to brightness because increasing the white content of a saturated color increases its energy content making it brighter.

Color television receivers are able to adjust the degree of saturation, brightness, and hue by means of suitable front panel controls. The saturation control is often referred to as the *color control.* Hue is adjusted by a control usually labeled *tint control.* Brightness is controlled by the *contrast* and *brightness controls.*

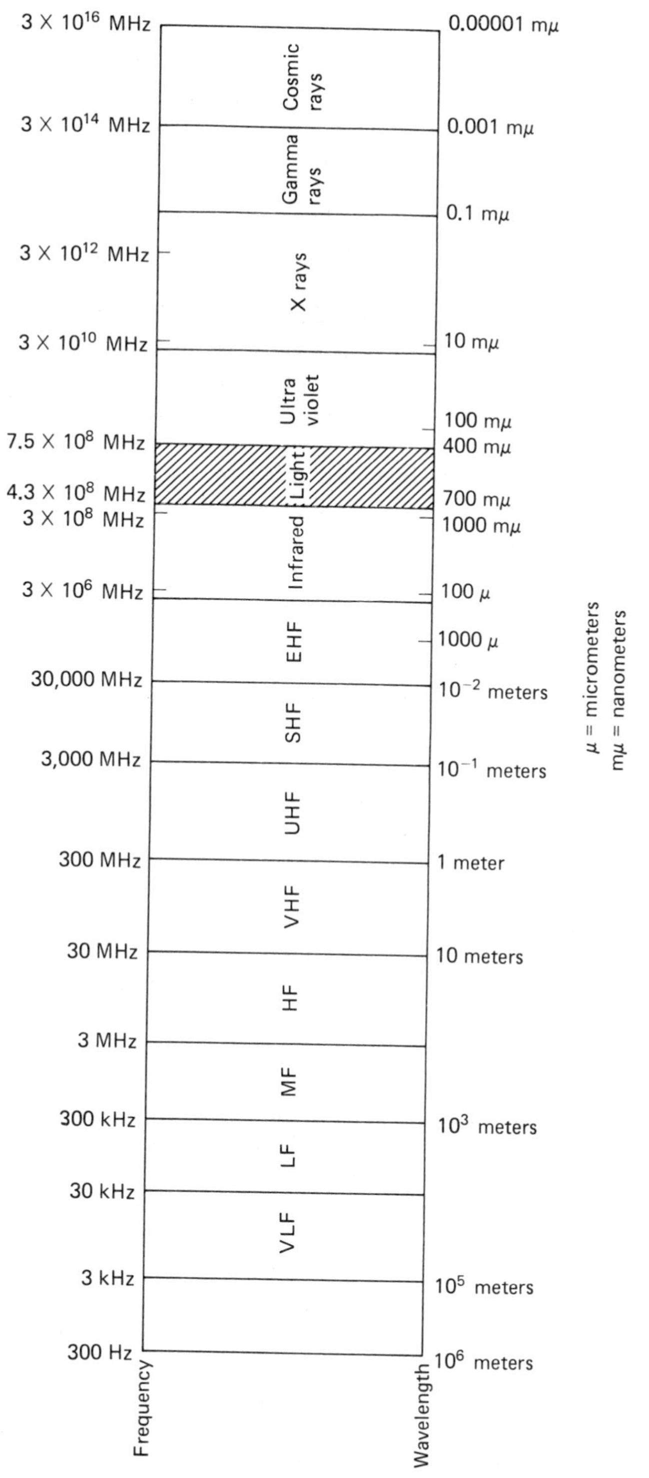

Figure 3-1 The electromagnetic spectrum.

Color mixing. Sir Isaac Newton was one of the first to note that white light (sunlight) when passed through a glass prism will separate into a multitude of colored components. This display is called a *spectrum,* and under proper atmospheric conditions it occurs naturally in the form of a rainbow. If the spectrum such as the one shown in Figure 3-2(a) is examined, it will be seen that the colors follow a sequence similar to that of the resistor color code that is used to identify the value of a carbon resistor. Thus, red is followed by orange, yellow, green, blue, and deep blue.

It is possible to reverse Newton's experiment and make white light from a spectrum of colors. Fortunately, it is not necessary to use all the colors of the spectrum to produce white light. As few as two or three spectral colors may be used for this purpose. However, the colors that are used to produce white light must be of equal energy. White is not the only color that may be made by mixing two or more spectral colors together. As a matter of fact, a very wide range of spectral and nonspectral colors may be obtained in this manner.

Subtractive color mixing. Colors may be mixed either by: *subtractive* mixing or by *additive* mixing. Subtractive mixing makes use of the filtering action that dyes, pigments, and inks have on reflected light. For example, if white light falls on a red dress, the dyes that color the dress will absorb all color in the white light except red. The red light component of the white light will be reflected and the observer will see a red dress. Similarly, if red paint were mixed with blue and green paint, all the spectral components of white light would be absorbed and black (the absence of light) would result. In short, all color printing, color painting, color photographs, and color movies use subtractive color mixing to obtain a wide range of color reproduction.

Additive color mixing. Color television makes use of the *additive system* of color mixing. In this system colored *light sources* are mixed to produce a wide variety of different colors. Additive mixing is superior to subtractive mixing because additive mixing is able to reproduce a much wider range of colors.

The colored light sources that are used in additive mixing are called *primary colors.* A primary color is one that can combine with other colors to form new colors not including itself. In color television the primary colors are red, green, and blue. Figure 3-2(b) shows the effect of combining three equal energy color light sources by projecting them onto a screen. The colors are arranged so that they overlap one another by a small amount. In this way the results of additive mixing may be clearly seen. Notice that where the three colors superimpose the resultant color is white. This is opposite to that of subtractive mixing where the result of combining three primary colors would be black. Of course, in additive color mixing the absence of light corresponds to black just as it does in the subtractive system of mixing. Figure 3-2(b) clearly indicates that the combination of red and green light sources results in yellow. Blue mixed with red produces magenta, and blue mixed with green produces cyan.

If the relative intensities of the colors being mixed are changed, a variety of new colors will be generated. Figure 3-3(a) illustrates the effect of changing the intensity of red and green. In this figure the vertical cylinders represent the intensity of each color. First red is held constant and green is varied; then the procedure is reversed. Notice that as the amplitude of either of the primaries is varied, the resultant color that the observer sees also changes. Especially note that when red and green are equal in intensity, the resultant color

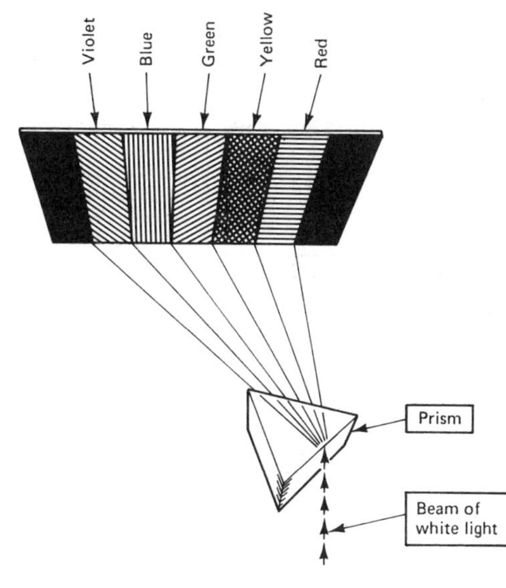

(a)

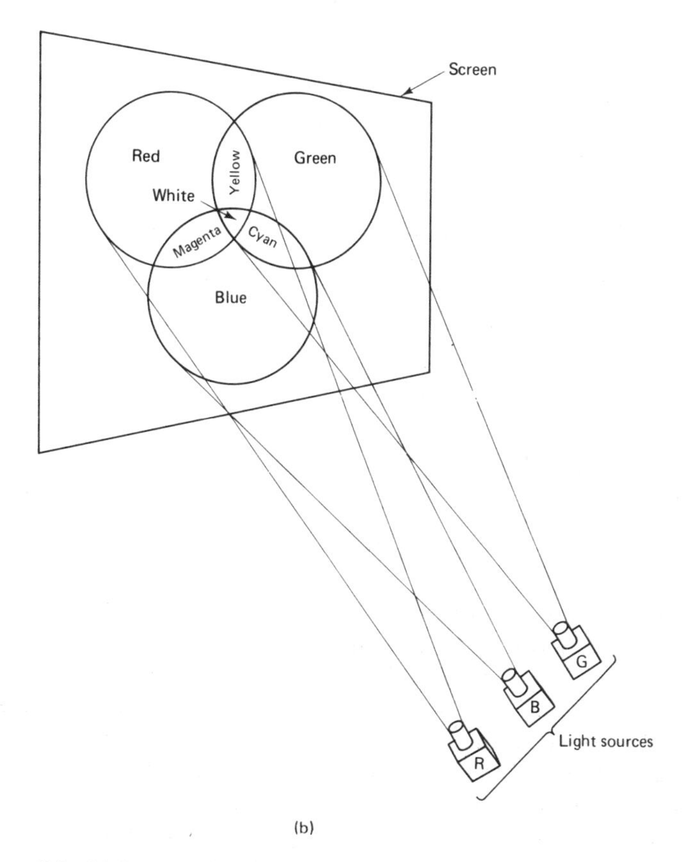

(b)

Figure 3-2 (a) Spectrum of white light; (b) result of additive mixing of three light sources.

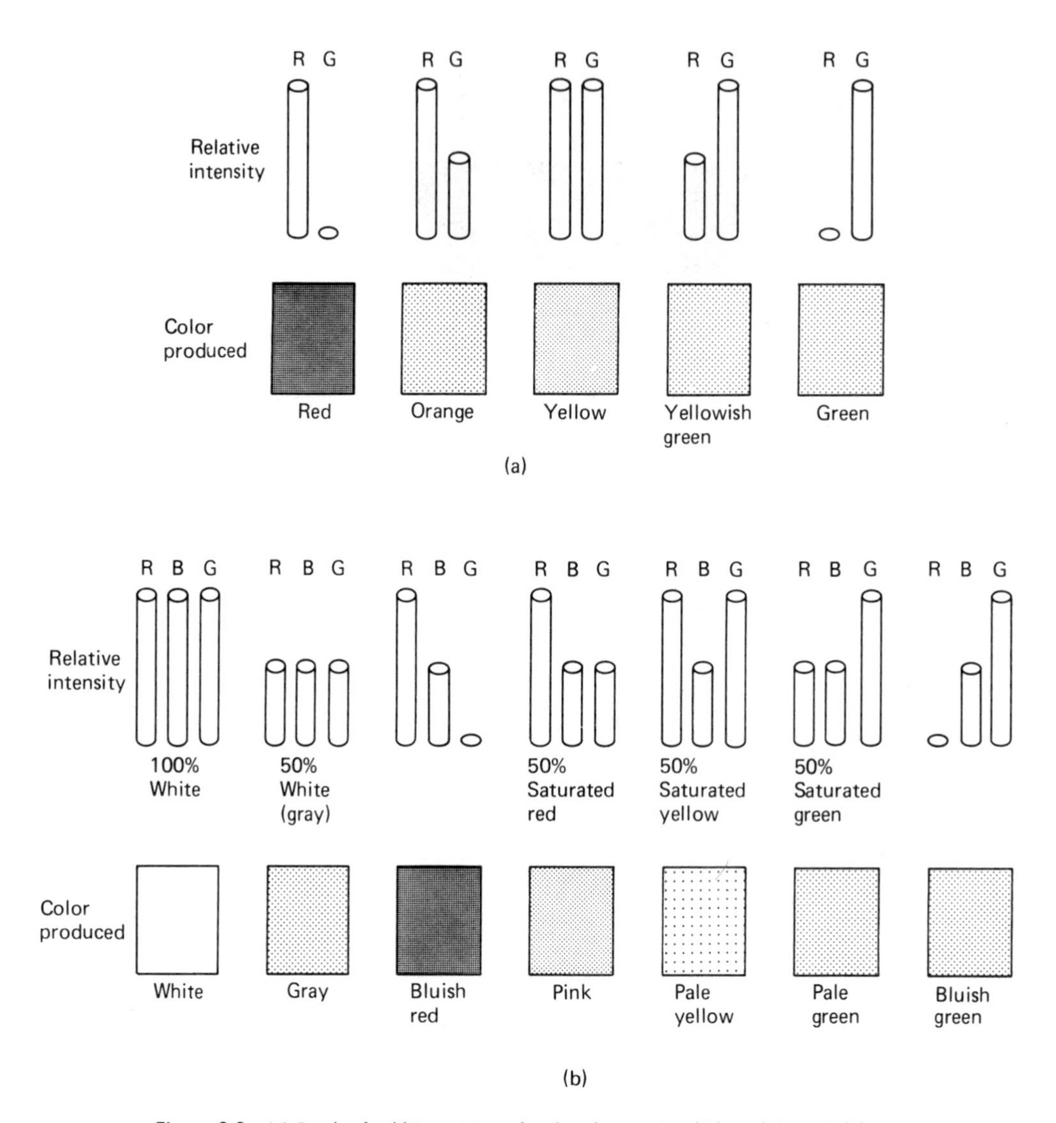

Figure 3-3 (a) Result of additive mixing of red and green in which each is varied from zero to maximum; (b) result of three-color additive mixing in which blue is held constant and red and green are both varied from maximum to zero.

is a saturated yellow. But when green is made weaker than red, orange is the result. Similarly, when red is made weaker than green, the observer sees a yellowish green.

In Figure 3-3(b) the vertical cylinders are again used to represent the relative amplitude of each of three primaries (red, green, and blue) used in the mixing process. Under each of these combinations is shown the colors that they generate. The first two cases illustrate the conditions needed to produce white and gray. In all of the following cases, blue is held constant at one-half the intensity of the other two primaries. In effect, for the cases in which all three primaries are present, white has been added to the mixing process. Since blue is held constant at the 50% level, the result may be termed 50% white or gray. One hundred percent white would be the condition in which all three primaries are at maximum intensity. Referring to Figure 3-3(b) and reading from left to right, we see that a mixture of red and 50% blue produces bluish-red. Increasing green to 50% results in a 50% white. When this is mixed with 100% red it produces a washed out red or pink. Increasing green to maximum (100%) results in the mixture of yellow (R and G) and 50% white which the observer sees as a pale yellow. Reducing red to 50% causes a pale green since the fully saturated green is diluted with 50% white. When red has been reduced to zero, the mixing process combines the 50% blue with the 100% green and produces a bluish-green.

Thus, through the additive mixing technique almost any color can be produced in terms of hue, saturation, and brightness by simply varying the relative amplitudes of the three primary colors used.

The addition of more primaries would increase the complexity of the color TV system as well as the range of colors that could be reproduced. However, the three colors that are used in commercial color television reproduce a range of colors greater than any other reproducing system, while providing a relatively simple system of color TV.

Luminance level. Color mixing produces colors that have different brightness levels. Thus, it is possible to have either a bright red or a dull red. In terms of light intensity levels, brightness can be compared to shades of gray. A bright yellow would correspond to a shade of gray approximating white, and a dark yellow would produce a shade of gray approaching black. The brightness or *luminance* level that corresponds to any given color can be determined by the following formula:

$$Y = 0.59G + 0.30R + 0.11B$$

For example, if equal energy 100% green, red, and blue were combined, the luminance level produced would be

$$\begin{aligned} Y &= 0.59(1) + 0.30(1) + 0.11(1) \\ &= 1.00 \end{aligned}$$

This combination produces the brightest shade of gray possible: *white.* Take another example. Assume that there is a shade of pale cyan whose component colors are 50% red, 100% blue, and 100% green. The shade of *gray* that this color corresponds to is

$$\begin{aligned} Y &= 0.59G + 0.30R + 0.11B \\ &= 0.59(1) + 0.30(0.5) + 0.11(1) \\ &= 0.59 + 0.15 + 0.11 \\ &= 0.85 \end{aligned}$$

This means that the shade of gray corresponding to pale cyan is 15% less intense than white. In the same way, saturated magenta would have a gray scale of 0.41 and would appear less than one-half as intense as white. Saturated blue would correspond to a shade of gray (0.11) that would almost be black. Black is the absence of all light and, therefore, would have a luminance level of zero.

The fact that it is possible to determine the corresponding level of gray for any color from the color levels themselves is important in color television because it is then possible to obtain a black & white video signal (Y signal) from the three primary color signals. More about this in a later section.

Object size and color. A characteristic of the eye that has important applications in the color television system is that the eye's color response changes with object size. The practical effect of this is to reduce the amount of color information that the television transmitter must transmit for a given scene. Expressed in terms of a 23-in. television picture tube, it may be shown that video signals of 0.5 MHz or less correspond to horizontal picture tube object sizes ranging from 0.34 in. or more. For these video frequencies, all three primary colors will be needed in order to provide complete color reproduction.

For picture object sizes ranging from 0.34 in. to a minimum of 0.12 in. corresponding to a half-cycle of video frequency of 1.5 MHz, only two colors, *orange* and *cyan*, are needed to provide complete color reproduction.

Since the eye is colorblind for very small objects, the color television system need only supply brightness information (black & white) when the video signal exceeds 1.5 MHz. This means that when the object size of a 23-in. television picture tube is less than 0.12 in. in width, no color information need be transmitted. This characteristic of the eye is very important to the color television system because it restricts all color video information to a range of signals extending from 0 to 1.5 MHz, whereas the brightness or *luminance* signal information has a required bandpass from 0 to 4.2 MHz. Obviously, the narrower the bandwidth of the color information, the less complex the color TV system need be.

3-3 color television transmission

The signal radiated by the color TV transmitter must be compatible with black & white receivers but at the same time it must provide the information needed to produce good color reception. Essentially, this is done by combining the color video information with the ordinary black & white video signal. However, the addition of this new information must not extend the video bandpass beyond its previously established limit of 4.2 MHz or introduce any interference or distortion to the existing black & white television transmission.

The means by which this can be accomplished relies on the principle that there is less likelihood of interference between color and black & white television signals if high-frequency video (corresponding to small picture information) instead of low-frequency video (large picture information) is added to the existing black & white video signal.

Unfortunately, the video signal generated by the color television camera is primarily low-frequency video. Therefore, the main function of the color circuitry used in the color transmitter is to *change* these low-frequency signals into higher-frequency signals. At the same time, the information originally generated by the camera must be preserved. The addition of this high-frequency color video to the black & white video results in a relatively interference-free composite video signal.

At the color receiver the process is reversed. First, the high-frequency color signal is separated from the composite video signal. Then this signal is changed back into its original low-frequency form. In this form it is fed into the color picture tube where it intensity modulates the three electron guns that in turn paint the colored pictures on the face of the picture tube. In the following sections we examine in detail how this system works.

A simple color television system. A simple yet cumbersome color television system might be constructed as shown in Figure 3-4. In this system four separate television cameras are used to generate video signals corresponding to black & white, red, green, and blue picture information. Each of these video signals is then amplified and used to modulate a complete television transmitter. The high-frequency television signals are then transmitted and are received by four complete television receiver systems.

The black & white camera and transmitter equipment are necessary to make the system compatible so that those people who own black & white television receivers can view the transmitted program material. Since this arrangement is no different from any other existing black & white system, it will require a full television channel allocation.

Each of the color television cameras is identical to the black & white camera with one exception. The color cameras use filters that make the camera colorblind to all colors

Figure 3-4 Complete transmitter and receiver color TV system.

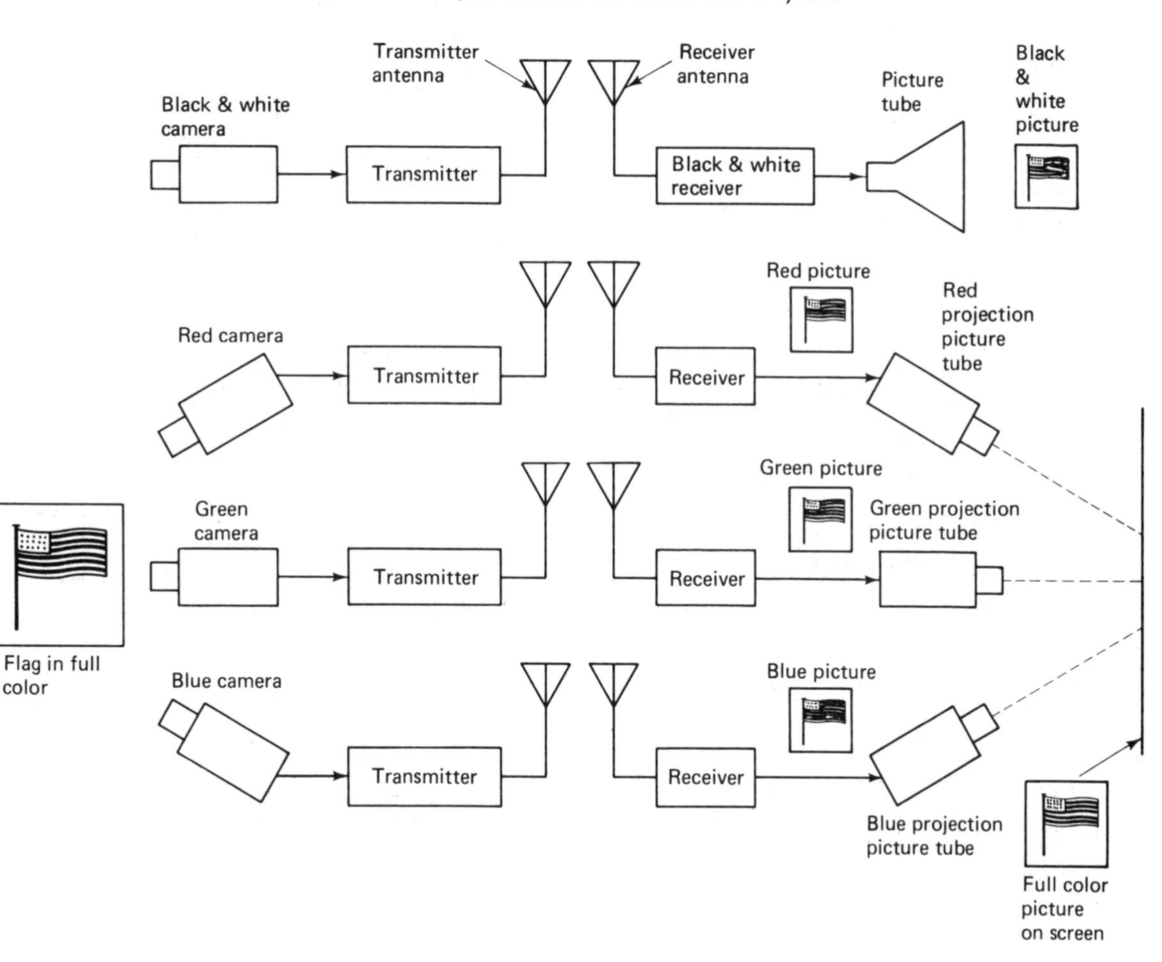

except the primary color for which it is designed. The video output frequency range of each camera will still extend from 0 to 4.2 MHz, and as a result each transmitter will require a full 6-MHz television channel allocation for a total of 24 MHz.

At each of the receivers the video information is used to excite a projection type picture tube whose light output corresponds to one of the primary colors. Each projection picture tube generates its own color raster that falls on the viewing screen. The projection tubes are so positioned that their rasters are made to superimpose exactly one on the other. Projection television systems of this type are commercially available. Under these conditions, the observer will see the television picture reproduced in full color. Unfortunately, this success was accomplished at enormous expense both in equipment (four complete television transmitters) and in channel allocations (four channels at 6 MHz each for a total of 24 MHz).

The matrix. Simplification of the transmitter can immediately be achieved by generating the brightness levels of the black & white signal (*Y signal*) from the three color cameras, thereby eliminating the black & white camera.* The Y signal can be made by adding 59% of the green signal to 30% of the red signal and 11% of the blue signal. The circuit in which this is done is called a *matrix*.

A matrix is a computerlike circuit that is able to combine a number of signals in a resistive voltage divider to produce a desired output signal. In Figure 3-5(a) the Y signal is obtained by adding the outputs of three voltage dividers that constitute the matrix. Each voltage divider produces a fixed percentage output for each camera. Thus, the red camera output feeds into a divider whose output is 30% of its input. The green camera feeds a divider whose output is 59% of its input. The blue camera output is also divided and the output of its resistive voltage divider is 11% of its input. The simple addition of these divider outputs results in the formation of the Y signal. Although in practice the computer-like matrix may be more complex than indicated in Figure 3-5(a), the principles of its operation are not changed.

Color difference signals. Additional transmitter simplification can be achieved by converting the color signals (R, G, and B) into special signals called $R - Y$ (red minus Y) and $B - Y$ (blue minus Y) signals. These color difference signals are video signals obtained from the matrix by inverting the Y signal and adding it to the color signals. The color difference signals are used solely for color reproduction. Therefore, by taking advantage of the eye's color characteristics, their bandwidth may be reduced to 1.5 MHz each. As a result, the three channels may have a total bandpass of about 12 MHz.† Changing the color signals into color difference signals has the advantage of reducing the three color signals into two signals.

$G - Y$ matrix. At the receiver it is possible to use a special matrix circuit to extract a $G - Y$ signal from the $B - Y$ and $R - Y$ signals. See Figure 3-5(b). This may be done by adding the outputs of two inverting amplifiers ($-0.51(R—Y)$ and $-0.19(B—Y)$) whose gain has been adjusted to the required values. The output of this matrix is a video signal corresponding to G—Y.

*In practice, four cameras are sometimes used, one of which generates the Y signal.
†The color channels would make use of double sideband transmission giving each a 3-MHz bandwidth.

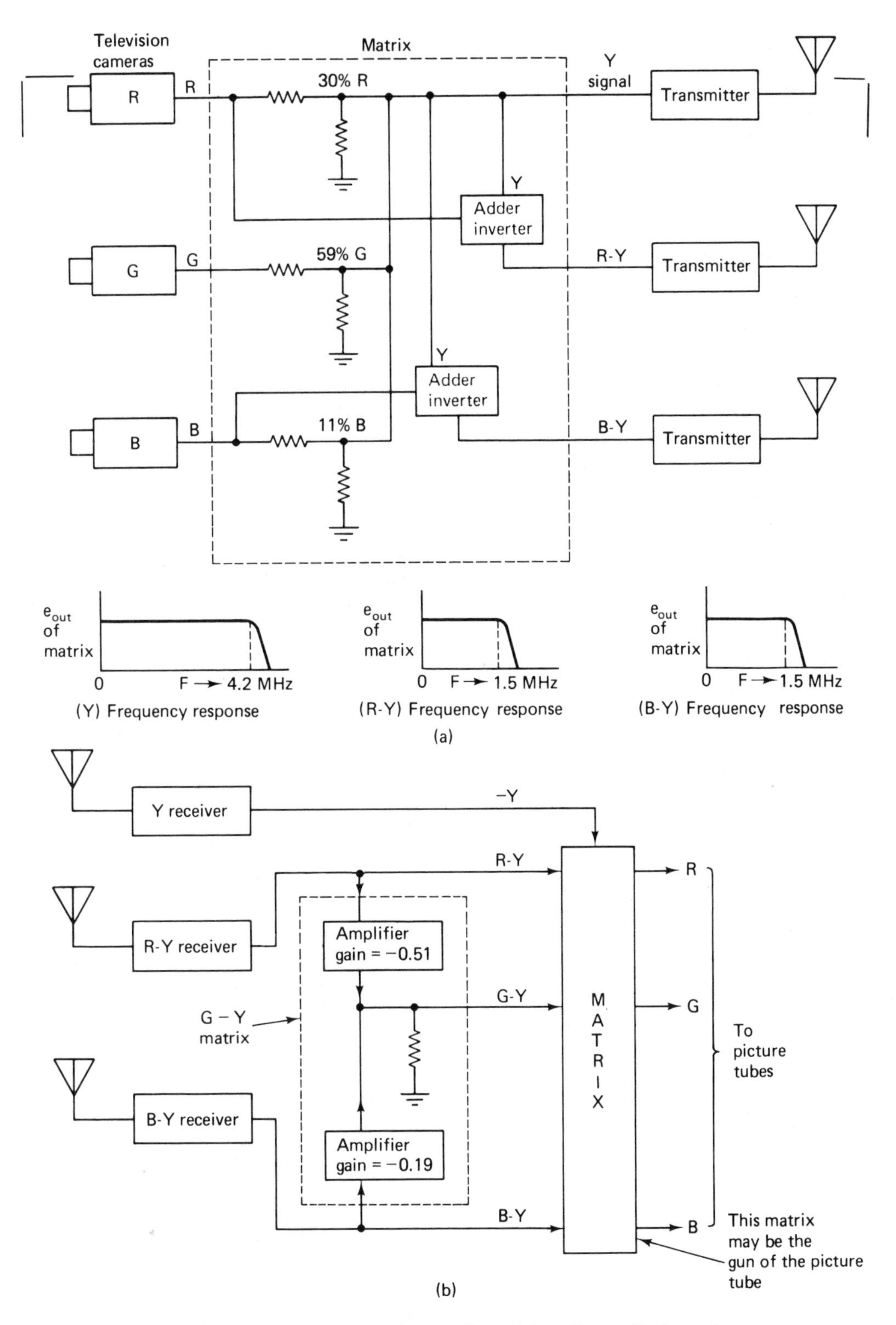

Figure 3-5 (a) *R — Y, B — Y* color transmitter; (b) *R — Y, B — Y* color receiver.

At the receiver, decoding the $R - Y$, $B - Y$, and $G - Y$ signals into their original red, blue, and green video signal levels is accomplished by simply adding, in a receiver matrix, the Y signal to each of the color difference signals. Thus,

$$(R - Y) + Y = \mathrm{R} \qquad (B - Y) + Y = B \qquad (G - Y) + Y = G$$

System simplification. The overall result of these simplification methods has been to reduce the number of cameras from four to three and to reduce the number of transmitters by the same amount. Another result has been the reduction of bandwidth requirements from 24 MHz in the first case to approximately 12 MHz as things stand now. But the simplification thus far achieved has not been enough. It is necessary to concentrate the system still further so that eventually only *one* transmitter having a bandpass of 6 MHz remains.

Obviously, to combine the Y signal with both the $R - Y$ and the $B - Y$ signals so that they fit into one 6-MHz channel requires that these signals be added together. The question arises as to how they should be added. At first glance it might appear that all one has to do is add the signals directly. If this were actually done, severe interference would result, making the system unusable. This is because the color information is primarily *low*-frequency video and as such corresponds to large picture areas. Hence, the interference that will result from the addition of the video signals will affect the large picture areas and will cause objectionable interference. To avoid this problem, it is possible to increase the frequency of the color (chrominance) signals. Thus, any interference that might result from the addition of the Y signal to the chrominance signals will appear only in the fine detail areas of the picture. As such, the interference will be barely visible.

Balanced modulator. The manner by which the $R - Y$ and $B - Y$ signals are increased in frequency is illustrated in Figure 3-6. Notice that each color difference signal along with a 3.58-MHz signal from a local oscillator is fed into a *balanced modulator* that functions as a frequency changer. The balanced modulator is similar in operation to the mixer stage of a radio since its output contains the sum and difference sideband frequencies of the input signals. But it differs from the ordinary mixer stage of a radio because the high-frequency local oscillator signal, called the *subcarrier,* is suppressed and does not appear at the output of the balanced modulator. The suppression of the subcarrier is necessary in order to minimize the possibility of a fixed interference pattern that would be seen when viewing a color telecast. This pattern would be the result of a beat equal to 920 kHz, the difference frequency between the sound carrier of the transmitter and the color subcarrier.

Color burst. The suppression of the color subcarrier at the transmitter requires that the *color receiver* have a 3.58-MHz oscillator which is used during demodulation to reinsert the subcarrier and restore the color signal to its original form. The frequency and phase of this reinserted carrier are critical for color reproduction. Therefore, it is necessary to synchronize the color receiver's local 3.58-MHz oscillator so that its frequency and phase are in step with the subcarrier signal at the transmitter. Synchronization is accomplished by transmitting a small sample of the transmitter's 3.58-MHz subcarrier during the back porch interval of the horizontal blanking pulse. Figure 3-7 shows one horizontal blanking and sync interval that is used to modulate the television transmitter during a color broadcast. It should be noticed that the sync pulse and the front porch and blanking intervals are essentially the

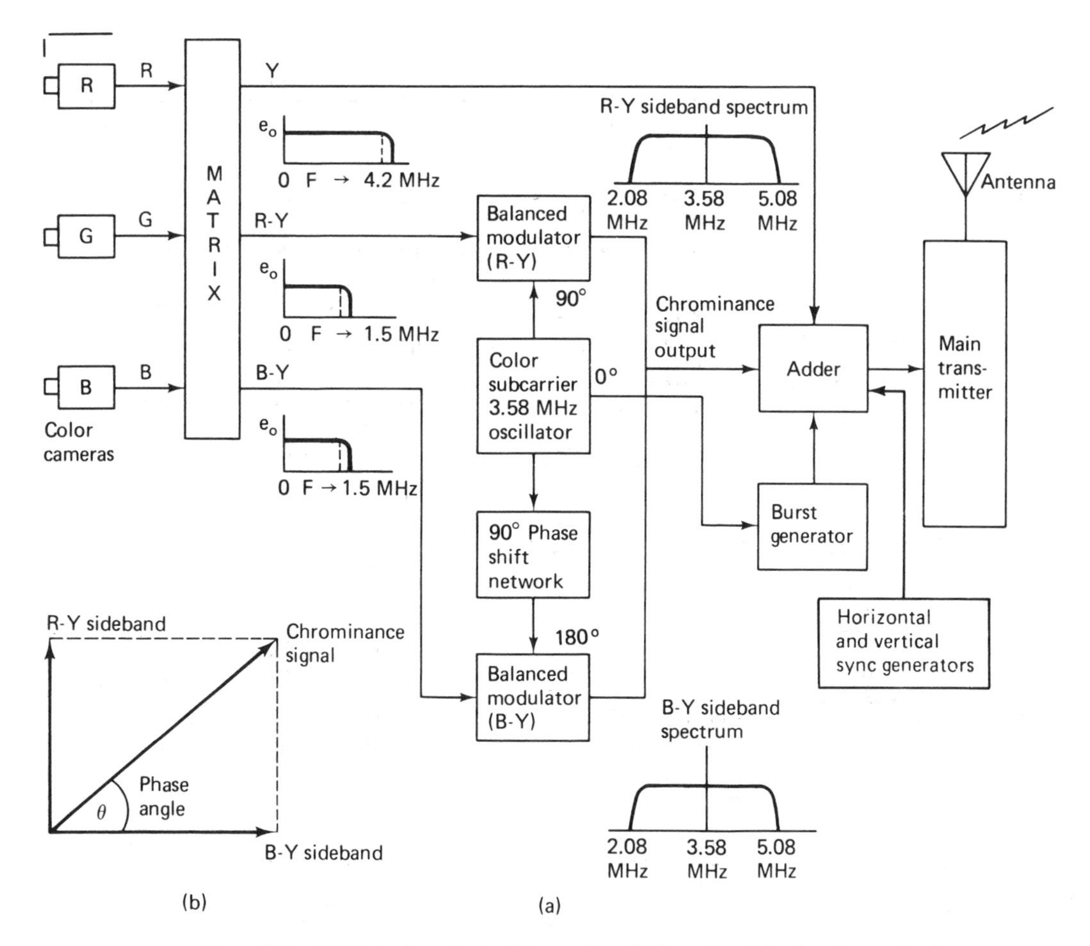

Figure 3-6 (a) Single-channel color TV transmitter; (b) formation of the chrominance signal.

Figure 3-7 (a) Color synchronizing signal; (b) vertical interval reference (VIR) signal.

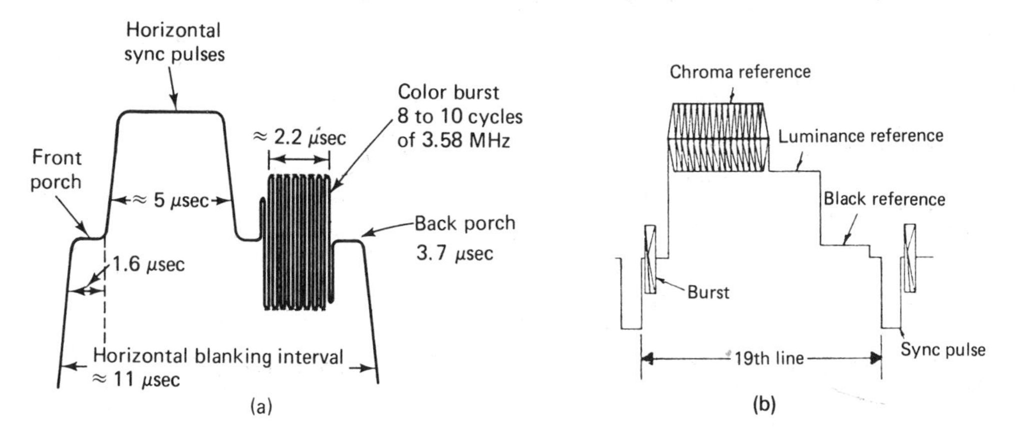

same as that used during black & white transmission. However, during color broadcasts there is superimposed on the back porch between 8 and 10 cycles of the 3.58-MHz subcarrier that is to be used as the color sync. This color sync interval is often referred to as the *color burst.* It should be further noticed that the color sync amplitude is approximately the same peak-to-peak amplitude as the horizontal sync pulse. Therefore, the color burst is approximately 25% of the amplitude of the entire composite video signal. During the color television vertical blanking and sync intervals a color burst is transmitted following each horizontal sync pulse. The color burst is removed during black & white transmission.

VIR. In 1975 the FCC designated line 19 of each vertical field of video information as reserved for a specific vertical interval reference (VIR) signal. See Figure 3-7(b). The FCC does not require that this signal actually be transmitted. The VIR signal contains three reference signals: (1) a chrominance reference signal of 3.579545 MHz having the same phase as the color burst; (2) a black level reference; and (3) a luminance (white) level reference. Some television receiver manufacturers have used the VIR signal for automatic tint control purposes.

3-4 phase relationships

As indicated in Figure 3-6(a), the local 3.58-MHz oscillator is fed to both balanced modulators, but the 3.58-MHz oscillator does not deliver the same signal to each modulator. The signal delivered to the $B - Y$ modulator is 90° out of phase with that feeding the $R - Y$ modulator. As a result, the $R - Y$ and $B - Y$ sidebands appearing at the output of the modulators also differ from each other by 90°. Vectorially combining these two output signals into one signal results in a *composite chrominance signal* that has two important characteristics: *amplitude* and *phase.* See Figure 3-6(b).

It should be noted that the number of signals that now must be transmitted to convey complete picture information has been reduced to two: the luminance (Y) and the composite chrominance signal. By using balanced modulators the frequency of the chrominance signal has been changed so that it is clustered about the 3.58-MHz subcarrier frequency. The increase in chrominance frequency allows the Y signal to be combined with the color signal with little likelihood of interference between them.

The *amplitude* of the composite chrominance signal represents the color *saturation* information, and its *phase* corresponds to the actual *color* or hue information of the picture being transmitted.

Since phase relationships play such an important role in color television transmission and reception, it is worthwhile to recall some of the important characteristics of these relationships.

Phasors. It should be noted that any sinusoid (sine and cosine waves) can be represented by a straight-line arrow called a *phasor.* A phasor has three properties: length, speed of counter clockwise rotation, and relative angular position. These properties correspond to amplitude, frequency, and phase of a sinusoid.

The technique by which the resultant amplitude and phase are determined when two sinusoids are added together is called *phasor addition.* In this method of phasor addition the two sine-wave voltages to be added are represented by phasors of proper length that are positioned so that the angle between them corresponds to the phase difference between the voltages to be added. Next, a parallelogram is constructed by using the phasors to be added as two of the sides of the parallelogram. The other two sides are constructed by drawing lines parallel and equal in length to the original two phasors. The phasor addition is completed by drawing a line between the intersection of both the original two phasors and their two parallel sides. See Figure 3-6(b).

3-5 formation of the chrominance signal

The $R - Y$ and $B - Y$ signals coming from the camera matrix and feeding the input of the balanced modulators are relatively low-frequency signals that determine the instantaneous *polarity* input of the balanced modulator. It is the characteristic of the balanced modulator that its output amplitude is directly related to its input amplitude and that its output phase depends upon the *polarity* of the input color difference signals. Thus, a positive going color difference signal at the input will produce a 3.58-MHz sine-wave output that can be considered 0° phase. See Figure 3-8(a). A negative going color difference signal will then produce a 3.58-MHz sine-wave output that will be 180° out of phase. If the subcarrier phase is shifted 90°, the output of the balanced modulator will also be shifted by 90°, as is also shown in Figure 3-8(a).

In Figure 3-6(a) the chrominance signal is shown to be made up of the sideband outputs of both the $R - Y$ balanced modulator and the $B - Y$ balanced modulator. The output of each balanced modulator may be either 0° or 180°, depending upon its input polarity. However, one of the balanced modulator outputs must differ from the other by 90° because of the 90° phase shift of the subcarrier oscillator. Therefore, the phasor representation of the sideband outputs of both balanced modulators may be represented as shown in Figure 3-8(b). In this diagram the $-(B - Y)$ sidebands are designated 0° and the $B - Y$ sidebands are located at 180°. $R - Y$ sidebands are positioned at 90° and the $-(R - Y)$ sidebands are located at 270°. Rotation in this diagram has been assumed to be clockwise.

It is now possible to examine the manner by which the $R - Y$ and $B - Y$ sidebands are combined into a composite chrominance signal. Referring to Figure 3-6(a) and assuming that a 100% saturated magenta is being scanned by the cameras, we see that the following will occur: (1) the output of the red camera will have an amplitude of 1.0 unit, (2) the output of the blue camera will also be 1.0 unit, and (3) the output of the green camera will be zero. The matrix output levels are determined as follows:

$$Y = 0.59G + 0.30R + 0.11B = 0.59(0) + 0.30(1) + 0.11(1) = 0.41$$
$$R - Y = 0.70R - 0.11B - 0.59G = 0.70(1) - 0.11(1) - 0.59(0) = 0.59$$
$$B - Y = 0.89B - 0.30R - 0.59G = 0.89(1) - 0.30(1) - 0.59(0) = 0.59$$

The Y signal is fed to the adder on its way to the main transmitter. The $R - Y$ and $B - Y$ signals are both positive and, therefore, produce a balanced modulator sideband output level of $+ (R - Y) = 0.59$ and $+ (B - Y) = 0.59$. This is indicated in Figure 3-9(a). The resultant composite chrominance signal produced by the phasor addition of the two sideband components will have a relative amplitude of approximate 0.83 units and will be at a frequency of 3.58 MHz.

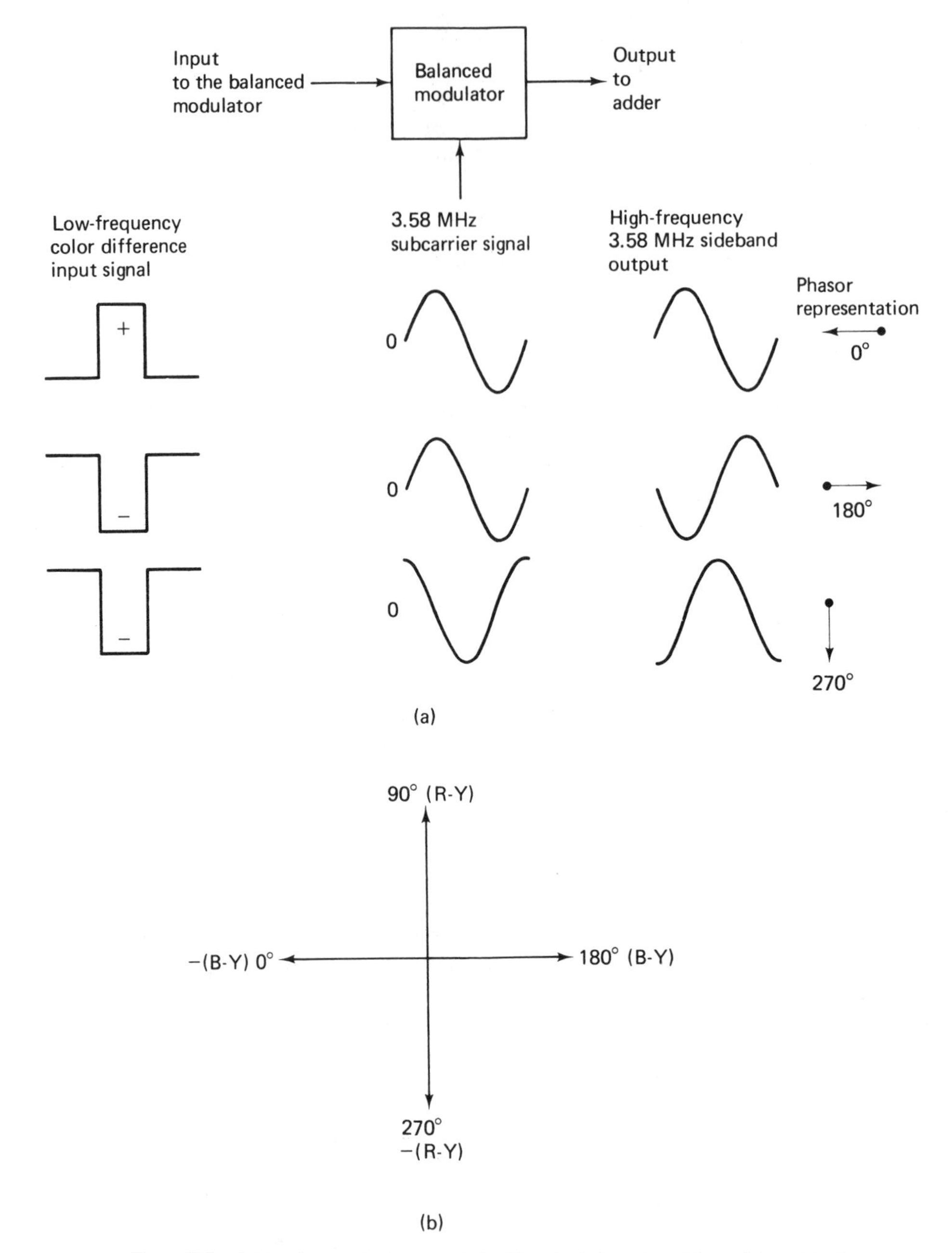

Figure 3-8 (a) Input and output phase relationships of a balanced modulator; (b) combined balanced modulator output phase relationships.

Overmodulation compensation. In actual practice, the $R - Y$ and $B - Y$ signals are compensated to prevent overmodulation of the transmitter. This is done by reducing their amplitudes to $0.877(R - Y)$ and $0.493(B - Y)$. This in turn causes changes in the amplitude

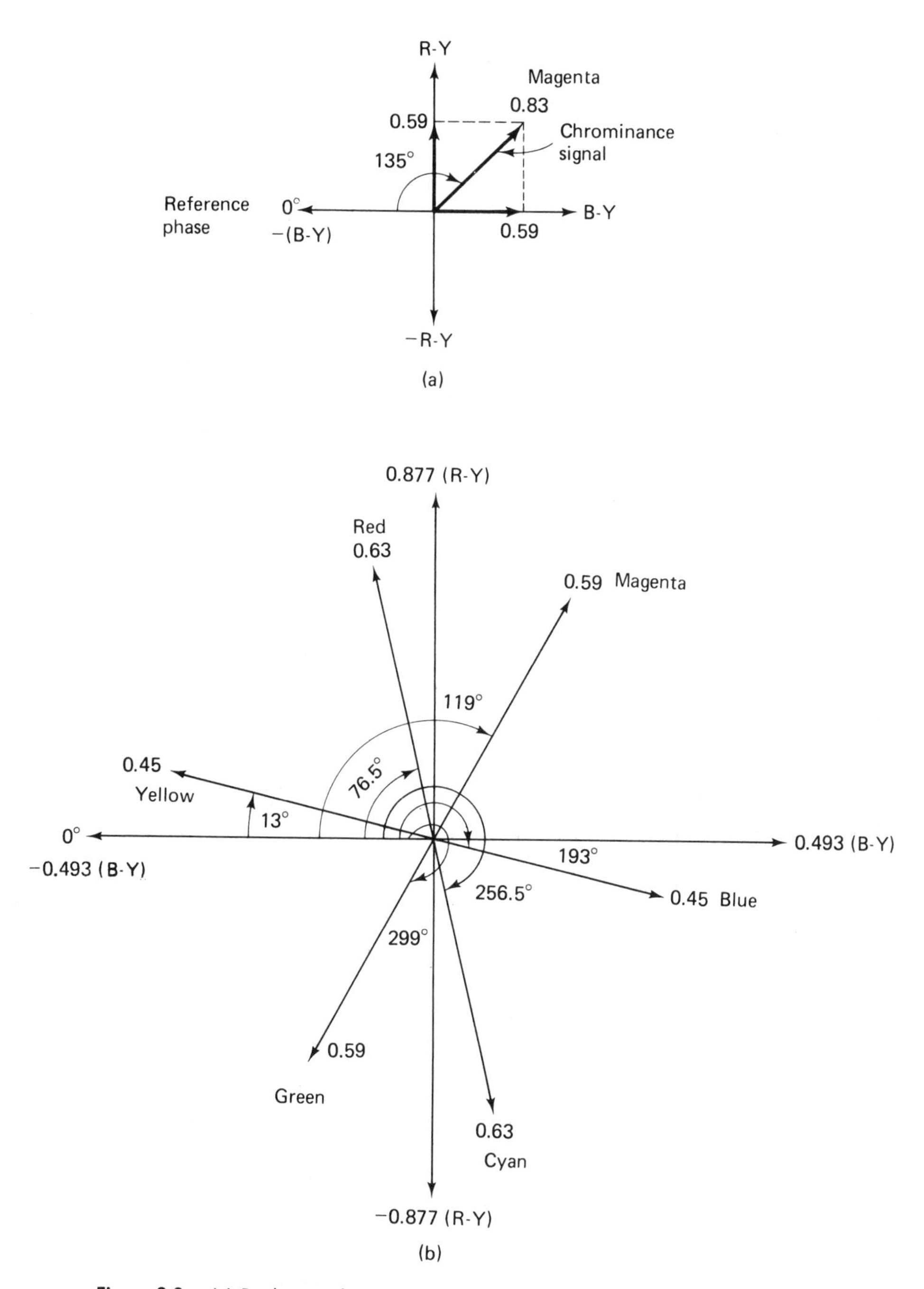

Figure 3-9 (a) Production of an uncompensated chrominance signal; (b) magnitude and phase relationships of compensated chrominance signals corresponding to the primary and complementary colors.

TABLE 3-1

Color	Uncompensated Chrominance		Compensated Chrominance	
	Amplitude	Phase	Amplitude	Phase
Magenta	0.83	135°	0.59	119°
Red	0.76	66.8°	0.63	76.5°
Blue	0.9	187°	0.45	193°
Green	0.83	315°	0.59	299.9°
Yellow	0.9	7°	0.45	13°
Cyan	0.76	246.8°	0.63	256.5°

of the chrominance signals and their phase angles. The phasor diagram showing the compensated values of the chrominance signal is shown in Figure 3-9(b).

At the receiver the chrominance signal must be adjusted to take this compensation into account. Table 3-1 is a list of colors showing their compensated and uncompensated chrominance amplitudes and phase angles. The compensated values are those which are actually transmitted.

Reference phase. It must be remembered that since phasor addition deals with sinusoids, the resultant voltage is also a sinusoid. As such, the resultant chrominance signal will be at some phase angle with respect to the $R - Y$ and $B - Y$ sideband signals. This *angle* represents the information about the *color* of the scene being televised.

In the color television system it has been decided that all color phase angles are to be measured from the phase position of a $-(B - Y)$ signal that corresponds to a greenish-yellow color. This phase angle has been designated 0° or the *reference phase* position on the phasor diagrams indicated in Figure 3-9(b). This is also the phase of the color burst that is transmitted on the back porch of each horizontal blanking pulse.

Color phase angle. Referring to Figure 3-9(b), the compensated color magenta is represented by a phasor at an angle of 119°. Similarly, the color red is shown to be represented by a phasor at an angle of 76.5°. The remaining phasor diagram indicates the phase and amplitude of the chrominance signal for blue, cyan, green, and yellow, respectively. In each case, a change in color is represented by a change in phase.

From the composite diagram it is possible to see that complementary colors are equal in amplitude and are 180° out of phase. Thus, yellow mixed with blue produces white. Also notice that the primary colors (*R, G,* and *B*) are 120° apart.

The amplitude of the chrominance signal is determined by the saturation of the color. The phasor diagram of Figure 3-9(b) is for fully saturated colors. If the scene being televised has colors that are of low saturation, the phasor diagrams that represent the color information being transmitted will indicate this by a reduction of the *lengths* of the $R - Y$, $B - Y$, and resultant chrominance phasors. During black & white transmission the color signals are zero and, therefore all the phasors vanish.

To complete the discussion of Figure 3-6(a), we see that the $R - Y$ and $B - Y$ sidebands that have been combined into the chrominance signal are then added to the Y signal and the color burst in an adder. This forms the composite video signal which is now sent to the main transmitter. The RF output of the transmitter forms a sideband spectrum like that shown in Figure 2-22. Notice that the color subcarrier is placed 3.58 MHz above the picture carrier.

Summary. In summary, the color cameras convert the color picture information into three electrical signals: red, green, and blue. These signals are then combined in a matrix that generates the luminance (Y) or black & white signal. The color signals are also combined in the matrix to produce two color difference signals: the $R - Y$ and $B - Y$ signals. These signals are increased in frequency and are placed 90° apart by the balanced modulators. This chroma signal is then added to the Y signal and the color burst to form the composite video signal. The composite video signal may then be used to modulate the main transmitter. Unfortunately, this is not the end of the story. One additional factor must be added in order to place the signal in the desired condition for actual transmission.

3-6 I and Q signals

In the early section of this chapter we discussed the color characteristics of the eye. One such feature of the eye that has been found to have important consequences is that the eye is blind to the color characteristics of very small objects. Furthermore, the eye only requires two colors (orange and cyan) to be satisfied by the color reproduction of medium-sized objects. However, three primary colors are necessary to provide full-color reproduction of large objects. These characteristics are taken into account at the transmitter as a means of conserving bandpass.

It is to be recalled that medium-sized objects generate a video signal whose frequency falls in the range 0.5 to 1.5 MHz. Thus, if both the $R - Y$ and $B - Y$ signals are to carry all this color signal information, the double-sideband bandpass at the output of each balanced modulator would be an undesirable total of 3 MHz. See Figure 3-6.

The transmission of the *double*-sideband information appearing at the output of the balanced modulators is not necessary when transmitting chrominance information for frequencies above 0.5 MHz because *either* upper or lower sidebands contain all of the information being transmitted. The only effect of double-sideband transmission at the receiver is to double the output amplitude of the received signal as compared to single-sideband reception. In the case of color television, it is more important to minimize bandwidth than to provide a strong signal; therefore, as a means of restricting bandwidth a form of vestigial sideband transmission for the chrominance signals is resorted to.

In addition, the bandwidth of the chrominance signals may be further reduced by taking into account the fact that only orange and cyan need be transmitted to reproduce the colors of objects generating a video frequency of 0.5 to 1.5 MHz.

To help understand how this is done, it must be recalled that the $R - Y$ and $B - Y$ signals represent colors that can be indicated on a phasor diagram similar to those of Figure 3-9(b). The following question might be asked: What would be the result of transmitting *only* the $R - Y$ signal and suppressing the $B - Y$ signal? This situation would occur if a range of only two colors are to be transmitted. Under these conditions, the $(R - Y)$ signal would correspond to a bluish-red because the $R - Y$ phasor falls close to red but between it and blue. By the same reasoning, the $-(R - Y)$ signal would correspond to a bluish-green color.

If the $B - Y$ signal were transmitted and the $R - Y$ signal were suppressed, the $-(B - Y)$ signal would correspond to a greenish-yellow and the $(B - Y)$ signal would correspond to a reddish-blue. Unfortunately, neither the $R - Y$ nor the $B - Y$ signals when transmitted alone are able to produce the desired color range of orange and cyan. The only way to get orange-cyan color reproduction during transmission of only one axis of modulation is to *change* the angle of modulation from $R - Y$ to a new angle called I. This new axis is designed to produce the orange-cyan color range needed to reproduce the color characteristics of medium-sized objects. The I color signal is derived from $R - Y$ and $B - Y$ signals in the matrix and then is placed at an angle of 57° with respect to the color burst in the balanced modulators. Along this axis the $+I$ signal corresponds to orange and the $-I$ signal represents cyan. The I sideband phase angle of 57° is determined by the phase of the subcarrier signal that feeds the I balanced modulator.

In order to maintain the 90° phase relationship between the two color signals that combine to produce the chrominance signal, a Q (for quadrature) signal is derived from the $R - Y$ and $B - Y$ signals in the matrix, and by means of the balanced modulator it is placed at right angles to the I signal sidebands. The $+Q$ signal roughly corresponds to the color magenta, and the $-Q$ signal represents a color that is almost green. Figure 3-10 illustrates these relationships. *The I and Q signals carry all of the color information that the R − Y and B − Y signals carried but they have the advantage that during the transmission of color information concerning small objects (0.5 to 1.5 MHz) the Q signal vanishes.* This leaves only the I signal and the luminance signals to provide full-color reproduction.

Figure 3-10 Phasor diagram of the I and Q signals superimposed on the $R - Y$ and $B - Y$ phasors.

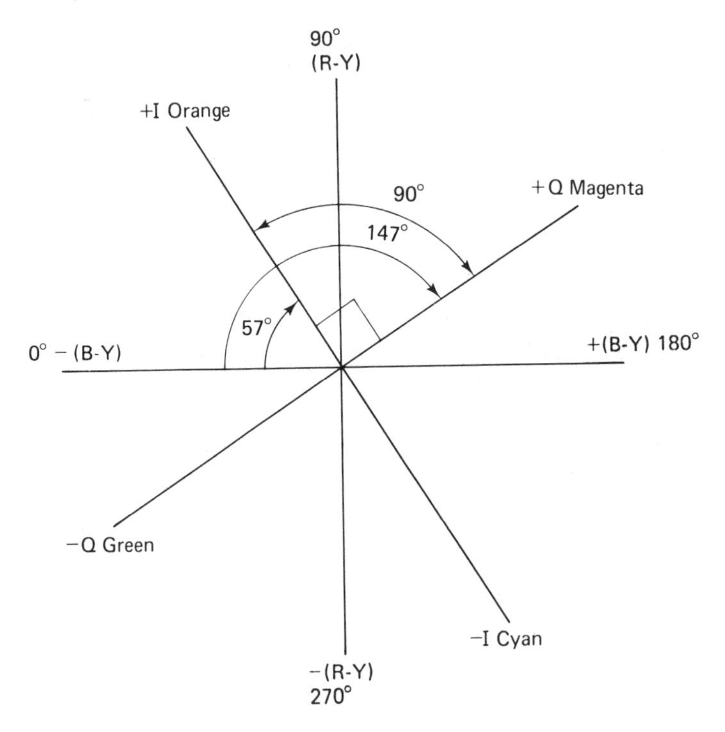

Bandwidth reduction. We will now examine how the change from $R - Y, B - Y$ transmission to I and Q as well as the use of vestigial sideband transmission can result in an overall reduction in bandwidth. First of all, the fact that the Q signal is not necessary when reproducing color information above 0.5 MHz means that the double-sideband Q channel bandwidth need only be 1 MHz. This represents a saving of 2 MHz as compared to the $R - Y, B - Y$ bandwidth requirements indicated in Figure 3-6. However, the I channel must carry video chrominance signals that extend to 1.5 MHz on each side of the subcarrier. Since the subcarrier frequency is 3.58 MHz, the upper sidebands of the I signal would extend to 5.08 MHz, which is well beyond the upper-frequency limit of the Y signal (4.2 MHz).

The object of all of the preceding efforts was to get the chrominance signals to combine with the Y signal so that they will both occupy the same space at the same time and yet not interfere with one another. To fit the I signal into the Y signal bandpass necessitates that its upper sideband be restricted to a maximum of 0.5 MHz above the 3.58-MHz subcarrier. As such, it is a form of vestigial sideband transmission since all of the lower sideband and only a small part of the upper sideband are available for transmission. Reducing the upper sideband by 1 MHz means that the total I-signal bandpass is 2 MHz. This is a saving of 1 MHz as compared to the bandpass requirements of the $R - Y$ and $B - Y$ channels shown in Figure 3-6.

The *I* and *Q* transmitter. A simplified block diagram of a practical color television transmitter using I and Q signals is shown in Figure 3-11. It should be noted that it is almost identical to the $R - Y, B - Y$ color transmitter block diagram of Figure 3-6. The major differences between the two systems occur in the matrix and subcarrier generator's phase shift networks. In the I and Q transmitter the I-balanced modulator is fed by a subcarrier signal that is phase shifted 57° with respect to the color burst and an I signal is obtained from the matrix. The Q signal is shifted an additional 90° from the I signal, thus placing it at 147° with respect to the color burst. Except for the phase angles, this is similar to what is done to the $B - Y$ signal in the color difference system.

The other major difference that exists is the bandpass characteristics of the I and Q modulators. The I modulator has a bandwidth of 2 MHz, and the Q modulator has a bandwidth of 1 MHz. This compares to a bandwidth of 3 MHz for both the $R - Y$ and $B - Y$ modulators in the color difference transmitter.

Both the I and Q and $(R - Y), (B - Y)$ systems are similar to each other in that the composite chrominance signal is added to the luminance signal in an adder. This circuit combines the two signals and then feeds them to the modulators of the main transmitter completing the system. The composite bandpass characteristic of the adder is shown in Figure 3-11. Notice that the Q signal is centered at 3.58 MHz and that the upper and lower sidebands are equal in bandwidth. Superimposed on the Q signal is the I signal. The I signal's upper sideband is reduced to 0.5 MHz, whereas the complete lower sideband is maintained for a full 1.5 MHz.

The bandwidth and frequency spectrum of each stage of the I and Q color transmitter, shown in Figure 3-11, is drawn in ladder diagram form below the block diagram. In this diagram the camera signals (R, G, and B) and the Y, I, and Q signals all start at zero frequency and extend to 0.5 MHz for the Q signal and to a maximum of 4.2 MHz for the Y and camera output signals. On the output side of the balanced modulators the

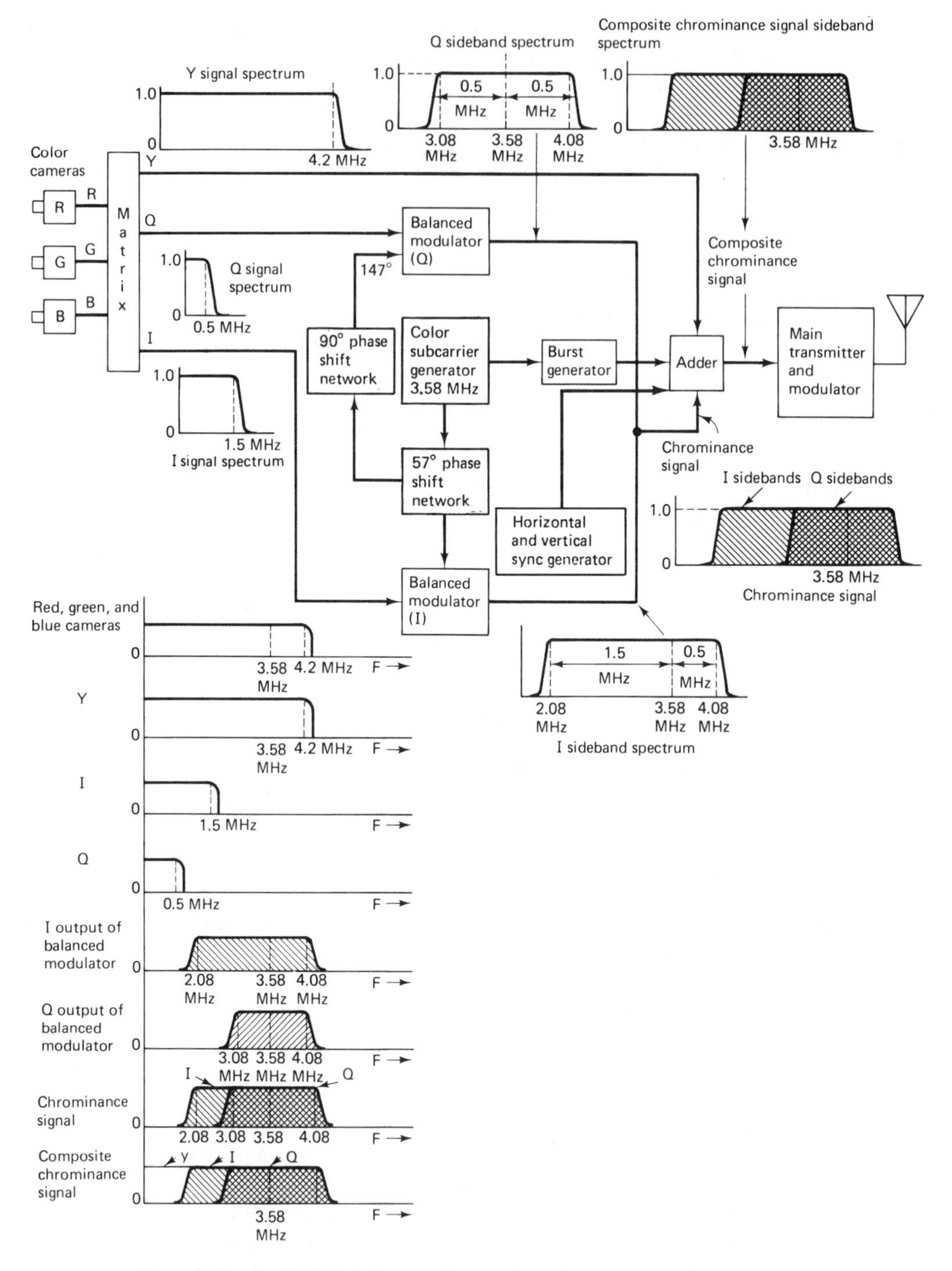

Figure 3-11 Simplified block diagram of an *I* and Q color transmitter. Also indicated are the bandwidth characteristics of each stage.

frequency of the I and Q signals has been increased so that both signals are centered at approximately 3.58 MHz. In the case of the Q signal, the sideband spectrum extends 0.5 MHz above and below the subcarrier frequency. The upper sidebands have a maximum frequency of 4.08 MHz and the lower sideband frequency limit is 3.08 MHz. The I signal has an unequal sideband spectrum, the upper sideband reaching a maximum frequency of 4.08 MHz which is 0.5 MHz higher than the color subcarrier frequency of 3.58 MHz, and the lower sideband extends from 1.5 MHz to a low frequency of 2.08 MHz. The addition of the I and Q signals takes place at the output of the balanced modulators forming the chrominance signal. An adder is used to combine the Y signal with the chrominance signal producing the composite chrominance signal which is then used to modulate the main transmitter. The overall bandwidth of this signal extends from 0 to 4.2 MHz. At this point all of the color and black & white video signal information has been concentrated into the same channel allocation as for a black & white signal alone.

3-7 frequency interlacing

The bulk of this chapter has been concerned with methods by which two signals (luminance and chrominance) of approximately the same frequency range can be combined and yet not mutually interfere with one another. Two methods by which interference was minimized was to increase the frequency of the chrominance signal and to suppress the subcarrier. Another method for reducing interference is to make the subcarrier frequency an *odd multiple of one-half the horizontal scanning frequency.* This is done by making the subcarrier frequency exactly 3.579545 MHz. The reasons for this choice of subcarrier frequency are outlined below.

Analysis of both the Y signal and of the chrominance signal has determined that the signal energy in each is distributed in discrete concentrations. In the case of the Y signal, the concentration of signal energy occurs at multiples of the horizontal line frequency (15,750 Hz). Surrounding each of the harmonics of the horizontal scanning rate (15,750 Hz) are clusters of signals that are harmonics of the field (60 Hz) and of the frame scanning rate (30 Hz). Figure 3-12(a) illustrates (in a simplified way) what this means. Notice that the figure is arranged in the form of a graph where frequency increases from left to right in discrete steps. Each step is separated from the next by an amount equal to the horizontal scanning frequency. The first step at the left is designated H and the next higher steps are designated as whole-number multiples of the first step. These higher-order steps are referred to as *harmonics* of the horizontal line frequency. As a general rule, the higher the frequency of the harmonic, the weaker it becomes. Theoretically, the number of harmonics needed to reconstruct the luminance signal is infinite. Figure 3-12(a) shows only a small number of them extending to the 286th harmonic; nevertheless, enough are shown so that it is obvious that there are large spaces between each cluster of harmonics. All one need do to minimize interference between the luminance and chrominance signals is to place the chrominance signal in these empty spaces.

Figure 3-12(b) shows the frequency position of the spectral components (harmonics) that make up the chrominance signal. Notice that the chrominance information peaks at a frequency of about 3.58 MHz and its harmonic components fall between the Y-signal harmonic components. To achieve this, the chrominance signal components are all made

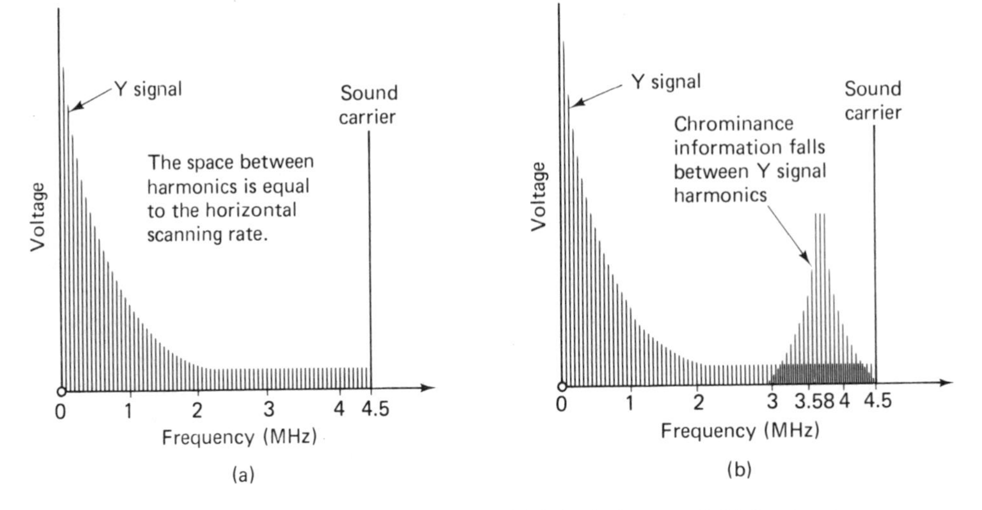

Figure 3-12 Harmonic signal distribution of (a) the luminance and (b) the chrominance signals.

into odd harmonics of one-half the line frequency. This system of interference avoidance is called *frequency interlace.*

Scanning frequencies. The subcarrier and the sound carrier of the color television signal differ in frequency from one another by approximately 920 kHz. In the receiver this difference frequency may cause picture interference. It has been found experimentally that the beat signal between the subcarrier and the sound carrier is much less objectionable if the sound carrier is made an odd multiple (117) of one-half the horizontal line frequency.

If the horizontal line frequency is *15,750 Hz,* the 285th and 286th harmonics correspond to frequencies of 4.48875 MHz and 4.5045 MHz, respectively. Unfortunately, the sound IF system of almost all television receivers is centered at 4.5 MHz. This means that if interference between the subcarrier and the sound IF is to be avoided and if the horizontal line frequency of 15,750 Hz is used, almost all existing receivers must be realigned to 4.5045 MHz in order to receive sound information during color transmission.

Another more practical solution is to change the horizontal scanning rate so that its harmonic (286th) falls exactly at 4.5 MHz. In practice, this is done by making the horizontal scanning rate 15,734.26 Hz (4.5 MHz/286 = 15734.26 Hz). It is this scanning rate that then determines the subcarrier frequency of 3.579545 MHz [455 × (15,734.26/2) = 3.579545 MHz]. Furthermore, a change in the horizontal scanning rate requires a change in the vertical scanning rate if interlace scanning is to be preserved. Therefore, the vertical scanning rate is made 59.94 Hz (15734.26/262.5 = 59.94 Hz). Compatibility with black & white television receivers is maintained because both scanning rate changes are so small that the receivers' scanning generators are easily able to synchronize with them.

3-8 composite chrominance signal

Up to this point we have examined the hows and whys of producing a compatible color television signal. We have examined the methods by which the color signal is modified to make the combination of luminance and chrominance signals free from mutual interference and at the same time allow them to simultaneously occupy the same 6-MHz channel bandwidth. All of the previous discussion revolved around phase relationships and the bandwidth consideration necessary to achieve the goal of compatible color television. In this section we will examine the overall *wave shape* as well as the wave shapes produced by the various color sections of the television transmitter.

The overall output wave shape of the color television transmitter is called the *composite chrominance signal.* The formation of this signal is detailed in the block diagram of Figure 3-11 and in Figure 3-13. Both of these figures are similar to each other in that they both indicate the manner in which the composite signal is formed. Figure 3-11 shows the bandwidth relationships for each stage and Figure 3-13 shows the output voltage wave shapes throughout the transmitter. These wave shapes represent the output voltages of each stage during one line of horizontal scan of a color bar pattern by the color television cameras.

Figure 3-13 is arranged as a ladder diagram in which each voltage waveform is positioned one over the other so that the waveform changes that take place in each circuit at any instant can be easily followed.

The color bar pattern at the top of the diagram is an arrangement of colors "seen" by the color cameras. For purposes of this discussion, each color is asssumed to have a brightness level of 100%. The color bar pattern has eight bars ranging from white through red, magenta, blue, green, yellow, cyan, and ending in black.

The voltage wave shapes seen below the color bar pattern are generated during one horizontal scanning line. The red camera output voltage will be maximum for colors that contain 100% red. Thus, the red camera produces maximum output for white, red, magenta, and yellow. The green camera output voltage is maximum for colors that contain green. Therefore, its output is maximum for white, green, and yellow. The blue camera "sees" blue in the white, magenta, and blue bars of the color bar pattern and produces a maximum output voltage as it scans through them.

The output of each camera is fed into a matrix that combines these signals in the proper proportions to produce the Y signal. The Y signal represents the brightness level of any given color as a shade of gray. A Y level of 0 is black and a level of 1 is white.

The matrix also feeds the I and Q balanced modulators. These circuits convert the input video frequency to higher-frequency color subcarrier sidebands centered at 3.58 MHz. The inputs of *each* modulator are shown in Figure 3-13.

It should be noted that the I and Q signal outputs are both zero for either of the black or white bars in the color bar pattern.

The 3.58-MHz I and Q sidebands appearing at the output of the balanced modulators are 90° out of phase. They are added together vectorially to produce the chrominance signal. The amplitude of the chrominance signal corresponds to the saturation of the color being scanned. The phase of the chrominance signal represents the color being scanned.

The chrominance signal is combined with the Y signal, the color burst, and the vertical and horizontal blanking and sync pulses to form the composite chrominance signal that is actually transmitted. Figure 3-14 shows a picture of an actual composite chrominance signal as seen on an oscilloscope.

Summary. At this point it is worthwhile to summarize the functions of both the color transmitter and the color receiver.

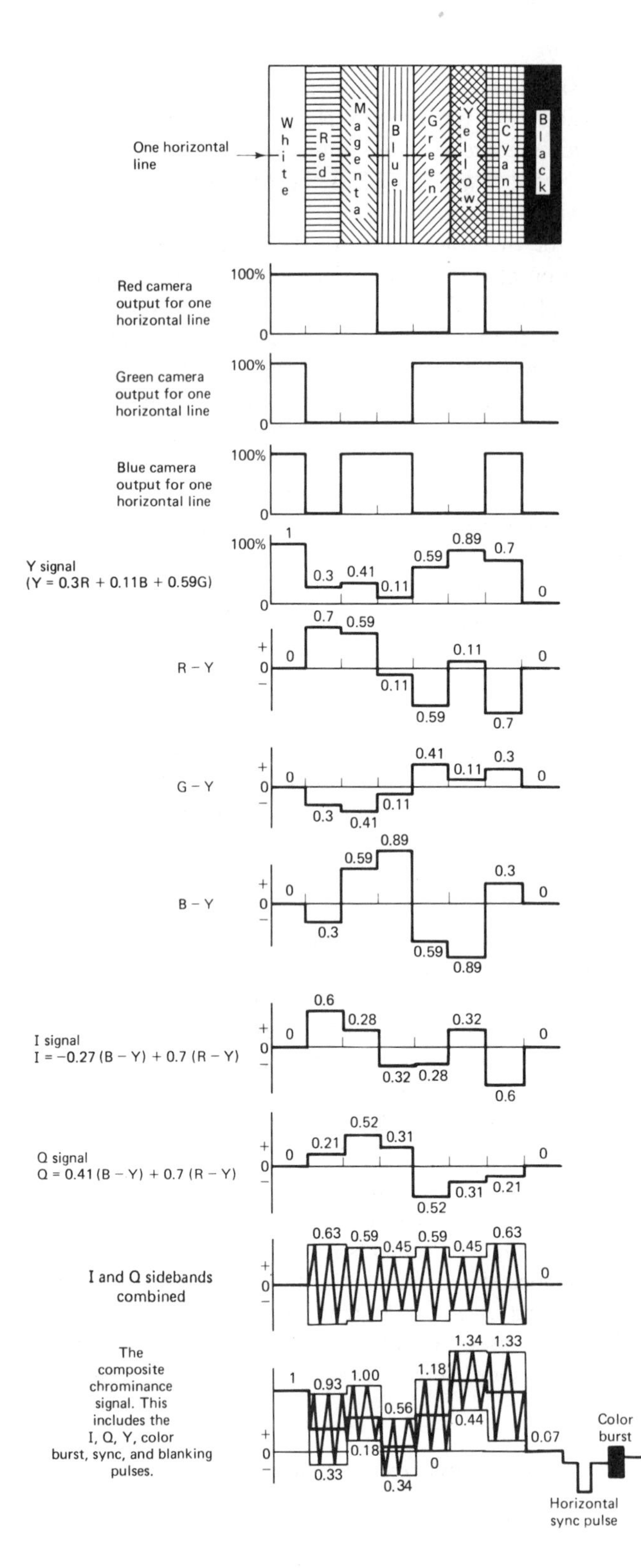

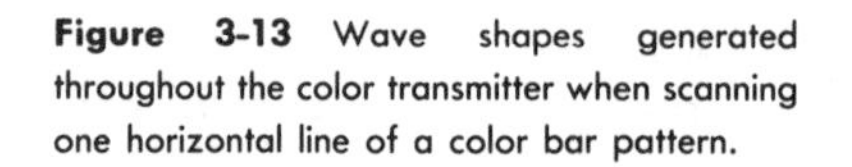

Figure 3-13 Wave shapes generated throughout the color transmitter when scanning one horizontal line of a color bar pattern.

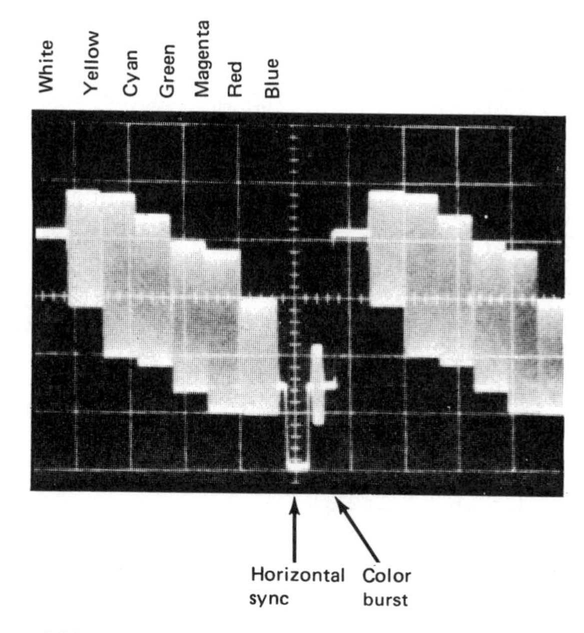

Figure 3-14 Oscilloscope presentation of the composite chrominance signal.

The functions of the *color transmitter* may be summarized as follows:

1. The color cameras convert the color picture into red, green, and blue video signals.
2. The color video signals are then converted by means of matrix into *Y, I,* and *Q* video signals.
3. The *I* and *Q* video signals are increased in frequency and are converted into *I* and *Q* chrominance signals by the balanced modulators.
4. The *Y* signal is added to the chrominance signal and the color burst to form the composite chrominance signal which then modulates the transmitter.

At the receiver the process is reversed.

1. The composite chrominance signal is separated into its components; the chrominance signal, the *Y* signal, and the color burst.
2. The chrominance signal is then converted, by means of the demodulators into $R - Y$ and $B - Y$ video signals.
3. The color difference video signals are then converted by means of a matrix into red, green, and blue video signals.
4. The red, green, and blue video signals drive the color picture tube to reproduce the colored picture.

3-9 block diagram of a color television receiver

The simplified block diagrams of both a black & white receiver and a color receiver are shown in Figure 3-15. In these diagrams only the essential circuits are indicated so that an overall comparison can be made. A survey of both systems will reveal that all the circuits

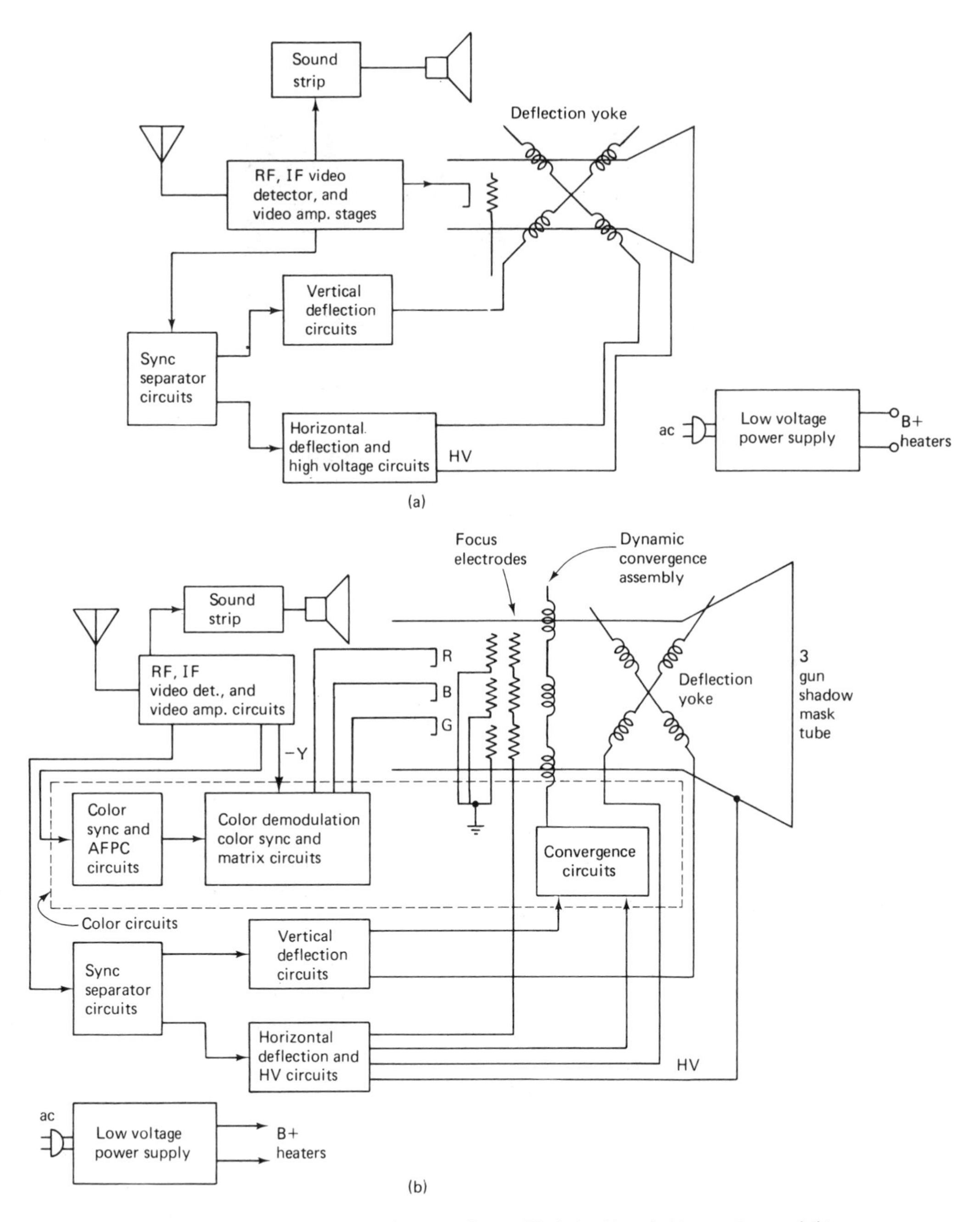

Figure 3-15 Simplified block diagrams of (a) a black & white television receiver, and (b) a color television receiver.

in the black & white receiver appear in the color receiver. However, the color receiver has three additional blocks that are not included in the black & white receiver. These are the convergence circuits and the two blocks found in the color circuit portion of the receiver.

Furthermore, those circuits that are common to both systems are expected to perform additional functions not required in the black & white receiver. For example, each of the deflection circuits is required to provide convergence outputs necessary for proper color picture tube operation. In addition, the horizontal deflection circuit is required to deliver to the picture tube high-voltage second anode potentials of approximately 25 kV, and in some receivers a focusing potential of 5 or 6 kV. The high voltage in the color receiver must be regulated. This is in contrast to the unregulated high voltage used in black & white receivers.

The color television picture tube is primarily responsible for many of the additional requirements placed on the television receiver circuitry. The most commonly used type of picture tube is a three-gun system in which each electron gun generates an electron beam that produces three complete rasters of red, green, and blue. The picture tube is so arranged that each colored raster is, in the ideal case, exactly superimposed one on the other. Unfortunately, because of the geometry of the tube the three rasters do not exactly superimpose, producing the condition of misconvergence. Misconvergence is especially noticeable as color fringing when viewing a color receiver during a black & white program. Figure 3-16(a) and (b) are photos of a color receiver that has an extreme case of misconvergence.

To compensate for the misconvergence normally produced by the picture tube, special circuits called *convergence circuits* may be used. These circuits are used in conjunction with three yokelike coils called the *dynamic convergence assembly* that is mounted on the neck of the picture tube. When properly adjusted these devices provide a picture that is fully converged.

The composite video and chrominance signal that modulates the transmitter's picture carrier is radiated by the transmitter and then after pick up by the receiver is processed by the RF, IF, and video detector stages of the receiver. At the output of the video detector the composite video and chrominance signals reappear in their original premodulation form (Figure 3-14.) There are four tasks required of the two blocks designated "color circuits" shown in Figure 3-15. These are (1) to separate the color information from

Figure 3-16 (a) Photo of a badly misconverged crosshatch pattern; (b) a misconverged dot pattern. See color insert.

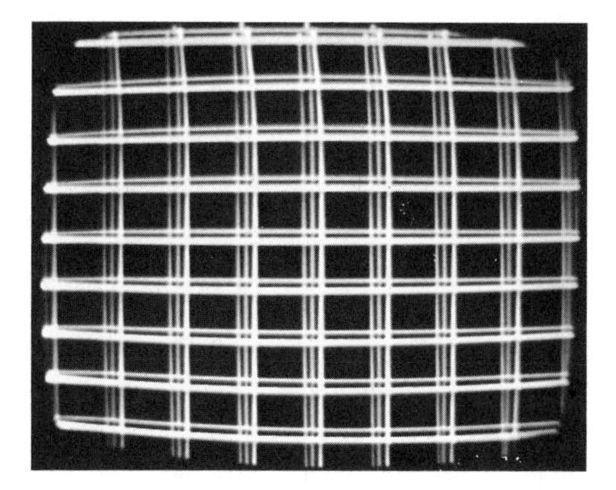

(a)

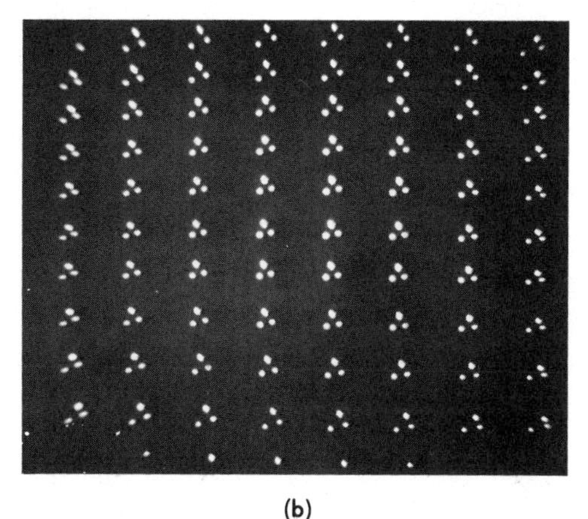

(b)

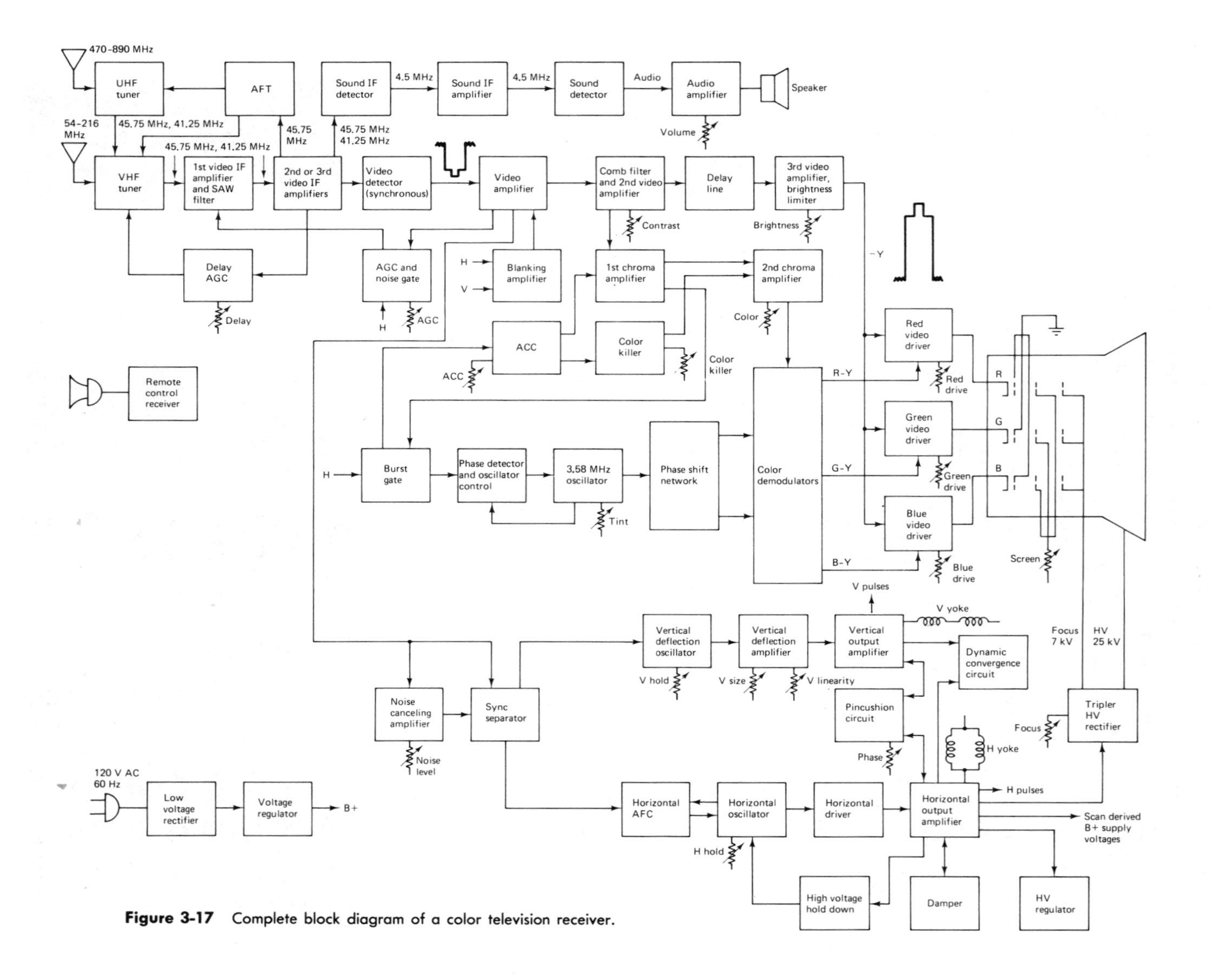

Figure 3-17 Complete block diagram of a color television receiver.

the black & white signal; (2) to reinsert a synchronized color subcarrier into the color signal; (3) to change the high-frequency chrominance signal to its original low-frequency form; and (4) to matrix the color and black & white signals such that the original color signals generated by the color cameras appears between grid and cathode of the color picture tube electron guns.

Current color receivers may use any one of a number of different matrix systems to reconstitute the original color signals. The system illustrated in Figure 3-15 is one of the more common types. In this system, the color demodulators produced $R - Y$, $B - Y$, and $G - Y$ signals that are matrixed with the $-Y$ signal. The outputs of this circuit are the original red, blue, and green signals. These signals are then fed to the corresponding cathodes of the picture tube.

Complete block diagram. A complete block diagram of a typical color television receiver is shown in Figure 3-17. This diagram includes all of the circuits normally found in a color receiver as well as all of the front and rear panel controls and adjustments that are usually associated with these circuits. This diagram depicts only one of several possible matrix system variations that are in use. This system is a commonly used arrangement and it, as well as the other systems, should be committed to memory. The other systems of matrixing will be discussed in detail in the chapter devoted to that topic.

Table 3-2 lists the stages and controls commonly found in color receivers and describes their functions. The table also indicates the effect on raster, picture, sound, and color if a stage becomes defective.

TABLE 3-2

Stage or Control	*Function*	*Raster*	*Picture*	*Sound*	*Color*
UHF and VHF tuners	This stage selects the desired TV channel from the signals fed to it by the antenna and converts it into the video IF.	O.K	No picture	No sound	No color
AFT (automatic fine tuning)	This stage is used to free the observer from making fine-tuning adjustments. This is especially important in color receivers because fine tuning is a critical adjustment.	O.K.	O.K.	O.K.	Manual fine tuning required for good color
Remote tuning	This stage is usually a low-frequency receiver that is used to activate from a distant point, various circuits that control volume, contrast, channel selection, tint and saturation controls.	O.K.	No channel selection	No sound control	No color control

TABLE 3-2 *(continued)*

Stage or Control	*Function*	*Raster*	*Picture*	*Sound*	*Color*
Video IF amplifiers and synchronous detector	These stages provide most of the signal amplification of the receiver. They also provide the necessary selectivity and bandpass characteristics required to compensate for vestigial sideband transmission and they also minimize interference caused by adjacent channel picture and sound carriers. The synchronous video detector changes the high-frequency IF into the composite video signal.	O.K.	No picture	No sound	No color
Sound strip	This block can actually be expanded into at least four stages. (1) A video IF to sound IF detector that converts the high-frequency video IF to the lower 4.5 MHz IF. This block is drawn separately. (2) The sound IF amplifier that operates at a frequency of 4.5 MHz. (3) The sound detector converts the sound IF into audio-frequency signals. (4) The audio output stage is an amplifier that boosts the level of the signal coming from the detector to that necessary to drive a loudspeaker. The sound strip differs from that in a black & white receiver in that its input is obtained from the video IF instead of from the video detector.	O.K.	O.K.	No sound	O.K.
Volume control	This control is located in the audio amplifier stage and controls the audio signal level.				
First video amplifier	This stage amplifies the composite video signal and acts like a signal dividing center. The output of this stage provides signals for (1) the second video ampli-	(a) O.K. (b) No raster	(a) Weak picture (b) No picture	O.K.	(a) O.K. (b) Weak or no color

TABLE 3-2 *(continued)*

Stage or Control	*Function*	*Raster*	*Picture*	*Sound*	*Color*
	fier, (2) the sync separator, (3) the AGC stage.				
Contrast control	This control adjusts the peak-to-peak amplitude of the composite video signal				
Comb filter and second video amplifier	Provides additional amplification for the video signal, and separates the color signal from the black & white signal (Y).	(a) O.K. (b) No raster (c) Blooming	(a) Weak picture (b) No picture	O.K.	O.K.
Brightness control	This control adjusts the overall light output of the picture tube. It may be located in the second video amplifier stage and it controls the bias of this stage. The second video amplifier may be directly coupled to the cathode of the picture tube. Hence, any change in bias of the second stage is communicated to the picture tube.				
Delay line	Strictly speaking, this is not a stage but a circuit element. Its purpose is to delay the Y signal so that it arrives at the picture tube at the same time as the color signal information.	(a) O.K. (b) No raster (c) Blooming	(a) No picture (b) Ringing in picture	O.K.	O.K.
Third video amplifier and brightness limiter	Provides sufficient additional amplification for the video signal so that it is able to drive the picture tube. The brightness limiter prevents excessive brightness that can lead to blooming.	(a) O.K. (b) No raster (c) Blooming	(a) No picture (b) Dim picture (c) Weak picture	O.K.	O.K.
AGC noise gate	This block is actually two stages that have the responsibility of generating a noise-free dc voltage that is proportional to the strength of the input RF signal. The dc voltage is used as an AGC voltage that controls the gain of the IF amplifiers.	O.K.	(a) Excessive AGC—weak or no picture (b) Insufficient AGC—overloaded picture or negative picture	O.K.	Same as picture conditions (a) and (b)

TABLE 3-2 (*continued*)

Stage or Control	*Function*	*Raster*	*Picture*	*Sound*	*Color*
AGC level control	This control is adjusted to prevent strong signals from overloading the RF and IF amplifiers.				
Noise-canceling amplifier and sync separator	These stages separate the sync pulses from the composite video signal with a minimum of noise interference.	O.K.	Defective sync—loss of both horizontal and vertical hold	O.K.	O.K.
Vertical deflection oscillator and output stages	These stages generate the vertical deflection sawtooth and drive the yoke.	No vertical deflection; one horizontal line in center of raster	Video signal reaches picture tube	O.K.	Color signal reaches the picture tube
Vertical hold control	This control adjusts the vertical oscillator frequency.				
Vertical height control	Adjusts the amplitude of the vertical deflection sawtooth.				
Vertical linearity control	Adjusts the linearity of the vertical deflection sawtooth.				
Horizontal AFC and oscillator	The AFC (automatic frequency control) portion of this block compares the horizontal sync pulses with the local horizontal oscillator output and generates a dc voltage that depends upon any frequency error between them. This error voltage is used to correct the error. The horizontal oscillator generates the horizontal deflection sawtooth.	O.K. (a) No raster (b) O.K.	Loss of horizontal sync	O.K. O.K.	O.K. Color signal reaches picture tube. No color
Horizontal hold control	Controls the frequency of the horizontal deflection oscillator.				
Horizontal driver	This stage matches the horizontal oscillator to the output amplifier stage and provides proper drive.	No raster	No picture	No sound	No color
Horizontal output	This stage is probably one of the most important circuits in the television receiver. It drives the deflection yoke to provide horizontal scanning. It also supplies pulses for AGC,	No raster	No picture	No sound	No color

TABLE 3-2 *(continued)*

Stage or Control	*Function*	*Raster*	*Picture*	*Sound*	*Color*
	blanking, and the burst gate as well as providing the ac voltages needed for the high-voltage second anode and focusing electrodes of the picture tube. It is responsible for the pulses driving the scan-derived $B+$ supply rectifiers				
Damper	This circuit works in conjunction with the horizontal output stage to provide horizontal scanning. The damper is primarily responsible for deflection on the left-hand side of the raster.	No raster	No picture	No sound	No color
High-voltage rectifier	This circuit rectifies the high-voltage pulses coming from the flyback transformer associated with the horizontal output amplifier into high-voltage dc. This circuit also provides focus voltage.	No raster	No picture	O.K.	No color
High-voltage regulator	This circuit keeps the high voltage that is fed to the picture tube constant under varying load conditions.	(a) No raster (b) Blooming	No picture	O.K.	No color
High-voltage hold-down circuit	To prevent excessive x-ray production this circuit disables the receiver if the high voltage exceeds its design rating. Defective high-voltage holddown circuits can cause the receiver to become defective by loss of raster, loss of picture, loss of horizontal or vertical sync or any other symptom that would force the receiver to be repaired.	No raster	No picture	No sound	No color
Pincushion circuit	This circuit is used to overcome a defect of deflection called *pincushioning.* This defect causes vertical deflection to decrease as horizontal deflection approaches the center of the	(a) Pincushion distortion of the raster (b) No raster (c) Ringing in raster	O.K.	O.K.	O.K.

TABLE 3-2 (continued)

Stage or Control	Function	Raster	Picture	Sound	Color
	screen, and it also causes horizontal deflection to decrease as vertical deflection approaches the center of the screen.				
Focus control	This control adjusts the dc voltage output of the focus supply, thereby adjusting the focus of the raster.				
Dynamic convergence circuit	Convergence circuits do not contain any transistors but they are essential to proper color television picture tube operation. These circuits take a sample of the vertical and horizontal deflection voltages and combine them to form the correction voltages and currents used by the convergence assembly mounted on the neck of the picture tube in order to compensate for misconvergence. Modern receivers may not have any convergence circuits or controls.	O.K.	Poor dynamic convergence	O.K.	O.K.
Dynamic convergence controls	Achieving proper convergence may require as many as 12 different controls that shape and determine amplitude and phase of the convergence correction voltages and currents fed to the dynamic convergence assembly.				
First and second chroma bandpass amplifiers	These stages amplify the high-frequency chrominance signal fed to it by the first video amplifier and comb filter. The output of the second video amplifier drives the color demodulators.	O.K.	O.K.	O.K.	(a) No color (b) Weak color (c) Poor color
Color control	The color control determines the output amplitude of the chroma bandpass amplifier. Its effect on color reproduction is to adjust the color saturation of the color picture.				

TABLE 3-2 *(continued)*

Stage or Control	*Function*	*Raster*	*Picture*	*Sound*	*Color*
Color killer	This stage acts like a switch that is able to turn on or turn off the bandpass amplifier. The "switch" turns on the bandpass amplifier when a color signal is transmitted and turns it off during black & white transmission.	O.K.	Color snow in picture	O.K.	(a) O.K. (b) No color (c) Weak color
Color killer control	This control adjusts the signal level required to activate the color killer.				
Burst amplifier or burst gate	The burst amplifier is a switch that is keyed on and off by a pulse coming from the horizontal output amplifier. The timing of this pulse is such that the burst gate is turned on during the time that the color burst appears at its input. Therefore, the only output of the stage is the eight cycles of color burst. During black & white transmission there is no color burst; hence, there is no output from the burst gate.	O.K.	O.K.	O.K.	Loss of color sync
Automatic color control (ACC)	This stage keeps the color level constant under varying signal conditions.	O.K.	O.K.	O.K.	(a) Weak or no color (b) Excessive color
Chroma phase detector	This stage compares the color burst coming from the burst gate with the 3.58-MHz signal generated by the local chroma oscillator. If any phase or frequency difference exists between these two signals, a dc error voltage is generated that tends to force the oscillator into synchronism with the color burst. This circuit also generates a dc voltage that disables the color killer and thereby activates the chroma bandpass amplifier.	O.K.	O.K.	O.K.	Loss of color sync

TABLE 3-2 *(continued)*

Stage or Control	*Function*	*Raster*	*Picture*	*Sound*	*Color*
Chroma oscillator control	This stage is fed by the dc output of the phase detector. Depending upon the magnitude and polarity of this dc voltage, this stage forces the chroma oscillator to synchronize with the color burst.	O.K.	O.K.	O.K.	(a) No color or weak color (b) Wrong color
Chroma oscillator	This stage generates the 3.58-MHz subcarrier signal that was suppressed at the transmitter.	O.K.	O.K.	O.K.	(a) No color, weak color (b) Loss of color sync (c) Wrong color
Tint control	This control adjusts the phase of the local 3.58-MHz oscillator and thereby controls the resultant picture color.				
Phase shift network	This passive circuit shifts the 3.58-MHz oscillator output phase by the amount necessary to allow the color demodulators to function properly.				
Color demodulators	This circuit is a frequency changer that converts the chrominance signal (3.58-MHz) into $R — Y$, $B — Y$, and $G — Y$ color signals (0 to 0.5 MHz).	O.K.	O.K.	O.K.	$R — Y$ demodulator (a) Loss of red (b) Excessive red raster $B — Y$ demodulator (a) Loss of blue (b) Excessive blue raster $G — Y$ demodulator (a) Loss of green (b) Excessive green raster
Red, green, and blue video drivers	These stages perform two functions. First, they act as a matrix combining the $-Y$ signal with the color difference signals producing red, green, and blue signals. Second, they are amplifiers	Red amplifier (a) Cyan raster (b) Red raster (c) Blooming Green ampli-	O.K.	O.K.	Red amplifier (a) Loss of red (b) Excessive red Green amplifier (a) Loss of green

TABLE 3-2 *(continued)*

Stage or Control	*Function*	*Raster*	*Picture*	*Sound*	*Color*
	that provide sufficient amplification to the difference signals to drive the color picture tube.	fier (a) Magenta raster (b) Green raster (c) Blooming Blue amplifier (a) Yellow raster (b) Blue raster (c) Blooming			(b) Excessive green Blue amplifier (a) Loss of blue (b) Excessive blue
Horizontal and vertical blanking amplifier	This stage obtains its input pulses from the vertical and horizontal output amplifiers. Its output is used to turn off the picture tube during vertical retrace, horizontal retrace, and the color burst intervals.	Retrace lines visible	O.K.	O.K.	O.K.
Master screen control	This control adjusts the screen grid (first anode) voltages of the picture tube as part of the adjustment procedures used to obtain gray scale tracking.				
Red, green, and blue drive controls	These three controls adjust the grid to cathode bias of each electron gun of the picture tube and help to obtain proper gray scale tracking.				
Low-voltage power supply	This stage converts the ac line voltage into the required ac and dc voltages needed to power the color television receiver.	No raster	No picture	No sound	No color
Voltage regulator	This circuit keeps the low-voltage supply voltage constant under varying load conditions.	No raster	No picture	No sound	No color

QUESTIONS

1. In what way is color television superior to black & white TV?
2. Define the following terms: (a) color, (b) brightness, (c) saturation, (d) nonspectral colors.
3. What is white light made of? What is black made of?
4. What is the difference between additive color mixing and subtractive color mixing?
5. Explain how a color picture tube produces the color cyan.
6. What are the frequency limits of the luminance and chrominance requirements of the color TV system based on the characteristics of the eye?
7. What are the differences in frequency and wave shape between the R, G, and B camera signals, the Y signal, the $R - Y$, $B - Y$, and $G - Y$ signals, and the chrominance signals?
8. How is the $G - Y$ signal extracted from the composite video signal?
9. What is the function of the balanced modulator in a color transmitter?
10. What is the function of the color burst?
11. What does the phase and amplitude of the chrominance signal correspond to?
12. Why does the color television transmitter transmit I and Q signals?
13. To what colors do the I and Q signals correspond?
14. What is frequency interlacing?
15. List two reasons for making the color subcarrier frequency exactly 3.579545 MHz.
16. Explain why the horizontal scanning frequency is 15,734.26 Hz and the vertical scanning frequency is 59.94 Hz in the color TV system.

chapter 4

The Front End

4-1 introduction

Figure 4-1 is a block diagram of an entire color television receiver. Those sections of the receiver that are to be discussed in this chapter are shaded. A number of conclusions may be drawn from a study of this block diagram. First, the RF signal coming from the transmitter enters the receiver via the "front end." The function of the front end is to select, amplify, and change this incoming signal to a lower frequency. Second, the front end consists of as many as four groups of circuits as follows:

1. The VHF tuner
2. The UHF tuner
3. Automatic fine-tuning (AFT) circuits
4. Remote control

The VHF and UHF tuners are usually mounted on a separate subassembly chassis, such as those shown in Figure 4-2. AFT circuits are more often found in color receivers and are usually mounted on separate printed circuit boards that are located physically near the tuner. Remote control is usually all electronic; however, in older receivers it was electro-mechanical; therefore, the motors, gears, and belts are mounted on and around the tuner itself. Third, the block diagram indicates that the front end is used in the same way for both color and black & white receiver operation. This means that defects in the tuner will show themselves during either color or black & white modes of operation. Similarly, a given defect in a tuner will result in the same kind of symptom on color as well as on black & white receivers.

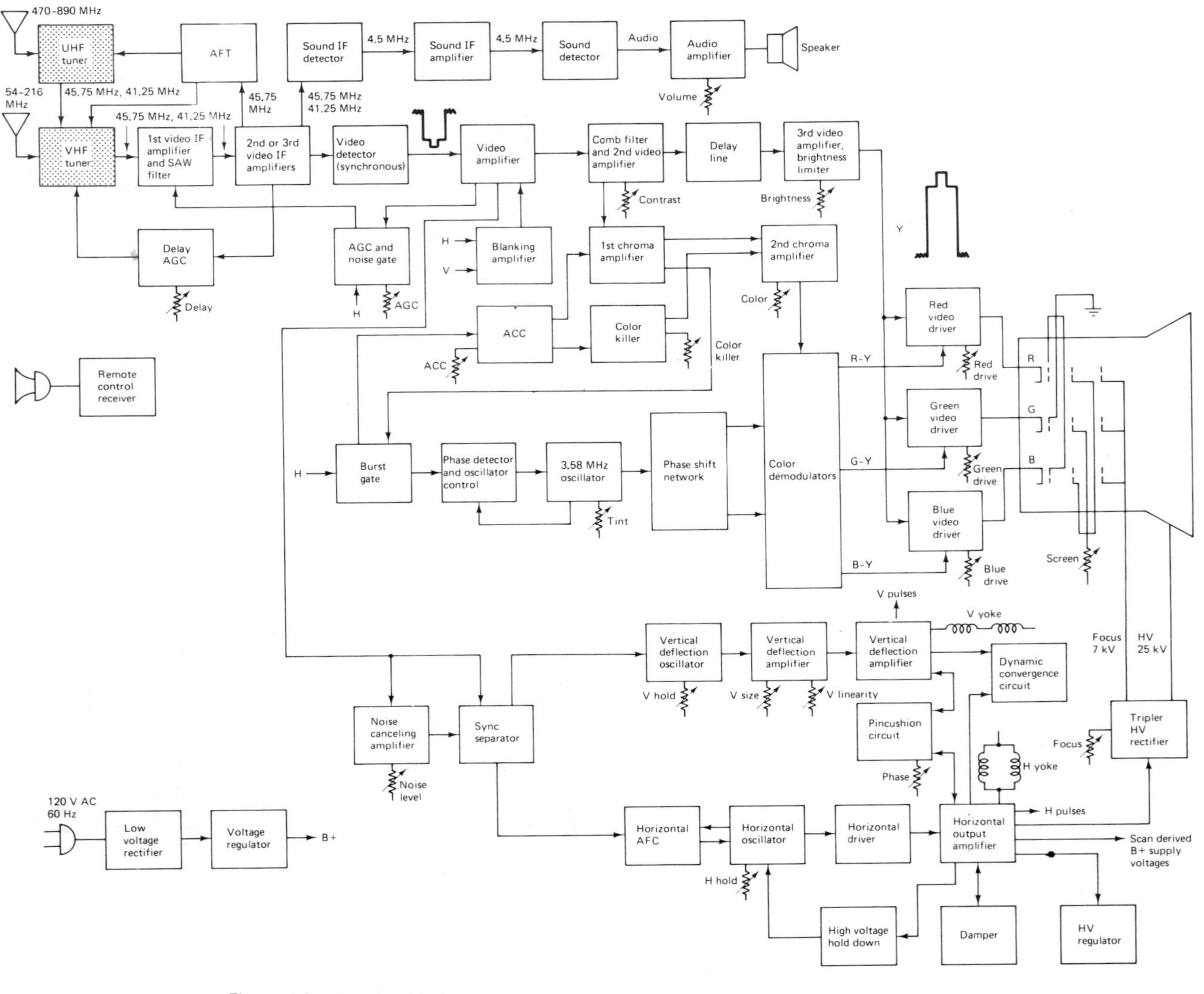

Figure 4-1 Complete block diagram of a color television receiver. The shaded blocks indicate the circuits to be discussed in this chapter.

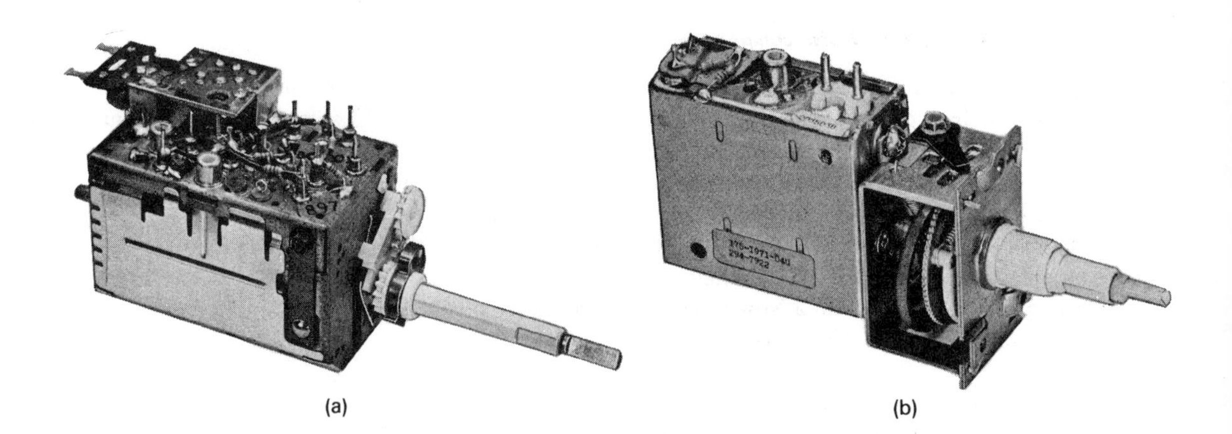

(a) (b)

(c)

Figure 4-2 (a) VHF tuner; (b) UHF tuner; (c) VHF/UHF integral varactor tuner. (Courtesy of Zenith.)

4-2 VHF and UHF comparable tuning

Because of the large differences in operating frequencies between VHF and UHF television channels, two separate tuners are used to receive them. This has led to a problem in that VHF and UHF tuners may select channels differently. The FCC ruled that after July 1, 1974, all new television receivers must make VHF and UHF tuning *comparable.* That is, the channel selecting mechanism should be able to select with equal ease a given channel whether it is VHF or UHF. This same FCC ruling directed that receivers should be able to receive 12 VHF channels and 6 UHF channels. This ruling and other earlier rulings have resulted in a situation in which there have been four different front-end combinations in use. These are:

1. VHF tuner only (prior to 1964)
2. VHF and UHF tuners mechanically and electronically linked to provide a full 82-channel reception capability
3. VHF and UHF tuners arranged so that tuning is comparable; reception may be limited to 18 channels: 12 VHF and 6 UHF
4. Random access tuning allowing the reception of as many as 135 VHF, UHF, and cable channels

4-3 VHF tuner

The complete block diagram of a VHF tuner is shown in Figure 4-3. The block diagram is the same for both color or black & white receivers and for vacuum tube or solid-state tuners. The basic functions of the tuner are to:

1. Select the desired channel from the many that may be transmitted.
2. Amplify the selected signal.
3. Change the frequency of the incoming RF (54 to 890 MHz) to the lower intermediate frequency (41.25 to 45.75 MHz).

The tuner block diagram indicates that the function of channel selection is accomplished by simultaneously adjusting the tuned circuits of all three stages that make up the tuner. In practice, this means that three or four tuned circuits must be changed when changing channels. The tuned circuits found in both vacuum tube and transistor tuners are as follows:

1. The input tuned circuit to the RF amplifier
2. The output tuned circuit of the RF amplifier
3. The input tuned circuit to the mixer
4. The oscillator tuned circuit

In tuners that have three tuned circuits it is usually the mixer input tuned circuit that is left out. Each tuned circuit must consist of a coil and a capacitor. The resonating capacitance consists of the *distributed capacitance* of the circuit plus small fixed or variable ceramic capacitors.

Tuners may be constructed in four different forms:

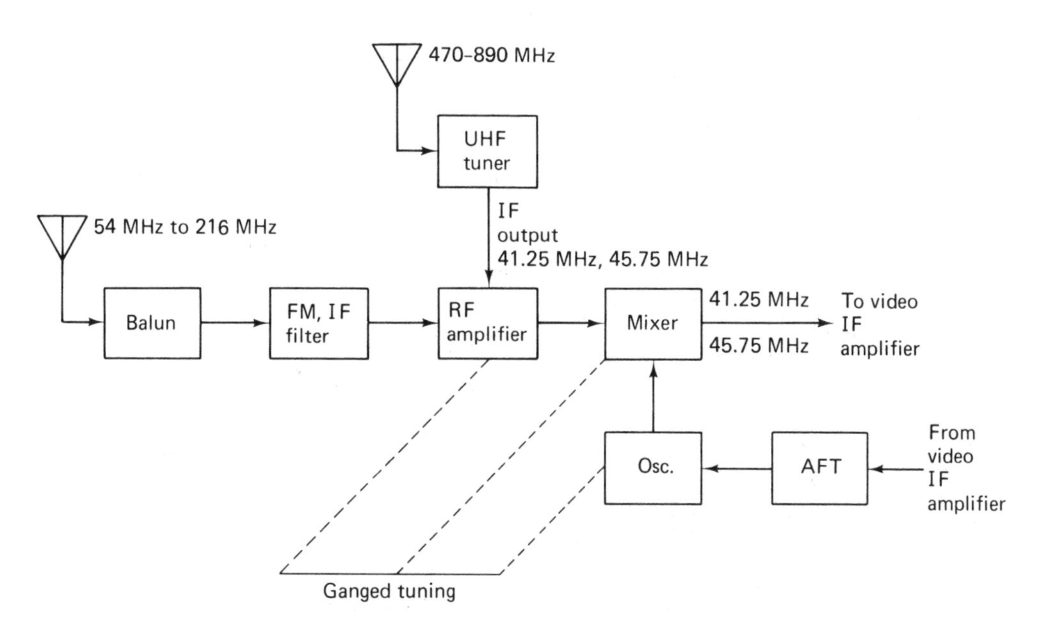

Figure 4-3 Complete block diagram of a VHF tuner.

1. Drum or turret
2. Wafer switch
3. Electronic tuning
4. Continuous tuning

The following paragraphs examine these types.

4-4 drum-type tuners

Changing channels may be accomplished as follows: either the inductance of the tuned circuit may be varied leaving the resonating capacitance fixed, or vice versa. The most common method is to mechanically vary the inductance of the tuned circuit. This may be done by physically changing all of the resonating inductances simultaneously by means of a *turret* or *drum*-type tuner. See Figure 4-4. In this tuner, channel selection is accomplished by rotating a drum made up of 13 (12 VHF and 1 UHF) individual strips that contain all four necessary tuned circuit coils. Each strip has a number of contact points associated with the coils that mate with contact points in the tuner and thereby are connected into the circuit. The tuner is locked into its proper position by a mechanical device called a *detent.* This is formed by a notched wheel that is part of the tuner shaft. A spring-loaded roller is placed on the edge of the wheel and it falls into a notch each time a channel is changed.

Many tuners provide fine tuning by means of a small gear-head metal screw that is inserted into one end of each oscillator coil strip. See Figure 4-4. The position in the end coil may be changed by rotating the gear head. The position of the screw adjusts the inductance of the oscillator section of the tuner, adjusting its frequency. In the usual

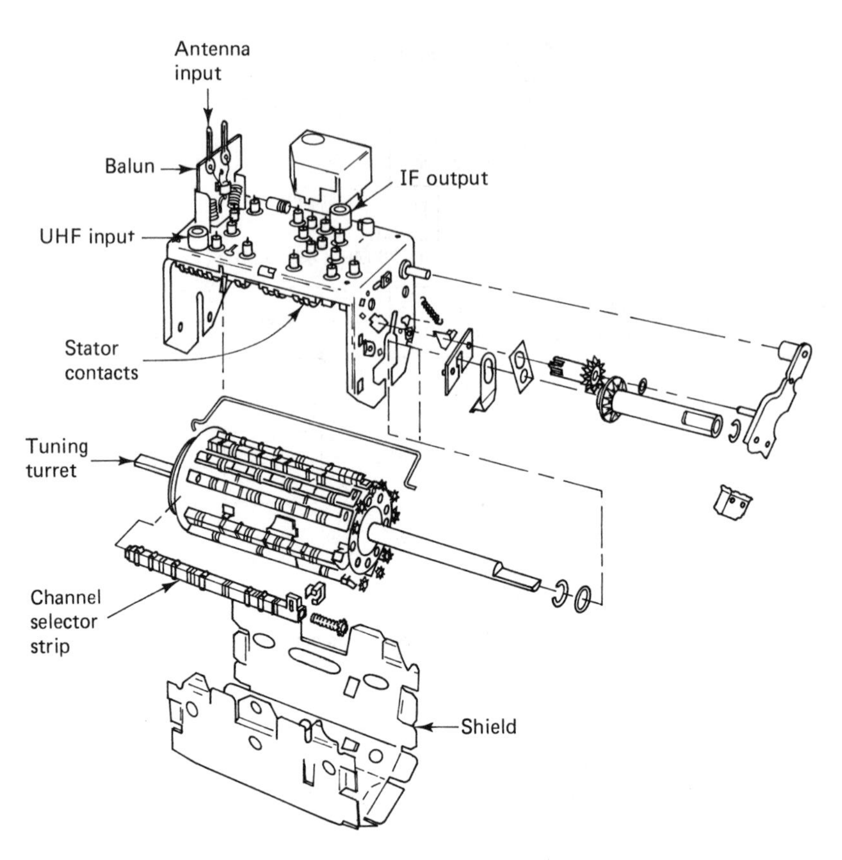

Figure 4-4 Turret-type tuner.

arrangement, the fine-tuning adjustment has a gear mechanism that meshes with the gear-head screw. Turning the fine tuning knob causes the screw in the oscillator coil of the strip to change its position, thereby changing the inductance of the coil and the frequency of the local oscillator. A number of different kinds of tuner strips have been in use.

Dirty tuners. One of the major electromechanical problems associated with the drum-type tuner is the accumulation of dirt and grease on the contact points. The trouble symptoms produced by the dirt accumulations fall into two categories:

1. Weak, snowy, or no picture or sound on one or more channels.
2. One or more channels are selected with difficulty. The channel selector knob must be wiggled or even placed at a critical point between channels in order to get reception. During the wiggling of the tuner knob loud scratchy noises are heard in the speaker and dark streaks appear in the picture.

Cleaning the tuner contacts will restore normal operation of the receiver. Cleaning is accomplished in one of two ways. Either each contact is abraded with something like the eraser end of a pencil, or a chemical tuner cleaner in the form of an aerosol spray, is applied to the internal and to the external contacts of the tuner. The disadvantage of friction cleaning of tuner contacts is that the tuner drum must be removed from the tuner chassis if the internal stator contacts are to be cleaned. Spray cleaning avoids this problem because it washes away the dirt and grease by chemical action. Care must be taken when using cleaning sprays that have lubricants because they may tend to detune various critical adjustments, such as neutralization, and may cause the tuner to oscillate.

4-5 switch-type tuners

Another commonly used method of channel selection is the wafer switch, or simply the switch-type tuner. See Figure 4-5. In this arrangement, the four necessary resonating inductances are each divided into 12 coils. These are mounted on four wafer switches that are all ganged together by a common shaft. Rotating the shaft rotates one or more contact fingers on each wafer that selects the necessary inductances to produce resonant circuits for the desired channel.

The coils may be arranged in series or they may be individually selected. This is also shown schematically in Figure 4-5. The antenna (E_1), RF wafer (E_2), and mixer wafer (E_3) coils are all of the series type. The oscillator coil (E_4) wafer is usually arranged as individual coils.

The usual arrangement found on most manufacturers' schematic diagrams shows the switch in the channel 13 position. In this position the coil (a single turn or two of wire) used to resonate the circuit to the frequency of channel 13 is the only coil not usually part of the wafer switch assembly. It should be noted from Figure 4-5(a) that when the switch is in the channel 13 position, the other coils are shorted out. As the switch position is moved to the lower channels, the coil necessary for the desired channel is produced by the series combination of the higher channel coil plus the inductance that is added to it by progressively removing the short placed across them by the switch. When channel 2 is reached, the total resonating inductance will be maximum because all of the coils mounted on the wafer will now be connected in series. This method of coil selection is most often used in the RF amplifier and mixer circuits of the tuner.

Shown in Figure 4-5(b) is the oscillator wafer switching arrangement. In contrast to the other tuned circuits, the individually tuned circuits in the oscillator section are all slug tuned. These adjustments are either screwdriver adjustments, accessible from the front of the receiver and intended for coarse adjustment of each channel, or part of the fine-tuning mechanical arrangement of the tuner.

As was the case with the turret tuner, dirt accumulations on the switching contacts can result in a weak picture or no picture or in noisy and erratic reception. In this type of tuner, spray chemical cleaning is the only method used to clean the tuner other than replacing the wafer.

4-6 electronic tuning

As a consequence of the FCC ruling on tuner comparability, the TV receiver industry has introduced UHF tuners with 70 position *detents* and all *electronic tuning* methods. See Figure 4-6. The tuning functions of the all electronic type of tuner are identical to those of the turret and switch types. That is, channel selection is accomplished by changing one of the reactive elements in all of the resonating circuits used in the tuner. In

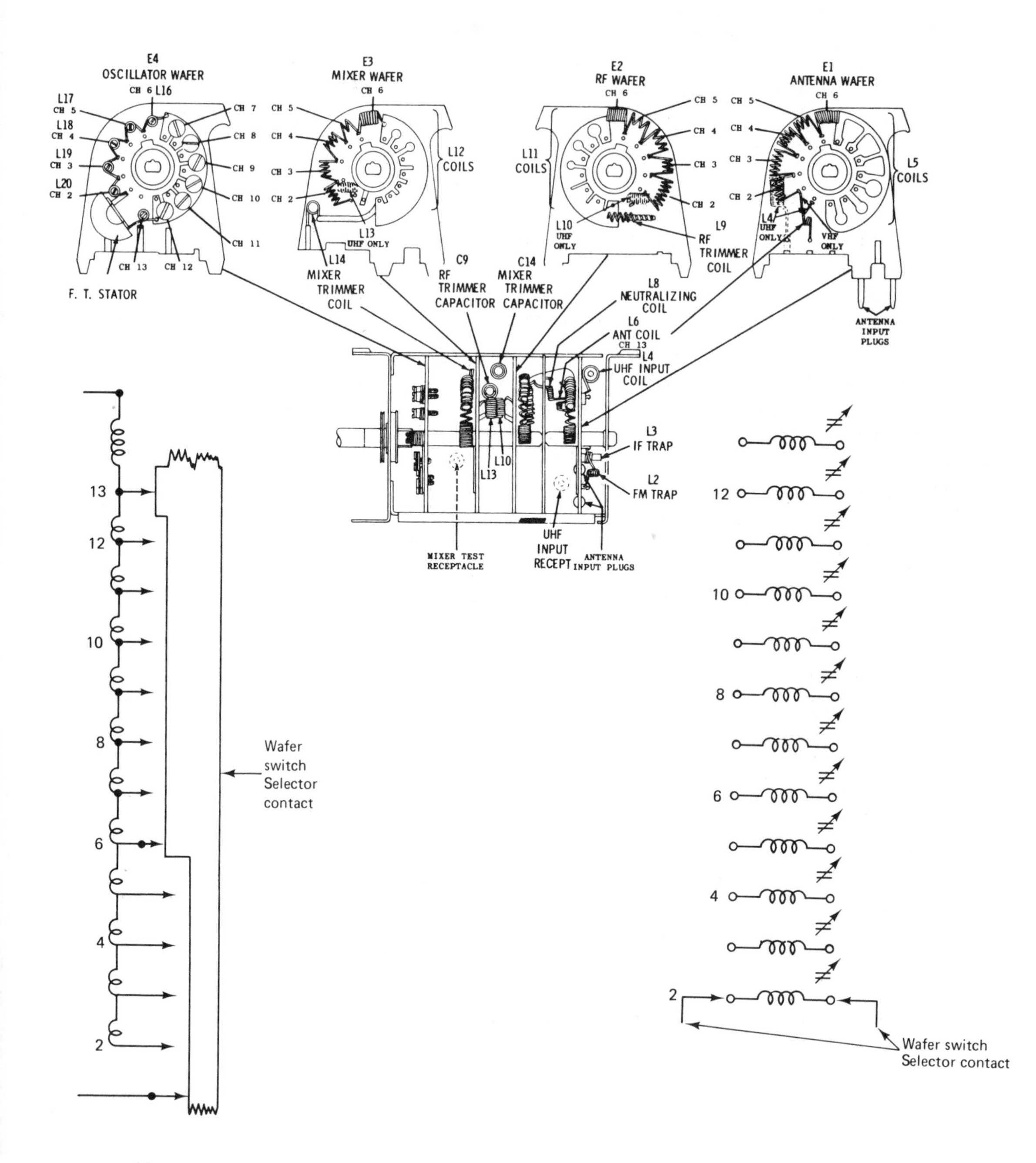

Figure 4-5 Wafer assemblies and their positions in the tuner: (a) schematic diagram of a series-connected coil selector wafer switch; (b) individual coil selector wafer switch.

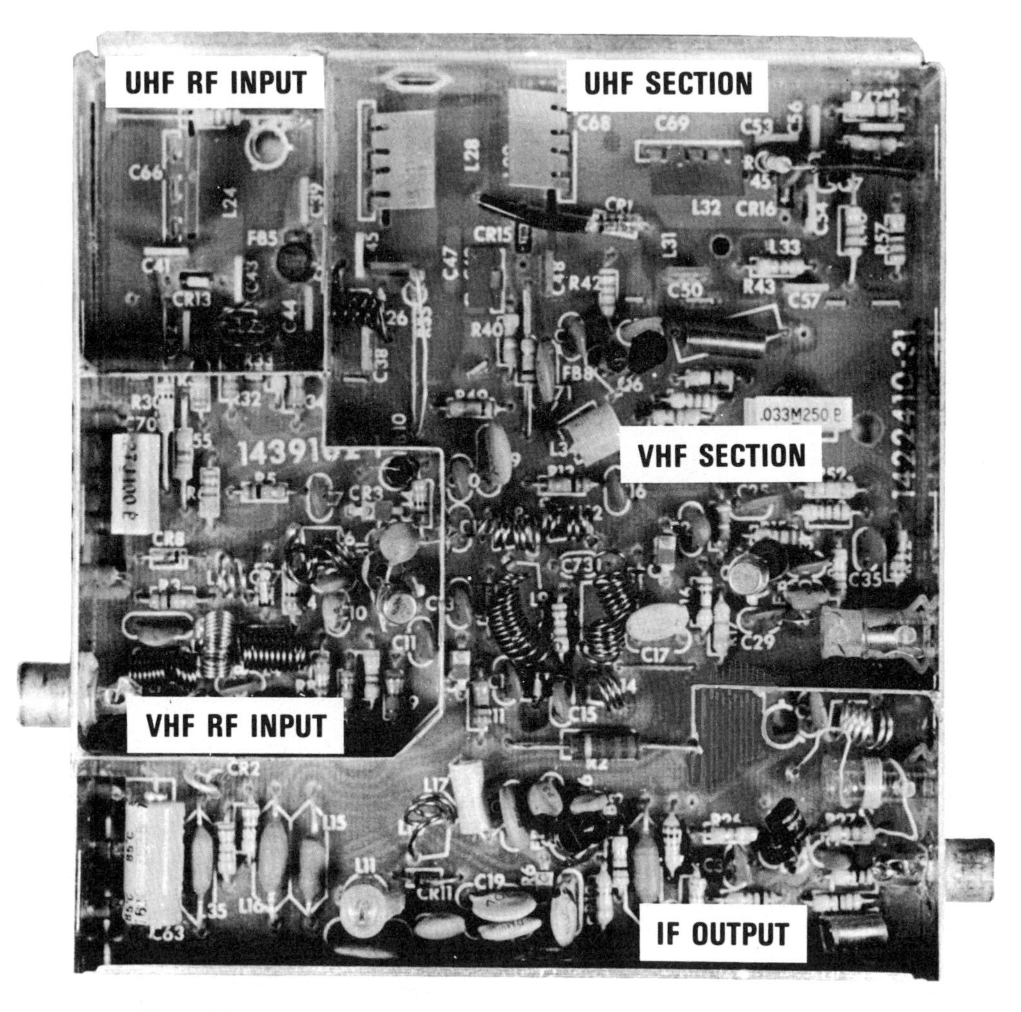

Figure 4-6 Internal construction of an electronic tuner. (Courtesy of RCA, Consumer Electronics.)

the electronic tuner, capacitance is the variable tuning element. The turret and switch tuners changed channels by mechanically changing the amount of inductance that the tuned circuits need for a given channel selection. In the electronic tuner a *dc voltage* is used to adjust the capacitance of a varactor and thereby change the resonant frequency of the tuned circuit. This method may be used in both VHF and UHF tuners.

In the VHF tuner the frequency range necessary to tune all 12 VHF channels is too large to be handled by one coil and its associated varactor. To overcome this difficulty, methods involving diode band switching are used to switch coils so that a large inductance is used for low-frequency channels and a smaller inductance is used with the high-frequency channels.

Not all electronic tuners have to be on a separate subchassis. They may be mounted as part of the main chassis. This is because the front panel channel selector controls are all dc operated; therefore, long cables can be used without difficulty.

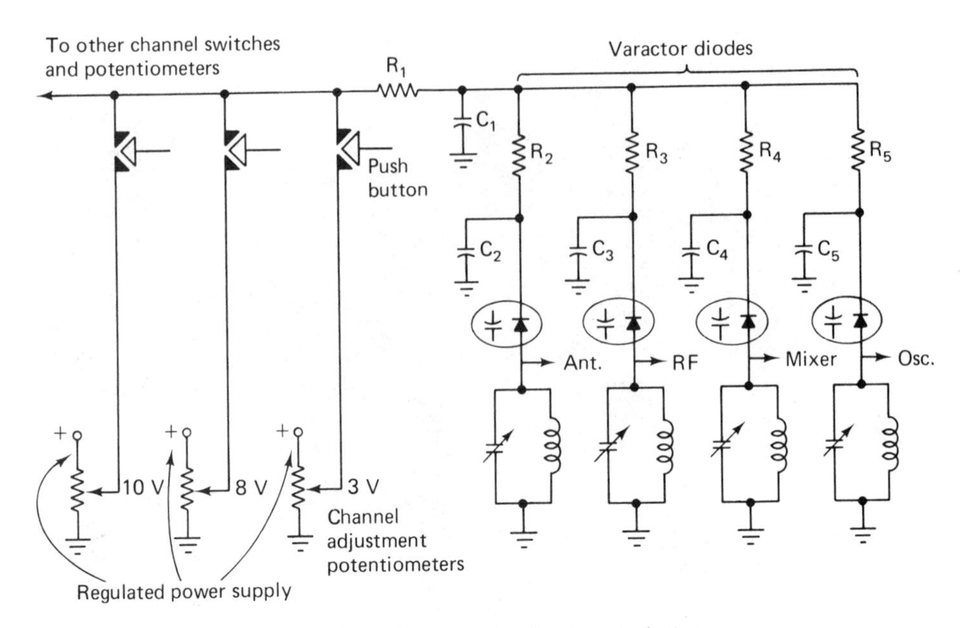

Figure 4-7 Pushbutton method for channel selection.

Channel selection. A simplified schematic diagram showing a common method by which the dc voltage across the varactor may be changed is shown in Figure 4-7. In this circuit, dc selection is made by pushbuttons. In each case, the dc voltage polarity that appears across the varactors located in each tuned circuit (antenna, RF, mixer, and oscillator) reverse biases them. The amount of reverse bias determines the amount of capacity developed by the diodes and thereby the resonant frequency of each tuned circuit. All of the tuned circuits are adjusted simultaneously in order to provide proper tracking. A channel adjustment potentiometer is required for each channel. This potentiometer is used to adjust all of the varactors throughout the tuner for any given channel. The potentiometers act as a kind of "fine tuning" that adjusts all the RF as well as the oscillator tuned circuits.

The dc tuning voltage source must be regulated in order to prevent line voltage changes, aging, temperature effects, and the like from affecting tuning. The dc output voltage of each potentiometer is fed to all of the varactors via a filter network consisting of R_1 and C_1. The resistors R_2, R_3, R_4, and R_5 are used to couple the dc voltage from the potentiometers to the varactors, while at the same time isolating one varactor from the other. The capacitors C_2, C_3, C_4, and C_5 are large and may therefore be considered a short circuit for RF. From an ac point of view, the varactor and the capacitors (C_2, C_3, C_4, and C_5) are in series and are connected across the tuned circuit. In effect, this places the varactor in parallel with the tuned circuit. If any of the filter capacitors were to open, the varactor would no longer be able to control the frequency of the tuned circuits and channel selection would not be possible.

Band switching. Band switching may be done in any number of different ways. One possible method is shown in Figure 4-8. In this circuit, separate diodes are used for band switching

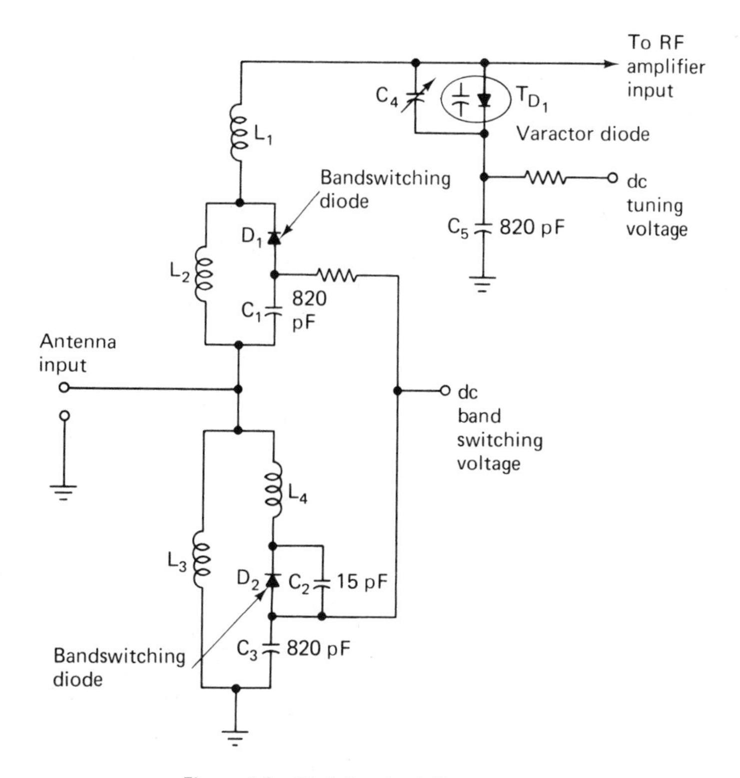

Figure 4-8 Diode band switching circuit.

and varactors are used for tuning. The varactor's (TD_1) capacity is adjusted by a dc voltage obtained from one of 12 potentiometers. Changing channels is accomplished by a two-pole pushbutton switch that selects both the necessary potentiometer that provides the required dc voltage for the varactor and the dc voltage for the switching diode. Since the low-numbered channels (2 through 6) are considerably lower in frequency than the high-numbered channels (7 through 13), the resonant circuits must resort to band switching in order to tune the required range of frequencies. Band switching is usually accomplished by introducing or removing coils from the tuned circuit. This is done in Figure 4-8 by the band-switching diodes D_1 and D_2. If the *low-frequency* channels are desired, the dc voltage obtained from the pushbutton channel selector switch *reverse* biases the diodes. This removes the diodes and any associated circuit elements that might be connected in series with them from the circuit. The result of reverse biasing the switching diodes is to connect L_1, L_2, and L_3 in series, thereby increasing the total inductance of the tuned circuit.

If it is desired to receive the *high-frequency* VHF channel, the push-button channel selector switch provides the required dc voltage to the varactor and a *forward* biasing voltage to the switching diodes D_1 and D_2 forcing them to conduct. Since the forward resistance of these diodes is very low, they can be considered short circuits. As a result, D_1 effectively shorts out L_2. At the same time D_2 connects L_4 in parallel with L_3. The effect of these connections is to reduce the total inductance of the resonant circuit and increase its resonant frequency, thereby tuning the receiver to a high-frequency TV channel.

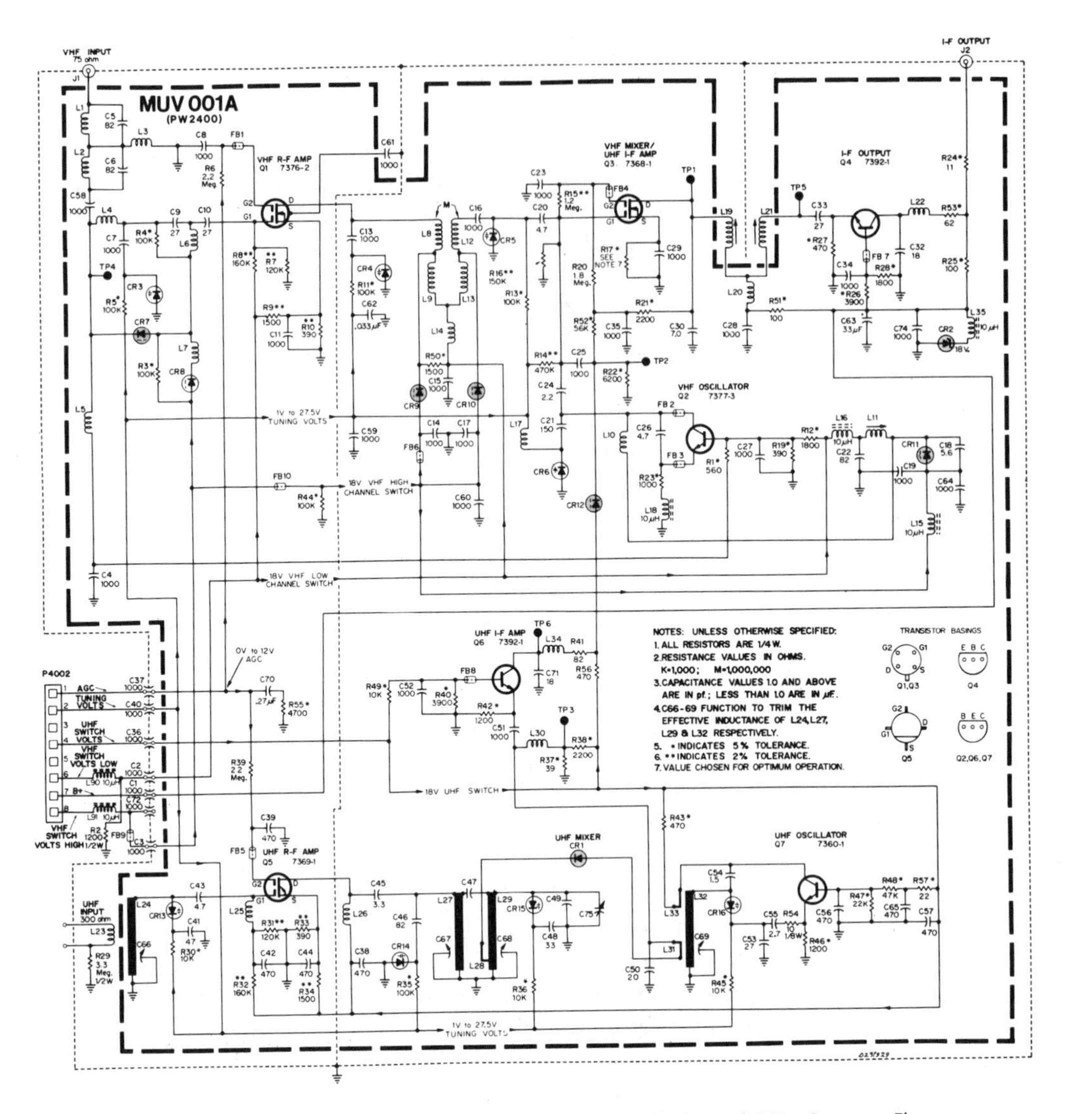

Figure 4-9 Complete schematic of a varactor tuner. (Courtesy of RCA, Consumer Electronics.)

Practical VHF tuner. When selecting any given channel in a varactor tuner, the tuned circuits for the antenna, RF amplifier, mixer, and oscillator circuits must all be simultaneously tuned to their proper frequencies. Therefore, varactor diodes and switching diodes will be found in each of these circuits. The complete schematic diagram of a tuner using varactor and switching diodes is shown in Figure 4-9.

Diodes CR_7 and CR_8 are the band-switching diodes for the RF amplifier input tuned circuit. Diodes CR_9 and CR_{10} are the band-switching diodes for the RF amplifier output and mixer input. The local oscillator uses diode CR_{11} to provide the band-switching function. Varactor diodes CR_3, CR_4, CR_5, and CR_6 provide simultaneous tuning of the antenna input tuned circuit of the RF amplifier, the output tuned circuit of the RF amplifier, the mixer input tuned circuit, and the oscillator tuned circuit, respectively. CR_{12} is used to couple the UHF IF signal into the UHF IF amplifier when the receiver is tuned to UHF.

Continuous tuning. Up to this point we have discussed three methods for channel selection: the turret, the wafer switch, and the varactor tuner. A fourth method of VHF channel selection that may be used is called *continuous tuning.* In this method it is possible to tune from the lowest channel to the highest channel as well as to everything in between. In older systems this was accomplished by either a variable inductance that had a movable tap or by variable capacitors that included a band-switching provision. These continuous tuners are no longer found in modern VHF television receivers. However, some modern receivers use a potentiometer to control the dc voltage feeding a varactor tuner to achieve a VHF continuous tuner. Continuous tuning is used in UHF tuners and will be discussed under that topic.

Channel-tuning systems. An ideal tuning system should incorporate the following characteristics:

1. It should be able to select the VHF, UHF, and cable channels for a total of 135 channels. VHF and UHF channels alone account for 82 channels. The remaining are the cable midband channels A to I (120 to 174 MHz), superband cable channels J through W (216 to 300 MHz), and the hyperband channels AA to BBB (300 to 776 MHz).
2. It should be "cable ready." This means that the TV tuner acts as a frequency converter that changes the cable channel frequency into a signal that is within the VHF band, usually channel 3.
3. No system alignment.
4. No mechanical switching.
5. Remote control capability.
6. The tuning system should be reliable and low in cost.

Varactor tuning in conjunction with digital and IC circuitry has permitted a close approximation of the ideal tuning system. However, in practice a variety of different tuning schemes have been introduced in both color and black & white receivers. Unfortunately, there is no common system in use. Listed below are some of these tuning methods.

1. In touch tuning, channels are changed by touching a touch-plate contact rather than rotating a channel selector knob. The touch plate can simply be a switch, or may be some other system such that the resistance of the touch plate is reduced by the operator's skin resistance. The switch or resistance change activates the circuitry necessary to develop the dc tuning voltage required by the varactor and switching diodes to change channels.

Touch tuning permits the operator to change channels sequentially up or down, in effect scanning the available channels. Also, the operator may select channels in any order that he or she wishes. Some manufacturers call this method "direct access." Others refer to it as "random access."

2. In another system that may employ microprocessors, the tuner starts from a parked position (perhaps channel 2) and then scans up through all the remaining channels, stopping when it finds an operating channel. These receivers are factory programmed for all the VHF channels, but must be operator programmed for UHF channels, video games,

and video cassette recorders. Up-and-down scanning buttons are also provided to allow a sequential selection of programmed channels.

3. In an alternative scheme, the operator manually programs the system to tune in each channel. The program is in the form of a digital code that is recalled when two channel selector buttons are depressed. This is a random access system.

4. The most common method of channel selection is *frequency synthesis* (FS) tuning. This type of tuning is often called *phase-locked-loop* (PLL) tuning. Modern FS systems incorporate microprocessors as part of the control circuit necessary to develop the dc voltage needed by the tuning varactors and switching diodes. The microprocessors may be used to operate digital clock and channel number TV screen displays. Channel numbers may also be displayed by LED seven-segment units.

Phase-locked-loop (PLL) basics. PLL circuits are used in a number of different stages of many modern television receivers. Two circuits that use PLLs are the horizontal AFC circuit that controls the frequency of the horizontal oscillator, and the color AFPC circuit that controls the phase of the color oscillator. In addition, frequency synthesis (FS) circuits, used in many television receivers for their tuning systems, incorporate PLL circuitry.

The basic phase-locked-loop circuit is shown in Figure 4-10(a) and consists of three basic circuits:

1. A controlled oscillator.
2. An error- or phase-detecting circuit that compares the oscillator output to a reference signal and develops a dc error voltage whose amplitude and polarity depends on the amount and direction of frequency error.
3. An oscillator control circuit that shifts the frequency of the oscillator in response to the dc error voltage. The direction of the oscillator frequency shift is always in the direction that will reduce the dc error voltage to a minimum. At this point the oscillator is at the same phase and frequency as the reference oscillator. The combination of the oscillator and the oscillator control stages is called a voltage-controlled oscillator (VCO).

The low-pass filter removes any residual ac or noise that may appear at the output of the phase detector. It also acts to stabilize the PLL and prevents hunting. See Section 13-10.

Frequency synthesis. Frequency synthesis is a technique for generating the oscillator signals needed for the mixers of the VHF and UHF tuners without the use of mechanical switching of coils or capacitors. The heart of this system is the phase-locked loop.

An example of a phase-locked loop used to generate a desired output frequency is shown in Figure 4-10(b). In this figure the VCO is the local oscillator of a tuner that is tuned to channel 2. As such, its output frequency will be 101 MHz. The output of the VCO is fed into a feedback programmable digital frequency divider which is set to divide its input signal frequency by 103,424, producing an output frequency of 976.5625 Hz. This signal is then fed into the phase detector.

This feedback frequency divider is made up of three parts. First is a prescaler, which is a high-frequency, high-speed divider that divides by a fixed factor of 256. The prescaler is usually a separate IC chip. The output of the prescaler feeds a divide-by-4 divider, which, in turn, feeds a programmable frequency divider. This divider is designed

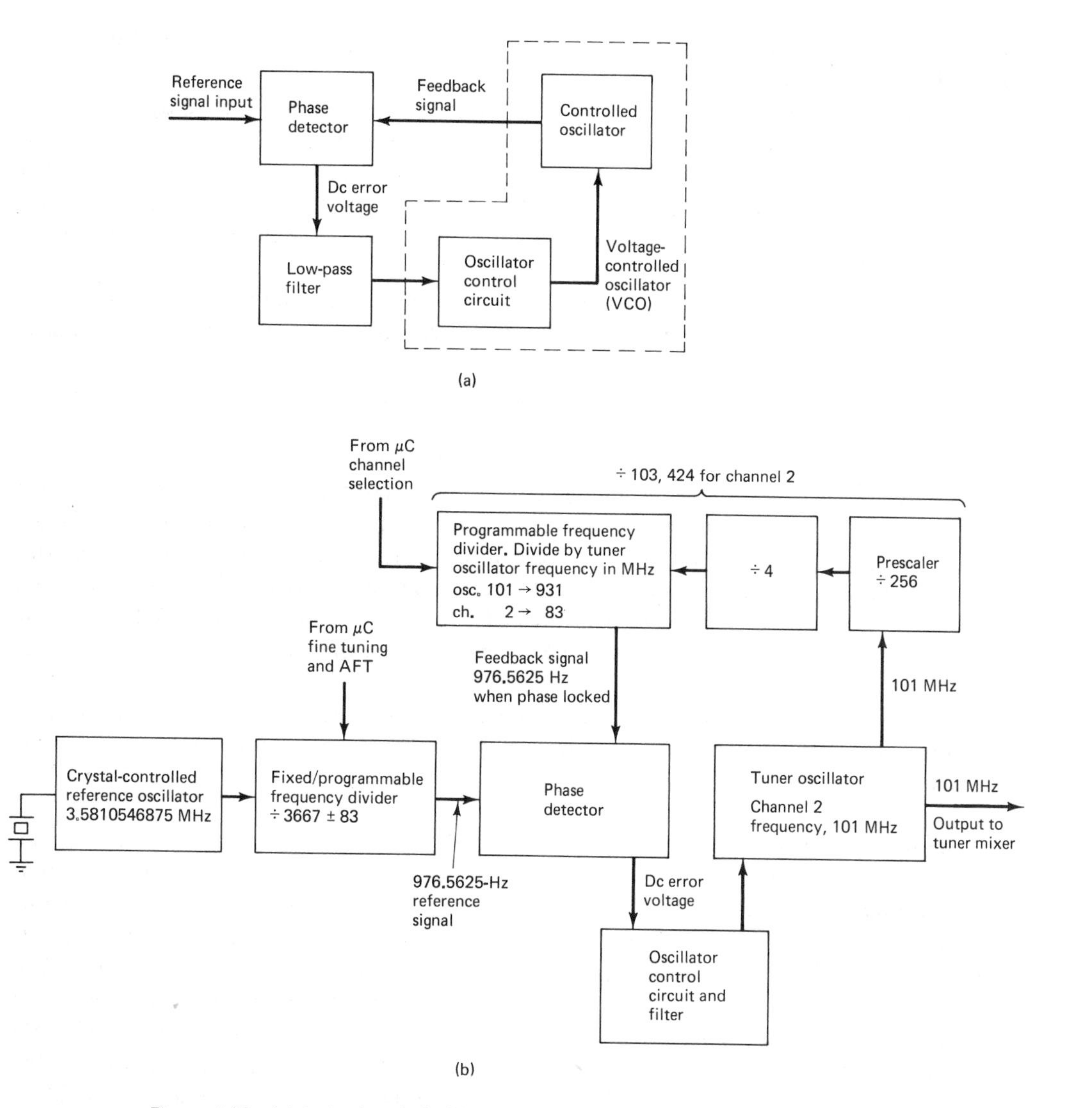

Figure 4-10 (a) Basic phase-locked loop; (b) phase-locked loop being used for frequency synthesis.

so that the divide factor is numerically equal to the local oscillator frequency in megahertz. Digital control signals obtained from a microcomputer can alter the divide factor depending upon the channel being selected. For the channel 2 example used here, the overall divide factor is the product of all the divide factors used in the divider chain, or $256 \times 4 \times 101 = 103{,}424$.

Also feeding the phase detector is the output of another programmable frequency divider which, in turn, is fed by a very stable crystal oscillator. For channel selection purposes this divider chain is fixed. However, it can be made programmable to provide manual fine tuning and automatic fine tuning for cable TV or video games that do not have the correct carrier frequencies.

The reference oscillator is set to a frequency of 3.5810546875 MHz by the crystal. For channel 2 operation the reference divider is set to divide by 3667, producing an output frequency of 976.5625 Hz at the input to the phase detector. The dc tuning voltage developed will be such that the VCO output frequency will be 101 MHz. For this to occur, both input frequencies to the phase detector must be identical. Typically, the dc voltage developed to tune the receiver to channel 2 will be about +2 V.

The reference frequency divider factor of 3667 can be increased or decreased by as much as ±83. Changes in divider factor are controlled by a microcomputer in response to the need for fine tuning.

If the channel desired were changed to channel 13, the VCO output frequency would be 257 MHz. The feedback programmable frequency divider would then be changed to divide by 263,168. This is done to produce the 976.5625-Hz signal at the input to the phase detector. The input to the phase detector coming from the reference oscillator remains fixed. Notice that in a PLL circuit, the frequencies of the feedback and reference signals driving the phase detector are always forced into being equal to one another, no matter what channel the receiver is tuned to. The dc voltage developed by the phase detector when the receiver is tuned to channel 13 is in the order of +16 V. This voltage is fed to all four varactor controlled tuned circuits in the VHF tuner, forcing them to be adjusted for channel 13 reception.

Figure 4-11 is the schematic diagram of a keyboard entry, direct access, PLL tuning system. This circuit makes use of a microcomputer IC to act as a central processor for transmitting and executing instructions.

Microcomputer. The microcomputer IC as well as all computers consist of three basic parts: input-output (I/O), memory, and a central processing unit (CPU). The "input" section feeds information into the CPU. The output of the CPU is delivered to the outside world via the "output" section. The "central processing unit" (CPU) is sometimes called a microprocessor and is programmed to perform sequences of step-by-step operations. The programs are stored in memories. Memories are used to store data and information necessary for computer operation. Two kinds of memories are used. A "read-only memory" (ROM) is used to store programs or other data which will remain unchanged for the life of the unit. "Random access memory" (RAM) is used to store data until it is needed by the CPU. The stored data may then be changed depending upon the CPU requirements. To perform all these operations the microcomputer is very complex. It is therefore not surprising that the microcomputer shown in the schematic diagram contains over 11,000 transistors in what is called a very large scale integration (VLSI) chip.

This microcomputer is used to:

1. Scan the digital keyboard for contact closures and to act on them
2. Control the operation of the LED channel display through LED drivers
3. Provide B+ switching
4. Control the PLL frequency synthesizer

The microcomputer used in Figure 4-11 is a low-end 4-bit device. It has 16 I/O lines and is available in a 20-pin dual-in-line package. One 8-bit port drives the LED channel

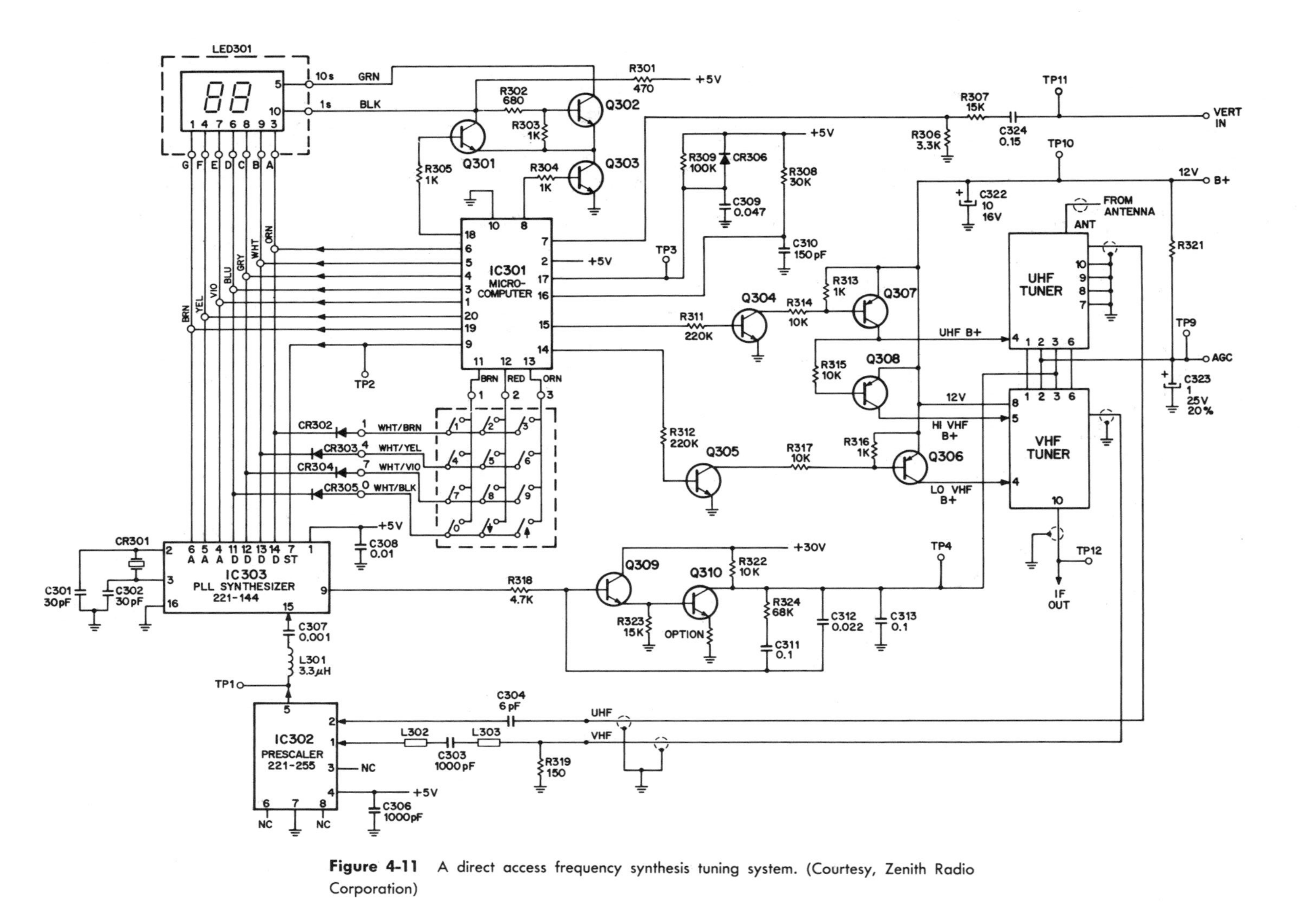

Figure 4-11 A direct access frequency synthesis tuning system. (Courtesy, Zenith Radio Corporation)

display segments, strobes the keyboard, and outputs the address and data information to the PLL IC. This is done in sequential order. The other lines read the keyboard, output the band information, give the data valid signal (strobe) to the PLL IC, and synchronize the entire system to the vertical retrace period. Noise produced by the logic and LED displays is eliminated by turning on and off the LEDs during the vertical retrace time.

Transistors Q_{301}, Q_{302}, and Q_{303} in conjunction with the microcomputer (pins 8 and 18) determine whether the LEDs are on or off. When a single-digit channel is called up, the right LED (1's) is turned on and the left (10's) is turned off. When a channel whose number is greater than 9 is called up, both LEDs are turned on.

The PLL synthesizer IC contains all of the blocks shown in Figure 4-10 except for the prescaler. When a channel is selected the microcomputer "informs" the PLL by means of the address (A), data (D), and sync (ST) lines that a channel has been selected and which channel number is being called up. The programmed PLL then adjusts the divide factors of the programmable dividers so that the phase detectors will generate the necessary dc voltage. In this circuit the dc voltage is obtained from pin 9. It is amplified by transistors Q_{309} and Q_{310} and then coupled to the varactor found in the VHF and UHF tuners (pin 3).

B supply switching is required to turn off the VHF tuner when going from VHF to UHF, and vice versa. In addition, the switching diodes must be properly energized when going from the low VHF channels to the high VHF channels, and vice versa. This is done by transistors Q_{304}, Q_{305}, Q_{306}, Q_{307}, and Q_{308} in conjunction with output commands of the microcomputer (pins 14 and 15).

4-7 the RF amplifier

The RF amplifier circuit may be considered to start at the antenna terminals of the receiver and end at the input of the mixer stage as shown in the block diagram of Figure 4-3. The RF amplifier performs at least four functions. These are:

1. Improves signal-to-noise ratio of receiver
2. Provides amplification
3. Provides selectivity
4. Reduces oscillator radiation

In the paragraphs that follow each of these functions is discussed in detail.

The RF amplifier improves the signal to noise ratio of the receiver. Electrical noise is defined as any electrical disturbance that tends to mask a desired signal. In terms of a television receiver, electrical noise most often appears on the picture tube as a random "salt and pepper" pattern called *snow*. See Figure 4-12.

Noise may be the result of two primary sources: noise sources external to the receiver and noise sources internal to the receiver. Noise produced by sources external to the receiver is caused by such things as auto ignition, atmospherics, cosmic noise, household appliances, and other man-made electrical disturbances. Unfortunately, there is not too much that can be done to eliminate these interfering noise sources. If the source of noise is man-made, locating it and introducing filters or changing brushes in motors may reduce its impact upon the receiver. Such noise sources usually produce dark horizontal streaks

Figure 4-12 Appearance of random noise (snow) on a TV picture.

through the picture. If the interference is bad enough, it may cause the picture to lose both horizontal and vertical synchronization. This type of noise interference lasts only as long as the noise source is operating.

Electrical noise internal to the receiver may either be referred to as *white noise* or *random noise*. It carries these names because it is generated by the random motion of electrons and holes in conductors, tubes, and transistors, and it has the same kind of spectral makeup as white light.

The voltage generated by random noise is in the order of a few millionths of a volt. This very small voltage may not seem to be large enough to cause any difficulty. Unfortunately, it and other internal noise contributors are the chief limiting factors on the sensitivity of a receiver because the weakest signal that a receiver can respond to must be larger than the noise voltage generated by the receiver itself.

All circuit elements generate electrical noise. For the noise to become important, the noise must be amplified sufficiently to have an effect on the output of the system. Since the voltages we are dealing with may be in the order of a few millionths of a volt, amplification of a million times or more is necessary before it has an effect on the output of the receiver. This means that those stages of a receiver that have the greatest overall amplification are most important in terms of noise. In the television receiver the RF amplifier and the mixer stages fall into this category. Hence, low-noise front ends are essential if good picture quality is to be achieved in weak signal fringe areas (80 to 200 miles from the receiver).

The noise generated by the mixer stage is the chief source of noise in the receiver. In fact, it is 5 to 10 times more noisy than that produced in the RF amplifier. It is the function of the RF amplifier to amplify a weak signal so that it can override the noise generated in the mixer.

Transistors and FETs generate random noise. To minimize the effect of this noise, transistors used in RF amplifiers are designed to have high beta, high F_t, and small-base spreading resistance.

Of all the devices discussed, FETs provide the best low-noise performance. FETs used in low-noise RF amplifiers are designed to have high transconductances (gm). FETs have another characteristic that makes them highly desirable for weak signal reception. They have an almost perfect square law characteristic. This means that the FET RF amplifier will be relatively free from cross-modulation (the transfer of modulation from one carrier to another).

The RF amplifier provides amplification. In the preceding section it was shown that in receivers designed for weak signals, RF amplifiers must generate less noise than the mixer stage and provide sufficient amplification to the desired signal to allow it to override the noise generated in the mixer. RF amplifiers that fill these demands may use transistors, FETs, and ICs. By and large, the circuit configuration most commonly used has been the common emitter for transistors, and the common source for the FET. Typical examples of these amplifiers are shown in Figures 4-13 and 4-14.

The balun. The transformer that provides an impedance match between the antenna and the input impedance of the amplifier is T_1. This transformer is called a *balun.* The name refers to the types of circuits that it is connected between. The input circuit is fed by an antenna that is *bal*anced with respect to ground, whereas the input to the amplifier is said to be *un*balanced with respect to ground. Figure 4-15 shows that the antenna transmission line has equal distributed capacity to ground; therefore, it is said to be balanced. The input circuit of the amplifier is said to be unbalanced because it returns directly to ground via

Figure 4-13 Neutralized common-emitter RF amplifier.

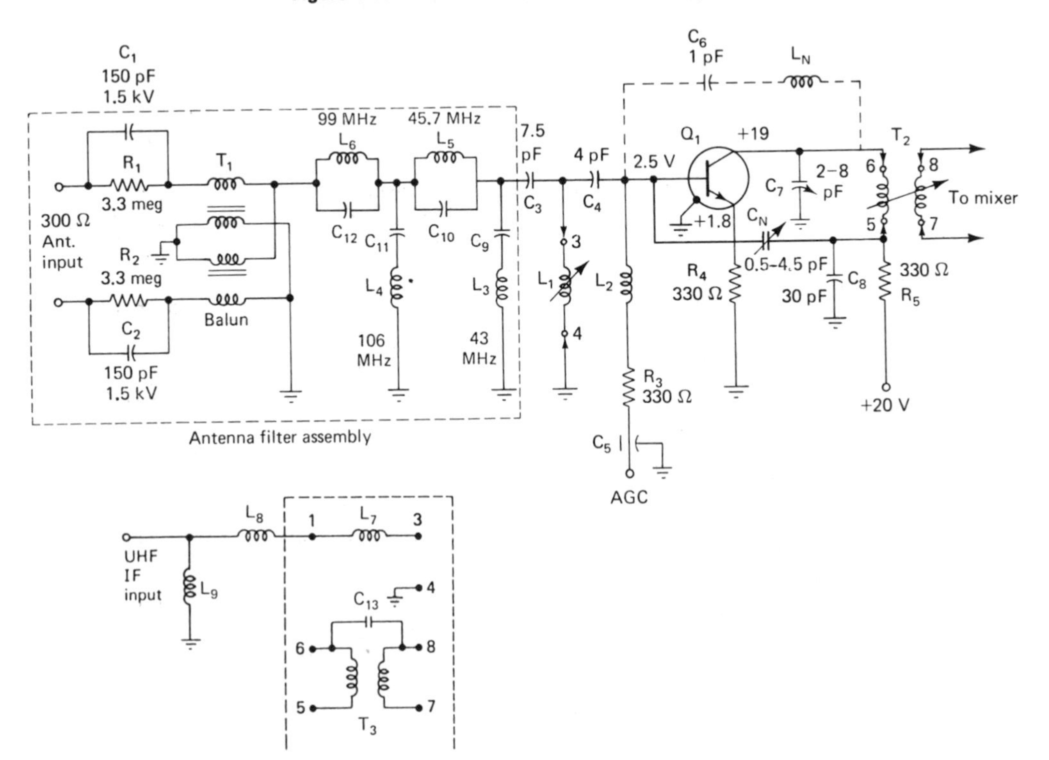

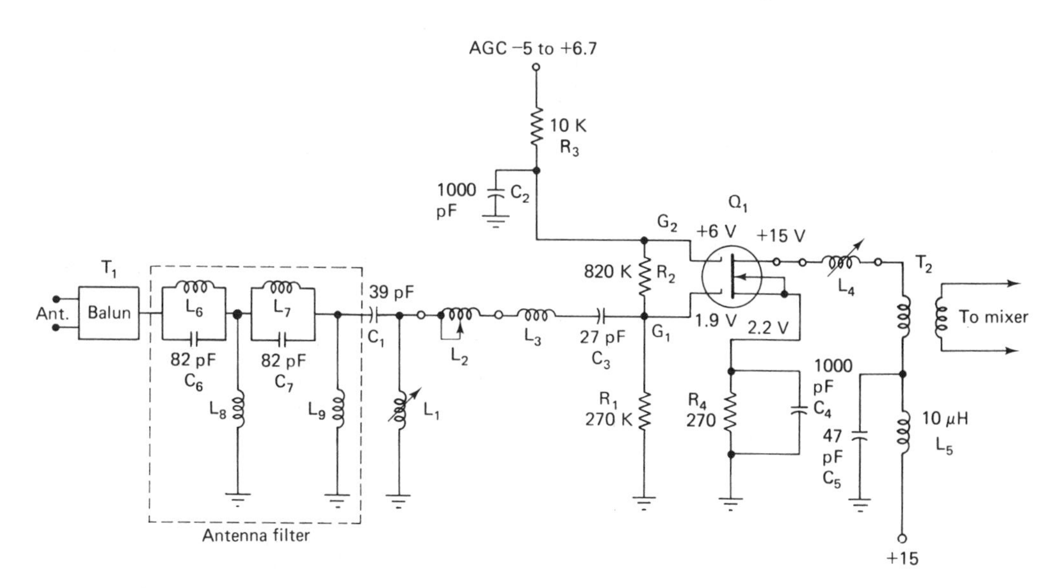

Figure 4-14 MOSFET RF amplifier.

the distributed capacitance. The balun plays an important role in reducing noise interference caused by pick up on the *transmission line.* Figure 4-15 also shows how this is accomplished. The electrical disturbance (auto ignition, motor brushes, and so on) sets up an electromagnetic field that passes through the transmission line. To ensure that the transmission line will have equal noise-induced currents, the transmission line is twisted as it is run from the antenna to the receiver. If it is assumed that the lines of force of the electromagnetic wave are such that its electrostatic field is vertical, the induced current in each wire will

Figure 4-15 Cancellation of noise pulses picked up on the transmission line.

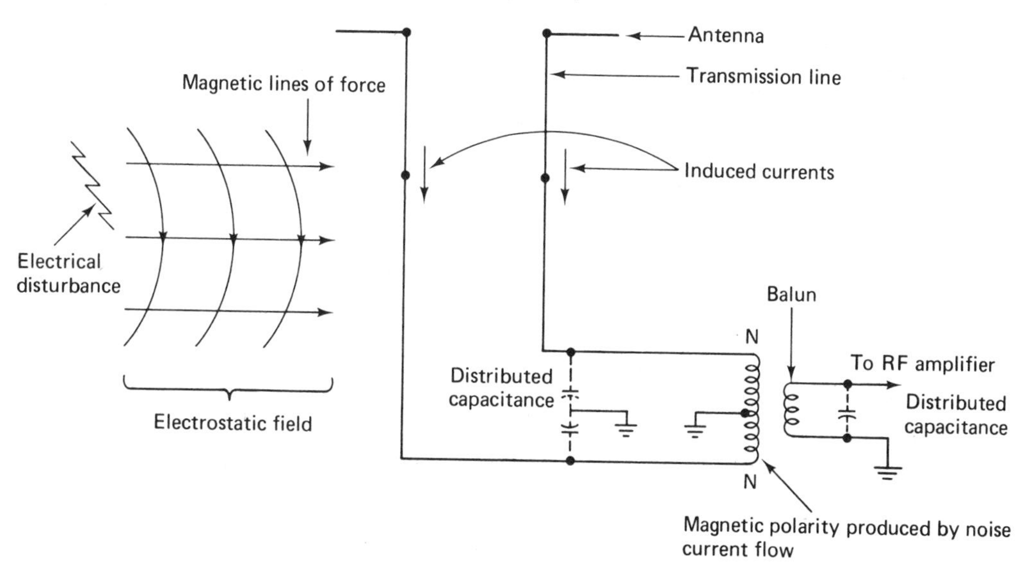

be in the direction shown. The current flow through the primary coil of the balun will produce changing magnetic fields that are equal and opposite with respect to the center tap of the transformer. These magnetic fields will cancel each other out, resulting in little or no induced voltage in the secondary, thereby greatly reducing this form of noise interference.

Normal signals are picked up by the antenna and produce signal currents that flow in opposite directions in each wire of the transmission line at any given time. Thus, the magnetic fields produced in the transformer are equal and reinforcing to one another, thus inducing a signal voltage into the secondary of the transformer.

Capristors. The parallel circuits R_1C_1 and R_2C_2 at the antenna input to the receiver shown in Figure 4-13 are safety devices called *capristors.* These high-voltage (1.5 kV) capacitors are used to isolate the antenna terminals from chassis ground and thereby prevent shock hazard in sets not using isolating power transformers. This isolation is at the 60-Hz line frequency. The reactance of the capacitors at 60 Hz is approximately 16 MΩ each and is in effect an open circuit. However, at a frequency of 60 MHz (channel 2) the capacitors have a reactance of approximately 16 Ω and result in negligible signal attenuation.

As a means of preventing these protective capacitors from being shorted because of the high-voltage electrostatic charges that may be developed during lightning storms, large valued resistors are placed across the capacitors. The resistors allow the electrostatic charges to leak off thereby protecting the capacitors. In addition, a *spark gap* is also provided across the capacitor. This allows rapid discharge of high-voltage buildup that might not be discharged fast enough by the shunt resistor. In practice, these units (R, C, and spark gap) are packaged together forming one unit that looks very much like a ceramic tubular or disk capacitor.

Preselection. The antenna filter assembly is a series of resonant circuits that are used to prevent undesired signals from reaching the input of the RF amplifier. These circuits are a preselector that substantially reduces interference and spurious responses. The tuned circuits are resonant to the FM band and the associated video IF. Frequencies of 43, 45.7, 99, and 106 MHz are typical. Examples of IF/FM trap assemblies are shown in Figures 4-13 and 4-14.

The isolating capacitors, balun, and antenna filter assembly are usually mounted on the top of the tuner subchassis and are often arranged on a small phenolic board of their own. See Figure 4-2. If the isolating capacitors or the balun become defective, the picture on all channels may become weak or snowy or it may be lost altogether.

Feed-through capacitors. Another component of interest in Figure 4-13 is C_5. Also see Figure 4-9 C_1, C_2, C_3, C_{36}, C_{37}, C_{40}, and C_{72}. These capacitors are called *feed-through capacitors* and are usually mounted on the chassis or one of its divider panels. The capacitor is mounted through a hole and is such that its leads are at right angles to both surfaces of the chassis. One plate of the capacitor is the chassis itself, and the other is the wire passing through the chassis. The dielectric is usually some form of ceramic material. Feed-through capacitors are generally used as simple RF bypass capacitors such as C_5 in Figure 4-13, or they may be used as part of tuning and phase shifting networks.

Transistors used at VHF. High-frequency transistors that are used in RF amplifiers differ from their low-frequency counterparts in a number of respects. High-frequency transistors have very narrow base and small base spreading resistance and small interelectrode capacitances as compared to low-frequency transistors. One of the major developments that allowed the production of high-frequency transistors was the introduction of an accelerating field in the base region of the transistor. This field has the effect of reducing the transit time between emitter and collector by increasing the speed of transit of the current carriers. The accelerating field is produced by doping the base so that its resistivity increases from emitter to collector. This type of base is called a *diffused base.* All high-frequency transistors use this base structure.

High-frequency transistors usually have four leads. Three of the leads correspond to the emitter, the base, and the collector. The fourth lead is tied to the transistor case and allows it to act as a shield. If this lead is open or is not provided in a replacement transistor, the circuit may oscillate.

The common-emitter RF amplifier. The common-emitter RF amplifier such as that shown in Figure 4-13 is a popular circuit configuration found in many modern solid-state tuners. It is characterized by high gain and low noise. Unfortunately, the common-emitter amplifier is susceptible to cross-modulation.

It is well known that when an amplifier operates at high frequencies, it tends to oscillate. This is due to positive feedback via the transistor's large interelectrode capacitance. One method used to prevent the amplifier from oscillating is capacitor neutralization. See Figure 4-13. The neutralizing capacitor C_N is usually mounted under the tuner chassis and may have an adjustable screw that is reached by an access hole on the side of the tuner chassis. In some tuners this adjustment is in the form of two small pieces of insulated wire that are twisted together and is called a *gimmick.* In Figure 4-13 the RF feedback voltage used for neutralization is obtained across R_5 and C_8. R_3 isolates the base from the ac short circuit that the AGC supply represents.

If the neutralizing adjustment is not right, perhaps because of an overdose of cleaning spray, the RF amplifier will oscillate on one or more channels. This will result in a tunable interference pattern that will be seen in the picture. The interference may be so bad that it completely obliterates the desired picture. Readjustment touch up of the neutralizing capacitor is done by observing the picture and adjusting the capacitor until the interference pattern is gone.

Channel selection is accomplished by changing L_1 and T_2. Capacitor C_7 is used to adjust the output tuned circuit to resonance.

The antenna filter assembly provides the hardware to (1) provide an impedance match (or slight mismatch for best noise characteristic) by means of a balun; (2) prevent shock hazard by means of C_1R_1, C_2R_2; and (3) reduce interference and spurious response by means of the resonant traps located between the balun and the RF amplifier.

The dc operating point of the amplifier is established by the AGC voltage and the voltage drops produced by R_3, R_4, and R_5.

An alternative method of neutralization is shown by dashed lines in Figure 4-13. In this case, feedback via the interelectrode capacitance between collector and emitter is minimized by making this capacitance part of a high-impedance parallel resonant circuit.

Thus, by placing L_N in parallel with the transistor's internal interelectrode capacitance, the relatively low-impedance path from collector to emitter is converted into a high impedance, thereby reducing or eliminating the feedback. The capacitor C_6 is a dc blocking capacitor that prevents a dc short between collector and emitter.

When the tuner is in the UHF position, the RF amplifier is converted into a 40-MHz IF amplifier that is fed by the IF output of the UHF tuner. This is done by changing the strip that L_1 and T_2 are mounted on as shown in the lower left of the diagram.

When replacing an RF amplifier transistor, a direct replacement must be used. If this is not possible, a substitute transistor that is recommeded by a transistor substitution guide may be used. Care in selecting a replacement transistor is necessary for two reasons. First, the replacement transistor must match the dc characteristics of the original. A poor dc match will also affect the operating point of the transistor, which in turn will affect stage gain and distortion characteristics of the amplifier. Second, the replacement transistor must have a good ac match to the original. The replacement transistor must be designed to operate at VHF frequencies.

The MOSFET RF amplifier. Many modern solid-state tuners use field-effect transistors for the RF amplifier. In many respects, this type of amplifier is superior to the other RF amplifiers because it features very low noise, high gain, high input impedance, and good stability, and because of its excellent low distortion characteristics interference and spurious signal response are lower than in any other device.

The RF amplifier shown in Figure 4-14 uses an N-channel dual gate MOSFET. The use of a dual gate serves two important purposes. First, it allows the MOSFET to be used in the cascode* mode of operation. Second, it provides separate means for applying AGC and RF signal voltages to the amplifier.

In the schematic diagram it is seen that the antenna delivers its signal to the usual arrangement of balun and antenna filter assembly, and then through the channel selecting tuned circuits (L_1, L_2, and the distributed capacitance of the circuit) to gate 1 of the MOSFET. The source is tied to ac ground by C_4. Therefore, G_1 and the source form the grounded source portion of the cascode amplifier. Gate 2 is returned to ground by means of the RF bypass capacitor C_2. The ac grounded gate, the channel, and the drain form the grounded gate portion of the cascode circuit arrangement. The use of this circuit results in a number of advantages. First, the input circuit (grounded source) is loaded by the low input impedance of the output circuit (grounded gate) and reduces the tendency for the circuit to oscillate. Therefore, neutralization is not necessary. Second, RF amplifier gain control may be obtained by varying the bias on G_2. A change of bias controls the drain current through both sections to the amplifier and produces a semiremote cutoff characteristic. The RF amplifier is isolated from the AGC source by R_3. The AGC voltage is coupled to gate 1 by means of R_2. By means of this resistor gate 1 voltage will "track" with the applied AGC, thereby helping to reduce input circuit capacitance variations that would affect the tuning characteristics of the amplifier.

The dc operating point of the amplifier is determined by the bias developed by resistor R_4. Capacitor C_4 is a bypass capacitor that prevents degeneration.

*A FET cascode amplifier is a two-stage amplifier consisting of a common source amplifier feeding a common gate amplifier. This combination provides high gain and low noise.

L_4, T_2, and the distributed capacitance of the circuit constitute the output tuned circuit that provides coupling to the mixer stage.

The RF amplifier provides selectivity. In addition to providing amplification and determining the noise characteristics of the receiver, the RF amplifier is also used to isolate the mixer from the antenna; thereby minimizing spurious responses and image interference. In a television receiver these undesired responses may appear as fine herringbone patterns in the picture or they may appear as a "windshield wiper" effect in which the hammerhead or sync pulse pattern of one channel passes back and forth superimposed on the desired signal. See Figure 6-6.

RF interference occurring on the VHF TV channels is mainly the result of intermodulation involving the harmonic of the oscillator plus or minus the IF, FM signals plus or minus a TV channel, or the second harmonic of FM signals.

This interference, as well as direct IF pickup, can be minimized if tuned filters are placed between the antenna and the RF amplifier. The RF amplifiers shown in Figures 4-13 and 4-14 indicate these filters as the *antenna filter assembly.*

RF amplifier frequency response. The input and output tuned circuits associated with the RF amplifier provide the selectivity required to minimize image and spurious responses. The overall bandpass of these circuits (antenna input to mixer input) must pass the 6 MHz of video information transmitted by each channel. An example of the ideal RF amplifier response curve, black & white receiver response, and a color TV response is shown in Figure 4-16. The ideal response has two important characteristics. First, it has a flat bandpass that amplifies all frequencies falling in it equally. Second, its vertical skirt response completely rejects any signals falling outside the bandpass. In comparing the responses of the black & white receiver and the color receiver to the ideal response, it is evident that the color receiver's RF amplifier bandpass is a closer approximation to the ideal response than is the "haystack" response of the black & white receiver. In particular, the color receiver's response should be relatively flat, thereby amplifying all signals equally. This is important because any shift in color signal amplitude can result in poor color reception.

The RF amplifier reduces oscillator radiation. The FCC regulations require that a TV receiver's local oscillator not act as a long-range transmitter. In fact, the receiver local oscillator is not to have meaningful radiation beyond 100 ft. The RF amplifier and its associated tuned circuits act to reduce the distributed capacitance between the oscillator and the antenna, thereby isolating one from the other minimizing radiation and possible interference problems.

4-8 the mixer

The purpose of the mixer stage is to change the VHF TV channel frequency into a much lower intermediate frequency. Therefore, the terms "mixer" and "frequency changer" will be used interchangeably throughout this text.

There are three advantages to be gained by lowering the frequency of the channel being received: (1) amplifier gain can be increased, (2) amplifier stability improves, and (3) tuned circuit selectivity improves.

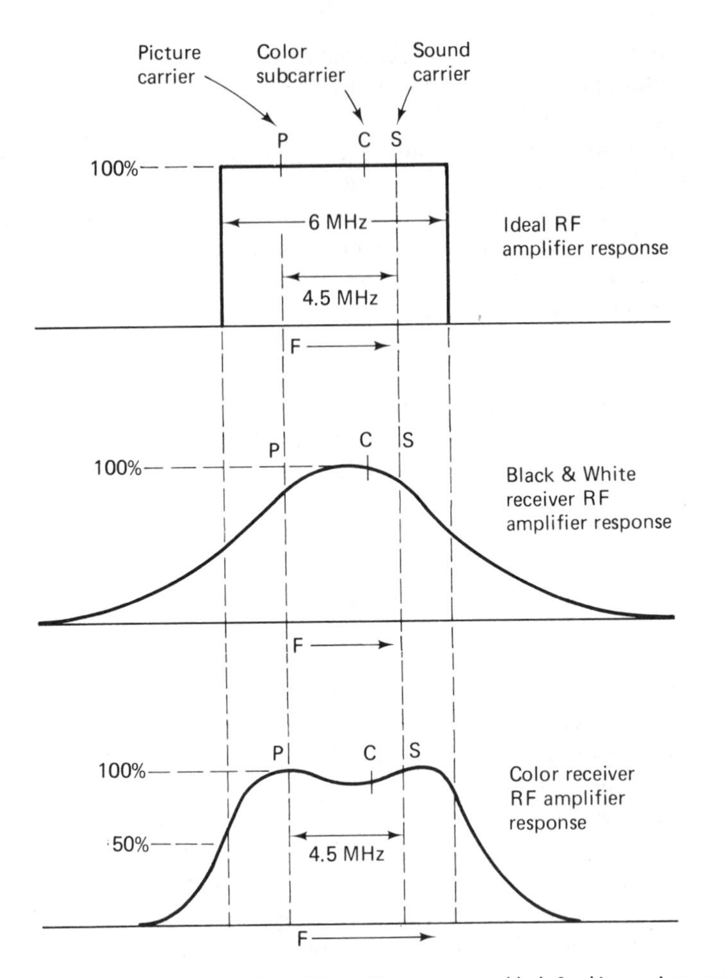

Figure 4-16 Comparison of an ideal RF amplifier response, a black & white receiver, and a color receiver.

Frequency changing. All frequency changers must operate nonlinearly in order to function. Nonlinear operation of a tube or transistor refers to the condition in which the output voltage or current is distorted in comparison to the input.

Nonlinearity of a tube or transistor is caused by any curvature or unilateral (current flow only in one direction) characteristics of the device's voltampere characteristic curve.

Practical frequency changers can use diodes, triodes, pentodes, pentagrid converters, transistors, and FETs. In some cases an integrated circuit may be used to provide the functions of mixer and oscillator. But depending upon the frequency of operation, certain frequency changers will function better than others.

The block diagram of a frequency changer is shown in Figure 4-3. Here the nonlinear device is called a mixer and is fed by two signal inputs. One of the inputs is the desired RF signal; the other is obtained from an oscillator that is part of the receiver front end. The oscillator frequency is usually placed above the incoming RF signals by an amount equal to the intermediate frequency used in the receiver. In television there are two input RF signals (video and sound) and therefore two output IF signals (45.75 MHz and 41.25 MHz).

One of the consequences of placing the oscillator above the input RF signal frequency is the frequency reversal of the intermediate frequencies. This is illustrated in Figure 4-17. The topmost portion of the drawing shows the spectral output of television station channel 3 as well as the upper and lower adjacent sound and picture carriers. Indicated below this diagram are the assigned frequencies for the lower adjacent sound carrier, the associated picture carrier, the associated color subcarrier, the associated sound carrier, the upper adjacent picture carrier, and the receiver's local oscillator frequency. Notice that in this diagram the lowest frequency is on the left side and that it reaches maximum on the right side.

Below the diagram of the spectral output of channel 3 is shown the overall IF output of the receiver. The shape of this curve indicates that the amplification of the amplifier changes with frequency. However, the important thing to notice is that the frequency is lowest on the right and reaches maximum on the left. This is because the intermediate frequency (IF) is the *difference* between the incoming RF and the local oscillator frequency. Notice that the *smallest* difference, and therefore the lowest IF frequency, is that occurring between the highest RF signals and the local oscillator. Thus, the difference between the local oscillator (107 MHz) and the upper adjacent picture carrier is only 39.75 MHz, while that between the lower adjacent sound carrier and the oscillator is 47.25 MHz. As a result, the IF output is reversed in frequency compared to the RF input.

Figure 4-17 Frequency reversal of the IF compared to the RF signal.

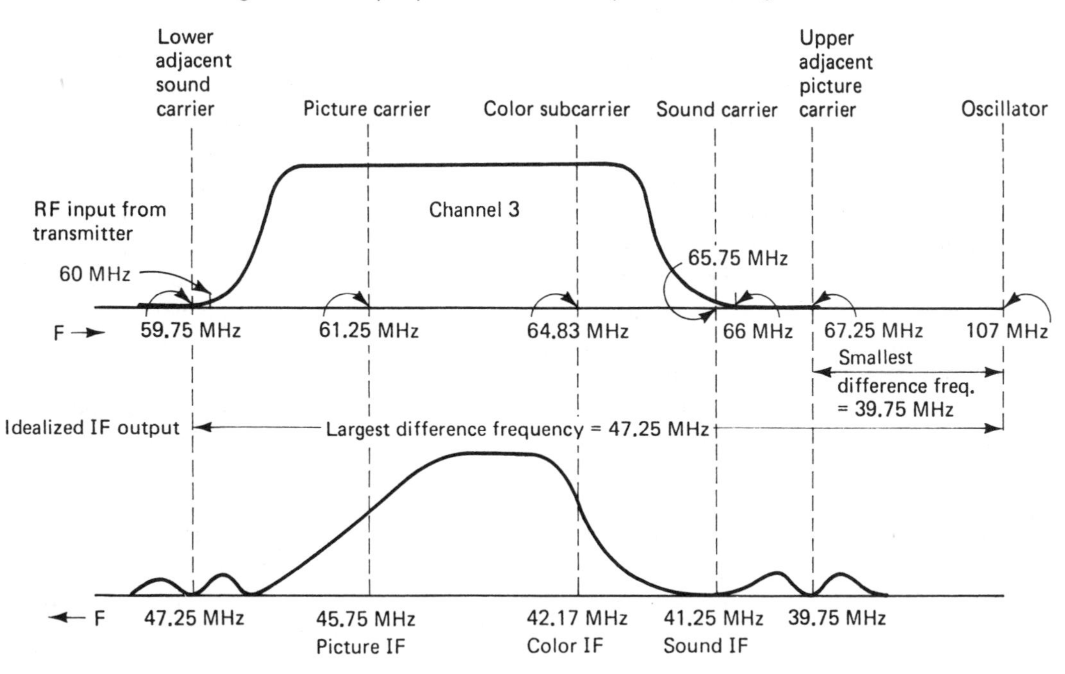

Oscillator pulling. At VHF frequencies it is very difficult to use any form of converter in which the oscillator and mixer are in one device. This is because of oscillator pulling. Oscillator pulling is the condition in which the local oscillator tends to synchronize its frequency with that of the incoming RF. This condition causes loss of conversion gain and under extreme conditions, when the oscillator and RF are synchronized, loss of reception altogether. To prevent this, it is necessary to reduce the amount of RF signal that gets into the oscillator. Thus, the mixer and oscillator stages are kept separate in all VHF and UHF tuners.

Oscillator level. Another condition that insures maximum frequency conversion gain is to make the oscillator signal voltage many times larger than the incoming RF signal. The exact value depends upon the characteristics of the mixer and is determined by the coupling mechanism used between the oscillator and mixer. Inductive or capacitive coupling methods may be used to control the amount of injected oscillator signal, as well as the amount of RF getting into the oscillator.

Typical mixer circuits. Two typical mixer circuits are shown in Figures 4-18 and 4-19. Transistor type tuners may use a common emitter type or a cascode arrangement of transistors. FET mixers are usually of common-source configuration.

The transistor mixer. The transistor mixer shown in Figure 4-18 is basically a common-emitter amplifier whose input base-to-emitter junction is used as a diode. The diode is nonlinear and therefore provides the mixing action; the remaining portion of the transistor is used as an IF amplifier.

The dc operating conditions of the mixer are established by the resistors R_1, R_2, R_3, and R_4. In addition to setting up the bias of the stage, R_3 also provides a degree of temperature stabilization.

The input RF tuned circuit feeding the mixer consists of the distributed capacitance, C_1, L_1, and L_2. L_1 and L_2 are changed for channel selection. Bypass capacitor C_4 keeps the emitter at ac ground potential and thereby ties the emitter to the input tuning capacitor C_1. Oscillator signal injection is accomplished by capacitor C_2. This capacitor couples the oscillator voltage into the mixer, and its size determines the level of the injected signal.

The output circuit is tuned to the IF by means of the output circuit distributed capacitance, C_3, C_5, and L_3. The output transformer that L_3 is part of is used to reduce the output impedance of the circuit so that long cables may be used between the tuner and the IF amplifier's printed circuit boards. This is also important because alignment of the tuner will not have important effects on the alignment of the IF amplifier. Hence, replacing tuners will not necessitate realignment.

TP_1 is a terminal that is connected to the collector of the mixer. This test point is used during sweep alignment as a means of observing the RF frequency response curve of the tuner.

The cascode transistor mixer. In an attempt to reduce feedback from the IF output of the mixer to the RF input, and thereby maintain mixer stability and high input impedance, two separate transistors are used as a cascode mixer. Two examples of this arrangement are shown in Figures 4-9 and 4-19. The circuit used in Figure 4-9 is part of a varactor

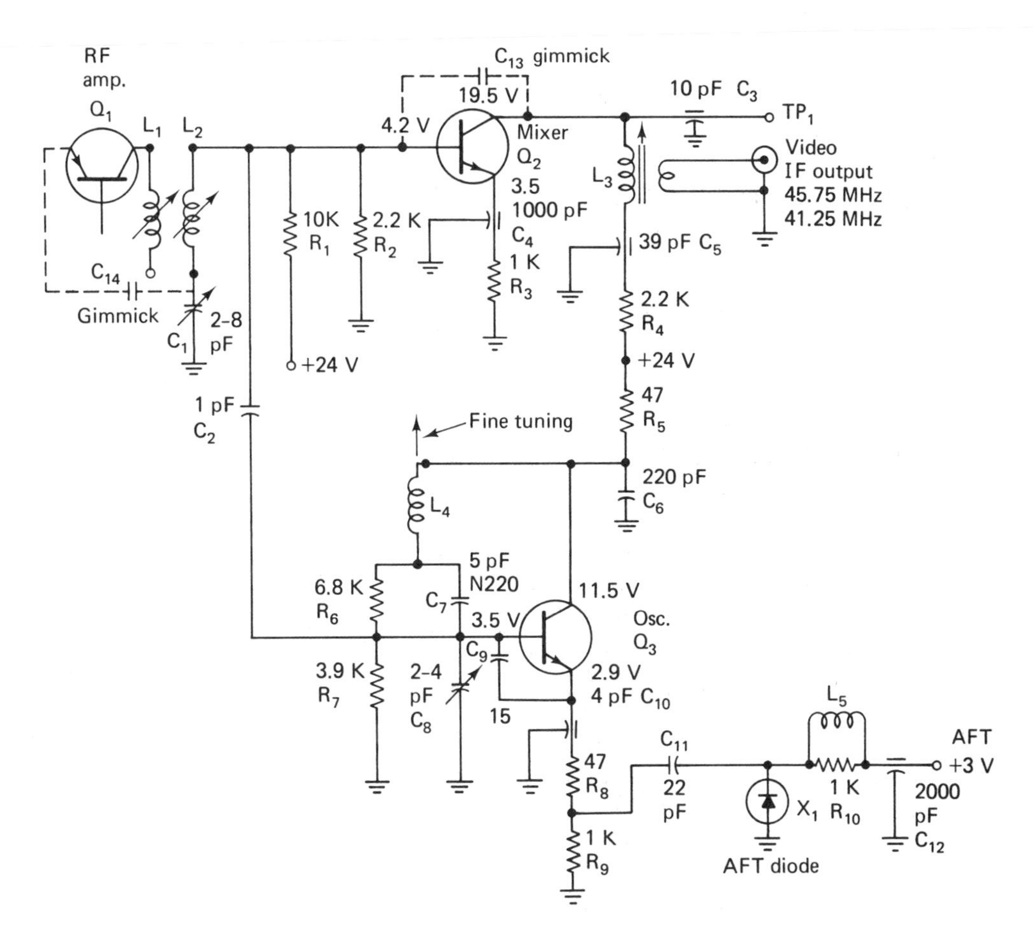

Figure 4-18 Typical transistor mixer.

electronic tuner, and that shown in Figure 4-19 is the general form to be discussed here and is found in many modern tuner designs.

The first stage (Q_1) is designed in the common emitter circuit configuration. A common base configuration is used in the second stage (Q_2). The low input impedance of the common base and the lack of tuned circuits between stages allow the two stages to be physically separated and mounted on a separate chassis, as is done in some RCA tuners. But when this is done, capacitive coupling instead of direct coupling is used between the stages as shown here.

Frequency conversion takes place because of the nonlinearity of the base to emitter junction of Q_1. The remaining portion of the mixer is used as an amplifier. The input to the first stage consists of the RF signal and the oscillator signal. Both inputs are capacitively coupled: the RF signal by C_1 and the oscillator signal by C_2. The size of the capacitors not only determines the levels of the input signals, but it also provides impedance match between these stages and the mixer.

The dc operating conditions of the mixer are established by R_1, R_2, R_3, and R_4. R_1, R_2, and R_4 are a voltage divider that establishes the voltage on the base of Q_1. Capacitor

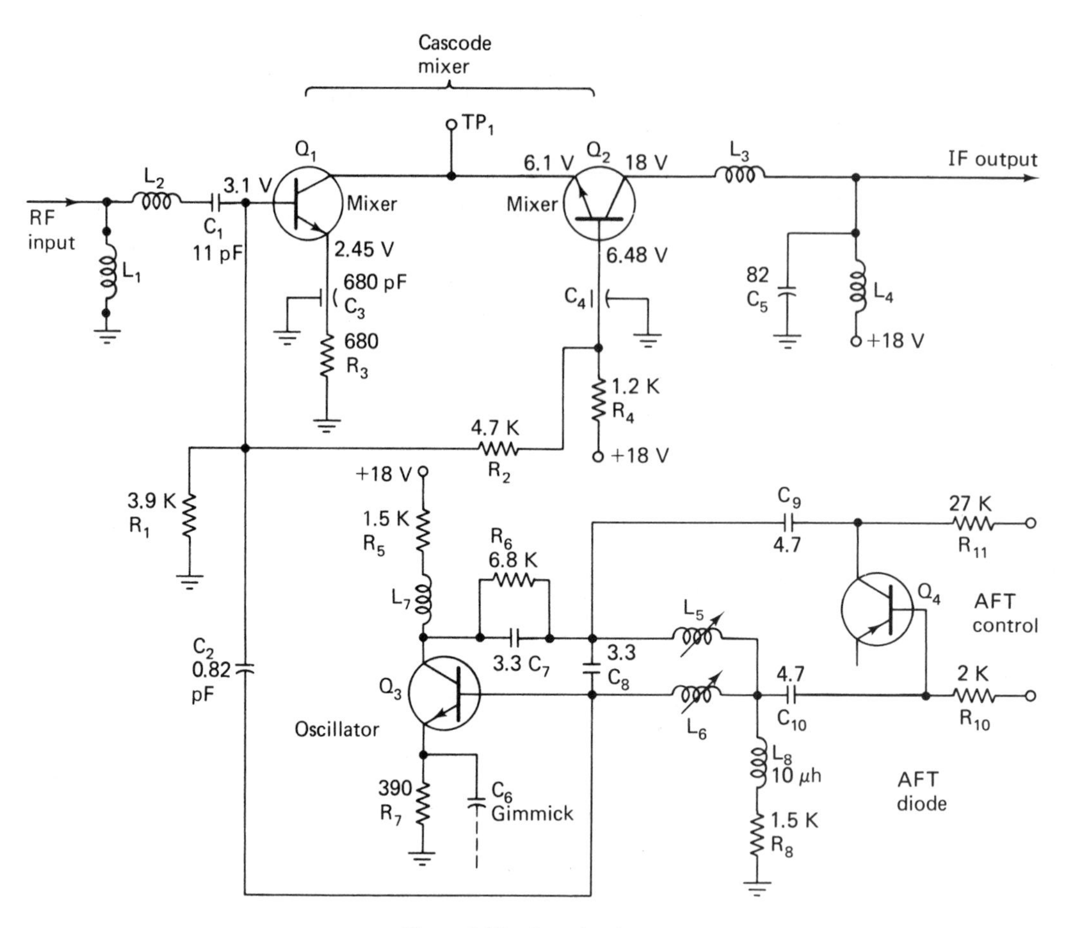

Figure 4-19 Cascode mixer.

C_4 ties the base to ac ground, thereby allowing all of the input signal to be applied to the emitter of Q_2.

The input circuit of Q_1 is tuned to the RF signal frequency by L_1, L_2, and the distributed capacitance of the circuit. L_1 is varied by means of a channel selector switch when changing channels. The output circuit is tuned to the IF frequency by L_3, L_4, and C_5.

The FET mixer. An example of a MOSFET mixer is shown in Figure 4-20. This stage is a VHF-to-IF frequency changer that is used as an IF amplifier when UHF is being received. The input tuned circuit is varactor tuned by CR_4 and CR_5. Band switching is accomplished by diodes CR_9 and CR_{10}.

Oscillator and RF voltages are coupled to gate G_1 of the mixer by capacitors C_{24}, C_{16}, and C_{20}. Resistors R_{14} and R_{15} are used to optimize conversion gain by causing the voltage on G_2 to track the voltage changes on G_1.

The drain of the mixer FET drives a double-tuned IF transformer which couples the IF signal to a buffer IF amplifier. This amplifier provides matching and gain for the mixer output signal.

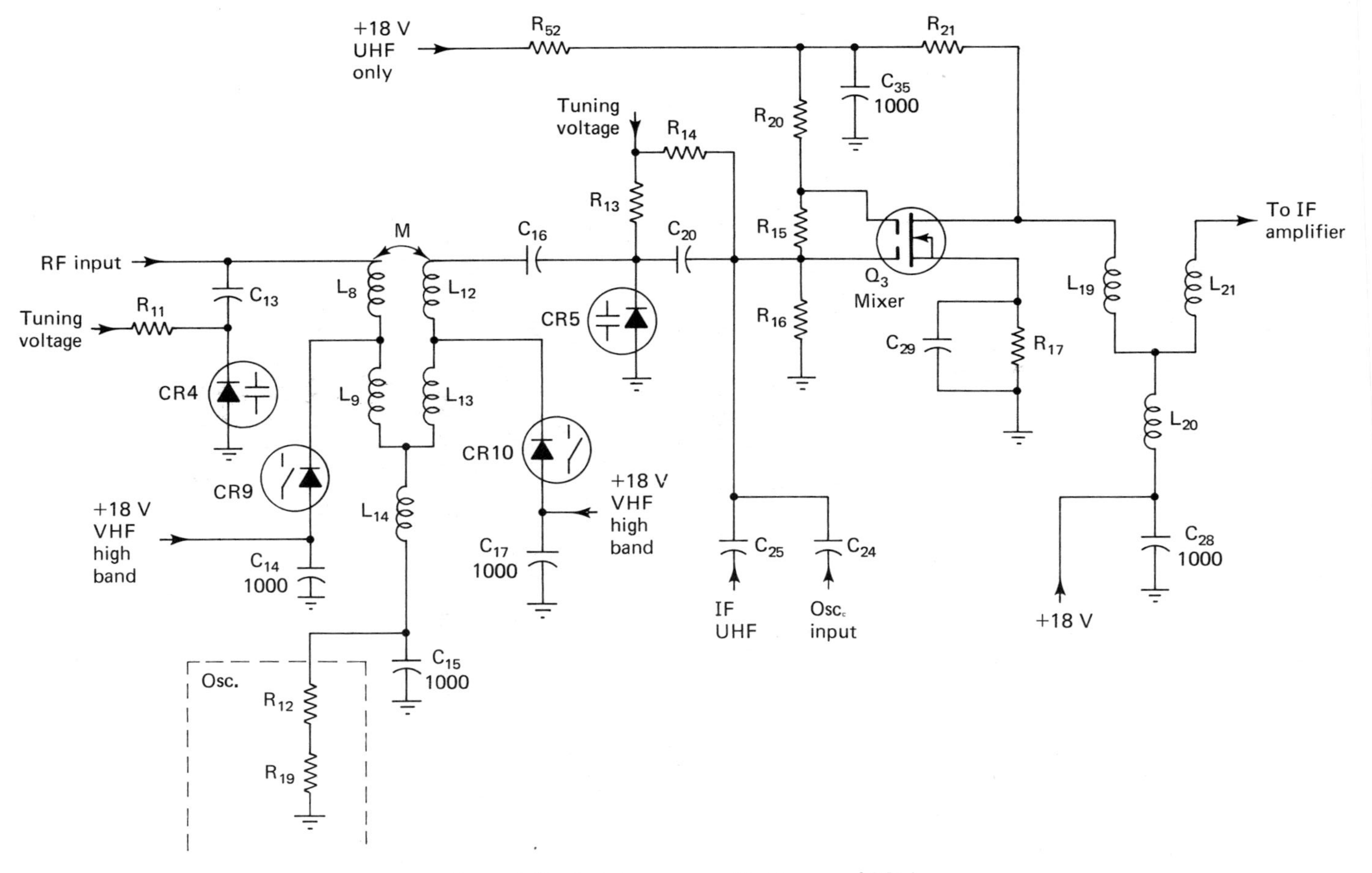

Figure 4-20 VHF mixer using a FET. (Courtesy of RCA.)

4-9 *the oscillator*

The local oscillator used in the front end of a television receiver is a high-frequency ac generator designed to produce a sine-wave output. The frequency of the oscillator is almost universally placed above the incoming RF signal such that the picture carrier produces a mixer output IF of 45.75 MHz and the sound carrier generates a mixer IF output of 41.25 MHz. The amplitude of the oscillator voltage injected into the mixer is adjustèd by the coupling method and is many times larger than the RF input signal in order to produce maximum conversion gain.

Transistor local oscillators. The Colpitts oscillator is the most commonly used oscillator in transistor VHF tuners. An example of a transistor Colpitts oscillator is shown in Figure 4-21. In many respects the transistor Colpitts is similar to the vacuum tube version. Thus, L_1, C_1 and the feedback capacitors C_3 and C_4 constitute the resonant tuned circuit and determine the frequency of operation. The feedback capacitors C_3 and C_4 include the transistor's interelectrode capacitance in a manner similar to the vacuum tube ultra-audion. Capacitor C_4 is often made tunable and is used to adjust the oscillator frequency during tuner alignment.

The dc operating conditions of the oscillator circuit are established by R_1, R_2, and R_3. These resistors provide a small forward bias that allows the oscillator to be self-starting. L_2 is a short circuit for dc and ties the collector to the B supply, but it is an open circuit for ac and isolates the collector from the ac ground of the power supply. In a similar way, C_1 is a short circuit for ac and eliminates the effect that R_1 might have had on the tuned circuit, but it is an open circuit for dc, thereby allowing R_1 and R_2 to provide the base voltage for the transistor.

Typical oscillator circuits. Practical examples of this type of oscillator are shown in Figure 4-19 and in the complete tuner schematic diagrams shown in Figure 4-9. In all of these circuits the output of the oscillator is coupled into the mixer by a capacitor. In each case, the oscillator frequency is changed mechanically during channel selection, with the exception of the varactor tuner of Figure 4-9. In this case, band switching is accomplished by diodes and channel selection is obtained by adjusting the dc voltage on a varactor.

Figure 4-21 Transistor Colpitts oscillator.

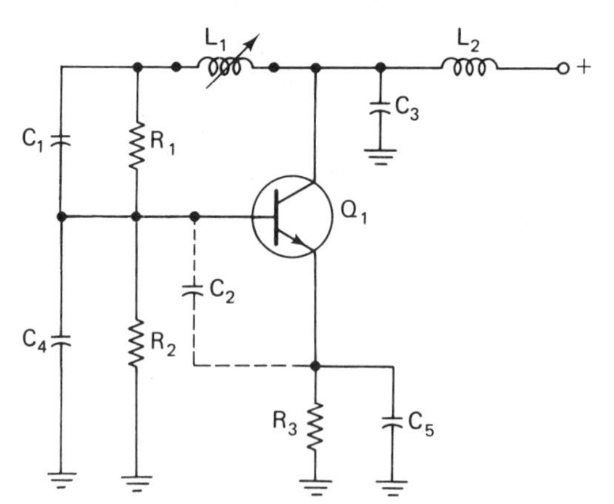

The diode X_1 in Figure 4-18 and the diode junction between collector and base of transistor Q_4 in Figure 4-19 are both part of an automatic fine-tuning circuit used with the tuners.

A defective oscillator or mixer will result in no picture, no color, no sound, and good raster conditions for all positions of the channel selector.

Practical VHF tuners. Figure 4-22 shows the schematic diagrams of a transistor VHF turret-type tuner, a FET/transistor wafer switch VHF tuner and an all transistor wafer switch tuner. In the turret-type tuner [Figure 4-22(a)] a common-base RF amplifier (Q_1) is inductively coupled (L_C and L_B) to a common-emitter mixer (Q_2). The Colpitts-type AFC-controlled oscillator is coupled to the mixer via capacitor C_{18}.

The RF amplifier transistor's metal case is returned to ground to provide shielding. The RF amplifier obtains its AGC control voltage through a 470-Ω isolation resistor R_2 and feed-through capacitor C_7.

Two RF antenna inputs are available. If a 300-Ω transmission line feeds the receiver, it is coupled to a balun (T_{001}), which in turn drives the IF/FM assembly. If a 75-Ω transmission line feeds the receiver, an input jack is provided for direct coupling to the IF/FM assembly. Channel selection is obtained by changing a strip mounted on a drum containing 13 such strips. Each strip contains four tuned circuit coils corresponding to the input of the RF amplifier, the output of the RF amplifier, the input to the mixer, and the local oscillator.

When the receiver is tuned to UHF (channel 1), the turret strip changes the resonant frequency of the RF amplifier and mixer-tuned circuits to the video IF (45.75 MHz), converting these stages into video IF amplifiers. The VHF oscillator is disabled by removing the oscillator coil. The UHF tuner is powered by a jumper that ties the VHF tuner B+ to the UHF tuner B+ line.

The FET/transistor wafer switch VHF tuner [Figure 4-22(b)] uses a common source RF amplifier (TR-1) that is inductively coupled, via the coils (L_{10}, L_{11}) mounted on the wafer switch, to the common source mixer (TR-2). The AFC-controlled Colpitts local oscillator (TR-3) is capacitively coupled by C_{18} into the mixer gate. The AGC voltage is fed to gate (G_2) of the RF amplifier.

The 75-Ω antenna delivers the VHF signal to an IF/FM filter assembly.

This transistor tuner has four tuned circuits (antenna, RF, mixer, and oscillator) mounted on wafer switches. Channel selection is obtained by changing the amount of inductance that is allowed to act in the circuit.

When the tuner is in its UHF position (channel U), the oscillator is disabled because L_{12} has no coil in that position. The RF amplifier and mixer stages are converted into video IF amplifiers, as was the case with the transistor tuner.

The all transistor wafer switch tuner of Figure 4-22(c) is similar, except for the switching system, to that of the turret tuner shown in Figure 4-22(a).

4-10 UHF tuners

Since 1964 all television receivers have been equipped to receive both VHF and UHF television channels. This is usually accomplished by completely separate tuners, one for VHF and the other for UHF. These tuners are mechanically mounted together so that

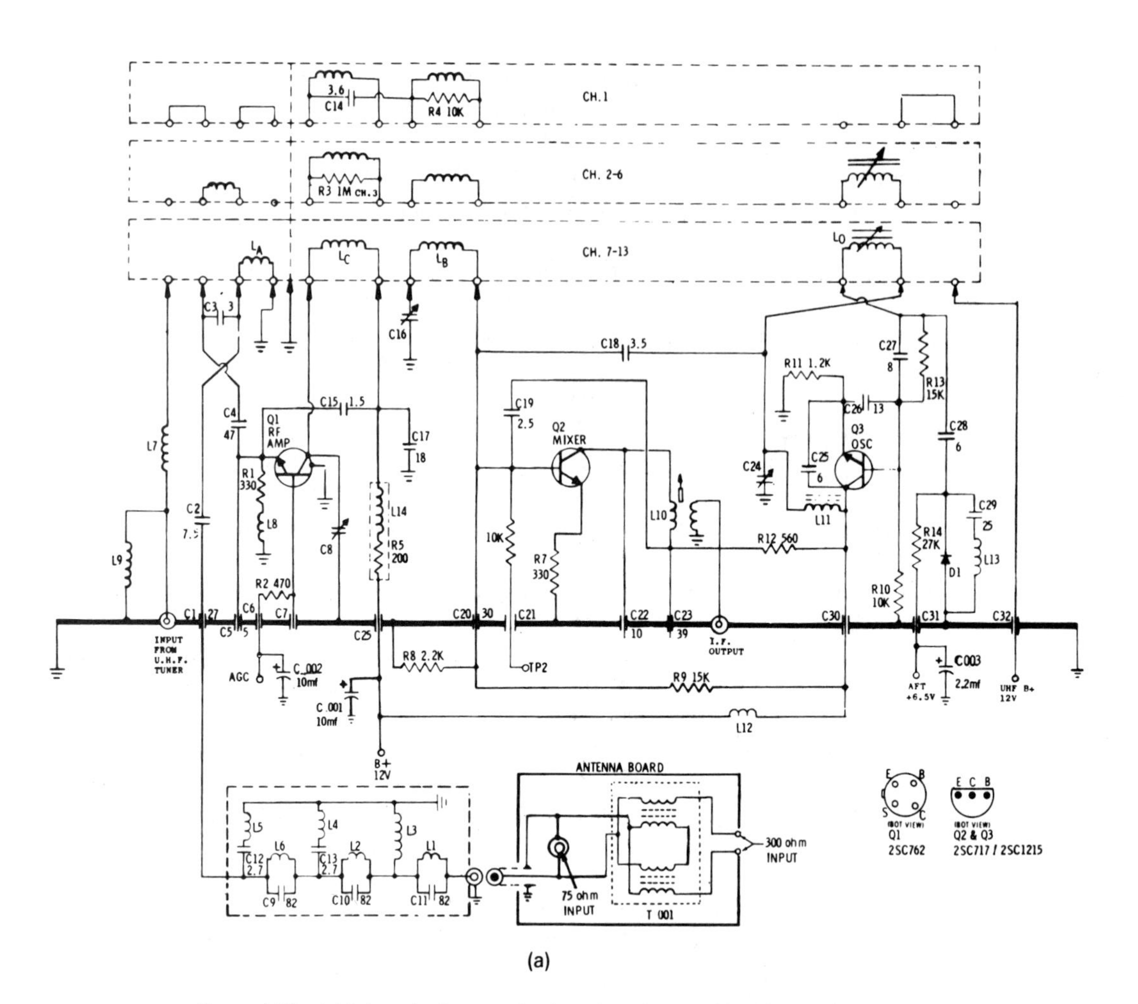

(a)

Figure 4-22 (a) Schematic diagram of a drum (turret) tuner; (b) wafer switch tuner. (c) a transistor wafer switch tuner. [(a) Courtesy of Quasar Industries, (b) courtesy of Panasonic.]

separate tuning shafts are used for VHF and UHF tuning. An example of these tuners is shown in Figure 4-2.

Transmission line losses. The selection of TV channels whose operating frequency ranges from 470 MHz (channel 14) to 890 MHz (channel 83) presents some special problems of its own. The very act of getting the UHF signal from the antenna to the UHF tuner is in itself a formidable problem. This is because of the high losses that occur in the transmission line. For example, channel 14 has almost twice the line attenuation loss as channel 13. If the transmission line is dirty or wet, the loss may increase six to eight times. Similarly, the upper UHF channels have still greater line attenuation loss. The ordinary 300-Ω ribbon line is not recommended for outdoor use with UHF tuners. The round polyfoam-filled cable or shielded twin lead give the best results. These types of transmission line combine low loss with stiffness that helps keep the cable from being shifted by the wind.

As a result of the small wavelengths at UHF, any change of as little as an inch

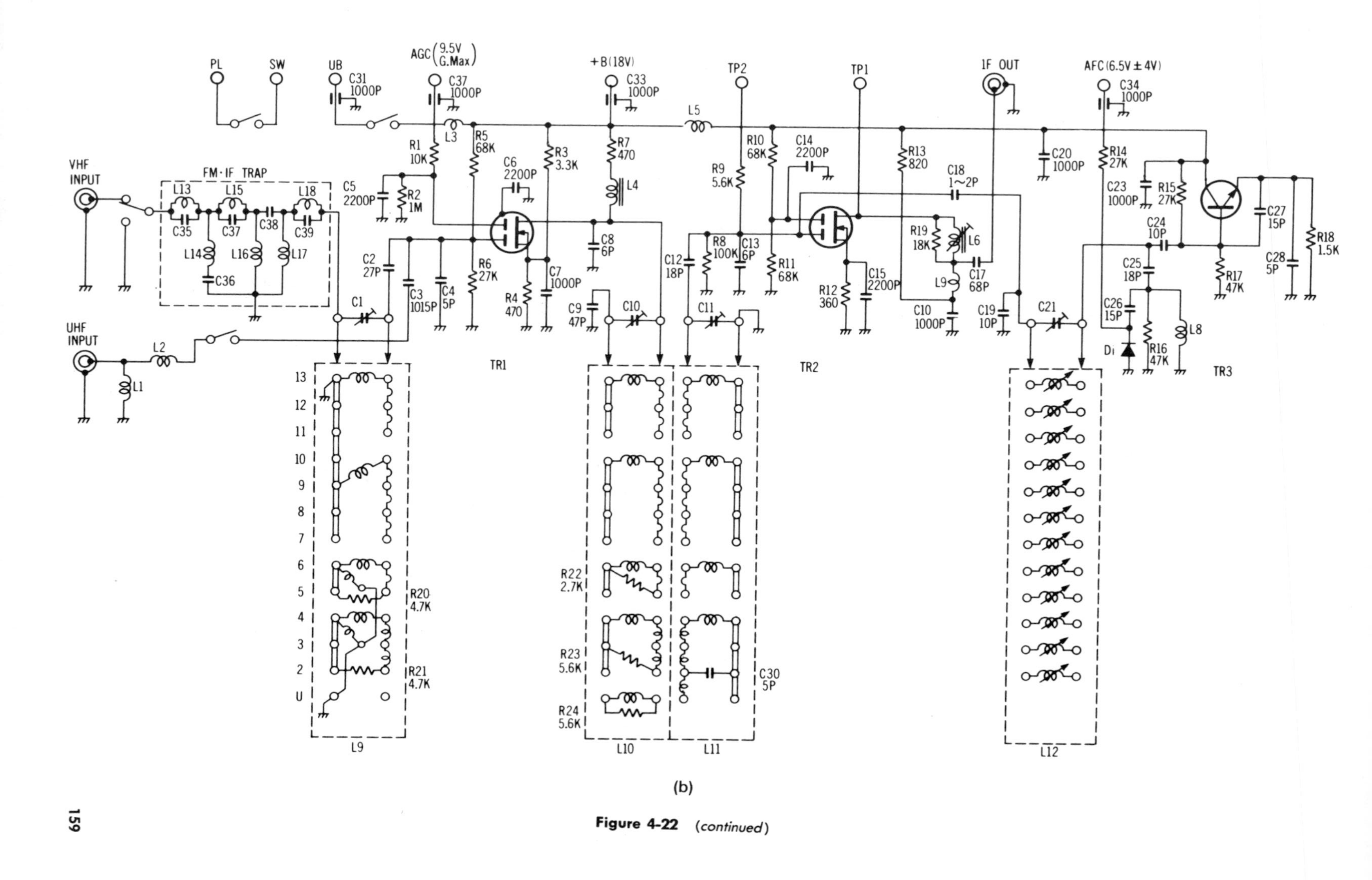

(b)

Figure 4-22 *(continued)*

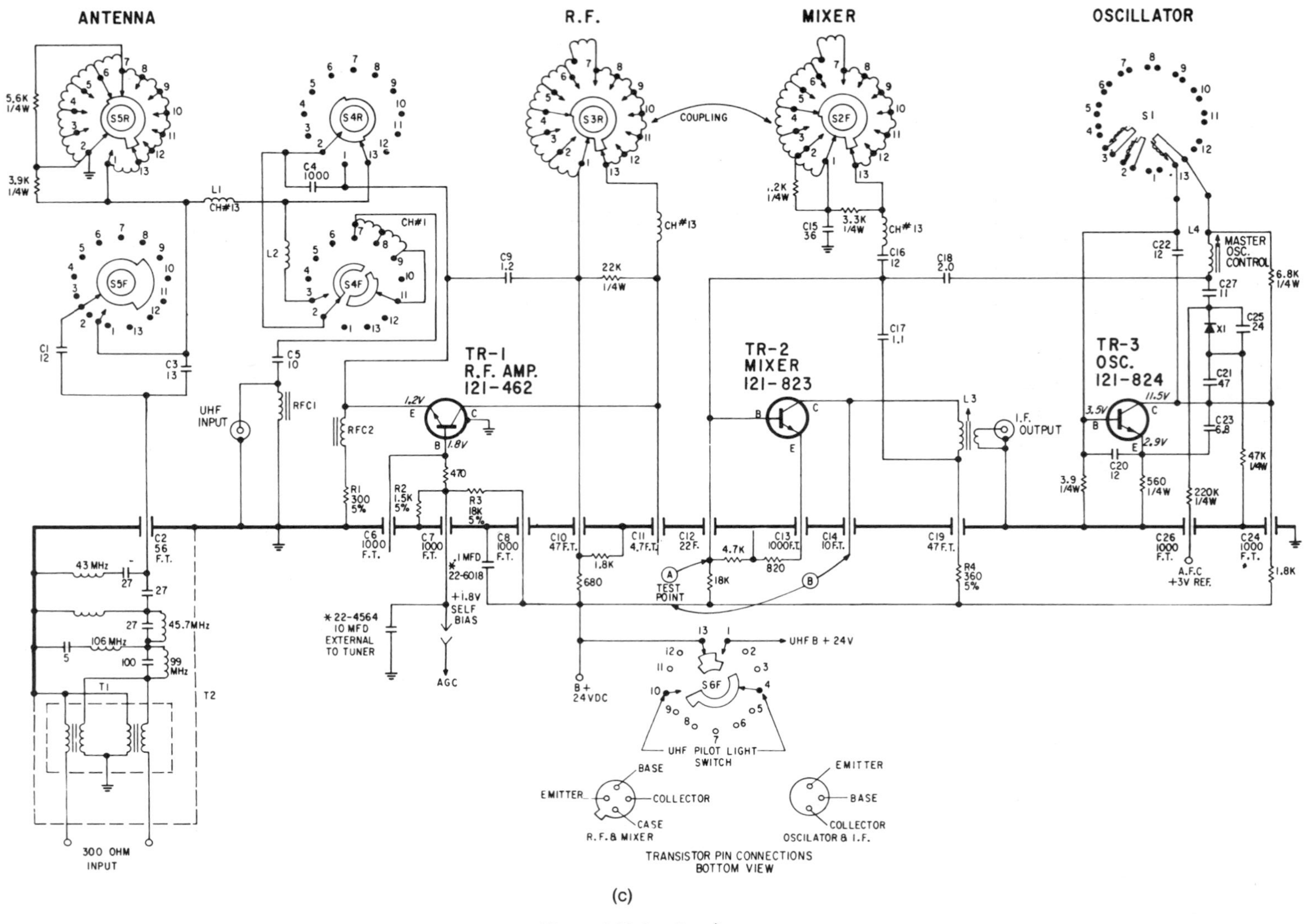

(c)

Figure 4-22 (*continued*)

or two in spacing, length, or position of wires carrying these signals may cause severe changes in the quality of reception. Indoor UHF antennas are very sensitive to position; a movement of a few inches can mean the difference between a good picture and no picture at all.

Tuned circuits. For technicians who have only worked at lower frequencies, the tuning techniques used at UHF may seem strange because at lower frequencies lumped (concentrated) electrical properties are used. Thus, familiar coils and capacitors are used for tuned circuits. At UHF frequencies it is no longer possible to use lumped coils or capacitors to form resonant circuits. The distributed electrical properties of L and C are used instead. This usually takes the form of sections of transmission lines that are cut to specific lengths and whose termination provide the desired impedance. Thus, a quarter-wave section of transmission line, called a *stub,* and terminated in a short circuit, appears to the generator as a parallel resonant circuit. For a frequency of 600 MHz, this would require a transmission line (two parallel wires or a coaxial cable) of approximately 5 in. long. If the shorted stub is made shorter, at the same frequency of operation, it appears to the input generator as an inductance. To return this quarter-wave stub to resonance, a lumped capacitor may be placed across the transmission line. If it is desired to make the resonant circuit variable, a variable capacitor may be used. Similarly, coaxial conductors may also be used. In this arrangement it is possible to put "windows" in the wall of the outer conductor, and by placing loops into this opening signal energy may be extracted. Coupling between two coaxial transmission lines may also be obtained by simply having a window in the common wall of the conductor separating the two inner conductors. Coupling is accomplished by an exchange of the electrostatic and magnetic fields passing through the window.

The UHF converter. The main function of a UHF converter is to change the frequency of the UHF channel being received to the video IF.

The block diagram of Figure 4-23 shows the circuit arrangement that has commonly been used for UHF reception. In this arrangement, the UHF signal is converted directly into the video IF and is fed into the UHF input jack on the VHF tuner. The VHF tuner is placed in its channel 1 position when UHF reception is desired. In this position both the RF and mixer stages are converted into video IF amplifiers.

Referring to the block diagram we see that UHF channel 24 feeds a preselector tuned circuit that (1) helps tune the desired signal, (2) rejects undesired signals that might cause image or other spurious responses, and (3) helps to isolate the oscillator from the antenna, thereby keeping oscillator radiation to the limits set by the FCC. In modern receivers, the signal then may feed a transistor (NPN silicon epitaxial planar type) used as an RF amplifier, or in older receivers the signal is fed directly into a crystal mixer. Crystal mixers such as the 1N82A have been commonly used at UHF frequencies because they have good conversion efficiency and they do not generate too much noise. In modern sets transistor mixers are used which also provide a conversion gain.

The output of the mixer is at the video IF frequency (45.75 MHz for the picture carrier and 41.25 MHz for the sound carrier) and is obtained by beating a signal of 577 MHz generated by the UHF oscillator (vacuum tube oscillators in older sets and transistor oscillators in modern receivers) with that of the picture and sound carriers of channel 24. In this case, the oscillator frequency is higher than the incoming UHF signal. This is done

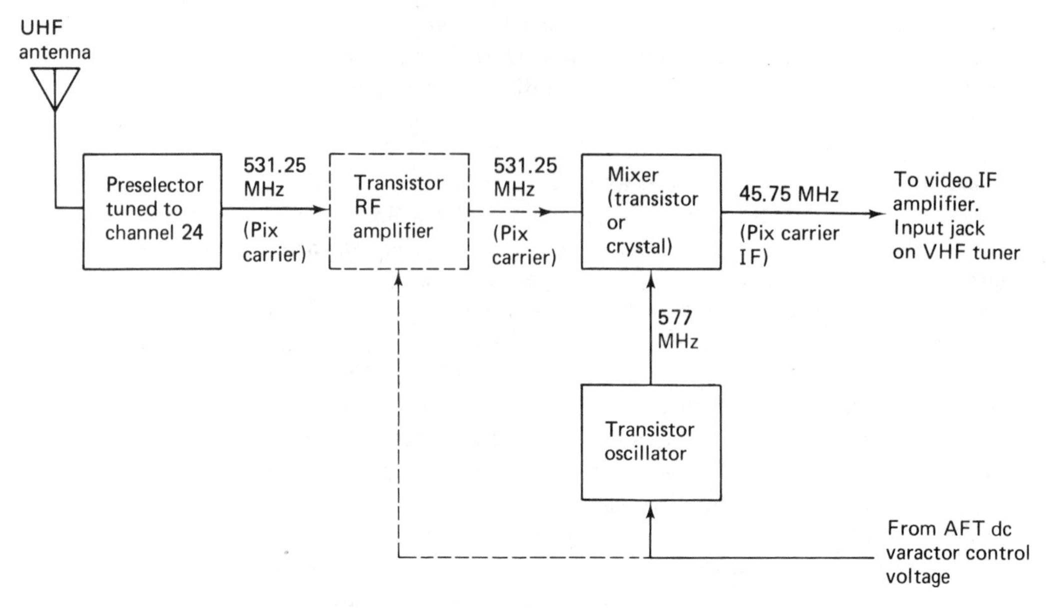

Figure 4-23 Block diagram of a UHF tuner.

because the output IF of the mixer must be frequency reversed, as is the case in VHF tuners. Modern receivers may use AFT systems for fine tuning the UHF tuner or they may resort to varactor channel selection methods entirely. The block diagram shows to which circuits the varactor control voltage is fed. UHF channel selection may be accomplished by detent tuning (70-position switch), continuous tuning, or varactor tuning.

Typical UHF tuners. The pictorial diagram of the internal structure of a UHF tuner, with its covers removed, is shown in Figure 4-24. The schematic diagram of this type of tuner is shown in Figure 4-25.

Examining the pictorial diagram we see that the tuner is divided into four thick-walled compartments. From left to right these are (1) the gear and shaft assembly, (2) the antenna input and tuned line, (3) the mixer tuned line, and (4) the oscillator tuned line and circuitry. The thick walls of each compartment and the massiveness of all the tuned circuit elements are the result of the need for frequency stability, reduction of radiation, and as a means of reducing losses.

The antibacklash gear and shaft assembly is used to allow continuous tuning over the entire UHF range.

If the tuner is used in a receiver that does not use an isolating power transformer, the shaft must be made of insulating material. This is done to minimize the possibility of shock.

The remaining three compartments help form shorted coaxial transmission line segments that are smaller then a quarter-wave in length. These are tuned by thick-plated variable capacitors located at their open end. These capacitors are slotted to allow proper tracking. Trimmer capacitors (not visible in this pictorial diagram) are placed across main tuning capacitors of some UHF tuners as an aid to tracking.

The antenna input coil looks something like a paper clip and matches the antenna

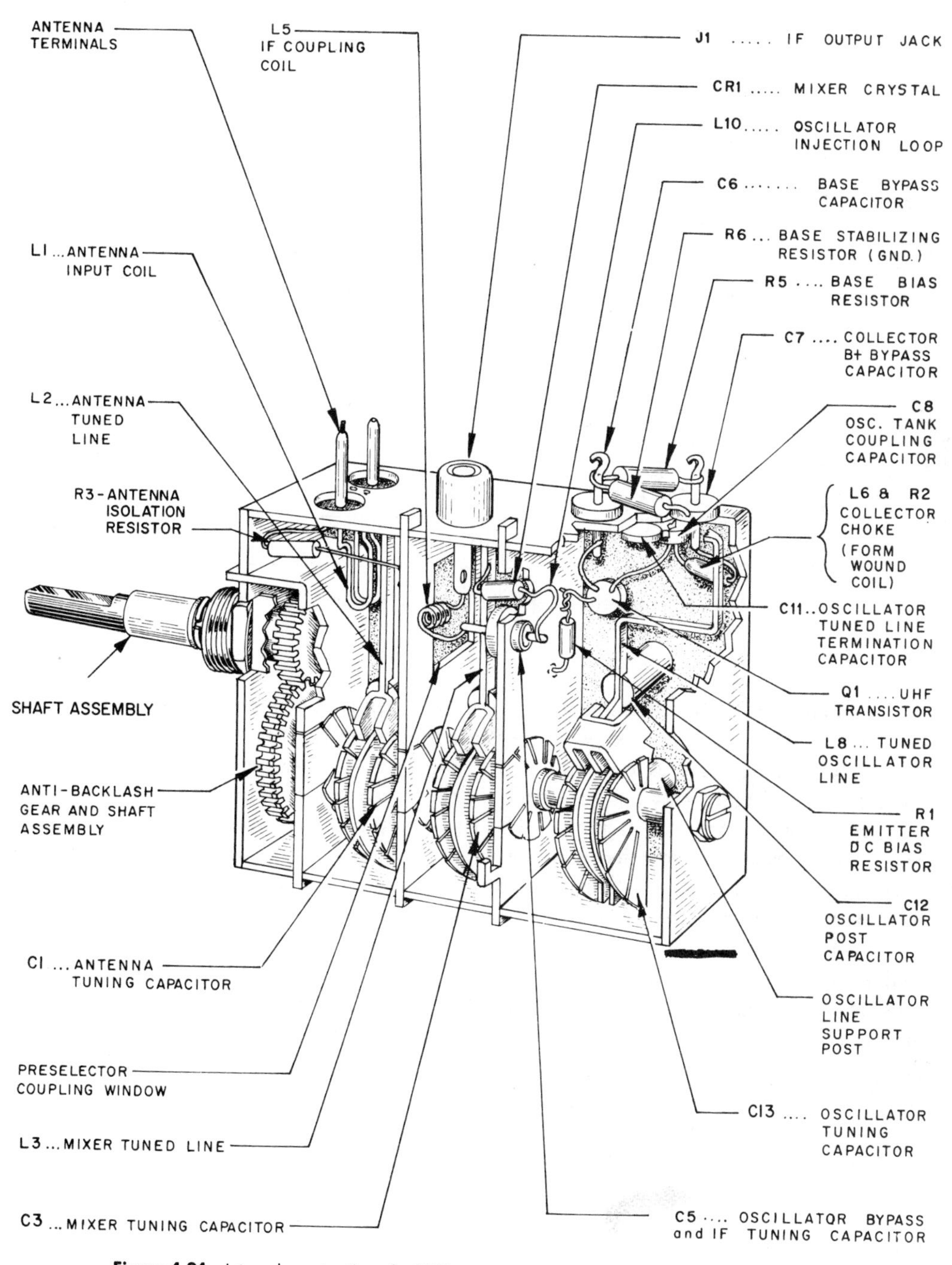

Figure 4-24 Internal construction of a UHF tuner. (Courtesy of RCA, Consumer Electronics.)

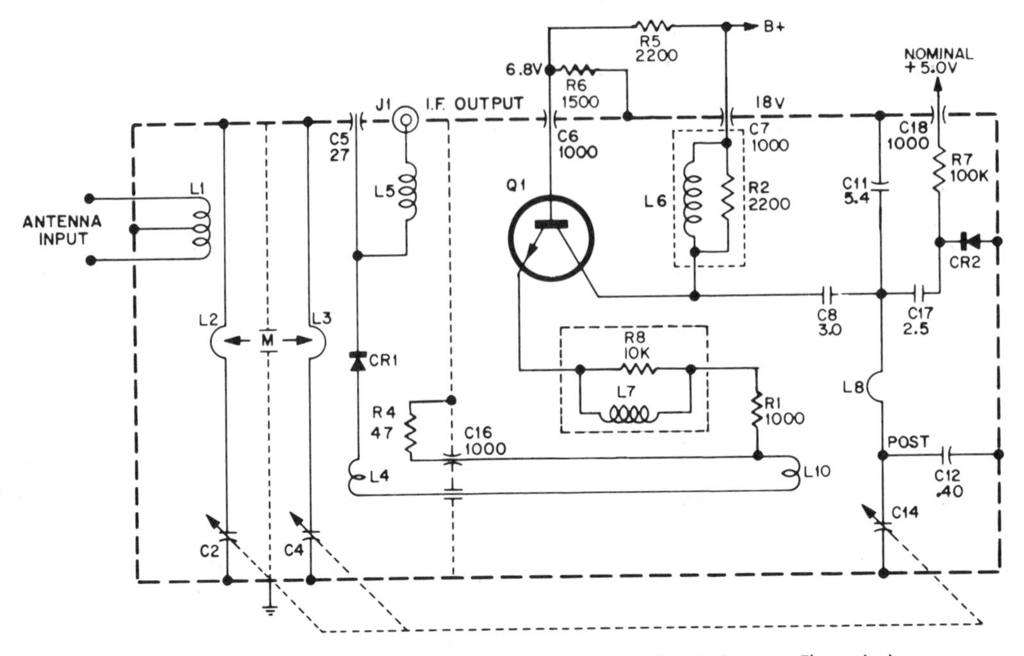

Figure 4-25 Schematic diagram of a UHF tuner. (Courtesy of RCA, Consumer Electronics.)

to the antenna tuned line. Coupling between the antenna tuned line and the mixer tuned line takes place through a window in the plate separating the two lines.

The mixer requires two inputs (RF and oscillator) to perform its function. The method used in Figure 4-24 has the mixer crystal placed in a hole in the compartment wall between the mixer and the oscillator. The leads of the crystal act as the coupling coils. Thus, on the oscillator side the oscillator injection loop (L_{10}) is used to couple the mixer to the oscillator line, and it consists of not quite one turn of wire. A similar loop (L_4) on the mixer side couples the mixer tuned line to the crystal mixer.

The fourth compartment houses a transistor Colpitts oscillator. The oscillator section is mechanically similar to the others. The tuned circuit as shown in Figure 4-25, consists of a tuned line (L_8 C_{14}). Two trimmer capacitors may be included in some UHF tuners. These are used to provide tracking adjustments at both the low and high ends of the tuning range of the oscillator. Dc operating voltages are provided by L_6, R_1, R_4, R_5, and R_6.

In this UHF tuner AFT is provided. The AFT varactor diode (CR_2) is located in the oscillator compartment and is placed across the tuned line. Terminals are provided for the AFT control voltage.

Fully transistorized UHF tuner. The schematic diagram of a fully transistorized UHF tuner is shown in Figure 4-26. The first thing that should be noticed is that the RF, mixer, and oscillator stages of the tuner use transistors. These transistors are silicon epitaxial planar transistors that can easily operate at UHF frequencies. The use of transistors increases the sensitivity and selectivity of the tuner and it further reduces oscillator radiation.

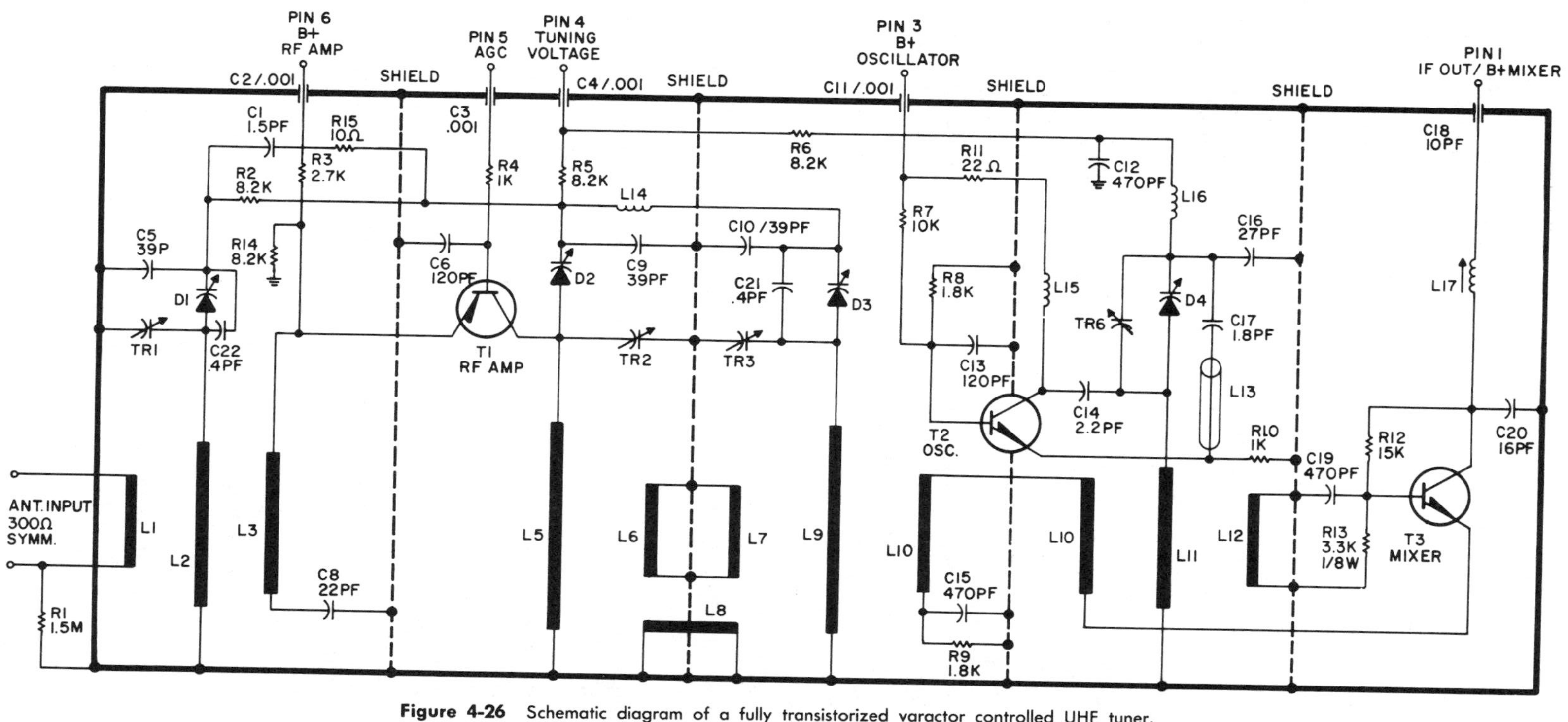

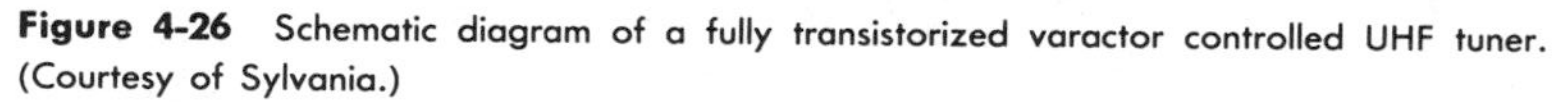

Figure 4-26 Schematic diagram of a fully transistorized varactor controlled UHF tuner. (Courtesy of Sylvania.)

The second thing that should be noticed is the use of varactors D_1, D_2, D_3, and D_4 for tuning each stage. The varactors allow the tuner to be completely free of all mechanical channel-selecting or fine-tuning arrangements. Tuning is accomplished by changing the dc voltage applied to the varactor. In this tuner the tuning voltage varies from 1.5 to 30 V, controlling all four varactors and tuning the tuner through the entire UHF range (channels 14 through 83).

The RF amplifier is a PNP used in a common base amplifier configuration. The base is kept at ac ground by C_6 and its dc voltage is varied by an AGC input. The input tuned circuit consists of L_1, L_2, L_3, D_1, TR_1, C_5, and C_{22}. The output tuned circuit consists of L_5, L_6, L_7, L_8, L_9, D_2, D_3, TR_2, TR_3, C_9, C_{10}, and C_{21}. The dc B supply feeds the emitter via R_3 and the collector is returned to ground through L_5. Many modern UHF RF amplifiers use MOSFETs, in place of bipolar transistors.

The oscillator is a Colpitts that uses an NPN transistor. The tuned circuit consists of L_{11}, TR_6, D_4, and C_{17}. The dc B supply for this transistor feeds the collector via R_{11} and L_{15}, and feeds the base by means of R_7 and R_8. The emitter returns to ground through R_{10}.

The mixer transistor has the oscillator signal injected into the base by inductive coupling between the oscillator tuned line L_{11} and the mixer tuned line L_{12}. The RF signal is injected into the mixer's emitter by inductive coupling between the RF amplifier output tuned line L_9 and L_{10}, the tuned line in the emitter of the mixer. The video IF output is obtained from the collector of the mixer. The dc operating conditions of the mixer are established by the B supply feeding the collector and base via L_{17} and R_{12}. The emitter returns to ground through L_{10} and R_9.

4-11 tuner troubleshooting

The visual effect (as seen on the picture tube) of tuner defects can be classified as follows:

1. Both black & white and color picture and sound are dead on all channels; raster is O.K.
2. No picture, no sound on some channels only; raster is O.K.
3. Weak, snowy pictures on all channels; sound may be O.K.
4. Weak, snowy pictures on some channels; sound may be O.K.
5. Intermittent operation on one or more channels.
6. Poor-quality picture.
7. Oscillation.
8. Either no color or weak or poor-quality color.
9. Color, sound, and black & white pictures occur at different fine-tuning settings.
10. No UHF reception; all VHF channels are O.K.

Trouble symptoms one and two are similar and therefore the faults that cause them may also be similar. Using standard troubleshooting methods and assuming all obvious defects have been eliminated, defective stage isolation is the next step.

For the receiver to be dead with a working raster indicates that the power supply is O.K. but the signal is not being processed between the antenna and the output of the video detector. This is so because if the signal interruption were in the video amplifier stages, sound and possibly color would still produce outputs.

The next troubleshooting step is to determine whether the trouble is in the tuner or in the IF amplifiers. This can be done by signal injection with a tuner subber or an RF signal generator. If the subber restores the picture and the sound, the IF strip is good and the receiver's tuner is at fault.

In electromechanical type tuners, fault isolation may be divided into two parts, mechanical and electrical. Mechanical faults that can cause a dead receiver on all channels would require some fault common to all positions of the turret or channel selecting switch. This means that the *stator* contacts in the turret type tuner or the *rotor* contacts in switch type tuner may not be making the electrical connections necessary for channel selection. Visual inspection of these contacts may reveal broken or bent contacts. If such contacts are found, part replacement of broken strips or switches, or repair of bent contacts is possible. However, taking labor costs into consideration, it is probably cheaper to replace the entire tuner.

1. Electrical defects that can cause a dead receiver can occur in the RF amplifier, the mixer, the oscillator, or AFT stages of the tuner.

A defective RF amplifier stage will not allow the signal to reach the mixer. Therefore in weak signal areas this defect will produce a snowy raster without a picture. If the mixer is defective, frequency conversion will not take place even though both the RF signal and the oscillator signal are being fed into it. The loss of the oscillator will disable the mixer and no picture or sound will be developed. The oscillator stage may become defective in one of two ways; a component in the stage fails, or the AFT stage fails in such a way as to cause the oscillator to cease functioning. For example, if the AFT varactor should short, the receiver's local oscillator tuned circuit would be shorted and the oscillator would be dead.

2. It is also possible for a tuner defect to be responsible for the loss of low or high frequency channels only. This can be due to a defective oscillator transistor, open or shorted capacitors in the antenna filter assembly, or defective coils or capacitors in the tuned circuits of the RF, mixer, or oscillator sections of the tuner. Misadjustment of the oscillator or other tuned circuits as well as defective switch contacts also can be responsible for loss of reception at high or low frequencies. Misadjustment of the neutralizing capacitor can cause severe oscillation on some channels and not others that may block out one or more channels.

3, 4. The third and fourth categories, weak, snowy pictures may be due to:

(a) Insufficient signal reaching the antenna terminals of the tuner or the input to the RF amplifier. This can be caused by such things as a defective antenna, an open transmission line between the antenna and the receiver, or defective input circuit components (balun, antenna filter assembly, and capristors).

(b) Insufficient amplification or inefficient frequency conversion. These defects can be caused by such things as weak tubes or transistors, dirty contacts in the channel selecting switches (particularly when only one or two channels are affected), excessive AGC voltage, changed value components causing misalignment, and insufficient oscillator voltage injection possibly due to defective coupling transformer or open coupling capacitor.

Many tuners provide a test point at the mixer, called a *looker point* that can be used to observe the response curve of the RF amplifier during alignment. This same point may be used to determine whether a signal can pass through the RF amplifier stage. Signal tracing after the mixer requires the use of a high-gain scope and a demodulator probe.

5. Intermittent operation on one or two channels may be due to different conditions from those that affect all channels. This is because defects affecting one or two channels must be related to the channel selecting tuned circuit elements associated with the channels. When all channels are affected the defects must be related to the dc operating conditions in any of the stages in the tuner. Hence, dirty contacts or defective channel selector strips or switches are the chief probable causes of intermittent operation on one or two channels. Intermittent operation on all channels can be caused by almost any other component in the tuner. Isolation of the defect begins by noting operation on channel 1 (UHF). If all is well, then trouble must be in the oscillator section of the tuner. The RF and mixer are suspect if intermittent operation continues.

6. Poor-quality black & white or color pictures are primarily due to misalignment of either the tuner or the IF amplifiers. This also may be due to circuit defects that cause changes in alignment.

7. Oscillation of the RF or mixer stages of the tuner may result in such symptoms as motorboating, herringbone pattern interference, loss of picture and sound, or completely erratic operation that is affected by the fine tuning. Oscillation can be due to such things as a misadjusted or defective RF amplifier neutralizing capacitor, poor lead dress, open bypass capacitors, open AGC isolating resistor or tuner misalignment.

8. The symptoms of no color and good black & white reception can be due to many different defective stages or adjustments. However, one of the simplest and easiest to check is whether the color receiver's fine tuning is improperly set. If misadjustment of the fine tuning is the cause of loss of color, this can be checked by simply turning up the color control and then adjusting the fine tuning until a good color picture is obtained. If a color picture still cannot be obtained the trouble lies elsewhere.

Another possible cause of misadjusted fine tuning is a defective AFT circuit. A defective AFT circuit is indicated if all channels are working normally and then they all simultaneously become badly detuned. This analysis is confirmed if normal manual fine tuning is restored when the AFT switch is turned off.

Weak or poor-quality color, such as poor color fit or smear, may be due to defects in many stages of the receiver. However, defective tuner alignment also can be responsible for this symptom. A quick check of the overall alignment of the receiver can be made by adjusting the fine tuning and observing the picture. In a normal receiver, changing the fine tuning should have little effect on picture quality. Large changes in picture quality or changes in the number of "ghosts" (ringing) associated with vertical lines in the picture indicate poor alignment, and warrant a sweep generator check of both the tuner and IF alignment of the receiver.

Ghosts. Ghosts or repeated pictures are usually due to causes external to the receiver. TV signals operate at frequencies where radio waves act like light and large buildings or other structures act like mirrors. It is therefore possible for the receiving antenna to receive the transmitted signal from two or more sources. The direct source is the transmitter; the other sources are the reflections that arrive at the receiving antenna some time after the direct signal. See Figure 4-27(a).

Assuming only one reflected signal, the observed picture will consist of the original direct picture plus a second picture to its right called a "ghost." The degree of separation

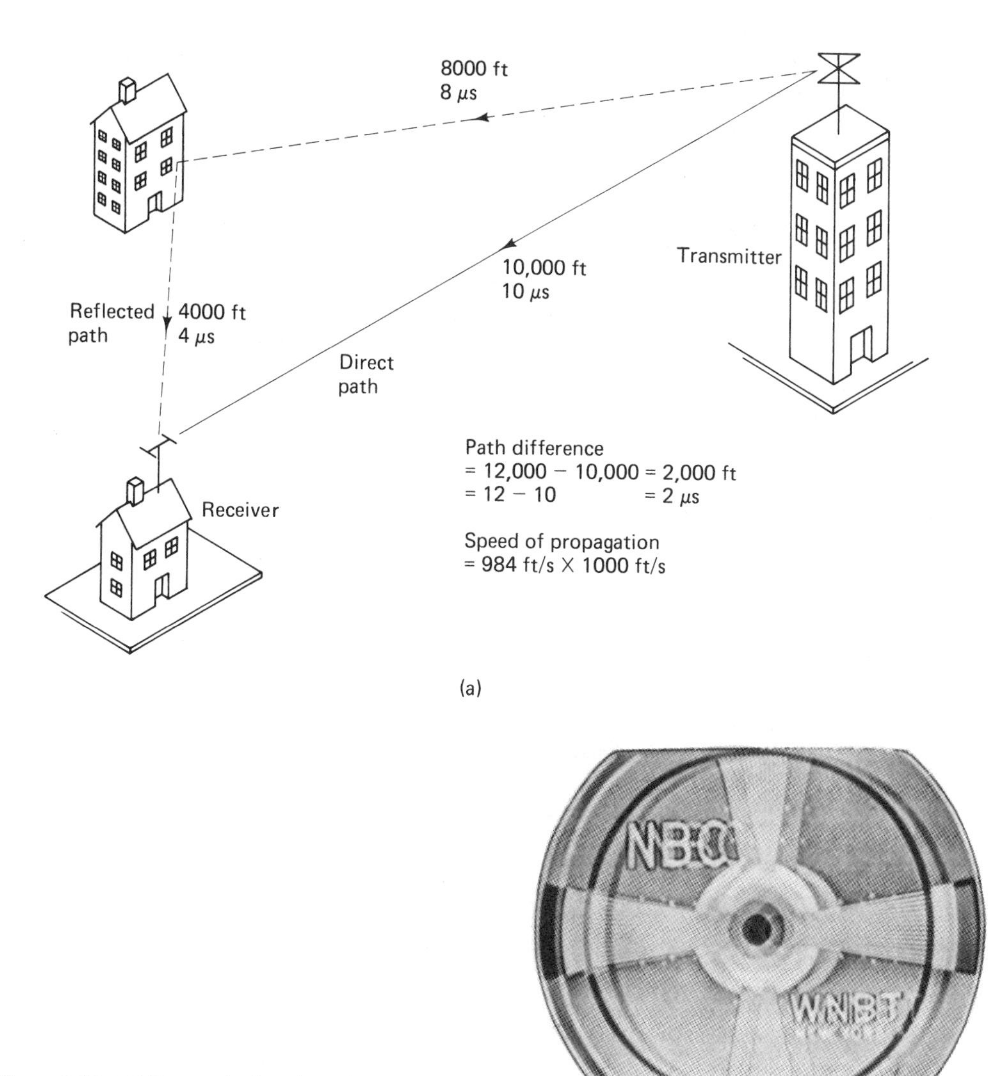

Figure 4-27 (a) Direct and reflected signals and their time relationships; (b) the effect of a reflected signal (ghost). (Courtesy of RCA, Consumer Electronics.)

between the picture and the ghost depends upon the time difference between the reception of the two signals. The ghost is to the right of the direct picture because horizontal scanning moves the deflection beam from the left to the right side of the raster. As a result, signals that arrive first will appear to the left of signals that are received later. See Figure 4-27(b).

Ghosts may be minimized by using highly directive antennas. Ghosts may also be minimized at the transmitter by transmitting the signal with circular polarization.

9. Color, sound, and black & white pictures occur at different fine tuning settings. These symptoms will usually be accompanied by poor picture quality, buzz, and color smear or poor color-fit. All of these symptoms indicate defective tuner, video IF, or sound

IF alignments or possibly IF regeneration. A sweep generator check of tuner and video IF alignment will quickly determine whether this is the cause of the difficulty.

10. No UHF reception. All VHF channels are O.K. These symptoms may be due to one of three possible general causes: (1) defective UHF antenna, (2) UHF strip in VHF tuner is defective, and (3) defective UHF tuner.

A defective antenna or antenna lead-in can be checked by visual inspection. Antenna substitution or modulated RF (UHF) generator signal substitution may be used to verify that the signal is not reaching the properly operating UHF tuner.

To check the UHF strip in the VHF tuner, tune receiver to UHF position on VHF tuner, disconnect UHF output coaxial cable from the VHF tuner and inject into the cable the output of a UHF subber. If a normal picture appears on the screen of the picture tube, the UHF strip is O.K. If not, the strip is defective.

The tuner subber. A good method for signal injection that has the advantage of testing the entire tuner is to completely substitute the tuner with a known good one. A commercially available battery-operated transistor tuner mounted in a carrying case is called a *tuner subber.* These units may contain both a VHF tuner and a UHF tuner.

The schematic diagram of a VHF or UHF tuner subber that can easily be constructed is shown in Figure 4-28. A transistor tuner is converted into a subber by simply supplying its B supply and AGC voltage requirements. In this circuit two 9-V batteries

Figure 4-28 Tuner subber.

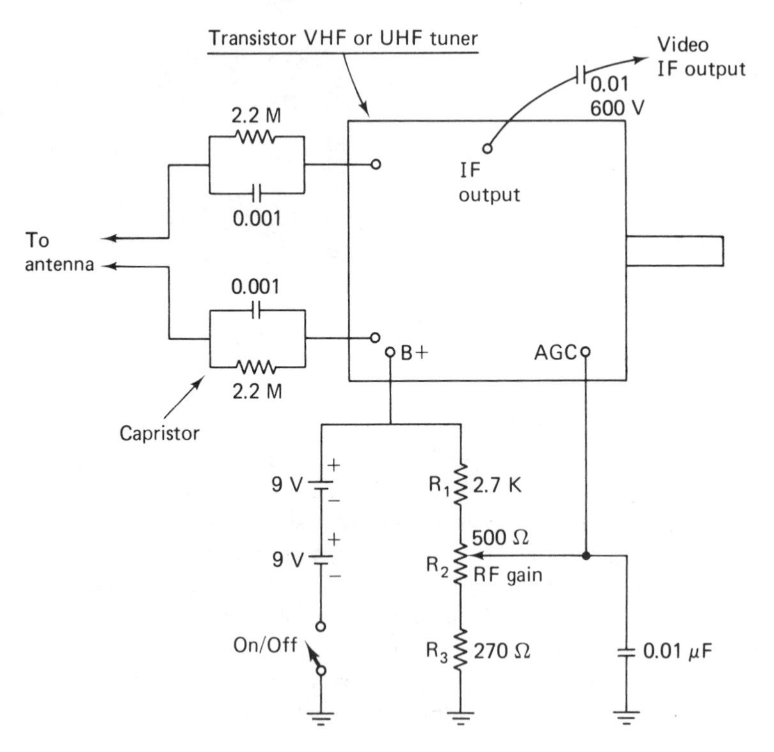

connected in series provide the B supply voltage. A variable voltage divider formed by R_1, R_2, and R_3 is placed across the B supply and supplies the tuner with an adjustable AGC voltage. The exact values of this voltage divider will vary depending upon the tuner used and must be determined experimentally. The location of the B+ and AGC tuner terminals varies from one tuner to another. Manufacturer service literature generally will indicate the position of these points. When such information is not available, wire tracing of the tuner can locate the B+ and AGC points.

The IF outputs of the subber are the sound IF (41.25 MHz) and the picture IF (45.75 MHz). Subbers cannot be used with older receivers having a 20-MHz IF.

4-12 VHF tuner alignment

The main objectives of tuner alignment are to:

1. Adjust the input and output RF amplifier and mixer input tuned circuits for the necessary 6-MHz bandpass for each selected channel. A typical RF amplifier to mixer frequency response curve is shown in Figure 4-29.
2. Adjust the oscillator frequency so that when it is combined with the RF signal in the mixer, the proper picture and sound IF will be produced. This should occur when the fine-tuning adjustment is in the center of its range.

Tuner alignment is a difficult procedure that should not be attempted unless the technician has the manufacturer's alignment instructions, as well as all of the matching pads, amplifiers, demodulator probes, sweep generators, marker generators, and bias clamp power supplies that the manufacturer calls for in its instructions. In practice, tuner alignment is seldom necessary. When it is, many service technicians exchange the defective tuner for a rebuilt properly aligned tuner. Rebuilt tuners are obtained at low cost from companies that specialize in rebuilding, repairing, and aligning tuners. As a matter of fact, some manufacturers seal their tuners and recommend that defective tuners be discarded and replaced with new ones.

Alignment procedures are almost identical for color and black & white as well as for vacuum tube and transistor tuners. The procedure may be modified, depending upon whether the tuner is a turret or wafer switch type. A typical equipment hookup for VHF tuner alignment is shown in Figure 4-30. The sweep generator is connected to the antenna through a matching pad. Proper matching is necessary in order to avoid the effects of

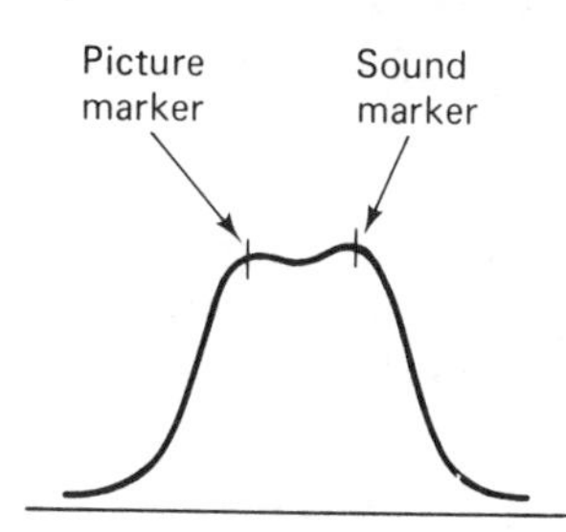

Figure 4-29 Typical RF response curve.

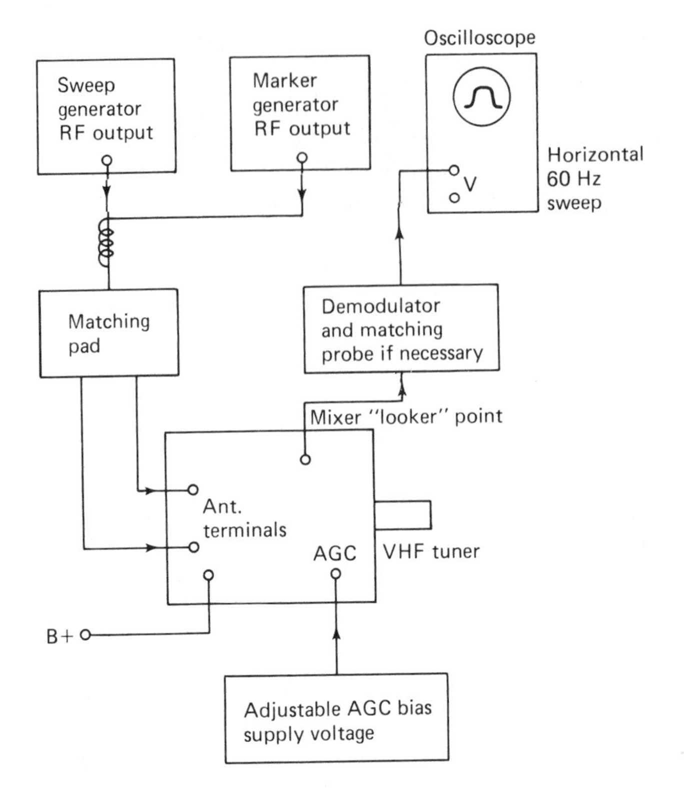

Figure 4-30 Typical VHF tuner alignment hookup.

standing waves. The oscilloscope is connected to a "looker" point in the mixer. Demodulator probes are used only when the mixer does not provide a demodulated output of its own.

RF amplifier alignment. If a turret-type tuner is being aligned, each tuned circuit strip must be adjusted to provide the desired bandpass. If the equipment hookup is proper and if all initial conditions, such as bias, have been met, alignment begins by adjusting both the sweep generator and the tuner to channel 13. The sweep generator output is kept low in order to prevent overloading. Then the spacings between the turns of wire making up the antenna input, RF amplifier output, and mixer input tuning strip inductors are adjusted for a response curve of maximum gain and best symmetry, as indicated by the marker positions on the curve. When channel 13 has been aligned, the remaining strips (channels 12 through 2) are then aligned in the same manner.

Alignment of the RF circuits of a wafer switch tuner is similar to that of the turret-type tuner. The only difference between them is the layout of the tuned circuit inductances. Here again, alignment starts on channel 13 and consists of changing the spacing between the windings of the inductances mounted on the wafer switches. These coils make up the tuned circuits of the tuner. After channel 13 has been aligned, the inductances of channels 12 through 2 are adjusted to obtain maximum gain and response curve symmetry.

Neutralization. In some tuners the RF amplifier is neutralized to prevent the amplifier from oscillating. The neutralization procedure is usually performed with tuner and sweep generator both tuned to channel 10. The output of the sweep generator is turned up to maximum and the AGC bias for the RF amplifier is increased to cutoff. Then the neutralizing capacitor, or *gimmick* as it is sometimes called, is adjusted for *minimum* output indication on the oscilloscope.

Oscillator adjustment. The object of the oscillator adjustment is to set the oscillator frequency for each channel so that the proper picture and sound IFs (45.75 MHz and 41.25 MHz) are produced at the output of the mixer. In general, there are two basic methods for doing this. The simplest method is to adjust the fine tuning of each channel for the best picture and sound that can be obtained. The second method is used when either the tuner being aligned is physically separated from a receiver or alignment is being done in an area where no stations are operating. In this method a frequency counter or an accurate beat frequency crystal oscillator marker generator is used to determine when the oscillator is adjusted properly. A beat frequency oscillator compares the tuner oscillator signal frequency to its own frequency, and it generates an audible tone whose pitch indicates the frequency difference between the two signals. When the two signals are equal in frequency, the output of the beat frequency generator will be zero. This sharply defined zero beat is used to indicate when the oscillator adjustment is correct.

In turret-type tuners each channel tuned circuit strip has its own oscillator fine-tuning screw adjustment. These are adjusted for proper oscillator frequency.

Wafer switch tuner oscillator adjustment usually consists of adjusting the spacing between turns of the oscillator coil for each channel. In some tuners, however, there may be a single slug tuned coil that adjusts the oscillator range of the high-frequency channels (7 through 13) and another slug tuned coil that adjusts the range of the low-frequency coils (2 through 6). In addition, the individual oscillator coils for each channel must be adjusted for correct oscillator frequency.

4-13 a vacuum tube VHF tuner

Figure 4-31 shows the schematic diagram of a vacuum tube VHF turret-type tuner.

A neutralized triode grounded cathode RF amplifier is capacitively coupled to a pentode mixer. The local oscillator is an ultra-audion and it capacitively injects its signal into the mixer via C_{12}.

The triode RF amplifier is neutralized by a feedback variable capacitor C_3. The RF amplifier obtains its AGC control voltage through a 47-kΩ isolation resistor and feed-through bypass capacitor C_2.

The 300-Ω antenna feeds the RF signal into a balun (T_2) that in turn feeds the signal through an IF/FM filter assembly. Channel selection is obtained by changing a strip mounted on a drum containing 13 such strips. Each strip contains four tuned circuit coils corresponding to the input of the RF amplifier, the output of the RF amplifier, the input to the mixer, and the local oscillator.

When the receiver is tuned to UHF (channel 1), the turret strip changes the resonant frequency of the RF amplifier and mixer-tuned circuits to the video IF (45.75 MHz), converting these stages into video IF amplifiers. The VHF oscillator is disabled by removing the oscillator coil.

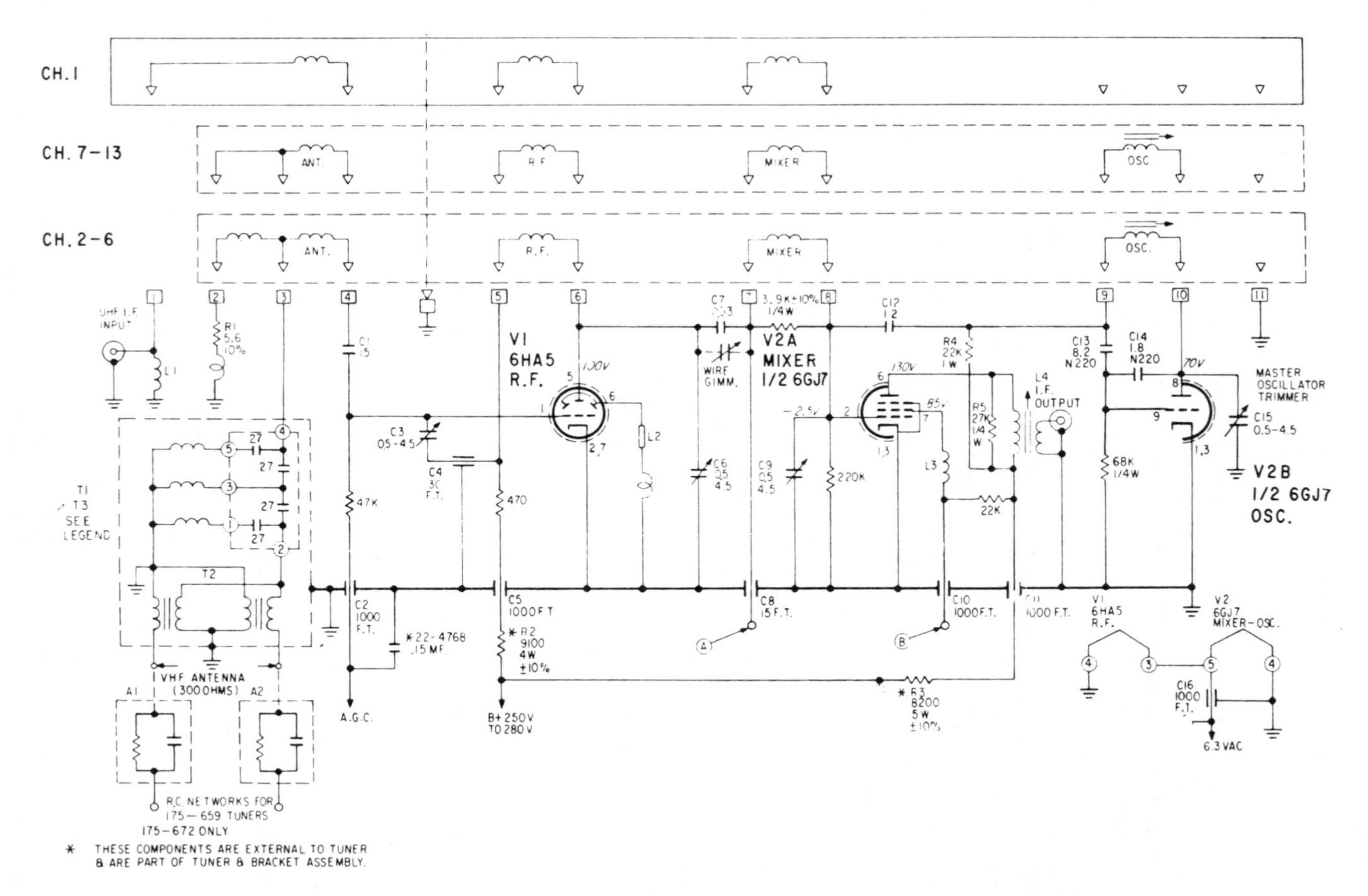

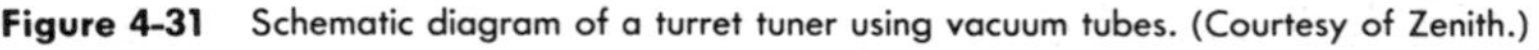

Figure 4-31 Schematic diagram of a turret tuner using vacuum tubes. (Courtesy of Zenith.)

QUESTIONS

1. List four groups of circuits that are part of the front end of a receiver.
2. Describe the symptoms and the method of repair of a dirty tuner.
3. Describe four methods of changing channels.
4. What is a varactor? What is another name for this device?
5. What is the effect of internal electrical noise on picture reproduction?
6. Which stage of a TV receiver generates the most noise?
7. List four functions of the RF amplifier.
8. Describe the circuit and characteristics of (a) the neutralized transistor RF amplifier and (b) the dual gate MOSFET RF amplifier.
9. Refer to Figure 4-13 and list three possible causes for the collector voltage to decrease.
10. What is the purpose of a balun?
11. What is the effect on receiver operation if (a) the RF amplifier is defective, (b) the mixer stage is defective, or (c) the oscillator stage is defective?
12. Refer to Figure 4-9 and determine (a) which components are used as filters, (b) which components are used to tune the tuner to the desired channels, (c) which components set the dc operating point, and (d) which components help maintain front-end stability.
13. Refer to Figure 4-19. What would be the probable effect of C_2 opening?
14. Refer to Figure 4-19 and list as many factors as you can that might cause the base voltage of Q_1, to decrease by a volt or more. What might make it increase?
15. What effect would a weak oscillator output have on the mixer?
16. What type of oscillator is most commonly used in vacuum tube and transistor tuners?
17. Why are transmission line segments used to tune UHF tuners?
18. Why do UHF converters place the oscillator frequency above the incoming RF?
19. What would be the effect on UHF operation if the VHF RF amplifier stage became inoperative or if the VHF oscillator stage became inoperative?
20. What are the two main objectives of tuner alignment?
21. Why are matching pads required between sweep generator and tuner during VHF tuner alignment?
22. What is frequency synthesis?
23. What is the function and the basic components of a PLL?
24. What are the basic parts of a computer?
25. If the tuning system of Figure 4-11 were tuned to channel 83, what would the feedback programmable divider factor be equal to?

chapter 5

AFT and Remote Control Circuits

5-1 introduction

This chapter is devoted to a discussion of the automatic fine tuning (AFT) and remote control circuits. Their positions in terms of the complete block diagram of a color receiver are shown in Figure 5-1.

Notice that the AFT circuit is fed by the video IF amplifier. The output of the amplifier is an ac signal corresponding to the picture IF (45.75 MHz). The AFT's dc output voltage is then fed to the tuner varactor-controlled oscillator. This correction voltage is able to change the oscillator frequency, thereby controlling the receiver's fine tuning.

The remote control circuit allows receiver functions such as on/off, volume, channel selection, color, and tint control to be controlled from a distant point. A handheld transmitter radiates either an ultrasonic or infrared signal that activates the remote control receiver. The frequency or the modulation of these signals determines the function activated.

5-2 automatic fine tuning

In the usual sense a television receiver is properly tuned when the best picture and best sound are obtained simultaneously. Fine tuning is not critical in black & white television receivers because the sound output is determined by the 4.5-MHz frequency *difference* between the picture carrier and the sound carrier. This frequency difference is established at the transmitter and is unaffected by receiver tuning. Since the sound will automatically be correct, proper tuning is judged by the sharpness of the picture.

Proper fine tuning of color receivers is judged by sharpness of picture *and* the quality of color. Hence, it is much more critical.

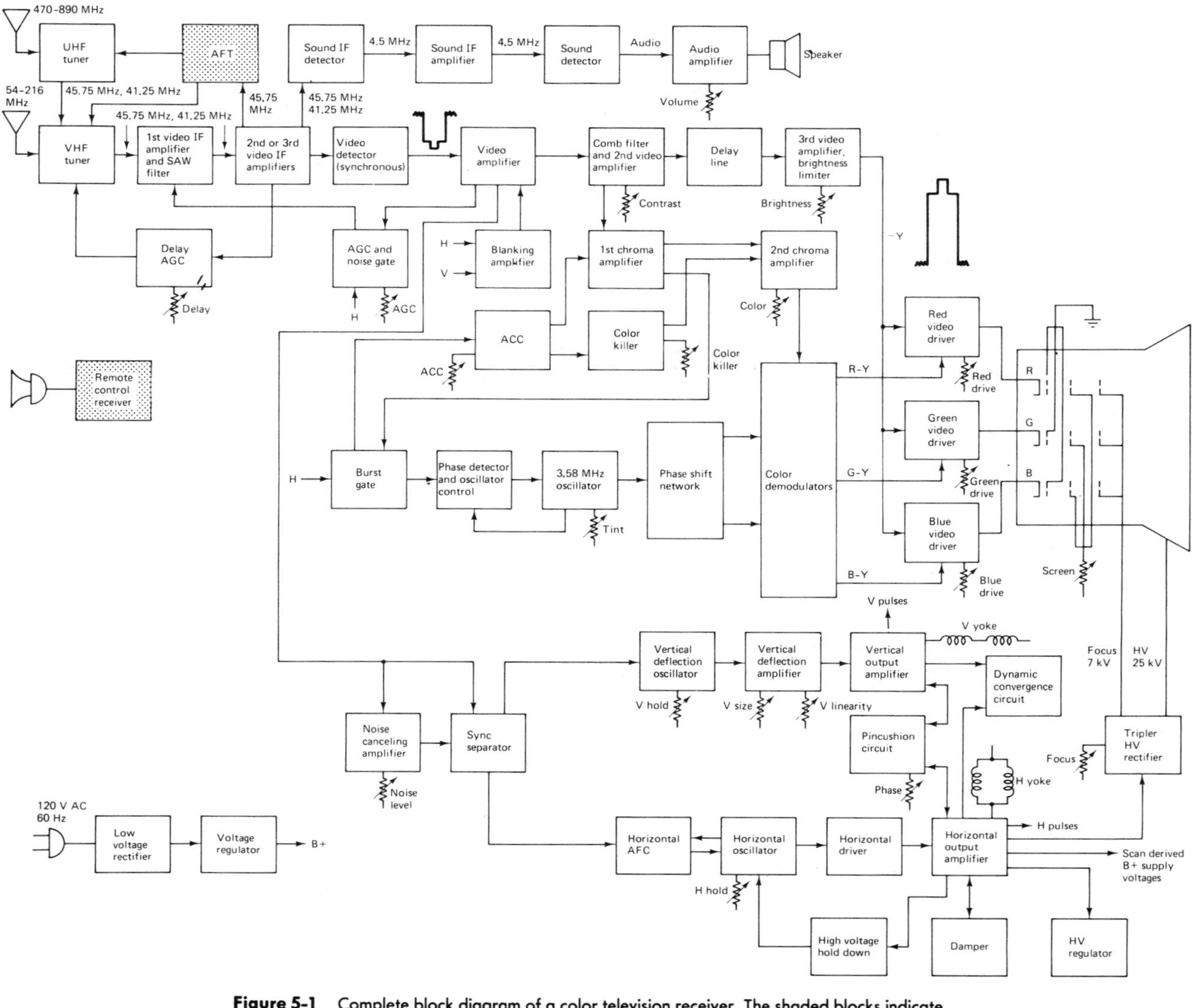

Figure 5-1 Complete block diagram of a color television receiver. The shaded blocks indicate the circuits to be discussed in this chapter.

Automatic fine tuning (AFT) is sometimes referred to as *automatic frequency control* (AFC) and is used in television receivers for two reasons:

1. Receivers that have provision for remote tuning have no provision for remote fine tuning.
2. Color receivers must be precisely tuned in order to obtain the best color reproduction.

Color receiver fine tuning. Proper tuning of a color receiver is critical because the color subcarrier must be properly positioned on the IF response curve if color reception is to be of high quality. If the IF amplifiers have been aligned properly and if the fine tuning has altered the local oscillator to its correct frequency, as is indicated in Figure 5-2(a), a proper color picture will be produced. This diagram shows an IF response that has been permanently adjusted during alignment to a particular shape over a given frequency range. When the fine tuning has adjusted the local oscillator frequency to the correct value (107 MHz) for a given channel (3), the IF signals that are generated by the mixer (sound IF 41.25 MHz; picture IF 45.75 MHz) will fall on the IF response in their design positions.

If the fine tuning is misadjusted to 108 MHz, as in Figure 5-2(b), the oscillator frequency will be 1 MHz too high. Thus, the picture IF, subcarrier IF, and the sound IF frequencies will also be 1 MHz too high, thereby falling on incorrect positions of the IF response curve. In this case, the picture carrier IF and the low-frequency video sidebands clustered around it are made weaker than they should be. At the same time, the color subcarrier and its associated chrominance sidebands are positioned at maximum IF response and are made stronger than is normal. Normally, very selective traps in the IF amplifier reject the sound IF and place it at zero level on the response curve. If however the oscillator frequency is too high, the sound IF output of the mixer will be placed higher up on the IF response curve and will result in substantial amplification. The most important result of these IF response position changes is to allow the sound carrier and color subcarrier to heterodyne producing a 920-kHz beat. As a video signal, the relatively low frequency 920-kHz beat represents fairly large picture information and will cover the color picture with a large undesired interference pattern as shown in Figure 5-2(b).

If the receiver fine tuning is adjusted too far in the opposite direction so that it is 1 MHz too low in frequency, the picture IF, subcarrier, and sound IF will all be 1 MHz lower in frequency than they should be. See Figure 5-2(c). Here again these IF signals appearing at the output of the mixer will fall on incorrect positions on the IF response curve. In this case, the picture IF will be shifted to the maximum level of the IF response and the subcarrier will be reduced in amplitude by being shifted to a low level of the IF response. As a result of the decrease in color level, and in particular the loss of the color burst, only a black & white picture will be seen as shown in Figure 5-2(c).

From the above discussion it should be clear that proper tuning is critical and absolutely necessary for high-quality color television reception. Improper fine tuning can result in loss of color, weak color, wrong colors, and color interference. Unfortunately, many color television receiver operators do not know how to tune a color receiver correctly. Hence, the need for a system of automatic fine tuning.

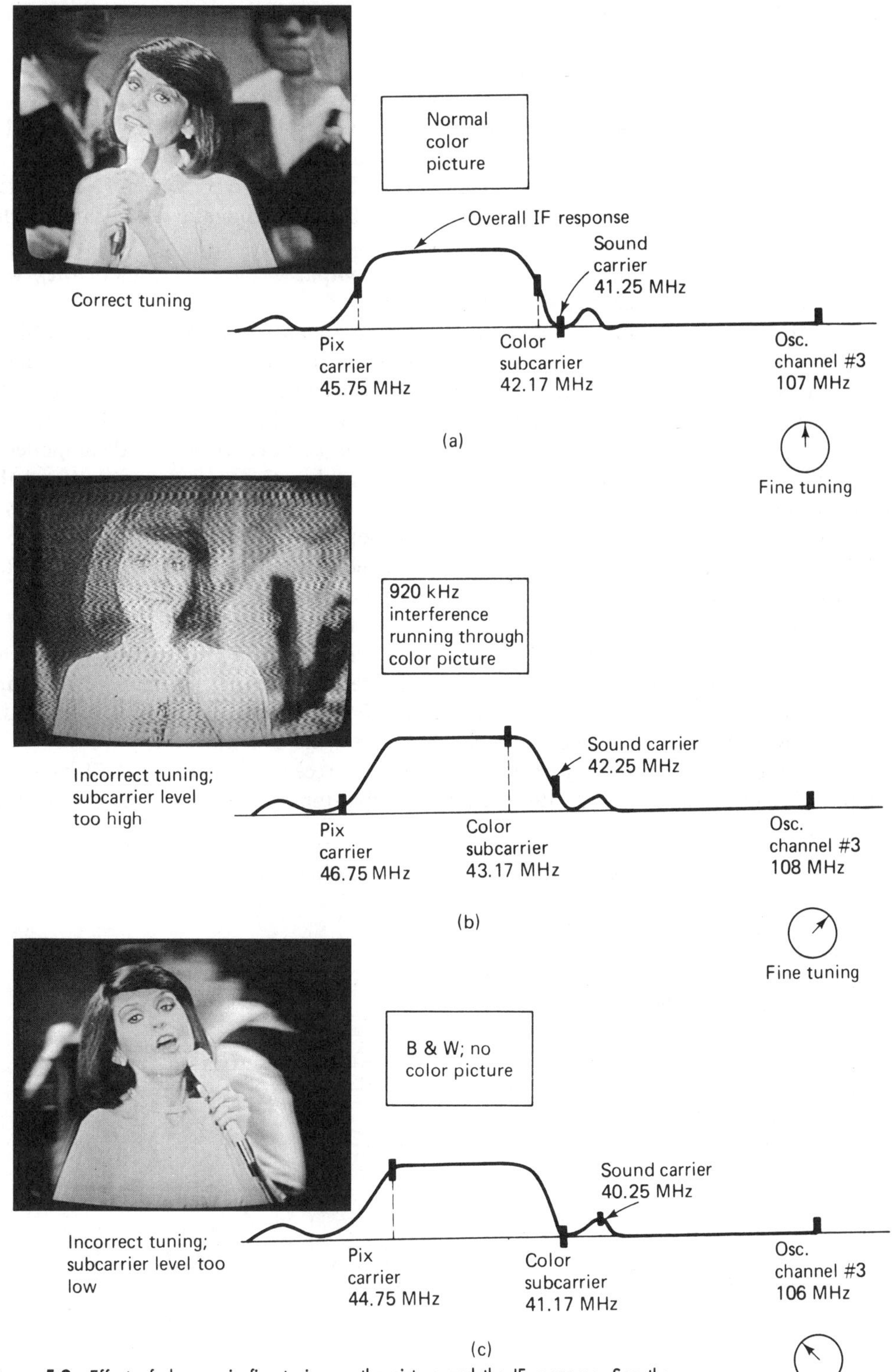

Figure 5-2 Effect of changes in fine tuning on the picture and the IF response. See the color plate.

5-3 AFT circuits

The block diagram of a commonly used arrangement of automatic fine tuning is shown in Figure 5-3. In this diagram a sample of the video picture carrier IF (45.75 MHz) is taken from the second or third video IF amplifier and fed into the tuned transformer of a Foster-Seeley discriminator. The discriminator is designed to convert a given IF frequency into a dc voltage. The S-curve output characteristic of the discriminator is indicated in the same figure.

If the tuned circuit is aligned properly, the output of the discriminator (dc_1) will be 0 V when the input is exactly 45.75 MHz.

If an error in tuning causes the IF to shift to a higher frequency, say 46.00 MHz, the output of the discriminator (dc_1) would be a negative dc voltage. This voltage is coupled into a dc amplifier that amplifies the error voltage and feeds it (dc_2) to a varactor. The varactor's characteristic curve is drawn below the varactor block diagram and shows that its capacity increases as its reverse bias control voltage (dc_2) decreases. The dc amplifier is designed to have a dc output voltage of, say, -6 V, when the tuning is correct and the input to the amplifier is zero. This establishes a bias on the varactor that sets its capacity and determines the normal oscillator frequency. It should be noticed that the output voltage (dc_2) of the dc amplifier is opposite in direction to the output voltage of the discriminator because of inverting properties of a common emitter amplifier.

A summary of the AFT action is illustrated in Figure 5-3 by means of arrows.

Of course, if the fine-tuning error were to shift the IF in the lower direction, the opposite reactions would take place in order to correct it. Alternate systems of AFT are possible. Some AFT systems amplify the picture IF (an ac signal) before it reaches the discriminator rather than provide dc amplification after the discriminator.

The block diagram of Figure 5-3 also shows that receivers may use an AFT indicator lamp. The lamp is used to indicate when proper fine tuning has been achieved. In this

Figure 5-3 Automatic fine-tuning circuit that uses a dc amplifier to drive the varactor.

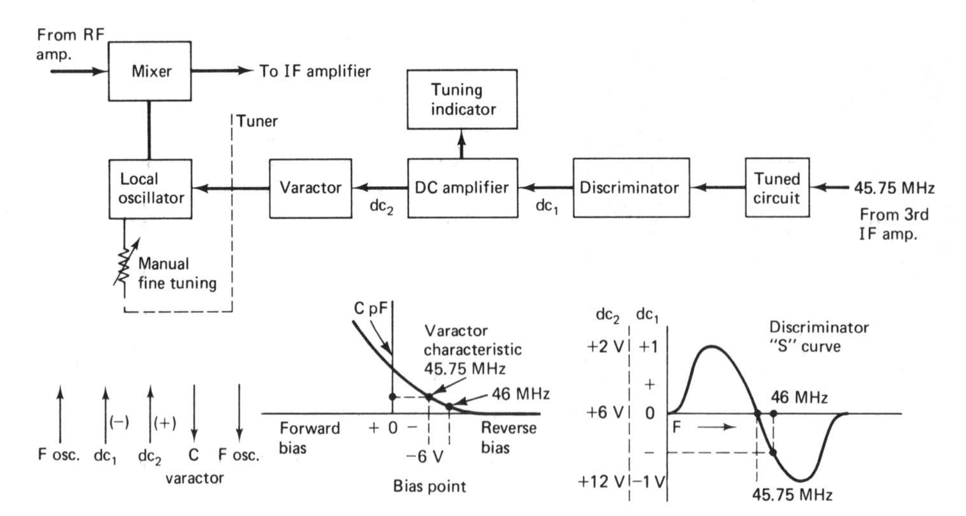

arrangement, the 45.75-MHz picture carrier is amplified and then rectified. The dc output is amplified and drives an indicator lamp that may turn on or off, depending upon the system design, when AFT has completed its function.

The AFT varactors used in actual TV receivers are usually mounted on the tuner itself and are physically close to the local oscillator. Varactors are silicon diodes; therefore, the base to collector junction of silicon transistors may be used in their place, as shown in Figure 4-19.

Varactor diodes or transistors are generally connected across part of the tuned circuit of the oscillator as shown in Figures 4-19 and 4-22. In Figure 4-19 the dc AFT control voltage is coupled to the varactor by a cable coming from the AFT chassis located some distance from the tuner. The base to collector junction of Q_4 is placed directly across the oscillator tuned circuit inductance L_5 by means of the coupling capacitors C_9 and C_{10}.

Practical AFT circuits. The schematic diagrams of two AFT circuits are shown in Figure 5-4. The top diagram [Figure 5-4(a)] is an example of a discriminator followed by a dc amplifier. The transformer T_1 converts frequency changes into amplitude changes. These voltage changes are rectified by diodes D_1 and D_2 and produce an output depending upon the tuned circuit S-curve response. The dc error voltage is coupled into the common emitter DC amplifier whose output feeds the varactor in the tuner. The bias adjust potentiometer in the emitter circuit of the transistor is used to set the zero error voltage bias for the transistor and varactor.

The bottom drawing [Figure 5-4(b)] is an AFT circuit that uses an integrated circuit to perform the functions of error detection and amplification. The 45.75-MHz input tuned circuits, the discriminator transformer, and decoupling *RC* networks are the only components that are mounted externally to the IC. All blocks drawn inside the heavy lines are the integrated circuit. The outer dashed lines indicate that the AFT circuit board is mounted inside a metal container that provides shielding.

5-4 AFT alignment

Proper automatic fine-tuning circuit operation requires correct AFT discriminator alignment. The only really correct alignment procedure is that given in the manufacturer's alignment instructions. This procedure will vary depending upon circuit type and design.

In general, the alignment procedure begins with the same equipment hookup and precautions as used in IF sweep alignment. First, dc bias clamps adjust the RF amplifier and IF amplifier bias to the manufacturer's specifications. Then, the sweep generator is connected to the mixer input and its output level is adjusted so that the response curve at the output of the video detector has the prescribed level (typically 2 to 3 V p-p). The oscilloscope is then moved to the output of the AFT circuit feeding the varactor. With the marker set to 45.75 MHz, the tuned circuits of the discriminator are adjusted so that the S-curve response (Figure 5-3) is symmetrical around the "zero" crossover frequency of 45.75 MHz. Final adjustment of the discriminator zero point is usually done with a VOM connected to the appropriate points in the AFT circuit [between points *A* and *B* in Figure 5-4(a)].

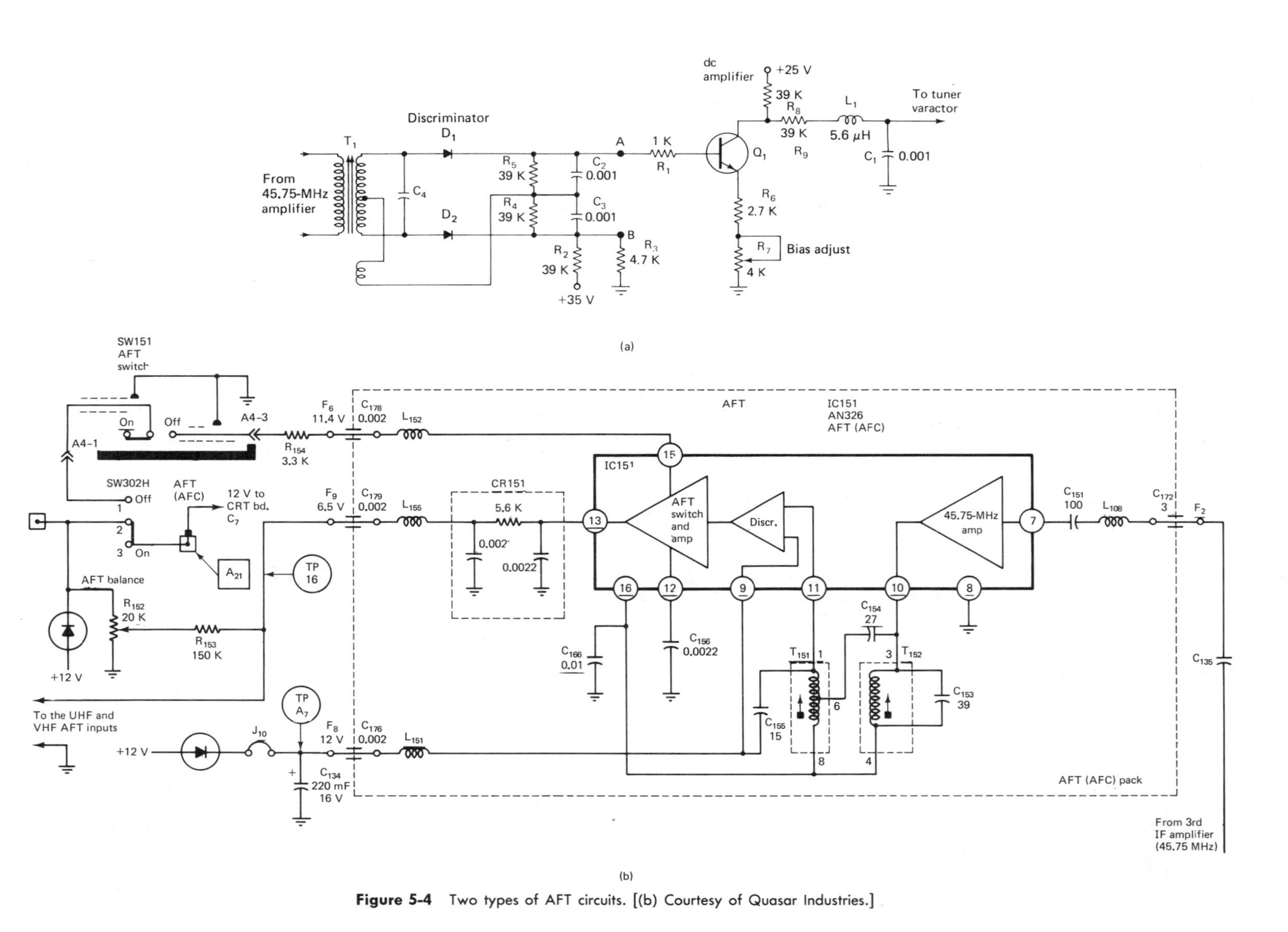

Figure 5-4 Two types of AFT circuits. [(b) Courtesy of Quasar Industries.]

5-5 troubleshooting AFT circuits

Defective AFT circuits can produce the following symptoms:

1. Weak or no AFT performance
2. No picture, no sound, raster O.K.
3. Large tuning error on all channels

Weak or no AFT performance can be caused by such things as the following:

1. A defective transistor or associated circuits such as Q_1 in Figure 5-4(a)
2. A defective transformer such as T_1 in Figure 5-4(a) and T151 or T152 in Figure 5-4(b)
3. A defective IC or associated circuits such as an open L151 in Figure 5-4(b)
4. Misalignment of the AFT tuned circuits

When there is no picture and no sound and the raster is O.K., it is generally the result of the local oscillator's being made inoperative by some defect in the AFT circuit. This can be caused by the following:

1. A shorted varactor
2. A varactor forced into forward bias by some defect in the dc amplifier, such as an open transistor Q_1 in Figure 5-4(a); another possible cause might be for R_2 to open and force Q_1 into cutoff, thereby turning the varactor diode on

In general, a large tuning error on all channels may be caused by a shift in varactor bias. This shift can be caused by the following:

1. Dc amplifier balance change caused by a change in the transistor or a change in resistor values anywhere in the dc amplifier
2. Defective varactor

5-6 remote control

The purpose of remote control is to allow the television viewer the luxury of controlling many, if not all of the controls situated on the front panel of the receiver without leaving his or her point of observation. Depending upon the make of the receiver, the viewer may be able to control many receiver functions. Some are volume, color, tint, on/off, and VHF/UHF channel selection. VHF/UHF channel selection may be accomplished by motor rotation, which allows up-and-down scanning, or by microcomputer-assisted direct access systems. All other receiver functions are usually all-electronic.

All remote control systems are basically similar. They all have a transmitter that is small enough to be held in the hand and operated by the viewer and they all have a remote receiver that is mounted in the receiver cabinet. All remote control functions are controlled by specific signals generated by the transmitter and acted upon by the receiver. A block diagram of an ultrasonic receiver is shown in Figure 5-5. In older receivers motors were used to control each function. Modern receivers using remote control, use all-electronic

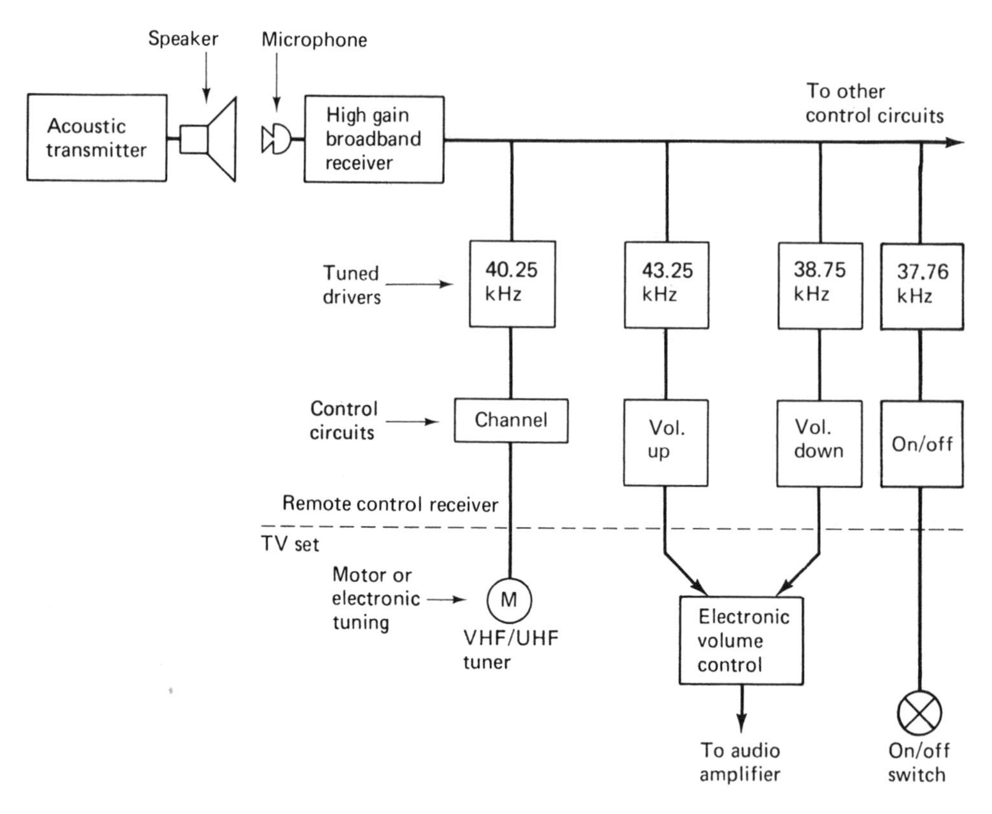

Figure 5-5 Block diagram of a remote control ultrasound system.

methods for channel selection, on/off, volume up and down, and mute functions. Some manufacturers still use motor control for channel selection alone. Automatic control circuits for such functions as color, tint, contrast, and brightness has resulted in the elimination of these functions from most modern remote control systems.

5-7 the ultrasonic transmitter

Acoustic transmitters. The acoustic transmitter is an ultrasonic sound generator. Its sound energy output usually falls in a frequency range of from 19 to 45 kHz and is radiated from a special high-frequency speaker. The exact frequencies used depends upon the manufacturer of the unit.

There are two methods for generating these tones. In one method a small hammer is made to strike a metal rod (as in chimes). An exploded view of such a transmitter is shown in Figure 5-6(a). In the other method a transistor oscillator is used to generate the desired tone. The number of specific frequencies generated by the transmitter determines the total number of functions that may be controlled. Manufacturers have produced units that are capable of generating three, four, six, seven, and as many as 18 tones. See Figure 5-6(b) for a seventeen function transmitter. In the case of the three-tone and four-tone units, one tone may perform several functions. Thus, in the three-tone system a 40-kHz tone is used to turn the volume down and may also be used to inform the receiver of a change in function of another tone. For example, in this system 41.5 kHz is used to turn the volume up and 38.5 kHz is used to select VHF/UHF channels. If, however, the 40-kHz tone is depressed preceding the 41.5-kHz tone, this combination will cause the tint control to be adjusted toward green tints, and the 38.5-kHz tone will reverse the rotation of the tint control and shift the colors toward the red.

The six- and seven-tone systems are capable of controlling six or seven separate receiver functions. The controlled functions are (1) volume up, (2) on/off, (3) volume down, (4) channel up, (5) channel down, (6) mute, and (7) recall, which causes the channel number and time to appear on the screen. Remote control direct access tuning systems (Section 4-6) require as many as 17 tones to perform all of the seven functions listed above, plus a keyboard of 10 more in order to select the desired channel from the 125 that are available.

It is possible to make one tone perform both volume and on/off functions. This is done by means of a stepping relay that changes resistors in the color receiver's volume control circuit each time the volume tone is generated. The first time the volume key of a transmitter is depressed it causes the receiver to be turned on, with low volume. When it is depressed a second time, the volume increases to medium. When it is again depressed, the volume increases to a preset maximum. If the transmitter key is depressed once more, the receiver is turned off.

The ultrasonic transmitter circuit. The schematic diagram of a three-tone acoustic transmitter is shown in Figure 5-7. The PNP transistor is used to form a low-frequency Armstrong oscillator. The frequency of its operation is determined by the inductance of the secondary winding of L_1 and the precision capacitors that are placed across it by depressing any one of the three transmitter's keys. The oscillator is operated at the same frequency as that being radiated by the sonic transducer. The diodes D_1 and D_2 and capacitors C_9 and C_{11} form a voltage doubler that is used to rectify the signal and produce a dc voltage that is used to bias the sonic transducer. This enables the sonic transducer to produce a fundamental output frequency. L_1 is used to set the channel function frequency of 41.5 kHz. Individual trimmer capacitors are used to adjust each of the remaining tones separately.

Alignment of the transmitter may be accomplished by means of a frequency counter or by establishing some kind of frequency standard with which to compare the frequency of the transmitter. This can be done either by using a known good hand transmitter as a signal generator or by using a properly adjusted and normally operating remote control receiver.

The transmitter can be responsible for at least three kinds of remote control faults:

1. Weak insensitive response on all functions
2. Dead all functions
3. Weak or dead on some functions but not all

Weak insensitive or dead response on all functions can be the result of a weak battery, a defective transistor, a defective sonic transducer, or misalignment of the oscillator. To make sure that the trouble is in the transmitter, the receiver may be checked with a known good transmitter or in some cases by simulating a transmitter by shaking a set of keys in front of the receiver microphone. If the trouble is definitely in the transmitter, the battery should be replaced and the oscillator checked for output with an oscilloscope. If

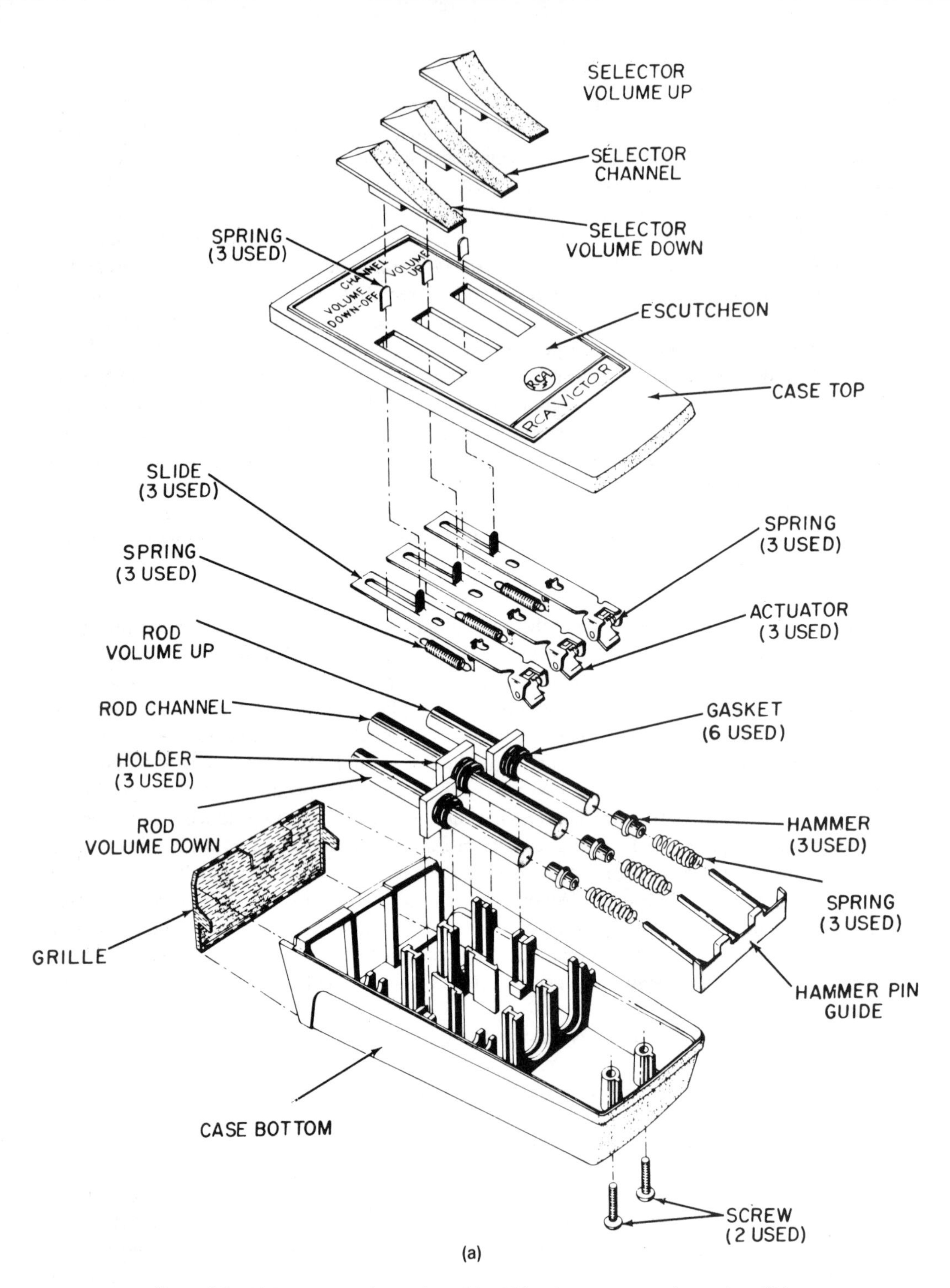

Figure 5-6 (a) Remote control transmitter; (b) a 17-function remote control transmitter. [(a) Courtesy of RCA, Consumer Electronics; (b) courtesy of Zenith.]

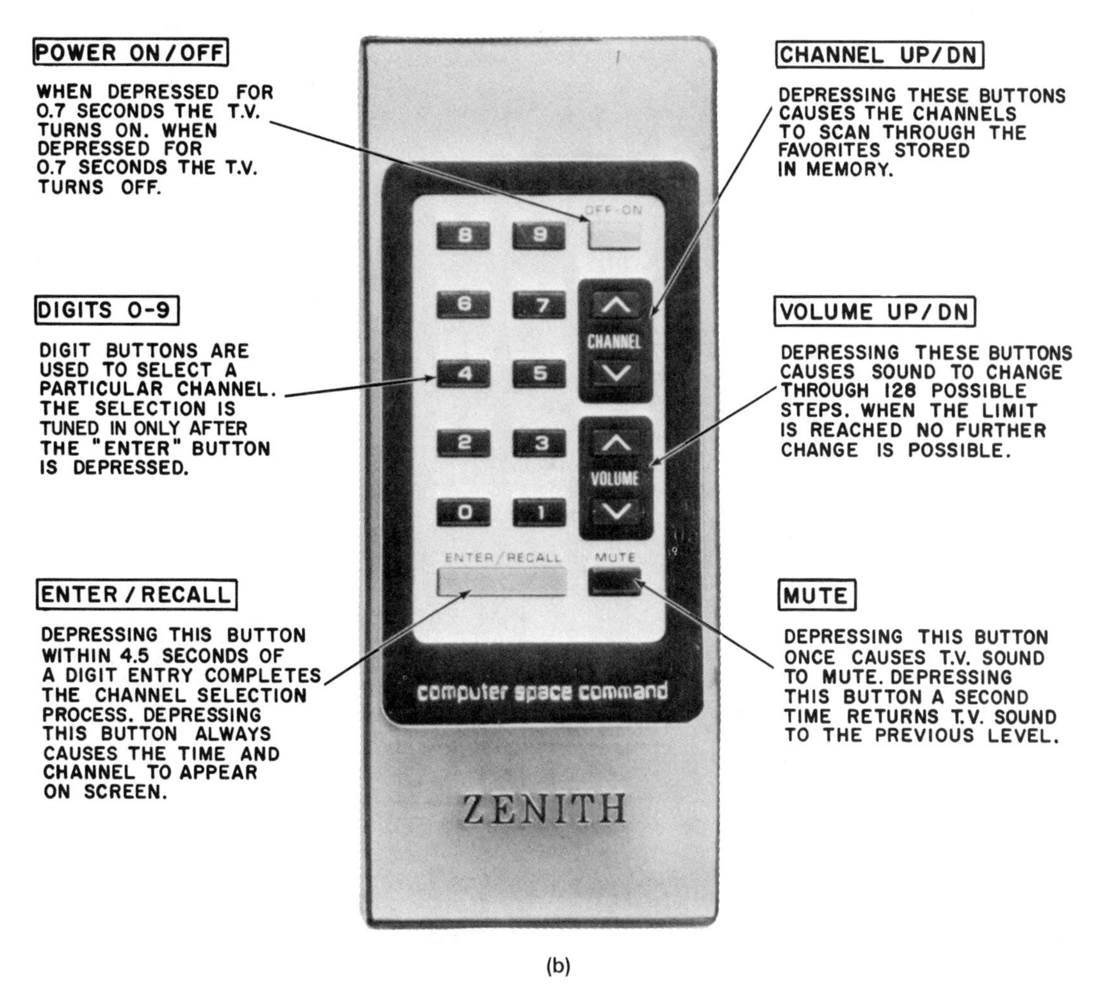

(b)

Figure 5-6 *(continued)*

the oscillator is functioning, its frequency should be checked by a frequency counter. If the frequency and amplitude are O.K., the sonic transducer should be checked by replacing it.

Weak or dead conditions on some but not all functions can be caused by dirty or broken switch contacts or a change in frequency of the tuned circuits corresponding to the inoperative functions.

5-8 the ultrasonic remote receiver

The block diagram of an ultrasonic remote control receiver is shown in Figure 5-5. For purposes of simplification, the receiver shown here is limited to only four functions, but the basic principle may be extended to any number of operations.

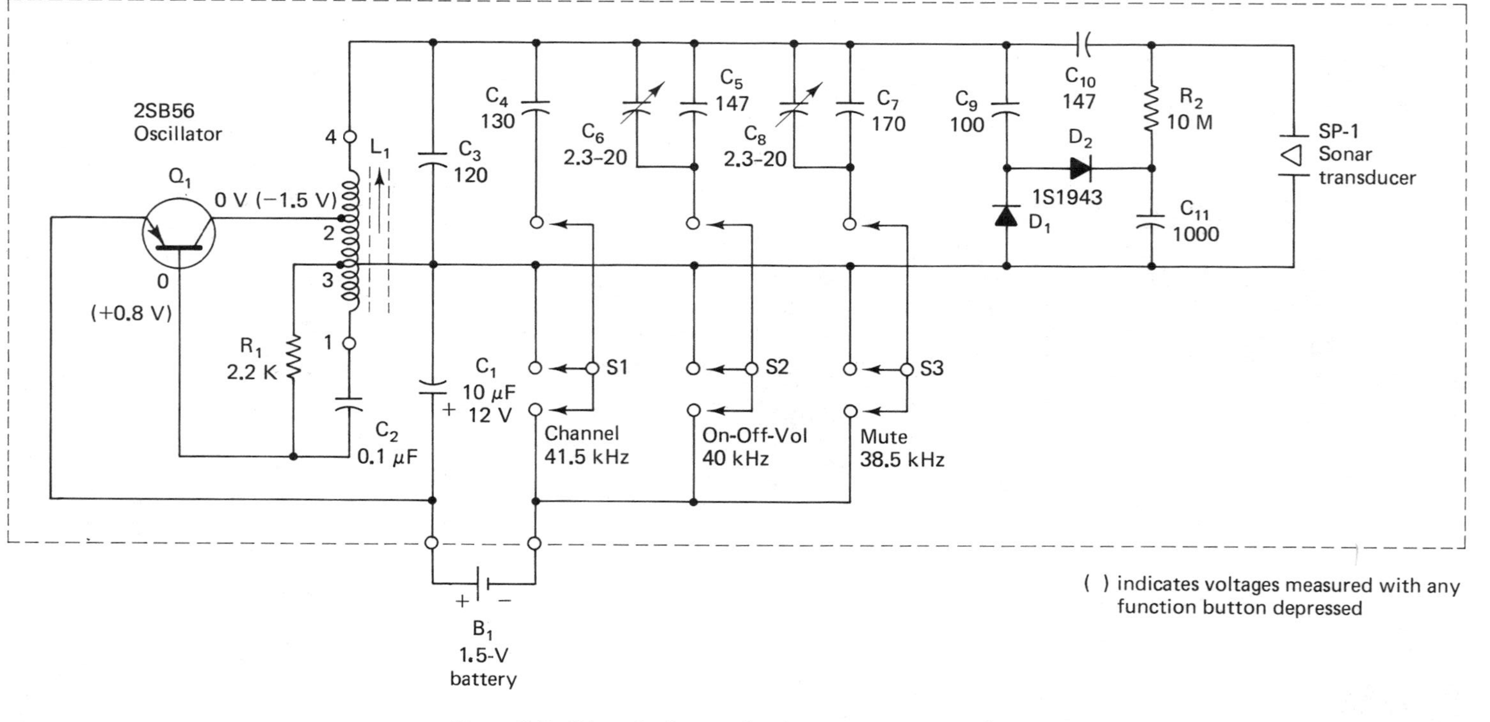

Figure 5-7 Schematic diagram of a three-tone remote transmitter.

The block diagram reveals that the remote receiver may be broken into the five following basic parts:

1. A microphone that picks up the ultrasonic tones of the transmitter and converts them into electrical signals.
2. A high-gain (power gain of 10^{12}) broadband (30 to 50 kHz) amplifier that amplifies any signal picked up by the microphone and delivers it to an array of tuned circuits.
3. Each function to be performed by the remote control system has a specific frequency that activates it. Each function in the receiver is separated from the other by a filter that is tuned to the frequency corresponding to that generated by the transmitter. Each tuned circuit is associated with a transistor power amplifier (driver) that is used to activate a relay or an electronic control circuit.
4. A relay. A relay is an electromechanical form of power amplifier that permits a relatively small current flow in its windings to close a switch that controls very much higher currents. In the remote control receiver the function relay activates a motor that completes the functions desired.
5. All of the components of the block diagram discussed so far are mounted on a separate remote control receiver chassis. The channel selector motor is a low-speed reversible type that is attached to the shaft of the tuner. In addition to a motor, some manufacturers use stepping or ratchet-type relays that may have two or more switching positions, depending upon the number of times the activity pulse is received. In more recent systems, motors and relays have by and large been replaced by electronic controls and systems.

The complete schematic diagram of a three-function remote control receiver is shown in Figure 5-8.

At the top of the diagram is shown a four-stage wideband transistor *RC* coupled amplifier. All of the common emitter stages are identical to each other. The input of the amplifier is fed by a capacitor microphone. The output of the amplifier is transformer coupled to the relay driver portion of the receiver. Modern receivers replace the four-stage transistor amplifier with one or two IC packages.

If the remote control amplifier or its microphone were to become defective, the remote control receiver would become insensitive or completely dead for *all* remote functions. The microphone can be checked by substituting a low-level audio signal generator for it and noting the effect on the remote control function by varying its frequency. If the remote control functions are not restored, it is possible that the amplifier is the cause of the trouble. This can be easily verified by injecting the input signal and then checking the output of the amplifier with an oscilloscope. A weak or no-output result means that the amplifier is at fault. The specific stage causing the trouble can be isolated by signal tracing back toward the input signal generator. Once the stage has been isolated, standard troubleshooting methods can be used to isolate the defective component.

The relay driver portion of the remote receiver (Figure 5-8) consists of three identical circuits. Each circuit has three basic parts: *a series resonant circuit* that feeds a *power transistor* that in turn drives a *relay*. The series resonant circuit is tuned to the frequency designated for a specific function. In this way, the signals are effectively separated and delivered to

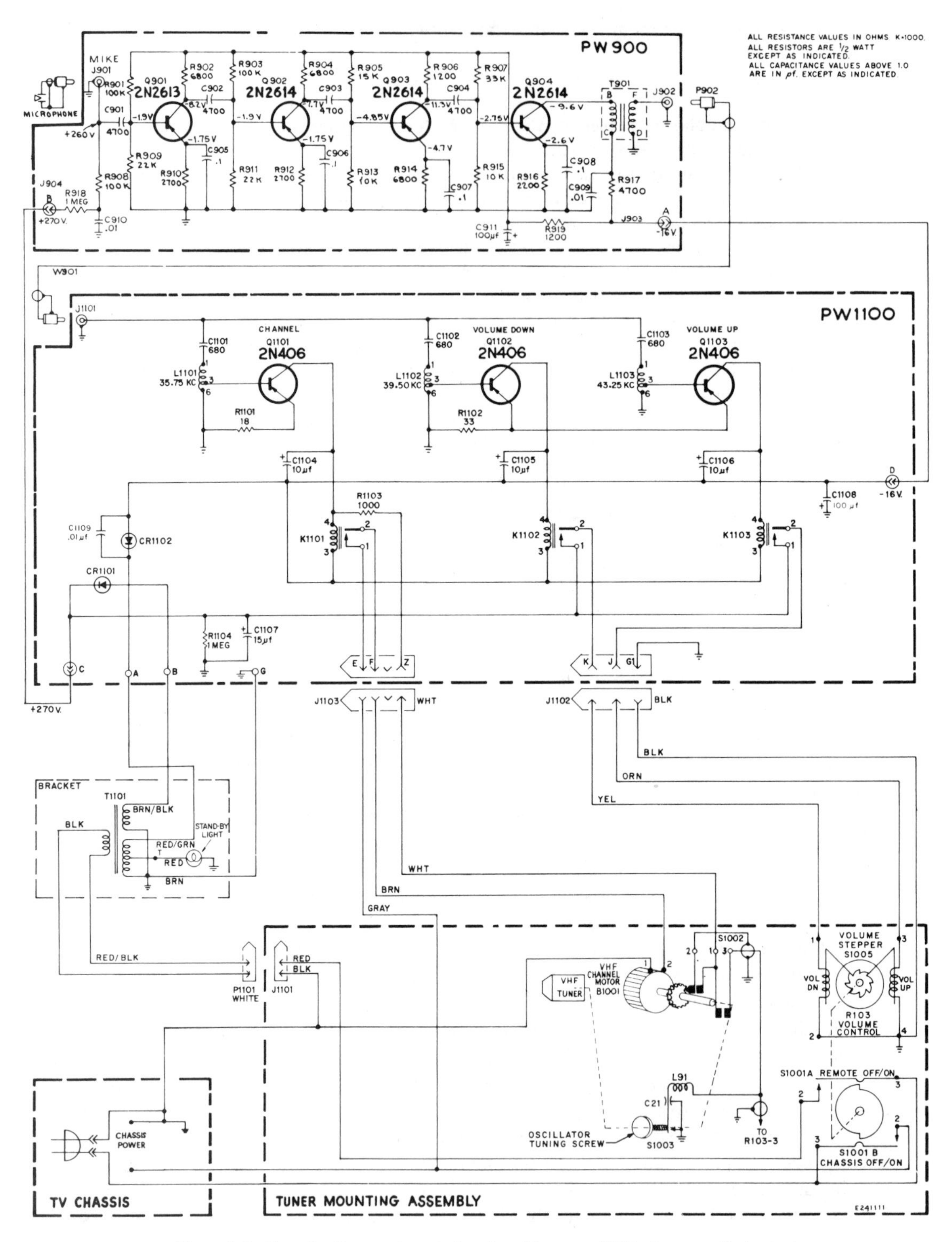

Figure 5-8 Three-function remote control receiver. (Courtesy of RCA, Consumer Electronics.)

the driver transistor. The coil of the resonant circuit is tapped in order to permit impedance matching to the low-input impedance of the transistor.

The driver transistor is a medium-power **PNP** whose normal no-signal base to emitter bias is zero. Therefore, it is cut off. When a signal is applied to the base, the base to emitter behaves like a diode, and the transistor is forced in and out of conduction at the input frequency rate. The dc collector current flows through the windings of a single-pole, single-throw relay closing it and activating the motor or stepping relay that controls the desired function.

A large capacitor is placed across the relay coil to prevent its inductive kick from damaging the transistor. The collector current also flows through an emitter resistor that is common to all of those driver transistors. The voltage drop produced across the resistor reverse biases the remaining transistor and prevents them from accidentally turning on.

A defect in any one of the relay driver circuits can result in the loss or improper operation of *one* of the control functions. Loss of a control function may be caused by such things as an open transistor, dirty relay contacts, or a detuned or defective tuned circuit.

Relay operation can be checked by simply shorting the collector of the driver transistor to the emitter. If relay closes, it is good. If it does not close, it should be checked both mechanically and electrically and replaced if found defective.

The transistor may be checked for an open by removing it from the circuit and testing it on a transistor tester. A shorted transistor (*C* to *B* or *C* to *E*) would keep the function relay closed all the time. In the case of channel selection, it will result in the tuner continuously changing channels.

A detuned or defective tuned circuit can be easily checked by signal tracing with an oscilloscope. The transmitter is first keyed to a working channel; then the peak to peak input voltage delivered to the base of its corresponding driver transistor is observed. If it is assumed that a fixed distance between the transmitter and receiver is maintained, a similar voltage should appear at the base of the inoperative circuit's transistor when the transmitter is keyed to the inoperative function. If not, the tuned circuits should be realigned and checked for defective components.

The electronic memory circuit. Television receiver manufacturers have developed remote control systems that are either completely free of motors or may use only one motor. In systems that do not use any motor, dc voltages are used to vary the controlled functions. Volume, tint, and color control are obtained by varying the operating bias of the amplifiers involved in these operations. This is the same technique that is used in AGC circuits. Dc control of channel selection makes use of varactor tuning.

In motor operated remote control systems the level of volume, color, tint, and the channel selected is "remembered" from one viewing session to the next by the mechanical position of the controls. In the motorless remote control systems a memory circuit is required to "remember" the control positions established at the last viewing session. If memory is not provided, turning off the receiver returns all control voltages to zero and the receiver operator has to readjust all controls each time the receiver is turned on.

The schematic diagram of a memory circuit used to control receiver volume is shown in Figure 5-9. This circuit is able to increase or decrease volume upon command of

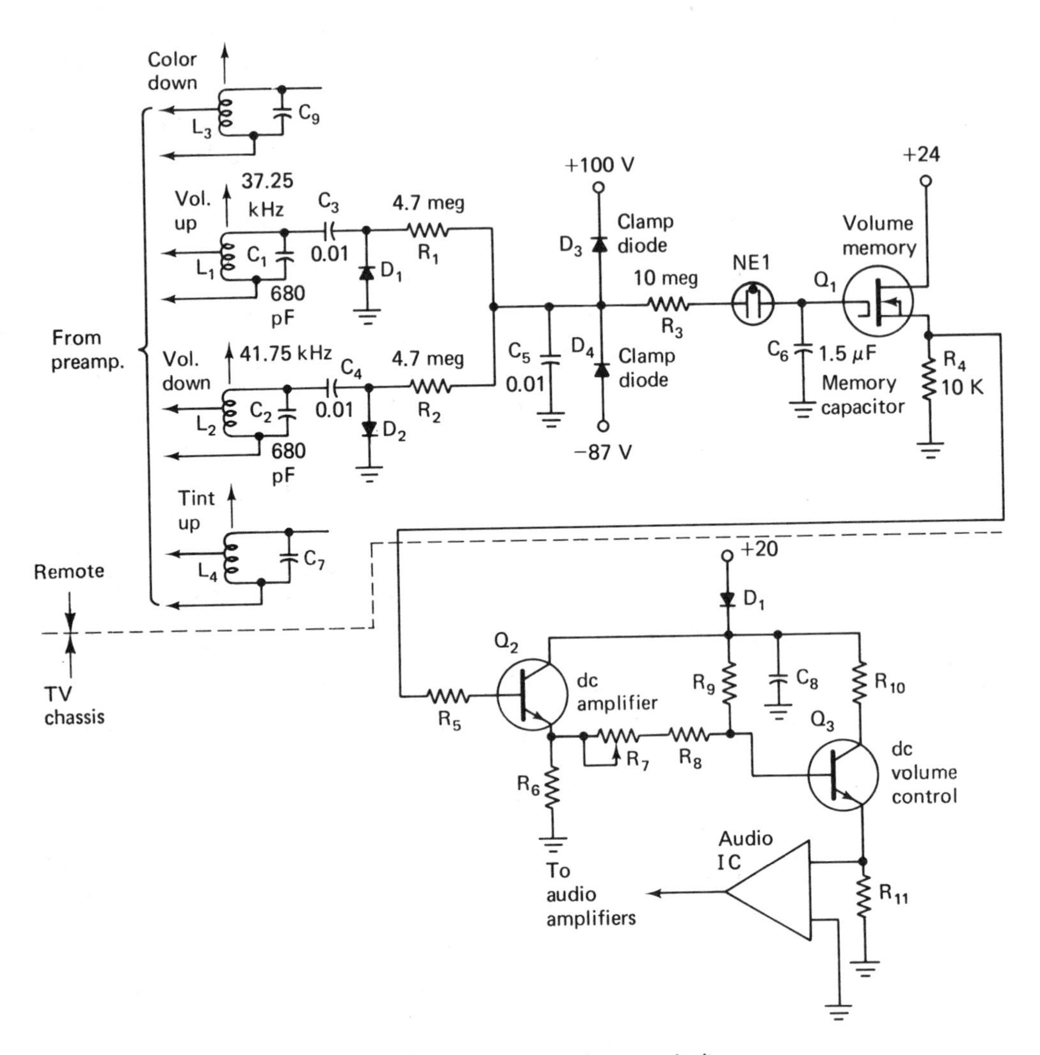

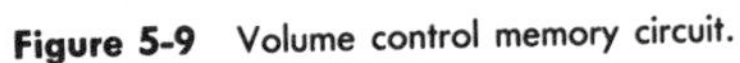
Figure 5-9 Volume control memory circuit.

the remote control transmitter. For example, volume increase is controlled when the transmitter is activated to produce a 37.25-kHz ultrasonic output. The remote receiver amplifies this signal and delivers it to a filter network. One of these filters (L_1, C_1) selects this signal, steps it up, and routes it to a rectifier (D_1) that converts the ac (37.25 kHz) signal into a pulsating dc voltage of positive polarity. The polarity is determined by the direction of rectifier conduction. Pulsations are filtered out by C_5, leaving a large positive dc voltage at the junction of D_3 and D_4. Should this voltage exceed the +100-V clamp voltage of the cathode of D_3, the diode will turn on. This in turn applies +100 V to the neon bulb NE1, turning it on. As a result, the high-Q memory capacitor C_6 begins to charge via R_3, in a *positive* direction. The memory capacitor is a special low-leakage capacitor that can hold a charge for several months after power has been removed. The amount of charge that is developed across C_6 depends upon how long the "volume up" key of the transmitter is depressed.

C_6 usually charges to only approximately 5 or 6 V because the neon bulb has a voltage drop of approximately 80 V across it when it conducts, and the very large time constant of R_3 and C_6 limits the voltage developed across C_6. Once C_6 has been charged to a given level, it will remain charged even though the receiver has been turned off and then on again.

The voltage across C_6 is the bias for Q_1, the MOSFET memory follower. The MOSFET does not discharge C_6 because of its very high input resistance. However, changes in bias provide a dc output voltage across R_4 that is proportional to the voltage across C_6. These voltage changes are fed to dc amplifiers (Q_2 and Q_3) that then force the audio IC to increase the volume.

To decrease volume, the transmitter is activated to produce an ultrasonic tone of 41.75 kHz. This is selected by filter L_2, C_2 and is rectified by D_2. Since D_2 is connected in the opposite direction from D_1, it produces a *negative* dc output voltage. This voltage exceeds the −87-V clamp voltage and D_4 conducts, clamping the junctions of D_3 and D_4 to −87 V. This voltage forces the neon bulb NE1 into conduction, charging C_6 in the negative direction. The amount of charge developed across C_6 is in direct proportion to the time the transmitter key is depressed. Here again, the voltage across C_6 acts as the bias for Q_1, which in turn forces the dc amplifiers (Q_2 and Q_3) to reduce the gain of the audio IC, lowering the volume.

5-9 infrared systems

The infrared transmitter. An alternative method for remote transmitter to receiver communication is by means of invisible infrared (IR) radiation. Ultrasonic remote transmitters do not have to be pointed directly at the remote receiver to function; they can activate remote receivers from another room. IR systems are "line of sight." They must be pointed directly at the remote receiver to be detected and acted upon. The infrared generators are LEDs that produce infrared radiation instead of visible light. Infrared sources can be checked by converting their invisible low frequency energy into higher frequency visible light. This may be done by means of a special fluorescent material coating a laminated plastic card. Exposing the card to infrared energy forces the coating to glow red. These cards are available from RCA (Part #153093).

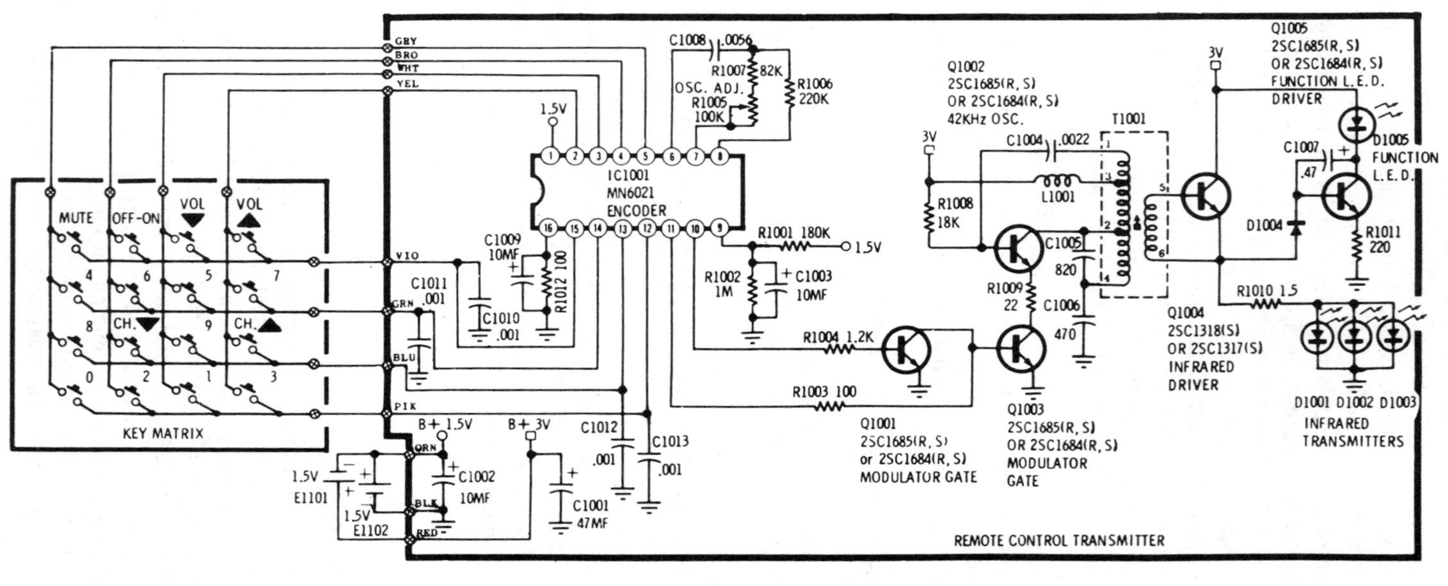

Figure 5-10 Schematic diagram of an infrared remote control transmitter. (Courtesy of Quasar Industries.)

The schematic diagram of an IR remote control transmitter is shown in Figure 5-10. This transmitter is part of a remote control direct access system having 16 functions. Depressing a keyboard button energizes the encoder IC (IC_{1001}) to generate a digital signal. This digital pulse code is different for each function selected and has a frequency of about 100 Hz. The digital pulse output of the encoder (pins 10 and 11) feed modulator gates Q_{1001} and Q_{1003}, which, in turn, modulates the 42-kHz oscillator Q_{1002}. The output of the oscillator feeds the infrared driver Q_{1004}, which pulse modulates three IR LED diodes (D_{1001}, D_{1002}, and D_{1003}) connected in parallel. A function LED (Q_{1005}) is activated by the function driver transistor Q_{1005} whenever the transmitter is energized. The transmitter is hand held and uses two 1.5-V batteries in series..

An infrared receiver. Figure 5-11 is a combination schematic and block diagram of an infrared remote control system. The remote control infrared sensor and its associated components, the low-voltage power supply for the remote receiver, and the solid-state switch used to turn on the main receiver, are shown as a schematic diagram. The remaining direct access microprocessor tuning system is shown as a block diagram.

The digitally modulated infrared signal (42-kHz carrier) from the transmitter is sensed by the photodiode D_{1101}. Its output is then coupled to the remote processor IC (IC_{1101}) which amplifies the coded signal. The signal is further amplified by transistor Q_{1101}, and demodulated by detector diode D_{1102}. A pulse-shaper circuit in the IC ensures that the digital pulses feeding the microprocessor (MPU) of the tuning system are of the proper amplitude and do not have any overshoots.

The remote receiver has its own independent power supply that is on all the time. An ac line is connected through a fuse to a step-down transformer T_{001} that converts the 120 V ac into 12 V ac. This voltage feeds through a 1A fuse to a bridge rectifier that converts the ac voltage into dc. The dc is filtered by C_{24}, and regulator Q_{11} provides a 5-V dc output.

Turning the main receiver on or off without the use of relays as well as isolating the low-voltage solid-state remote control circuit from the 120-V ac line is the function of triac D_{1104} and its associated photo-optical isolator (IC_{1102}). Triacs are very similar to SCRs except each triac is the equivalent of two SCRs wired in parallel, but with reverse polarity. As a result, triacs conduct both on positive and negative half-cycles of the ac line voltage acting like a switch. Triacs are turned on by applying a large enough voltage to the gate of the device. Removing the gate voltage will turn off the ac-operated triac.

The *photo-optical isolator* consists of a light-emitting diode that is optically coupled to a light-dependent resistor (LDR). An LDR has very low resistance when it is illuminated and has very large resistance when it is dark.

When the on/off remote transmitter button is depressed, it activates the infrared LEDs. The transmitted digital signal is picked up by the remote receiver and is passed on to the microprocessor, which in turn energizes the photo-optical isolator's LED. This lowers the resistance of the LDR, which couples a greater ac voltage to the triac gate, turning it on. In effect this closes the receiver on/off switch, turning on the main receiver.

When the "off" button is depressed, the process is reversed and the gate voltage of the triac is reduced, cutting off the triac and turning off the main receiver.

Figure 5-11 Infrared remote control system. (Courtesy of Quasar Industries.)

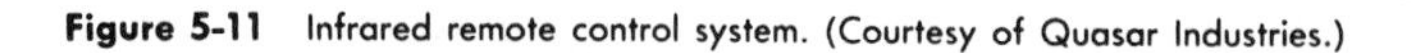

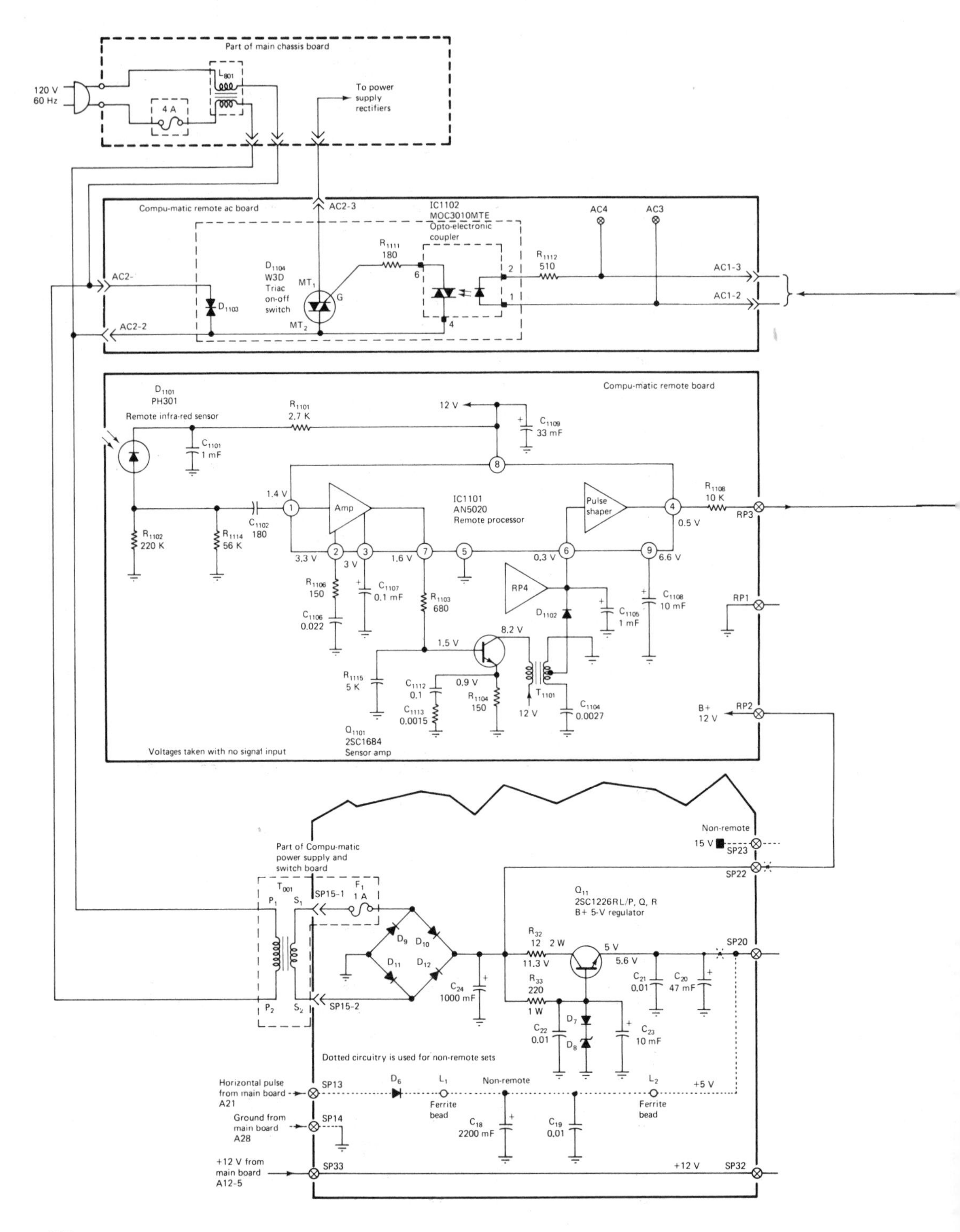

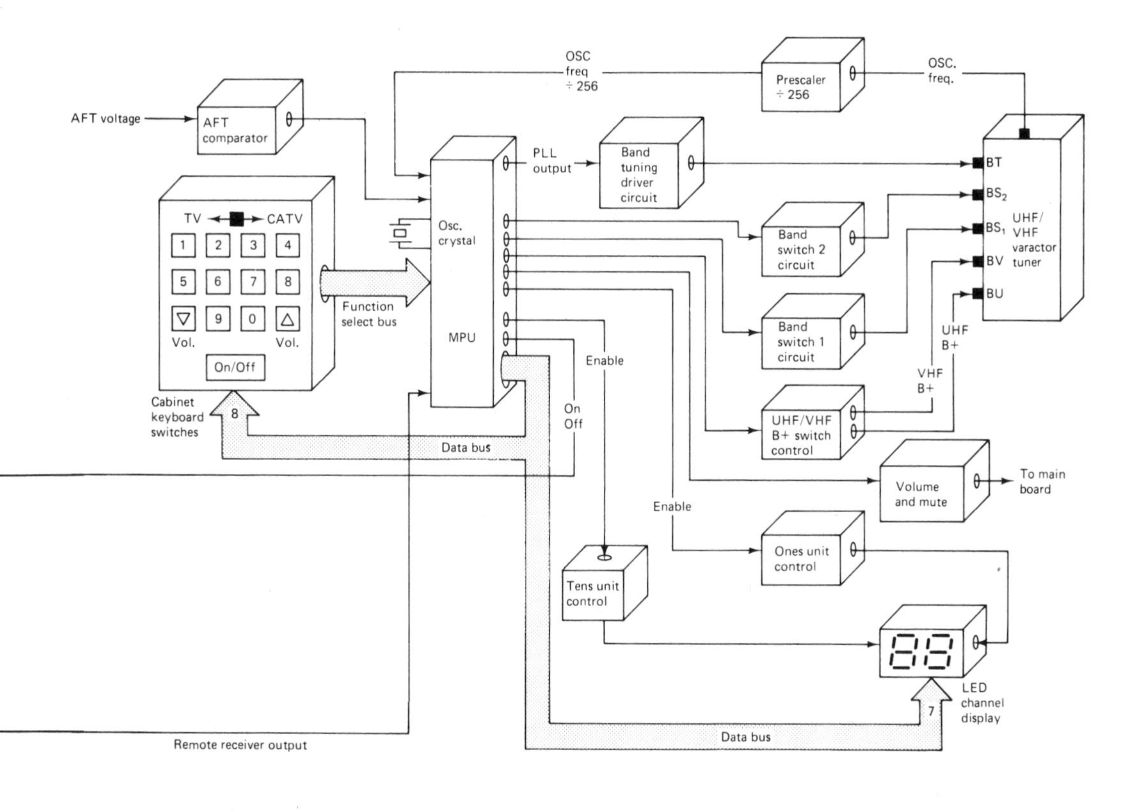

Figure 5-11 *(continued)*

QUESTIONS

1. To what frequency is the AFT discriminator transformer tuned?
2. Why is fine tuning critical in color receivers?
3. How does the AFT circuit control the frequency of the tuner oscillator?
4. What are three trouble symptoms that a defective AFT circuit could produce?
5. At what frequency do remote control transmitters operate?
6. What are two types of remote control transmitters?
7. In Figure 5-8 what would be the effect on receiver operation if Q_{1101} shorted?
8. What type of oscillator is being used in Figure 5-7, and why is its base voltage positive?
9. What type of capacitor is C_6 in Figure 5-9?
10. What is the purpose of the memory circuit such as that found in Figure 5-9?
11. What is meant by a "direct access" tuning system?
12. What is a photo-optical isolator, and what are its characteristics?
13. What is a triac, and what are its characteristics?
14. Explain why triacs are used in remote control systems.
15. What would be the effect on remote control operation if Q_{1101} (Figure 5-11), the sensor amplifier, opened?
16. List all of the inputs and outputs of the MPU in Figure 5-11.

chapter 6

Video IF Amplifiers

6-1 introduction

The complete block diagram of a color receiver is shown in Figure 6-1. The shaded blocks refer to those sections (video IF amplifiers and video detector) of the receiver that are discussed in this chapter. This block diagram indicates that the IF output of the mixer is fed into the video IF amplifier stage of the receiver. Since these sections provide common signal paths for both black & white and color signals, defects in these sections will affect *both* color and black & white reception.

6-2 intermediate frequency

The advantage of the superheterodyne receiver over other receiver systems lies in its ability to reduce its operating frequency. This reduction in frequency is accomplished in the mixer and allows the receiver amplifiers to have better selectivity and higher gain than they could if they were operated at the incoming RF frequency.

In the case of television receivers, narrow-band reception is not the desired characteristic. Wide-band flat-top frequency response curves are needed. Lower-frequency operation allows the skirt or side response of these circuits to be steeper. The means by which this type of response is obtained will be discussed later.

Operating a tuned amplifier at a lower frequency allows higher amplification levels to be obtained than might be achieved at high frequencies. This is primarily because of the fact that at low frequencies the reactance of the interelectrode capacitances increase; therefore, there is less stray and interelectrode coupling between output and input circuits of the amplifier. This reduction in undesired feedback makes the amplifier less likely to oscillate at high levels of amplification.

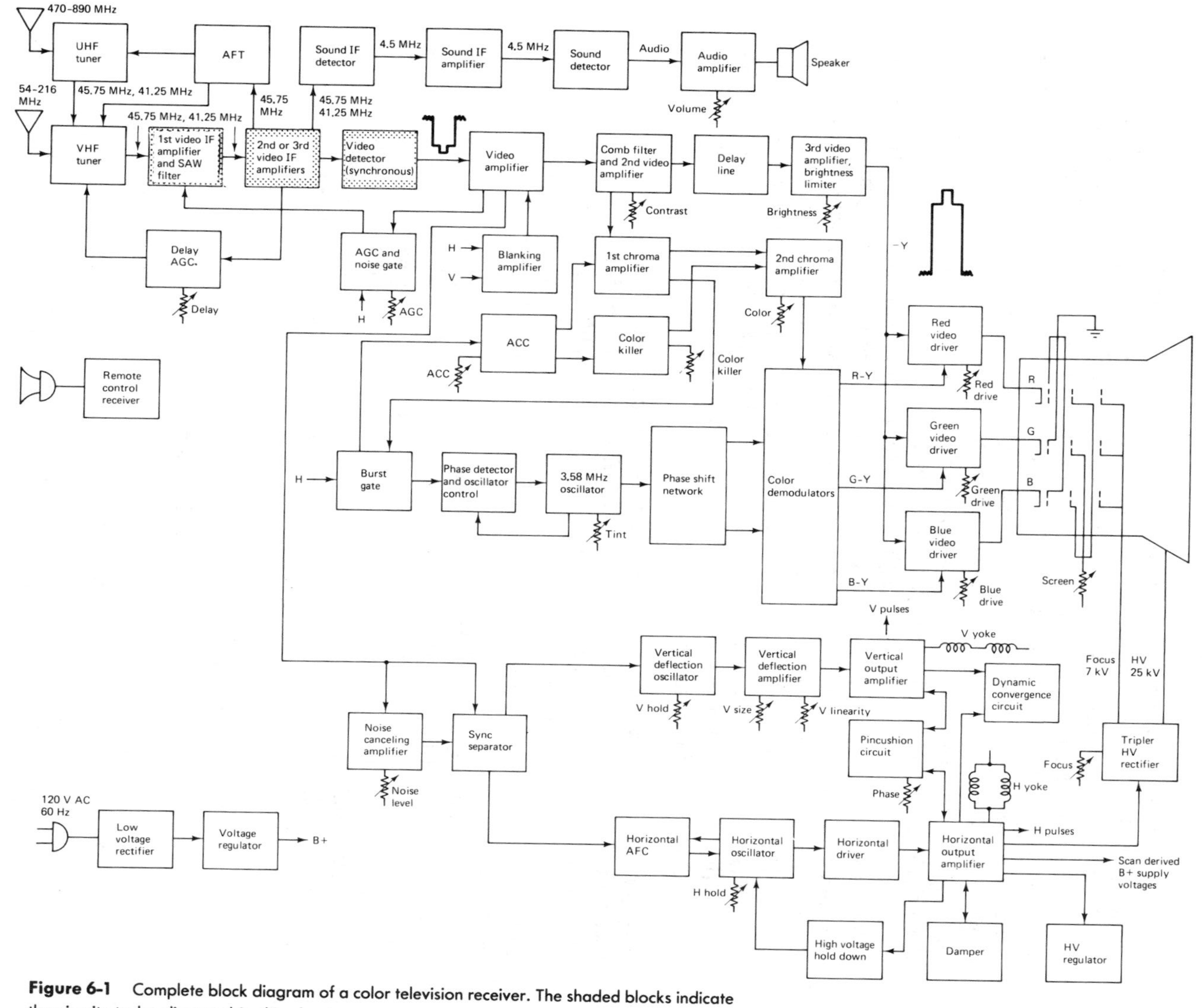

Figure 6-1 Complete block diagram of a color television receiver. The shaded blocks indicate the circuits to be discussed in this chapter.

The actual frequency selected for the video IF is complicated by other factors than selectivity and sensitivity. Such factors as image and local oscillator interference also play an important part.

From 1946 to the mid-1950s most television receivers were designed to reduce the frequency of the video picture carrier and sound carrier to an IF frequency of 25.75 MHz and 21.25 MHz, respectively. Unfortunately, these frequencies resulted in interference between receivers. This happened because a 25.75-MHz IF requires that the local oscillator of the receiver be tuned to a frequency 25.75 MHz above the incoming RF signal. This means that when the receiver is tuned between channels 2 and 9, the local oscillator frequencies will range from 80 to 212 MHz, which, unfortunately, falls in the frequency range of channels 5 to 13. Hence, it was possible for your TV receiver to be tuned to channel 13 (210 to 216 MHz) and your neighbor's TV receiver to be tuned to channel 9 (186 to 192 MHz). This placed your neighbor's receiver's local oscillator at 212 MHz, which falls in channel 13. Because of the close proximity of both rooftop antennas, your neighbor's local oscillator radiation was picked up by your receiver and interfered with its reception. The interference appeared as fine stripes that continually moved across the picture screen.

This form of interference was effectively eliminated by changing the video IF to 45.75 MHz, and the sound IF to 41.25 MHz. These higher IFs placed the local oscillator outside the VHF band of TV stations eliminating oscillator radiation interference. These IF frequencies are almost universally used today.

IF stages. An important difference exists in the block diagrams of the video IF amplifiers used for color receivers and those used for black & white receivers. See Figure 6-2. In color receivers the sound takeoff is obtained from the last IF amplifier and is fed to *a separate sound IF detector.* The black & white TV receiver uses the video detector as a sound IF detector, and sound takeoff is obtained either from the video detector or from the output of the video amplifier.

The number of video IF stages used depends upon the type of receiver and the manufacturer. Color receivers may have three or four stages of amplification; black & white receivers have two or three stages. Integrated circuits are used in many receivers to replace the transistor or vacuum tube amplifiers that have been used previously. One or two IC chips may completely replace all of the transistors, diodes, and resistors ordinarily used in discrete component receivers. The only components not replaced are the tuned circuit elements and various decoupling *RC* filters.

6-3 the video IF strip

The video IF strip has four major functions:

1. To compensate for vestigial sideband transmission
2. To provide most of the selectivity required by the receiver for interference-free reception
3. To provide most of the picture and sound signal amplification of the receiver
4. To frequency change the IF into the video signal at the video detector

The output IF signal voltage of the mixer can be viewed in two ways: in terms of *signal wave shape* or in terms of *signal frequencies.* Both approaches contain useful information for the technician. The *signal wave shapes* that appear at the input to each stage

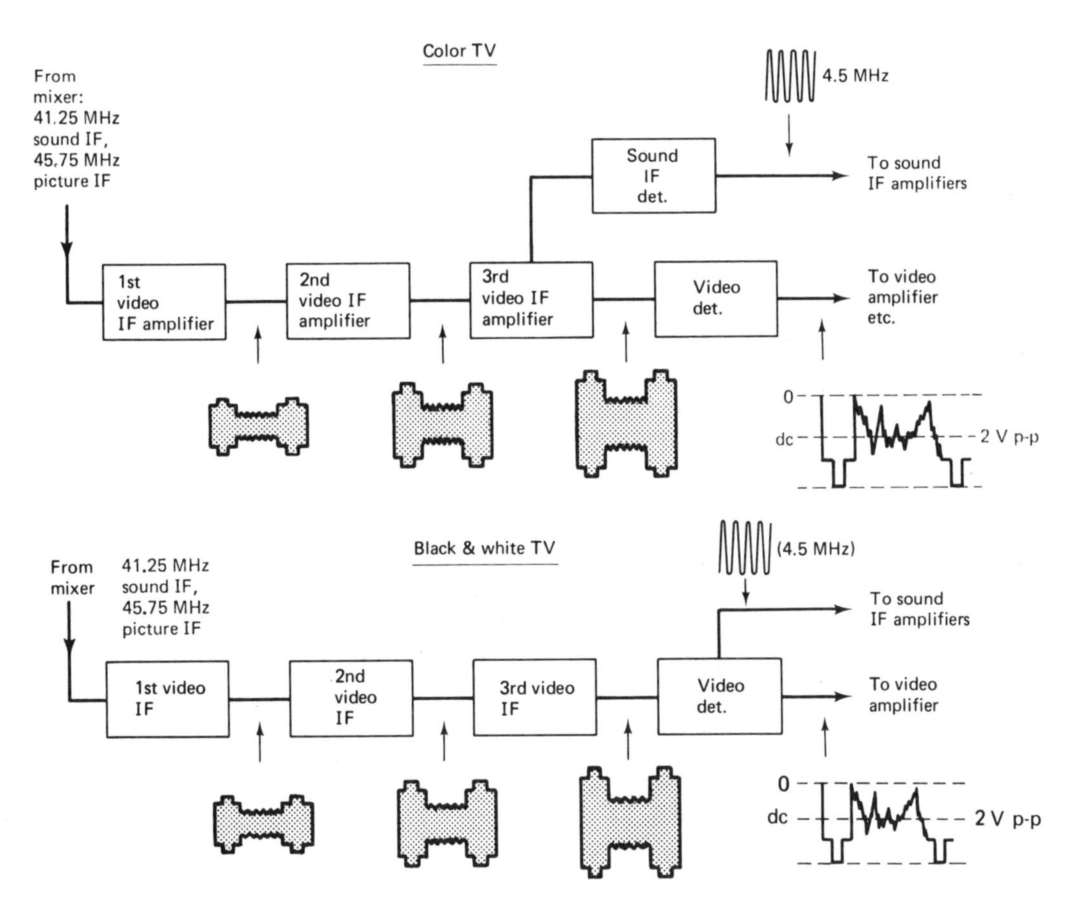

Figure 6-2 Sound takeoff points for color and black & white television receivers. Also indicated are the signal wave shapes and their respective frequencies.

are shown in Figure 6-2. The peak-to-peak output voltage of the mixer is in the order of 0.001 V and its wave shape is that of a modulated RF television signal indicating horizontal sync, blanking, and video for one horizontal line. This signal is amplified sufficiently so that the output voltage of the detector is 1 or 2 V p-p.

In terms of frequency, the mixer stage has converted the frequency of the transmitted picture, sound, and color subcarrier signals to lower frequencies. Typically, the picture carrier is 45.75 MHz, the sound carrier is 41.25 MHz, and the color subcarrier is 42.17 MHz. These frequencies are well beyond the high-frequency limits of service-type oscilloscopes and therefore direct *wave shape* observations of these signals are not usually possible. But it is possible to observe the *modulation envelope* of these signals if the oscilloscope is equipped with a demodulation probe.

6-4 intercarrier sound

In the early days of television the sound and picture information of the television signal were separated at the mixer, and each signal was processed independently in its own IF amplifiers. This was called the *split-sound IF system.* The main problems with this conventional system were the difficulty of tuning and the cost of having two complex IF amplifier systems. Tuning was a problem because the viewer adjusted the fine tuning of the receiver for best sound and hoped that best picture would also be present. Misalignment of the sound IF could result in a condition in which best picture and poor or no sound occurred at one point in the fine tuning, and best sound and poor or no picture occurred at another. Furthermore, fine tuning was necessary whenever channels were changed because the local oscillator was frequently 25 to 50 kHz above or below the value required to produce the proper sound IF.

The *intercarrier IF system* was introduced in the early 1950s. In this system *both* the sound and the picture IF outputs of the mixer are amplified in a common video IF amplifier chain. The sound IF is separated from the video signals at the video detector, or as is done in a color receiver, by means of a separate sound IF detector. The intermediate frequency used for the sound IF takes advantage of the 4.5-MHz frequency separation between the picture and sound carriers of the television signal. Since this frequency separation is established at the transmitter, it is fixed and very accurate. The sound and picture IFs are allowed to beat together in a detector, producing the 4.5-MHz intercarrier sound IF.

The advantages gained by using this system are low cost (since only one multistage IF amplifier is required) and ease of channel fine tuning. In a receiver using the intercarrier system fine tuning is adjusted for best picture. *Best sound is automatically obtained* because fine tuning has no effect on the frequency separation between picture and sound carriers since this is established at the transmitter. The sound IF will be produced at the video detector if adjustment of the fine tuning does not move either the picture or the sound carrier out of the IF bandpass of the receiver. If one carrier is lost, it would in effect eliminate the 4.5-MHz difference that is the sound IF.

Since the video detector is being used as a frequency changer in exactly the same way as the mixer in the tuner, the relative amplitudes of the sound and picture carriers become important. In the tuner the oscillator signal injected into the mixer is made many times greater than the RF signal in order to obtain maximum conversion efficiency. The same is true in the intercarrier IF system. During alignment the sound carrier is adjusted to be between one-tenth and one-twenty-fifth of the amplitude of the picture carrier in order to obtain high-conversion efficiency for the 4.5-MHz FM sound IF.

The television picture carrier is an amplitude modulated signal, and the sound carrier is frequency modulated. The 4.5-MHz sound IF produced by the video detector must retain its FM characteristics and at the same time have as little amplitude modulation as possible. This desired result can be achieved by making use of the fact that when two signals are heterodyned (mixed), the resultant difference signal will contain only the characteristics of the *weaker* of the two signals, if the weaker signal is considerably weaker. The frequency response curve of the IF amplifier is adjusted to achieve this by making the sound IF very much weaker than the picture IF.

6-5 IF response

In some respects the sound IF level is critical. If it is made too large, not only may intercarrier buzz occur, but slope detection of the frequency modulated sound IF may result in sound bars appearing in the picture that vary in width and position with the sound content of the signal. See Figure 6-6.

Figure 6-3 is a ladder diagram showing the relationship between the RF signal, the local oscillator, and the overall IF frequency response of both a black & white and a color receiver. The topmost diagram shows the frequency and amplitude relationships of three UHF channels (14, 15, and 16). Below this diagram is the relative frequency position of the local oscillator. Also identified are the various mixer-produced difference frequencies (IF) generated between the desired channel (15) and the local oscillator frequency. The desired IF signals produced are the picture IF carrier (45.75 MHz), the sound IF carrier (41.25 MHz), and the color subcarrier IF (42.17 MHz). The undesired IF signals are the upper adjacent picture carrier IF (39.75 MHz) and the lower adjacent sound carrier IF (47.25 MHz).

It is the function of the IF amplifier's tuned circuits to eliminate the undesired signals and to pass the signals that are desired. On the next lower rung of the ladder diagram are shown the mixer to video detector frequency characteristics of a black & white receiver. Notice that the associated sound is kept at a level of 2 to 5% and that the picture IF is at 50%.

The last diagram on the bottom rung of the ladder is the frequency response of a color receiver. Notice that the sound IF is held at zero level. This is done in order to minimize the 920-kHz picture interference that would occur if the sound IF and the color subcarrier IF were allowed to be detected together at the video detector. This interference may be produced by simple misadjustment of the fine tuning or by misalignment of the video IF. An example of such an interference pattern is shown in Figure 5-2(b) and in the color plate. Also note that the color subcarrier and the picture carrier are both at the 50% level.

In both black & white and color receivers the lower adjacent sound and upper adjacent picture IFs are undesirable and therefore they are kept at zero level.

Intercarrier buzz. One of the undesired results of the intercarrier system is intercarrier buzz. This may be heard when the receiver is tuned to a channel. It is a harsh, raw 60-Hz sound, not at all like the tone produced by a pure sine wave. The frequency of the buzz and the fact that it is only present when tuned to a channel tells us something about what causes it. The frequency indicates that the vertical sync pulse interval is the buzzlike sound that is heard. The fact that it is heard only when the receiver is tuned to a channel confirms this idea. The buzz gets into the sound by one of the three following mechanisms:

1. Misalignment can cause the relative amplitudes of the sound and picture carriers to change. The result is an excessive amplitude modulation of the 4.5-MHz sound IF by the video signal and in particular the vertical sync pulses.
2. Clipping in the video IF, video amplifiers, or sound IF can cause the loss of one or both of the sound and picture carriers and result in a loss of the 4.5-MHz sound IF. If this interruption is repeated at a rapid rate, a buzz is heard. The most likely portions of the video signal to be clipped are the sync pulses because they have the greatest amplitude. Therefore, it is not surprising that the buzz frequency is at the 60-Hz rate of the vertical sync pulses.
3. Ineffective amplitude limiting in the FM detector may not be able to remove the normal residual AM in the 4.5-MHz sound IF resulting in buzz.

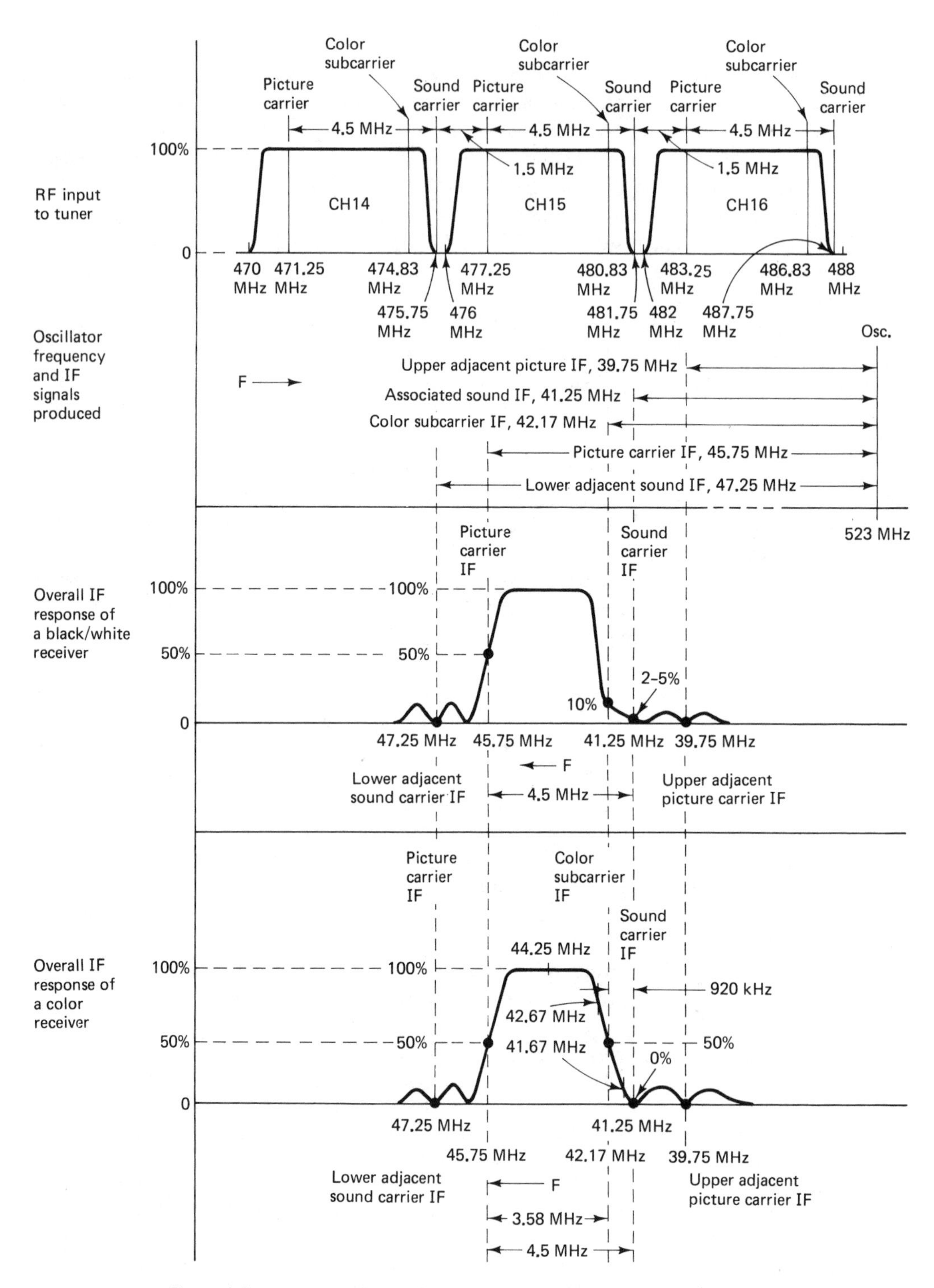

Figure 6-3 RF input, oscillator frequency, and overall IF response of both a black & white and a color receiver.

Compensation for vestigial sideband transmission. The transmitted black & white video signal consists of three signal components: the lower sideband, the carrier, and the upper sideband. The lower sideband extends 1.25 MHz below the carrier, and the upper sideband extends 4.2 MHz above the carrier. This method of partial sideband suppression is called *vestigial sideband transmission.* The sideband spectrum produced by the picture transmitter for one channel is shown in Figure 6-4. Notice that *double*-sideband transmission is used for video frequencies up to 0.75 MHz and that *single*-sideband transmission is used for frequencies above 0.75 MHz.

If it is assumed that a receiver is designed to amplify *all* sideband frequencies equally, the uncompensated video IF frequency response curve would have the same shape as the channel sideband spectrum of Figure 6-4. This means that if a video signal of 0.75 MHz or less were to modulate the transmitter, *two* sidebands would be generated: one above the carrier and the other below it. Both would be equally amplified by the video amplifier and fed to the video detector. If the video modulating signal is greater than 0.75 MHz, only *one* sideband would be generated and only *one* signal would be delivered to the video detector. Therefore, the output of the video detector would be twice as great for video frequencies below 0.75 MHz as for video signals above. This constitutes a form of frequency distortion since all video frequencies are not reproduced equally. The visual effect of this on the picture tube would be an increase in contrast for large picture areas and a relative loss in contrast for fine detail.

To avoid this, the video IF frequency response curve is altered so that the total *double*-sideband energy of a sideband *below* the carrier and a sideband *above* the carrier is equal to the *single*-sideband energy for those video frequencies higher than 0.75 MHz. This is called *vestigial sideband compensation* and is accomplished by placing the picture carrier at a 50% response point on the IF response curve, as shown in Figure 6-3. This response

Figure 6-4 Sideband spectrum for TV channel 4.

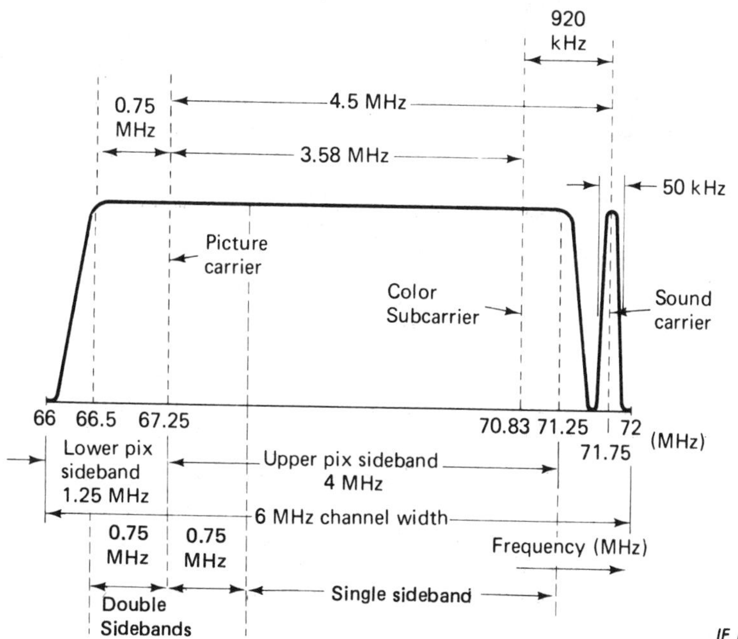

curve is the desired shape that both color or black & white television receivers are adjusted for. The slope of the video IF response curve is adjusted to produce the desired sideband amplitudes. A properly shaped response curve indicates that all video frequencies can be reproduced faithfully without distortion and can duplicate the transmitted video signal exactly.

6-6 color receiver IF response

The output of the mixer for color program material consists of:

1. The video sound IF (41.25 MHz).
2. The picture carrier IF (45.75 MHz).
3. Upper and lower sidebands associated with the black & white or luminance signal; these signals range from 0.75 MHz below the carrier to 4.2 MHz above the picture carrier.
4. Upper and lower sidebands associated with the color or chrominance portion of the signal centered at 42.17 MHz. The chrominance sidebands may extend to a maximum of 0.5 MHz above the subcarrier and 1.5 MHz (the I signal) below the subcarrier. In terms of video IF frequencies, the chrominance sidebands range from 41.67 to 43.67 MHz. Modern receivers no longer make use of the entire I sideband signal. In practice, only 0.5 MHz above and below the subcarrier is used. Thus, the chrominance video IF frequencies of practical use extend from 41.67 to 42.67 MHz.

The block diagram and the bandpass characteristics associated with the broad bandpass and narrow bandpass systems of color IF amplification are shown in Figure 6-5. Since both systems differ only in their bandpass characteristics, the block diagram represents both the *broad bandpass IF system* and the *narrow bandpass IF system.*

In the broad bandpass IF system *all* transmitted color sidebands are amplified uniformly. Hence, the IF response curve is flat out to 41.7 MHz. After detection the video IF frequencies are changed to their original lower video frequencies, amplified and are passed on to a color bandpass amplifier. The bandpass amplifier acts like a filter and separates the chrominance information from the composite video signal. Since its frequency response is flat, all color sidebands are amplified equally and provide a color signal output free from distortion.

The narrow-band IF system is the most universally used color IF system, primarily because its simplicity makes it cheaper to manufacture. In this system both the picture carrier and the color subcarrier IF (42.17 MHz) are placed on the 50% points of the video IF response. The shape of the IF response takes on the appearance of a haystack. Placing the color subcarrier at the 50% point causes the upper color sideband (41.67 MHz) to be attenuated to a greater extent than the lower sidebands (42.67 MHz) in a manner similar to vestigial sideband transmission compensation. In this case, however, such compensation is not necessary since only the double-sideband portion (0.5 MHz above and 0.5 MHz below the 42.17-MHz subcarrier frequency) of the chrominance signal is to be used by the color receiver. The simplification of the video IF bandpass characteristics that has been achieved by this means, has also introduced a high-frequency roll off in the video output of the video detector which must be taken into account. This distortion may be compensated

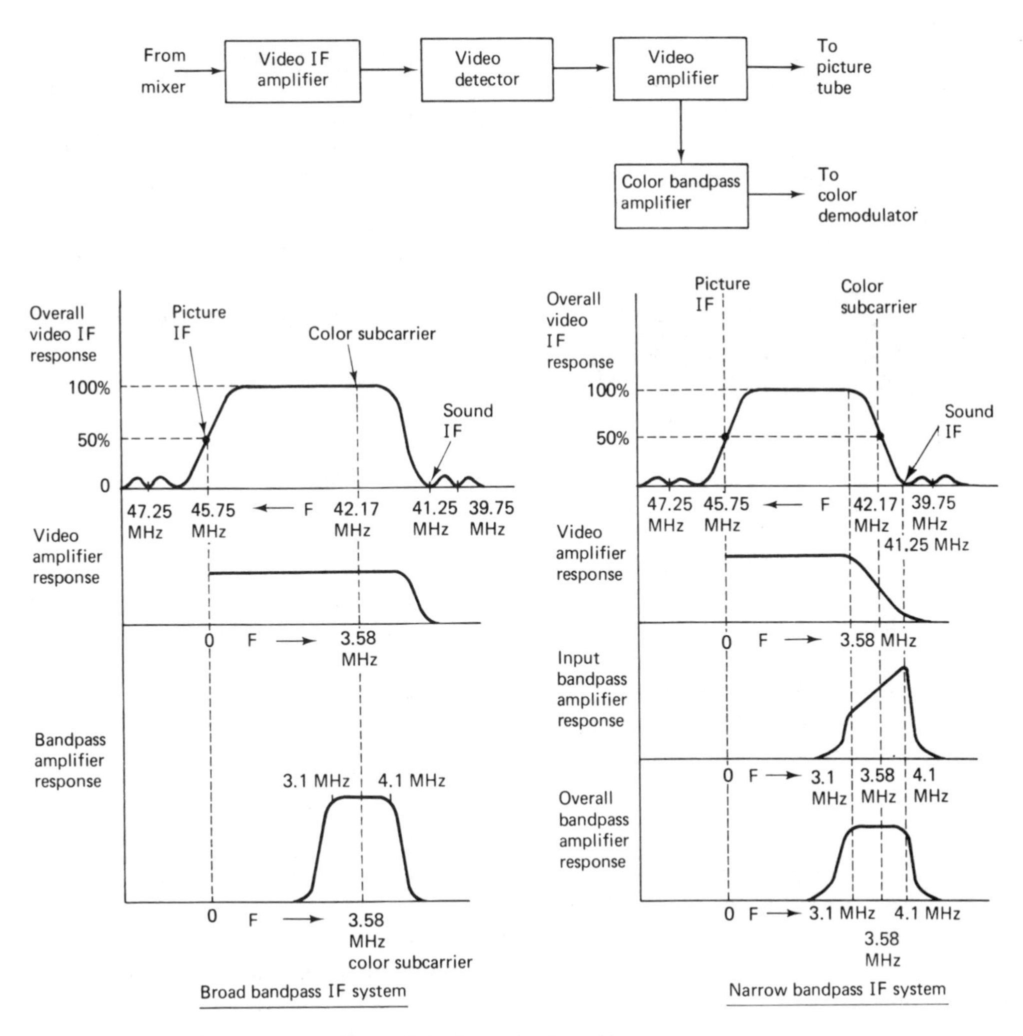

Figure 6-5 Two color IF amplifier systems.

for by the bandpass characteristics of the color bandpass amplifier. This is also shown in Figure 6-5. The frequency response of this stage is centered at 3.58 MHz, the color subcarrier frequency. The upper sidebands (4.1 MHz) received increased amplification to offset the decrease in output that takes place in the video detector. Similarly, the lower color sidebands (3.1 MHz) are not amplified as much in order to be compensated for their excess amplification by the video IF amplifier.

The *overall* effect of the video IF amplifier and the bandpass amplifier frequency response curves is to produce an output frequency response which is flat from 3.1 to 4.1 MHz.

Selectivity. The video IF amplifier is responsible for providing most of the selectivity of the television receiver. The desired channel, as well as any other channel near or distant that

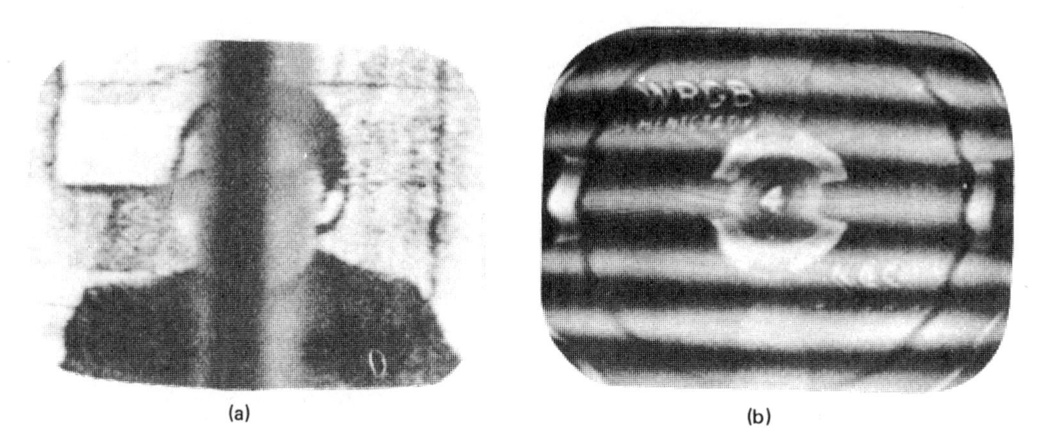

(a) (b)

Figure 6-6 Interference caused by poor IF selectivity: (a) windshield wiper interference; (b) sound interference. (Courtesy of General Electric.)

is picked up by the antenna, is changed to the IF frequency by the mixer. However, because of the relatively poor selectivity characteristics of the front end, signals such as the upper adjacent channel picture carrier and the lower adjacent channel sound carrier will also be converted into IF frequencies. The upper adjacent channel picture carrier will be converted into a video IF of 39.75 MHz and the lower adjacent sound carrier will be changed by the mixer into an IF of 47.25 MHz. If these signals are allowed to be amplified, they will interfere with the desired channel, as shown in Figure 6-6.

6-7 IF traps

To minimize interference from the upper adjacent picture carrier and the lower adjacent sound carrier, the video IF response for these frequencies is greatly reduced by means of highly selective resonant filters called *traps.* Video IF amplifiers in both black & white and color receivers may use traps for three different frequencies. One trap is tuned to the upper adjacent channel's picture IF (39.75 MHz), another is tuned to the lower adjacent channel's (47.25 MHz) sound IF, and the third trap is tuned to the associated sound (41.25 MHz) of the desired channel. The usual location and the frequency to which these traps are tuned are shown in the block diagram in Figure 6-7. Notice that in many receivers most of the traps are placed between the output of the mixer and the input of the first IF.

In a black & white receiver the associated sound trap (41.25 MHz) is used to adjust the level of the sound IF signal to the required 2 to 5% level rather than to get rid of it altogether. In color receivers one or two associated sound traps are used to attenuate this signal to as low a value as possible and to provide a sound takeoff point for the sound IF detector. Traps are not only used to get rid of an undesired signal: they may also be used to shape the overall response curve.

Many different types of traps have been used in color and black & white television receivers. Some of the more common types are as follows:

1. Series resonant shunt traps
2. Parallel resonant series traps

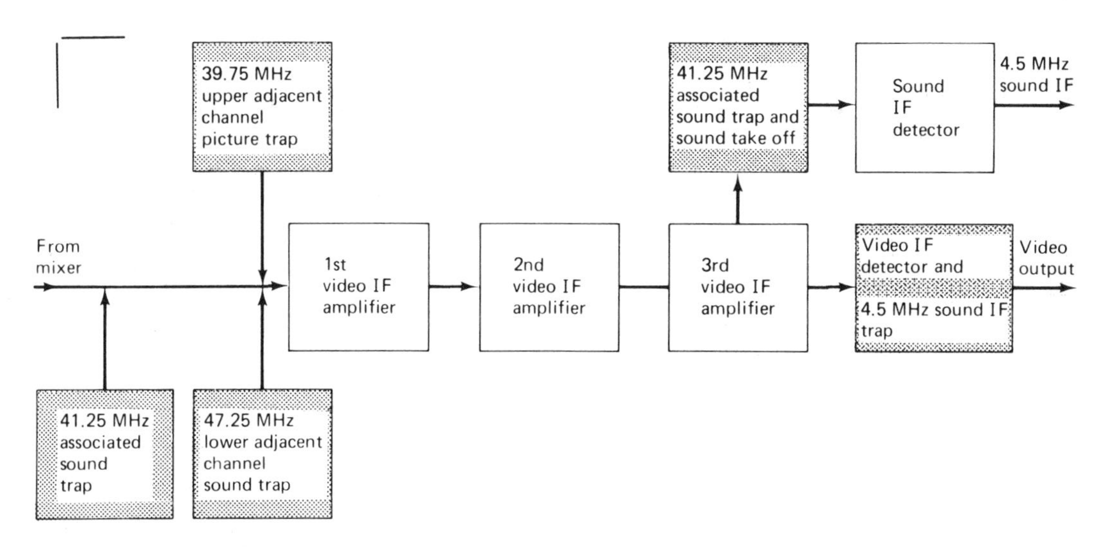

Figure 6-7 Typical video IF traps, their usual locations, and the frequencies to which they are tuned.

3. Transformer traps
4. Absorption traps
5. Bridged T traps

Shunt traps. A high Q series resonant circuit may be considered a *short circuit* at resonance and an open circuit at frequencies removed from resonance. When used as a *shunt trap*, a series resonant circuit is connected across the signal path. This parallel connection acts like a short circuit for the undesired signal and prevents it from passing on to the next stage.

An example of a circuit that uses shut traps is shown in Figure 6-8. L_3 and C_3 form a series resonant circuit that is tuned to 41.25 MHz, the associated sound IF. L_4 and C_4 also form a series resonant circuit, but it is tuned to 39.75 MHz, the upper adjacent picture carrier IF. Both of these shunt traps effectively short circuit the signals to which they are tuned and prevent them from reaching the IF amplifiers. The coil L_2 and the circuit stray capacitance are used to tune the input of the first IF amplifier.

Another application of a shunt trap used in color receivers is to act as a sound IF takeoff point (at the junction of C and L) that feeds a separate sound IF detector.

Series traps. When a parallel resonant circuit is placed in series with the signal path, it is termed a *series trap*. Figure 6-8 illustrates a circuit in which a parallel resonant circuit is used as a trap. In Figure 6-8 L_5 and C_9 form a series trap that isolates the base of the second IF amplifier from the first IF amplifier output at the trap frequency of 41.25 MHz.

Transformer traps. Also shown in Figure 6-8 is a parallel resonant circuit (C_6, L_6) that is combined with a transformer (T_1) to provide signal cancellation at the undesired frequency. This is accomplished by allowing the primary current of T_1 to flow through L_6 and thus develop a voltage drop across the tuned circuit at the trap frequency. The primary current

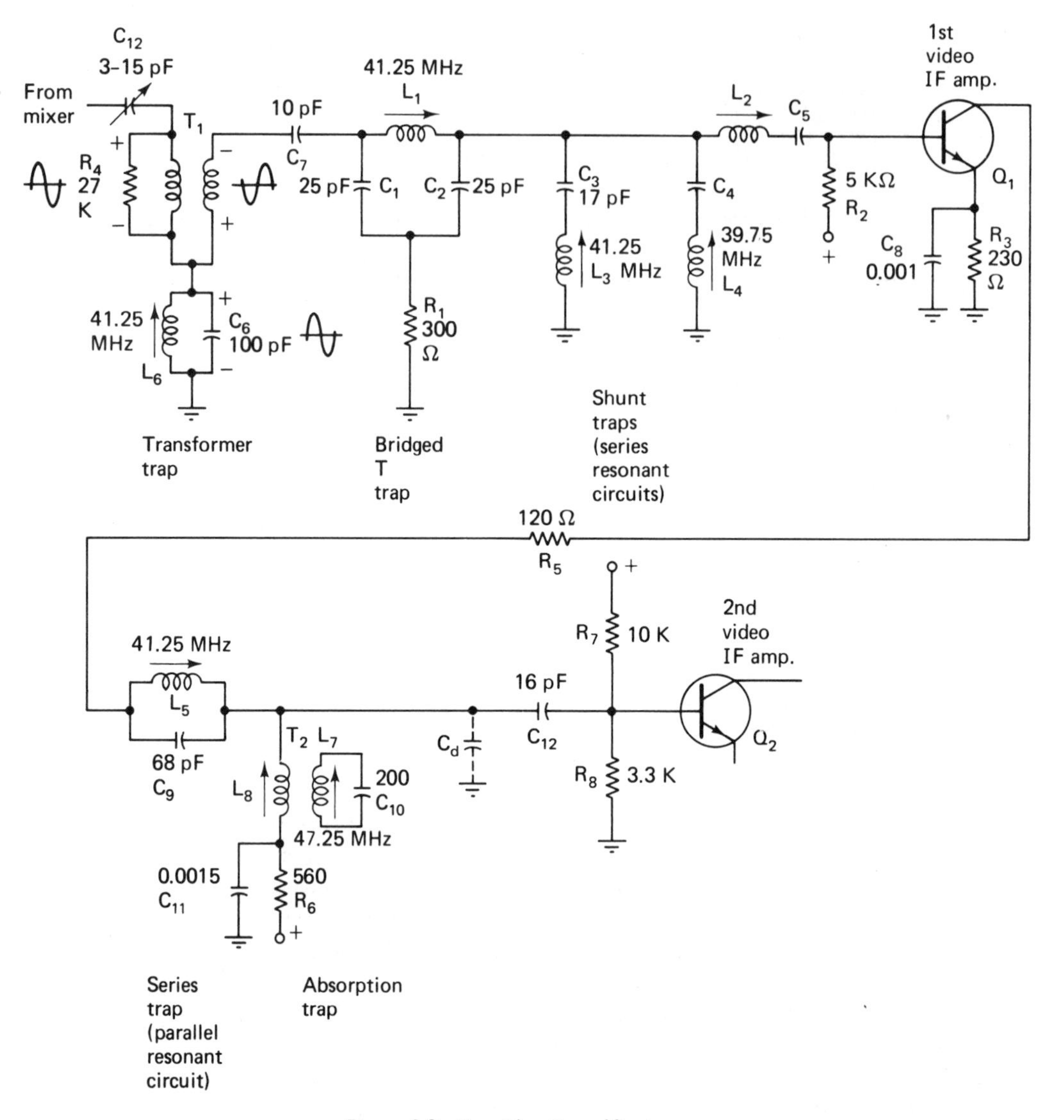

Figure 6-8 Five video IF amplifier traps.

of T_1 also induces a voltage into the secondary winding of the transformer. The transformer secondary is wound so that the induced secondary voltage is 180° out of phase with the primary voltage. This is indicated in the diagram by the plus and minus signs on either side of the transformer. The voltage drop across the tuned circuit (L_6, C_6) will have the same phase as the voltage across the primary. Therefore, both circuits will have the same instantaneous polarity signs. If a signal at the trap frequency (41.25 MHz) passes through this circuit, the output voltage feeding the base of the first IF amplifier will be the sum of the voltage across the tuned circuit (L_6, C_6) and the transformer (T_1) secondary voltage. Since these voltages are 180° out of phase, their sum will be zero provided that the voltages are equal. To achieve this, many receivers make R_4 a variable resistor which, when adjusted properly, will provide the desired attenuation.

Absorption traps. An absorption trap, such as the one shown in Figure 6-8, acts as a trap because it reduces the gain of the IF amplifier stage at the undesired frequency. The absorption trap is easily identified since it appears on the schematic diagram as the tuned secondary of a transformer that goes nowhere.

The absorption trap is able to eliminate the undesired signal as follows. The stage gain of the first IF amplifier in Figure 6-8 is determined by the characteristics of the transistor Q_1, and the impedance of the parallel resonant circuit formed by L_8 and the circuit distributed capacity C_d. At the trap frequency the tuned circuit impedance is made to drop, thereby reducing the stage gain.

At or near parallel resonance the ac voltage across the tuned circuit (L_8, C_d) will be maximum. The ac current flow in L_8 will be maximum. The high current flow in L_8 will induce a voltage into the secondary winding (L_7) that forms part of the absorption trap. Since there is only one path for current to flow in this circuit, L_7 and C_{10} form a *series resonant* circuit. At the trap frequency the impedance of this series circuit is resistive and very small. Therefore, its current flow is high. This current must come from the primary, (L_8) resulting in an increase in primary current. In effect, this causes the impedance of the primary tuned circuit to decrease, which in turn causes a large decrease in stage gain at the trap frequency.

Bridged T traps. The bridged T trap is a more complex trap than those discussed previously, but it is also more effective. An example of a typical bridged T trap used in color and black & white receivers is shown in Figure 6-8. The bridged T circuit is essentially a tuned phase shift network that at the trap frequency is in effect able to develop a negative resistance in series with the resistor R_1. This negative resistance is not accessible and it cannot be measured with test equipment. However, it effectively cancels all resistance in the circuit and makes the circuit an open circuit at the trap frequency, thereby preventing the undesired signal from passing on to the next stage.

The resistor R_1 has the function of determining the amount of attenuation received by the undesired signal. With an optimum value of R_1, as much as 40 dB (100 to 1) reduction is possible. This amount of attenuation occurs when the value of R_1 equals the effective negative resistance developed by the circuit. In many circuits R_1 is made variable as a means of adjusting the degree of attenuation. The bandwidth of the bridged T trap depends upon the Q of the inductor. The higher the Q, the narrower the bandwidth and the better the skirt selectivity. It is because of the high attenuation and narrow bandwidth of this type of filter that it is often referred to as a *notch filter.*

Proper adjustment of traps can only be made during alignment of the video IF amplifiers. Above all else, this requires the use of accurate markers to ensure that the traps are adjusted to their correct frequencies and do the job that they are designed to do.

6-8 coupling methods

The video IF amplifier is required to pass a very wide range of frequencies. Black & white receivers may have bandpass requirements of at least 3 MHz, and color receivers using the wide-band IF system may have bandwidth requirements of 4 MHz. To get an idea of how large this is, compare these requirements to the 200-kHz bandwidth requirement of an FM receiver and the 10-kHz bandwidth requirement of an AM receiver. Such large video IF bandwidth requirements that are necessary in television systems may be achieved by using one or more of the following techniques:

1. Overcoupled tuned transformers
2. Loaded tuned circuits
3. Stagger tuning
4. Surface acoustic wave (SAW) filters

Overcoupled tuned transformers. Transformers whose primary and secondary are both tuned and whose degree of mutual coupling has been adjusted to exceed critical coupling are termed *overcoupled transformers.* If the coupling continues to become tighter, unity coupling is eventually achieved. This is the condition in which all of the magnetic field generated in the primary links the secondary. Unity coupling of a double-tuned transformer results in essentially a single-peaked response.

Bifilar transformers. Tuned transformers that are designed to have unity coupling are called *bifilar transformers.* In this transformer the enamel-insulated wires used for the primary and secondary windings are placed parallel to one another throughout their entire length. The wires are mechanically banded together to ensure tight coupling; then this twin lead is wrapped around a coil form to make the transformer.

Two examples of interstage coupling commonly found in transistor video IF amplifiers are shown in Figures 6-9 and 6-10. The circuits are straightforward examples of single tuned (Figure 6-9) and impedance coupled (Figure 6-10) IF amplifiers that are used in television receivers and are similar to those found in AM and FM broadcast receivers. The only real difference lies in the value of the bypass capacitors and the interstage tuned circuits used. The television receiver circuits use smaller valued components because of the higher frequency (45.75 MHz) of operation.

Figure 6-9 Transistor IF amplifier using transformer coupling.

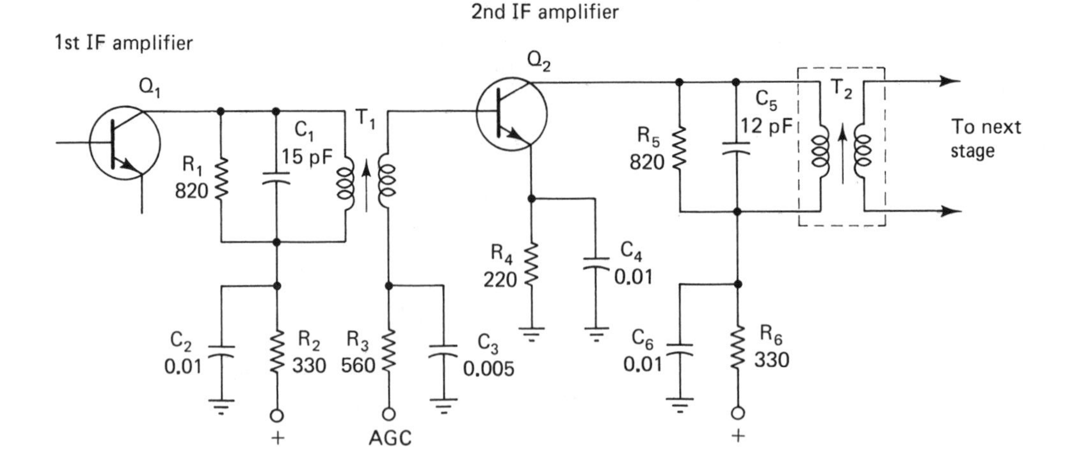

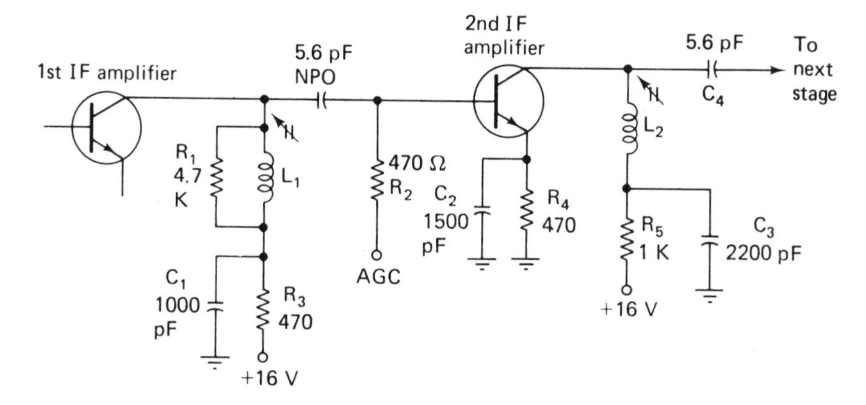

Figure 6-10 Transistor video IF amplifier using impedance coupling between stages.

Loaded tuned circuits. From basic theory we know that the bandpass of a parallel tuned circuit depends upon the resistance associated with it. If the resistance is in *series* with the coil or capacitor and if it is made larger, it will cause the bandpass of the circuit to increase. However, the same increase in bandpass can occur if a resistor is placed *across* the tuned circuit and is decreased in value. Therefore, one of the methods used to obtain the desired wide bandpass for the video IF amplifier is to place a resistor, called a *loading resistor,* across one or more of the amplifier's tuned circuits. An example of a loading resistor may be seen in Figure 6-9. In this transformer-coupled circuit, loading resistors R_1 and R_5 are placed across the primaries of the input and output transformers. The primaries of these transformers are tuned by capacitors C_1 and C_5 associated with the circuits. In the impedance coupled transistor circuit of Figure 6-10 a loading resistor (R_1) is placed across the tuned circuit inductance L_1. This is in addition to the loading imposed on L_1 by the transistor input resistance. The circuit is made resonant by the stray circuit capacitance.

Synchronous tuning. Tuned circuit loading may be all that is necessary to provide the proper video IF bandpass. In such receivers each tuned circuit is tuned to the same frequency. This is called *synchronous tuning.* Loading of the video IF tuned circuits is achieved by the parallel combination of the loading resistors and the low-input resistance of the transistor amplifier that it feeds. An example of such a circuit is seen in Figure 6-11. In this circuit three NPN transistors (Q_{200}, Q_{201}, and Q_{202}) are used as the video IF amplifier. The impedance coupled tuned circuits T_{201} and T_{202} and the tuned transformer T_{203} associated with each amplifier are all tuned to the same frequency (44 MHz). Proper bandpass is obtained by loading placed on the tuned circuits by the input circuit of the next stage. For example, T_{202} is ac shunted by the parallel combination of R_{214}, R_{215}, and the input resistance of the transistor.

Each of the coils is returned to B+ via a tap on the coil. From an ac point of view, this point is at ac ground because of the 0.001-μF bypass capacitor that ties it to ground. The ac voltages that appear at each end of the coil are 180° out of phase with one another as measured to ground. The ac voltage at the bottom end of each coil is used to neutralize the amplifier and is returned to the input via a neutralizing capacitor (C_{212}, C_{220}, and C_{226}).

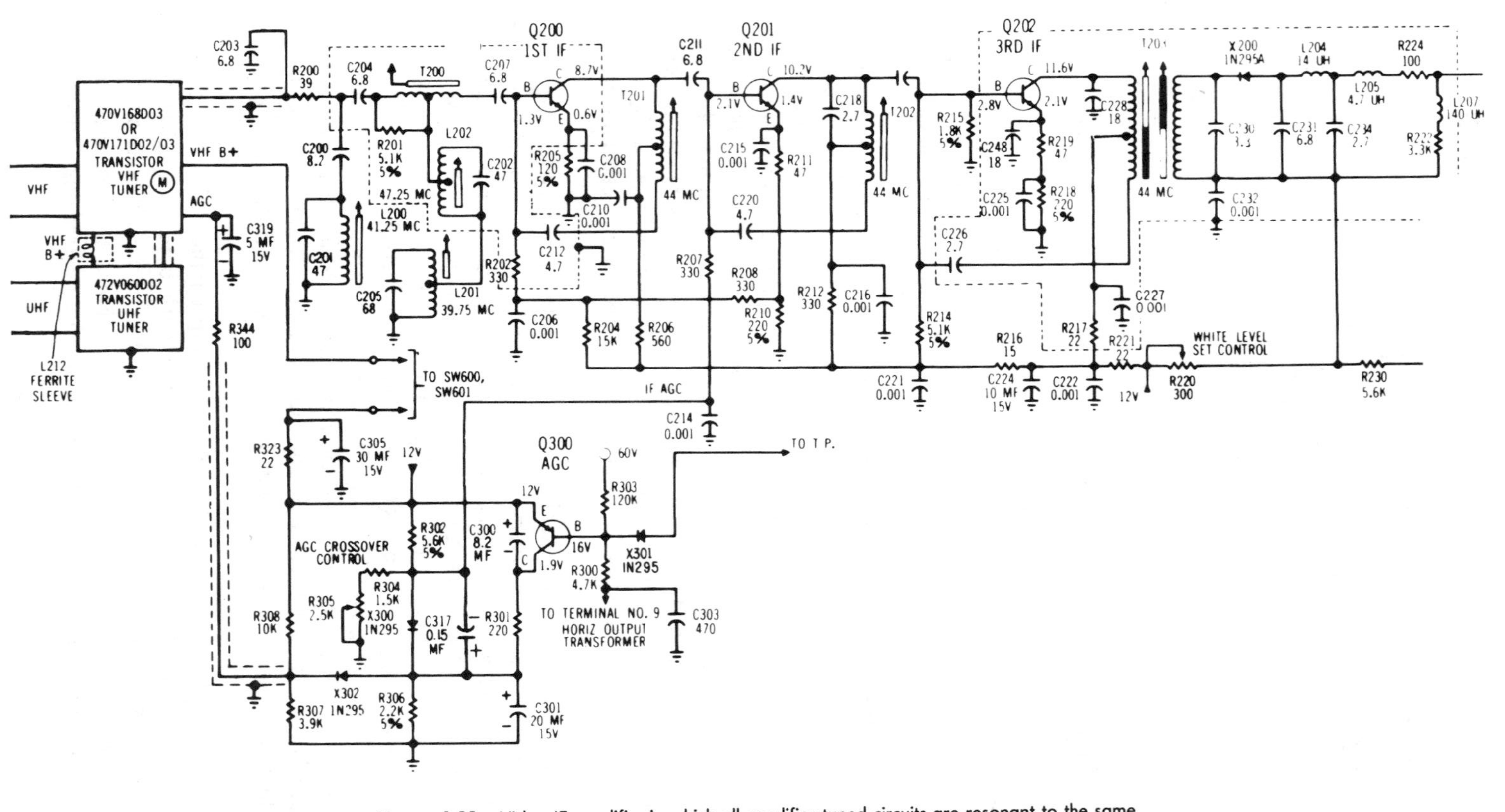

Figure 6-11 Video IF amplifier in which all amplifier tuned circuits are resonant to the same frequency. This arrangement is called synchronous tuning.

Three traps are used to adjust the associated sound carrier (41.25 MHz) amplitude and to eliminate the upper adjacent picture IF (39.75 MHz) and the lower adjacent sound IF (47.25 MHz). These traps are combinations of a shunt trap (C_{200}, C_{201}, and L_{200}) and a complex trap consists of T_{200}, L_{201}, L_{202}, C_{202}, and C_{205}.

Stagger tuning. Stagger tuning is a system for obtaining a wide bandpass amplifier. In this arrangement each amplifier of a multiple-stage IF system is tuned to a different frequency. The addition of each tuned response results in an overall bandpass that may be made very wide. This system is widely used in color and black & white TV receivers. The main advantage of the stagger tuned system is its ease of alignment. Alignment consists of adjusting each tuned circuit to a resonant peak at its assigned frequency. Although it is possible to peak these tuned circuits with a signal generator and a VTVM, this method is not recommended because small errors in each adjustment will result in an overall video IF response curve whose shape is not proper. For color receivers, such changes in response curve shape can mean the difference between good color reception and weak, poor, or no color reception. Therefore, video IF alignment should always be done with a sweep generator following the manufacturer's alignment procedure.

Another advantage of the stagger tuned system is that it is less likely to oscillate than are synchronously tuned amplifiers. As a result, stagger tuned systems are less likely to require neutralization.

A typical video IF amplifier using stagger tuning is shown in Figure 6-12. In this three-stage transistor IF amplifier each of five tuned circuits used to obtain the required bandpass is tuned to a different frequency. Thus, L_{301}, the first IF-input tuned circuit, is tuned to 42.8 MHz, and L_{302}, the first IF collector tuned circuit, is tuned to 44.2 MHz. L_{303}, the second IF tuned circuit, is also tuned to 44.2 MHz. The third IF tuned circuit, L_{305}, is tuned to 45.3 MHz, and L_{306}, the detector-input tuned circuit is tuned to 43.5 MHz. In this circuit both L_{301} and L_{303} are loaded by resistors R_{305} and R_{309}, which are placed across the tuned circuits as a means of broadening the bandpass of the system. In this circuit each tuned inductance is made resonant by means of the distributed and stray capacitance associated with the components found in each amplifier. Figure 6-13 is a ladder diagram showing the frequency response of each stage and the overall resultant bandpass of the stagger tuned IF amplifier shown in Figure 6-12.

Walking IF. In some of its television receivers Zenith has introduced an adjustable bandpass circuit that is called a "walking IF." In this arrangement, strong signals produce an IF response that is normal. Weak signals result in a distorted bandpass such that the picture IF and its associated lower-frequency sidebands are boosted, while the high-frequency sidebands are attenuated. This results in an improvement in picture quality and stability, since the low-frequency video components, representing large picture areas and sync, are made stronger. The high-frequency video signals, corresponding to fine detail and snow, are made weaker.

The automatic bandpass feature of this circuit is due to AGC control of the transistor and IC interelectrode capacitance. These capacitances change when the dc bias of the circuit is altered in accordance with signal strength. As a result of the design of this circuit, realignment of the IF response takes place.

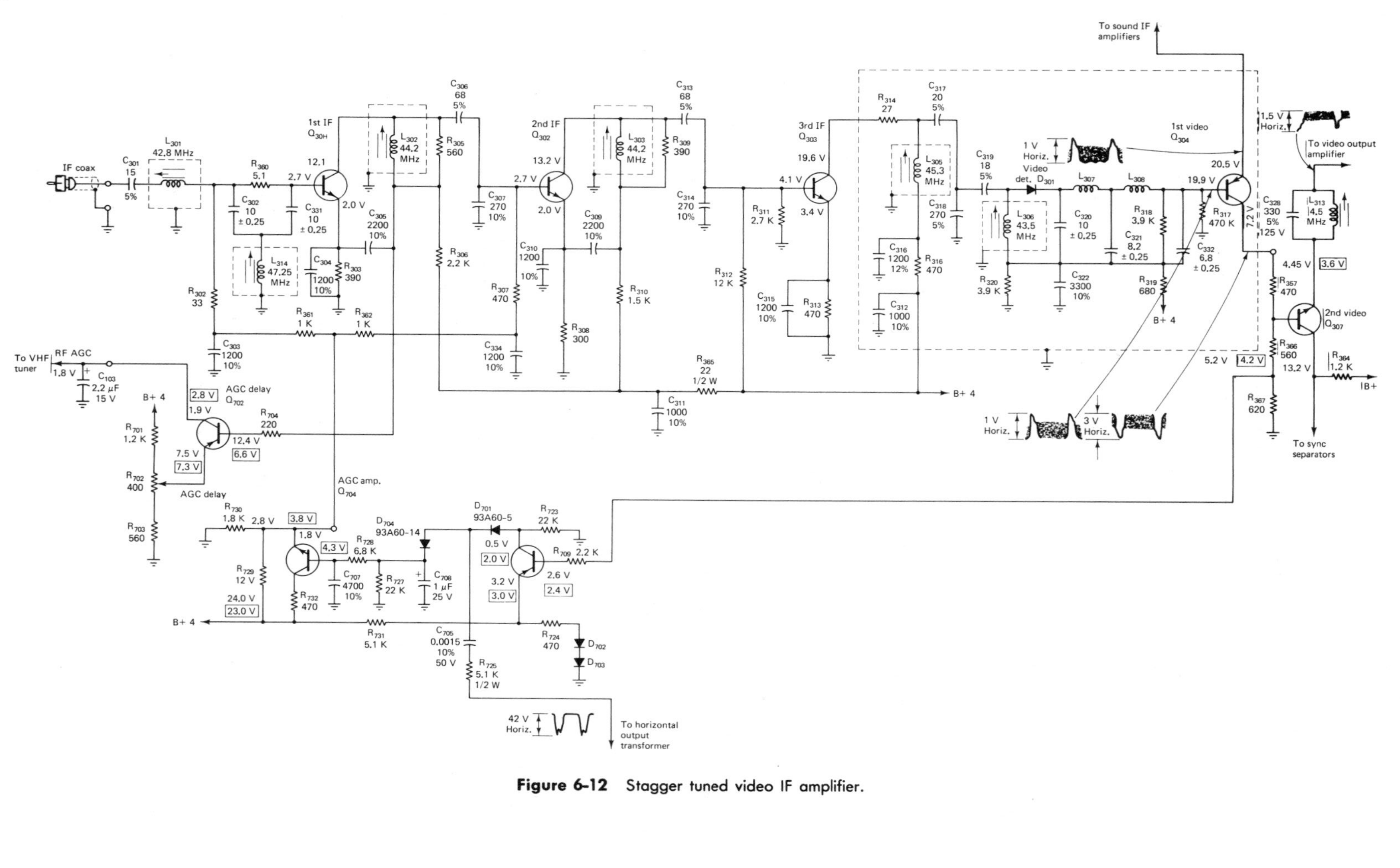

Figure 6-12 Stagger tuned video IF amplifier.

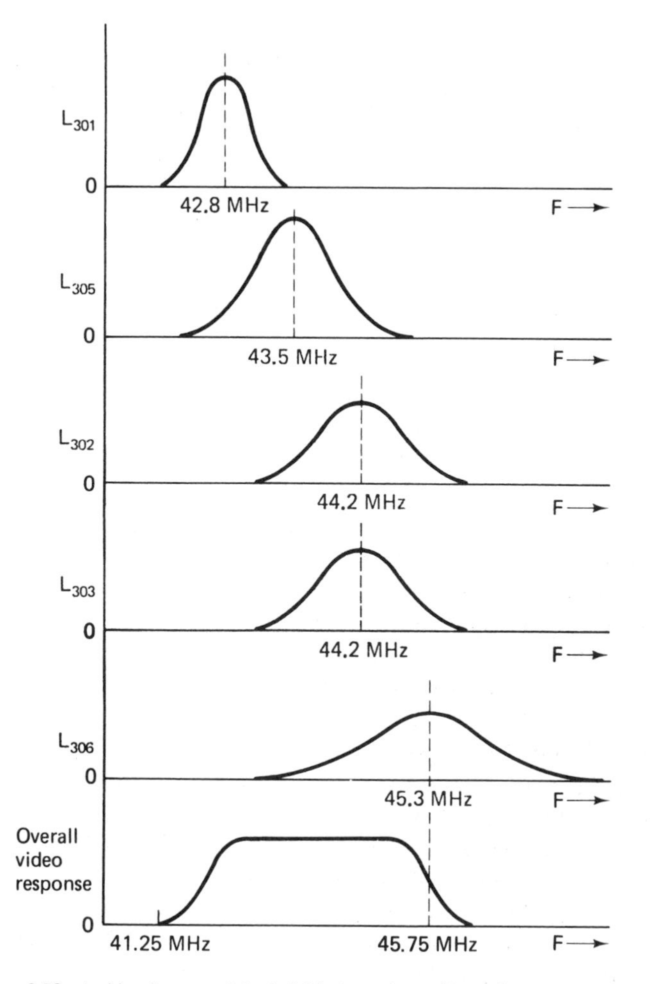

Figure 6-13 Ladder diagram of the individual tuned transformer frequency response curves and their overall results.

Surface acoustic wave filters (SAW). In the early 1970s a great deal of research and development was done on applying acoustic waves (Rayleigh waves) to processing signals in communication systems. This work resulted in the development of surface acoustic wave (SAW) filters for use in television receivers.

Acoustic waves are ripples on the surface of a solid which travel out from the point of excitation at the speed of sound. Surface acoustic waves can be generated on the face of the earth by forces as great as an earthquake. They can also be generated by applying an ac voltage to a thin slice of piezoelectric material such as quartz, lithium niobate, or lead zirconate.

Piezoelectric crystals generate a voltage when they are mechanically stressed, and conversely they will become distorted or compressed when a voltage is applied to them. The degree of stress is proportional to the applied voltage and the generated voltage is proportional to the applied stress.

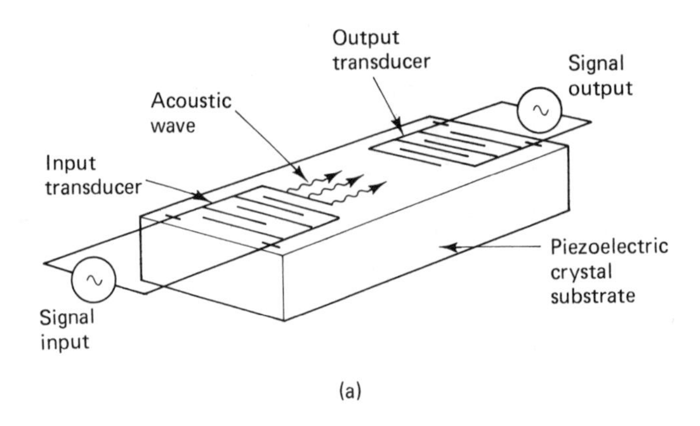

Figure 6-14 (a) Surface acoustic wave device; (b) schematic diagram of an IF amplifier using a SAW filter. [(b) Courtesy of RCA.]

A surface acoustic wave filter is a small, thin device only 0.15 in. square. It is made of piezoelectric material on which two pairs of electrodes are plated. One pair is the input transducer; the other is the output transducer. See Figure 6-14(a). The transducers are fingerlike projections whose length and spacings (called weighting) determine the frequency response of the filter. By adjusting the dimensions of the transducer, it is possible to construct a filter that has a frequency response characteristic that is similar to that required by a TV IF amplifier. See Figure 6-3.

The SAW filter is fixed tuned and cannot be adjusted. This eliminates practically all IF alignment adjustments except for a lower adjacent sound IF trap (47.25 MHz) and a SAW matching tuned circuit adjustment.

A schematic diagram of a typical IF amplifier using a SAW filter is shown in Figure 6-14(b). The IF output of the tuner is impedance coupled to the base of the IF amplifier Q_{301}. In the base of the amplifier is a lower adjacent sound trap. The IF amplifier is necessary to make up for the signal loss produced by the SAW filter. L_{302} is used for purposes of impedance matching. The output of the SAW filter feeds a three-stage IF amplifier found in IC U_{301}. The output of the IF amplifier then feeds a synchronous video detector.

6-9 the integrated circuit video IF amplifier

Figure 6-15 shows the schematic diagram of a color receiver video IF amplifier that used integrated circuits. The two ICs used in this circuit are contained in a single 20-pin dual-in-line package. This package in turn is mounted on a module plug-in board that contains all of the components for the video IF, video amplifier, and AFT circuitry.

The video IF circuit is very straightforward. IC1A and IC1B provide the required video IF amplification, sound IF (4.5 MHz) detection, and video detection. The transistor Q_1 is a voltage regulator that ensures a stable dc voltage source. Inductances L_1 through L_6 resonate with their associated distributed circuit capacity to provide the desired selectivity characteristics of the color receiver. Elimination of undesired upper and lower adjacent channel carrier IFs, the associated sound IF, as well as adjusting the skirt selectivity of the video IF, are accomplished by three traps. The upper adjacent picture carrier trap (39.75

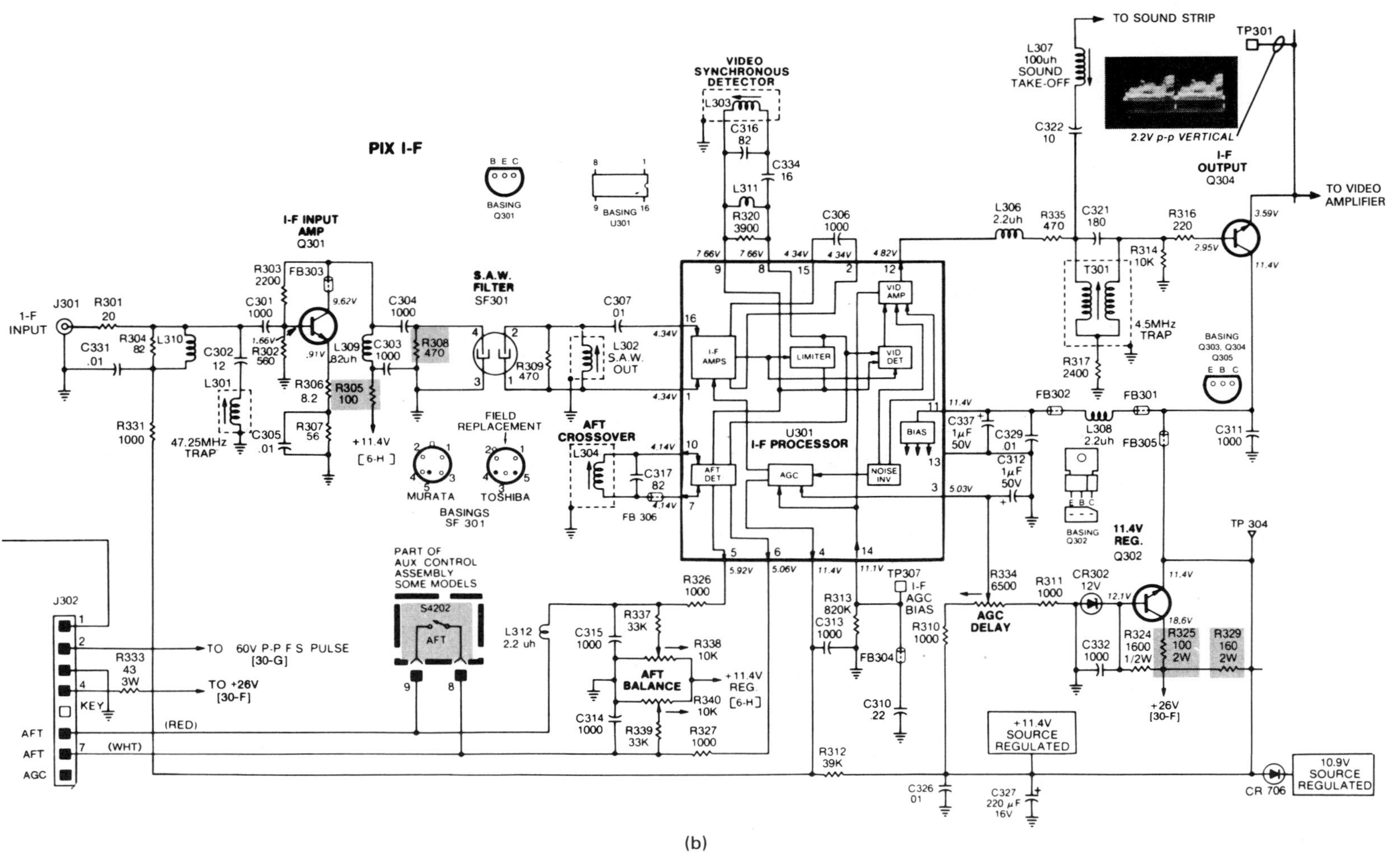

(b)

Figure 6-14 *(continued)*

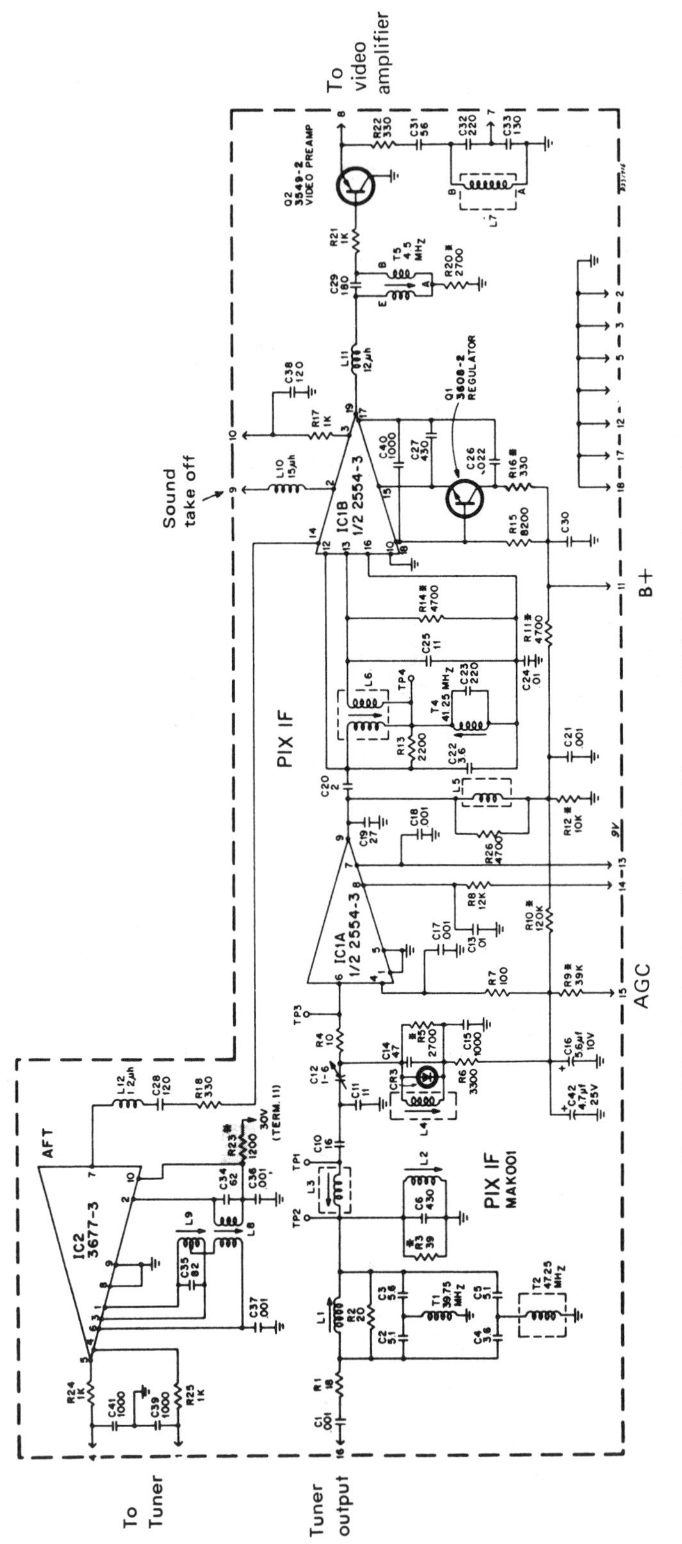

Figure 6-15 IC video IF amplifier module. (Courtesy of RCA, Consumer Electronics.)

MHz) is a bridged T trap formed by T_1, C_2, C_3, and R_2, L_1. The lower adjacent sound carrier trap (47.25 MHz) is also a bridged T type trap and is formed by T_2, C_4, C_5, and R_2, L_1. The associated sound trap (41.25 MHz) is formed by a complex trap that consists of T_4 and L_6. Resistors R_3, R_5, and R_{13} are loading resistors used to broaden the bandwidth characteristics of their associated tuned circuits. Resistors such as R_6 through R_{12} as well as R_{15} and R_{16} are used as voltage dividers that provide the proper dc and AGC voltages for the ICs. Capacitors C_1, C_{10}, C_{12}, and C_{20} are coupling capacitors. Capacitors C_6, C_{14}, C_{23}, and C_{25} are used to resonate the tuned circuits with which they are associated. Almost all of the remaining capacitors are used as bypass filter capacitors. The sound take-off is obtained from IC1B (pin 2) and is fed to the sound strip. The video output is also obtained from IC1B (pin 19) and is fed to Q_2, the video preamp.

6-10 video IF sweep alignment

The video IF amplifier chain is essentially a broadband amplifier filter system. The shape of the *color* video IF amplifier frequency response curve is critical. Ideally, the overall frequency response should appear to be similar to that shown in Figure 6-3. In a properly aligned color receiver the picture carrier (45.75 MHz) and the color subcarrier (42.17 MHz) should be at the 50% points of the overall IF response.

The slope of the response curve in the region of the color subcarrier IF is fairly steep. On the slope is the color sideband information extending from 41.67 to 42.67 MHz, for a bandwidth of 1 MHz. The sound IF (41.25 MHz) response is zero. The opposite slope is not as steep and the picture carrier is positioned at the 50% point. The slope rises linearly for best low-frequency video signal recovery. The lower adjacent channel sound carrier (47.25 MHz) and the upper adjacent channel picture carrier (39.75 MHz) are trapped out to prevent interference. At these frequencies the amplifier has minimum or zero gain. These traps also help shape the response curve at opposite ends of the response curve.

The overall video IF frequency response curve for a *black & white* receiver differs from that of a color receiver in two important characteristics. First, the sound IF (41.25 MHz) is not attenuated to zero; it has a response of 2 to 5%. Second, the color subcarrier response is greatly attenuated to minimize interference.

Circuit defects or faulty adjustment can cause misalignment of the tuned circuits forming the IF chain and will cause the frequency response curve to change shape. This in turn can cause such trouble symptoms as (1) no color, (2) ringing in picture, (3) poor picture quality, (4) interference, (5) buzz in sound, (6) sound in picture, and (7) poor horizontal and vertical synchronization.

Alignment equipment. Video IF amplifier alignment must begin with the proper equipment and technique. The essential equipment needed for sweep alignment is as follows:

1. Manufacturer's service and alignment instructions. Without these, proper alignment is not possible.
2. A sweep generator with an accurate marker generator combination. In modern equipment the sweep generator and the marker generator are combined into one unit. The most important characteristic of the sweep generator is that it have a constant output (360 mV) over a deviation wide enough to cover the entire IF response (15 MHz). The chief characteristic of the marker generator is its accuracy. Without accuracy good alignment is not possible.

3. An oscilloscope. The low-frequency response of the oscilloscope must be very good if distortion of the response curve is to be avoided.
4. Matching pads between the sweep generator and the receiver. These pads terminate the sweep generator's 75-Ω transmission line output cable with its characteristic impedance. This prevents standing waves from being generated on the line. Standing waves will distort the response curve. The presence of standing waves is indicated when running a hand along the transmission line causes the response curve to change in amplitude and shape. The matching pad circuit is usually obtained from the manufacturer's service and alignment literature.
5. Bias supplies. Depending upon the manufacturer, as many as three AGC bias clamps may be needed to fix the operating point of the video IF amplifier. In some cases, this voltage may be as high as 40 V. The exact values and the point of bias injection are obtained from the manufacturer's service and alignment literature. Bias clamps are needed to prevent changes in response curve shape. These changes result when the gain of the IF amplifiers is changed because of shifts in AGC bias that occur during alignment.
6. Demodulator probes. The manufacturer's alignment instructions often call for observing the frequency response at points before the video detector. This can only be done by placing a demodulator probe between the point being measured and the vertical input of the oscilloscope. In some cases, the gain of the scope is not sufficient to produce a usable pattern. To increase the response curve level, amplifiers, voltage doubler or quadrupler demodulator probes may be used. Circuit diagrams for these amplifiers and special probes are usually given in the manufacturer's alignment data.

Alignment hookup. A typical video IF equipment hookup is shown in Figure 6-16. The sweep generator may have three outputs: an RF output, a 60-Hz horizontal deflection voltage, and an intensity post-injection marker.

The RF frequency modulated output is set to a midfrequency of 44 MHz and the deviation is set to approximately ± 10 MHz. This signal is fed via a short properly terminated transmission line to the input of the tuner mixer.

As a means of synchronizing the oscilloscope horizontal deflection with the sweep generator frequency deviation, a sample of the 60-Hz modulating voltage is used as the horizontal deflection voltage for the oscilloscope. In this application the horizontal selector switch of the oscilloscope is switched to the horizontal input position. This disconnects the internal sawtooth generator and allows an external signal to be applied to the horizontal amplifiers.

Modern sweep generators have marker generators built into them. These marker generators may be designed to simultaneously add as many as eight markers to the response curve, as shown in Figure 6-16, or they may be designed to produce pulses which are then applied to the Z-axis of the oscilloscope. This introduces a bright spot or interruption into the response curve at the frequency being marked.

The vertical input of the oscilloscope is connected to the output of the video detector. If the manufacturer requires that the frequency response be checked at the input to the IF amplifier or at some other point, a demodulator probe must be used.

Alignment procedure. Alignment may begin with a number of preliminary receiver modifications. The horizontal output amplifier and vertical deflection circuit may be disabled to

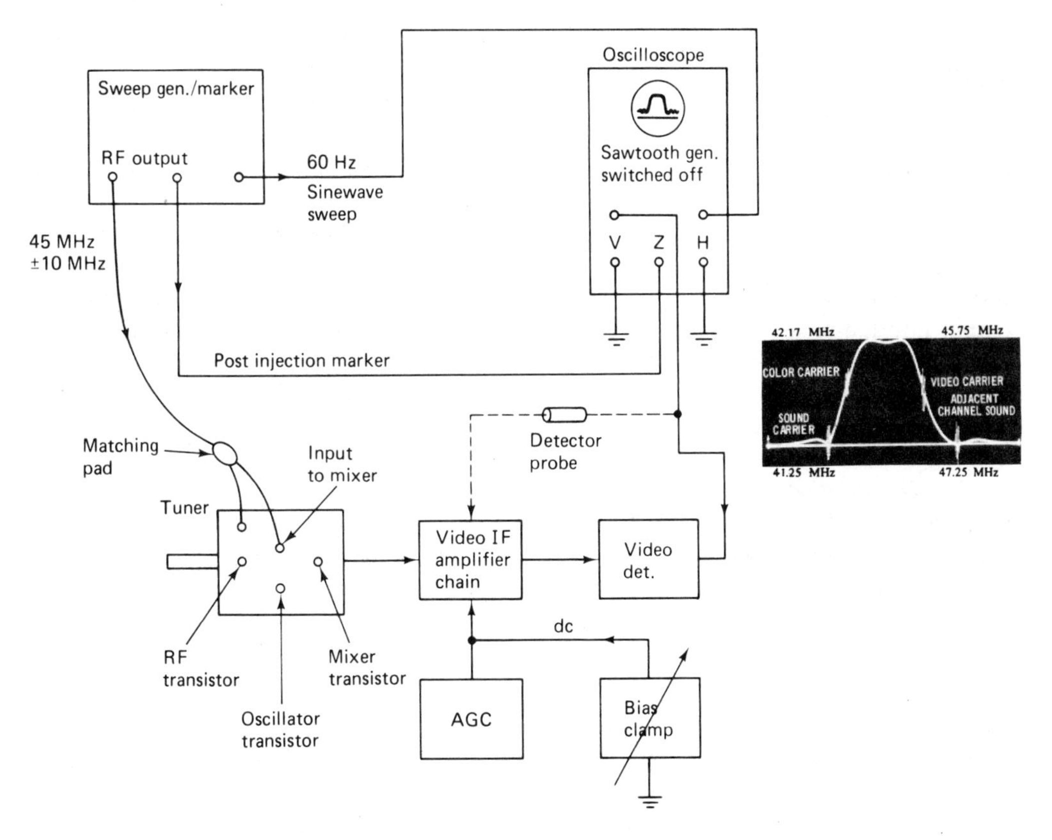

Figure 6-16 Typical video IF alignment equipment hookup.

eliminate "hash" or "grass," as well as spike interference of the response curve that may make alignment more difficult. In some sets the manufacturer may remove the IC module from the set and supply its B supply needs by a separate supply. The required conditions are obtained from the manufacturer's service literature.

During alignment certain precautions and operating procedures should be followed. As a general rule, the oscilloscope vertical gain control should be set to its most sensitive position. This ensures that the sweep generator output will be kept at a low enough value so that IF overloading will be avoided. Overloading the IF amplifiers will produce a flat-topped response curve regardless of adjustment.

Lead dress and the use of short cables are very important. The sweep generator output leads or demodulator output leads should not be brought near or over the IF PC board because it can result in detuning or oscillation. Regeneration and oscillation appear as a sudden reversal and change of shape of the response curve.

In general, sweep alignment of black & white or color receiver IF amplifiers follows the same basic procedures, but the color receiver alignment is much more critical than that of the black & white receiver. In either case, the procedure to follow is that outlined in the manufacturer's service notes.

This procedure may be divided into three basic parts. First, the traps must be aligned, then link alignment must be performed, and finally the remaining tuned circuits for the overall response (both amplitude and shape) given by the manufacturer must be adjusted.

Trap alignment is very important and should be done first so that any misadjustment does not affect the shape of the overall response. Misalignment of the 41.25-MHz trap can cause sound bars in the picture or excessive 920-kHz beat interference. An improperly adjusted lower adjacent sound trap (47.25 MHz) can cause smear in the picture.

Link alignment adjusts the mixer output coil in the tuner, the coaxial cable between the tuner and the chassis, and the input circuits to the first IF stage so that they are properly tuned together. In general, this alignment is performed by injecting the sweep generator output into the mixer input via a 75-Ω matching pad. The link response and the effect of tuned circuit adjustments are observed at the output of the first IF amplifier. Since the frequency at this point is above the frequency limits of the oscilloscope, a demodulator probe is needed to perform this part of the alignment. Link alignment is usually all that is needed after a tuner has been replaced.

Overall alignment is observed at the output of the video detector. The remaining tuned circuits are then adjusted to obtain the desired amplitude and shape. In some receivers the manufacturer's procedure may include signal injection of the sweep generator at the first or second IF amplifier and demodulator probe observation of the individual stage response.

6-11 the video detector

The output of the video IF amplifiers feeds into the video detector stage. The video detector is often referred to as a *second detector* because it is a frequency changer in exactly the same way that the mixer stage is in the tuner. The function of the video detector is to lower the frequency of the video IF to that of the video signal generated by the camera at the transmitter. The output of the crystal diode video detector in a black & white receiver consists of the following:

1. The composite video signal (0 to 4.2 MHz)
2. 4.5-MHz sound IF
3. A dc component corresponding to the average illumination in the picture
4. Residual IF and higher-frequency harmonic components

In a color receiver the output of the video detector is similar to that of the black & white receiver, except that the video signal contains the chrominance signal and the 4.5-MHz sound IF is greatly attenuated.

The output signal voltages and circuit position of the video detector are shown in the block diagrams of Figure 6-2.

The basic circuit of a diode video detector is shown in Figure 6-17. The circuit is essentially a half-wave rectifier and consists of three basic parts; the tuned circuit input, the diode detector, which is almost always a crystal germanium diode that is often mounted inside the transformer housing of the tuned circuit, and the detector load. The tuned circuit input is fed by the last video IF stage. Its wave shape and frequency are indicated in the diagram. The input signal is a high-frequency ac signal that is rectified by the diode. The results of this rectification are indicated in Figure 6-18 as the input voltage to the video amplifier.

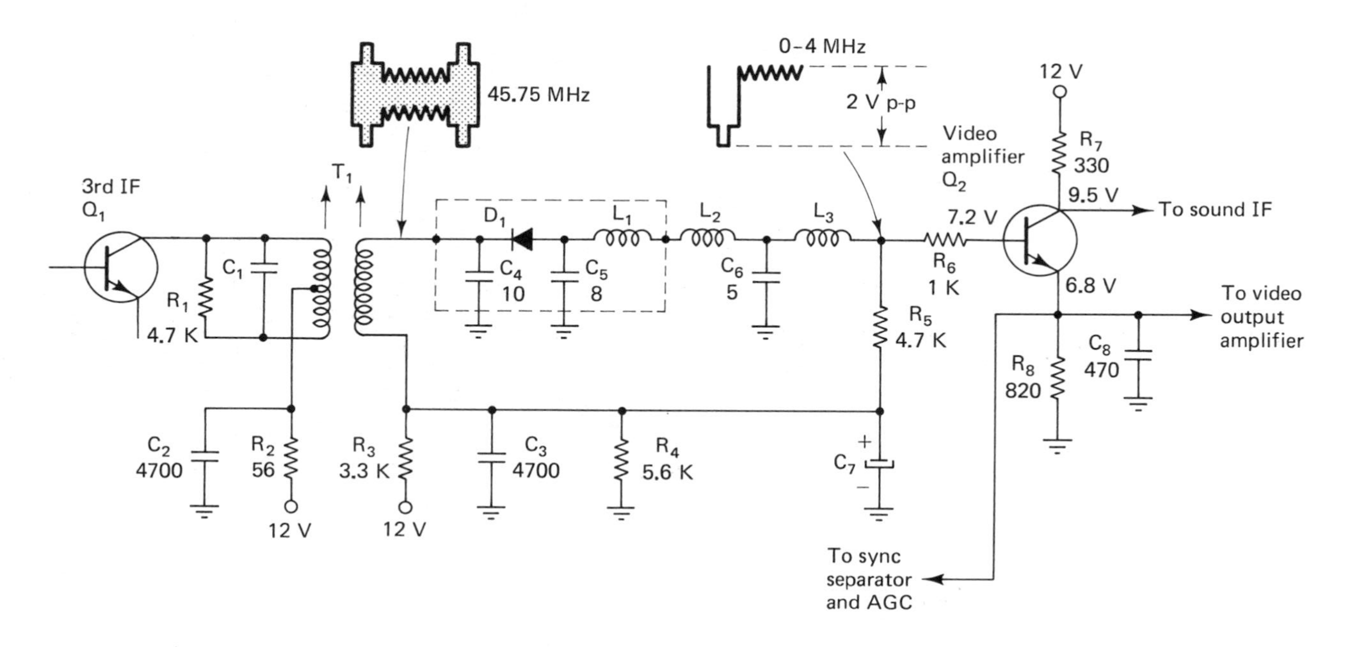

Figure 6-17 Typical video detector used in black & white receivers.

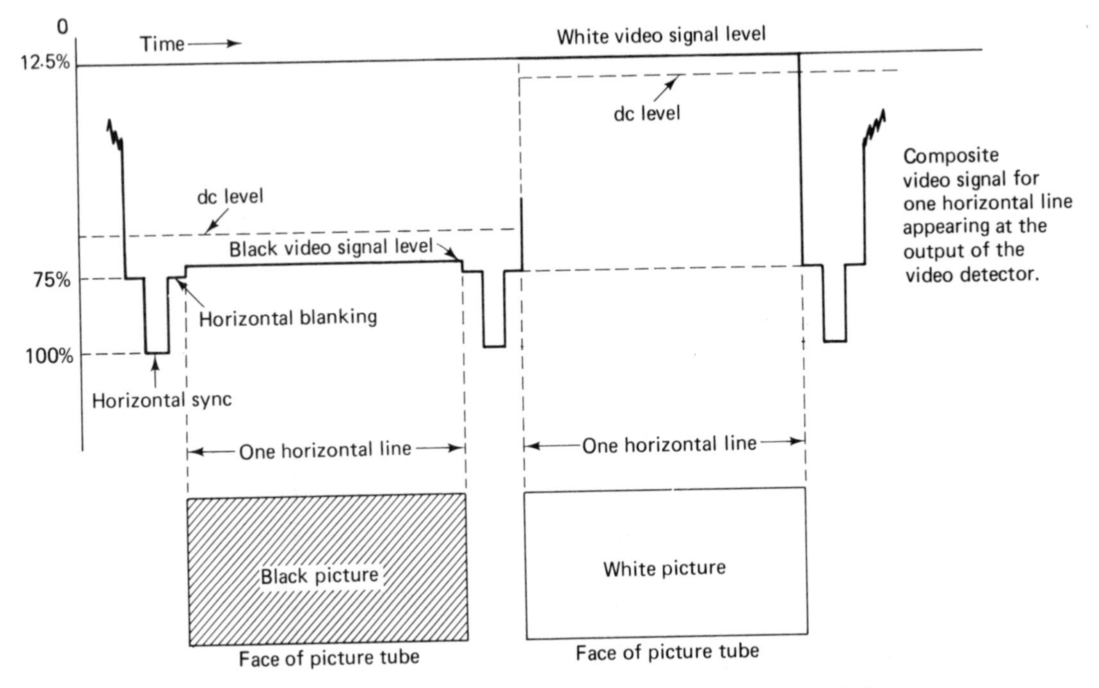

Figure 6-18 Change in video dc level with a change in picture content.

For all practical purposes, a diode may be considered a switch. When the voltage across it is positive on the anode and negative on the cathode, the diode is on and its resistance is almost zero. Hence, all of the input voltage appears across the load (R_5). On the positive half cycle of input voltage the diode is reverse biased making the diode an open circuit. Therefore, during this time the output voltage is zero. Over a great many cycles of input IF signal, the output video signal will consist of *only* the negative half-cycles. Hence, the output of the detector is pulsating dc. The ac component of this signal is the composite video signal. The magnitude of the dc level is determined by the input signal strength and the nature of the modulating signal. If the signal is weak, the output dc level will be small. At zero input the output will be zero. As the signal strength increases, the output dc level increases. This fact is important in automatic gain control (AGC) systems.

The dc component. The dc level is affected by the picture content. This is illustrated in Figure 6-18. If a completely *black picture* were to be transmitted, the video signal for each horizontal line would have almost the same level as the horizontal and vertical blanking pulses. Since these pulses represent 75% of the peak video signal, a black picture will produce a large dc output voltage.

If, however, a completely *white picture* is being transmitted, the video signal level will be almost zero for each horizontal line. Therefore, the dc level will decrease to almost zero. The dc level associated with the video signal corresponds to the average brightness of the picture and is important in black & white receivers for reasons to be discussed in the chapter on video amplifiers. In color receivers loss or change in the dc level will result in color changes that must be avoided if proper color reproduction is to be achieved. Therefore, many color receivers are designed to maintain the dc level through the video amplifiers.

In actual practice, depending upon input signal strength, the peak to peak video output to be expected from a video detector will vary from 1 to 4 V, and the dc components will be from 0.5 to 3 V.

The ac components that appear at the detector output consist of four signals: (1) the composite video signal, (2) the video IF signal, (3) the 4.5-MHz sound IF, and (4) harmonics and sum and difference frequencies of the video IF, sound IF, and video signal.

Polarity. The composite video signal is the desired output of the detector. One important aspect of the composite video signal is its polarity. The output of the video detector may be arranged to produce a composite video signal whose horizontal and vertical sync pulses go either in the negative direction or in the positive direction. The polarity output of the video detector depends upon the direction of the video detector diode. If the diode is connected into the circuit as shown in Figure 6-17, the sync pulse output will have a negative going polarity with respect to ground. Reversing the diode will reverse the direction of current flow. Hence, the video output signal will have positive-going sync pulses.

In both black & white and color receivers the video signal is amplified and then fed to the *cathode* of the picture tube. Since the blanking and sync levels of the video signal correspond to black, the picture tube must be cut off if it is to operate normally. This is accomplished by feeding the cathode of the picture tube with a video signal that has a positive going sync pulse of sufficient amplitude to cut off the picture tube (50 to 200 V p-p). This arrangement is used in preference to a negative-going video signal fed to the control grid as a means of avoiding white compression of the video signal. White compression is due to the video amplifiers nonlinear characteristics.

In general, both black & white and color receivers use video detectors that provide negative going sync pulses. This is primarily for two reasons: (1) Noise pulses will be negative going and tend to drive NPN transistor video amplifiers into cut off, thereby making them act as noise limiters. (2) The dc component at the output of the video detector can act as the bias for the video amplifier. However, bias for transistor video amplifiers is usually a combination derived from the B+ supply and the signal generated dc component developed by the video detector.

Frequency response. Another factor that affects the composite video signal is the frequency response of the video amplifier. The output of the detector will have desired signal components whose frequencies will extend from 0 to 4.5 MHz. However, because of the stray shunt capacity the high-frequency components will tend to be bypassed and lost. In a black & white receiver this would cause loss of detail and sharpness in the picture. In a color receiver the loss of high-frequency signals would result in weak or even complete loss of color. This undesired condition can be compensated for by decreasing the value of the load resistor and by using series or shunt peaking coils in conjunction with the load resistor. A typical solid-state receiver may use diode detector load resistors of 2 to 5.6 kΩ.

The black & white receiver detector. An example of a typical video detector used in a black & white receiver is shown in Figure 6-17. In this circuit T_1 and C_1 are part of the IF amplifier tuned circuit that helps to form the IF response curve. D_1 is the diode detector. R_5 is the diode load resistor.

In this detector the 4.5-MHz sound IF is a desired output and is passed on to the video amplifier. The sound IF takeoff is obtained from the collector of the video amplifier. In other receivers it may be taken directly from the detector. Also, appearing at the output of the detector are the high-frequency video IF (45.75 MHz), its harmonics, and various sum and difference frequencies which if left unattenuated can be radiated, picked up on the antenna lead in, and produce picture interference. This form of interference called *birdies* or *tweets* can be on one or more channels. Since, the second harmonic of the video IF falls in channel 6 (82 to 88 MHz), it is the most affected. Video detectors incorporate filters that attenuate these undesired signals. The video detector may also be shielded as a means of reducing radiated interference. In the typical circuit of Figure 6-17, C_5, C_6, L_1, L_2, and L_3 attenuate these undesired output components of the video detector.

Color receiver detector. The video detector used in a color receiver is more complex then that used in a black & white receiver. A typical video detector used in a color receiver is shown in Figure 6-19. As is indicated in the block diagram of Figure 6-2, color receivers locate the sound takeoff point in the last IF amplifier. This is done to minimize any 4.5-MHz beat that may occur at the output of the video detector.

The diode D_1 is fed by a double-tuned transformer T_1. Coil L_1 and capacitor C_4 form a filter that minimizes tweets. A bridged T trap tuned to 4.5 MHz is formed by C_6, T_2, and R_4. This trap prevents any residual 4.5-MHz beat that may be produced by the difference between the sound and picture IF from reaching the video amplifiers. The trap is insurance against the production of a 920-kHz beat that may be formed if any 4.5-MHz signal were to be mixed at the video amplifier with the chrominance signal (3.58 MHz). The video high-frequency response is maintained by the shunt and series peaking coils L_2 and L_3. The resistor R_6 shunts L_3 and helps to provide the required video amplifier response.

The video detector feeds a negative going video signal to the video amplifier. The video detector and the video amplifier are dc coupled. Since the transistor being fed is PNP, the dc component of the video signal developed by the video detector must be the proper polarity (negative) to provide proper transistor bias. However, since the input signal may greatly vary from channel to channel, the bias of the stage would also tend to shift from zero (transistor cut off) for no signal to a value of forward bias determined by signal strength.

Operating a transistor near or at cut off will result in distortion caused by clipping and low gain. To avoid this, a fixed bias is provided for the video amplifier by means of the 18-V B supply and the resistors R_2, R_3, R_4, and R_5. Capacitors C_1 and C_5 are bypass capacitors that tie the points to which they are connected to ac ground.

The composite video signal may be observed at the output of the video detector. In the circuit shown in Figure 6-19 test point *TP* may be used to gain access to the video detector output. When making this type of observation a low-capacity probe should be used with the oscilloscope. The observed waveforms should appear as shown in Figure 6-19(b). The wave shapes will differ, depending upon the sweep rate of the oscilloscope. If the sweep rate is 30 Hz, the observed pattern will consist of two complete fields totaling 525 horizontal lines. If the oscilloscope sweep rate is increased to 7875 Hz, one oscilloscope sweep across the face of the CRT will be exactly equal to the time duration of two horizontal lines. Therefore, the scope will display two horizontal sync and blanking pulses.

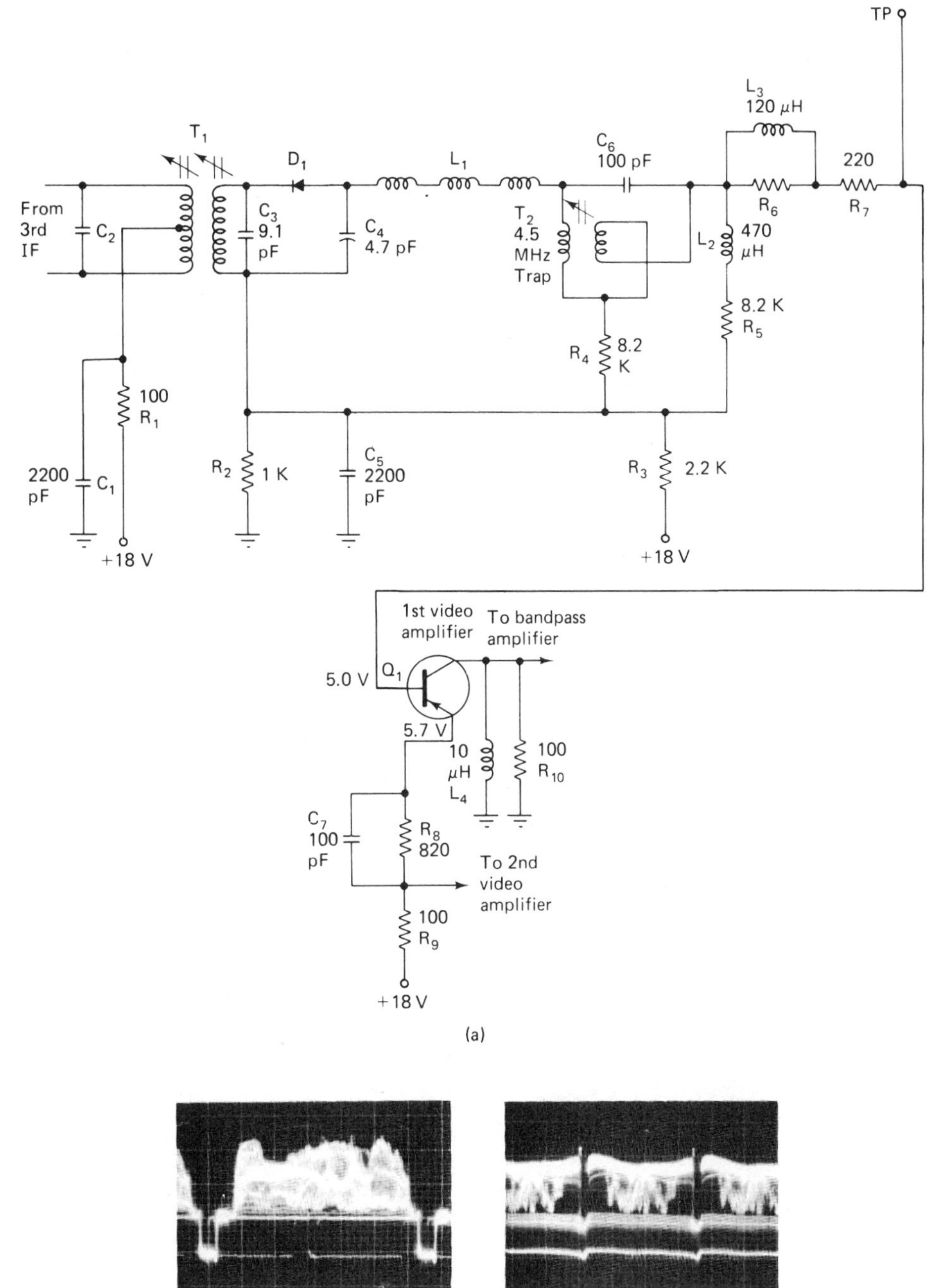

Figure 6-19 (a) Video detector used in a color receiver; (b) composite video signal at the output of the detector.

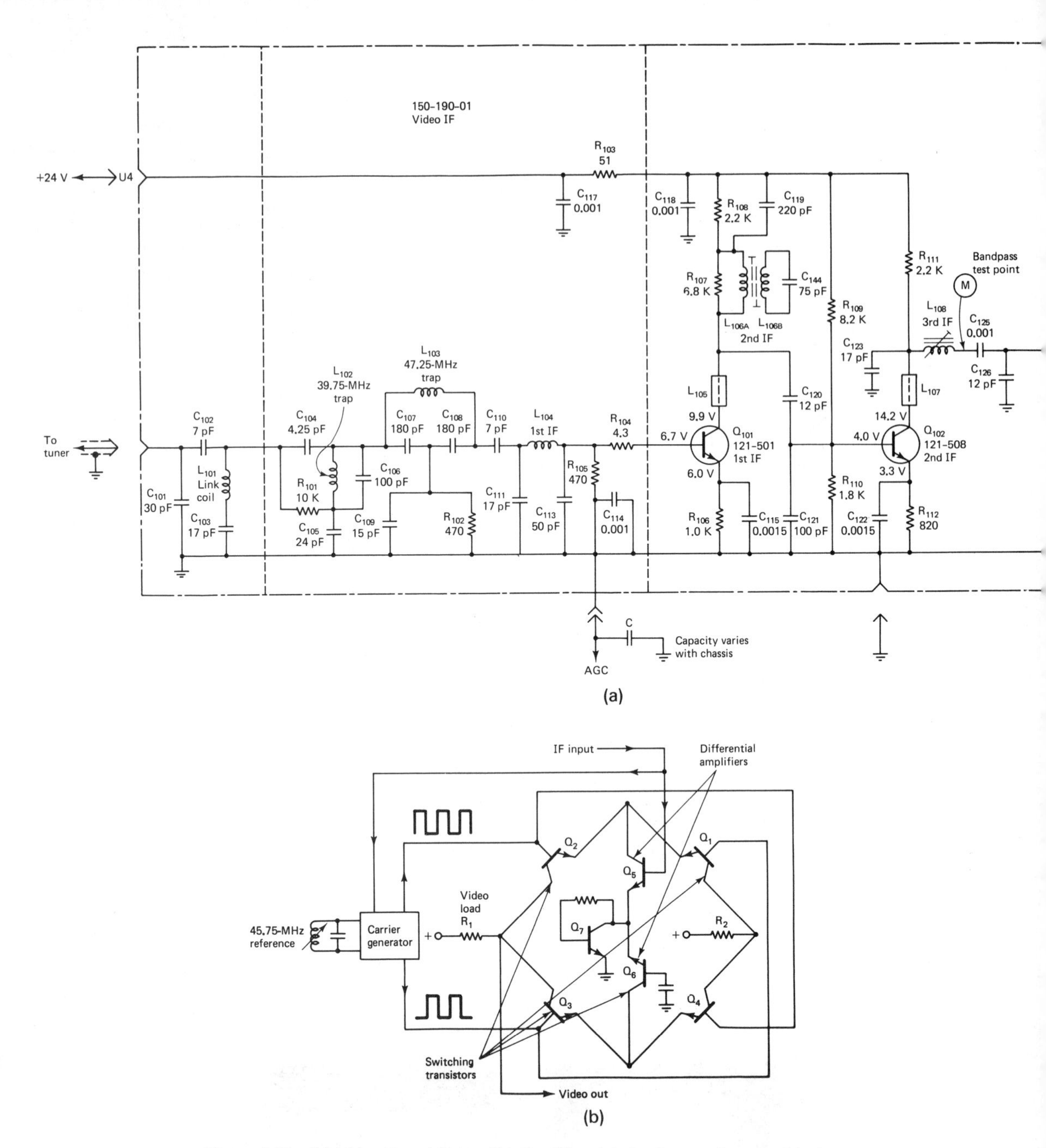

Figure 6-20 (a) Video IF module in which the video detector is a synchronous detector; (b) simplified diagram of a synchronous detector. [(a) Courtesy of Zenith; (b) Courtesy of Sencore.]

Synchronous detector. The germanium diode detector has been used as the video detector primarily because of its low cost. However, the diode (or envelope) detector introduced a number of problems, the primary one being that the diode detector is nonlinear. This means that different input signal levels and modulation levels will produce different outputs. Thus,

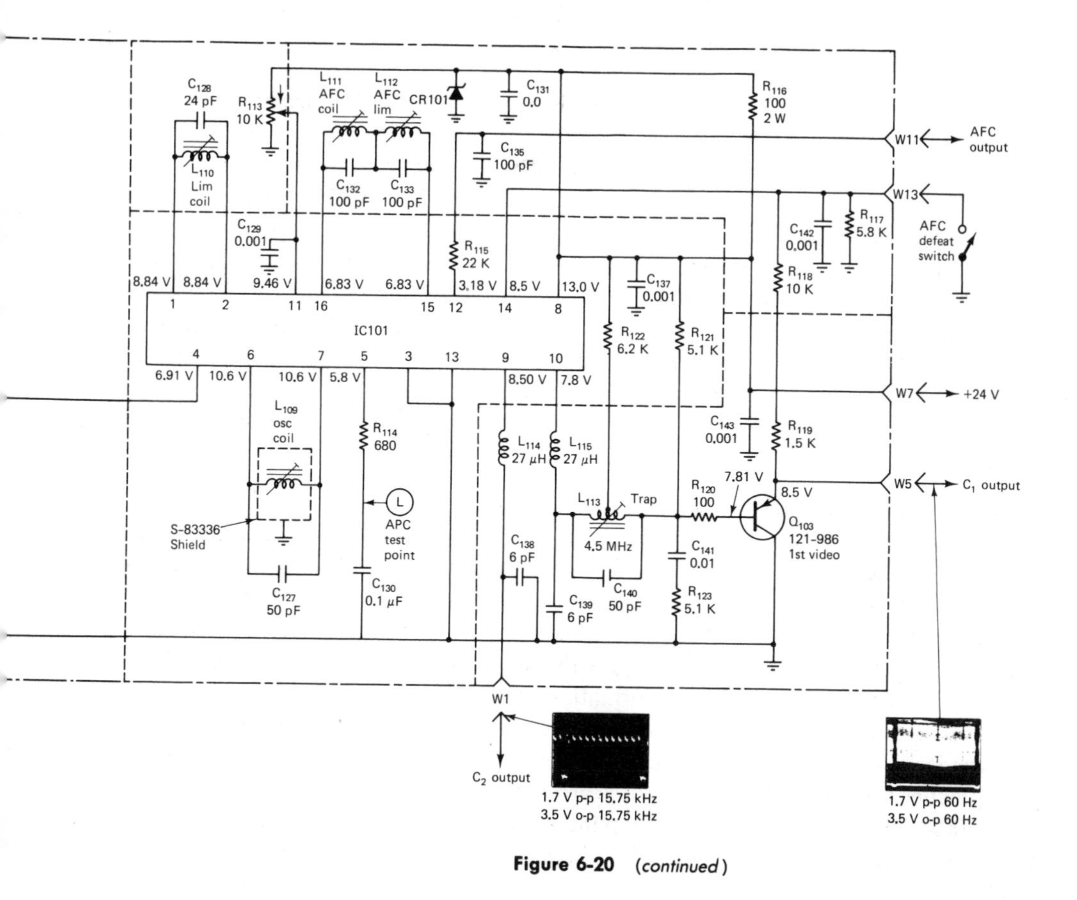

Figure 6-20 (*continued*)

if an equally spaced staircase waveform were to be transmitted, it might appear at the output of the detector as having unequal step spacings. In addition, the nonlinear detector characteristics result in sum and difference beat frequencies of the IF carriers and associated sideband components. These beat frequencies can be radiated back to the antenna and IF amplifiers and result in interference called "tweets." To minimize this problem, extensive filtering and shielding of the detector is resorted to. The diode detector will respond to any input signal and therefore will produce undesired output beats corresponding to adjacent channel sound and video IF signals as well as the desired video and sound signals. Finally, the diode detector has no gain and, in fact, introduces a signal loss.

Modern integrated circuit technology has allowed complex circuits to be constructed at low cost. One such development is the synchronous video detector. This type of video detector is of the same type that has been used for many years in the color demodulator section of the color receiver. The synchronous detector overcomes all of the defects of the germanium diode video detector. Since it provides gain, fewer IF and video amplifier stages are required. Improved linearity characteristics result in fewer traps required, less shielding and filtering, and much improved picture contrast and brightness quality. Fewer traps are required because the synchronous detector does not detect adjacent picture and sound carriers. This results in better reception on cable TV systems.

An example of a typical circuit using a synchronous detector is shown in Figure

6-20(a). A similar circuit, made by RCA, is shown in Figure 6-14. In this circuit the internal block diagram of the circuit is shown.

A major difference between diode detectors and synchronous detectors is that synchronous detection requires that an IF carrier reference signal of 45.75 MHz be generated against which the transmitted video IF signal is compared. In Figure 6-14 the internal circuit generator is identified by an adjustable coil labeled "video synchronous detector." In Figure 6-20(a) the adjustment is called the oscillator coil. This circuit also includes a zero carrier adjustment potentiometer R 113 (sometimes referred to as dc level or centering adjustment) and a limiter coil adjustment. The potentiometer is used to set the dc level of the signal from the detector to the video amplifier stages, thus determining the black level of the video. The limiter adjustment is set for the best balance between good linearity and minimum overshoot.

Synchronous detector operation. The synchronous detector can be thought of as a mixer (frequency changer) which has two inputs; the signal, and an oscillator operating at the same frequency as the signal. This is the reason it is called a synchronous detector. The net result of this arrangement is to generate sum and difference frequency outputs. The sum frequency is in the order of 90 MHz and is easily filtered out. The difference frequency for the carrier is zero. The sidebands associated with the modulated carrier produce difference frequencies equal to the original video information. Thus, the output of the synchronous detector is the original composite video signal. Signals that are not synchronized to the desired IF carrier will not produce a detected output.

In practice the synchronous detector is arranged in a bridge configuration as shown in Figure 6-20(b). This is done to reduce coupling of IF signal energy from the 45.75-MHz generator to the output of the detector. The bridge consists of four switching transistors (Q_1, Q_2, Q_3, and Q_4) and a differential amplifier (Q_5, Q_6, and Q_7). The carrier generator drives the bases of the switching transistors. Q_7 forces the differential amplifier circuit to behave as a constant-current device. This means that if the IF input to the base of Q_5 increases in the positive direction, the current through Q_5 would also increase and the current through Q_6 would decrease. On the negative half-cycle of the IF input signal the reverse would occur. However, the total current flowing through Q_7 would remain constant on both half-cycles.

A differential amplifier produces two outputs equal in amplitude but opposite in phase. In this circuit only the output associated with R_1 is used.

The switching transistors act like commutator switches. In response to the carrier generator the transistors act like on/off switches that automatically switch the load resistors from one collector of the differential amplifier to the other. The end result of this is to full-wave rectify the video IF carrier signal. This pulsating dc signal is then filtered and fed to the input of the video amplifier.

Alignment. Alignment of the synchronous detector may be done in at least three different methods. RCA (Figure 6-14) uses a marker generator and a digital voltmeter. The marker generator is set to 45.75 MHz and is applied to the input of the IF strip. A dc voltage is measured at the IF output amplifier (TP_{301}). L_{303} is then adjusted for minimum output voltage. This completes the alignment.

Alignment of a Zenith receiver using a synchronous detector such as shown in Figure 6-20(a) requires a sweep/marker generator, an oscilloscope, a DVM, an IF detector probe, and an AGC bias supply. The general procedure consists of first applying an AGC bias clamp to the first IF amplifier. Then the zero carrier control is adjusted to set the dc video amplifier conditions. The sweep/marker generator is injected into the input of the IF strip and the response of the first and second IF amplifiers is observed at the output of the second IF amplifier by means of a special demodulator probe. The traps and the IF tuned circuits are adjusted to the manufacturer's response curve specifications. The synchronous detector is adjusted by observing the output response curve and tuning the oscillator coil for a locked signal (no oscillation) over the right half of the bandpass. The limiter coil is adjusted for maximum gain at the 45.75-MHz marker, which should be at 30 to 75% of the peak response.

The third method of alignment makes use of a special signal generator, the VA48, manufactured by Sencore. See Figure 1-10. This generator produces a composite video stairstep multiburst signal as shown in Figure 6-21(a). The signal is fed into the UHF input or the VHF tuner of the receiver and is modified by the tuned circuits in the mixer, IF amplifier, and synchronous detector circuits to a shape that may look similar to Figure 6-21(b). Basically, alignment consists of simply adjusting the IF and detector tuned circuits while observing the waveform at the detector output so that (1) the stairstep portion of the waveform has equal spacings (good linearity), (2) the burst signals should all have approximately the same amplitude, and (3) when the synchronous detector limiter coil is adjusted, overshoots and undershoots of the low-frequency bursts should be minimized. See Figure 6-21(c).

When observed on the face of the receiver picture tube, this video signal produces a pattern as shown in Figure 6-21(d). After alignment has been completed, these bars will all appear to have equal brightness levels. IF alignment may also be made by simply observing the picture tube pattern while making the tuned circuit adjustments.

Traps are aligned by switching the VA48 to a crystal-controlled 1-kHz modulated signal at the trap frequency. This produces a horizontal zebra-like bar pattern of about 15 black and 15 white bars on the screen of the picture tube. Proper trap adjustment occurs when the zebra bar pattern is minimized.

6-12 troubleshooting the video IF amplifier and detector

Typical symptoms. Since the video IF amplifier and detector stages process both color and black & white signal information, defects in these sections of a color or black & white receiver will produce similar symptoms in either receiver. These symptoms are listed below. Pictures showing the visual appearance of some of these symptoms are shown in Figure 6-22.

1. No picture, no sound, raster O.K.
2. No picture, sound O.K., raster O.K.
3. Weak color and black & white picture, weak sound, raster O.K.
4. Overloaded or negative picture, buzz in sound, raster O.K.
5. Intercarrier buzz in sound
6. Poor vertical or horizontal synchronization
7. Tunable ringing or oscillation in the picture
8. Smeared picture
9. Sound bars in picture
10. Poor color fit

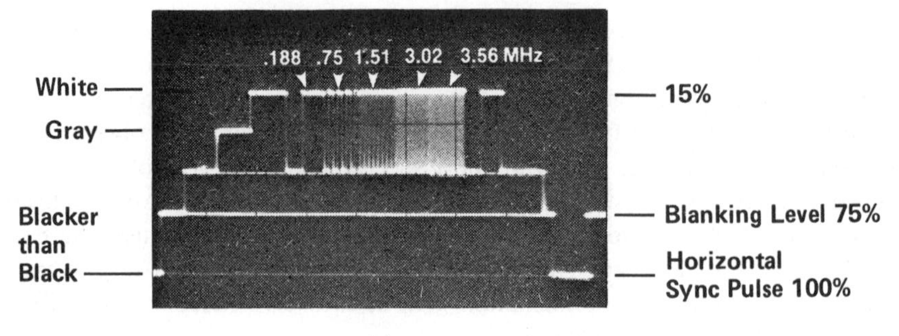

(a)

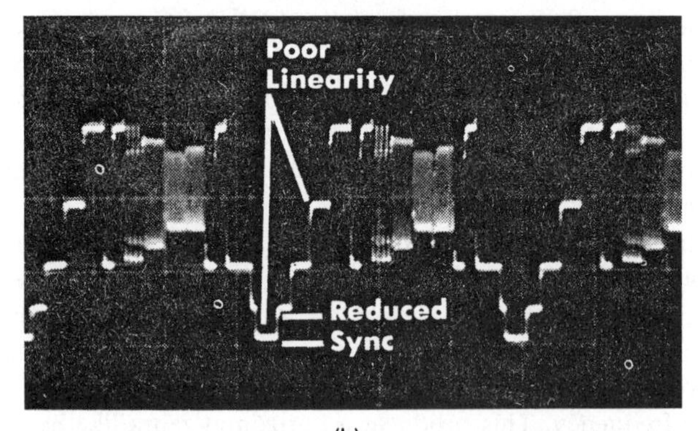

(b)

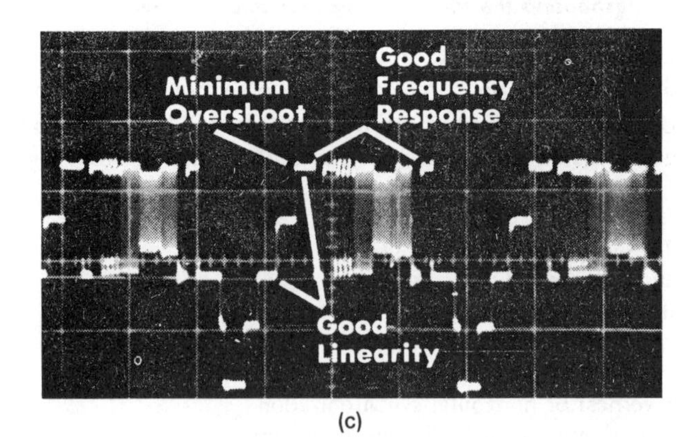

(c)

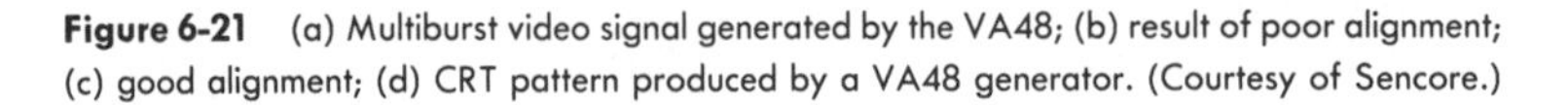

Figure 6-21 (a) Multiburst video signal generated by the VA48; (b) result of poor alignment; (c) good alignment; (d) CRT pattern produced by a VA48 generator. (Courtesy of Sencore.)

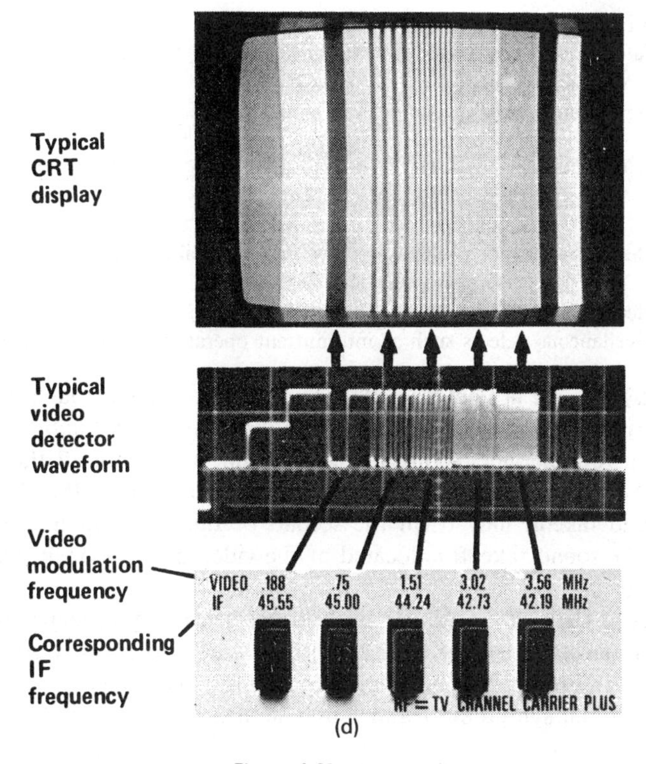

(d)

Figure 6-21 *(continued)*

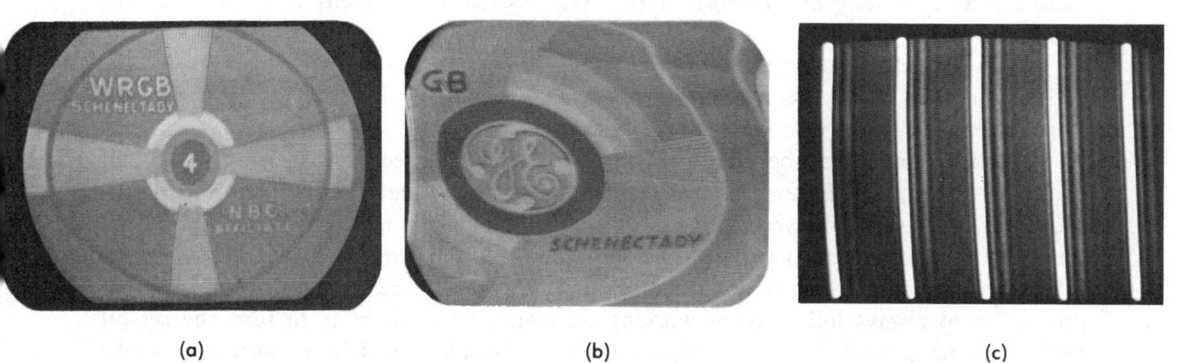

(a) (b) (c)

Figure 6-22 Trouble symptoms caused by defects in the IF amplifier section of a TV receiver: (a) weak picture; (b) negative picture (overloading); (c) ringing in picture. (Courtesy of General Electric.)

11. No color
12. Best color and best black & white have different fine-tuning positions
13. Hum in picture
14. Intermittent defects

All of the troubles listed above may be classified according to the three following types of *causes*:

1. Defects caused by an increase or decrease in total amplification of the video IF
2. Defects caused by changes in alignment
3. Miscellaneous defects such as intermittent operation or hum modulation

Defects caused by changes in total IF amplification. Symptoms 1 through 4 are examples of defects caused by changes in overall IF amplification. Symptoms 1 and 2 differ only in that in symptom 1 there is neither picture nor sound and in symptom 2 there is sound. This difference is caused by the position of the sound takeoff point in the IF strip. In a color receiver the sound takeoff may be in the second or third IF, but in almost all black & white receivers the sound takeoff is located at the video detector or in the input circuit of the video amplifier.

Because of stray coupling and the extreme sensitivity of some sound detectors it is possible for a defective receiver with symptoms 1 or 2 to produce no sound output in weak signal areas and to produce a sound output in a strong signal area. Hence, depending upon receiver location, symptoms 1 and 2 may actually be the same.

Bias clamp. Complete loss of picture and sound in a color receiver must be the result of an open in the signal flow path through the video tuner or IF amplifier. If the tuner has been found to be good, these symptoms generally mean that one or more stages in the video IF amplifier are dead. A dead stage, especially the first or second IF amplifiers, may be dead because it has been driven into extreme conditions of cutoff or saturation by a defective AGC system. Therefore, one of the first things that should be done, after such things as bad transistors, broken wires, and so on, have been eliminated as possible causes, is to substitute one or more *bias clamps* for the AGC system and thereby eliminate it as the possible cause of trouble.

A bias clamp is a variable power supply that is placed across the outputs of the AGC system. In effect, this procedure is the same as signal substitution, except that the "signal" voltage in this case is dc.

Bias clamps may be placed in transistor television receivers such as the circuit shown in Figure 6-23. Here again bias clamps are placed at the RF AGC bus and the IF AGC bus. In this case, however, the polarity of the bias clamp is positive to ground. Care must be taken to be sure that the bias clamp voltage is the same as that indicated in the schematic diagram. If this is not done, damage to the transistors may result. Another precaution to always follow when working on transistor receivers is to turn the set off before making jumper or bias supply connections. In addition, it is important to make ground connections first. Some of the transistors in these receivers are easily damaged by making circuit connection while the receiver is on. In this circuit reverse AGC is used and a decrease of AGC voltage will cause a drop in gain. It should also be noticed that a change of less than 1 V is required to shift the operating point of the IF amplifier from maximum

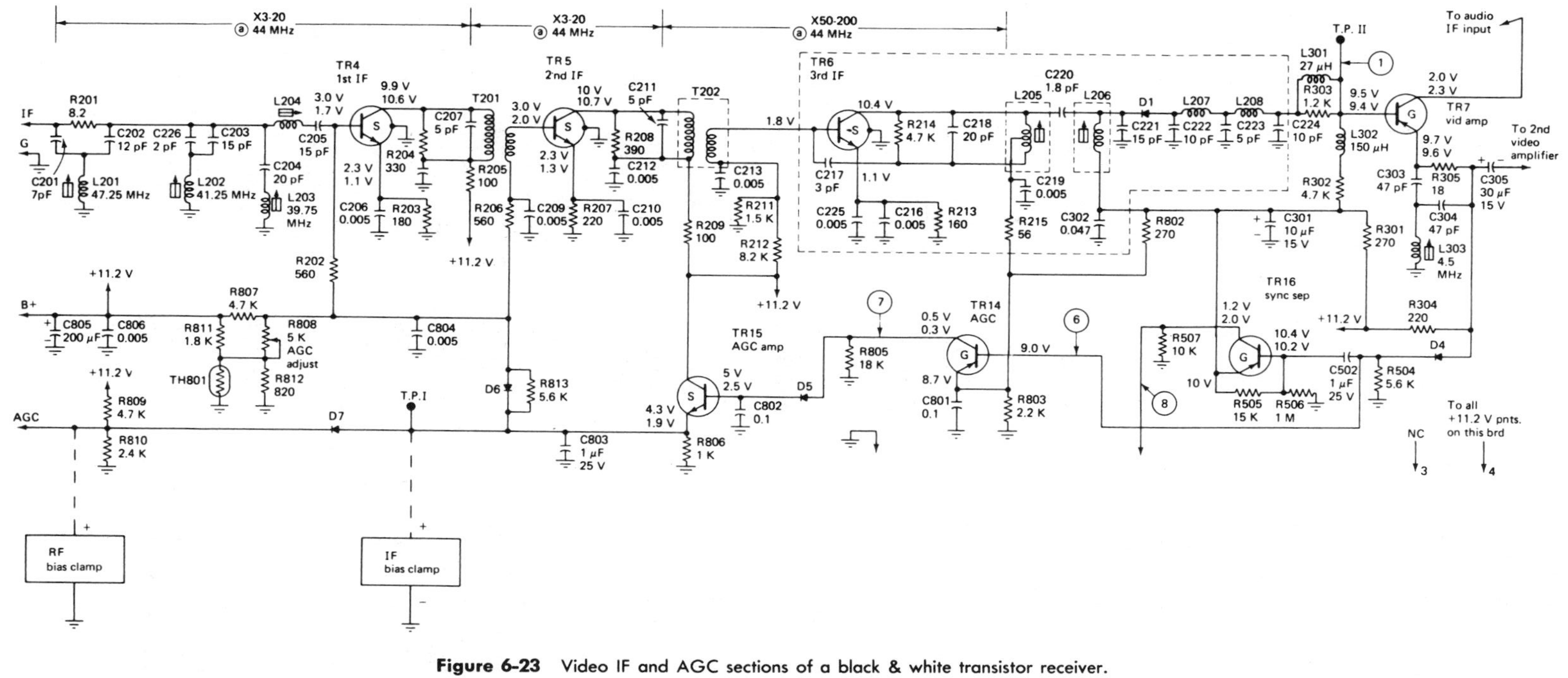

Figure 6-23 Video IF and AGC sections of a black & white transistor receiver.

gain (+3V) to lower values (2V). Thus, the bias clamp voltage should not be varied by more than is necessary to force the base voltages of the controlled IF stages back to their normal range. A return to normal receiver operation means that the AGC circuit is defective. If the receiver remains without picture or sound, the defect must be in the IF or video detector stages.

Once it has been decided that the trouble is in the video IF or detector stages, signal tracing or signal substitution can be used to isolate the defective stage.

Defective component isolation is the next step. This is done primarily with the voltmeter and ohmmeter.

Symptoms 3 and 4 (weak picture and weak sound, no snow or overloaded picture, and possible buzz in the sound) are those cases in which the gain of the video IF amplifier has decreased resulting in a weak picture or has increased resulting in a high-contrast overloaded picture. Under extreme conditions an overloaded picture may become negative, that is, blacks will become white, and vice versa. See Figure 6-19. Colors will also shift and become badly distorted or may be lost altogether. The picture will have the appearance of a photographic negative. This is usually the result of inoperative AGC that allows the final IF amplifier to be overdriven. An overdriven final IF amplifier will act as rectifier and will distort the 40-MHz IF signal. When the distorted (clipped) signal is passed on to the video detector, it is again rectified and the phase of the video signal driving the first video amplifier will be opposite to that normally produced. As a result, the video signal feeding the cathode of the picture tube will be reversed and a negative picture will be produced.

Defects caused by changes in alignment. Symptoms 5 through 12 are examples of defects caused by changes in video IF alignment.

Defective alignment of a color or black & white receiver may be recognized by adjusting the receiver's fine tuning. In a properly aligned receiver the quality of the picture will not change too drastically. In a receiver whose alignment is defective, changing the receiver's fine tuning will cause large changes in picture quality. The picture will smear, become weak, or overloaded. Ringing may appear and interference patterns may also appear indicating that the circuit is oscillating.

Changes in alignment can cause defects in the black & white picture, in the sound, and in the color portion of the picture. The specific defect depends upon which portion of the IF response curve has altered. Thus, changes in the relative amplitude of the picture carrier, and sound carrier, may give rise to intercarrier buzz. Changes in the relative amplitudes of the picture carrier and the color carrier may result in weak color, poor color fit, and fine-tuning problems where best color and best black & white do not correspond. Such changes in relative amplitude may be caused by misadjustment of one or more traps or misalignment of the video IF strip.

Misalignment that causes changes in IF response shape can be responsible for smeared pictures, loss of color, loss of horizontal or vertical sync, and ringing. Both smeared pictures and loss of sync indicate changes in IF response curve shape that have resulted in poor low-frequency response. This may be seen during alignment as a narrow bandpass response. This response may also be responsible for loss of color since those portions of the composite signal necessary for color reproduction will not be present at the output of the video IF amplifier.

Ringing is an indication that the IF amplifiers are regenerating at some frequency in the bandpass of the IF amplifier and are causing excessively high gain at these frequencies. In this case, the IF response curve would appear to have one or more large peaks at the frequency of regeneration. In extreme cases of regeneration the IF amplifiers will oscillate. When observing this kind of IF response curve during alignment it will become inverted as the IF amplifiers begin to oscillate.

Intercarrier buzz (symptom 5) is often due to defects in the sound IF amplifier or detector stages. It is possible, however, that defects in the video IF and detector stages may cause changes in the relative picture and sound IF amplitudes that may also cause it.

Common causes of intercarrier buzz are:

1. Defective transistors or ICs.
2. Overloaded signal.
3. Misadjusted traps.
4. Improper alignment of the video IF stages. Here again a bias clamp on the AGC bus will eliminate the possibility of overload due to a defective AGC system. The other possibilities (3 and 4) require an overall alignment check to determine whether this is actually where the fault lies.

Poor vertical and/or horizontal sync (symptom 6) is usually due to defects in the sync separator or video amplifier stages of the receiver. However, it may also be caused by such things as:

1. Overloaded IF amplifiers, defective AGC.
2. Hum modulation of the signal. B supply ripple level and lead dress should be checked.
3. Misalignment of the video IF, causing poor low frequency response. Misalignment can be due to defective components, or the result of touch-up tuning by someone.
4. Interference pickup due to improper shielding of the IF strip.

Ringing (symptom 7) is a symptom of excessive IF regenerative feedback. See Figure 6-22(c). This defect appears in the picture as two or more ghosts whose number, intensity, and relative position will vary with fine tuning. For example, if a crosshatch pattern obtained from a color bar generator is used to obtain a fixed pattern on the screen, a normal receiver's fine tuning control should be adjusted so that the vertical and horizontal lines are of equal intensity. This indicates that the low-frequency video signal content, as represented by the *horizontal lines,* is being amplified by the same amount as the high video frequencies, represented by the *vertical lines.* For these conditions the picture carrier must be positioned at its proper 50% point. In a receiver whose IF response is defective, it may not be possible to adjust the fine tuning to obtain a crosshatch pattern whose vertical and horizontal lines are of equal intensity. If regeneration and ringing are present the vertical lines in the crosshatch pattern will be followed (right side) by a number of repeating lines that will be affected by the fine tuning. This is because the fine tuning shifts the position of the picture carrier on the IF response curve. If the response curve has large peaks and valleys due to misalignment, large changes in picture content will take place as the picture carrier is moved through them.

Regeneration may not be apparent on strong signals. This is because the television receiver's AGC system tends to reduce the system's gain, and minimizes the tendency for the circuit to regenerate. If the input signal strength is reduced, the IF amplifier gain will increase and any regenerative tendencies will show themselves as ringing in the picture.

Isolating the cause of the regeneration requires that the feedback loop be identified and then broken. As a first step, a bias clamp on the AGC line will prevent the AGC circuit from confusing the situation. Then a VTVM is placed across the video detector load resistor. If the circuit is oscillating, the VTVM will read several volts without any input signal. Moving the hand near or even touching the input of each stage will cause large changes in picture content for regenerative IF amplifiers. It also will cause large changes in VTVM readings if the circuit is oscillating. Those stages that react strongly to the hand test are part of the feedback loop and should receive more attention.

The most common causes of regeneration and oscillation are:

1. Defective transistor or IC
2. Open decoupling and bypass capacitors
3. Open or changed value neutralizing capacitors
4. Open damping resistors
5. Open or shorted AGC filter capacitors
6. Defective tuned circuits
7. Shields that have been removed or disconnected
8. Poor ground connections
9. Improper lead dress or component placement
10. Misalignment of the video IF

A *smeared picture* (symptom 8) represents the visual effects of poor frequency response and phase shift. This is usually associated with loss of picture detail. The most common cause of this trouble is a defect in the video amplifier or video detector stages. However, receiver alignment also may be the source of this symptom. Before attempting receiver alignment a complete voltage and resistance check of the circuit should be made.

Sound bars (symptom 9). The shape of the IF response curve and the amplitude of the picture and sound IF signals are critical to proper intercarrier receiver operation. Normally the sound IF is placed at a level of about 5% of maximum on the response curve. If the sound IF level is increased, due to misalignment, slope detection will convert the FM sound into AM whose amplitude varies at an audio rate. This in turn will modulate the IF and will appear at the output of the video detector as a low frequency audio signal. After video amplification the audio signal will appear on the TV screen as black & white sound bars that vary in position and shading as the sound changes.

Poor color fit (symptom 10), *no color* (symptom 11) and *best color and best black & white have different fine tuning positions* (symptom 12), may all be symptoms of the same defect: misalignment of the video IF amplifier. Poor color fit also may be due to defects or misalignment of the chroma bandpass amplifier. This and many other circuits also may be the cause of no color symptoms. If symptom number (11) occurs on all channels, the defect is definitely at the receiver and is not a transmitter defect. This type of trouble is associated with a distorted or regenerative IF response curve. Therefore, it is best checked by a visual sweep check of the alignment.

Hum in the color or black & white picture (symptom 13) may be caused by hum modulation originating in the RF or IF portions of the receiver. Hum getting into the video amplifier stages will be visible as one black and one white horizontal hum bar with or without a picture. In either case if the hum bars are caused by 60 Hz the picture would consist of two or three alternating black and white horizontal bars. Two horizontal black and two horizontal white bars would indicate that the 120-Hz B supply ripple is reaching the cathode of the picture tube.

Hum bars caused by *hum modulation* in the IF stages will only appear on the screen in the presence of a television picture. This is because the 60- or 120-Hz hum voltage is only able to pass through the IF filters if it is able to become a part of the signal passing through the IF amplifiers. This can only occur if one of the stages is operated nonlinearly or is overdriven.

From a servicing point of view, if the hum bars are at either a 60-Hz rate or a 120-Hz rate and are only present when a station is tuned in, the probable trouble is poor power supply filtering in that part of the power supply that feeds the IF amplifiers.

Intermittent defects are defects that come and go, often without any apparent reason. Sometimes temporary or even permanent repairs of these defects can be made by shaking or jarring the receiver. These results are a sure indication of a loose connection, or a cold solder joint. Troubleshooting these defects is made more difficult because when the receiver is working normally all voltages and resistances are also normal.

6-13 vacuum tube video IF amplifier

A typical video IF amplifier using stagger tuning is shown in Figure 6-24. In this three-stage vacuum tube IF amplifier each of the four tuned circuits used to obtain the required bandpass is tuned to a different frequency. Thus, T_{301}, the first IF grid transformer, is tuned to 44.0 MHz; T_{302}, the first IF plate transformer, is tuned to the picture IF 45.75 MHz; T_{303}, the second IF transformer, is tuned to 42.5 MHz; and T_{304}, the third IF transformer, is tuned to 43.8 MHz. In this case both T_{301} and T_{303} are loaded by resistors R_{302} and R_{319}, respectively, which are placed across the tuned circuits as a means for broadening the bandpass of the system. T_{301} also serves the dual purpose of a tuned IF transformer and part of a complex trap formed in conjunction with L_{301} and C_{302}.

An interesting feature of this circuit is the dc current path of the first and second IF amplifiers. These two stages are connected in series, because current will flow from the cathode to the plate of V_{301} and then to the cathode and plate of V_{302} on its way to the B+ supply. This circuit is sometimes referred to as a *stacked* or *totem pole* arrangement. Notice that the supply voltage is 280 V and that the two tubes divide the voltage so that 154 V appear at the cathode of V_{302}. Hence, in order to provide proper bias for this tube, the grid voltage is made 148 V.

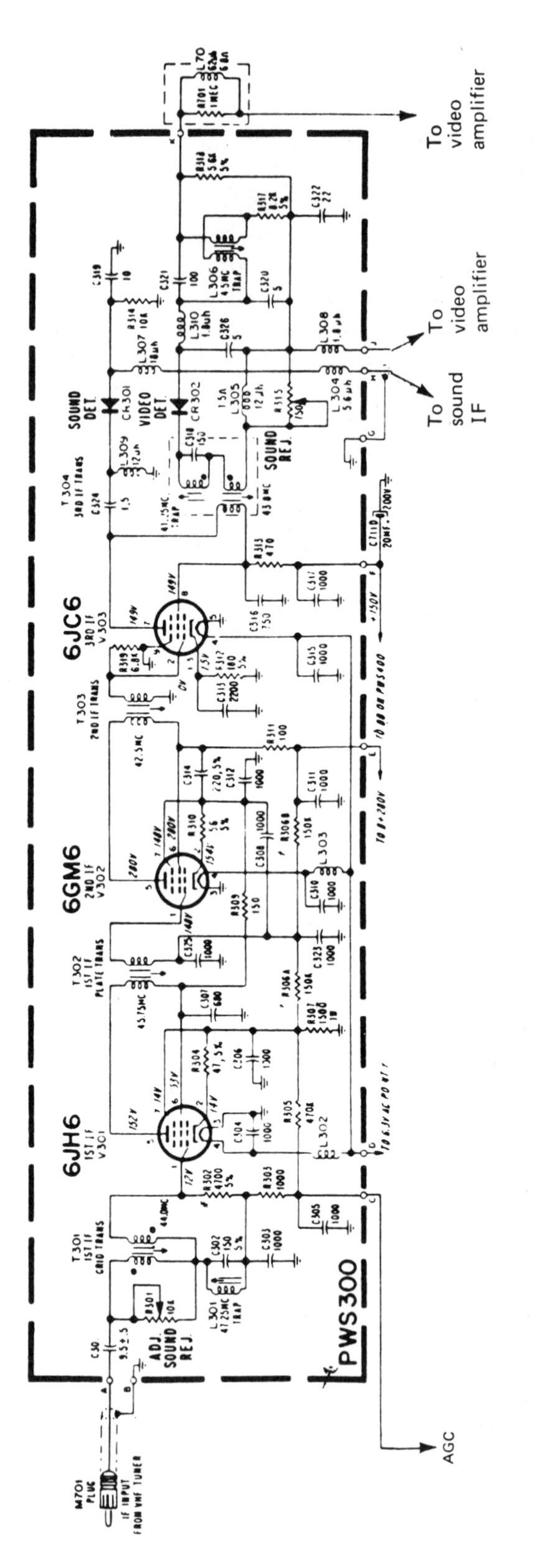

Figure 6-24 Stagger tuned video IF amplifier. (Courtesy of RCA, Consumer Electronics.)

QUESTIONS

1. In what way does the block diagram of the video IF amplifier for a black & white receiver differ from that of a color receiver?
2. What is the maximum number of video IF stages in modern TV receivers? What is the minimum number?
3. What are the four functions of the video IF strip?
4. Under what conditions can the video IF be directly observed on an oscilloscope?
5. Why does the output phase of the video detector usually consist of negative going sync pulses?
6. What are the advantages of the intercarrier sound IF system? What are its disadvantages?
7. Why is the video sound IF amplitude made much smaller than the picture IF?
8. Name three things that might cause intercarrier buzz.
9. Why must vestigial sideband transmission be compensated for, and how is it done?
10. What causes sound bars in the picture?
11. How does the video IF response for black & white receivers differ from color receivers?
12. What is a bifilar transformer and why is it used?
13. What are the three frequencies that IF traps are tuned to?
14. Explain the operation of four kinds of IF traps.
15. How does stagger tuning increase the bandpass of an amplifier?
16. Why are matching pads required for sweep alignment of tuner and video IF amplifiers?
17. Where and why are bias clamps used during sweep alignment?
18. What is the function of a video detector and what output signals does it produce?
19. What picture information does the dc component of the video detector output correspond to and why is it important in color TV?
20. What would be the effect on receiver operation if the video detector were reversed? Explain.
21. What is a SAW filter made of, and where is it used?
22. What is the cause and cure of 920-kHz beat interference?
23. What are four disadvantages of the diode video detector?
24. What are the important advantages of the synchronous detector?
25. Explain the operation of the synchronous detector.
26. Describe three methods for aligning the synchronous detector.
27. What are the causes of changes in total IF amplification?
28. What trouble symptoms might be produced by changes in IF alignment?

chapter 7

Automatic Gain Control and Noise-Canceling Circuits

7-1 introduction

The block diagram of a complete color television receiver is shown in Figure 7-1. The automatic gain control and noise-canceling stages discussed in this chapter are shaded. The automatic gain control (AGC) stage has two inputs: a horizontal retrace pulse and the composite video signal. Its output is a dc control voltage that feeds the RF and IF amplifiers.

The television receiver must be able to present a clear picture over a rather large range of input signal levels. Usable signal strengths at the receiver input may vary from as little as 50 μV to as much as 100,000 μV and more, depending upon the channel being received and the distance between the receiver and the transmitter. This is especially noticeable in areas where some of the channels being received have transmitters that are nearby and some that are located many miles away. If a receiver had no automatic gain control, channels with moderately strong signals would provide pictures that were clear, noise-free, and of proper contrast. Switching the receiver to a weak channel would produce a weak picture of low contrast.

Airplane flutter. Because of atmospheric conditions, the time of day, or the season of the year, television signals may erratically fade and become stronger or weaker. Television signals usually become stronger at night and in the winter.

Slow changes in signal level are not as noticeable as are the rapid changes in signal strength that may occur when aircraft pass nearby. To explain the rapid changes in signal strength consider that at television frequencies the transmitter's radiated signal normally reaches the receiver's antenna by two paths: the direct wave and the ground reflected wave. This is shown in Figure 7-2. When an airplane passes between the transmitter and the receiver, a third component of the radiated signal, the sky wave, now can reach the receiver's

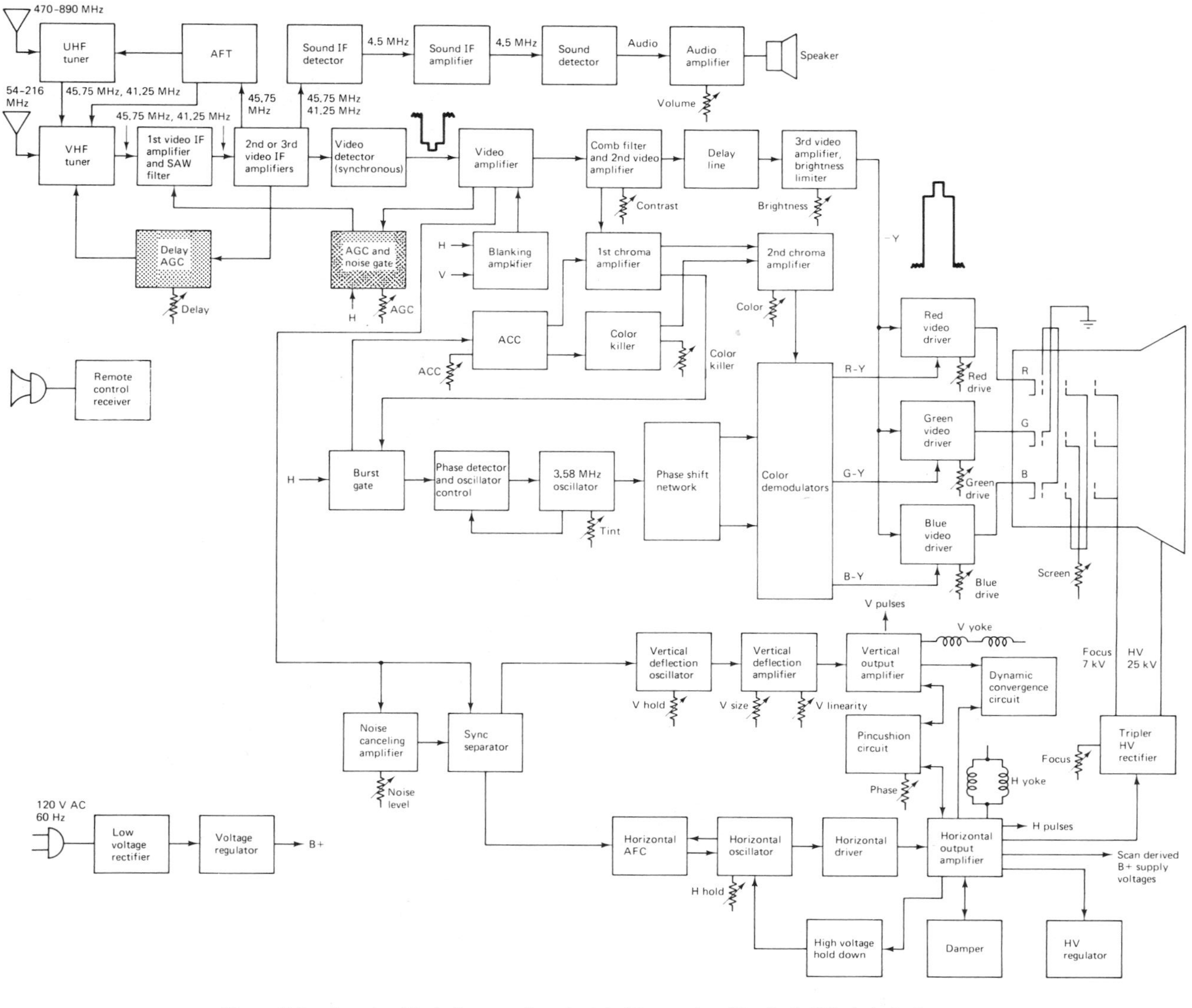

Figure 7-1 Complete block diagram of a color television receiver. The shaded blocks indicate the circuits discussed in this chapter.

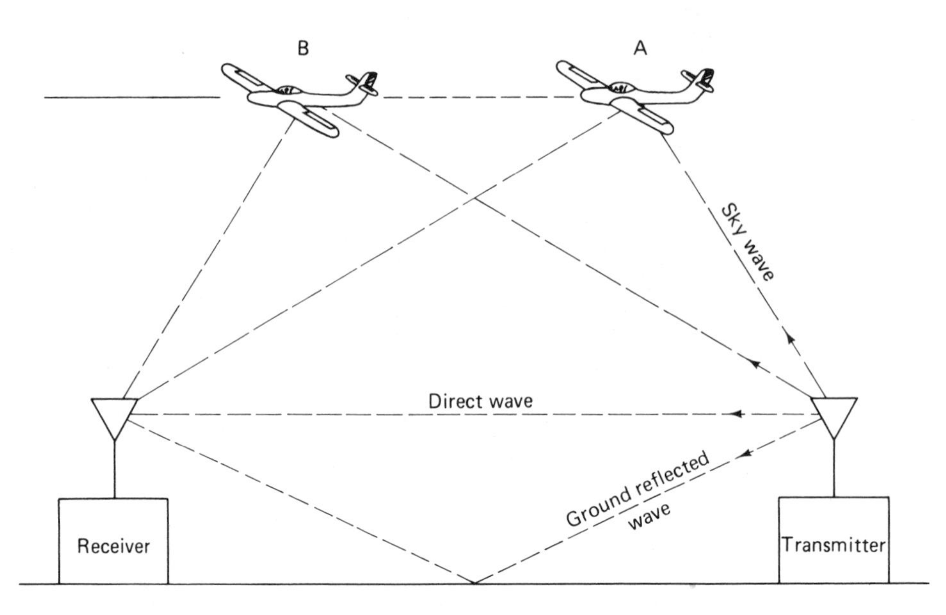

Figure 7-2 Electromagnetic propagation paths between the transmitter and receiver when an airplane flies overhead.

antenna by reflection from the airplane. The vector sum of the signals arriving at the receiving antenna determines signal strength and is related to the length of the path that the signals take. Because of the movement of the airplane, the length of the sky reflected wave is constantly changing. Therefore, the signal strength at the receiver antenna will also rapidly change. Such signal variations may last as much as 15 or 30 seconds. This gives rise to a picture that appears to flutter, that is, its contrast changes as rapidly as 20 or 30 times per second. In metropolitan areas where there is heavy airplane traffic, airplane flutter is a serious problem.

To the observer changes in picture contrast are an annoyance whether they are fast or slow changes. Such contrast changes require that the television receiver operator readjust the receiver contrast control* each time he or she changes channels or whenever signal levels change. Thus, if the signal were to get stronger, the contrast control would have to be adjusted for less contrast in order to restore the receiver to its original contrast level. If the signal were to become weaker, the contrast control would be adjusted for more contrast. Automatic gain control (AGC) may be considered an automatic contrast control since these circuits perform this function automatically.

7-2 the ideal AGC system

If a graph is drawn showing the output voltage of the video detector versus the input RF signal to the television receiver for a receiver that does not have any AGC system, it will appear as shown in Figure 7-3. In this case, an increase in input signal causes a corresponding increase in output voltage. Of course, this cannot continue indefinitely since

*The contrast control is located in the video amplifier and controls the peak-to-peak signal that is delivered to the picture tube. It may be compared to the volume control in the audio portion of the receiver.

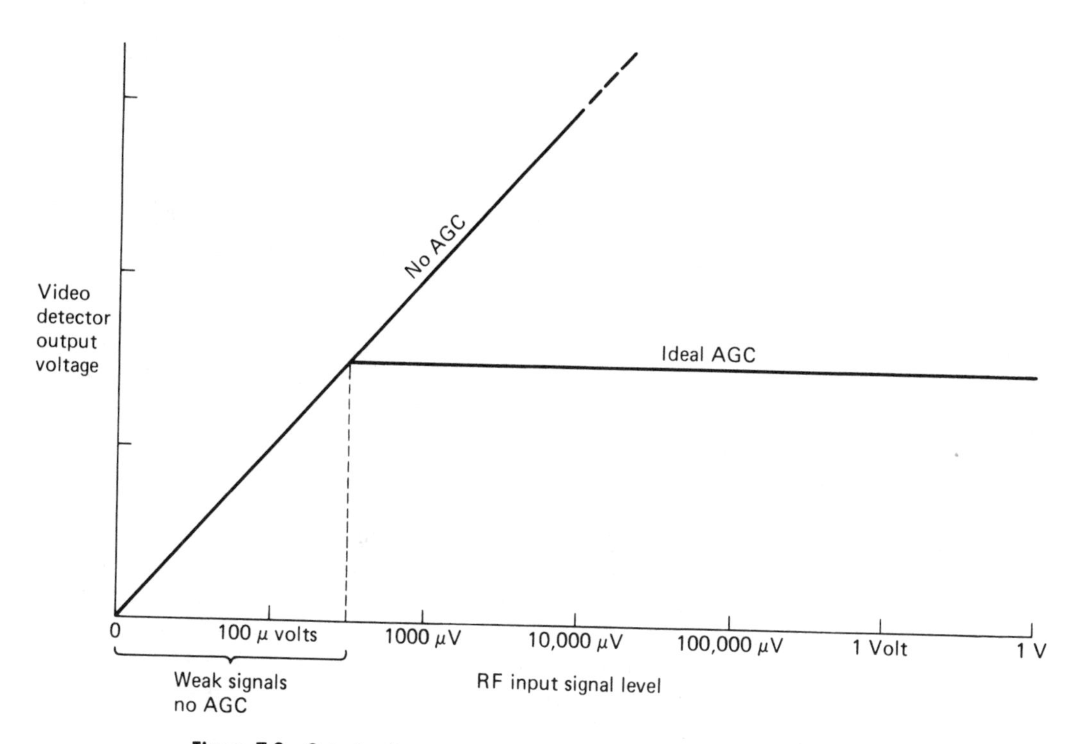

Figure 7-3 Output voltage versus input signal level for a receiver with no AGC and for an "ideal" AGC system.

at higher input signal levels nonlinearity will cause clipping in the RF and IF stages and would limit the output obtainable. In addition, nonlinear operation (cutoff and saturation) of the RF amplifier stage will cause cross-modulation. This is the condition in which the modulation of an interfering signal is transferred to the desired signal. Cross-modulation as seen on the picture tube can result in a form of windshield wiper interference. Cross-modulation can be minimized by preventing the RF and the first IF amplifiers from overloading. This is one of the functions of an AGC system.

The characteristic of an *ideal* AGC system is also shown in Figure 7-3. In this case, the output voltage of the video detector is held constant over a large range of input voltages. The output voltage referred to is the peak-to-peak output voltage of the composite video signal. Put another way, the ideal AGC system keeps the sync pulse peaks at a constant level regardless of changes in the input signal. For weaker signals, the ideal AGC system has exactly the same characteristics as the receiver without any AGC at all because all AGC systems reduce the gain of the receiver in the presence of a signal, no matter how weak it may be.

Delayed AGC. A reduction in RF amplifier gain means a smaller RF signal feeding the mixer. The result is greater noise output. Therefore, in an attempt to improve the signal to noise characteristics of the receiver, the ideal AGC system should *delay* the application of automatic gain control to the RF amplifier until the signal becomes stronger. In this way maximum amplification is available for weak signals when it is needed. Because of the delay characteristics built into the system, the receiver will have the best signal-to-noise ratio possible. When the signal becomes stronger, AGC will then prevent the RF amplifier from overloading, thereby preventing cross-modulation.

7-3 gain control methods

Since most of the television receiver's amplification is provided in the video IF amplifiers, effective automatic gain control of the receiver must use these stages. The RF amplifier must also have AGC applied to it if cross-modulation is to be minimized.

Assume that a television signal has been received and that the contrast control of the receiver has been adjusted for a pleasing picture. If the signal strength at the antenna were to decrease suddenly, the action of the automatic gain control system would be to increase the gain of the receiver so that the picture contrast would remain unchanged. Picture contrast will also remain constant for increases in signal strength since in this case the AGC will decrease the overall amplification of the receiver.

FET amplifiers. The gain of a FET amplifier may be determined from the equation $A_V = g_m Z_L$, where g_m is the transconductance of the device and Z_L is the impedance of the load. Z_L is determined by the components used in the tuned circuit. The load impedance does not lend itself to simple manipulation, and therefore cannot be used for AGC control.

The other factor that determines amplifier gain is the g_m or transconductance of a FET. The g_m of FETs may be controlled by varying their bias. Hence, all AGC systems vary the bias of the RF and IF stages to control their gain.

In a FET, the g_m is smallest near cutoff and it increases as the bias decreases toward zero. The g_m also decreases if the operation of the FET is brought close to saturation. In short, as the operating point of a FET is made to approach a region of nonlinearity (cutoff or saturation), the g_m and the gain of the amplifier will decrease. In FET AGC systems the region of operation that is used for AGC applications is the region near cutoff. Hence, in FET AGC systems the bias of the IF and RF stages is shifted toward or away from cutoff, depending upon signal strength. For example, if the signal strength at the antenna of a receiver suddenly increased, tending to increase picture contrast, the AGC system will act to counteract this by increasing its negative dc output voltage. This in turn is fed to the gates of the controlled FETs causing the bias of the IF and RF amplifiers to increase. The transconductance decreases and causes the amplification of these amplifiers to decrease. The decrease in amplification returns the picture contrast to the condition that existed before the signal strength increased.

Transistor amplifiers. The power gain of a transistor amplifier may be determined by the equation

$$G = \beta^2 \frac{R_L}{R_{in}}$$

where β is beta (hfe) the current gain of the transistor, R_L is the effective load fed by the transistor, and R_{in} is the input resistance of the transistor. Here again the most convenient "handle" for varying the overall amplification of the receiver is to vary one of the parameters of the transistor instead of the load or input resistance of the amplifier.

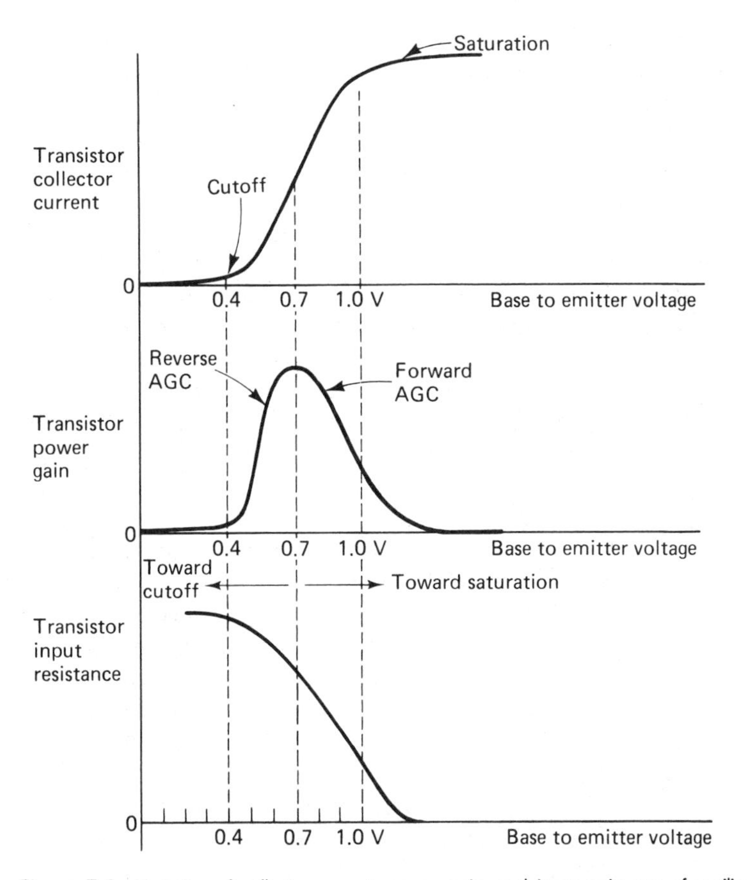

Figure 7-4 Variation of collector current, power gain, and input resistance of a silicon transistor amplifier as its base-to-emitter voltage is varied.

As was the case with transconductance in FETs, the magnitude of beta depends upon the operating point of the transistor which is established by the base to emitter forward bias. Shifting the operating point toward collector current cutoff (zero bias) or collector current saturation causes a decrease in beta with a corresponding decrease in power gain.

Figure 7-4 summarizes the effects that a change in base-to-emitter voltage of an NPN silicon transistor would have on the transistor collector current, the amplifier's power gain, and the input resistance of the transistor. A number of conclusions may be drawn from these curves. First, the amount of base to emitter voltage change that is necessary to shift the silicon transistor's operating point from cutoff to saturation is small, only 0.4 or 0.5 V. If the transistor is germanium only 0.2 or 0.3 V is needed. Second, at some optimum value of forward bias (0.7 V in this diagram) the power gain of the amplifier is maximum and then decreases as the bias is either increased, shifting the operating point toward saturation, or decreased, moving it toward cutoff.

Forward and reverse AGC. AGC systems that control receiver amplification by shifting the operating point toward saturation are referred to as using *forward AGC.* AGC systems that shift the operating point toward cutoff are referred to as using *reverse AGC.* Television receivers may exclusively use either forward AGC or reverse AGC. In some receivers, combinations of forward and reverse AGC may be used simultaneously for different parts of the RF and IF amplifier chain. Receivers that use either reverse or forward AGC do not operate the stages at peak gain; they place the no-signal operating point on the slope of the power gain curve so that they may increase or decrease the stage gain without moving to the other side of the power gain peak.

The power gain curve in Figure 7-4 is not symmetrical, that is, the reverse AGC region of the curve falls off much more rapidly than does the forward AGC region. This means that reverse AGC will require smaller voltage changes to make the amplifier go from maximum gain to minimum than will forward AGC. Forward AGC is often used for controlling video IF amplifiers because it is more linear in its control action. It can also be seen in Figure 7-4 that the input resistance of a transistor decreases with an increase in forward bias. This change in input resistance causes a power mismatch between the tuned IF transformers and the transistor, thereby providing an additional control on power gain. In summary, the power gain of a transistor is determined by the beta and the degree of mismatch between the signal source and the transistor. By changing the transistor bias, both of these "handles" can be used to provide automatic gain control.

If a receiver uses a reverse AGC system in conjunction with NPN transistors used in the RF and IF stages, the AGC system would have to decrease the transistors' bias toward zero in order to decrease the amplifier's gain for strong signals. The reverse would be true for weakened signals. This means that the AGC system should deliver a negative going voltage to the base of the transistor being controlled that is directly proportional to the signal strength. If the RF and IF transistors are PNP types, the AGC control voltage should be positive going.

Figure 7-5 shows a single-stage NPN transistor IF amplifier that is controlled by reverse AGC. The arrows indicate the changes that take place when the RF signal increases. The arrows are reversed when the signal input decreases.

Figure 7-6 shows another single-stage tuned IF amplifier that uses an NPN transistor. In this case, however, forward AGC is used. This means that for increased signal strength, the base-to-emitter forward bias must increase, shifting the operating point of the transistor toward saturation. A decrease in signal strength would require a decrease in forward bias. To achieve this, the AGC system must deliver a positive-going voltage to the base of the amplifier. If a PNP transistor is used, the forward AGC system would develop a negative-going voltage proportional to signal strength. Amplifiers that use forward AGC often use a large resistor (R_4) in series with the collector. This resistor causes the effective collector supply voltage of the transistor to shift whenever a bias change takes place, thereby allowing the transistor to approach saturation more easily.

The arrows at the bottom of Figure 7-6 indicate the changes in the circuit conditions necessary to keep the output constant for an increase in RF signal.

7-4 AGC methods: simple AGC

The simplest AGC system makes use of the signal derived dc voltage generated at the output of the video detector or some other AGC diode. This voltage is proportional to the input signal and may be positive or negative in polarity. This makes it suitable for reverse AGC application. An example of this system is shown in Figure 7-7. The circuit

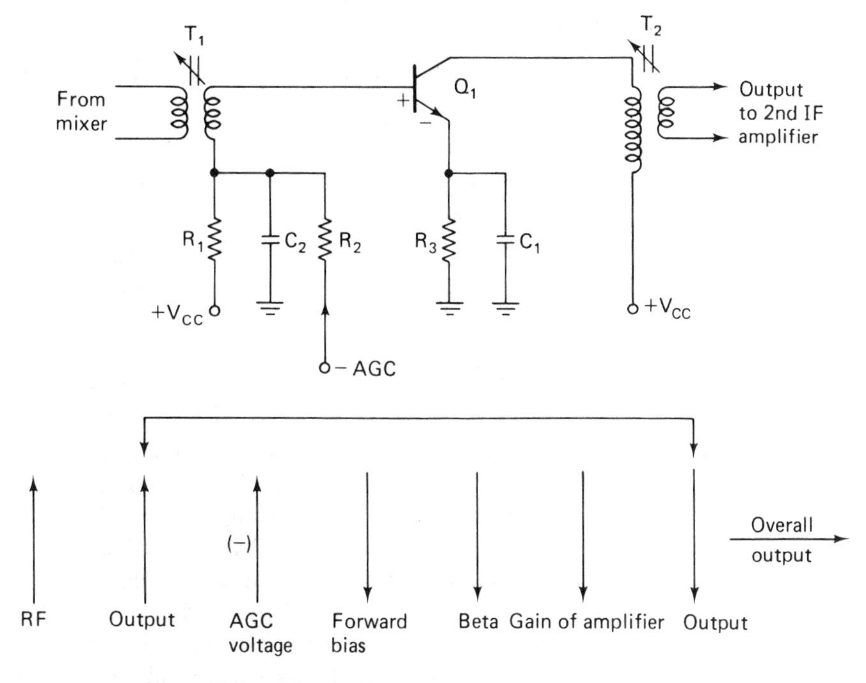

Figure 7-5 Tuned amplifier that uses reverse AGC.

Figure 7-6 Tuned amplifier that uses forward AGC.

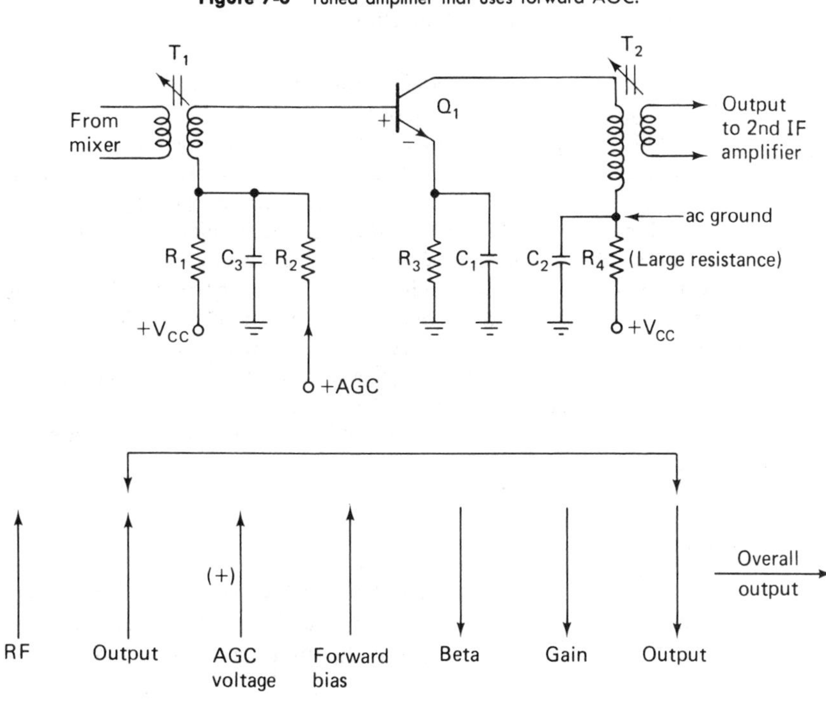

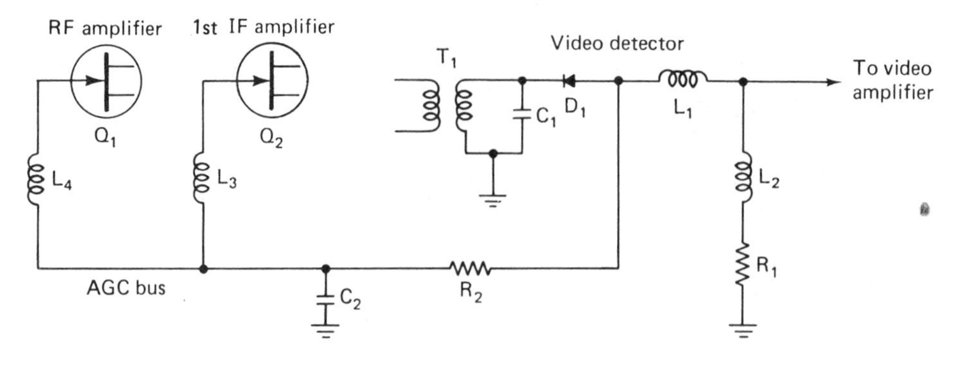

Figure 7-7 Simple AGC circuit.

consists of three basic parts: the RF and IF stages to be controlled, the video detector that develops the dc AGC voltage, and an AGC filter (R_2, C_2) that removes the video ac components from the desired dc that feeds the controlled stages.

This filter is in fact a frequency-sensitive voltage divider. At high frequencies the reactance of C_2 is very small. Therefore, most of the ac input to the filter is dropped across R_2. As the frequency gets lower, the reactance of C_2 gets larger and more ac voltage appears across C_2, and less appears across R_2. If, however, the filter is designed correctly, the ac voltage developed across C_2 at the lowest video frequencies should not be more than approximately 0.1Vp-p.

If the filter became defective because C_2 opened, some of the IF signal appearing in the detector output would be returned to the input and would cause the circuit to oscillate.

Simple AGC has a number of limitations:

1. Weak signals develop AGC voltages that tend to reduce RF amplifier gain and cause poor signal to noise characteristics.
2. The level of dc voltage appearing at the output of the video detector is small and is related to picture content: therefore, the range of amplifier control is also small and may vary as light or dark picture scenes are portrayed. Amplified AGC such as shown in Figure 6-23 may be used to minimize this problem.
3. Noise pulses that accompany the desired signal increase the dc output of the video detector. This increase in AGC voltage cuts down the gain of the receiver, resulting in still poorer signal to noise characteristics. This is called *noise setup.*
4. The video output of the video detector may have frequencies as low as 30 Hz. To effectively remove these low frequencies from the AGC output, long-time constant filters must be used. This makes simple AGC slow-acting. Therefore, it cannot follow and eliminate the results of the rapid signal changes characteristic of airplane flutter.

7-5 keyed AGC

All of the limitations of simple AGC may be overcome by the use of keyed AGC. The block diagram of Figure 7-8 shows the input and output signals and voltages of a keyed AGC system. In the basic system one of the inputs is a video signal with a positive

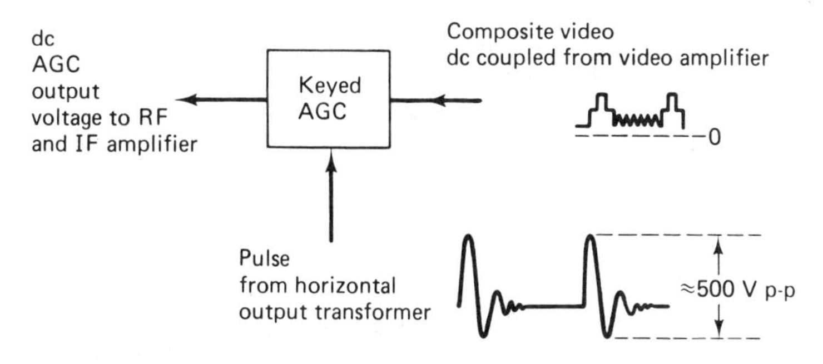

Figure 7-8 Input and output signals and voltages in a keyed AGC system.

going sync pulse that is taken from the output of a dc-coupled video amplifier. The other input is obtained from the horizontal output transformers and is a large constant amplitude positive going pulse that is generated during horizontal retrace. Thus, this pulse and the horizontal sync pulse occur at the same time. The keyer circuit is arranged to be at cutoff between horizontal sync pulse intervals and to conduct when both pulses are present. The amount of conduction is controlled by the amplitude of the horizontal sync pulse tips which in turn determines the output dc voltage used for AGC. To ensure that the sync pulse level is not affected by picture content, the dc component of the video signal is preserved by using dc coupling between the video detector and the keyer.

Noise-canceling keyed AGC systems. The need for minimizing noise setup in AGC systems has given rise to numerous systems of noise cancellation. The block diagrams of two systems that have been used in color and black & white receivers are shown in Figure 7-9. In Figure 7-9(a) a diode is used as a switch that opens in the presence of noise preventing it from reaching the keyed AGC circuit. The block diagram of the noise-canceling circuit shown in Figure 7-9(b) eliminates noise by using a separate noise gate that separates the noise from the composite video signal, amplifies it, and then adds it to the inverted composite video signal at the output of the video amplifier. Since the noise pulses are equal and opposite, they cancel. This system is frequently used in solid-state receivers.

The diode noise gate. Figure 7-10 shows a simplified schematic diagram of a diode noise gate. The crystal diode D_1 and its associated components (T_1, L_1, L_2, and R_1) constitute a video detector, the output of which is a composite video signal whose sync pulses are negative going. The diode load resistors and the input transformer T_1 return to a potentiometer (R_2) that is tied to B+. This connection allows the diode detector to function normally. It lifts the output voltage from its normal negative dc output to a positive value that places the horizontal sync pulses slightly above zero. Under these conditions, the diode noise gate is conducting and may be considered a short circuit. Therefore, the output voltage across R_3 will be exactly the same as the input. Noise pulses that have amplitudes larger than the sync pulses will make the anode of D_2 negative. The diode will stop conducting and the noise pulse will not appear across R_3 and it will not appear at the input to the video amplifier. Consequently, keyed AGC and sync circuits will have noise-free operation.

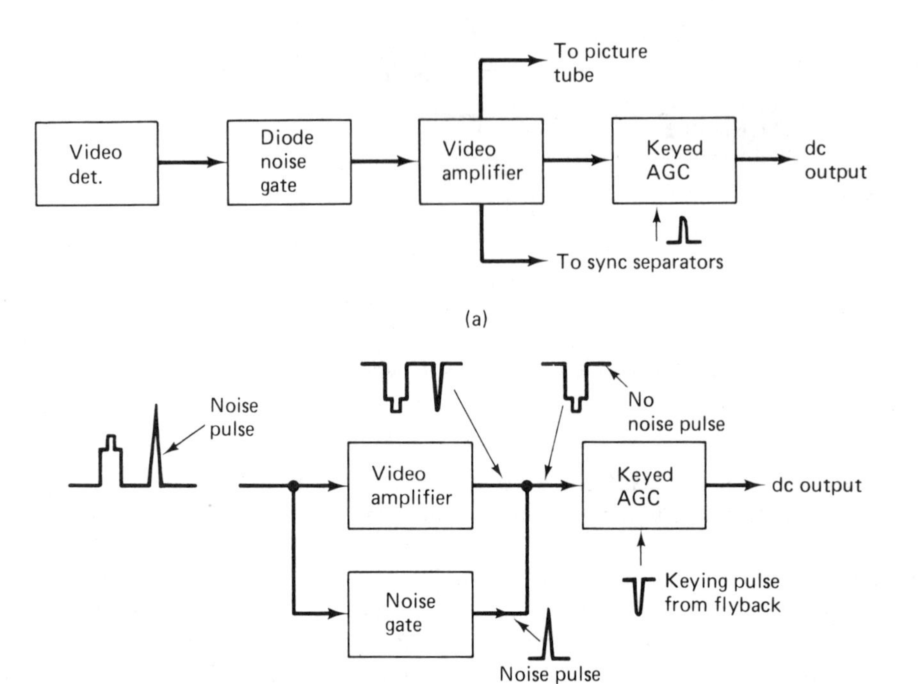

Figure 7-9 Two systems of noise cancellation used with keyed AGC.

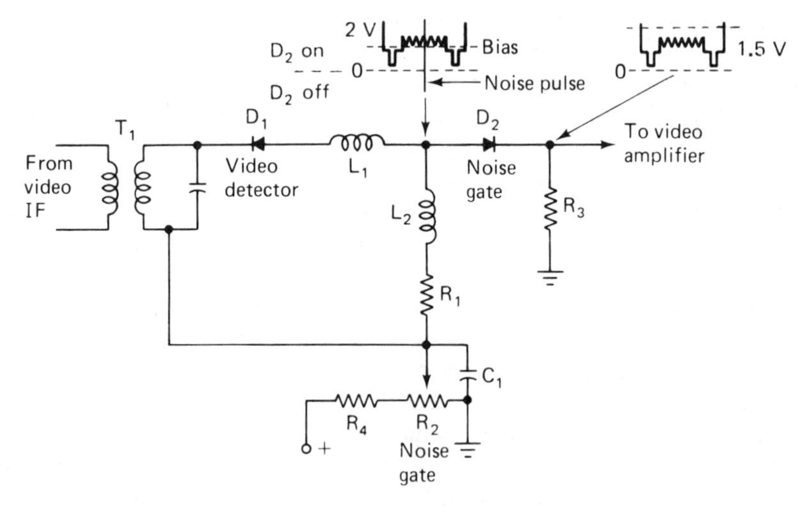

Figure 7-10 Diode noise gate.

The amplified noise-canceling circuit. An example of an amplified noise-canceling circuit is shown in Figure 7-11. Q_1 is a straightforward keyed AGC circuit. The video signal obtained from the video amplifier is a positive-going sync pulse and is fed to the base of the keyer

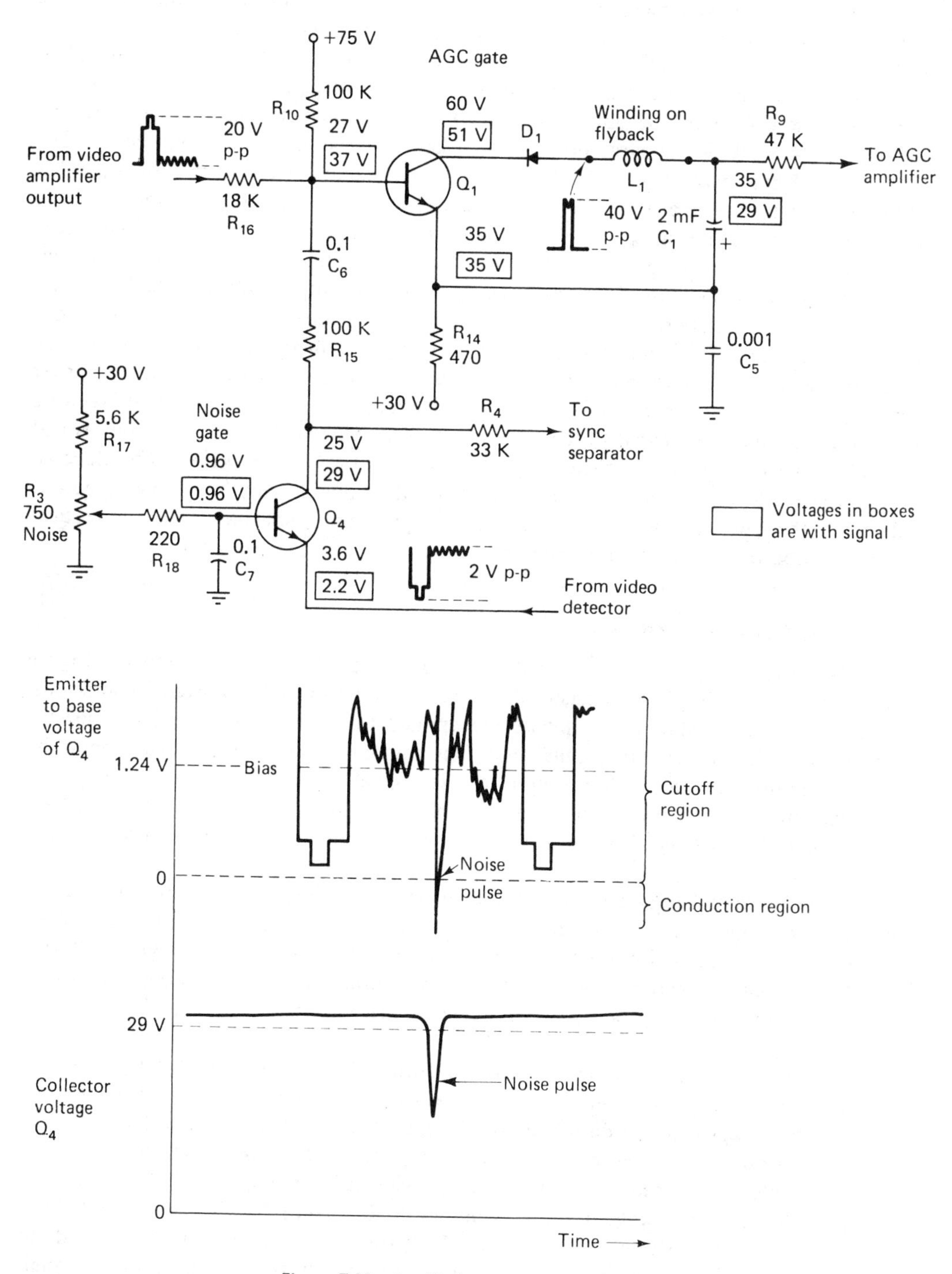

Figure 7-11 Amplified noise-canceling circuit.

transistor via R_{16}. Also feeding the base is the output of Q_4, the noise amplifier. This transistor is arranged as a common-base amplifier. Its negative-going video input signal feeds the emitter and is obtained from the video detector output. The base is returned to ac ground by means of the bypass capacitor C_7. The dc voltage on the base with signal is 0.96 V and the voltage on the emitter is 2.2 V. This means that the bias between base and emitter is 1.24 V. With this large bias, the transistor (Q_2) is operated beyond cutoff. The exact value of bias may be adjusted by the noise gate control (R_3). The noise control is adjusted so that all normal signals are in the cutoff region of the transistor and therefore do not appear at the output of Q_2. This is indicated in the ladder diagram drawn below the schematic diagram. However, noise pulses that exceed the sync pulse amplitude will drive the noise amplifier into conduction and produce an output voltage. The input and output noise pulses are both negative going because it is the characteristic of the common-base amplifier that it does not invert the signal being amplified. The noise pulse is then added to the positive-going video signal that has been coupled to the base of the keyer from the output of the video amplifier. If the noise amplifier's gain is such that its noise pulse output amplitude is equal to that in the composite video signal, the noise pulses will cancel. The noise-free composite video signal is then fed to the AGC keyer to produce an AGC voltage independent of noise. This circuit also does double duty by having its output delivered to the sync separator stages via R_4, thereby providing stable horizontal and vertical synchronization of the deflection oscillators.

7-6 transistor keyed AGC

An example of a commonly used transistor keyed AGC circuit is shown in Figure 7-12. In this circuit the PNP transistor Q_1 is biased to cutoff under no-signal conditions. Negative going horizontal retrace pulses are applied to the collector circuit. Transistor conduction can only occur during pulse time. The amount of conduction depends upon the negative sync tip amplitude of the composite video signal fed to the base of the transistor. If the transistor were an NPN, both pulses would be positive going. The horizontal keying pulse is in the order of 25 V p-p and the composite video signal is in the order of 2 V p-p.

The horizontal retrace pulse is obtained from a winding on the horizontal output (flyback) transformer and is fed to the collector of the keyer transistor via diode D_1. When the transistor is turned on by a horizontal sync pulse, current will flow through R_1, R_2, Q_1, D_1, and L_1 charging C_1 and complete the circuit by returning to R_1. The amount of current flow, and therefore the charge on C_1, is determined by the sync tip amplitude of the composite video signal.

Diode D_1 prevents the discharge of C_1 during the time between keying pulses. If D_1 were not used, the charge on C_1 would forward bias the collector-to-base junction of Q_1 and allow the capacitor to discharge, resulting in the loss of the AGC voltage.

In many receivers the dc operating conditions of the AGC keyer transistor are made adjustable so that the circuit is able to provide AGC voltages without causing IF overloading for a greater range of input signal levels. In Figure 7-12, R_2 is the AGC level control. As the control is moved toward the +10-V source, the keyer transistor is made to conduct more heavily, generating a greater AGC output voltage for a given input signal. Moving the control toward ground shifts the transistor's (Q_1) operating point further into

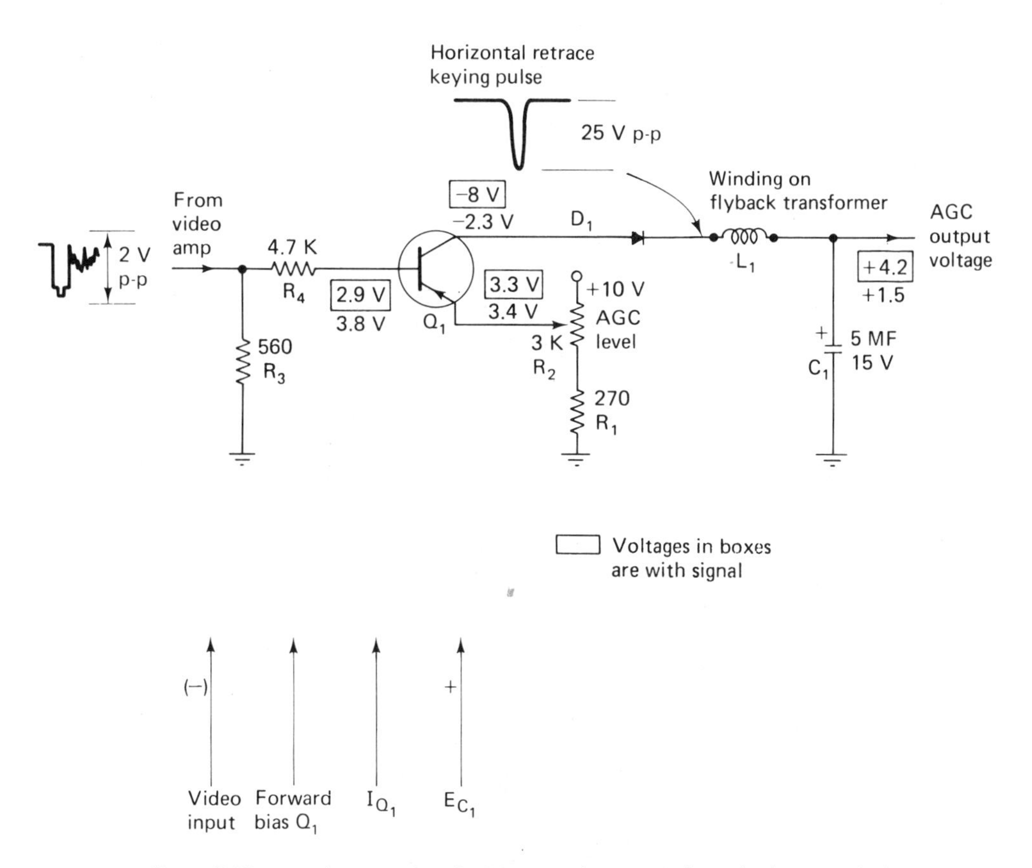

Figure 7-12 Typical transistor keyed AGC circuit. The arrows indicate the changes in circuit operation that occur as a result of an increase in signal level.

cutoff and makes it more difficult for the transistor to conduct. This in turn means that the AGC output voltage will be reduced. This control is adjusted for minimum picture overload when the receiver is tuned to a strong channel.

In the same diagram arrows are used to indicate the changes that would take place in the circuit for an increase in signal strength. Reading from left to right, an increase in the negative going video signal will cause an increase in Q_1 forward bias which in turn will cause an increase in collector current. This results in an increased positive voltage across C_1. Thus, this keyed AGC circuit produces an output dc voltage that is suitable for use with a forward AGC IF NPN amplifier system. If a reverse AGC system using NPN IF amplifiers is to be driven, the keyer transistor must be changed to an NPN, the diode must be reversed, and the polarity of the pulses must be inverted.

Transistor AGC systems. Transistor keyed AGC systems may be very complex. In some circuits five or more transistors may be used to develop the required AGC voltages. Transistors are used as amplifiers, inverters, and for delay purposes. Some receivers use only forward AGC; other receivers make use of both forward and reverse AGC in different parts of the RF IF chain. There is no standard approach used by all manufacturers. Three

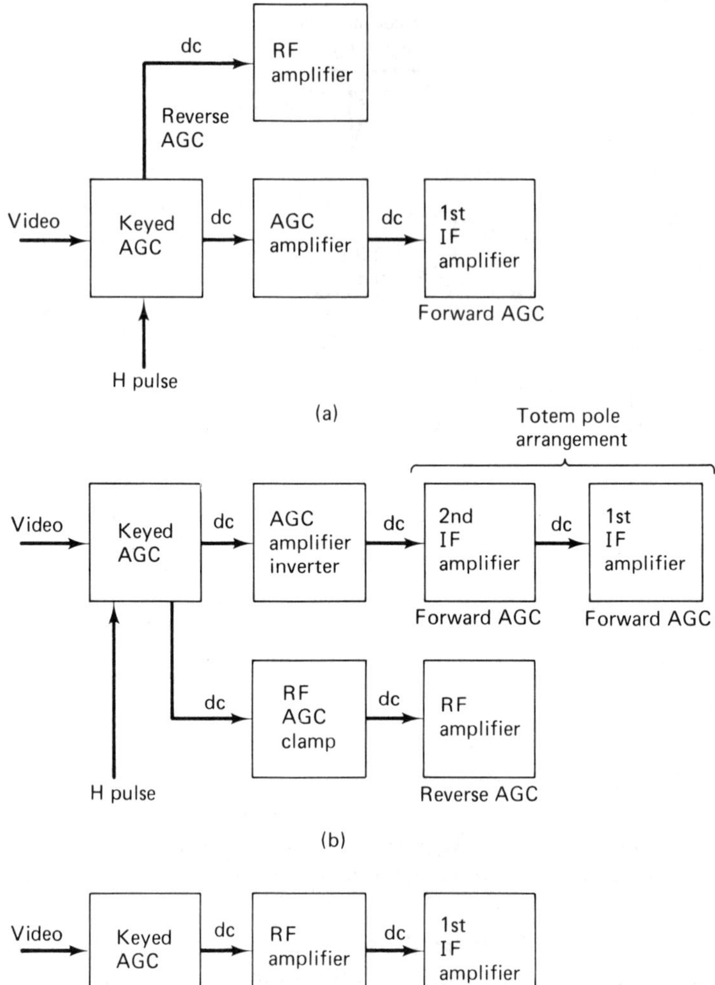

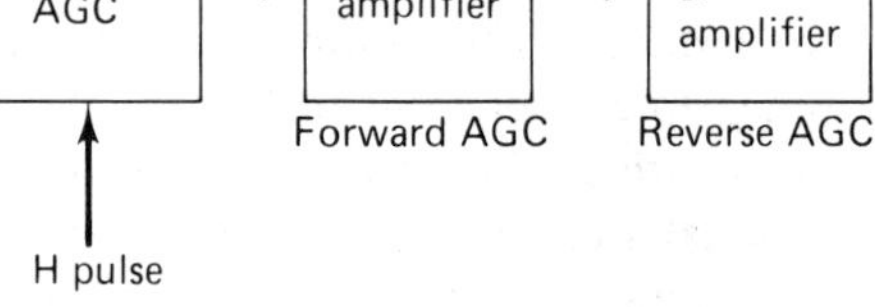

Figure 7-13 Different systems of transistor keyed AGC.

possible arrangements that have been used are shown in the block diagrams of Figure 7-13.

In Figure 7-13(a) the keyer transistor develops a dc voltage that is suitable for reverse AGC and is used as such by the RF amplifier. The IF amplifiers are designed to use forward AGC. Therefore, a transistor placed between the keyer transistor and the IF amplifiers is used as an amplifier providing greater control and as an inverter providing the proper dc voltage for use with forward AGC.

In Figure 7-13(b) the keyer develops a voltage suitable for reverse AGC. This is fed through a delay diode (called an *AGC clamp* in the circuit) to the RF amplifier. The keyer output is also fed to a dc amplifier that inverts the direction of voltage change and makes the voltage suitable for use with forward AGC. This dc voltage is fed to the base of the *second* IF amplifier that is connected in a totem pole arrangement with the first IF amplifier. Hence, the dc control voltage used to adjust the gain of one stage is able to control the gain of another stage as well. In terms of dc the second video IF acts like an emitter follower. Therefore, the control voltage is not inverted and both stages use forward AGC.

The keyed AGC system used in Figure 7-13(c) develops a positive-going AGC control voltage that provides forward AGC for the RF amplifier. The RF amplifier is then used as an inverter and applies a negative-going dc voltage to the IF amplifier. This voltage is used to control a reverse AGC system used by the IF amplifiers.

A complete transistor AGC system. The schematic diagram of Figure 7-14 shows a modern solid-state AGC system. This circuit uses four transistors that are used to develop the dc AGC voltage, to amplify and invert the control voltage, to provide an AGC delay, and to provide noise cancellation.

The keyer Q_1 is an NPN transistor that develops a negative AGC voltage across C_1. This voltage is proportional to signal strength. In this circuit the voltage across C_1 is in series with B+ supply. Therefore, an increase in signal strength will cause the charge across C_1 to *increase* and the voltage from the negative terminal of C_1 to ground to *decrease.* This causes the AGC amplifier Q_2 to conduct more heavily. Collector (Q_2) current flows through R_1 and R_2 and causes an *increased* positive voltage to be developed across them that is then delivered to the first video IF amplifier. This is the polarity of AGC voltage needed to provide forward AGC.

Under no-signal conditions, the collector voltage of the AGC amplifier Q_2 is 2.2 V, and the base of the AGC delay transistor Q_3 is set to 11.5 V by the delay potentiometer. Therefore, the difference of potential between these two points is 9.3 V, the base of Q_3 being positive with respect to the collector of Q_2. For these conditions, both D_2 and the base emitter junction of the PNP transistor Q_3 are reverse biased. They will remain reverse biased until an increased signal strength causes Q_2 to conduct more heavily and increase its collector voltage. When the collector voltage of Q_2 exceeds the delay voltage (11.5 V) both the diode D_2 and the emitter-base junction of Q_3 are forced into conduction. Since Q_3 is arranged in the common-base configuration, no voltage inversion takes place and the output of the delay AGC transistor provides forward AGC to the RF amplifier. Changing the delay voltage by adjusting the delay potentiometer determines the amount of signal strength required before AGC voltages are made available to the RF amplifier.

The noise immunity of the AGC system is improved by means of the noise gate Q_4. This transistor is connected in the common base configuration and is biased by the noise level control (R_3) to be at cutoff for all normal composite video signals that are fed to its emitter input. It is driven into conduction by negative going noise pulses that exceed the sync pulse amplitude. These noise pulses then appear without inversion in the collector circuit of Q_4 and the base circuit of Q_1. At the same time a positive-going composite video signal obtained from the output of the video amplifier feeds the base of the AGC gate.

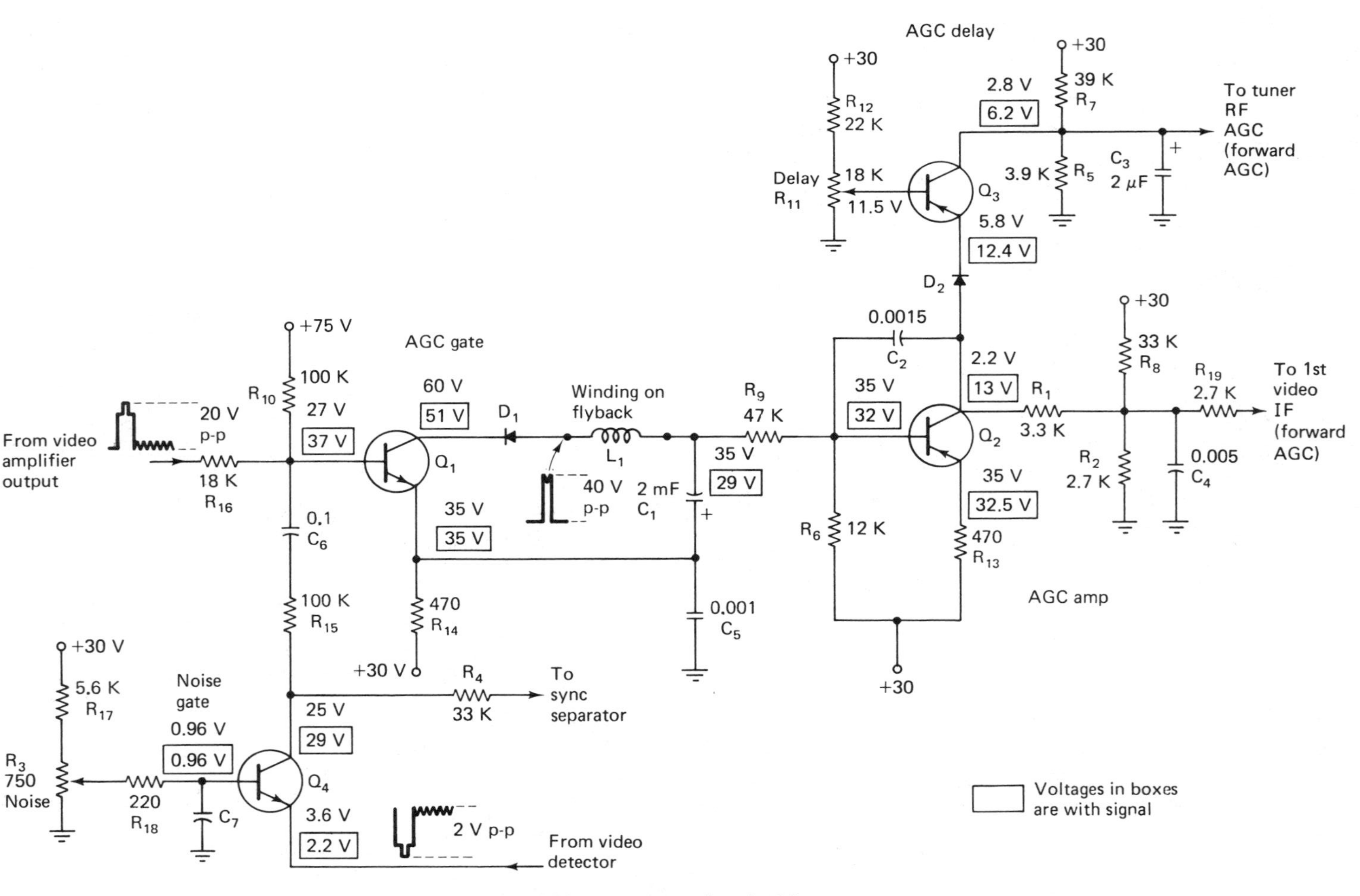

Figure 7-14 Four-transistor keyed AGC system.

Since there are two noise pulses appearing at the base of the keyer transistor at the same time of opposite polarity, they cancel and provide a noise-free composite video signal.

The noise output of the noise gate is also used to prevent noise setup in the sync separator circuits. This is accomplished by coupling the noise output from the collector of Q_4 to the sync separator circuit via R_4.

7-7 AGC level controls

Manufacturers of television receivers must design their receivers' AGC system with great latitude because they cannot possibly know in what kind of signal strength area their product will eventually be operated. If the receiver is located in a strong signal area, the signal may exceed the signal-handling capacity of the AGC system and the receiver may overload. The same receiver in a weak signal area may not be able to increase the gain of the RF and IF stages sufficiently, causing a weak and snowy picture. As a means of extending the range of signal strengths that an AGC system can handle, many manufacturers include IF AGC level and RF AGC delay controls as back panel or PC board service adjustments.

The usual AGC level control adjustment procedure begins by tuning the receiver to a strong station and adjusting the AGC level control until the picture overloads. Then the control is backed off until a normal, stable picture is obtained. If the receiver has an RF delay adjustment, it is usually made after the AGC level adjustment has been completed.

The delay adjustment is made for minimum snow while the receiver is tuned to a weak channel. In some cases, adjustment is made by setting the RF AGC voltage to some specific value while tuned to a weak channel. Since the actual procedures may vary somewhat, depending upon receiver make and model, it is always best to use the manufacturer's recommended adjustment procedure. This information is usually found in the service literature for the receiver.

Transistor AGC level controls are usually located in the emitter circuit of the keyer transistor. An example of this is shown in Figure 7-12. Since the basic function of the control is to vary the no-signal bias conditions in the circuit, it is also possible to place this control in the base circuit of the keyer. Transistor AGC circuits have not become standardized; each manufacturer has its own system.

Many transistor AGC circuits include some type of AGC delay control. An example of a delay control is shown in Figure 7-14. In this circuit R_{11} determines when Q_3 will conduct. If the control is at the ground end of the potentiometer, the transistor will conduct and the RF amplifier will be AGC controlled. This would be the condition in which no weak signals are to be received. This condition is most probable in a large city in which all the channels are strong and may cause overloading of the IF amplifier. If weak signals are to be received, high RF amplifier gain is desired. Therefore, the control is moved up toward the +30-V source. This cuts off Q_3 and prevents AGC control of the RF amplifier until a strong signal overcomes the reverse bias on Q_3 and normal AGC control of the RF amplifier is resumed.

7-8 the complete video IF strip integrated circuit

Modern color and black & white television receivers use integrated circuits to perform almost every function in the entire television receiver. Weak signal sections of the receiver that commonly use integrated circuits are the video IF, video detector, AGC, video amplifier, and sound IF strip. All of these circuits, with the exception of tuned circuit filters and decoupling resistors and capacitors, may be fabricated on a single IC chip such as the RCA type CA3068. This IC is an in-line flat pack having 19 active terminals.

Figure 7-15 is a block diagram showing the internal construction of this IC. The signal flow path and the stages used are similar to conventional circuits. Thus, the output of the tuner feeds pin 6, the input to the first IF. The output of this stage is delivered to an external tuned circuit filter that helps provide the required IF response. This circuit then delivers its output to the second and third video IF stages as well as the sound IF amplifier and detector. The 4.5-MHz sound IF output is then amplified by the 4.5-MHz sound IF amplifier and its output is delivered to terminal 2 for further processing. The output of the second and third video IF is fed into the video detector where the signal frequency is lowered to a range between 0 and 4.2 MHz. This signal is then amplified by the video amplifier limiter stage and then delivered to output terminal number 19 for additional amplification. The output of the video amplifier and a horizontal pulse are fed to a keyed AGC system. The dc control voltage developed by this stage is then used to control the stage gain of the first video IF. In addition to its normal signal-handling function, the video

Figure 7-15 Block diagram of the internal circuits found in the RCA CA3068 integrated circuit. (Courtesy of RCA.)

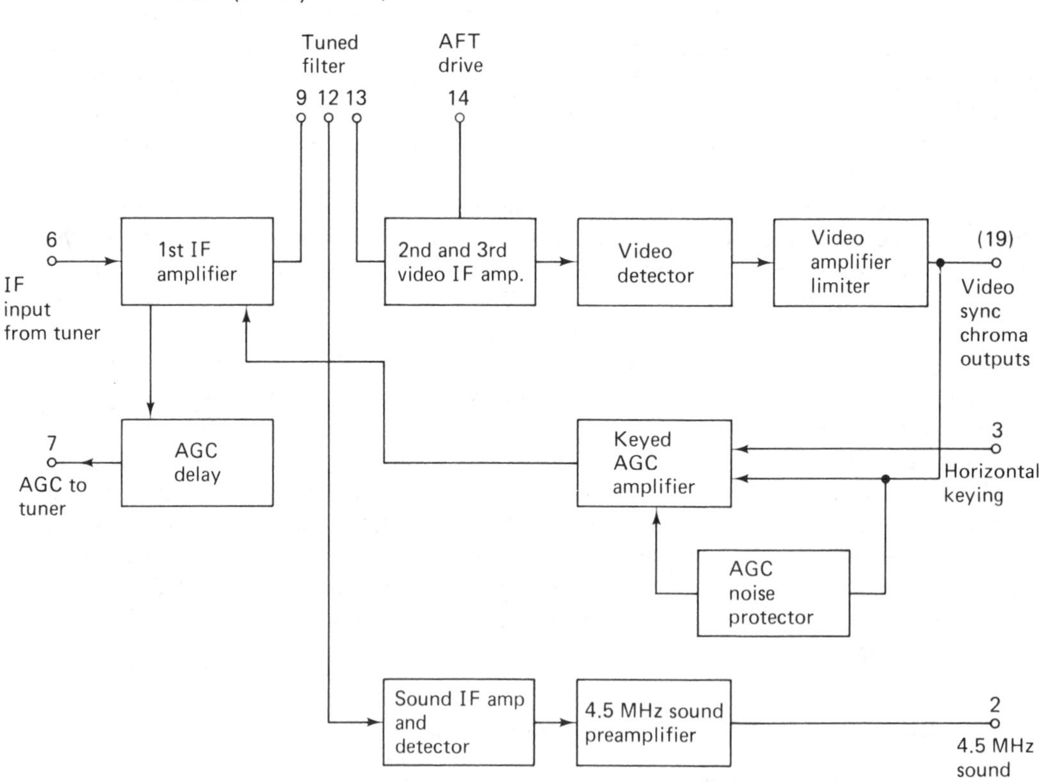

IF is used as a dc amplifier that feeds an AGC delay circuit that provides a delayed tuner AGC.

The CA3068 integrated circuit consists of 39 transistors, 10 diodes, 67 resistors, and 18 capacitors. These circuit elements are combined to form the complex circuit that performs the functions indicated in the block diagram discussed earlier. From a servicing point of view, the internal circuitry of an IC is only of limited interest since the service technician is not able to repair or modify the IC.

7-9 troubleshooting AGC systems

Defects in the AGC system of a television receiver produce one of three symptoms:

1. Excessive AGC output
2. Insufficient AGC output. Depending upon whether the AGC is forward or reverse, either condition can result in a weak or nonexistant picture and sound, or an overloaded or even a negative picture, possibly accompanied with buzz in the sound.
3. Poor AGC filtering resulting in symptoms of regeneration or oscillation.

These symptoms are similar to those discussed previously for video IF troubleshooting. The first step in servicing the AGC system is to decide whether the symptoms listed above are caused by the AGC circuit or the video IF amplifiers. This is primarily done by means of a bias clamp (refer to IF troubleshooting). If the bias clamp restores normal operation, the trouble is probably in the AGC system. If normal operation is not restored, the defect is probably in the IF amplifiers. Exceptions to this general rule may occur in black & white receivers when a defective video amplifier is the cause of the trouble.

In color receivers clamping the AGC bus may cause the sound and color to reappear without any black & white picture because the sound takeoff is in the IF amplifier and the defective video amplifier is blocking the *Y* signal from reaching the picture tube as well as shifting the operating point of the keyed AGC.

In some color receivers the AGC takeoff is obtained from a separate detector that is used to supply the composite video signal to the sync separator and AGC keyer. In other sets the AGC takeoff may be found in the output circuit of the video amplifier. In any event, defects in the takeoff point of the keyed AGC system will affect both the video stages and the IF stages. Therefore, to help isolate the defective stage, it is good practice when dealing with troubles in the video strip (RF, IF, detector, and video amplifier) to (1) clamp the AGC bus and leave it there, (2) check the operation of the last video IF stages with an oscilloscope equipped with a demodulator probe, and (3) check the operation of the video detector and video amplifier with an oscilloscope using an LCP.

If the defect has been narrowed down to the AGC system, troubleshooting consists of determining which section of the AGC system is defective.

The AGC system consists of three basic sections: the input system from which the AGC voltage is derived, the output network that delivers the AGC and delay voltages to the IF amplifiers and RF amplifiers, and the pulse generator and coupling system in keyed AGC. Defects in these sections may result in one of three conditions: (1) excessive AGC voltage, (2) low AGC voltage, or (3) lack of filtering in the AGC line.

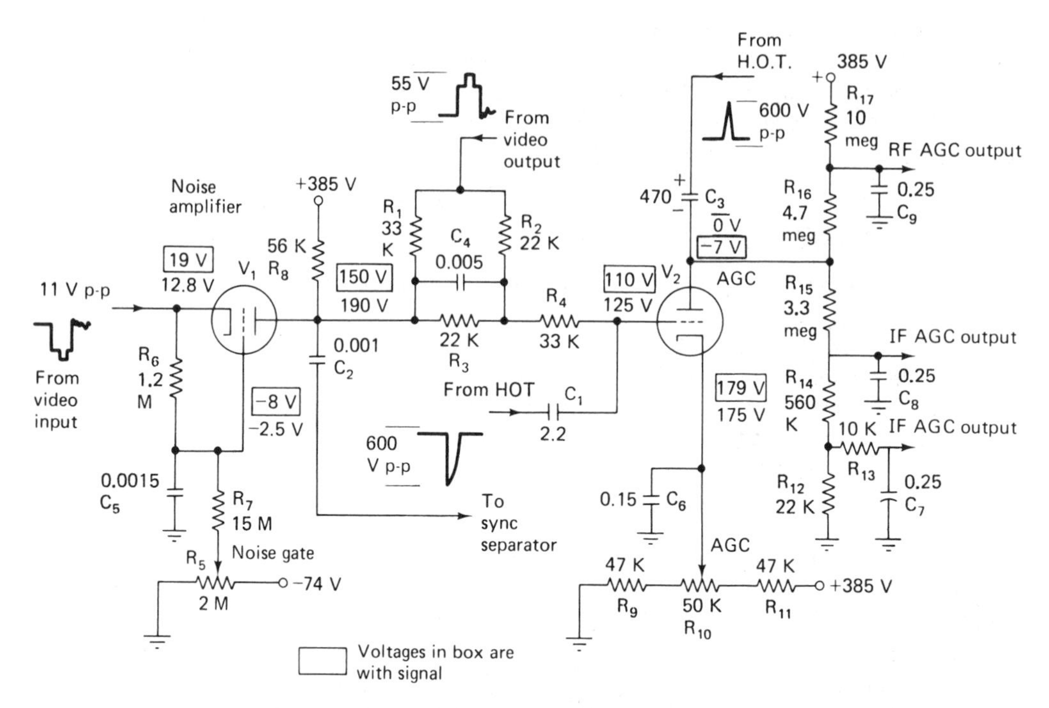

Figure 7-16 Keyed AGC and Amplified noise-canceling circuit.

In troubleshooting a complex circuit such as the transistor keyed AGC circuit of Figure 7-14, a bias clamp would be extremely useful for isolating the defective portion of the circuit. The bias clamp would be used in a way similar to the way a signal generator is used for signal substitution. Thus, starting with the AGC amplifier transistor Q_2, the bias clamp would be carefully adjusted to approximately 33 V and then placed between the base and ground. If the receiver is restored to normal operation, the defect is located in the keyed AGC circuits associated with Q_4. If the defect remains, the trouble is in the circuits associated with Q_2 and Q_3. To isolate which of these two circuits is at fault, the bias clamp is removed from the base of Q_2 and its output voltage is adjusted to approximately 6 V and then placed across C_3. If normal operation is obtained, then Q_3 or its associated circuitry are at fault. If normal operation is not obtained, then the defect is in the circuit associated with Q_2.

7-10 a vacuum tube amplified noise-canceling circuit

An example of an amplified noise-canceling circuit is shown in Figure 7-16. V_2 is a straightforward delayed keyed AGC circuit. The video signal obtained from the plate of the video amplifier is a positive-going sync pulse and is fed to the grid of the keyer tube via R_1, R_2, R_3, and R_4. Also feeding the grid is the output of V_1, the noise amplifier. This tube is arranged as a grounded grid amplifier. Its negative-going video input signal feeds the cathode and is obtained from the video detector output. The grid is returned to ac ground by means of the bypass capacitor C_5. The dc voltage on the grid with signal is -8 V and the voltage on the cathode is 19 V. This means that the bias between grid and cathode is 27 V. With this large bias, the tube (V_1) is operated beyond cutoff. The exact value of bias may be adjusted by the noise gate control (R_5). The noise control is adjusted so that all normal signals are in the cut off region of the tube and therefore do not appear at the output of V_1. This is indicated in the ladder diagram drawn below the schematic diagram. However, noise pulses that exceed the sync pulse amplitude will drive the noise amplifier into conduction and produce an output voltage. The input and output noise pulses are both negative-going because it is the characteristic of the grounded grid amplifiers that it does not invert the signal being amplified. The noise pulse is then added to the positive-going video signal that has been coupled to the grid of the keyer from the output of the video amplifier. If the noise amplifier's gain is such that its noise pulse output amplitude is equal to that in the composite video signal, the noise pulses will cancel. The noise-free composite video signal is then fed to the AGC keyer to produce an AGC voltage independent of noise. This circuit also does double duty by having its output delivered to the sync separator stages via C_2, thereby providing stable horizontal and vertical synchronization of the deflection oscillators.

The diagram shows that the large positive going horizontal retrace pulse at the plate of the keyer is coupled, via C_3, from a winding or tap on the horizontal output transformer. This voltage is a constant amplitude pulse 600 V peak to peak and has a duration of 7 to 10 μs. The horizontal retrace pulse at the plate occurs at the same time as the horizontal sync pulse on the grid. Since both pulses are positive going, the tube will be turned on during this time. Between pulses horizontal scanning (53 μs) is taking place. During this time the keyer tube is cut off. Hence video signal noise pulses that occur during scanning time are prevented from affecting the AGC level. Noise pulses that occur during sync pulse peaks will affect the AGC level and may be eliminated by the use of noise-canceling circuits.

The dc voltage between grid and cathode establish the keyer tube's bias and determines the operating conditions of the circuit under signal conditions. The grid is $+$ 125 V and the cathode is $+$ 175 V; therefore, the grid is minus 50 V with respect to the cathode. This is more than enough voltage to keep the tube cut off between retrace pulses. The tube bias may be controlled by adjusting R_{10} in the voltage divider consisting of R_9, R_{10}, and R_{11}. R_{10} is called the AGC level control and is used to determine the strongest video signal that the receiver can handle before it overloads.

During horizontal retrace the keyer tube conducts. Current flows from the cathode of the keyer to the plate through C_3 charging it with the polarity indicated. The current then flows through the horizontal output transformer winding to ground, and finally it returns to the cathode of the keyer to complete the circuit. The amount of the current flow, and therefore the amount of charge on C_3, is determined by the amplitude of the sync pulsse tips.

During the time when the tube is cut off (scanning) C_3 discharges through R_{12}, and R_{15} to ground and returns to the positive side of C_3 via the horizontal output transformer. The discharge of C_3 develops the negative AGC voltage required for the RF and IF stages. It is during this time that the plate voltage becomes negative and is responsible for the -7

V that is developed on the plate of the keyer when the receiver is tuned to a signal. The plate is 0 V when no signal is applied because it returns to ground via $R_{12,14}$ and R_{15}. Little or no current flows through these resistors because the RF AGC take-off point is clamped to zero. C_7, C_8 and C_9 are part of the AGC filter that removes any horizontal retrace pulse, noise, or video that may appear at the plate of the keyer tube. The time constants of these filters are relatively short and allow the circuit to be fast acting, thereby minimizing airplane flutter.

QUESTIONS

1. What are the characteristics of an ideal AGC system?
2. Explain the operation of the keyed AGC circuit of Figure 7-12.
3. What are the main characteristics of forward AGC and reverse AGC?
4. By examining a schematic diagram how can a service technician tell the difference between forward AGC and reverse AGC?
5. Describe the operation of two kinds of noise-canceling circuits.
6. What is the function of the AGC level control?
7. What is the function of delayed AGC and why is a delay control necessary?
8. Why do some receivers have a noise level control? What does this control do?
9. What is the purpose of D_1 in Figure 7-12?
10. In Figure 7-14 what is the probable effect on circuit operation if Q_2 opens?
11. Explain how a bias clamp may be used when troubleshooting an AGC/IF circuit.
12. Discuss how to troubleshoot a circuit that has an IC IF amplifier and AGC circuit.
13. Explain the symptoms produced when a defective AGC system produces excessive IF amplification. Insufficient amplification.
14. Referring to Figure 7-12, explain why a negative going horizontal retrace pulse is applied to the collector of Q_1.
15. Assuming that the RF signal has increased, what is the required polarity of the AGC voltage and what is its direction (increasing or decreasing) if reverse AGC were applied to the emitter of an NPN IF amplifier? What if it were applied to the base? What if forward AGC was used? Note that the starting conditions are such that the base voltage is 5 V and the emitter voltage is 4.5 V.
16. Repeat Question 15 assuming a PNP IF amplifier. Note the starting conditions are such that the base is 5 V and the emitter is 5.5 V.

chapter 8

The Sound Strip

8-1 introduction

The block diagram of a complete color receiver is shown in Figure 8-1. The four shaded blocks refer to that portion of the receiver called the *sound strip.* This section of the receiver is discussed in this chapter.

The sound system of a television receiver is essentially the same as that of a commercial FM receiver. In terms of the input signal, both receivers are similar because they process frequency-modulated signals. Thus, both receivers contain such circuits as limiters and frequency discriminators. In both cases, the audio frequency range that is transmitted extends from 50 Hz to 15 kHz. As a result, the TV sound system is capable of high-fidelity sound reproduction, but in practice it has been seldom used to its fullest extent. Both systems also share the freedom from noise that is inherent in frequency modulation systems. Also, in recent years both systems have utilized stereo sound reproduction.

Noise is caused by man-made or natural electrical disturbances, and it may be added to the desired RF signal in two ways: (1) the noise amplitude modulates the desired signal and (2) it introduces a small amount of phase modulation (indirect FM). Of the two, amplitude modulation is the more important means of noise addition. Since all FM receivers are equipped with some form of amplitude limiter, almost all of the noise is removed from the signal before it can appear at the audio output of the receiver's detector. The small amount of noise that does reach the audio amplifier is caused by the noise contributed by indirect frequency modulation of a signal that the limiters cannot remove.

A number of differences also exist between the commercial FM receiver and the sound system of the television receiver. First, the commercial FM system is permitted a maximum deviation of 75 kHz. This means that each commercial FM station occupies a

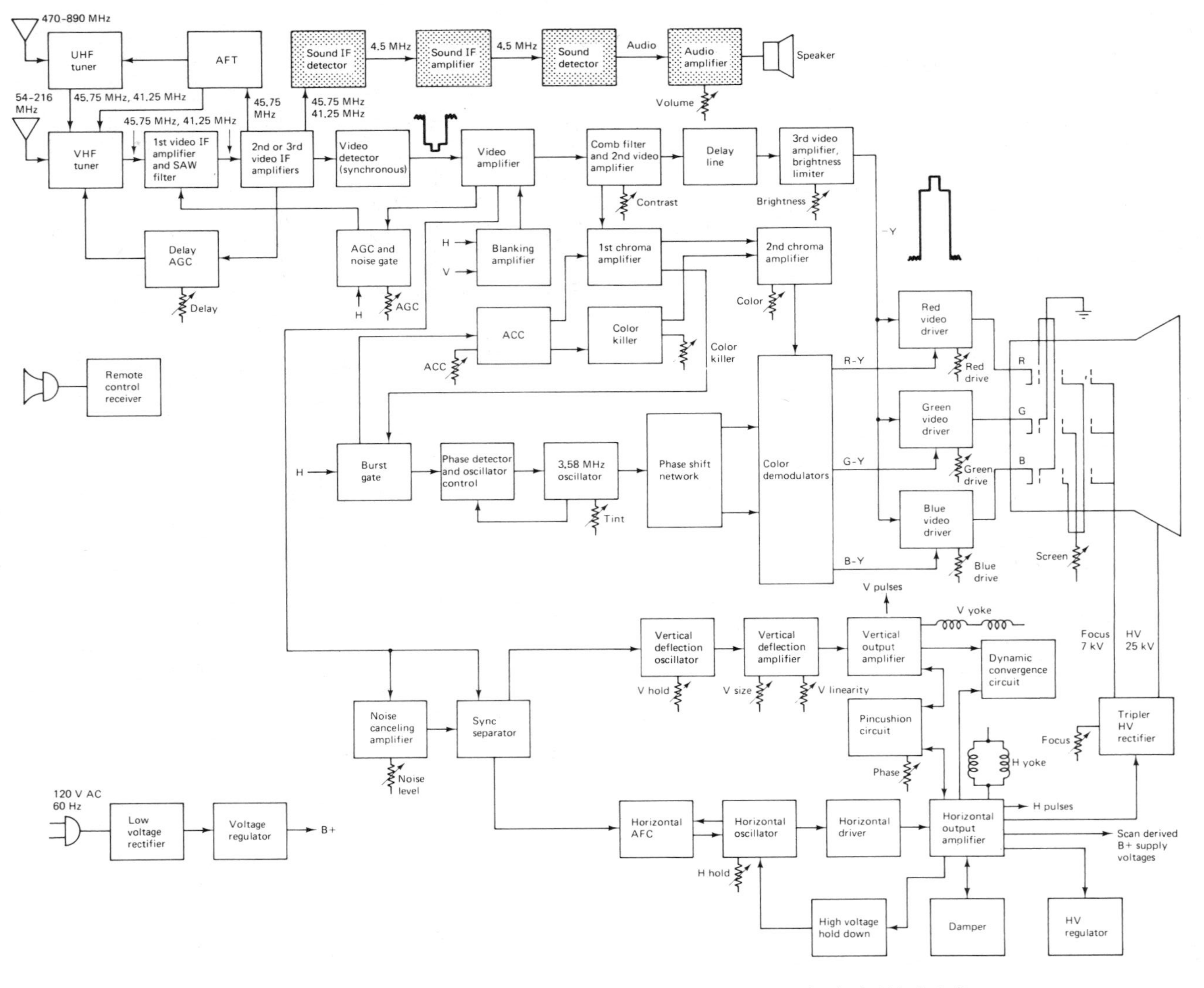

Figure 8-1 Complete block diagram of a color television receiver. The shaded blocks indicate the circuits to be discussed in this chapter.

bandwidth of 150 kHz. In contrast, the television sound carrier has a deviation of only 25 kHz. As such, it occupies a bandwidth of only 50 kHz. An advantage of this narrow bandpass is that it allows the television receiver sound IF amplifiers to operate at a higher gain than is possible for the same stage in a commercial FM receiver.

In addition, the commercial FM receiver has a tuned front end that is used to select the desired station from the great number being transmitted. In effect, the television sound system is tuned to only one "station," the 4.5-MHz IF output of the sound IF detector. In the intercarrier IF system, channel selection is the function of the tuner. The picture and sound IF outputs are then amplified together in the video IF amplifier. The sound IF (4.5 MHz) is the difference frequency produced when the picture IF (45.75 MHz) and the sound IF (41.25 MHz) are mixed together in the video detector.

Deemphasis and preemphasis. The deemphasis network used in television receivers serves the same noise reduction function as those found in FM receivers. In both cases the audio signal is given a high frequency boost at the transmitter. This audio preemphasis improves the signal to noise ratio and is accomplished by passing the audio signal through an *LR* network having a 75-μs time constant. The receiver compensates for preemphasis by passing the audio output of the detector through a 75-μs *CR* network called a deemphasis network.

Color and black & white sound strips. A comparison of the block diagram of the sound strips used in color receivers and black & white receivers is shown in Figure 8-2.

The sound strip in a color receiver usually consists of four stages: the sound IF detector, the sound IF amplifier, the sound detector, and the audio amplifier.

In a color receiver these stages perform the four following separate functions:

1. The *sound IF detector* changes the video IF (41.25-MHz sound IF and 45.75-MHz picture IF) into the sound IF (4.5 MHz).
2. The *sound IF amplifier* provides amplification, and in some receivers it may also act as a limiter for the sound IF (4.5 MHz).
3. The *sound detector* changes the frequency modulated sound IF (4.5 MHz) into audio.

Figure 8-2 Comparison of the sound strip used in a color receiver and a black & white receiver.

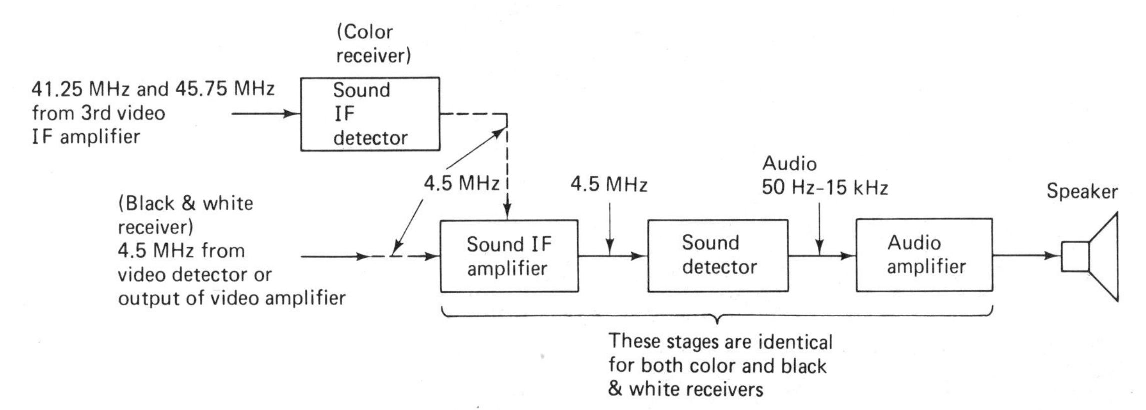

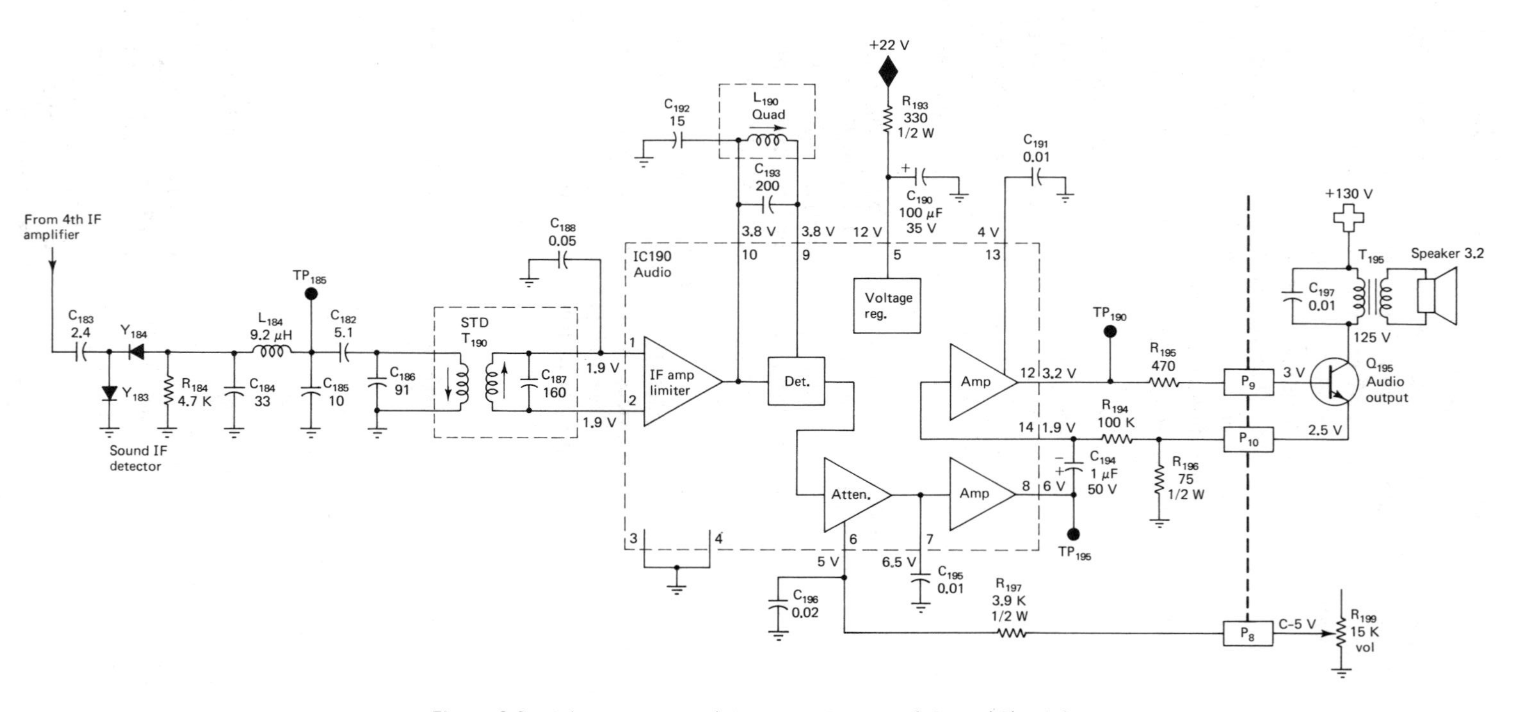

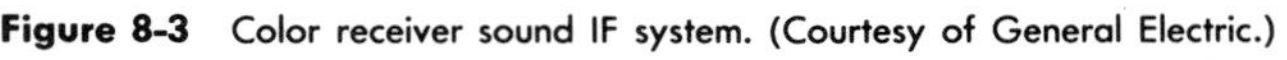
Figure 8-3 Color receiver sound IF system. (Courtesy of General Electric.)

4. The *audio amplifier* provides sufficient amplification of the audio output of the detector so that it is able to drive the loudspeaker.

These same functions are performed in black & white television receivers with the exception that converting the video IF (41.25 MHz and 45.75 MHz) into the sound IF (4.5 MHz) takes place in the *video detector.* Therefore, a black & white receiver does not require a separate *sound IF detector.* As a result, the sound strip consists of only three stages.

The sound takeoff point that feeds the sound strip of the black & white receiver is obtained at either the output of the video detector or at the output of the video amplifier. In color receivers the sound takeoff is located in the second or third video IF amplifier. A separate diode is used as a sound IF detector which converts the video IF into the sound IF.

8-2 sound IF amplifiers

The functions of the black & white sound IF amplifier and its associated input circuits are to separate the 4.5-MHz sound IF from the composite video signal and to provide sufficient amplification of the 4.5-MHz sound IF to permit proper operation of the FM detector. In some cases, the sound IF amplifier may also act as a limiter. A typical sound IF amplifier used in color receivers is shown in Figure 8-3. Examination of this circuit will reveal that it may be broken into four sections; the sound takeoff IF detector input circuit, the sound IF amplifier, detector circuits, and the audio amplifier circuits.

Sound takeoff circuits. The input circuit of the sound IF in a black & white receiver always contains a sound takeoff circuit that couples the 4.5-MHz difference signal from the point of its origin to the sound IF amplifier. In practice, this circuit usually performs two functions simultaneously. It separates the 4.5-MHz sound IF from the composite video signal and delivers it to the input of the sound IF. It also acts as a trap that prevents the 4.5-MHz sound IF from reaching the picture tube. If this were not done and the sound IF were to reach the picture tube cathode, a highly undesirable "herringbone" pattern similar to that shown in Figure 8-4 would appear on all channels. Proper adjustment of this trap will eliminate this defect.

Figure 8-4 A 4.5-MHz herringbone interference pattern. (Courtesy of General Electric.)

Transistor sound takeoff circuits. Figures 8-5 and 8-6 illustrate two sound takeoff circuits in receivers using transistor and integrated circuits. The usual sound takeoff point in a black & white transistor receiver is at the output of the video driver stage. This stage is usually a buffer stage, typically an emitter follower that is used to minimize loading of the video detector by the video amplifier stage.

In Figure 8-5 the video driver is acting as a phase splitter that provides *two* outputs. The emitter circuit output feeds the video output amplifier. Since the emitter output contains the full composite video signal, including the 4.5-MHz sound IF, a shunt trap consisting of C_1 and L_1 is used to eliminate this source of interference. The collector circuit contains a tuned transformer (T_1) that is used to separate the sound IF from the composite video signal and deliver it to the base of Q_2, the sound IF amplifier. This transformer is tapped to provide proper impedance match and maintain a high tuned circuit Q.

The coil in the base of Q_1 is a tweet filter that is part of the video detector. Note that the manufacturer gives the frequency (82 MHz) to which it is tuned. This happens to be the frequency of channel 6, and it is also the second harmonic of the video IF.

The sound takeoff arrangement used in a receiver using integrated circuits is shown in Figure 8-6. The fact that the sound IF amplifier is an integrated circuit does not necessarily mean that any different method of sound takeoff is necessary. The double tuned transformer T_{301} provides 4.5-MHz selectivity and an impedance match between the high collector resistance of the transistor Q_{304} and the relatively low input resistance of the IC. This arrangement also allows the Q of the tuned circuit to remain high and thereby ensures good selectivity and effective sound IF separation.

Figure 8-5 Transistor sound takeoff circuit found in a black & white receiver.

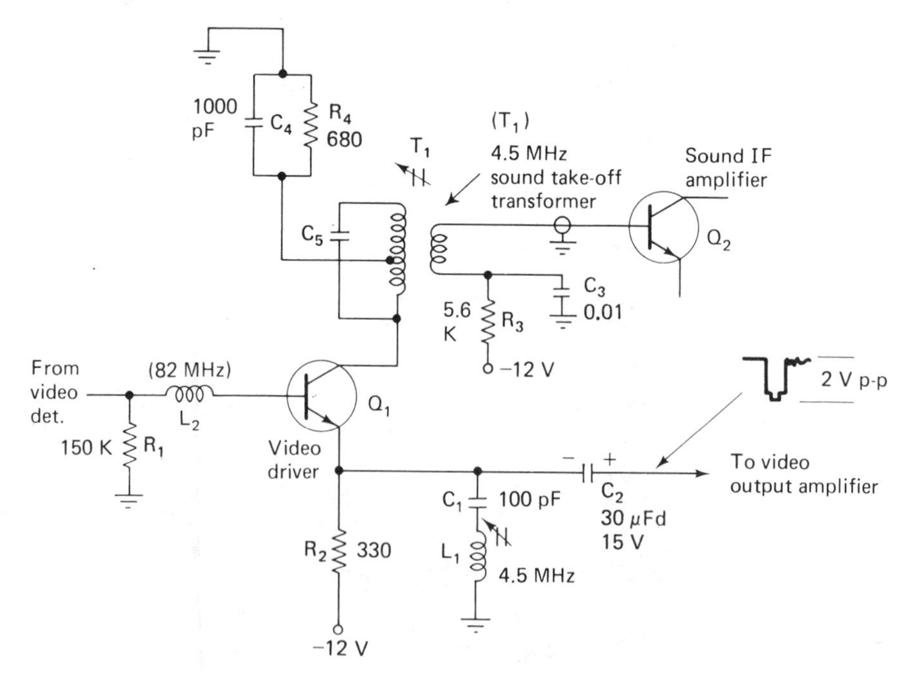

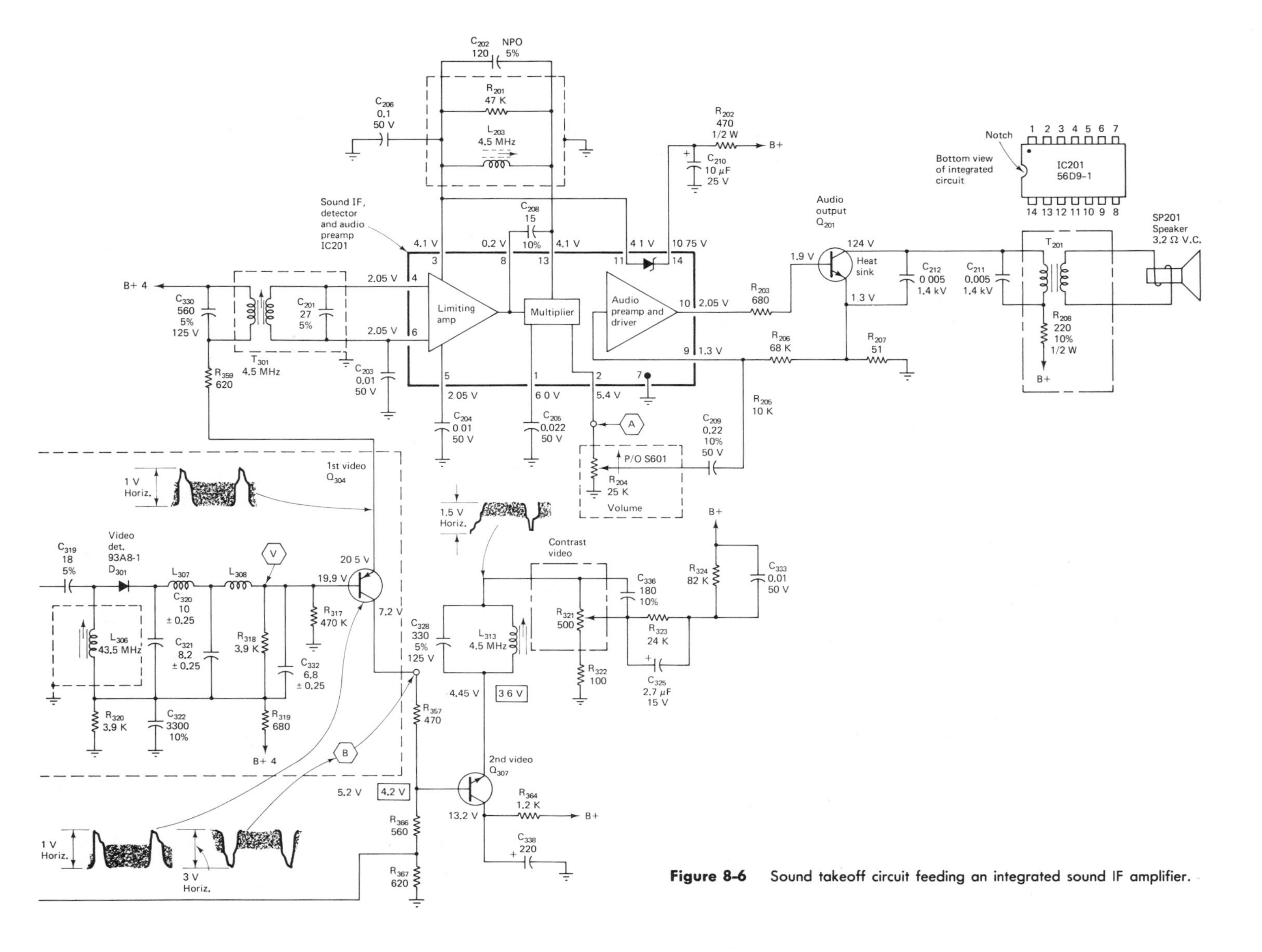

Figure 8-6 Sound takeoff circuit feeding an integrated sound IF amplifier.

Color sound takeoff circuits. A typical sound takeoff circuit used in color television receivers is shown in Figure 8-3. The object of this circuit is to convert the picture (45.75 MHz) and the sound (41.25 MHz) IF signals into the intercarrier sound IF of 4.5 MHz, and to separate it from the composite signal which is simultaneously produced. Some manufacturers may also make use of the composite video signal obtained at this point to drive the sync and AGC stages of the receiver. From Figure 8-3 it can be seen that the sound takeoff point is located in the fourth video IF amplifier stage. The sound takeoff is placed before the video detector to minimize the 920-kHz video interference that would be produced if the sound IF (4.5 MHz) is permitted to beat with the 3.58-MHz color subcarrier. When the sound takeoff is obtained before the video detector, greater attenuation of the 4.5-MHz sound IF is possible after the video detector. This is done by a bridge type trap located after the video detector.

Conversion of the video IF to the sound IF (4.5 MHz) is accomplished by diodes Y_{183} and Y_{184} arranged in a voltage doubler detector configuration. In some circuits the detector may use series or shunt detector diode arrangements. In most modern television receivers the sound IF detector as well as the entire sound strip is encapsulated in a single IC chip.

R_{184} and C_{184} are the detector load resistor and IF filter. L_{184}, C_{185}, C_{182}, and tuned transformer T_{190} are the tweet filters and the 4.5-MHz sound IF tuned circuits that are used to separate the sound IF (4.5 MHz) from the composite video that appears at the output of the sound IF detector. The secondary winding of T_{190} drives the IF amplifier limiter section of the integrated circuit IC_{190}.

8-3 sound IF amplifier circuits

The sound IF amplifier has two functions: It is a tuned amplifier that is designed to deliver adequate sound IF signal to the sound detector and it may also provide amplitude limiting. Defects in these stages may cause loss of sound, weak sound, and distortion and buzz if misalignment is present.

Solid-state arrangements of the sound IF amplifier in both color and black & white receivers usually do not contain more than one stage of amplification. In general, this circuit is very similar to the tuned amplifier used in both the RF and video IF stages of the receiver. The major difference between these amplifiers is their frequency of operation. Since this is a function of the tuned circuits, all three (RF, video IF, and sound IF) amplifiers will appear schematically the same.

Another difference between these amplifiers is that the sound IF amplifier does not use AGC because the sound signal is frequency modulated and the use of limiters in the sound strip tends to maintain a constant signal level without resorting to automatic gain control methods.

Solid-state sound IF amplifiers may use transistors, FETs, or ICs as the amplifier. If oscillation is to be avoided when transistors or FETS are used, the amplifier must be neutralized.

Modern television receivers use either transistor or integrated circuit sound IF amplifiers. The function of these amplifiers is to provide sufficient amplification to drive the FM detector and provide amplitude limiting which ensures noise-free sound.

The transistor circuit shown in Figure 8-7 uses an NPN transistor arranged in a

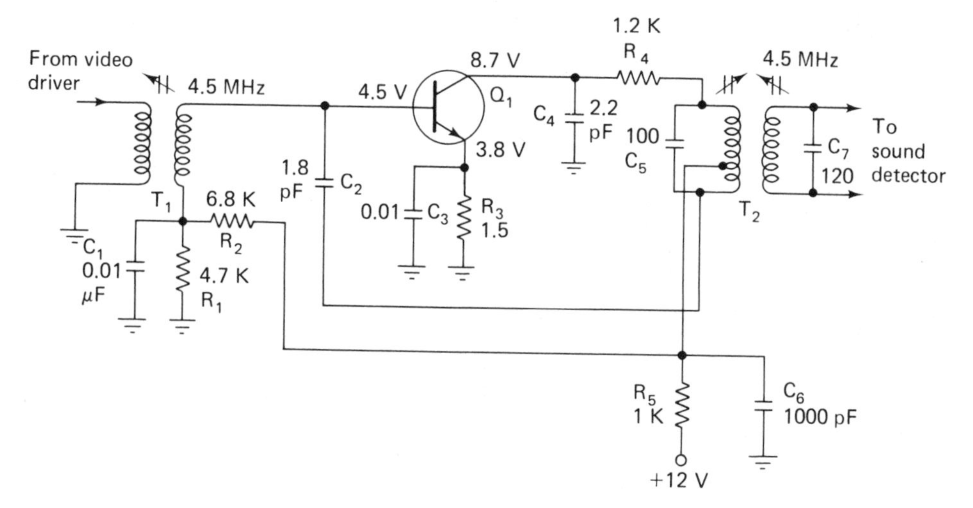

Figure 8-7 Transistor sound IF amplifier.

standard IF amplifier common-emitter circuit configuration. Resistors R_1, R_2, R_3, and R_5 are used to set the no-signal forward bias for the transistor. Capacitors C_1, C_3, and C_6 are RF bypass capacitors that tie their points of connection to ac ground. If either C_1 or C_3 were to open, the amplifier would go dead because the input signal is applied between base and emitter via the series connection of C_1 and C_3. Therefore, if either C_1 or C_3 were to open, the input signal base current would be greatly attenuated by R_1 and R_3. The result would be a loss of sound IF amplification.

Tuned transformers T_1 and T_2 restrict the bandpass of the amplifiers to the sound IF. The primary of transformer T_2 is tapped in order to provide for neutralization via C_2. A transistor may be used as a limiter by simply overdriving it.

To ensure that small input signals can cause saturation, the collector supply voltage may be reduced. The cutoff point of the transistor is a function only of the transistor and is not appreciably affected by circuit conditions.

One of the important effects that takes place when a transistor is driven into saturation is that the instantaneous collector voltage will fall below the base voltage. This means, in the case of the NPN transistor, that the collector voltage may be +5.5 V when the base voltage is 6 V. As a result, when the collector to base junction is forward biased, the collector may be considered to be tied to the base. In effect, 100% feedback takes place during this time, since all of the ac voltage across T_2 will be shunted across T_1. During this time the circuit will oscillate and possibly cause interference. To prevent this from happening, an isolating resistor R_4 is used to prevent the primary of T_2 from being tied to the secondary of T_1 when saturation causes the collector to be shorted to the base.

8-4 an IC sound IF amplifier

The single integrated circuit of Figure 8-8 contains all of the stages of the entire sound strip. This chip, called the "sound processor," contains the IF amplifier and limiter stages, an FM detector, an electronic attenuator, an audio power amplifier, and other power regulating and controlling stages.

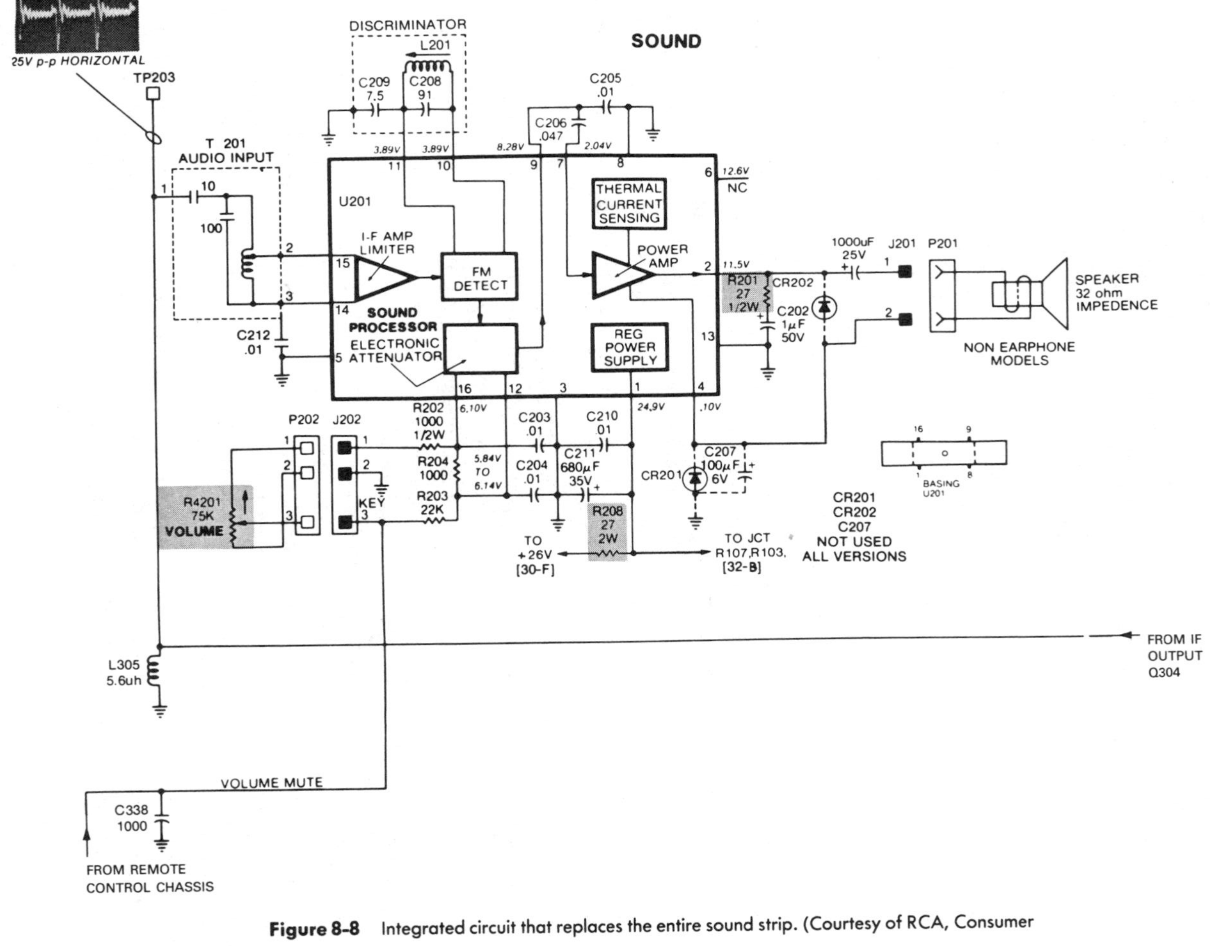

Figure 8-8 Integrated circuit that replaces the entire sound strip. (Courtesy of RCA, Consumer Electronics.)

The 4.5-MHz audio IF input signal is obtained from the output of the video amplifier and is fed through the audio input tuned circuit T_{201} to the IF inputs of the IC. This signal is then amplified and demodulated by the internal circuits of the chip. The audio output of the demodulator is fed into an electronic attenuator. Volume control R_{4201} provides a variable dc bias to the gain-controlled stages in the electronic attenuator, which permits control of the audio level. The use of a dc voltage to control volume level makes remote control of volume less complex and economical. The output of the attenuator is coupled to the power output stage by capacitor C_{206}. Tone is controlled by C_{205}. The output of the power amplifier is coupled via capacitor C_{201} to a 32-Ω speaker or earphone.

Alignment of this circuit is simple. First, tune the receiver to a strong local station and adjust the volume for normal-level sound. Then adjust L_{201}, the detector tuned circuit, for maximum volume. Finally, adjust the input tuned circuit, T_{201}, for maximum volume with minimum noise.

8-5 sound detectors

The output of the sound IF amplifier is a constant amplitude, frequency modulated, 4.5-MHz signal. As such, it cannot be directly used to produce an audio output. It is the function of the sound detector to act as a limiter and to convert the frequency modulated sound IF into audio. In general, the two following types of sound detectors are being used:

1. The ratio detector
2. The integrated circuit sound detector

8-6 the ratio detector

Many transistor receivers use the ratio detector as the sound detector. Its main advantage is that it is self-limiting, but this is obtained at the expense of the audio output which is approximately one-half that of the Foster-Seely discriminator for the same amount of frequency deviation. Therefore, the audio output of this detector requires substantial audio amplification before it is able to drive the loudspeaker.

Typical circuit. A typical schematic diagram of an unbalanced ratio detector is shown in Figure 8-9. The function of the detector is to act as a frequency changer, that is, to change the high-frequency sound IF (4.5 MHz) into low-frequency audio. This is done in two steps. First, the frequency modulated sound IF is converted into an equivalent amplitude modulated IF signal. Second, the amplitude modulation is detected and converted into audio. Frequency modulation to amplitude modulation conversion is done by the tuned input transformer T_1. Detection is provided by the series connected diodes D_1 and D_2.

Shown in the schematic diagram of the detector are the wave shapes at its input and output. Notice that the input is a constant amplitude frequency modulated signal and that the output of the transformer is an amplitude modulated signal. The amplitude changes are directly related to the amount of frequency change. Thus, a large frequency change (corresponding to a large audio amplitude) will produce a high percentage of amplitude modulation. Similarly, the rate of frequency deviation (corresponding to audio frequency) will appear at the output of the transformer as the rate of amplitude change.

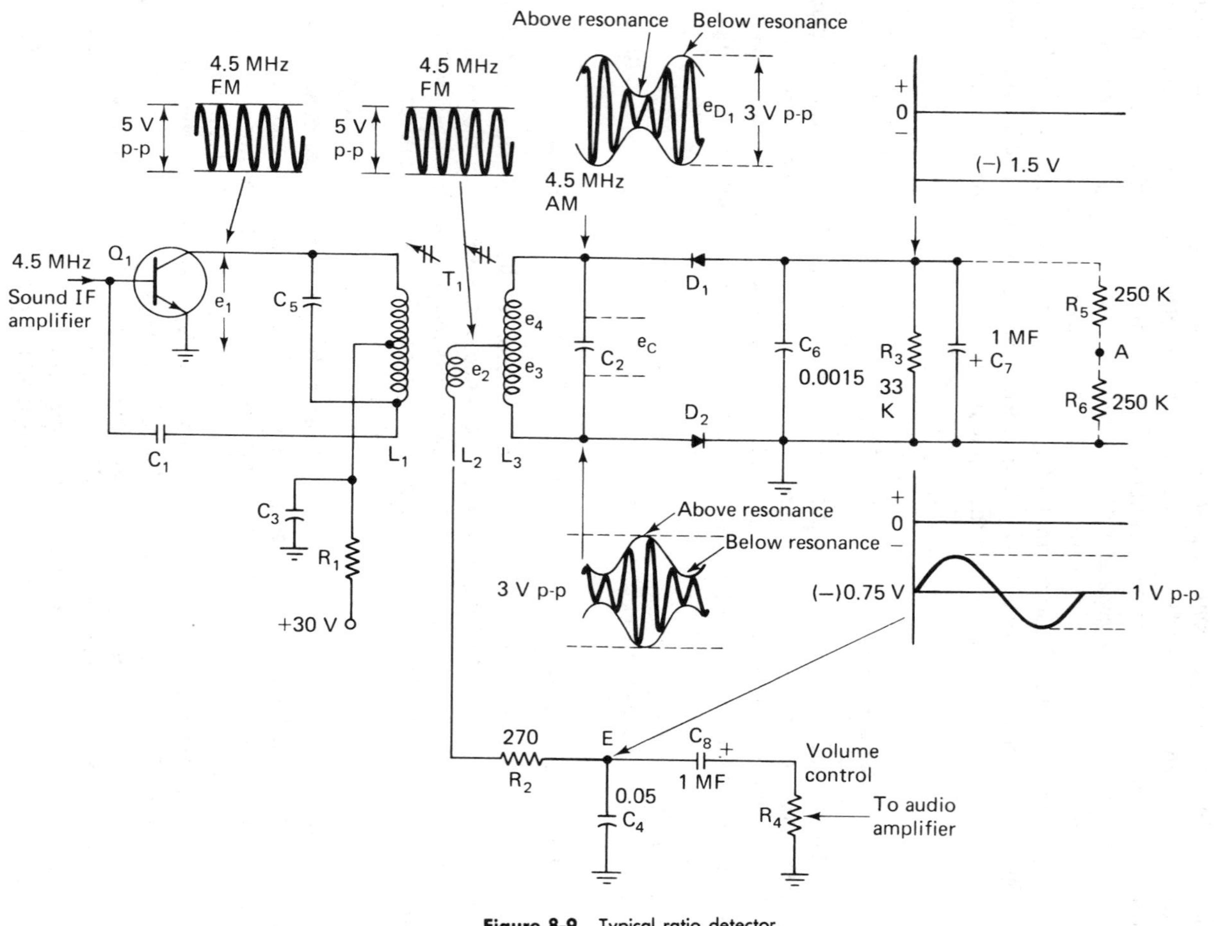

Figure 8-9 Typical ratio detector.

Transformer action. To understand how frequency modulation to amplitude modulation transformation takes place, it is necessary to examine the phase relationships that occur in a tuned transformer as the frequency of the signal applied to it changes. Three conditions will be examined: at resonance, above resonance, and below resonance.

At resonance. When the input signal is at 4.5 MHz, both primary and secondary tuned circuits of T_1 are at resonance. As shown in Figure 8-9, the primary is a *parallel* resonant circuit because it has two paths for ac current flow, and the secondary is a *series* resonant circuit in terms of its induced voltage because there is only one path for current flow. Since both circuits are at resonance, they are both resistive. L_2 is another secondary winding called the *tertiary winding.* Because of the direction of the turns in L_2, the voltage induced into it (e_2) is in phase with the primary voltage (e_1). Figure 8-10 shows the phasor diagram illustrating the phase relationships between the voltages and currents in the circuit.

The primary voltage (e_1) forces a current (i_L) through the primary inductance L_1. The current lags the voltage by 90°. This current produces a changing magnetic field (ϕ) which induces a voltage (e_2) into the tertiary winding L_2 that has the same phase as the primary voltage e_1. Another voltage (e_{induced}) is also induced into the tuned secondary L_3. The induced voltage may be considered an ac generator that is effectively in series with the secondary and does not appear as a terminal voltage.

The induced voltage (e_{induced}) acting in the secondary, "sees" a resistive circuit at resonance. Therefore, the secondary current (i_c) is in phase with the voltage. This current passes through the secondary tuned circuit capacitor C_2 and produces a voltage drop (e_c) across it which lags the current by 90°. The voltage across the capacitor is also the voltage across the secondary winding and is not the same as the induced voltage (e_{induced}).

Notice that the output secondary terminal voltage (e_c) of the transformer is 90° out of phase with (e_1) and (e_2), the voltages across the primary and tertiary windings, respectively.

Above resonance. If the frequency of operation is increased above 4.5 MHz, the secondary series circuit impedance will appear to be inductive.

The phasor diagram for the tuned transformer operating above resonance is shown in Figure 8-10. Because of the inductive secondary impedance, the secondary current (i_c) now lags the induced voltage (e_{induced}) by some angle that depends upon the Q of the circuit

Figure 8-10 Phase characteristics associated with a ratio detector.

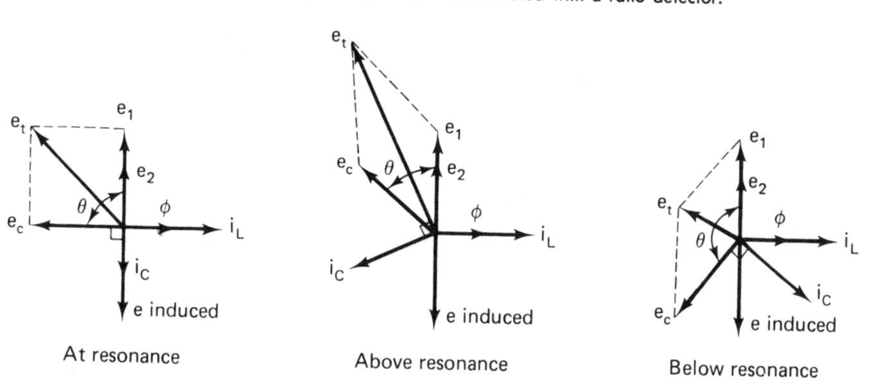

and the amount of input signal frequency deviation. The secondary current (i_c) passes through the resonating capacitor C_2 and produces a voltage drop (e_c) across it that lags the current by 90°. As a result, the phase angle (ϕ) between the primary voltage (e_1) and the secondary terminal voltage (e_c) is now less than 90°.

Below resonance. When the input signal shifts below resonance, the secondary tuned circuit becomes capacitive, and the phasor diagram shown in Figure 8-10 results. The secondary current (i_c) now leads the induced voltage (e_{induced}) by an angle that depends upon the tuned circuit Q and the input frequency. This current produces a voltage drop across C_2 that lags the current by 90°. In this case, the secondary terminal voltage (e_c) and the primary voltage (e_1) are now more than 90° apart.

Summary. In summary, it can be seen from the phasor diagrams that a change in frequency will cause a change in phase between the primary and secondary voltage. *If these voltages are added together vectorially as shown in Figure 8-10, their vector sum (e_1) will be seen to change in amplitude.*

How is this addition of voltages done in the practical circuit? In Figure 8-9 it can be seen that the secondary of T_1 is center tapped and that the tertiary winding L_2 is connected to the center tap. The voltage induced into L_2 (e_2) is of the same phase as the primary voltage (e_1); therefore, it may be considered the primary voltage. The addition of primary and secondary voltages is effectively obtained by adding the tertiary voltage to the secondary center point of the transformer.

The secondary terminal voltage (e_c) is divided into two voltages e_3 and e_4. With respect to the center tap, these voltages are 180° out of phase. The vector addition of the tertiary voltage (e_2) and the secondary voltages (e_3 and e_4) are shown in Figure 8-11. The vector sum of the tertiary voltage e_2 and one-half the secondary voltage e_4 produce the voltage (e_{D_1}) which is fed to the diode D_1. The addition of voltages e_2 and e_3 produce the diode voltage e_{D_2} which is fed to the diode D_2. These two voltages (e_{D_1} and e_{D_2}) are equal in amplitude when the primary and secondary voltages (e_2, e_3, and e_4) are 90° out of phase. This will occur when the input signal is at the resonant frequency of 4.5 MHz.

When the frequency modulated signal increases above resonance, the phase relationships change, and the vector addition of the tertiary voltage (e_2) and the secondary voltages e_3 and e_4 result in a situation in which the voltage amplitude e_{D_2} is larger than e_{D_1}.

Below resonance the relative phase between primary and secondary voltages

Figure 8-11 Vector addition of the ratio detector secondary and tertiary voltages.

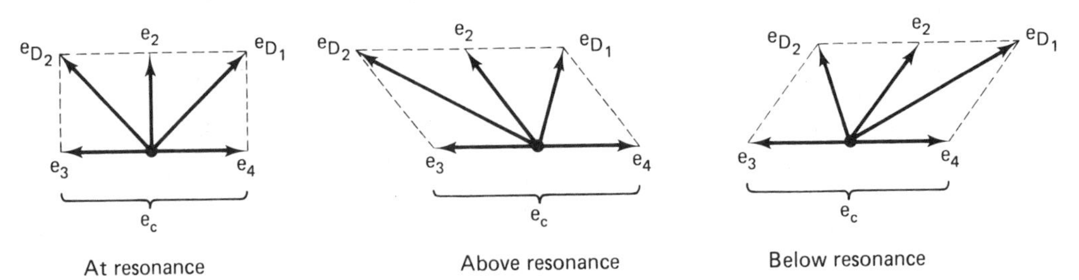

changes, resulting in a situation in which e_{D_1} is now larger than e_{D_2}. In short, frequency modulation is converted into amplitude modulation as shown in Figure 8-9.

The audio output. The next step in the detection process is to convert the high-frequency (4.5 MHz) amplitude modulated signal into audio. Notice in Figure 8-9 that when the frequency of the incoming signal is *below resonance,* diode D_1 receives a larger voltage than does D_2. Also notice that both diodes are connected into the circuit so that they both conduct on the same half-cycle of input signal (e_3 and e_4). Thus, when the top diode D_1 instantaneously receives a negative voltage, the lower diode D_2 is receiving a voltage which is 180° out of phase and is therefore a positive voltage. The degree of conduction of each diode depends upon its driving voltage and the charges on C_4 and C_7. Therefore, for the below resonance condition, D_1 will conduct more heavily than D_2. The electron current flow path for D_1 is as follows: current flows from D_1 to C_7 to ground *up* through C_4 through L_2 and L_3 and back to D_1. Diode D_2 also conducts and its path will be from D_2 to L_3 and L_2 *down* through C_4 to ground and then returning to D_2. The currents flowing through C_4 are in the opposite direction, but they are not equal. Therefore, C_4 will tend to charge in the direction of the greater current. In this case, the current resulting from D_1 is greater than that of D_2 and the top of C_4 will shift toward a positive polarity.

When the frequency shifts to the *above resonance* conditions, the magnitudes of the voltages feeding D_1 and D_2 will reverse, making e_{D_2} larger than e_{D_1}. This in turn will force D_2 to conduct more heavily than D_1, charging C_4 in a negative direction. Since the frequency modulated input signal changes frequency at an audio rate, the voltage across C_4 will vary at an audio rate, thereby producing the audio output. In the practical circuit of Figure 8-9 resistor R_2 and capacitor C_4 form a deemphasis network. C_8 is the audio coupling capacitor used to pass the audio and block the dc component developed across C_4. Volume is controlled by R_4 which is a voltage divider. When the variable tap is at the top of the control, all of the signal is fed to the next stage. When it is at the bottom end of the control, the tap is tied to ground and no signal reaches the next stage.

If a constant-amplitude, constant-frequency (4.5 MHz) signal were applied to the ratio detector, both diodes would receive equal ac voltages and it would appear that they would conduct equally, forcing a net current of zero through C_4 (Figure 8-9). Hence, it might be assumed that the voltage across C_4 is zero at resonance. Unfortunately, under these conditions, actual voltage measurements in the circuit would reveal that a dc voltage that is equal to one-half that developed across C_7 appears across C_4 (see Figure 8-9); this is because of the unequal conduction paths for each diode.

Amplitude limiting. The ratio detector is noted for its amplitude limiting ability. Because of this characteristic the preceding sound IF stage need not be a limiter, although in practice it often is.

The ratio detector is designed to act as a form of dynamic limiter. A dynamic limiter is able to keep the amplitude of the detector input signal constant by automatically varying the Q of the tuned circuit feeding the diodes. The Q is varied in such a direction that it opposes the change in signal level. The ac voltage across both the inductance and capacitance that make up the series resonant secondary winding of the input transformer of the detector is directly related to the Q of the tuned circuit. Thus, a high Q will produce a large output voltage and a low Q will produce a smaller output signal.

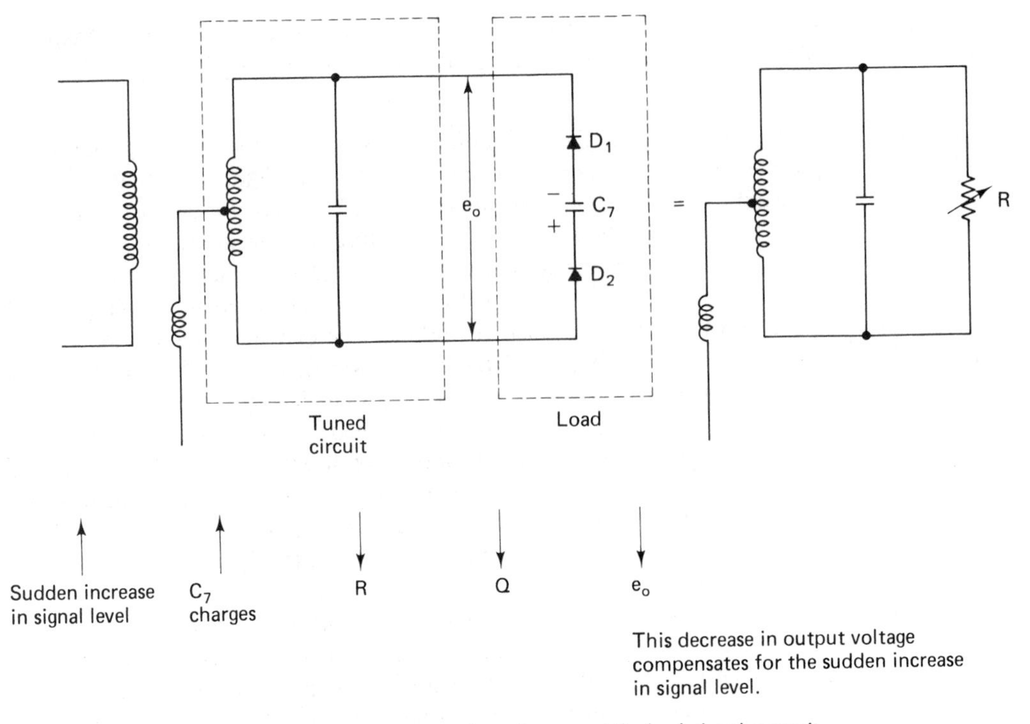

Figure 8-12 Ratio detector tuned transformer and the load placed across it.

The Q of the tuned circuit can be varied by changing the loading across it. Thus, if the tuned circuit were shunted by a small resistance (heavy load), the Q would decrease and cause the output voltage to also decrease. The opposite would occur under light loading.

In the practical circuit the tuned circuit is shunted by the series combination of the diodes (D_1 and D_2) and the electrolytic capacitor C_7. This is shown in Figure 8-12. The diodes and C_7 may be replaced by a variable resistor (R) because the amount of diode conduction, and therefore the effective load across the tuned circuit, depends upon the signal derived dc voltage across C_7. For example, the arrows in Figure 8-12 show that if the signal were to *suddenly* increase in level because of a noise pulse, both diodes would conduct very heavily since the input signal would exceed the reverse bias of C_7. In effect, the high charging current may be said to be the result of a small resistance R placed across the tuned circuit. This would reduce the Q of the tuned circuit and cause the output voltage (e_o) to decrease, thus compensating for the original increase in signal level.

A decrease in signal level would cause the diodes to reduce their conduction which in effect would be the same as increasing the tuned circuit shunt resistance. This in turn would force the Q of the tuned circuit to rise and the output voltage would increase, thus compensating for the decrease in signal level.

Alignment. The alignment procedure for the radio detector of Figure 8-9 is straightforward. First, an unmodulated 4.5-MHz signal generator is connected to the input of the first sound IF amplifier and a dc VOM is placed across C_7. Then the input IF tuned transformer and

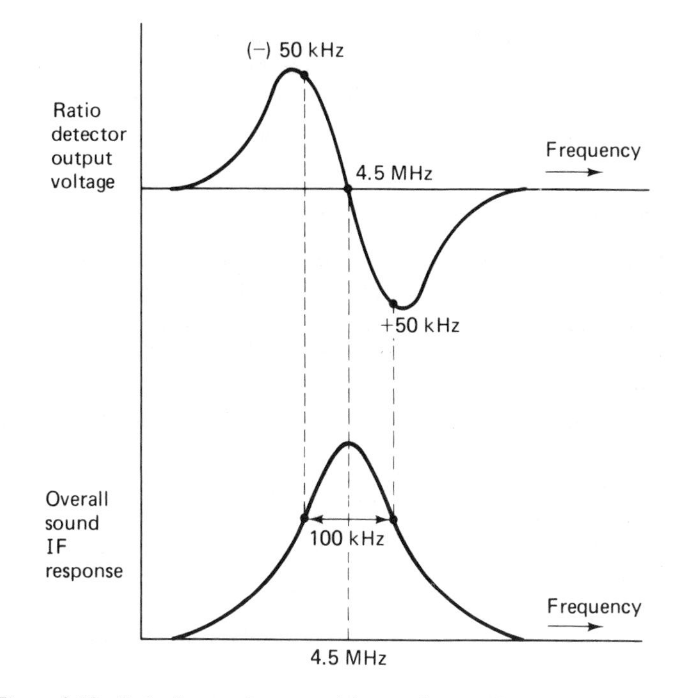

Figure 8-13 Ratio detector S curve and the sound IF amplifier response curves.

the primary of the detector transformer are adjusted for maximum dc output voltage across C_7. This means that the input IF transformer and the primary of the detector transformer are at resonance. Next, the VOM is placed between points A and E and the secondary of the detector transformer is adjusted to its resonant frequency. This is indicated by a zero dc voltage that is obtained between the audio output (point E across C_4) and the junction (point A) of two series connected resistors, of approximately 250 kΩ each, placed across C_7.

These adjustments may also be made using a sweep generator and an oscilloscope. The oscilloscope display obtained at the audio output (across C_4) of the ratio detector during sweep alignment is the familiar S-curve response shown in Figure 8-13. Also shown is the overall sound IF response. The bandpass of the tuned IF amplifiers is approximately 100 kHz. The IF response is observed across R_3, with C_7 open.

Ratio detector troubleshooting. From the discussion above it can be seen that if the electrolytic capacitor or either of the diodes were to open or short, limiting action of the detector would cease and the audio output of the detector would decrease and be interfered with by a loud buzz. The buzz is the result of the 60-Hz vertical blanking pulse that amplitude modulates the sound IF and appears in the audio if limiting is not effective. Another possible defect that causes buzz is a misadjustment of the secondary tuned circuit. Therefore, the first thing that should be tried before any other troubleshooting procedure, is to attempt a touch-up adjustment of the tuned transformer secondary. If the buzz cannot be eliminated and good distortion-free sound produced, then the other possibilities should be checked out.

Diodes may be checked in-circuit by an ohmmeter. The resistance should change by a factor of 5 to 1 when the ohmmeter leads are reversed. If identical readings are obtained in both directions, the diode is defective. Electrolytic capacitors should be checked by direct substitution.

8-7 integrated circuit detectors

The FM detector was one of the first TV receiver circuits to be made into an integrated circuit. Since that time a number of different detector IC circuits have been developed. One of the aims of this development has been to eliminate the need for tuned circuit alignment. Other factors stimulating IC development have been system performance, cost, and a desire to reduce IC complexity.

Four integrated circuit FM detector systems that may be used are as follows:

1. Discriminator
2. Differential peak
3. Quadrature
4. Phase-locked loop

The block diagrams of each system are shown in Figure 8-14.

The discriminator detector. The topmost block diagram in Figure 8-14(a) is that of the basic discriminator form of FM detector. This arrangement may use either the ratio detector or the Foster-Seely discriminator. The Foster-Seeley discriminator has twice the sensitivity of the ratio detector and does not require a large stabilizing capacitor. These advantages are achieved at the expense of good AM rejection characteristics, but this may be overcome by preceding the discriminator with a good amplitude limiter. Receivers using this IC detector will have tuned circuits external to the IC that will require alignment.

The differential peak detector. Figure 8-14(b) illustrates a differential peak detector. This detector features a single-tuned circuit; therefore, it combines good sensitivity with ease of alignment. In this system the detector compares the peak voltages detected on each side of a single-tuned circuit by applying them to a differential amplifier. Amplitude limiting is provided by the differential amplifiers preceding the detector.

The output of the differential amplifier feeds an electronic attenuator. This portion of the IC allows a dc voltage to control the audio amplitude and therefore permits the use of a single-wire volume control. The volume control is used as a rheostat and is part of a voltage divider circuit that controls the bias level of one of the audio amplifier transistors used in the IC. In effect, gain is controlled in the same way as the gain of a video IF amplifier that uses an AGC system. The only difference is that in the electronic attenuator the gain control is manual, whereas AGC systems control gain automatically. The use of a single-wire volume control that controls a dc voltage has the advantage of being able to be placed at almost any distance from the IC without fear of hum pickup. Therefore, shielded cables are not needed. Another advantage of the single-wire system is the ease with which it may be used in some remote control systems. In these systems a dc voltage may be used to replace the geared motor and motor relay used in conventional remote control systems.

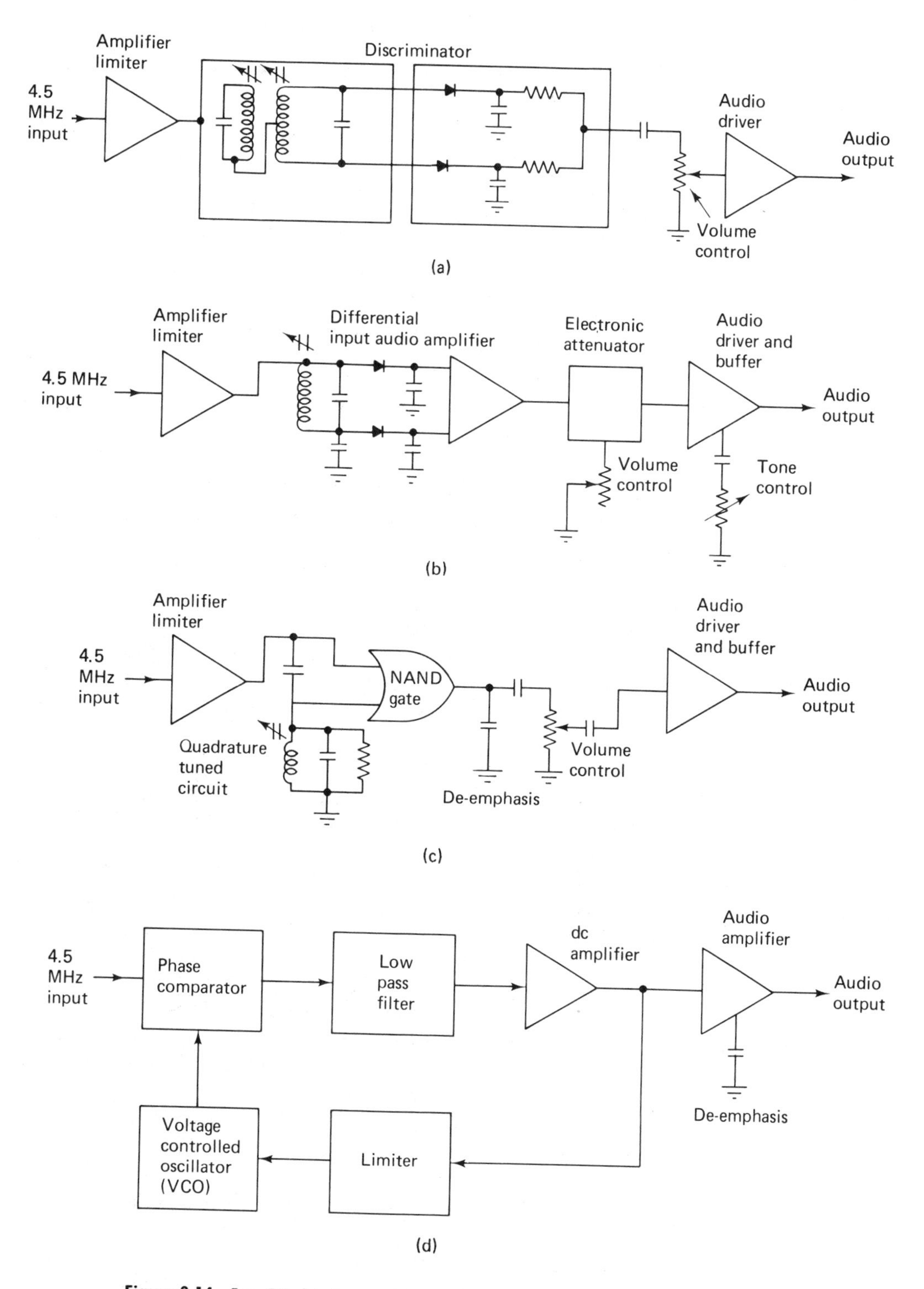

Figure 8-14 Four FM detectors used in integrated circuit sound systems: (a) discriminator detector; (b) differential peak detector; (c) quadrature detector; (d) phase-locked-loop detector.

The quadrature detector. The third IC method of FM detection [Figure 8-14(c)] is a quadrature detector. The detector is fed by an amplifier limiter that removes all noise and amplitude variations from the FM signal. In this system the relative phases of the signal on either side of a single-tuned circuit are compared. This is done by means of a balanced phased detector which in terms of computer logic terminology, is the equivalent of an *equivalence* logic function. In this arrangement, NAND logic blocks are used so that their output current is gated on only when the two input signals are either both negative or positive. The output pulses vary in width at an audio rate that depends upon the deviation of the input FM signal. Here again, the advantages of this system are its low cost and easy alignment. The usual alignment procedure is to tune the receiver to a station and then adjust the coil for maximum undistorted sound output.

The phase-locked loop. The last system [shown in Figure 8-14(d)] is the phase-locked-loop detector. This detector circuit arrangement is very similar to the AFT circuit discussed earlier. Thus, a dc error voltage is generated by a phase detector that compares the phase of the input FM signal with that of a locally generated oscillator of the same frequency. The dc voltage is directly related to the degree of phase difference between the two signals. This voltage is used as a control voltage that is returned to the voltage controlled oscillator (VCO) and forces the oscillator to reduce the phase error. When the free-running frequency of the VCO is sufficiently close to the input signal frequency, the system locks onto this frequency and the VCO tracks the input FM signal. Since the dc control voltage will change in step with the changes in input signal frequency modulation, it constitutes the demodulated audio output. To ensure noise-free reception, this system requires a good limiter preceding the phase detector. This detector has the advantage that it may be classed as an inductorless detector requiring no alignment.

A practical IC FM detector. The schematic diagram of an integrated circuit detector is shown in Figure 8-15. This circuit may be divided into two parts: (1) the three-transistor audio driver and power amplifiers and (2) the IC package.

The IC package has 14 leads and performs the function of limiting, differential peak detecting, and audio preamplification. The input circuit consisting of L_1, C_1, L_2, C_2, and C_3 couples the sound IF to the IC and also constitutes a 4.5-MHz tuned circuit. The other tuned circuit (L_3, C_4) is used in the differential peak detector and is tuned for maximum undistorted audio output. The B supply feeds the IC via R_1 and R_2. Capacitor C_7 is a bypass capacitor that keeps terminal 5 at ac ground potential for high-frequencies. The tone control (R_4) is part of a low-pass filter that determines the high-frequency content of the audio signal. When the movable arm of the potentiometer is adjusted so that the resistance of the potentiometer is zero, capacitor C_9 is tied directly to ground and bypasses the high-frequency tones. When the potentiometer is adjusted for maximum resistance, C_9 is less able to shunt the high-frequency tones to ground; therefore, the sound output of the receiver will be richer in high-frequency content. The volume control (R_6) is part of an electronic attenuator circuit in the IC that allows a dc voltage to control the gain of the IC.

The detector deemphasis network is formed by capacitor C_{11}. Capacitor C_8 couples the audio signal from the internal IC buffer stage to the first audio amplifier portion of the

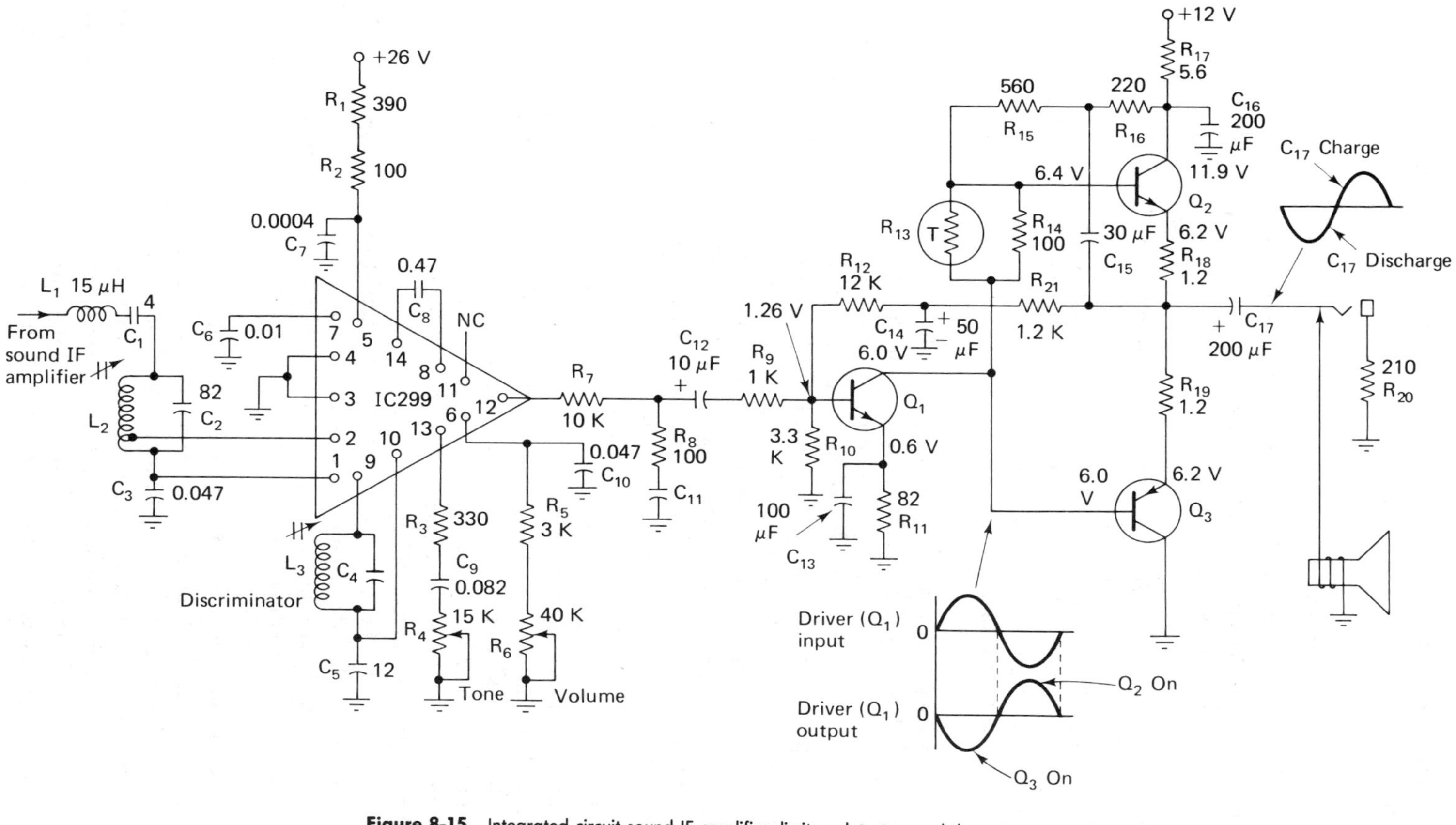

Figure 8-15 Integrated circuit sound IF amplifier, limiter, detector, and three-stage complementary symmetry audio amplifier.

IC circuitry. The audio output of the IC is coupled to the audio driver transistor Q_1 by R_9 and C_{12}.

8-8 audio amplifiers

For a given input FM signal, the audio output level of the FM detector depends upon the type of detector used in the receiver. This in turn determines the number and type of audio amplifiers required by the receiver.

Receivers that use a ratio detector must use a stage or two of audio preamplification in order to increase the typical 1-V p-p audio output level of this detector to that needed to drive the audio power amplifier.

Integrated circuit detectors also have relatively small audio output levels. Therefore, they are usually followed by one or more stages of audio preamplification prior to driving the audio power output stage. These audio amplifier stages may or may not be included as part of the IC package.

Transistor audio power amplifier. A typical example of a form of television transistor power amplifier that has one input signal and does not require an output matching transformer is shown in Figure 8-15. This amplifier is referred to as a *complementary symmetry amplifier* because one of the series-connected output transistors is PNP and the other is NPN. In this arrangement the dc current (electron flow) flows from the emitter of the PNP power transistor Q_3 to the emitter of the NPN power transistor Q_2. The emitter current of the PNP transistor (Q_3) consists of the collector current coming from the B− terminal of the power supply and the base current obtained from the collector of the driver transistor Q_1. The emitter current of the NPN transistor (Q_2) divides into the collector current, which flows directly to B+, and the base current, which returns to the B supply via R_{15} and R_{16}.

The base current of a transistor sets the transistor's operating point and thereby establishes the conditions under which the transistor will function. For this circuit, the operating points of Q_2 and Q_3 are determined by the combined effects of the collector current of the driver transistor Q_1 flowing through R_{13}, R_{14}, R_{15}, R_{16}, and R_{17}, as well as the voltage drops across the emitter resistors R_{18} and R_{19}. The proper dc collector current flowing through Q_2 may be controlled by R_{14}. The manufacturer's service literature will often indicate the value of this current, and R_{14} must be adjusted to obtain the recommended current. A typical value of idling current is 4 mA.

In circuits that use germanium power transistors a thermistor (R_{13}) is used to temperature stabilize the circuit and prevent thermal run away. When silicon power transistors are used, the thermistor is replaced by a forward biased silicon diode, which provides temperature compensation but with less voltage drop than a resistor would.

Class B operation. The most efficient method of audio power amplification is Class B. In this arrangement two transistors are both operated at cutoff and are alternately driven into conduction by an input signal. The input signal during the first half-cycle simultaneously drives one transistor on and the other off, thus providing an output for that half-cycle. On the next half-cycle the process is reversed. The end result is an output signal that contains both half-cycles of the input signal. If the transistors were perfectly linear (output current is directly proportioned to input base current), the amplified output signal would be undistorted. Unfortunately, the output characteristics of the transistor are not linear, and they become very nonlinear as the operating conditions approach cutoff. Therefore, when operating transistors are Class B power amplifiers, severe distortion will occur as the signal drives the transistor in and out of cutoff. This distortion is called *crossover distortion.* Crossover distortion may be minimized by providing a small forward bias for each transistor. It is for this reason that the complementary symmetry audio power amplifier of Figure 8-15 has such an elaborate biasing arrangement.

Output conditions. The dc voltage appearing at the emitter junction of Q_2 and Q_3 is approximately one-half of the B supply voltage. This means that both transistors must have matched characteristics even though they are of different types. Furthermore, C_{17}, the large 200-μF capacitor, will become charged to this voltage, since it is connected by means of the voice coil of the speaker between the junction and ground. In this circuit the junction voltage is also made use of as a 6-V power source for the low-level amplifiers elsewhere in the receiver. As such, it is used to provide the base bias voltage of the driver transistor. Filter network R_{21} and C_{14} are used to remove all audio variations before the dc voltage is allowed to reach the driver. R_{10} and R_{12} are voltage dividers that adjust the driver Q_1 base voltage to the proper value to provide the necessary bias for Class A operation. R_{11} provides thermal stability and also helps to establish the driver transistor bias.

When a signal is applied to the base of the NPN driver transistor (Q_1), the first positive half-cycle will cause the driver collector current to *increase.* This in turn will cause the base and collector currents of Q_3 to rise and the base and collector currents of Q_2 to decrease. The heavy conduction of Q_3 causes C_{17} to discharge. The discharge current passing through the speaker voice coil winding produces an audio output that corresponds to the first half-cycle of the audio signal input.

On the negative half-cycle of input signal the driver transistor's forward bias is decreased reducing both the driver collector current and the voltage drops across R_{13}, R_{14}, R_{15}, and R_{16}. This in turn tends to increase the base voltages of both Q_2 and Q_3. As a result, the forward bias of Q_3 is decreased and tends to turn it off, and the forward bias of Q_2 is increased and tends to turn it on. Consequently, C_{17} now begins to *charge.* Electron current flows up through the speaker and Q_2 to B+. The direction of current flow through the speaker voice coil is opposite to that which occurred when C_{17} was discharging. The direction of current through the voice coil and the movement of the speaker cone will produce a sound output that corresponds to the second half-cycle of the audio input signal.That is, if the first half-cycle of audio were to produce a compression of air in front of the speaker cone, the second half-cycle would produce a rarification of air pressure.

One of the problems associated with audio power amplifiers that use speaker matching output transformers is the tendency for the power transistors or the output transformer to be damaged if the speaker voice coil opens because the removal of the transformer load effectively increases the primary inductance, thereby allowing any sudden changes in the audio signal to develop very large transients across the primary. To prevent this from occurring, capacitors and varistors are placed across the primary, and load resistors are used to place a load across the secondary during that instant that the speaker is disconnected and an earphone is used to replace it.

The situation is somewhat different when dealing with a complementary symmetry, power amplifier. If the speaker voice coil were to open, the charge and discharge currents of C_{17} would be reduced to zero, thereby reducing the demands placed on Q_2 and Q_3. Therefore, no damage to the amplifier should be expected when the speaker load is removed. But, the situation changes if the speaker is shorted. Under these conditions, Q_2 will conduct very heavily and it or its associated components may burn up.

8-9 heat sinks

The audio power output stage and the vertical and horizontal output transistors are usually operated with sufficient power dissipation to make them very hot. To help dissipate this heat and thereby keep the transistor cool, large metal fins called *heat sinks* may be attached to the transistor.

Heat sinks come in many different shapes and forms. They may physically look like a gear, the center hole of which is filled by the transistor case. They may be strips of heavy aluminum or the chassis on which the transistor is mounted. Or they may look like clothespins, with the clamping ends placed around the transistors. Any number of different designs are possible and are in use. See Figure 8-16.

Heat sinks dissipate the heat generated by power transistors by three methods: (1) thermal conduction, (2) convection, and (3) radiation.

Heat may be carried away from its source by materials that carry heat in the same way that electrical conductors carry electrical energy. *Heat* is defined as the degree of motion of the atoms in a material. This motion is most readily transferred from one atom to another in metals. In fact, the best electrical conductors are also the best heat conductors. Therefore, heat sinks are almost always made of such metals as aluminum or copper.

Although air is not the best thermal conductor, air will become heated when it is in contact with a hot object such as a power transistor. The warm air will rise and take some of the heat with it. Then the air will be replaced by cooler air. This process, called *convection,* will continue as long as the heated surface is warmer than the surrounding air. The effectiveness of convection for removing heat depends upon the surface area of the heat sink. Most of the heat sink shapes (gear tooth, radiating fins, and so on) are all attempts at increasing the heat-dissipating surface area of the heat sink, thereby increasing the amount of heat carried away by convection.

The third method of heat dissipation, radiation, makes use of the fact that heat is

Figure 8-16 Gear-shaped fin-type heat sink. (Courtesy of RCA, Consumer Electronics.)

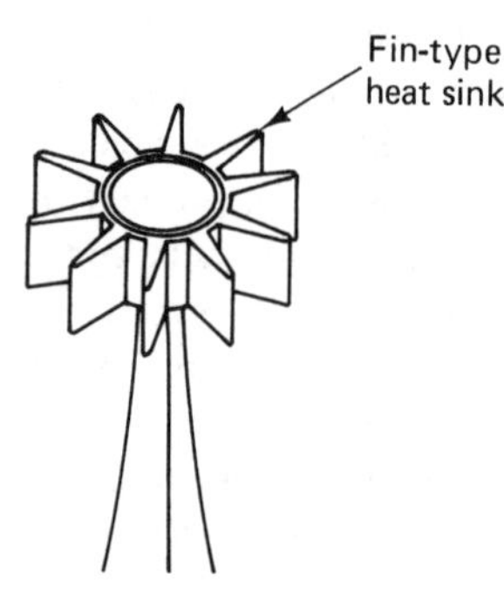

a form of electromagnetic energy similar to radio waves and light. This radiant characteristic of heat is evident when one considers how the heat generated by the sun is able to reach the earth. Certainly, the vacuum of space will not permit heat transfer by conduction or convection. Therefore, it must be concluded that heat may also be transferred by radiation.

As is well known, light can be reflected by mirrored or shiny surfaces. Since black surfaces do not reflect light, all of the light energy striking a black surface is absorbed by black material. Similarly, heat is also reflected by mirrored or shiny surfaces. Therefore, if the surface of a heat sink is made highly reflective, the heat generated at the interior of the heat sink would be reflected back into the heat sink and the cooling ability of the heat sink would be somewhat impaired. To improve the radiating efficiency of the heat sink, many heat sinks are anodized (a colored aluminum oxide coating) black.

Silicone grease. Some power transistors are arranged so that the collector of the transistor is tied to the metal case of the transistor. This is done to make the transistor case a heat sink. The collector is the element in the transistor that dissipates most of the transistor power; therefore, it must be adequately cooled. If it is desired to increase the size of the heat sink by clamping the transistor to the chassis of the receiver or some other large metal surface, a problem may arise because the collector voltage and its polarity may not be the same as that of the chassis. How does one make a good thermal bond between the transistor and the chassis while at the same time electrically insulate one from the other? This is done as shown in Figure 8-17 by placing between the power transistor and the heat sink a thin sheet of insulating mica or plastic that is coated with a film of silicone thermal grease. The mica sheet is perforated so that the transistor leads and the mounting screws may pass through it. The silicone grease is an excellent electrical insulator that has good thermal conductive properties, thereby providing the desired conditions. Whenever a power transistor

Figure 8-17 Heat sink mounting hardware used with a power transistor. (Courtesy of RCA, Consumer Electronics.)

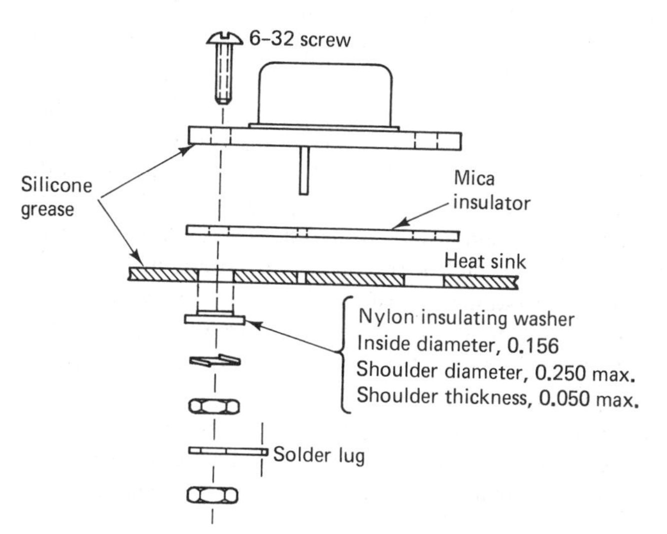

has to be replaced, it is good practice to renew the thermal grease and to check the mica sheet for shorts before power is applied.

8-10 troubleshooting the sound strip

The five most common troubles encountered in the sound strip are listed below:

1. No sound; picture and raster O.K.
2. Weak sound; picture and raster O.K.
3. Distorted sound; picture and raster O.K.
4. Buzz in sound; picture and raster O.K.
5. Intermittent sound; picture and raster O.K.

The fact that in each case the picture and raster are not affected indicates that the trouble must be located in the sound strip.

One exception is symptom 4 (buzz in sound; picture and raster O.K.). This symptom can be caused by defects in the tuner, video IF amplifiers, video amplifier, and the sound strip. Almost all picture circuit troubles that result in buzz also cause visible defects. Therefore, if the picture is of good quality and exhibits no sync problems, the difficulty is probably in the sound strip. Typical defects might be overloaded tuner or video IF amplifier caused by a defective AGC system, misalignment of tuner, video IF, or sound IF, and buzz pickup from the vertical deflection circuits.

8-11 vacuum tube sound IF amplifier

Figure 8-18 shows a typical schematic of a sound IF strip using a pentode sound IF amplifier. In this circuit tuned transformer L_4, L_5 separates the 4.5-MHz sound IF from the composite video signal and feeds it to the pentode tuned sound IF amplifier (V_5A). *Neutralization* (at the relatively low frequency of 4.5 MHz) is not necessary because the screen grid of the pentode reduces the interelectrode capacitance to a very small value, thereby reducing the undesired feedback to a negligible amount.

Coupling capacitor C_{23} and L_{11} form a series resonant circuit tuned to 4.5 MHz. The sound IF voltage developed across L_{11} feeds the grid of the pentode IF amplifier.

The pentode is made into a limiter by reducing the screen and plate voltages. This may be done by increasing the size of the decoupling resistor R_{30}, which, in effect, reduces the plate and screen supply voltages, shifting the tube's operating point toward saturation. Under these conditions, a relatively small input signal can drive the tube between cutoff and saturation, removing all amplitude variations. C_{25} is an RF bypass capacitor that ties the screen grid and the lower end of the output tuned circuit to ac ground.

The output double-tuned transformer performs two functions. Its most obvious function is to restrict the bandpass of the stage to those frequency-modulated signals clustered (± 25 kHz) about the 4.5-MHz IF carrier. C_{26} and C_{27} resonate the transformer to this band of frequencies.

The quadrature grid detector (V_{6A}) performs the same limiting and detecting functions as the gated beam detector, but it also operates as a "lock-in oscillator" which improves its limiting capabilities and makes it an extremely sensitive detector. As little as 0.15 V rms

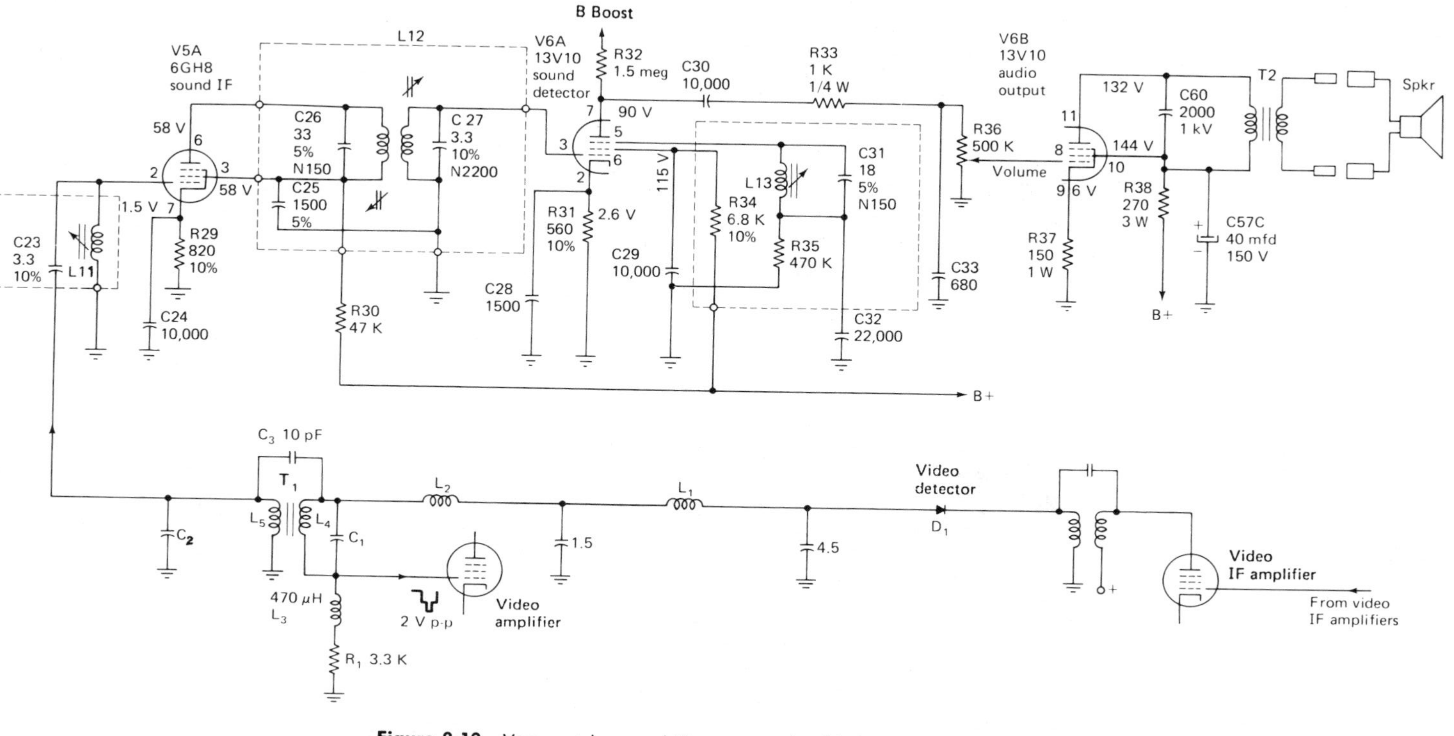

Figure 8-18 Vacuum tube sound IF system used in black & white receivers.

input can produce an output of approximately 18 V rms. This means that in weak signal areas this detector can produce strong noise-free audio although the picture may be weak and snowy.

The audio power amplifier V_{6B} is of conventional Class A_1 design that provides enough power amplification to drive the speaker.

QUESTIONS

1. Why are FM receivers relatively insensitive to electrical noise?
2. In what way does the sound strip of a color receiver differ from that of a black & white receiver?
3. Where is the sound IF take-off placed in a black & white television receiver that uses transistors?
4. Why do color receivers use a separate sound IF detector, and why is the sound takeoff placed in the second or third video IF amplifier?
5. Explain how signal bias may be used for amplitude limiting.
6. Why is a small resistor placed in series with the collector of the sound IF amplifier shown in Figure 8-7?
7. What are the two types of FM sound detectors used in television receivers and what are their general characteristics?
8. Explain how a phase-locked loop (PLL) can be used as an FM detector.
9. In what other television receiver circuits are PLLs found?
10. What would be the probable effect on receiver operation if the sound IF detector became defective?
11. What are the bandwidth requirements of the TV IF amplifier? Why?
12. How does the ratio detector convert frequency changes into amplitude changes?
13. Explain how the ratio detector provides amplitude limiting.
14. Explain how the phase-locked-loop detector works.
15. What would be the effect on receiver operation if C_7, the stabilizing capacitor in the ratio detector circuit of Figure 8-9, opened?
16. Explain how a heat sink is able to keep a transistor from overheating. Also explain the purpose of silicone grease.
17. What is crossover distortion, and how is it avoided?

---

chapter 9

Picture Tubes and Associated Circuits

9-1 basic structure

The end result of the television system is the picture of the televised scene. As such, it should not be surprising that the entire design of the television receiver is based upon the characteristics and demands of the television picture tube, or *kinescope* as it is sometimes called. Thus, the nature of the composite video signal, the required amount of receiver amplification, the method of raster formation, and the power supply requirements are all based upon the needs of the picture tube.

Figure 9-1 is the block diagram of the complete color television receiver. The shaded areas represent those sections of the receiver that are discussed in this chapter. The color receiver picture tube requires the use of many auxiliary circuits that compensate for beam deflection errors that unavoidably occur. An example of one of these is the automatic degaussing circuit located in the low-voltage power supply that eliminates stray magnetic fields that have detrimental effects on color picture tube purity. Automatic pincushioning is another circuit that minimizes a form of deflection distortion that tends to stretch the deflection in the corners of the raster. Also discussed are the dynamic convergence circuits and setup procedures used to converge the three electron beams throughout the color picture tube screen.

The beam deflection problems found in a color receiver are far more complex than those found in a black & white picture tube. Therefore, black & white picture tubes do not usually have auxiliary picture tube circuits.

The television picture tube in both color and black & white receivers consists of one or more electron guns, a thick glass envelope and faceplate combination, and a fluorescent screen. The basic features of these tubes are shown in Figure 9-2.

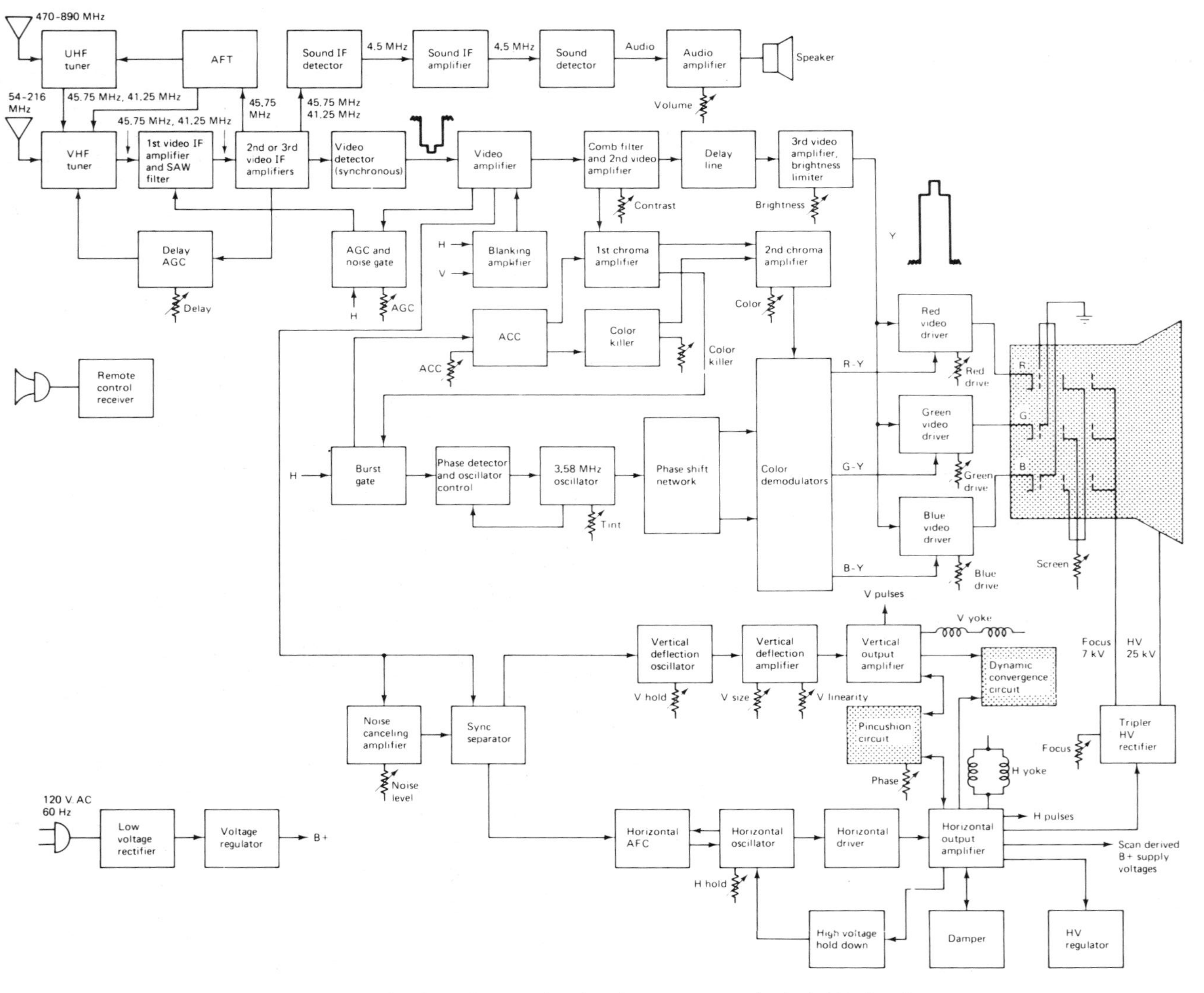

Figure 9-1 Complete block diagram of a color television receiver. The shaded blocks indicate the circuits to be discussed in this chapter.

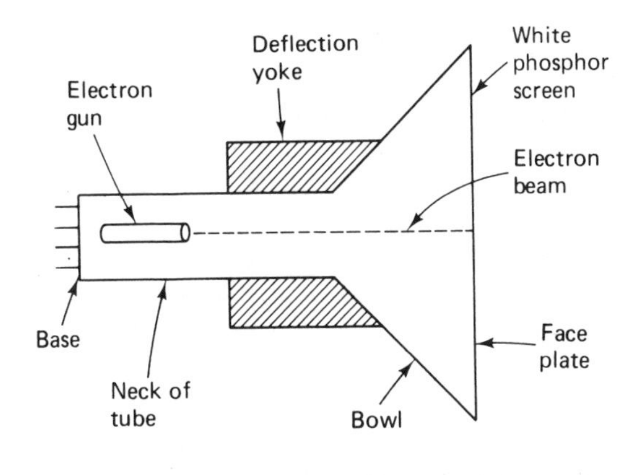

(a)

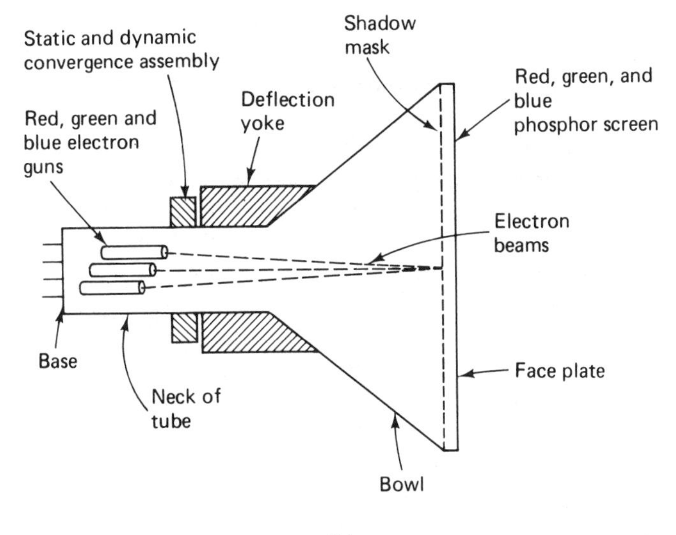

(b)

Figure 9-2 Essential features of (a) a black & white picture tube and (b) a color picture tube.

Electron guns are used in each tube to provide high velocity, well-focused, electron beams. In a black & white picture tube there is only one gun; in a color picture tube there are three guns.

In both color and in black & white receivers the electron beam is accelerated and made to strike a phosphor screen. The impact of the high-velocity electron beam causes the phosphor to fluoresce and to emit light. The brightness of the light output depends upon the velocity of the electron beam, the number of electrons in the beam, and the type of phosphor material used. The color of the light output depends upon the type of phosphor used.

Placed directly above the phosphor screen in a color picture tube is a shadow mask or aperture grille. This sieve-like plate contains thousands of holes or slots and is used to help form, shape, and direct the electron beams to the proper color phosphors.

The raster is generated by the simultaneous horizontal and vertical deflection of the electron beam. Deflection in both color and black & white picture tubes is obtained by the interaction of the magnetic field surrounding the electron beam and a magnetic field generated by the deflection yoke mounted on the neck of the picture tube.

Also mounted on the neck of the color picture tube is the dynamic convergence assembly. This yoke-like structure provides increasing correction of the deflection of the electron beams as they move further away from the center of the picture tube screen, thereby preventing color fringing near the edges of the picture tube screen.

The glass envelope. All modern picture tubes are made of glass. The glass may be new or old, depending upon whether the tube is rebuilt or not. A rebuilt tube is a tube that was once defective and has had its gun and phosphor renewed. Many tubes in use today use old glass for the envelope.

The face of the picture tube is almost 1/2 in. thick. The neck and bowl are somewhat thinner. Such heavy construction is necessary to support the enormous atmospheric pressure exerted on the picture tube caused by the extremely high vacuum inside it. A high vacuum is needed to ensure proper picture tube operation. Unfortunately, this combination of high vacuum and glass results in a dangerous condition. It is possible that careless handling or an accidental blow can shatter the glass envelope and an extremely dangerous implosion can be produced. If this happens, the glass envelope collapses inward and heavy pieces of glass may fly in all directions with considerable force. It is highly recommended that safety goggles be worn whenever picture tubes are being handled.

Service technicians often have to discard defective picture tubes. For this to be done safely, these tubes must have their vacuum destroyed. All picture tubes are manufactured with a vacuum seal located in the base of the picture tube. This seal is pointed and is the thinnest part of the picture tube. When a picture tube must be disabled, this seal should be broken.

The usual procedure is to place the picture tube in a carton. The bowl and as much of the tube's neck as possible should be covered with cardboard. This limits flying glass in case of an implosion. Then the vacuum seal is broken. The seal may be broken by snipping it with a pair of diagonal pliers or by placing a long metal rod against it and striking the end of the rod with a hammer.

Face plate. The danger of picture tube implosion has been recognized by the picture tube manufacturers as well as by receiver manufacturers. Older receivers were manufactured with a laminated safety glass window placed in front of the picture tube. Unfortunately, this resulted in undesired reflections and accumulations of dust coating the inside of the protective window. To prevent this, modern picture tubes have the safety glass bonded to the faceplate of the tube. This eliminates the need for a protective window. Bonding may be accomplished by cement or by means of a welded tension band placed around the rim of the tube.

As a means of increasing picture contrast some black & white picture tubes have a tinted faceplate. The tint makes the faceplate gray. Unfortunately, light transmission is also cut by about 22%. Faceplates may also be frosted in the manner of light bulbs. This is done in order to diffuse and reduce annoying faceplate reflections of lamps or other light sources in the room.

The faceplate thickness and its lead or strontium content also ensure that the X-ray radiation produced by the picture tube will be reduced to a safe level.

Some General Electric picture tubes called Neo-Vision use a faceplate that contains neodymimin oxide. This material gives the faceplate a blue color. The blue-tube glass has two advantages. First, the blue glass absorbs the yellowish room light associated with ordinary house lighting. This is said to produce richer, more natural colors. The second advantage is the appearence of the blue tube when it is turned off. This is in contrast to the gray or green appearance of ordinary tubes when they are turned off.

Picture tube size and shape. In the early history of both black & white and color television the face of all picture tubes was round. This was primarily due to the type of glass available and the technical manufacturing know-how of the time. Round picture tubes have at least two important disadvantages. First, because the television picture is a rectangle having a 3:4 ratio, a good deal of picture tube area is not used. See Figure 9-3. As a result, parts of the picture do not fall on the face of the tube. Second, the unused portions of the picture tube cause the cabinet size of the receiver using the picture tube to be larger than it might be if the tube were rectangular.

Figure 9-3 Round-face picture tube and the television raster. The rectangular picture tube accommodates the entire raster. Also illustrated is the method for determining picture tube size.

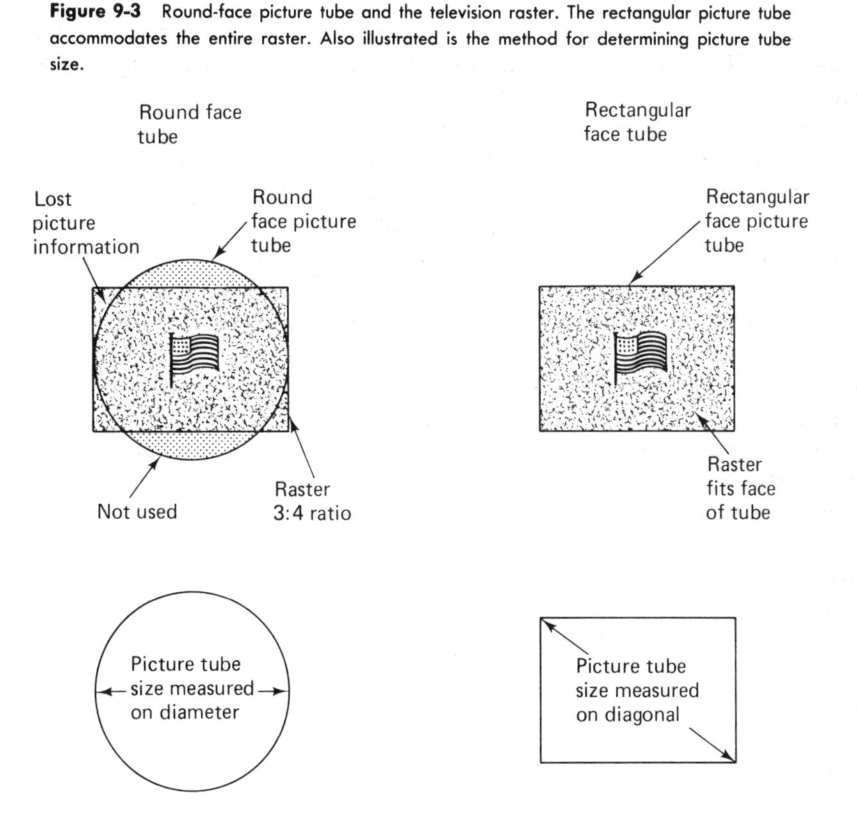

Figure 9-4 Color television picture tube. (Courtesy of Electronic Components Group, GTE Sylvania.)

Almost all modern black & white and color picture tubes have rectangular faces. See Figure 9-4. This shape permits the full 3:4 raster to be displayed with no loss of corner information.

Picture tube size is rated in either the diameter of an equivalent round tube or in terms of surface area. Thus, a tube rated as a 10-in. round picture tube has a *diameter* of 10 in. A rectangular picture tube rated as 25 in. has a *diagonal* length of 25 in. The diagonal length is used because this would equal the diameter of a round tube, and it makes it possible to make a comparison between both tubes. A 25-in. rectangular tube would have a width of 20 in. and a height of 15 in. making a surface area of approximately 300 sq in. Picture tubes used for television have varied in size from 1 in. to 30 in.

Picture tube numbering system. The first number in the television picture tube numbering system used by American manufacturers represents the picture tube face size within a 1½-in. approximation. For example, the first number in a 23EGP22 color picture tube indicates that this tube has a 22.375-in. rectangular faceplate. A 21FBP22A is a round color picture tube whose diameter is 19.25 in. Unfortunately, there is no way of knowing from the picture tube number whether the tube is round or rectangular. This information is listed in the manufacturer's tube data or in a tube manual. Picture tubes may also use metric dimensions to indicate their diagonal size. For example, a 470DLB22 Trinitron color picture tube indicates that its diagonal size is 470 mm in length.

The letters following the picture tube size are EIA* alphabetical designations given in the order of registration that differentiate one tube from another. Picture tubes of the

*Electronic Industries Association.

same size but of different letters are not always directly interchangeable. Interchanging is possible, but it should only be done after consulting available tube substitution guides.

The letter V following the faceplate size number indicates that this tube is using a system of faceplate identification that more accurately indicates picture tube size. For example, a color picture tube that has a diagonal rectangular dimension of 22.995 in. might have been designated a 25ABP22 under the old system. Under the new system the designation would be 23VABP22, accurately reflecting the actual picture diagonal measurement.

Phosphor numbers. The letter P and the number following it refer to the type of fluorescent phosphor that the tube uses. A color tube is designated by P22 and a black & white tube by P4. Picture tubes using metric nomenclature use the letter B instead of P to designate the phosphor number.

The phosphor number indicates two things: the duration of light output after excitation (persistence) and the color of the light. A P4 phosphor has medium persistence and produces a gray-white light output. The P22 designation indicates that the color picture tube uses three medium-persistence phosphors: one for red, one for green, and one for blue.

Older color tubes had phosphors of unequal efficiency. That is, for equal amounts of electron beam current the light output of each phosphor would be different. Red was the phosphor that had the poorest efficiency and therefore required the highest beam current. In modern color tubes the phosphors have been made to have approximately the same efficiency; therefore, all three beam currents may be made equal.

Other P numbers are possible, but they are found in cathode ray tubes that have special applications other than television. For example, radar display tubes may require long persistence phosphors that have a greenish-yellow light output. These are designated P7. Oscilloscope cathode ray tubes may have a green light output of medium persistence and be designated P1. Other types of cathode ray tubes used for photographic applications may have short persistence phosphors that emit blue light and are designated P11.

Some picture tube designations have letters following the P number, for example, 21HP4A. The letter A refers to a change that has been introduced into the tube compared to tubes of the same type that are not so designated. Details about the differences may be obtained from the manufacturer's tube manual.

Worldwide type designation system. A new system of picture tube numbering called the Worldwide Type Designation System or WTDS is replacing the numbering system previously used by U.S., European, and Japanese manufacturers. The WTDS has 5 parts reading from left to right. The first part is a single letter indicating whether the tube is a picture tube (A) or a monitor (M). Next is a 2 digit number indicating screen size in centimeters. The 3rd part is a 3 letter family code assigned in sequence from AAA to DZZ for US tubes, from EAA to HZZ for European tubes and JAA to MZZ for Japanese tubes. The fourth part is a code of one or two digits indicating such things as different mounting hardware, or neck size, which makes the tube a specific member within a particular family. The 5th part of the number is a one or two letter code indicating phosphor type. Color tubes are represented by the letter X and black & white tubes use the letters WW. An example of a picture tube number using the WTDS might be A62AAAOOX.

Picture tube length and deflection angle. The length of the picture tube from base to picture tube faceplate determines the cabinet depth. Picture tubes may have lengths as small as 9 in. (11GP4) or as much as 25½ in. (21FJP22). Picture tube length depends upon two factors: picture tube faceplate size and yoke deflection angle. The deflection angle is defined as the total angle that the electron beam can move as the result of the magnetic field created by the yoke. Actually, there are three different deflection angles: horizontal, vertical, and diagonal. As a result of the 3 to 4 aspect ratio, the vertical deflection angle is smaller than the horizontal deflection angle, and both of these are smaller than the diagonal deflection angle. The nominal deflection angle that is listed for picture tubes is usually the diagonal deflection angle. For example, a 20SP4 black & white picture tube is rated as having a 114° deflection angle. Its diagonal deflection angle is 114°, its horizontal deflection angle is 105°, and its vertical deflection angle is 87°.

Picture tubes may have deflection angles ranging from 50° to 114°. Color picture tubes have deflection angles of as much as 114°. Experimental black & white tubes have been constructed that have 180° deflection angles.

Figure 9-5 shows how deflection angle and faceplate size determine picture tube

Figure 9-5 Effect of deflection angle on picture tube length and faceplate size.

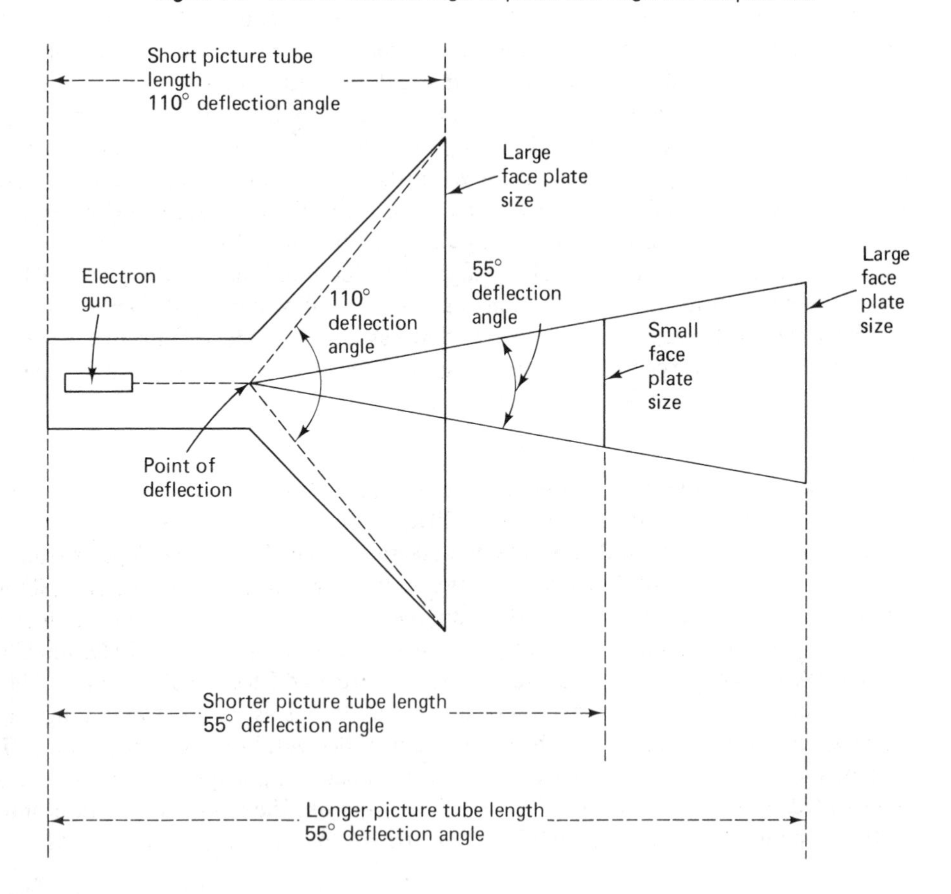

length. In the figure is shown a picture tube with two deflection angles, one of 55° and the other of 110°.

When the electron beam is deflected to its maximum angle, the distance between the faceplate and point of deflection determines the length of the picture tube. An example of this is shown in Figure 9-5. Note that when the distance between the faceplate and point of deflection for a tube with the same deflection angle as before has been decreased, it results in a shorter picture tube of decreased faceplate size. Also indicated in Figure 9-5 is the result of increasing picture tube deflection angle. Notice that increasing the deflection angle allows a shorter picture tube to accommodate a large faceplate.

Neck size. Large deflection angles introduce problems in yoke design. Stronger magnetic fields are required, that in turn force the diameter of the picture tube neck to be reduced. Black & white picture tubes that have deflection angles of 90° or less have neck diameters of 1 7/16 in. (36.5 mm). Picture tubes that have deflection angles of 110° and 114° require stronger magnetic fields and therefore have a smaller [1 1/8 in. (29.1 mm)]-diameter neck for the same-size faceplate picture tube.

Picture tubes used in many small black & white portable transistor receivers may have extremely small neck diameters of less than 1 in. Color picture tubes have larger neck diameters because they must accommodate three electron guns. The neck diameter for a 70° tube is 2 in. (50.8 mm) and for 90° or 110° color tubes it is 1 7/16 in. (36.5 mm) in diameter.

Base and socket considerations. Black & white picture tubes of modern manufacture have 8 or 12 pin terminals that connect the tube electrodes to the receiver circuit. Color picture tubes use bases that have either 13 or 14 pins. The basing diagrams for these tubes are shown in Figure 9-6. Black & white picture tubes rated for deflection angles of 90° or less use 12-pin bases. Tubes rated for 110° and 114° use 8-pin bases. Most modern color picture tubes have 13-pin bases; the 7-in. Trinitron color picture tube, however, has an 8-pin base. The larger 17-in. Trinitron uses a 14-pin base.

Tubes using a 12-pin base have leads coming out of the glass envelope that are soldered to terminal pins mounted on a plastic form that is cemented to the neck of the tube. Eight and 14 pin tubes use heavy-gauge steel wire for the leads that pass through the glass envelope of the tube as is done in miniature tubes. These same leads are used as the base pins of the tube. All picture tubes have a plastic key placed in the center of the

Figure 9-6 Picture tube bases.

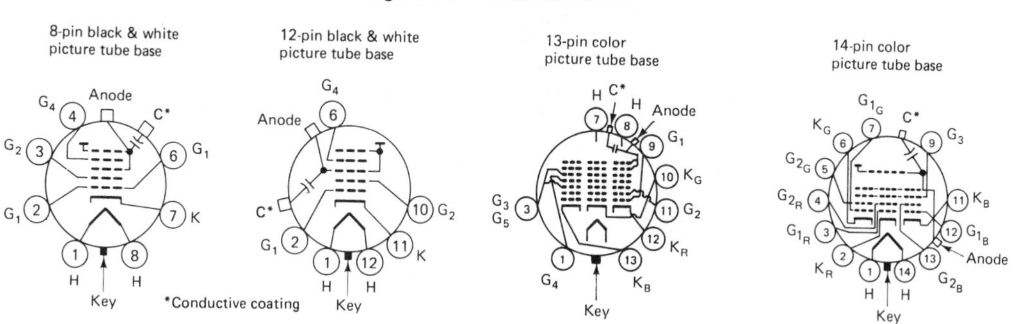

ring formed by the pins. This key is used to ensure that the socket and the base pins mate properly. The plastic key also serves as a protective cover for the fragile vacuum seal located under it. The pins of the picture tube are numbered starting with the left edge of the key way and proceeding in a clockwise direction toward the right side of the key. The base diagram and the actual picture tube pin positions are the same. When the front of the *socket,* is being viewed, however, pin positions are counted in a counterclockwise direction.

9-2 the electron gun

All picture tubes currently available base their operation on a beam of electrons generated by the electron gun. All practical color picture tubes have three electron beams and hence *three electron guns.* Black & white picture tubes have one electron gun. A cross-sectional view of a black & white picture tube showing its internal gun structure as well as the yoke and centering controls mounted on the neck of the tube is shown in Figure 9-7. This arrangement of a heater, a cathode, and five grids and anodes is called the *unipotential focus electron gun* and is used in all modern black & white picture tubes. Other systems using fewer grids are possible. One system using only four grids is commonly used in color picture tube guns. A cross-sectional view of this gun is shown in Figure 9-8(a). This gun is called the *bipotential focus electron gun.* In an attempt to reduce spot size (a 30% reduction) and improve focus, in-line picture tubes have used six grids and anodes in an arrangement called a *tripotential electron gun.* Three of the six grids are used for focus. This system requires the use of two focus supplies, one of 7 kV and the other of 13 kV in addition to the second anode supply voltage of 25 kV. See Figure 9-8(b). Another high-resolution electron gun is the *hi-bipotential* (hi-bi) gun. In this scheme only one focus electrode is used as in the bipotential gun. However, its voltage is made 28.2% of the second anode voltage, making its voltage about 7 kV compared to 5 kV for the bipotential gun.

Figure 9-7 Cross-sectional view of a black & white picture tube showing its electron gun. (Courtesy of RCA, Consumer Electronics.)

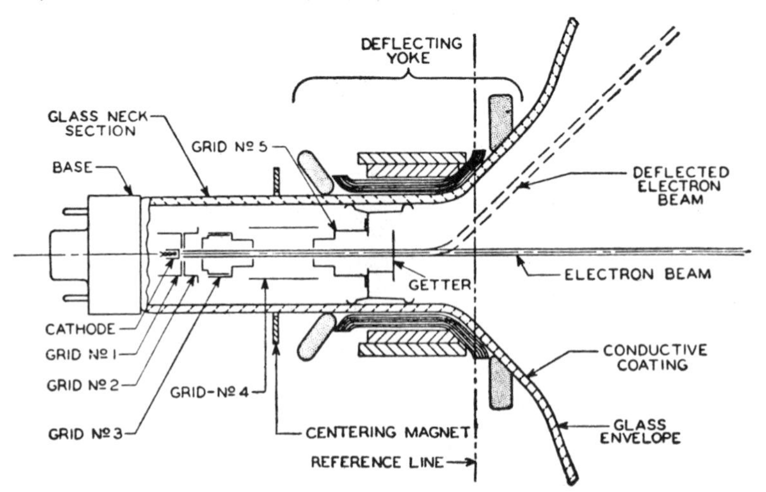

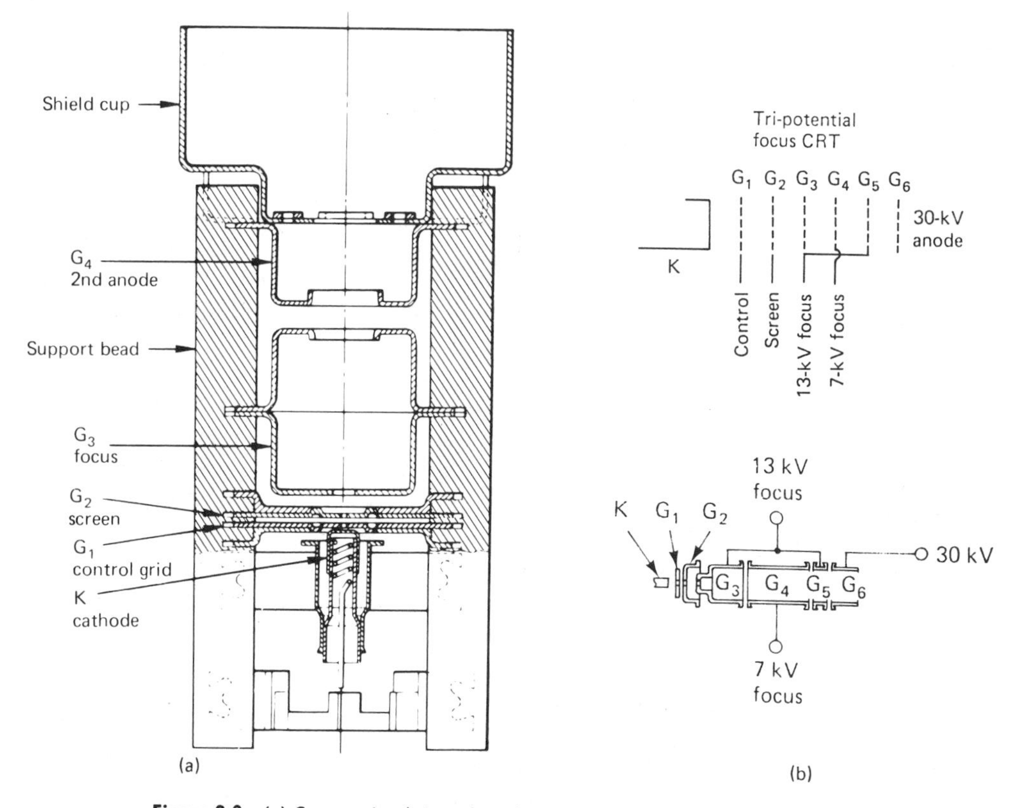

Figure 9-8 (a) Cross-sectional view of a color picture tube gun; (b) schematic diagram and cross-sectional view of a tripotential gun. [(a) courtesy of RCA, Consumer Electronics; (b) courtesy of Magnavox.]

All electron guns, no matter what their arrangement, are constructed as flat plates or cylindrical metal tubes of various lengths that may be closed at each end with an aperture disk that has a small hole in its center. See Figure 9-8(a). Each section of the electron gun is tied to the other sections of the gun by means of glass supporting beads that are fastened to the cylinders. These cylinders establish the required electrostatic fields necessary to produce an electron beam. The aperture disks are used to restrict the number of electrons that can reach the phosphor screen to those that are moving along the axis of the tube toward the screen. This tends to make the spot of light produced by the electron beam on the screen as small as possible and it also prevents stray electrons from illuminating the screen.

The heater. Indirectly heated cathodes are used in all television picture tubes for the emission of electrons. The heater consists of a tungsten wire that is twisted into a helix of sufficiently small diameter to fit into a cathode sleeve. See Figure 9-8(a). The heater wire is coated with several layers of aluminum oxide that forms an insulating barrier between the heater and the cathode. This insulation must be sufficient to prevent heater to cathode shorts from occuring. Typically, this insulation is rated at 300 V dc in black & white picture tubes and 1000 V in large color tubes.

Picture tube heaters are frequently rated for 6.3 V, but a number of small picture tubes do have higher as well as lower voltages ratings such as 2.35 V, 4.2 V, 9.45 V/300 mA, 11 V/85 mA, 12.6 V/150 mA for use in small-sized battery operated portable television receivers. These tubes are usually used in solid-state receivers and obtain their heater voltage from the low-voltage dc supply.

The current ratings of black & white tubes range from 85 to 600 mA. Tubes with a 600-mA rating are used in parallel heater arrangements and 450-mA tubes are used in series string heaters. Color picture tubes have higher current ratings because they have three electron guns and are in effect three picture tubes in one. Typical heater current ratings range from 580 to 1800 mA. The three heaters are usually tied in parallel inside the picture tube.

The cathode. The cathode sleeve is a cylinder that is closed on one end. Inserted into the sleeve is the heater. The closed flat portion of the sleeve is coated with an oxide material that is capable of generating high electron emission at relatively low temperatures. The cathode is usually kept at a positive potential, with respect to ground, of between 40 V and 250 V. The higher voltage is usually found in color picture tubes. In addition to the dc voltage found on the cathode, the usual practice is for the composite video signal to be fed to the cathode of the picture tube. Typically, this signal should have an amplitude ranging between 40 and 150 V peak to peak.

Grid 1 (the control grid). The control grid electrode (G_1) is adjacent to the cathode emitting disk. See Figures 9-7 and 9-8. It may be constructed as a short closed cylinder or a metal disk. In both cases, the electron beam exits the grid through a small aperature hole in its center. This hole helps shape the electron beam into a small-diameter circular beam. The control grid may have almost any potential: It may be zero, positive, or negative. The potential on the grid must be such that it establishes a difference of potential (bias) between the control grid and the cathode and makes the control grid negative with respect to the cathode. This difference of potential also creates an electrostatic field between the grid and cathode that forces divergent electrons to come together at a *first crossover point* in front of the grid. At the same time this electrostatic field determines the number of electrons that are able to reach the crossover point, and subsequently the phosphor screen of the picture tube. Increasing the bias voltage by making the grid more negative than the cathode reduces the number of electrons that can reach the screen and thereby reduces the light output of the picture tube. The intensity or brightness of the picture tube light output depends upon the number of electrons that strike the screen and upon their velocity. Increasing the bias sufficiently will cause the light output to drop to zero because few if any electrons will be able to reach the phosphor screen. Decreasing the bias will increase picture tube brightness by making more electrons available.

The brightness control of a television receiver may be connected in the grid or cathode circuit of the picture tube and is able to vary the bias of the tube and hence the brightness of the picture tube light output.

Grid 2 (first anode). The preaccelerating electrode (G_2) is located between the control grid and the accelerating anode (G_3). This electrode is called the first anode in black & white picture tubes and the *screen* in color picture tubes. The purpose of this electrode is not only to accelerate the electrons toward the phosphor screen but also to provide focusing action. Acceleration is produced by a positive voltage placed on this electrode. Increasing this voltage will cause an increase in picture tube brightness. Typically, the voltage found on this electrode in black & white picture tubes may be as little as 45 V or as much as 350 V. In color receivers the first anode may have voltages of 800 or 900 V. This voltage is adjustable; each of the three screens of a color picture tube gun usually has a potentiometer that allows the screen voltage to be varied.

Grid 3 (the focus electrode). Following the first anode is an electrode called the *focus electrode.* See Figures 9-7 and 9-8. The electrode is necessary because the divergent electrons leaving the first crossover point mutually repel each other and tend to spread out. If this were allowed to continue, the electrons would strike the phosphor screen as a large blob instead of the small diameter spot that is necessary for good picture definition. To force the divergent electrons into a fine point at the screen requires some form of focusing action. Two basic methods are used: *magnetic focusing* and *electrostatic focusing.* All modern picture tubes use electrostatic focus. Black & white tubes use a form of low-voltage focus; color picture tubes use either low-voltage focus or high-voltage focus.

Electrostatic focus. Electrostatic focus is the result of one or more electrostatic fields that are set up in the electron gun. These fields act like an optical lens that forms an image of the first crossover point on the screen of the tube.

An example of an electrostatic lens consisting of two anodes is shown in Figure 9-9. This high-voltage focusing arrangement is primarily found in color tubes. On the extreme left of the drawing is the source of electrons, the crossover point. The lines radiating to the right represent the divergent paths taken by electrons as they are accelerated by the positive potential on the electrodes toward the picture tube screen. The first electrode is called a *focusing anode* and may be at a potential of as high as 7000 V. The accelerating anode is at a potential of 20,000 V. Because of the 13,000-V difference of potential that exists between the two anodes, a strong electrostatic field is created between the two anodes. Because the accelerating anode is more positive than the focusing anode, the direction of these lines of force is from the accelerating anode to the focusing anode. It is to be remembered that electrons tend to be accelerated in a direction opposite to the direction of the arrows. Therefore, the diverging electrons coming from the cathode will be acted on by three forces: (1) As a result of their negative charge they will be repelled from one another and will tend

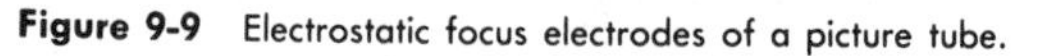

Figure 9-9 Electrostatic focus electrodes of a picture tube.

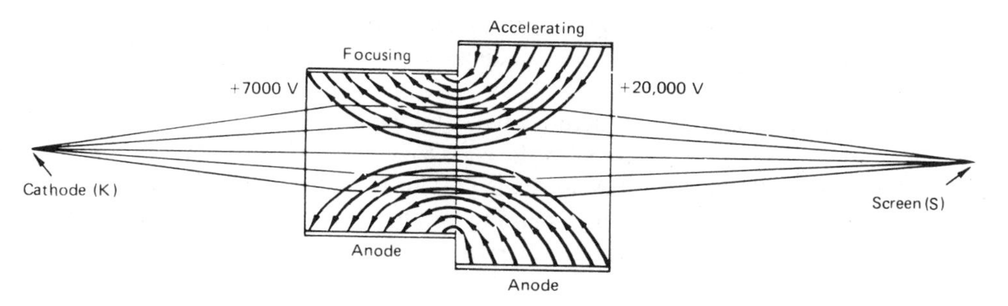

to spread out. (2) They will move forward at high velocity because of the large accelerating voltage. (3) They will attempt to follow the path laid out for them by the electrostatic field.

Since the shape of the electrostatic field is convex toward the axis of the tube, the entering divergent electrons will be forced toward the axis of the tube as they travel toward the screen. The entering velocity of the electrons into the lens system is much less than the exit velocity; thus, the electrostatic field at the input will have a greater effect on electron direction than will the electrostatic field on the output side of the electrostatic focusing lens.

If the difference of potential between the focusing anode and the accelerating anode is proper, the divergent electrons will be forced to come together at the screen. If the potential difference between the anodes is too high, excessive bending of the divergent electrons will take place and focus will take place at a point in front of the screen. If the potential difference is too low, the converging beam of electrons will strike the screen before coming to a focus and again the beam will appear on the screen as a large spot of light that will not produce a high definition picture.

In actual practice, two types of electrostatic focus have been used in both color and black & white picture tubes: *high-voltage electrostatic focus* (focus voltage: 4 to 13 kV) and *low-voltage electrostatic focus* (focus voltage: −200 to +300 V). Almost all modern black & white picture tubes use a system of low-voltage focus called *unipotential focus* using the *Einzel lens.* Older color tubes by and large use high-voltage focus, called *bipotential focus.* Modern color tubes are manufactured with electron guns that either use low-voltage focus or *tripotential focus* systems.

Magnetic focus. Older black & white picture tubes (1946 to 1959) used magnetic focus. In this form of focusing a strong magnetic field produced by an electromagnet or a permanent magnet mounted behind the yoke on the neck of the tube forces divergent electrons to take a helical (coil-like) path to reach the picture tube screen. The helical path is usually only one turn long. With the proper magnetic field strength oppositely divergent electrons will move forward along coil-like paths that rotate in opposite directions and meet at the phosphor screen, thereby bringing the electrons together again.

The accelerating anode (second anode). The accelerating anode (G_4) consists of a cylinder in combination with a conductive coating on the inner surface of the picture tube envelope. See Figure 9-7. A disk with a small hole in the center closes the cylinder and prevents stray electrons from reaching the picture tube screen. The conductive coating is a form of graphite called *aquadag* that is sprayed or painted onto the inner surface of the envelope. The coating is connected to an anode cylinder by means of metal clips extending from the second anode cylinder to the walls of the tube. The aquadag coating then extends up to the phosphor screen, but it does not touch it.

The outer surface of the picture tube envelope also has an aquadag coating. The inner and outer coatings are separated by the glass envelope; therefore, they form a capacitor of approximately 200 pF. This capacitance is used to filter the high-voltage power supply. The outer aquadag coating is tied to ground by means of spring clips that press up against the tube when the tube is installed in the receiver. The inner conductive coating is connected to the high-voltage power supply by means of a circular metal cup imbedded in the wall of the bulb. This connecting point is called the *ultor.*

The accelerating anode is connected to the positive terminal of a high-voltage supply. Depending on the type of tube used, this supply may range in voltage output from 7 to 80 kV. The higher voltage is necessary in some projection picture tubes. Typically, the second anode voltage for a color tube is 25 kV and that for a black & white tube is 20 kV.

Path of current flow. Figure 9-10 illustrates the path of current flow in a television picture tube. Notice that electrons flow from the cathode of the picture tube where they are emitted to the screen of the tube. At this point the high-velocity electrons may dislodge secondary electrons that are collected by the positive second anode aquadag coating. In modern tubes the phosphor screen is coated with a thin coat of aluminum that conducts the electrons from the screen to the positive terminal of the second anode. From this point on, current flow is external to the tube and flows through the high-voltage power supply and completes the circuit by returning to the cathode of the tube. The high-voltage power supply may be considered to be made up of a battery and an internal resistance. The effect of the internal resistance is to lower the available accelerating voltage that appears between cathode and second anode. It is also responsible for changes in second anode potential that may occur when beam current levels change. For example, if a defect has caused the internal resistance to increase and if the brightness control has been adjusted for more brightness, the following action will take place. Beam current will increase and cause the second anode voltage to decrease. This in turn will cause a reduction in brightness, poor focus, and an increase in beam deflection. This condition is called *blooming.* The increase in deflection is the result of slower-moving electrons that are acted upon for a longer time by the deflection magnetic fields.

Figure 9-10 Path taken by electron current flow in a picture tube.

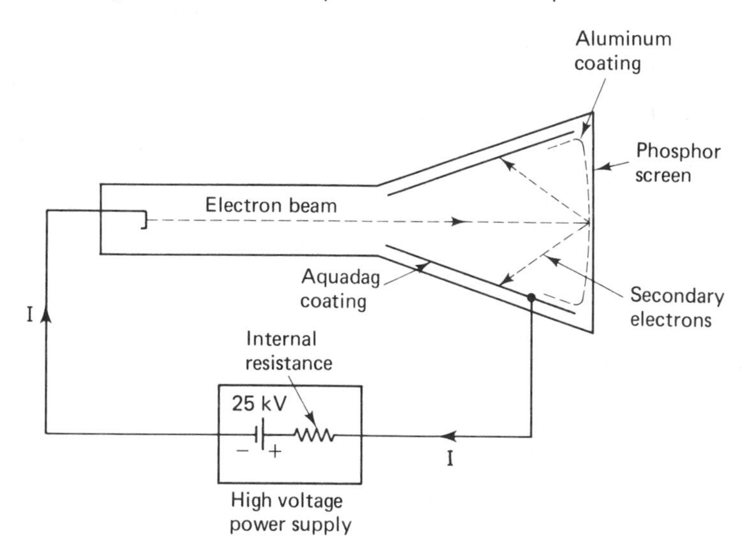

Beam currents of 100 μA are typical for black & white picture tubes. Color picture tubes have typical beam currents of 1000 μA. This higher beam current results from the use of a shadow mask and the use of three electron beams in color tubes. This places a greater demand on the high-voltage power supply of the receiver than does the black & white tube's beam current.

9-3 ion burns

In the early days of television it was discovered that after a few weeks of operation a permanent dark spot approximately the size of a silver dollar would form in the center of the picture tube screen. As time went by, the spot would get darker and in some cases the phosphor would literally flake off the face of the tube and leave bare glass. Upon investigation it was determined that the cathode was emitting negatively charged ions (molecules or atoms) as well as electrons. These ions are many times more massive than the electrons. Both ions and electrons are accelerated toward the phosphor screen by the accelerating voltages. When the electrons strike the screen, light is emitted, but when ions strike the screen, they have enough energy to burn the phosphor. This accounts for the darkened screen. Ordinarily it might be expected that the burn would be uniformly spread over the entire raster. A uniform burn requires that both electrons and ions are deflected by the same amount. Such burns in fact, do occur in tubes that use electrostatic deflection to produce the raster. In most modern television picture tubes, however, magnetic deflection is used and the more massive charged particles receive considerably less deflection than do the less massive electrons. As a result, the ions fall on the center of the screen and burn it.

Types of ion traps. Elimination of ion burns takes two forms: offset ion traps used in sets prior to 1965, and aluminized phosphor screens.

Many different ion traps have been used. One is called the *offset ion trap.* The principles used in the offset ion trap are similar to those used in all other ion traps. In general, the gun is constructed in such a way that it creates a diagonal electrostatic field between the first anode (G_2) and the focus electrode (G_3) that forces *both* ions and electrons to be deflected toward the wall of the G_3 cylinder. This action prevents the ions from reaching the phosphor screen. Unfortunately, this also prevents the electrons from reaching the screen. This is overcome by taking advantage of the fact that magnetic fields deflect electrons to a greater extent than do ions. Thus, one or more neck-mounted permanent magnets are used to deflect the electron beam so that it returns to the axis of the tube and is able to reach the phosphor screen.

Proper adjustment of the ion trap magnet was obtained by simultaneously rotating and moving forward and back the ion trap magnet on the neck of the tube. At the same time, the raster is observed and ion trap magnet adjustment continues until the raster is at maximum brightness.

Aluminum-coated screens. Modern picture tubes do not use ion traps to prevent ion burn because these tubes have been built with a thin aluminum coating placed over the phosphor screen. This coating is used in both color and black & white picture tubes, as shown in Figure 9-10. The aluminum coating is thick enough to act as a net that prevents the massive ions from reaching the phosphor screen while being coarse enough to allow the less massive electrons to pass through it. Furthermore, since the aluminum coating is a good heat conductor, it tends to carry away and dissipate any heat developed by the ion bombardment. Aluminized tubes provide several other advantages. It was pointed out earlier that in nonaluminized tubes the electron beam returns to the high-voltage second anode supply via secondary emission from the phosphor screen. There is a limit to how much secondary emission can be produced by the phosphor screen. Increasing the second anode voltage will increase the amount of secondary emission until this limit or sticking potential is reached. Further increasing in the second anode voltage will not increase secondary emission. Therefore, beam current is limited. This limit also means that the light output of the tube is fixed because it is directly related to the beam current. Because an aluminum coating is a good electrical conductor, it provides a direct return path for the electron beam current to the high-voltage supply. Therefore, the sticking potential limit is removed and higher second anode voltages may be used to obtain brighter pictures.

When the electron beam strikes the phosphor screen it produces a light output. Ordinarily, the light output radiates in all directions. As a result, approximately one-half of the light output is lost because it radiates to the rear of the picture tube. If an aluminized picture tube is used, a large percentage of the "lost" light can be reflected by the mirror-like aluminum coating out of the front of the tube. This greatly increases the brightness and contrast of the television picture tube.

9-4 flat picture tubes

Over the past few years a great deal of research has gone into the development of a flat panel picture display device to replace the electron beam picture tube. The major advantage of such a picture display panel is the reduction in television cabinet depth due to the elimination of the picture tube neck.

Flat panel picture displays will permit television receivers to be hung on the wall like an oil painting. In addition, portable television receivers of very small size become possible.

Flat panel TV displays are being developed along several different lines. The most promising schemes for black & white TV use are the flat CRT and the liquid crystal display (LCD). Color flat panel displays are also being developed.

Flat CRT. In the flat CRT system the electron gun is placed parallel and to the side of the phosphor screen. Two problems that had to be overcome in this arrangement were the distortion of the electron beam into an elliptical spot shape, and the difficulty of making the raster fill the corners and edges of the screen.

The tube produced by Sony is paddle-shaped and measures 5 × 2.1 × 0.65 in. The screen measures 2 in. diagonally. A cross-sectional view of this tube is shown in Figure 9-11(a). A unique feature of this tube is that the observer looks through the transparent front window and the electron beam at the light generated from the phosphor screen placed in the rear of the tube. This arrangement is said to produce greater brightness with less power consumption.

Electrostatic deflection is used for vertical scanning and magnetic deflection is used for horizontal scanning. Focusing and beam correction of the elliptical spot are provided

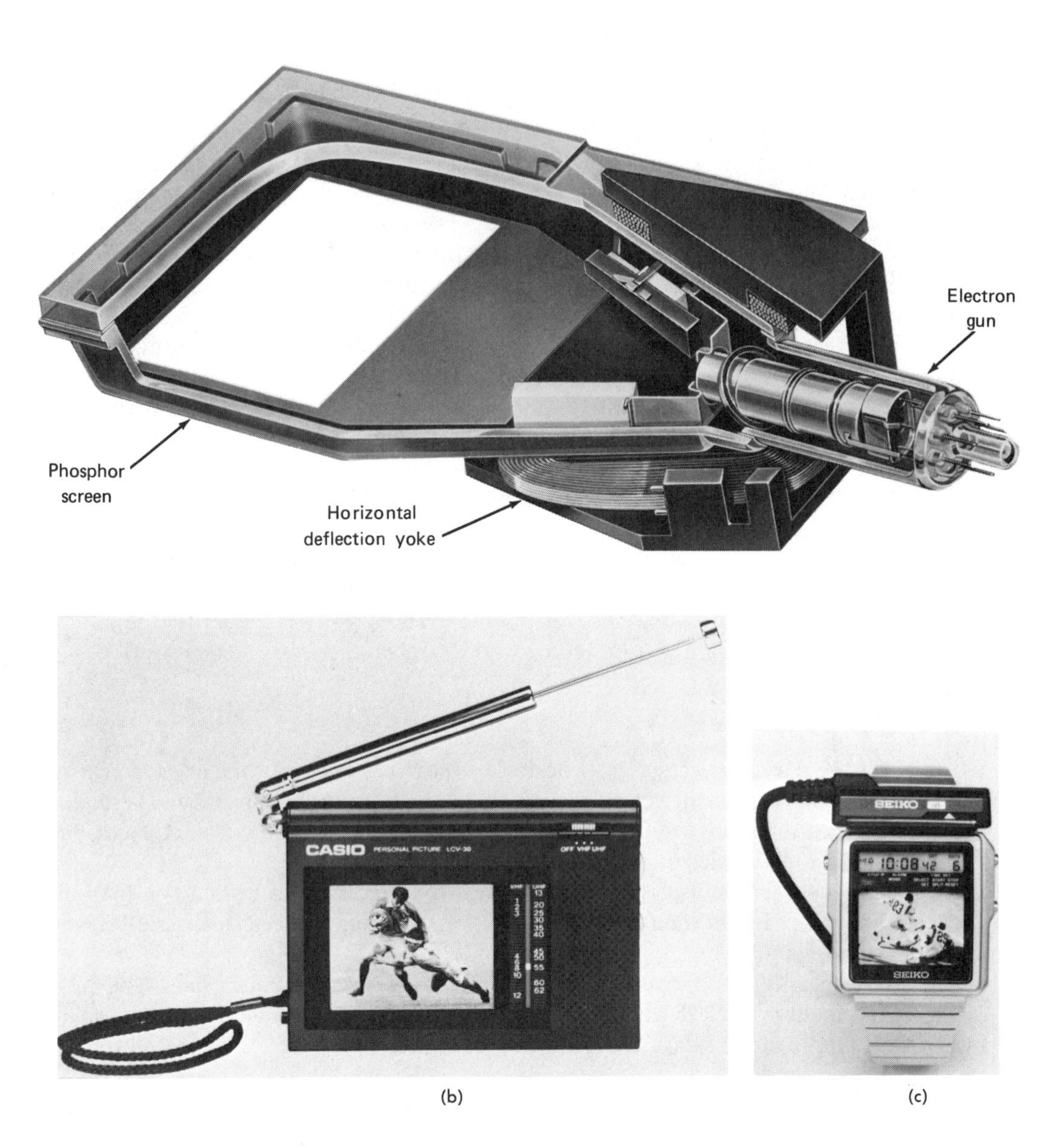

(b)

(c)

Figure 9-11 (a) Flat picture tube; (b) TV receiver using an LCD TV panel for display; (c) wristwatch television receiver using LCD display. [(a) Courtesy of Sony; (b) courtesy of Casio; (c) courtesy of Seiko.]

by a focusing field created by a transparent electrode on the inside surface of the front window.

Color and 50-in. CRTs of this type are said to be possible.

LCD TV panels. Liquid crystal displays of the type used in watches and calculators use ambient light which is selectively absorbed and reflected. The areas of absorption produce black. The reflected areas appear as white areas. Two examples of TV receivers using LCD TV panels are shown in Figure 9-11(b) and (c).

The typical *field-effect LCD* is constructed as a six-layer sandwich of metal plates, etched and plain glass plates, polarizer plates, and nematic fluid (liquid crystal). The liquid crystal is placed between two transparent metalized glass plates. When a voltage is applied across the nematic fluid, its orderly molecular structure is randomized, so that reflected light is blocked, making the energized area black. Removing the voltage restores the molecular structure of the LCD to its original orderly form, allowing light to be easily reflected. The exciting voltage is typically a 30-Hz square wave of about 8 V p-p. Ac voltage is used to prevent plating of the electrodes that would occur if dc were used.

Flat panel TV display schemes using LCDs make use of matrix scanning. A matrix display is an orderly mosaic of picture elements. Picture resolution improves as greater numbers of picture elements are used in the mosaic. The electronics necessary to scan the matrix display are shown in Figure 9-12(a).

The mosaic panel is constructed as a grid. Horizontal rows (220) of electrodes are run along the back plate of the device, one electrode for each row of picture elements. The vertical column electrodes (240) of the grid are deposited on the inside surface of the glass plate of the device. Coating both horizontal and vertical electrodes are transparent insulator layers. Placed between these plates is a square-shaped layer of material that is electrically connected to the electrodes. The material at the intersection of the row and column electrodes is the picture element that is excited when both electrodes are energized. In effect the pixel acts like an AND gate, with a number of additional requirements. The degree of excitement (light output) must be proportional to the video signal, resulting in a light output ranging from white through shades of gray to black. The second requirement is that during matrix addressing, the pixel holds its shade of gray until the next frame (any one line is scanned only once for each frame). Also, the pixel must be able to change its brightness level very quickly to avoid smear as the picture content changes. An example of a display matrix pixel is shown in Figure 9-12(b). Each pixel consists of a liquid crystal display element which conceals a thin-film switching transistor and storage capacitor fabricated directly on the panel. The storage capacitor keeps the display element "on" between adressing pulses.

As outlined above the array of pixels will require 220 row electrodes and 240 column electrodes for a total of 460 wires interfacing between the display matrix and the receiver. One method for reducing the enormous number of connections between the matrix and the receiver is to incorporate the driving electronics for the row and column electrodes directly on the display matrix panel. This arrangement can reduce the number of connections necessary to activate the display unit to as few as six per side.

9-5 the color picture tube

There are three types of color picture tubes currently being used. Shown in Figure 9-13 are simplified drawings of the three color picture tube electron gun arrangements. These are the three-gun delta configuration, the three-gun in-line system, and the Trinitron single-gun system.

In the delta system the three electron guns are positioned at 120° angles to one another and are tilted inward in relation to the axis of the tube so that they form a triangular configuration. Each gun corresponds to a specific color. The blue gun is usually positioned

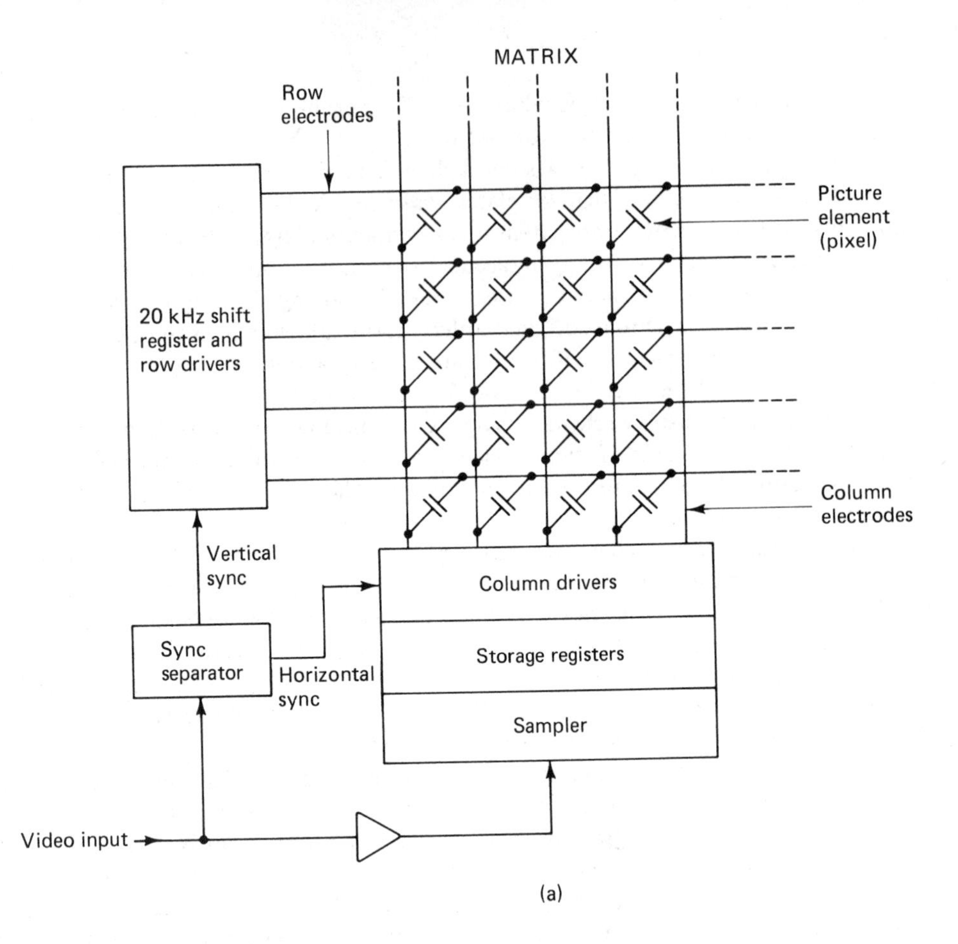

(a)

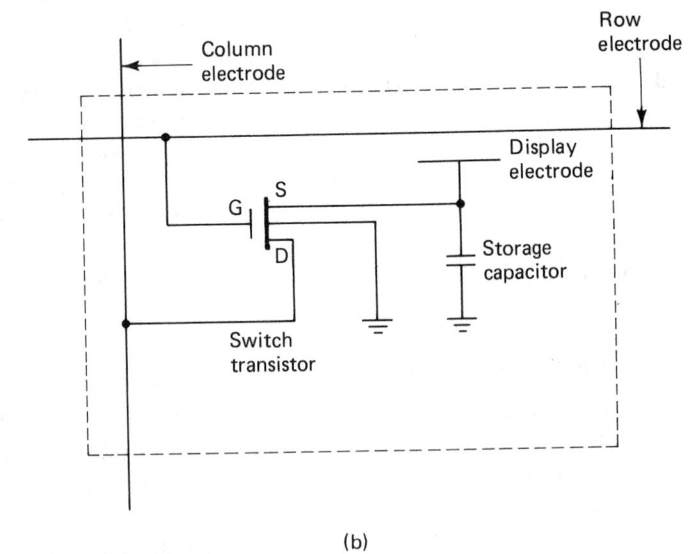

(b)

Figure 9-12 (a) Simplified matrix display indicating the method of pixel address; (b) picture element circuit.

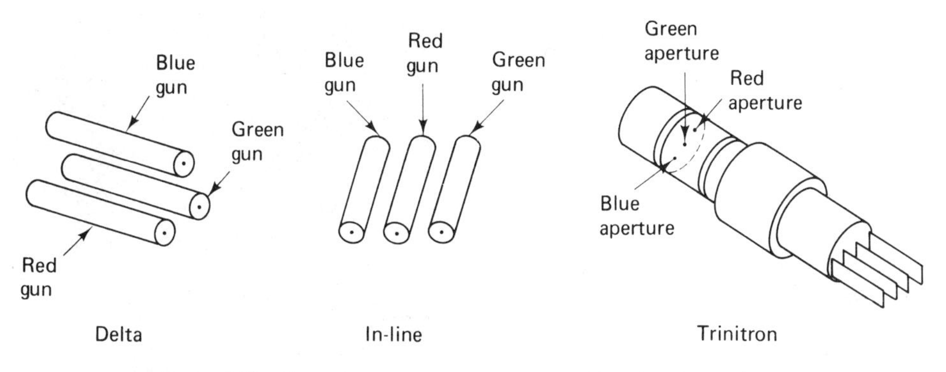

Figure 9-13 Three electron gun arrangements used in color picture tubes.

as the upward vertical apex of the triangle. Some manufacturers, however, position the apex downward.

In the in-line tube the three electron guns are positioned parallel to one another. The blue gun is on the left, the red gun is in the center, and the green gun is on the right. The Trinitron is an in-line gun structure for the three beams.

Pictures of both a delta-type and in-line electron guns are shown in Figure 9-14.

The Trinitron tube uses three electron beams generated by their own cathodes. All three beams, however, are accelerated, focused, and adjusted for convergence by a single gun structure. The Trinitron is similar to the in-line tube because the electron beam apertures are positioned so that the electron beams are on the same plane, with the outer two beams tilted inward so that they meet with the center beam at the screen. In this tube the center electron beam is responsible for green output, the right beam corresponds to red, and the left is blue. This arrangement is similar to that used by GE in-line tubes. However, it differs from the order used in RCA in-line tubes, which is similar to that shown in Figure 9-13.

Color picture tubes differ from black & white picture tubes in at least two other important ways. First, the luminescent phosphor screen is capable of emitting red, green, and blue light. Second, located between the electron gun and the phosphor screen is a sheet steel shadow mask of one type or other.

9-6 the phosphor screen

The physical arrangement of the phosphor screen depends upon which type of electron gun is used and upon the manufacturer. Thus, the three-gun delta tube uses a phosphor screen in which a group of three phosphor dots are arranged in a repeated form of a triad such as that shown in Figure 9-15(a). Depending upon the size of the picture tube, approximately 1,000,000 such dots forming approximately 333,000 triads are deposited on the glass faceplate. Each phosphor dot is responsible for its own color. From a front view the bottom apex of the triad is blue, the right apex is red, and the left apex is green.

In an attempt to increase the light output of color tubes many manufacturers blacken the space between phosphor dots. This tends to reduce scattered light and it allows the use of larger diameter electron beams without overlapping one another. Larger diameter electron beams mean more of the phosphor dot surface is used and therefore greater light output is possible. These tubes are referred to as *black matrix*-type picture tubes. In addition

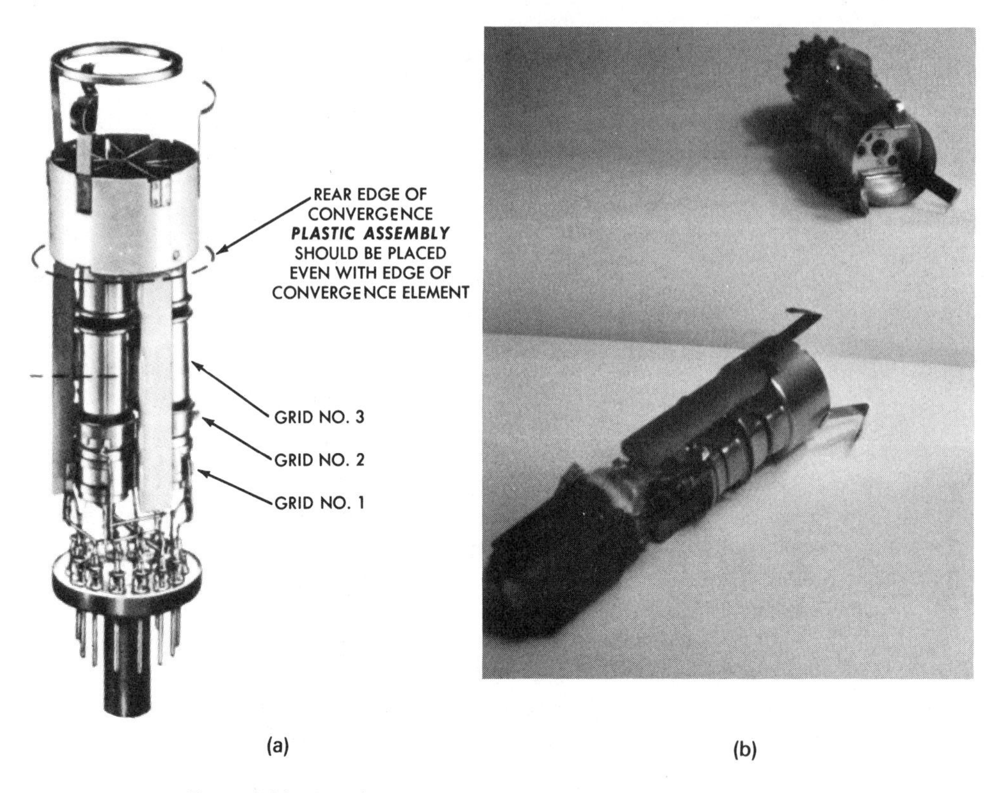

Figure 9-14 (a) Delta-type picture tube gun; (b) in-line picture tube gun.

to the delta-type tubes, some General Electric in-line tubes use screens that employ a phosphor dot structure.

As shown in Figure 9-15(b), RCA in-line tubes use a phosphor screen that consists of vertical stripes of alternating green, red, and blue phosphors. Each stripe is only a few thousandths of an inch wide.

The Sony Trinitron, shown in Figure 9-15(c), also uses vertical stripes, but they alternate in a red, green, and blue sequence.

9-7 the shadow mask

Situated directly behind the phosphor screen of each type of tube is a shadow mask. This is also illustrated in Figure 9-15. The physical arrangement of each mask depends upon whether phosphor dots or stripes are used, and upon the manufacturer. Thus, the delta electron gun system uses a sheet steel shadow mask that has one hole for each triad. This mask has been used in GE in-line tubes. RCA in-line type tubes use slot-shaped apertures, one vertical line of slots for one group of green, red, and blue phosphor stripes. In comparison, the Trinitron shadow mask, called an *aperture grille,* consists of vertical metal stripes. One space between metal stripes corresponds to one group of red, green, and blue phosphor stripes.

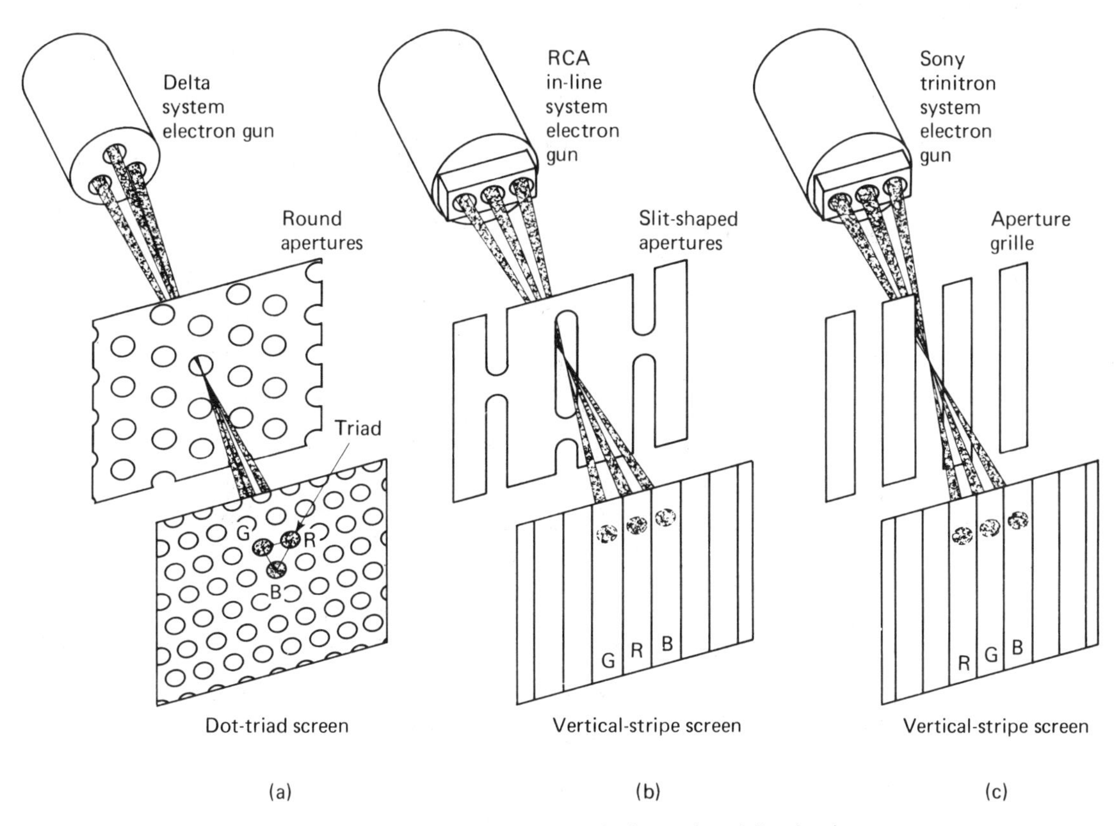

Figure 9-15 (a) Delta system indicating gun, shadow mask, and phosphor dot arrangements; (b) RCA in-line system; (c) Sony Trinitron system.

In each type of color picture tube the purpose of the shadow mask is to ensure that a specific electron beam emitted from each gun strikes only its corresponding phosphor. This is accomplished by the shadow mask because each electron beam enters the shadow mask at a different angle. This is most easily seen in the case of the in-line tubes. See Figure 9-16. In these tubes the electron beams are on the same plane. As a result, the center beam will move along the axis of the tube. The guns on the right and left will emit electron beams that because of their inward tilt will meet the center beam at the aperture grill. The diameter of the electron beams is wide enough to cover at least two slots in the aperture grille. As shown in Figure 9-16, the green beam will pass through the slots in the aperture grille and strike only the green phosphor. The aperture grille effectively prevents the green electron beam from striking the red or blue phosphors.

The electron beam corresponding to blue is designed to arrive at the aperture grille at an angle to the axis of the tube. If it enters the slot of the aperture grille as shown in Figure 9-16, it will only strike the blue phosphor. Here again, the aperture grille effectively prevents the blue electron beam from striking either red or green phosphors. Similarly, the red electron beam enters the aperture grille from the right and therefore is prevented from striking any phosphor other than red.

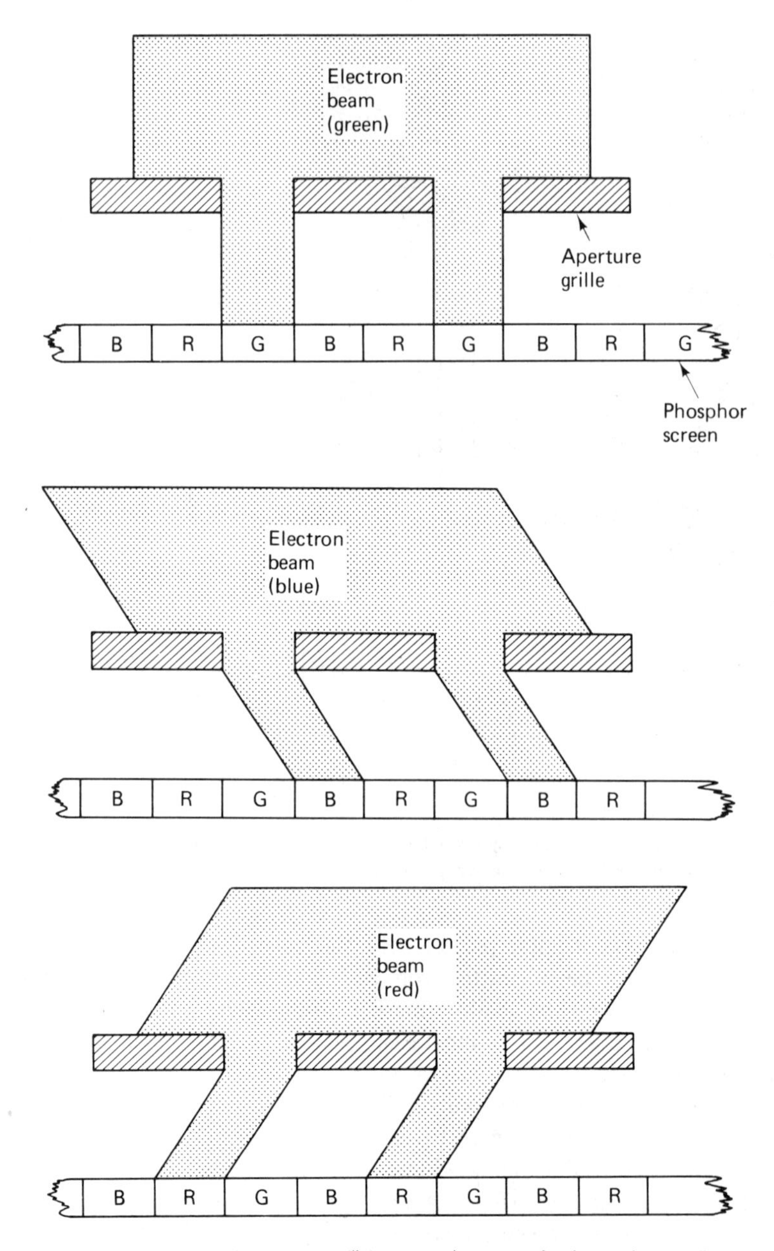

Figure 9-16 Shadow mask (aperture grille) ensures that a specific electron beam will strike only its corresponding phosphor. Drawings represent tube as seen from above.

The shadow mask performs a similar function in the delta-type tube. Here again, the electron beams enter the shadow mask holes at different angles [see Figure 9-15(a)] and are therefore prevented from striking any phosphor dots other than the ones they are identified with.

9-8 purity

Ideally, the entry angles of each electron beam should be unaffected by deflection or other factors throughout the entire raster. Under these conditions, turning off the green and blue guns, by reducing the electron gun screen voltage, should produce a completely red raster. Similarly, a completely green or blue raster may also be obtained. These rasters would be said to have good *purity.* It is often convenient to consider the color televison picture as being made up of three superimposed red, green, and blue rasters. If these three rasters all have good purity and if their light levels are proper (gray scale), the black & white picture they produce will be free of color.

Unfortunately, the entry angles of the electron beams are affected by the *shape of the shadow mask,* the *mechanical centering* of the electron guns, *stray magnetic fields,* and the *point of deflection.* The result of defects in any one or all of these things is to cause areas of discoloration in any one or in all three color rasters. This results in a black & white picture that may have blotches of color in one or more areas of the picture. Poor purity is more of a problem in black & white reception than in color reception because poor purity is somewhat camouflaged by the color program material.

Magnetic fields. Mechanical considerations, such as the shape of the shadow mask and the tilt and position of the guns, can and are solved by careful precision construction of the tube. In addition, proper adjustment of beam centering and yoke position are necessary to overcome any remaining defects.

The chief cause of poor purity is the susceptibility of the shadow mask and its mounting frame to become magnetized by the earth's magnetic field or by any other stray magnetic field. Stray magnetic fields can be caused by such things as turning off a vacuum cleaner near a color receiver or placing an electric clock or radio on top of the color receiver's cabinet. These localized magnetic fields deflect the electrons entering the shadow mask and cause the electron beams to strike the wrong color phosphor. This problem is common to all color picture tubes. To minimize the effects of both the earth's magnetic field and stray magnetic fields, color picture tubes are magnetically shielded. This is done by placing a thin silicon steel housing around the bowl of the tube. Unfortunately, this housing and the shadow mask structure both have high magnetic retentivity and over a period of time will become weakly magnetized by the earth's magnetic field. Whenever the color receiver is moved from one location to another, the field changes and this in turn will cause a change in picture purity. To overcome this, all modern color receivers, including the in-line types, incorporate some form of automatic degaussing to destroy the magnetic field. Figure 9-17 shows cross-sectional views of both the in-line and delta picture tubes showing the location of both the magnetic shields and the degaussing coils. Some modern tubes incorporate the magnetic shield inside the picture tube bowl.

9-9 automatic degaussing circuits

A magnetized object may be demagnetized by placing it in an alternating magnetic field that becomes weaker over a period of time. The magnetized object will be forced to assume the magnetic strength of the external degaussing field and will therefore become

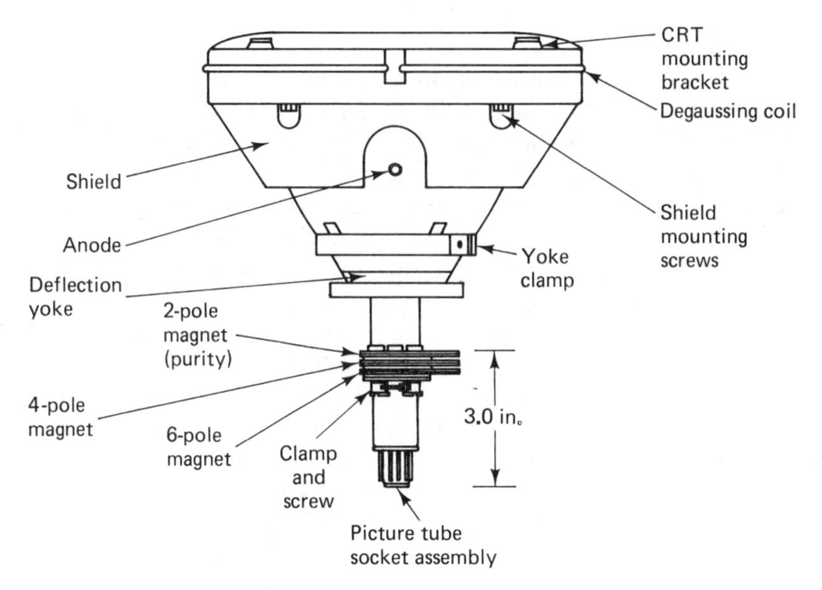

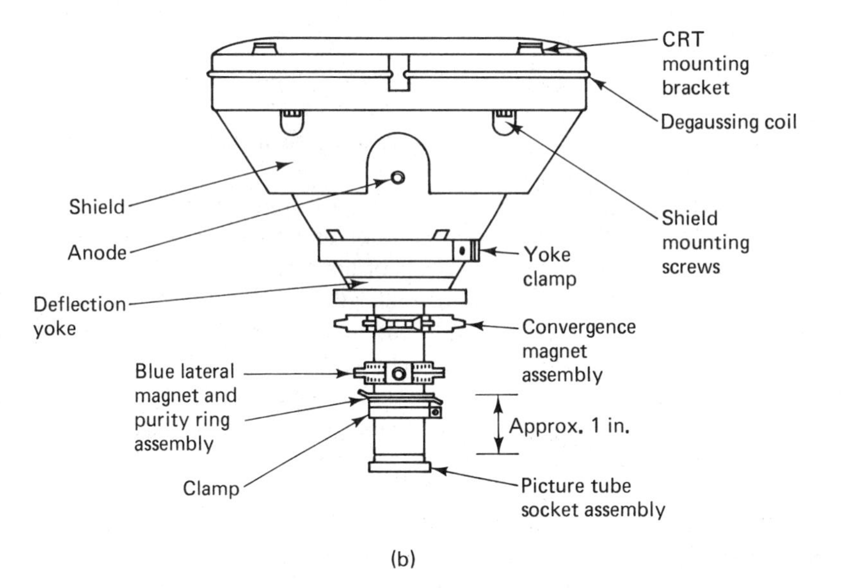

Figure 9-17 Color picture tubes and their associated components; (a) in-line; (b) delta.

weaker as the degaussing magnetic field becomes weaker. In the same way, a degaussing coil wrapped around the bowl of the color picture tube is provided with an ac current whose first cycle peak may be 5 or 6 A. Over a period of four-tenths of a second the current may drop to only 25 mA peak. During this time the magnetic field created by the degaussing coil will begin with a very strong field and will decay into a very weak one, forcing all nearby magnetic objects to assume the same terminal field strength as that of the degaussing coil.

Degaussing coils. Portable degaussing coils are available that may be used to degauss stubborn cases of poor purity. These coils are approximately 12 in. in diameter and consist of 425 turns of No. 20 enameled wire forming a bundle of wire approximately 3/4 in. thick. Degaussing is accomplished by connecting the degaussing coil to a 120-V 60-Hz source and then moving the coil over the faceplate, sides, top, and bottom of the tube. Care should be taken not to bring the coil too close to the neck of the tube or to the speaker where permanent magnets are mounted. After a few seconds the coils should be moved back approximately 10 ft from the tube. Then the coil should be turned at right angles to the set before power is turned off. If the degaussing coil is not moved away from the picture tube before the power is turned off, a large strange-looking colored impurity will be put into the raster.

ADG circuits. Three types of automatic degaussing (ADG) circuits may be found in many receivers. These are the *manual DG* in which a switch is used to excite the ADG coil, the *thermistor varistor* ADG, and the *positive temperature coefficient resistor* ADG in which a *P*-type resistor is placed in series with the degaussing coil and the combination is placed across the ac power source. See Figure 9-18(a). When the (*P*) resistor type of ADG is first

Figure 9-18 (a) P-type thermistor ADG; (b) thermistor and varistor automatic degaussing circuit.

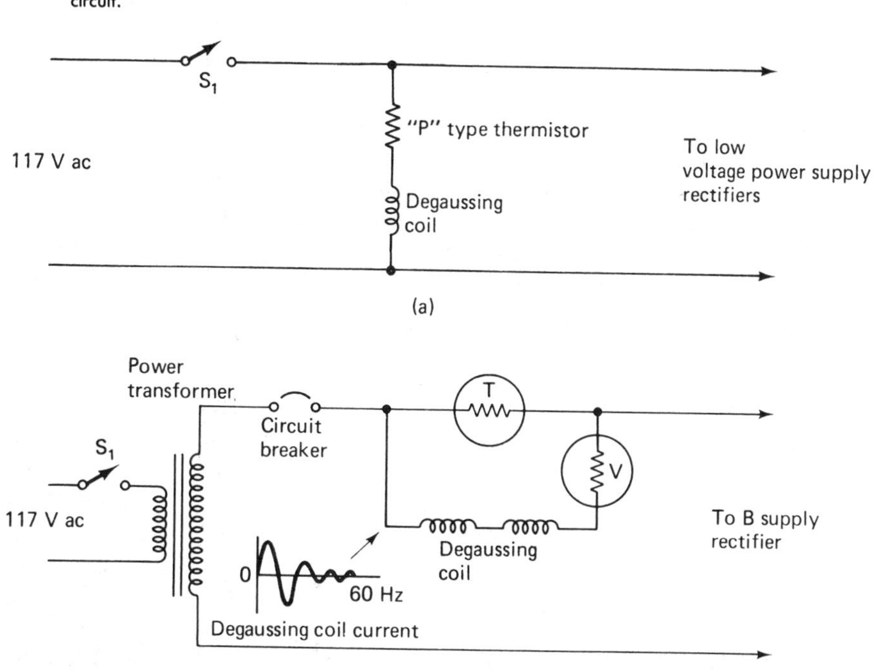

energized, its resistance is very small (25Ω) and a high current flows through to the ADG coil. When the (*P*) resistor gets hot, its resistance is high (1 MΩ or more) and the ac current flow through the ADG coil drops to almost zero, completing the degaussing action.

Thermistor/varistor ADG. A popular automatic degaussing circuit (ADG) is shown in Figure 9-18(b). This circuit incorporates two unusual resistors, a thermistor and a varistor.

The thermistor is a resistor that is temperature-sensitive; its resistance decreases with an increase in temperature. Such negative temperature coefficient resistors can be expected to have their resistance change from 100 Ω to 1 Ω as their temperature increases from room temperature (25°C) to operating temperature (145°C).

The varistor is a voltage-dependent resistor. In this type of component an increase in voltage of 35 V causes a decrease in resistance of approximately 10,000 Ω (from 10 kΩ to 10 Ω).

In this circuit automatic degaussing takes place before the set warms up sufficiently to produce a picture. The degaussing action in this circuit is as follows. When the operator closes the on/off switch power is applied to both the thermistor and the varistor. Since the thermistor is cold, its resistance is high, and most of the ac secondary voltage appears across it. Since the series combination of the varistor and degaussing coil is in parallel with the thermistor, all of its voltage divides across both of these components. The resistance of the degaussing coil is approximately 6 or 7 Ω, and with a voltage across the varistor of approximately 40 V, its resistance is approximately 100 Ω. Therefore, a high ac surge current of approximately 4 A flows through the degaussing coil. Because of the power supply current, the thermistor heats up, its resistance falls, and the voltage drop across it decreases. As a result, the voltage across the varactor decreases, increasing its resistance. This in turn forces the ac current flowing through the degaussing coil to drop to a low of approximately 25 mA. Degaussing is ended at this point, and normal receiver operation begins.

In this ADG circuit it is not possible to get repeated degaussing action without turning off the set and allowing the thermistor to cool. If either the varactor or the degaussing coil opens, automatic degaussing will stop and picture purity will become progressively worse. Degaussing with an external degaussing coil will only give temporary relief. If the thermistor opens, it will open the low-voltage supply and kill reception. If one of the rectifier diodes in the low-voltage power supply opens, the degaussing circuit will actually magnetize the tube mask and frame and cause severe impurity.

9-10 purity adjustments

Basically, proper purity is obtained by making five adjustments. The order and manner of making these adjustments may vary from one manufacturer to another. Therefore, exact procedures should be obtained from the manufacturer's service manual. The five adjustments are (1) making preliminary receiver adjustments such as gray scale, brightness limiter, high voltage, horizontal and vertical centering; (2) obtaining good static convergence (to be discussed in a later section); (3) manually degaussing; (4) adjusting the position of deflection yoke for the proper point of deflection; and (5) obtaining proper beam centering with the centering magnet tabs mounted on the neck of the picture tube. These steps are essentially the same for both the delta-type color picture tube and the in-line color picture tube.

Blank raster. One requirement for purity adjustment is that a blank snow-free raster be obtainable. This raster is needed so that any areas of impurity that might exist on the screen can be easily seen. This can be accomplished in at least four different ways:

1. Many receivers have service switches mounted on the rear apron of the receiver. One of the positions of the switch will produce a blank raster and is intended for purity adjustment.
2. A color bar generator may be used that is able to generate an unmodulated RF carrier (usually channel 3 or 4) that is injected at the antenna. This high-level signal will produce a noise-free blank raster.
3. The AGC level control can be adjusted for a blank raster.
4. The tuner to IF PC board cable usually has a jack on the tuner that can be disconnected, thus producing a clean blank raster.

Purity rings. Mounted on the neck of both the delta and in-line picture tubes is an adjustment device referred to as *purity rings.* See Figure 9-19. This type of adjustment is found on both delta and in-line tubes and is used in both for the same purpose. Purity rings consist of two flat washerlike magnets held together in such a way that both are able to rotate freely. On the outside edge of the rings are tabs that allow the operator to rotate the rings during purity adjustment. Purity rings are similar in physical shape and purpose to the *centering rings* used on black & white picture tubes. In both cases, the magnetic field of the ring magnets passes through the neck of the tube and interacts with the magnetic field of the electron beams. This interaction forces all three electron beams and their associated rasters to be deflected. Rotating both rings at the same time rotates the magnetic field and causes the beams to rotate in a circular path around the axis of the tube. Spreading the tabs increases or decreases the magnetic field strength and causes the electron beams to be deflected at right angles to the rings' magnetic field. In effect, this makes the electron beams move up and down, right or left, or diagonally, depending upon the positions of the ring magnets.

The purpose of the ring magnets is to compensate for any tilt in the electron gun assembly that would cause the electron beams to be off-center when they arrive at the phosphor screen. The purity rings (centering rings) are able to force the electron beams back to the mechanical axis of the tube. This places the electron beams in the proper position for deflection. In some RCA in-line receivers purity and static convergence adjustment assemblies are replaced by magnetic tape cemented to the neck of the picture tube.

In-line tubes also employ purity rings for the purposes outlined above. However, in addition, they use four- and six-pole ring magnets as indicated in Figure 9-19(b). These adjustments are designed to provide static convergence. Mounted on each magnet assembly is a knob that can be rotated, varying the strength of the magnetic field passing through the neck of the tube. Shifting the knob around the axis of the neck of the picture tube varies the position of the resultant magnetic field. The purity rings are constructed in the same manner and may be placed near the socket end of the assembly or at the opposite end of the assembly. The position of the purity static convergence assembly on the neck of the tube is critical.

Yoke position. The position of the yoke on the neck of the tube determines the point of deflection of the electron beams. If this point of deflection is forward or back from the

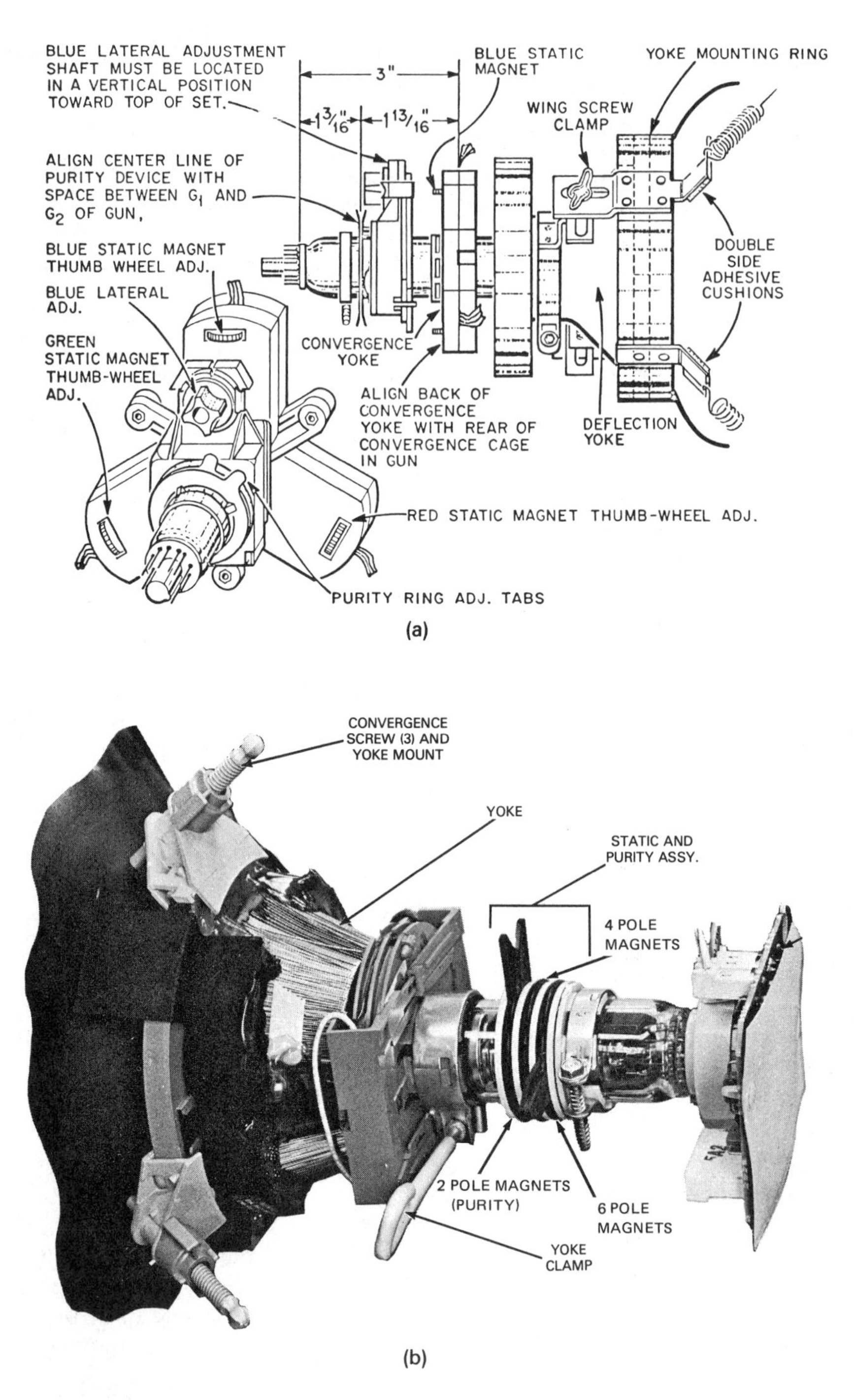

Figure 9-19 (a) Location of the purity and static and dynamic convergence controls mounted on the neck of a delta-type picture tube; (b) location of purity and static convergence adjustments in an in-line picture tube. [(a) Courtesy of Admiral Group/Rockwell International; (b) courtesy of Zenith.]

proper point of deflection, the entry angle of the beams into the shadow mask will be improper and poor purity will result. Therefore, one of the more important purity adjustments is finding the yoke position that corresponds to the proper point of deflection. In some modern in-line tubes this position has been found by the manufacturer and the yoke is cemented in place.

Typical purity adjustment procedure. For best results, purity adjustment should be made in the receiver's final location. If the receiver is to be moved, make the adjustment with the receiver pointing north. The receiver should be operating for 15 minutes and the CRT faceplate must be at room temperature before any purity adjustments are made.

1. Position the yoke, convergence assembly, purity, and blue lateral adjust on the neck of the tube. See Figure 9-19. Make all preliminary receiver adjustments, such as high-voltage, horizontal and vertical centering, gray scale, static convergence, and pincushion adjustments.
2. Manually degauss the front, sides, top, and bottom of the receiver cabinet.
3. Using one of the methods discussed earlier, obtain a blank raster.
4. Turn off the green and blue guns of the picture tube, leaving a red raster only. In delta-type tubes the red raster has been the raster that purity adjustments are usually made on. In-line tubes may use green or red rasters for purity adjustments.
5. Loosen the yoke and slide it back as far as it will go toward the base of the tube. This should produce a multicolored raster. Somewhere in the splash of colors there should be a red blotch. Adjust the purity magnets so that the red blotch is in the center of the screen. This will ensure that the electron beams are centered. A similar procedure is used in in-line tubes; however, instead of a red blotch, a red or green vertical bar is obtained.
6. Slide the deflection yoke forward until a uniform red or green raster is obtained. If small areas of impurity remain, adjust the purity rings for best purity. If good purity is not obtained, repeat the purity procedure from step 1 on.
7. Check green and blue rasters for uniformity by turning off the red gun and then turning on the green or blue guns. If impurity exists in either raster, repeat the entire purity procedure.
8. Make sure that the yoke is not rotated and then tighten it in place.

This completes a typical purity adjustment for delta and in-line tubes.

9-11 convergence

Because the picture tube face is essentially flat, the distance from the electron guns to the center of the phosphor screen is less than the distance from the guns to the edges of the picture tube. The amount of electron beam deflection depends upon gun to screen distance. This means that if the electron beams are adjusted so that they converge at the center of the screen, they will be divergent along the edges of the picture tube. This is illustrated in Figure 9-20. The visual effect of misconvergence is color fringing. This is most noticeable when viewing black & white pictures in which all vertical or horizontal lines in the picture are outlined with a red, green, or blue fringe. When viewing color program

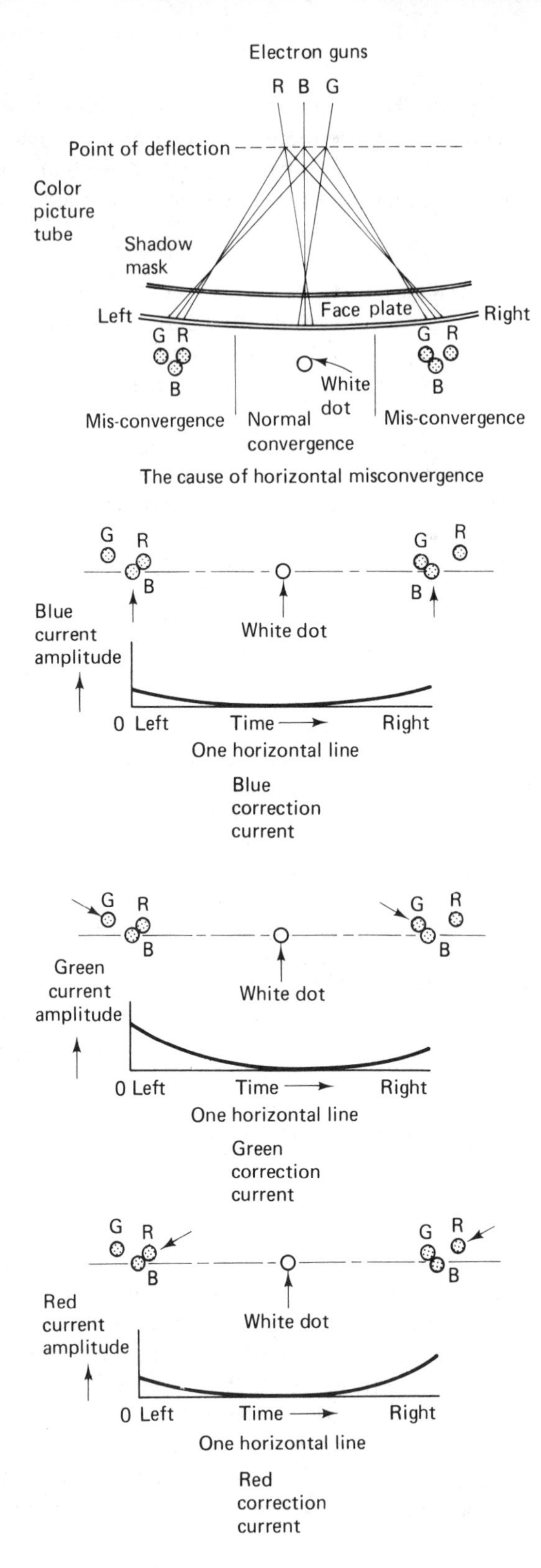

Figure 9-20 Cause of horizontal misconvergence and the parabolic correction currents required to overcome the defect.

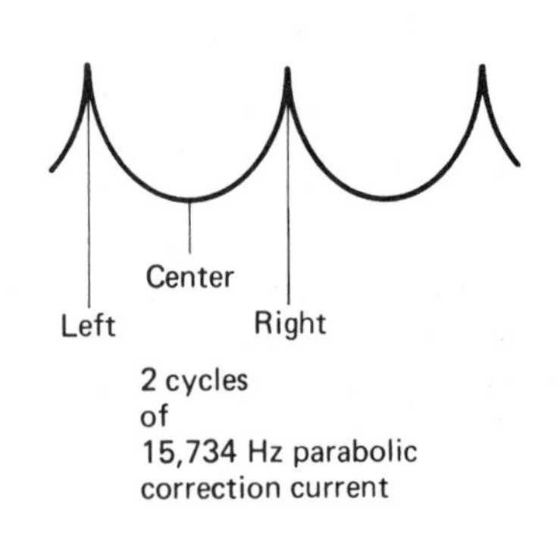

material, misconvergence is not as noticeable because color fringing is covered up by the color content of the picture.

A horizontal cross section of the delta-type picture tube is shown in Figure 9-20. Assume that there is no deflection. The three electron beams meet at the center of the shadow mask. They will then pass through it and illuminate their associated red, green, and blue phosphor dots of the same group of triads. The superposition of the three spots produces the visual appearance of a single white dot of light in the center of the screen. Convergence in the center of the color tube is called *static convergence.* When the electron beams are moved to the right or the left, the three beams converge before they reach the shadow mask. After passing through the shadow mask, the three beams produce illuminated dots that are separated from one another and form a triangular pattern of colored dots.

Similarly, a vertical cross-sectional view of the picture tube, shown in Figure 9-21(a), indicates that if center convergence is proper, top and bottom convergence is poor. Here again, the positions of the triangular colored dots depend upon the relative position and distance between the gun and the screen. The resultant pattern of horizontal and vertical misconvergence is shown in Figure 9-21(b).

In-line tubes. Because in-line tubes place their electron guns on the same plane, the distance between the guns and the screen will be the same for *vertical deflection,* but gun to screen distance will differ for *horizontal deflection.* Therefore, each electron beam will receive the same amount of vertical deflection and unequal amounts of horizontal deflection. As a consequence, vertical misconvergence *does not* occur in in-line tubes. This results in considerable reduction in hardware and circuitry needed to correct misconvergence. In delta-type tubes, four static convergence magnets, a dynamic convergence assembly mounted on the picture tube neck, and twelve controls are needed in order to achieve proper horizontal and vertical convergence. Some in-line picture tubes are available with few if any convergence adjustments at all. All adjustments that are required are done at the factory. Purity and static convergence adjustment magnets are mounted on the neck of the tube. Dynamic convergence may be accomplished by means of tilting a specially designed yoke or by adjusting two or three dynamic convergence controls. See Figures 9-17, 9-19, and 9-22.

9-12 static convergence

Static convergence is defined as convergence in the center of the screen. In delta-type tubes static convergence is obtained by adjusting four magnets (three static convergence and one blue lateral magnet) that are mounted on the neck of the tube. Three of these magnets are located in the dynamic convergence assembly mounted directly behind the deflection yoke. See Figure 9-19. A cross-sectional view of the neck of the picture tube, the dynamic convergence assembly, and the static convergence magnets are shown in Figure 9-23. This diagram discloses that each electron gun includes magnetic pole pieces that are placed on either side of each electron beam. Also see Figure 9-14. Above each pair of pole pieces is the core of the dynamic convergence electromagnet. Mounted in each of the three cores is a rodlike adjustable static convergence magnet. The rod is magnetized along its length so that one-half of its diameter is north and one-half is south. The exact location and manner of adjustment of these magnets vary from one manufacturer to another. In all cases, however, the static convergence permanent magnet's field is conducted by the dynamic

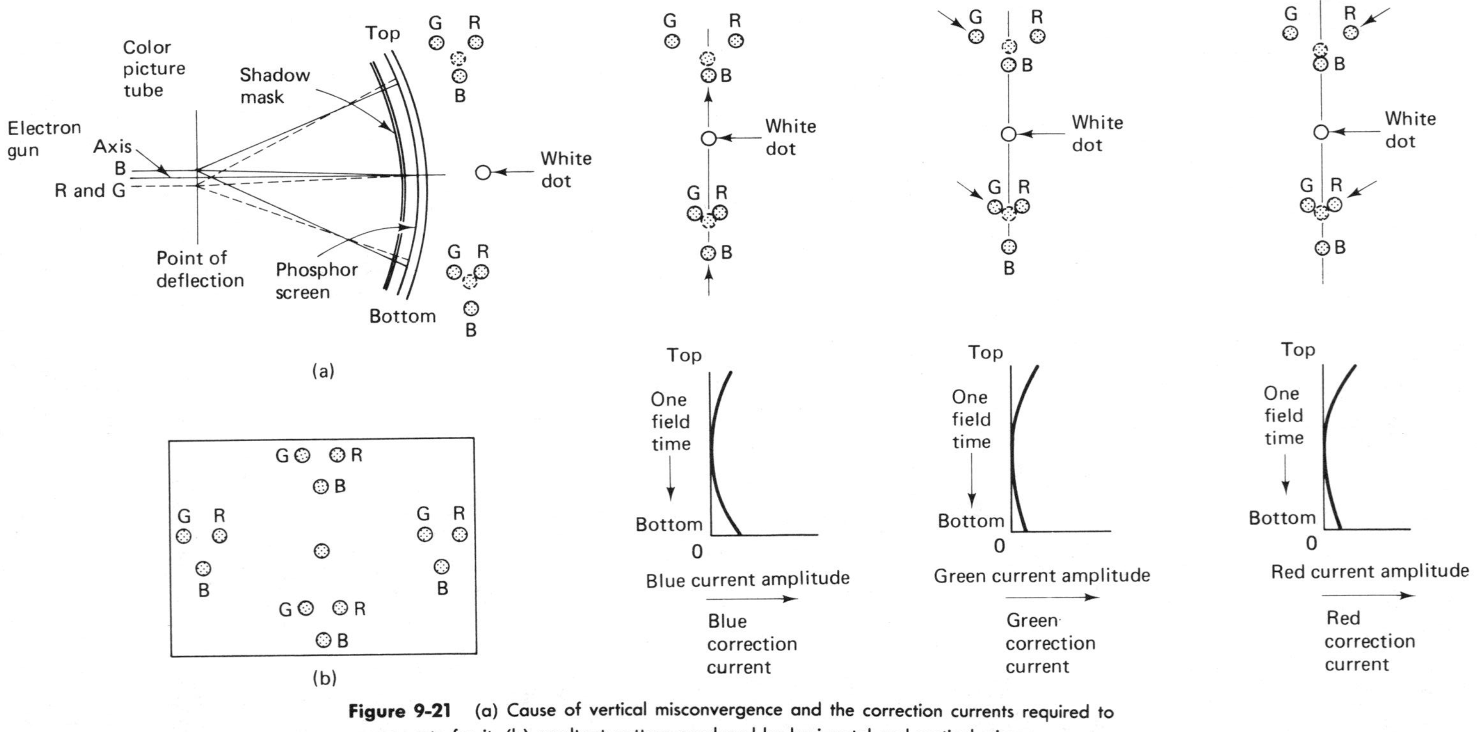

Figure 9-21 (a) Cause of vertical misconvergence and the correction currents required to compensate for it; (b) resultant pattern produced by horizontal and vertical misconvergence.

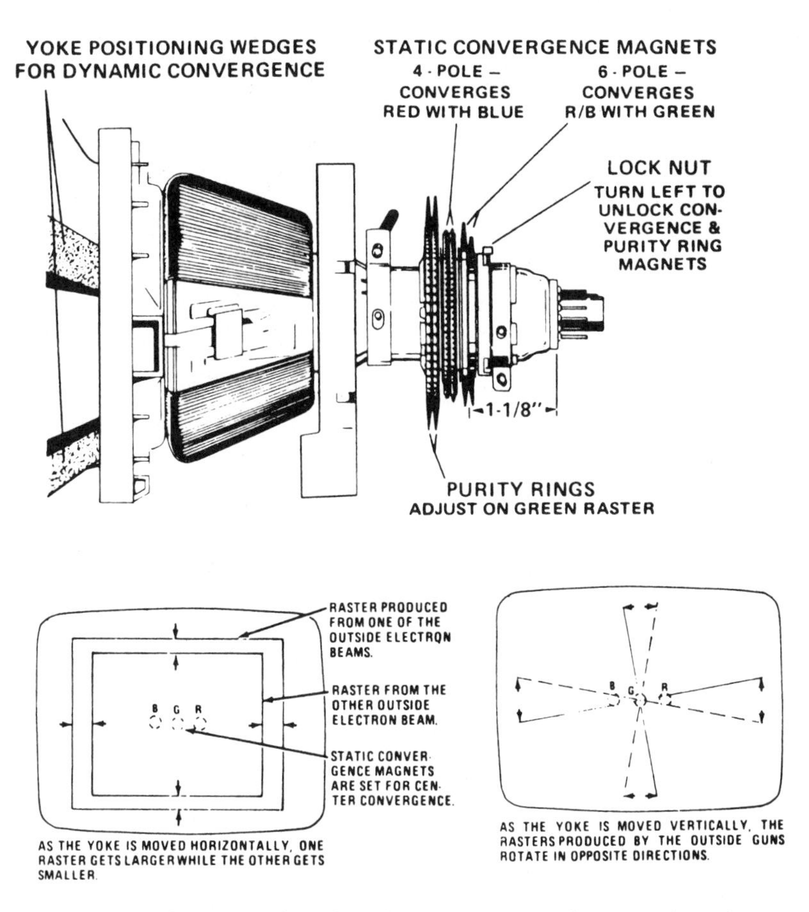

Figure 9-22 In-line picture tube and its adjustments. (Courtesy of Quasar Industries.)

convergence assembly core (through the glass of the tube to the magnetic pole pieces). The pole pieces are made of a magnetic material that conducts and directs the magnetic field so that it is at right angles to the path of the electron beam.

Electron beams are deflected at right angles to the magnetic fields that they enter. In the case of the blue gun, this forces the blue electron beam to move vertically. The red and green beams are also deflected at right angles to the magnetic fields produced by their static convergence magnets and pole pieces. However, because of the position of these electron guns, these electron beams are deflected diagonally. By rotating the static convergence magnet the amount and direction of beam deflection depends upon the strength and direction of the magnetic field between the pole pieces. This is controlled by rotating the static convergence magnet.

The end result of these adjustments is to force the three electron beams to converge at the center of the screen. As shown in Figure 9-24, the red and green beams move along diagonal paths and therefore must intersect at some point. The blue beam moves vertically and may or may not intersect the other two electron beams. The topmost drawing shows

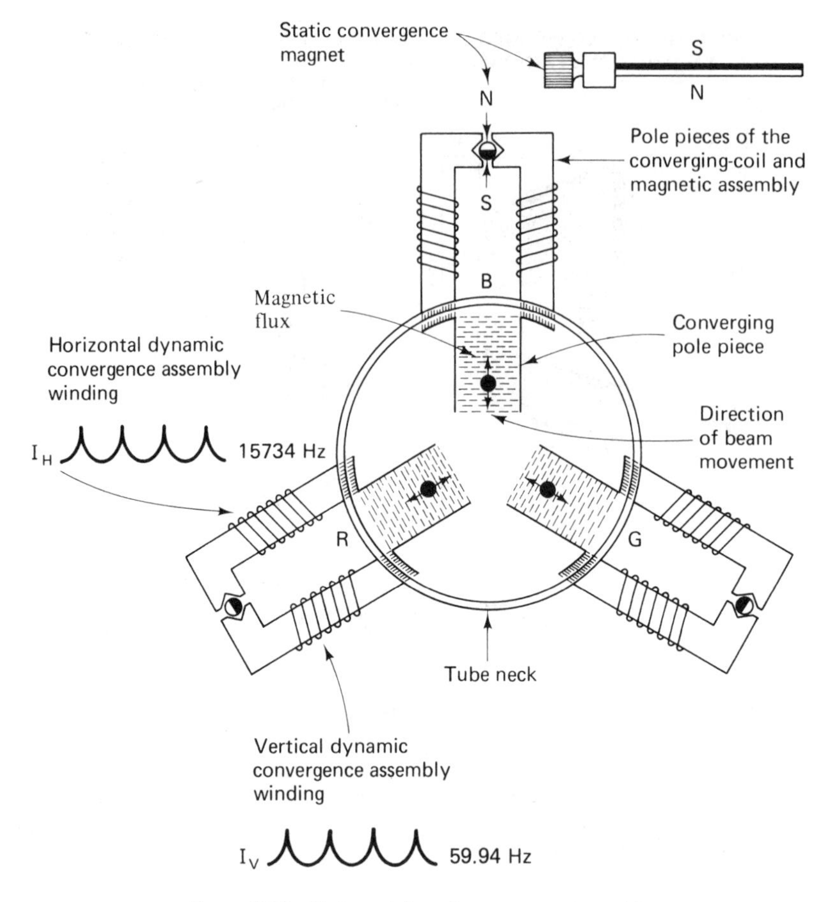

Figure 9-23 Static and dynamic convergence assembly.

the ideal case in which all three electron beams intersect. The center drawing shows the more usual case in which the red and green beams, intersect but the blue beam is off to one side. To correct this requires that the blue beam be deflected horizontally. This is accomplished by an adjustable magnet, called the *blue lateral adjust,* placed on the neck of the tube directly behind the dynamic convergence assembly. See Figure 9-19(a). The blue lateral adjust magnet produces a magnetic field that passes through the blue gun in a vertical direction. As a result, the blue electron beam is deflected at right angles to the field and therefore is forced to move to the right or left, depending upon the direction of the field and its strength.

Static convergence adjustment. Before proceeding with the convergence adjustment make sure that the following basic receiver adjustments have been made:

1. Picture tube neck components properly placed (Figure 9-19)
2. High voltage and efficiency coil adjustments (only available on older receivers)
3. Focus

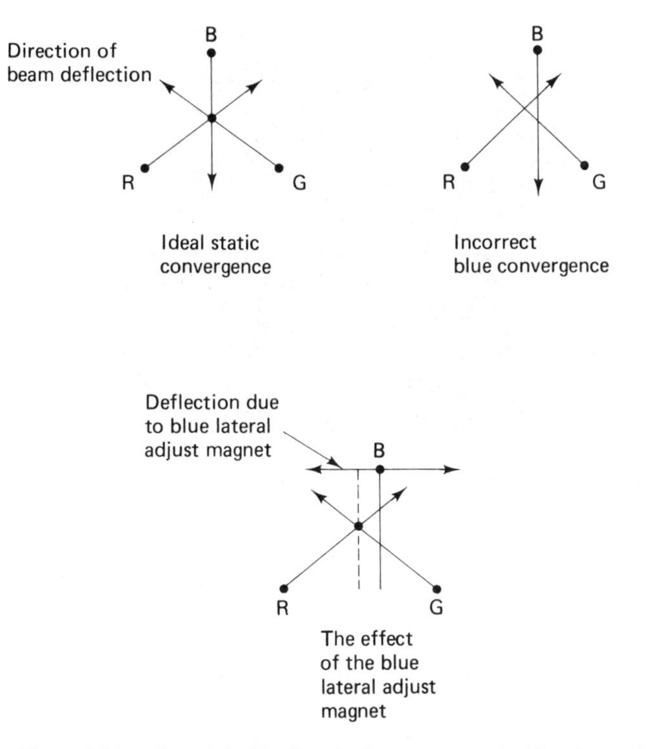

Figure 9-24 Effect of the blue lateral adjust magnet on the blue electron beam.

4. Brightness limiter and gray scale
5. Horizontal and vertical centering (very important for proper convergence)
6. Purity
7. Vertical size and linearity (has great effect on convergence and should not be touched after it has been made)
8. Pincushion adjustment
9. All other adjustments that produce a good black & white picture

A mirror mounted on a stand in front of the set so that it can be viewed from the rear can make purity and convergence adjustments somewhat easier.

1. Connect a dot/crosshatch generator to the receiver. Adjust the generator for a white dot pattern. Then set the brightness control for a dark background and small white dots. Adjust the fine tuning for best dot pattern. Turn off any residual color by turning down the color control.
2. Adjust the red, green, blue, and blue lateral adjust static convergence magnets so that the colored dots superimpose (converge) at the center of the screen to form a single white dot. An example of proper static convergence is shown in Figure 9-25. If static convergence is difficult to obtain, check the position of the convergence yoke assembly. The blue convergence electromagnet should be placed directly over the blue gun. Weak magnets are another cause of inability to obtain static convergence.

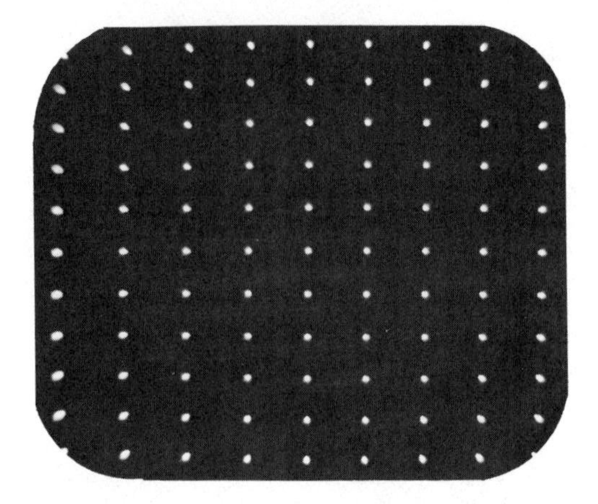

Figure 9-25 Dot pattern showing proper static convergence. See the color plate.

This completes the static convergence adjustment for the delta-type tube. The static convergence procedure for an in-line picture tube of the type shown in Figure 9-19(b) is outlined below.

1. Connect a color bar generator to the TV receiver and obtain a crosshatch pattern on the picture tube screen.
2. The four-pole magnet (center ring) is adjusted by simultaneously separating the tabs and rotating the device to converge blue with red (some sets blue with green) horizontal and vertical lines in the center of the screen.
3. Adjust the six-pole magnet (rear rings) by simultaneously separating the tabs and rotating the device to converge the blue and red lines with green.

9-13 dynamic convergence

It was pointed out earlier that if the three electron beams converge in the center of the screen, they will be out of convergence along the edges of the screen. This defect could be remedied if the tube were built with a spherical faceplate so that all points on the surface of the screen were the same distance from the beam's point of deflection. Clearly, this is an impossible, or at least a highly impractical way to build tubes. Since flat-face tubes are highly desirable and spherical face tubes are not possible, an electronic system for forcing the misconverged beam into convergence is necessary.

It should be recalled that each individual electron beam can be positioned by means of the static convergence magnets. If this is possible, would it not also be possible to force the beams to converge along the edges of the screen by varying the convergence magnetic field strength? In practice, this is exactly what is done. The convergence magnetic field for each electron beam is varied by passing a current of the proper shape and frequency through the coils wound around the cores of the dynamic convergence assembly. See Figure 9-23. Since there are three electron beams and each one must be corrected for both horizontal and vertical misconvergence, a total of six correction currents are needed. To see what wave-form and frequency these currents must have, refer to Figure 9-20. At the top of this drawing there is a horizontal cross section of the picture tube. Also indicated are the three electron beams and their resultant horizontal misconvergence. Drawn below this diagram are three diagrams showing the current requirements for the blue, green, and red dynamic convergence electromagnets. These currents are needed to produce a magnetic field that is in step with horizontal deflection. Notice that on the left side of the screen all three beams need different amounts of correction. Thus, the amount of current needed to force the three beams into convergence will differ from one horizontal winding of the dynamic convergence assembly to the other. When horizontal deflection has brought the three beams to the center of the raster, the three beams will require no correction if proper static convergence has been obtained. Therefore, at the center of the screen the correction currents will be zero. On the right side of the screen corresponding to the end of one horizontal line of scan, correction currents of different amplitudes are again necessary. The current required to produce dynamic convergence has a parabolic waveform and has a horizontal scanning frequency of 15734.26 Hz. An illustration of two cycles of such a current waveform is shown in Figure 9-20.

Figure 9-21 illustrates a vertical cross-sectional view of a color picture tube and the currents required by the three vertical dynamic convergence electromagnets to force the electron beams into proper convergence. Here again, as was the case in horizontal convergence, each current has a different amplitude requirement at both the top and bottom of the raster. Similarly, at the center of the raster no correction current is necessary. The waveform is therefore parabolic, and its frequency is that of the color television vertical scanning rate, 59.94 Hz.

In practice, each electromagnet of the dynamic convergence assembly has two separate windings: one for the horizontal correction current and the other for the vertical. See Figure 9-23. The currents are obtained from a twelve-control dynamic convergence circuit which in turn is fed by the horizontal and vertical deflection circuits. See the block diagram of Figure 9-1. This means that any changes in vertical or horizontal size or linearity can have an effect on dynamic convergence.

The dynamic convergence circuit is usually found on a separate circuit board. It is mounted on the neck of the picture tube in conjunction with the dynamic convergence assembly (as shown in Figure 9-26), or it is detachable from the chassis.

An examination of the convergence board will reveal that there are a total of twelve controls necessary for proper dynamic convergence. These controls are designed to adjust the amplitude and tilt of the parabolic currents passing through the coils. See Figure 9-27. Adjusting the amplitude of the parabolic waveform simply means making it larger or smaller.

Adding a sawtooth to a parabolic waveform tends to accentuate one peak or the other producing tilted waveform. The direction of the tilt depends upon the phase of the sawtooth. Some of the convergence adjustments on the board are able to change the direction of parabolic tilt by changing the phase of the sawtooth. The effect of amplitude and tilt on the picture is to force the red, green, and blue rasters into convergence.

9-14 convergence circuits

Before 1965 the dynamic convergence assembly for each gun had only one winding. Both vertical and horizontal parabolic correction currents flowed through this winding. Static convergence was achieved by a small permanent magnet mounted in the core in the manner already discussed.

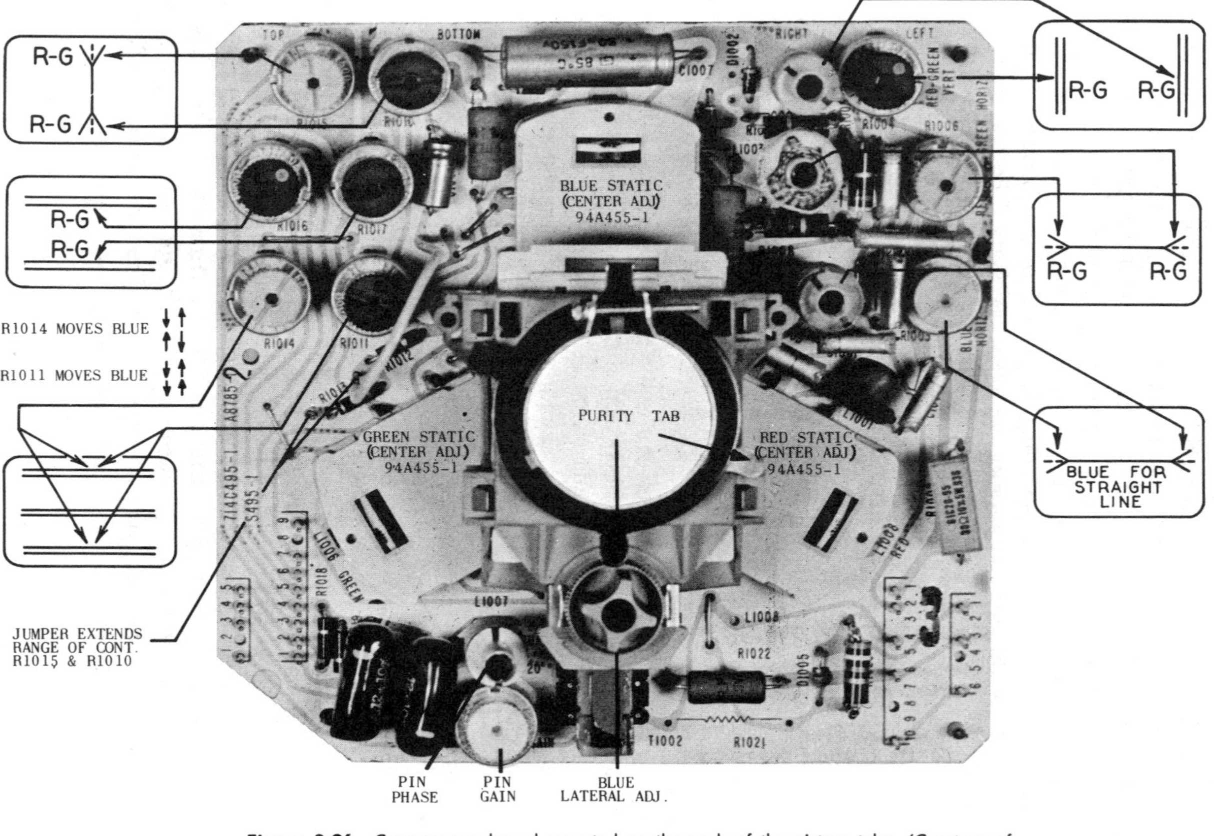

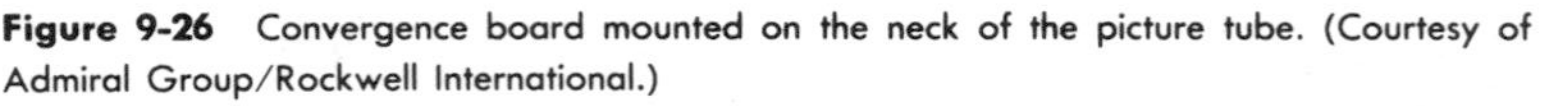
Figure 9-26 Convergence board mounted on the neck of the picture tube. (Courtesy of Admiral Group/Rockwell International.)

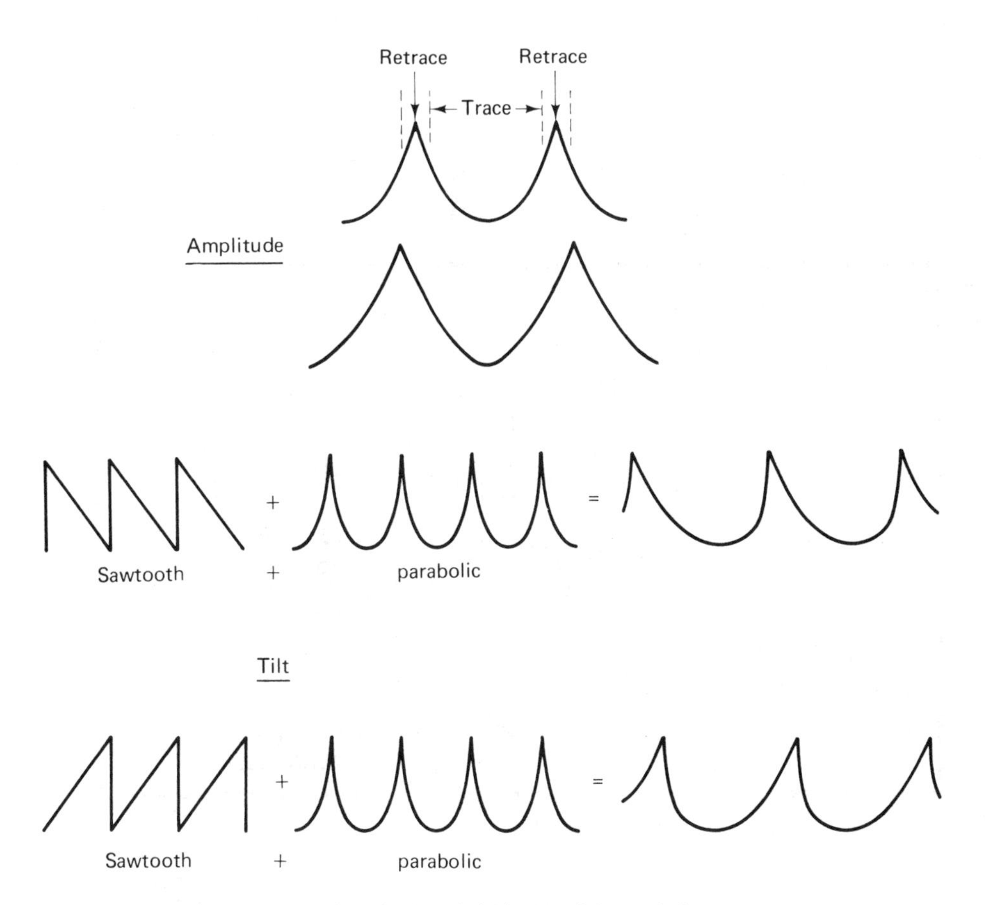

Figure 9-27 Effect of amplitude and tilt changes of the parabolic correction current.

Since 1965 almost all delta-type color television receivers have been using circuits similar to the dynamic convergence circuits shown in Figures 9-28 and 9-29. These circuits use dynamic convergence electromagnets that have separate vertical and horizontal windings on the same cores.

Transformerless convergence circuits. Manufacturers of solid-state receivers are using vertical deflection circuits that do not use output transformers. These circuits are very similar to the transformerless audio amplifier circuits discussed in Chapter 8.

Figure 9-28(a) shows the complete schematic diagram of the convergence board used with this receiver. The horizontal convergence circuit is essentially similar to that already discussed, but the vertical convergence circuitry is different. In this circuit the convergence circuit and the vertical deflection yoke are in series. Refer to the simplified circuit of Figure 9-28(b) and notice that the yoke sawtooth current passes through the low-resistance resistors R_1 and R_2. Some yoke current, however, does flow through the convergence coil L_1, via diodes D_1 and D_4, when the yoke current is positive. The conduction through L_1 reverses its direction and flows through D_2 and D_3 when the yoke current is negative. In effect, the diodes have converted a sawtooth into a parabolic waveform. Because

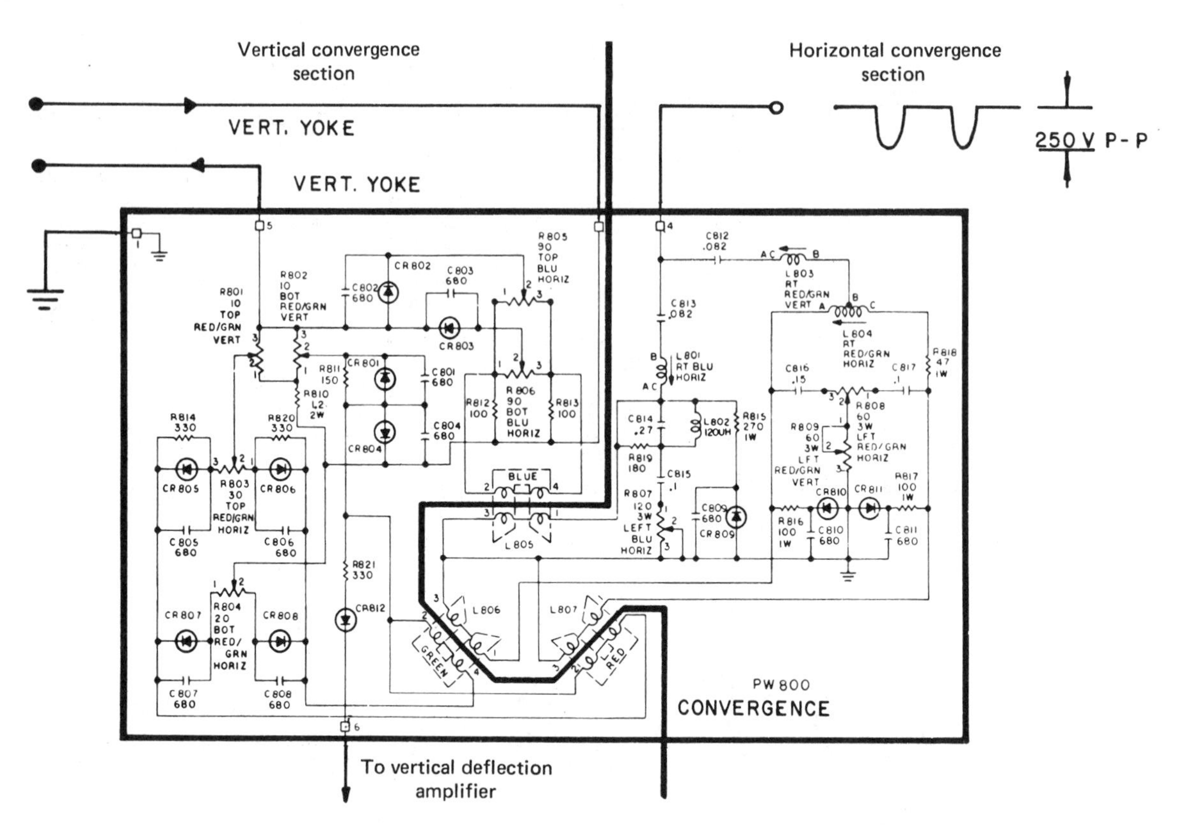

(a)

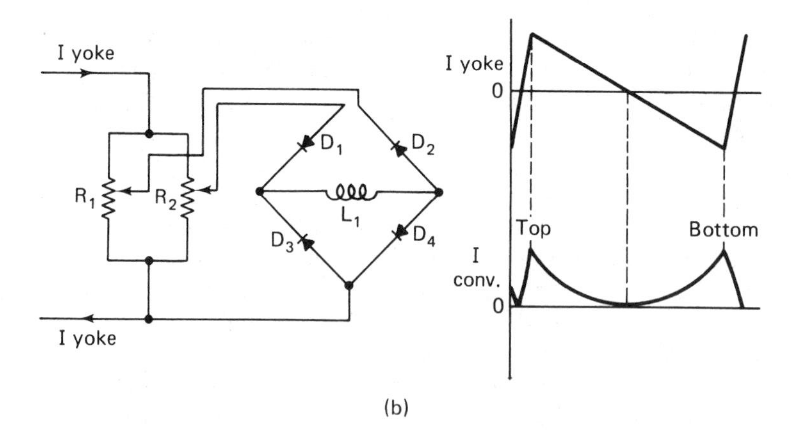

(b)

Figure 9-28 (a) Complete schematic diagram of a convergence circuit used in receivers that do not use vertical output transformers; (b) simplified diagram of the vertical convergence circuit.

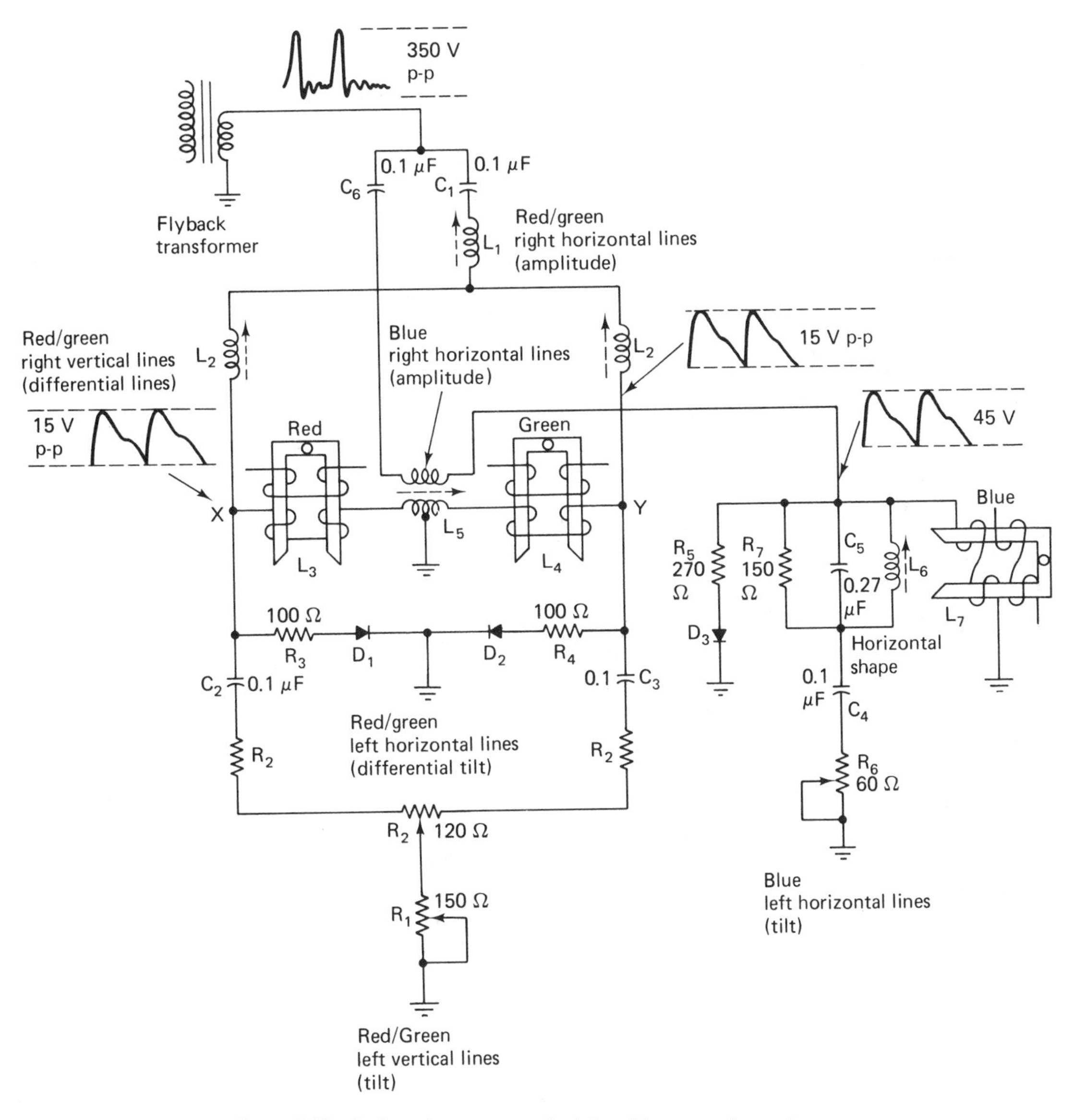

Figure 9-29 Horizontal convergence circuit found in many color receivers.

of the inductive impedance of the convergence coil, this current is almost a true parabolic. The amount and direction of tilt can be changed by varying the value of R_1 or R_2.

Horizontal dynamic convergence. The horizontal dynamic convergence circuit is mounted on the same printed circuit board as the vertical dynamic convergence circuit. See Figure 9-26. A typical circuit used for horizontal dynamic convergence is shown in Figure 9-29. In this circuit a positive going pulse is obtained from a special winding on the flyback transformer and is fed to the convergence electromagnets via C_1 and C_6. The circuit consisting of C_6, L_5 and the parallel combination of the blue convergence coil (L_7), the diode clamp (R_5, D_3), and the horizontal shape circuit (L_6, C_5, R_7, C_4, and R_6) change the pulse into a parabolic current. The amplitude of this current is controlled by L_5. This control affects the blue horizontal lines of a crosshatch pattern on the right side of the screen. The horizontal shape inductance (L_6) forms a parallel resonant circuit with C_5 and the damping resistor R_7. This tuned circuit is shock excited by the input pulse and is designed to resonate at twice the horizontal line frequency. In some circuits the shape coil is adjustable, in others it is fixed tuned. In any event, this circuit is not usually adjusted during dynamic convergence setup.

Adjusting R_6 determines the effect (tilt) that the horizontal shape circuit has on the parabolic correction current. This potentiometer has its greatest effect on blue horizontal lines on the left side of the screen. Diode D_3 is a clamp that ties the parabolic wave to ground when the electron beams pass through the center line of the picture screen.

Capacitor C_1 couples the positive going flyback pulse to the red/green section of the dynamic convergence board. Coil L_1 provides a series impedance that shapes and controls the amplitude of convergence currents fed to the red/green convergence coils. This coil is adjustable and has its greatest effect on the red/green crosshatch lines on the right side of the screen. The current flowing out of L_1 divides at the center tap of L_2 (for convenience L_2 is divided into two sections) into two branch currents of a Wheatstone bridge. These currents flow through C_2, C_3, R_2, and R_1 to ground. When R_2 is properly positioned, the bridge is balanced, and the voltage from X to Y is zero. Under these conditions, no current flows through the crossarm of the bridge. The crossarm consists of the two convergence electromagnets (L_3 and L_4) and the inductance L_5. However, when R_2 is adjusted so that the bridge is no longer balanced, current will flow through L_3 and L_4. The amount and direction of current flowing in either coil depend upon the degree and direction of unbalance. On this basis, R_2 is referred to as a *differential tilt control* that has its greatest effect on red/green horizontal lines on the left side of the crosshatch pattern. R_1 controls the magnitude of total current flow and is primarily responsible for the convergence of the red/green vertical lines of the crosshatch pattern on the left side of the screen. D_1 and D_2 are diode clamps of the same type as D_3. R_3 and R_4 limit the peak currents that flow through the diodes to a safe level.

9-15 convergence adjustments

In general, dynamic convergence setup is a "hunting" procedure. That is, the adjustment zeros in on proper convergence by trial and error. Of course, experience reduces the amount of trial and error to a minimum. The procedure outlined in service literature varies from one manufacturer to another. The procedure given below is a summary of the basic steps that are usually followed.

Before proceeding the following adjustments must be made:

1. Check and adjust the picture tube neck components as shown in Figure 9-19(a). If the neck components are not in their proper positions or locations, convergence adjustments may not be possible.
2. Also check and adjust, if necessary, width, high voltage, efficiency coil, focus, vertical size, vertical linearity, centering, gray scale, and brightness limiters.

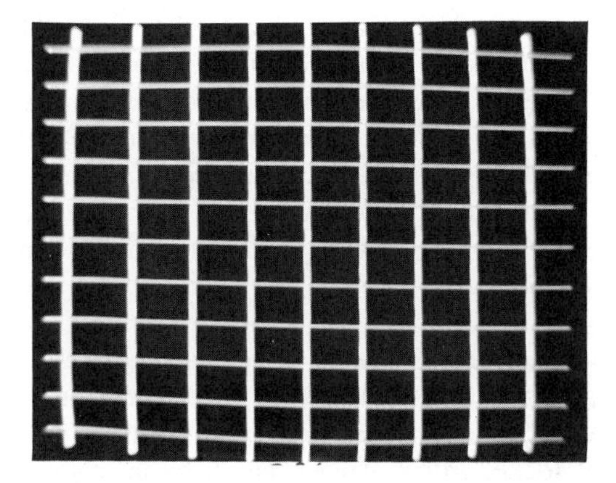

Figure 9-30 Crosshatch pattern of proper convergence.

3. Perform static convergence.
4. Connect a crosshatch generator to the receiver and adjust channel selector, fine tuning, AGC, contrast control, and color killer for a sharp, stable crosshatch pattern free of color interference (worms). See Figure 9-30. Connect the gun killer switches to the red, green, and blue grids of the picture tube. Gun killer switches are used to disable one or more picture tube guns so that gray scale (black & white raster) adjustments are unaffected during purity and static convergence adjustments. The switches are usually part of the color bar generator. Gun killer switches are only used in those older color receiver systems where the *Y* signal feeds the cathode of the picture tube and the color signals feed the grids.
5. Refer to Figure 9-26. Notice that each control has associated with it a diagram showing the region and colors of the crosshatch pattern affected by a given control. Adjust each control so that convergence along the edges of the picture is as good as possible. Frequent static convergence adjustments may be necessary. In general, 85% convergence is considered adequate. If some misconvergence is unavoidable, try to have blue fringing because this is least noticeable.

Troubleshooting. Sets that refuse to converge may have trouble in (1) the dynamic convergence assembly, (2) the dynamic convergence board, (3) the horizontal or vertical circuits, (4) the deflection yoke, or (5) they may have a bad picture tube. Typically, open or shorted turns in either the vertical or horizontal windings of the convergence electromagnets may be responsible for the inability to obtain dynamic convergence. Almost any defective component on the convergence board can cause misconvergence. One of the most common defects is an open or shorted diode. However, changed value resistors, open capacitors, and shorted turns in the adjustment coils are also common. If the picture linearity and size is normal, and if the waveforms coming from the vertical and horizontal output circuits are normal, the trouble is not likely to be in the deflection circuits. If, however, the waveform feeding the convergence board is improper, it may be the result of defects in the deflection circuits. Poor convergence may also be caused by a defective deflection yoke or a defective picture tube.

Convergence of in-line tubes. In-line picture tubes have considerably less convergence problems than do the delta-type tubes, because during vertical deflection the distance from the electron guns to the screen remains essentially the same for all three electron beams. Therefore, vertical misconvergence is considerably reduced. Horizontal misconvergence is still a problem because of the changing distance from gun to screen that occurs during horizontal scanning. Vertical misconvergence is still further reduced by the use of vertical phosphor strips rather than triads because any vertical displacement of an electron beam will fall on the same color.

Dynamic convergence in-line tubes. Dynamic (edge) convergence of an in-line picture tube of the type shown in Figure 9-19(b) is obtained by tilting the front of the yoke up and down and/or left and right. The yoke tilt is first positioned by hand or by screw for best convergence, then it is locked into a fixed position by means of a wedge placed between the yoke and the bowl of the picture tube.

Convergence, the Trinitron. In the Trinitron color picture tube the three electron beams are arranged in an in-line configuration. See Figure 9-31. The electron gun structure depends upon picture tube size. Each electron gun uses three individual cathodes each with its own heater. The control grid (G_1) in small Trinitrons is a single electrode with three electron beam apertures that enclose the cathodes. In larger tubes all cathodes and control grids are separate electrodes. In tubes in which control grid voltage is common to all three beams, varying the individual cathode voltages causes changes in beam intensity which then results in light output variations. In those tubes, using separate grids, changing either cathode or grid voltages will control beam current. A common screen grid (G_2) and a common focus electrode are used to accelerate and focus the electron beams. As a result of using a common focus electrode for all three electron beams, uniform focus, spot shape, and small spot size are obtained throughout the entire screen.

The three electron beams leave the focus electrode and pass through a set of four electrostatic convergence plates that provide horizontal static convergence adjustment. Since there is no difference of potential between the two center plates, the green beam passes through the convergence plates unaffected. If a dc voltage (almost equal to the second anode voltage) is applied to both of the outside convergence plates, the electron beams that pass through the electrodes will be deflected. Horizontal static convergence is obtained when the adjustable dc voltage on these plates is sufficient to produce beam convergence

Figure 9-31 Trinitron electron gun as viewed from above.

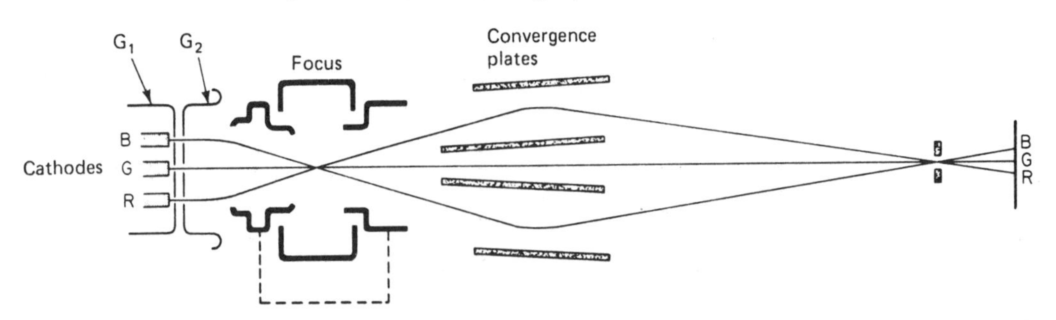

at the aperture grille. In small-screen Trinitrons a parabolic waveform obtained from the horizontal output stage is added to the dc convergence voltage to obtain dynamic horizontal convergence. At the center of the screen no correction is needed and the parabolic voltage is zero, leaving only the dc voltage necessary for static convergence. As the electron beams are deflected toward the edges of the screen, the parabolic voltage makes the convergence electrodes more positive. This forces the electron beams to diverge slightly and move their point of convergence forward to the phosphor screen. In larger screen Trinitrons no parabolic voltage is applied to the convergence plates.

The Trinitron convergence circuit. An example of the horizontal dynamic convergence circuit used with the Trinitron is shown in Figure 9-32. In this circuit the horizontal output transistor directly feeds the horizontal windings of the yoke. The sawtooth current flowing through the yoke is wave-shaped by L_3 and C_1 and produces a parabolic voltage across C_1. Capacitors C_2 and C_3 act as an ac voltage divider that determines the amount of parabolic voltage fed to the primary of the horizontal convergence transformer (T_2) via the tilt control R_1. The amount and polarity of the sawtooth voltage developed across R_1 and added to the parabolic waveform are determined by the position of the movable arm of R_1. This combined ac voltage appears at the secondary of the horizontal convergence transformer (T_2) and is added to the high voltage developed by the high-voltage rectifier. The sum of these two voltages is then fed to the convergence anode.

Vertical convergence is built into the picture tube system. Because of its in-line gun structure and its phosphor strips, vertical misconvergence is minimal. All additional compensation is obtained by winding the deflection yoke in the form of a toroid (doughnut-shaped) instead of the conventional saddle windings usually used. There are no adjustments

Figure 9-32 Simplified form of the horizontal dynamic convergence circuit used with the Trinitron color picture tube.

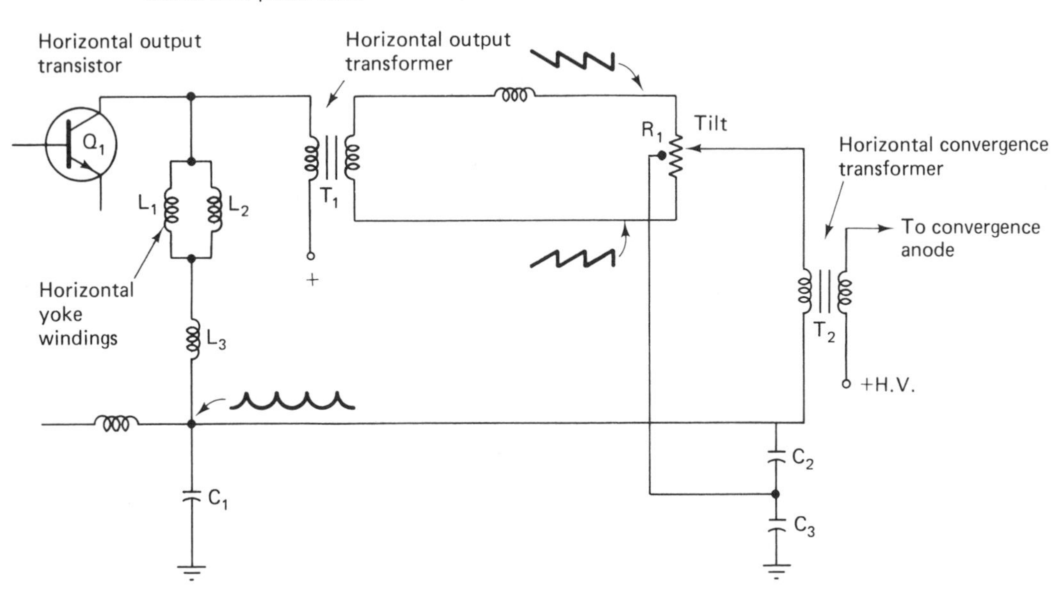

to be made to obtain vertical convergence in small-screen Trinitrons, but some adjustment is necessary in the large-screen tubes.

Purity. Purity adjustments for the Trinitron are similar to those of delta and other in-line tubes. Thus, centering magnets are used to position the beams along the tube's axis and yoke position determines the point of beam deflection. In addition to these adjustments, the Trinitron uses a special yoke called the *neck-twist coil.* This coil is positioned on the neck of the tube close to the socket. The magnetic field that it generates is such that it twists (rotates) the outer beams around the green beam. When the current flowing through the coil is adjusted for proper magnitude and direction, the three electron beams will be positioned in a horizontal plane relative to the electron gun assembly. In small-screen tubes the dc current flowing through the neck-twist coil is controlled by a potentiometer; in large-screen tubes the current flowing through the coil is the horizontal sawtooth current of the yoke. In the large tubes the neck-twist coil is called the *beam alignment coil* and is considered part of the dynamic convergence system.

Convergence procedure. The purity and convergence procedures for the Trinitron are similar to those used in other color tubes. Exact purity and convergence should be obtained from the manufacturer's service literature because they vary somewhat, depending upon the size of the picture tube.

9-16 pincushion cause and correction

When larger black & white or color picture tubes are used, the top, bottom, left, and right edges of the raster tend to bow inward toward the center of the screen. This condition is known as *pincushioning.* Pincushion distortion is the result of the type of yoke used and the fact that the amount of electron beam deflection depends upon the distance between the point of deflection and the phosphor screen. Thus, the amount of beam deflection in the corners of a rectangular-faced picture tube is greater than along the edges. This gives rise to a pincushion raster that looks like that shown in Figure 9-33.

Figure 9-33 Pincushion distortion.

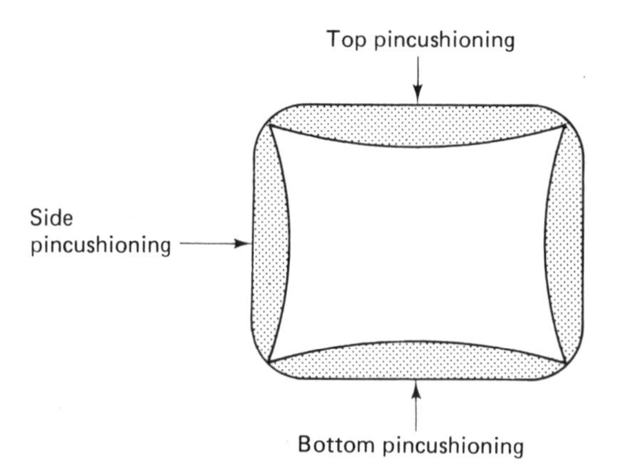

Black & white receivers. In black & white receivers, pincushion distortion is eliminated by the design of the yoke and the use of four small permanent magnets that are mounted in the yoke housing near the bowl of the picture tube. These magnets are positioned so that they stretch the raster along the edges of the picture and compensate for the pincushion distortion. In some receivers the pincushion magnets are adjustable and may be rotated or positioned closer or farther from the bowl of the picture tube. Correct adjustment occurs when the pincushioning is minimum.

Color receivers. The use of permanent magnets for the elimination of pincushion distortion is not feasible for color receivers because the magnets would tend to introduce purity problems. Modern color receivers eliminate pincushion distortion by careful design of the yoke and picture tube, thereby requiring no pincushion adjustments. Older color television receivers used antipincushion circuits that automatically increased horizontal width and vertical size in those regions of the raster that are shrunken because of pincushion distortion. The deflection waveforms required to correct pincushion distortion are shown in Figure 9-34. To correct side pincushioning, the horizontal deflection sawtooth current must be amplitude modulated at a vertical rate so that when the electron beam is at the top or bottom of the picture, the horizontal sawtooth amplitude is minimum and when the electron beam is at the center of the vertical deflection interval, the horizontal sawtooth amplitude must be maximum.

Figure 9-34 Horizontal and vertical waveforms required for pincushion correction.

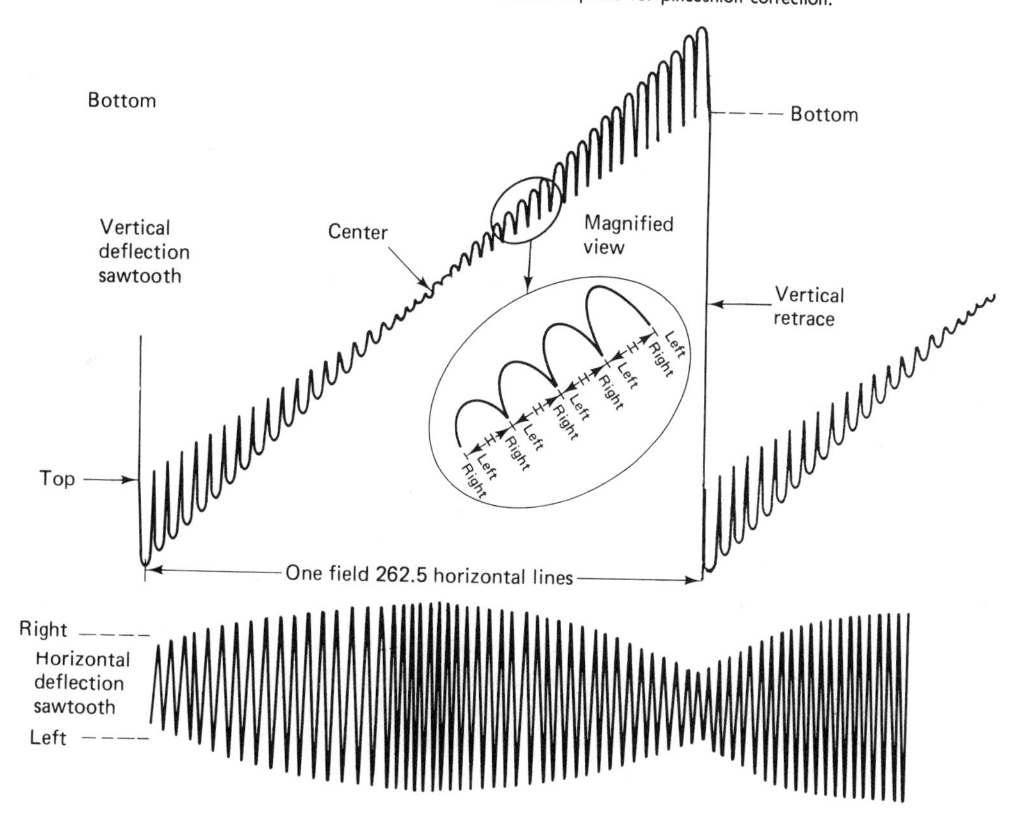

The top and bottom pincushioning is corrected by forcing the vertical deflection sawtooth current to pulsate in amplitude at the horizontal scanning rate. At the top and bottom of raster scan, a parabolic waveform at the horizontal rate is superimposed on the vertical deflection sawtooth. In effect, this increases vertical size during the time that the electron beam is moving through the midpoint of its horizontal scan. The parabolic waveform at the top of the raster is of opposite polarity to that at the bottom since the raster stretch required at the top is opposite to that needed at the bottom of the raster. The amplitude of the parabolic waveform required for top and bottom pincushion correction decreases to zero as vertical deflection passes through the center of the raster.

Antipincushion circuits. Below are listed the two basic methods used to eliminate pincushion distortion:

1. Antipincushion circuits that make use of special saturable core reactors and resonating circuits. Side correction and top and bottom correction may be obtained in one or two circuits, depending upon manufacturer.
2. Dynamic antipincushioning circuits that use transistors to produce the desired amplitude variations in the deflection sawtooth.

The saturable reactor antipincushioning circuit. An example of a circuit that uses saturable reactors for pincushion correction is shown in Figure 9-35. This circuit combines both side and top and bottom pincushion correction in one circuit.

Top and bottom pincushion correction. Top and bottom pincushion correction is obtained by adding the proper amplitude parabolic waveform at the horizontal rate to the vertical deflection sawtooth. In order to do this, a special transformer called a *saturable reactor* is used. A saturable reactor is a coil or transformer that uses a solid no-air gap ferrite core and whose inductance can be varied by changing the magnetic flux density passing through the core of the coil. The change in core magnetic field strength is accomplished by passing a current through a control winding on the coil. L_1 and L_2 of the saturable reactor (T_1) of Figure 9-35 are the control windings of this reactor.

Some manufacturers place permanent magnets on their saturable reactors as a form of magnetic bias. The magnets set the magnetic operating point of the core near saturation and therefore less control current is needed to shift the core into magnetic saturation.

In this circuit the saturable transformer (T_1) acts to couple horizontal pulses obtained from the horizontal flyback windings into the vertical deflection yoke. The horizontal pulses are filtered by the resonant circuit consisting of L_5 and C_1 into sine waves. Since a sine wave has approximately the same shape as a parabolic waveform, it may be used in place of it. The amplitude of the induced horizontal pulse appearing in the secondary (L_3 and L_4) of the saturable transformer depends upon the amount of horizontal deflection current flowing through the series connected control windings L_1 and L_2. The amplitude and phase of the total secondary voltage are the sum of the series connected secondary coils. These coils are connected in series opposition. Therefore, when the secondary (L_3 and L_4) voltages are equal, the sum voltage will be zero. This occurs when the vertical yoke current is zero and neither coil is saturated. At this time the electron beam is passing through the center of the raster and no horizontal correction current is applied to the

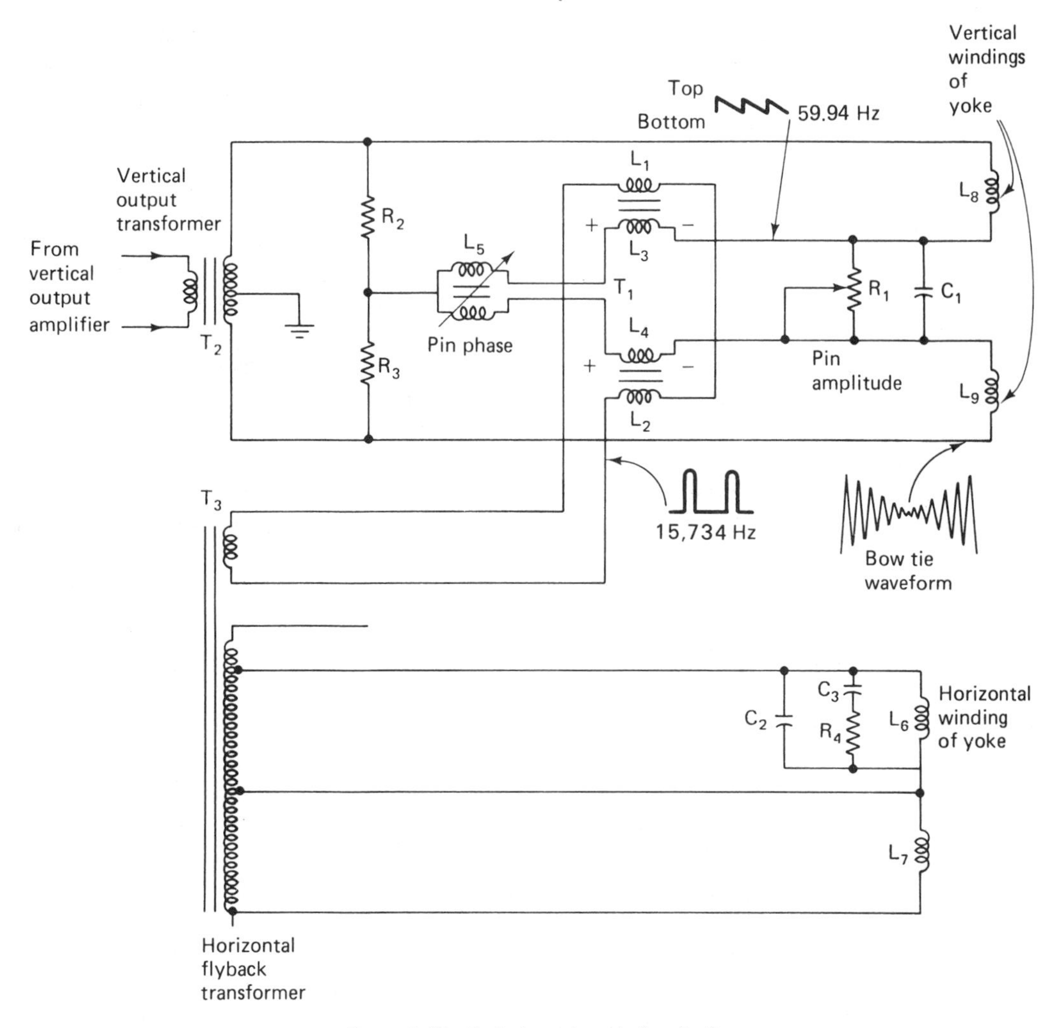

Figure 9-35 Typical antipincushioning circuit.

vertical yoke. When the vertical deflection current is maximum and the electron beam is at the top of the raster, one of the saturable reactors (L_1, L_3) becomes saturated and the induced secondary voltage in this reactor drops to zero. The induced voltage in the other reactor is now maximum. Therefore, the sum voltage is maximum, and maximum correction current is applied to the vertical yoke windings. When the electron beam is at the bottom of the raster, the vertical yoke current is maximum in the reverse direction, taking L_1 and L_3 out of saturation and forcing L_2, L_4 into saturation. This results in a sum voltage that is maximum but of reversed phase from the previous condition. This introduces maximum vertical correction current in the reverse direction.

The amplitude of the horizontal correction current flowing through the vertical yoke windings is adjusted by means of R_1. Minimum correction amplitude occurs when the movable arm shorts out R_1 and prevents any correction current from flowing through the vertical windings of the yoke. The phase of the parabolic correction current is adjusted by slug tuning L_5, the pin phase control. In some circuits the position of the permanent magnet mounted on the saturable reactor is variable. Moving the magnet will shift the position of the crossover point of the "bowtie" parabolic correction current.

Side pincushion correction. In this circuit, side pincushion distortion is corrected by placing a variable load across the horizontal output transformer. When the electron beams are in the center of the screen, the vertical deflection current flowing through L_3 and L_4 is zero and transformer T_1 is not saturated. Therefore, the inductances L_1 and L_2 are maximum and minimum horizontal current flows through these windings. This allows all of the available deflection energy in the horizontal output transformer to provide maximum horizontal deflection. This action increases horizontal width and compensates for side pincushioning.

When the electron beam is at the top or bottom of the raster, maximum vertical deflection current flows through L_3 and L_4 of the saturable reactor, forcing the transformer into saturation. The inductance of transformer windings L_1 and L_2 are minimum and allow maximum current flow. The current that flows through these windings represents a loss of horizontal deflection energy and forces the raster to become narrower. If the amount of variable loading is proper, the change in horizontal raster size will counteract the side pincushioning that is ordinarily present.

Transistor antipincushion circuit. An example of a circuit that uses a transistor to correct top and bottom pincushion distortion is shown in Figure 9-36. In this circuit, as in the previously discussed circuit a saturable reactor (T_1) is used to modulate the vertical deflection yoke current at the horizontal rate. The base of the transistor (Q_1) is fed by a signal obtained from the vertical output stage. This voltage controls the amount of collector current flowing through the primary or control winding of T_1 and, therefore, determines when it becomes saturated. At some time during the vertical deflection interval, the electron beam is in the center of the raster. At this time, T_1 is forced into saturation and the horizontal pulses fed to the emitter of Q_1 do not appear at the secondary of T_1. When the electron beam is at the top or bottom of the raster the vertical pulse at the base of Q_1 is such that the saturable reactor is unsaturated and the horizontal pulses are induced into the secondary winding of T_1. These pulses are filtered by L_1 and C_1 into sine wave currents that are added to the vertical deflection sawtooth. Adjusting L_1 controls the phase of the pincushion correction currents. The amplitude of the horizontal pulses fed to the emitter of Q_1 is controlled by the variable resistor R_1.

Pincushion adjustment. To check the operation of the pincushion correction circuits accurately and to adjust them properly, a crosshatch pattern should be displayed on the screen of the receiver. The adjustment procedure for top and bottom pincushion adjustment consists of first increasing the pin amplitude control until the top and bottom horizontal lines are curved. Then the pin phase adjustment is used to move the curvature to the center of the screen. The pin amplitude control is then adjusted for straight horizontal lines on the top and bottom of the raster. This completes the pincushioning adjustment.

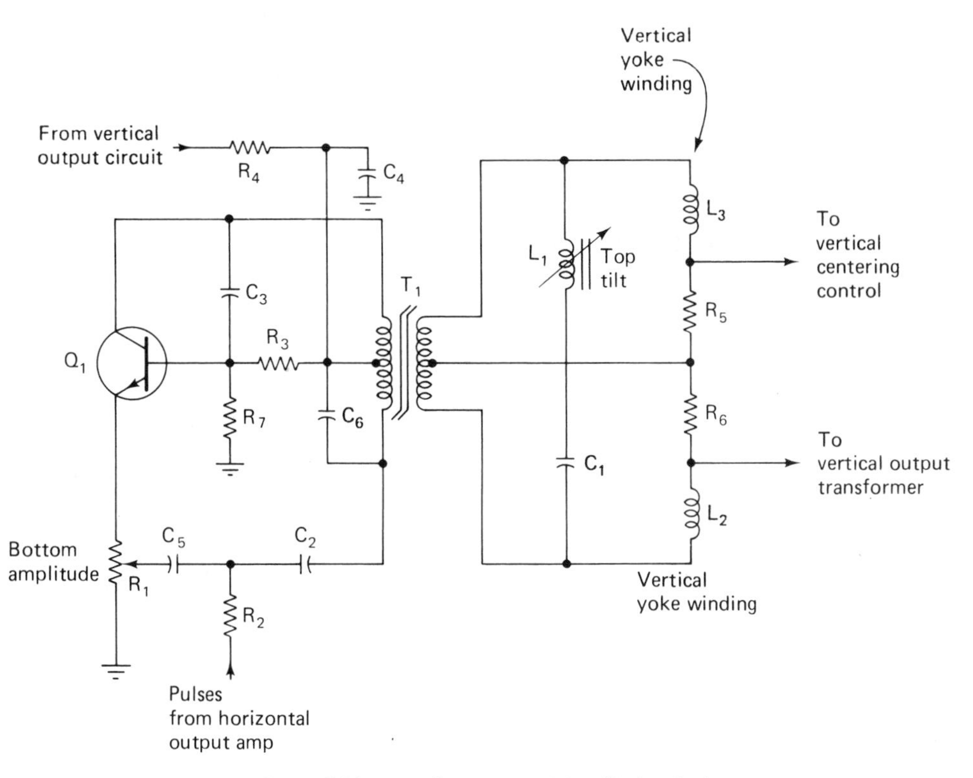

Figure 9-36 Typical transistor antipincushioning circuit.

9-17 the gray scale

Gray scale, or *color temperature* as it is sometimes called, refers to the shade of white generated by the color picture tube. Color temperature refers to the fact that a black object will turn different colors when it is heated. At a temperature of 6800°K a black object turns to a shade of white that corresponds to a standard white raster as defined by the NTSC. Proper adjustment of a raster for neutral white will result in black & white pictures of sharp contrast.

The sensation of white light is produced by combining equal energy red, green, and blue light. The gray scale adjustment procedure is concerned with setting the beam currents of each gun so that a white light output is obtained. This does not necessarily mean that the beam currents are simply made equal to one another. This is only possible in picture tubes that use equal efficiency phosphors. In older picture tubes in which the light efficiency of each phosphor was different, equal beam currents would not produce a neutral white raster. The best way to adjust gray scale is by looking at the raster and judging its color during the adjustment procedure.

Figure 9-37 shows a schematic diagram of the electron guns of a color picture tube and their associated gray scale adjustment control circuits. These controls represent those that might be found in any set. The purpose of each control is outlined below.

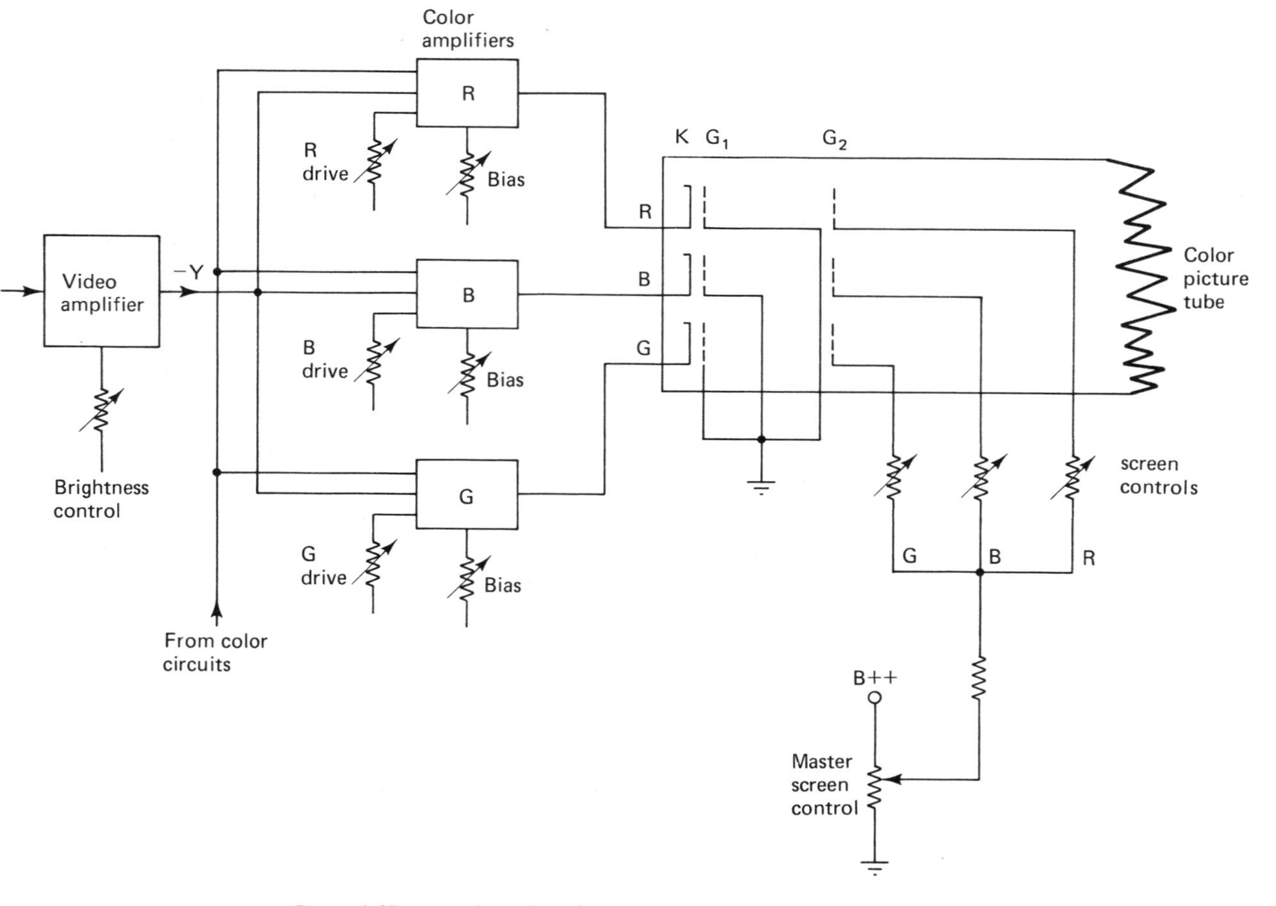

Figure 9-37 Controls used in adjusting a color picture tube for gray scale.

The dc picture tube bias voltage developed on the cathodes of the color tube is determined by *brightness control.* This control may be located in the first or second dc coupled video amplifier section of the receiver.

Also connected to the CRT cathodes, through their associated color amplifiers, are the drive and CRT *bias controls. Drive controls* determine how much signal is applied to the cathode of each gun. The bias controls are located in the emitter circuit of the dc-coupled color driver amplifier. This control adjusts the bias of the driver amplifier, which in turn controls the bias of the picture tube.

Many older color receivers have three *screen controls.* These controls adjust the dc voltage applied to the first anode electrode of each color electron gun. Modern receivers have master screen controls that simultaneously adjust the dc voltages and therefore raster brightness on all of the screens.

Gray scale adjustment. Unfortunately, there is no universally used gray scale setup procedure. Specific instructions for any given set can only be obtained from the manufacturer. In every case, the gray scale controls are adjusted to obtain a good white raster.

9-18 picture tube testing and troubleshooting

The picture tube is the most expensive component in either black & white or color receivers. Therefore, when the picture tube becomes defective, replacing it represents a sizable investment for the owner and considerable income for the service technician.

Both black & white and color picture tube troubles fall into the four following categories:

1. Open heaters
2. Low-emission or gassy tubes
3. Shorts and opens between tube electrodes
4. Poor color tube gun tracking

Many if not all of these defects may be checked and sometimes repaired by the use of a good-quality picture tube tester.

Open heaters. In black & white picture tubes the symptoms produced by an open heater are no raster, no picture, and sound O.K. There is no remedy for this defect except replacement of the picture tube. Before condemning the picture tube, the heater wiring and the tube socket should be checked. If there is nothing wrong with these, the open heater may be confirmed by means of an ohmmeter.

Color picture tubes have three heaters that are usually connected in parallel. The parallel heater connections are made inside the picture tube envelope and cannot be reached by the service technician. If one of the heaters opens, the color associated with the open heater will be missing.

Low emission. As a picture tube (either black & white or color) ages, the cathode-emitting material becomes "poisoned" with an overcoating of ions collected from the residual gas in the tube. This overcoat reduces the amount of electron emission available from the cathode and in time results in a weak, low brightness picture. In some cases, the low brightness picture is accompanied by silvery highlights. The picture tube tester can be used to confirm this by an emission test similar to that used in ordinary tubes. Gassy tubes generally display the same kinds of symptoms as a tube of low emission and should be serviced as such.

Low emission can be remedied either by picture tube boosters or by rejuvenation.

Picture tube boosters. A picture tube booster is useful primarily in older TV receivers and is able to restore a low-emission black & white or color picture tube to its normal levels of light output by increasing the ac heater voltage of the tube. This in turn increases cathode emission and restores the light output of the picture tube to normal. When the heater voltage is dc, boosters cannot be used and picture tube rejuvenation may be used.

A booster consists of a small step-up transformer (6.3 to 8.1 V) that is attached to a picture tube base on one end (secondary) and a picture tube socket (primary) on the other. All other pin connections are transferred from the receiver socket to the picture tube base by means of wires that are part of the booster. Since picture tubes have different kinds of base, the service technician must have a supply of boosters that will fit the different types of tubes. When a booster is to be used, the socket connected to the base of the picture tube is removed from the tube. The booster socket is attached to the tube and the booster picture tube base is inserted into the receiver socket.

Picture tubes found in solid state receivers often obtain their heater voltage from windings on the horizontal output transformer. Such picture tubes can be supplied with special boosters that are designed to be placed between the flyback and the picture tube heater, increasing the heater voltage and restoring emission to normal.

Picture tube rejuvenation. Another technique that has been successfully used to restore low-emission picture tubes to normal operation is *rejuvenation.* In this process the picture tube tester is adjusted so that the CRT heater voltage is increased by 50% and a high voltage (700 V) is applied to the control grid for an instant. This causes a sudden increase in cathode current that strips away the overcoat of impurities that limits cathode emission. If emission is not restored to normal, some rejuvenation units have provision for increasing the high voltage applied to the grid. Others extend the time and increase the heater voltage. In some types all three (heater voltage, grid voltage, and time) are increased. If the boosted rejuvenation does not work, the tube is considered beyond repair and is replaced.

Shorts. Picture tubes can develop short circuits between their electrodes as a result of flakes of conductive material that lodge between them or because the electron gun becomes warped. Shorts between heater and cathode and between grid and cathode will result in an increased raster brightness and loss of picture information. In color tubes the short will affect only one color.

Shorts between screen grid and the first anode or screen and focus electrode or shorts to the second anode will all result in loss of the raster.

A short caused by a flake of conductive material very often can be remedied by burning the flake out. This is done by adjusting the picture tube tester to apply an ac or dc voltage of approximately 600 V between the shorted electrodes. If a high-enough current is passed through the flake, it will burn up and open the short.

Opens. Open picture tube electrodes usually result in no raster. An open cathode or control grid may produce a dim raster. Opens can be produced by breaks at the gun electrode or at the point where the base pins are welded to the electrode wires. Opens may be checked for by tapping the neck of the tube and noting the effect on the picture. If the picture is momentarily returned to normal, the trouble is an open electrode.

In some cases, notably an open cathode, it may be possible to weld the open cathode together again. The usual procedure is to apply a high voltage (1000 V) between the cathode and the open electrode. At the same time, the neck of the tube is gently tapped. Because of vibration of the open leads, a momentary contact between them may be sufficient to spot weld them together.

Tracking. Color picture tube electron guns are designed to have similar electron emission and control characteristics. This is necessary in order to ensure that gray scale tracking remains constant as the brightness control is varied from low to high levels. Occasionally, the emission characteristics of the guns change and are no longer matched. As a result, gray scale tracking cannot be maintained from low to high brightness levels. Some picture tube testers are equipped to measure the tracking characteristics of the three guns. If tracking is defective, rejuvenation of each gun may be used to restore tube operation to normal.

9-19 test jigs

A color test jig is a device that is used to directly substitute for the color picture tube of a defective receiver.

The color TV test jig consists of a color picture tube, the deflection yoke, a high-voltage meter, the convergence assembly, and other cables and sockets needed to substitute for and match a picture tube used in a specific color receiver. An example of such a test jig is shown in Figure 9-38. The major reasons for using a test jig are as follows:

1. The use of a shop test jig means that, in many cases, the entire television set need not be brought to the shop. If the defect is not related to the sweep, convergence, or picture tube sections of the receiver, the chassis and tuner are the only parts of the defective receiver that must be brought to the shop. Leaving the CRT in the home has the advantage of minimizing potential damage to bulky, heavy, and expensive cabinets as well as avoiding the need for a purity and convergence setup when the receiver is returned.
2. Direct substitution of a defective receiver picture tube with a test jig will leave no doubt about whether or not the receiver picture tube should be replaced. A portable test jig permits tests to be done in the customer's home, thus reducing bench work and assuring the customer that he or she needs a new picture tube.

The color test jig must be matched both physically and electrically to the receiver for which it is substituting the picture tube. To do this for all of the vacuum tube and solid-state receivers on the market, requires special connecting cables such as CRT base leads, CRT grounding, and CRT high-voltage connectors. In addition, yoke matching devices and dynamic convergence assembly substitution adapters are also required. Test jig manufacturers supply the cables and adapters that are needed.

Figure 9-38 Color test jig with its accessory cables. (Courtesy of Electronic Components Group, GTE Sylvania.)

Focus voltage. Many test jigs are equipped with a picture tube that requires 4000 to 6000 V for its focus electrode. This tube will not produce a raster when it is used with a chassis that uses a low-voltage focus system. In order to provide a universal test jig, the required focus voltage must be obtained from a separate high-voltage power supply. Such power supplies may not be available. A simpler solution used in many test jigs is to place a high-resistance (80 MΩ), high-wattage (20 W) variable voltage divider across the second anode voltage of the chassis and tap off the required focus voltage. This is the method used by many color television manufacturers to obtain focus voltage. Therefore, the components that the manufacturers use for their voltage divider may be obtained and added to test jigs that do not have the required focus voltage.

9-20 the color bar generator

One of the most important pieces of test equipment used by the color television service technician is the color bar generator. See Figure 9-39(a). Essentially, this generator is a simulated TV transmitter that has a fixed number of selectable pictures (patterns) that are able to modulate the RF signal. Some examples of the patterns generated are shown

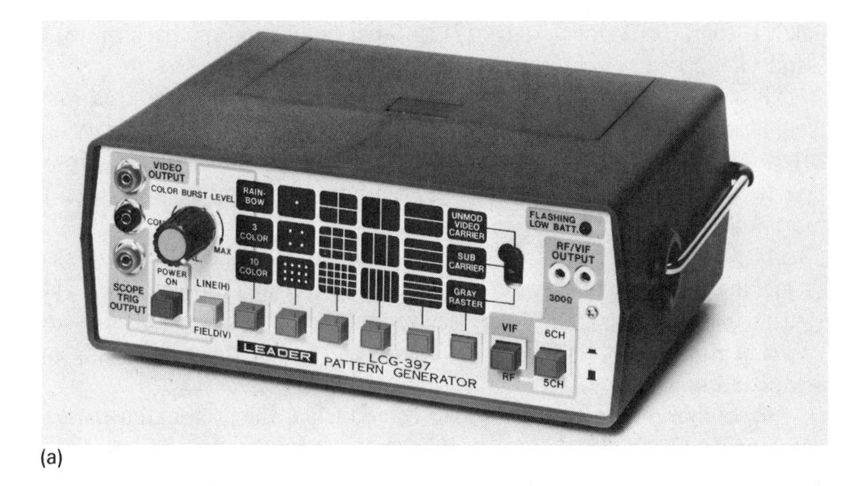

(a)

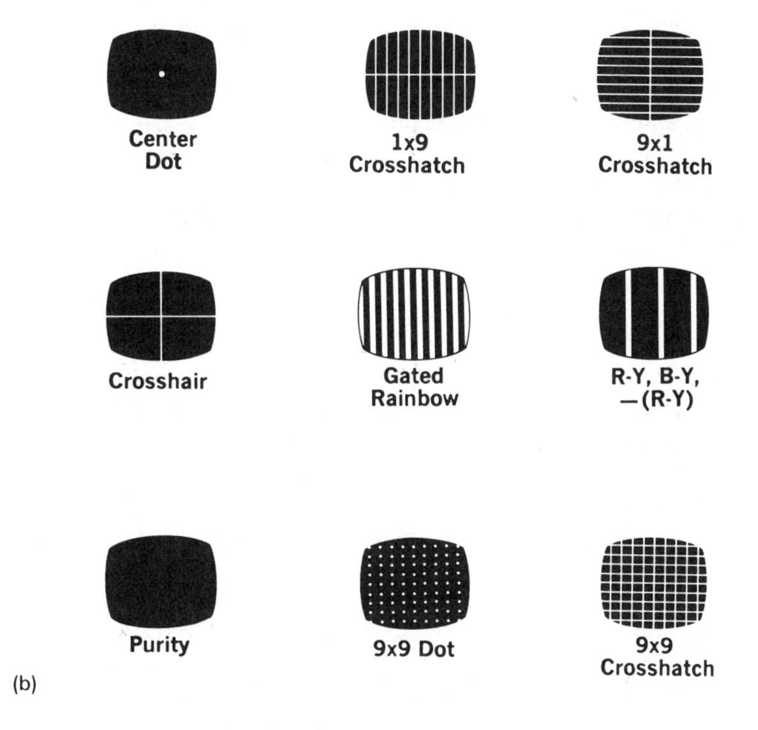

(b)

Figure 9-39 (a) Photo of a gated rainbow color bar generator; (b) some of the patterns that may be generated by a color bar generator. [(a) Courtesy of Leader Instrument Corp.]

in Figure 9-39(b). Most color bar generators provide an RF output at channel 3 or 4 that is coupled to the antenna terminals of the receiver. Depending upon the position of the generator's pattern selector switch, the picture produced on the screen of a normally operating color receiver can be any one of the following patterns: (1) vertical lines, (2) horizontal lines, (3) a single dot in the center of the screen, (4) a cross-hair pattern, (5) a crosshatch pattern, (6) a dot pattern, (7) a 10-bar color rainbow pattern, (8) a 3-bar $R-Y$, $B-Y$, and $(R-Y)$ color pattern, and (9) a blank white raster.

All of these patterns, except for the color bar patterns, are used for purity, static, and dynamic convergence adjustments. The blank raster is used for purity adjustment, the dot patterns for static convergence, and the crosshatch and other lines for dynamic convergence. Because of the equal spacing between horizontal and vertical lines, the crosshatch pattern may also be used for linearity adjustments.

The color patterns are used for identifying faults in the color sections of the receiver and for making automatic frequency phase control (AFPC) adjustments. The wave shapes produced in the color sections of the receiver by these color bar patterns are essential for signal tracing throughout these circuits. These aspects of the color bar generator will be discussed in Section 16-2.

Proper fine tuning of the receiver occurs when the crosshatch pattern of the color bar generator produced on the screen of the picture tube has horizontal lines and vertical lines that are of equal brightness. When this occurs, the low video frequency (horizontal lines) and the high video frequency (vertical lines) response of the receiver are equal.

The simplified block diagram of a color bar generator is shown in Figure 9-40. The color bar generator makes use of digital circuits to produce stable test patterns. The heart of the sync and pattern-forming section of the generator is a 189-kHz crystal oscillator. This frequency is the twelfth harmonic of the horizontal scanning line frequency used with this generator. This means that 12 cycles of this signal can occupy the time of one horizontal scanning line. One of these cycles is used as the horizontal sync pulse; another adjacent cycle occupies the time interval set aside for the color burst. This leaves 10 cycles for the production of 10 vertical bars.

Horizontal lines are obtained from the 189-kHz clock by using a chain of flip-flops as counters. Each flip-flop intrinsically divides by 2. By means of feedback it is possible to make a number of series or parallel connected flip-flops divide by any desired amount. Thus, if 189,000 is divided by a chain of flip-flops designed to produce a division of 210, the output will produce 900 pulses per second. This means that in $^{1}/_{60}$s, the time of one vertical field, a total of 15 of these pulses is produced ($^{1}/_{60} \times 900 = 15$). If the time duration of each pulse is one or two horizontal lines, 14 equally spaced horizontal lines one or two scanning lines thick, will appear on the screen of the picture tube. The fifteenth pulse occurs during vertical retrace and is not seen.

Horizontal sync pulses of 15,750 Hz and vertical sync pulses at 60 Hz may also be obtained from the 189-kHz clock by dividing its output by 12 and 3150 respectively.

Each of the desired outputs of the divider chain is passed through a pulse shaper that sharpens the pulses so that distinct lines will be produced. The desired pattern that is to modulate the RF carrier is selected by the function selector switch. When the switch is in the horizontal lines position, only the 900-Hz signal reaches the modulator. Similarly, in the vertical lines position, only the 189-kHz clock pulse reaches the modulator.

If a crosshatch pattern is desired, the function selector switch selects the output of an adder that simply combines the 189-kHz and the 900-Hz signals.

To obtain a dot pattern, the 189-kHz and 900-Hz pulses are fed into an AND gate. An AND gate is equivalent to a number of switches connected in series. In this arrangement the current can only flow when all the switches are closed at the same time. Thus, the only time the AND gate will produce an output is when both the 189-kHz clock

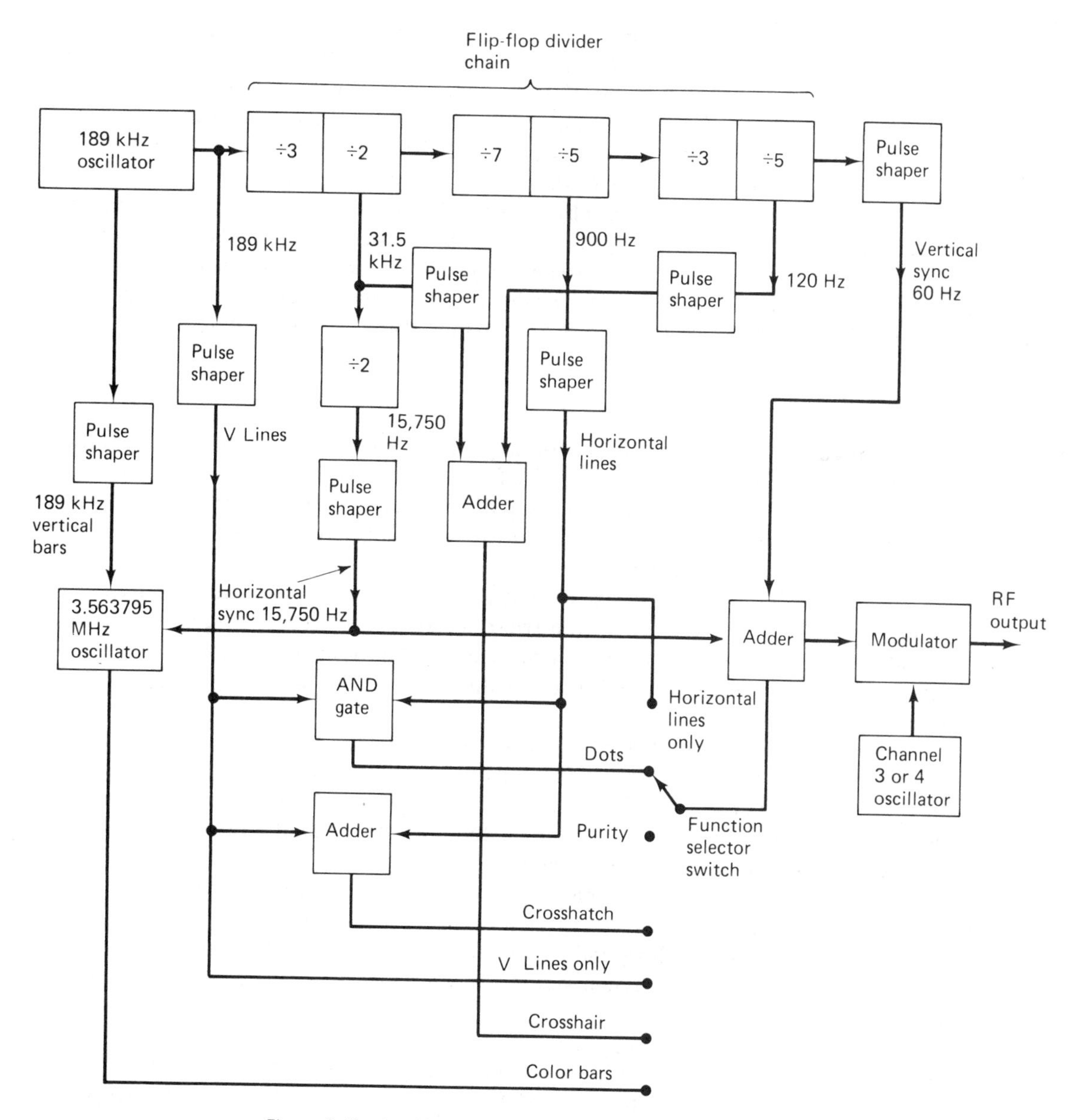

Figure 9-40 Simplified block diagram of a color bar generator.

pulse and the 900-Hz pulse simultaneously present an input pulse to the AND gate. This only occurs at the points of intersection in the crosshatch pattern.

A blank raster for purity adjustments is obtained by not modulating the RF oscillator at all.

To produce a cross-hair pattern, the flip-flop dividers are designed to produce outputs at 31.5 kHz and at 120 Hz. Each of these frequencies is twice the scanning rates. Therefore, they will both produce two pulses during one scanning period. If the timing is correct, one of the pulses will occur during the sync interval and the other will occur midway between the sync intervals. Thus, the 31.5-kHz pulse will produce a single vertical line in the center of the raster and the 120-Hz pulse will produce a single horizontal line in the center of the raster. If these two pulses are combined in an adder, a signal that can be used to generate a cross-hair pattern will result. If the divider outputs are fed to an AND gate, a single dot in the center of the screen will be the result.

Other patterns may be obtained by proper frequency division and gating techniques. Unfortunately, there are no universally accepted standard patterns used by all manufacturers. Furthermore, for a variety of reasons, some manufacturers of color bar generators select clock frequencies of 189.8 or 189.6 kHz. The divider chains and numbers of horizontal lines and vertical lines produced will also vary from manufacturer to manufacturer.

9-21 projection TV

Projection TV is an optical method for obtaining large-screen television pictures. Depending upon the manufacturer, screen size for home noncommercial use can vary from 40 in. (1.016 m) to 120 in. (3.05 m) as measured diagonally. Auditorium-type systems are also available that can produce 42 × 55 ft (13 × 17 m) displays. Projection TV may be obtained in either color or black & white format.

Projection TV systems may be obtained in a two-piece or single-unit arrangement as shown in Figure 9-41(a) to (c). In the two-piece system a television receiver that has been suitably modified is used to project an enlarged television picture onto a separate screen that is placed some distance in front of the receiver. This scheme is similar to that used in a movie theater.

The single-unit projection TV system may be designed in a number of different arrangements, as shown in Figure 9-41(b) and (c). In the self-contained front projection method, the light output from the TV projector strikes a flat mirror whose reflective surface is on the front surface of the glass. This mirror then reflects the light onto a screen, which in turn reflects the television picture to the viewer.

In the rear screen projection method, two front surface mirrors are used to project the TV picture onto a viewing screen that is textured on both sides. The projection side is a Fresnel lens [see Figure 9-41(d)] and the observer's side is lenticular. The screen has a focal length of 20 ft, making optimum viewing at a distance of about 15 ft from the front of the receiver.

Projection systems. There are three main categories in projection TV schemes: refraction, Schmidt, and light valves. Each produces a projected television picture with its own particular characteristics and trade-offs.

Refraction systems. As is commonly known, convex lenses have magnifying characteristics. If the lens is brought close to a TV screen in a darkened room, a projected inverted and reversed image of the TV picture can be projected onto a nearby wall. The lens is able to provide this magnified image because it can bend or refract the light coming off the TV screen. Lens systems may be divided into two types: single-lens systems and multiple-lens systems.

In the single-lens system a large-diameter lens is mounted approximately 1 ft in

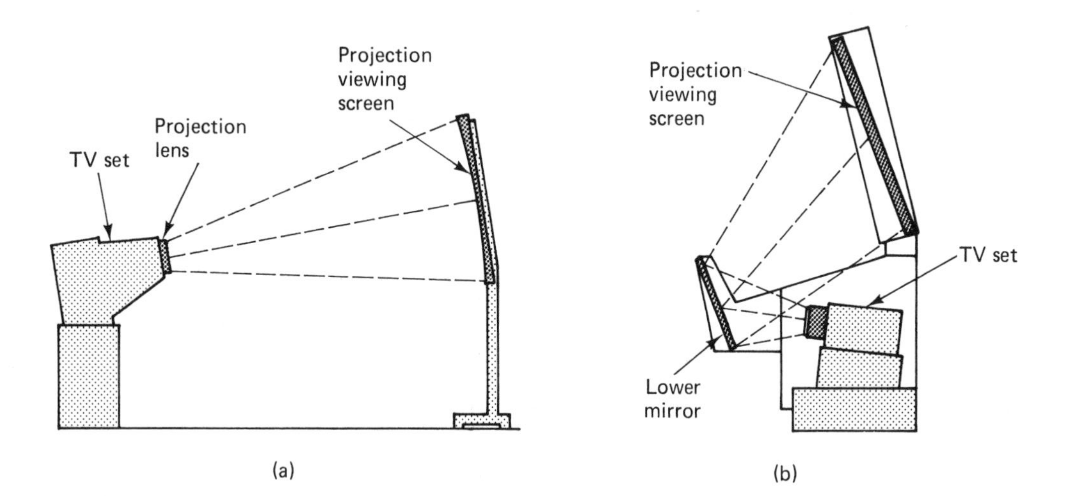

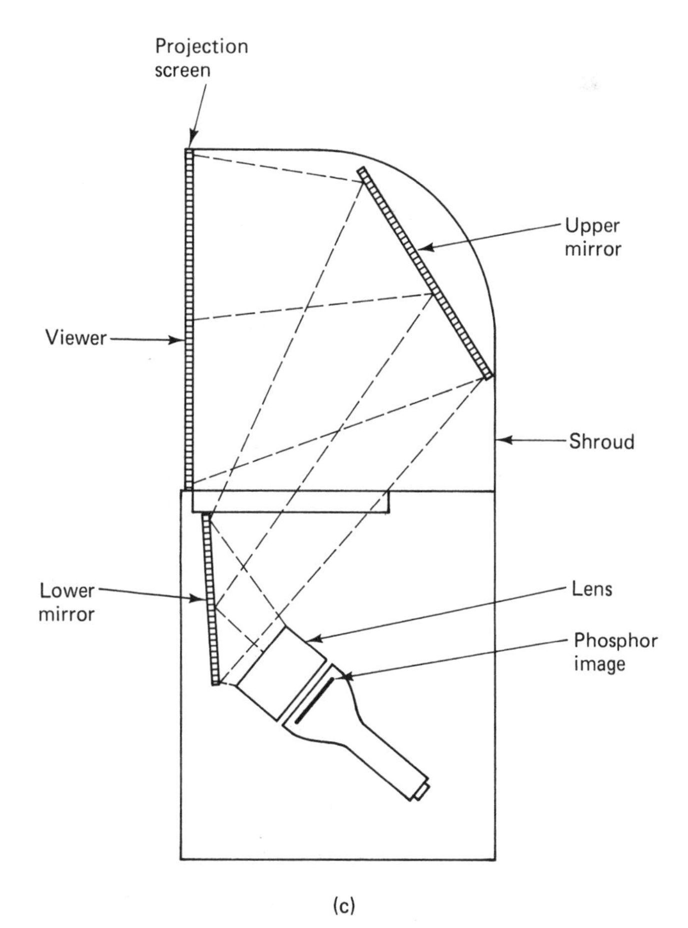

Figure 9-41 (a) Two-piece projection television system; (b) one-piece front projection television system; (c) rear projection television system; (d) construction of a rear projection viewer screen; (e) Schmidt-type projection tube.

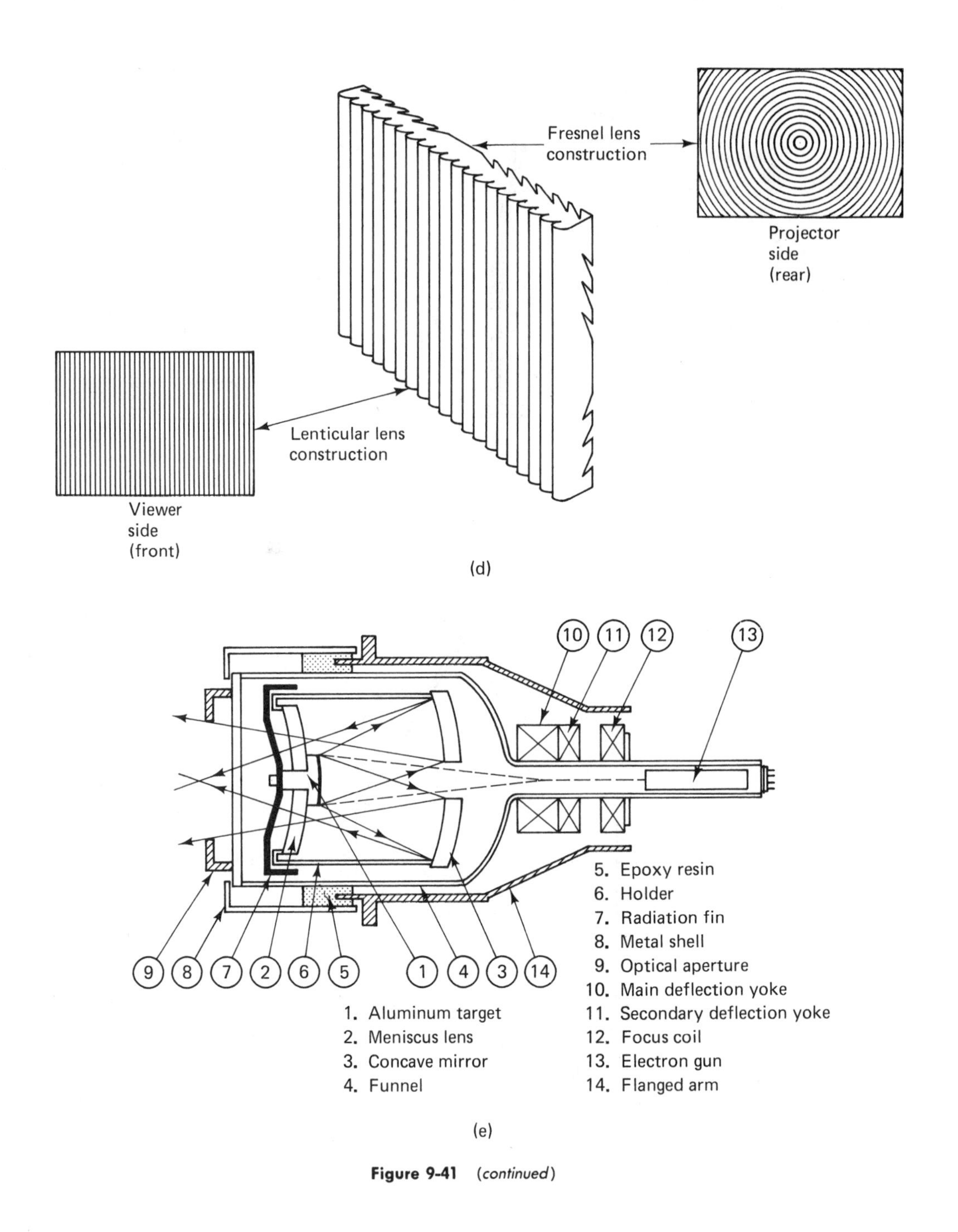

Figure 9-41 *(continued)*

front of the television receiver's picture tube screen. The lens then gathers the light from the picture tube face and projects it out of the front of the lens onto a reflecting screen for viewing.

To compensate for lens reversal and inversion of the image, the TV receiver yoke wiring must be modified so that vertical scanning inverts the TV picture and horizontal scanning reverses the picture so that it is wrong-reading.

The main problem with the projection scheme is its low light levels. This means that viewing must be done in a darkened room. There are two reasons for the low light levels. First, the lenses are not perfect transmitters of light. Some light energy is lost by scattering and absorption of parts of the light spectrum. Lens efficiency is measured by F numbers. The lower the F number, the greater the light transmission of the lens.

Second, the greatest problem with low light levels is the picture tube itself. The shadow mask in a color receiver picture tube cuts the energy from the electron guns by 80%. Thus, only 20% is available to produce usable light. In addition, magnification of the light source image reduces the brightness of the final projected image. For example, magnifying an image by two times can reduce the resultant picture brightness by 10 times.

To overcome these problems requires very bright TV pictures. In a single-tube system the light output of a picture tube may be increased by increasing the second anode voltage. This approach is limited by a reduction in contrast and x-ray production.

Three-tube projection television systems produce as much as five times the light output of a single tube. Such brightness levels are twice that of a movie theater. This is because each projection tube (red, green, and blue) in conjunction with its own lens does not require a shadow mask. Compared to single-tube systems, picture quality is improved in the three-tube system. Convergence defects in a single-tube projection system are magnified. Multiple-tube systems do not have this problem. However, they do require three yokes, in some cases three focus coils, heavy-duty low- and high-voltage power supplies, and multiple adjustable controls for linearity and amplitude. In addition, special safety circuits may be required to protect the picture tubes from phosphor burn if the vertical deflection circuit becomes defective and a high-intensity horizontal line is produced.

Schmidt system. The Schmidt projection system is a three-tube arrangement in which each tube, one red, one green, and one blue, has its own built in catadioptic lens system of very low F number. See Figure 9-41(e). The tube is about 5 in. (12.7 cm) in diameter and operates its small phosphor target at approximately 28 kV. This produces an intense light output. The light is first directed toward a spherical concave mirror which reflects the light outward toward the projection screen. Unfortunately, a spherical mirror introduces optical distortion. To compensate for this, a Schmidt corrector meniscus lens is placed in front of the system, giving the mirror-lens combination its name. The result is a picture with a brightness of 10 to 15 times that of a single-tube system; this is about four times the brightness of a movie theater.

Light valve systems. In this arrangement three special units (red, green, and blue) using projection lenses produce a picture whose light level depends only upon an external light source. No phosphor screen is used. The light source used is usually a very high intensity xenon lamp.

The light valve units are constructed with a spherical mirror that is coated with a thin layer of oil. Above the oil is an electron gun scanning in a conventional manner. The electron beam charges the oil and distorts it into grooves that are proportional to the video information. The light passing through the oil is reflected to the projector lens and out to the screen in a standard television pattern. Light valves have sufficient light output to illuminate a 42 × 55 ft (13 × 17 m) TV screen.

Reflector screens. The total brightness of the TV picture the viewer observes depends upon three factors: source brightness, picture size, and viewing screen *gain.* Compared to a matte screen (a chalk-like flat reflective screen), projection television screens have high light gain. They are beaded or of lenticular design and reflect more light or even concentrate light rays for a brighter picture. Screens can have a gain of as much as 6 to 10 times that of a matte screen. Screens are curved and inclined to ensure greater center illumination and to reflect extraneous light downward. Moving 20 or 30° off center-screen horizontally or vertically may cause loss of the viewed picture.

QUESTIONS

1. Describe the function of each electrode in an electron gun used in a black & white picture tube.
2. Describe the major differences between the delta, GE in-line, RCA in-line, and Trinitron color picture tube.
3. What are three methods of focus, and how do they affect picture tube requirements?
4. What are three reasons for aluminizing the phosphor screen?
5. List six factors that determine the light output of the picture tube.
6. What effect will an increase in the internal resistance of the high-voltage power supply have on picture tube operation?
7. Give a brief description of the purity procedure used in delta and in-line tubes.
8. What is static convergence, and how is it obtained?
9. What is dynamic convergence, and how is it obtained?
10. What is the waveform of current that passes through the windings of the dynamic convergence assembly?
11. What is done to the dynamic convergence current to provide tilt?
12. Why is the dynamic convergence circuit of the in-line tubes much simpler than those used in delta tubes?
13. Why is the Trinitron considered a single gun tube?
14. What is the cause of pincushioning?
15. What is meant by gray scale tracking?
16. What is meant by picture tube rejuvenation, and how is it done?
17. What are the advantages and limitations of using a picture tube test jig?
18. Describe the construction of a flat CRT.
19. Describe the operation of a field-effect LCD.
20. What are three types of projection TV systems?
21. What must be done to the television receiver in a refraction projection system?
22. What does a catadioptic lens system consist of?
23. Explain how a light-valve projection system works.
24. Where is Fresnel/lenticular lens construction used?
25. What type of reflector screen has the highest gain?
26. What is the main advantage of a Schmidt projection tube?

chapter 10

The Video Amplifiers

10-1 introduction

The video amplifier stage of the television receiver is located between the video detector and the cathode of the picture tube. This is illustrated in the block diagram of a complete color receiver shown in Figure 10-1. Video amplifiers used in black & white receivers are located in the same relative position and serve the same functions as those found in color receivers.

The basic function of the video amplifier is to amplify the output of the video detector to a level sufficient to fully intensity modulate the picture tube. That is, the video signal must be able to vary the picture tube bias over a range large enough to vary the light output of the tube.

The signal output of the video detector for a moderately strong signal is approximately 2 V p-p. Depending upon the type of picture tube used, a video signal of 80 to 160 V p-p may be required to drive it between maximum beam current (white) and beam current cutoff (black). This means that the video amplifier must be able to provide distortion-free voltage amplification of between 40 and 80 times. In black & white sets only one stage of amplification may be used; color receivers may have from two to five stages of amplification.

10-2 picture tube light output characteristics

The light output characteristic of a television picture tube is such that making its cathode less positive with respect to the control grid will cause an increase in light output, and vice versa. An example of a picture tube characteristic curve is shown in Figure 10-2. Notice that when the grid to cathode voltage (E_{KG}) of the picture tube's electron gun is zero, the light output will be maximum. This corresponds to white in a black & white

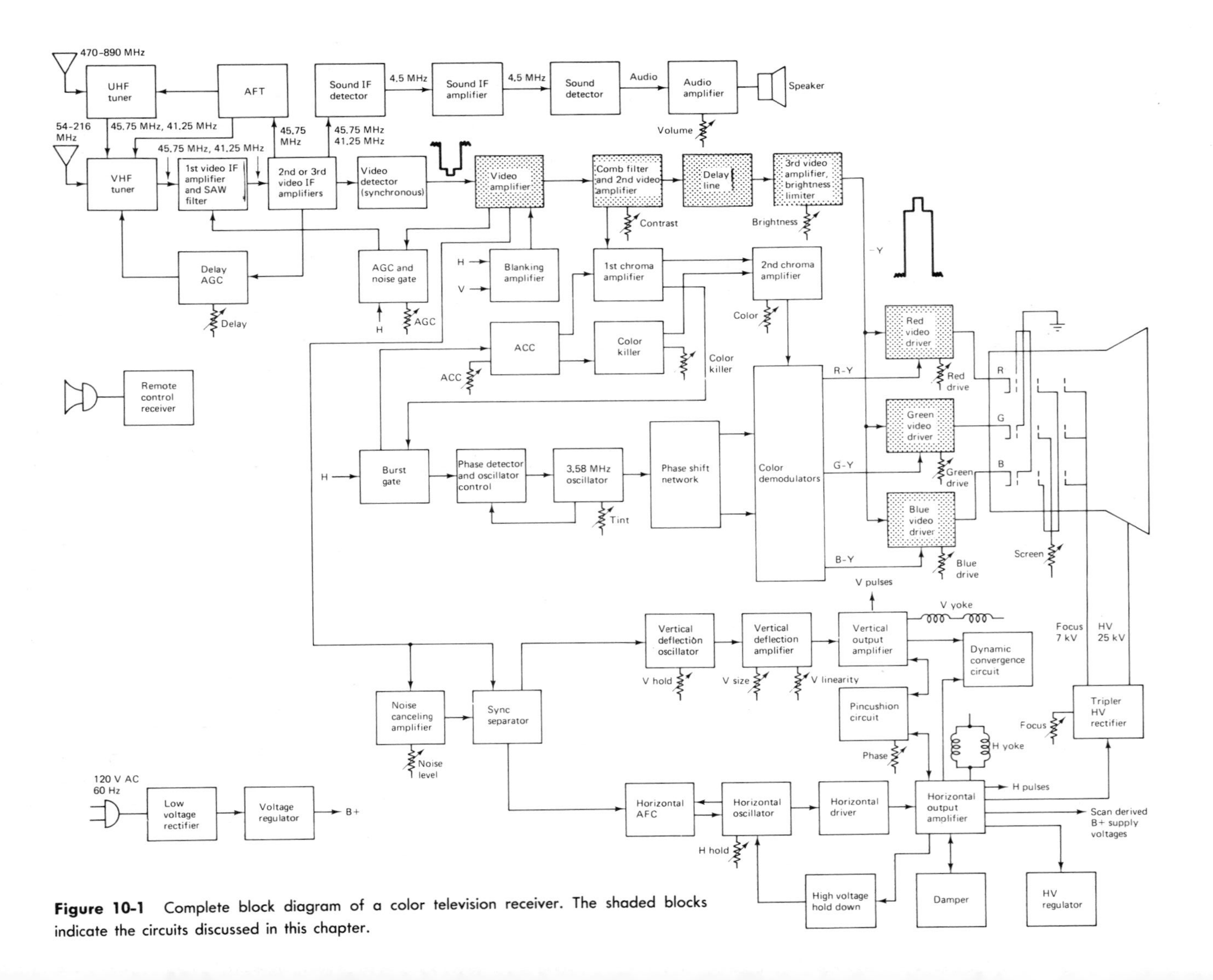

Figure 10-1 Complete block diagram of a color television receiver. The shaded blocks indicate the circuits discussed in this chapter.

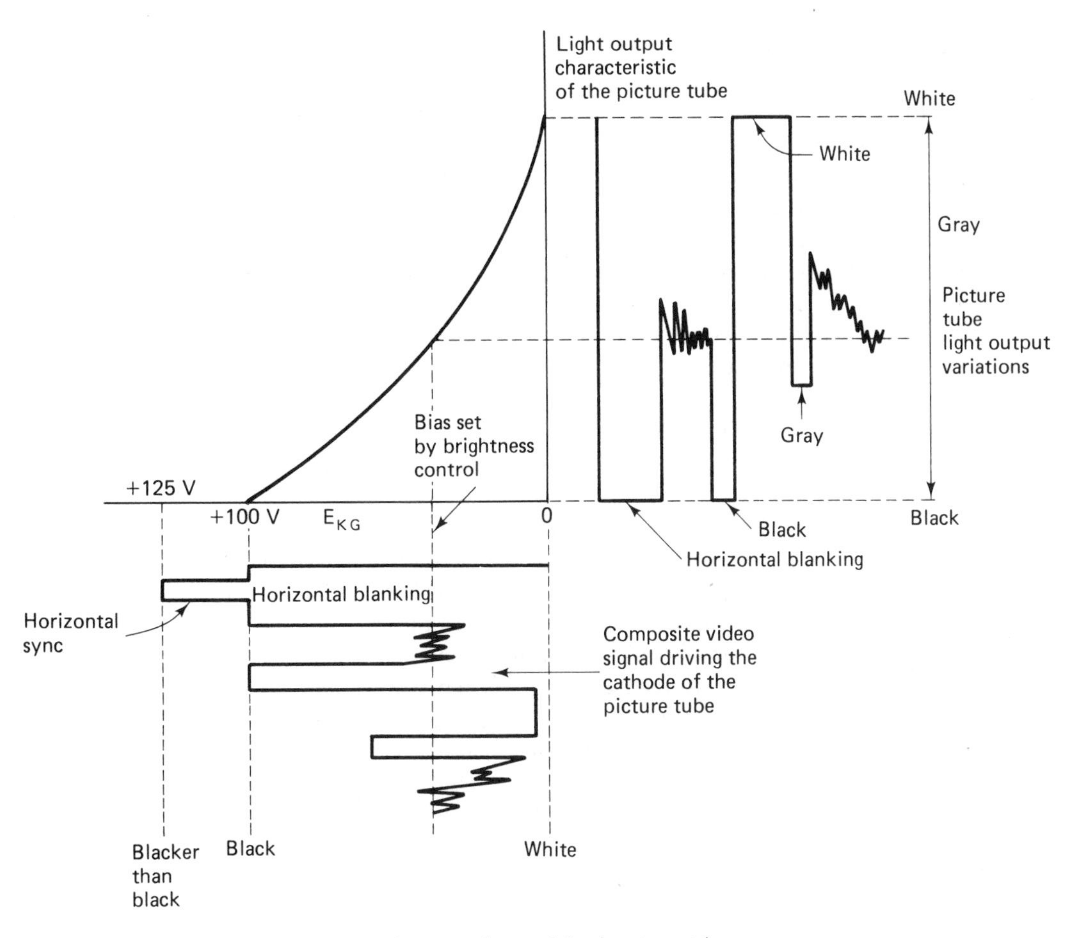

Figure 10-2 Light output characteristic of a picture tube.

picture tube and to bright red, green, or blue in a color picture tube. When the grid is made sufficiently negative with respect to the cathode, the electron beam current will be reduced to the point where no light is emitted from the picture tube, corresponding to black. Values of grid to cathode voltage between these extremes will result in corresponding shades of gray or color brightness.

When the composite video signal is used to drive the picture tube, it is usually fed to the cathode of the tube, and the grid is held at a constant potential. The phase of the video signal is such that the sync and blanking portions of the signal are able to drive the picture tube into beam current cutoff. This means that the wave shape at the cathode of the picture tube must have a positive going sync pulse. In Figure 10-2 the video signal is positioned on the light transfer curve by the brightness control that sets the operating bias of the tube. When the bias is correct, the video signal is positioned so that the blanking pulse is at cutoff, corresponding to black, and the sync pulse level is at a level that is blacker than black. In this bias position the video signal is able to vary the light output of the picture tube over a wide range from white to black.

The brightness control. The brightness control is a front panel control in both color and black & white receivers that electrically adjusts the picture tube bias. It may be located in either the grid or cathode circuit of the picture tube. If the brightness control is misadjusted so that the operating point of the picture tube is shifted toward cutoff (+100 V), the average brightness of the picture will become darker. Those levels of the video signal that would normally correspond to gray now are black and those levels of the video signal that normally correspond to white now appear as a shade of gray. Of course, those levels of the video signal that are almost black will be shifted into the black region and will not be visible on the screen of the tube.

If the brightness control setting is now moved toward zero, the average brightness of the picture will increase.

The video signal phase. The choice as to which electrode of the picture tube the video signal is to be applied is based upon the effect that the video amplifier has on picture quality and noise.

If the *grid* of the picture tube is to be driven by the video signal, the sync and blanking pulses that are impressed on it must be negative-going. Consequently, amplifier inversion will cause the composite video signal delivered to the input or base of the driver NPN video amplifier to have positive-going sync pulses. As such, this signal would be positioned on the video amplifier's nonlinear transfer curve such that the sync and biasing pulses drive the amplifier toward collector current saturation. As a result of the tube's nonlinear characteristics, those levels of the video signal corresponding to white are compressed and those levels of the video signal that are not normally seen because they are black become visible because they have been expanded. The visual result of this is to have a picture that has lost its ability to reproduce all shades of gray.

In addition to this undesirable result, strong noise pulses that accompany the video signal will be rectified and produce signal bias. This in turn will increase the bias of the video amplifier shifting part of the video signal into cutoff; resulting in some of the white picture information being removed from the output. In effect, this *noise set-up* of the video amplifier bias further aggravates white compression of the video signal.

If the video signal is applied to the cathode of the picture tube the polarity of the video signals at both the input and output of the video amplifier reverses from that used for grid drive. The effect of this on video amplifier operation is to cause the sync pulses to be compressed and the white portions of the video signal to be expanded. Furthermore, the noise pulses are no longer rectified but are amplitude limited by cutoff. It is for these reasons that the picture tube cathode is driven by the video signal in all modern color and black & white receivers.

10-3 amplifier distortion

The nature of the input signal to the video amplifier determines the required characteristics of the video amplifier. The composite signal is obtained at the output of the video detector and consists of the following signals:

1. The video (Y) signal (0 to 4.2 MHz)
2. The chrominance signal (2.08 to 4.08 MHz)

3. Horizontal and vertical blanking pulses (15,734.26 Hz and 59.94 Hz)
4. Horizontal and vertical sync pulses (15,734.26 Hz and 59.94 Hz)
5. Color burst (3.58 MHz)
6. A dc level that is directly proportional to signal strength and picture content
7. A sound IF of 4.5 MHz (not present in color receivers)

In short, the video amplifier must be able to reproduce without distortion a range of signal frequencies extending from dc to 4.2 MHz and whose wave shapes are extremely complex.

In black & white receivers sound IF take-off traps tuned to 4.5 MHz are provided in the output circuit of the video amplifier that remove this signal from the video amplifier, and deliver it to the sound IF amplifiers. It is important that the sound IF does not reach the picture tube. If it does, a severe herringbone interference pattern will appear on the picture tube screen. See Figure 10-3(a).

In color receivers the sound IF (4.5 MHz) must be prevented from reaching the video amplifiers. This is done to avoid a 920-kHz picture interference beat produced by mixing the sound IF with the color subcarrier. This form of interference [see Figure 10-3(b)] is prevented by traps tuned to 4.5 MHz that are placed at the output of the video detector.

Types of distortion. In electronics the term *distortion* refers to any undesired change that occurs in a signal when it is passed through an electronic circuit. Video amplifiers must be relatively distortion-free in order to provide proper picture reproduction. The four types of distortion that may occur in video amplifiers are as follows:

1. Harmonic distortion, also referred to as amplitude distortion
2. Intermodulation distortion
3. Frequency distortion
4. Phase distortion

Figure 10-3 (a) Herringbone interference due to 4.5-MHz sound IF reaching the picture tube; (b) 920-kHz interference due to a beat between the 3.58-MHz color subcarrier and the 4.5-MHz sound IF.

(a)

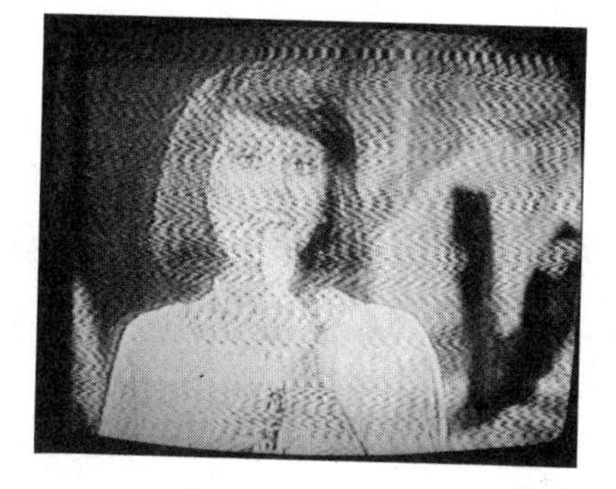

(b)

Harmonic and intermodulation distortion. Harmonic and intermodulation distortion are both caused by the nonlinear operation of the amplifier. That is, the amplifier is either overdriven and therefore is forced to operate in cutoff or saturation, or the bias has been shifted so that the amplifier's operating point is now in a nonlinear portion of its dynamic characteristics.

Harmonic distortion obtains its name from the fact that when a sine wave is applied to the input of an amplifier that is operated in a nonlinear fashion, the output signal wave shape will be found to have changed. The output wave shape may be then analyzed into fundamental and *harmonic* sine-wave components that did not exist at the amplifier's input.

The same may be said for intermodulation distortion. The exception is that instead of only one input signal, two or more input signals are applied to the input of the amplifier. Under these circumstances, the output wave shape of the amplifier may be analyzed into fundamental components, harmonic components, and *sum* and *difference* frequency components. In short the amplifier acts like a mixer.

The hammerhead test. In a practical sense, the effect of harmonic distortion on video amplifier operation is to cause clipping. That is, if the bias of the transistor or FET has been changed or if the amplifier is being overdriven, the peak portions of the composite video signal will be amplitude limited (clipped) by the cutoff characteristics of the amplifier. Since the peak portions of the video signal are the sync and blanking pulses, their removal will cause erratic or possibly complete loss of synchronization. This type of clipping may be made visible by observing the hammerhead pattern.

Under normal conditions, the hammerhead pattern should appear as shown in Figure 10-4(a). Here the brightness control has been increased to the point where the black level has become gray and the vertical sync and equalizing pulses are now black. If nonlinear circuit operation has caused clipping to occur and if the sync pulses have been clipped, the hammerhead will appear as shown in Figure 10-4(b). Here the vertical blanking interval has been made gray and the clipped vertical sync and equalizing pulses that constitute the hammerhead are now missing.

Figure 10-4 (a) Normal appearance of the hammerhead pattern; (b) hammerhead that results when the video amplifier is clipping the sync pulse level. (Courtesy of General Electric.)

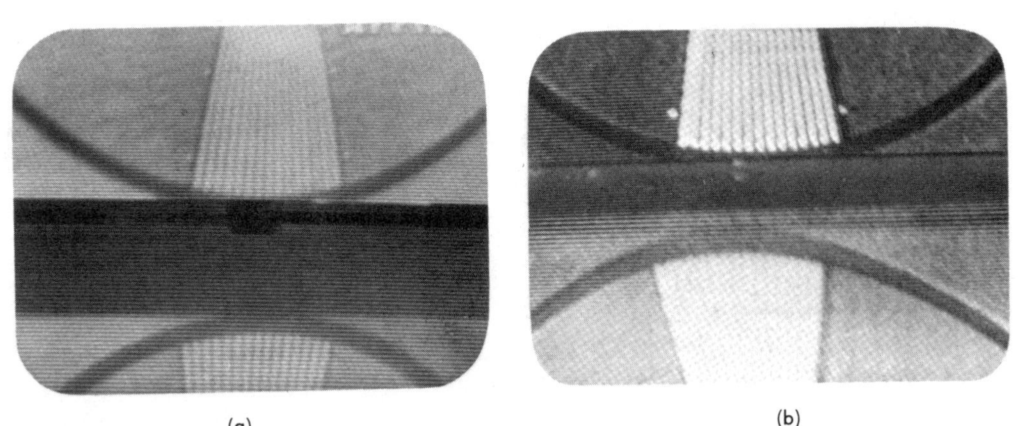

(a) (b)

Frequency and phase distortion. Frequency and phase distortion are both primarily caused by the effects of the interstage coupling networks of the amplifier. The visual effect of poor frequency and phase response on the TV picture is to cause loss of fine detail when the high frequencies are affected and to cause smear and loss of contrast when the low-frequency response is defective.

Frequency distortion is the condition in which the gain of an amplifier changes with frequency. Figure 10-5(a) shows the effects of frequency changes on the response of an ideal amplifier. Notice that the input voltage to the amplifier is kept constant while its frequency is varied from zero to infinity. In this example the input voltage at all frequencies is 1 V p-p, and the voltage amplification is assumed to be 40; therefore, the output voltage will be 40 V p-p throughout the entire frequency range of the amplifier. This amplifier has a perfectly flat response, which means that it is free of frequency distortion.

Directly below the frequency response curve of the ideal amplifier is drawn the phase response curve for this amplifier. The phase of a sine wave refers to either its starting position or to its time difference measured in degrees as compared to another sine wave. For example, if a sinusoidal generator's output voltage is maximum at the instant it starts, the phase of this signal would be said to be 90°. If the starting voltage of another such generator having the same peak voltage were half that of the first generator, it would have a phase of 30°. The phase or time difference between these two signals would be 60°. The phase relationship that will be referred to in this discussion is that between two signals. In the ideal amplifier there is no phase shift at all, and the phase difference between input and output voltages, neglecting normal amplifier inversion, is zero for all frequencies.

In actual practice, the reactances associated with the coupling circuits change with frequency and therefore affect both the amplification and the phase shift of the amplifier. The results of this are shown in Figure 10-5(b). In this case, the input voltage is kept constant in amplitude but is varied in frequency, as was the case in Figure 10-5(a). However, the output frequency and phase response of the amplifier changes drastically from that of the ideal case. Notice that the amplification is no longer constant for all input frequencies and that the frequency and phase response curves of the actual amplifier may be divided into three regions: low frequency, midfrequency, and high frequency. The midfrequency region is the region of amplification where both the frequency and phase response of the amplifier are the same as that of the ideal amplifier. The low-frequency portion of the frequency and phase response curves show that the voltage amplification *increases* with an increase in frequency and that the phase shift *decreases* with an increase in input frequency. The phase shift is said to be leading because e_o leads e_{in} by the angles indicated on the curve.

The high-frequency portion of the response curve indicates that the voltage amplification *decreases* with an increase in frequency and that the phase shift *increases* with frequency. In this case, the phase shift is lagging because e_o lags e_{in}.

The bandpass of an amplifier is defined as that range of frequencies that fall between the one-half power points on the frequency response curve. In terms of voltage, the one-half power points are the points on the response curve that correspond to 70.7% of the midfrequency output voltage. The low-frequency one-half power point is designated F_1 and that at the high frequency is F_2. The phase shift at the midfrequency portion of the curve is zero and that at the one-half power frequencies is 45°. Unfortunately, the effects of as little as 1° phase shift can be seen in the television picture. Hence, if the combined effects

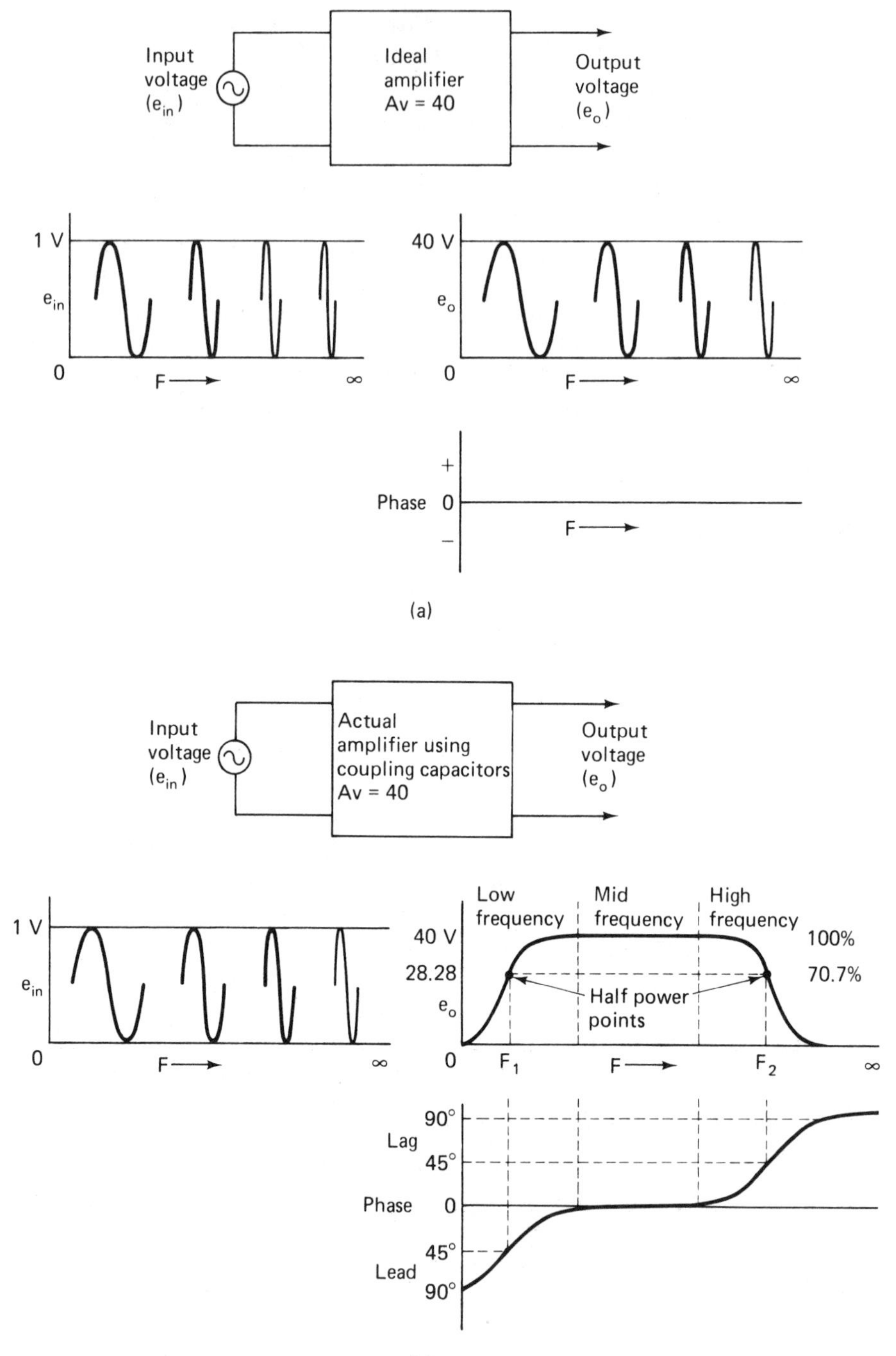

Figure 10-5 (a) Frequency and phase response of a perfect amplifier; (b) frequency and phase response of a practical amplifier.

of frequency and phase distortion are to be eliminated from the reproduced television picture, the bandpass of the video amplifier must be flat from 0 to 4.2 MHz. Of course, when the frequency response is flat, there is no frequency or phase distortion.

10-4 video frequency and picture content

High-frequency video signals correspond to the fine detail in the television picture and low-frequency video signal information is concerned with large picture areas. To understand this, consider the effect of simultaneously applying to the cathode of the picture tube a square-wave signal whose frequency is 15,750 Hz and to the yoke a deflection sawtooth current whose frequency is 15,750 Hz. This means that the grid to cathode voltage will be able to turn on and then turn off the picture tube beam current for one-half of a horizontal line. Since each horizontal line is identical and appears on the raster one below the other, the visual effect is that of a white vertical bar next to a black vertical bar.

If the frequency of the square-wave signal applied to the cathode is increased to 157,500 Hz, or exactly 10 times the horizontal scanning rate, a pattern of 10 black and 10 white vertical bars is displayed on the face of the picture tube, assuming zero retrace time. If the video frequency is increased to 1,570,000 Hz, 100 black and 100 white bars will be displayed. Increasing the video frequency to 4.2 MHz, the limit set by the FCC standards, the number of black & white bars will increase to approximately 266 each. These bars will be stationary and vertical for video frequencies that are an exact multiple of the horizontal scanning rate. Any other frequency relationship between the video signal and the horizontal scanning rate will cause the bars to move.

If the video frequency is reduced so that it is lower than the horizontal scanning rate (15,750 Hz), the picture will change from vertical bars to horizontal bars. This comes about because one-half cycle of a low-frequency video signal may have a time duration of more than one horizontal line. Therefore, more than one full horizontal line will be black and more than one full horizontal line will be white. For example, if the video frequency was made 300 Hz, each one-half cycle would correspond to a time duration of 1666 μs. Since one horizontal line has a time duration of 63.5 μs, approximately 26 horizontal lines will be scanned during this time. When the picture tube is turned on, these lines will be white; when the picture tube is turned off, these lines will be black. A video frequency of 300 Hz is five times the vertical scanning rate of 60 Hz. Therefore, in one vertical scan from the top of the picture to the bottom five cycles of video signal will appear between grid and cathode of the picture tube. This means that the picture should display five alternate horizontal black and white bars. Here again, these bars will remain stationary only if the video signal is an exact multiple of the vertical scanning rate.

In summary, narrow black or white bars whose duration is less than that of one horizontal line represent frequencies higher than the horizontal scanning rate. In contrast, long-duration black or white bars that occupy more than one horizontal line must represent frequencies lower than the horizontal scanning rate. It is therefore possible to relate any video signal to black & white areas whose widths depend upon the video frequency. Figure 10-6 shows the visual result of applying positive-going pulses of different time durations to the cathode of a picture tube. In each case, the pulse width is given in microseconds and is considered to be one-half cycle of a video signal whose frequency is also given.

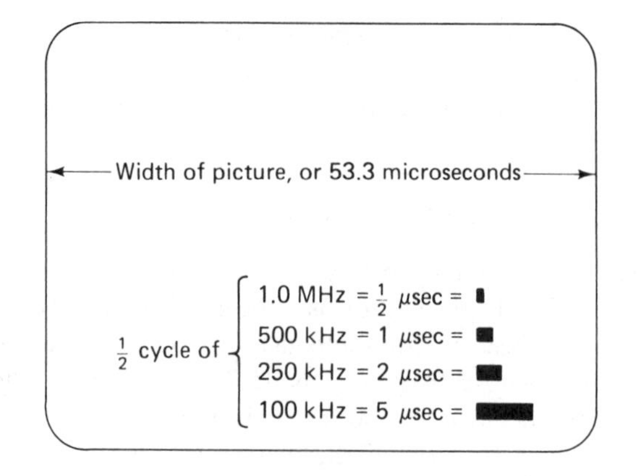

Figure 10-6 Relationship between picture size and video frequency.

The test pattern. An example of one type of television test pattern is shown in Figure 10-7. This pattern has been used at the television transmitter to test the TV system for various defects. Such test pattern generators have also been available to television service technicians. These generators use the flying spot scanning method for developing the video signal. This video signal then modulates an RF generator whose frequency of operation is adjustable over the entire VHF TV band. In short, this type of signal generator is a low-power television station.

The test pattern RF signal is injected at the receiver antenna and then is processed by the receiver and displayed on the picture tube. As indicated, this test pattern will reveal the following about the TV system:

1. It allows the aspect ratio, centering, and the vertical and horizontal linearity to be checked
2. The range of contrast possible (gray scale)
3. The low- and high-frequency and phase response of the system
4. Focus
5. Interlace
6. The horizontal and vertical resolution

Resolution. The sharpness of a picture formed on the screen of a television picture tube is called resolution. Resolution can be defined vertically or horizontally, and is governed by such factors as the diameter of the spot formed by the electron beam, the number of horizontal lines used to construct the raster, and the bandwidth of the video amplifier driving the picture tube.

Vertical resolution is the ability of the video system and picture tube to sharply display *horizontal* lines. At first thought it would seem that a TV system should yield 525 horizontal lines of resolution. However, the number of horizontal lines that can actually be seen is limited by the number of lines lost during vertical retrace, the pairing of horizontal lines due to faulty interlace, and the number of lines forming the raster. A small-diameter electron beam spot can generate a raster having a greater number of lines than a larger spot. Therefore, spot size is an important factor for high-resolution systems. The vertical

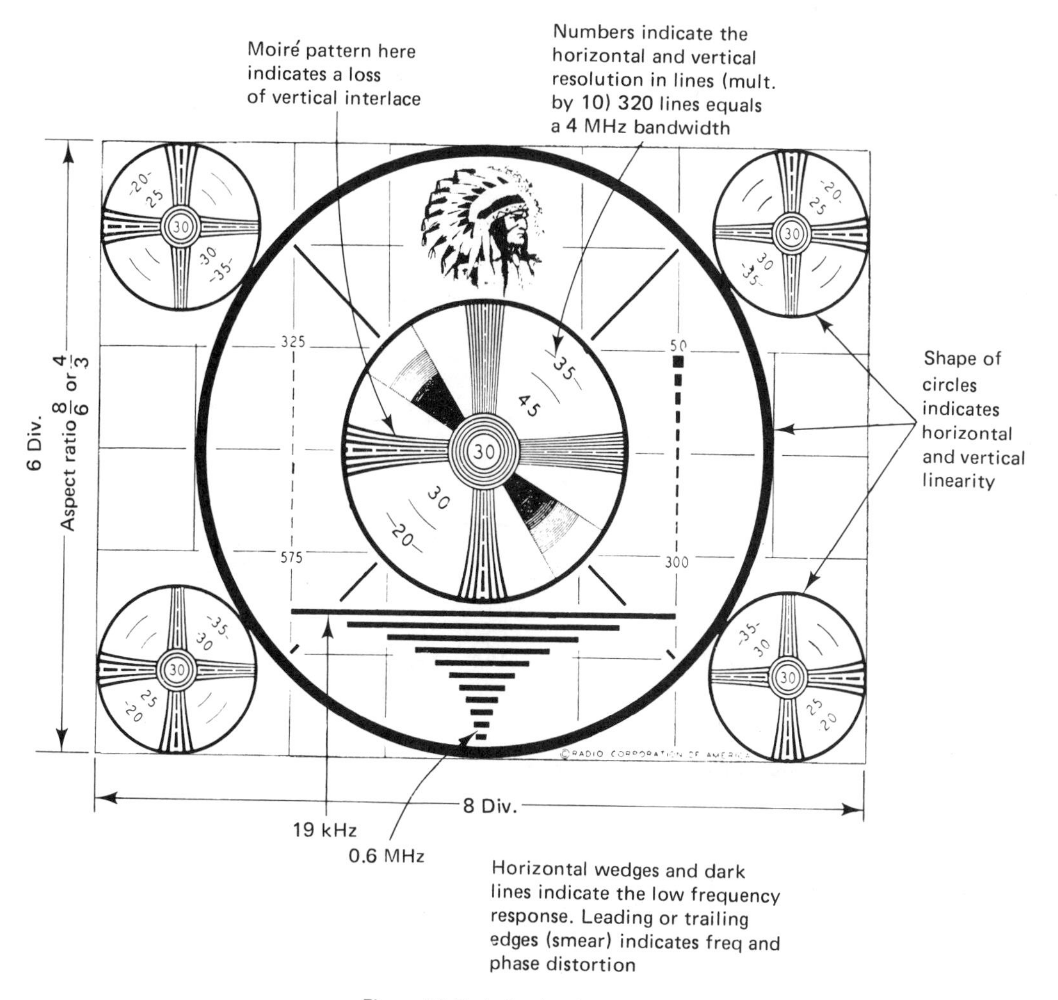

Figure 10-7 Indian head test pattern.

resolution of a TV receiver can be obtained by multiplying the number of lines in the system (525) by a utilization factor of 0.7. The utilization factor consolidates all of the system's limitations into a single quantity. For the conventional TV system using 525 lines, the maximum vertical resolution will be 343 lines.

Horizontal resolution is defined as the ability of the system to display sharply defined *vertical* lines. The horizontal resolution is determined by multiplying the active time of each line by the video bandwidth, and then multiplying the result by 2. For example, if we assume an ideal system bandwidth of 4.2 MHz, and an active time of 53.5 μs, the horizontal resolution will be 4,200,000 × 0.0000535 × 2 = 449 lines of resolution. In actual practice, due to blanking, overscan, and spot size the active time is usually taken as 40 μs resulting in a practical maximum of 336 lines of resolution. Television receivers often limit the video response to 3.8 MHz or less and have resolutions ranging from 240 to 335 horizontal lines.

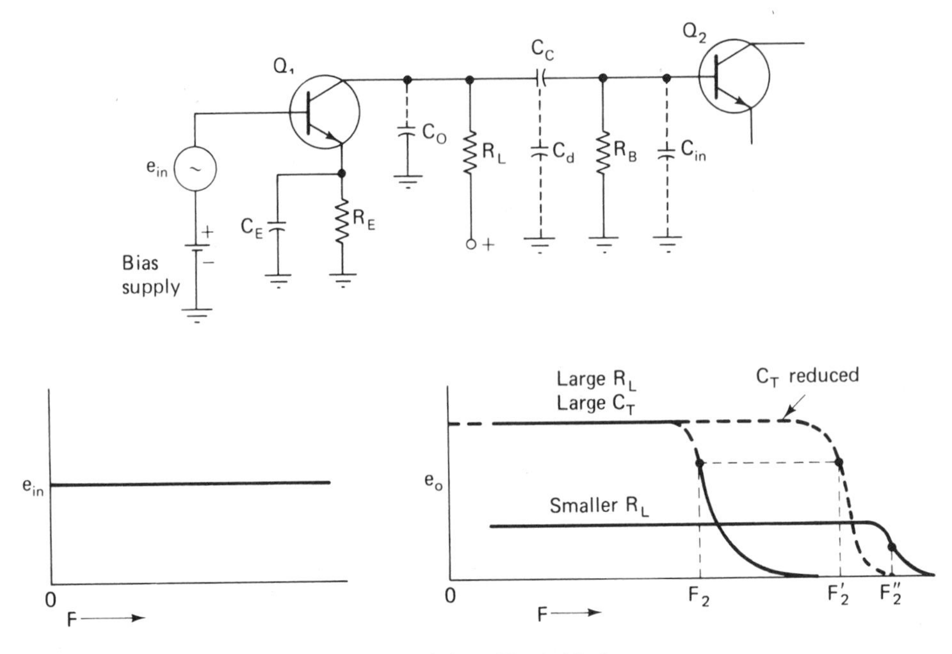

Figure 10-8 *RC-coupled amplifier and its frequency response.*

10-5 methods of frequency compensation

Most video amplifier circuits are either *RC*-coupled or direct-coupled amplifiers. Therefore, the following discussion will be concerned with the characteristics of these amplifiers. It was pointed out earlier that the perfect amplifier has frequency and phase response curves that are flat and therefore do not introduce either frequency or phase distortion. The practical video amplifier does not have a flat response and therefore will introduce both frequency and phase distortion, causing a deterioration of picture quality. The object of the following discussion is to determine the causes of frequency and phase distortion and to determine the means for extending the flat response of the video amplifier so that these forms of distortion are minimized over the video frequency band.

The schematic of a basic transistor *RC*-coupled amplifier is shown in Figure 10-8. To simplify the circuit, the input generator is connected in series with a bias supply. Drawn below the generator is the response curve of the generator. The response curve is flat, indicating that the output voltage of the generator is constant for all frequencies. The output response curve is drawn below the output of the amplifier and it indicates that frequency distortion is taking place at high frequencies. The effect of this on the picture is loss of fine detail, producing a fuzzy picture.

10-6 high-frequency compensation

The output circuit of the transistor consists of three discrete components: R_L, the collector load resistor; R_B, the transistor input resistor for the next stage; and C_c, the interstage coupling capacitor. The circuit also contains a number of invisible components that are due to the circuit's distributed capacity. The capacity labeled C_o is due to the transistor's output capacity, that labeled C_d is due to the circuit wiring and the physical size of the components, and C_{in} is due to the input capacity of the next stage. Since the reactance of the coupling capacitor (X_{Cc}) is very small at high frequencies, it may be considered a short circuit, thereby effectively placing all three distributed components of capacity in parallel. The total distributed capacity is referred to as C_T and is equal to $C_o + C_d + C_{in}$. It is the distributed capacity of the circuit that is primarily responsible for the loss in high-frequency response. To understand this, remember that the output voltage of an amplifier is directly related to the size of the amplifier load impedance. Thus, if the impedance were to decrease, the amplification of the stage would decrease and the output voltage would also decrease. Of course, the opposite would occur if the load impedance were to increase.

The load impedance of the basic amplifier consists of the parallel combination of R_{aL}, R_B, and C_T as well as the output impedance of Q_1 and input impedance of Q_2. As the frequency *increases,* the reactance of the total distributed capacitance C_T *decreases,* lowering the output impedance of the amplifier, thereby making the gain of the amplifier frequency dependent. At the frequency where the shunt distributed capacitive reactance equals the effective load resistance the response of the amplifier will have decreased to the half power point (F_2).

C_T reduction. To extend the high-frequency response of the amplifier requires that the shunt capacity be reduced in size. This may be done by using physically small components, careful component layout, short lead lengths, and proper choice of transistors. The effect of reducing the total shunt capacity on frequency response is also shown in the response curve of Figure 10-8. Notice that when C_T is reduced, the frequency response is extended and the gain of the amplifier remains high.

Unfortunately, the distributed capacity cannot be reduced to zero. Therefore, it would appear that an upper limit to the frequency response is reached when C_T cannot be reduced in size any further. Fortunately, there are at least five other methods for improving the high-frequency characteristics of a video amplifier. These are:

1. Reduce R_L
2. Shunt compensation (peaking)
3. Series compensation (peaking)
4. Combination compensation (peaking)
5. Emitter or cathode peaking

R_L reduction. The high-frequency half power point (F_2) of an amplifier occurs when R_L equals the capacitive reactance of the shunt capacity (X_{C_T}). Therefore, *reducing* R_L will require that the operating *frequency be increased* in order to *reduce* X_{C_T} to the new value of R_L. Consequently, this will extend the high-frequency response of the amplifier. The effect of reducing R_L on the response curve of an amplifier is shown in Figure 10-8. Notice that a reduction of R_L causes a reduction in gain while providing an extended bandpass.

This method of extending the video amplifier's bandpass is commonly used in transistor amplifiers. Unfortunately, this method alone is not sufficient to provide the video amplifier with the necessary gain and bandpass required by the television receiver. Typical values for R_L used in transistor amplifiers range from 5.6 to 10 kΩ.

Shunt peaking. The second method by which the high-frequency response of the amplifier may be extended is by placing a small coil in series with the load resistor R_L, as shown in Figure 10-9(a). In this method the coil (L_S), the shunt capacity (C_T), and R_L form a low Q parallel resonant circuit. The effect of using the high-impedance characteristics of a parallel circuit is to make the load impedance of the circuit at F_2, equal to the value of R_L at midfrequencies. This in turn makes the gain of the amplifier at F_2 equal to that at midfrequencies and extends the *flat* portion of the response curve out to the upper half power point (F_2).

The proper value of inductance that L_S must have in order to provide high-frequency peaking is such that it and the distributed capacity of the circuit resonate at a frequency of 1.41 times F_2. See Figure 10-9(a). In practice, typical values of L_S range from 300 to 700 μH. This applies to both color and black & white receivers.

Series peaking. Another method often used to extend the high-frequency response of the video amplifier is to use the characteristics of series resonance. In this method of frequency compensation, called *series peaking,* a small coil is placed in series with the coupling capacitor as shown in Figure 10-9(b). In some cases, the coil may be placed between the collector of the transistor and the load resistor R_L. In either arrangement the peaking coil L_C is made to resonate with the input capacity (C_{in}) of the next stage.

Two characteristics of series resonance are important to series peaking. First, at the resonant frequency of the circuit the impedance is very small and a high current will flow. Second, because of the high current flow, a large voltage drop will appear across each reactive element of the circuit. This effect is sometimes referred to as *voltage magnification.* These reactive voltages are Q times larger than the applied voltage.

The effect of series peaking on the response curves of the amplifier is shown in Figure 10-9(b). Notice that the frequency response as measured at the collector of the transistor exhibits a severe drop in output at F_2. This occurs because of the large drop in amplifier gain resulting from the very substantial reduction in load impedance occurring at the circuit's resonant frequency. The frequency response measured at the input of the next stage is also shown in Figure 10-9(b). Notice that at this point the frequency response has been extended and is flat out to F_2. This dramatic change in response from one side of the peaking coil to the other is caused by the series resonant magnification that appears across the input capacity of the following stage.

Peaking coil Q. The choice of peaking coil Q is important. If the Q is too small, incomplete compensation will occur. If the Q is too large, excessive peaking will result and the output frequency response curve will display excessive high-frequency peaking such as that shown by the dashed lines. The result of a mild form of this kind of response on picture quality is to provide an apparent crispness to the picture. Some television receivers are equipped with *peaking controls* that vary the high-frequency response and provide an adjustable crispness to the picture.

Ringing. A high-Q coil is subject to ringing oscillation when pulsed by a sudden change in current through it because the coil and its distributed capacitance form a parallel resonant circuit. Ringing occurs on the leading and trailing edges of the video signal because of the sudden changes that occur here. These variations in video signal amplitude will cause the

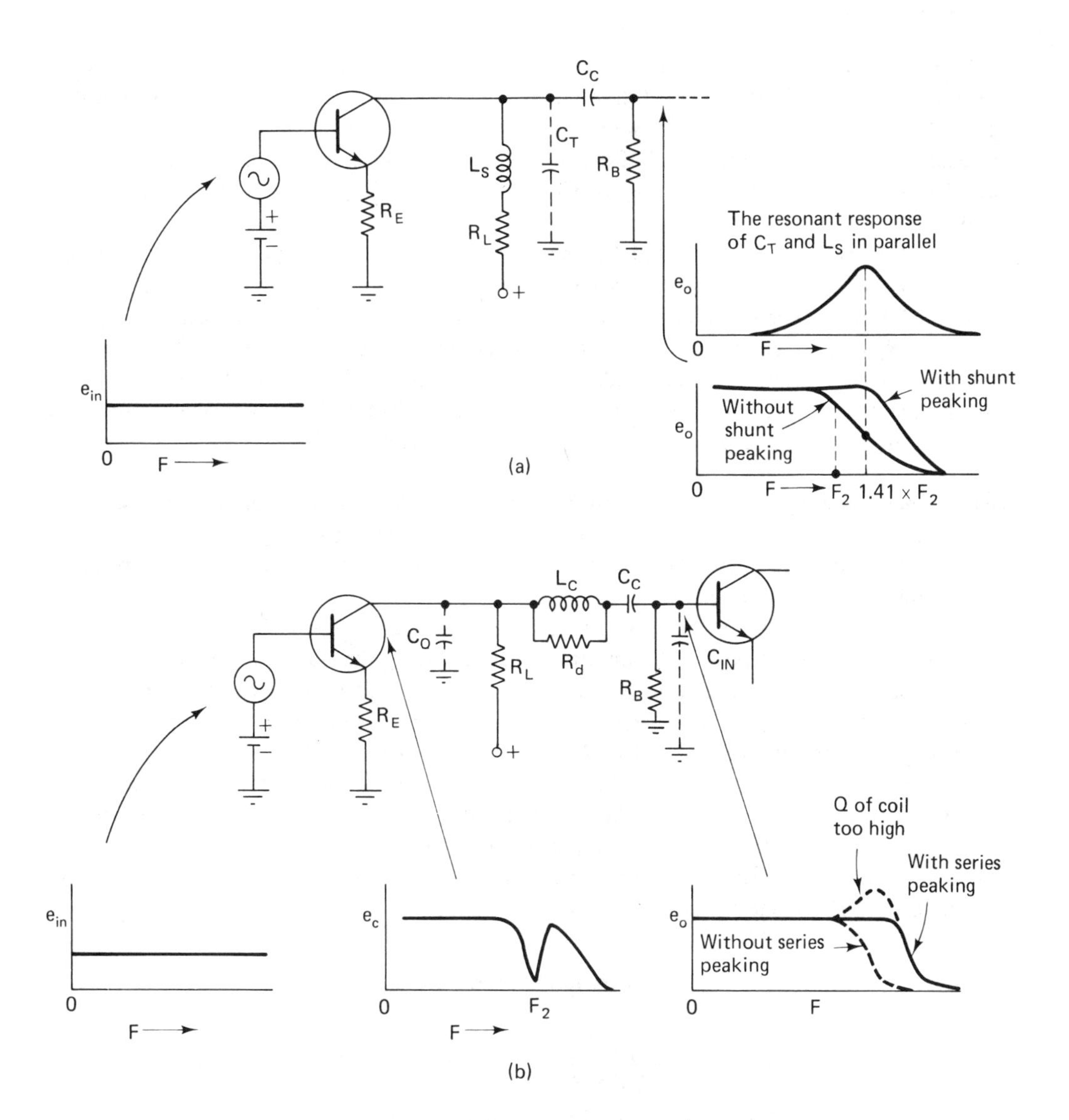

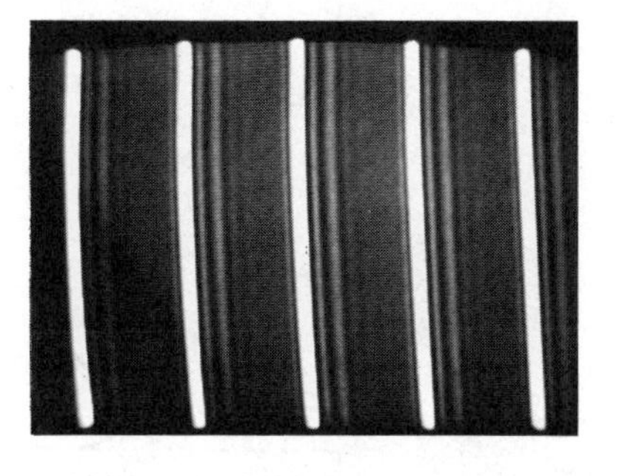

(c)

Figure 10-9 (a) Shunt peaking circuit and frequency response; (b) series peaking; (c) ringing in picture.

picture to become periodically lighter and darker. On positive alternations of the ringing the electron beam will be increased in intensity causing greater light output. On the negative alternations the reverse will occur. Since this action repeats itself at the same relative position for each horizontal line, the visual effect of ringing is to cause vertical lines adjacent to the original picture information. See Figure 10-9(c). These lines repeat themselves as they grow weaker.

Ringing may be prevented by using the proper Q for the peaking coils. Peaking coil Q may be adjusted by placing a damping resistor (R_d) in parallel with the coil. See Figure 10-9(b). In some cases, the peaking coil may be wound on the resistor. Typical values of R_d range from 8 to 100 kΩ. The inductance of a series peaking coil is typically smaller than that used for shunt peaking and ranges from 60 to 400 μH. In both cases, the ohmmeter resistance of the coil is typically less than 10 Ω. Most good service literature will give the value of the peaking coil as well as its resistance.

Degenerative peaking. Another method for improving the high-frequency response of an amplifier is illustrated in Figure 10-10. In this arrangement, frequency dependent negative feedback is used to increase the gain at high frequencies. This is done at the expense of gain at mid and low frequencies.

An unbypassed emitter resistor will introduce negative feedback for all input signals, regardless of their frequency, while simultaneously reducing the stage gain because the input signal causes variations in the amplifier's output current that flows through the emitter resistor and develops an ac voltage drop across it. This ac component of voltage is of the same frequency as the input signal and is of opposite phase. In terms of the emitter, the ac voltage developed across the emitter resistor R_E is in series with the input signal. Since the two voltages are out of phase, they tend to cancel each other and reduce the effective ac signal voltage that is actually developed between base and emitter. A reduction of the effective input voltage between emitter and base of the transistor amplifier means that the amplified output voltage is reduced.

Figure 10-10 Emitter peaking.

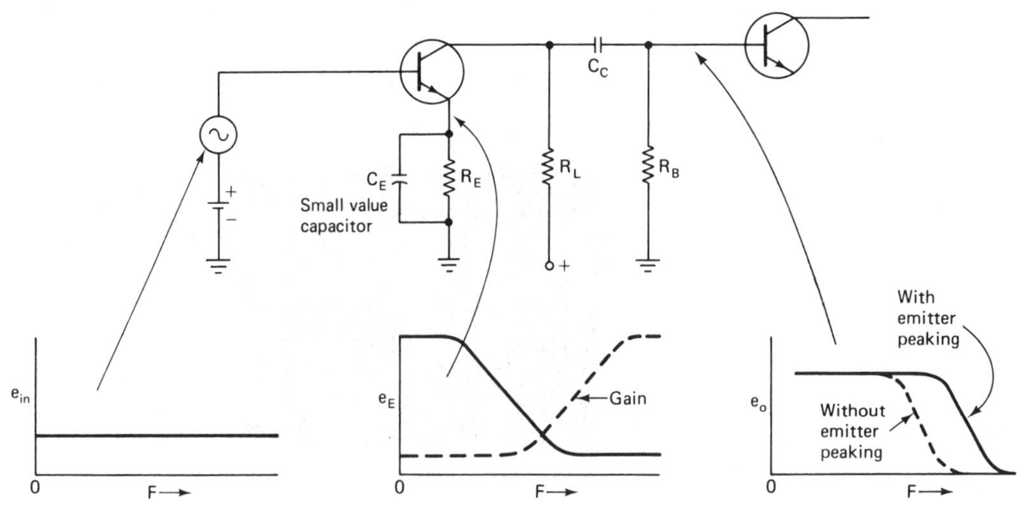

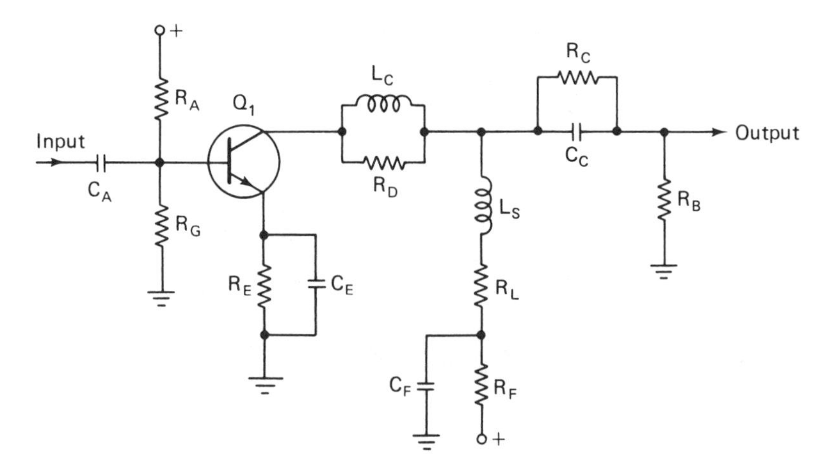

Figure 10–11 Combination peaking.

Placing a large bypass capacitor (100 μF or larger) across the emitter resistor will eliminate the negative feedback for all but the very lowest frequencies because the capacitor short circuits the ac component of voltage developed across the resistor. At low frequencies the reactance of the capacitor will have increased, preventing it from being an effective bypass capacitor.

If the large capacitor is removed and a small capacitor (400 pF) is placed across the emitter resistor, it would have no effect on amplifier gain at low frequencies because of its high reactance, but it would eliminate the negative feedback at high frequencies (above 1.5 MHz). This, in turn, would *increase* the gain of the amplifier in the frequency range where the high-frequency gain normally drops off.

The frequency response curves associated with the schematic diagram illustrates that because of the small size of the bypass capacitor (C_E) the ac voltage across R_E decreases with increasing frequency. As a result, over the same frequency range the gain of the stage increases, compensating for the poor high-frequency response of the amplifier.

Combination peaking. Shunt and series peaking may be combined in order to provide a higher gain for the same bandwidth. An example of combination peaking is shown in Figure 10-11.

10-7 low-frequency distortion

At low video frequencies both phase and frequency distortion must be eliminated if proper reproduction of the television picture is to be obtained. A loss of low-frequency response without phase distortion will result in a picture in which the large black & white areas of the picture appear to be washed out and have poor contrast. Low-frequency phase distortion introduced by the video amplifier will cause severe smear in the picture. See Figure 10-12.

Phase distortion is introduced at both the low- and the high-frequency portions of the amplifier's frequency response curve. It is associated with any departure of the frequency

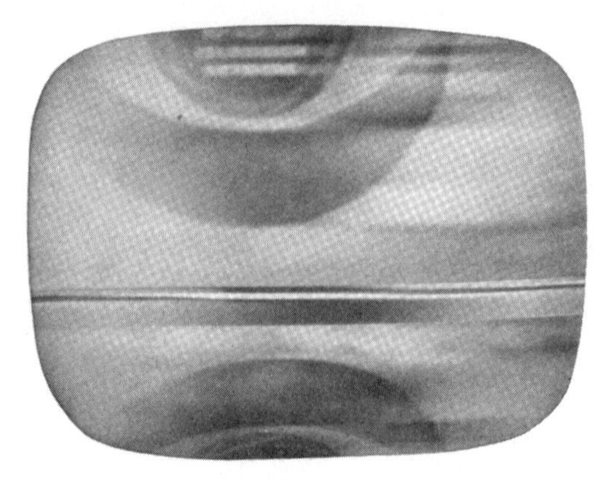

Figure 10-12 Picture smear produced by poor low-frequency response with phase distortion. (Courtesy of General Electric.)

response curve from flatness. See Figure 10-5(b). Unfortunately, phase distortion is more of a problem at low frequencies than at high frequencies because *phase distortion does not occur if the phase shift is proportional to frequency,* and it will be very severe if it is inversely proportional to frequency. These are precisely the conditions that exist at the respective high- and low-frequency regions of the amplifier.

The square wave. The simplest way to understand this is to observe the effect of phase distortion on a voltage that has a square-wave wave shape. A square wave is a complex wave and as such it may be shown to consist of an infinite number of sinusoidal components. The lowest-frequency component is called the *fundamental,* and the higher-frequency components are called *harmonics.* It should be noted that if the fundamental is assumed to have an amplitude of unity, the third harmonic will be one-third the fundamental amplitude, the fifth harmonic will be one-fifth the fundamental amplitude, and so on. Thus, the higher the frequency of the harmonic, the weaker it becomes. In theory, an infinite number of these components must be used to reproduce a square wave. In practice, a satisfactory square wave may be obtained with as few as 10 harmonic components. The square-wave components not only have specific amplitudes but they also have definite phase relationships. Other periodic complex wave shapes, such as the sawtooth clipped sine wave, etc., have different harmonic and phase relationships than those found in the square wave. Consequently, it may be concluded that any change in relative amplitude or phase relationships between the fundamental and harmonic components of the square wave will cause a change in the wave shape. This in turn will cause a change in picture content.

Phase distortion and picture quality. Figure 10-13(a) illustrates the effect of a normal 15,750-Hz square-wave video signal on the television picture. This video signal produces a picture in which the left side of the picture is black and the right side is white. The transition from black to white is very sharp because of the rapid fall time of the square-wave video signal.

The effect of a lagging phase shift (high-frequency phase distortion) of the fundamental component of the square wave is shown in Figure 10-13(b). In this case, phase

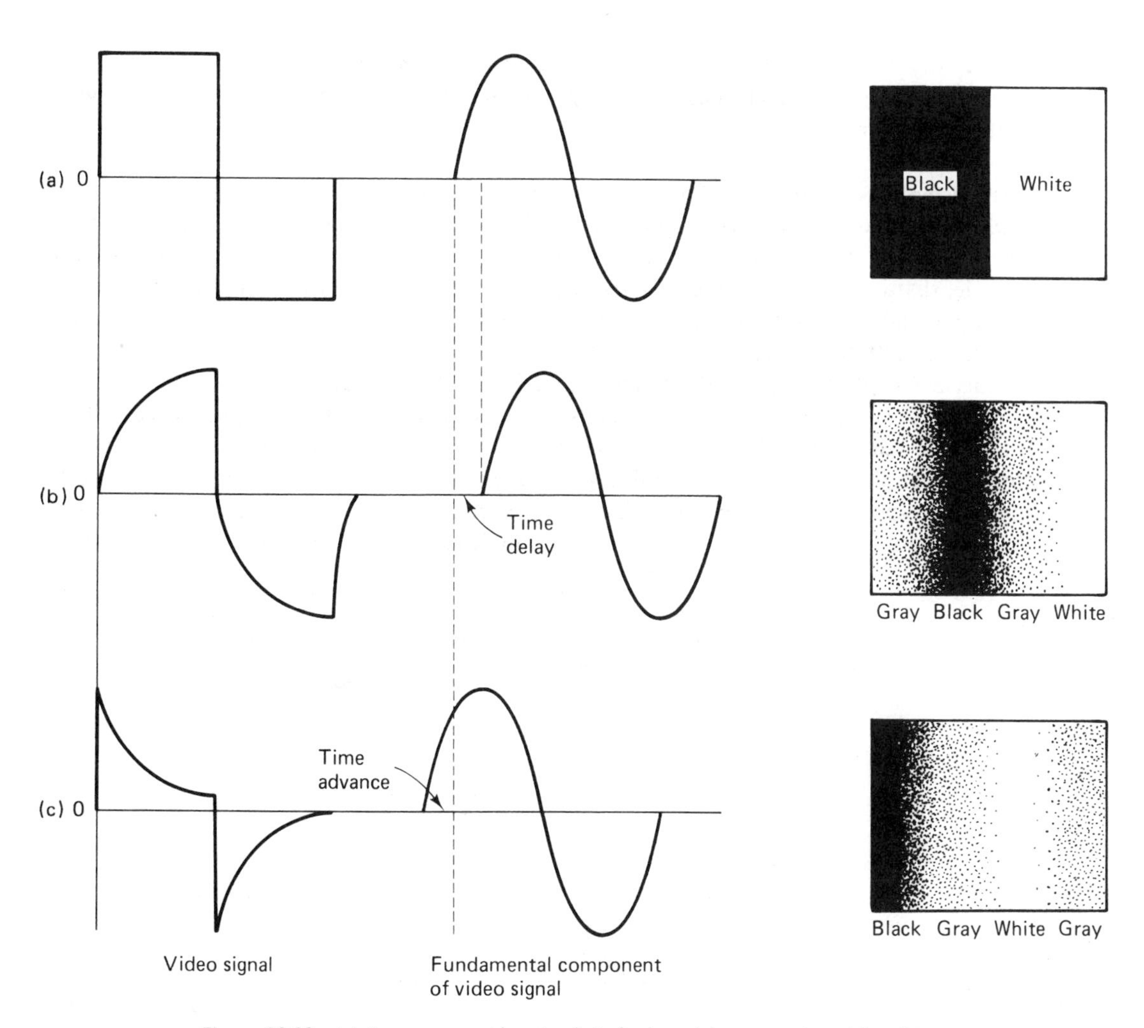

Figure 10-13 (a) Square-wave video signal, its fundamental component, and the picture obtained; (b) distorted square wave, its fundamental component lagging that of drawing (a), and the resulting picture; (c) wave shape of a leading phase shift and the resulting picture.

distortion has caused the fundamental component of the square-wave signal to be delayed from its normal position, thereby changing the resultant video wave shape. This in turn changes the picture content. Here, the gradual rise in signal voltage causes the picture to go from a gray, on the extreme left, to black at the center. The transition from black to white is also gradual, in contrast to the normal condition.

If the phase shift causes the fundamental component of the square wave to be advanced (low-frequency phase distortion) relative to the harmonic components, the result will be as shown in Figure 10-13(c). Here, phase distortion has caused the square wave to change into a narrow double-pointed wave shape that produces a different picture from that of the other two conditions. The picture now begins on the left with a dark black area that rapidly becomes a light gray at the center of the picture. At this point, a sudden transition take place which makes the screen white. This rapidly changes to a light gray on the extreme right.

Phase distortion compensation. Phase distortion may be minimized by either providing the amplifier with a flat frequency response or by arranging it so that the phase shift is proportional to frequency. A phase shift that is proportional to frequency means, for example, that if the fundamental is shifted 20° the third harmonic will be shifted three times as much or 60°, the fifth harmonic five times as much, and so on. Figure 10-14 illustrates how phase shift that is proportional to frequency avoids distortion. Notice that in terms of time, a 60° phase shift of the fundamental is equal to a 180° phase shift of the third harmonic. This means that all of the sinusoidal components of the complex wave will be in the same relative position as in an undistorted wave, and no distortion will take place. Proportional phase shift normally takes place at the high-frequency end of the amplifier's frequency response curve. Therefore, phase distortion is less of a problem at high frequencies. Unfortunately, the low-frequency region of the amplifier frequency response exhibits a phase

Figure 10-14 Relationship between the fundamental and the third harmonic in a normal square wave; (b) relationship between the fundamental and the third harmonic when each has had a phase shift that is proportional to frequency.

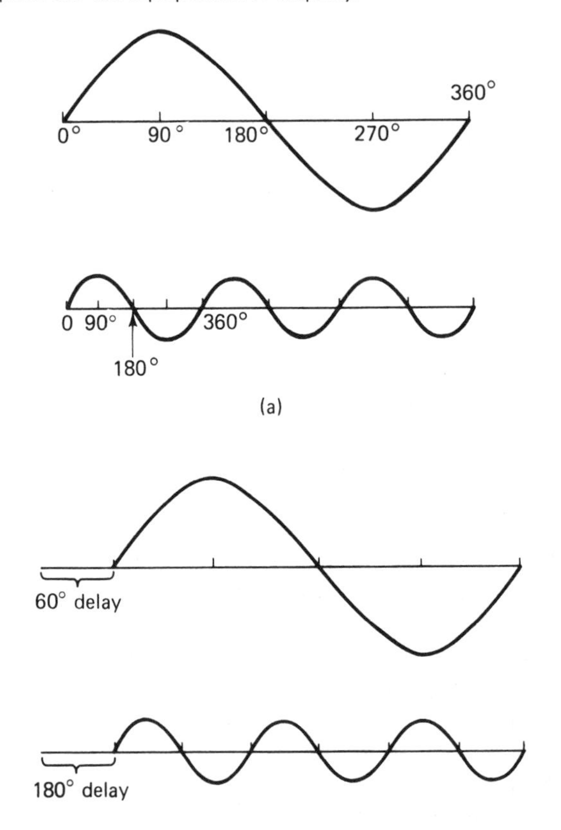

shift that is *inversely proportional* to frequency. In this case, if the fundamental component is shifted 120°, the third harmonic will be shifted 40°, the fifth harmonic by 24°, and so on. This form of phase distortion tends to make a bad situation even worse. It may be avoided by compensating the amplifier so that it has a flat response and therefore introduces no phase distortion at all.

Low-frequency response of a basic amplifier. Figure 10-15 shows the frequency response curves for a simple transistor amplifier. In this circuit the input voltage e_{in} is shown to have a flat response at all frequencies. The output frequency response (ignoring the high-frequency response) measured at the collector of the amplifier is also flat, indicating that neither frequency nor phase distortion exists at these points. But, when the frequency response is measured at the output of the amplifier after passing through the coupling capacitor (C_c), the low-frequency response is seen to drop off very rapidly. It may then be concluded that the loss of low-frequency response is due to the coupling capacitor (C_c) and output resistor (R_o).

The coupling capacitor and output resistor act as an ac voltage divider. As the input frequency is decreased, the capacitor's reactance increases, reducing the available output voltage across the output resistor, as well as introducing a phase shift.

Coupling capacitor and the output resistor. The low-frequency response of the amplifier is determined by the size of the coupling capacitor and output resistor. If either the capacitor C_c or resistor R_o is increased in value, the low-frequency response will improve. See Figure 10-15. This is because of the smaller reactance of C_c and the reduced voltage drop developed across C_c at any given frequency. Of course, if C_c or R_o is made smaller, the low-frequency response will become poorer.

The bypass capacitor. Other circuit elements in addition to the coupling capacitor may be responsible for poor low-frequency response. In transistor amplifiers the emitter bypass capacitor can also be responsible for poor low-frequency response.

Another form of frequency-dependent negative feedback occurs as a result of in-

Figure 10-15 Low-frequency response of a transistor *RC*-coupled amplifier and the result of increasing or decreasing the value of C_c or R_o.

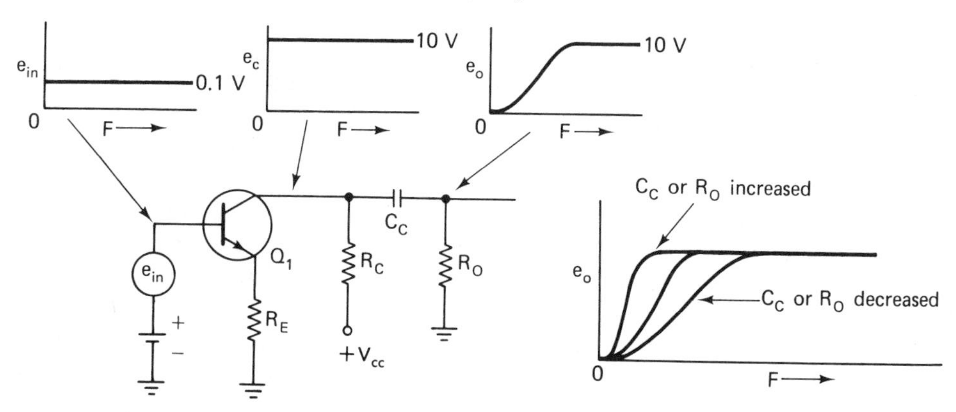

complete bypassing of the emitter resistor in transistor amplifiers. The ac voltage developed across the emitter resistor (R_E) increases as the frequency is lowered because at low frequencies the reactance of the bypass capacitor (C_E) increases and the capacitor is not able to effectively short the ac signal component developed across the emitter resistor to ground. At low enough frequencies C_E is effectively an open circuit and is in effect out of the circuit.

The ac signal voltage developed across R_E introduces negative feedback that lowers the gain of the stage at low frequencies.

Low-frequency compensation. Two methods may be used to minimize low-frequency distortion introduced by C_E, R_E. First, R_E may be left completely unbypassed, thereby reducing stage gain. This means that the negative feedback developed by R_E is uniform for all frequencies and therefore no frequency or phase distortion is introduced. This is the method most modern television receivers use. Second, C_E may be made very large (1000 μF or more) so that the low video frequencies are effectively bypassed. Unfortunately, no matter how large C_E is made, it is ineffective for extremely low video frequencies and dc level changes that are important in color television.

Low-frequency compensation of the frequency distortion caused by the coupling capacitor C_c and the input resistance R_S may take the three following basic forms:

1. Decoupling filter compensation
2. Modified direct coupling
3. Direct coupling

Examples of decoupling and direct coupling circuit arrangements are shown in Figure 10-16.

Decoupling filter compensation. In Figure 10-16(a) low-frequency compensation is obtained by making the gain of the amplifier increase as the frequency is decreased in order to overcome the loss in output due to C_c and R_o. For an amplifier to have a gain characteristic which is inversely proportional to frequency requires that its load impedance must increase as the input frequency decreases. This kind of load impedance can be obtained by placing a capacitor (C_F) in series with the load resistor (R_L). As the frequency of operation decreases, the reactance of C_F increases and is added to the resistance of the load resistor. Therefore, the frequency response curve at the collector of the transistor will exhibit an increased output voltage at low frequencies. If the increase in ac collector voltage matches the loss in output voltage resulting from the ac voltage division between C_c and R_o, the output frequency response curve will be flat down to the lowest frequencies. However, if C_F is made too small (X_c becomes larger), the low-frequency voltage amplification will increase faster than the loss of output voltage, and the output frequency response will exhibit a peak at the low frequencies. The reverse will occur if C_F is made too large. Low-frequency compensation occurs when the time constant of C_F, R_L is equal to that of C_c, R_o.

In this circuit the purpose of R_F is to return the collector of the amplifier to the power supply (B+). From an ac point of view, R_F and C_F are in parallel and therefore the maximum impedance that this combination can have is the value of R_F. This means that perfect compensation is not possible because at very low frequencies the gain of the amplifiers will no longer increase to compensate for the voltage division of C_c and R_o. Therefore, the

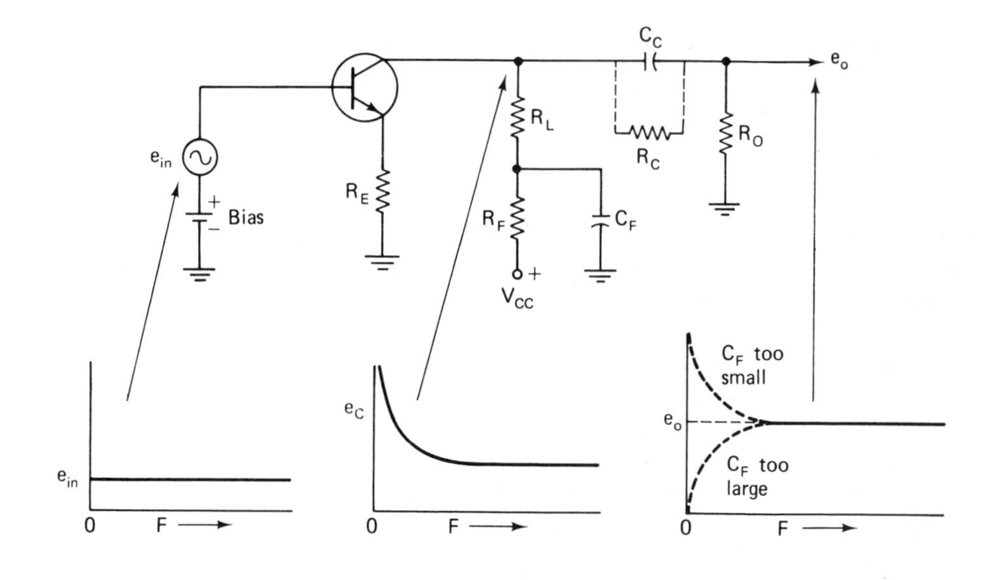

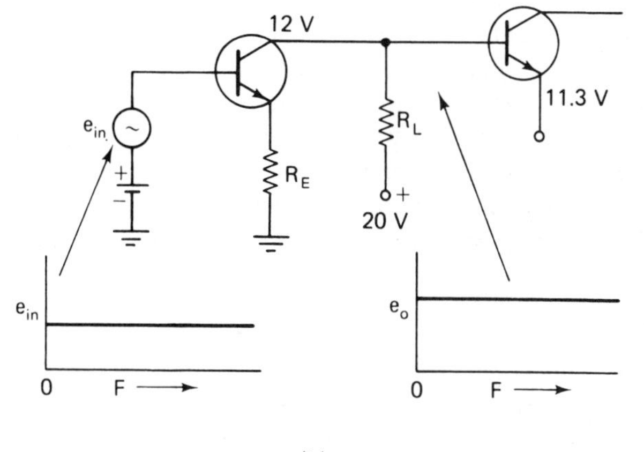

Figure 10-16 Two methods of low-frequency compensation: (a) decoupling capacitor; (b) direct coupling.

value of R_F is made as large as possible consistent with supplying the transistor its required dc collector voltage.

Modified direct coupling. To overcome the limitations of the previous method of compensation, and to allow low-frequency response that includes dc level changes, the *RC*-coupled amplifier may be modified. In this case, a resistor (R_c), typically 0.5 MΩ, is placed across the coupling capacitor. See Figure 10-16(a). As a result, the dc signal levels that are developed on the collector side of C_c are then communicated to the other side of C_c. Proper low-frequency compensation may be obtained in this circuit when

$$\frac{R_L}{R_o} = \frac{R_F}{R_c} = \frac{C_c}{C_F}$$

In this system of compensation C_F is used as a peaking capacitor as it was in the previous method of compensation and is typically 0.05 μF. Under these conditions, the low-frequency response of the amplifier will be flat down to dc.

Direct coupling. An amplifier system that maintains the dc level and provides perfect low-frequency response without any compensation is the dc-coupled amplifier. An example of a simple dc-coupled transistor amplifier is shown in Figure 10-16(b). In this arrangement the collector of the first stage is tied directly to the base of the second stage. This means that if the dc collector voltage is 12 V, the base of the second stage is also 12 V. Proper bias for the second NPN transistor stage is obtained by making the emitter voltage of the second stage less positive than the base.

Signals developed at the collector of the first stage are then directly connected to the base of the second stage. Since there is no coupling capacitor between the two stages, the signal voltage developed at the base of the output transistor will be identical to the collector signal voltage of the first stage for all signal frequencies including dc.

Direct-coupled amplifiers find widespread application in color television receivers where they are used in video amplifiers and color difference amplifiers. One of the problems associated with the direct-coupled amplifier is its dc operating point stability. Operating point stability is primarily a problem when the dc amplifier is cascaded with two or more stages. Under these circumstances, small changes in the low-level stages can, because of amplification, cause larger changes in operating levels in the output stage. To minimize this problem cascaded direct-coupled amplifiers seldom have more than two stages.

10-8 dc restorers

The composite video signal that appears at the output of the video detector consists of an ac and a dc component. The ac signal component represents black & white picture information (Y signal), color signal information (chrominance), color sync, and horizontal and vertical deflection sync and blanking pulses. The dc component of the composite video signal represents the average brightness of the scene. Thus, if the scene were completely black, the dc level associated with the composite video signal would be maximum. It could either be positive or negative, depending upon the output phase of the detector. If the scene were all white, the dc level would be minimum (close to zero).

If the composite signal is passed through a coupling capacitor, the dc component will be blocked by the capacitor and will not be coupled to the input of the next stage. This is illustrated in Figure 10-17. In this circuit the ac component of the video signal for one horizontal line is represented by a generator and the dc component is represented by a battery. The output wave shape of the combination is drawn below the diagram. Also shown is the resultant picture produced by this video signal.

If the scene being televised is a black background with a white vertical bar passing

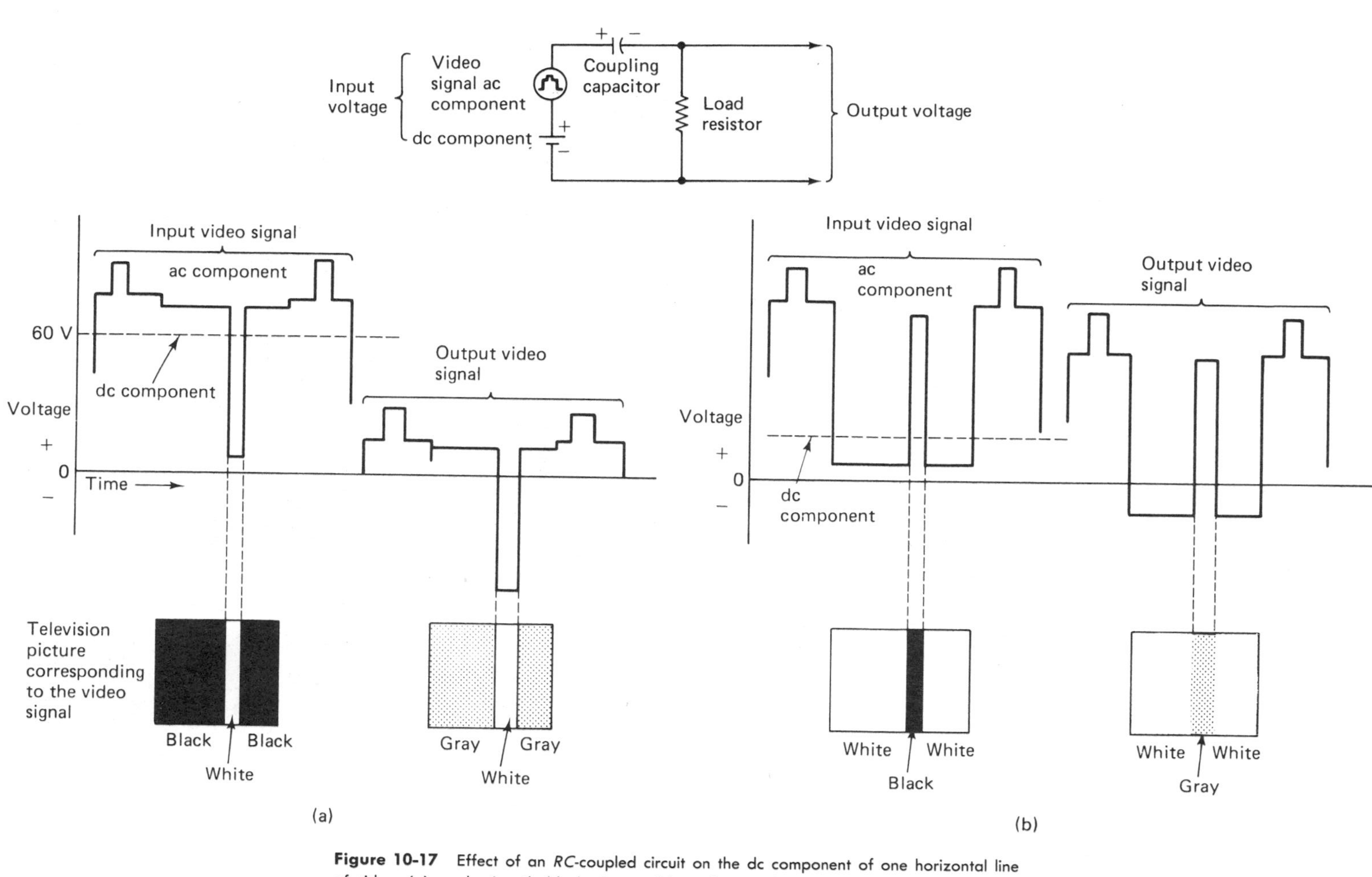

Figure 10-17 Effect of an *RC*-coupled circuit on the dc component of one horizontal line of video: (a) predominantly black picture; (b) predominantly white picture.

through the picture, the dc component of the video signal will be almost equal to the black level of the horizontal blanking pulses. See Figure 10-17(a). Let us assume that this voltage is 60 V. When this signal is applied to the coupling capacitor-load resistor combination, the capacitor charges to the dc voltage. Since the applied dc voltage and the dc voltage across the capacitor are equal and are series opposing to one another, the dc voltage that appears across the load resistor is their total, which is zero. In other words, the coupling capacitor blocks the dc voltage from reaching the load resistor. If only a dc voltage were to be applied to the input of the circuit, the voltage across the load resistor (input voltage to the next stage) would remain at zero. But the applied voltage also contains an ac component (video signal) that causes the applied voltage to increase and decrease, depending upon the nature of the signal. This variation in input signal causes the coupling capacitor to charge in one direction and then discharge in the opposite direction as the applied voltage exceeds or falls below the voltage across the capacitor. The charge and discharge current flows through the output load resistor and produces an ac voltage drop across it that is almost identical in amplitude and wave shape to the applied ac voltage. See Figure 10-17(a). It should be noticed that the output wave shape is identical to that of the input except that the sync and blanking pulses are now at a much lower level (closer to zero) than at the input and that the portion of the signal corresponding to the white bar is now negative. Notice that as a result of losing the dc component the black areas of the picture have become gray.

Figure 10-17(b) illustrates the same loss of dc component but here the input video signal is for a white picture with a black bar. Again, the loss of dc level at the output has caused the sync and blanking pulses to shift to a lower level and has caused the white portion of the video signal to become negative. Notice that in each case the output zero level occupies the same relative position on the video signal as did the dc level in the input voltage. Here again the loss of the dc component causes the picture quality to change. The black region is degraded into gray.

In summary note that in a video signal that contains its dc component, all of the sync pulse tips have the same amplitude and are "lined up" regardless of picture content. Loss of the dc level results in a video signal whose sync pulse tips no longer line up and vary with picture content.

Three undesirable conditions that may result from the loss of the video signal dc level are listed below:

1. Retrace lines become visible. This is because the blanking pulses do not remain at a constant level and may not be able to provide proper retrace blanking. Most modern receivers incorporate special vertical and horizontal retrace blanking circuits as a means of preventing retrace lines from becoming visible.
2. Possible loss of sync. This is because loss of the dc component will cause sync pulse amplitudes to vary with picture content. If synchronization is to remain stable, all television receivers must restore the dc level (line up sync pulses) of the video signal before it is passed through the sync separator stages.
3. Loss of average scene brightness. This means that dark and light scenes may not be easily distinguished.

Color and the dc component. In color television a change in the (Y) signal will cause a change in the brightness of a color. Therefore, the loss of the video signal dc component will result in poor color reproduction. The color signals may also have dc levels associated with them which if lost as the result of *RC* coupling would cause poor color reproduction.

Ordinarily there is no measurement that can be made of the ac output voltage of an *RC*-coupled circuit that will indicate the original dc voltage level. In effect, the dc level associated with the input signal is lost when the signal is passed through a coupling capacitor.

Methods of dc restoration. The dc level may be preserved by using direct coupled amplifiers or it may be *restored.* Restoring the dc level after it has been lost is accomplished by dc level clamping. This may be achieved by either one of the following:

1. The video signal (ac component) has a characteristic such that the original dc level is indicated by the signal itself. In practice, this is possible because the video signal is transmitted with all sync and blanking pulses *lined up.* If the dc component is lost, the sync pulse tips will vary in amplitude with respect to zero level. Restoring the dc level is accomplished by simply lining up the sync pulse tips. Dc restorers of this type are sometimes called *automatic brightness controls.*
2. A reference pulse or an offset bias may be used to establish a *dc level* which may not be identical to the *original dc level* but is proportional to it.

The diode clamp. Restoring the dc level by lining up the sync pulses is most easily accomplished by a simple diode clamp circuit such as shown in Figure 10-18. The circuit operation of the diode clamp is identical to the development of signal bias in transistor amplifiers.

Figure 10-18 Diode dc restorer. The circuit in the dashed lines is an alternate arrangement that may be used to replace D_1. The boxes indicate the voltages and polarities that occur when the offset clamping diode (D_2) is used.

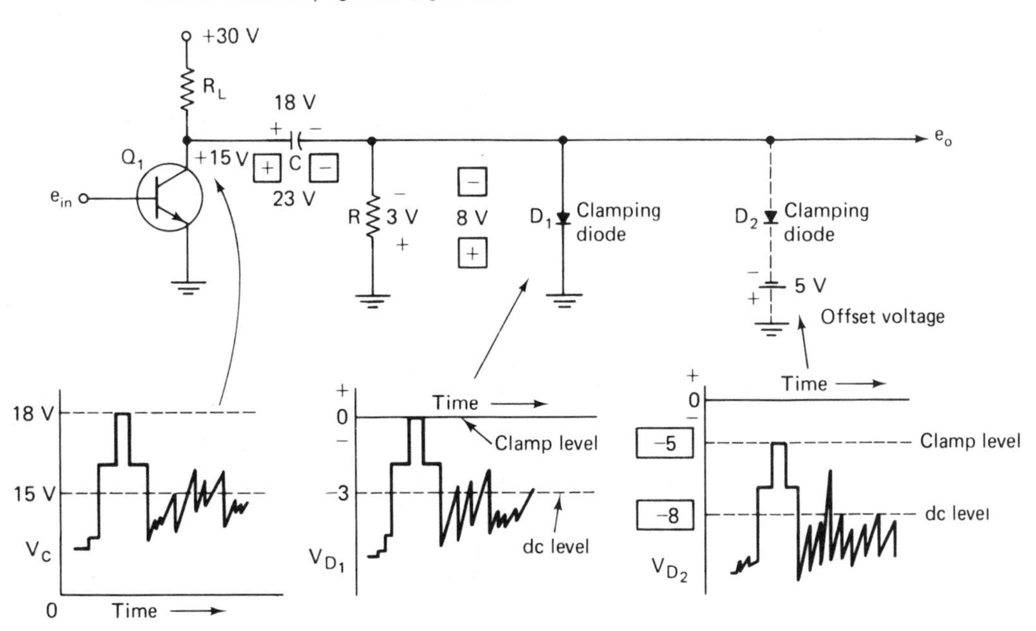

In these circuits the base and emitter of the next stage act as the diode clamp. Signal bias is in fact a form of dc restoration commonly used.

In the Class A amplifier transistor circuit of Figure 10-18 notice that the collector circuit contains a dc component of 15 V around which a composite video signal rises and falls. Under no-signal conditions, the capacitor (*C*) will charge to 15 V and the voltage across the parallel combination of the diode and the resistor (*R*) will be zero. If the input signal is assumed to be 6 V p-p, the transistor collector voltage may rise to a peak of 18 V. The positive increase in collector voltage is coupled through the coupling capacitor to the anode of the diode (D_1), turning it on. Since a conducting diode may be considered a short circuit, it effectively ties (clamps) the output circuit to ground (zero level). In effect, each positive sync pulse tip will be clamped to zero level, thereby lining them up and restoring the dc level of the video signal. The dc voltage developed across the diode will be -3 V because the conduction of the diode will cause the coupling capacitor to charge to the peak voltage of 18 V. The voltage across the capacitor and across the transistor (*C* to *E*) is series opposing. The sum of these voltages is the dc voltage across the diode and is -3 V. Notice that restoring the dc level of a signal does not mean that the dc level on both sides of the capacitor becomes the same.

Reversing the diode in the dc restorer circuit will result in the negative peak of the input signal being clamped at zero. This would mean that the dc output voltage of the circuit will be positive. If the video signal has a negative sync pulse phase, the sync pulse tips will be clamped to zero. If the video signal maintains a positive sync pulse phase, the white peaks will be clamped to zero.

Offset voltage. The basic diode clamp dc restorer circuit may be modified by placing an offset voltage in series with the diode clamp as shown by the alternate dashed circuit of Figure 10-18. In this circuit under no-signal conditions the capacitor (*C*) will charge to the sum of the collector and offset voltages, or 20 V. Charging will stop when the voltage across the coupling capacitor reaches this value because then the net voltage across the diode (D_2) will be zero.

When a signal whose positive peak amplitude at the collector of the transistor is 18 V is applied, the coupling capacitor will charge to a maximum of 23 V. This in turn means that the dc voltage across the output resistor will be 8 V ($23 - 15 = 8$) and the positive peak of the input voltage will be clamped at the offset voltage of 5 V.

10-9 the comb filter

In recent years many video cassette recorders and color television receivers have incorporated a circuit which is able to separate the chrominance and luminance signals from the composite video signal. This circuit, called a comb filter, can do away with the need for a chroma bandpass amplifier stage.

Before the use of the comb filter the high-frequency response of the video amplifier chain (luminance channel) was limited to about 3 MHz out of a possible response of 4.2 MHz. As a result, receiver performance suffered due to the loss of fine-detail information. There were two reasons for degrading receiver luminance performance:

1. High-frequency luminance (Y) signals between 3.08 and 4.08 MHz can simulate color information and can be processed by the color circuits. This results in false colors that appear as swirls of color on such things as clothing that have herringbone weave or on striped ties.
2. The overlapping chrominance and luminance signals tend to produce beats that add a grainy roughness to the picture. The comb filter eliminates both of these defects and does not introduce any of its own.

The operation of the comb filter is based on four important characteristics of the NTSC composite video signal:

1. The harmonic components of the luminance signal are spaced at intervals equal to the horizontal scanning rate. See Figure 3-12.
2. The color harmonic components fill the space between the luminance harmonics by being an odd multiple of half the horizontal scanning rate. This results in the chrominance signal reversing phase from one horizontal line to the next. It is this comblike relationship of these harmonics that gives rise to the name of the comb filter.
3. Adjacent successive horizontal lines of luminance information are almost identical.
4. All components of the composite video signal have definite phase or timing relationships at the start of each horizontal line.

The comb filter is able to separate the chrominance and luminance signals by delaying the composite video signal by one horizontal line and then adding and subtracting this signal from the composite video signal. A block diagram showing this arrangement is shown in Figure 10-19(a). Notice that the composite video signal is applied simultaneously to three circuits: a noninverting amplifier, a one-horizontal-line (63.5 μs) delay circuit, and an inverting amplifier. The output of the delay circuit feeds two adders. Each amplifier drives only one adder.

Assuming that two horizontal lines have been processed by this circuit, line 2 will be arriving at the input of the adders at the same time as the delayed composite video on line 1. See Figure 10-19(b).

At this point the nature of these signals becomes important. If, for example, the Y signal (luminance) is at a frequency which is the 133rd harmonic of the horizontal line frequency, and the chrominance signal is 1.5 times higher in frequency at the 399th odd harmonic of half the horizontal rate, the two signals will appear as shown in Figure 10-19(c). Notice that during both horizontal scanning lines 1 and 2, the phase of the Y signal remains the same, but that the color (C) signal phase reverses from one line to the other. This means that when the presently arriving video associated with line 2 appears at the input of adder 1, (see Figure 10–19(b)) it will combine with the video on horizontal line 1 that has been stored in the delay circuit. Since the Y components in both of these signals are in phase, the output of the adder will be $2Y$. Similarly, since the two color signals are 180° out of phase, the adder color output will be zero. The inputs to adder 2 consist of the delayed video on horizontal line 1 and the inverted video on horizontal line 2. Since the Y signals are now 180° out of phase and the color signals are in phase, the output of the second adder is $0Y$ and 2C. In this manner the color and luminance signals

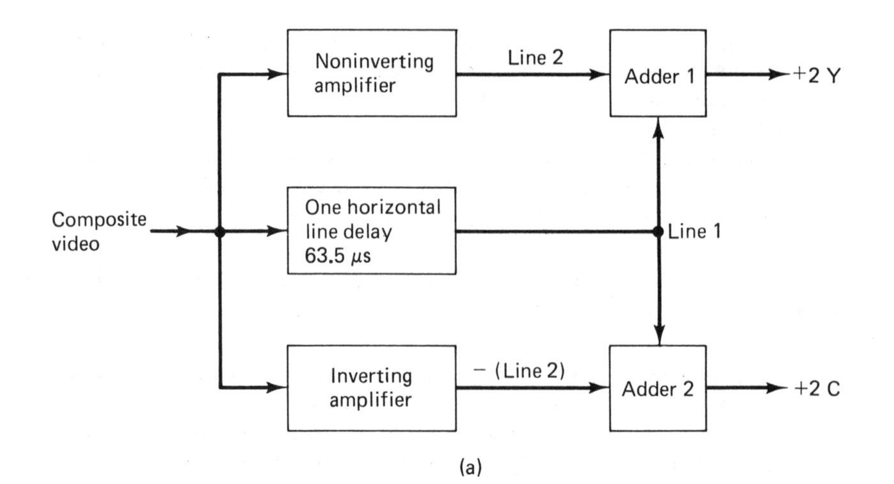

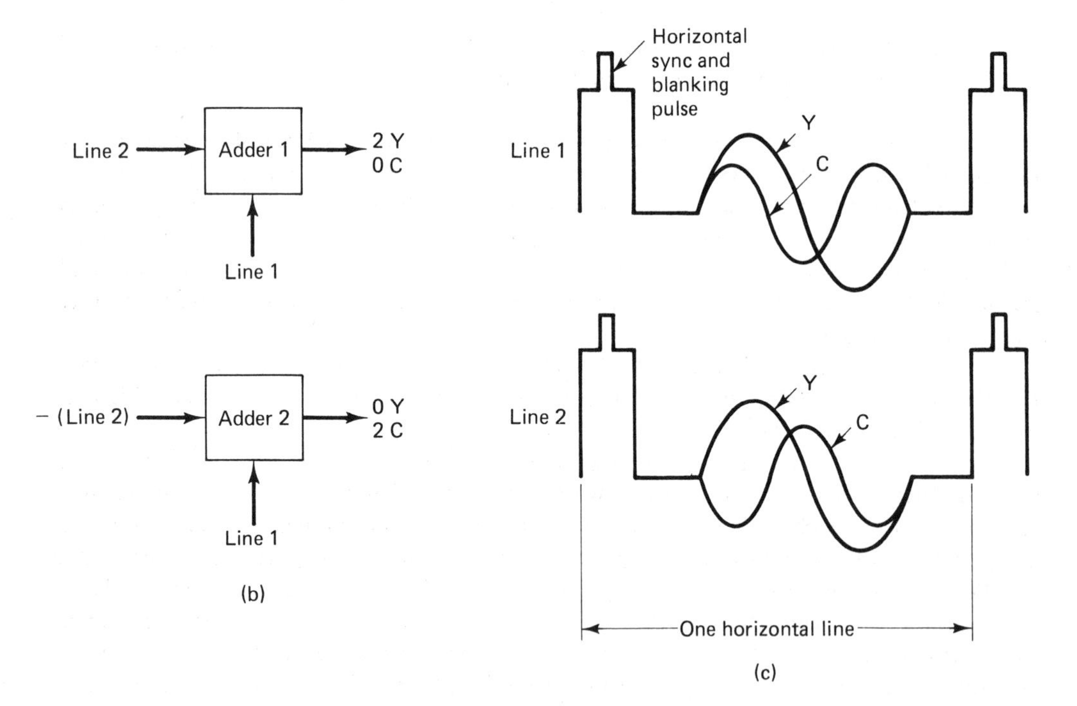

Figure 10–19 (a) Basic comb filter arrangement; (b) action of the adder circuits; (c) phase relationships of the Y and color (C) signals for two adjacent lines.

are very effectively separated. The net effect of this filtering technique is a 25% increase in horizontal resolution from 260 lines to 330 lines, and the reduction of dot patterns and barber pole cross-color effects.

Practical comb filters. Two comb filter systems are in use. The major difference between them concerns the method of obtaining the necessary 63.5-μs time delay. Magnavox and Zenith use a glass acoustical delay device; RCA uses a charge-coupled device (CCD) to obtain the necessary one-horizontal-line composite video signal delay. The RCA CCD system delays the entire video spectrum from 0 to 4.2 MHz. The acoustical delay system is frequency sensitive, only delaying the chrominance signal spectrum falling between 3.08 and 4.08 MHz. The RCA system also compensates for the loss of vertical resolution due to the merging of Y-signal information on adjacent horizontal lines. The block and schematic diagrams of both systems are shown in Figures 10-20 and 10-21.

Comparing the block and schematic diagrams of the Zenith comb filter, Figure 10-20 reveals that the composite video signal is fed into the delay line driver Q_{2404}, which has two outputs, (B) and (C). The inverted output (C) at the collector goes to the delay line, and the uninverted output (B) appearing at the emitter is added to the output of the delay circuit (E), producing a video signal having a strong chroma component which is fed to the output buffer (Q_{2404}). See the waveform ladder diagram, (D), (F), and (G). The output of this stage (F) is taken from its emitter and is sent to the chroma circuits, which filter out the remaining low-frequency luminance signals.

The output buffer also feeds a luminance-removing bandpass filter (C_{2403}, L_{2426}, C_{2404}, and C_{2405}) that feeds the base of Q_{2401} (waveforms G and J). Also feeding the emitter of this stage is a composite video signal (waveform H) that has been attenuated and delayed by the network consisting of R_{2413}, C_{2402}, R_{2409}, C_{2406}, and R_{2412}. Since the inputs to the base and emitter (H and J) of Q_{2401} are in phase, it is their difference which appears between base and emitter (waveform J-H); hence this circuit acts like a subtractor. The chroma inputs having been made equal are canceled, leaving only the luminance signal (I), which is amplified and appears at the collector. Q_{2405} and Q_{2403} are video amplifiers that drive the delay line and feed the video amplifier chain.

Zenith comb filter adjustments. The Zenith comb filter has four adjustments, two of which (R_{2426} and L_{2427}) are factory set and should not be field adjusted. The procedure for the remaining two adjustments, R_{4201} and L_{2426}, is as follows:

1. Tune the receiver to a color bar pattern.
2. Adjust the color level to minimum, and the picture and sharpness controls to maximum, to produce maximum visibility of the 3.58-MHz dot crawl.
3. Adjust R_{2401}, luminance amplitude, and L_{2426}, luminance phase, for minimum dot crawl and brightness.

CCD comb filter. The RCA comb filter block and schematic diagrams are shown in Figure 10-21. The block diagram (Figure 10–21(a)) reveals that the system of filtering is similar to that discussed earlier concerning Figure 10-19(a). However, the comb filter incorporates three novel features: (1) a charge-coupled device (CCD) of 682.5 elements providing one horizontal line (63.5 μs) delay, (2) a vertical resolution enhancement circuit, and (3) a vertical peaking circuit.

CCD. The charge-coupled device is not frequency sensitive and can delay one horizontal line of video information without degrading the quality of the video signal. The action of the CCD is similar to a shift register. Each of the 683.5 CCD elements acts like capacitors

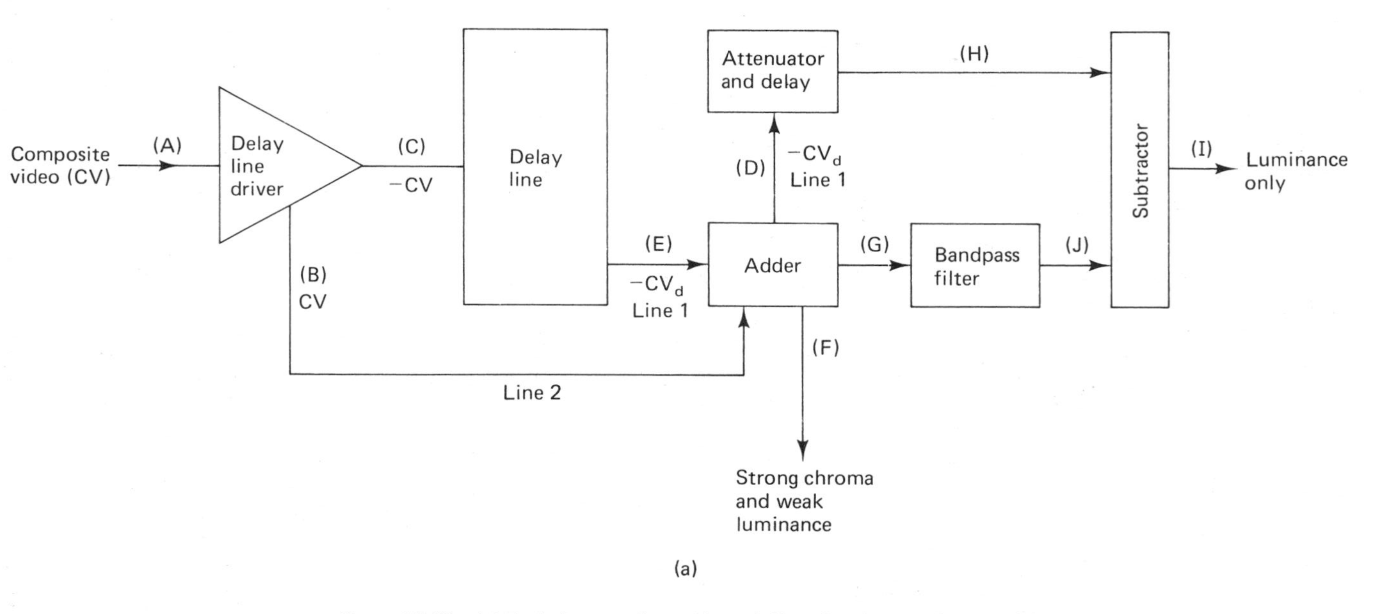

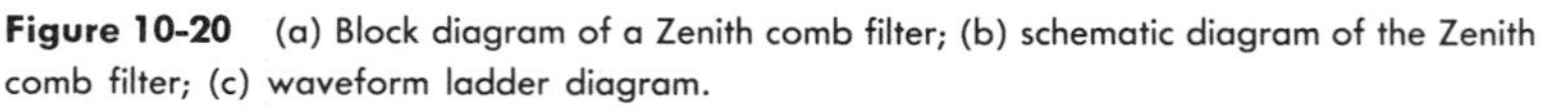

Figure 10-20 (a) Block diagram of a Zenith comb filter; (b) schematic diagram of the Zenith comb filter; (c) waveform ladder diagram.

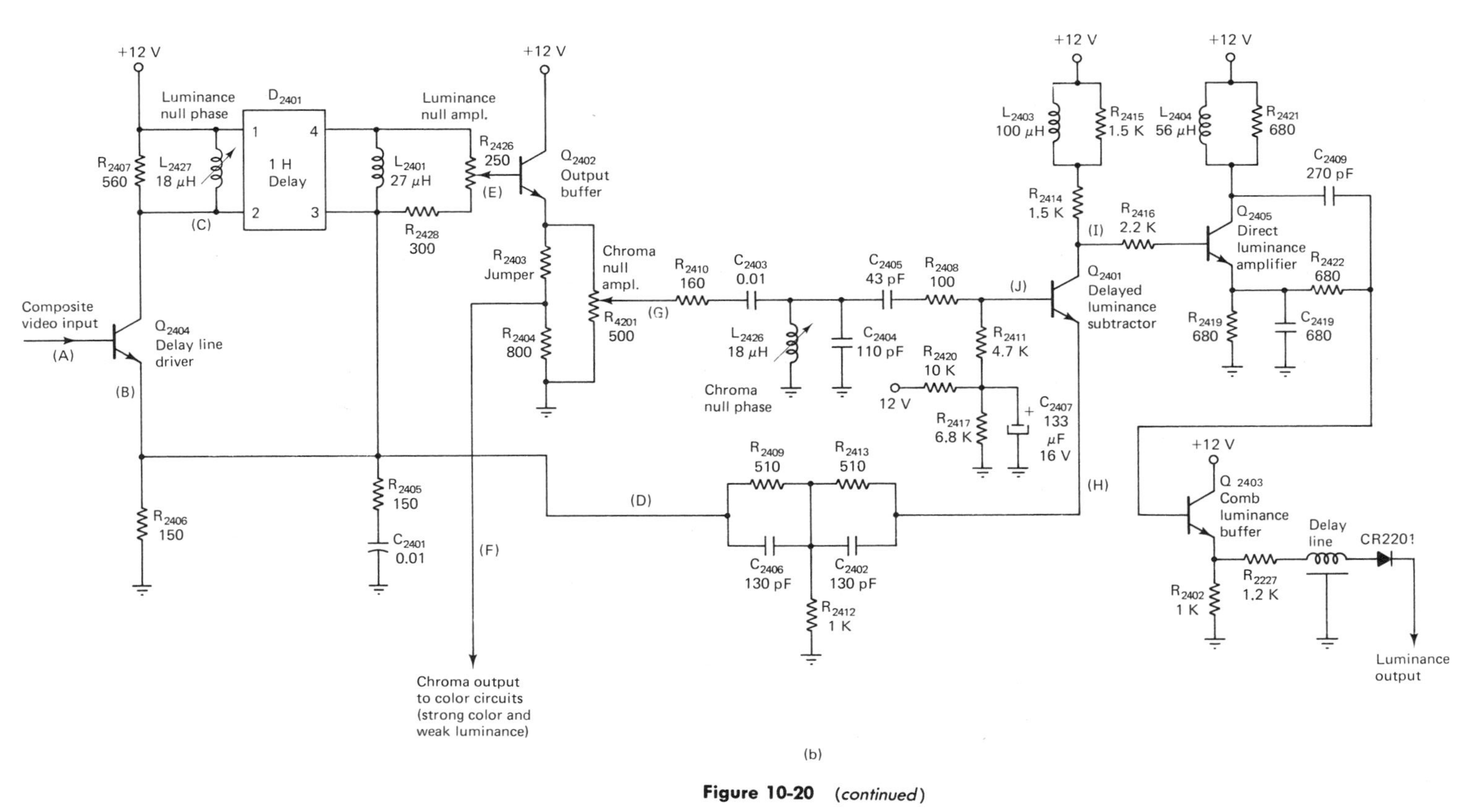

(b)

Figure 10-20 *(continued)*

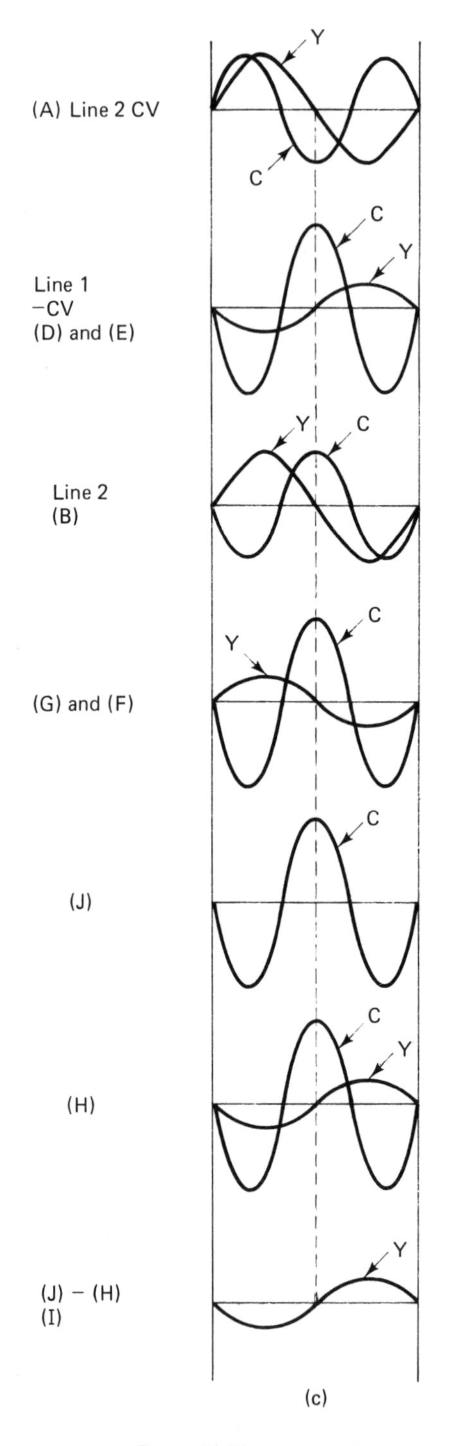

Figure 10-20 *(continued)*

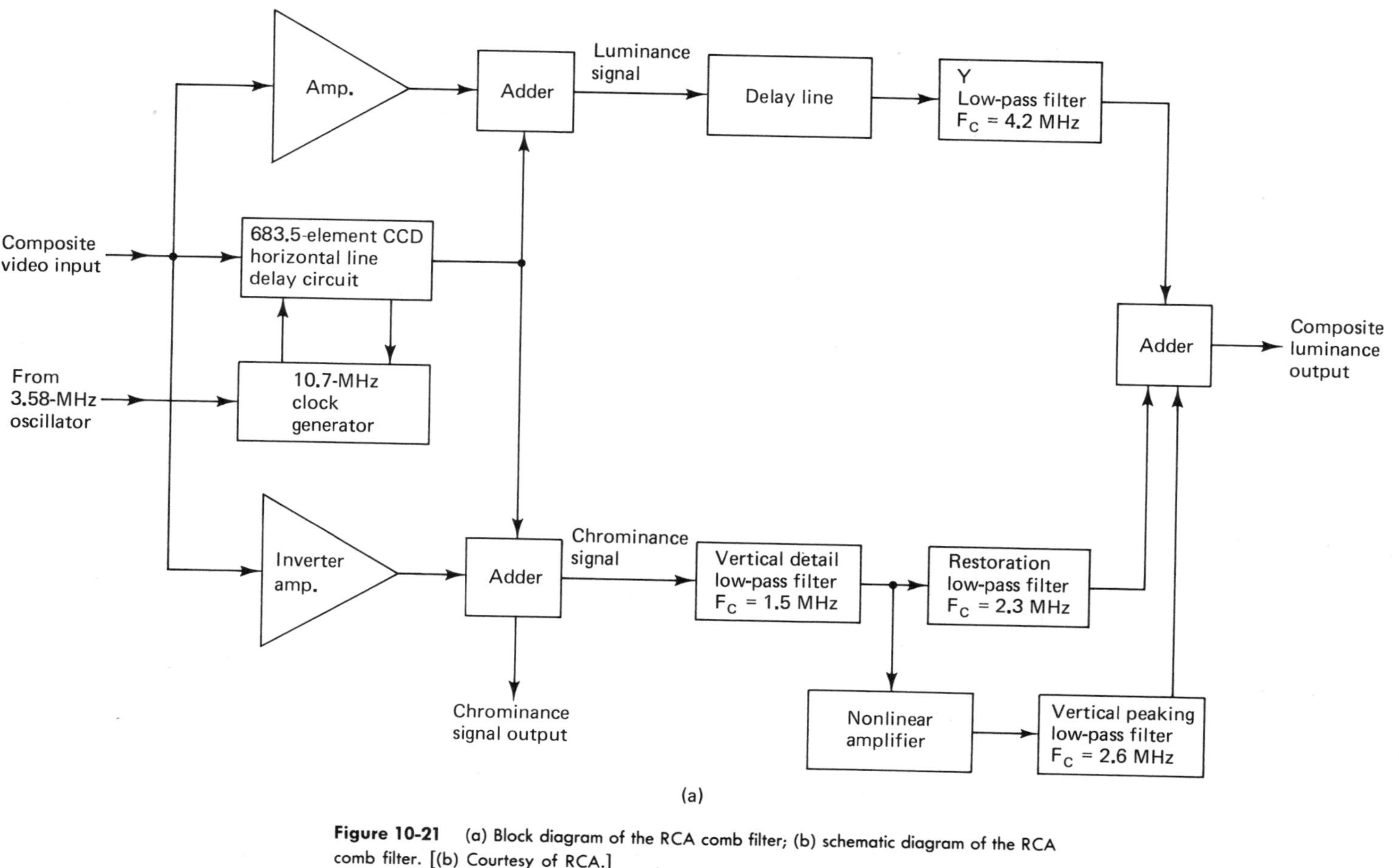

Figure 10-21 (a) Block diagram of the RCA comb filter; (b) schematic diagram of the RCA comb filter. [(b) Courtesy of RCA.]

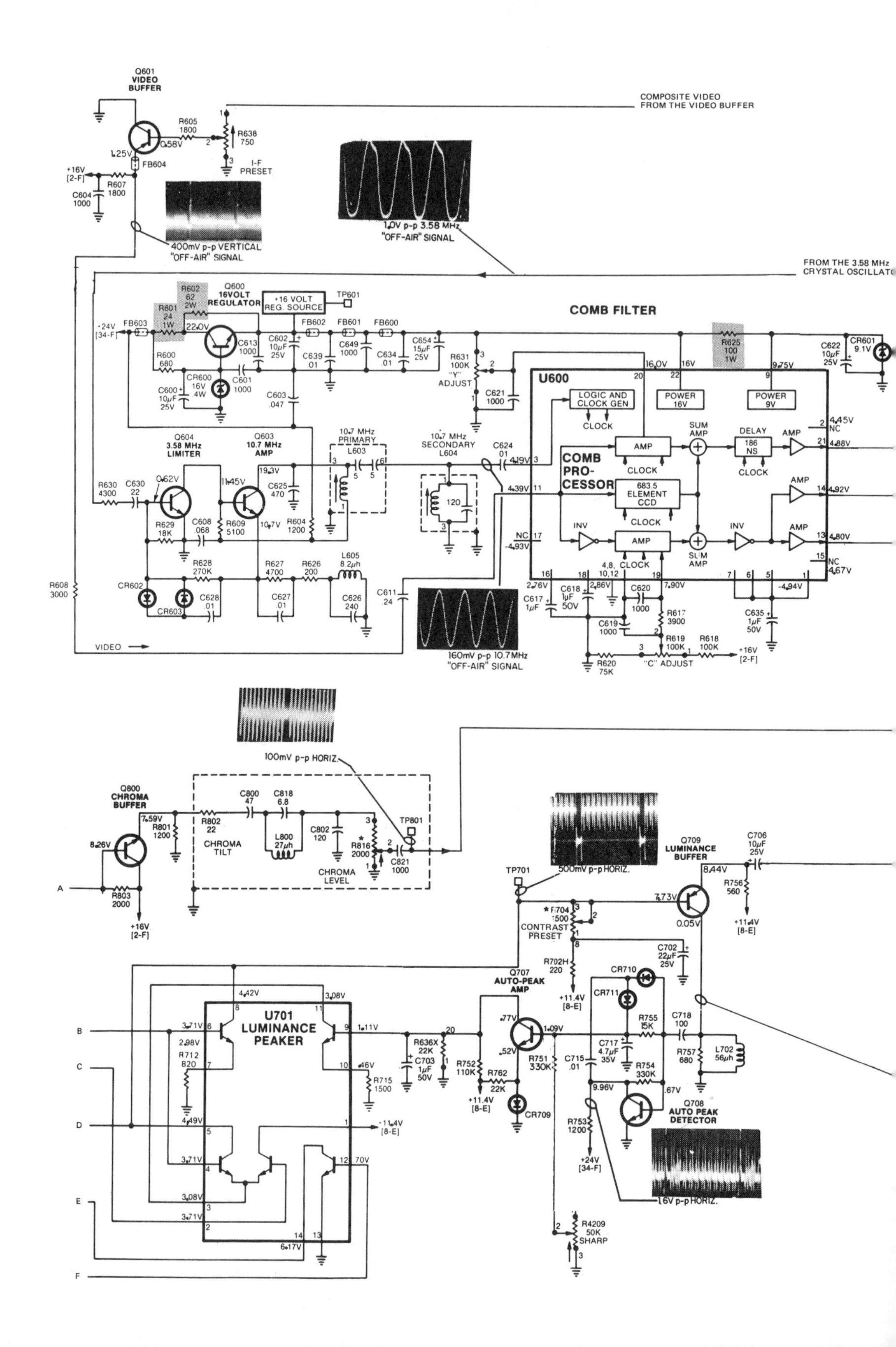

Q601 VIDEO BUFFER
COMPOSITE VIDEO FROM THE VIDEO BUFFER
R605 1800
R638 750
I-F PRESET
FB604
R607 1800
C604 1000
400mV p-p VERTICAL "OFF-AIR" SIGNAL
1.0V p-p 3.58 MHz "OFF-AIR" SIGNAL
FROM THE 3.58 MHz CRYSTAL OSCILLATOR
COMB FILTER
Q600 16VOLT REGULATOR
+16 VOLT REG. SOURCE
TP601
R602 62 2W
R601 24 1W
R600 680
CR600 16V 4W
C600 10μF 25V
C601 1000
C613 1000
C602 10μF 25V
C603 .047
C639 .01
C649 1000
C634 .01
C654 15μF 25V
R631 100K "Y" ADJUST
C621 1000
R625 100 1W
C622 10μF 25V
CR601 9.1V
U600
LOGIC AND CLOCK GEN
POWER 16V
POWER 9V
CLOCK
AMP
SUM AMP
DELAY 186 NS
COMB PROCESSOR
683.5 ELEMENT CCD
INV
Q604 3.58 MHz LIMITER
Q603 10.7 MHz AMP
10.7 MHz PRIMARY L603
10.7 MHz SECONDARY L604
C624 .01
R630 4300
C630 22
R629 18K
C608 .068
R609 5100
C625 470
R604 1200
R628 270K
R627 4700
R626 200
L605 8.2μh
CR602
CR603
C628 .01
C627 .01
C626 240
C611 .24
R608 3000
VIDEO
160mV p-p 10.7MHz "OFF-AIR" SIGNAL
C617 1μF
C618 1μF 50V
C620 1000
C619 1000
R617 3900
R619 100K
R618 100K
R620 75K
"C" ADJUST
C635 1μF 50V
100mV p-p HORIZ.
Q800 CHROMA BUFFER
R801 1200
R802 22
C800 47
C818 6.8
L800 27μh
C802 120
CHROMA TILT
R816 2000
CHROMA LEVEL
TP801
C821 1000
R803 2000
500mV p-p HORIZ.
TP701
Q709 LUMINANCE BUFFER
C706 10μF 25V
R756 560
R704 500 CONTRAST PRESET
R702H 220
C702 22μF 25V
Q707 AUTO-PEAK AMP
CR710
CR711
R755 15K
C718 100
U701 LUMINANCE PEAKER
R712 820
R715 1500
R636X 22K
C703 1μF 50V
R752 110K
R762 22K
CR709
R751 330K
C715 .01
C717 4.7μF 35V
R754 330K
R757 680
L702 56μh
Q708 AUTO PEAK DETECTOR
R753 1200
1.6V p-p HORIZ.
R4209 50K SHARP

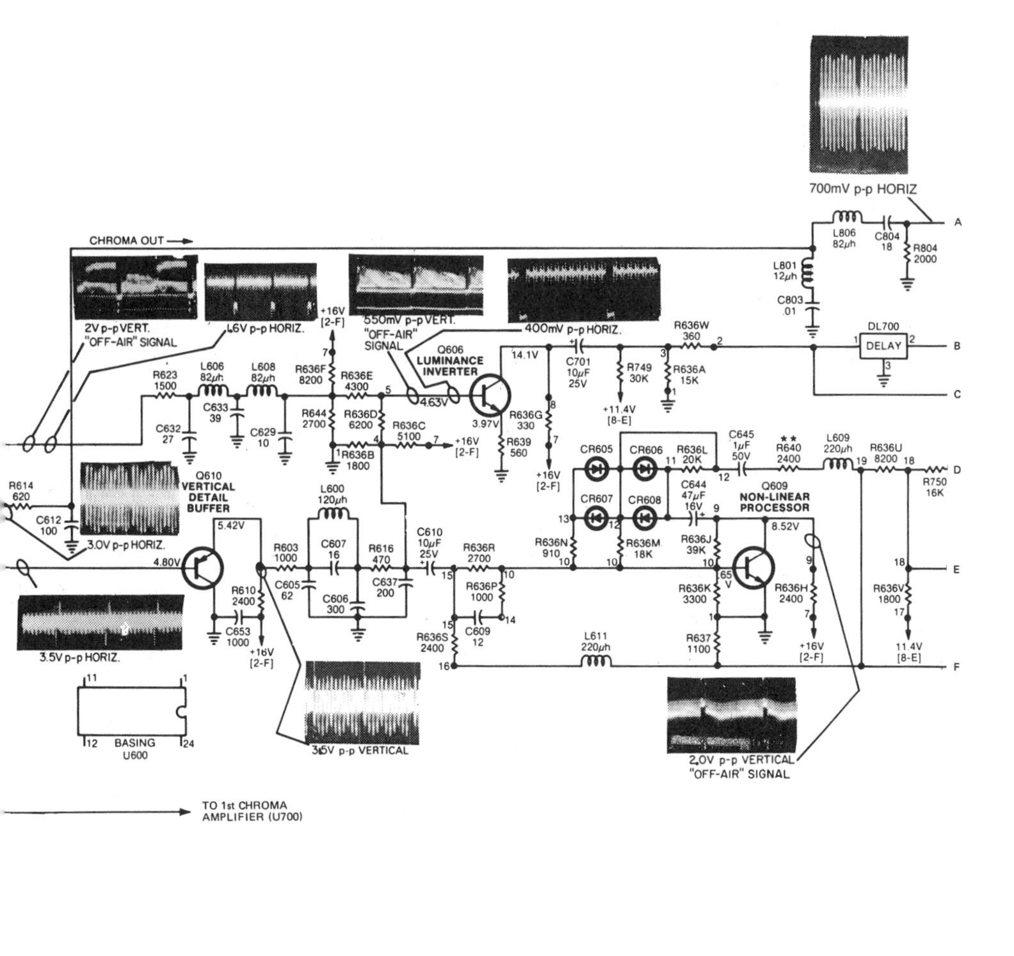

TO LUMINANCE AMPLIFIER (U700)

180mV p-p HORIZ.

(b)

Figure 10-21 *(continued)*

that can store a given amount of charge. The video signal is sampled at a 10.7-MHz rate (three times the chroma subcarrier frequency) and a charge equal to the video signal amplitude is applied to a CCD element. This charge information is shifted from element to element at a 10.7-MHz rate with each shift corresponding to a 93-ns delay. The video information is shifted through 683.5 elements to achieve a 63.5-μs delay.

The 10.7-MHz timing signal is obtained by passing the 3.58-MHz color oscillator signal through a frequency tripler circuit consisting of Q_{604} and Q_{603} and their associated circuits. The very high Q transformer L_{603} and L_{604} is tuned to 10.7 MHz. The output of this circuit is then capacitively coupled to the CCD comb filter processor IC chip.

Vertical enhancement circuit. Since the comb filter technique adds signals together after a one-horizontal-line delay, the signals contained in adjacent horizontal lines tend to merge, reducing the distinction between lines. This results in a reduction in vertical resolution corresponding to video frequencies of under 1 MHz. The lost information can be recovered by passing the combed chroma information through a bandpass filter consisting of Q_{610}, L_{600}, C_{605}, C_{607}, C_{606}, R_{616}, C_{637}, and R_{636R}, which extracts signals below 1.5 MHz. This signal is called the vertical detail output (VDO). Vertical resolution can be restored by adding the VDO to the luminance signal (emitter of Q_{606}). In addition, by selectively increasing the amount of VDO added, the vertical resolution is "peaked." This results in a crisp, sharp picture in which leading edges of picture details are made whiter-than-white and the trailing edges are made blacker-than-black. A nonlinear processor amplifier Q_{609} and its associated diodes (CR_{605}, CR_{606}, CR_{607}, CR_{608}) provides peaking and prevents overpeaking that would exaggerate the effects of noise, co-channel interference, and low-level field-to-field shift in video setup (black) transmitted by some television stations.

At low modulation levels (5 IRE units) the VDO information is added at a level only sufficient to restore the original information. Above this level (5 to 40 IRE) the VDO is peaked by about 30%. Between 40 and 100 IRE units the peaking is decreased to about 3% peaking at 100% IRE.

The final luminance signal is the sum of (1) the luminance signal output of the comb filter, (2) the vertical detail information (VDO), and (3) the peaked vertical detail signal.

This signal is obtained by adding the peaked vertical detail to the restored vertical detail output (junction of L_{611} and L_{609}, also pin 12 ICU_{701}). This signal is amplified and its output (pin 14 U_{701}) is added to the luminance signal (Y) at pin 8 of U_{701}. The combed luminance signal is developed at pin 21 of U_{600} is then amplified and inverted by Q_{606}. The Y signal is then passed through delay line DL_{700}, which in turn drives two amplifiers in the luminance peaker U_{701}. These amplifier outputs (pins 8 and 5) combine with the vertical detail signals to form the composite luminance signal driving the luminance buffer Q_{709}. The gain of the differential luminance amplifier in U_{700} is automatically controlled by a dc feedback loop consisting of Q_{708} (auto-peak detector) and Q_{707} (auto-peak amplifier).

CCD comb filter adjustment. The RCA comb filter (U600 IC)has three adjustments: (1) the "Y adjust" (R_{631}), (2) the "C adjust" (R_{619}), and (3) the IF preset adjustment (R_{638}). These adjustments must be done in the order listed. RCA's procedure calls for an off-the-air signal and the use of a dual channel oscilloscope. Channel 2 of the oscilloscope is used to provide trigger sync. In the case of the "Y and C adjust" the trigger pulse is obtained from the collector of the horizontal driver transistor Q_{403}. During IF preset adjustment the trigger is obtained from the collector of the bottom vertical output transistor Q_{502}. When making the "Y adjust," channel 1 of the oscilloscope is connected to the base of the luminance inverter transistor Q_{606}, and Y adjust R_{631} is adjusted to the point where the chroma content of the signal is reduced to minimum (nulled).

Chroma comb filter "C adjust" is made by connecting channel 1 of the oscilloscope to the emitter of the vertical detail buffer transistor Q_{610} and adjusting R_{619} the C adjust for minimum horizontal sync amplitude in the VDO signal.

IF preset adjustment is made by placing the input to channel 1 of the oscilloscope to the emitter of the vertical detail buffer transistor Q_{610} and adjusting the IF preset control R_{638} to obtain a symmetrical VDO output signal during the vertical retrace interval.

10-10 video amplifier controls and typical circuits

Contrast control. Contrast refers to the relative brightness that exists between the brightest and darkest regions in the television picture. This is determined by the peak-to-peak video signal that is applied between the grid and cathode of the picture tube. Typical video signals range from 90 to 150 V p-p. If the video signal amplitude is very small, the picture will appear to be washed out and is said to have poor contrast. If the video signal is of proper amplitude, the television picture will range from a maximum white through several shades of gray to black.

It is the function of the contrast control to allow the operator of the television receiver a means for adjusting the peak-to-peak amplitude of the video signal. This control is a front panel adjustment found on all types of television receivers.

Contrast controls may use one of two basic circuits. They are either used as input or output circuit ac voltage dividers or as degenerative feedback controls. Three examples of commonly used circuits found in solid-state television IC receivers are shown in Figures 10-22 to 10-24.

The voltage divider contrast control. In Figure 10-22 the contrast control (R_2) is located in the emitter circuit of the first video amplifier. The contrast control is ac coupled through C_4 to the second video amplifier which is coupled to the picture tube. When the movable arm of the potentiometer (R_2) is placed at the top of the control, maximum video signal will be delivered to the output video amplifier. If the movable arm of the potentiometer is positioned at the bottom of the control, the video signal delivered to the next stage will be very small because the arm will be almost shorted to ground. Positions of the control between these extremes will provide video signal levels of any desired intermediate value.

This contrast control is similar in operation to the volume control found in the audio amplifier section of the receiver.

Another type of contrast control is seen in Figure 10-23. Here the contrast control (R_6) is placed in the collector circuit of the output video amplifier. Notice that the contrast control potentiometer is placed between the collector and the B supply voltage. As a result, the full output video signal appears across it. At the top end of the control (collector end) the video signal is maximum and at the bottom end of the control the video signal is minimum. Here again, the control is acting as a voltage divider. Moving the potentiometer arm allows the receiver operator to select the most pleasing level of video signal between these two extremes.

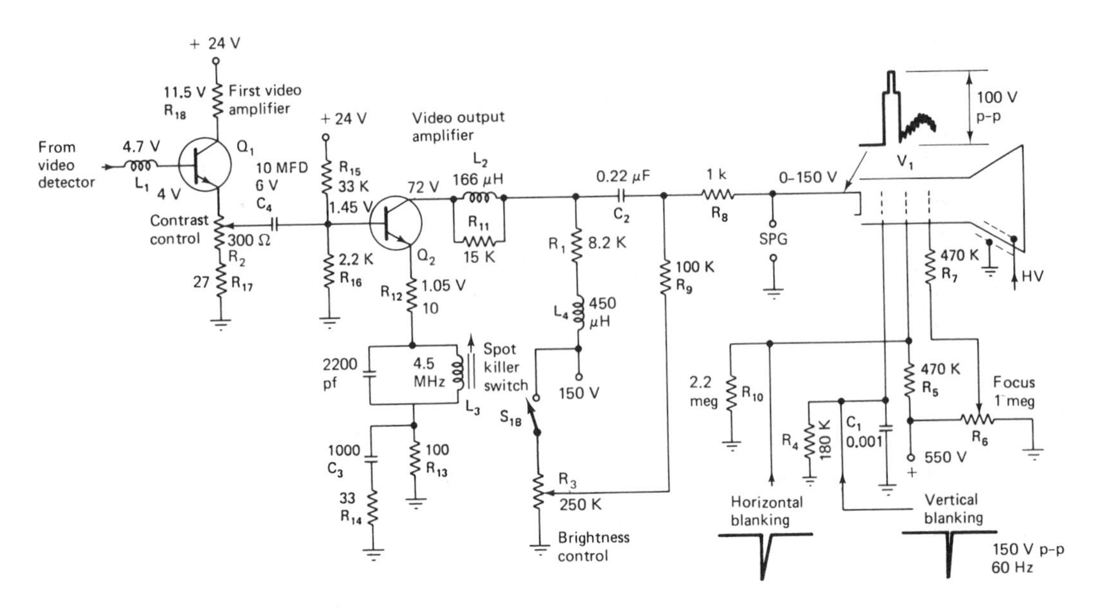

Figure 10-22 Transistor video amplifier used in a black & white receiver.

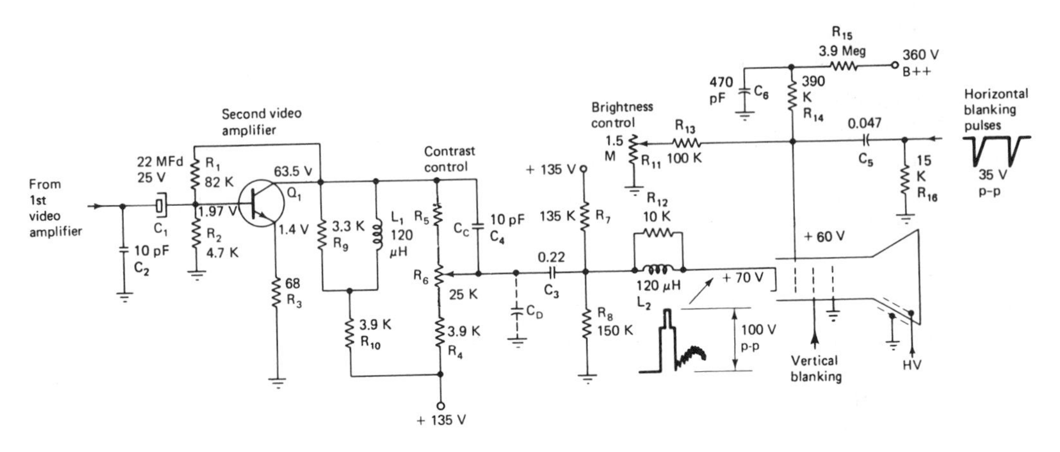

Figure 10-23 Transistor video amplifier.

Voltage divider compensation. Voltage divider contrast controls are subject to changing high-frequency response when they are used in high-impedance circuits. To use a contrast control in the high-impedance collector of a video amplifier requires that the circuit be frequency compensated.

Frequency distortion does not occur when voltage division takes place across like impedances. That is, if the circuit of Figure 10-23 is to be frequency compensated, resistor R_5 must be shunted with a capacitor C_c forming Z_1. If the proper value of C_c is selected, the impedances Z_1, and Z_2 (formed by C_D and R_6 plus R_4) will change at the same rate as the input frequency is increased. Therefore, the voltage division between them will be constant for all frequencies, and frequency distortion will be eliminated. The correct value of C_C is that which makes the time constant of Z_1 equal to the time constant of Z_2. This frequency compensation technique is also used in oscilloscope step attenuators and low-capacity probes.

The degenerative contrast control. Another type of contrast control is shown in Figure 10-24. In this color receiver circuit the peak to peak amplitude of the video signal (Y signal) is controlled by varying the stage gain of the video output amplifier. This contrast control (R_9) is placed in the emitter circuit of a transistor video amplifier. The ac component of the emitter current will pass through the contrast control and produce a form of negative feedback. The larger the contrast control resistance, the greater the degree of negative feedback and the greater the reduction in stage gain. Thus, in Figure 10-24 when the contrast control movable arm is at the top of the control, the contrast control R_9 is shorted out and C_3 provides minimum degeneration by effectively shorting out any signal ac at the emitter of Q_2. Under these conditions, the gain and contrast are maximum. When the movable arm of the contrast control is moved to the bottom of the control, the impedance of the electrolytic capacitor (C_3) is increased by the addition of R_9. The bypass action of C_3 having been reduced, the ac voltage at the emitter of Q_2 will be maximum. Consequently, negative feedback will cause the gain of the stage to be reduced.

The brightness control. Brightness refers to the average illumination of a scene. Thus, in a bright daytime scene the brightness level would be high. In terms of the video signal, this would correspond to a dc level approaching zero. See Figure 10-17 (b). Similarly, a nighttime scene would be an example of a low brightness scene. In this case, the video signal would develop a large dc component. See Figure 10-17 (a).

The brightness control is a front panel adjustment found in both color and black & white television receivers that controls the bias of the picture tube. As such, it may be placed in one of the following circuits:

1. The cathode of the picture tube
2. The grid of the picture tube
3. The base circuit of the dc-coupled output video amplifier

Examples of these arrangements may be seen in Figures 10-22 to 10-24.

Cathode brightness control. The most commonly used arrangement found in black & white receivers is shown in Figure 10-22. In this arrangement the cathode of the picture tube is made positive by returning it through the brightness control (R_3) to the B supply. The picture tube control grid is returned to ground through a 180-kΩ resistor (R_4) and is also fed by a negative-going vertical retrace blanking pulse. The grid is thereby made negative (dc) with respect to the cathode. Varying the brightness control potentiometer will cause the cathode voltage of the picture tube to change from 0 V to 150 V. When the cathode is at 0 V, the bias between grid and cathode of the picture tube is almost zero and the beam current will be maximum, resulting in maximum light output. When the cathode is +150 V, the picture tube will be cut off and there will be no light output (black) from the picture tube.

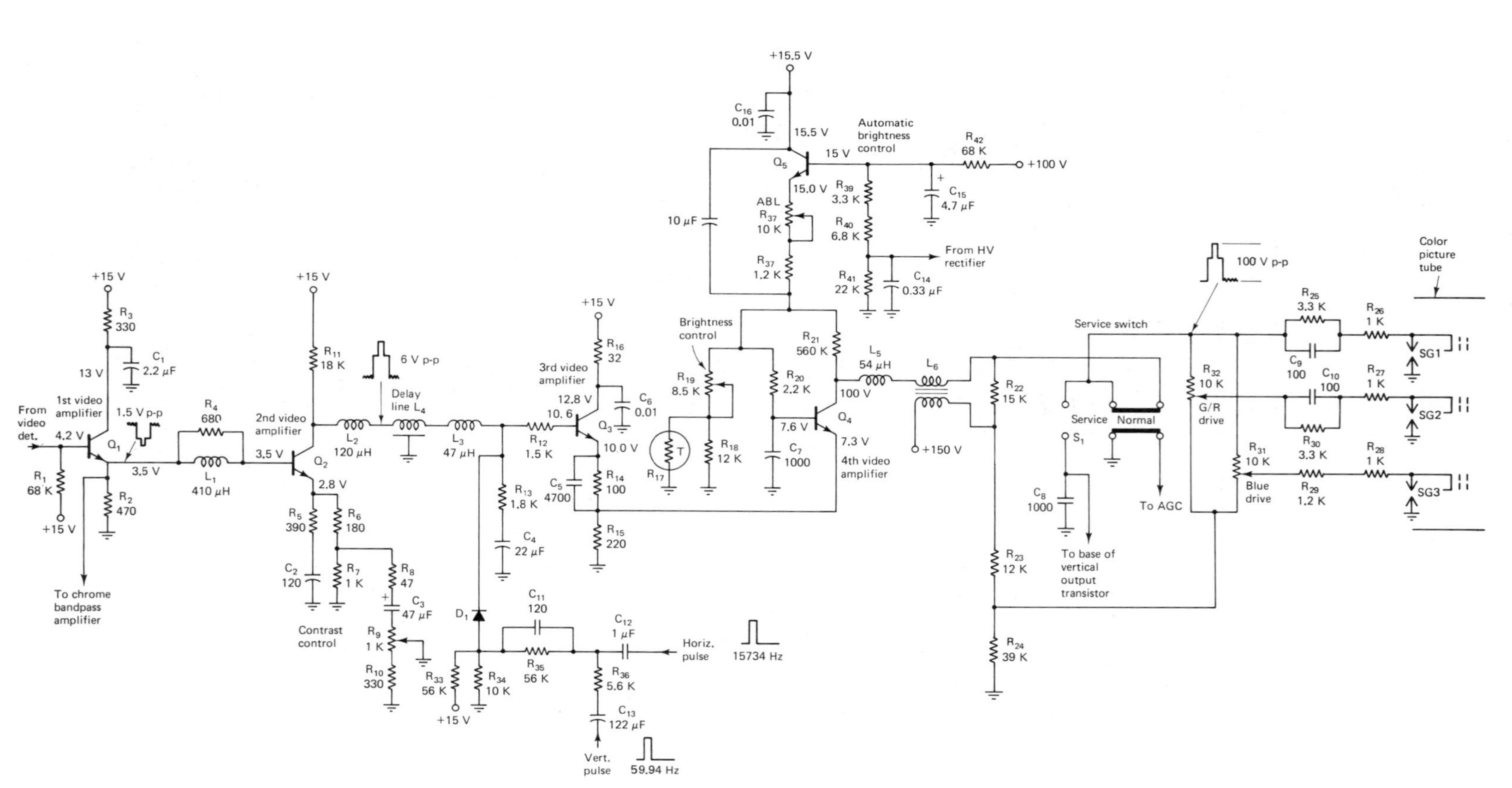

Figure 10-24 Video amplifier used in a color receiver.

The grid circuit brightness control. In Figure 10-23 the brightness control is placed in the grid circuit of the picture tube. Here the brightness control acts as part of a voltage divider for the B+ boost voltage. Also feeding the grid is a negative-going pulse obtained from the horizontal deflection circuit. This pulse is used to provide retrace blanking. The cathode of the picture tube is kept at +70 V by the dc voltage divider formed by R_7 and R_8. Changing the position of the brightness control changes the positive voltage on the grid from 0 V to +65 V, thereby changing the bias and the brightness of the picture tube.

The dc-coupled brightness control. Color receivers that use dc coupling between video amplifier stages often place the brightness control in the base circuit of the output video amplifier. An example of such a circuit is shown in Figure 10-24. In this circuit the brightness control (R_{19}) acts as a dc voltage divider. This control adjusts the dc base voltage which in conjunction with the emitter voltage establishes the bias for the video output amplifier. When the brightness control is rotated, the base voltage of Q_4 will become more or less positive. Making the base more positive means that the collector current in the transistor must increase, causing the collector voltage to fall. The collector of the video amplifier is direct coupled to the cathodes of the picture tube. Therefore, any changes of video amplifier collector voltage are communicated directly to the cathode of the picture tube. The control grids of the picture tube are all kept at a fixed voltage. Hence, the grid-to-cathode voltage (bias) and the brightness of the picture tube depend upon the degree of video amplifier collector current, which is controlled by R_{19}. In summary, when the base voltage of Q_4 is made maximum positive, the collector and picture tube cathode voltages will fall increasing the brightness, and vice versa.

The service switch. Included in Figure 10-24 is an example of a *service switch.* This double-pole, double-throw switch is used during gray scale brightness or color temperature setup procedures of the color picture tube. In this procedure it is desired to obtain equal light output from each of the red, green, and blue phosphors of the picture tube. If purity is good, this will result in a white raster.

When the switch is moved from normal to the service position, three things happen: (1) The picture tube cathode is disconnected from the video amplifier, removing the video signal. (2) The AGC is altered such that it disables the RF and video IF amplifier preventing noise and color information from reaching the speaker and the picture tube grids, respectively. (3) The vertical deflection amplifier is disabled. This results in a collapse of the raster into three horizontal lines of red, green, and blue. It has been found that gray scale adjustments are made more easily if three lines are used rather than the whole raster. When the service switch is returned to normal the raster and picture content will return to the screen.

In some receivers a three-position service switch may be used. In these circuits the third position produces a blank raster by disconnecting the video amplifier, and is used for purity and gray scale adjustment.

10-11 video picture tube driver circuits

The color video amplifier circuit of Figure 10-24 is an example of an older system of picture tube drive. In this arrangement, the cathodes of the picture tube were tied together and driven by a special video driver transistor that was able to withstand collector voltages of several hundred volts. In addition, each of the picture tube grids was driven by a color signal obtained from a similar type of transistor. Thus, this arrangement required the use of four expensive driver transistors.

An example of a modern type of picture tube driver circuit is shown in Figure 10-25. In this arrangement the picture tube control grids are kept essentially at ground potential, and each picture tube cathode is driven by a separate driver transistor (Q_{5001}, Q_{5002}, and Q_{5003}). There are at least three important advantages to the use of this system:

1. The number of picture tube driver transistors is reduced from four to three.
2. Since the cathodes of the picture tube are not tied together, the total distributed capacitance shunting the video circuit is reduced by one-third, thereby improving the driver amplifier high-frequency response.
3. In this arrangement the picture tube control grid is grounded; therefore, in the event of internal tube arcing, it provides additional high-voltage arc protection to the cathode driver transistors.

The red, green, and blue driver transistors (Q_{5001}, Q_{5002}, and Q_{5003}) are arranged in a common-base configuration. The collectors are dc coupled to the cathodes of the picture tube. Three bias controls (R_{750}, R_{752}, and R_{754}) control the dc emitter voltage of each transistor. These bias controls are used to adjust the dc operating point of each of the driver transistors, which in turn controls the dc cathode voltage of the picture tube, establishing raster temperature (white raster). Color bias controls adjust low-light (dark) areas of the color picture.

The signal voltages feeding the emitters of the driver transistors are of the standard form. These waveforms are obtained when a gated rainbow color bar generator is used as an RF signal source applied to the antenna terminals of the receiver.

The red, green, and blue transistors are fed by the red, green, and blue bias transistors, (Q_{704}, Q_{705}, and Q_{706}), arranged in a common-emitter configuration. The emitters of each of these transistors are tied to the drive control resistors (R_{756} and R_{758}). These resistors adjust the emitter voltage of all three bias transistors. Since dc coupling is used throughout the picture tube driver circuits, these controls also affect the color temperature of the raster. Color drive controls are used to adjust the highlight (white) areas of the color picture. The emitter voltage of the bias transistor is also controlled by a dc luminance voltage that automatically adjusts brightness levels as scene content changes.

The bases of the bias transistors are driven by video signals corresponding to red, green, and blue picture information, and are obtained from a chroma processor integrated circuit. In this IC, color difference signals are matrixed with the Y signal, producing the red, green, and blue video signals. In addition, retrace blanking pulses are added to the signal to ensure that the picture tube is turned off during retrace.

Other arrangements of this type of picture tube driver circuits are possible. One such circuit is shown in Figure 10-33. Here the base of each driver transistor is fed by a color difference signal ($R - Y$, $B - Y$, and $G - Y$). The emitter of each driver is driven by a black & white $-Y$ signal. Thus matrixing to produce red, green, and blue signal information is done in the driver transistor directly. Again, the color temperature of the picture tube is adjusted by low-light and drive controls associated with setting the bias of the driver transistors.

10-12 miscellaneous circuits

Brightness limiters. The video amplifier stage has associated with it a number of useful circuits that will be discussed in the following paragraphs.

Black & white television receivers have picture tube electron beam currents of approximately 100 μA. Color picture tubes may have beam currents that are as high as 1000 μA, or approximately 10 times that of black & white picture tubes. Color receivers are more susceptible to blooming because of their high picture tube beam currents. Selective blooming can also occur. This causes a fuzziness in the bright sections of the picture. Blooming is also caused by high-level video signals that overdrive the picture tube whenever the brightness or white signal level is high.

Automatic brightness limiters (ABL) are circuits used to prevent excessive beam current in the color picture tube, thereby preventing blooming or selective picture fuzziness.

Typical circuit. An example of a brightness limiter circuit is shown in simplified form in Figure 10-26. It is the function of the brightness limiter transistor to sense a change in the average picture tube beam current and then to shift the bias of the picture tube in a direction in order to prevent the change. Control of picture tube bias can most easily be accomplished in dc coupled video amplifiers. Hence, the video amplifiers in these circuits are dc coupled from the video detector to the picture tube cathodes.

Figure 10-26 picture tube beam current (electrons) flows from ground up through the brightness limiter potentiometer R_1 through Q_3 the third video amplifier and the driver transistors Q_4, Q_5, and Q_6 to the cathodes of the picture tube. The voltage drop developed across R_1 is the beam current sensing voltage used to drive the base of the brightness limiter transistor (Q_1). The brightness limiter (Q_1), in conjunction with bias resistors R_2 and R_3, determines the bias of the second video amplifier Q_2. A change in the operating point of Q_2 causes its collector voltage to change. As a result of the direct coupling between this stage and all the following stages, including the cathode of the picture tube, this change will cause a shift in picture tube brightness. The exact means by which this takes place is illustrated by the arrows drawn below the schematic diagram. These arrows show the directions of voltage and current changes that take place throughout the circuit (it is assumed that the picture tube beam current has increased). Note that by this action the original increase in beam current is effectively counteracted.

In this circuit the brightness control and the brightness range control are located in the second video amplifier. These controls vary the emitter bias of the second video amplifier, and because of the dc coupling used, this change in bias is coupled directly to the CRT cathodes. Brightness is minimum when the brightness control arm is nearest the +24-V source. The brightness range control determines the maximum level of brightness by setting the amount of positive voltage that appears on the low side of the brightness control.

Proper adjustment of the brightness limiter requires that it be adjusted in conjunction with the brightness range control. The adjustment procedure is as follows:

1. Turn the brightness limiter control fully clockwise.
2. Turn the brightness control fully clockwise.
3. Adjust the brightness range until picture blooms approximately 1 in.
4. Adjust brightness limiter until blooming is only 1/4 in.

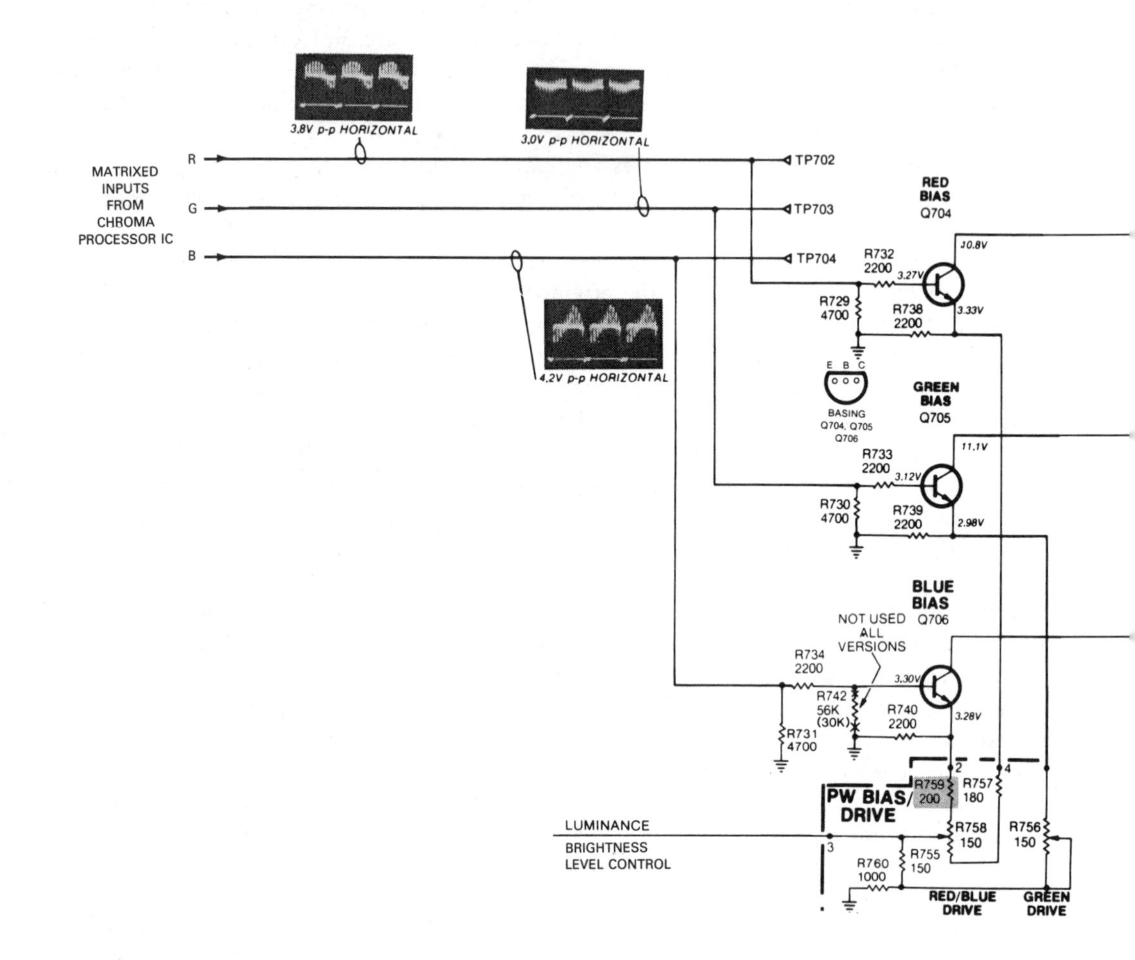

Figure 10–25 The picture tube driver circuits in a modern color television receiver. (Courtesy RCA, Consumer Electronics.)

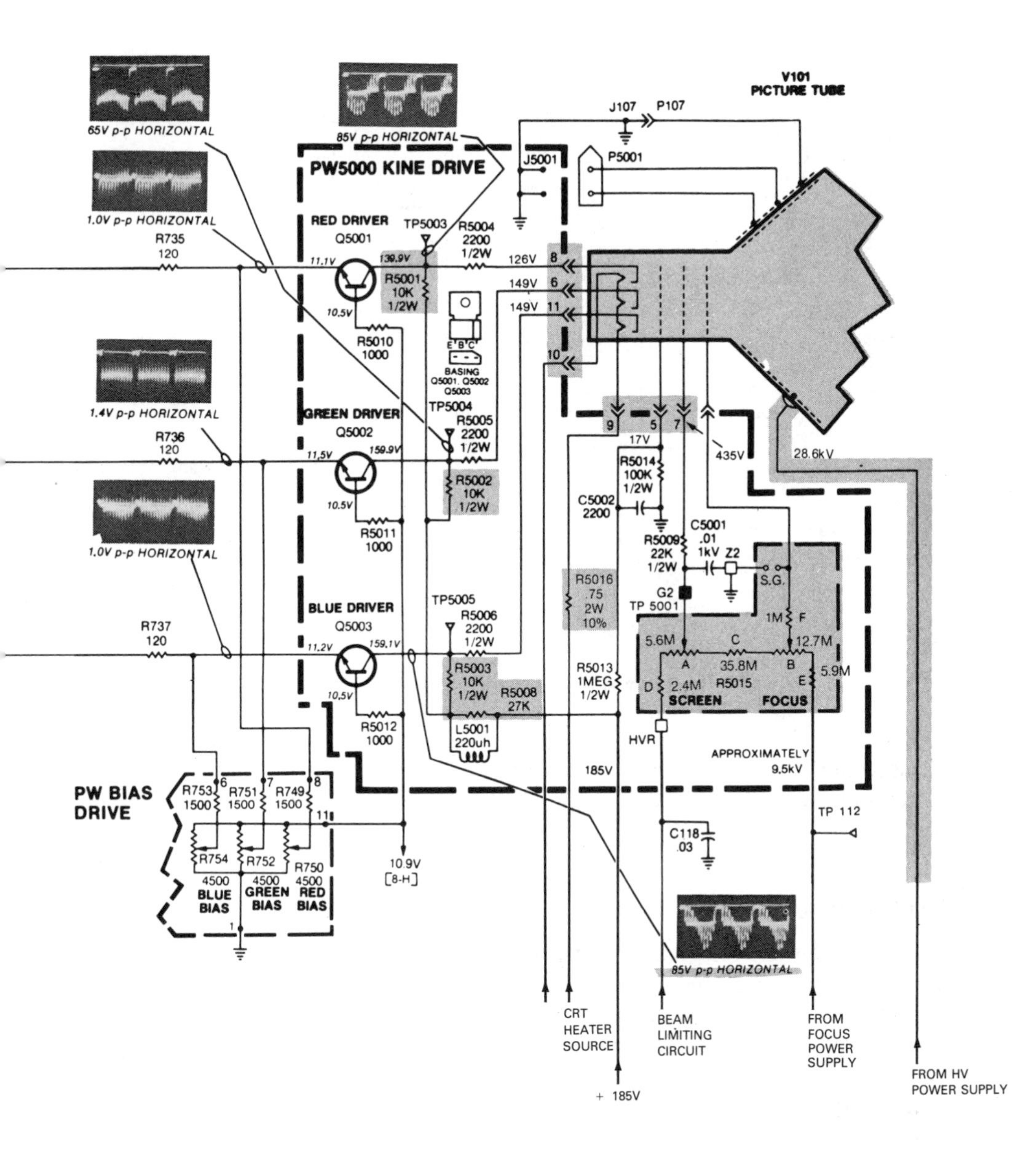
V101
PICTURE TUBE
J107
P107
J5001
P5001
PW5000 KINE DRIVE
65V p-p HORIZONTAL
1.0V p-p HORIZONTAL
85V p-p HORIZONTAL
1.4V p-p HORIZONTAL
1.0V p-p HORIZONTAL
RED DRIVER
Q5001
TP5003
R5004
2200
1/2W
R735
120
11.1V
139.9V
126V
R5001
10K
1/2W
149V
149V
10.5V
R5010
1000
BASING
Q5001, Q5002
Q5003
GREEN DRIVER
Q5002
TP5004
R5005
2200
1/2W
R736
120
11.5V
159.9V
R5002
10K
1/2W
R5011
1000
BLUE DRIVER
Q5003
TP5005
R5006
2200
1/2W
R737
120
11.2V
159.1V
R5003
10K
1/2W
R5008
27K
R5012
1000
L5001
220uh
17V
435V
28.6kV
R5014
100K
1/2W
C5002
2200
R5009
22K
1/2W
C5001
.01
1kV
Z2
S.G.
G2
TP 5001
R5016
.75
2W
10%
1M
F
5.6M
C
12.7M
A
35.8M
B
5.9M
R5013
1MEG
1/2W
D
2.4M
R5015
E
SCREEN
FOCUS
HVR
APPROXIMATELY
9.5kV
185V
PW BIAS
DRIVE
R753
1500
R751
1500
R749
1500
R754
R752
R750
4500
4500
4500
BLUE
BIAS
GREEN
BIAS
RED
BIAS
10.9V
[8-H]
TP 112
C118
.03
85V p-p HORIZONTAL
CRT
HEATER
SOURCE
BEAM
LIMITING
CIRCUIT
FROM
FOCUS
POWER
SUPPLY
FROM HV
POWER SUPPLY
+ 185V

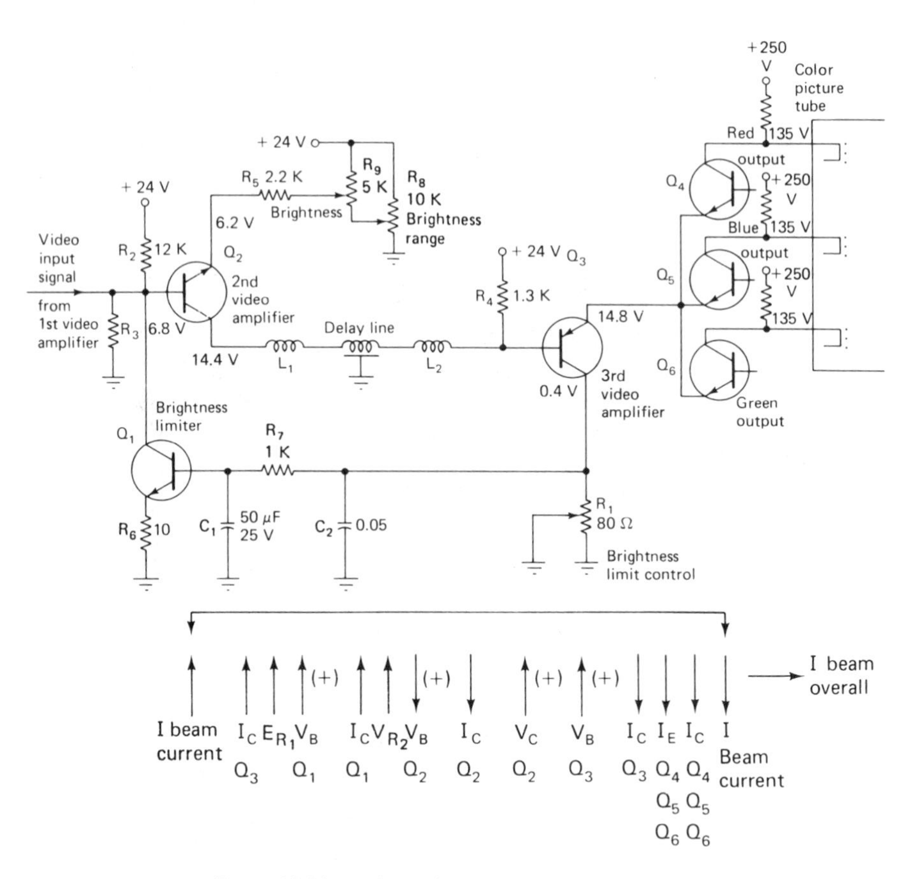

Figure 10-26 Brightness limiter controlled by beam current.

In general, adjustment procedures will vary from one manufacturer to another. Proper procedures for other circuits will be found in the manufacturer's service literature or in Sams Photofact.

10-13 peaking controls

The high-frequency response of the video amplifier of a color receiver is of necessity limited in order to minimize objectionable interference patterns that may result from interaction between the chrominance signal (3.58 ± 0.5 MHz) and the higher frequencies (3 to 4 MHz) of the video signal (Y). Unfortunately, loss of high-frequency response tends to degrade image definition. In order to compensate for this, many color receivers use video peaking switches or controls. These controls are located on the back panel of the color receivers and allow the operator to vary the degree of video peaking. These circuits are designed to enable the operator to adjust the high-frequency response of the video amplifier so that either a sharp, clearly defined picture or a somewhat "softer" picture may be obtained. A sharply defined picture has the disadvantage of sharpening the appearance of noise pulses. This is especially troublesome when receiving weak signal or high local noise

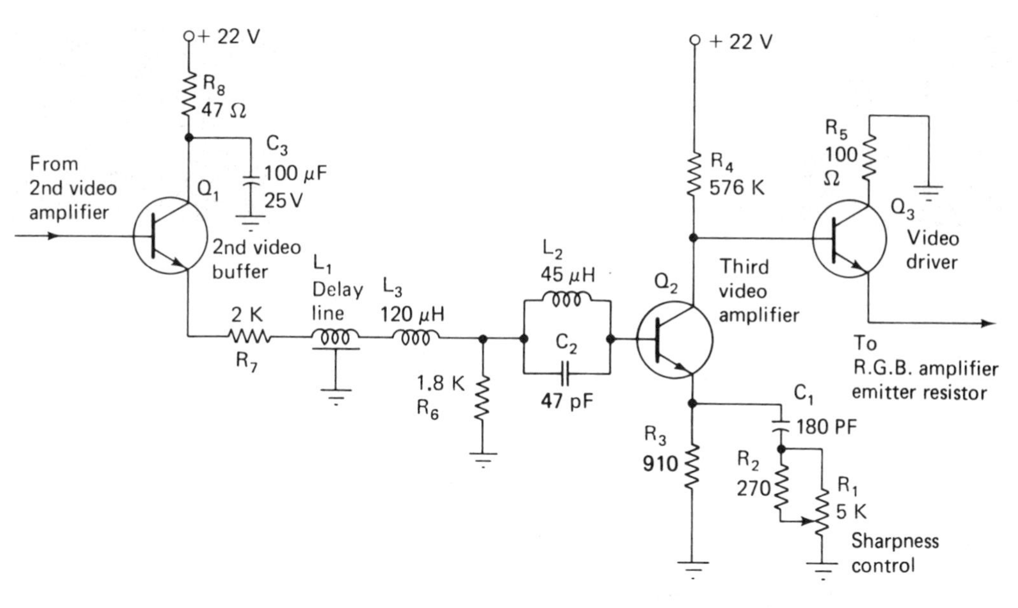

Figure 10-27 Sharpness (peaking) control.

content signals. A high degree of video peaking tends to accentuate the appearance of the noise, making it desirable for the viewer to have a control that can soften the impact of the noise on picture quality.

Typical circuit. An example of a video peaking circuit is shown in Figure 10-27. In this figure the video peaking control is called a *sharpness control* and is located in the emitter of the third video amplifier transistor. The function of the control is to vary the amount of high-frequency degeneration. When the control arm is at the top of the control maximum resistance is inserted in series with the high-frequency 180-pF peaking capacitor (C_1). Hence, the capacitor is less able to bypass the high frequencies and maximum degeneration takes place. In other words, the high-frequency gain is minimum. When the control arm is at the ground end of the potentiometer (R_1), the capacitor is more able to bypass the high frequencies and therefore little or no high-frequency degeneration occurs in the circuit and the high-frequency response is peaked.

10-14 the automatic brightness control

A common problem encountered by the television viewer is the effect of changes in the ambient (surrounding) light levels upon the brightness and contrast of the television picture. In the usual situation the television receiver may be turned on and adjusted for a proper picture during daylight hours, when the receiver's light output should be high to overcome the ambient light, and left that way into the evening hours, when the light level output of the receiver can or should be lowered. Excessive or insufficient picture brightness can result in a poor-quality picture. To minimize the effect of these changes, it is possible to use a light dependent resistor (LDR) that is mounted on the front panel of the receiver

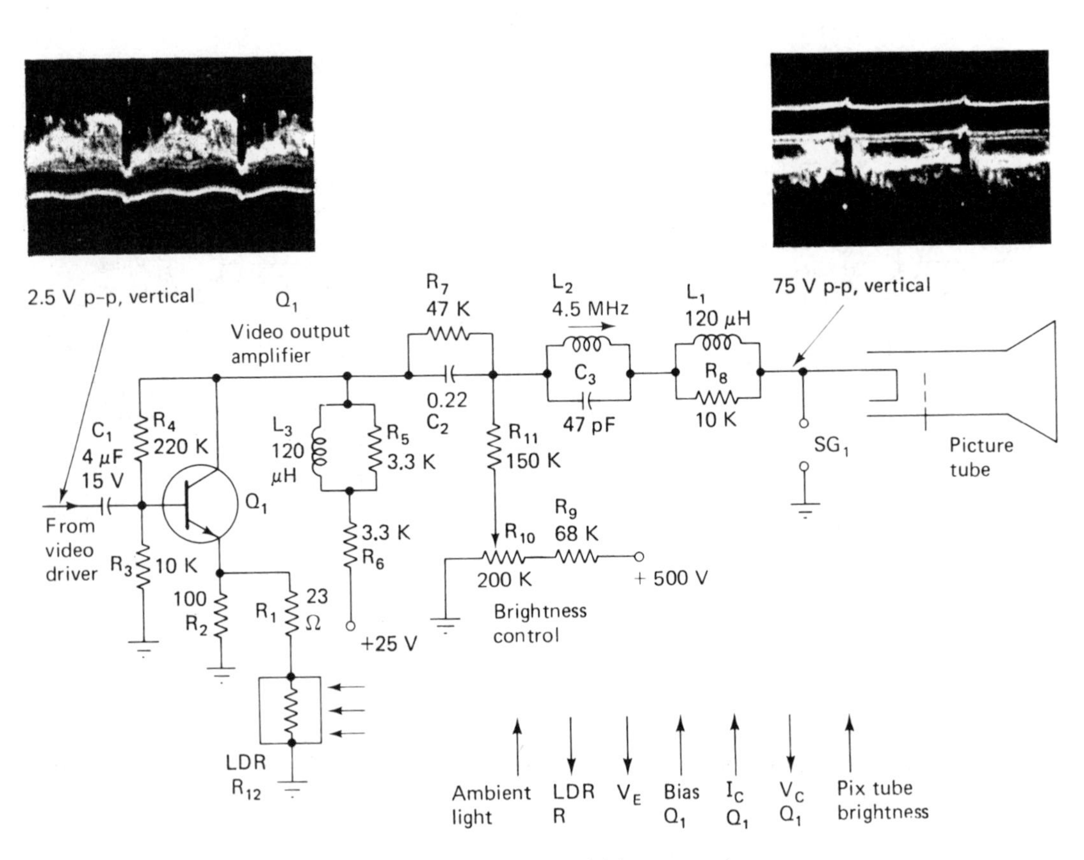

Figure 10-28 Automatic brightness control.

to automatically adjust the brightness and contrast of the receiver in response to changes in the ambient lighting. An LDR is a photoconductor whose resistance is very high in the dark and becomes very much lower in the presence of light.

Typical circuit. A system of automatic brightness control is shown in Figure 10-28. In this circuit the LDR is placed in the emitter circuit of the video output amplifier and affects both the brightness and contrast of the picture. Changes in LDR resistance will affect the bias of the video amplifier transistor. Such shifts in bias change the dc and ac characteristics of the amplifier. For example, if the ambient light level is high, the resistance of the LDR will be small, resulting in reduced emitter voltage. This in turn will increase the forward bias of the amplifier, causing greater collector current flow and a decreased collector voltage. Since dc coupling exists between the collector and the cathode of the picture tube, a decreased collector voltage will also force the bias of the picture tube to decrease, resulting in more light output from the CRT.

In addition, lowered LDR resistance will result in less total emitter resistance. Since the emitter resistor is not bypassed, the reduction in emitter resistance means that there will be less degeneration and the stage gain will increase, driving the CRT with a higher-amplitude video signal to produce more contrast.

As the room lighting decreases, the LDR resistance increases, produces more degeneration and less gain thereby reducing contrast. The output stage conducts less heavily due to the decreased forward bias of the transistor and allows the collector dc voltage to become more positive increasing the bias of the CRT and reducing picture brightness. A 23-Ω resistor is placed in series with the LDR to limit the variations of resistance in the circuit so that overcorrection of brightness and contrast does not occur.

10-15 spot killer circuits

When a television receiver is turned off, the filament of the picture tube, as well as other tubes, remains hot for a period of time and does not instantly stop emitting electrons. Similarly, the charge developed across the dc filter capacitor of the high-voltage supply and the various other power supply circuits does not drop to zero the instant the receiver is turned off. In some receivers these conditions result in a bright spot of light that appears at the center of the picture tube screen and lingers for a considerable length of time after the receiver has been turned off. This spot usually does no damage to the picture tube but it is annoying to look at. Consequently, many manufacturers incorporate special circuits, called *spot killers,* whose purpose is to eliminate the lingering spot on the CRT screen when the set is turned off.

Typical circuit. An example of such a circuit is shown in Figure 10-22. In this circuit a switch (S_{1B}) is ganged to the on/off switch of the receiver. When the receiver is turned on, switch S_{1B} is closed and the 150-V dc supply is used to establish proper picture tube bias. This bias is set by the brightness control.

When the receiver is turned off, switch S_{1B} is opened and the 150-V dc supply is removed from the brightness control causing the picture tube cathode voltage to drop to zero. Since both cathode and grid are at zero potential, the bias on the picture tube is zero and maximum beam current flows. This in turn quickly discharges the high-voltage filter capacitor and rapidly extinguishes the spot.

The spark gap. As shown in Figure 10-28, a spark gap (SG_1) is connected from the cathode of the picture tube to ground. This component is used as a safety device that allows high-voltage picture tube transients to discharge, thereby protecting the relatively low voltage rated circuit components. These high-voltage transients may be caused by internal arcing in the picture tube or by electrostatic induction caused by the picture tube's second anode high voltage. Spark gaps may be found in a number of different packages. They may simply be two electrodes held rigidly at a fixed separation in a ceramic container. They are often molded into the picture tube socket, or they may be found as an integral part of a disk-type ceramic capacitor. Many types are available, depending upon the manufacturer.

Defective spark gaps are seldom found. If, however, the gap should become too small, repeated arcing will occur and cause intermittent operation. If the gap should become too wide, excessive voltages that build up in the circuit may cause repeated breakdown of transistors or other low-voltage components in the circuit.

Spark gaps are usually rated at approximately 5000 V, but spark gaps found in the socket of picture tubes may have ratings of only 1500 V. Spark gaps may also be found in horizontal deflection circuits, color amplifiers, and other sections of the receiver.

10-16 the delay line

An important component found in color receiver video amplifiers but not in black & white receivers is the delay line. In the block diagram of Figure 10-1 it can be seen that the composite video signal leaving the first video amplifier divides into a luminance (brightness) or *Y* signal and a chrominance signal. These signals are acted upon by the circuits that they pass through and then both arrive at the picture tube. The *Y* signal has fewer stages to pass through than does the chrominance signal. It should be noted that each stage tends to introduce a time delay in inverse proportion to its bandpass. Therefore, the wide-band *Y* signal tends to arrive at the picture tube cathodes before the related narrow-band chrominance signal can reach the grids of the color picture tube. The result of this is to produce a picture in which the color information appears slightly to the right of the black & white information. This is illustrated in Figure 10-29(a). Also shown are the conditions of proper arrival times and excessive *Y* signal delay.

To ensure that all signals arrive at the picture tube at the same time, a delay line is introduced into the *Y* channel. See Figures 10-24, 10-26, and 10-27. Its purpose is to delay the *Y* signal by approximately 1 μs so that the *Y* signal will arrive at the picture tube at the same time as the color picture information. The delay line is essentially a long coil (3 to 6 in.) that has been wound on a form that is constructed with a metal foil throughout its length. See Figure 10-29(b). The metal foil is used as a ground plane and it has a terminal tied to ground. The ground plane ensures that there will be capacity from every turn of the coil to ground. The capacity in conjunction with the inductance introduces a time delay because it takes time for the capacity to build up a charge and the coil to produce its magnetic field.

A defective delay line or its termination can cause at least the following symptoms:

1. No raster
2. Raster O.K.; color picture without brightness information (*Y* signal) [see Figure 10-29(c)]
3. Ringing in picture

An open delay line is in series with the signal, and therefore it interrupts the signal path in the *Y* channel. In dc coupled video amplifiers the delay line also interrupts the dc conditions in the video amplifier and may cause the video output stage to go into cutoff. Under these conditions, the video amplifier plate or collector voltage rises to B+ and the picture tube cathodes become very positive, cutting off the picture tube. Since beam current is reduced to zero, no raster and no picture result.

If the circuit is ac coupled, an open delay line prevents the *Y* signal from reaching the picture tube without disturbing the dc conditions of the picture tube. This defect results in the very strange condition of a colored picture without any blacks & whites.

Improper line termination or open ground plane connections result in picture ringing that appears very similar to ghosts, but these ghosts have distinctive characteristics. They remain unchanged regardless of channel or fine-tuning position and they are evenly spaced. Improper delay line termination may be the result of a changed value resistor or a defective or open peaking coil.

A good way to eliminate the antenna derived ghost as a possible cause of the ringing is to disconnect the antenna from the receiver and replace it with a known good

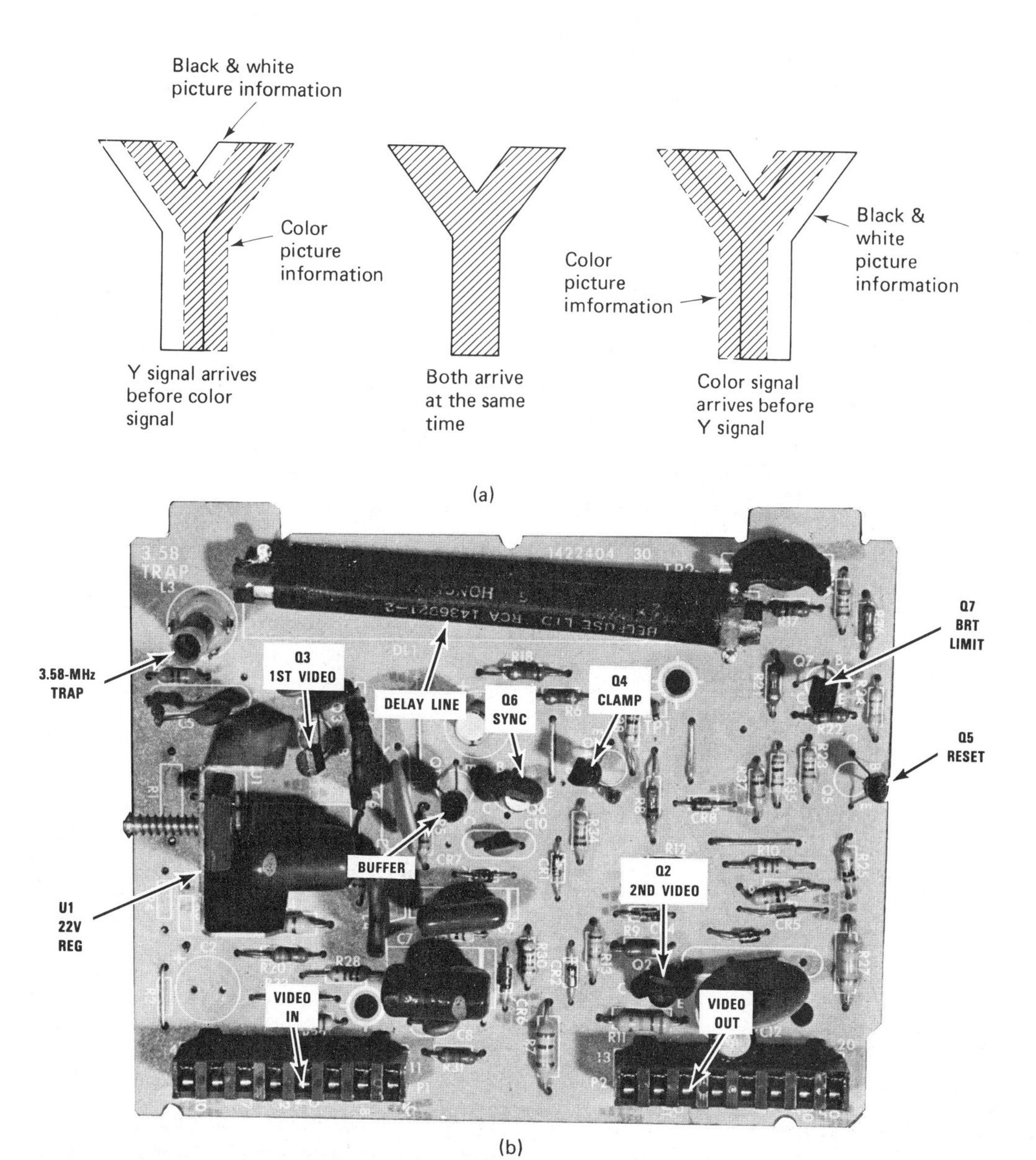

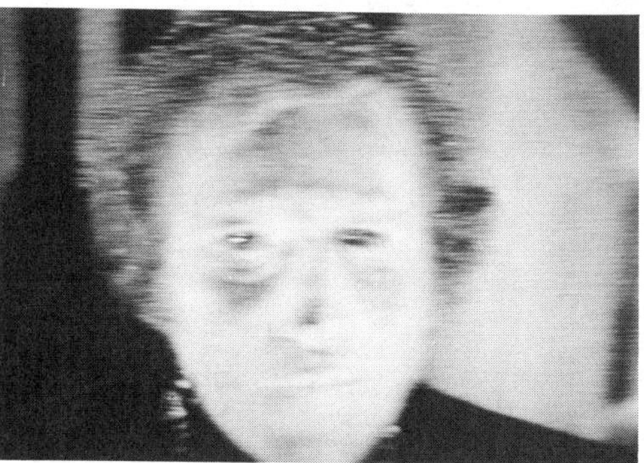

Figure 10-29 (a) Effect of *Y*-channel time delay on picture reproduction; (b) A typical delay line mounted on a video amplifier module; (c) color picture without brightness information (see color plate). [(b) Courtesy of RCA, Consumer Electronics.]

crosshatch generator. If the ghosts remain, then the defect must be in the receiver. If fine tuning and changing channels have little effect on the ringing, then alignment and regeneration are not likely causes. This leaves the delay line as a probable cause.

10-17 retrace blanking circuits

The purpose of retrace blanking is to prevent the effects of horizontal and vertical retrace from being seen. This is done by forcing the picture tube to turn off during periods of horizontal and vertical retrace. Black & white receivers may use vertical retrace blanking alone, but many black & white receivers do use both horizontal and vertical blanking. In color receivers both horizontal and vertical retrace blanking are almost always used.

Vertical retrace occurs at the end of vertical scanning when the electron beam is at the bottom of the picture. When the beam returns to the top of the screen, the beam moves at a relatively slower rate. Therefore, vertical retrace produces heavy diagonal white lines that are superimposed on the raster such as those shown in Figure 2-9.

Horizontal retrace occurs at the end of the horizontal scanning period when the electron beam is on the right side of the screen. During retrace the beam returns to the left side of the screen at a somewhat faster rate than the scanning lines. These retrace lines are not as bright as vertical retrace lines and appear as a haze covering the raster.

The effects of horizontal retrace can be somewhat more annoying in color receivers. During color transmission the color burst is transmitted. The color burst rides on the back porch of the horizontal blanking pulse. If the horizontal blanking is not used, the color burst will produce a greenish-yellow bar smeared through the picture.

Theoretically, the blanking pulses of the composite video signal should be able to turn off the picture tube during retrace periods. Unfortunately, changes that take place in the video signals dc level during changes in scene brightness can affect the blanking pulse effectiveness and make retrace lines visible. Also, without retrace blanking, brightness control adjustment is important, if not critical, in setting the proper conditions for retrace-free television viewing. In the case of color transmission, there is no real provision in the composite video signal for blanking out the color burst; it is left to the receiver designer to provide the necessary circuitry. Retrace blanking circuits ensure that the television receiver will be free from annoying retrace lines or color burst smear independently of the operator or signal level changes.

Picture tube requirements. The television picture tube must be turned off during retrace time if the effects of retrace are to be removed. This may be done by applying a pulse that occurs during horizontal and vertical retrace time to the proper electrode of the picture tube. The pulse must also be of proper amplitude and polarity. Figure 10-30 shows color and black & white picture tubes, their electrodes, and the polarity of the retrace blanking pulses. Notice that in both tubes the control grid and screen grids may be fed negative going pulses and that the cathodes are fed positive going pulses. These pulse polarities are necessary to force the picture tube to reduce its electron beam current to cutoff.

Typical circuit: black & white receiver. The pulses are obtained from the vertical and horizontal output stages of the receiver and are coupled via voltage dividers and pulse-shaping networks to the electrodes of the picture tube. Unfortunately, there are no universally used circuit

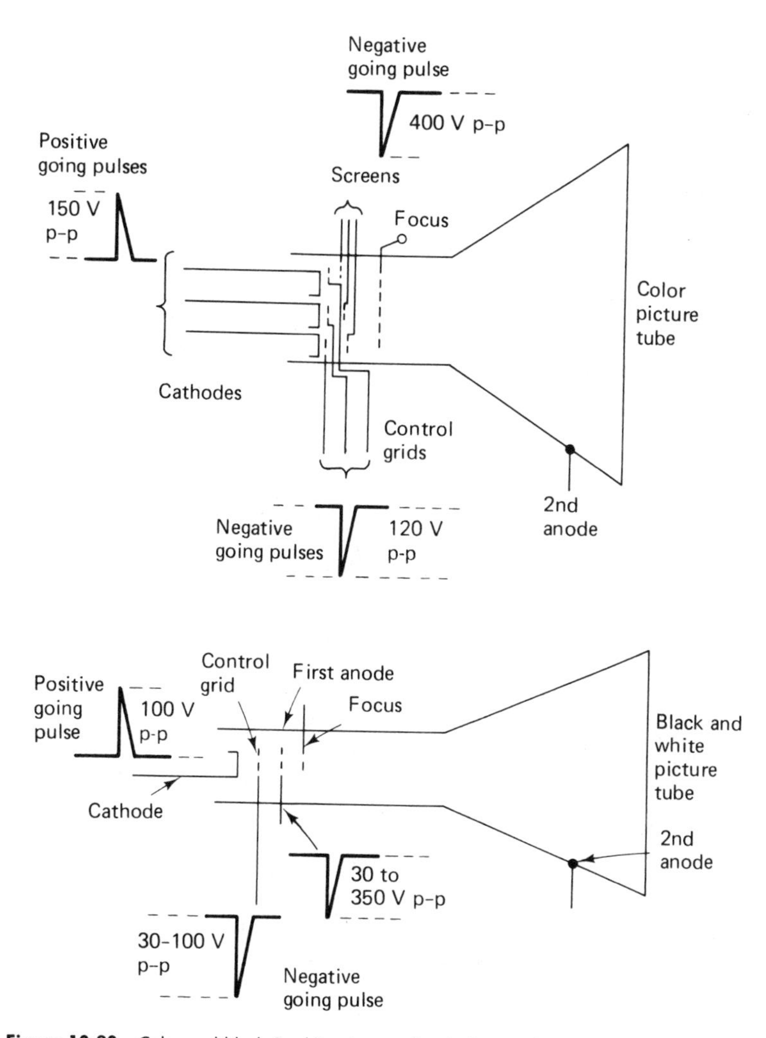

Figure 10-30 Color and black & white picture tubes indicating the polarity, amplitude, and electrode to which the blanking pulses may be applied.

arrangements. A typical example of a blanking scheme used in a black & white receiver is shown in Figure 10-31.

In this circuit vertical retrace blanking is produced by a negative going pulse that feeds the control grid of the picture tube and is obtained from the vertical output transformer. This pulse is generated during the retrace period of sawtooth generation. The coupling circuit between the transformer and the grid of the tube is used to block B+ from reaching the picture tube grid this is done by C_1. R_1 and R_2 act as a voltage divider that reduces the amplitude of the pulse to usable levels. The combination of R_1, R_2, and C_2 forms a pulse-shaping network that broadens the retrace blanking pulse and ensures that the entire retrace period is blanked out.

The amplitude of the vertical blanking pulse is 90 V p-p, but this value will vary from one receiver to another, depending upon the type and size of the picture tube. It

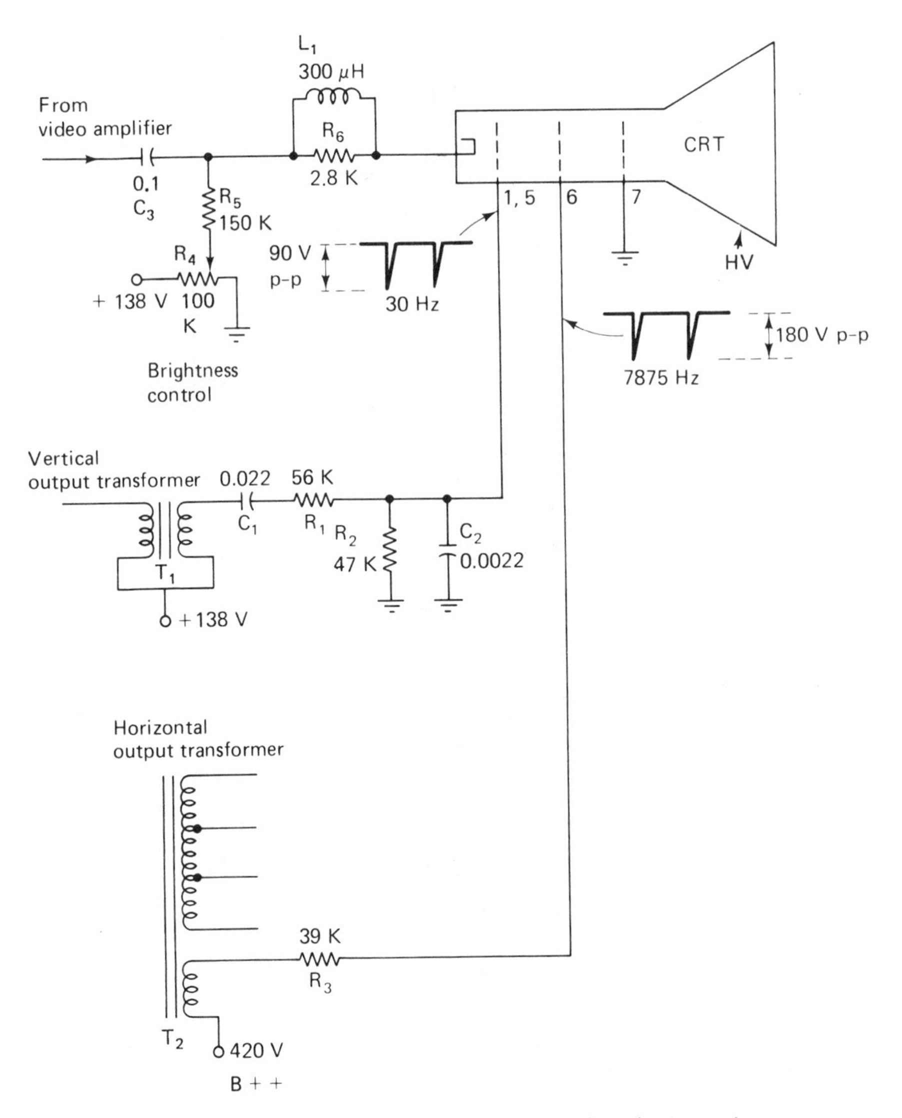

Figure 10-31 Typical blanking circuit used in a black & white television receiver.

should also be noted that some manufacturers feed the vertical retrace pulse to the first anode of the picture tube instead of to the control grid. Since the control grid is more sensitive than the first anode, the first anode pulse amplitude should be at least two to four times larger than that needed on the control grid.

In this circuit, horizontal retrace blanking is produced by a large negative going pulse that is fed to the first anode of the picture tube and is obtained from a special winding on the horizontal output transformer. This pulse is generated by the horizontal deflection circuit during the retrace portion of the deflection sawtooth.

The pulse is coupled via R_3 to the first anode of the picture tube. This resistor, in conjunction with the distributed capacitance of the picture tube, forms a wave-shaping circuit that broadens the pulse and ensures that blanking will be effective. The amplitude of horizontal blanking voltage required is twice that required for vertical blanking.

Retrace blanking: color receiver. Retrace blanking is usually more complex in color receivers. Furthermore, servicing is made more difficult because each manufacturer has its own circuit for achieving blanking. Many color receivers use separate IC or transistor stages that are used as blanking buffers, amplifiers, and/or inverters. These circuits may be used either for horizontal or vertical blanking alone or they may be fed both pulses and be used to provide both types of retrace blanking pulse output. In general, the blanker stage feeds the video amplifier. In older receivers it may also have an output that feeds the color difference amplifiers. The video amplifier then acts as an inverter and amplifier for the blanking pulses. Figure 10-32 shows a partial block diagram of a color receiver showing one possible arrangement of the stages involved in the blanking process and the polarity of the blanking pulses at each point in the circuit. Notice that the output pulses of the deflection amplifiers are positive going and that the blanker inverts them producing two negative-going pulses. Both pulses feed the video amplifier stage. The negative-going pulses driving the video

Figure 10-32 Block diagram of the blanking circuits that may be found in a color receiver.

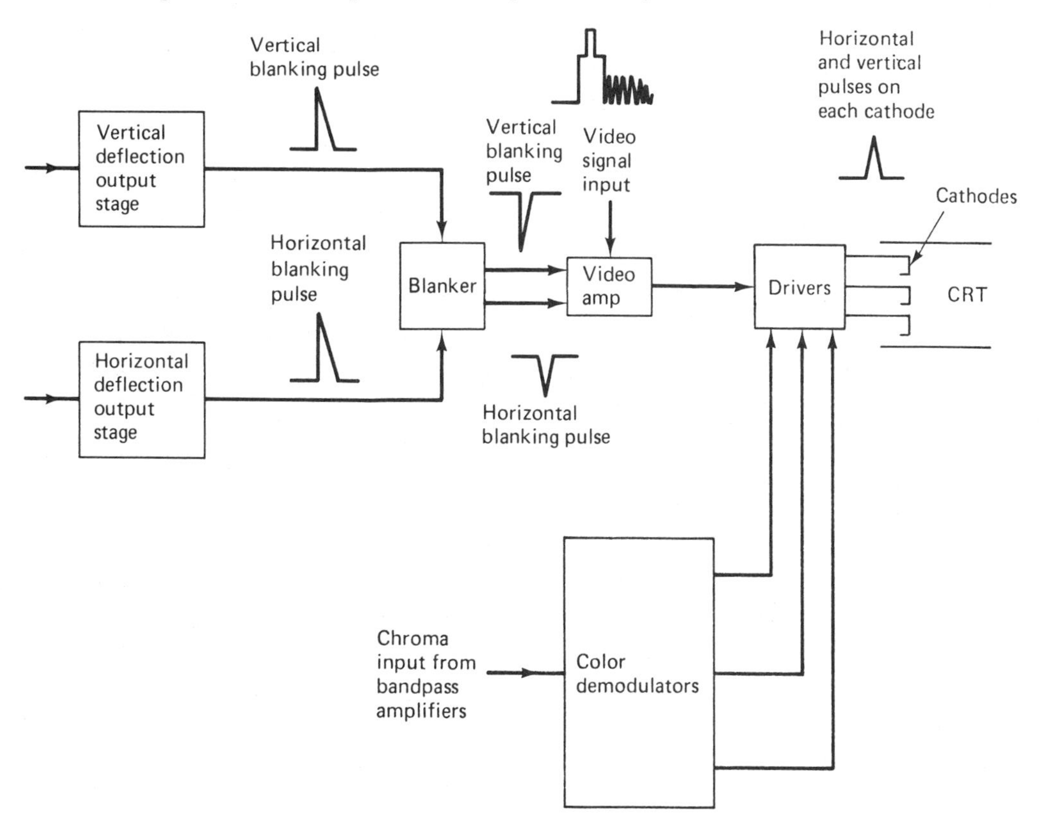

amplifier are again inverted and feed the noninverting driver stages which in turn drive the cathodes of the picture tube, driving them into cutoff during horizontal and vertical retrace periods. It should be borne in mind that this block diagram is only representative and that other arrangements are possible. In almost every case, however, a color receiver will incorporate vertical retrace blanking, horizontal retrace blanking, and some may also include color burst blanking in its design.

Typical circuit: color receiver. A partial schematic diagram of the blanking circuits used in a color receiver is shown in Figure 10-33. In this circuit a single blanker circuit is used for both horizontal and vertical blanking. The vertical blanker (Q_{305}) obtains a positive-going

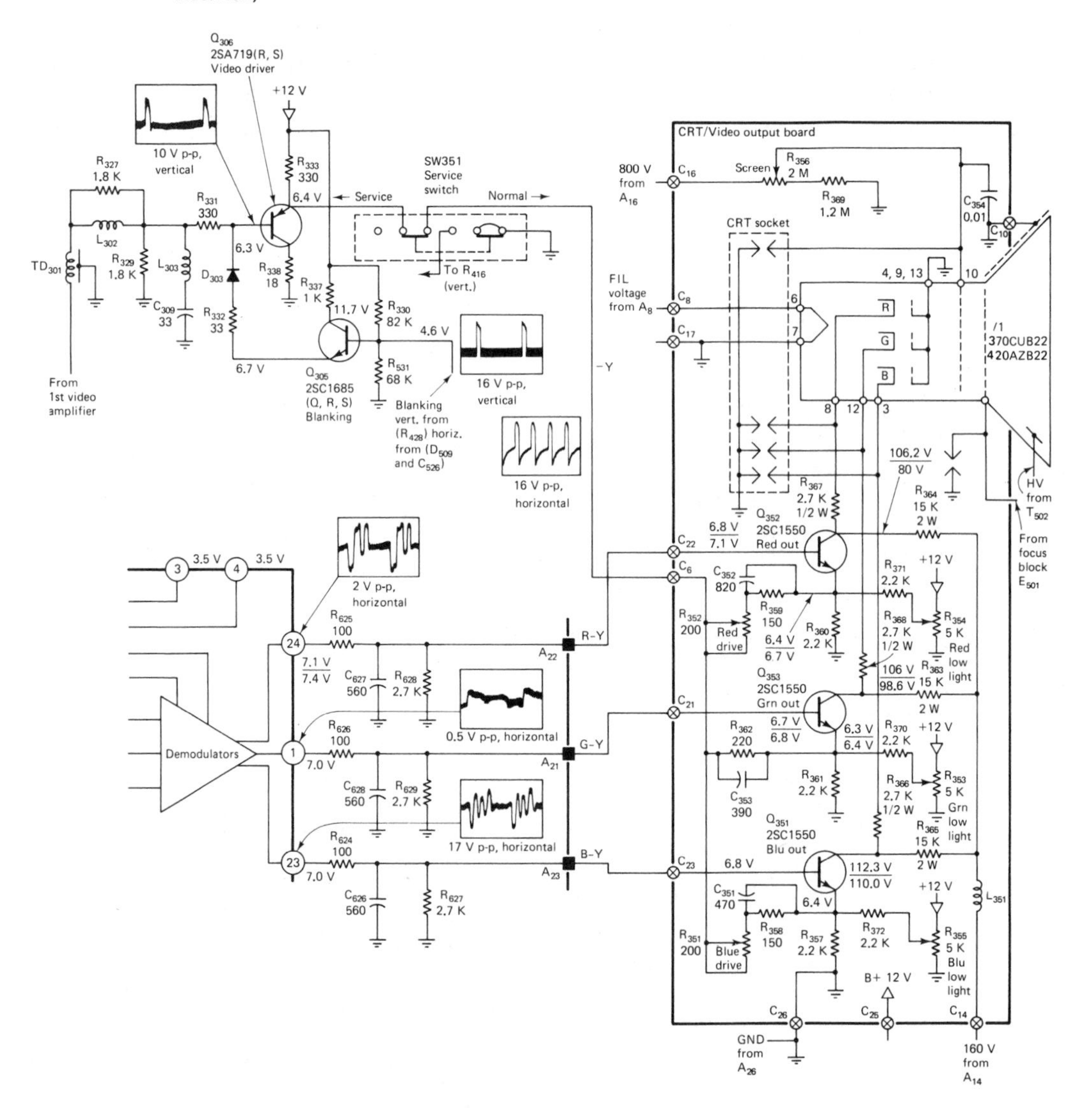

Figure 10-33 Blanking circuit used in a color television receiver. (Courtesy of Quasar Industries.)

vertical pulse from the vertical deflection oscillator, and a positive-going horizontal pulse from the horizontal flyback transformer.

The blanker transistor is biased so that it conducts on the most positive portions of the input pulses. The noninverted output of Q_{305} is taken from the emitter and is coupled through diode D_{303} to the base of the video driver Q_{306}. Diode D_{303} conducts only on the most positive portions of the blanker output pulses, preventing any negative excursions of the blanking pulses from reaching the base of the video driver Q_{306}.

The video driver is an emitter follower and its noninverted emitter output consists of the luminance video plus positive-going blanking pulses. These are coupled through the service switch to the emitters of the red output driver Q_{352}, the green output driver Q_{353}, and the blue output driver Q_{351}. In terms of the blanking and luminance video signals, the drivers act like common-base amplifiers. The collector output is noninverted and consists of positive-going blanking pulses and luminance information ($-Y$). Since the blanking pulses are positive and applied to the cathodes of the picture tube, they are the correct polarity to drive the picture tube into cutoff during both horizontal and vertical retrace intervals.

10-18 video amplifier troubleshooting

Common trouble symptoms caused by the video amplifier of a color or black & white television receiver are listed in Table 10-1. The trouble symptoms may be divided into four categories: gain, frequency response, brightness, and interference. The first seven defects are concerned with gain and brightness, the next two are related to frequency response, and the remaining four are interference problems of one sort or another.

In a black & white receiver similar symptoms would appear, except, of course, that no color would be seen.

TABLE 10-1

Color	*Picture*	*Sound*	*Raster*
1. No	No	O.K./No	O.K.
2. No	No	O.K.	No
3. O.K.	O.K.; loss of horizontal or vertical sync	O.K.	O.K.
4. O.K.	Dark or negative	O.K.; may have buzz in B&W	O.K.
5. O.K.	Dim	O.K.	O.K.
6. O.K.	No	O.K.	O.K.
7. Present	Present	O.K.	Blooming; excessive brightness
8. O.K.	Smeared picture	O.K.	O.K.
9. O.K.	Loss of fine detail	O.K.	O.K.
10. O.K.	Ringing in picture	O.K.	O.K.
11. O.K.	Hum bars	O.K.	O.K.
12. O.K.	Retrace lines	O.K.	O.K.
13. O.K.	4.5-MHz or 920-kHz interference beat	O.K.	O.K.

Video amplifier output circuits. To understand how these symptoms may be caused by defects in the video output stage of the receiver, it is necessary to understand that the video amplifier feeds at least five different circuits. This is illustrated in the block diagram shown in Figure 10-34. This block diagram is for a color receiver.

As shown in this diagram, the video amplifier has one input and as many as six outputs. The composite video signal obtained from the video detector may only be 2 V p-p and may be amplified to a level of 100 V or more. In a color receiver this may mean as many as five stages of video amplification. The chroma sync and AGC takeoff points may be located in any of these stages. For example, the sync and AGC takeoff point may be located at the collector of the first video amplifier, and the chroma takeoff point may be located at the emitter of the same amplifier. This would mean that defects in the second or third video output stages would have no effect on the color or sync characteristics of the picture. Defects in the first video amplifier could affect sync, color, and AGC portions of the receiver.

The block diagram for a black & white receiver would differ from the color receiver in two ways. First, the sound takeoff for a color receiver usually occurs before the video detector, whereas in the black & white receiver the sound takeoff is usually found in the video amplifier. Second, the black & white receiver would not have any color bandpass

Figure 10-34 Signal outputs of the video amplifier.

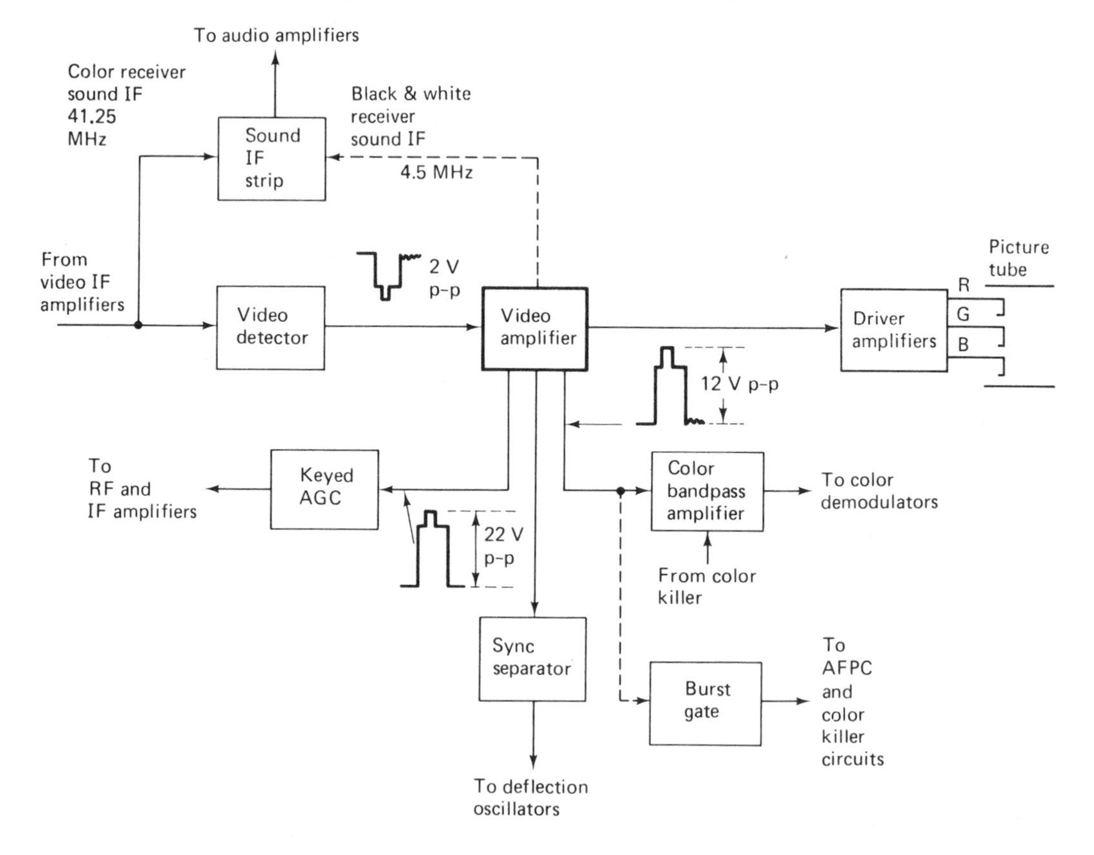

amplifier or burst gate circuits to feed. In some color receivers the burst gate is fed by the bandpass amplifier instead of by the video amplifier. However, both color and black & white receivers use the video amplifier as the takeoff point for the sync separator and keyed AGC circuits.

10-19 vacuum tube video amplifiers

Figure 10-35 is an example of a vacuum tube video output amplifier. This stage operates Class A and drives the cathodes of the picture tube with a $-Y$ signal.

The degenerative contrast control. In this color receiver circuit the peak-to-peak amplitude of the video signal (Y signal) is controlled by varying the stage gain of the vacuum tube video output amplifier. This contrast control (R_3) is placed in the cathode circuit of a vacuum tube video amplifier. The ac component of the tube current will pass through the contrast control and produce a form of negative feedback. The larger the contrast control resistance, the greater the degree of negative feedback and the greater the reduction in stage gain. Thus, when the contrast control movable arm is at the top of the control, the contrast control R_3 is shorted out for ac by the large 200-μF electrolytic capacitor C_3 placed across it. Under these conditions, the gain and contrast are maximum. When the movable arm of the contrast control is moved to the bottom of the control, the electrolytic capacitor is shorted out. The ac voltage drop across R_3 will be maximum and because of negative feedback the gain of the stage will be reduced.

The dc coupled brightness control. Color receivers that use dc coupling between video amplifier stages often place the brightness control in the grid circuit of the output video amplifier. In this circuit the brightness control (R_2) acts as a dc voltage divider. This control adjusts the dc grid voltage which in conjunction with the cathode voltage establishes the bias for the video output amplifier. When the brightness control is rotated, the grid voltage changes from +1 V to +3.5 V. Making the grid more positive means that the plate current in the tube must increase, causing the plate voltage to fall from +175 V to +150 V. The plate of the video amplifier is direct coupled to the cathodes of the picture tube. Therefore, any

Figure 10-35 Hybrid video amplifier used in color receivers.

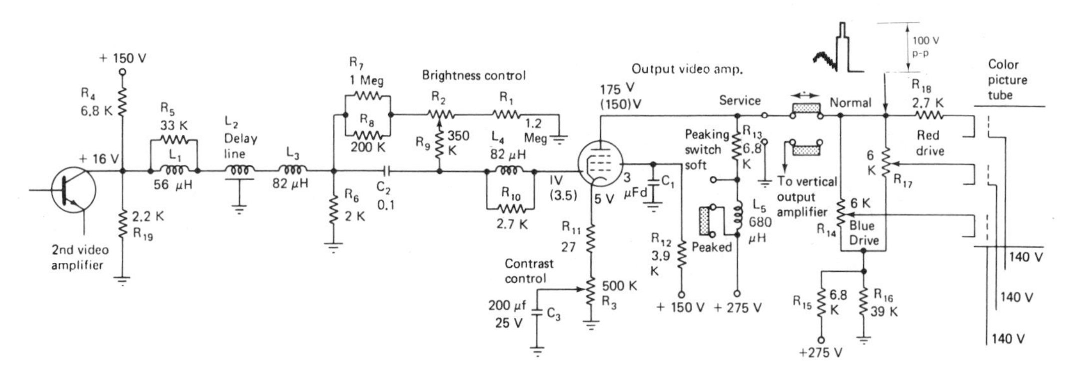

changes of video amplifier plate voltage are communicated directly to the cathode of the picture tube. The control grids are all kept at a fixed voltage. Hence, the grid-to-cathode voltage (bias) and the brightness of the picture tube depends upon the degree of video amplifier plate current.

The service switch. Included in Figure 10-35 is an example of a *service switch.* This double-pole, double-throw switch is used during gray scale brightness or color temperature setup procedures of the color picture tube. In this procedure it is desired to obtain equal light output from each of the red, green, and blue phosphors of the picture tube. If purity is good, this will result in a white raster.

When the switch is moved from normal to the service position, two things happen. The picture tube cathode is disconnected from the video amplifier (removing the video signal), and the vertical deflection amplifier is disabled. This results in a collapse of the raster into three horizontal lines of red, green, and blue. It has been found that gray scale adjustments are made more easily if three lines are used rather than the whole raster. When the service switch is returned to normal the raster and picture content will return to the screen.

In some receivers a three-position service switch may be used. In these circuits the third position produces a blank raster by disconnecting the video amplifier, and it is used for purity and gray scale adjustment.

Peaking control. In this circuit the peaking control is a two-position switch placed across the shunt peaking coil of the video amplifier. In the normal or peaked position of the switch the peaking coil is part of the video amplifier circuit and maximum high-frequency response is obtained. When the switch is moved to the "soft" position, the shunt peaking coil is removed (shorted) from the circuit and the high-frequency response of the amplifier is attenuated.

QUESTIONS

1. What is the basic function of the video amplifier?
2. Explain why the video signal is usually fed to the cathode of the picture tube instead of to the grid.
3. Explain what is meant by frequency distortion, phase distortion, harmonic distortion, and intermodulation distortion. How would each of these forms of distortion appear on the television screen?
4. What would be the visual appearance on the television screen of a video signal that was a square wave and had a frequency of 1,575,000 Hz?
5. List five ways to extend the high-frequency response of an amplifier.
6. What is the effect of phase distortion on the television picture?
7. List four causes of poor low-frequency response.
8. List three methods for extending the low-frequency response of an *RC*-coupled amplifier.
9. What effect on picture reproduction will the loss of the dc component have in a black & white receiver and in a color receiver?
10. List two kinds of contrast controls. Explain their functions and how they work.

11. What does the brightness control do, and in what circuits may it be located?
12. What is the function of an automatic brightness limiter? Why is this circuit used only in color receivers?
13. What is the purpose of peaking or sharpness controls? Explain how the one shown in Figure 10-27 works.
14. Explain how an automatic brightness control works, and discuss the characteristics of an LDR.
15. What is the purpose of spot elimination circuits? Explain how the one shown in Figure 10-22 works.
16. What is a delay line and why is it used in color receivers?
17. What are the possible results of a defective delay line?
18. Which elements of color and black & white picture tubes are blanking pulses fed and what are their polarities?
19. List four basic trouble symptoms for which the video amplifier strip can be responsible.
20. What is the purpose of the comb filter?
21. What are two advantages to be gained by using the comb filter?
22. What are the two methods of time delay used in comb filters?
23. Referring to Figure 10-19(a), explain how the comb filter works.
24. In the RCA comb filter, vertical enhancement and vertical peaking circuits are used. Explain their purpose.

---

chapter 11

Low-Voltage Power Supplies

11-1 introduction

As shown in Figure 11-1, the low-voltage power supply is common to almost all stages in both color and black & white television receivers. The power supply stage is responsible for providing the dc (B supply) requirements of each circuit and for providing the heater needs of the picture tube. Such circuits as the video detector, sound detector, horizontal AFC, color phase detector, and in some cases color demodulators do not have any dc power requirements because these circuits obtain their power needs by rectifying their input signal.

Many modern television receivers obtain most of their low-voltage dc power from the horizontal output circuit. In these circuits taps are placed on the winding of the horizontal output transformer, which provides horizontal retrace pulses of different voltages. These pulses are then rectified, providing the required dc supplies. This type of power supply is sometimes referred to as a *scan-derived power supply.*

Solid-state receivers may have dc power supply current demands ranging from 400 to 1200 mA. Here again, the current needs of black & white receivers are somewhat less than the current needs of color receivers.

With the exception of the high-voltage needs of the picture tube, the low-voltage power supply used in solid-state receivers provides positive voltages with respect to ground that range from 4 to 350 V.

In general, the dc current demands of the television receiver will increase when a signal is applied. This change in current demand can increase by as much as ten times.

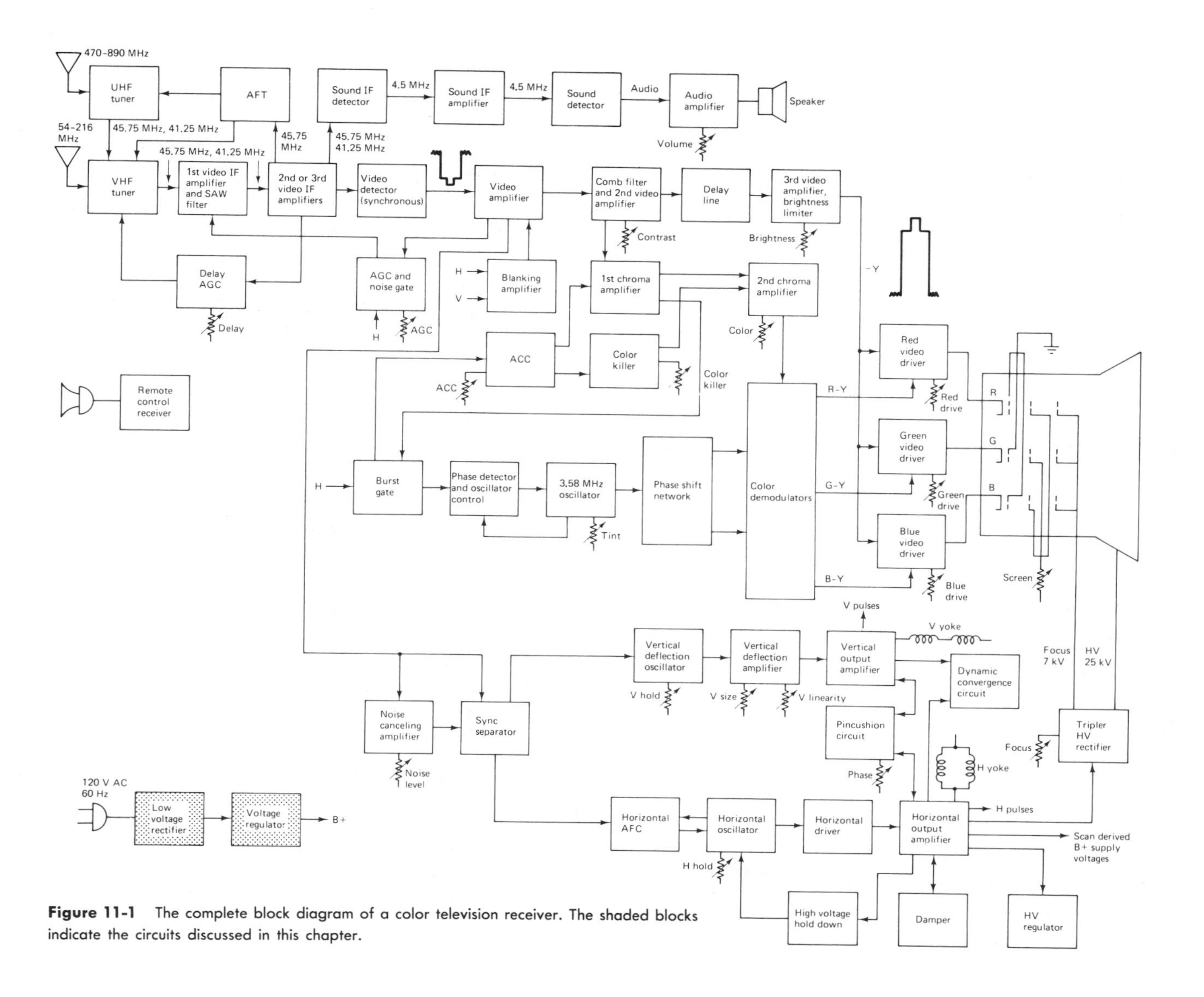

Figure 11-1 The complete block diagram of a color television receiver. The shaded blocks indicate the circuits discussed in this chapter.

11-2 shock hazard

Under certain conditions, touching the chassis of a television receiver can result in a severe shock. This could happen because one side of the ac power line is returned to earth ground. See Figure 11-2(a). Notice that it is the practice of power companies to supply the electrical consumer with a step-down transformer that produces a center-tapped secondary voltage of 220 V rms. The usual practice is for the center tap to be returned to *earth* ground. It is also the usual practice for the ac voltage provided at any outlet to be taken between the grounded wire and one side of the transformer. This means that the available ac outlet voltage is approximately 110 V. Because of the changing load placed on the power company's facilities, the output line voltage may vary from as little as 95 V to as high as 130 V. The nominal line voltage that is assumed for design of television receivers is 117 V rms.

Most modern ac outlets have three terminals. Two terminals, *A,* and *B* in Figure 11-2(a), provide the outlet voltage to the load and the third terminal *C* is internally tied to earth ground. This third wire is a safety device used to protect electrical equipment by grounding their metal cases. This means that when the third terminal is properly connected, it is impossible for the metal frame of any hand tool or piece of equipment to assume any potential other than ground. Thus, shock hazard from internal equipment shorts is minimized. For example, if the field winding of a hand drill shorted to its case and the third wire is used, the worst that could happen is that the house fuse would blow out. If the third wire were not used, and if there were a field winding to case short, the operator's hand would be tied to one side of the ac line. If any other part of the operator's body were grounded, the operator would receive a potentially lethal shock.

Unfortunately, since television receivers seldom use three-terminal plugs, it is possible for shock hazard to exist. How this may come about is illustrated in Figure 11-2(a) to (c). Let us assume that a simple electrical load consisting of an electric lamp surrounded by a metal box is to be connected to the ac outlet. Notice that in this example one side of the lamp is connected directly to the metal box. Therefore, when the lamp is plugged into the ac outlet, the metal box will be either tied to earth ground or to the high side of the transformer secondary. Let us further assume that a person places one hand on the metal box and the other hand on a nearby water pipe. If the ac plug has been inserted into the outlet so that the metal box is tied to ground, the effect on the experimenter is shown in Figure 11-2(b). Notice that by placing one hand on the water pipe and the other on the metal box the person is in effect holding his or her own hand and tying both hands to ground. Since no difference of potential exists, no shock will be experienced.

If, however, the plug were reversed when placed in the ac outlet, the unfortunate experimenter would be placing one hand on ground and the other on the high side of the transformer. In effect, the person would be placed directly across the ac line as shown in Figure 11-2(c). As a result, the person would experience a severe shock.

Shock hazard protection. To avoid shock hazards, one of the three methods discussed below can be used.

Isolation transformers. A very common method is to use an *isolation transformer* (separate primary and secondary windings) that can be placed between the ac outlet and the load.

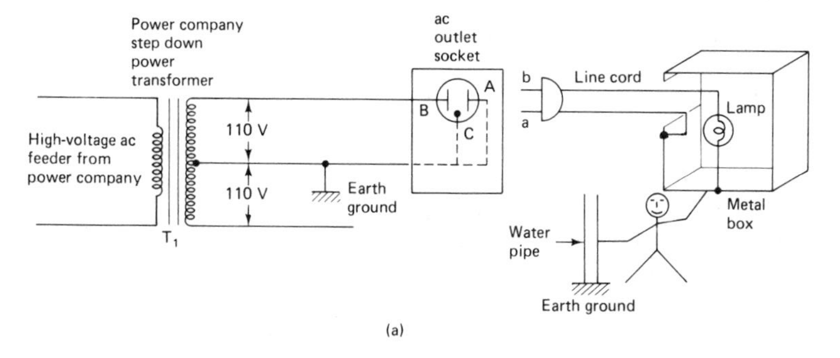

(a)

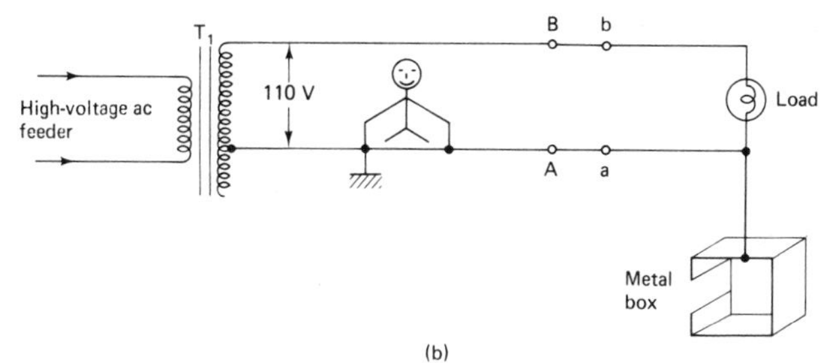

(b)

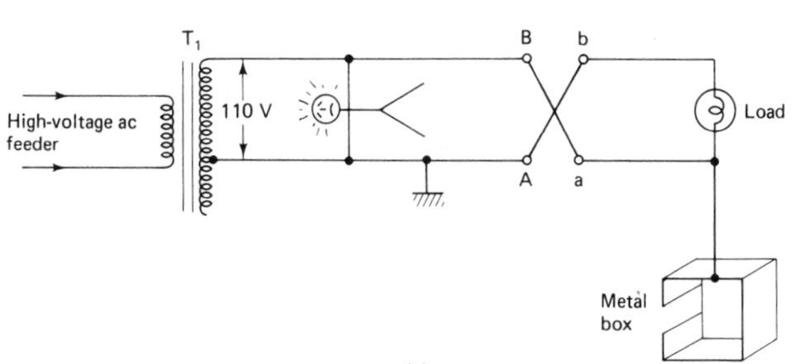

(c)

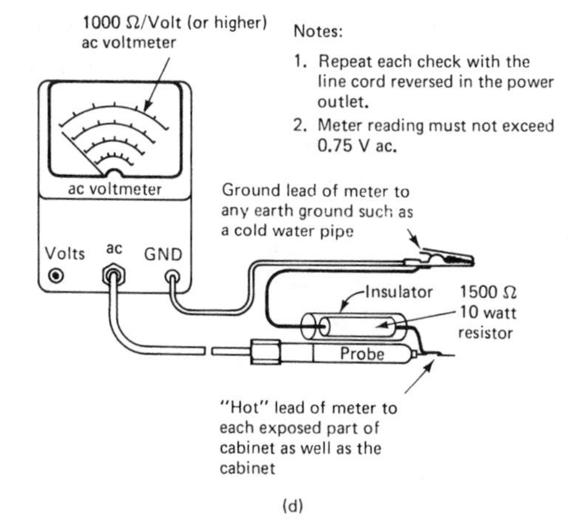

(d)

Figure 11-2 (a) Typical ac outlet and load that has one side of the line returned to chassis; (b) proper connection for no shock hazard; (c) improper connection of ac plug results in a shock hazard; (d) shock hazard test circuit. [(d) Courtesy of Quasar Industries.]

This effectively insulates the load and its chassis from earth ground and removes any possibility of shock hazard.

Floating chassis. Another simple method is to float the chassis, that is, have no electrical connection to the chassis at all. Under these conditions, no shock hazard exists, but other problems arise. The chassis actually has two functions: it is the structure on which the components that comprise the television set are mounted and it may behave as an electrostatic and/or magnetic shield.

To be an effective shield, the chassis must be ac grounded. This is done by returning the chassis to ground via a bypass capacitor whose value is such that it is an open for 60-Hz and a short for RF and IF frequencies. This capacitor is shunted with a large resistor (1 MΩ or more) that discharges any accumulated charge across it that might cause the capacitor to short.

Insulated cabinet. Still another method for preventing shock hazard is to tie the chassis to one side of the ac line and enclose the entire receiver in an insulated cabinet. In this arrangement all knob mountings, screws, and cabinet fixtures are made of plastic, wood, or some other insulating material. Since the operator has no direct access to the chassis, shock hazard cannot exist. The only place that bare wires reach the outside of the cabinet is the antenna lead-in terminals. To prevent shock hazard, capristors are placed in series with the antenna leads. Of course, the service technician is not protected when he has removed the set from its cabinet.

It is imperative that when the service technician is working on a television receiver that has no isolation transformer of its own that the technician connect a 1-to-1 isolation transformer between the ac outlet and the receiver's plug. This is important not only from a safety point of view, but also for very practical servicing reasons. For example, most isolation transformers are adjustable and allow low and high line voltage checks on receiver operation.

Shock hazard test. Most manufacturers recommend that after a repair on a transformerless television receiver has been completed, a leakage check should be made to determine whether or not the set is dangerous. First, plug the ac line cord of the receiver directly into a 120-V ac receptacle, then, using two clip leads, connect a 1500-Ω 10-W resistor paralleled by a 0.15-μF capacitor between all exposed metal cabinet parts and a known earth ground, such as a water pipe or conduit. Next, using an ac VTVM or a VOM with 1000 Ω per volt or higher sensitivity, measure the ac voltage drop across the resistor. While continuously monitering the ac voltage drop across the resistor, move the ungrounded lead of the 1500-Ω resistor to all exposed metal parts (antenna, metal cabinet screws, knobs, control shafts, and so on). Any reading of 0.75 V rms or more is excessive and indicates a potential shock hazard that must be corrected before returning the receiver to the owner. See Figure 11-2(d) for one possible arrangement of the test circuit hookup.

11-3 overload protective devices

Television receiver power supplies are powered by being connected to the ac power line. This power source may be capable of delivering short-circuit currents in excess of 400 A (as high as 10,000 A in some cases). As such, defects in the receiver power supply can cause very large currents to flow into the power supply. If safety precautions are not taken to limit the maximum current flow, the television receiver can catch fire. It is for this reason that all television receivers are equipped with one or more overload protection devices.

The most commonly used protective devices are fuses and circuit breakers. In some of the solid-state circuits special *shutdown* protective circuits are used to minimize damage to the receiver if the power supply becomes overloaded.

Fuses. The simplest method of circuit protection is the use of a fuse. A fuse is a piece of wire whose length, diameter, and composition are such that when the current flow through the wire exceeds a certain rating, the wire melts and opens the circuit.

From a mechanical point of view, fuses used in television receivers are usually contained in glass cylinders approximately 1 in. long and 1/4 in. in diameter. The ends of the cylinder are capped with metal caps that are the connection points for the fuse wire inside the cylinder. These leadless fuses require some form of fuse holder to hold them.

Some fuses have wire pigtails attached to the end caps. These pigtails are used to wire the fuse directly into the protected circuit. If this fuse blows, it may be replaced by either cutting it out and replacing it with a good pigtail fuse or by using special clips that permit a good leadless fuse to be attached to the bad one. These work-saving clips may be obtained in a number of different styles. They are either individual clips (two are required, one for each cap) shaped like the letter "S" and made of thin, flat metal 1/4 in. wide or they are made like fuse holders with a pair of electrically connected back-to-back clips at each end of the holder. One set of clips attaches the holder to the defective fuse. The other set of clips holds the good fuse that shunts the defective fuse.

Not all fuses are mounted in housings. Some fuses have no housings at all. These fuses are lengths of thin wire that are used as fuses. For example a No. 34 copper wire can be used as a 5-A fuse; that is, if the current in the circuit containing this wire exceeds 5 A, the wire will open. Other wire sizes often used and their associated fusing currents are as follows:

Copper Wire Size Number	*Fusing Current (A)*
36	3.6
34	5
32	7
30	10
28	14

Fuses used in television receivers may be rated as either *fast-acting* or *time delay.* The fast-acting (or normal blow) fuse is usually applied in circuits that do not have surge or transient currents. The fuse is constructed of a single strand of wire that has no heat sinks to dissipate momentary overloads. These are normally rated at 200 to 300% of the full-load current. Fast-acting fuses can be obtained in current ratings ranging from 1/4 to 20 A.

Time delay fuses (sometimes referred to as *slow blow*) are designed to pass transients and surges without opening. They will, however, blow with sustained overloads or short circuits. Physically, these fuses are usually constructed of a fuse wire that is under spring tension. The spring is embedded in a solder alloy that acts like a heat sink for momentary current surges. When a sustained overload occurs, the solder melts and the fuse opens. Current ratings of these fuses may range from $1\frac{1}{4}$ A to as much as 30 A.

Voltage rating. Fuses also have voltage ratings. These ratings indicate the maximum voltage that can appear across the fuse when it has blown and is an open circuit without arcing across the fuse gap. Since the voltage across an open circuit is equal to the fully applied voltage, the fuse should have a voltage rating equal to or greater than the voltage on the circuit. Standard voltage ratings used in television receivers are 125 V and 250 V.

Circuit breakers. Circuit breakers are overload protection devices that are based upon the fact that a bimetallic strip will bend when heated. If this strip is made one of the contacts of a switch, it can be used as an overload protective device. During normal operation the circuit breaker is placed in series with the circuit to be protected and is in a closed position. The current flowing through the switch will not warm the bimetallic strip sufficiently to cause it to open. This level of current is called the *hold current.* Typical circuit breaker hold current ratings of 2 or 3 A are common.

If there is an overload, the current flow through the bimetallic switch will be sufficient to bend it, and the switch contacts will snap open. Springs associated with the bimetallic strip cause the contacts to snap to open rapidly. Once the circuit breaker is open, it will remain open until it is reset by pushing a red button protruding from the circuit breaker. The current level required to open the circuit breaker is called the *tripping current.* Typical tripping current ratings are 4 or 5 A.

11-4 basic television power supply arrangements

The low-voltage power supply used in television receivers may be found in a number of different configurations. In general, color solid-state receivers use power supply circuits that are more complex than those used in black & white receivers.

A composite block diagram showing the essential features found in most types of power supplies is shown in Figure 11-3. An on/off switch is placed between the ac line voltage and the line filter. The switch determines whether or not the receiver is energized. The line filter consists of a number of coils and capacitors that prevent high-frequency signals from going into or coming out of the receiver, thereby minimizing interference. In some receivers the ac output of the line filter feeds a power transformer. In other types

Figure 11-3 Essential features of the low-voltage power supply.

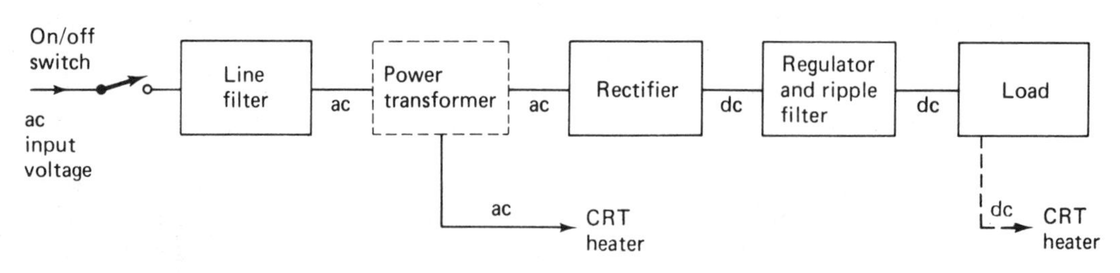

there is no transformer and the line filter feeds directly into the rectifier. The transformer is used to provide isolation. It also either steps up or steps down the line voltage as required. The rectifier conducts only in one direction and therefore is able to convert ac into dc. The dc output of the rectifier has a considerable ac component that must be removed before the dc can be used by the receiver. This is done by the filter portion of the power supply. This circuit consists of large electrolytic capacitors, coils, and in some cases active filters that use transistors. After the pulsating dc has been filtered, it is fed to the load. The load consists of the circuits that make up the rest of the receiver. In many modern television receivers the low-voltage power supply is regulated. This results in a power supply whose output voltage does not change under varying load or line voltage variations.

Picture tube heaters. All television receivers must use a picture tube. Picture tube heater voltages are usually 6.3 or 12 V. Therefore, all television receivers whether they are hybrid or completely solid-state must have some provision for energizing the heater. In solid-state receivers in which only the picture tube heater is to be supplied, a dc power supply is often used. This power source is obtained from the B supply or a separate dc supply is provided. Other picture tube heater sources that have been used are (1) separate 60-Hz step-down power transformers and (2) separate picture tube heater winding on the horizontal output transformer (flyback). See Figures 11-5 and 11-13.

11-5 basic B supply circuits

Television receiver circuits are powered by direct current obtained by rectifying the ac line voltage. The specific circuit used for rectification depends upon cost, number of receiver stages, degree of filtering required, regulation, voltage and current needs. As a result, television receivers incorporate a wide variety of rectification systems. In this section we will discuss the following types:

1. The half-wave rectifier
2. The full-wave rectifier
3. The bridge rectifier

11-6 the half-wave rectifier

The basic half-wave rectifier as shown in Figure 11-4 is a series circuit consisting of the ac source, the rectifier diode, and the load. The ac source provides a sine-wave voltage and in practice can be the ac line voltage or the secondary winding of a power transformer. In modern receivers the rectifier diode is invariably a silicon diode. Silicon diodes are characterized by very low forward resistance when they conduct.

In this basic circuit two load conditions are possible, depending upon the position of S_1. In each position the load represents the total power needs of the receiver and is represented by a resistor R_L. In one position of S_1 the load is purely resistive; in the other position the load resistor is shunted by a large capacitor.

The operation of the half-wave rectifier with a resistive load (S_1 open) is indicated by the waveform ladder diagram accompanying the circuit. The circuit and the waveforms are so arranged that the waveform is adjacent to the component across which it is found.

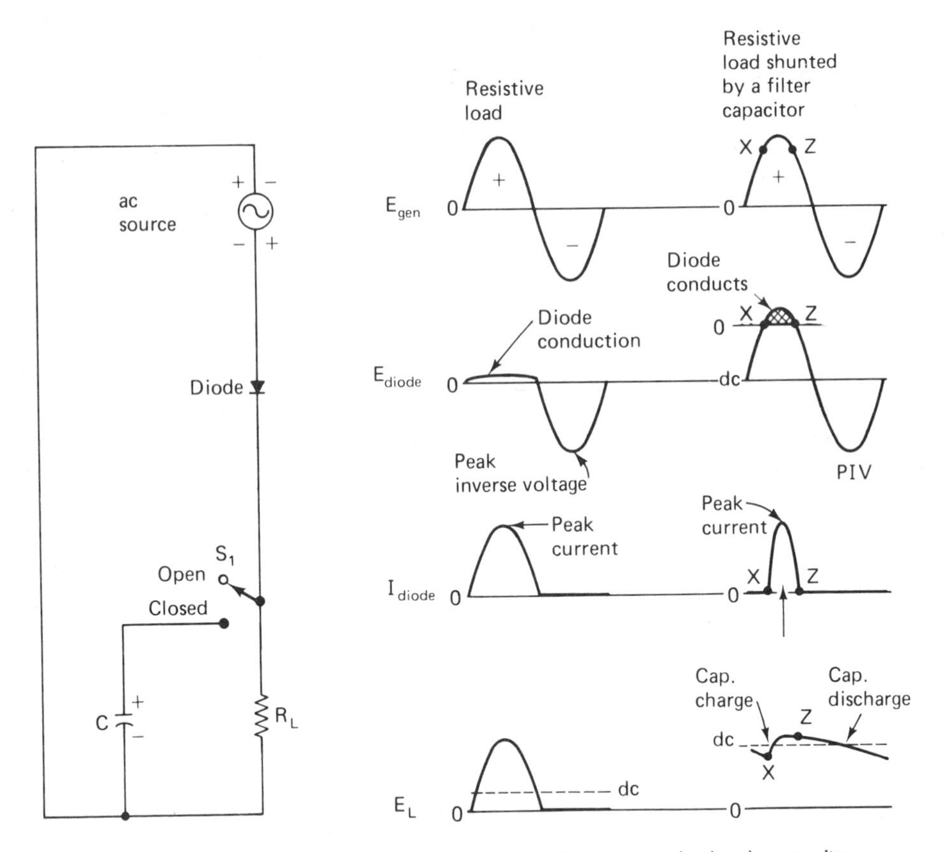

Figure 11-4 Basic half-wave rectifier and waveforms for a resistive load and a capacitor input filter load.

The topmost waveform is the applied ac signal. On the positive half-cycle the diode is turned on and current flows through the load. During this time the diode may be thought of as a short circuit, with the result that almost all of the input voltage is developed across the load resistor R_L. At the same time, the voltage drop across the diode is almost zero.

On the negative half-cycle the diode is reverse biased and is an open circuit. Therefore, no current flows through the load. During this time the voltage across the load is zero and all of the applied voltage is across the diode. It is because current only flows for one cycle of the input voltage that this circuit is called a half-wave rectifier.

Peak inverse voltage (PIV). The maximum voltage developed across the diode during the time it is reverse biased is called the *peak inverse voltage* (PIV). The PIV developed in this circuit for an input of 120 V rms is 169 V (120 × 1.414 = 169 V). The diode used in this circuit must have a PIV rating that is at least equal to this voltage. The PIV developed in a given rectifier circuit depends upon the peak input voltage and the type of load.

This simple rectifier circuit has a number of disadvantages. An ideal rectifier would be one that developed a pure dc output voltage with no ac component (ripple). In this circuit the dc voltage is very small, only 32% of the peak, and the 60-Hz ripple has a peak-to-peak swing equal to the peak of the applied voltage. This is clearly a long way from the desired ideal.

Capacitive load. If S_1 is closed, placing a large electrolytic capacitor (C) across the load (R_L), the operating conditions of the half-wave rectifier change. If the rectifier has been in operation for a considerable period of time, and if the capacitor has some charge across it at all times, the operation of the circuit is as follows.

The voltage across the rectifying diode is always the sum of the source voltage and the voltage across the capacitor (C). When the source voltage is zero, the voltage across the capacitor is such that it reverse biases the diode and prevents it from conducting. As the source voltage increases in the positive direction, it counteracts the voltage across the capacitor because they are series opposing to one another. When the two voltages are equal, the voltage across the diode is zero. This point is indicated on the ladder diagram as point X. When the source voltage increases above this point, the diode conducts and C charges to the peak of the applied voltage. Current flows for a short period of time and stops at point Z when the decreasing source voltage is less than or series aiding to the voltage across the capacitor. During this time the diode is reverse biased. The diode will remain reverse biased until the next peak positive alternation of the source. While the diode is nonconducting the capacitor is discharging through the load, maintaining a relatively constant flow of current. In short, when the diode is not conducting, the charged capacitor acts as the power supply.

The longer the time constant of C and R_L, the slower the discharge of C, and the nearer the dc voltage will be to the peak of the applied voltage. Similarly, the sawtooth-shaped peak-to-peak ripple voltage will be smaller. In practice, this means that the capacitor must be fairly large, 100 to 4500 μF. In semiconductor receivers or circuits having many stages the value of R_L may be very small (as little as 20 Ω), resulting in a rapid discharge of C. To counteract this, C is made very large, that is 1000 μF or more. The addition of a large capacitor across the load smoothes the excessive ripple of the resistive load to an acceptable level. This circuit is referred to as a *capacitor input power supply filter.*

Peak inverse voltage capacitive load. The voltage across the rectifier diode in a half-wave capacitor input rectifier system is the sum of the ac source voltage and the dc voltage across the capacitor. When the source voltage is positive going on the anode of the diode, the two voltages are series opposing. On the negative half-cycle the two voltages are series aiding and the total peak inverse voltage across the diode is the sum of the dc voltage across the capacitor and the ac source voltage. For a 120-V rms ac line voltage, the dc voltage across the capacitor may be as much as 150 V and the peak voltage of the ac signal may be as much as 169 V, resulting in a peak inverse voltage of approximately 319 V. This is approximately equal to the peak-to-peak voltage of the applied voltage. This means that replacement diodes used in this circuit must have PIV ratings of at least 350 V if they are to remain trouble-free. In addition, the replacement diodes must have a high enough peak current rating to accommodate the current demands of the circuit.

Typical half-wave rectifier circuit. An example of a half-wave rectifier circuit similar to those found in many TV receivers is shown in Figure 11-5. The ac input voltage is fed through a safety interlock (cheater cord) and an RF choke T_1 to the on/off switch. The safety

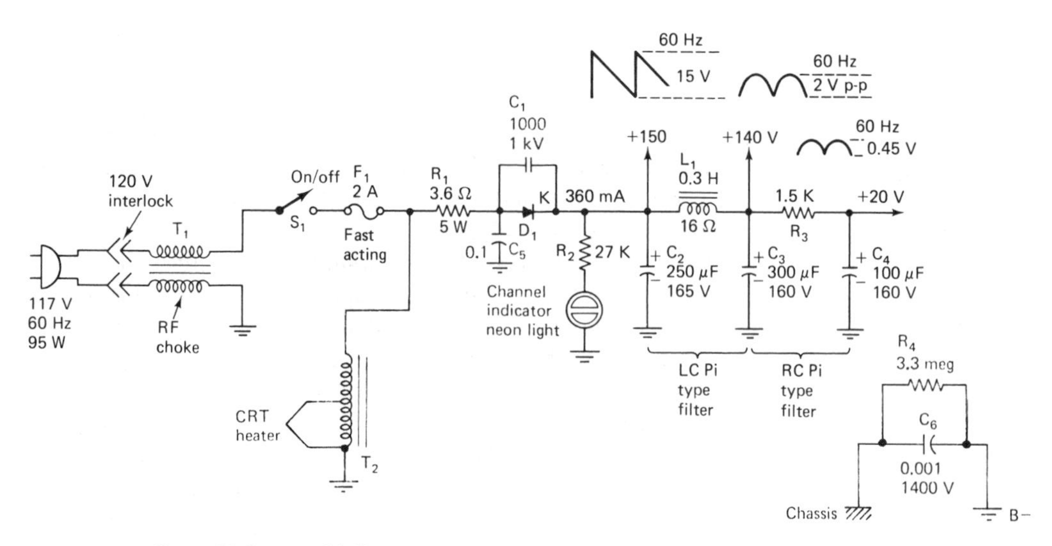

Figure 11-5 Typical half-wave rectifier used in a hybrid black & white television receiver.

interlock is part of the ac line cord and automatically disconnects the receiver from the ac power source if the back cover of the receiver is removed. The on/off switch is usually ganged to the volume control of the receiver. Once the switch is closed, power is applied to the CRT heater, autotransformer T_2 and the rectifier diode D_1 through a 2-A fast-acting fuse. In series with the diode is a fusible surge limiting resistor (R_1) of 3.6 Ω.

Surge resistor. When power is first applied to a half-wave rectifier that uses a capacitor input filter, a very high initial surge current (35 A or more) flows. This current lasts for a fraction of a cycle of the input source voltage and is due to the charging of the filter capacitor. Once the capacitor is charged, the current demands drop to their normal levels. The amount of surge current that flows depends upon the size of the capacitor. Large capacitors have higher surge currents than smaller capacitors. High surge currents place a stress on the rectifier diode. Rectifier diodes must have current ratings that can accommodate the surge current. Many receiver manufacturers limit the high surge currents by placing a fusible resistor in series with the diode. This surge resistor is typically a 5-W resistor of 3 to 5 Ω in value and is designed to act as a fuse if the continuous current should exceed the design ratings.

The diode (D_1) is connected into the circuit such that a positive polarity dc B supply voltage is provided. Reversing the diode reverses the polarity of the dc output voltage. If the diode is inadvertently reversed when it is replaced the polarized electrolytic capacitors will overheat, excessive current will flow, and one or both of the fuses will blow.

The capacitor (C_1) across the diode is used to minimize RF radiation that results from the sudden turn on and turn off of the diode.

Filter circuit. The large electrolytic capacitor C_2 develops the dc output voltage of 150 V and provides substantial filtering. At this point the ripple voltage is approximately 10% of the dc volts. Additional filtering is obtained by means of the ac voltage divider formed by L_1 and C_3. The combination of C_2, L_1 and C_3 forms an LC pi filter. The resistance of L_1 is usually small, approximately 16 Ω. As a result, little dc voltage is dropped across it, and the dc voltage across C_3 is almost the same as the voltage across C_2.

R_3 and C_4 combine to form another ac voltage divider. Because of the small reactance of C_4, most of the ac ripple voltage is dropped across R_3, leaving less than 1 V peak-to-peak ripple across C_4. At the same time, R_3 and the transistor load of the receiver act as a dc voltage divider. The dc current demands of the load determine the current flowing through R_3 and establishes the voltage drop across it. In this case, 120 V are dropped across R_3 leaving 20 V across the load. This voltage is intended for the transistor circuits in the receiver. The combination of C_3, R_3, and C_4 may be referred to as an *RC pi-type filter.*

The floating chassis. As a means of reducing shock hazard in this receiver the chassis is floating. R_4 and C_6 act as a high-frequency short circuit between B− and the chassis.

11-7 the full-wave rectifier

The half-wave rectifier has the disadvantage of producing a ripple frequency of 60 Hz. As a result, a considerable degree of filtering is required to minimize the effects of hum and picture pulling that may result from this ac component. A 60-Hz power supply ripple also means that the filter capacitors must act as the power source for the receiver for a considerable period of time (almost a full cycle) between charging pulses. Therefore, the average voltage (dc) is less than it might be if the charging periods occurred more frequently.

In contrast, the full-wave rectifier is a circuit that permits current to flow in the same direction through the load on both half-cycles of the ac input and therefore it produces a 120-Hz ripple. This has the advantage of higher output voltage and more easily filtered ripple.

An example of a typical full-wave rectifier circuit is shown in Figure 11-6. The full-wave rectifier is characterized by an isolating power transformer that has a center-tapped secondary winding. The secondary winding may either be step up or step down, depending upon the desired application. The purpose of the secondary winding tap is to provide two ac voltages that are equal to one another but opposite in phase with respect to the center tap. In effect, the center tap cuts the full secondary voltage in half. In this circuit the full secondary voltage is 120 V p-p, but each half of the secondary has only 60 V p-p.

Another characteristic feature of the full-wave rectifier is the use of two diodes (D_1 and D_2), one for each half of the secondary winding. The anodes of each diode are connected to the ends of the secondary and the cathodes are tied together. If these connections are reversed, the dc output polarity of the rectifier will also reverse.

For the phases of ac voltage indicated, D_1 will conduct and D_2 will be nonconducting on the positive half-cycle of the primary voltage, because the instantaneous polarity of the secondary makes the anode of D_1 positive and the anode of D_2 negative. The path of D_1 electron conduction current begins at the negative center tap terminal of the secondary. It then flows down through the fuse and ground (B−) and up into the filter capacitors (C_4 and C_5) and the load (R_L), charging the capacitors to the peak of the applied voltages. It then returns to the positive terminal of the secondary through the conducting diode D_1.

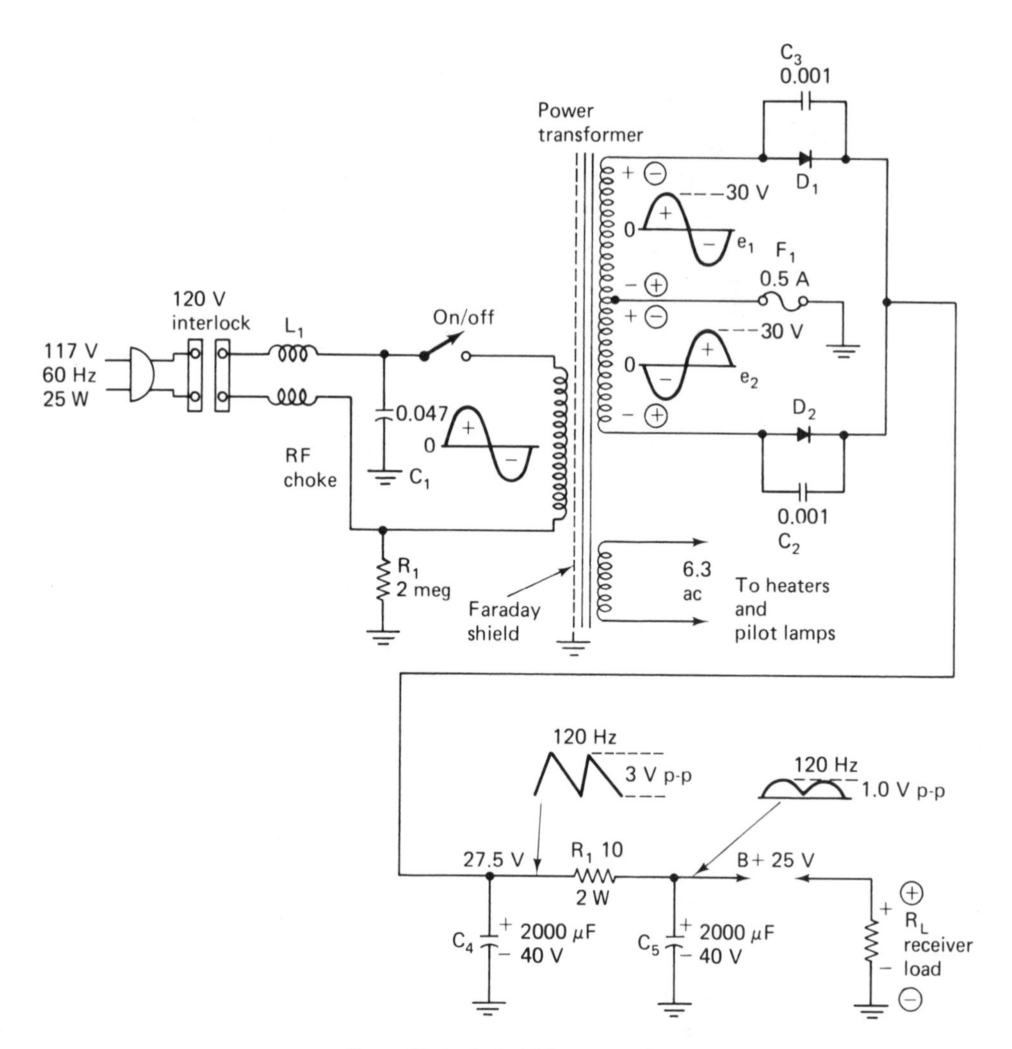

Figure 11-6 Typical full-wave rectifier circuit.

Peak inverse voltage. During the time that D_1 is conducting D_2 is reverse biased and is cut off. The total voltage across D_2 is the sum of the peak secondary voltage (e_2) and the dc voltage across C_4. This peak inverse voltage is approximately equal to the total peak secondary voltage. For this circuit, this is approximately 60 V. This means that replacement diodes must have the proper PIV ratings or they will fail.

On the negative half-cycle of the input ac voltage (indicated by polarity signs enclosed in circles) D_1 is reverse biased and D_2 is turned on. The path of current flow under these conditions is similar to that of the previous case. Current leaves the negative secondary center tap and flows through the fuse, B−, capacitors (C_4 and C_5), and the load, and then it returns to the positive terminal of the secondary through the conducting diode D_2. Since the direction of current flow is the same for both half-cycles of input, the capacitors charge in the same direction for both half-cycles. Since charging of the filter capacitors takes place twice for each full cycle of input voltage, the ripple frequency will be 120 Hz.

The PIV across D_1 is determined as before and will be found to be the same as for D_2. Therefore, the same replacement precautions must be followed.

RF filters. As was the case in the half-wave rectifier, radiation of high-frequency interference resulting from the switchlike operation of the diodes is supressed by means of capacitors (C_2 and C_3) shunting the diodes.

The primary circuit of the power transformer contains a number of important circuit elements. The RF choke L_1 in conjunction with the bypass capacitor C_1 prevent transient and other undesired RF signals from entering the receiver via the ac power line and also prevents any RF radiation from the receiver from getting into the power line. Radiation and interference problems are also minimized by using a *Faraday shield* placed between the primary and secondary windings of the transformer. In effect, this shield reduces the distributed capacitance between primary and secondary, thereby reducing high-frequency coupling between them.

A 2,000,000-Ω resistor (R_1) is connected from one side of the ac line to B− (ground). It may seem a bit strange that after all the expense and trouble of isolating B− from the ac line by using an isolation transformer, a resistor is placed between these two points. But this resistor is necessary to prevent a static build up of voltage on the chassis from causing an insulation breakdown between windings of the power transformer. The static dc voltage buildup is similar to the charging of a capacitor and is a result of the close proximity of the picture tube to the chassis and of electrostatic pickup from the antenna. R_1 discharges the "capacitor" by being placed across it. In this case, the capacitor's plates are the chassis and the primary winding of the transformer. The capacitor's dielectric is the insulation between the primary and secondary windings.

Ripple filters. The filter capacitors (C_4 and C_5) are very large low-voltage electrolytic capacitors that combine with R_1 to form a pi-type ripple filter. Because of the considerably heavier current drains found in transistor receivers, the filter capacitors found in these receivers will always be much larger than those found in vacuum tube sets. In both cases, the purpose of this filter is to reduce hum levels to the point where they do not have any effect on the raster, picture, or sound quality of the receiver.

The ripple wave shapes that accompany Figure 11-6 are those that would be found in the circuit provided no signal is applied to the receiver. Under signal conditions, some of the ac signal voltage is coupled into the power supply. When this voltage is viewed with an oscilloscope, it appears across the output filter capacitor C_5. If C_5 should open or decrease in value, the level of this coupled signal voltage would increase enormously. This could result in feedback between signal processing stages of the receiver and cause such trouble symptoms as motorboating, picture pulling, and sound bars in the picture.

11-8 the bridge rectifier

Another type of full-wave rectifier commonly used in solid-state and color television receivers is the bridge rectifier. This rectifier may use the ac line voltage directly, but it is usually characterized by an isolation transformer whose secondary is not center tapped and by four silicon rectifier diodes. An example of a bridge rectifier is shown in Figure 11-7.

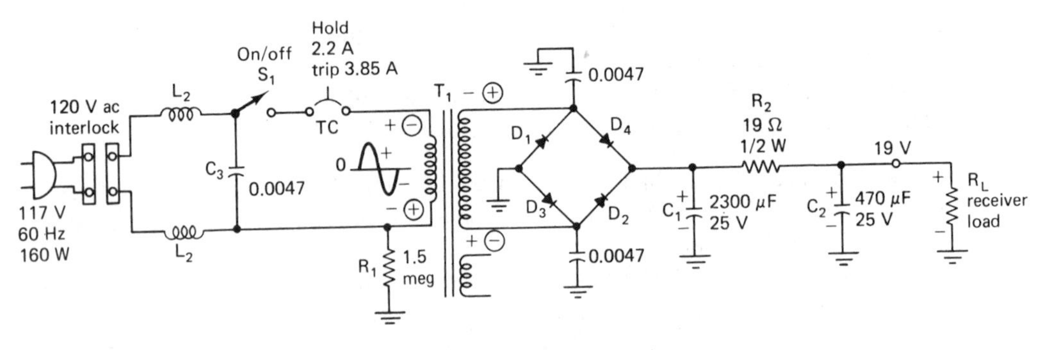

Figure 11-7 Typical bridge rectifier.

The operation of this full-wave rectifier may be outlined as follows. On the positive half-cycle of the input ac voltage the polarity of the secondary is such that only two diodes (D_1 and D_2) are turned on. If the diodes are considered to be short circuits when they conduct, it can be seen that the load (R_L) will in effect be placed directly across the secondary winding of the transformer. This will produce a voltage drop across R_L that is positive with respect to ground.

On the negative half-cycle of input voltage the instantaneous polarity of voltage across the secondary is such that D_1 and D_2 turn off and D_3 and D_4 turn on. Here again, neglecting the small voltage drops across the diodes, the load (R_L) is effectively placed across the secondary of the transformer. As a result, the voltage across the load is of the same polarity as during the first half-cycle of operation. Since current flows through the load on both half-cycles of the input ac voltage, the bridge rectifier is a full-wave rectifier. If one of the diodes should open, the circuit would become a half-wave rectifier and the ripple frequency would be 60 Hz, which would make filtering more difficult.

The filtered dc output of this rectifier is approximately equal to the peak of the full secondary voltage instead of half the full secondary voltage, as was the case in the previous full-wave rectifier. This means that for a given output voltage, the power transformer used with a bridge rectifier can have fewer turns and therefore can be smaller and less expensive. The peak inverse voltage across the diodes that are not conducting is approximately equal to the peak of the secondary voltage. The nonconducting diodes are effectively placed in shunt with the load.

Filters. The ripple filter consists of two large electrolytic capacitors C_1 and C_2 and a resistor R_2 arranged in a pi-type configuration. Tied between each end of the T_1 secondary and ground are two RF filter capacitors. These capacitors help minimize RF radiation produced by the rapid turn on and turn off of the diodes. Additional RF filtering is obtained by the input line filter L_2 and C_3 located in the primary of the transformer.

Overload protection. Overload protection in this circuit is obtained by means of a thermal overload circuit breaker (*TC*) placed in series with the on/off switch S_1. In some circuits it may be placed in the secondary winding of the power transformer.

11-9 regulator circuits

The ability of a power supply to maintain a constant output voltage despite changes in input voltage or output circuit conditions is referred to as the *voltage regulation* of the system. Power supplies that have a low effective internal resistance are said to have *good regulation*. The ideal television power supply should have an effective internal resistance of zero. In practice, the internal power supply resistance is considerably larger. It is determined by such things as the forward resistance of the rectifier diodes, the transformer winding resistance, the size of the filter capacitors, the filter resistance, whether the rectifier is full-wave or half-wave, and the frequency of operation.

Transistor television receivers operate at relatively low dc voltages (4 to 60 V) but relatively high currents (400 to 1200 mA). The high current demands of the receiver are responsible for the need for special regulator circuits that keep the output voltage constant. Regulation also protects the receiver transistors if the line voltage becomes too high or too low or is subject to surges. In addition, fast-acting regulators can be used to provide additional filtering for ripple reduction. Low-voltage regulation also prevents changes in B supply or ac line voltage from causing high-voltage output variations.

Four types of regulators that are commonly used in television receivers of all types are as follows:

1. Transformer regulator
2. Zener diode regulator
3. Transistor-operated regulator
4. Switch-type regulator

Transformer regulator. Constant-voltage power transformers (CVT) are used in many color television receivers to provide voltage regulation. These transformers can be designed to regulate the low-voltage power supply for both line voltage and load variations. In addition, the CVT is also able to suppress transient pulses which may appear on the ac line. If not suppressed, these short-duration high-amplitude pulses can result in apparently mysterious transistor and diode breakdown.

The CVT transformer is able to keep its output voltage constant by making the secondary winding a resonant circuit. In addition, the primary and secondary coils of the transformer are wound on a specially constructed core that allows the primary to operate normally while the secondary is operated in magnetic saturation. The secondary winding is forced into saturation by the large series resonant circuit current flowing through the winding.

A saturated transformer will produce a square-wave secondary output voltage. Changes in line voltage or load that are not sufficient to bring the transformer out of saturation will produce a constant output voltage. In practice, television receivers equipped with these transformers produce a good full-size picture with line voltages that have dropped to as little as half of their normal 120-V value.

An example of a circuit that uses a CVT is shown in Figure 11-8. Notice that the circuit appears to be exactly the same as any other transformer-operated full-wave rectifier except for a 1.5-μF ac electrolytic resonating capacitor (C221) placed across the secondary windings.

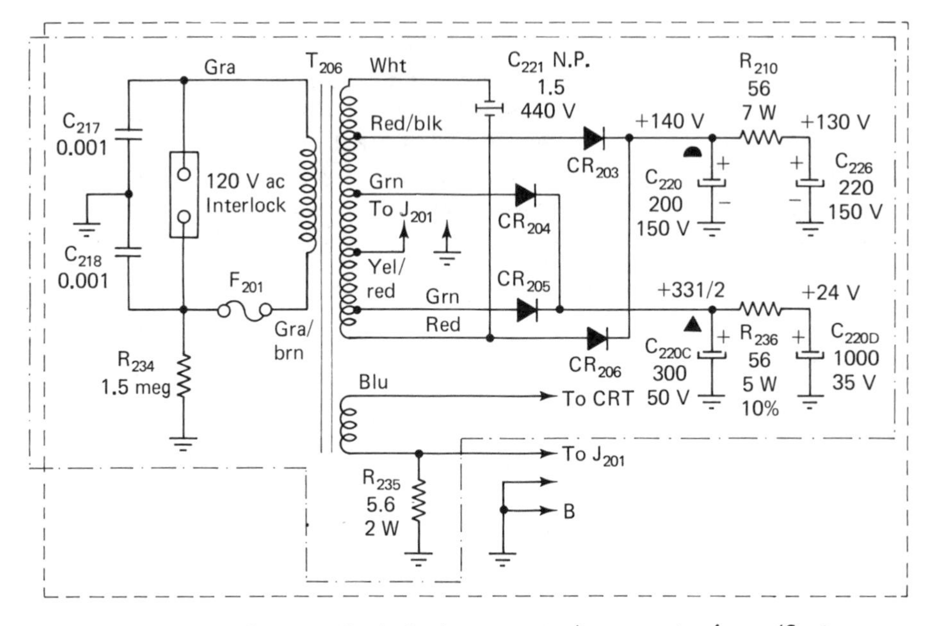

Figure 11-8 Full-wave rectifier circuit using a constant-voltage power transformer. (Courtesy of Zenith.)

The zener diode voltage regulator. The zener diode regulator is widely used and is a relatively simple shunt type voltage regulator. The zener diode is unusual in that, under normal operating conditions, it is reverse biased and conducting because the diode has exceeded the reverse bias breakdown voltage of the diode. The breakdown voltage of the diode is the output voltage of the regulator. Diodes are available over a wide range of breakdown voltages, making regulation possible at almost any desired voltage. The voltampere characteristics of the diode in the breakdown region are such that a large change in current through the diode will be accompanied by a very small change in the voltage across the diode. It is this feature that makes the zener diode useful for voltage regulator applications.

Typical circuit. The circuit shown in Figure 11-9 is an example of a typical zener diode voltage regulator. The unregulated output of the B supply is fed to the series combination of R_1 and the zener diode (D_1). In parallel with the diode is the load fed by the regulated power supply. Since the applied voltage of +40 V is considerably larger than the rated breakdown voltage of the zener diode (24 V), the diode will be forced into its breakdown region and reverse diode conduction will take place. Current (electrons) will flow from B− (ground) up through both the zener diode and the load resistance R_L. Their combined current will then flow through R_1 and complete the circuit by returning to the power supply. This current (91.6 mA) will produce a voltage drop of 16 V across R_1.

The manner by which voltage regulation is obtained is indicated by means of arrows drawn near the diagram. In fact, ripple voltage variations may also be smoothed out by the use of a voltage regulator to improve filtering action. As a result, smaller filter capacitors can be used than would be necessary if voltage regulation were not used.

In this circuit C_1 is used to provide ripple filtering and to provide a low ac impedance

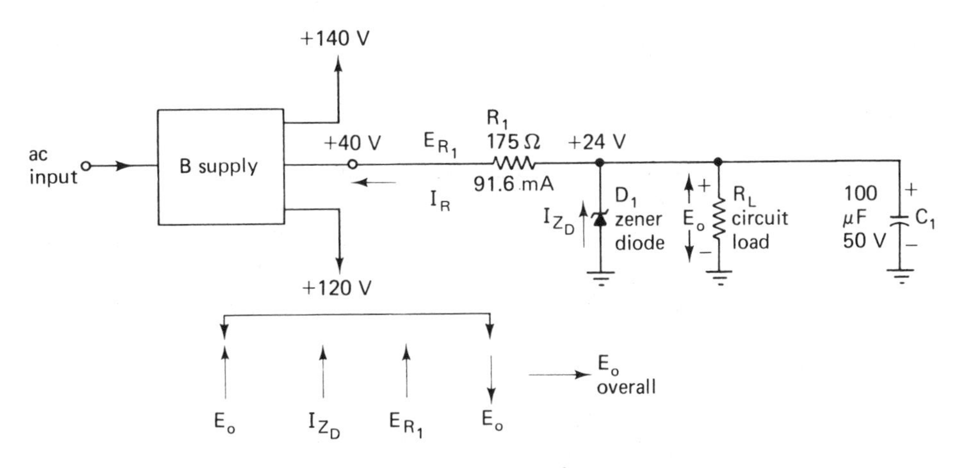

Figure 11-9 Zener diode regulator.

across the receiver load. This ensures that undesired signal feedback via the power supply will not occur between stages of the receiver. If this capacitor were not used, the receiver might experience oscillation, motorboating, or crosstalk between stages.

Transistor-operated regulated power supply. Very effective regulation of the dc output voltage of a power supply can be obtained by means of a regulator circuit similar to that shown in Figure 11-10. This circuit tends to maintain a constant output voltage despite changes in ac line voltage or changes in load conditions. Regulation in this circuit is obtained by placing a power transistor (Q_1) in series with the load. Because of the high current flow through Q_1, it is usually provided with a heat sink. Voltage regulation is accomplished by varying the effective resistance of this transistor. The voltage division between it and the load causes the output terminal voltage of the power supply to increase or decrease, thus counteracting any undesired change in output voltage. The effective resistance of Q_1 is controlled by an amplifier Q_2. This amplifier senses any change in output voltage by comparing it to a standard voltage set by a zener diode (D_5).

In this circuit the input ac line voltage is stepped down to approximately 14.5 V rms by the power transformer T_1. The ac voltage is rectified by the bridge rectifier consisting of D_1, D_2, D_3, and D_4, and it develops a dc voltage of 16.5 V across the 2000-μF filter capacitor C_1. The 120-Hz ripple content at this point is approximately 1 V p-p.

The regulator circuit consists basically of three circuits: the regulator transistor Q_1, the error amplifier Q_2, and the zener diode D_5. The conditions of the circuit are such that Q_1 is normally conducting. Because of the current flow through it, a drop of 5 V is developed across Q_1, leaving 11.5 V as the output B supply voltage. Zener diode D_5 and the emitter resistor R_1 of the error amplifier Q_2 are connected between B+ and ground. The zener diode is reverse biased (6.9 V) and is conducting. Therefore, the voltage across R_1 is 4.6 V ($11.5 - 6.9 = 4.6$). Since the voltage across the zener diode is constant, any variation in load output voltage appears across R_1. Also connected from B+ to ground is a voltage divider consisting of R_2, R_3, and R_4. R_3 is a potentiometer whose movable arm is connected to the base of Q_2. In this way, a percentage of the B+ output voltage variations is fed to the base of Q_2 while at the same time the emitter is made to vary directly with any output voltage variations. The difference of potential between these two voltages forms the bias for the error amplifier. The collector of the error amplifier (Q_2) is dc coupled via R_5 to the base of the regulator transistor.

Arrow diagram. Directly below the schematic drawing in Figure 11-10 arrows are used to outline the operation of the regulator. If an increase in output voltage (E_0) occurs as a result of a load change or an increase in ac line voltage, the base voltage of Q_2 (E_{BQ2}) will increase in a positive direction. The emitter voltage of Q_2 will increase by a greater amount. Therefore, the forward bias of this NPN transistor will decrease, causing a decrease in its collector current (I_{CQ2}). This in turn will cause the base and collector currents of Q_1 to decrease. In effect, the resistance of Q_1 (R_{eff} Q_1) has increased. Because of the voltage division between Q_1 and the load, the output voltage (E_0B+) decreases. Since the original voltage change was an increase in output voltage, and the end result of regulator action is to reduce the output voltage, the overall change in output voltage (E_0 overall) will be small enough to consider the output voltage constant.

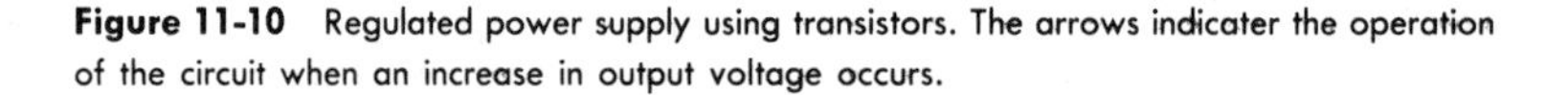

Figure 11-10 Regulated power supply using transistors. The arrows indicater the operation of the circuit when an increase in output voltage occurs.

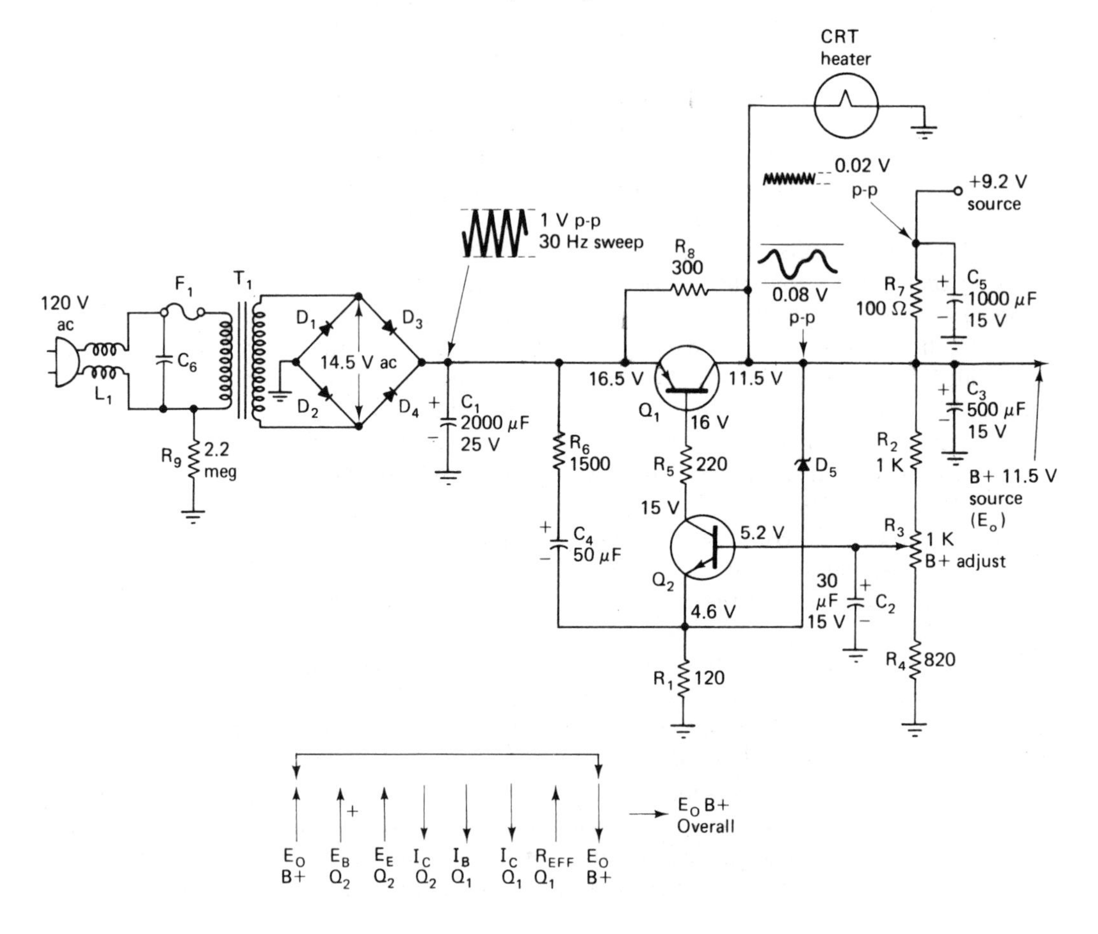

If the output voltage of the power supply decreases, the regulator will act in an opposite manner to that already discussed.

Output voltage. The output dc voltage of this supply is adjustable. This is accomplished by means of potentiometer R_3. Changing the position of the arm of R_3 causes a change in error amplifier bias, which in turn changes the effective resistance of Q_1, resulting in a change in B supply voltage.

Filtering. This type of regulator circuit is sometimes referred to as an *active filter* or as a *capacitor multiplier circuit* because the circuit is extremely effective in reducing the output ripple voltage. Typically, the output ripple voltage may only be 50 mV p-p. Ripple reduction is accomplished by feeding a portion of the high-level ripple voltage found across C_1 to the emitter of the error amplifier Q_2. R_6 and C_4 provide coupling between these two points. The error amplifier now acts as a common base amplifier for the ripple voltage and delivers an amplified ripple voltage to the base of the regulator (Q_1) that is *in phase* with the ripple voltage found on the emitter of Q_1. If these two voltages are equal, the net change in regulator (Q_1) bias will be zero and no 120-Hz ripple will appear in the collector circuit of the regulator. Additional filtering may be obtained in the conventional manner by means of resistor capacitor combinations such as R_7 and C_5.

CRT heater supply. The regulated dc output voltage of this circuit is also used as the heater voltage of the picture tube (CRT). In this way, a special heater winding on the power transformer is not necessary for the CRT heater.

Switch regulator. Silicon-controlled rectifiers (SCRs) are used in many television receivers to perform the function of low-voltage regulation. This type of regulator has three advantages:

1. Load current and line voltage variation have little or no effect on raster width or high voltage.
2. Switching regulators have high efficiency.
3. Low cost.

A simplified schematic diagram of an SCR switch-type regulator is shown in Figure 11-11. The circuit may be divided into four parts: (1) the bridge rectifier circuit, (2) the startup circuit, (3) the SCR gating circuit, and (4) the SCR regulator. Associated with these circuits are horizontal retrace pulses produced by windings on the horizontal output transformer (flyback transformer).

The bridge rectifier is powered directly by the ac line voltage and produces a dc output voltage of about 165 V. This pulsating dc voltage is filtered by a resonant pi-type filter formed by C_1, C_2, and C_3 in shunt with L_1. C_3 and L_1 form a parallel resonant circuit that provides increased filtering at the ripple frequency. The output of the filter feeds the anode of the SCR regulator through a winding (L_2) of the horizontal output transformer. In addition, the pi-filter output also feeds the startup circuit. This circuit is formed by R_1, D_1, D_2, and D_3 and delivers the unregulated bridge rectifier voltage to the gate of the SCR.

When the set is first turned on, the SCR does not conduct because it lacks a proper

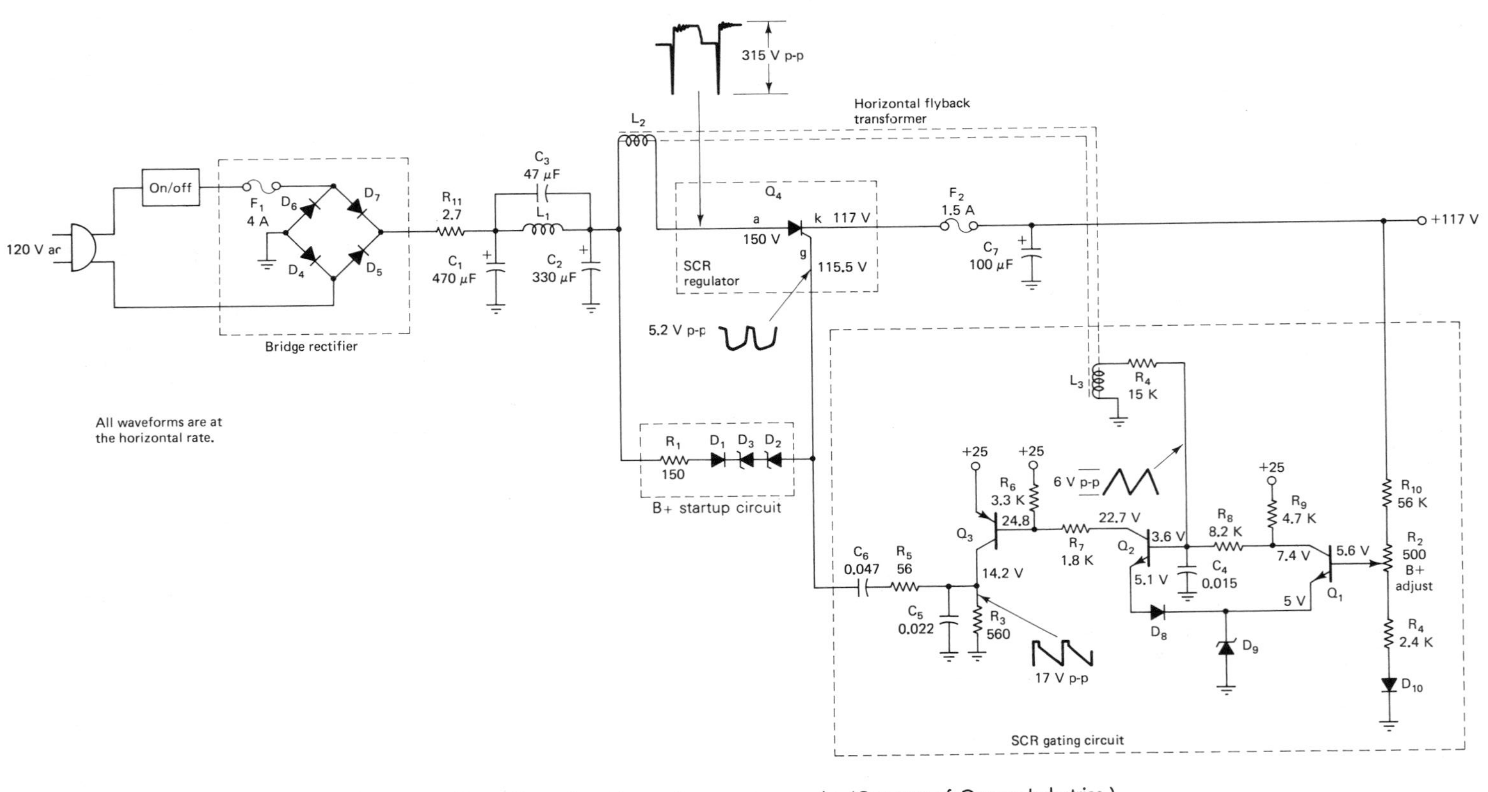

Figure 11-11 SCR-regulated low-voltage power supply. (Courtesy of Quasar Industries.)

gate-to-cathode voltage (gate more positive than the cathode). However, the startup circuit forward biases the gate-to-cathode SCR junction, allowing approximately 94 V to be applied to the horizontal oscillator, horizontal driver, and horizontal output stages, permitting them to begin operation. The horizontal circuit startup produces the proper gating voltage for the SCR, turning it on, and increases the B+ to 117 V. The startup circuit now becomes inactive.

An SCR, once turned on, will remain on unless it is turned off (reset) by momentarily reverse biasing the anode of the SCR. This reset operation is performed by a horizontal pulse (315 V p-p) that is applied to the anode of the SCR by L_2. To turn the SCR on again requires placing a positive-going pulse on the gate of the SCR. The width of the gating pulse determines the conduction time of the SCR. The longer the SCR conducts, the higher the output voltage, and vice versa. If the B+ supply voltage should fall, the gating circuit must form a gating pulse whose positive width has increased in order to counteract the fall in B+ voltage. This is done by the SCR gating circuit, consisting of Q_1, Q_2, and Q_3.

In the gating circuit, a horizontal retrace pulse obtained from the horizontal flyback winding L_3 is passed through an integrator consisting of R_4 and C_4. The resulting sawtooth waveform is coupled to the base of Q_2. The dc bias for Q_2 is established by the conduction of Q_1. The level of Q_1 conduction is determined by the setting of the B+ adjustment R_2, which in turn sets the B+ power supply output level to 117 V.

Transistor Q_2 is driven from cutoff to saturation by its input sawtooth waveform. The period of time Q_2 is in conduction is determined by the amplitude of the horizontally derived sawtooth waveform and the dc bias applied by Q_1.

The resultant waveform at the collector of Q_2 is inverted by PNP transistor Q_3, shaped by R_3, R_5, and C_5, and coupled by C_6 to the gate of the SCR regulator Q_4. This positive-going pulse determines the length of time that the SCR is on.

If the amplitude of the horizontal retrace pulse (high voltage) or B+ varies in amplitude, the conduction time of Q_2 is changed. Thus, the width of the gating pulse applied to the gate of the SCR is varied to turn the SCR on for the period of time required to maintain the +117 V B+. Longer conduction time tends to increase the B+ voltage, and vice versa.

11-10 high-frequency power supplies

Power transformers that are designed to operate at 60 Hz use large, relatively expensive laminated iron cores. Laminated cores are used to minimize power losses caused by eddy currents induced into the core. Hysteresis loss is determined by the type of silicon steel used in the core. The lower these losses, the less power is wasted in heating the transformer. Both of these losses are frequency sensitive and become greater at high frequencies.

The amount of wire used in winding the primary and secondary of the transformer depends upon the *frequency* of operation. If the inductive reactance of the primary is approximately 50 Ω, the ac current drawn from the power supply will be 2.4 A. If the ac line frequency is increased, the primary reactance ($X_L = 2\pi FL$) of the power transformer can be kept at 50 Ω by *reducing* the inductance of the primary in a manner inversely proportional to the increase in frequency. This is done by either reducing the number of turns in the primary or reducing the amount of iron in the core. Of course, if the number of turns in the primary is reduced, the number of turns in the secondary windings can also be reduced without changing the turns ratio. If both the number of turns and the core dimensions of a transformer are reduced, a considerably smaller power transformer can be used. Unfortunately, core losses increase sharply at higher frequencies. By using ferrite core materials, however, eddy current and hysteresis losses can be kept to tolerable levels.

Block diagrams. As a result of the advantages obtained by high-frequency operation, many television receiver manufacturers have introduced high-frequency, low-voltage power supplies. In general, there are two types of circuits being used. These are shown in block diagram form in Figure 11-12. One type, called a scan-derived power supply, makes use of the existing horizontal deflection output circuit, and the other is designed around a special self-contained oscillator. This oscillator is synchronized to the horizontal scanning rate.

In Figure 11-12(a) the horizontal output stage is designed to have an output flyback transformer that replaces the power transformer and provides four ac pulse outputs. These are rectified by suitable half-wave rectifier diodes providing 20,000 V for use as the accelerating voltage for the picture tube; 145 V, 27.5 V, and 17.5 V dc are for the other stages on the receiver. Detailed discussion of this type of low-voltage power supply will be found in the chapter devoted to horizontal deflection circuits.

The block diagram for the second type of high-frequency power supply circuit is shown in Figure 11-12(b). In this circuit an oscillator is synchronized to the horizontal line frequency and drives a switching transistor that activates a power transformer. This in turn produces seven ac outputs. These outputs are rectified by suitable diodes and produce six dc output voltages ranging from 6 to 200 V. The remaining 6.3-V ac voltage is used for pilot lamps.

Typical circuit. The circuit for a separate type of power supply is shown in Figure 11-13. The oscillator (Q_6) is powered by a full-wave voltage doubler that is supplied from the ac line and provides dc voltages of +280 V, +50 V, and +33 V. These voltages are not connected directly to the receiver circuits. The schematic shows that the chassis and the ac input to the power supply are isolated from one another. As a result, shock hazard is minimized. The heater of the CRT is supplied by a small filament transformer (T_{801}) that prewarms the heaters when the receiver is off. When the receiver is switched on, the filament transformer is opened and the CRT heater is supplied by the regulated 6-V dc output of the power supply.

The oscillator (Q_6) is synchronized with the horizontal deflection oscillator and feeds the shaper (Q_3), and the driver (Q_2) stages of the power supply. These stages deliver a square-wave voltage at the horizontal line frequency to the switching transistor (Q_8). During the positive half cycle of input voltage the switching transistor (Q_8) is turned on. This allows a large current to flow through the primary winding of the power transformer (T_3), inducing voltages into the many secondary windings. On the negative half cycle of input voltage the switching transistor is turned off and primary current stops flowing. The collapsing magnetic field accompanying the reduction in current induces a voltage of opposite polarity into the secondary windings. The use of a switching transistor improves circuit efficiency. This is because the transistor is either on or off, thus transistor power dissipation is at an absolute minimum.

The ac voltage induced into each secondary winding of T_3 is rectified by a diode

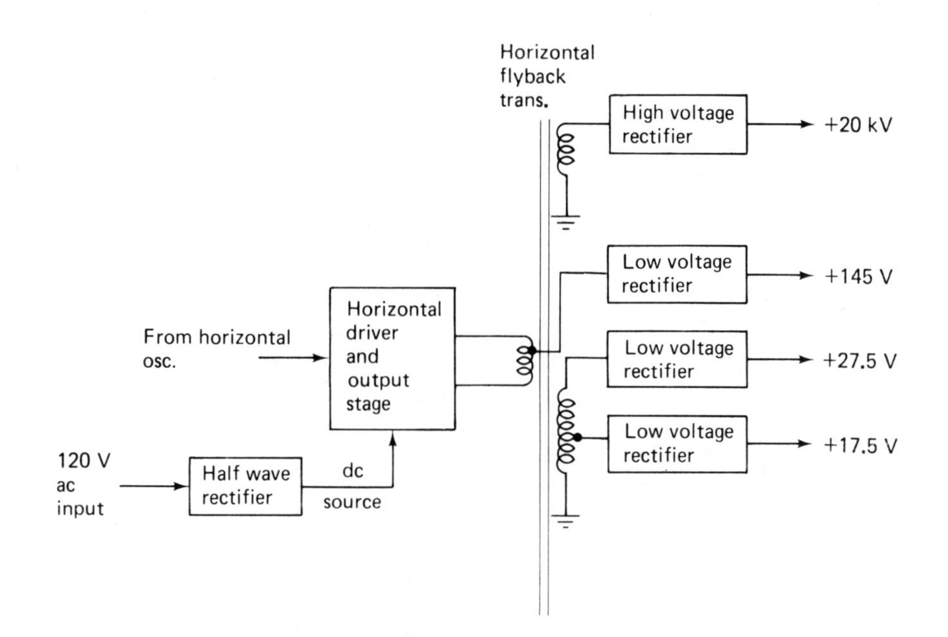

(a)

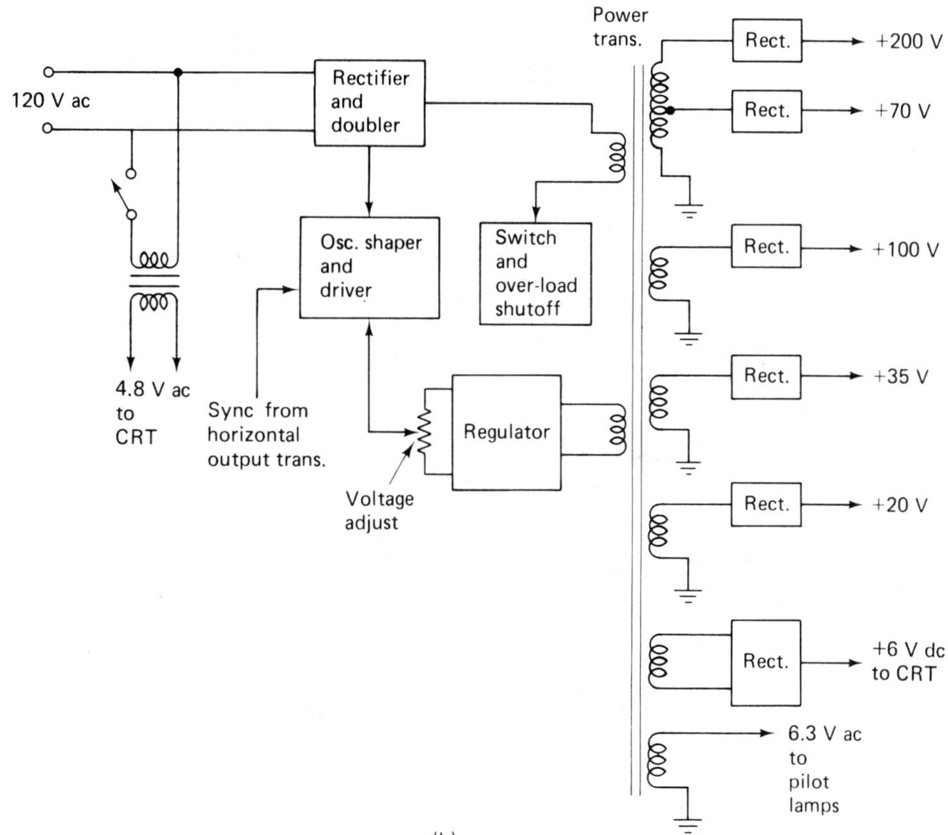

(b)

Figure 11-12 (a) Circuit in which dc power is obtained from the horizontal output circuit; (b) high-frequency ac generator that is synchronized to the horizontal line frequency and provides six dc outputs and one ac output.

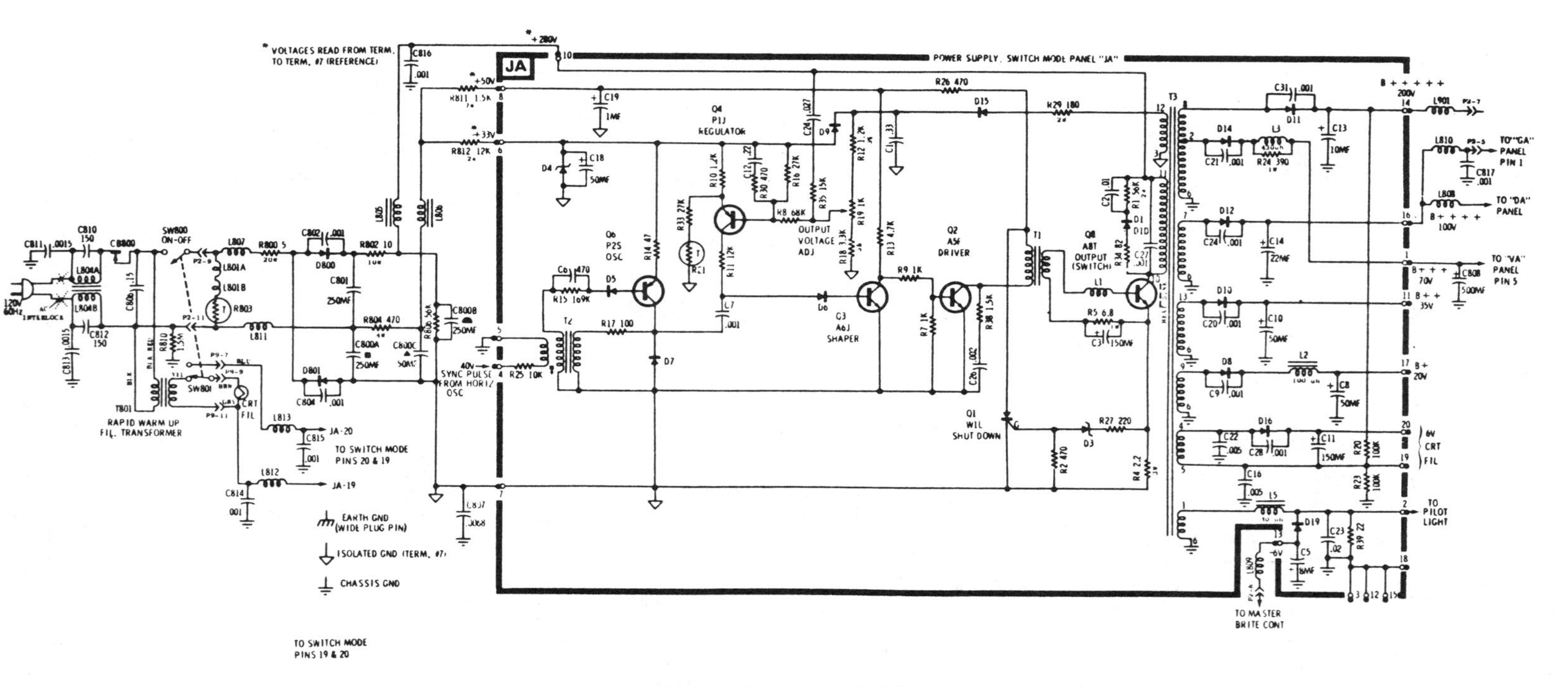

Figure 11-13 Schematic diagram of a regulated high-frequency power supply.

arranged in a half-wave rectifier configuration. Each diode is shunted by a small capacitor to reduce RF radiation. These rectifier circuits also contain RF bypass capacitors (C_{817}, C_{23}) and small RF chokes (L_{901}, L_3, L_{810}, L_{808}, L_2, and L_5) that are used to prevent undesired RF radiation. The output of each rectifier is filtered by means of electrolytic capacitors (C_{13}, C_{14}, C_{808}, C_{10}, C_8, C_{11}, and C_5). These capacitors are smaller than they would have been if the circuits were operated at 60 Hz.

This power supply circuit is also regulated. Regulation of the ac voltage output of the power transformer is accomplished by Q_4. This transistor is in series with the base of Q_3 and is able to regulate the power supply output because the collector current of Q_4 is the base current of the shaper transistor Q_3. This means that if the conduction of Q_4 should decrease, the conduction of Q_3 must also decrease. However, since Q_3 is driven by the oscillator in a switchlike manner, being either on or off, the reduction of the Q_3 shaper collector current results in a reduction of pulse width. If regulator (Q_4) collector current should increase, the pulse width of the collector current of the shaper transistor (Q_3) would also increase. Thus, the pulse width feeding both the driver (Q_2) and the output switch (Q_8) would vary, depending upon the conduction of the regulator. A narrow pulse width driving the output transformer would mean less energy and therefore an output voltage reduction. A wide pulse width would mean more energy available and the output voltages of the transformer would increase.

The conduction of the regulator transistor (Q_4) is determined by its base to emitter bias. The emitter voltage is obtained from the +33-V dc source that is regulated by a zener diode (D_4). The base voltage is obtained from a secondary winding on the power transformer. This winding provides a sample of the ac output voltage that is rectified by diode D_{15} producing a dc voltage across C_1. This voltage is fed to the base of the regulator transistor via the output voltage adjustment potentiometer (R_{19}).

Summary. Regulator action may be summarized as follows. If for some reason the ac output voltage were to increase, the dc voltage across C_1 would increase, causing a decrease in regulator conduction. This in turn would reduce the conduction time of the shaper Q_3 and would force the switching transistor to conduct for a shorter period of time. As a result, the ac output voltage would tend to decrease, keeping the output voltage constant. Of course, the opposite would occur if the ac output voltage were to decrease.

Overload protection. The power supply and especially the relatively expensive switching transistor are protected by an SCR (silicon-controlled rectifier) shutdown circuit. If for some reason the switching transistor were overloaded, its emitter current would increase and cause an increase in voltage drop across R_4. This would reverse bias zener diode D_3 to the point of breakdown. If this were to occur, a positive pulse would develop across R_2 and be applied to the gate of the SCR, thus turning it on. The conduction of the SCR would short out the dc supply for the driver transistor (Q_2). Loss of the driver pulses would reduce the switching transistor bias to zero, turning it off.

The characteristic of the SCR is such that once it conducts, it remains conducting until the anode voltage is removed. Therefore, once the shutdown circuit has been activated, it will remain activated until the receiver is turned off and then on again. If the overload were temporary, the receiver would again operate normally; if not, the shutdown circuit would activate once more, indicating a circuit defect.

11-11 scan-derived low-voltage power supplies

Low-voltage power supplies which are energized by pulses obtained from the horizontal output flyback transformer are commonly used in modern television receivers. The main reason for using such scan-derived power sources is increased efficiency resulting in less power consumption. Scan-derived power supplies may be divided into four basic parts.

First, the ac line voltage is rectified and converted into an unregulated B supply of about 150 V. This rectifier may use an isolation transformer and therefore have an isolated "cold" ground, or if no isolation transformer is used the power supply will have a nonisolated "hot" ground. In addition, grounds associated with the isolated horizontal output scan derived B supply sources are often called "cold" grounds. Those grounds associated with nonisolated ac line voltage power supplies are termed "hot." Since this circuit is unregulated, it cannot be used to directly power other circuits in the receiver.

Second, the output of the unregulated power source is fed to a dc regulator. This regulator may use a pass-transistor and a zener diode. Other circuits may use an SCR regulator. In both cases their purpose is to keep the B supply feeding the *horizontal output amplifier* constant, thereby providing good high-voltage regulation.

Third, the flyback transformer provides the high voltage needed to operate the picture tube, the pulses for the AGC and other circuits, the picture tube heater voltage, and it also generates all of the dc voltages for all the other circuits in the receiver. These dc sources may have their own regulation, although the regulator supplying the horizontal output does a good job of regulating these supplies, too.

Fourth, as indicated above the horizontal oscillator obtains its dc power from the horizontal output stage. However, it would appear that an impossible situation has arisen, since the horizontal output circuit cannot supply power without it being driven by an energized horizontal oscillator.

Kick- and trickle-start circuits. To overcome the problem of having the horizontal output amplifier supplied with B supply power and the horizontal oscillator lacking its B supply voltage, a starting voltage for the horizontal oscillator is required. This is done in one of two ways: a kick-start or a trickle-start circuit is used.

The kick-start circuit used in many RCA chassis applies a small amount of voltage to the horizontal oscillator for a few seconds after the receiver has been turned on. See Figure 11-14(a). In this circuit the primary of a start transformer (T_1) is connected in series with a large capacitor C_1. When the receiver is first turned on, C_1 charges. During the short period when C_1 is charging the charging pulses energize the transformer (T_1) and deliver ac pulses to diode D_1. This diode rectifies the pulses and develops a dc voltage that powers the horizontal oscillator starting the system. The dc voltage developed by the horizontal scan-derived power source passes through D_2 and reverse biases D_1, turning it off. If the horizontal oscillator has not started by the time the capacitor is fully charged, the receiver will remain inoperative.

In the trickle-start system, a large-valued resistor, a diode, or a pass-type transistor is connected between the unregulated B+ and the horizontal oscillator. The resistor will not supply enough current to operate the receiver but will allow the oscillator to operate. The flyback transformer output supplies power to the horizontal oscillator as soon as the horizontal oscillator has started, closing the power loop.

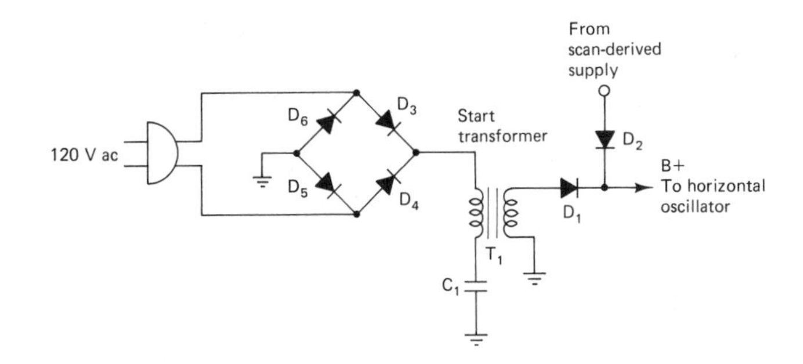

(a)

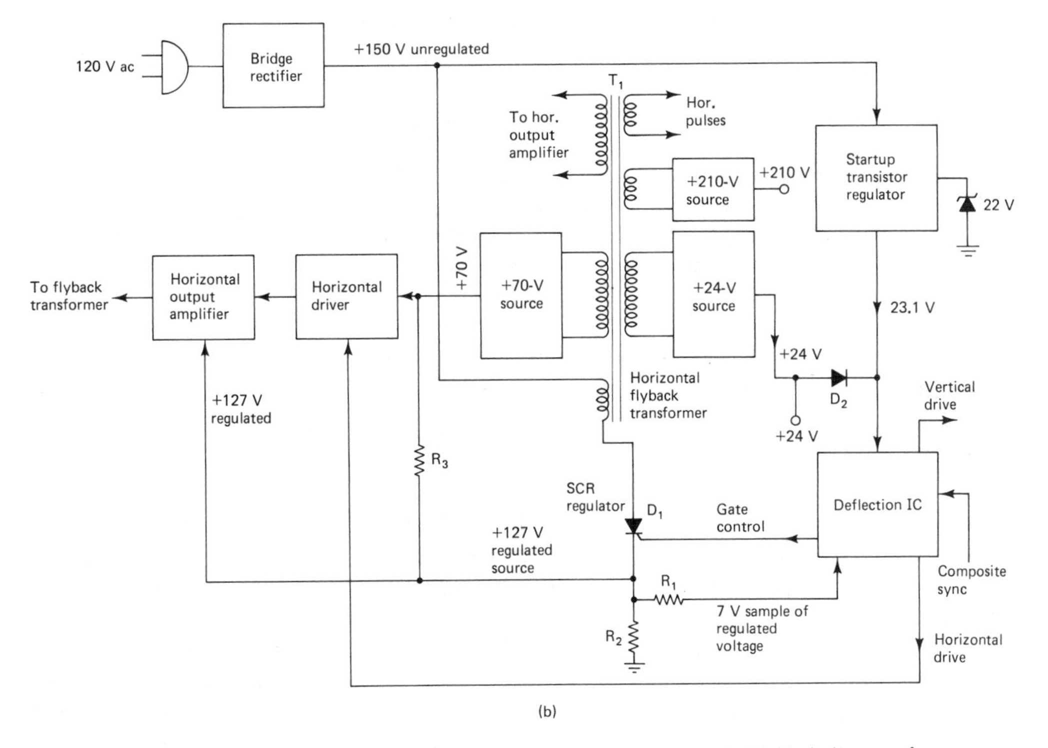

(b)

Figure 11-14 (a) Simplified schematic diagram of a kick-start circuit; (b) block diagram of an RCA low-voltage power supply.

The block diagram of a recent RCA television receiver using a scan-derived power supply is shown in Figure 11-14(b). In this circuit a bridge rectifier converts 120 V ac into +150 V of unregulated dc. This voltage feeds the anode of an SCR (D_1) and a startup pass-type regulator. This voltage is used to energize the deflection integrated circuit (IC) chip. The deflection IC produces a pulse that drives the horizontal driver stage as well as a pulse that activates the gate of the SCR, turning it on. Turning the SCR on develops the +127 V regulated output voltage, which is then used to power the collector circuits of both the horizontal driver and output stages. R_1 and R_2 provide a sample of the regulated 127 V supply, which is then processed by the deflection IC and determines the pulse width of the gate control voltage. If for any reason the regulated voltage should decrease, the gate pulse width would increase, allowing the SCR to conduct longer, forcing the output to increase, compensating for the original decrease in voltage.

Once the scan-derived voltages are generated, D_2 conducts, providing the 23.1 V necessary for proper deflection IC operation. Also, the 70-V source in conjunction with R_3 overrides the 127-V source and fixes the horizontal driver collector source voltage at 70 V.

11-12 troubleshooting

Television receivers, whether they are color or black & white, exhibit similar trouble symptoms when the power supply becomes defective. Defects may be classified as those that are caused by a reduction or loss of B supply voltages or those that are the result of poor filtering. Reduction or loss of B supply voltages may result in the following symptoms:

1. Loss of raster; no sound or weak sound [see Figure 11-15(a)]
2. Dim raster; hum in sound
3. Shrunken raster; hum in sound [see Figure 11-15(b)]
4. Poor picture linearity (bent and stretched); hum in sound
5. Overheating of the transformer or some other component in the power supply accompanied by one of the above symptoms
6. Repeated fuse or circuit breaker failure

Loss or reduction in B supply voltage can be caused by open, leaky, shorted, or changed value components. Defects that do not cause overheating or open protective devices are usually caused by an open circuit or component. Defects that result in overheating or blown fuses are caused by shorts or excessive component leakage.

Loss of raster; no sound. Loss of raster and sound [Figure 11-15(a)] may be the result of a loss of power caused by a break in a printed circuit board or some open component. In Figure 11-5 these symptoms may be caused by such defective components as the on/off switch, fuse F_1, R_1, D_1, and L_1. The quickest way to isolate the fault is by means of a VTVM. Starting with dc measurements at the cathode of the diode and then using ac voltage measurements at the anode work back toward the ac plug. The open is isolated at the point where a voltage is first measured. For example, if R_1 were open, an ac voltage would not be measured at the anode of the diode but it would be measured at the junction of the fuse and the CRT heater.

Loss of raster; weak sound. Loss of raster but not of sound may be the result of defects in the filter network, for example, a resistor that has increased in value (R_3 in Figure 11-5). This symptom may also be caused by a leaky or open filter capacitor (C_2 in Figure 11-5).

Dim, shrunken, and poor linearity rasters. Dim raster, shrunken raster, and poor picture linearity, accompanied by a hum in sound, [Figure 11-15(b)] can all be the result of a reduction in B supply voltage. They can be caused by such things as defective rectifier diodes, leaky or open input filter capacitors, defective regulator transistors, and defective zener diodes.

(a) (b)

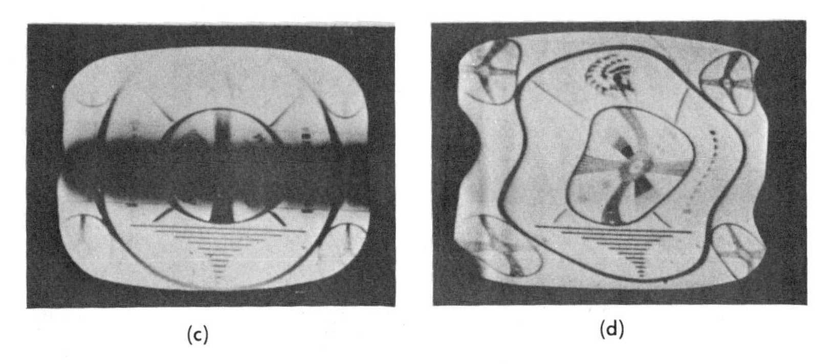

(c) (d)

Figure 11-15 (a) Blank raster, no sound; (b) shrunken raster, hum in sound; (c) hum bars in picture, hum in sound; (d) picture and raster pulling. (Courtesy of General Electric.)

Overheating. Overheating of power transformers or other rectifier circuit components, or in the extreme case repeated fuse or circuit breaker failure, is an indication of excessive current demand. Current overloads in a circuit, such as that of Figure 11-7, can be caused by shorted turns in the power transformers (T_1) or a short circuit in one of the loads fed by the power supply.

Isolation of a component that causes repeated fuse failure is made somewhat difficult because it is impossible to make voltage measurements. One way to overcome this problem is to place a 25-W lamp in series with one of the input power leads of the defective receiver. Under the short-circuit conditions, the lamp will glow brightly. To find the defective component, disconnect one load at a time from the power supply and note its effect on the lamp. If the lamp glow decreases markedly after a load has been removed, the possibility is that the short will be found in that load.

Loads may be disconnected in printed circuit wiring by slicing the printed circuit foil with a razor blade. Once the defective branch of the load has been found, an ohmmeter may be used to isolate the defective component. The same method may be used if the short circuit is in the rectifier circuit proper. Repair of the circuit board cut is accomplished by soldering the break closed.

Precautions. A number of precautions should always be observed when servicing television receivers. When working with transformerless power supplies, such as that of Figures 11-5 and 11-11 an isolation transformer should be used to avoid shock hazard.

In many television receivers the chassis and B− are not identical. An example of this type of receiver is shown in Figure 11-5. In this receiver, if the negative lead of the VTVM is placed on chassis ground, erroneous voltage readings will result. To avoid this difficulty, refer to the schematic diagram notes that will tell between which two points in the receiver voltages are measured.

Poor filtering. Poor filtering can result in large B supply ripple content or crosstalk between receiver stages via the common impedance of the power supply. These conditions can result in such trouble symptoms as the following:

1. Hum bars in picture and hum in sound [see Figure 11-15(c)]
2. Picture and raster pulling and garbled sound [see Figure 11-15(d)]
3. No raster and no sound
4. Dim raster and hum in sound
5. Shrunken raster and hum in sound
6. Pulsating or erratic picture and motorboating sound

These defects are usually caused by such things as open or leaky filter capacitors or defects in rectifier circuits that cause the ripple frequency to shift from 120 Hz to 60 Hz. This condition could, for example, be caused by an open diode in a bridge rectifier.

Excessive ripple and low voltage may also be caused by excessive current demands placed on the power supply by the other circuits of the receiver. This defect can often be checked by inserting milliammeter in series with each load. To aid in making this type of measurement many manufacturers provide the normal current values in their schematic diagrams.

In general, excessive ripple and crosstalk between stages can be observed across the output filter capacitor with an oscilloscope. Here again, reference to the waveforms associated with the receiver's schematic diagram will allow the service technician to judge whether or not the peak to peak amplitude is excessive.

Open filter capacitors can be quickly checked by jumping them with known good capacitors. If receiver operation returns to normal, the circuit filter capacitor is bad. A leaky capacitor can only be checked by removing it from the circuit and then substituting a known good one.

Defective diodes can be checked by an ohmmeter. A normal diode will read a high resistance with one polarity of the ohmmeter; reversing the leads of the ohmmeter should cause a low reading. If the readings are both high or low, the diode is defective.

Hum bars and picture pulling. Such symptoms as hum bars in the picture, garbled sound, and picture pulling [Figure 11-15(c) and (d)] may be the result of excessive (60 Hz or 120 Hz) ripple that is affecting the video amplifiers, audio amplifiers, and sync circuits, respectively. Hum in the sound that cannot be altered by adjusting the vertical hold control (indicating that vertical sawtooth fluctuations are not getting into the sound amplifiers) is an indication that excessive ripple voltage is being injected into the audio amplifiers.

Loss of raster. A reduction in B supply voltage can cause the loss of raster. This may be caused by the failure of the horizontal oscillator or by severe reduction in second anode potential. If the second anode voltage does not drop too low, a dim blooming raster will result.

Shrunken raster. Low B supply voltages can be the cause of insufficient vertical and horizontal deflection because both vertical and horizontal output stages require considerable power to provide deflection. If the B supply cannot provide the energy requirements of these deflection stages, insufficient deflection results and the raster will be reduced in size.

Low B supply voltages can be caused by an open input filter capacitor (C_1 in Figure 11-10), defective diodes, defective regulator transistor, or any other associated regulator circuits that might cause the effective regulator resistance to increase (base-to-emitter voltage decreasing toward zero).

Motorboating and oscillation. Open output filter capacitors (C_5 in Figure 11-10) can be responsible for crosstalk between stages and possible oscillation because the large discharging current requirements of output stages cause considerable voltage fluctuation across the filter capacitor. These voltage fluctuations are at the frequency rate of the signals passing through the amplifier stages. Since the output of the power supply feeds all the other low-level stages of the receiver, some of this signal voltage developed in the power supply will be fed into these stages. If the decoupling filters associated with the stages are not adequate to remove this undesired feedback voltage, oscillation may occur.

QUESTIONS

1. What is meant by shock hazard? How can it occur?
2. Explain how an isolation transformer can minimize shock hazard.
3. What is the purpose of returning the chassis to the ac line by means of a bypass capacitor in those circuits where the chassis is supposed to be floating?
4. Explain how a shock hazard test circuit is used.
5. List three methods for avoiding shock hazard.
6. A fuse is rated at 2 A 250 V. What does the 250-V rating refer to?
7. In a half-wave rectifier using a capacitor input filter, the diode conducts for a short interval once each cycle. Explain why this happens.
8. What is the surge current? Why is it so large?
9. Explain why the peak inverse voltage in a half-wave rectifier using a capacitor input filter is equal to the peak-to-peak voltage of the applied.
10. What would be the likely effect on the circuit and what would the trouble symptoms be if the following components in Figure 11-5 opened? (a)C_2, (b)D_1, (c)C_5, (d)R_1, (e)F_1.
11. What would be the effect on the circuit if the following components in Figure 11-5 shorted? (a)C_2 (b)L_1, (c)D_1, (d)C_4, (e)C_5.
12. What is the purpose of L_1, C_1, C_2, and C_3 in Figure 11-6?
13. What would be the effect on circuit operation of the bridge rectifier of Figure 11-7 if D_1 opened?
14. Explain the functions of R_1 in Figure 11-7.

15. What are the functions of R_4 and C_6 in Figure 11-5?
16. Explain what would happen if the B+ adjust (R_3) in Figure 11-10 were adjusted so that the arm of the potentiometer was moved to the bottom of the control.
17. Explain how the circuit of Figure 11-10 reduces the ripple content of the circuit.
18. What would be the effect on the regulator circuit operation of Figure 11-10 if the zener diode D_5 were to open?
19. What are two reasons for operating power transformers at frequencies of 15,750 Hz?
20. Explain how a constant-voltage power transformer works.
21. What kind of waveform appears across the secondary winding of a CVT?
22. What are the advantages of a switch-type regulator?
23. What is the purpose of the "startup" circuit in Figure 11-11?
24. Explain how the kick-start circuit of Figure 11-14(a) works.
25. What is the effect upon the gate control pulse width if the regulated voltage of Figure 11-14(b) were to increase?

chapter 12

Sync Separators

12-1 introduction

This chapter is devoted to a discussion of those sections of the television receiver responsible for synchronizing the deflection oscillators. The synchronizing circuits are responsible for keeping the receiver sawtooth deflection oscillators in step with the transmitter sawtooth deflection oscillators. Both the deflection oscillators and sync separator circuits are fundamental to the operation of all types of receivers.

The receiver deflection oscillators are synchronized to the transmitter deflection oscillators by means of the horizontal and vertical sync pulses that are part of the composite video signal. These pulses force the free-running receiver deflection oscillators to assume the frequencies of the pulses.

In the block diagram of the color receiver shown in Figure 12-1 those circuits to be discussed in this chapter are shaded.

12-2 sync separator

A more detailed block diagram of the sync and deflection oscillator sections of the television receiver is shown in Figure 12-2. This block diagram represents the typical circuit configurations found in both color and black & white receivers.

A composite video signal having positive going sync pulses is obtained from the output of the video amplifier and is fed to the input of the sync separator. It is the function of the sync separator stage to remove the blanking pulses and video signal from composite video signal, leaving only the vertical and horizontal sync pulses at the output of the stage. This process is called *amplitude separation.*

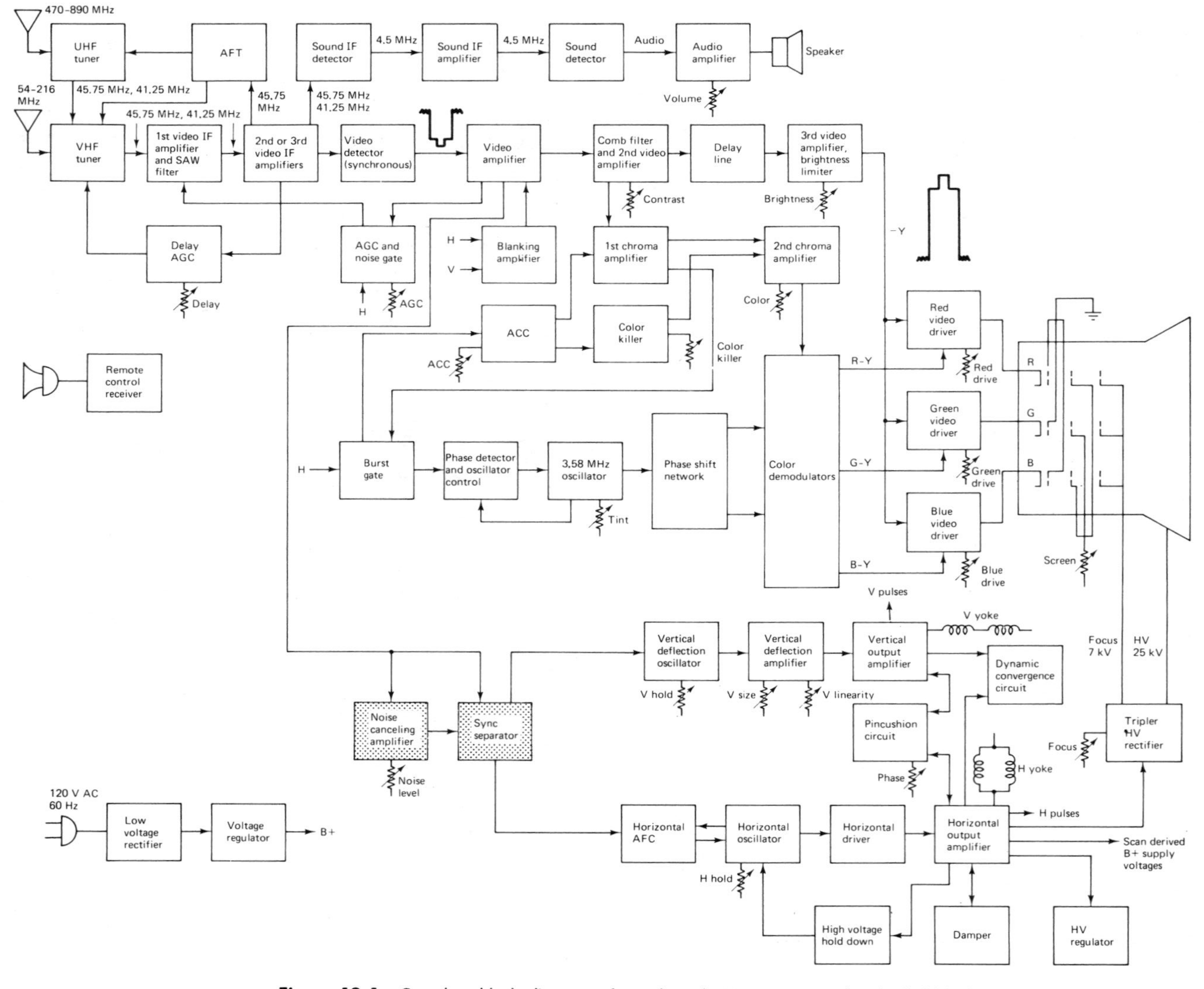

Figure 12–1 Complete block diagram of a color television receiver. The shaded blocks indicate the circuits discussed in this chapter.

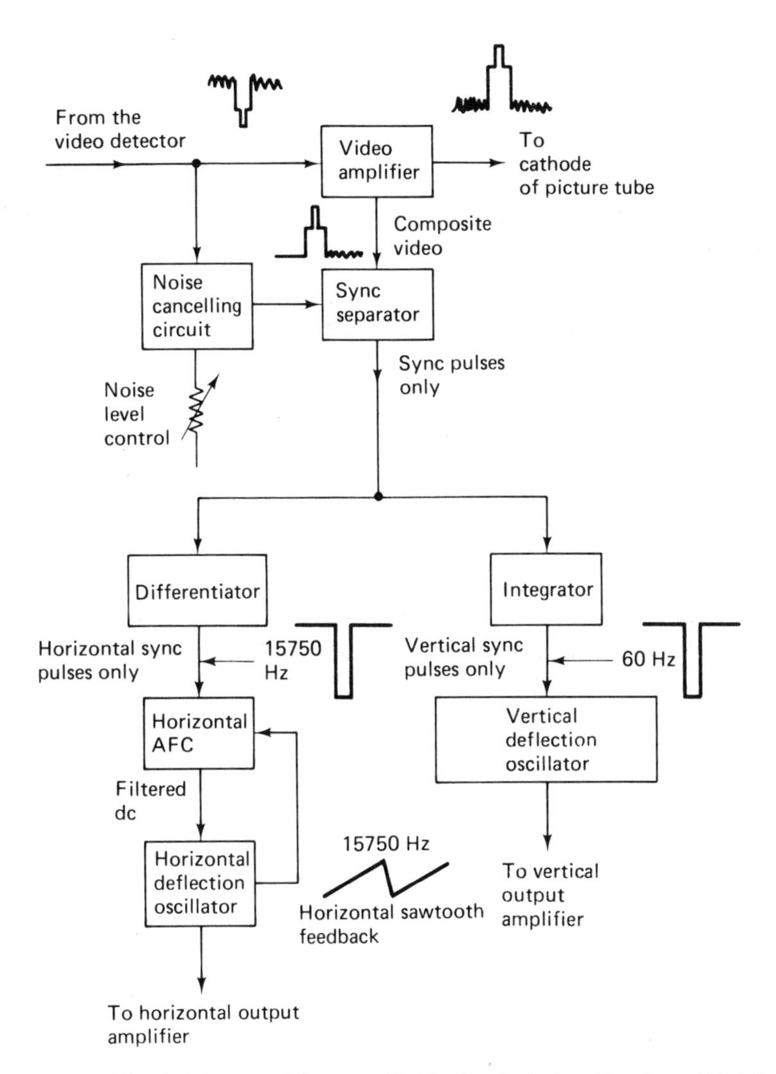

Figure 12–2 Block diagram of the sync and deflection circuits found in color and black & white television receivers.

To ensure that the random noise pulses that accompany the composite video signal have little effect on oscillator synchronization, a *noise-canceling circuit* is usually associated with the sync separator circuit. A control called a *noise level control* is used to adjust this circuit for minimum noise and maximum sync stability.

The output of the sync separator consists of both vertical and horizontal sync pulses of the same amplitude. These pulses are separated from one another by means of filter circuits called *integrators* and *differentiators*. The outputs of each of these filters is then fed to its associated deflection oscillator. The integrator is placed between the sync separator and the vertical deflection oscillator. In effect, it blocks the passage of the horizontal sync pulse but it passes the vertical sync pulses. The differentiator is placed between the sync separator and the horizontal AFC. Its function is to prevent the vertical sync pulses from reaching the horizontal deflection oscillator. This technique of separating the vertical sync pulses from the horizontal sync pulses is called *waveform separation.*

12-3 amplitude separation

The typical sync separator stage of a television receiver performs the five following important functions:

1. Amplitude separation of the sync pulses from the composite video signal
2. Amplification of the sync pulses
3. Dc restoration
4. Improvement of noise immunity preventing loss of sync because of noise
5. Waveform separation prevents horizontal and vertical sync pulses from interfering with one another

Transistor sync separators. Amplitude separation is achieved by using large amplitude video signals that drive the sync separator stage deeply into cutoff. In effect, the stage is operated as a Class C amplifier. Figure 12-3 illustrates how this is done by using the dynamic transfer curve of a transistor sync separator. The transfer curve shows both the regions of current cutoff and the region of collector saturation. The base to emitter input voltage is the

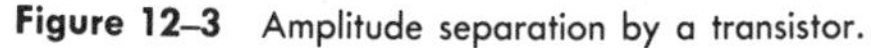

Figure 12–3 Amplitude separation by a transistor.

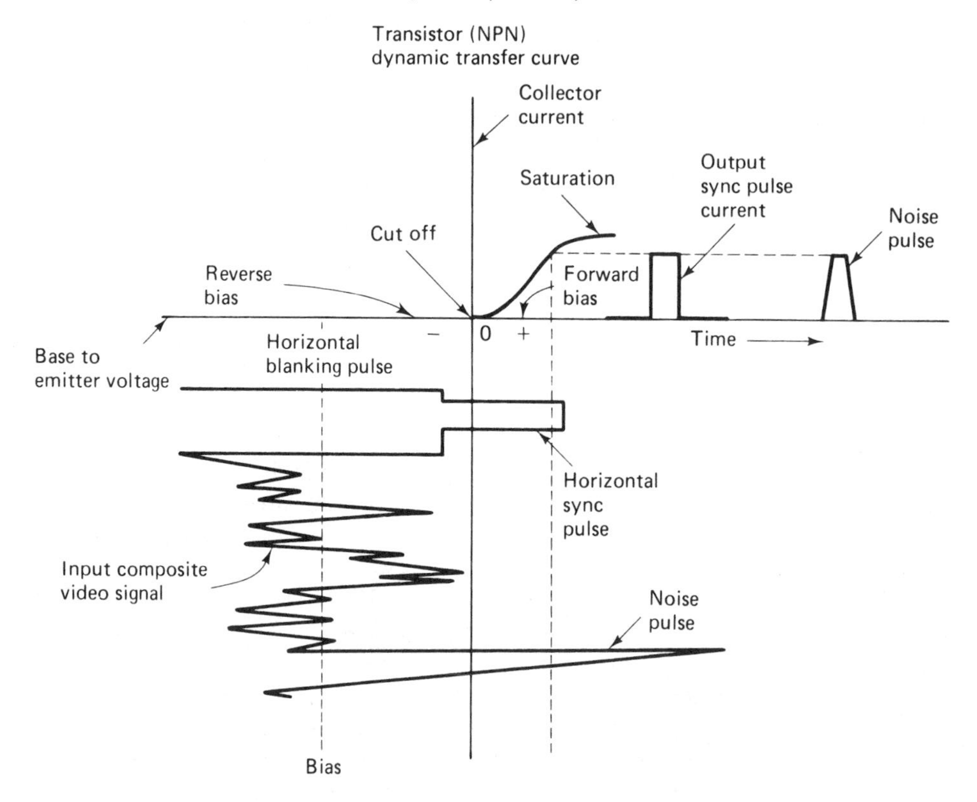

composite video signal. Collector current cutoff in a transistor occurs when the base-to-emitter voltage is approximately zero. Collector current flows only when the transistor is forward biased. As indicated, the composite video signal is applied to a transistor that is biased into the region of cutoff. If the signal is large enough, only the sync pulse will have sufficient amplitude to drive the transistor into conduction. Therefore, the output current of the transistor will consist only of horizontal and vertical sync pulses.

When a transistor amplifier is driven into the extreme forward bias region of its dynamic characteristics, it becomes saturated. A further increase in transistor forward bias in this region, will not cause an increase in collector current. Sync separators are designed by proper selection of load resistor and bias point so that the sync pulse tips drive the transistor into saturation. This has the advantage of keeping the output sync pulse amplitude constant for almost all input signal strength variations. In addition, noise immunity is improved because noise pulses that are stronger than the sync pulses are limited to the same amplitude as the sync pulses. This can be seen in Figure 12-3.

In practice, transistor sync separator stages are driven by video signals having peak-to-peak voltage ranging from 0.5 to 6 V. The amplified output voltage of the transistor sync separator is approximately 40 V p-p.

Dc restoration. Sync separators always use *signal bias* to set their operating point. Signal bias serves two important functions: (1) it restores the dc component of the video signal, and (2) it is a form of automatic bias that adjusts itself as the signal strength changes so that only the sync pulse tips are responsible for sync separator conduction.

Dc restoration is necessary because the composite video signal is often coupled from the video amplifier to the input to the sync separator through a coupling capacitor. As a result, the dc component is lost. The practical effect of this is to cause the sync pulse tips to change in relative amplitude as the scene content changes. If this were not corrected, the varying sync amplitudes would cause erratic synchronization of the deflection oscillators. Signal bias forces the sync pulse tips to *line up,* restoring the dc component of the video signal. For a more detailed discussion, see Section 10-8.

12-4 waveform separation

The standards governing the horizontal and vertical sync pulses are such that both types of pulses have the same amplitude but differ in pulse duration. How does the receiver know which pulses are to synchronize the horizontal deflection oscillator and which pulses are to synchronize the vertical deflection sawtooth oscillator? The method by which horizontal and vertical sync pulses are separated from one another is called *waveform separation* and is accomplished by filters at the output of the sync separator.

The differentiator. A series RC circuit can be made into a high-pass filter that is used to block the vertical sync pulses and pass and shape the horizontal sync pulses into trigger voltages for the horizontal oscillator. This circuit is shown in Figure 12-4 and is called a *differentiator.* Differentiators are designed to have a short time constant compared to the period (width) of the horizontal sync pulses. In this circuit the output voltage is developed across the *resistor* R.

The leading edge of the input causes C to rapidly charge through R to the peak

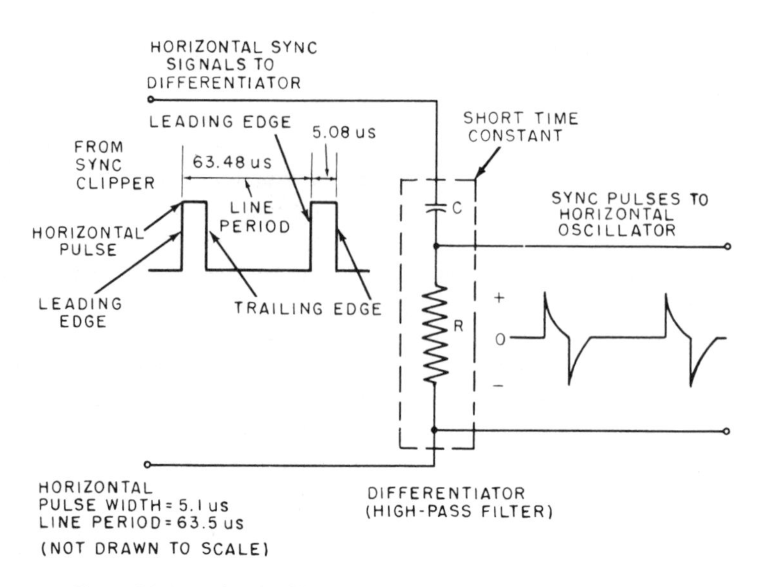

Figure 12-4 High-pass differentiator used for horizontal sync pulses.

of the applied voltage. On the trailing edge of the input pulse the capacitor rapidly discharges through R. The flow of charge and discharge current through R causes sharp voltage spikes to appear across the resistor equal to twice the peak-to-peak voltage of the applied pulses. Either the positive or negative spikes can be used to trigger the oscillator.

The integrator. The integrator is a low-pass filter used to block the horizontal pulses and pass and shape the vertical sync pulses for use as vertical deflection oscillator trigger voltages. An integrator circuit is shown in Figure 12-5. In this series RC circuit the output is taken across the *capacitor (C)*. The RC time constant is made long with respect to the width of the vertical sync pulse interval. Because of the long time constant of the integrator circuit, C can only charge to a small fraction of the peak voltage of a short-duration pulse during the time the applied voltage rises to maximum. During the time that the applied voltage is zero the capacitor discharges toward zero. The capacitor is forced to repeat this sequence whenever the applied voltage rises to maximum once again. If the *duration* of the applied voltage is increased, the voltage across the capacitor will also increase because the long time constant circuit has more time to charge.

This action can be seen in Figure 12-5. The vertical blanking portion of the composite video signal input waveform fed to the integrator consists of two horizontal sync pulses (5 μs), six leading equalizing pulses (2.5 μs), six vertical sync pulses (27.3 μs), six serrations (4.4 μs), and six lagging equalizing pulses. Each of the leading short-duration equalizing pulses causes the integrator output voltage to rise slightly, and then because of the long period between pulses, the output falls to zero because the long time constant integrator is only able to charge to a small fraction of the applied voltage during the time the equalizing pulses are present.

Each of the long-duration serrated vertical sync pulses causes the integrator to progressively charge toward the peak of the applied voltage. During each short-duration

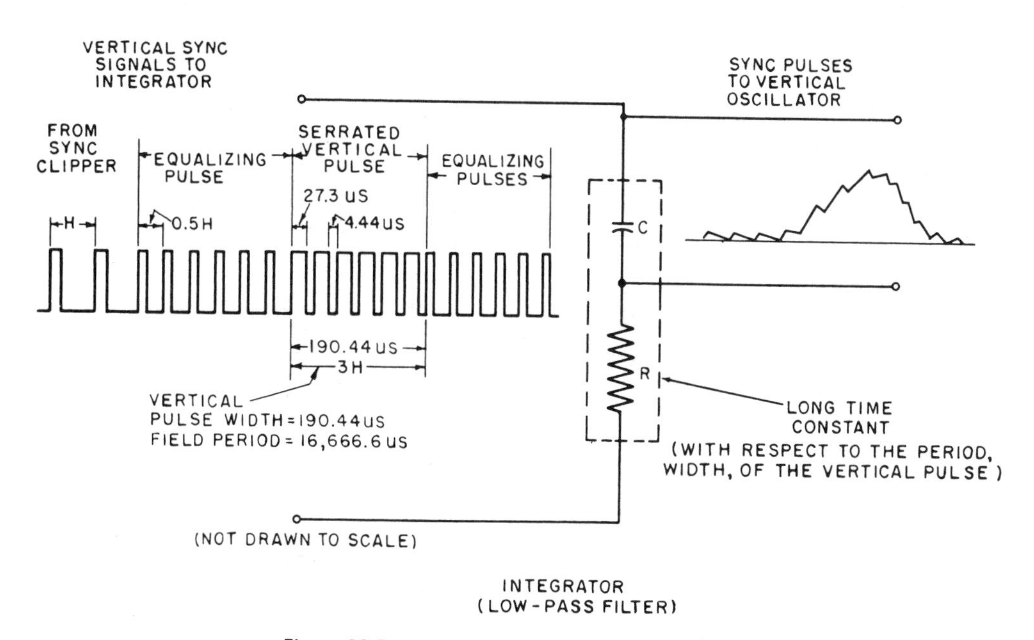

Figure 12-5 Low-pass filter (integrator) used for vertical sync pulses.

serration the capacitor discharges by a small amount, but the next vertical sync pulse continues to build the charge on the integrator capacitor. At some point during the increase in vertical sync pulse amplitude the vertical sawtooth oscillator will be forced to initiate vertical retrace and a new vertical scanning field will begin.

Summary. A summary of the action of both integrator and differentiator circuits is shown in Figure 12-6. The top row shows all of the pulses (equalizing, vertical sync, horizontal sync, and serrations) that are transmitted during the vertical blanking interval. Below this row are two rows of waveforms showing the output waveforms for the differentiator and the integrator circuits. Those differentiator spikes that are used to synchronize the horizontal oscillator are labeled *A*. Notice that every other equalizing pulse and vertical serration is used as a horizontal sync pulse. These pulses maintain horizontal synchronization during vertical retrace blanking.

The fourth row of waveforms corresponds to the vertical blanking interval for the second field of the frame. The only difference between the sync pulse waveforms of the first row and those of the fourth row is that the separation between the first equalizing pulse of each vertical blanking period and the last horizontal sync pulse of the field differs by one-half line. This one-half line difference also occurs between the last equalizing pulse and the first horizontal sync pulse of the vertical blanking interval. The one-half line difference between fields is the result of the use of odd-line interlace scanning to produce a raster. Unfortunately, this half line difference can cause loss of interlace scanning. Equalizing pulses are included in the vertical blanking interval as a means of avoiding the resulting pairing of lines and consequent loss of picture detail produced by a loss of interlace scanning.

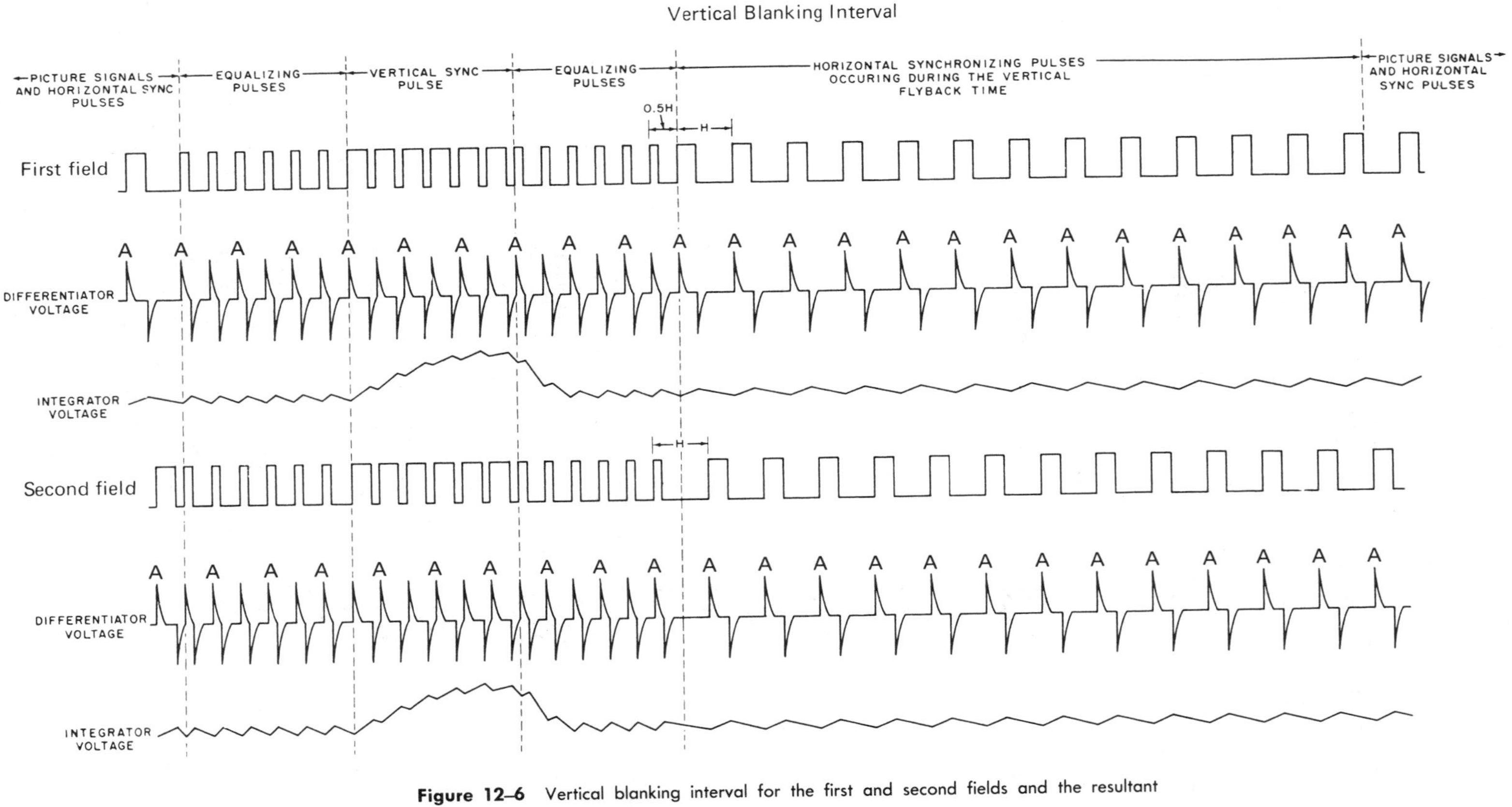

Figure 12–6 Vertical blanking interval for the first and second fields and the resultant differentiator and integrator outputs.

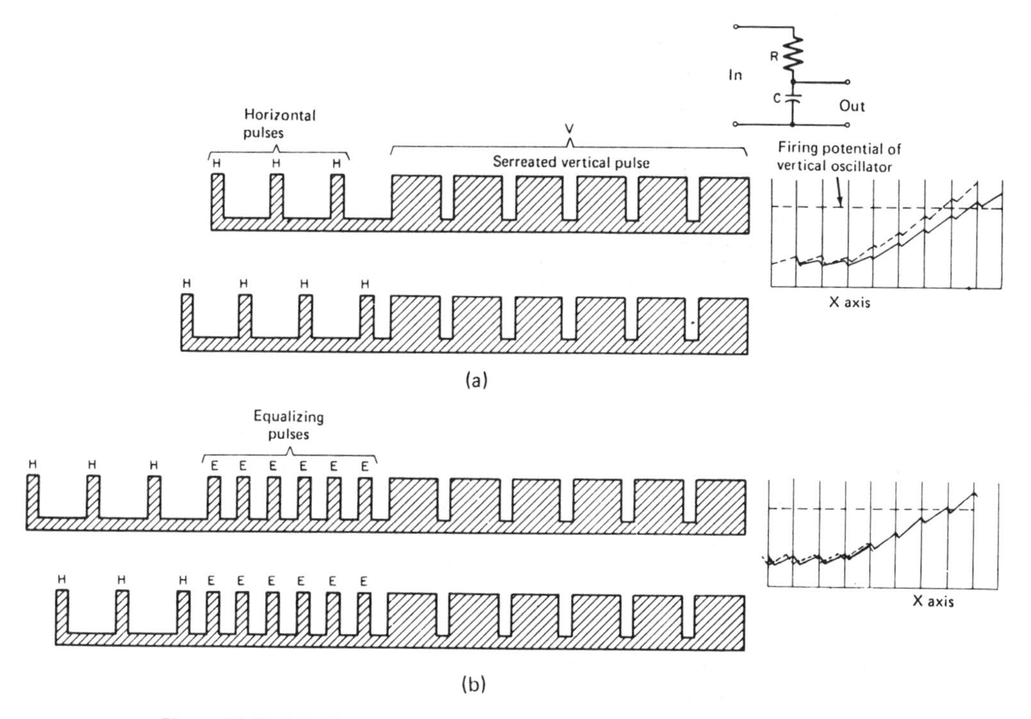

Figure 12-7 (a) Effect on vertical synchronization of no equalizing pulses; (b) effect on vertical synchronization with equalizing pulses.

Equalizing pulses. How equalizing pulses prevent loss of interlace scanning is shown in Figure 12-7. First, let it be assumed that no equalizing pulses are used, as indicated in Figure 12-7(a). In this case, the one-half line difference between fields causes the output voltage of the integrator to build up to the firing potential of the vertical oscillator in less time at the end of one field than at the other. This results from the fact that the slight charge on C, caused by the horizontal pulse, does not have time to leak off before the vertical pulse arrives. The residual voltage across C, plus the voltage caused by the vertical sync pulse cause the vertical oscillator to fire sooner on one field than on the other.

The situation is corrected as shown in Figure 12-7(b) by the use of equalizing pulses. The integrator output voltage buildup across C now begins at the same point, regardless of whether the vertical pulse occurs at the end of a one-half line or at the end of a full line.

12-5 the transistor sync separator

A typical transistor sync separator circuit is shown in Figure 12-8. This circuit uses signal bias. C_1 and R_1 form the signal bias network.

Under no-signal conditions, the base voltage is slightly positive because R_1 returns to a positive source. This type of connection ensures that the transistor is driven into saturation when the positive going sync pulse is applied. The input signal driving this sync separator is only 10 V p-p. This signal is rectified by the base to emitter junction of the

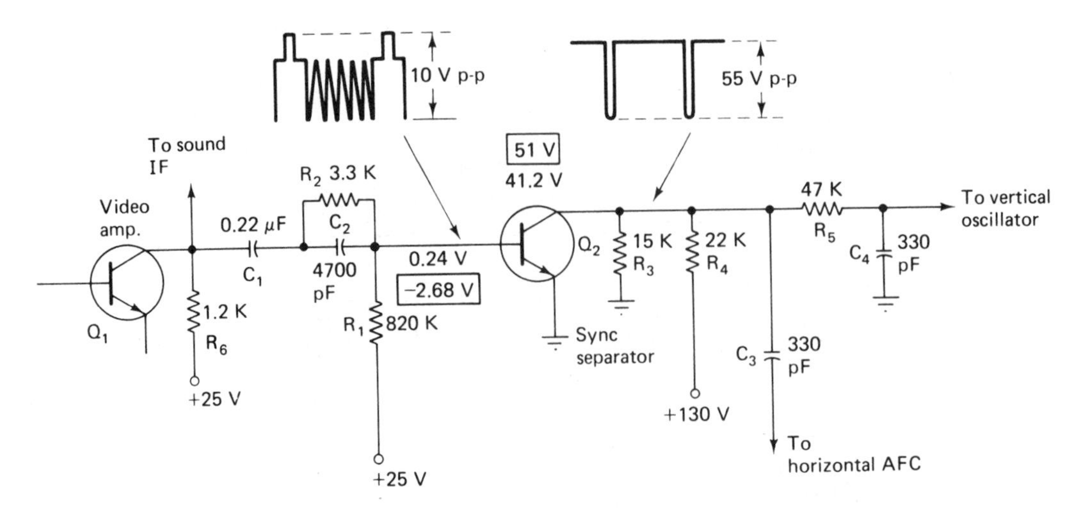

Figure 12-8 Typical transistor sync separator. The voltages in the boxes are the dc voltages with an input signal applied.

transistor and develops a negative voltage from base to ground of approximately 2.7 V. This voltage reverse biases the transistor and pushes all of the composite video signal except the sync pulses below collector current cutoff. This provides the required amplitude separation. As a result, only amplified sync pulses appear at the output of the sync separator.

R_2 and C_2 provide the pulse-shaping and antinoise setup characteristics needed to reduce the effects of noise.

The collector load consists of two resistors R_3 and R_4. R_3 is in parallel with the emitter to collector portion of the transistor. This combination is in series with R_4, which in turn is connected to B+. The purpose of R_3 is to prevent the collector voltage from becoming equal to the B supply voltage. If this were allowed to happen, the transistor collector to emitter voltage rating would be exceeded and it might become damaged. This could occur during the time the transistor is cut off. It is during this time that the collector voltage tends to increase and become equal to the source voltage (130 V). Because of the voltage division between R_3 and R_4, the collector voltage is limited to a maximum of approximately 51 V. During no-signal operation the transistor conducts because of the small positive base voltage, and the collector voltage falls to approximately 41 V.

The output horizontal sync pulses are coupled to the horizontal AFC circuit via C_3, and the vertical sync pulses pass through the integrator circuit R_5 and C_4 on their way to the vertical deflection oscillator.

The collector load resistors R_3 and R_4 determine the total peak-to-peak output voltage of the sync separator. The collector load resistor cannot be made too large since this will result in a loss of the sync separator high-frequency response.

The output sync pulse amplitude will remain constant for most video input signal variations because transistor saturation and the base to emitter diode action keep the sync pulses squared off and lined up. If the signal is very weak, it is possible that amplitude separation of the composite video signal will include part of the blanking and video signal as well as the sync pulses. This can result in erratic sync.

Noise set-up. Noise may be considered an undesired signal. As such, it contributes to the signal bias produced by the desired signal. This has the effect of changing the transistor bias and pushing the composite video signal deeper into cutoff. As a result, only a small part of the sync pulse is able to force the transistor into conduction reducing the amplitude of the output sync pulse. This condition is called *noise setup.* The reduction of sync pulse amplitude resulting from presence of noise may cause erratic horizontal and vertical picture synchronization.

As shown in Figure 12-8 the addition of C_2 and R_2 in series with the signal bias capacitor C_1 tends to minimize noise setup because the parallel combination of C_2 and R_2 forms a short-time-constant circuit. Positive going noise pulses applied to the base of the transistor cause base current flow that charges both C_1 and C_2. Since C_2 is very much smaller than C_1, it receives the *greatest* voltage. The charge across C_2 is promptly discharged through R_2 and does not contribute to the signal bias of the transistor. This noise protection is only suitable for nonrecurring noise. If the noise pulses are periodic, the small charge on C_1, contributed by each noise pulse, will be cumulative and noise setup will still occur.

12-6 the transistor noise-canceling amplifier

The circuit shown in Figure 12-9 makes use of a special transistor (Q_1) that is used as a noninverting noise amplifier. The function of this stage is to amplify noise pulses obtained from the video detector and combine them with oppositely phased noise pulses at input of the sync separator (Q_3) canceling the noise.

Figure 12-9 All-transistor noise-canceling circuit.

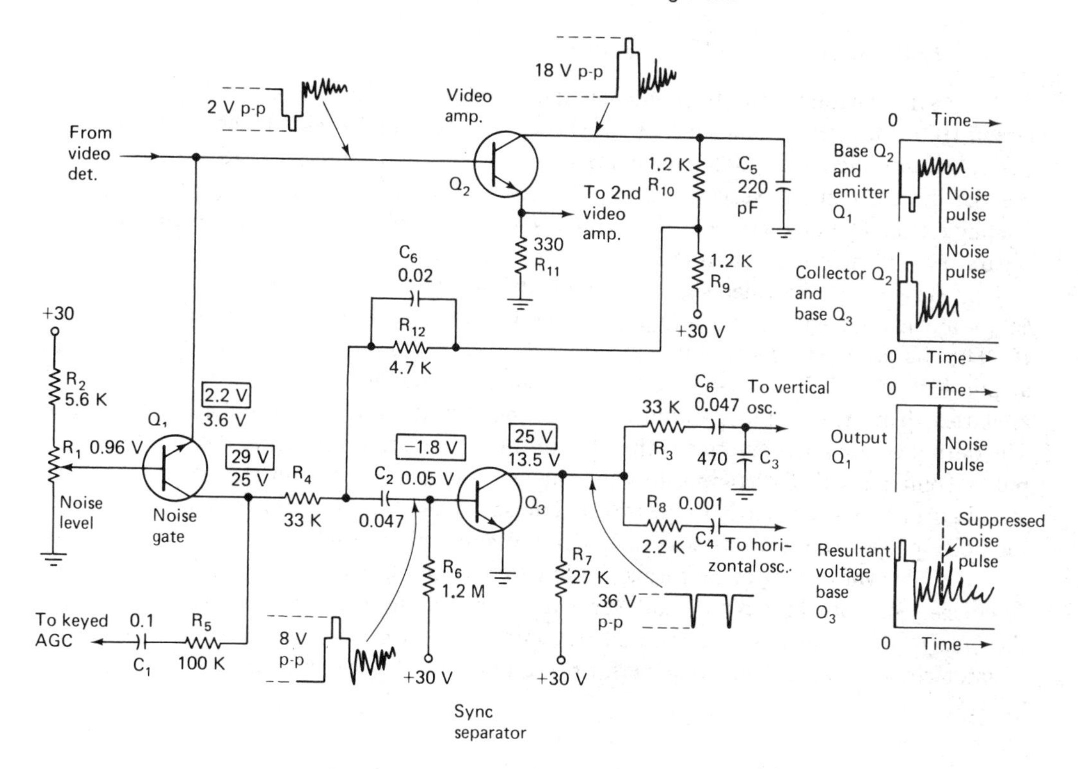

In this circuit the video detector feeds a negative-going video signal to both the video amplifier and the emitter of a noninverting common base noise amplifier Q_1. This transistor's bias is established by R_1, the noise level control, and is normally biased beyond cutoff.

The normal level of the video signal is insufficient to drive the noise amplifier into conduction. Therefore, under normal conditions, Q_1 may be considered as being out of the circuit.

The sync separator (Q_3) is fed by both the output circuit of the video amplifier (Q_2) and the noise amplifier (Q_1). Under normal conditions, however, the composite video signal at the input to the sync separator point is solely the result of the video amplifier and has a positive going sync pulse.

When a sufficiently strong negative going noise pulse appears at the input to the noise amplifier, it conducts and produces an amplified *negative-going* noise pulse at the input of the sync separator. Simultaneously, the noise pulse has been amplified by the video amplifier and appears at the input of the sync separator as a *positive-going* pulse. Since both of these pulses are opposite in direction and are equal in amplitude, the noise pulse is effectively eliminated from the input of the sync separator. This noise-free signal is also fed to the input of the keyed AGC circuit via C_1. The action of this circuit is summarized in the ladder diagram of Figure 12-9.

To adjust this circuit, first tune the receiver to the strongest channel. Then, starting with the noise gate control (R_1) in its off position (CCW, where the base is at its minimum voltage) slowly rotate it clockwise until the picture starts to bend or fall out of sync. At this point, back off the control until the picture is stable. This completes the adjustment.

12-7 integrated circuit sync separator

Sync separators and their associated circuits are often found as part of an integrated circuit (IC). An example of such an IC is shown in Figure 12-10. This IC chip also contains the video amplifier, the video gain control, and dc restorer circuits. External to the chip are the picture control, service switch, sharpness control, automatic brightness control, coupling circuit, and their related circuits. Two stages of the chip are devoted to sync separation and noise cancellation.

The composite video signal is obtained from the base of the second video amplifier (Q_{302}) and is coupled via R_{309} and C_{305} to the input of the sync separator at pin 1 of the IC. The composite video signal is also fed via L_{305}/C_{317}, and C_{306} to the noise inverter circuit at pin 4 of the IC. The noise inverter is designed to conduct only when the noise pulses are larger than the sync pulse amplitude. These noise pulses are then amplified and inverted. The output of the noise inverter is then fed into the sync separator. Here the inverted noise pulses combine with the composite video signal. Any noise pulses in the composite video signal are of opposite polarity to the inverted noise pulses, canceling them out and preventing the noise pulses from causing loss of horizontal and vertical sync.

The sync separator performs the function of amplitude separation. Its output (pin 3) consists of both horizontal- and vertical-negative-going sync pulses. These pulses are then fed to the vertical and horizontal deflection oscillators via the integrator and differentiator circuits (not shown) that perform the function of waveform separation.

12-8 vertical countdown circuits

Many recent television receivers have incorporated as part of their sync separators a system for deriving both the horizontal and vertical deflection drives from one master oscillator. This arrangement is called a vertical countdown circuit and has a number of important advantages:

1. Perfect interlace scanning. This means that vertical picture resolution is improved because there cannot be any pairing of horizontal scanning lines.
2. The vertical countdown circuit does not require any vertical hold control.
3. Changes in vertical height and vertical linearity control settings cannot have any effect on the vertical sawtooth frequency. In receivers not using vertical countdown circuits, changing these controls usually also affects the vertical sawtooth frequency, causing the picture to roll.
4. This circuit uses one master oscillator to generate both the horizontal and the vertical deflection sawteeth. This is in contrast to the previous methods, where two separate deflection oscillators were necessary.

A number of different schemes for vertical countdown are being used. One of the chief differences is the frequency of the master oscillator. For example, Zenith receivers use a master oscillator whose frequency is 503.49 kHz, which is the 32nd harmonic of the horizontal scanning rate. RCA's circuit uses a frequency of 31.46852 kHz, which is twice the horizontal scanning rate. As a result, RCA's circuit can use a shorter frequency divider chain ($\div 2$) than that required by the Zenith circuit ($\div 32$) to obtain the horizontal scanning frequency. Figure 12-11 is the block diagram of the Zenith system and Figure 12-12 shows the circuit arrangement used in the RCA system.

The Zenith vertical countdown system. The schematic diagram of the two integrated circuit chips used to provide vertical and horizontal deflection drive is shown in Figure 12-11. The master oscillator chip 221-105 provides six functions:

1. *Video amplification.*
2. *AGC processing* provides AGC voltages for the IF, VHF tuner and the UHF tuner.
3. *Sync separation* generates a vertical sync pulse that is fed to the countdown IC and a horizontal sync pulse that is fed to the horizontal phase detector.
4. *Noise-canceling amplifiers,* in conjunction with the video amplifier, improve system noise immunity by eliminating noise pulses whose amplitude exceeds that of the horizontal sync pulse amplitude.
5. The *horizontal AFC* section of the IC compares the horizontal sync pulse frequency with the horizontal sawtooth frequency obtained from a pulse generated by the horizontal flyback transformer. This pulse is integrated by an *RC* network forming a sawtooth which then feeds the phase detector. If the comparison shows that the two frequencies are equal, the dc error voltage generated is zero. Any frequency error will generate a positive or a negative voltage depending on whether the frequency of the oscillator is too high or too low. The dc error voltage then feeds the 503.5-kHz master scan oscillator,

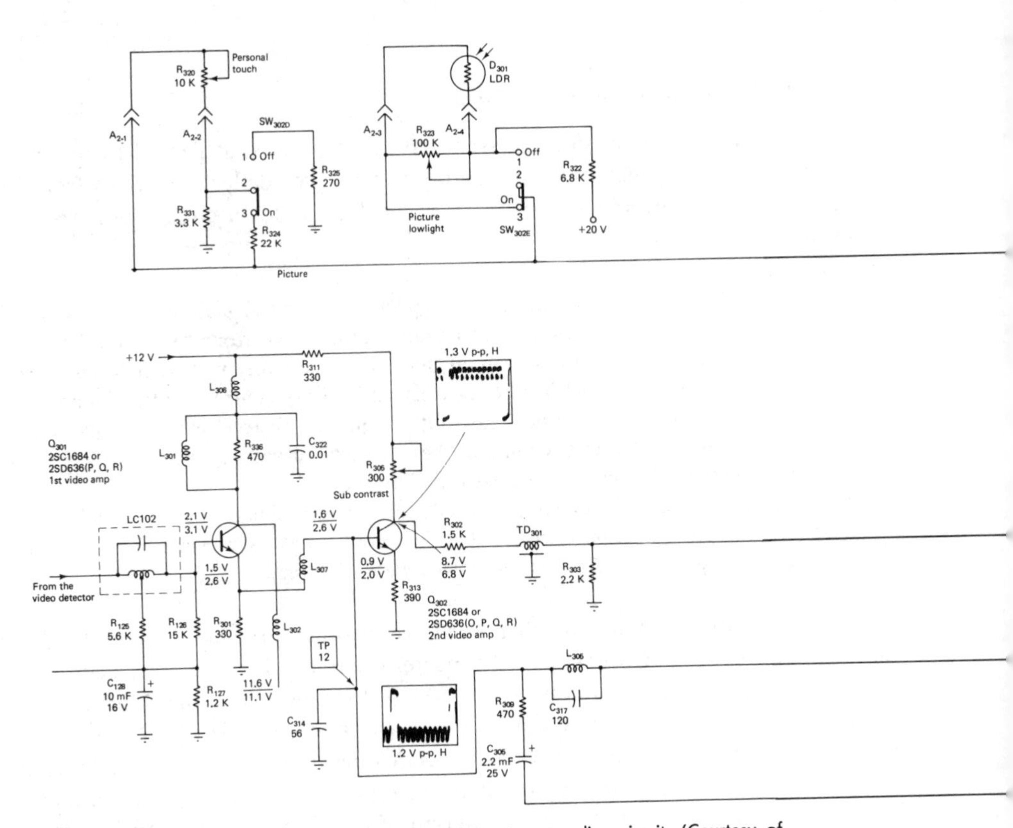

Figure 12-10 Integrated circuit sync separator and noise canceling circuit. (Courtesy of Quasar Industries.)

correcting any frequency error. This type of circuit arrangement is often called a phase-locked loop (PLL).

6. The *master scan oscillator* circuit operates at a frequency of 32 times the color horizontal scanning rate or 503.5 kHz. Its frequency is controlled by the horizontal AFC section of the chip. However, its frequency is adjustable by means of the "frequency adjust" coil. This coil is adjusted by first shorting pin 3, thereby removing horizontal sync, and then adjusting the coil for a picture that has minimum roll. The output (pin 7) of this oscillator is then used by the countdown IC chip to develop both the horizontal and vertical drive signals.

Vertical countdown chip. The block diagram of the vertical countdown chip and module is shown in Figure 12-11. This chip performs three functions:

1. *Provides a blanking pulse.* This pulse is amplified by a buffer stage and is coupled to the low-level-luminance circuits.
2. *Provides horizontal drive.* This is accomplished by dividing the 503.5-kHz master oscillator sine-wave input signal at pin 9 by 16 and then by 2, for a total of 32. The output of this divider chain produces a square-wave output whose frequency is 15,734.26 Hz. This output is then used to drive the horizontal deflection system. The output of the divide-by-16 section of the IC also feeds the logic section of the IC.

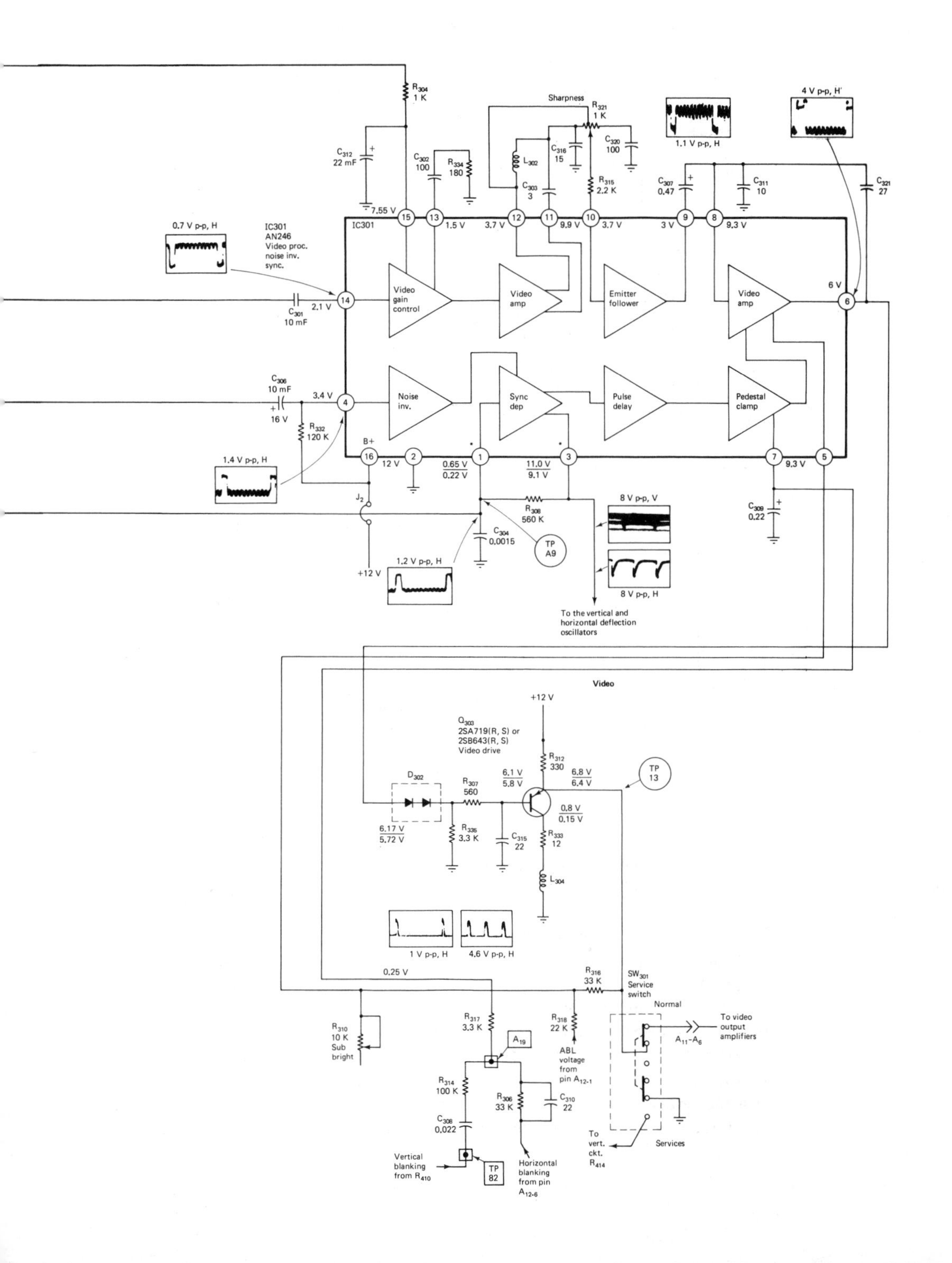

R304
1 K
Sharpness
R321
1 K
C312
22 mF
C302
100
R334
180
L302
C316
15
C320
100
1.1 V p-p, H
4 V p-p, H
C303
3
R315
2.2 K
C307
0.47
C311
10
C321
27
7.55 V
1.5 V
3.7 V
9.9 V
3.7 V
3 V
9.3 V
0.7 V p-p, H
IC301
AN246
Video proc.
noise inv.
sync.
IC301
Video
gain
control
Video
amp
Emitter
follower
Video
amp
6 V
2.1 V
C301
10 mF
C306
10 mF
3.4 V
16 V
Noise
inv.
Sync
dep
Pulse
delay
Pedestal
clamp
R332
120 K
B+
1.4 V p-p, H
12 V
0.65 V
0.22 V
11.0 V
9.1 V
9.3 V
C309
0.22
J2
R308
560 K
8 V p-p, V
C304
0.0015
TP
A9
+12 V
1.2 V p-p, H
8 V p-p, H
To the vertical and
horizontal deflection
oscillators
Video
+12 V
Q303
2SA719(R, S) or
2SB643(R, S)
Video drive
R312
330
D302
R307
560
6.1 V
5.8 V
6.8 V
6.4 V
TP
13
0.8 V
0.15 V
6.17 V
5.72 V
R335
3.3 K
C315
22
R333
12
L304
1 V p-p, H
4.6 V p-p, H
0.25 V
R316
33 K
SW301
Service
switch
Normal
R310
10 K
Sub
bright
R317
3.3 K
A19
R318
22 K
ABL
voltage
from
pin A12-1
To video
output
amplifiers
A11–A6
R314
100 K
R306
33 K
C310
22
C308
0.022
To
vert.
ckt.
R414
Services
Vertical
blanking
from R410
TP
82
Horizontal
blanking
from pin
A12-6

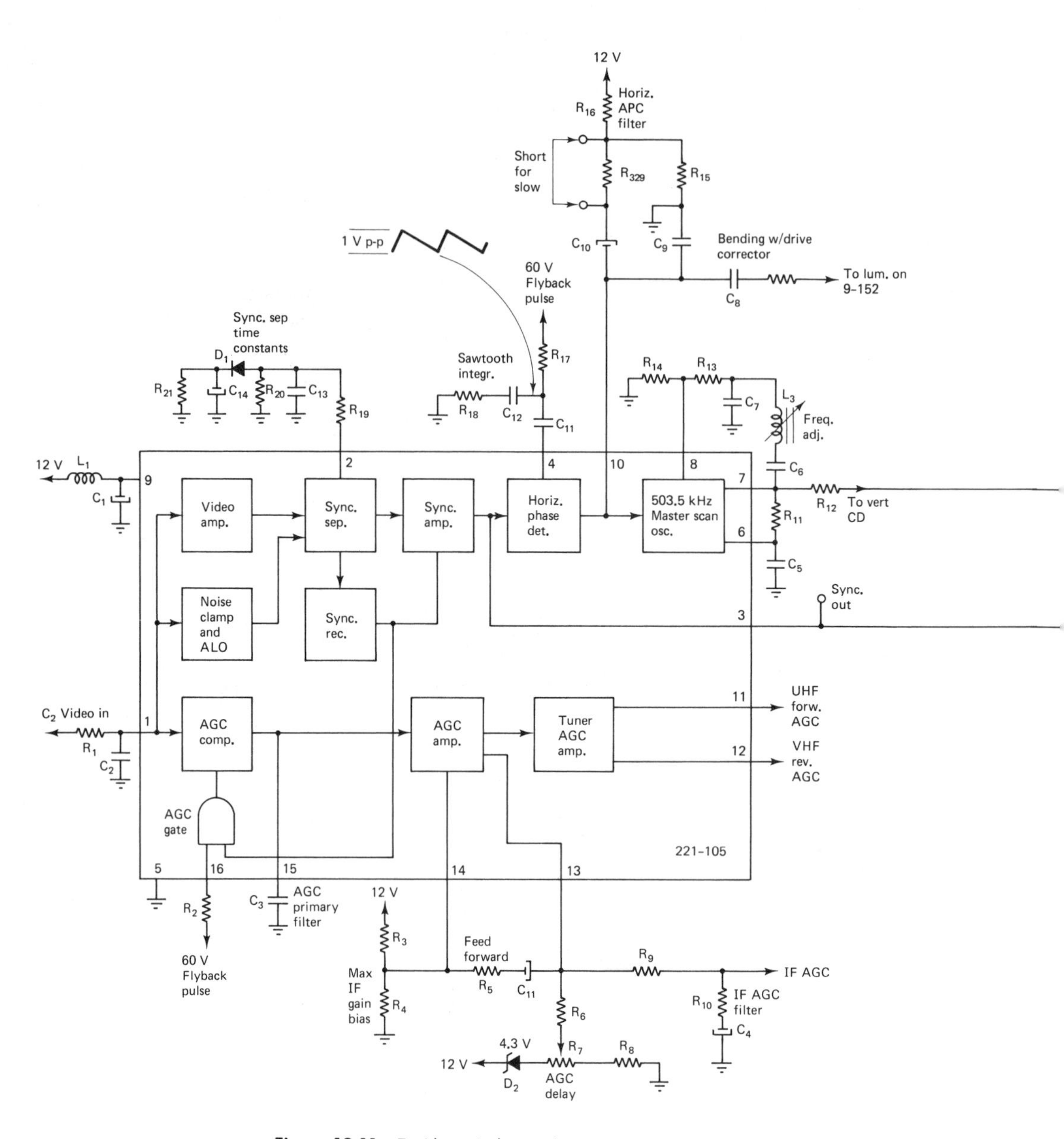

Figure 12-11 Zenith vertical countdown circuit. (Courtesy of Zenith.)

3. *Provides drive for the vertical deflection amplifier.* The vertical drive is obtained from the logic portion of the countdown IC chip. Composite sync and a sample of the master oscillator (31,468.52 Hz) taken from the divide-by-16 section of the chip are the inputs to the logic circuit. During the countdown mode of operation, the logic circuit divides its input signal by 525 producing a vertical output rate of 59.94 Hz. An *RC* integrator injects a vertical sync pulse at pin 12. This is done to ensure that the vertical retrace interval coincides with the vertical sync interval of the composite sync. The logic circuitry also uses these inputs to decide whether or not the signal being viewed has 525

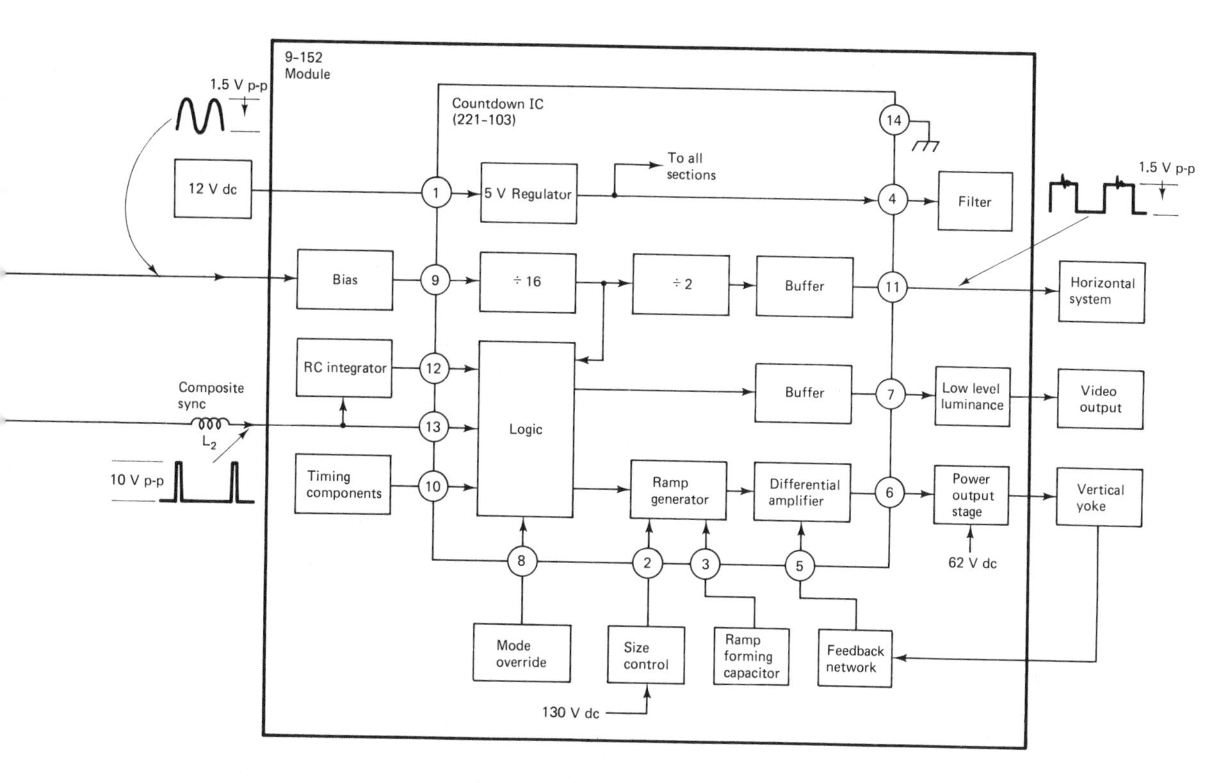

Figure 12-11 *(Cont.)*

lines per frame. If it does, the logic circuit delivers a pulse to the ramp generator which develops the deflection sawtooth. Associated with the circuit are a size control and a ramp-forming capacitor. Both of these components are external to the chip and are mounted on the module. The output of the ramp generator feeds a differential amplifier, which in turn drives the power output stage and yoke. Feedback is used to improve the linearity of the deflection sawtooth current.

If nonstandard signals such as that which might be generated by cable systems, some pattern generators, and some closed-circuit TV camera systems are applied to the receiver, the logic circuit identifies these signals and compensates for them by switching to a direct sync mode of operation.

The RCA countdown system. Figure 12-12 shows the schematic diagram of the countdown chip and its associated circuitry. The chip's internal structure is indicated in block diagram form. This chip performs two functions: it generates the horizontal drive pulses and it

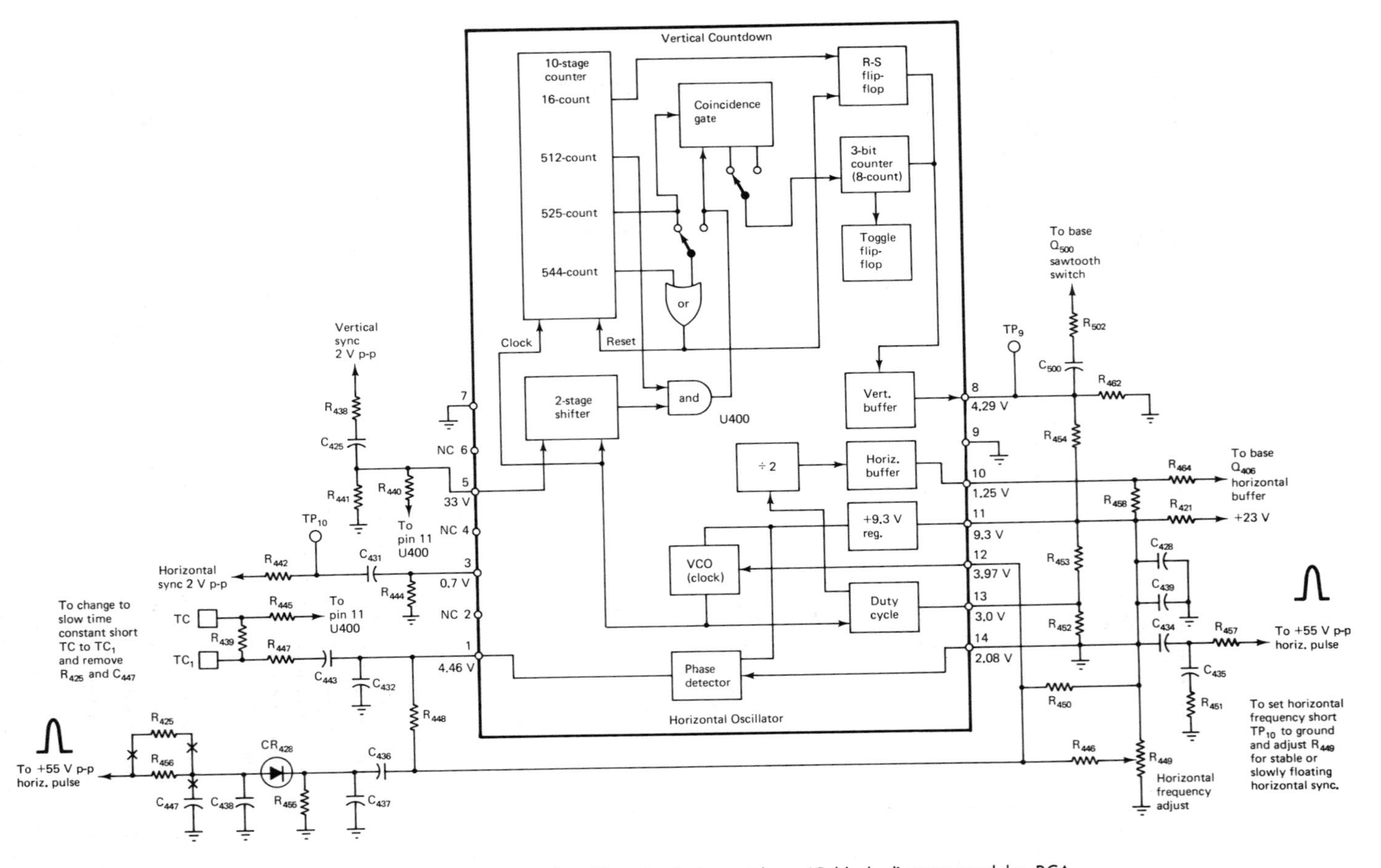

Figure 12-12 Horizontal oscillator/vertical countdown IC block diagram used by RCA. (Courtesy of RCA, Consumer Electronics.)

generates the vertical drive pulses. Both of these outputs are obtained from a master voltage-controlled oscillator (VCO) that operates at a frequency of 31,468.52 Hz, which is twice the horizontal scanning rate.

Differentiated horizontal sync pulses are fed to pin 3 and into the phase detector. Also feeding the phase detector is a sawtooth derived from a horizontal pulse obtained from the flyback transformer. The phase detector dc output (pin 1) is filtered, then dc coupled to the VCO (pin 12). To correct for picture bending (hook) the voltage at pin 12 also has a sawtooth voltage derived by the application of a flyback pulse to C_{436} and R_{450}.

The phase detector output achieves frequency lock to the broadcast signal horizontal sync by modifying the VCO frequency as required. The horizontal oscillator "time constant" is designed to be compatible with the requirements of most home video cassette recorders. Under weak-signal or high-impulse-noise conditions, it may be desirable to slow the time constant and prevent picture hook. This is done by shorting stake TC to TC_1 and removing C_{447} and R_{425}. The output of the VCO is coupled into a duty cycle adjustment circuit. The duty cycle is determined by the dc voltage at pin 13. The output of the duty cycle block is then fed to a divide-by-2 counter, producing a square-wave output whose frequency equals the horizontal scanning rate (15,734.26 Hz). This output (pin 10) is then used to drive the horizontal buffer.

The vertical countdown system. The vertical output drive pulses are obtained from pin 8 of the countdown chip. To obtain the 59.94-Hz vertical drive from the master oscillator (31,468.52 Hz) requires a 10-stage divider chain that divides the master oscillator signal by 525. In addition, a complex logic circuit consisting of AND gates, OR gates, coincidence gates, RS flip-flops, shift registers, and counters is used to determine whether or not there is an input signal, and whether the input signal is using a standard NTSC vertical sync pulse or some kind of nonstandard vertical sync. When an NTSC vertical sync pulse is present, the logic circuit uses the countdown circuit to generate the vertical output pulse. However, under no-signal or nonstandard sync conditions the logic circuit switches to "sync mode" and the circuit behaves as a synchronized oscillator.

12-9 sync separator troubleshooting

The defective sync separator stage may exhibit the following trouble symptoms:

1. Loss of vertical synchronization or picture roll [see Figure 12-13(a)]
2. Loss of horizontal synchronization or picture tearing [see Figure 12-13(b)]
3. Loss of both horizontal and vertical synchronization [see Figure 12-13(c)]
4. Horizontal pulling or bending [see Figure 12-13(d)]
5. Vertical jitter

Many of these trouble symptoms can be caused by clipping in the video amplifier stage. Such clipping can be caused by AGC or video IF amplifier alignment defects that result in very large video signals that force the video amplifier to overload. These trouble symptoms may also be the result of changes in the video amplifier such that sync pulse clipping takes place. It is therefore essential to be able to determine quickly whether the trouble is in the sync separator stages or in the video strip. This may be done by observing the "hammerhead" pattern. If the hammerhead is normal, the trouble must be in the sync circuit. If the hammerhead is washed out or missing, the trouble must be in the video strip.

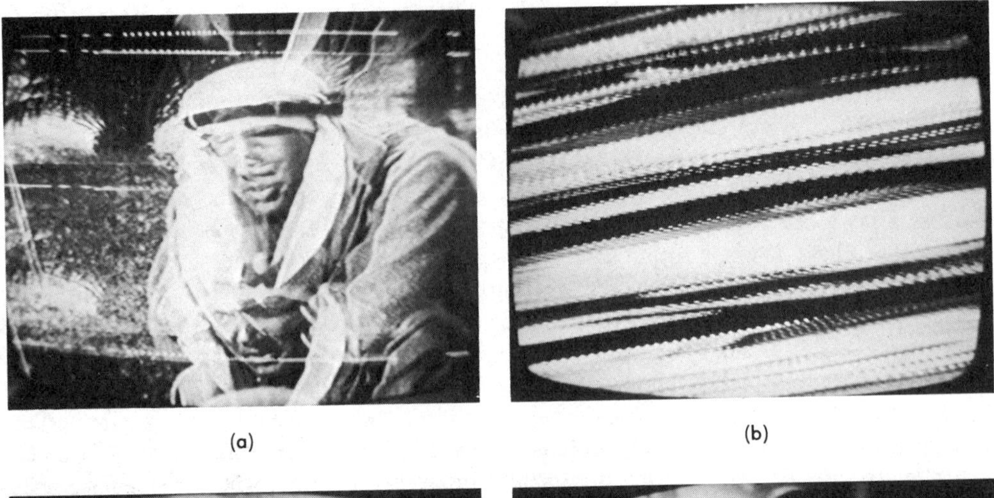

(a) (b)

(c) (d)

Figure 12–13 (a) Loss of vertical sync; (b) loss of horizontal sync; (c) loss of both horizontal and vertical sync; (d) horizontal pulling. (Courtesy of General Electric.)

Loss of vertical sync. If it has been established that the trouble is in the sync separator circuit, the trouble symptom can tell much about where the defect is located. For example, the loss of vertical sync [see Figure 12-13(a)] with normal horizontal synchronization indicates that the sync separator is functioning, but the vertical sync pulses are either distorted or are not reaching the vertical oscillator. The most probable causes of this defect are:

1. Defective tube or transistor
2. An open integrator resistor such as R_5 in Figure 12-8
3. Shorted integrator capacitor C_4 in Figure 12-8
4. An open resistor between the sync separator and the vertical oscillator, such as R_5 in Figure 12-8

5. A 60-Hz hum present in the sync signal; this may be caused by poor B supply filtering

In some receivers more than one stage of sync separation is used. These circuits may be arranged so that the sync separator stage feeds separate horizontal and vertical sync pulse amplifiers. In such circuits the loss of vertical sync can be caused by defects in the vertical sync amplifier.

Loss of horizontal sync. Figure 12-13(b) shows a picture of a receiver that has lost horizontal sync. If this condition is not accompanied by a loss of vertical sync, it may be assumed that the sync separator stage is functioning and that the horizontal sync pulses are not reaching the horizontal AFC circuit. This trouble symptom can be caused by such defects as the following:

1. Defective horizontal sync amplifier tube or transistor
2. Open or leaky coupling capacitor to horizontal AFC, C_3 in Figure 12-8
3. Increased value of plate or collector load resistor such as R_4 in Figure 12-8

A defective horizontal AFC circuit can also be responsible for the loss of horizontal sync. This will be discussed under the topic of automatic frequency control (AFC).

Loss of both horizontal and vertical synchronization. The complete loss of both horizontal and vertical synchronization is illustrated by the photograph in Figure 12-13(c). This trouble symptom indicates that both deflection oscillators are free-running and that either the composite video signal is not reaching the sync separator or the sync separator itself is not functioning. Signal tracing with the scope is very useful in determining why the horizontal and vertical sync pulses are not reaching their respective oscillators. Once the general area has been isolated, the voltmeter and ohmmeter may be used to isolate the defective component.

Possible causes of this trouble symptom are as follows:

1. Defective transistor in the sync or noise limiter stages, for example, if Q_3 of Figure 12-9 is open
2. Video coupling capacitors, open, shorted, or leaky (see C_1 in Figure 12-8 and C_2 in Figure 12-9)
3. Video sync isolation resistors open or increased in value (see R_{10} in Figure 12-9)

Horizontal bending or pulling. An example of horizontal pulling is shown in Figure 12-12(d). This trouble symptom may be caused by defects in such stages as the following:

1. The video strip (tuner, video IF, and video amplifier)
2. The AGC circuit
3. The sync separator stage
4. The horizontal AFC circuit
5. The low-voltage power supply filter

Hum modulation or overloading in the video strip are important causes of horizontal bending. Hum modulation can be the result of poor power supply filtering. The first thing to do in servicing this is to test by direct substitution the B supply filter capacitors. If these checks do not locate the fault, check the hammerhead pattern for overloading. If the hammerhead is normal, break the circuit between the sync separator and the horizontal AFC. This should cause the picture to lose horizontal sync. Adjust the horizontal hold control until the picture momentarily stops tearing and note if the picture still is pulling. If it is, the trouble is in the horizontal AFC circuit; if it is not, then the trouble is in the sync separator circuit.

Possible causes for this trouble symptom might be the following:

1. Defective transistor or IC
2. Leaky coupling capacitor (see C_1 in Figure 12-8)
3. Open capacitor in compensating network (see C_2 in Figure 12-8)
4. Decreased plate or collector load resistance (see R_3 or R_4 in Figure 12-8)
5. Defective filtering of power supply

Vertical jitter. In this trouble symptom the picture appears to be about to lose vertical synchronization and it is usually the result of poor waveform separation. This gives the picture the appearance of bounce. This symptom may also be accompanied by poor interlace.

Some possible causes for vertical jitter are as follows:

1. Defective IC or transistors in sync separator stage
2. Defective component in vertical integrator network (see R_5 and C_4 in Figure 12-8 and R_3 and C_3 in Figure 12-9)

Vertical jitter may also be caused by horizontal sweep pulses that get into the AGC as a result of insufficient filtering. This can cause alternate vertical sync pulses to vary in amplitude, causing the erratic vertical jitter. Other causes of jitter may be traced to the integrator circuits found in the vertical deflection multivibrator. These circuits are used to prevent horizontal pulses developed in the yoke from reaching the vertical oscillator.

12-10 triode sync separator

Figure 12-14 shows the schematic diagrams of a vacuum tube sync separator circuit. In vacuum tube sync separators, grid leak bias (signal bias) is used to establish the operating point of the tube. Capacitor C_1 and resistor R_1 form the grid leak bias circuit. Without any signal, the grid voltage is -1.4 V because of the rectification of random noise. When a signal such as that shown in the figure is applied to the grid of the tube, the grid acts like the anode of a rectifying diode and develops a large negative bias for the tube. This bias depends upon the time constant of R_1 and C_1 as well as the signal amplitude. Large signals will develop large bias voltages and weak signals will develop smaller bias voltages. The action of signal bias is to act as a dc restorer, thereby "lining up" the tips of the sync pulses.

For the circuit under consideration, the grid bias voltage with a normal signal is -40 V. Such a large bias means that the tube is being operated far into its cutoff region.

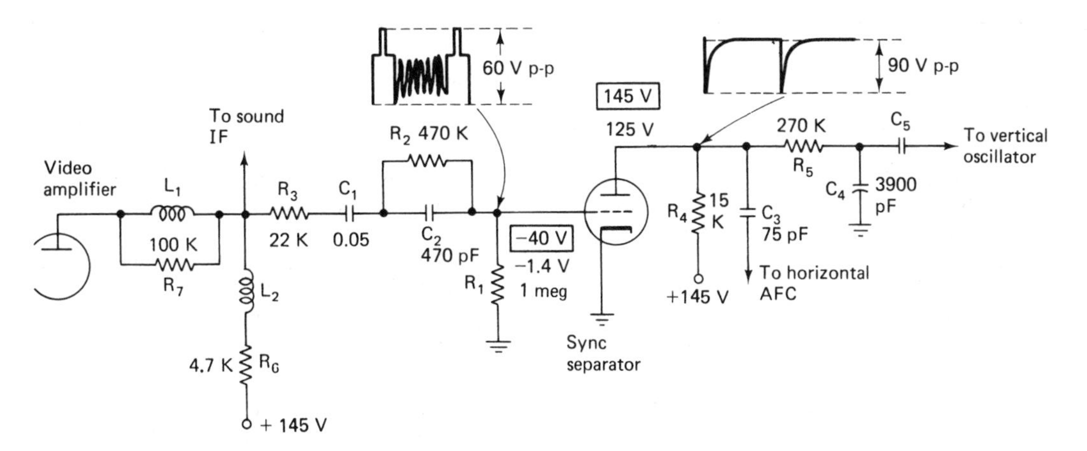

Figure 12-14 Vacuum tube sync separator circuit.

As a result, only the sync pulses are able to force the tube into conduction producing amplified output sync pulses of 90 V p-p.

If C_1 should become leaky (15 MΩ), the grid leak bias will be reduced because C_1 will be discharging itself as well as coupling the positive plate voltage from the previous stage to the grid. This reduction in sync separator bias will permit some of the blanking and video signal information to appear at the output of the sync separator. As a result, erratic horizontal and vertical picture synchronization will occur.

In some receivers the grid leak resistor R_1 may return to B+ instead of ground. This is done to improve the weak signal sync stability of the deflection oscillators. This connection ensures that the tube will be easily driven into saturation and will clip noise pulses that may accompany the video signal.

C_2 and R_2 form a high-pass filter that is placed in series with the grid. This network performs two functions. First, it is a frequency response compensator similar to that used in the step attenuator of the oscilloscope or that used with the contrast control of the video amplifier. This type of network effectively extends the frequency response of the circuit and maintains the sharp rise time of the input sync pulses. Second, this circuit is able to improve the noise immunity of the deflection oscillators.

The plate load resistor R_4 determines the total peak to peak output voltage of the sync separator. The plate load resistor cannot be made too large since this will result in a loss of the sync separator high-frequency response.

The output sync pulse amplitude will remain constant for most video input signal variations because tube saturation and the grid to cathode diode action keep the sync pulses squared off and lined up. If the signal is very weak, it is possible that amplitude separation of the composite video signal will include part of the blanking and video signal as well as the sync pulses. This can result in erratic sync.

In the plate circuit of the sync separator are found parts of the waveform separating circuits. These components are C_3, which is part of a differentiator circuit feeding the horizontal AFC, and R_5 and C_4 which form an integrating network. This circuit feeds the vertical sync pulses to the vertical deflection oscillator.

QUESTIONS

1. What is meant by amplitude separation? By waveform separation?
2. What is the function of the integrator circuit? The differentiator circuit?
3. Why is an integrator circuit called a low-pass filter, and why is a differentiator called a high-pass filter?
4. Explain why sync separators use signal bias.
5. What is noise setup? How can it be minimized?
6. What is the effect on the composite video signal if it loses its dc component?
7. How can the dc component be restored?
8. Explain how the equalizing pulses prevent a loss of interlace scanning.
9. Explain how R_2 and C_2 in Figure 12-8 are used to minimize the effects of impulse noise.
10. What would be the result if C_1 of Figure 12-8 became leaky?
11. Explain how the noise gate transistor of Figure 12-9 is able to minimize the effects of impulse noise.
12. What would be the effect on receiver operation if C_{304} in Figure 12-10 shorts?
13. What is the purpose of the vertical countdown circuit?
14. What are the advantages of the vertical countdown circuit?
15. What is one of the chief differences between the Zenith and RCA vertical countdown circuits?
16. What is the frequency of the master oscillator in the Zenith and in the RCA vertical countdown systems?
17. What are the three functions performed by the Zenith vertical countdown chip?
18. What stages might be responsible for the following trouble symptoms? (a) loss of both vertical and horizontal sync, (b) loss of horizontal sync only, (c) loss of vertical sync only, (d) horizontal pulling, (e) vertical jitter.
19. In what ways does the transistor sync separator of Figure 12-9 differ from the vacuum tube circuit of Figure 12-14?

chapter 13

Deflection Oscillators

13-1 introduction

This chapter is devoted to a discussion of those sections of the television receiver that generate the deflection sawtooth for both horizontal and vertical scanning. The deflection oscillators and sync separator circuits are fundamental to the operation of all receivers and are found in vacuum tube, solid-state, color, and black & white television receivers.

The receiver deflection oscillators are synchronized to the transmitter deflection oscillators by means of the horizontal and vertical sync pulses that are part of the composite video signal. These pulses force the free-running receiver deflection oscillators to assume the frequencies of the pulses.

In the block diagram of the color receiver shown in Figure 13-1 those circuits to be discussed in this chapter are shaded. The horizontal *automatic frequency control* (AFC) may be thought of as part of the synchronizing circuit since it directly controls the frequency of the horizontal oscillator. However, its operation is so closely related to the operation of the horizontal oscillator that its theory of operation will be discussed in this chapter.

13-2 the deflection oscillator and sawtooth formation

The scanning process at both the camera and the picture tube that is responsible for producing a raster requires simultaneously synchronized horizontal and vertical deflection. Horizontal and vertical deflection of the picture tube electron beam is accomplished by ac magnetic fields that are generated by the yoke. These magnetic fields are created by ac sawtooth currents that pass through the horizontal and vertical windings of the yoke.

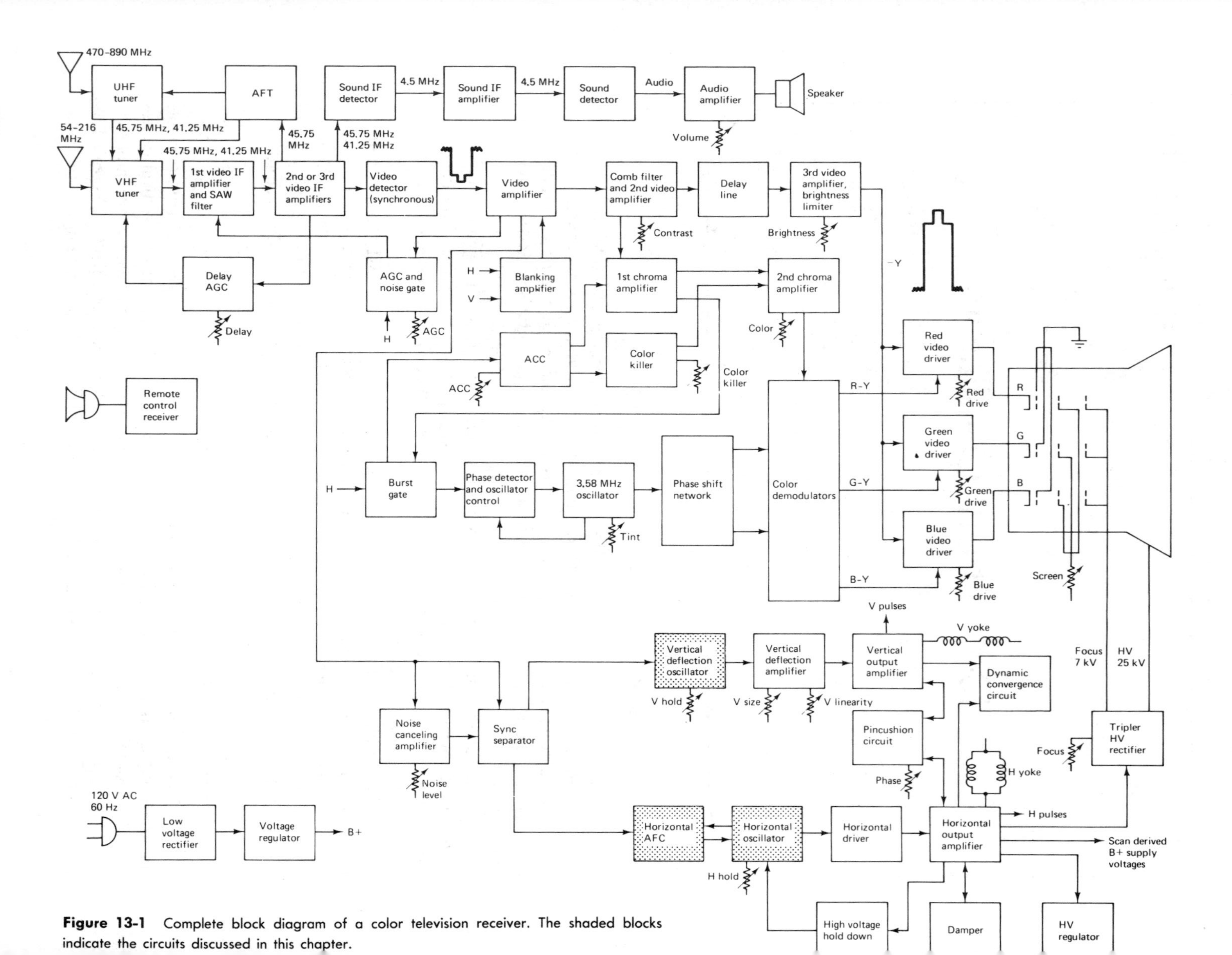

Figure 13-1 Complete block diagram of a color television receiver. The shaded blocks indicate the circuits discussed in this chapter.

The sawtooth deflection current has three important characteristics that affect the television picture:

1. Amplitude
2. Linearity
3. Frequency

Sawtooth amplitude characteristics. The peak-to-peak amplitude swing of the sawtooth determines the size of the raster. Figure 13-2 illustrates the effect of changing the amplitude of the horizontal deflection sawtooth. The top drawing shows a normal size raster with a test pattern of a circle. In addition, both the vertical and the horizontal deflection sawteeth

Figure 13-2 Effect on the raster of reducing the deflection sawtooth amplitude.

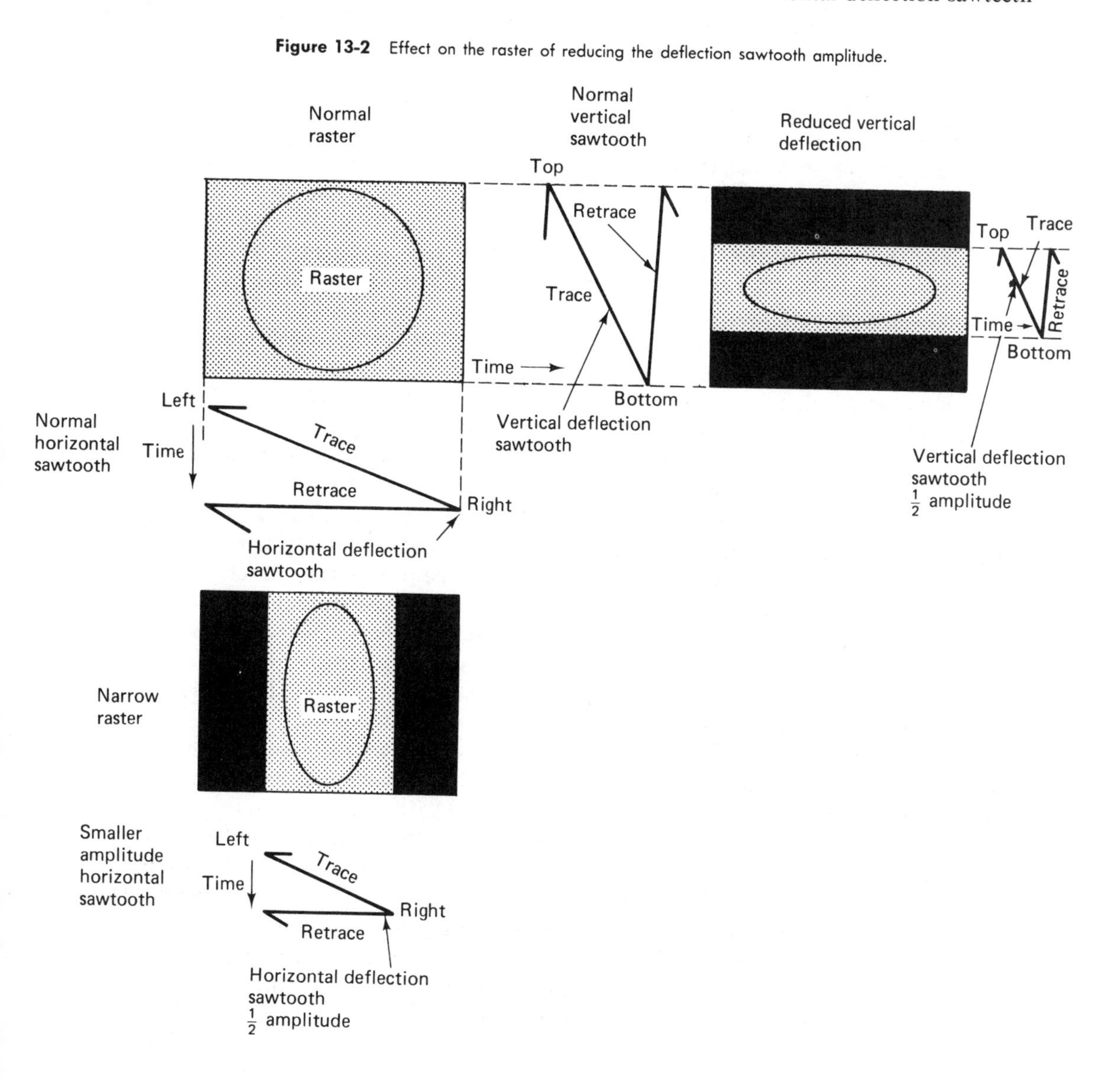

responsible for it are also shown. The normal-size raster fills the picture tube screen and the television picture of a circle is clearly indicated. In the lower drawing the horizontal deflection sawtooth has been reduced to one-half its normal amplitude. As a result, the raster does not fill the screen horizontally and the circle becomes distorted into a vertically positioned ellipse.

If the vertical deflection sawtooth amplitude is varied, the normal circular test pattern will be distorted into an ellipse oriented in the horizontal direction. This is also shown in Figure 13-2. The diagram on the right shows the effect on raster height of reducing the vertical deflection sawtooth to one-half of its full raster amplitude. The picture of the test pattern is flattened into a blimp-like shape.

In summary, the amplitude of the horizontal deflection sawtooth determines the *width* of the raster, and the amplitude of the vertical sawtooth determines the *height* of the raster.

Sawtooth linearity characteristics. The ideal deflection sawtooth is characterized by a trace time that forces the electron beam to move at a uniform rate across the screen. When viewed on an oscilloscope such a sawtooth will have a trace period that is a straight line. The straightness of the trace portion of the sawtooth is referred to as its *linearity.* If the linearity of the TV *camera* deflection sawtooth is perfect, any curvature or bending of the trace portion of the *receiver* deflection sawtooth will result in a picture that is badly distorted. In addition, the raster brightness will change in step with raster nonlinearity because of the variation in electron beam writing speed.

Figure 13-3 illustrates how sawtooth linearity affects picture quality. In the upper left-hand portion of the diagram is shown a raster and the linear horizontal and vertical deflection sawteeth that formed it. On the raster is a picture of a stick figure. This is how the raster and picture appear at the transmitter. This picture must be reproduced at the receiver without distortion. To the right of this drawing are shown the results at the receiver of a nonlinear horizontal deflection sawtooth.

In the center top drawing the horizontal deflection sawtooth is such that the electron beam remains on the extreme left side of the raster for a longer period of time than it does at the transmitter. This means that from the center of the picture tube raster to the right side of the screen almost the entire transmitted picture must be reproduced. As a result, the left side of the picture will appear compressed and the right side will appear stretched. In the picture, the stick figure's left shoulder appears larger than the right shoulder. In addition, the left side appears to be brighter than the right side. This is because of the slower writing speed of the electron beam on the left side of the screen.

The drawing on the far right of Figure 13-3, shows the results of horizontal nonlinearity opposite to that just discussed. As a result, the raster will be stretched on the left side and compressed on the right.

The effect of vertical deflection nonlinearity is illustrated in Figure 13-3 in two drawings located below the drawing of the transmitter's raster. In the first drawing the vertical nonlinearity is such that the electron beam is forced to move rapidly from the top of the raster to the center and slowly from the center to the top of the raster. The result of this nonlinearity is to cause the picture to be stretched to the top and compressed at the bottom. The stick figure drawing shows this nonlinearity in that the head and neck are stretched and the legs and lower torso are compressed.

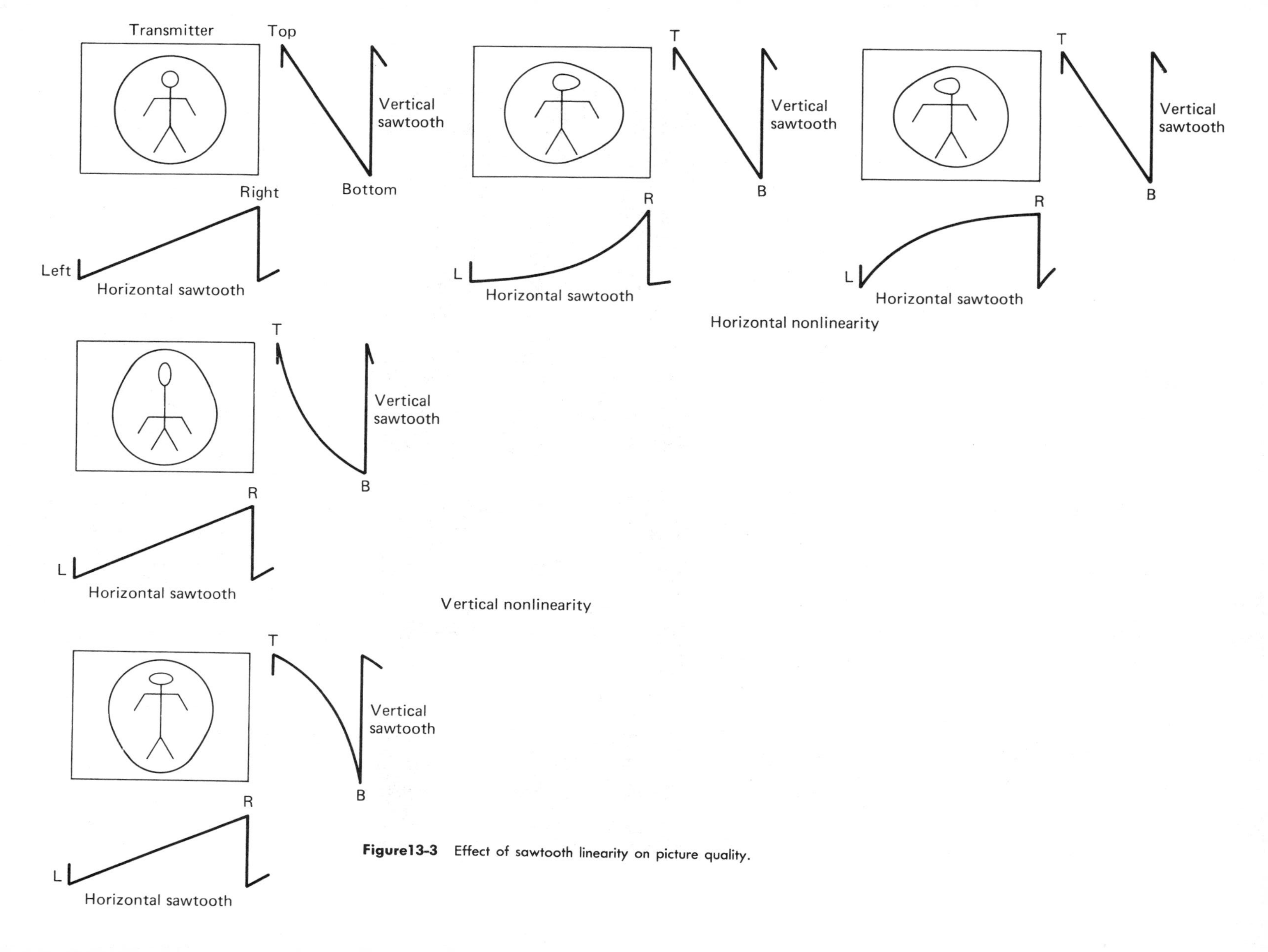

Figure13-3 Effect of sawtooth linearity on picture quality.

If the vertical nonlinearity is reversed, as indicated in the lower drawing, the top of the raster will be compressed and the bottom will be stretched. The effect on the stick figure is to flatten the head and to stretch the legs.

If either of the transmitter's sawteeth is nonlinear, a linear picture may only be obtained at the receiver by distorting the receiver sawteeth to match the transmitter's sawteeth. In this case, raster brightness will change because of the change in electron beam writing speed.

Receiver sawteeth amplitude and linearity may be adjusted by four back panel controls. The horizontal sawtooth in some receivers may be adjusted by the width and horizontal linearity adjustments. The vertical sawtooth is usually adjusted by the height and vertical linearity controls.

Sawtooth frequency. The frequencies of the horizontal and vertical deflection sawteeth are set by the FCC transmission standards. The vertical deflection sawtooth has a frequency of 60 Hz for black & white transmission and 59.94 Hz for color. The horizontal deflection sawtooth frequency is 15,750 Hz for black & white transmission and 15,734.26 Hz for color transmission. The transmitter and receiver deflection sawtooth frequencies must be identical. If they are not, the resultant television picture will be out of synchronization and will appear as shown in Figure 13-4.

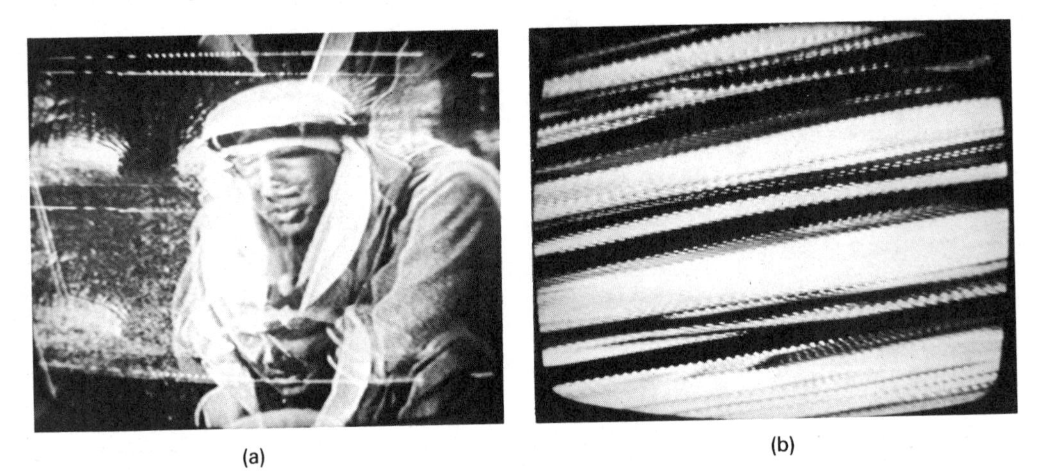

(a) (b)

Figure 13-4 (a) Loss of vertical synchronization; (b) loss of horizontal synchronization.

Sawtooth generation. All methods for generating a sawtooth are based on the charging and discharging characteristics of a capacitor.

The basic circuit for generating a sawtooth and its associated waveforms are shown in Figure 13-5. This circuit consists of the series connection of a battery (E_A), a switch (S_1), a resistor (R), and a capacitor (C). Across the capacitor is placed a second switch (S_2).

At the start, both S_1 and S_2 are open, and the voltages across R and C are zero. At some time T_1 switch S_1 is closed. This completes the circuit and allows the capacitor C to charge. The voltage across the capacitor rises in accordance with the universal charging curve for capacitors. If the time constant (TC) of R and C is long ($TC = R \times C$), the voltage across C will gradually increase toward the source voltage. At the same time the

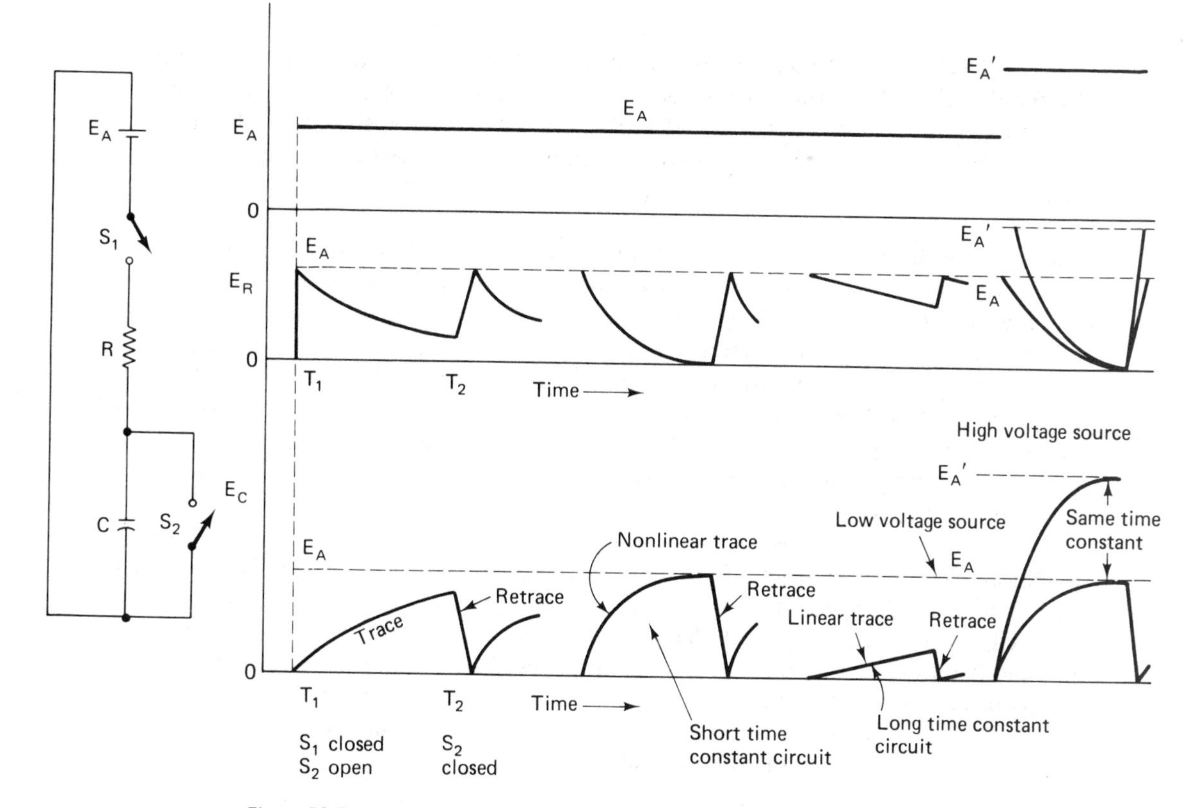

Figure 13-5 Basic circuit for generating a sawtooth and the effect of changing time constants and source voltages.

voltage across the resistor will drop toward zero. The voltage developed across the capacitor during this time can be considered the *trace* portion of the sawtooth.

If S_2 is now closed (T_2), the capacitor (C) will be short circuited and will discharge rapidly to zero. This corresponds to the *retrace* portion of the sawtooth. If this action is made to repeat itself, a train of sawtooth waves will be generated.

Sawtooth linearity. The *linearity* of the sawtooth is affected by the time constant of the circuit and the total time the capacitor is allowed to charge. If the time constant of the RC circuit is *short* compared to the period of trace, the capacitor will charge to the full source voltage before retrace begins. The sawtooth wave shape of the voltage (E_C) across the capacitor will be highly nonlinear and will appear as shown in Figure 13-5.

If the short-time-constant RC circuit is replaced by a long-time-constant circuit, the resultant sawtooth will be linear. The effect of this may be seen in Figure 13-5. Improved linearity is the result of the fact that the capacitor does not have sufficient time to charge to the source voltage before retrace begins. In other words, only a small portion of the entire capacitor charging curve is used to develop the trace portion of the sawtooth. As a general rule, the smaller the segment of the charging curve used, the better the linearity of the sawtooth.

The length of the segment of the capacitor charging curve that is used is determined by the duration of the trace period or by the time constant of the *RC* circuit. In actual practice, the period of charge time (trace) is fixed by the TV standards. Therefore, the primary means for determining sawtooth linearity is by selecting the proper *RC* time constant.

Sawtooth amplitude. The output *amplitude* of the sawtooth depends on the time constant of the *RC* network and the source voltage. This is also shown in Figure 13-5. In the drawing for E_C a short-time-constant *RC* circuit (small values of *R* or *C*) is seen to allow the capacitor to quickly charge to the dc source voltage (E_A). In this case, the peak-to-peak amplitude of the sawtooth becomes almost equal to the source voltage and will be maximum. Increasing the value of *R* or *C* will increase the time constant of the circuit and the capacitor will charge more slowly. This is again shown in Figure 13-5. As a result, at the point that S_2 closes and retrace begins, the peak-to-peak amplitude of the sawtooth will be less than the dc supply voltage (E_A).

As shown in Figure 13-5, the magnitude of the source voltage also affects sawtooth amplitude. This is because the charge developed across the capacitor is a fixed percentage of the source voltage for a given period of time. Therefore, if the source voltage E_A is doubled, (E_A') the capacitor will charge to twice its former value during the same period of time. Put another way, if the source voltage is only 20 V, and if the capacitor is allowed to completely charge, the sawtooth amplitude cannot exceed 20 V. If, however, the source voltage is increased to 300 V, and if the conditions are the same as before, the peak-to-peak amplitude of the sawtooth can be as much as 300 V.

13-3 the discharge transistor

In the preceding section it was seen that a sawtooth can be generated by allowing a capacitor to charge and then to discharge it rapidly by means of a switch placed across the capacitor. In actual practice, the switch placed across the capacitor can be replaced by a transistor or some other low-resistance device that can be turned on or off. An example of a sawtooth forming circuit using a transistor is shown in Figure 13-6. This circuit is called a *discharge transistor.*

As shown in both the schematic diagram and the ladder diagram below the circuit, a pulse generator drives the base of the transistor with a large narrow positive going pulse. The base to emitter voltage of the transistor is zero. Therefore, in the absence of pulses, the transistor is cut off. During this time the sawtooth forming capacitor C_2, found in the collector circuit, charges. Charging current flows from B− (ground) into the lower plate of C_2, out of the upper plate, and through R_2 to the positive terminal of the B supply, completing the circuit. This long time constant charging period corresponds to the trace portion of the sawtooth and will continue until retrace is initiated by the input pulse that momentarily forward biases the transistor and turns it on.

The schematic shows that the collector and emitter of the transistor are tied directly across the sawtooth forming capacitor C_2. When the transistor is turned on, the capacitor rapidly discharges through the low emitter to collector resistance of the conducting transistor. Since the time constant of the discharge circuit is very short, discharge occurs very rapidly. This period of time corresponds to electron beam retrace in the picture tube.

The discharge transistor is made to act like a switch by the pulses that are fed to

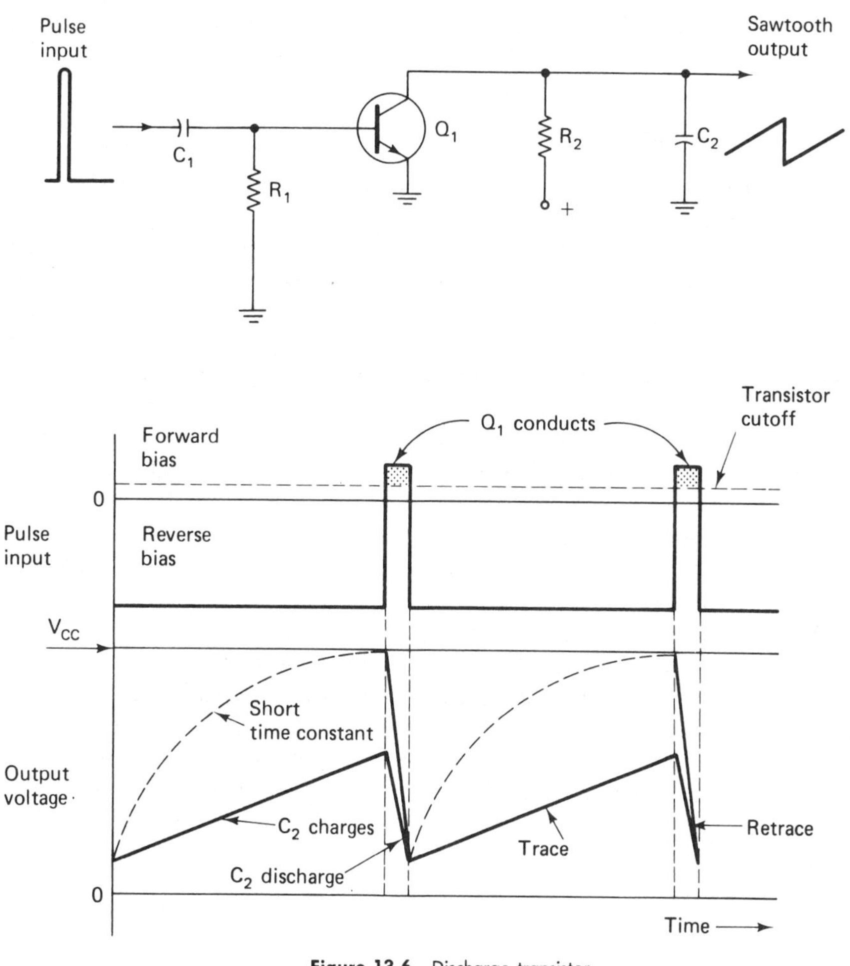

Figure 13-6 Discharge transistor.

the base of the transistor. These pulses may be obtained from a separate generator, as is done in this circuit, or they can be generated in an oscillator that contains the discharge transistor.

Sawtooth amplitude. Changing the time constant of the charging circuit of the discharge transistor results in important changes in the sawtooth waveform. For example, if R_2 of Figure 13-6 is reduced in value, the charging time constant will decrease, allowing C_2 to rapidly charge toward the B supply voltage. The net effect of this is to increase the peak-to-peak amplitude of the sawtooth. However, since a greater percentage of the charging curve is used, the sawtooth linearity becomes poorer. Of course, increasing the time constant of the charging circuit by increasing the value of R_2 will produce the opposite effect, a small amplitude, good linearity sawtooth.

In practice, R_2 may be made a variable resistor and is used to control the amplitude of the deflection sawtooth. This method of controlling sawtooth amplitude is most commonly used with the vertical deflection circuit. The control is called the vertical *height* control and is usually found on the back apron of the television receiver.

13-4 the trapezoid

All modern television receivers use vertical and horizontal magnetic deflection of the picture tube electron beams to generate a raster. These magnetic fields are produced by sawtooth currents that pass through their respective yoke windings. The important thing to note is that the *voltage* waveform across the yoke inductance that is associated with the *sawtooth of current* flowing through it is not a sawtooth. It is a trapezoid waveform.

In the television receiver a sawtooth of current is made to pass through the winding of the deflection yoke. These windings consist of both resistance and inductance, and they may be represented by the series combination of *R* and *L* shown in Figure 13-7. As shown in this diagram, a sawtooth of current passing through the resistor will produce a sawtooth voltage drop across it. The same current passing through the coil will produce a negative going pulse across the coil. The total voltage appearing across the *RL* combination must be the sum of the individual voltage drops appearing across the resistor and coil. This waveform is shown at the bottom of the waveform ladder diagram and is called a *trapezoid.* If the resistance is very much greater than the inductance, the trapezoid will be more like a sawtooth. If the inductive reactance is larger than the resistance, the resultant waveform will look less like a sawtooth and more like a negative-going pulse.

In practice, the voltage feeding the output power amplifier that drives the yoke may have the appearance of a sawtooth, a trapezoid, or a square wave. This is largely determined by the internal resistance of the output amplifier driving the yoke. If it is a low-resistance device, the driving voltage will be a trapezoid. If the output power amplifier is a high-resistance device the driving voltage may be a sawtooth or a square wave. In general,

Figure 13-7 Resultant voltage and current produced by passing a sawtooth of current through an inductive circuit.

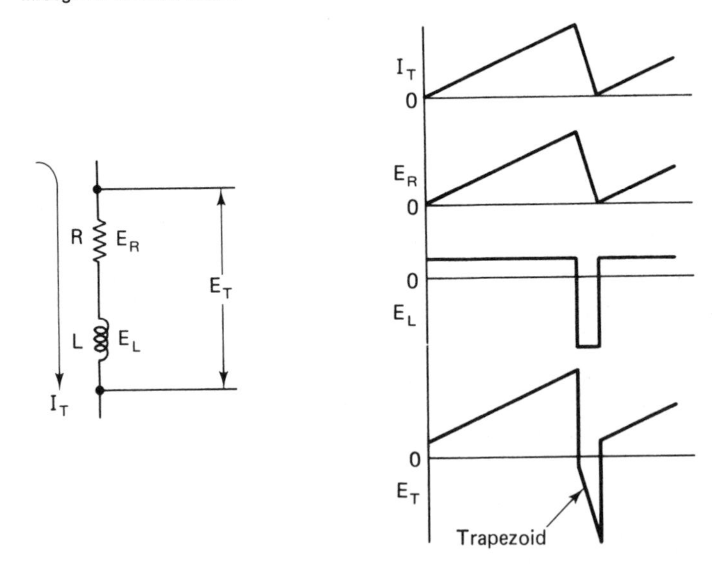

the greater the effective resistance of the yoke circuit, the more likely that the input driving voltage will look like a sawtooth.

The trapezoid generator. Whenever a sawtooth is required to drive the output power amplifier, a sawtooth generator such as the transistor discharge circuit can be used directly. If a trapezoid is required, it must be produced by the sawtooth generator. The usual way of doing this is to place a small resistor called a *peaking resistor* in series with the sawtooth-forming capacitor (C_2), as shown in Figure 13-8.

This circuit is essentially a discharge transistor circuit. Except for the addition of the small peaking resistor (R_p), this circuit and its operation are identical to that discussed previously. Because of its small size, the peaking resistor has little effect on the charging current passing through the capacitor. Therefore, the voltage developed across the capacitor will be a linear sawtooth.

It is a basic electrical fact that when the voltage across a capacitor increases at a linear rate, the charging current is constant. Therefore, the current flowing through the peaking resistor during the trace portion of the sawtooth will produce a constant voltage drop across it. During retrace the sawtooth-forming capacitor discharges. The direction of current flow flowing through the peaking resistor reverses direction and produces a voltage drop across the peaking resistor that is opposite in polarity to that produced during trace. This voltage is also constant because it is assumed that the voltage across the capacitor decays in a linear manner.

Adding the voltage drops across the sawtooth capacitor and the peaking resistor will provide the output trapezoidal waveform of voltage used to drive the output deflection amplifier.

13-5 the sawtooth oscillator

The purpose of the sawtooth generator used in a television receiver is to generate a sawtooth (or trapezoid) of sufficient amplitude to drive the output deflection amplifier. The sawtooth must also be of the proper frequency and have good linearity. In both color

Figure 13-8 Trapezoid generator.

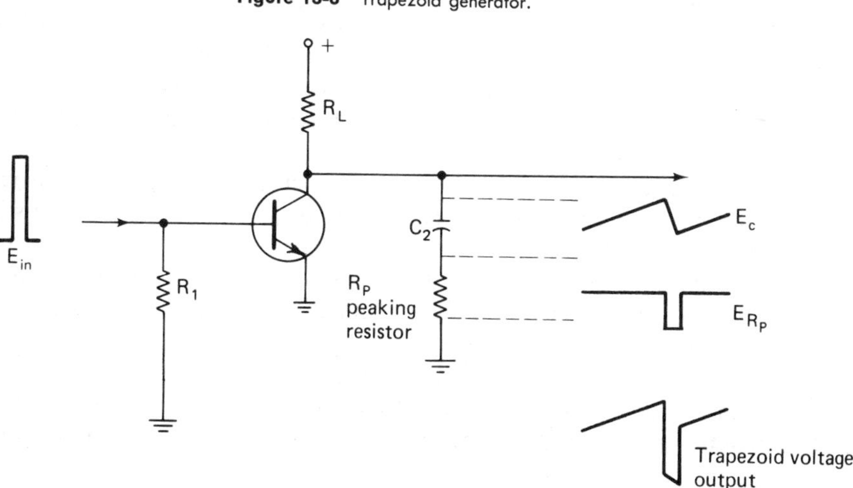

and black & white receivers synchronized free-running deflection oscillators of similar types are used. The vertical deflection oscillator is operated at 60 Hz (59.94 Hz for color) and the horizontal deflection oscillator operates at a frequency of 15,750 Hz (15,734 Hz for color). In every case, the oscillator provides the pulses that are necessary to drive a discharge transistor. The discharge circuit may be part of the oscillator itself or it may be a separate circuit.

The four deflection oscillators commonly used in both vertical and horizontal deflection circuits are as follows:

1. The blocking oscillator
2. The multivibrator
3. The overdriven Hartley oscillator
4. The IC oscillator

These oscillators may use either vacuum tubes or transistors for their operation.

13-6 the blocking oscillator

All feedback oscillators, whether they generate sine or sawtooth waveforms, have three basic electrical characteristics in common. They all must have an *amplifier* such as a FET or transistor. Oscillators must also have *frequency determining circuit elements* such as an *LC* or *RC* network, and they must also have sufficient *positive feedback* to sustain oscillation. If the oscillation uses an *LC* frequency determining circuit, it usually becomes a sine-wave generating oscillator.

An example of a transistor blocking oscillator is shown in Figure 13-9. The blocking oscillator, is essentially an Armstrong sine-wave generator. This is because positive feedback is obtained by means of a transformer. The transformer is designed so that changes in collector current in the primary induce voltages into the base-connected secondary of such polarity that they reinforce the original changes of collector current. The black dots drawn next to the primary and secondary windings of the feedback transformer indicate those terminals of the transformer that have the same instantaneous polarity.

The frequency of sinusoidal oscillation is determined by the inductance of the transformer and the lumped and distributed capacity shunting the transformer. In the case of a blocking oscillator, one-half cycle of oscillation takes place and then the oscillator stops, waits a considerable period of time, and then again generates one-half cycle of oscillation. In short, the blocking oscillator oscillates at two frequencies: the blocking rate and the ringing rate.

If the schematic diagram of an ordinary Armstrong oscillator is compared to that of a blocking oscillator, little or no differences would be noted. The main difference between them is the *Q* or damping factor of the tuned feedback transformer. In the blocking oscillator this *Q* is very small. This means that if the transformer should be pulsed, it would ring for only two or three cycles. In addition, as a result of the rapid loss of energy each cycle would be very much smaller than the cycle before it. In contrast, a pulsed high *Q* tuned circuit would ring for many cycles before its amplitude decayed to a negligible value.

This means that if the first cycle of oscillation in a blocking oscillator develops signal bias of sufficient amplitude, the second cycle will not have enough amplitude to drive

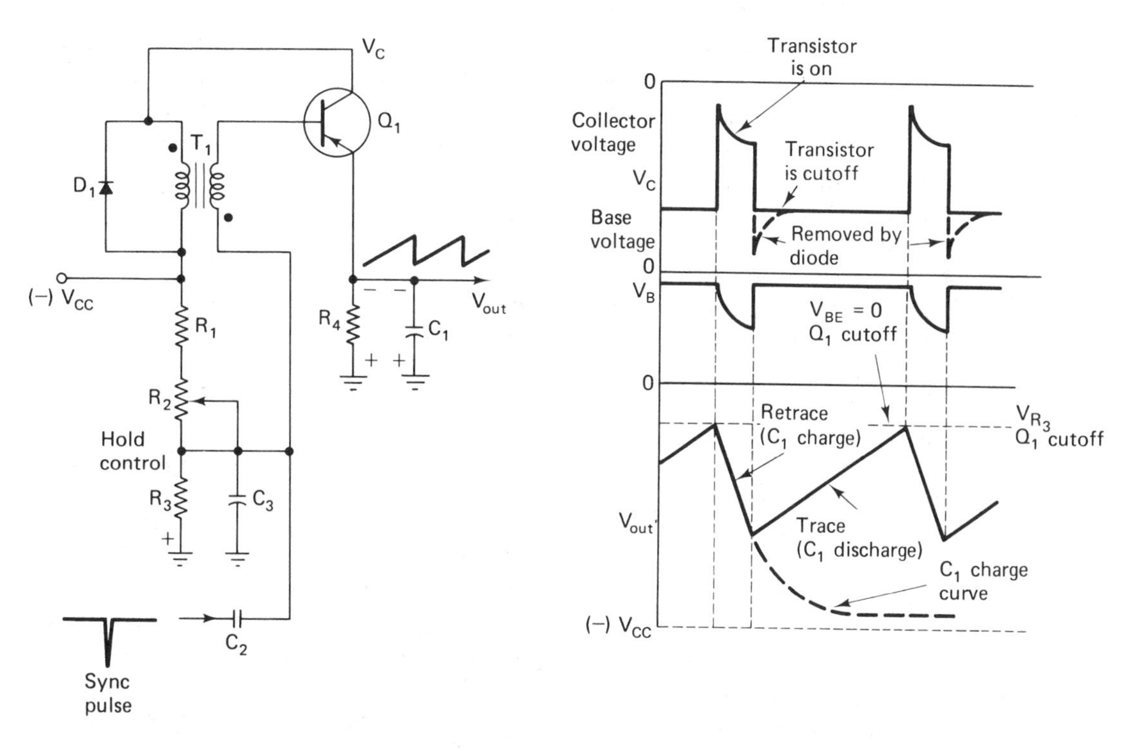

Figure 13-9 Transistor blocking oscillator with its associated waveforms.

the transistor or tube into conduction and it will block because reverse bias has developed across the signal bias *RC* network. The transistor will remain blocked until the *RC* network discharges sufficiently to allow conduction once again. The end result of this is that the transistor conducts for a short period of time and is cut off for a long period. During the time the transistor is cut off a sawtooth-forming capacitor is allowed to charge, forming the trace portion of the sawtooth. The retrace portion of the sawtooth is formed when the transistor is allowed to conduct.

Typical blocking oscillator circuit. Specifically, the operation of the circuit shown in Figure 13-9 is as follows. First, let it be assumed that the Armstrong oscillator has been operating for some time and that the circuit has reached steady-state operation. During the retrace time of the blocking oscillator's operation, the transistor will conduct and one cycle of damped oscillation (ringing) will take place. The duration of this period depends upon the amount of transformer inductance and the stray capacity. Transistor conduction places the parallel combination of R_4 and C_1 across the negative power supply and allows the capacitor to charge toward $-V_{cc}$. This can be seen in the waveform labeled V_{out}. The transistor bias is defined as the voltage between base and emitter. Therefore, the voltage difference between the voltage across R_3 in the base circuit and the charge across C_1 is the bias for the transistor. R_3 is part of a voltage divider consisting of R_1, R_2, and R_3 that is placed across $-V_{cc}$. During transistor conduction the voltage across C_1 will continue to increase if its voltage is smaller than the sum of the ac-induced voltage across the secondary of the transformer and the dc voltage drop across R_3. When the secondary voltage at the base of the transistor decreases, the voltage across C_1 will become larger than the base voltage and the transistor will go into cutoff. As a means of damping out any further oscillation, the diode D_1 is placed across the primary of the transformer. It conducts immediately after Q_1 is cut off and it shorts out any oscillation that may tend to continue. During Q_1 cutoff time the capacitor (C_1) will begin to discharge through R_4, forming the trace portion of the sawtooth. This will continue until the voltage across C_1 becomes slightly smaller than the voltage across R_2. At this point Q_1 conducts and retrace begins.

In effect, the blocking oscillator is a self-activated switch that is closed for a short time (retrace) allowing the sawtooth capacitor (C_1) to charge and is then open for a long time (trace), permitting the capacitor to discharge.

Blocking oscillator frequency. The time of Q_1 conduction (switch closed) is governed by the oscillation of the LC circuit formed by the transformer and its distributed capacity. The length of time that the transistor is blocked (switch open), allowing the trace portion of the sawtooth to be formed, is primarily determined by the time constant of R_4 and C_1. Increasing this time constant will lower the sawtooth frequency; decreasing it will increase the frequency.

The frequency of the blocking oscillator is also determined by the dc bias voltage between the base and emitter of Q_1. If the voltage across R_3 is decreased by making R_2 (hold control) larger, it will take more time for the discharge of C_1 to reach Q_1 cutoff; therefore, the frequency of the oscillator will decrease. Of course, the opposite will happen if the voltage across R_3 increases. Controlling the frequency of a sawtooth oscillator by means of dc bias variations is important in automatic frequency control applications.

Negative-going sync pulses fed to the base of the blocking oscillator transistor will momentarily increase the base voltage and force the oscillator to a higher frequency than its normal free-running frequency. In this way, the oscillator is forced to assume the frequency of the sync pulse.

13-7 multivibrators

Color and black & white television receivers may use two types of sawtooth generating multivibrators. In transistor receivers these are the collector-coupled multivibrators and emitter-coupled multivibrators. The horizontal deflection oscillator used in solid-state receivers commonly uses the emitter-coupled multivibrator for its deflection oscillator. Vertical deflection sawtooth oscillators are very often formed by combining three or four amplifier stages with the vertical output stage to form a collector-coupled multivibrator.

The collector-coupled multivibrator. The schematic diagram of a collector-coupled multivibrator is shown in Figure 13-10. Examination of the schematic will show that the multivibrator consists of two RC-coupled amplifiers connected in cascade, with the output of the last stage returning to the input of the first stage. As is the case with any oscillator, this oscillator requires sufficient positive feedback, sufficient amplification, and some type of frequency determining components for it to operate. The means of obtaining sufficient

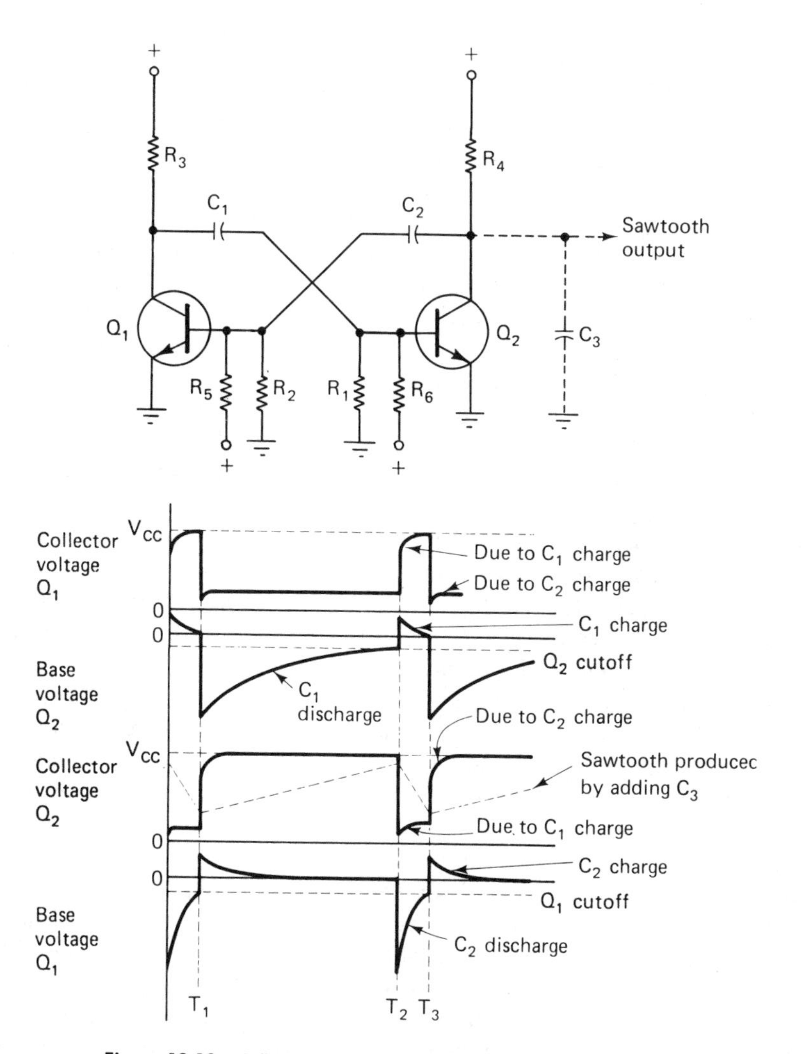

Figure 13-10 Collector-coupled multivibrator and its associated waveforms.

amplification is evident from the use of two transistor amplifiers connected in cascade. Positive feedback is obtained by taking advantage of the fact that a grounded emitter amplifier has a 180° phase reversal between input and output. Since two stages are used, a total of 360° phase shift is introduced between the input to Q_1 and the output of Q_2. Returning the output of Q_2 to the input of Q_1 provides the necessary feedback to sustain oscillation. The frequency and shape of the waveforms produced are determined by the time constants of *RC* coupling networks between the stages. In television receivers one of the networks has a short time constant and the other network has a long time constant. This results in a circuit that produces asymmetrical waveforms. This means that the positive and negative half-cycles have unequal time durations and have dissimilar shapes. Notice that R_5 and R_6 provide a small forward bias for each transistor to ensure that they can conduct and allow the circuit to self-start.

Waveform analysis. In the circuit of Figure 13-10 it is assumed that C_1 and R_1 have a long time constant and that C_2 and R_2 have a short time constant. Below the schematic diagram is a ladder diagram showing the waveforms that will be found on the base and collector of the multivibrator. The waveforms are arranged so that the collector voltage of one transistor is followed by the next stage base voltage.

The key to understanding how a multivibrator works is the idea that a given *change* in collector voltage will be capacitively coupled into the input of the next stage with both its peak-to-peak voltage change and the direction of change preserved. This can be seen by noting the changes that take place at time T_1 in Figure 13-10. Starting at the top of the ladder diagram, notice that the collector voltage of Q_1 drops to a value approaching zero. This large change in collector voltage is coupled to the base of Q_2 by coupling capacitor C_1. The decreasing collector voltage forces the base voltage of Q_2 to change in the same direction driving the transistor far into cutoff. As the collector current of Q_2 decreases toward cutoff, the collector voltage of Q_2 increases toward the B supply voltage. The large sudden increase in Q_2 collector voltage forces the base of Q_1 positive to drive Q_1 into conduction. This in turn forces the collector voltage of Q_1 to drop. At the same time, the positive base voltage (Q_1) draws base current that charges C_2.

The charging path for C_2 is as follows: hole current flows from the B supply through R_4 to C_2 to the base and emitter of Q_1 to ground. Because of the relatively small values of the base to emitter resistance of Q_1, and R_4, C_2 charges rapidly to the B supply voltage. The charging current affects the waveforms in two ways. First, the change in base voltage Q_1 due to the charging of C_2 is amplified and appears in the collector circuit of Q_1 as a negative-going overshoot. Second, the charging current passes through R_4, causing a voltage drop across it, thereby preventing the collector voltage of Q_2 from becoming equal to the B supply voltage at the time it was cut off.

Q_2 is held at cutoff by the charge developed across C_1 during a previous cycle. The length of time Q_2 is kept cut off determines the trace time of the deflection sawtooth and depends upon the time needed for C_1 to discharge to the cutoff level of the transistor (Q_2). The discharge time of C_1 is determined by the time constant of C_1 and R_1 and the dc emitter to collector resistance of Q_1. This time constant is made long by making C_1 and R_1 large. With the exception of the base voltage of Q_2, all of the voltages on the other transistor elements remain constant during the time Q_2 is cut off. Q_2 is forced into conduction when the base voltage of Q_2 becomes equal to the cutoff value of Q_2 (time T_2). This causes the collector voltage of Q_2 to drop, driving the base of Q_1 negative and forcing it into cutoff. C_2 now rapidly discharges through R_2 and the emitter to collector dc resistance of the transistor Q_2. When C_2 has discharged sufficiently, the base voltage of Q_1 will reach the cutoff level of the transistor and it will begin to conduct again. The time of C_2 R_2 discharge determines the retrace time of the deflection sawtooth.

During the time T_2 to T_3 the collector voltage of Q_1 rises toward B+. This change in collector voltage is coupled into the base of Q_2 by C_1, forcing Q_2 to conduct very heavily. At the same time C_1 begins to charge toward the B supply voltage. Charging current flows from the B+ source through R_3 into C_1 to the base and emitter of Q_2, and then returns to B– (ground). The voltage drop developed across R_3 subtracts from the collector voltage of Q_1 (now in cutoff) preventing it from rising directly to B+.

The cycle will repeat itself when C_2 had discharged to the cutoff point of Q_1, allowing it to conduct.

Summary. An examination of the ladder diagram reveals a number of important facts. First, it can be seen that one of the transistors (Q_1) conducts for a long period of time and the other (Q_2) conducts for a short period of time. These transistors are referred to as the *normally on* and the *normally off* transistors, respectively. Second, it should be noted that the normally off transistor behaves like a switch that is open for a long time and closed for a short time. This switching is exactly the type needed for generating a sawtooth as was previously discussed concerning both the blocking oscillator and discharge transistor. The collector coupled multivibrator can be converted into a sawtooth generator by placing the sawtooth forming capacitor (C_3) across the normally off transistor. This is shown in Figure 13-10 and by the dotted lines indicating the position of the capacitor and the change in waveform produced.

The size and linearity of the sawtooth may be controlled by varying R_4. In vertical deflection circuits this control is called the *height control.*

Frequency of operation. The frequency of the multivibrator is controlled by making R_1 variable. It is useful to remember that the location of the frequency control identifies the normally off transistor. The frequency of the collector coupled multivibrator may also be controlled by changing the bias of the transistors. An increase in the bias of the normally off transistor will cause the frequency of the sawtooth generator to decrease. Decreasing the bias will cause its frequency to increase. This is because an increase in bias will move the base waveform farther away (more negative) from cutoff. Therefore, it will take a longer time for the discharge of C_1 to reach cutoff, resulting in a lower sawtooth frequency. Increasing the bias of the normally on transistor will cause the frequency of the oscillator to increase.

Collector-coupled multivibrator. An example of a typical collector-coupled multivibrator used as a vertical deflection oscillator is shown in Figure 13-11.

It should be noted that the time constant of the coupling circuit C_1, R_1, and R_2 is very much longer than that of C_2 and R_3. This fact, plus the location of the hold control R_2, indicates that transistor Q_2 is normally off and that Q_1 is normally on. This conclusion is further reinforced by the waveforms found on the base of Q_1 and Q_2. The waveform at the base of Q_2 indicates that the normally off transistor has a relatively long period during which C_1 discharges. During this time Q_2 is cut off. Changing the time constant of C_1, R_1 and R_2 by varying the hold control R_2, causes a change in the discharge time of C_1, thus causing the frequency of the oscillator to change.

Synchronization of the oscillator is obtained by feeding a negative-going pulse into the emitter of the normally off transistor Q_2. Because of the discharge of C_1 these pulses force the transistor into conduction before it normally would. In this way, the oscillator is forced to assume the frequency of the synchronizing pulses.

The emitter-coupled multivibrator. An example of an emitter-coupled multivibrator is shown in Figure 13-12. In this circuit Q_2 is the normally off transistor and transistor Q_1 is normally on.

Notice that R_6 provides a small amount of forward bias to Q_2 to ensure that the oscillation is self-starting. If R_6 were not present, the base-to-emitter bias of each transistor would be zero and they would be cut off. After operation has started, Q_2 is kept cut off

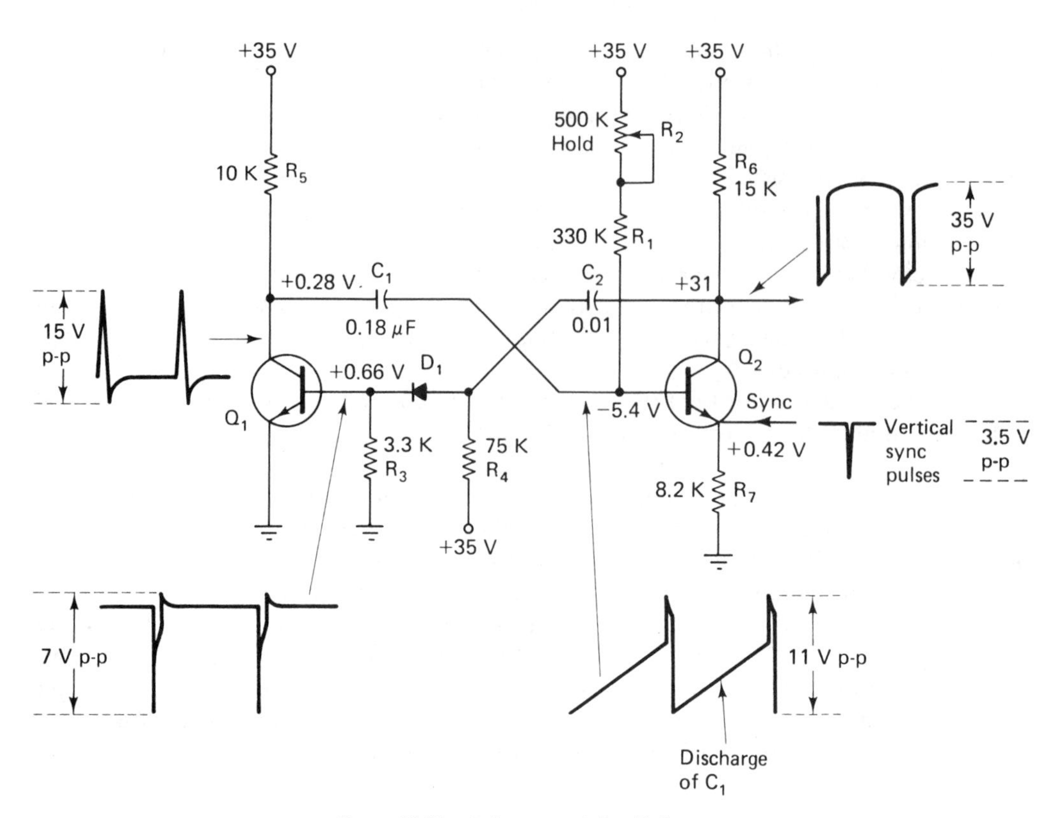

Figure 13-11 Collector-coupled multivibrator.

Figure 13-12 Emitter-coupled multivibrator.

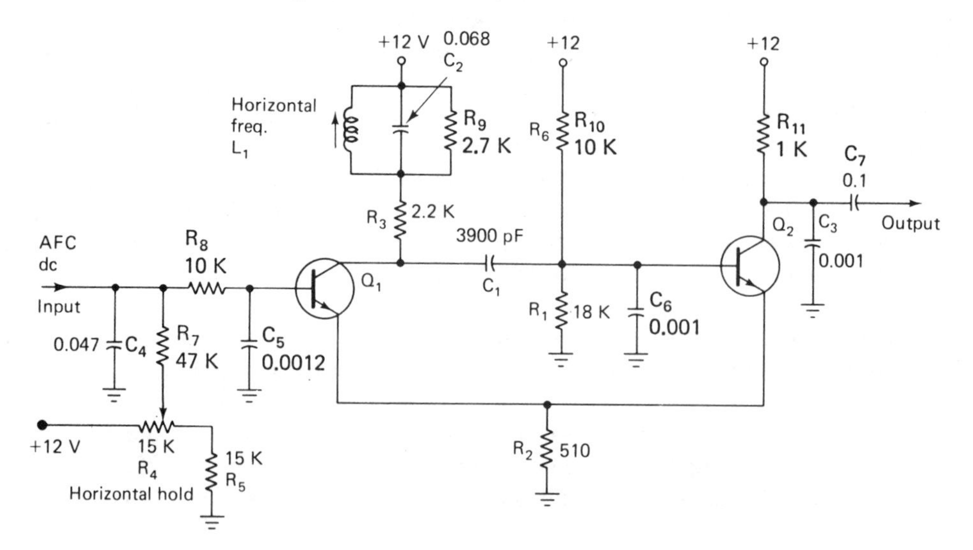

by the discharge of C_1 through R_1, R_2, and Q_1. The long time constant of this circuit determines the time duration of trace and consequently the frequency of the deflection sawtooth. When C_1 has discharged to the point where Q_2 conducts, a large positive voltage drop appears across R_2, biasing Q_1 into cutoff. The collector voltage of Q_1 then rises toward cutoff. This positive going increase in voltage is coupled to the base of Q_2, forcing it to conduct harder. Because of the base current flow, C_1 charges. The charging time constant of C_1 is short and is determined by C_1, Q_2, R_2, and R_3. This period corresponds to the horizontal retrace period of the sawtooth. As C_1 charges, the voltage across R_2 falls because of the reduced conduction of Q_2. When this voltage drops low enough, Q_1 begins to conduct. The resultant drop in collector voltage is coupled to the base of Q_2, forcing it into cutoff. Thus, the cycle is brought back to its starting point. As a result of the above action, Q_2 acts like a switch that allows C_3 to charge for a long time and discharge for a short time, forming the deflection sawtooth.

The ringing coil. As a means of increasing the frequency stability of the oscillator shown in Figure 13-12, a tuned circuit (L_1 and C_2) is placed in the collector circuit of Q_1. This tuned circuit is shock excited by the periodic switching of the transistor from an on to an off condition. As a result, a sine wave of oscillation at a frequency of 15,750 Hz is developed across the tuned circuit. The sine wave is added to the collector voltage of Q_1 and the base voltage of Q_2. The effect of this is to make the discharge curve of C_1 rise toward the cutoff level of Q_2 at a much faster rate. Noise immunity and frequency stability are improved because larger noise pulses are necessary to force the normally off transistor of the oscillator into premature conduction.

The frequency of this multivibrator is controlled by the horizontal hold control (R_4) and by the horizontal frequency control (L_1). The horizontal hold control (R_4) and R_5 form a dc voltage divider circuit that provides the bias for Q_1. Moving the adjustable arm of R_4 toward the +12-V end of the potentiometer increases the forward bias of Q_1 and tends to make it conduct harder when it does conduct. This in turn causes the collector voltage of Q_1 to drop to a lower voltage. The change in voltage is coupled to the base of Q_2, driving it deeper into cutoff than before. This in turn means that it will take C_1 longer to discharge to the point where Q_2 begins to conduct again, in effect decreasing the frequency of the oscillator. By similar reasoning, it can be seen that decreasing the forward bias of the normally on transistor Q_1, will cause the frequency of the oscillator to increase. Dc control of the frequency of oscillation is important for automatic frequency control of the oscillator.

Synchronization of deflection oscillators. For a stable television picture to be produced, the vertical and horizontal deflection oscillators must be forced to assume the same frequency and have the same timing as the transmitted vertical and horizontal sync pulses. This condition is referred to as *synchronization.* In practice, only the vertical deflection oscillator is directly synchronized by pulses. The horizontal oscillator frequency is indirectly controlled by an automatic frequency control (AFC) circuit placed between the oscillator and the sync separator.

In order for a deflection oscillator of the blocking oscillator or multivibrator type to be directly synchronized by pulses, the three basic conditions listed below and illustrated in Figure 13-13 must be met.

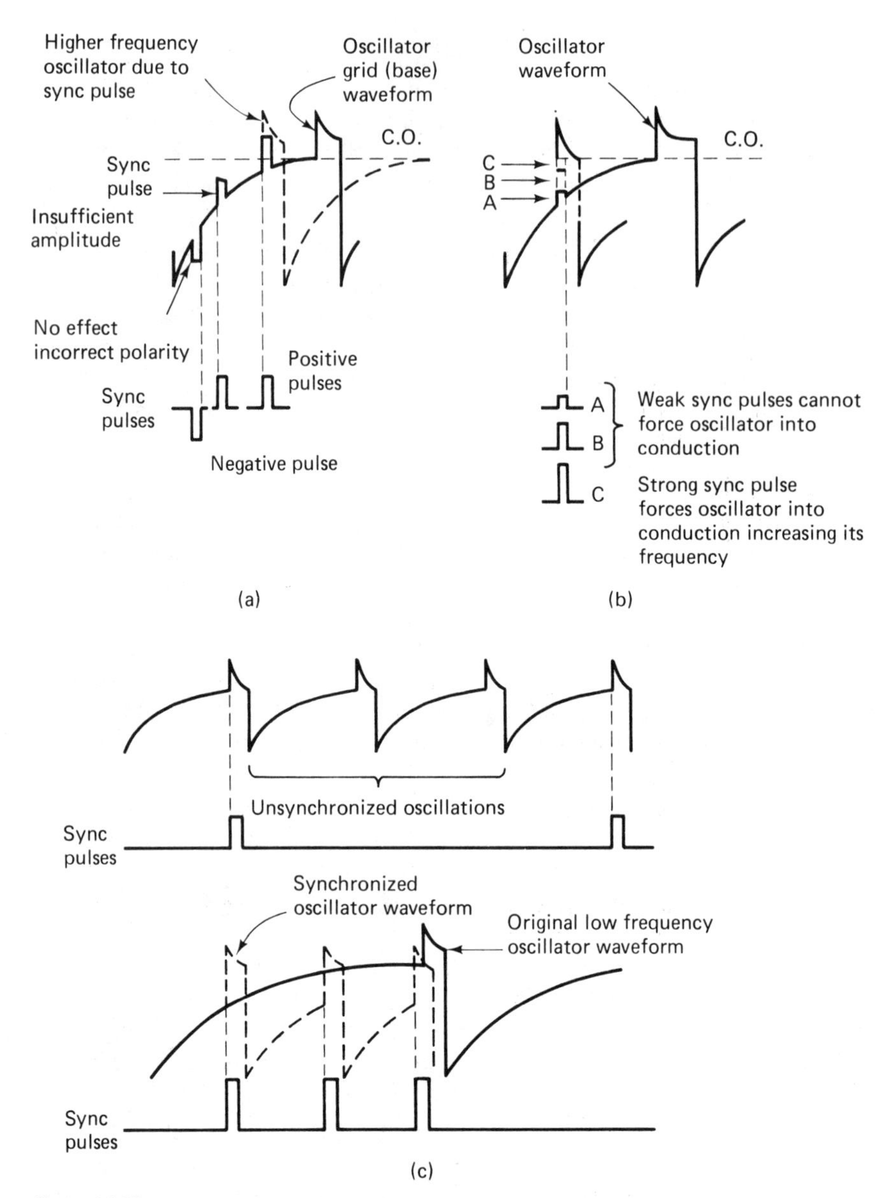

Figure 13-13 (a) How polarity affects synchronization; (b) how amplitude affects synchronization; (c) how sync frequency affects synchronization.

1. The sync pulses must have the correct polarity.
2. The sync pulses must have sufficient amplitude.
3. The frequency of the free-running oscillator must be slightly lower than the frequency of the sync pulses.

Synchronization of an NPN transistor multivibrator or blocking oscillator sawtooth generator is usually accomplished by feeding a positive-going pulse into the base of the *normally off* half of the multivibrator or into the base circuit of the blocking oscillator. Negative-going sync pulses are required if the transistors are PNP type. The action of the sync pulse shown in Figure 13-13(a) is to momentarily increase the base voltage of the oscillator above the cutoff value and force the normally off transistor into conduction. This begins retrace and starts a new sawtooth cycle. In Figure 13-11 a negative sync pulse is applied to the emitter of the normally off transistor Q_2 forcing it into conduction, thereby starting retrace.

If the sync pulse is the wrong polarity, it will force the transistor further into cutoff, which will inhibit synchronization.

If the sync pulse amplitude is insufficient [Figure 13-13(b)], it will not be able to force the normally off transistor into conduction and synchronization will not be achieved.

If it were desired to use negative-going sync pulses in a transistor multivibrator circuit such as the NPN transistor circuit of Figure 13-11, the pulses might be applied to the base of the normally on transistor Q_1. The circuit would then act as an overdriven amplifier and would invert the sync pulses so that positive-going sync pulses are applied to the base of Q_2. This arrangement has the additional advantages of amplifying weak sync pulses and shaping them into rectangular pulses.

In order to understand why the free-running frequency of the oscillator must be lower than that of the sync pulses, consider the situation shown in Figure 13-13(c) in which the oscillator frequency is three times higher than the sync pulse. Under these conditions, the oscillator produces three sawteeth for each sync pulse. As a result, two sawteeth are uncontrolled and free-running for every third sawtooth that is forced to assume the frequency and phase of the sync pulse.

Now consider the reverse condition in which the sync pulses are three times higher in frequency than the sawtooth output of the oscillator. In this case, each sync pulse having sufficient amplitude will force the oscillator into conduction, driving the sawtooth oscillator into its retrace period. As a result, the output frequency of the oscillator is completely controlled by the sync pulses. In practice, the deflection oscillator is only slightly lower in frequency than the sync pulses, but the basic principle is the same.

13-8 sine-wave horizontal deflection oscillators

Sine-wave oscillators are often used as horizontal deflection oscillators. This is possible, provided that the oscillator is overdriven so that the transistor is made to act like a switch that allows a sawtooth-forming capacitor to charge and discharge.

The frequency of the horizontal deflection oscillator is always controlled in one way or other by the AFC circuit feeding it. In blocking oscillators and multivibrators a dc error voltage generated by the AFC circuit controls the frequency of the oscillator. Frequency control of the sine-wave oscillator may use the dc error voltage developed by the AFC circuit directly, as shown in the block diagram of Figure 13-14, or the frequency of the oscillator may be controlled by a reactance transistor placed between the AFC circuit and the oscillator as shown by the dashed lines in Figure 13-14.

Directly driven sine-wave oscillators. An example of a directly driven sine-wave deflection oscillator is shown in Figure 13-15. In this circuit the AFC diodes (D_1 and D_2) compare the horizontal sync pulse and the horizontal oscillator signals. If there is a frequency

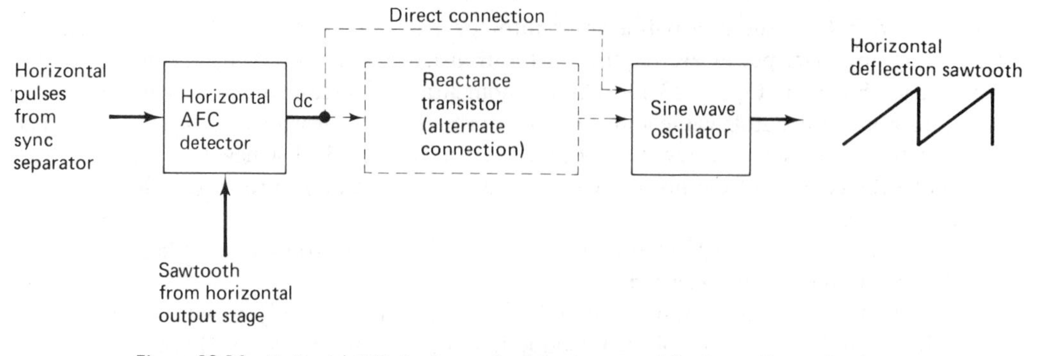

Figure 13-14 Horizontal AFC circuits may feed the sine-wave deflection oscillators directly or through a reactance transistor.

difference between them, they develop a dc error voltage. This is fed via R_1, R_2, and R_{14} directly to the base of the horizontal deflection oscillator. The oscillator (Q_1) is identified as an electron-coupled Hartley oscillator by the tap on the tuned circuit inductance L_1. This inductance is variable and is used as the horizontal hold or frequency control. The oscillator output is taken from the emitter of Q_1. The frequency of the oscillator is 15,750 Hz for black & white and 15,734 Hz for color. The oscillator operates the transistor in Class B. Thus, collector current flows for a half-cycle pulse during the positive peak of the base voltage. These periods of conduction roughly correspond to the time of horizontal retrace. The output of the oscillator Q_1 is taken from the emitter and is used to energize the driver Q_2. This stage is also operated Class B. It produces a square-wave output that is coupled by T_1 to the base of the horizontal output amplifier Q_3.

Dc frequency control. The dc error voltage developed across C_4 by the AFC diodes is able to directly change the frequency of the oscillator without the use of a reactance device because of the relatively short time constant of C_1, R_1, and R_2. These components are responsible for developing the signal bias for the oscillator. As is the case in any rectifier power supply, the bias voltage will contain a sawtooth-shaped ripple component. The peak-to-peak amplitude of the ripple depends upon the time constant of the rectifier filter C_1, R_1, and R_2. If the time constant is long, the ripple will be small. If the time constant is short, as is the case in this circuit, the ripple voltage will have a fairly large amplitude.

Ordinarily, changing the bias level of the oscillator by adding the dc error voltage developed by the AFC circuit to that of a long-time-constant filter signal bias network will simply move the bias level to a higher or lower value. This in turn will move the oscillator base signal voltage to a higher or lower bias level and will have little effect on oscillator frequency.

If, however, a *positive* AFC error voltage is added to the bias of a *short time constant* circuit, it will move the oscillator bias, its sizable sawtooth ripple voltage, and the oscillator sine-wave voltage closer to zero bias. The addition of the sawtooth and the sine wave results in a distorted base voltage that reaches the cut-off point of the transistor sooner than it would have if only a sine wave were present. Since the transistor is forced into conduction sooner than it should have been, *its frequency of operation increases.* Of course, reversing the polarity of the AFC error voltage will cause the frequency of the oscillator to decrease.

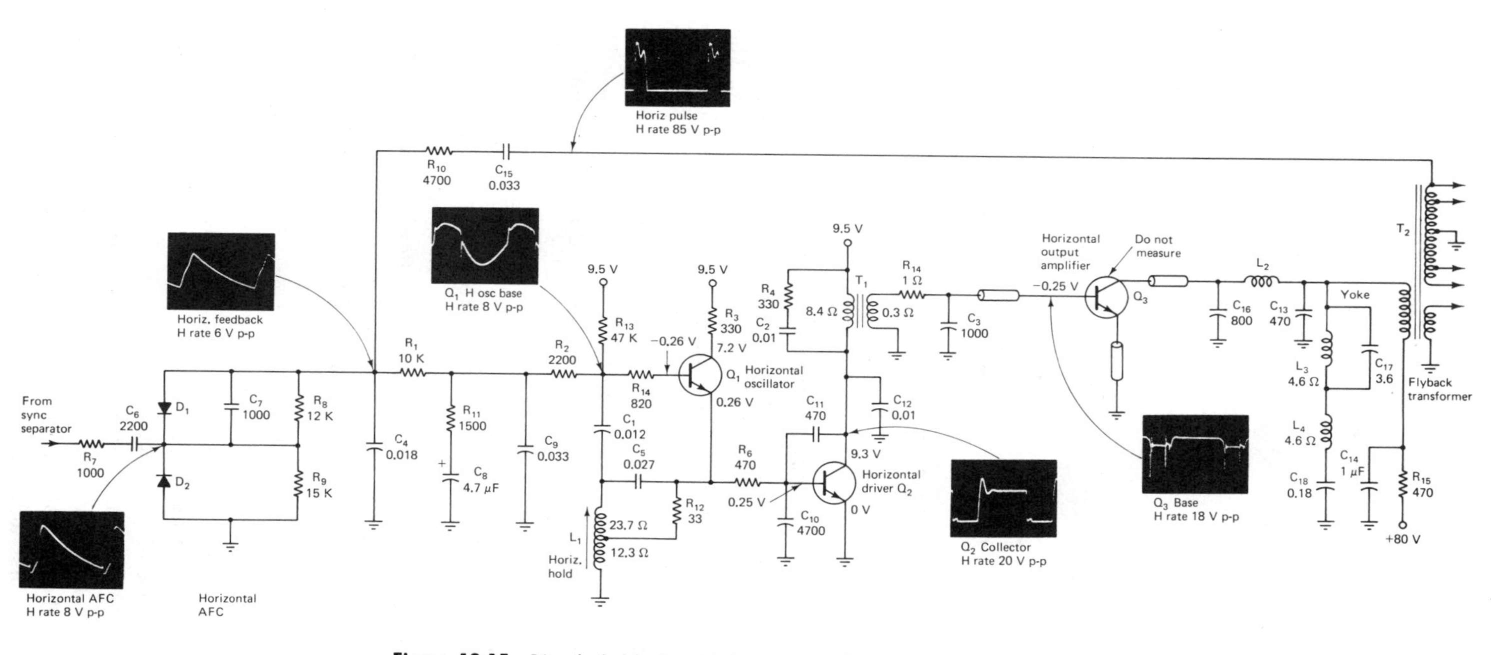

Figure 13-15 Directly fed horizontal sine-wave oscillator and output circuit. (Courtesy of RCA.)

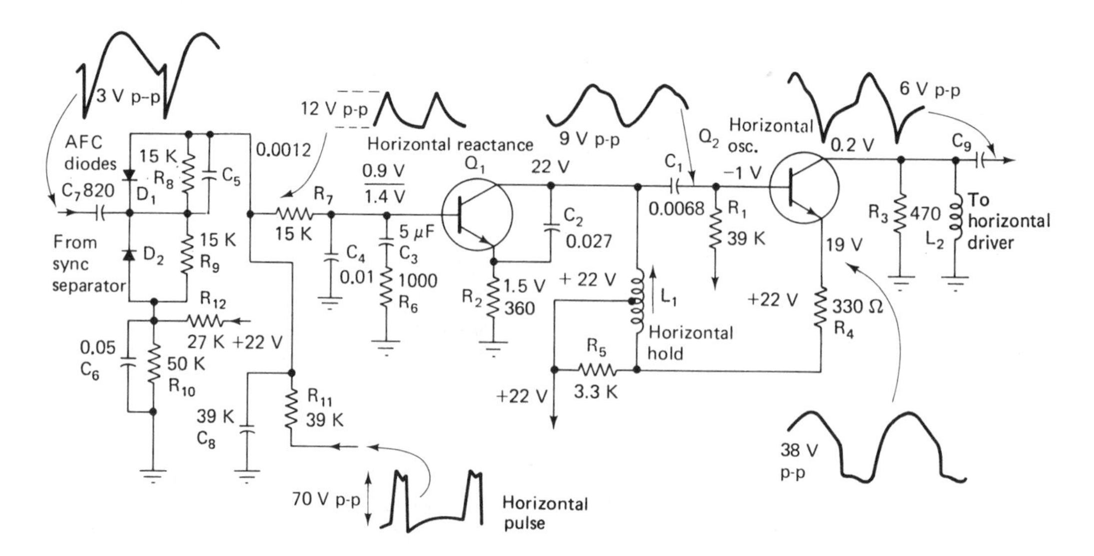

Figure 13-16 Transistor sine-wave horizontal deflection oscillator and its associated reactance transistor.

Controlled transistor sine-wave oscillator. Figure 13-16 shows a popular variation of a transistor sine-wave oscillator. In this circuit an Armstrong oscillator is used. The oscillator is overdriven so that the output current will flow in short-duration pulses. The transistor acts like a switch that is closed for a short time and open for a longer time. Signal bias is used in the oscillator to establish the transistor operating point. C_1 and R_1 form this bias network.

The output of the sine-wave transistor oscillator is usually fed to a separate driver stage. This stage isolates the oscillator from the horizontal output transistor amplifier and provides the required power amplification necessary to drive the output amplifier.

The reactance transistor. A reactance transistor circuit may be defined as a voltage controlled reactance. Reactance transistors may be designed to appear to the oscillator that they are shunting as either an inductance or as a capacitance. In either case, a change in bias of the Class A operated reactance transistor will cause the effective reactance of the transistor to change, which in turn will force the horizontal oscillator to change frequency. In this circuit, the bias of the reactance transistor is the combined result of the AFC derived base voltage and its emitter voltage.

A reactance transistor shunts the oscillator tank circuit and appears to the oscillator as a reactance because its ac collector current and ac collector voltage are 90° out of phase. In an ordinary amplifier the collector current and collector voltage are 180° out of phase. To achieve this phase shift, a feedback circuit (C_2 and R_2 in Figure 13-16) is used between the collector and the base of the reactance transistor. C_2 may be a capacitor or the interelectrode capacitance of the transistor, and R_2 is returned to ground.

The circuit shown in Figure 13-16 uses a reactance transistor (Q_1) between the AFC diodes and the horizontal oscillator. The theory of its operation shows that the reactance transistor acts like an *inductor.*

If the oscillator frequency should change, the AFC circuit would develop a dc error voltage that would be directly coupled into the base of the reactance transistor. If the transistor bias is forced to increase (base more positive), the collector current will increase. This in turn acts to decrease the effective reactance of the transistor which in effect causes the total inductance shunting the oscillator tuned circuit to decrease. As a result, the frequency of the oscillator will increase. The reverse would occur if the bias were to decrease.

Integrated circuit horizontal oscillator. Deflection oscillators may be made as part of an integrated circuit chip. Figure 13-17 is an example of such a horizontal deflection oscillator and its associated circuits. In this chip the B supply voltage (pin 4) is fed to a regulator, which in turn provides power to the remaining portions of the IC. The sync separator feeds a 6-V p-p horizontal pulse to pin 11 of the IC, which corresponds to the horizontal phase detector. Also feeding the phase detector at pin 9 is a horizontal sawtooth obtained by integrating (C_{702} and R_{703}) a positive-going horizontal pulse of 110 V p-p obtained from the flyback transformer T_{702}. The phase detector compares the horizontal sync pulse and the horizontal sawtooth and generates a dc error voltage (pin 8), which is then passed through a filter (C_{707}, R_{708}, C_{708}, R_{705}, and C_{706}) to remove any noise and correct for oscillator hunting. This dc voltage is also made adjustable by means of R_{707}, the horizontal hold control, and R_{709}, the horizontal subhold control. The horizontal oscillator is a voltage-controlled oscillator (VCO) which permits its frequency to be varied by changing its dc control voltage. This closed loop consisting of a phase detector, a VCO, and feedback, where the oscillator output returns to the phase detector, is called a phase-locked loop (PLL).

The horizontal oscillator feeds its output to the horizontal driver stage of the IC, which in turn develops an output voltage of 25 V p-p across the primary of the horizontal driver transformer. The output of the transformer then drives the base of the horizontal output amplifier.

13-9 troubleshooting deflection oscillators

Defects in the deflection oscillator can cause a number of raster symptoms. For example:

1. Oscillator stops functioning, resulting in no vertical or no horizontal deflection, depending upon which oscillator becomes defective.
2. Oscillator changes frequency. This forces the picture to be out of vertical or horizontal sync, depending upon which oscillator changes frequency.
3. Oscillator output waveform changes shape, resulting in poor picture linearity.
4. Oscillator output is reduced, reducing the size of the raster.
5. Oscillator is intermittent. Any one of the symptoms above suddenly appears and then disappears.

Oscillator stops functioning. Failure of the vertical oscillator produces different symptoms from those the horizontal deflection oscillator produces. Complete failure of the vertical oscillator will cause the loss of vertical deflection. This will result in a raster that consists of one horizontal line across the center of the screen. See Figure 13-18(a).

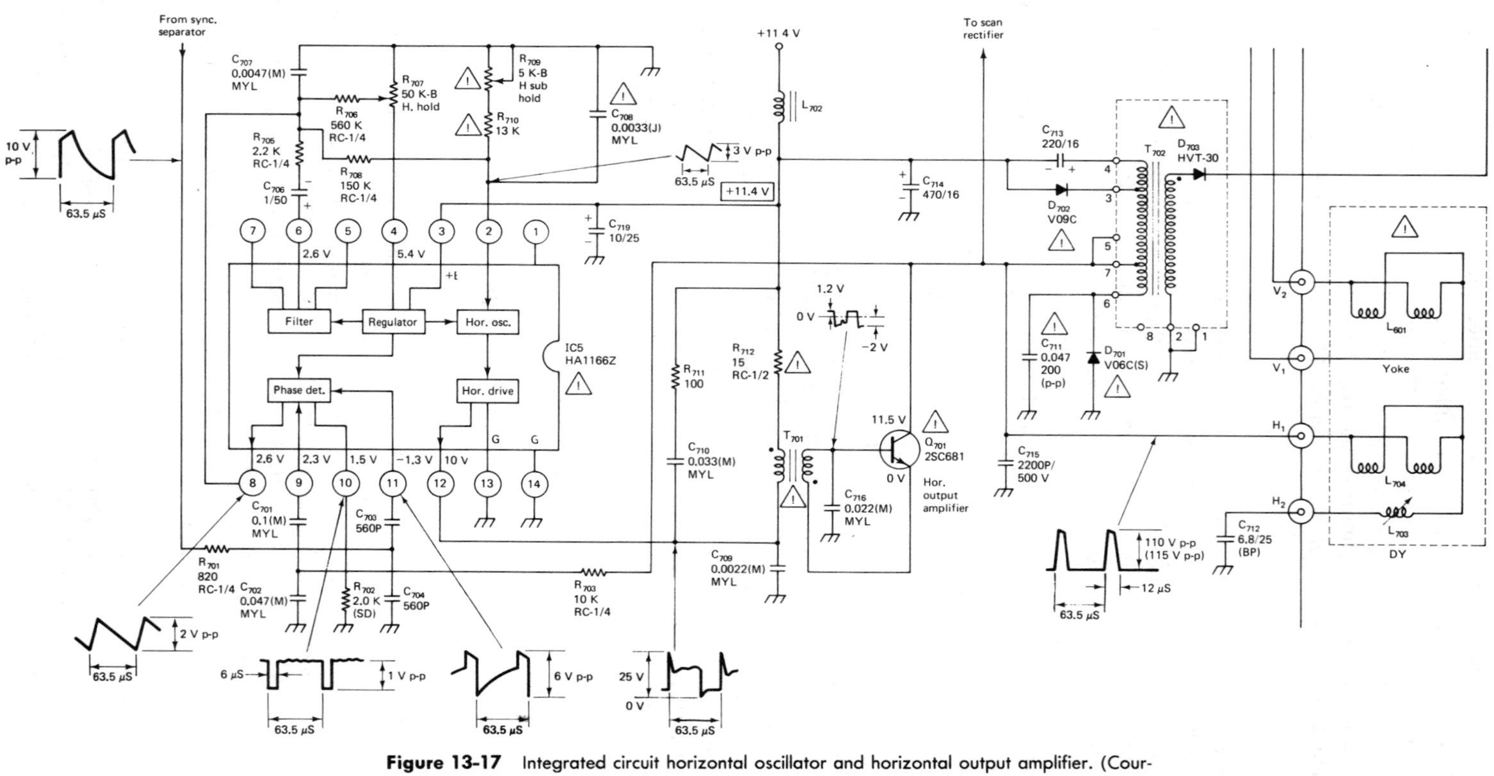

Figure 13-17 Integrated circuit horizontal oscillator and horizontal output amplifier. (Courtesy of Magnavox.)

If the horizontal oscillator ceases to operate, the raster will be lost because the output of the oscillator drives the horizontal output stage, which in turn develops the picture tube second anode voltage. See Figure 13-18(b). Loss of the second anode voltage will cause the picture tube to go dark. The loss of horizontal drive in a transistor circuit causes a loss of raster and will not cause the output stage to overheat.

The fastest way to verify that the deflection oscillator is not functioning is to observe the output and input waveforms of the oscillator. If they are not present, the oscillator is definitely defective. Voltage and resistance measurements can then be used to isolate the defective part.

Oscillator changes frequency. Changes in oscillator frequency must be caused by defects that are sufficiently large to cause loss of sync despite the horizontal AFC or vertical sync pulses. Examples of this defect are found in Figure 13-18(c) and (d). The most likely components that might cause such symptoms are the frequency determining components of the oscillator.

In horizontal oscillator circuits that use ringing coils, defects in the tuned circuit such as open or leaky capacitors or defective or misadjusted coils may be responsible for loss of picture sync. In this oscillator, it is possible to check the tuned circuits by simply shorting it out with a jumper. If normal operation is restored, the trouble is likely to be in the tuned circuit. If normal operation is not restored, the other frequency determining components must be checked.

Oscillator output waveform changes shape. The results of this defect are nonlinear pictures (circles become egg-shaped) foldover, and compression. See Figure 13-18(e) and (f).

In vertical deflection circuits in which the oscillator is a separate circuit, defects that tend to increase retrace time are responsible for foldover at the top of the picture. Foldover might be caused by a defective blocking oscillator transformer.

Poor waveform linearity in the horizontal oscillator can cause poor picture linearity on the right or left sides of the picture or can cause vertical white lines called *drive lines* in the center and to the right of center of the raster.

Oscillator output is reduced. The peak-to-peak output of a transistor horizontal oscillator ranges from 2 to 20 V p-p. If this voltage is reduced for any reason, the picture width and high-voltage output of the horizontal stage would be reduced. This would result in a narrow, dim picture. See Figure 13-18(g).

A reduction of vertical deflection sawtooth oscillator output would result in a raster that does not have sufficient height. See Figure 13-18(h). In either case, a reduction of sawtooth amplitude can be caused by such defects as low B+, increased value of collector load, open sawtooth forming capacitor, and a defective coupling capacitor.

Oscillator is intermittent. Any of the trouble symptoms described above can appear intermittently. In some cases, it may be a momentary disruption; in other cases, the trouble symptoms may last longer. The symptoms should be analyzed to determine the probable areas of the oscillator circuit that might cause this. The suspected group of components should be checked by means of chiller spray and inspection for cold joints, cracked PC

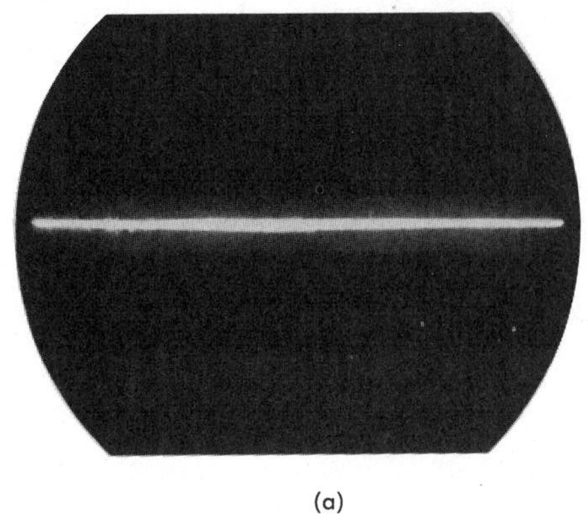

(a)

(b)

(c)

(d)

(e)

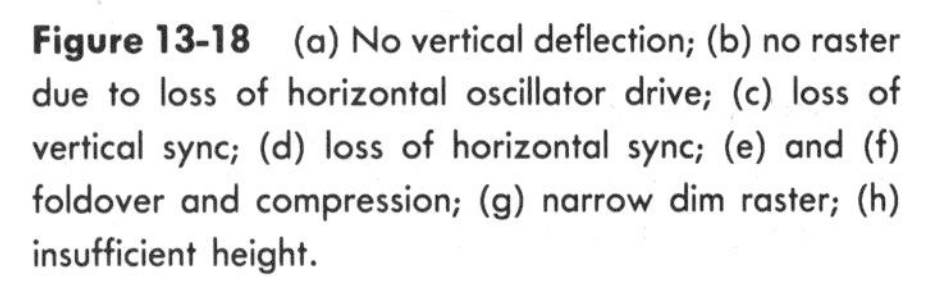

Figure 13-18 (a) No vertical deflection; (b) no raster due to loss of horizontal oscillator drive; (c) loss of vertical sync; (d) loss of horizontal sync; (e) and (f) foldover and compression; (g) narrow dim raster; (h) insufficient height.

(f)

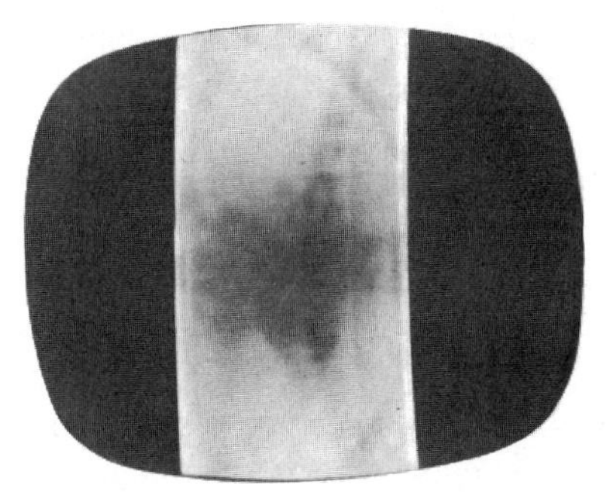

(g)

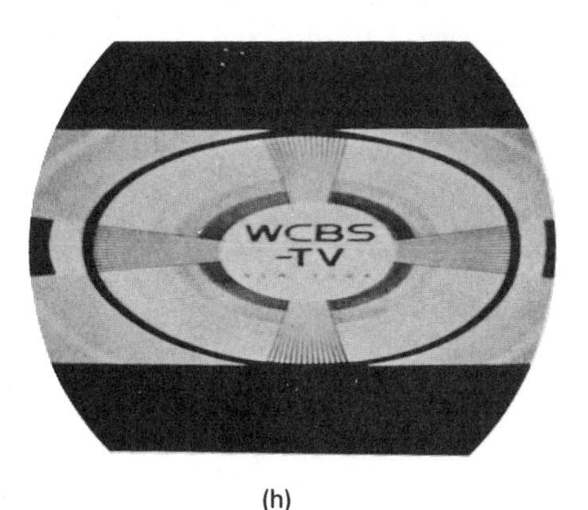

(h)

Figure 13-18 (*Cont.*)

boards, and solder drips that may have been left in the chassis during manufacturing. If these checks fail, and if the circuit area has been definitely established, change all of the components in the circuit. This usually means four or five capacitors, five or six resistors, and one or two transistors.

13-10 automatic frequency control

All modern color and black & white receivers incorporate some type of horizontal AFC. In general, the vertical deflection oscillators do not include any AFC circuit.

Horizontal automatic frequency control is used to *improve the noise immunity* of the horizontal deflection oscillator. This is necessary because the sync separator output contains noise pulses as well as vertical and horizontal sync pulses. The horizontal and vertical sync pulses and their accompanying noise are fed into the differentiator circuit that separates the horizontal sync pulses from the vertical sync pulses and then couples the horizontal sync pulses into the horizontal oscillator. The differentiator is essentially a high-pass filter. Since noise pulses have most of their energy in the high-frequency range, they are easily passed from the sync separator into the horizontal oscillator. Unfortunately, the frequency of the horizontal oscillator is very sensitive to pulses. Noise pulses that reach the horizontal oscillators act like sync pulses and can easily force the oscillator to change frequency. If direct horizontal synchronization were used, the picture would tear out in erratic groups of lines, sometimes at the top, center, or bottom of the raster, depending upon what point in the vertical deflection cycle the noise pulses appeared.

Vertical synchronization is not affected by the noise pulses to the same extent as the horizontal synchronization is, because an integrator is placed between the sync separator and the vertical deflection oscillator. The integrator is a low-pass filter that is able to effectively remove the noise pulses before they can reach the deflection oscillator.

AFC circuits. In the past many different systems of horizontal AFC have been used. Many years ago such systems as the Synchroguide and the Synchrolock AFC circuits were popular, but these systems have been generally replaced by two other types of AFC, the Gruen AFC system and the Phase Detector AFC system. Both systems are commonly used in transistor-type receivers. The essential features of both systems are shown in Figure 13-19. In both cases, the horizontal sync pulses are obtained from the sync separator and then they are compared to a sawtooth obtained from the horizontal oscillator or from the horizontal output circuit. The comparison between these two voltages is made in the *phase detector* portion of the circuit. In effect, this circuit converts the horizontal sync pulses into a *dc error voltage.* This voltage is then fed into a filter circuit that removes any noise that may be present. The dc error voltage is then fed to the deflection oscillator and corrects the timing of the oscillator.

The block diagram does not indicate two differences that exist between the Gruen and Phase Detector AFC systems. First, the Phase Detector AFC is driven by a phase inverter amplifier that provides two sync pulses of opposite polarity. The Gruen AFC is fed directly by a negative-going sync pulse obtained from the sync separator stage. The Gruen and Phase Detector horizontal feedback sawteeth are generally of opposite phase.

Dc error voltage. The output dc error voltage of the phase detector is approximately zero when the sync pulses and the horizontal sawtooth feedback voltage are the same frequency and of proper phase (the sync pulse occurs during horizontal retrace). If the oscillator frequency is higher than the sync pulse, the dc output of the phase detector may be positive; if the frequency of the oscillator is lower, the dc error voltage may be negative. The required polarity of the error voltage produced for a given change in oscillator frequency depends upon the type of deflection oscillator used and upon the AFC error voltage injection point.

If the oscillator is an emitter-coupled multivibrator such as that shown in Figure 13-12, the injection point is the base of the normally on transistor (Q_1). In this case, an increase in oscillator frequency will produce a positive error voltage that decreases the transistor bias.

The dc error voltage produced by the AFC circuit may also feed the base of a reactance transistor (Q_1) as shown in Figure 13-16. The polarity of the error voltage will always be such that it causes the horizontal oscillator frequency to shift back toward that condition in which the error voltage is zero. If the circuit is functioning properly, the error voltage will be zero when the frequency of the oscillator is equal to the horizontal sync rate. In some cases, an unbalanced phase detector is used and the error voltage may be a small positive or negative voltage when no timing error exists between the sync and horizontal oscillator.

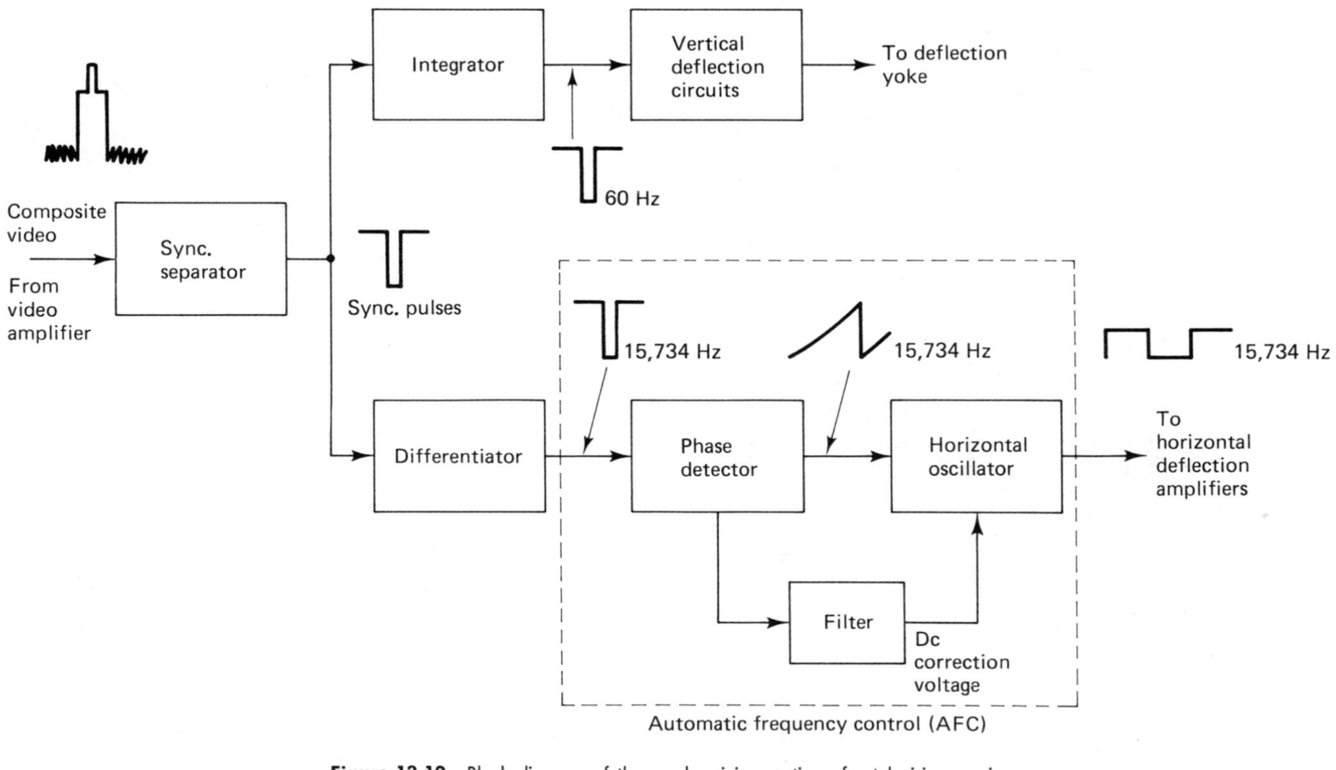

Figure 13-19 Block diagram of the synchronizing section of a television receiver.

13-11 the Gruen AFC detector

The schematic diagram of the Gruen phase detector circuit is shown in Figure 13-20. Other examples of this circuit and their associated horizontal deflection oscillators can be found in Figures 13-15 and 13-16. The waveforms found in these circuits are also indicated.

In the Gruen phase detector, a pair of selenium diodes are arranged so that their cathodes are tied together. In practice, the two diodes are often encapsulated in a small three-lead plastic package. The center lead of the package is the common cathode.

In Figure 13-20, it can be seen that a negative-going sync pulse of 17 V p-p is fed from the sync separator via coupling capacitor C_1 to the common cathodes of the diode phase detector. The phase detector diodes are also fed by a pulse obtained from a winding on the horizontal flyback transformer. The pulse is passed through an integrator circuit (R_6 and C_7) and is converted into a sawtooth of 12.4 V p-p. The horizontal pulse polarity determines the phase of the sawtooth. Positive pulses produce sawteeth whose retrace interval increase in amplitude; negative pulses develop sawteeth of opposite direction. See Figures 13-15 and 13-20. The polarity of the sawtooth is one factor in determining the polarity of the error voltage used to correct the frequency of the oscillator. Hence in any given circuit the sawtooth may have either polarity depending upon the design of the horizontal oscillator.

Circuit operation. Referring to the wave shape at the output of Q_1, let us *first assume that only the horizontal sync pulses are being applied* to the phase detector. During the time (T_1), before the sync pulse is applied, the sync separator is cut off and its collector voltage is

Figure 13-20 Gruen AFC phase detector.

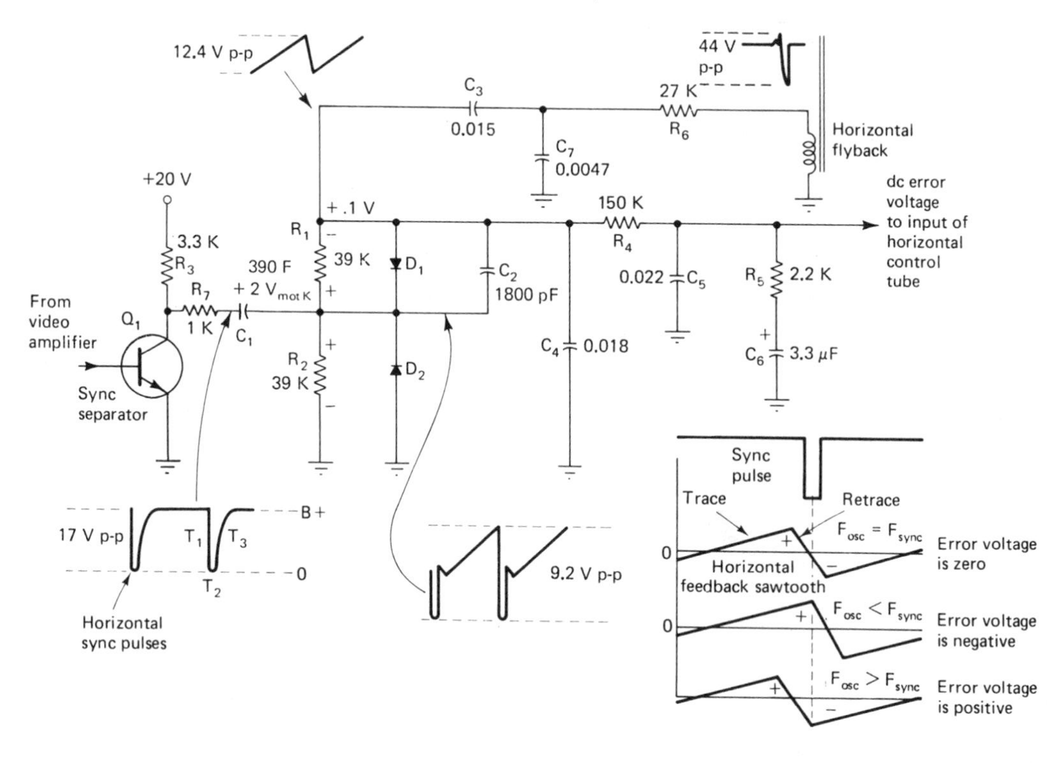

equal to the B+ supply voltage. As a result, the coupling capacitor C_1 is fully charged. When the sync pulse appears (T_2), Q_1 is conducting and the diodes D_1 and D_2 are also driven into conduction by the discharge of C_1. C_1 discharges through two paths. One path is from C_1 through D_2 to the emitter of Q_1 and back through the collector of Q_1 returning to C_1. The other path is from C_1 through D_1, to C_3, and C_7 back to the emitter of Q_1 and returning to C_1 via the collector current of Q_1. The important result of C_1's discharging is the charge of C_3.

Immediately following the short-duration (5 μs) sync pulse (T_2), the sync separator is driven into cutoff and its collector voltage rises to B+ (T_3). During this long period (58 μs) between sync pulses, C_1 charges and both diodes are turned off. Charging current flows through two paths. First, current flows from ground up through R_2 to C_1 then through R_3 and back to the B supply. This produces a dc voltage drop of approximately 2 V across R_2 such that the cathodes of the diodes are positive with respect to ground. In the second charging path, current flows from ground through C_7 and C_3, down through R_1, into C_1, and then through the sync separator load resistor R_3 to the B supply. The voltage drop across R_1 resulting from this charging current is approximately equal to the voltage drop across R_2. The polarity of the voltage drop across R_1 is such that the end tied to the anode of D_1 is negative. The voltage from the anode of D_1 to ground is the sum of the voltage drops across R_1 and R_2. Since these voltage drops are almost equal and are opposite, the net voltage from the anode of D_1 to ground is almost zero. In practice, this voltage may be a few tenths of a volt because of the mismatch between components that comprise the phase detector. In Gruen AFC circuits that are designed so that R_1 is not equal to R_2, an unbalanced condition occurs and the net voltage across the diodes will be greater than zero.

Sawtooth action. Now let us consider the effect of the sawtooth on the phase detector (we are *assuming that no sync pulses* are present). The sawtooth is an ac voltage that first forces one diode into conduction and then when it reverses its polarity it forces the other diode into conduction. When the sawtooth polarity is positive to ground, D_1 is turned on and current flows through R_2 and D_1 and charges C_3. The voltage drop produced across R_2 is 2 or 3 V positive at the cathode of D_1 and negative at the ground end of the resistor. On the next half-cycle the polarity of the sawtooth is negative to ground. This turns D_1 off and D_2 on. Current now flows through D_2 and R_1 and discharges C_3. The polarity of the voltage drop across R_1 is such that the end of the resistor tied to the cathode of D_2 is positive. If R_1 and R_2 are equal, the voltage drops across them will also be equal and opposite. Therefore, the net dc voltage across the diodes will be zero. In a practical circuit the voltage will be a few tenths of a volt.

Testing the Gruen AFC. From a servicing point of view this fact makes for a rapid method for testing the Gruen AFC circuit. First, adjust the receiver tuner for a blank channel so that the AFC circuit will not receive any horizontal sync pulses. Next, measure the voltage at the cathodes of the diodes and then the total voltage across both diodes. If the cathode voltage is 1 to 6 V and if the total voltage across the diodes is only a few tenths of a volt, the AFC circuit can be assumed to be functioning properly.

The sawtooth plus sync pulse. Now let us consider what happens when both the sync pulse and the sawtooth are applied to the circuit simultaneously. Shown in the ladder diagram of Figure 13-20 are the three possible conditions that can occur. First, the sync pulse may occur during the horizontal retrace portion of the feedback sawtooth. This is the desired or *normal* condition for the AFC circuit. Notice that the sync pulse occurs during the time that the sawtooth is passing through *zero.* It is as though the sync pulse were applied without the sawtooth, or vice versa. Consequently, the dc error voltage produced will be zero for a balanced system in which the diodes and their shunting resistors are matched.

Second, the frequency of the oscillator may be lower than the sync pulse rate. Under these conditions, the sync pulse will occur during the time that the sawtooth is *positive* with respect to ground. This means that the peak voltage across diode D_1 will be larger than the peak voltage developed across D_2 during the negative half-cycle of the sawtooth. As a result, D_1 conducts harder than D_2 which charges C_3 to a larger voltage in one direction than in the other. In this case, the conduction of D_1 charges C_3 so that the terminal of C_3 that is tied to D_1 is negative. When C_3 discharges, it will develop a voltage across R_1 that will be larger than the voltage drop across R_2, and the net voltage from the anode of D_1 will be negative to ground. This voltage is then fed to the base of the oscillator and forces the oscillator to increase in frequency, returning it to normal operation.

In the third case, the frequency of the oscillator may be higher than the sync pulse rate. See Figure 13-20. When this happens, the horizontal sync pulse will occur during the *negative* half-cycle of the feedback sawtooth. In this event, D_2 conducts harder than D_1 and C_3 charges a greater amount in one direction than in the other direction. The net result of this is that the positive terminal of C_3 is tied to the anode of D_1. When C_3 is discharged, it develops a net *positive* voltage across the diodes. This voltage is fed to the base of the normally off transistor of the horizontal oscillator and forces the frequency of the oscillator to decrease.

If either R_1 or R_2 should change value or if D_1 or D_2 should become defective, the phase detector could cease functioning and would cause the horizontal oscillator to go out of synchronization.

Shunting D_1 is a small capacitor C_2. This capacitor is a frequency compensating capacitor that ensures that the sawtooth voltage drops across D_1 and D_2 are similar in magnitude and wave shape. If this capacitor opens, the range of horizontal locking will be affected. In general, tighter locking in one direction of the horizontal hold control and poorer locking in the other will be experienced.

13-12 antihunt circuits

When a frequency error occurs, the AFC circuit will react to compensate for the error. See Figure 13-21(a). Unfortunately, the AFC circuit tends to overreact and therefore the error voltage forces the oscillator too far in the opposite direction. This new error then forces the AFC circuit to react and it now generates an error voltage of opposite polarity and forces the horizontal oscillator in the direction of the original frequency error. As this process continues the error will become smaller and smaller until it is completely corrected. The action is similar to that of an inexperienced elevator operator who is trying to bring an elevator level to the landing. First the operator overshoots and then the operator undershoots, each time bringing the elevator and the landing closer together. Eventually, the operator is able to match elevator and landing. This process is called *hunting.* In a television receiver, horizontal AFC hunting can cause such picture defects as horizontal

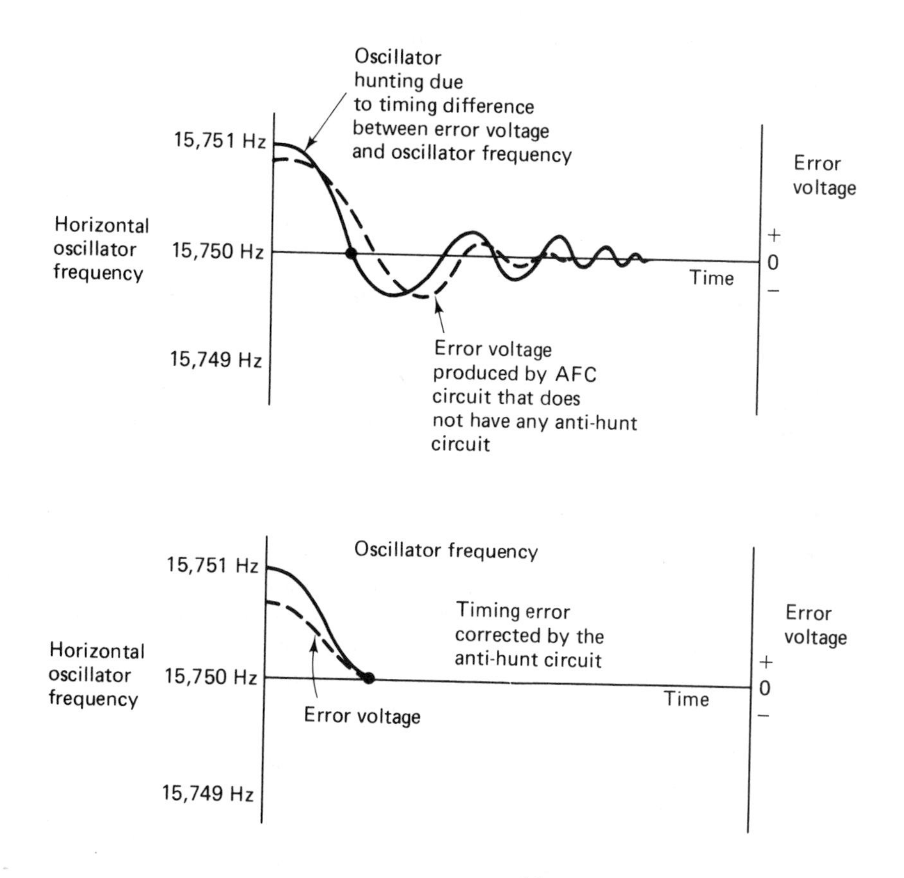

(a)

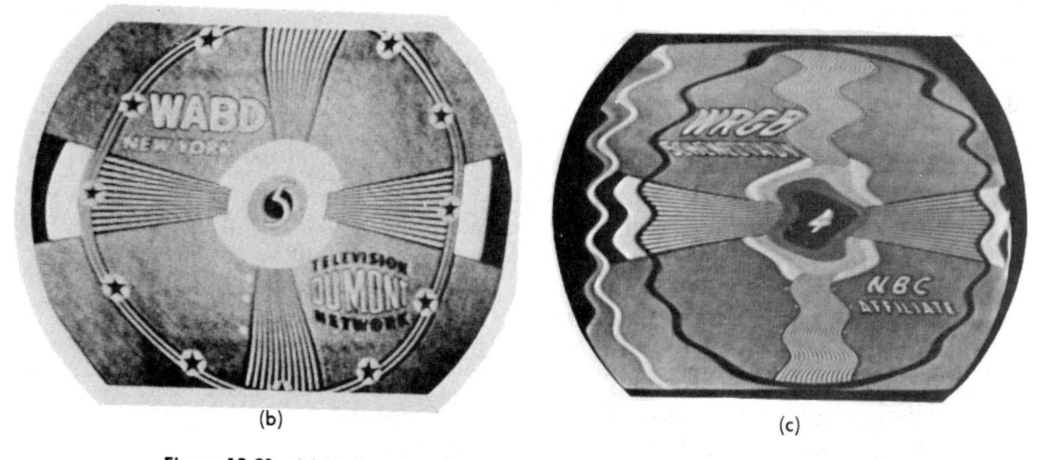

(b) (c)

Figure 13-21 (a) Horizontal oscillator hunting and its correction; (b) horizontal pulling; (c) piecrust picture distortion. (Courtesy of General Electric.)

hook, horizontal pulling, S bending, or "piecrusting." Examples of these defects are found in Figure 13-21(b) and (c). Horizontal hook and piecrusting are really the same defect in differing degrees. Both symptoms are the result of small periodic changes in the phase of the horizontal deflection sawtooth. When a group of horizontal lines is shifted in phase with respect to the remaining picture, the lines will be displaced to the right or left and a hook will be produced.

It is possible to minimize hunting by placing a phase shift network such as that shown in Figure 13-20, consisting of C_5, R_4, R_5, and C_6 across the output of the phase detector. This network effectively damps out the oscillating error voltage produced by the hunting. This is shown in Figure 13-21a. Any change of value of these critical components or leakage of the capacitors will produce horizontal bending of varying degrees or in severe cases, piecrusting.

The phase shift network also serves as a noise filter that prevents any impulse noise that has passed through the AFC circuit from reaching the deflection oscillator.

13-13 phase detector AFC

An example of a phase detector AFC system used in a transistor receiver is shown in Figure 13-22. The function of this circuit is to develop a dc voltage that is proportional to any timing error that might exist between the horizontal sync pulses and the sawtooth output of the horizontal oscillator. In terms of its function, this circuit is identical to the Gruen AFC, but the means of developing the error voltage is different. Also to be noticed from a comparison of schematics is that the duodiodes are connected somewhat differently. In the Gruen system the cathodes of the diodes are tied together and in the Phase Detector the diodes are connected in series, with the anode of one diode tied to the cathode of the other.

This circuit will be studied under three conditions: with only the sync pulses applied; with the sawtooth feedback voltage alone; and with both sync pulses and sawtooth present.

Sync pulses only. The phase inverter (Q_1) is an amplifier that provides two sync pulse outputs that are equal in amplitude and opposite in phase. The input of the phase inverter is a negative-going horizontal sync pulse obtained from the sync separator stage. Acting as an emitter follower, the output at the emitter is in phase with the input pulses. The collector output pulse is inverted with respect to the input, as in the case for an ordinary common emitter amplifier. When the input sync pulse (10 V p-p) is present, the transistor is driven into cutoff. Between pulses the transistor is conducting.

The positive-going sync pulses are coupled from the collector of the sync inverter (Q_1) through C_1 to the anode of D_1. At the same time the negative-going horizontal sync pulses are coupled via C_2 to the cathode of D_2. As a result of these pulses, electron current flows along the path of R_1, C_2, D_2, D_1, C_1, and R_4 and then returns to ground via the B supply. This flow of electrons charges C_1 and C_2 to approximately the peak of the sync pulses.

During the time between sync pulses the capacitors discharge, C_1 discharges through R_3, R_5, R_8, B−, B+, and R_4. C_2 discharges through R_1, R_8, R_5, and R_2. Two voltages are developed across R_8, one negative and the other positive with respect to ground. Since each voltage drop was caused by an equal capacitor discharge, the two voltages across R_8 are

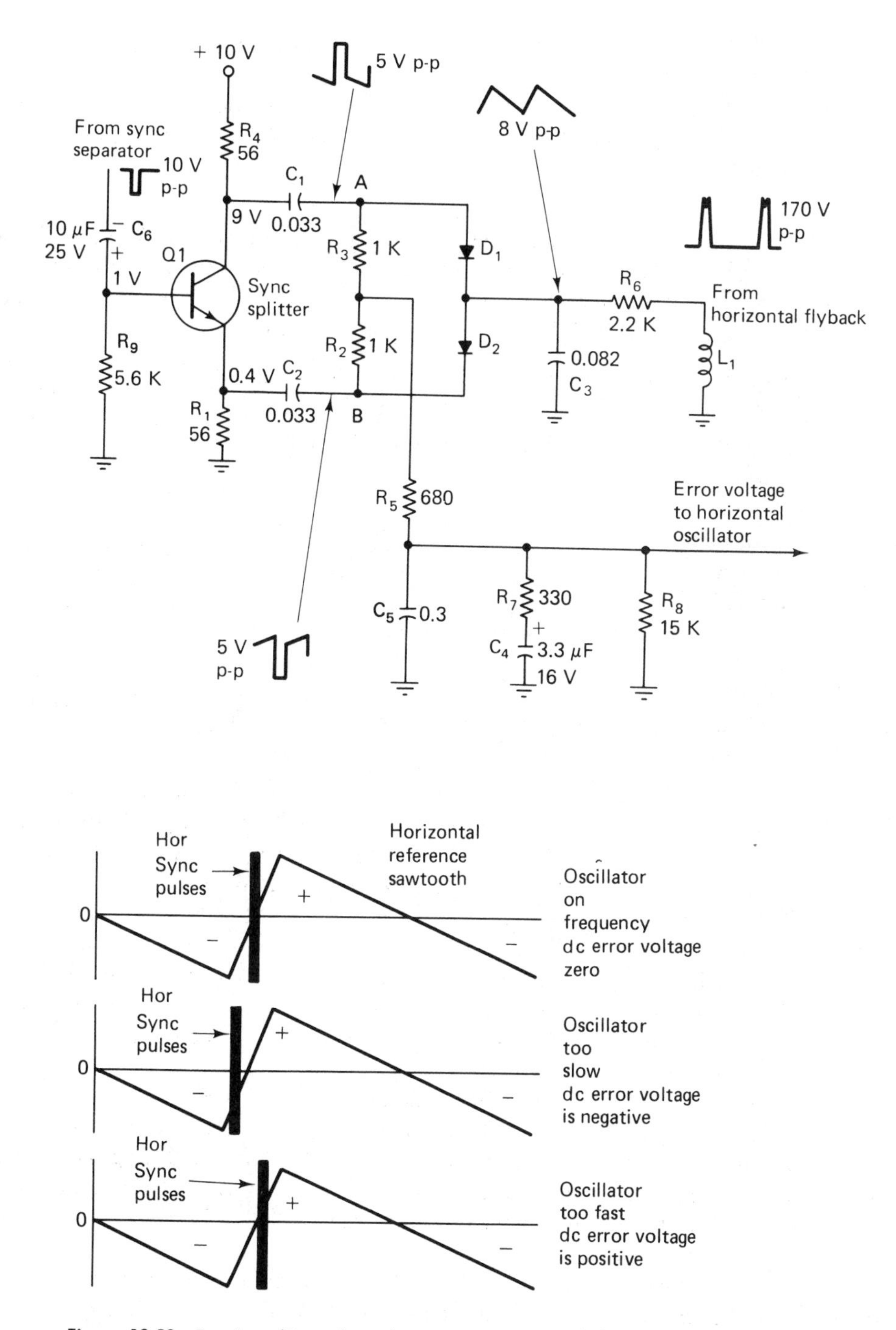

Figure 13-22 Transistor-driven phase detector AFC system and the timing relationships between the reference feedback sawtooth and the sync pulses at the output of the phase inverter.

equal and of opposite polarity. These voltages cancel each other, leaving a net voltage drop of zero across R_8.

The discharge of C_1 and C_2 produces voltage drops across the matched resistors R_1 and R_3 that are of such polarity as to reverse bias diodes D_1 and D_2 during the time between sync pulses.

Sawtooth only. The sawtooth reference signal is obtained from either the horizontal deflection oscillator or a winding on the horizontal flyback transformer. This winding supplies a positive-going pulse that is integrated by C_3 and R_6 and forms the ac reference sawtooth that feeds the junction of D_1 and D_2. If no sync pulses are present, the positive half-cycle of the sawtooth allows D_2 to conduct through path R_8, R_5, R_2, D_2, and R_6 and return to ground via the horizontal flyback winding L_1. On the negative half-cycle, diode D_1 conducts through path R_8, R_5, R_3, D_1, R_6, and L_1 to ground. The direction of current through R_8 produces opposite voltage drops for each half-cycle; therefore, the net voltage drop across R_8 is zero.

Sync pulse and sawtooth. When both signals are applied to the phase detector, three possible conditions can exist. These are illustrated in Figure 13-22. In the *normal* case, the frequency of the oscillator and the frequency of the sync pulses are the same and the sync pulse occurs when the sawtooth is passing through its zero point during retrace. As a result, the circuit behaves as though each signal were applied to the phase detector independently of each other. A *zero* error voltage is produced and delivered to the deflection oscillator. Therefore, the frequency of the horizontal deflection oscillator remains unchanged.

If the oscillator were to drift so that the oscillator frequency is *lower* than that of the sync pulses, the sync pulses would occur during the time that the sawtooth is in its negative half-cycle. See Figure 13-22. This is because a single cycle of oscillator sawtooth waveform will take longer to complete, and the sawtooth waveform will still be below zero when the sync pulses arrive. Since the sawtooth places a negative voltage on the cathode of D_1 and the anode of D_2, the conduction of D_1 caused by the sync pulses will make C_1 charge to a higher voltage than it did before. The negative voltage on the anode of D_2 reduces its conduction and C_2 charges to a smaller voltage. The net result is that the discharge currents through R_8 are not equal and a *negative* error voltage is developed across it. This voltage is then filtered by C_4, R_7, and C_5 and is applied to the horizontal oscillator, forcing its frequency to return to its normal value.

If the frequency of the horizontal oscillator were to *increase,* the sawtooth would be positive during the sync pulse interval. See Figure 13-22. As a result, the conduction of the diodes will be such that C_2 will charge to a higher value than C_1. During the discharge period of these capacitors their unequal charges will develop a *positive* error voltage across R_8. This voltage will be filtered and applied to the horizontal oscillator causing its frequency to decrease to its normal value.

The phase detector operation may be checked by first measuring the voltage at both points A and B without any input signal and then with a signal. Under normal conditions these voltages should be approximately equal and of opposite polarity. They should both get larger in the presence of a signal.

Horizontal oscillator off frequency or horizontal tearing. If the frequency of the oscillator is far off frequency, the picture will be out of sync and will appear to be broken up into many diagonal segments as shown in Figure 13-18(d). If the frequency is correct but is drifting in phase, the horizontal blanking bar will appear to slide slowly across the screen. These symptoms may be caused by the following defects:

1. Defective phase inverter or diodes.
2. Leaky or open coupling capacitors such as C_1 in Figure 13-20 and C_1 and C_2 in Figure 13-22.
3. Change values of matched resistors R_1 and R_2 in Figure 13-20 and R_2 and R_3 in Figure 13-22.
4. Defective feedback components C_3 and R_6 in either Figure 13-20 or 13-22.
5. Defective filter circuit components C_4 and C_5 in Figure 13-20.

Bias clamps may also be used to help determine whether the defect is in the AFC system or in the horizontal oscillator itself. The bias clamp is tied from the output of the AFC to ground and is adjusted for the normal no-signal bias indicated in the schematic diagram. If changing the dc voltage can momentarily force the oscillator to a normal in-sync picture, it can then be assumed that the oscillator is normal and that the trouble is in the AFC circuit.

Piecrusting. See Figure 13-21(c). This symptom is an indication of improper filtering of the error voltage. The cause lies in the *RC* filter circuit at the output of the phase detector. Components that would be suspect are C_6 and R_5 in Figure 13-20, and C_4, C_5, and R_7 in Figure 13-22.

Picture hook. This symptom is illustrated in Figure 13-21(b). It may indicate that vertical sync is affecting the horizontal deflection oscillator. A secondary cause might be a change in the filter components or in the integration feedback network (check C_3 and R_6, and C_4, C_5, and R_7 in Figure 13-22). Also to be suspected are open or leaky power supply electrolytic filter capacitors that may allow changes in current drawn during vertical retrace to affect the B+ supply and thereby affect the horizontal oscillator.

Curvature of vertical lines in a picture can be the result of *picture pulling* (hook) or *raster pulling.* At first glance both of these defects look alike. Closer examination, however, will reveal important differences.

Picture pulling. Picture pulling or hook is the condition in which horizontal scanning frequency shifts in phase above and below its proper value, causing the picture information associated with these horizontal lines to shift to the left and right of their normal raster positions. In this trouble symptom the width of the raster is unaffected and remains constant from the top to the bottom of the picture.

Picture pulling can be a difficult trouble to isolate, for it can be the result of defects in the tuner, video IF, video amplifiers, AGC, low-voltage power supply, sync separator, and AFC circuits. Picture pulling resulting from defects in the tuner, video IF, AGC, and power supply are caused by poor power supply filtering that hum modulates the video signal passing through the defective stage.

Picture pulling may also be the result of overdriving the IF or video amplifier stages. An excessively large signal can force the amplifier (tube or transistor) into cutoff or saturation, clipping off the sync pulses. Insufficient AGC can cause excessive IF gain and result in clipping. Bias clamping the AGC will quickly indicate whether the trouble is in the AGC circuit or in the video IF amplifiers.

The results of clipping can be quickly determined by observing the hammerhead pattern (vertical sync and blanking interval). If the sync pulse is blacker than any black in the picture, clipping has not occurred and the trouble is either in the sync separator or in the AFC circuit.

Poor video IF alignment can cause loss of low video frequencies and can result in picture pulling. In this case, however, the quality of the picture and the effect of fine tuning on the picture would tend to identify this problem as being bad alignment.

If it has been determined that video signal clipping or poor power supply filtering is not responsible for picture pulling, then the defect is probably in the sync separator or the AFC circuit. To determine which one of these circuits is at fault, break the connection between the sync separator and the AFC circuit and observe the picture. The picture will have lost sync, but adjusting the horizontal hold control can cause the picture to stop momentarily. If the hook is gone, the trouble is in the sync separator. If not, it is in the AFC circuit.

Raster pulling. Raster pulling is the result of *picture width changing* or of changes in raster centering as vertical scanning occurs. This trouble symptom is usually caused by the 60-Hz or 120-Hz modulation of the horizontal deflection sawtooth. The quickest way to determine which fault is present is to shift the horizontal centering of the raster so that the extreme edge is visible. If the edge of the raster is straight, *picture pulling* is present. If the edge of the raster is curved, the trouble symptom is classified as *raster pulling.*

Raster pulling is usually the result of defective filtering in the low-voltage power supply.

Phasing ghost. Another trouble symptom closely related to poor horizontal synchronization is improper phasing between the horizontal sawtooth and the sync pulse. Such an out of phase condition may result in a *split phase* picture [see Figure 13-23(a)] in which the horizontal sync and blanking pulses appear in the center of the picture. A somewhat less extreme example of improper phasing is shown in Figure 13-23(b). This is called a *phasing ghost* and usually takes the form of a white mist or cloud on the left side of the screen. The phasing ghost may be accompanied by picture pulling.

Possible causes for improper phasing are such things as defective feedback network components (C_3, C_7, and R_6 in Figure 13-20), leaky or open AFC filter capacitors and filter resistors (C_5, C_6, and R_4 or R_5 in Figure 13-20).

No raster. A no-raster condition can be caused by a defect in the AFC circuit, for example, referring to Figure 13-22, a shorted coupling capacitor C_1 would force a large positive

(a) (b)

Figure 13-23 (a) Split phase; (b) phasing ghost. (Courtesy of General Electric.)

voltage drop across R_8. This in turn will be coupled to the base of the horizontal multivibrator, forcing it to shift its frequency radically. This in turn may force the horizontal oscillator to cease functioning. If it continues to oscillate at a very low frequency, the efficiency of the output circuit will be reduced to the point where no high voltage is produced. In some sets, shorted diodes will produce the same effect.

13-15 reactance tube sine-wave oscillator

Figure 13-24 shows a typical sine-wave horizontal deflection oscillator that used a reactance tube between the Gruen AFC circuit and the oscillator. The oscillator (V_1) is identified as an electron coupled Hartley oscillator by the tap on the tuned circuit inductance L_1. This inductance is variable and is used as the horizontal frequency control. The oscillator uses the screen of the pentode as the plate of the oscillator and uses the plate of the tube to develop the output voltage. This arrangement is called an *electron-coupled oscillator* and is used to ensure that the oscillator is isolated from the large plate voltage variations. The frequency of the oscillator is 15,750 Hz for black & white and 15,734 Hz for color. The oscillator uses grid leak bias that develops a grid bias of −45 V. For a pentode, this voltage is more than enough to operate the tube in Class C. Thus, plate current flows in short-duration pulses only during the extreme positive peaks of the grid voltage. These periods of tube conduction are responsible for horizontal retrace. In order to ensure a large peak-to-peak output voltage swing (185 V p-p), the plate load resistance (R_{16}) is made very large. R_{17} and C_7 are responsible for producing the trapezoidal output voltage. C_7 is the sawtooth-forming capacitor and R_{17} is the peaking resistor. The ac voltage drop across C_9 and R_{12} is used to provide a sawtooth necessary for proper AFC operation.

A reactance tube (V_2) shunts the oscillator tank circuit and appears to the oscillator as a reactance because its ac plate current and ac plate voltage are 90° out of phase. To achieve this phase shift, a feedback circuit (C_1 and R_1 in Figure 13-24) is used between the plate and the grid of the reactance tube. C_1 may be a capacitor or the interelectrode capacitance of the tube, and R_1 is returned to ac ground via C_2.

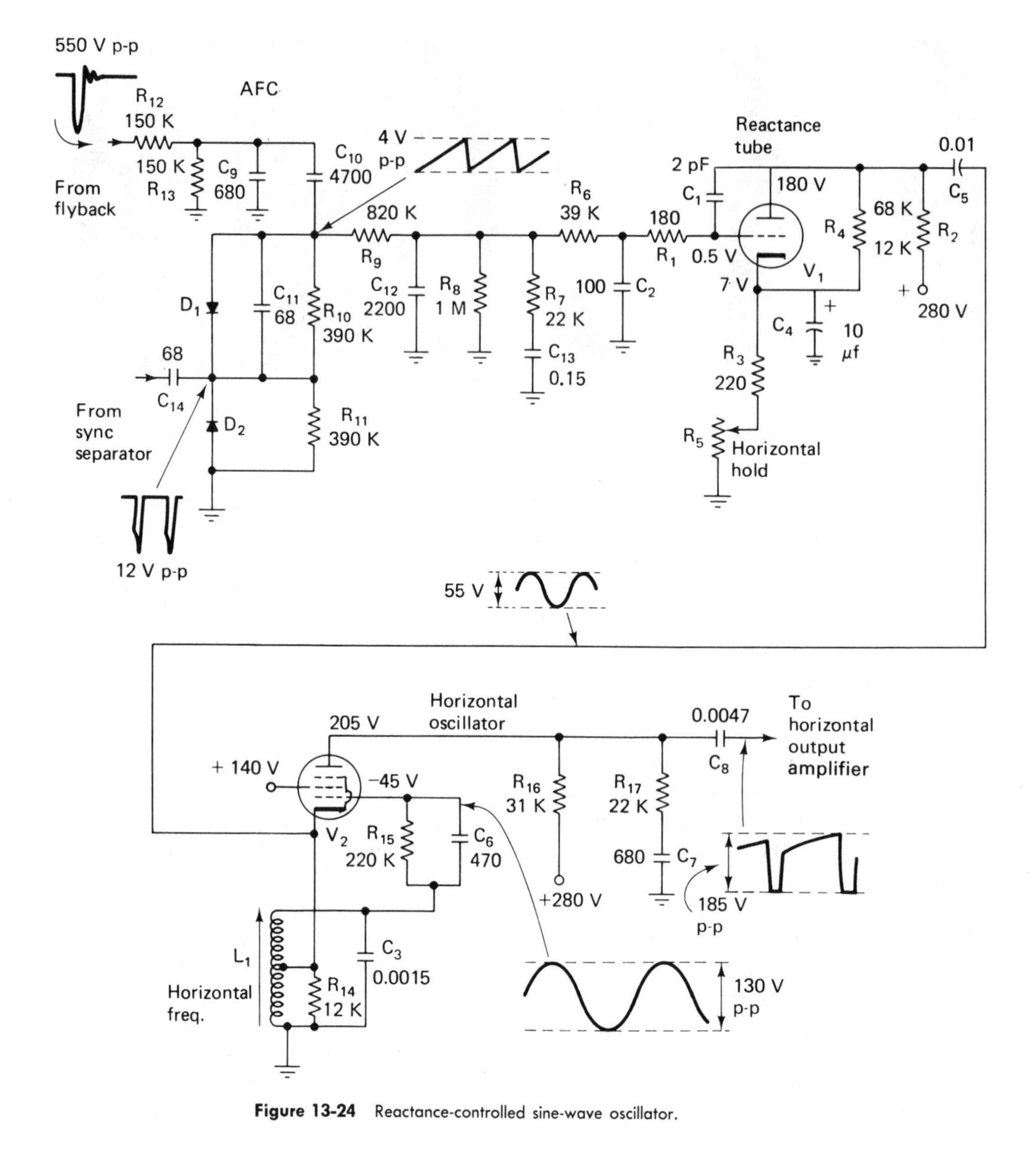

Figure 13-24 Reactance-controlled sine-wave oscillator.

QUESTIONS

1. List three important characteristics of a sawtooth and the effect of each on the raster.
2. What factors determine sawtooth linearity?
3. What factors determine sawtooth amplitude?
4. What effect does incorrect scanning frequency have upon picture reproduction?

5. What is the basic circuit used to develop a sawtooth waveform?
6. What defect might cause a bright compressed region at the top of the raster?
7. Why is a trapezoid waveform necessary? How is it formed?
8. The blocking oscillator is basically a sine-wave generating Armstrong oscillator. How does it produce a sawtooth?
9. What is the effect of increasing the reverse bias of the transistor blocking oscillator of Figure 13-9?
10. What are three characteristics of a feedback oscillator?
11. What is the effect of changing the bias in the blocking oscillator of Figure 13-9?
12. What are the two types of multivibrators used as deflection oscillators, and how do they differ from one another?
13. Across which transistor of a multivibrator is the sawtooth forming capacitor placed?
14. What are the three conditions necessary for proper synchronization?
15. How are horizontal deflection oscillators and vertical deflection oscillators synchronized?
16. How is a dc voltage able to vary the frequency of a horizontal sine-wave deflection oscillator?
17. What will happen to the oscillator frequency if the bias of the reactance tube in Figure 13-24 decreases?
18. What are three differences between the AFC circuits of Figures 13-20 and 13-22?
19. In the Gruen AFC system of Figure 13-20, what would be the effect of C_1 opening, of C_3 opening, and of C_6 opening?
20. Which components in Figure 13-22 are most likely to cause the following symptoms? piecrust, horizontal pulling, phasing ghost, loss of horizontal sync, and loss of horizontal deflection.

chapter 14

Vertical Deflection Circuits

14-1 introduction

Figure 14-1 is the complete block diagram of a color receiver. The shaded blocks correspond to the vertical deflection stages of the receiver. It is this section of the receiver that is the topic of this chapter.

Vertical deflection circuits are necessary in both color and black & white television receivers and perform two main functions: they generate the vertical deflection sawtooth and they provide the necessary power amplification to drive the vertical deflection yoke. The vertical deflection amplifiers may also provide important outputs for vertical retrace blanking, and in color receivers they may provide vertical convergence and top and bottom pincushion correction.

The frequency of the vertical deflection sawtooth for black & white picture information is 60 Hz and for color it is 59.94 Hz. The vertical deflection sawtooth current flows through the vertical windings of the yoke and creates a changing magnetic field that forces the electron beam to deflect from the top to the bottom of the raster.

Deflection circuit classification. The vertical deflection circuits may be classified according to the device used, circuit configuration, and output considerations.

Circuit arrangement. In terms of circuitry, the vertical deflection circuit may be arranged as shown in the three block diagrams of Figure 14-2. A very popular method used in many modern solid-state designs, is that indicated in Figure 14-2(a). In this method a separate oscillator develops the sawtooth that drives the output amplifier stage, which in turn energizes the yoke. A blocking oscillator is often used as the oscillator, but multivibrators and other types of oscillators have also been used. The output amplifier may use transformer

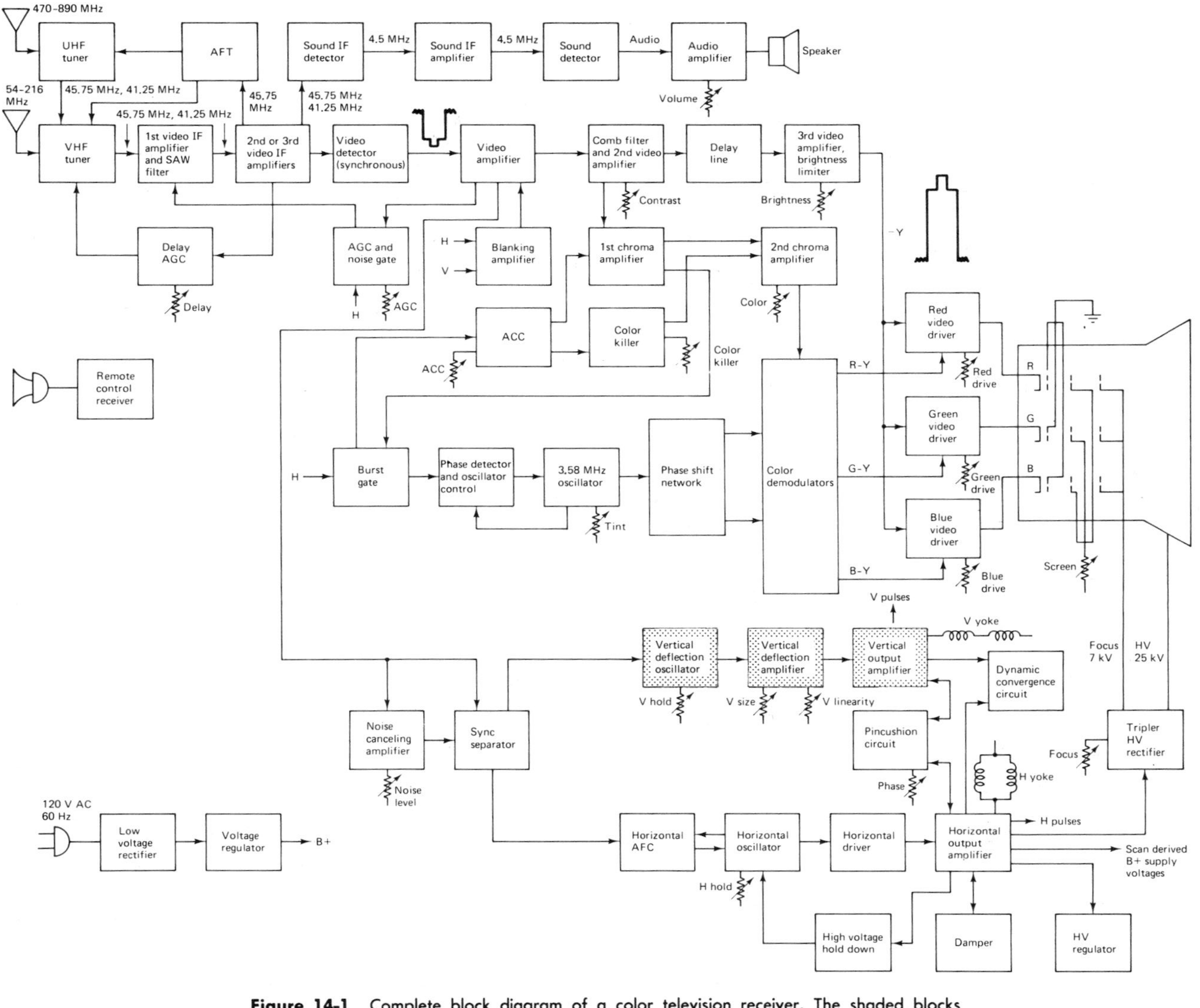

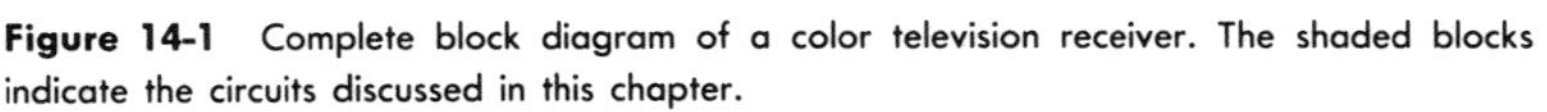

Figure 14-1 Complete block diagram of a color television receiver. The shaded blocks indicate the circuits discussed in this chapter.

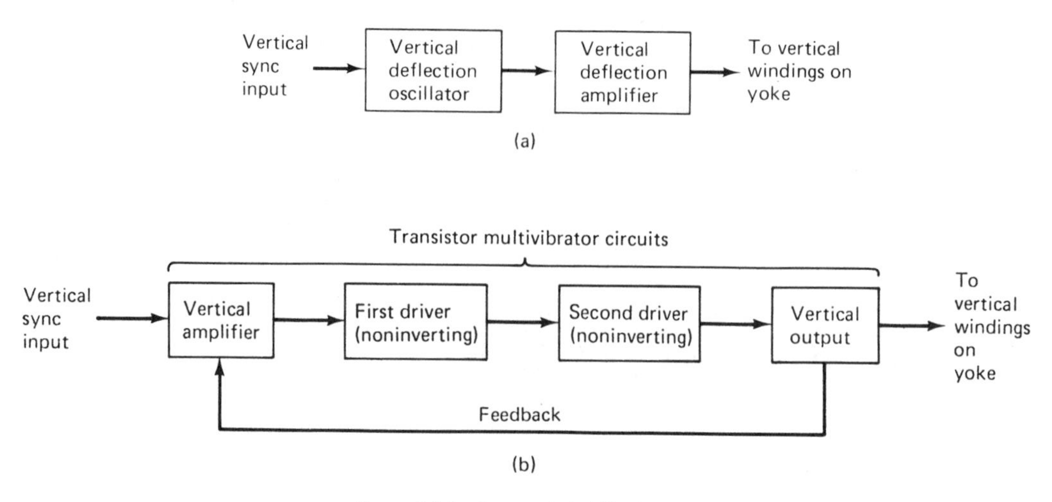

Figure 14-2 Two vertical deflection systems.

coupling to the yoke or it may use transformerless amplifiers of the type used in high-fidelity audio amplifiers.

Modern solid-state vertical deflection circuits may contain as few as two or as many as nine transistors. A block diagram of one possible system using four transistors in a multivibrator configuration is shown in Figure 14-2(b). In this arrangement positive feedback to sustain oscillation is obtained from the output stage and is returned to the input. In addition, negative feedback may also be used to improve sawtooth linearity.

Yoke connection. Vertical output amplifiers may also be classified according to the manner by which the amplifier is coupled to the yoke. In transistor circuits a transformer may be used between the amplifier and the yoke. The transformer may be of the isolated primary and isolated secondary type or it may be an auto transformer. In color receivers these transformers may have two or three additional secondary windings whose outputs are used in convergence and pincushion circuits. Most modern solid-state vertical deflection circuits are transformerless because the low yoke impedance can be easily matched directly to the output impedance of the amplifier without resorting to a transformer.

14-2 the vertical output stage

The purpose of the vertical output stage is to drive the vertical deflection winding of the yoke with an ac sawtooth current of sufficient amplitude to produce full vertical deflection of the picture tube. In practice, this current may be as large as 3 to 5 A p-p.

Output transformers. Because of the large internal resistance of some transistor vertical deflection amplifiers, proper power match can only be achieved if transformer matching exists between the yoke and the amplifier. As shown in Figure 14-3, an isolated primary and secondary transformer may be used for this purpose. In Figure 14-3 the yoke is isolated from the primary winding. This arrangement has the advantage of preventing a vertical shift in picture centering caused by dc collector current passing through the yoke. In some

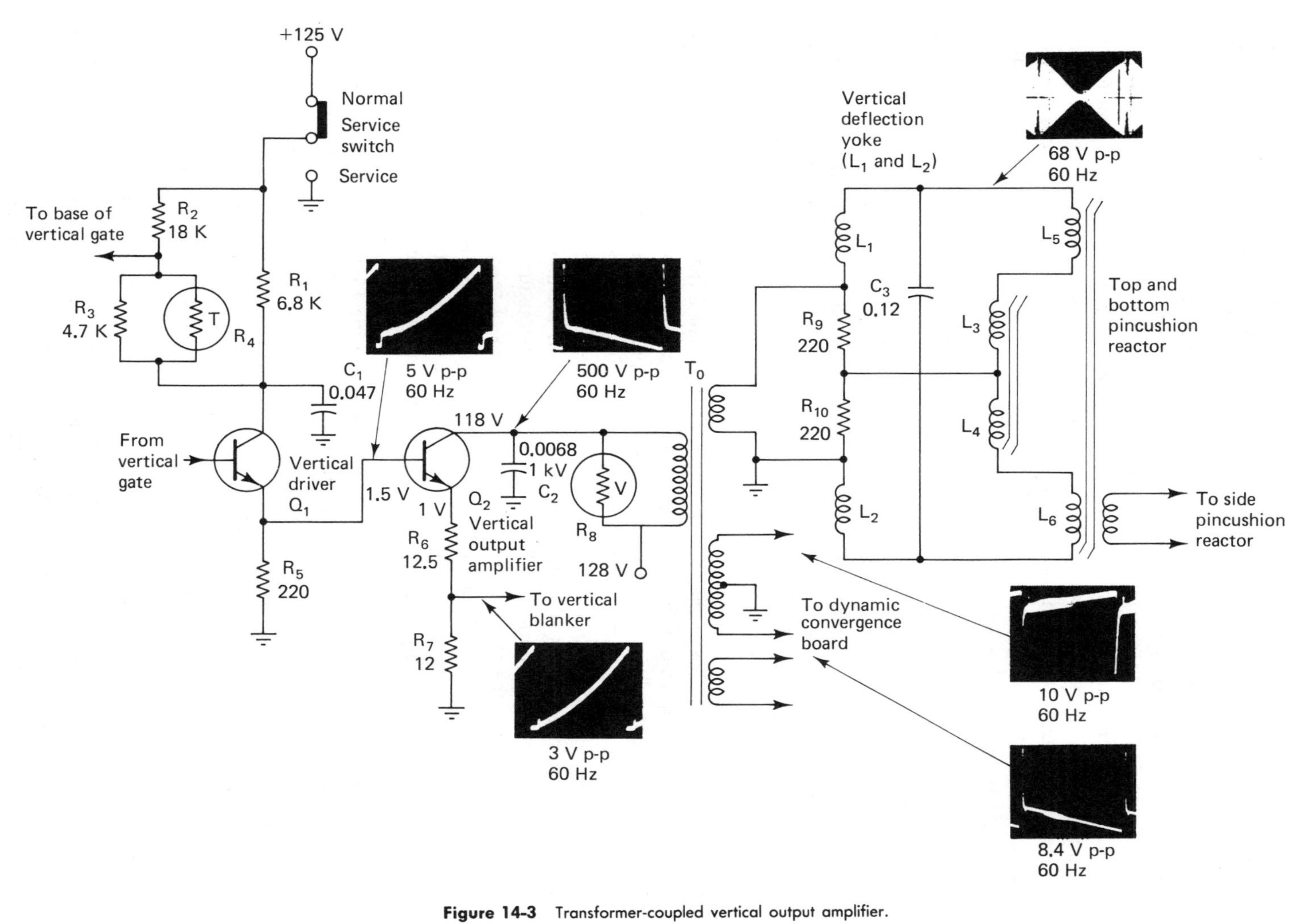

Figure 14-3 Transformer-coupled vertical output amplifier.

receivers, however, control of vertical centering may be desired. In such receivers a vertical centering control may be used to control a dc current that is allowed to flow through the windings of the yoke.

Output transformers are physically large. The inductance of the primary winding of the output transformer may range from 7 to 40 H, and its impedance can have a value of 2000 to 35,000 Ω. To provide the proper impedance match between the source and the yoke, a turns ratio ranging from 5:1 to 12:1 may be necessary. Vertical output transformers used in color receivers may have three or more secondary windings for use in the dynamic convergence circuits. When an output transformer or yoke has to be replaced, an exact replacement must be used if proper matching is to be obtained. Mismatching can result in such defects as a small raster, ringing, neck shadow, and other defects.

14-3 the vertical output amplifier

The vertical output amplifier is essentially a Class A or AB low-frequency power amplifier. In order to amplify the 60-Hz sawtooth with little distortion, the frequency response of the amplifier must extend down to 1 Hz. This is achieved by using large coupling capacitors, large inductance transformers and in most modern solid-state designs, dc coupling and negative feedback are used.

Input wave shape. The wave shape at the input of the vertical deflection amplifier may be a sawtooth or a trapezoid as in Figure 14-3. In some cases the trapezoidal driving voltage is designed to have a large negative-going pulse. This pulse occurs during vertical retrace; therefore it ensures that the transistor will remain cut off during retrace. Forcing the transistor to go deeply into cutoff is necessary because during retrace the conduction of the vertical output stage is suddenly reduced. As a result the magnetic field generated in the vertical output transformer will rapidly collapse. This induces a large positive going peak voltage across the transformer that can be in excess of several thousand volts. Such a large collector voltage could be sufficient to force the transistor into conduction, lengthening retrace and causing compression or foldover at the top of the picture. This is prevented by the large negative-going trapezoidal pulse at the input of the amplifier.

Transistor protection. When a transistor is used as the vertical output amplifier, the large peak voltages that appear aross the output transformer can damage the transistor. To prevent this, such components as a VDR (voltage-dependent resistor), a capacitor, or a diode may be placed across the primary of the transformer. See Figure 14-3. These components act to limit the peak voltages that the transformer primary can produce.

Thermistors. In some circuits such as Figure 14-5(a) a thermistor (R_3) having a normal cold resistance of from 4 to 75 Ω may be placed in series with the yoke. The thermistor is physically mounted on the yoke so that the temperatures of the thermistor and the yoke are identical. The sawtooth current passing through the yoke will heat the resistance of the copper wire that makes up the yoke. Copper has the characteristic of increasing its resistance as it gets hotter. As a result, less current can flow through the yoke, thus reducing the size of the raster. This is compensated for by the thermistor whose resistance decreases as the temperature increases. Since the thermistor is in series with the yoke, their total resistance remains constant, keeping the raster size constant.

Thermistors may also be found in the base circuit of the vertical output amplifier and are used to compensate for and prevent thermal runaway.

Damping resistors. The two vertical deflection yoke windings may be connected either in series or in parallel. See Figures 14-5 and 14-9. Connecting the windings in series increases their impedance and allows a smaller turns ratio transformer. In series with each of the windings in Figure 14-3 is a damping resistor of approximately 220 Ω. In some circuits a single damping resistor is placed across both of the windings. In Figure 14-8 the damping resistors are placed across each winding of the yoke. Damping resistors are needed to minimize the effects of crosstalk from the horizontal windings of the yoke.

The horizontal and vertical windings of the yoke act like a transformer. During horizontal retrace a great deal of energy is coupled (by transformer action) into the vertical windings. This energy tends to shock excite the vertical circuit into oscillation. As a result, each horizontal line exhibits a vertical undulation on the left side of the screen as shown in Figure 14-4. The damping resistors dissipate these oscillations and remove the undesired distortion of the raster. If these resistors were to increase in value, they would cause the reappearance of ringing on the left side of the raster.

Figure 14-4 Effect of vertical yoke ringing on the raster.

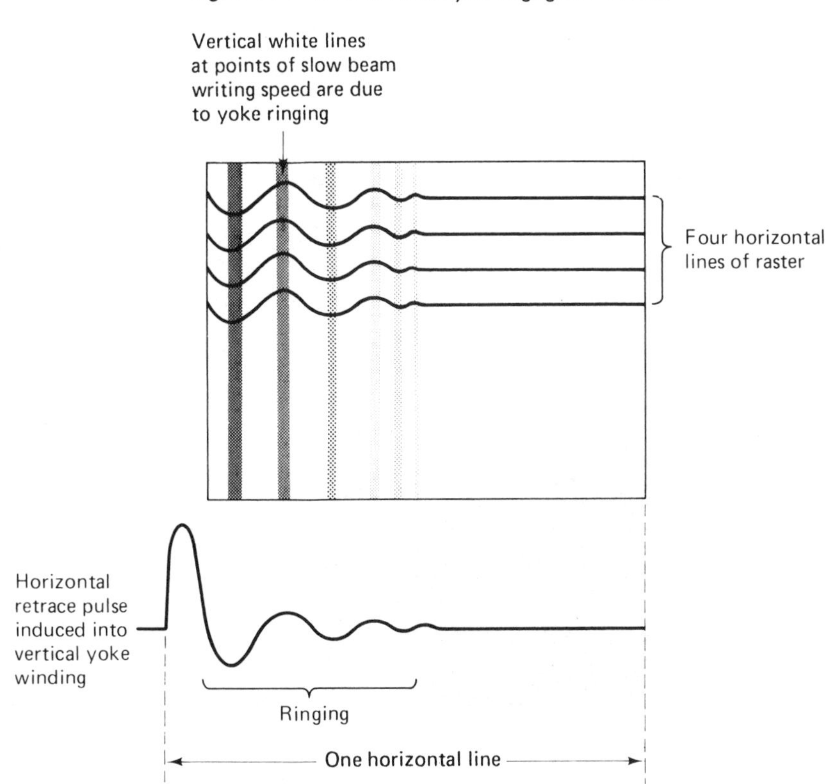

Connecting the vertical yoke windings in parallel reduces their circuit impedance to a value such that ringing is minimized.

14-4 transistor power amplifiers

Transistor power amplifiers are characterized by low internal resistance and high current capability. This amplifier is perfectly suited to drive directly either a series connected or parallel connected yoke without the need of an output matching transformer. Figure

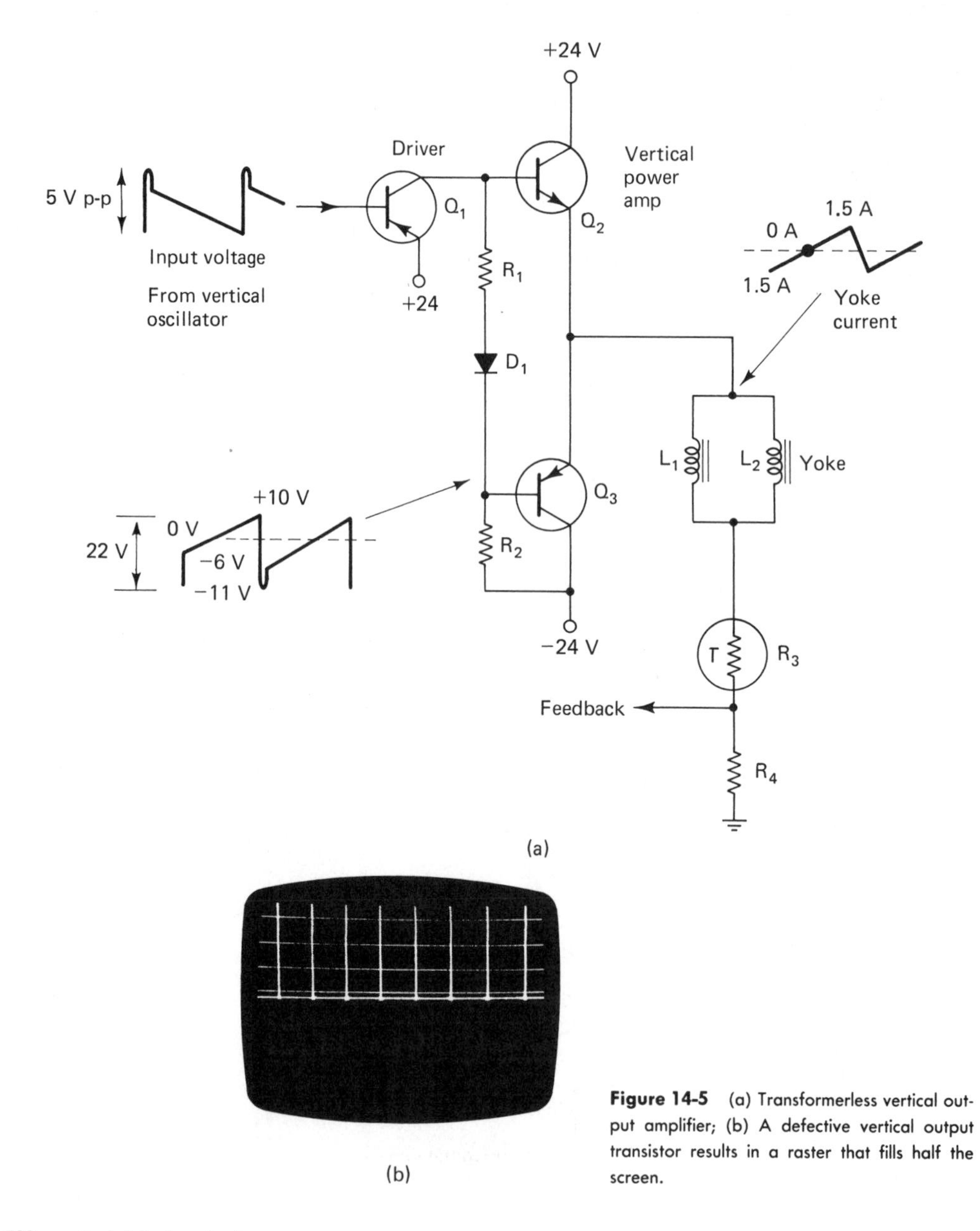

Figure 14-5 (a) Transformerless vertical output amplifier; (b) A defective vertical output transistor results in a raster that fills half the screen.

14-5 is an example of such a circuit. The output transistors Q_2 and Q_3 are arranged as a complementary symmetry power amplifier that is almost identical to that used in many solid-state audio power amplifiers. The output transistors are operated Class B and are alternately driven into conduction by a common trapezoidal input signal. When Q_2 is on and Q_3 is off, electron current flows *up* from ground through the yoke through Q_2 to the positive 24-V supply. On the alternate half of the input signal Q_3 is on and Q_2 is off. Current now flows from the negative 24-V power supply through Q_3 and *down* through the yoke to ground. Notice that an ac current flows through the yoke. Also notice that because of the equal and opposite power supplies, this circuit does not use the large electrolytic coupling capacitor that is often found in series with the yoke.

In this type of circuit if one of the output transistors becomes defective, vertical deflection will only occur for half the raster. For example, if a crosshatch test pattern were transmitted, the defective receiver would display half the pattern from center to top and a blank raster from center to bottom, and vice versa. See Figure 14-5(b).

Diode D_1 is forward biased and its voltage drop is used to prevent crossover distortion by providing bias for the power amplifier stages. The conduction of the diode effectively ties the bases of the power transistors Q_2 and Q_3 together, allowing the output signal of the driver to feed both of them simultaneously.

14-5 vertical linearity

The deflection current passing through the vertical windings of the yoke has a sawtooth wave shape. An example of such a wave shape is shown in Figure 14-6(a). The sawtooth is divided into two regions; trace and retrace. The trace portion of the sawtooth is responsible for forcing the electron beam to move from the top to the bottom of the screen. The retrace period is shorter in time duration and returns the electron beam from the bottom of the raster to the top. A sawtooth is said to be linear if the trace portion of the sawtooth forms a straight line. A linear sawtooth will force the electron beam to scan at a constant rate from the top to the bottom of the raster producing an equal number of horizontal lines for each vertical inch of the picture tube face. A constant rate of scanning means that the writing speed of the electron beam will also be constant. This will result in a raster of uniform brightness. If a crosshatch generator is used as a signal source, the crosshatch display will permit a rapid check of vertical linearity. Equal spacing between the horizontal bars at the top and bottom of the raster indicates good vertical sawtooth linearity. If the sawtooth is nonlinear, as indicated in Figure 14-6(b) and (c), the crosshatch horizontal line spacings will be crowded at either the top or bottom of the raster.

Nonlinearity may be caused by a short-time-constant sawtooth-forming *RC* circuit in the vertical oscillator. This defect produces a sawtooth like that shown in Figure 14-6(b). In this case, too much of the exponential charging curve has been used to form the sawtooth. Similar nonlinear compression at the bottom of the raster may be caused by positive peak clipping of the voltage driving the vertical output stage. Another possible cause might be collector current that can saturate the output transformer, effectively flattening the region of the sawtooth corresponding to the bottom of the raster. Compression at the top of the raster may be the result of defects that cause excessive retrace time and may be associated with the vertical deflection oscillator.

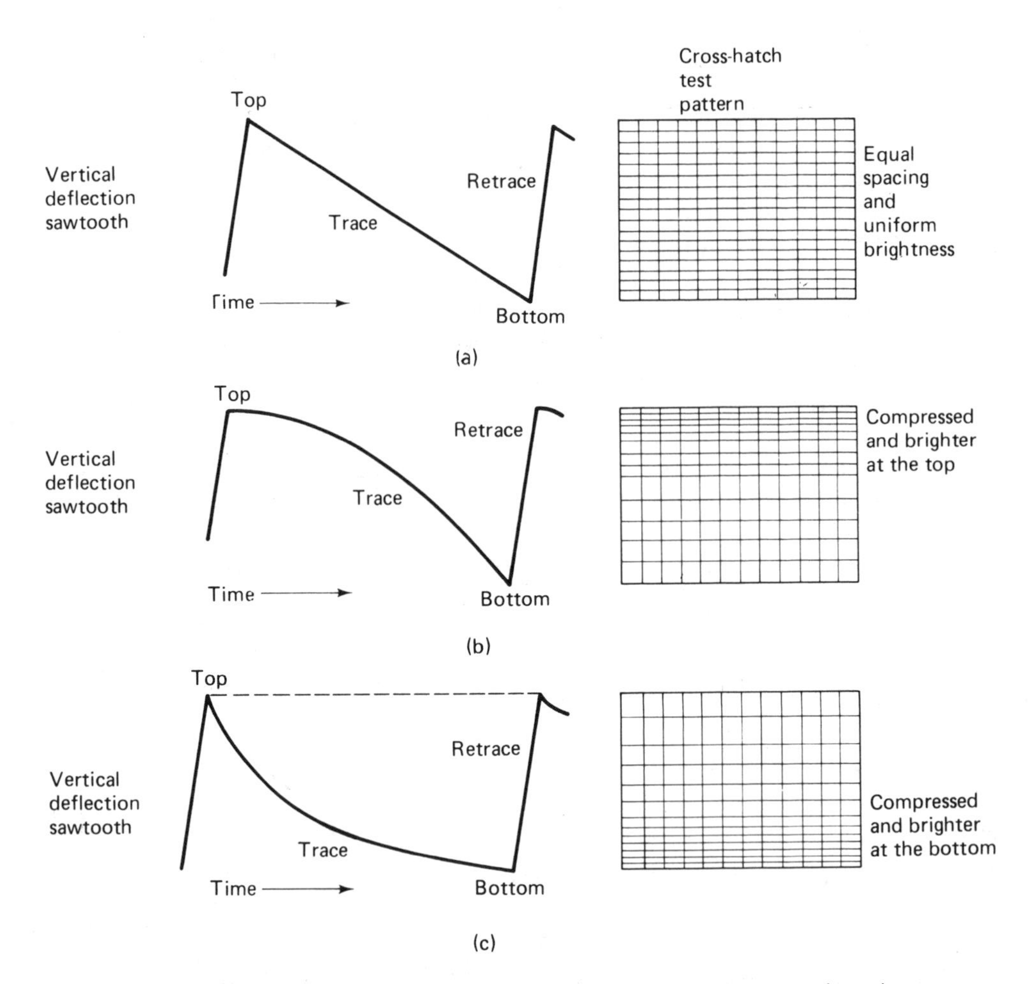

Figure 14-6 (a) Linear vertical deflection sawtooth and the resultant raster; (b) nonlinear sawtooth and its effect on the raster; (c) nonlinear sawtooth of opposite direction and its effect on the raster.

"S" shaping. Because of the curvature of some color picture tube faceplates and nonlinearity produced by the yoke resistance, a linear sawtooth of current in the yoke will result in a stretch at the top and bottom of the raster. In receivers using such tubes this is often compensated for by purposely distorting the sawtooth into one with an S shape. "S" shaping is usually accomplished by special feedback circuits.

The linearity control. Nonlinearity of the sawtooth is a normal byproduct of sawtooth generation. This is because of the nonlinear *RC* charging characteristic upon which sawtooth generation is based. Linearity control in transistor vertical deflection circuits is usually provided by means of feedback combined with a wave-shaping circuit. Unfortunately, there is no universal circuit arrangement. One example of this type of control is shown in Figure 14-7. In this circuit the sawtooth forming capacitors are C_1 and C_2. A transistor that acts like a low-resistance switch during retrace is placed across them. The sawtooth output is

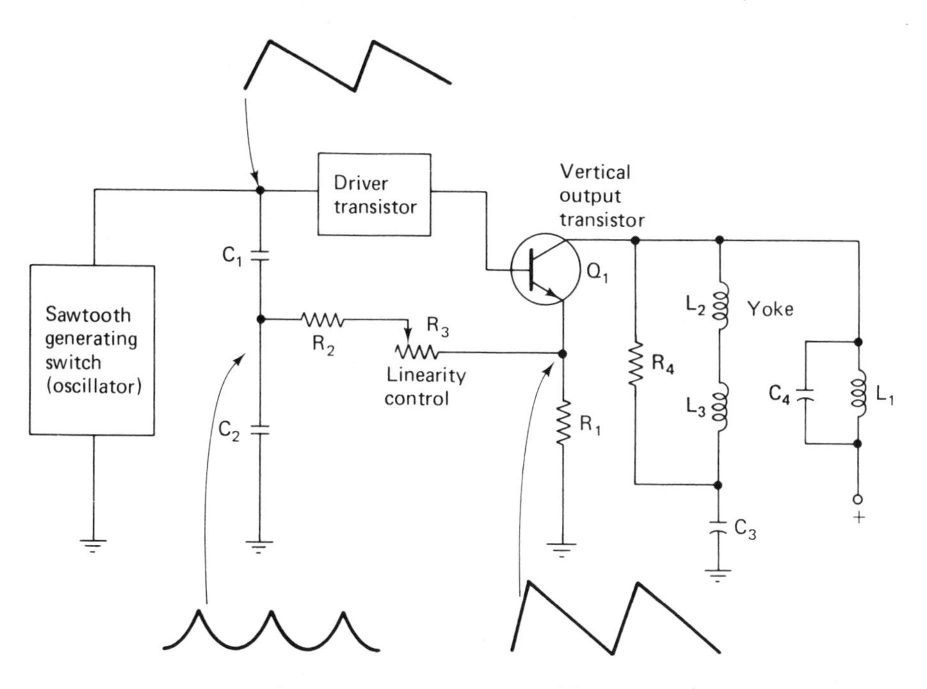

Figure 14-7 Feedback type of vertical linearity control.

then amplified and is used to drive the yoke. The ac sawtooth yoke current flows through R_1 and develops a sawtooth of voltage across it. This voltage is fed back to the junction of C_1 and C_2 through R_2 and the linearity control R_3. The combination of R_2, R_3, and C_2 forms an integrating circuit that converts the sawtooth (across R_1) into a parabolic waveform. This voltage is now added to the sawtooth ordinarily developed across C_1 and C_2 and corrects for any nonlinearity. The magnitude and shape of the feedback voltage are controlled by the linearity control R_3.

14-6 commonly used vertical deflection circuits

In this section the vertical output circuits that are commonly found in both black & white and color television receivers will be discussed.

A practical solid-state circuit consisting of a blocking oscillator and a two-stage amplifier is shown in Figure 14-8. This circuit is identified as part of a color receiver by the multiple windings on the vertical output transformer that are used to feed the dynamic convergence board.

Circuit action. The blocking oscillator transistor Q_1 is placed across the sawtooth forming capacitor C_1 and acts like a switch. When Q_1 is off, C_1 charges through R_1, forming the *trace* portion of the sawtooth. This voltage decreases in amplitude as C_1 charges because C_1 is in series with the 12.5-V source, and the voltage developed across C_1 is series opposing to the source voltage. During retrace Q_1 is turned on, shorting C_1 discharging it, and forcing the voltage across R_1 to return to 12.5 V.

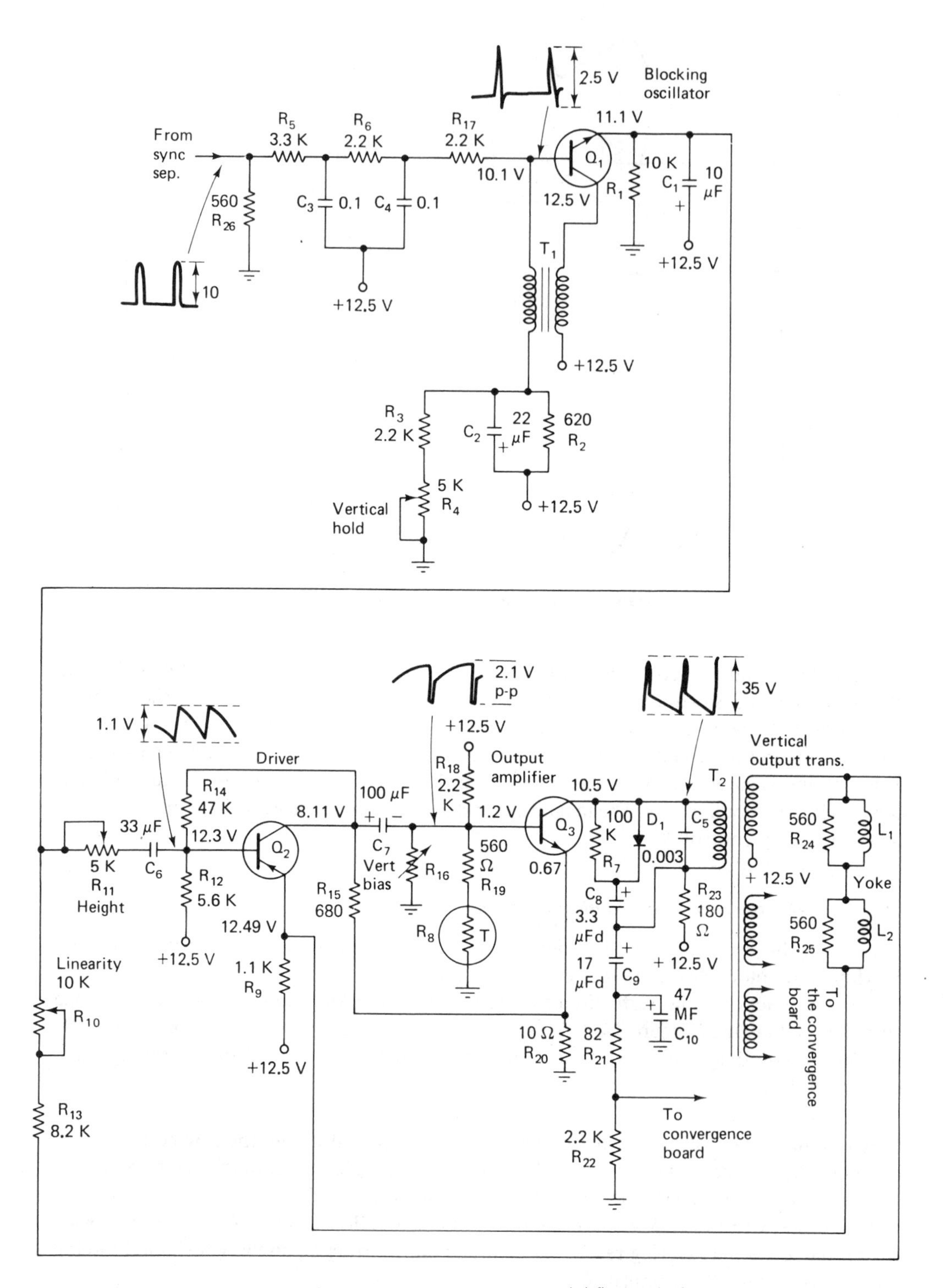

Figure 14-8 Three-stage transistor vertical deflection circuit.

Transformer T_1 provides the positive feedback that is necessary to sustain oscillation. During *retrace* the discharge current of C_1 flows through the primary winding of T_1 and induces a voltage into the secondary winding that makes the base of Q_1 positive. This polarity ensures that the NPN transistor will conduct. The base current that flows, charges C_2, forming the signal bias necessary to block the transistor. The frequency of the oscillator is determined by the length of time the transistor is blocked. This in turn is determined by the time constant of C_2 and R_2, as well as the reverse bias between base and emitter. The hold control R_4 is able to vary the frequency of the oscillator by changing the bias of Q_1.

Positive-going vertical sync pulses obtained from the sync separator stage are integrated by R_5, C_3, R_6, and C_4. The output pulse of the integrator forces the blocking oscillator to assume the same frequency as the sync pulse by driving the blocking oscillator into conduction starting retrace.

The sawtooth output of the blocking oscillator is inverted and amplified by the driver transistor Q_2 to a level sufficient to drive fully the output transistor Q_3. The output amplifier in turn drives the vertical output transformer (T_2) which provides the necessary impedance matching for the yoke.

During *retrace* the NPN output transistor Q_3 is driven into cutoff. The sudden decrease in collector current causes the magnetic field in the output transformer and yoke to collapse, inducing a large positive-going voltage across the primary winding of the transformer. To prevent this voltage from damaging the output transistor, C_5, C_8, D_1, and R_7 act to limit the peak primary voltage to a safe level.

In the base circuit of Q_3 is a thermistor R_8. This component prevents thermal runaway and ensures the thermal stability of Q_3. This is accomplished as follows. During normal operation the vertical output amplifier Q_3 heats up, causing a corresponding increase in its collector current. The thermistor R_8 also gets hot and its resistance decreases. As a result of voltage divider action between R_{18}, R_{19}, and R_8 the base dc voltage decreases, reducing Q_3's forward bias. This in turn reduces the collector current, compensating for the original temperature-caused increase in collector current.

The vertical deflection yoke windings are connected in series. Across each winding is a damping resistor of 560 Ω that prevents ringing. See Figure 14-8. One side of one of the three secondary windings of the vertical output transformer feeds the yoke. The other side of the winding returns to the 12.5-V B supply. Since both ends of the yoke return to the 12.5-V supply, the dc difference of potential across the yoke is zero and no dc current flows. The yoke is part of one of the two feedback circuits used in this circuit. It feeds back a sawtooth of current to the emitter resistor R_9 of the driver transistor Q_2. The other feedback path is from the top of the output winding of the secondary through the linearity (R_{10}) and height controls (R_{11}) to the base of Q_2. The linearity control R_{10} and the sawtooth-forming capacitor C_1 act to integrate the feedback voltage into a parabola. Both negative feedback paths are used to control the waveform linearity and stability of the amplifier. The linearity control R_{10} adjusts the amplitude and waveform of the feedback, thereby controlling the output sawtooth linearity. The height control in conjunction with R_{12} acts as a voltage divider that determines the amplitude of the sawtooth reaching the base of the driver transistor Q_2.

The service switch. In many color receivers a service switch is found on the rear apron of the television chassis. This switch may have two or three positions: The two-position switch has a "normal picture" position and "blank raster" position used for purity adjustments. The three-position service switch includes a position for gray scale adjustment. In this position vertical deflection is disabled and a red, green, and blue horizontal line appears on the screen. The details of gray scale adjustment are discussed in the chapter devoted to picture tubes and will not be repeated here.

Many different methods are available for disabling the vertical deflection circuit. Some receivers such as the one that uses the vertical deflection circuit of Figure 14-8 do not have any service switch. In such cases, the technician may have to improvise. For example, shorting the primary or secondary winding of the blocking oscillator transformers would safely disable the vertical deflection circuit. Very often the manufacturer's service literature will indicate other methods that might be used.

Service switches that are used in transistor-operated vertical deflection circuits may be placed in series with the base of a driver or output transistor. Opening the switch removes the driver's input deflection sawtooth and causes the raster to collapse into three horizontal lines. It may also bias the transistor into cutoff and prevent overheating of the transistor.

An example of a service switch is shown in Figure 14-3. In this system the service switch is placed in series with the collector of the vertical deflection driver Q_1. During normal operation the switch is closed, as shown in Figure 14-3. When the switch is placed in the service position, it opens the collector power supply, disabling the circuit and forcing the raster to collapse into three colored horizontal lines.

14-7 transformerless vertical output amplifier

Figure 14-9 shows the schematic diagram of a vertical output deflection circuit used by RCA. The output circuit is driven by an integrated circuit chip U_{401}. In this figure a partial block diagram of those sections of the chip devoted to the sawtooth generator and linearity correction are shown.

Power is supplied to the output transistors by the 24-V B+ source and the 127-V regulated source. The regulated source supplies energy through R_1 to assist retrace.

In this circuit fewer transistors are required in the output stage because the "bottom" transistor Q_2 acts as a driver for the "top" transistor. Vertical drive is obtained by the charging of C_1 through the vertical height control R_2. The "ramp reset switch" inside the IC initiates retrace by discharging C_1 through a switch transistor. The sawtooth developed across C_1 is coupled by C_2 to the vertical linearity and vertical drive portions of the IC, where it is processed and amplified. The output of the vertical drive is applied to the base of Q_2.

Vertical scan may be divided into two parts: *top scan,* where the electron beam moves from the top of the raster to the center, and *bottom scan,* where it moves from the center to the bottom of the raster. During the first half of scan, Q_2 amplifies the drive signal and drives the base of Q_1 through diode D_1. Q_1 then forces current through the yoke windings moving the electron beam from the top to the center of the screen. During the second half of scan, Q_2 conducts and drives the yoke through D_2. At the beginning of vertical trace, Q_1 is turned on, and maximum current flows from the 24-V source through D_3, Q_1, the yoke, and C_3, charging the capacitor. This charging current moves the beam from the top of the raster toward the center. Q_1 is turned on by current flowing through D_4, R_4, and R_5. During this same time Q_2 is not completely off and it shunts some of Q_1's base current by

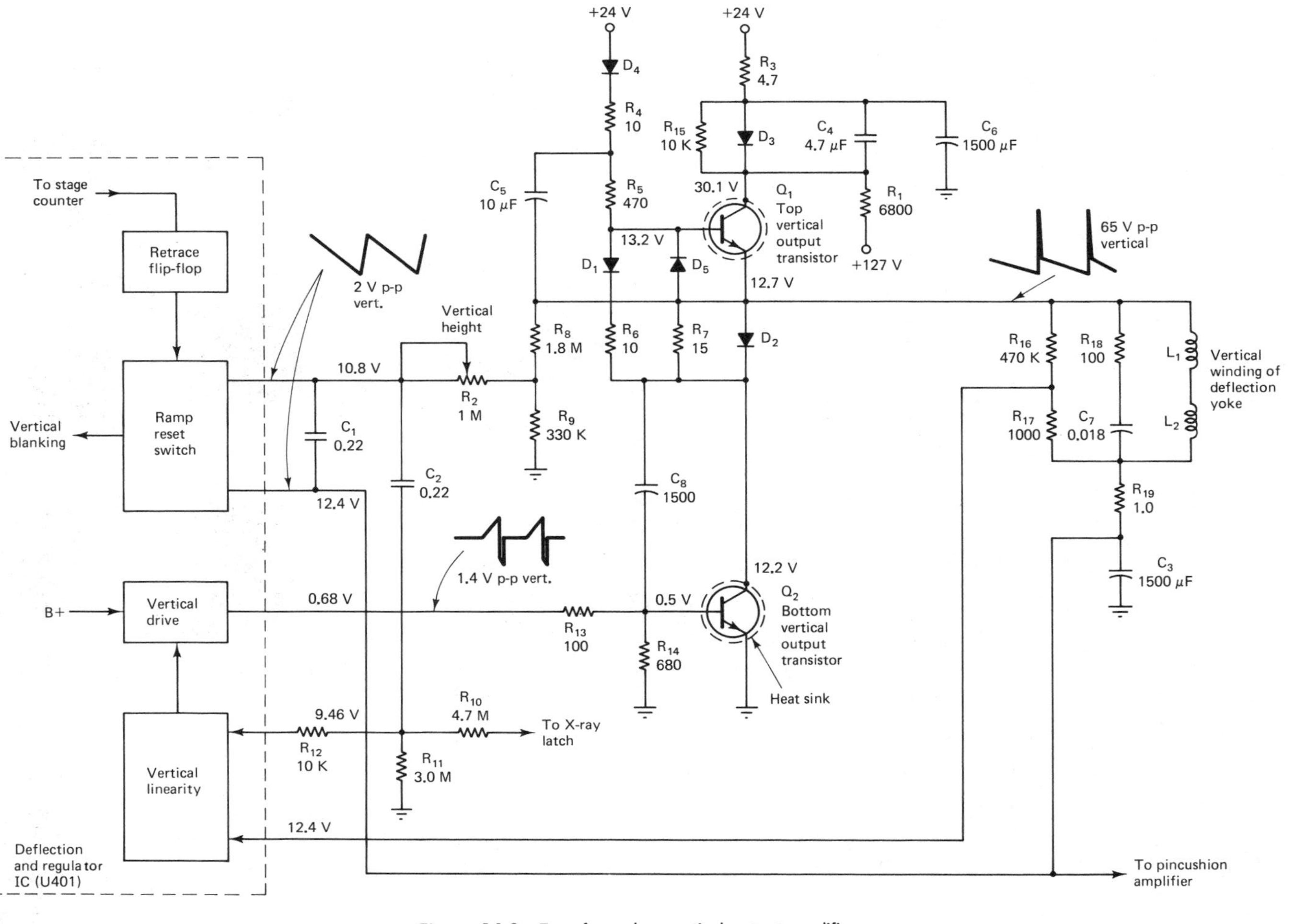

Figure 14-9 Transformerless vertical output amplifier.

drawing current through D_1 and R_6. Yoke current is not affected because D_2 is off and prevents Q_2 from influencing it. As a result, the current flow through Q_1 and the yoke decreases to the point where Q_1 is turned off (the beam is at the center of the screen). The increasing current flowing through Q_2 reverses the current through the yoke, moving the electron beam down from the center toward the bottom of the raster. The source of this current is the discharge current of C_3 that flows from C_3 up through the yoke, through D_2, Q_2, and then back to C_3.

Retrace is initiated by the negative-going pulse of the drive voltage. This suddenly forces Q_2 to go from maximum current flow (beam is at the bottom of the screen) to zero. However, the yoke current continues to flow through D_5 into the *base-collector* junction, forcing the transistor to conduct in the reverse mode from emitter to collector, charging C_4. Diode D_3 prevents the charging current from going to the 24-V supply.

The energy stored in C_5 supplies the base drive of Q_1 during vertical retrace and the first half of retrace. At the time that C_4 is fully charged and the beam is near the center of the screen, the charge on C_5 turns Q_1 on, discharging C_4 through the yoke in the reverse direction, moving the beam back up to the top of the screen. Retrace time is shortened by increasing the collector voltage of Q_1 by connecting it via R_1 to the 127-V source. At this time yoke current is maximum, the beam is at the top of the screen and Q_2 begins to conduct, shunting base current from Q_1, thus starting the cycle over again.

14-8 the three-stage transistor multivibrator

In the circuit shown in Figure 14-10 three transistors, Q_{18}, Q_{19}, and Q_{20}, form a collector-coupled multivibrator. The output of the last stage (Q_{20}) is fed back to the first stage (Q_{18}) via C_{600} and R_{601}. The first stage Q_{18} inverts the signal and feeds it to Q_{19}, a noninverting emitter follower driver amplifier. This stage is used to isolate and match the first stage to the third. The output of the third stage Q_{20} is again inverted by amplifier action and provides the phase shift needed for positive feedback.

Capacitors C_{602} and C_{603} make up the sawtooth-forming capacitors and are shunted

Figure 14-10 Three-stage multivibrator.

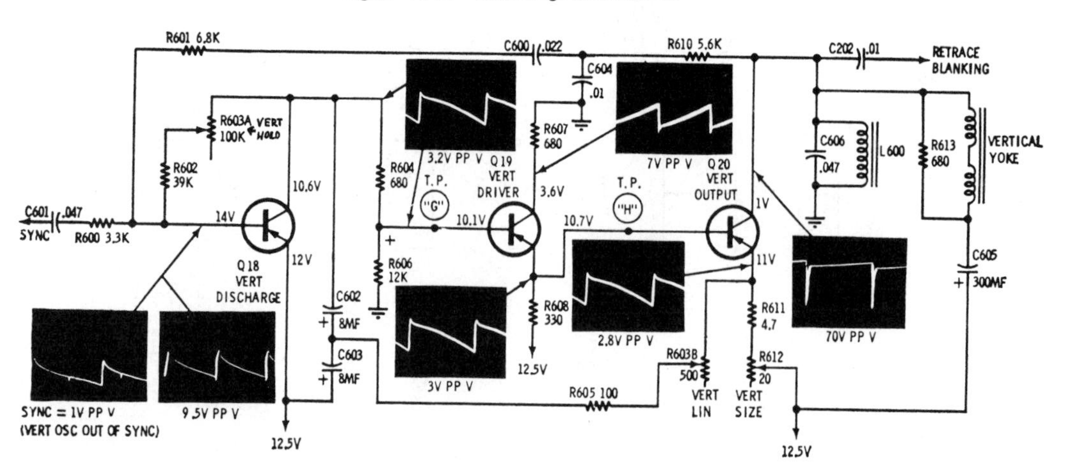

by the vertical discharge transistor Q_{18}. At the beginning of trace the capacitors are completely discharged. As a result, the voltage across Q_{18} is zero and it is cut off. As the capacitors charge, the forward bias on Q_{19} and Q_{20} also increases and forces the yoke current to increase and moves the electron beam from the top to the bottom of the screen.

The base voltage of Q_{18} is developed by a form of combination bias. The large negative pulse at the collector of the output deflection amplifier (Q_{20}) is fed back and rectified by the base-to-emitter junction of the vertical discharge transistor (Q_{18}). This charges C_{600}, producing signal bias. This bias voltage is reduced by the charge of C_{602} and C_{603}. The voltage across these capacitors is coupled via the hold control R_{602} and R_{603A} to the base. As these capacitors charge, the base voltage falls. When the base voltage drops below the emitter voltage (12 V), Q_{18} conducts, discharging the sawtooth-forming capacitors and initiating retrace. R_{603A} is the hold control because it determines how fast the base voltage of Q_{18} falls. When it is maximum, the base voltage falls more slowly and retrace is delayed, thus lowering the oscillator frequency.

The sudden rise in the collector voltage of Q_{18} rapidly decreases the forward bias of the vertical driver Q_{19} (an emitter follower) and that of the vertical output transistor Q_{20}, driving them into cutoff. During trace these transistors are conducting and building up the yoke's deflection magnetic field. When they are driven into cutoff, the yoke's magnetic field collapses and develops a large negative pulse. This negative pulse is coupled by R_{610}, C_{600}, and R_{601} back to the base of the discharge transistor Q_{18}, driving it into saturation, discharging C_{602} and C_{603}, and marking the beginning of a new cycle.

Synchronizing of the oscillator is accomplished by negative-going sync pulses that are applied to the base of the vertical discharge transistor Q_{18} forcing it to turn on and thereby initiating retrace.

In this circuit, the linearity control (R_{603B}) is tied between the emitter of the vertical output amplifier and the sawtooth forming capacitor C_{603}. This combination forms an integrator that converts the sawtooth into a parabolic wave shape that is added to the sawtooth developed across the vertical discharge transistor. The linearity control varies the shape and amplitude of the parabolic feedback waveform and is therefore able to change the shape of the resultant deflection sawtooth, thus compensating for any nonlinearity.

The size (height) control (R_{612}) is in the emitter circuit of the vertical output stage and determines the bias (operating point) and amount of degeneration introduced into the circuit. As such, it not only determines the peak-to-peak deflection sawtooth amplitude but it also affects its linearity.

Displacement of the raster is prevented by C_{605}, which is placed in series with the vertical windings on the yoke. This blocks any dc current from passing through the yoke, but it allows the ac deflection sawtooth to pass through without any attenuation.

Horizontal pulses are coupled into the vertical circuit by transformer action between the horizontal and vertical windings of the deflection yoke. If allowed to reach the base of the vertical discharge transistor, these pulses can cause vertical jitter. C_{604} and R_{610} form an integrator that filters out these pulses and prevents this from happening.

During retrace the collapsing magnetic field surrounding the vertical windings of the yoke is capable of inducing extremely high voltages that can damage the output transistor (Q_{20}) or other components. R_{613} and C_{606} prevent these negative pulses from exceeding the breakdown voltage of Q_{20}. These components load the yoke and the output load inductance L_{600} and limit the peak voltages that can be developed across them.

14-9 the Miller rundown vertical deflection circuit

In recent years a vertical deflection circuit called the *Miller rundown* or the *Miller feedback amplifier* has been introduced. This circuit is capable of generating a sawtooth of exceptionally good linearity. As a result, no linearity control is necessary in either color or black & white receivers that use this circuit.

The Miller rundown circuit makes use of the Miller effect to enormously increase the effective size of the sawtooth forming capacitor. Consequently, the charging time constant of the sawtooth forming *RC* network becomes very large and only a very small portion of the *RC* charging curve is used to form the trace portion of the sawtooth. In this manner a sawtooth of excellent linearity is obtained.

The Miller effect is the change in amplifier input impedance that the addition of feedback has on the amplifier. To understand this statement, consider the following. An ac generator is said to be connected across a capacitive circuit whenever the voltage across a circuit lags the current through it by 90°. If for any reason this current is increased, the generator will "see" a larger effective capacitor than the one that is physically there. In an amplifier the input generator can be made to have an effectively higher input capacitive current by introducing feedback.

The effective capacitance introduced into an amplifier circuit by the Miller effect is directly proportional to the amplifier gain. Therefore, if the overall no-feedback amplification of a multistage amplifier using capacitor feedback is 1 million, the input capacitance of the amplifier will appear to be 1 million times larger than the feedback capacitor. It is this principle that is used to increase the effective size of the sawtooth forming capacitor in the Miller rundown vertical deflection circuit.

Typical circuit. Figure 14-11 is a simplified schematic diagram of the Miller rundown vertical deflection circuit. The circuit contains five transistors that form a high-gain amplifier, four diodes, and four feedback paths. The feedback circuits convert the amplifier into a multivibrator-type oscillator. Examination of the circuit shows that the circuit is very similar to a transformerless high-quality audio amplifier. The output stage (Q_4 and Q_5) is a Class B complementary symmetry power amplifier that drives a sawtooth current through the vertical deflection winding of the yoke. This is done via C_1, a large dc blocking electrolytic capacitor. The sawtooth yoke current is also used to provide vertical dynamic convergence and dynamic pincushion correction.

Diode D_1 is a bias diode used to reduce crossover distortion. If it were to become shorted, a white horizontal line would be produced at the center of the raster. The zener diode D_2 conducts only during vertical retrace and combines with output transistor Q_4 to act as a series regulator that limits the yoke current during retrace. If this diode were to open, the base voltage of Q_4 would tend to increase and cause a corresponding increase in vertical output current which may result in premature failure of the vertical output transistors.

During the trace portion of vertical scan the driver transistor Q_3 develops a linear negative going signal at the bases of the output transistors Q_4 and Q_5. The portion of this drive signal that corresponds to the top to the center of the raster, turns off Q_5 and forces Q_4 to decrease its conduction linearly while charging C_1. When the scanning beam has reached the center of the screen, the drive voltage will be such that it turns Q_4 off and linearly increases the conduction of Q_5.

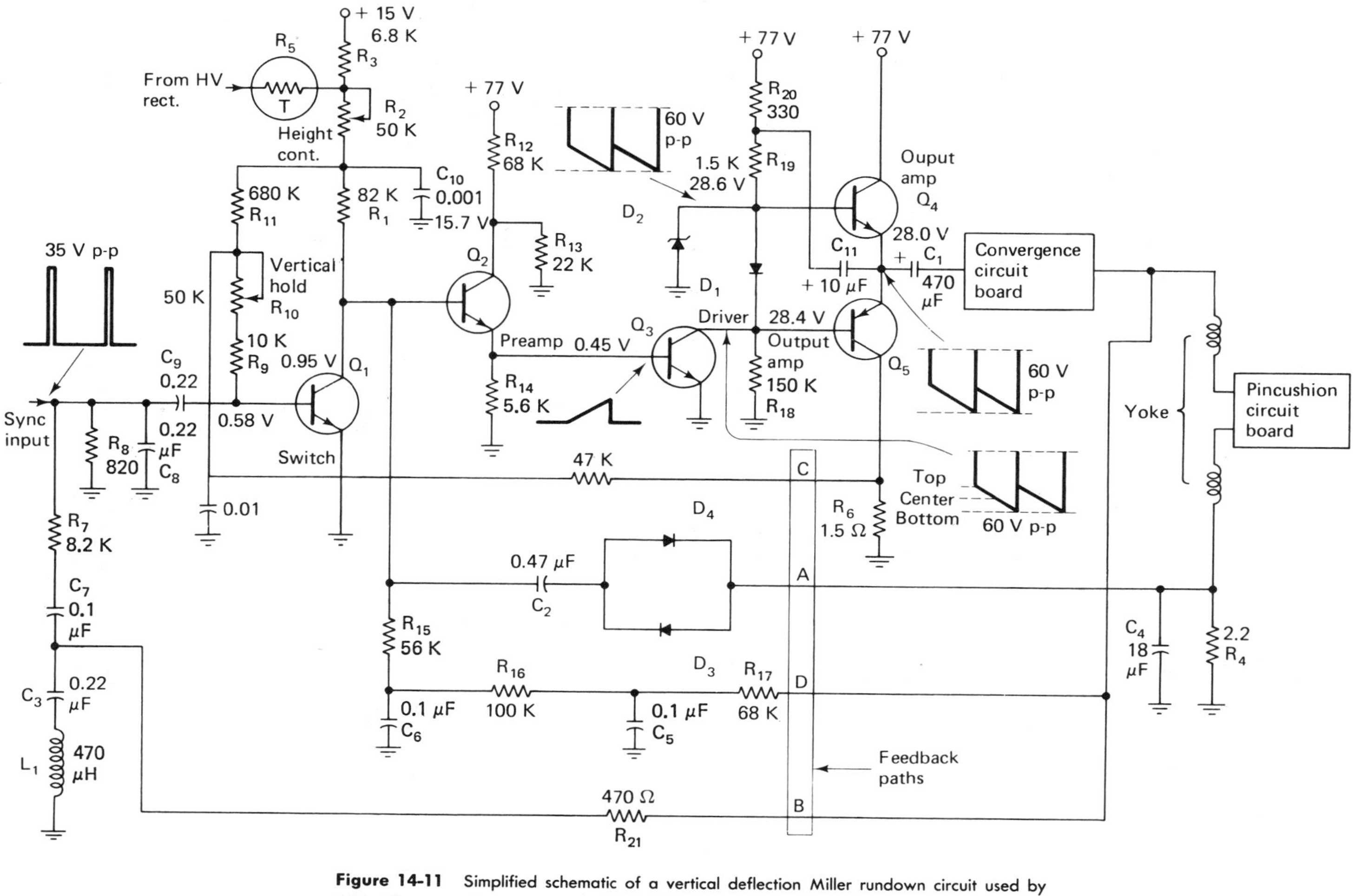

Figure 14-11 Simplified schematic of a vertical deflection Miller rundown circuit used by RCA. (Courtesy of RCA.)

Deflection from the center of the screen to the bottom of the raster is caused by the discharge of C_1. When the beam has reached the bottom of the raster, the driver Q_3 goes into cutoff and the output drive voltage suddenly becomes very positive and turns off Q_5 and turns on Q_4. The heavy conduction of Q_4 forces the picture tube electron beam to retrace quickly from the bottom of the raster to the top. Zener diode D_2 limits the maximum base voltage of Q_4 and therefore prevents the collector current of Q_4 from becoming excessive.

The preamp Q_2 is an emitter follower that acts as a noninverting buffer power amplifier between switch transistor Q_1 and the driver Q_3. The switch transistor (Q_1) generates the sawtooth. When it is off, the sawtooth-forming capacitor C_2 charges (through R_1, R_2, R_3, R_4 and D_3) forming the trace portion of the sawtooth. C_4 is a bypass capacitor that eliminates any horizontal pulses that might be introduced from the yoke or pincushion circuit. When the switching transistor comes on, C_2 discharges (via Q_1, R_4, and D_3) and forms the retrace portion of the sawtooth.

Because of the Miller effect, *feedback path A* effectively increases the size of C_2 by approximately 1 million times (the open loop gain of Q_2, Q_3, Q_4, and Q_5) ensuring excellent sawtooth linearity. The height control R_2 provides a means of controlling the peak-to-peak amplitude of the sawtooth by changing the RC time constant of the charging circuit.

Feedback path B provides the positive feedback needed to convert the high-gain amplifier into a sawtooth generating multivibrator. The free-running frequency of the oscillator can be made to vary from 40 to 80 Hz, depending upon the position of the hold control. The switch transistor Q_1 is the normally off transistor in the multivibrator and its associated circuit performs three functions: It controls the free-running frequency of the multivibrator, allows synchronization with the received signal, and determines retrace time.

Included in the feedback path is a series resonant trap C_3 and L_1. It is tuned to the horizontal scanning rate and is used to bypass any horizontal pulses that might have been coupled into the vertical output circuit by the pincushion circuit, or crosstalk in the yoke.

Positive-going vertical sync pulses are coupled to the base of the switch transistor Q_1 that force it into conduction. This results in the discharge of C_2, which initiates retrace.

Feedback path C is used to improve the frequency stability of the oscillator circuit. The feedback voltage is developed across R_6 only when Q_5 conducts. This occurs when the electron beam is moving from the center to the bottom of the screen. When it reaches the bottom of the raster, Q_5 rapidly turns off. The voltage developed across R_6 will be a positive going ramp that suddenly drops to zero during retrace. The addition of this voltage to that of the normal exponentially rising voltage at the base of Q_1 sharpens the voltage rise as it moves the transistor's operating point from cutoff to saturation. In addition, the sudden drop in voltage across R_6 when retrace occurs enhances the cutoff characteristic of the switch Q_1 circuit.

The Miller rundown circuit provides a very linear sawtooth current through the yoke. This current must be modified because the point of electron beam deflection is not located at the radius of curvature of the picture tube faceplate. As a result, pincushion distortion of the raster occurs. In addition, the raster is stretched at the top and bottom. To compensate for these difficulties, a pincushion correction circuit is connected in series with the yoke and *feedback path D* is used to introduce "S" shaping or correction to the yoke current to compensate for the stretch at the top and bottom of the raster. "S" compensation is achieved by generating a parabola of current and coupling it into the Miller feedback path. The Miller circuit combines this feedback component with that of the Miller feedback and modifies the yoke current in the proper fashion for "S" shaping.

The yoke. The deflection yoke consists of two pairs of coils wound at right angles to each other. One pair of coils is used for vertical deflection and the other pair is used for horizontal deflection. The coils are mounted around the neck of the picture tube such that the coils that provide horizontal deflection are mounted vertically and those used for vertical deflection are mounted horizontally. The yoke windings may be constructed on a ferrite core in a toroidal shape or a saddle shape. Figure 14-12(a) shows an example of the windings of a saddle-type yoke and Figure 14-12(b) shows an example of the windings found in a toroidal-type yoke.

To ensure uniform focus, the cross-sectional thickness of the windings are so organized that the distribution is in proportion to the cosine of the angle between the deflection axis of the coil and the position of the windings on the neck of the picture tube. Cosine yoke construction creates a uniform magnetic field that prevents elliptical distortion of the deflection beam that would result in nonuniform focus.

Deflection yokes are rated according to their deflection angle, neck diameter, winding inductance, and peak deflection current required for full screen deflection.

Yokes designed for modern black & white vacuum tube receivers have deflection angles of 90°, 110°, and 114°. Picture tube neck diameters range from 29 to 39 mm.The smaller the neck diameter, the smaller the yoke's inductance for the same deflection. The inductance of the vertical deflection windings may range from approximately 1 to 80 mH, and typical peak-to-peak currents range from 0.4 to 4 A.

Transistor-type yokes used in black & white receivers have vertical winding inductance that is typically 70 mH in black & white receivers and as little as 1 mH in color receivers. The peak-to-peak sawtooth current for full vertical deflection is typically approximately 3.0 A.

Color receivers use yokes that are often physically larger than those used in black & white receivers and do not have antipincushioning permanent magnets embedded in their

Figure 14-12 (a) Saddle-type yoke winding; (b) toroidal-type yoke winding. (Courtesy of Motorola.)

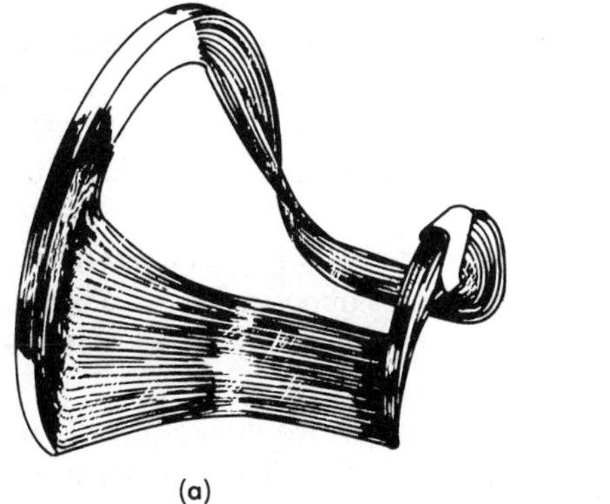

(a)

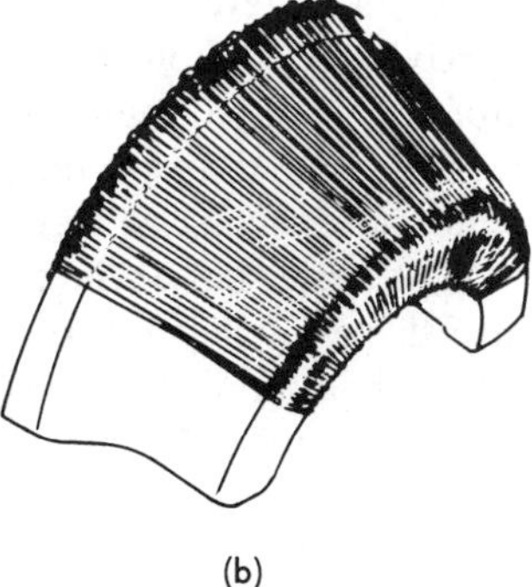

(b)

plastic housing as do black & white receiver yokes. The physical arrangement of the coil winding may be saddle, toroidal, or a combination of saddle and toroidal. The older color receiver yokes were primarily of the saddle type. Modern color receiver yokes are often toroidal in construction. One example of this type is the *precision static toroid* (PST) deflection yoke. The combination of a closely spaced in-line precision picture tube electron gun and the PST yoke makes practical an inherent self-converging color system that requires few if any dynamic convergence adjustments. The PST yoke is used on 90° and 110° narrow-neck picture tubes. The PST yoke consists of comparatively few turns of heavy wire wound into the winding grooves of molded plastic rings that are cemented to each end of the core. This yoke allows precise winding that tightly controls the magnetic field produced. The position of the yoke on the neck of the picture tube is extremely critical. At the time of manufacture it is properly positioned and then may be cemented into place with a thermosetting plastic. If either the yoke or picture tube were to fail, the entire unit may, depending on manufacturing, be replaced. The impedance of the PST yoke is very low and is well suited for use with solid-state deflection circuits. The vertical deflection yoke inductance for series-connected winding is 1.15 mH and that of the horizontal series-connected windings 0.158 mH.

14-10 a vacuum tube vertical deflection multivibrator

The vacuum tube vertical deflection multivibrator has been one of the most commonly used vertical deflection circuits and is found in both black & white and color vacuum tube receivers. In this circuit two tubes are usually combined in one glass envelope and act as both a plate-coupled multivibrator and the output power amplifier. The tubes may both be triodes, but in some cases one tube is a triode and the other is a power pentode.

Typical circuit. A typical schematic of a vertical deflection multivibrator used in a black & white receiver is shown in Figure 14-13. Notice that the plate of V_1 is coupled to the grid of V_2 by means of capacitor C_1 and that the plate of V_2 returns to the grid of V_1 via a feedback path consisting of C_2, R_1, C_3, R_2, C_4, and C_5. In this circuit V_1 is identified as the normally off tube because the vertical hold control (R_3) is located in its grid circuit. This is further confirmed by the wave shape found at the grid.

The normally on tube V_2 is operated as a power amplifier and uses an autotransformer (T_1) to provide an impedance match to the yoke. Damping resistors (R_4 and R_5) across the yoke prevent ringing. Capacitor C_6 acts to bypass horizontal pulses that are capacitively coupled from the horizontal yoke windings into the vertical windings. This capacitor also tends to reduce the peak voltages developed across the transformer during retrace. If C_6 were to open, the picture would tend to have a vertical jitter because of the horizontal pulses acting as vertical sync pulses.

The center tap of the transformer returns to B+, which is ac ground. As a result, the wave shapes at the ends of the transformer are of opposite phase. The negative-going pulse at the bottom of the transformer is coupled by C_7 and R_6 to the blanking circuits of the receiver and is used to produce vertical retrace blanking.

Negative-going sync pulses are coupled through two integrators R_7 and C_8 and R_8 and C_9 and coupling capacitors C_{10} and C_1 to the grid of V_2. These pulses are amplifed and inverted by V_2 and then return to the grid of V_1 through the feedback network R_{11}, C_2, R_1,

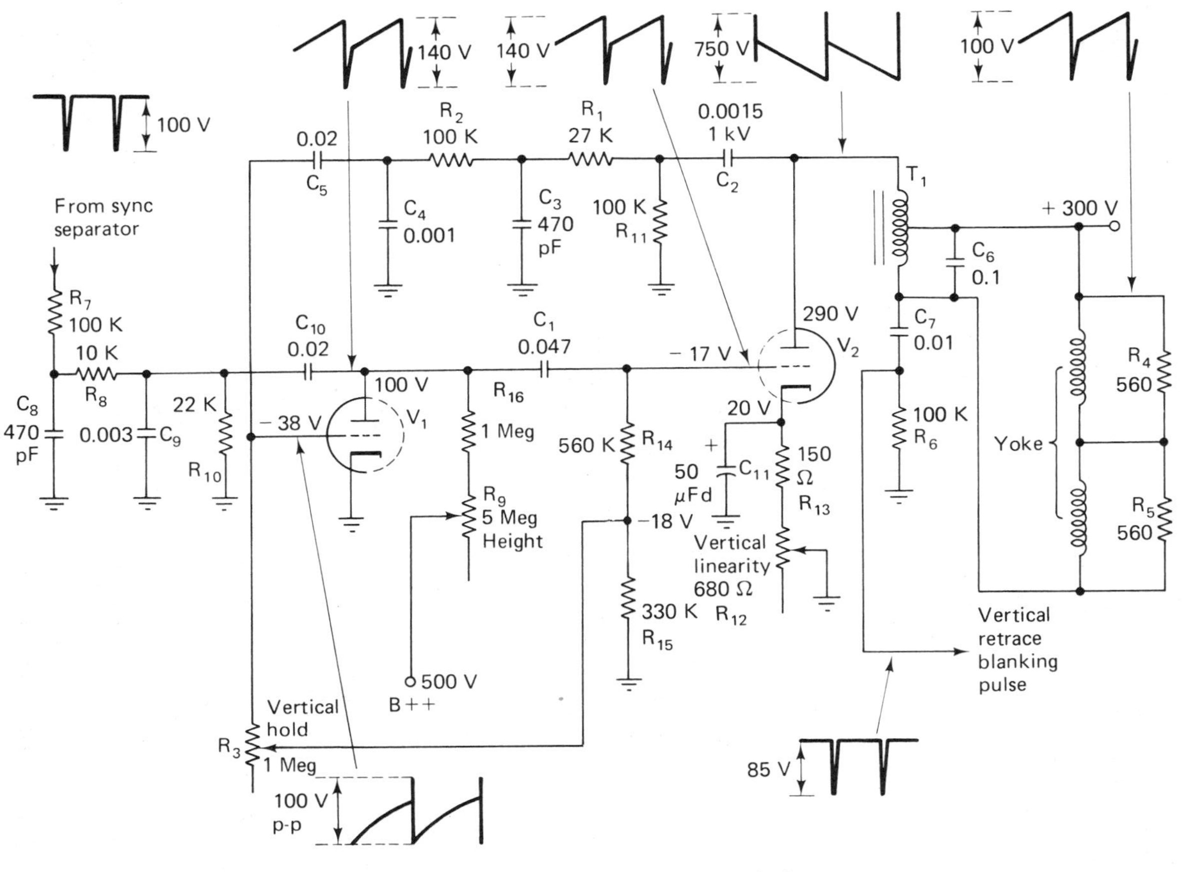

Figure 14-13 Vacuum tube vertical deflection multivibrator.

C_3, R_2, C_4, and C_5. The sync pulses are integrated by R_1, C_3 and R_2, C_4. In addition, these components have two other functions: They tend to eliminate any horizontal crosstalk pulses that may appear at the plate of V_2 via the yoke, and they act as a voltage divider that reduces the 750-V p-p plate voltage of V_2 to a more manageable size.

The sawtooth-forming capacitor is C_{10}. The amplitude of the sawtooth is varied by the height control R_9. R_{10} is the peaking resistor that converts the sawtooth into a trapezoid. The height control has its greatest effect on the bottom of the raster. The linearity control has its greatest effect on the top of the raster. The linearity control R_{12} is in the cathode circuit of V_2 and varies the tube's bias. Electrolytic capacitor C_{11} bypasses the cathode circuit and prevents degeneration that could reduce vertical deflection.

14-11 vertical deflection circuit troubleshooting

Defects in the vertical deflection circuit can be classified into the five following major categories:

1. Loss of vertical sweep [see Figure 14-14(a)]
2. Insufficient vertical height with good linearity [see Figure 14-14(b)]
3. Poor vertical linearity [see Figure 14-14(c)]
4. Loss of vertical sync [see Figure 14-14(d)]
5. Keystone raster [see Figure 14-14(e)]

Loss of vertical sweep. When the vertical deflection sawtooth is lost, the raster will collapse into a single horizontal line. If the brightness control is turned up too high, it is possible for this line to burn the phosphor screen and damage the picture tube.

If the vertical deflection system consists of a separate oscillator and separate amplifier such as shown in Figures 14-2(a) and 14-8, loss of vertical deflection can be the result of an inoperative deflection oscillator, defective amplifier, or open yoke. Isolating the defective stage may be accomplished by signal tracing with an oscilloscope or by signal injecting with a 60-Hz signal source.

Signal tracing may start at the output of the oscillator and work through the amplifiers to the yoke. The trouble is located between the point of no wave shape and the preceding point of normal wave shape. For example, in Figure 14-8 a normal wave shape at the base of the driver transistor Q_2 but no wave shape at the base of the output transistor Q_3 would indicate that the fault lies between the two points. If the defect were an open coupling capacitor (C_7), it can easily be isolated by moving the oscilloscope to the collector of Q_2. A normal wave shape tends to indicate that C_7 is open. Once the defective stage has been isolated, voltage and resistance checks of the circuit will isolate the defective component.

If signal injection is used to isolate the defective stage, the signal is injected into the input of the output stage first and the resultant effect on the raster is observed. If an expanded raster is observed there the stage is probably good. The signal source is then moved back to the driver and oscillator stages. The defective stage is located between the point of observed raster expansion and the point of no expansion. Isolation of the defective part is done by voltage and resistance checks as well as parts substitution.

The signal source used for signal substitution in transistor television receivers must be selected carefully. The signal source should have a low internal resistance and sufficient output so that it can supply the input current needs of the transistor. It should also have

(a)

(c)

(b)

(e)

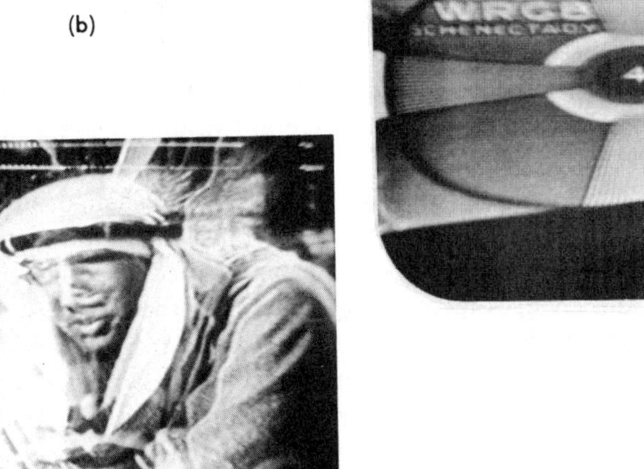

(d)

Figure 14-14 (a) Loss of vertical sweep; (b) insufficient height; (c) poor vertical linearity with foldover; (d) loss of vertical sync; (e) keystone raster.

a dc blocking capacitor in order to avoid dc circuit changes that might occur if any dc paths exist.

No vertical sweep conditions that occur in a receiver that uses a multivibrator-type circuit such as that shown in Figure 14-13 indicates that the circuit is probably not oscillating. Signal tracing with a scope will not be useful since no wave shapes associated with the oscillator will be present. In addition, all of the dc voltages will be in error since the signal bias of both stages will be zero. Signal injection can quickly isolate the defective section of the circuit. Starting at the grid of the output amplifier and then injecting at the grid of the first stage of the multivibrator will determine if the defect is in either of these two stages. If raster expansion occurs at both grid injection points, the trouble is likely to be in the feedback section of the circuit (R_1, R_2, R_{11}, C_2, C_3, C_4, and C_5 in Figure 14-13). The actual defective component may then be identified by means of an ohmmeter or by direct parts substitution.

Insufficient height with good linearity. A defect that causes a reduction in sawtooth peak-to-peak amplitude but does not affect its linearity results in a picture like the one shown in Figure 14-14(b). This defect can be the result of misadjusted height and linearity controls, defective tubes or transistors, increased time constant in the sawtooth-forming circuits, and lowered amplification in the amplifiers following the sawtooth oscillator. Signal tracing will only be meaningful in those circuits that have separate sawtooth oscillators and amplifiers. If the peak-to-peak amplitude of the sawtooth at the output of the oscillator is found to be proper, then the trouble must be in the amplifier section, and vice versa.

Multivibrator and other feedback circuits are more difficult to troubleshoot by means of signal tracing than are separate oscillator amplifier circuits because in a feedback circuit the output affects the input and makes it very difficult to interpret the observed oscilloscope wave shapes. In such circuits it is often faster to check each component with an ohmmeter to determine which one is defective.

Poor vertical linearity. This symptom, shown in Figure 14-14(c), indicates that the vertical deflection sawtooth has become nonlinear. There are various degrees of nonlinearity that can occur in vertical deflection systems. The most common types of nonlinearity are compression on the bottom, compression on the top, and foldover. Changes in sawtooth linearity can be caused by such things as defective tubes or transistors, improperly adjusted height and linearity controls, shifts in tube or transistor bias caused by leaky capacitors, defective output transformers, and changes in the time constant of the sawtooth forming capacitor.

Leaky coupling capacitors such as C_1 in Figure 14-13 or leaky cathode bypass capacitors such as C_{11} in the same circuit can cause the bias of the output tube V_2 to be reduced. As a result, the peak positive portion of the input trapezoid may be clipped. Since this portion of the input trapezoid corresponds to the bottom of the raster, it can cause the bottom of the raster to be compressed. In extreme cases, foldover will occur.

Loss of vertical sync. Loss of vertical synchronization is shown in Figure 14-14(d). This symptom may be caused by an insufficient vertical sync pulse amplitude or by a change in the free-running frequency of the vertical oscillator. If the vertical sawtooth frequency is lower than the frequency of the vertical sync pulse, the vertical retrace blanking bar (hammerhead pattern) will appear to move from the bottom of the picture to the top.

If the vertical hold control is able to momentarily stop the picture from rolling, it is likely that the oscillator is good and that the trouble is in the integrator circuit between the sync separator and the vertical deflection circuit. If the hold control is not able to stop the picture from rolling, it is likely that the trouble is in the deflection oscillator. Probably the fastest way to isolate the defective component is to check each component with an ohmmeter.

The keystone raster. An example of this trouble is shown in Figure 14-14(e). A keystone raster is caused by a defective yoke. This may be explained as follows: The vertical deflection windings of the yoke are placed horizontally on either side of the neck of the picture tube. These windings are responsible for creating a uniform horizontal magnetic field inside the neck of the picture tube. If a shorted turn were to develop in one of the windings, the magnetic field closest to it would be weakened. The magnetic field associated with the good winding would remain strong. As a result, the electron beam would receive little vertical deflection when horizontal deflection brings it near the weak vertical magnetic field and normal deflection when it is in the region of the strong vertical magnetic field. The height of the raster in the region of the weakened magnetic field would be less than that on the opposite side, consequently producing the keystone raster.

Shorted turns in the yoke are not the only cause of keystoning. In circuits in which the yoke windings are in parallel an open winding will produce keystoning. In series connected windings an open yoke can also produce a keystone raster if the winding is shunted by a damping resistor. The damping resistor provides a path for the sawtooth current to bypass the open winding and pass through the good one.

QUESTIONS

1. What are the two functions of the vertical deflection stage?
2. Describe the characteristics of the two vertical deflection systems shown in Figure 14-2.
3. Explain why it is possible for transistor vertical output amplifiers not to use vertical output transformers.
4. What is the advantage of an autotransformer?
5. What is the effect on the raster if a dc current passes through the yoke?
6. What class of operation do vertical output amplifiers operate at?
7. Why is the driving voltage of the vertical output stage often a trapezoid?
8. How are transistor vertical output power amplifiers protected from high peak voltages developed by an output transformer?
9. What is the function of a thermistor placed in series with the vertical windings of the yoke?
10. What is the effect on the raster if the damping resistors across the yoke windings open? Why?
11. What is the advantage of connecting the vertical deflection windings of the yoke in parallel?
12. What is the advantage of the complementary symmetry type of transistor output amplifier?
13. What is the effect of poor sawtooth linearity on picture performance?
14. Explain how vertical linearity is adjusted in transistor circuits.
15. What is "S' shaping? Why is it needed?
16. What is the Miller effect?
17. Explain how the Miller effect is able to

improve the linearity of the vertical deflection circuit shown in Figure 14-11.

18. Explain why clipping the positive peak of the drive voltage that feeds the vertical output stage will cause compression or foldover on the bottom of the picture.
19. What is the main advantage of PST yokes?
20. If a PST yoke failed, why might it be necessary to replace the picture tube?
21. Explain how a defective yoke can cause a keystone raster.
22. Explain why all of the dc voltages change in a defective vertical deflection circuit such as that of Figure 14-10.
23. What trouble symptoms would develop if the following defects occurred in the circuit of Figure 14-8. (a)C_7 open (b)D_1 shorted (c)L_1 open (d)L_2 shorted turns (e)R_5 open.
24. Referring to Figure 14-3, what is the purpose of R_8?
25. What class of operation is the vertical output stage of Figure 14-5(a) and Figure 14-8, operated at?

chapter 15

Horizontal Output Amplifier Systems

15-1 introduction

The block diagram of a typical color television receiver is shown in Figure 15-1. the shaded blocks are the subjects of this chapter. These stages comprise six basic circuits: the horizontal driver, the horizontal output amplifier, the damper, the tripler high-voltage rectifier, the high-voltage regulator, and the high-voltage hold-down circuit.

Black & white receiver horizontal deflection circuits are similar to those of color receivers except that they do not require a focus rectifier or a high-voltage regulator.

Black & white solid-state receiver horizontal output amplifiers consume approximately 4 W of power and color transistor receivers may consume 85 W. As a result, the electronic components used in color receivers are usually of heavier construction than similar components found in black & white receivers. Because of the high power dissipation of these stages, these components are subject to frequent breakdown. It is therefore not surprising that this section of the receiver is the one most often serviced.

A block diagram of a horizontal output stage indicating those stages used in black & white and color television receivers is shown in Figure 15-2. A comparison of the two systems leads to the following conclusions.

1. Both color and black & white receivers drive a sawtooth of current through the horizontal deflection winding of the yoke, but the peak-to-peak sawtooth current needed in the color receiver may be larger than that in the black & white receiver.

2. Both receivers need a high-voltage rectifier to convert the high ac voltage developed by the horizontal output amplifier into a dc high voltage. Color receivers may require second anode dc voltages as high as 30,000 V, but black & white receivers seldom exceed 20,000 V. In addition, color receivers may use special high-voltage regulators to keep the dc high voltage constant.

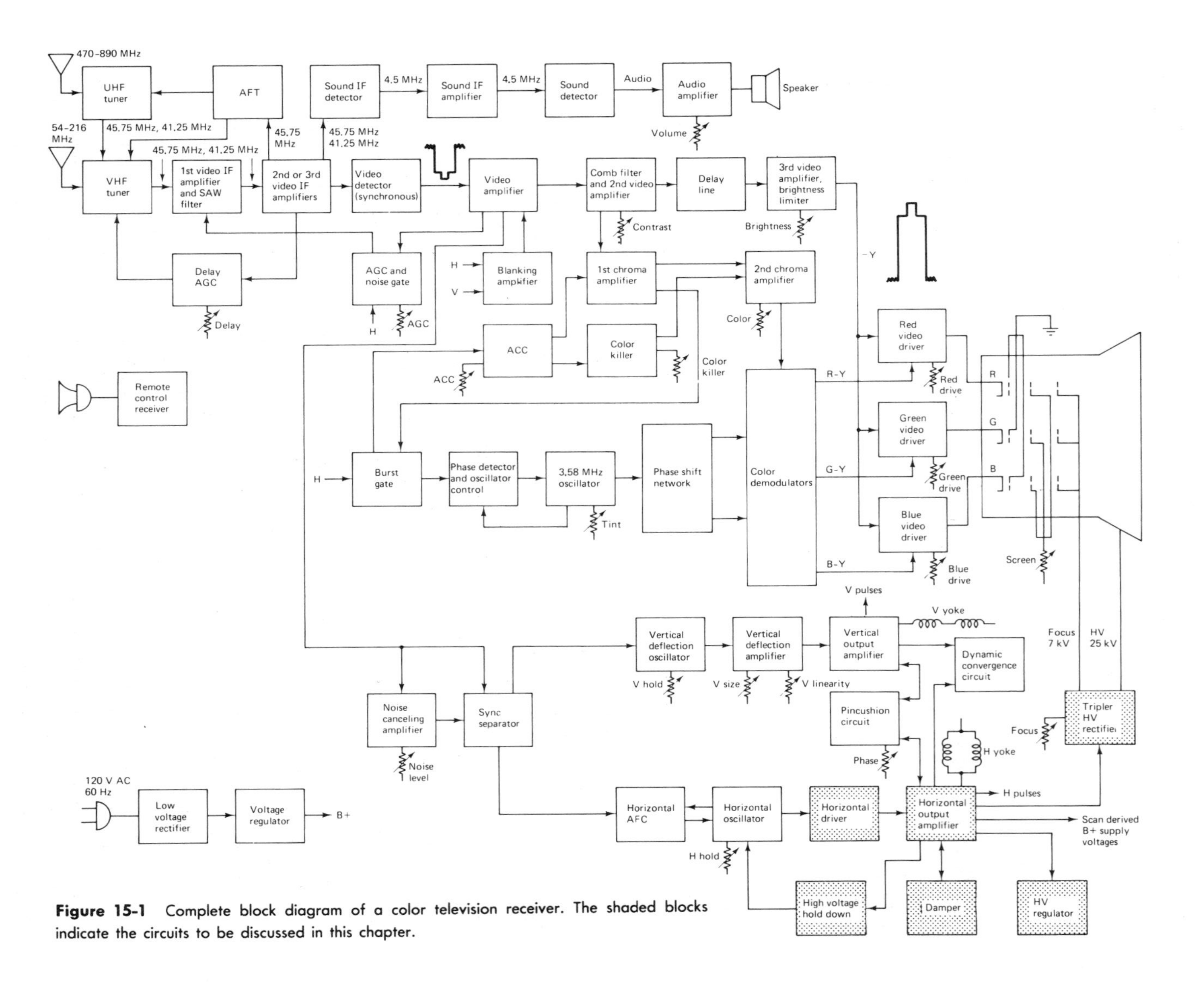

Figure 15-1 Complete block diagram of a color television receiver. The shaded blocks indicate the circuits to be discussed in this chapter.

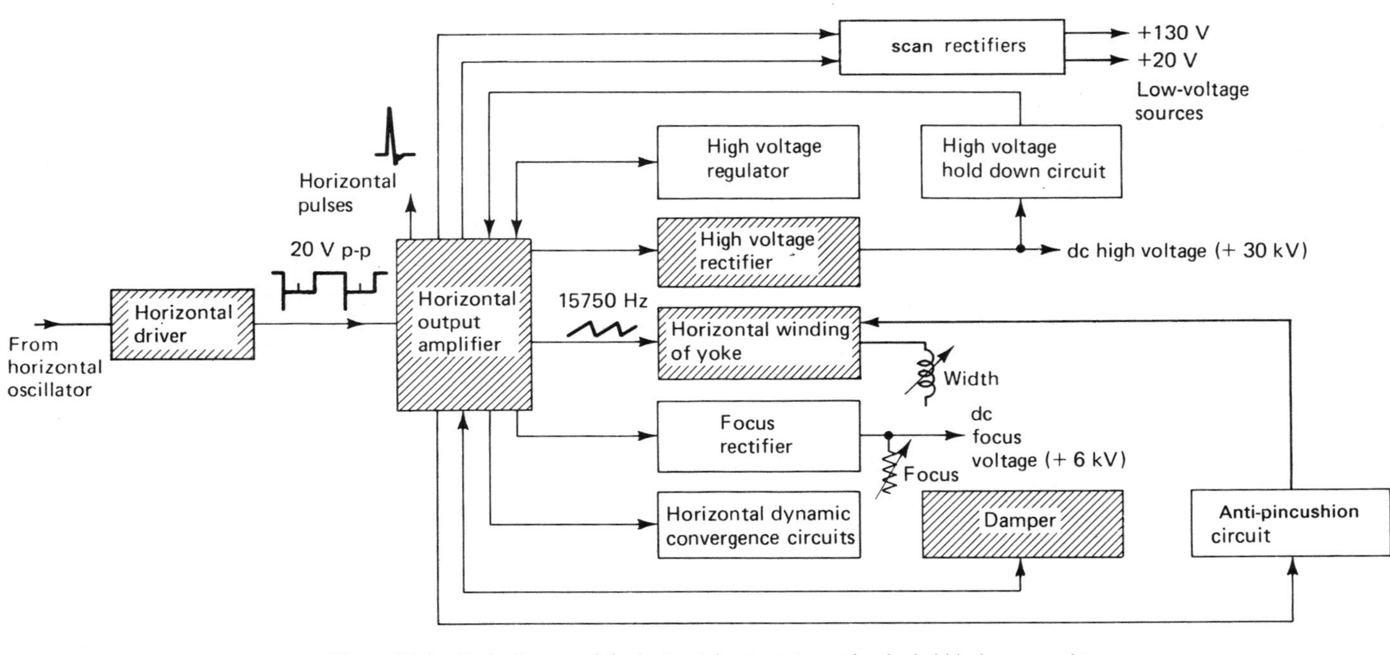

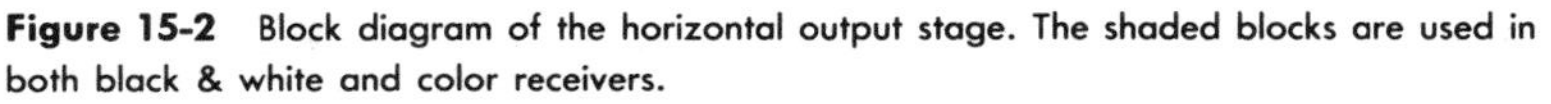

Figure 15-2 Block diagram of the horizontal output stage. The shaded blocks are used in both black & white and color receivers.

3. Many color receivers use picture tubes that use high-voltage electrostatic focus requiring approximately 6000 V or more for proper operation. These receivers may have a separate focus rectifier for this purpose. Except for some receivers built in the 1950s, all black & white television receivers use low-voltage electrostatic focus and do not require any special focus rectifier.

4. Both black & white and color receivers make use of the horizontal output stage to develop additional B supply voltages. These voltages are sometimes called B+ *boost* and may be higher or lower than voltages obtained from the conventional low-voltage power supplies. In most modern receivers the horizontal output stage supplies all of the B supply and CRT heater requirements for the entire receiver.

5. Color receivers may require dynamic convergence and anti-pincushioning circuits. These circuits are not found in black & white receivers. They are in part activated by pulses obtained from the horizontal output amplifier. Black & white and color receivers also use horizontal pulses obtained from the horizontal output amplifier for such things as keyed AGC, horizontal AFC, and horizontal retrace blanking. Color receivers may also use these pulses for other circuits such as the burst gate and color killer.

6. The Department of Health, Education and Welfare has established rules for reducing x-ray radiation from the television receiver under breakdown conditions. These rules have forced manufacturers to introduce special circuits, called *hold-down circuits,* that automatically disable the receiver if a fault should occur that would tend to increase the high-voltage excessively.

7. The input drive voltage of a transistor horizontal output amplifier may vary from 5 to 30 V p-p, depending upon output power considerations. The drive voltage is usually smaller in black & white receivers than in color.

Horizontal output amplifiers may be classified into three categories: vacuum tube, transistor, and SCR types. The characteristics of these circuits have many similarities as well as many differences. The remaining sections of this chapter will discuss these characteristics in detail.

15-2 the horizontal output amplifier

The schematic diagram of a typical horizontal output amplifier and its associated circuits commonly found in black & white receivers is shown in Figure 15-3. An examination of the circuits will show that basically this circuit consists of six major components: the horizontal output amplifier Q_1, the damper diode D_2, the high-voltage rectifier D_3, the horizontal output transformer T_1 (often called the *flyback transformer*), scan rectifiers D_4 and D_6, and the yoke Y_1. The horizontal output amplifier operating Class B is designed to act as a switch that supplies power to the flyback transformer and drives a sawtooth current through the yoke.

The flyback transformer is an autotransformer wound on a rectangular ferrite core and has three essential functions:

1. In some cases it acts as a step-down matching transformer between the horizontal output amplifier and the yoke. However, in most transistor horizontal output circuits, impedance matching is not required.

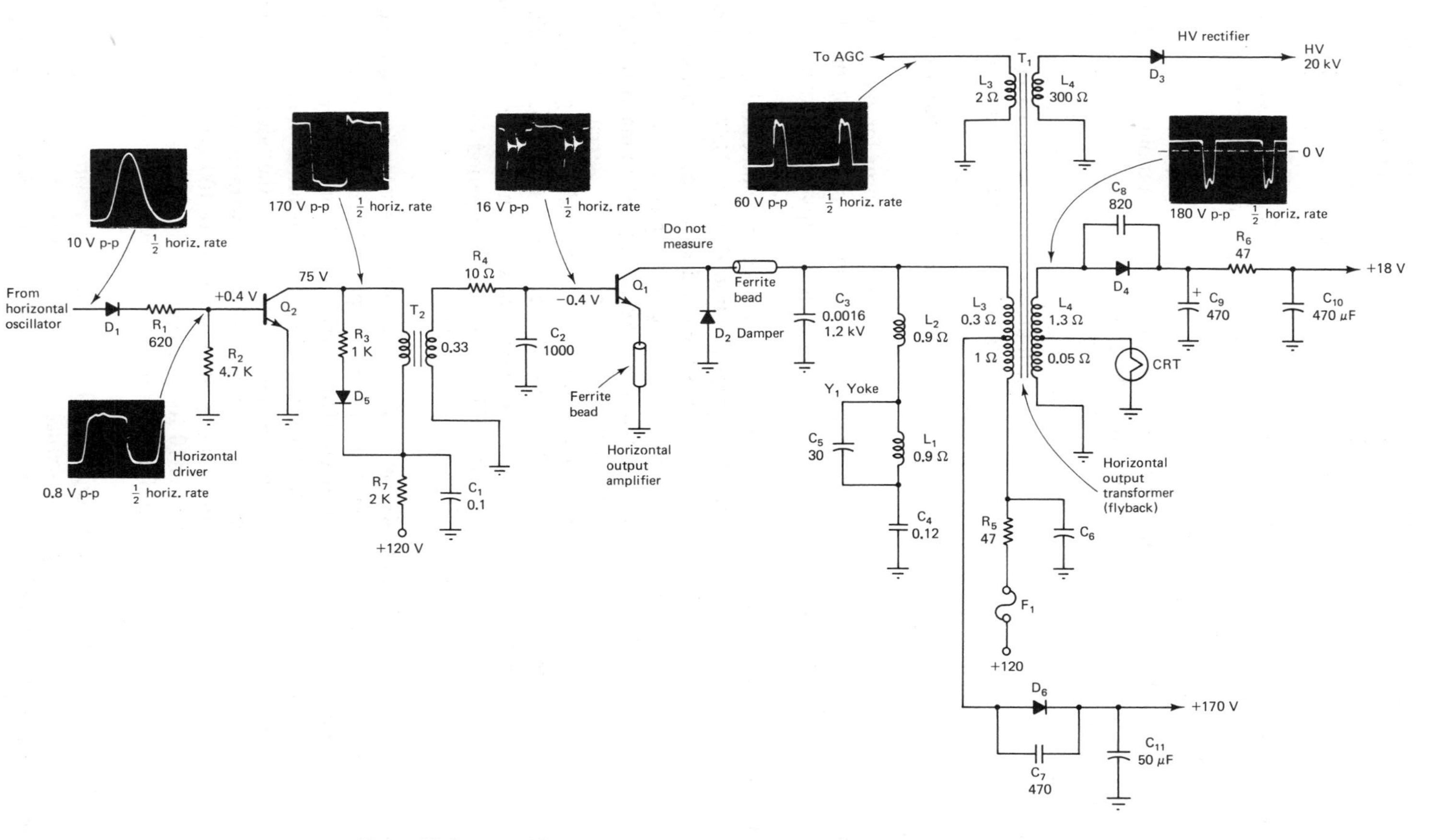

Figure 15-3 Typical horizontal output amplifier circuit used in black & white receivers.

2. It acts like a step-down transformer supplying ac voltages for the picture tube heater and rectifier circuits that supply most of the B+ needs of the receiver.
3. The flyback transformer also supplies horizontal retrace pulses for use in other circuits such as AGC, and AFC, and horizontal retrace blanking.

The damper D_2 acts like a switch that suppresses shock excited oscillations that occur immediately after retrace. It is also responsible for horizontal scanning on the left side of the picture tube screen.

Equivalent circuit. To help understand the operation of the horizontal output circuit of Figure 15-3, it is convenient to represent it as an equivalent circuit. The equivalent circuit of the horizontal output amplifier stage is shown in Figure 15-4 and is constructed by combining the lumped inductance of the flyback and yoke into a single equivalent inductance (L_1) that is shunted by the circuit's lumped (C_3) and distributed capacitance represented by an equivalent capacitance (C_1). A switch (S_1) representing the horizontal output amplifier is placed in series with the power source and the parallel resonant circuit formed by the parallel combination of C_1 and L_1. Another switch (S_2) representing the damper is placed in parallel with the tuned circuit.

Also shown in Figure 15-4 is the yoke's sawtooth waveform for one horizontal line. Below that is shown a sequence of diagrams indicating the position of the electron beam on the picture tube screen and the electrical state of the equivalent circuit.

Horizontal scanning, center to the right. Horizontal scanning begins when the horizontal output amplifier is first turned on (see Figure 15-4, beam position diagram 1). At this time the electron beam is in the center of the picture tube screen and switch S_1 has just closed and applied a constant voltage to L_1. At the same time the damper switch S_2 is open and is out of the circuit. As electron current flows into the yoke, an expanding magnetic field (see diagram 2) forces the electron beam toward the right side of the screen. The magnetic polarity associated with the yoke at this time is arbitrarily indicated as being north on top of the coil and south on the bottom. When the sawtooth current is maximum (see diagram 3), the magnetic field in the yoke has reached the maximum strength and the electron beam has been forced to the far right of the screen. This ends the trace period for the right side of the raster.

Retrace. At this time (see diagram 4) the horizontal output transistor stops conducting, which in terms of the equivalent circuit means that S_1 opens. The damper is so arranged that it also remains nonconducting (S_2 open). Under these conditions, the energy that has been stored in the magnetic field of the yoke and flyback transformer forces the tuned circuit into oscillation. The frequency of this ringing is determined by the values of equivalent inductance L_1 and the total distributed capacitance (C_1) associated with the yoke and flyback transformer. In practice, the circuit tends to ring at approximately 100 kHz. The period of time occupied by one cycle of 100 kHz is 10 μs and that of a half-cycle is 5 μs. It is the first half-cycle of yoke and flyback transformer ringing that provides horizontal retrace, or as it is commonly called *flyback.*

At the moment that S_1 opens (diagram 4) the magnetic field surrounding L_1 begins to collapse, the tuned circuit rings and *retrace begins.* This makes L_1 act as a generator whose peak voltage depends upon the rate of magnetic field collapse. Since the ringing rate

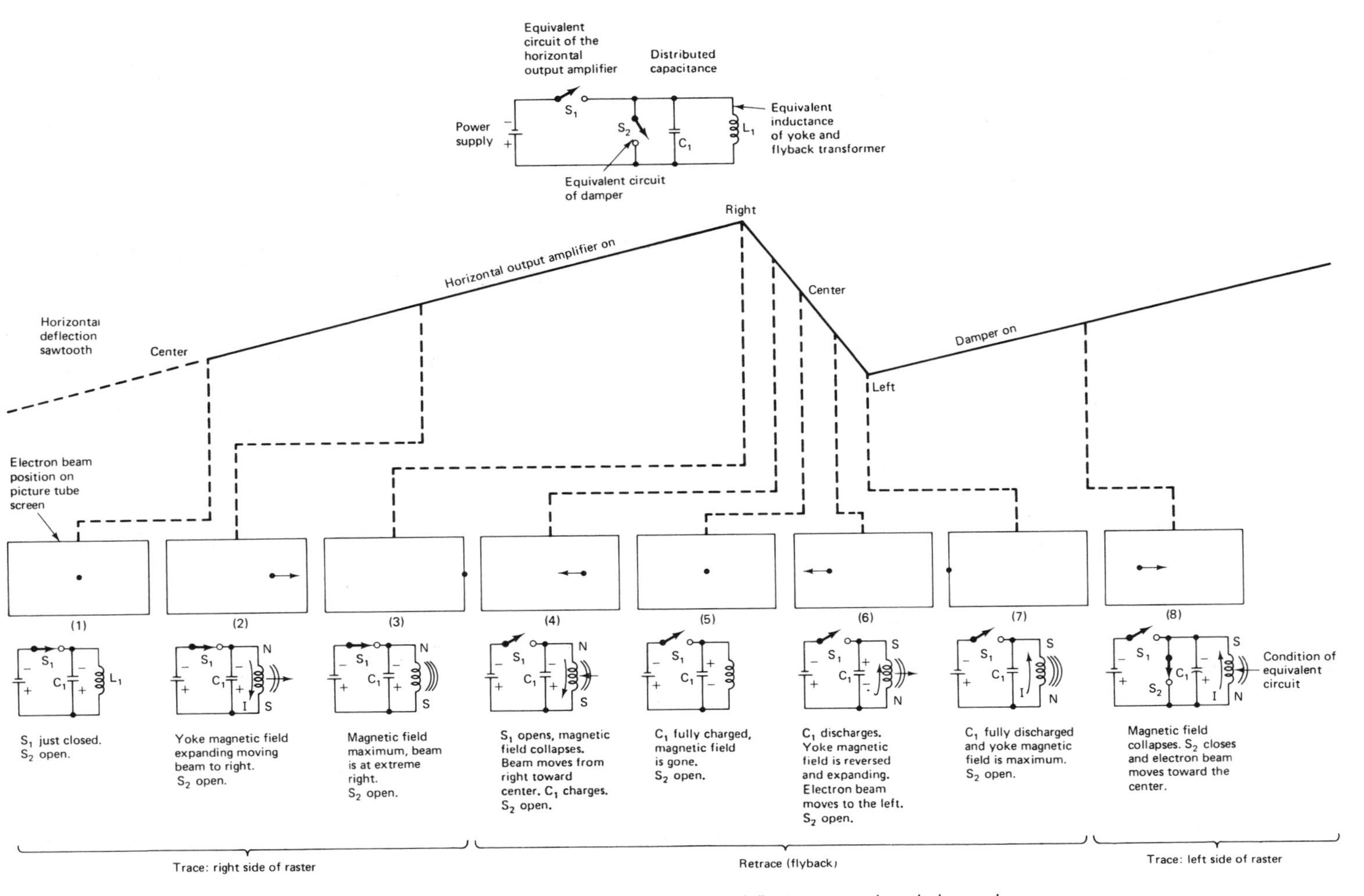

Figure 15-4 Relationship between the horizontal deflection sawtooth and electron beam position.

is 100 kHz, the voltage developed across L_1 will have a positive peak voltage of several thousand volts. If a step-up transformer is used this peak voltage can be increased to 20,000 or 30,000 V. After high-voltage rectification this voltage is used as the second anode voltage of the picture tube. During this same period of time the voltage induced across L_1 charges C_1. The polarity the voltage developed across C_1 is opposite to that which existed across it when the yoke magnetic field was expanding. This is because the collapsing magnetic field induces a voltage such that it prevents any change in current direction or amplitude. Since the decreasing current flowing through the yoke is in the same direction as before (diagram 2), the polarity of the collapsing magnetic field is the same as before. As a result, the electron beam moves from the extreme right toward the center of the raster.

When the electron beam has reached the center of the screen (see diagram 5), the magnetic field surrounding the yoke and flyback transformer has completely collapsed, and C_1 has become fully charged (the flyback pulse is maximum positive); S_1 and S_2 are still open.

C_1 now begins to discharge through L_1 (see diagram 6). Because of its reversed polarity, the discharge current direction is opposite to that flowing through the yoke during the time the beam was on the right. The expanding magnetic field produced by the discharge current of C_1 is also reversed, forcing the beam to move from the center of the screen to the left side.

Trace, left to center. When the beam has reached the extreme left side of the screen (see diagram 7) C_1 is fully discharged and current through the yoke is maximum. Note that in an oscillating *LC* circuit when the sine-wave voltage is zero, the sine-wave current is maximum. At this point retrace (flyback) has been completed and is the result of one-half cycle of oscillation of the tuned circuit formed by L_1 and C_1. If the oscillation is allowed to continue until it dies out, it will force the yoke current to oscillate at approximately a 100-kHz rate. This is turn will force the electron beam to retrace quickly toward the right side of the screen and then back toward the left. Each successive swing grows smaller until the oscillator dies out and the beam is in the center of the screen. Because of the varying writing speed of the electron beam, the left and right extreme excursions of the beam would appear brighter than the more rapid flyback periods. This action occurs for each horizontal line forming the raster. The resultant raster will exhibit a number of white vertical lines on each side of the screen and will exhibit picture foldover. The picture foldover is the result of the video signal intensity modulating the electron beam during the time the beam is being forced to oscillate back and forth across the screen. Each of these scans will superimpose a portion of the picture one on top of the other.

Damper. To avoid the continued oscillation of L_1 and C_1, the tuned circuit must be *critically damped*. This condition is one in which the tuned circuit oscillation is forced to rapidly decay through a small shunt resistance without reversing polarity. Converting the undamped equivalent circuit of the horizontal output amplifier stage into a critically damped *LC* circuit is accomplished by the conduction of the damper diode. In terms of the equivalent circuit, this means that immediately following retrace S_2 is closed (see diagram 8). At this point the magnetic field that forced the electron beam to be at the extreme right now collapses and allows the electron beam to return to the center of the screen. The deflection of the electron beam from the left to the center of the screen is linear because the circuit is critically damped. When the beam has reached the center of the screen, the damper ceases to conduct, S_2 opens, and the horizontal output amplifier turns on (S_1 closes). At this point horizontal deflection for one line has been completed and the next line repeats the process.

Composite sawtooth. The addition of the yoke current during critical damping (damper conduction) to the yoke current during the time that the horizontal output amplifier is conducting results in the required horizontal deflection sawtooth current. Figure 15-5 is a ladder diagram showing various wave shapes throughout the circuit and also illustrates the composite sawtooth and the raster produced by it. Notice that the left side of the raster is formed by conduction of the damper and the right side is formed by conduction of the horizontal output stage. During retrace neither one conducts. It should be noted, however, that in actual practice the high-voltage rectifier conducts during retrace.

The point at which the damper yoke current stops and the horizontal output yoke current begins may be called the *transition point.* It can be seen that if the damper nonlinear decay is exactly matched by the oppositely curved nonlinear rise in horizontal output amplifier current, the resultant sawtooth current will be linear. If for any reason (horizontal output amplifier bias change) the currents are not properly matched at the transition point, horizontal nonlinearity will appear in the raster. For example, excessive input signal to the horizontal output amplifier will cause the deflection sawtooth to develop a kink at the transition point. This appears in the raster as one or more white vertical lines in the center of the raster. The increased brightness is caused by the reduced electron beam writing speed at the time of the transition kink.

Depending upon load conditions, the time of horizontal output transistor conduction may vary from one manufacturer to another. Thus, in some systems the horizontal output transistor may be on for as much as 90% of the active time of one horizontal line. This is controlled by the duty cycle of the base voltage driving the horizontal output amplifier. The duty cycle of the base voltage shown in Figure 15-5 is about 60%. The high-voltage pulse (flyback) that appears across the damper corresponds to one half-cycle of circuit oscillation.

15-3 the PNP horizontal output stage

The horizontal output stage requires a switching device to energize the yoke. This switching device may be a vacuum tube, a transistor, or an SCR. In this section the transistor horizontal output circuit used in both black & white and color receivers will be discussed.

The two major disadvantages of the older vacuum tube horizontal output stages are that (1) the vacuum tube requires a power wasting heater and (2) the vacuum tube has a relatively high plate resistance that necessitates a matching transformer (flyback) between the tube and the yoke.

Transistor horizontal output stages do not require a heater and they do not need output matching transformers to match the low-resistance horizontal output transistor to the low-resistance yoke. The horizontal yoke windings are usually placed in parallel and are connected in shunt with the horizontal output transistor.

Transistors used in horizontal output stage applications may be either PNP or NPN, and they may be of the germanium or silicon type. The most common type used in modern sets is the silicon NPN transistor. This power transistor is designed to dissipate

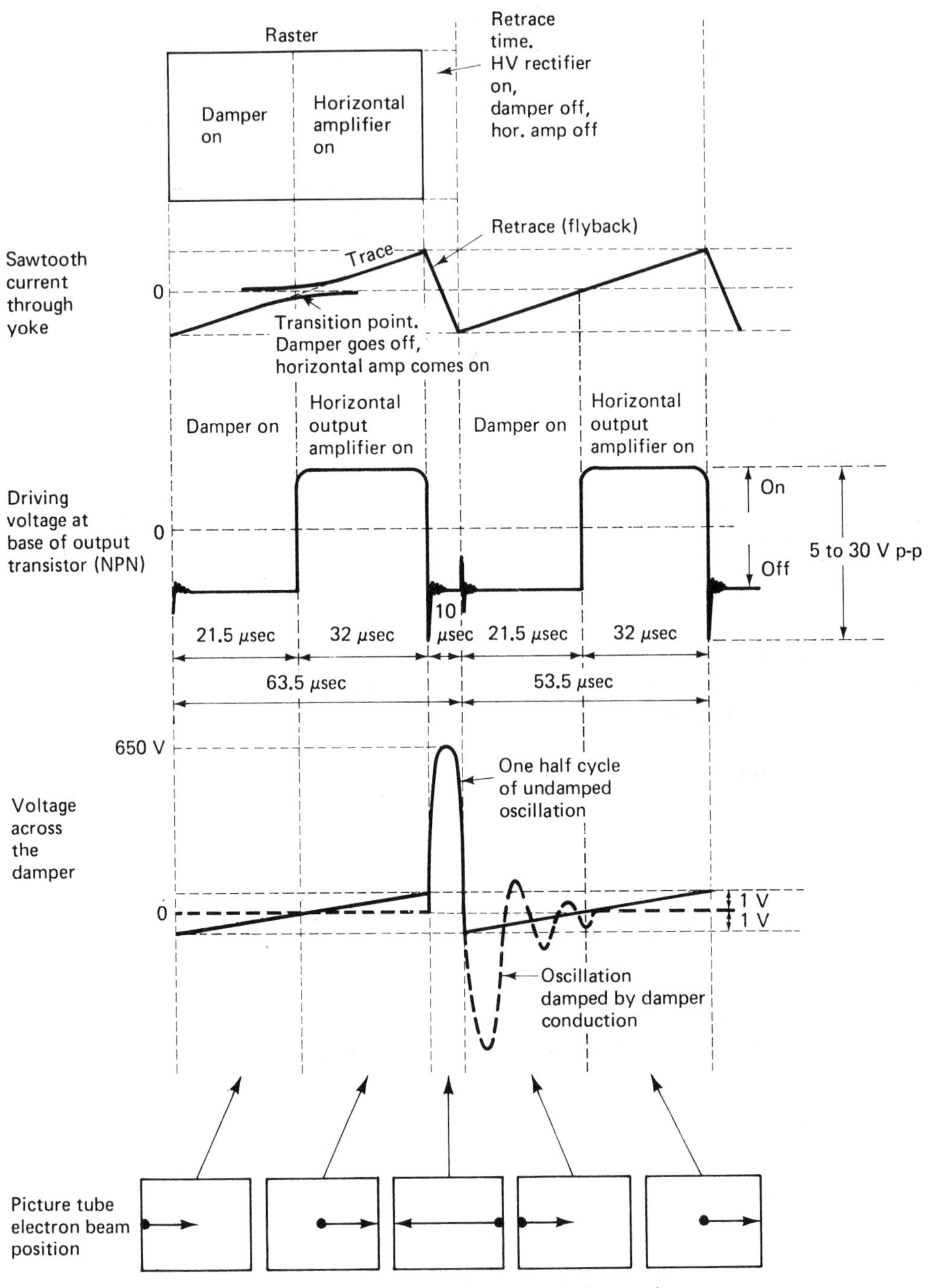

Figure 15-5 Voltage and current relationships in the horizontal output stage.

as much as 65 W continuously and 400 W on peaks. It can handle peak currents of as much as 10 A. Its major shortcoming is the relatively low peak voltage (1500 V maximum) that it can withstand between collector and emitter or between collector and base.

An example of a horizontal output stage using a PNP transistor is shown in Figure 15-6. The horizontal output transistor Q_2 is connected as a Class B common collector amplifier. The input drive signal is obtained from the secondary winding of the drive transformer (T_1), which is connected between the base and emitter of Q_2. Wave shapes taken from collector to ground and from base to ground appear to be identical because the relatively high-voltage pulse developed on the collector is coupled via the secondary of T_1 to the base. The actual drive voltage is the difference between the base and collector voltages. If the base to emitter voltage were observed on an oscilloscope, it would look like a distorted square wave of between 5 and 10 V p-p. This voltage is of sufficient amplitude to drive the output transistor between cut-off and saturation. The bias from base to emitter is zero and sets the operating point of the transistor point at cutoff; consequently, the transistor is operated Class B.

The emitter of the horizontal output transistor Q_2 is connected to the damper diode (D_1), the yoke (L_2), and the horizontal output transformer (T_2). In this circuit the horizontal yoke windings are connected in parallel and are coupled via C_1 to the emitter of the horizontal output transistor.

The horizontal output transformer has a somewhat different function in transistor horizontal output stages than that found in vacuum tube circuits. In vacuum tube deflection circuits the output transformer is primarily responsible for ensuring rapid retrace because of its self-oscillation. In addition, it is used to provide impedance matching between the tube and the yoke and to provide the ac high voltage. Special windings are used to provide pulses for AGC, AFC, and horizontal blanking.

In transistor horizontal deflection circuits such as that of Figure 15-6, retrace time is primarily determined by the oscillation of the yoke and capacitor C_2. The output transformer (T_2) is tuned with its distributed capacitance to the third harmonic of the yoke oscillation and is only used to modify the wave shapes in the circuit. In other respects, such as high-voltage generation and pulse production, transistor horizontal output transformers are similar to vacuum tube output transformers.

In transistor circuits the B supply needs of the receiver are provided by taps on the output transformer. See Figure 15-6. These ac pulse sources are then rectified by special fast-recovery diodes D_2 and D_3 that produce dc voltages of 260 V and 70 V, respectively. C_3 and C_4 are the filter capacitors that remove the 15,750-Hz ripple from the dc output of the B supplies. Resistors R_1 and R_2 act as bleeder resistors that discharge the capacitors when the receiver is turned off. D_4 is the high-voltage rectifier.

Fast-recovery diodes are silicon diodes that are designed to turn off very rapidly (0.5 μs). Conventional diodes, when used with scan rectifiers, turn on during the pulse time and then are not able to turn off fast enough after the pulse. This results in very high reverse current, which causes them to burn out.

Third harmonic tuning. To prevent transistor breakdown, the emitter peak voltage cannot be allowed to exceed the maximum reverse voltage ratings of the transistor. At the same time it is desirable to increase the high-voltage peak voltage developed by the output transformer. Both of these requirements can be obtained by tuning the leakage inductance of the output

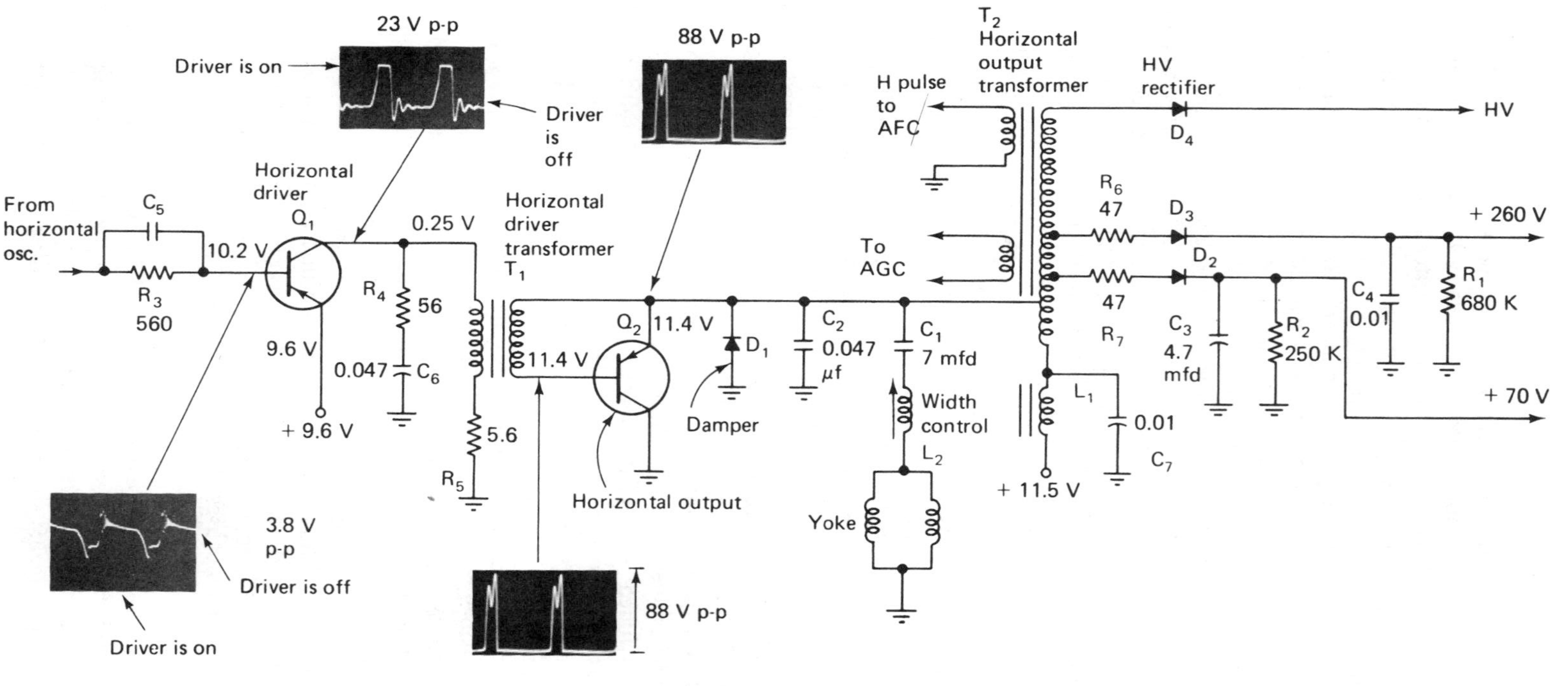

Figure 15-6 PNP transistor horizontal output amplifier.

transformer to the third harmonic of the yoke ringing frequency. When tuned properly, the phase of the third harmonic cancels part of the peak voltage across the yoke, resulting in a noticeable dip in the emitter voltage. See Figure 15-6. At the same time the phase of the output transformer's third harmonic ringing tends to increase the high-voltage pulse delivered to the high-voltage rectifier. In some sets third harmonic tuning is made adjustable by means of a variable coil placed in shunt with the flyback.

The damper. When retrace has been completed, the base drive voltage is still positive and keeps the output transistor off. At the same time, the yoke oscillation has reversed its polarity and has forced the damper diode (D_1) to turn on. The conduction of the diode ends the oscillation of the yoke and C_2, and it allows the energy stored in the yoke's magnetic field to decay at a controlled rate producing the negative half-cycle of sawtooth current. This portion of the sawtooth corresponds to the left side of the picture tube screen. During this time the horizontal output transistor remains cut off even though the input drive voltage has forward biased the transistor. This is because the emitter-to-collector voltage is negative and therefore prevents emitter to collector current flow. Damper conduction ends when the emitter voltage of the output transistor reverses polarity and allows the base to emitter forward bias of Q_2 to turn it on. This action starts a new deflection cycle.

An open damper forces the base-to-collector junction of the horizontal output amplifier to take over the damper function. Therefore, no effect will be noticed in the raster. However, the life of the horizontal output stage may be shortened since it will be doing all the work, and may not be designed to handle the additional load.

Coupling capacitor C_1 has two functions. First, it is a dc blocking capacitor that prevents any dc current from flowing through the yoke. Any such current would cause decentering of the raster and picture. Second, it modifies the shape of the yoke deflection current from a linear ramp into a wave shape that takes on the shape of a gentle S. A linear deflection sawtooth is not desirable because the curvature of the face of the CRT has a radius greater than the distance from the electron gun to the face. In addition, the small yoke resistance (1 Ω) tends to create a nonlinear sweep. It causes a left-hand stretch and a right-hand compression of the raster. C_1 and the yoke form a series resonant circuit tuned to about 5kHz. A portion of the sine wave produced provides the S-shaped current. "S" shaping of the sawtooth current is necessary for obtaining a linear visual sweep. If C_1 changes value, it can result in changes of raster width linearity, and possible blooming.

Width control. Raster width may be controlled in several ways. Some manufacturers use an adjustable strip of brass foil between the neck of the tube and the yoke, in the same way as was done in vacuum tube receivers. This method is limited to black & white receivers. Other manufacturers change width by either adding or removing capacitance placed in shunt with the resonating capacitor, C_2. Reducing C_2 makes the raster narrower by increasing the high voltage. The method of width control shown in Figure 15-6 is simply to place a variable coil in series with the yoke. Increasing the inductance of the coil reduces the current flowing through the yoke windings and reduces picture width.

15-4 the horizontal driver stage

In vacuum tube receivers the output trapezoid of the horizontal deflection oscillator is of sufficient amplitude to drive the horizontal output stage directly. In transistor circuits a driver stage operating Class B or C, is placed between the horizontal oscillator and the output stage. This stage is a buffer power amplifier that isolates the oscillator from the output stage and provides sufficient power gain to supply the input power requirements of the horizontal output stage. These needs may be very large. For example, during the time that the horizontal output transistor is on, its base current may be as much as 400 mA.

The driver transistor such as that shown in Figure 15-6 is driven by an asymmetrical rectangular waveform obtained from the horizontal oscillator. The negative-going half of the wave shape lasts for approximately 21μs and occurs during the retrace period of scan. C_5 and R_3 are wave-shaping components that tend to increase the rise time of the input pulse. During the negative half-cycle the PNP driver transistor (Q_1) is turned on and forces the transistor into saturation. Since the collector is returned to ground through the driver transformer, the collector voltage is approximately equal to the emitter voltage (9.6 V). The collector current passing through the transformer (T_1) builds up a magnetic field. On the positive half-cycle of input voltage the driver transistor Q_1 is cut off. The current flowing through the driver transistor ceases and the collapsing magnetic field induces a voltage across the primary winding of the driver transformer which forces the collector voltage to become highly negative (approximately 23 V). To ensure rapid turn-off of the horizontal output stage, the driver output wave shape is modified by R_4 and C_6 to produce a rise time overshoot in the base-to-emitter drive voltage of Q_2. The driver transformer T_1 has a 4:1 step-down turns ratio that matches the driver transistor output resistance to the low-input resistance of the output transistor.

15-5 the NPN transistor horizontal output stage

A circuit of a horizontal output amplifier that uses an NPN power transistor is shown in Figure 15-7. This basic circuit arrangement may be used in either black & white or color receivers. The circuit shown in this figure is a circuit found in a color receiver.

Essentially, the PNP and the NPN horizontal output circuits are similar. Both circuits drive the yoke directly and both circuits use the output transformer to provide the high- and low-voltage needs of the receiver. Both systems connect the damper diode across the yoke with the anode side grounded, and both systems provide retrace and scanning of the left side of the raster by means of yoke ringing and damper conduction.

The PNP and NPN transistor horizontal output circuits differ from one another in circuit configuration. The PNP transistor is used as an emitter follower, and the NPN horizontal output transistor is connected as a common emitter. This means that the waveform seen at the base of the NPN output transistor is the actual drive voltage instead of the retrace pulse coupled into the base circuit, as in the case in the PNP output transistor.

In the NPN transistor the fall time of the drive voltage corresponds to the start of horizontal retrace. This can be seen in Figure 15-7(b) which shows the relationship between the drive voltage, the yoke voltage, and the yoke current.

Examination of the schematic diagram of Figure 15-7 shows that the dc base voltage of the output transistor Q_1 is approximately 1 V negative. This reverse bias is developed by rectification of the input drive voltage. Resistor R_1 and capacitor C_1 form the bias network. If R_1 increases in value, the bias of Q_1 changes, forcing a shift in transistor operating point. As a result, the raster exhibits foldover or drive lines. If C_1 opens the effective drive to the output transistor decreases, lowering the high voltage and possibly causing complete loss of raster.

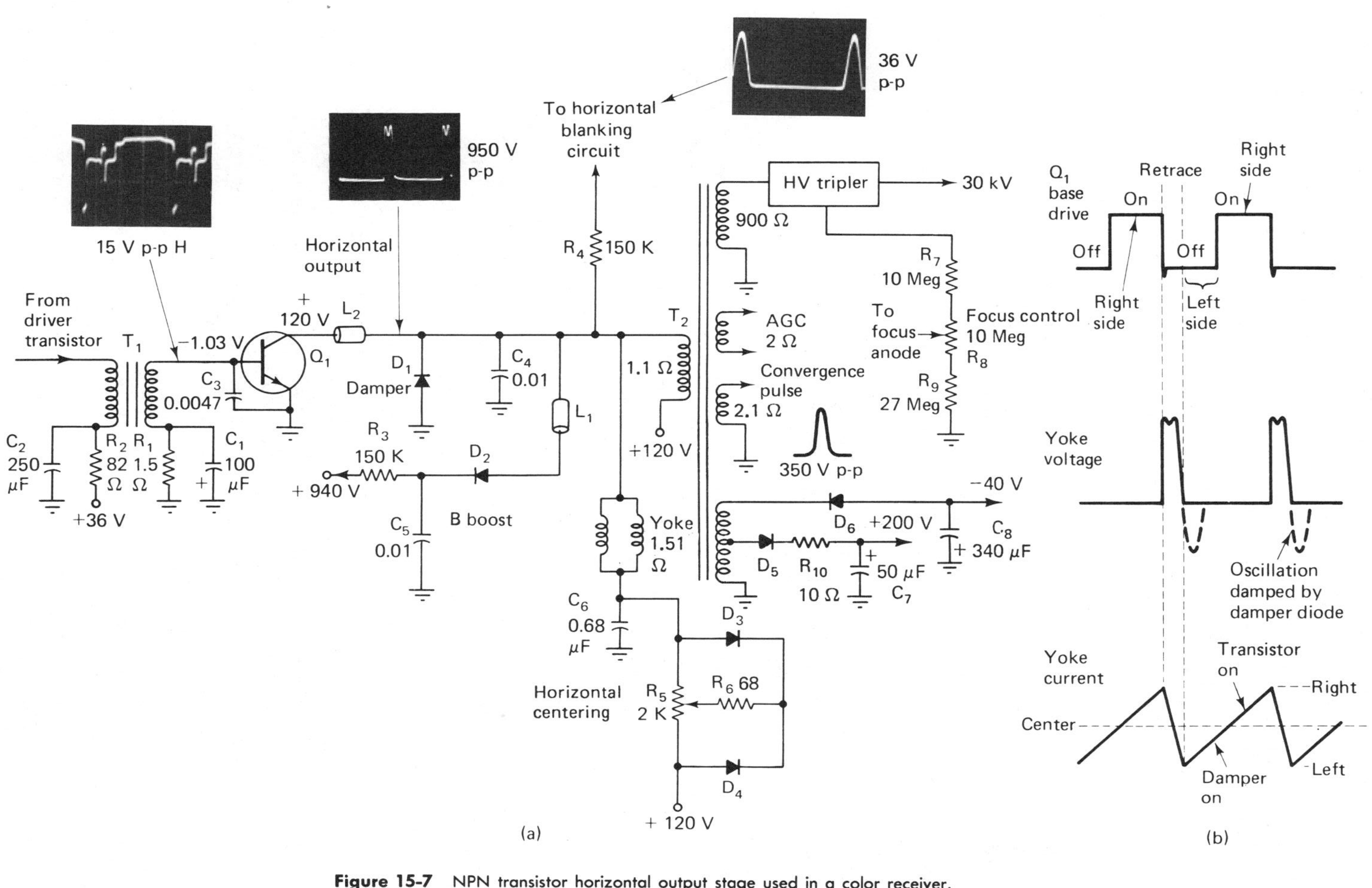

Figure 15-7 NPN transistor horizontal output stage used in a color receiver.

Centering control. In the collector circuit the yoke is returned to ac ground through capacitor C_6 and a Wheatstone bridge centering control consisting of D_3, D_4, and the two halves of R_5. Resistor R_6 is used to limit current flow when the bridge is unbalanced. A dc blocking capacitor is not needed because both sides of the yoke return to the positive 120-V supply, leaving a net dc voltage across the yoke of zero. If the centering control is moved from its center position, the bridge becomes unbalanced and the sawtooth deflection current is rectified, resulting in a net dc yoke current. The direction of the current depends upon which side of center the control is moved to and determines whether the raster moves to the right or to the left.

Damper. Shunting the yoke are C_4 and the damper diode D_1. The yoke and C_4 form a resonant circuit that is allowed to oscillate for one positive half-cycle when the output transistor turns off. During this time horizontal retrace takes place. Immediately following retrace the damper is turned on by the negative half-cycle of oscillation. Conduction of the damper eliminates the oscillation and converts this energy into that part of the deflection sawtooth responsible for horizontal deflection on the left side of the screen.

An open damper diode will cause no immediate noticeable effect because the collector to base junction of the horizontal output transistor (Q_1) will act as the damper. Since Q_1 is not rated for this kind of operation, it will quickly burn out. Some receiver manufacturers take advantage of this to eliminate the damper diode. In such receivers the horizontal output transistor is designed to also act as the damper. A leaky damper diode will cause a narrow picture with compression on the left side of the raster.

The value of the resonating capacitor (C_4) is extremely critical and should be replaced only with a capacitor that has the same specifications. These capacitors are low-loss high Q capacitors. If they are replaced by ordinary capacitors, they will overheat.

A small change of C_4 in either direction will result in noticeable changes in raster width and the amplitude of the high voltage. If the amplitude of the retrace pulse is increased too much, the output transistor may be damaged.

Many modern receivers divide the resonating capacitor (C_4) into two or three capacitors in parallel. This is done in accordance with HEW regulations that limit x-ray generation. The idea is that if one capacitor were used and if it were to open, the high voltage and x-ray radiation would both increase enormously. If three capacitors were used and one were to open the high voltage and x-ray radiation increase would be minimal.

Also connected to the collector of the output transistor is a diode D_2. This diode rectifies the 950-V retrace pulse that is developed at the collector and provides the 940-V dc that is used for the screen electrodes (first anode) of the color picture tube. C_5 and R_3 are filters that remove the 15,734-Hz ripple from the dc.

Diode D_5 is used to rectify a positive-going horizontal pulse and produce a +200-V source. R_{10} is a surge-limiting resistor and C_7 provides filtering.

The positive-going horizontal pulse feeding D_6 is an ac voltage which may have a positive peak of 300 or 400 V, and a negative peak of 40 or 50 V. Diode D_6 is polarized so that it passes the negative peak and blocks the positive peak. As a result, the dc output of the rectifier is a negative 40 V.

L_1 and L_2 are ferrite beads that are slipped onto the connecting wires that increase the distributed inductance of the wire. As such, these inductances act as *RF* chokes that tend to reduce undesired radiation from the circuit.

The horizontal output transformer is used primarily to step up the retrace pulse to approximately 10,000 V. This is then boosted by a tripler to approximately 30,000 V dc for use at the second anode of the picture tube. The dc voltage required by the focus electrode is also developed by the tripler.

The horizontal output transformer has several windings used to supply horizontal pulses to the AGC and convergence circuits. Horizontal blanking pulses are obtained from the collector of Q_1 via R_4.

A transistor is cut off when its bias is zero. The horizontal output stage (PNP or NPN) operates under zero or slightly reverse bias conditions. Therefore, if for any reason the horizontal oscillator or driver should fail, removing the horizontal drive signal from the base of the horizontal output amplifier, the horizontal output stage would cease functioning. As a result, high-voltage and low-voltage supplies would drop to zero, forcing the receiver to exhibit no raster and no sound symptoms. The horizontal output stage itself would not be damaged.

15-6 the flyback transformer and yoke

A photograph of the flyback transformer used in color receivers is shown in Figure 15-8. Aside from the larger physical size of the color receiver flyback transformer, both color and black & white transformers have similar appearance. The large cartwheel-like winding is the high-voltage winding of the transformer and the smaller core upon which it is wound comprises the remaining windings of the transformer. The winding is coated with a heavy insulation to prevent high-voltage arcing.

The flyback transformer is essentially a tuned transformer that is used to generate the ac high-voltage needs of the picture tube. Additional windings are usually included to provide pulses for other stages of the receiver. The transformer may be designed as an autotransformer to improve transformer efficiency.

Figure 15-8 Television flyback transformer. (Courtesy of Thordarson Meissner, Inc.)

The yoke. The two horizontal windings of the deflection yoke are positioned vertically on the neck of the picture tube. This produces a vertical magnetic field through the neck of the tube that forces the electron beam to deflect horizontally. The horizontal winding inductance of a yoke used in a transistor black & white receiver is typically 150 mH and may have a winding resistance of only 1 or 2 Ω. Transistor color receivers have horizontal yoke windings that may have inductances of 150 to 750 mH and winding resistances of 1 to 4 Ω.

Horizontal yoke windings may be connected in either series (see Figure 15-3) or parallel (see Figure 15-6). The series-connected yoke windings have a higher impedance than do the parallel-connected windings. The low impedance of the parallel arrangement allows it to be connected directly to the output transistor without the need for a matching transformer. Because of the large difference in series connected yoke and horizontal output amplifier impedances a step-down matching transformer (flyback) may be used.

The yoke balancing capacitor. In television receivers that use series connected yokes a small capacitor (30 pF) is placed across one of the horizontal yoke windings. See Figure 15-3. This capacitor (C_5) is called a *balancing capacitor*. To understand what C_5 does, let us first assume that it has been removed from the circuit. The high-potential side of the flyback transformer is tied to one end of yoke winding L_1. This winding is physically located above the neck of the picture tube, and L_2 is on the opposite side of the tube. During horizontal retrace a pulse of approximately 1000 V is developed across the horizontal yoke windings. The yoke windings can be considered as plates of a capacitor, and the neck of the picture tube is in effect the dielectric of the capacitor. The high-voltage pulse developed on the windings acts to produce vertical *electrostatic deflection* of the electron beam. The horizontal retrace pulse will shock excite the inductance of the yoke and the very small stray capacitance of the circuits. This causes harmonically related high-frequency ringing that lasts for some time after horizontal retrace has ended. As a result, each horizontal line will have a series of small vertical undulations on the left side of the screen. See Figure 15-9(a). The net visual effect is a series of vertical white and black bars on the left side of the screen. The variation in brightness is caused by change in writing speed of the electron beam.

Adding the balancing capacitor across L_1 tunes C_5 and L_2 to series resonance at the ringing frequency. At this frequency (100 kHz) the impedance across the yoke is very small and is resistive. This small resistance effectively damps out the yoke ringing and effectively stops the rippling of the left side of the raster. The value of C_5 is critical; too large or small a value will not tune the circuit to resonance and the ringing will not be eliminated. C_5 is a high-voltage ceramic capacitor whose voltage rating is typically 1000 to 3000 V.

Flyback and yoke defects. Defective flyback transformers usually take two forms. They either have an open winding or they have leakage resistance or shorts between turns. An open winding in a flyback will result in the loss of high voltage, and with it the loss of raster. Leakage or shorts between turns can result in a narrow picture, such as that shown in Figure 15-9(b). A greater number of shorted turns will cause the loss of high voltage, and with it the raster. In addition, the excessive load on the horizontal output transistor will cause increased current flow and the transistor will burn up or a fuse will open.

The manufacturer's service literature usually gives the resistance found between taps of the flyback transformer. See Figure 15-3. These resistance values are useful only

(a)

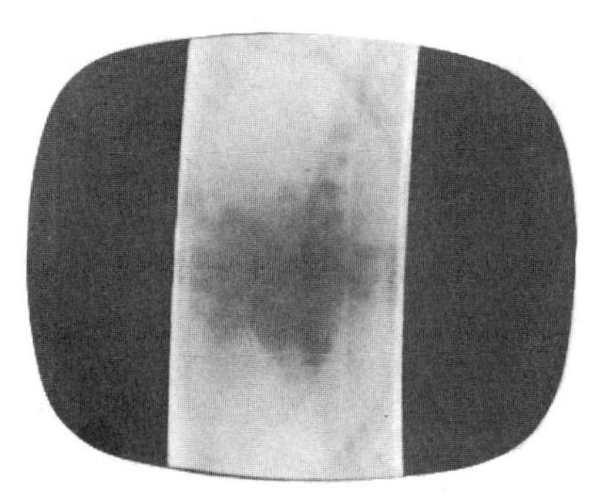

(b)

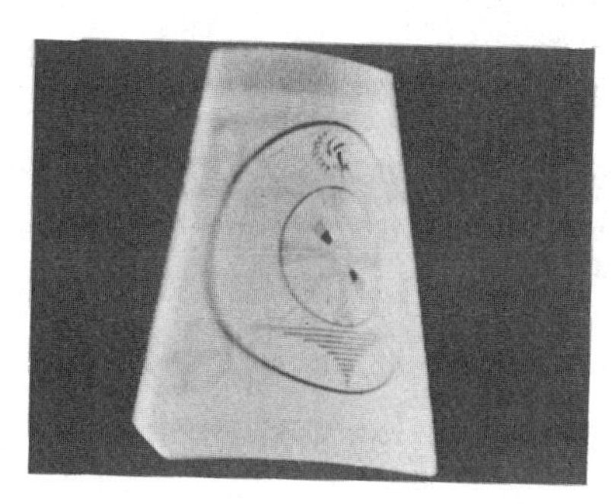

(c)

(d)

Figure 15-9 (a) Effect of an open balancing capacitor; (b) narrow raster due to shorted turns in flyback; (c) horizontal keystone; (d) yoke/flyback tester. [(a)–(c) Courtesy of General Electric; (d) courtesy of Heath Zenith.]

when there are gross changes in resistance resulting from open coil windings or shorts between layers of coil windings. Unfortunately, a single shorted turn cannot be detected with an ohmmeter, but it can cause a major reduction in winding inductance. Shorted turns also will cause the horizontal output transistor, the flyback transformer, yoke, or other coils to become hot. Overheated coils, especially hot spots, are useful in localizing transformers that have shorted turns.

Horizontal yoke winding may also develop open windings, shorts, or leakage between turns. An open yoke can cause a complete loss of raster. This loss of raster is caused by the reduction or loss of high voltage. In some cases, a single vertical line may be produced. Leakage or shorts between turns of the yoke may cause keystoning of the raster [see Figure 15-9(c)] or in extreme cases complete loss of raster may result.

Yoke and flyback testing. The loss of raster is most often caused by the loss or reduction in high voltage. Unfortunately, loss of high voltage can be caused by a defective yoke or flyback transformer, which in turn may cause the horizontal output transistor to blow out. If all other circuit elements have been checked and found to be good, a test is needed to determine whether or not the yoke or flyback is defective.

There are at least three practical ways of making this determination, for example:

1. *Direct substitution* of the yoke and the flyback. Since these components are relatively expensive, this procedure is most easily performed by service facilities that carry a large stock of parts.
2. *High-voltage substitution.* In this procedure the high voltage is restored and the raster is observed. If the raster exhibits keystoning, the yoke is defective. If the raster is narrow with straight edges, the flyback is to be replaced. This procedure is primarily useful in black & white receivers. Color picture tubes have additional voltage requirements such as focus voltage and screen voltages that are also derived from the horizontal output stage. In order to observe the raster, these voltages must also be supplied.
3. *Flyback and yoke testers.* These commercially available units check the Q of the flyback or yoke by means of a shock excited oscillator circuit. The meter indicates whether or not the component should be replaced. See Figure 15-9(d).

15-7 horizontal output amplifier controls

The horizontal output amplifier of a television receiver may have a width control, a linearity control, a third harmonic tuning control, and a centering control.

The third harmonic tuning control adjusts the resonent frequency of the flyback to the third harmonic of the horizontal retrace ringing rate. This causes the flyback pulse and the third harmonic to combine in such a way that the horizontal peak voltage is reduced. Proper tuning occurs when the dip in the horizontal retrace pulse is maximum. See Figure 15-10.

Width controls. Figure 15-10 illustrates the circuit connections for three widely used width controls found in color and black & white receivers. These controls are used to vary the amplitude of the horizontal deflection sawtooth. The sawtooth amplitude must be sufficient to produce a raster that "overscans" slightly. In this way, the front and back porches of the horizontal blanking pulse cannot produce a picture that appears to be narrow, even though the blank raster fills the screen. A width control allows the technician to adjust raster size to that point where the picture fills the screen.

One type of width control (A) shown in Figure 15-10 is a small variable coil placed in series with the yoke. In this case, the width control limits the current that can flow through the yoke. Shorting it will produce maximum deflection.

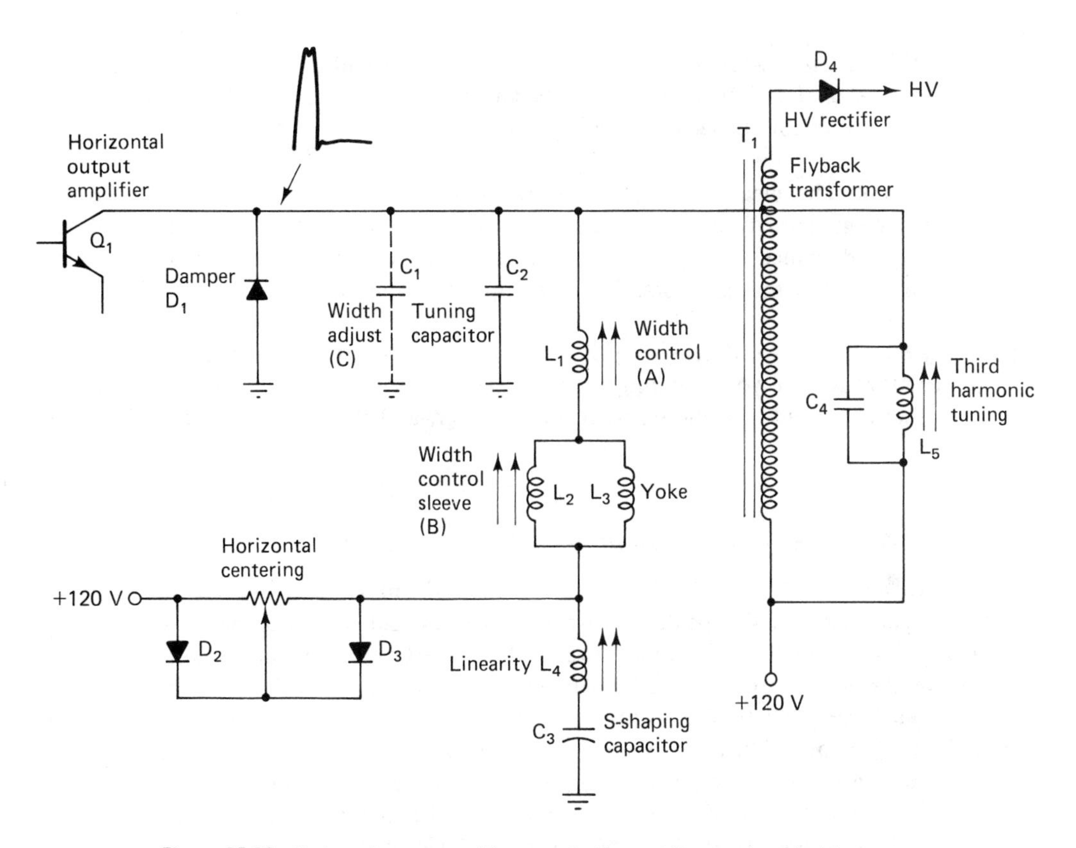

Figure 15-10 Horizontal output amplifier controls (three width controls, a third harmonic tuning circuit, and one linearity control). Only one method of width control is used in any given receiver.

Using a coil rather than a resistor as a width control has the advantage of controlling the yoke current without sizable power dissipation.

Also shown in Figure 15-10 is the schematic representation of a width control sleeve (B) that is mounted between the yoke and the neck of a black & white picture tube. The sleeve is made of thin-gauge brass and its position on the neck of the tube can be varied by sliding the sleeve in or out of the yoke. The sleeve acts as an ac magnetic shield. Weakening the horizontal deflection magnetic field by inserting the brass sleeve reduces horizontal deflection. Removing the brass sleeve increases the magnetic field and increases deflection. Vertical deflection is relatively unaffected because the magnetic shield's effectiveness is related to the frequency of the ac magnetic field.

Figure 15-10 shows an additional method (C) for adjusting horizontal width. In this case, a small tuning capacitor (C_1) (600 to 1500 pF) is used.

The addition of C_1 in parallel with the tuning capacitor C_2 lowers the resonant frequency of the output tuned circuit, increases retrace time, and causes the high voltage to decrease. This causes an increase in picture width. The value of C_1 and C_2 is critical. If the capacitors are made too large, foldover and nonlinearity will be introduced into the raster. If the capacitor is taken out of the circuit, the raster will be reduced to its narrowest width. Because of the high-voltage peaks occurring in this circuit, both C_1 and C_2 must have voltage ratings of between 1000 and 3000 V.

Linearity control. The linearity control shown in Figure 15-10 is able to adjust the horizontal sawtooth current linearity by adjusting the effectiveness of the "S" capacitor C_3. Making the linearity inductance L_4 larger reduces the total equivalent capacitive reactance in the circuit. This in effect makes C_3 smaller. The opposite occurs if L_3 is made smaller.

Centering control. Electrical centering of a television picture is accomplished by passing a dc current through the yoke windings. The centering control shown in Figure 15-10 is the same as that explained in section 15-5 of this chapter, and therefore will not be dealt with here.

15-8 the gate-controlled switch

In addition to PNP and NPN types of horizontal output amplifiers, horizontal output circuits using SCR-type devices are uses by RCA and Sony. Compared to transistors, SCRs are able to switch considerable power without overheating, exhibiting higher efficiency. SCRs would seem perfectly suited for use in wide-deflection-angle horizontal output amplifier application, except that they can only be turned on by gate current, and they can only be turned off by removing the anode voltage. As a result, SCR circuits as used by RCA require separate devices for trace and retrace functions. In practice precise switching and conduction timing of four devices (two SCRs and two damper diodes) require a very complicated circuit.

Sony's solution is to use an SCR type of device called a "gate-controlled switch." The GCS is a four-layer device similar to that used in an SCR. It may be considered as the equivalent of a PNP transistor connected to an NPN, where the base of the PNP is connected to the collector of the NPN, and vice versa. See Figure 15-11(a). This configuration allows the gate to turn the GCS *on and off.* When the gate is negative with respect to the cathode, the GCS is off. When the gate is driven positive, the GCS is turned on. This device has all of the advantages of a transistor except one. Losing drive in a transistor circuit shifts the operating conditions to zero bias, cutting off the transistor but causing it no damage. In the case of the GCS a loss of drive will turn it on to maximum current and burn it out. As such, the GCS is similar in its operation to the vacuum tube horizontal output amplifier.

The GCS is also sensitive to overload conditions, as are transistors. Overloads can be caused by such things as high-voltage arcs, defective scan rectifier circuits, shorted high-voltage circuits, and shorts in such circuits as convergence, pincushion, and AGC.

Examining Figure 15-11(b) reveals that the GCS circuit is very similar to the NPN horizontal output amplifier circuit. Q_1 is a drive transistor that provides the GCS Q_2 with an 18-V p-p square-wave gate signal. Driver transformer T_1 provides impedance matching. D_1 is the damper diode, which eliminates yoke-flyback ringing immediately after retrace. C_4 and C_5 are the tuning capacitors that resonate with the yoke to determine the retrace ringing frequency. If this frequency is too high, excessive high voltage will result; if it is too low, blooming will result. The horizontal deflection sawtooth of current passes through the yoke (Y_1) and the pincushion transformer (T_3). Horizontal centering of the raster is

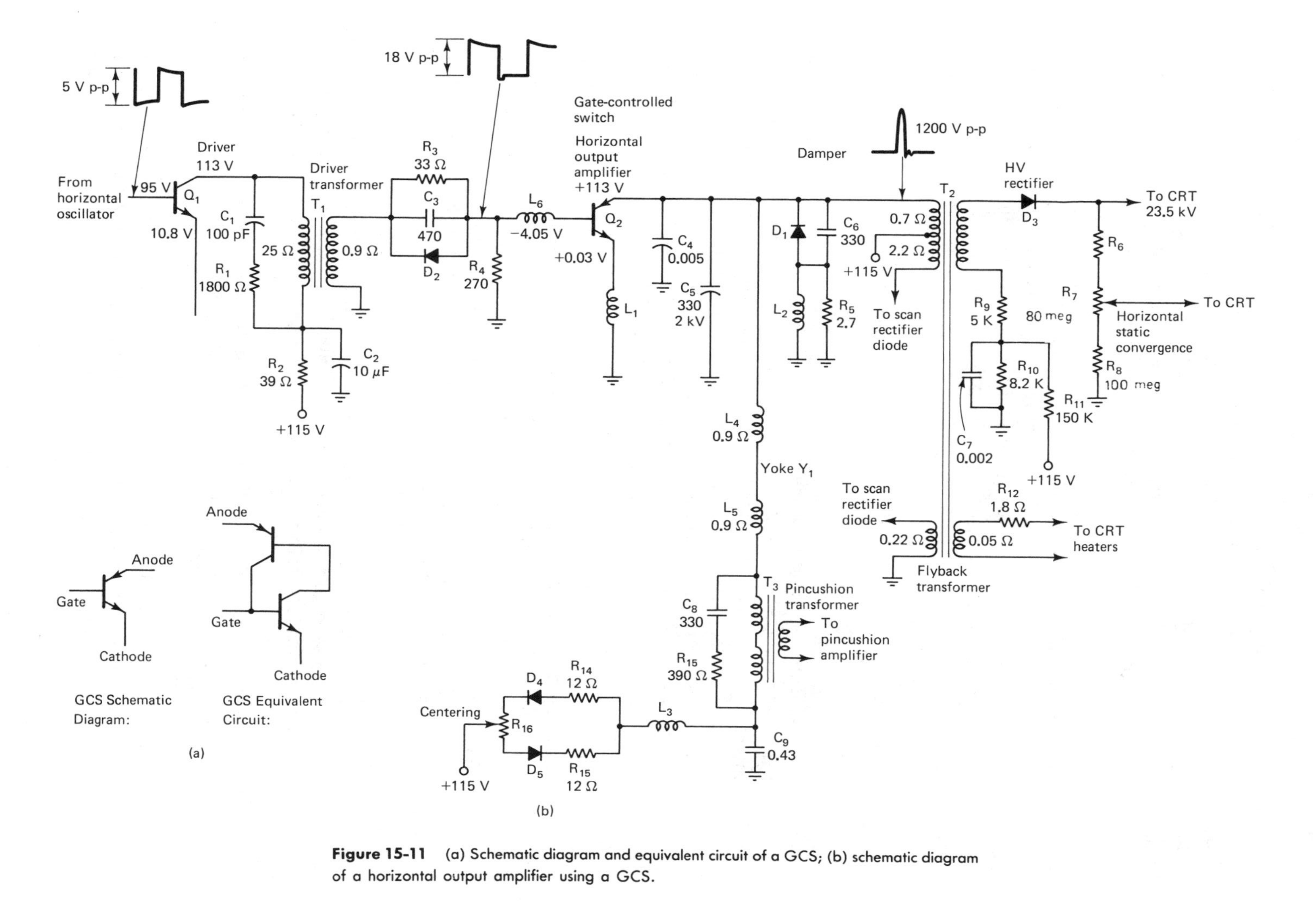

Figure 15-11 (a) Schematic diagram and equivalent circuit of a GCS; (b) schematic diagram of a horizontal output amplifier using a GCS.

provided by D_4, D_5, and R_{16}, which control the amount of dc current flowing through the yoke. C_9 is the S-shaping capacitor. The flyback transformer (T_2) develops ac high voltage which is rectified by D_3, producing 23.5 kV dc high voltage for use by the second anode of the picture tube. Some of this voltage is tapped off by R_6, R_7, and R_8 to provide the Trinitron picture tube with horizontal static convergence voltage. The flyback transformer also has windings for the CRT heater, as well as for two scan rectifiers.

Horizontal output amplifier troubleshooting. The horizontal output stage is under considerable stress due to high voltages and large deflection currents. It is therefore not suprising that this stage has the greatest frequency of breakdown of any of the stages in the TV receiver.

Transistor horizontal output circuits usually exhibit three types of defect symptoms:

1. No raster, no sound
2. Insufficient or excessive raster width; sound O.K., color O.K.
3. Horizontal foldover; sound O.K., color O.K.

No raster, no sound. This condition is usually caused by the loss of high voltage and all scan-derived B supply voltages. The loss of the high voltage accounts for the loss of the raster and the loss of the B supply voltages results in the loss of sound. All of these symptoms point to an inoperative horizontal output transistor (HOT).

An inoperative HOT can be caused by loss of drive from the horizontal oscillator. In the case of the transistor, there will not be any damage, but a GCS will burn out. Transistors may burn out due to excessive current flow through them. If a protective fuse is placed in series with the HOT, it may blow out first, saving the HOT from damage. Defects that could cause a fuse to blow out are a shorted HOT, a shorted damper diode, a shorted tuning capacitor, and a shorted S capacitor. Defects that could cause either the HOT or a fuse to blow out are such things as shorted turns in the yoke, a shorted flyback transformer, a shorted high-voltage rectifier, a shorted pincushion transformer, an open tuning capacitor, a shorted scan rectifier, incorrect horizontal drive frequency, and high-voltage arcing. In addition, the horizontal output stage may be disabled without damage by an activated high-voltage hold-down circuit.

Insufficient or excessive width, sound O.K., color O.K. Insufficient width can be caused by such things as reduced B supply voltage, a leaky S capacitor, excessive loading due to shorted or leaky components, insufficient drive, leaky damper, or a tuning capacitor that is too small.

Excessive width can be due to the S capacitor being too small in value, increased B supply voltage, or the tuning capacitor being too large.

Foldover, sound O.K., color O.K. This trouble symptom (see Figure 15-12) is usually caused by changes in value of bias components (C_1, R_1 in Figure 15-7) in the base circuit of the HOT or changes in the values of the driver decoupling filter circuit (R_7, C_1 in Figure 15-3).

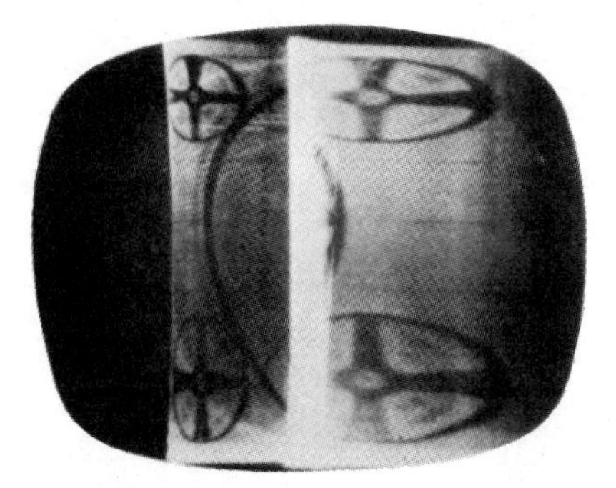

Figure 15-12 Horizontal foldover. (Courtesy of General Electric.)

15-9 high-voltage rectifier circuits

Both color and black & white picture tube operation requires that the second anode of the tube be energized with dc voltages that may be as high as 30,000 V. To obtain this voltage, the ac high-voltage retrace pulse developed by the flyback transformer is converted into high-voltage dc by a selenium high-voltage half-wave rectifier diode, as shown in Figures 15-3, 15-6, and 15-13(a).

The high-voltage ac retrace pulse is applied to the anode of the rectifier diode. The positive peak forces it to conduct, charging the picture tube capacitance formed by the inner and outer graphite conductors that coat the glass bulb of the tube. The high-voltage charge across this capacitor ensures that the diode will only conduct on the peak of the positive retrace pulse.

Figure 15-13 (a) Basic high-voltage rectifier circuit;

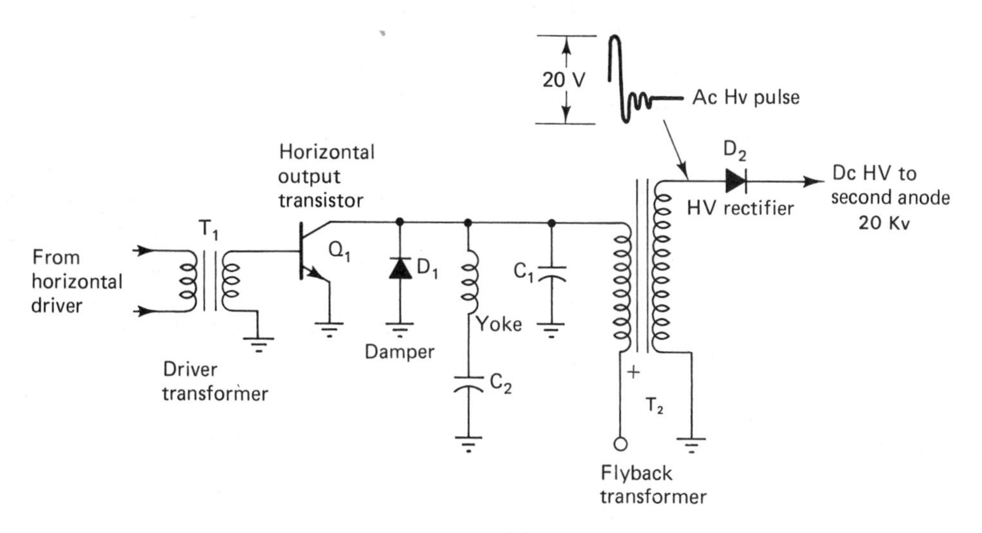

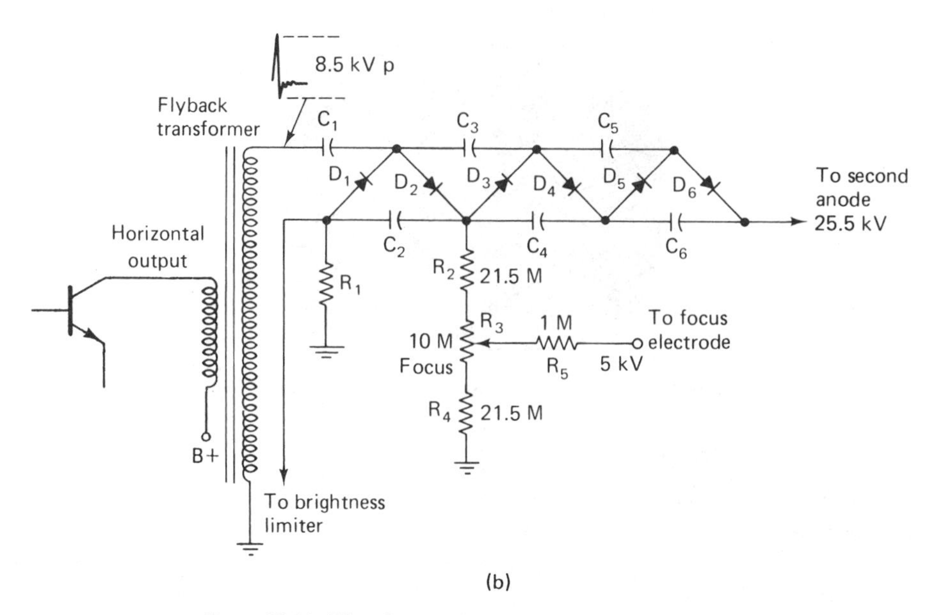

Figure 15-13 (*Cont.*) (b) solid-state tripler type of focus supply.

High-voltage multipliers. Many modern color receivers incorporate high-voltage tripler or quadrupler rectifier systems as a means of reducing x-ray production as well as reducing the peak voltage requirements of the flyback transformer. In a typical tripler arrangement such as that shown in Figure 15-13(b) the peak input voltage developed by the flyback transformer is 8.5 kV and the output dc voltage of the tripler is 25,500 V.

This high-voltage multiplier is similar to the half-wave voltage doubler used in some low-voltage dc power supplies.

The input pulse to the tripler has a large positive-going spike and a somewhat smaller negative-going peak. On the negative portion of the pulse diode D_1 conducts and charges C_1 to the peak voltage. The positive portion of the pulse now adds to the voltage across C_1 and causes D_2 to conduct, thus charging C_2. The voltage developed across C_2 is equal to the *peak-to-peak voltage* of the input pulse.

After a number of pulses the remaining diodes in the chain have produced similar effects. The capacitors C_3 and C_5 each charge to the negative peak of the input pulse, and then they add their charges to C_4 and C_6. Capacitors C_2, C_4, and C_6 are connected in series and each has a charge of 8.5 kV; these voltages add and produce an output terminal voltage of approximately 25 kV. Since the circuit time constant is very large, the accumulated dc voltages across each capacitor remain fairly constant. If this process is continued one step further by adding another pair of diodes and two additional capacitors, the circuit becomes a quadrupler and the output voltage will be four times the input. In practical quadruplers input voltage has a peak of approximately 7 kV. Therefore, the output should be approximately 28 kV.

High-voltage tripler and quadrupler circuits are totally encapsulated and therefore cannot be repaired other than by direct replacement.

Focus voltage and automatic brightness limiter control voltages may be obtained from the tripler and quadrupler high-voltage multiplier circuits. The focus voltage is obtained by placing a tap across C_2. This provides approximately 7 kV dc, which is then passed through a voltage divider to obtain the required focus voltage of 4 to 5 kV.

In some receivers the ground return for the tripler may be passed through a brightness limiter transistor. Thus, the beam current of the picture tube is used to control and limit maximum brightness automatically and prevent blooming.

The split high-voltage transformer. In many color television receivers a diode split high-voltage transformer similar to that shown in Figure 15-14 is used to eliminate the need for a tripler circuit.

The primary of the flyback transformer is fed by the collector of the horizontal output transistor. Four separate secondary windings each develop 6-kV pulses. Built into the transformer and in series with each winding is a high-voltage rectifier diode. These 6-kV horizontal pulses force the diodes (D_1, D_2, D_3, and D_4) to conduct and charge their interlayer distributed capacitances (C_2, C_3, C_4, and C_5). Each winding pulse is associated with the dc produced by its diode. Since the windings and the diodes are connected in a series-aiding configuration, each winding raises the dc voltage by 6 kV, until a maximum dc voltage of 25 kV is reached. The CRT capacitor formed by the inner and outer Aquadag coatings filters the high-voltage output of the transformer.

Figure 15-14 Typical diode split high-voltage flyback transformer.

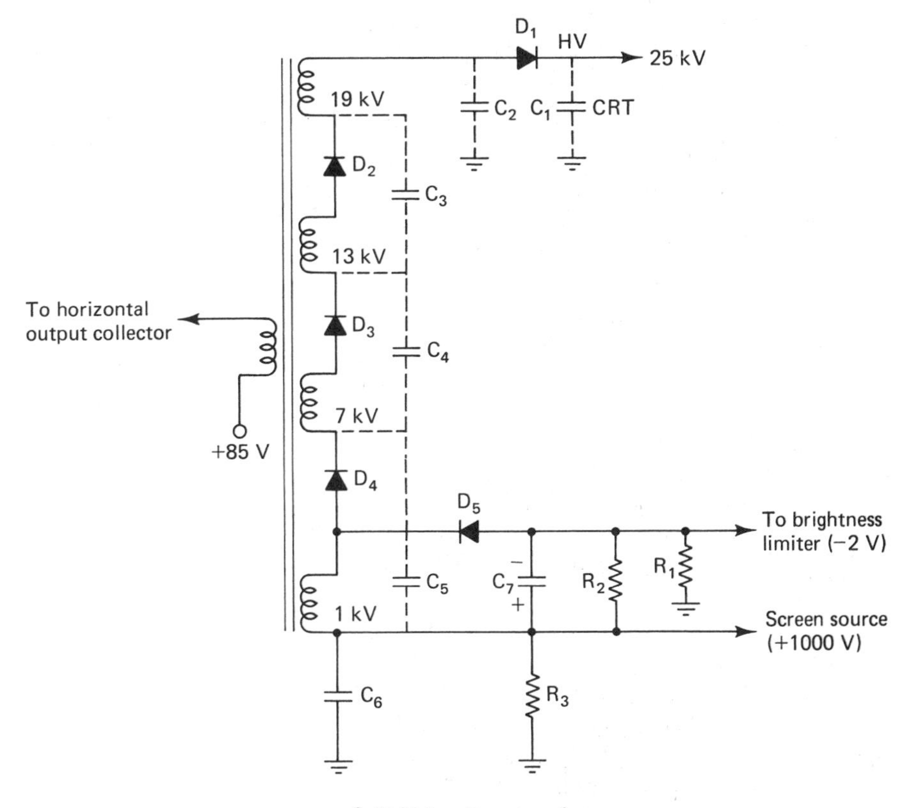

Split high-voltage transformer

D_5 conducts on the negative portion of the horizontal retrace pulse developed across the lowest winding of the flyback, charging C_7 to almost 1000 V. The voltage division across R_1 and R_3 is such that about 2 V is developed across R_1 and is used to control the brightness limiter circuit. The voltage developed across R_3 is used to provide 1000 V to the picture tube screens.

If any of the diodes in the transformer become defective, the entire transformer must be replaced.

Troubleshooting HV circuits. Defective high-voltage rectifier circuits can produce at least the five following important trouble symptoms:

1. No raster, sound O.K.; loss of high voltage
2. Blooming
3. Overloading of the horizontal input amplifier
4. Corona
5. Arcing

No raster, sound O.K. The trouble symptom of no raster but sound O.K. can be caused by a defective picture tube or improper picture tube electrode voltages. Even though the picture tube heater is illuminated and the picture tube cathode, control grid, and screen grid voltages are proper, the loss of second anode high voltage will result in a no-raster condition.

Loss of high voltage may be either an ac or a dc problem. A quick way to determine whether or not the horizontal output amplifier is working is to place a small neon bulb (mounted on an insulating rod) near the flyback transformer. If the bulb glows brightly, the horizontal output amplifier is functioning. If the neon bulb also glows when it is brought near the anode of the high-voltage rectifier, ac high voltage is reaching the rectifier.

If a neon bulb is not available, bringing the blade of an insulated screwdriver near the anode of the high-voltage rectifier will produce a high-frequency arc between the anode and the screwdriver blade if the ac high voltage is present.

To detect the presence of high-voltage dc, a high-voltage probe should be used. To measure the dc high voltage of vacuum tube receivers a momentary arc may be drawn between the second anode lead and ground. A long bright and cracking spark indicates the presence of dc high voltage. A weak arc or no arc at all indicates insufficient dc high voltage.

Arcing should not be done in hybrid or all transistor receivers because it can cause one or more ICs or transistors to become defective.

If ac high voltage is present but there is no dc high voltage, the trouble may be a defective high-voltage rectifier, an open filter resistor sometimes found in series with the high-voltage lead, a defective high-voltage regulator, and an internal picture tube short.

Blooming. Blooming is the condition in which an increase in brightness causes simultaneous raster expansion, loss of brightness, and a loss of focus. This trouble symptom is caused by poor high voltage regulation. This simply means that the high voltage decreases as the picture tube beam current increases. In general, blooming is caused by an increase in the internal resistance of the high-voltage power supply. Such things as a bad rectifier diode, a bad picture tube, or an increased series filter resistance are possible causes of blooming. Other causes of blooming are defects in the horizontal output amplifier, damper, horizontal oscillator stage, flyback transformer, high-voltage regulator, or high-voltage hold-down circuits.

Overloading of the horizontal output amplifier. Loss of raster, accompanied by a blown out horizontal output amplifier transistor, may be caused by an increased ac load on the horizontal output stage. Increased ac loading may be caused by shorts in the flyback, pincushion transformer defects, or shorts in other loads connected to the flyback transformer. A shorted scan rectifier is another possible cause of this trouble symptom.

Corona. Corona is the result of ionization of the air. It is caused by a very strong electrostatic field generated by a high voltage.

Corona makes itself known by a hissing sound, the smell of ozone, a bluish glow surrounding the point of discharge, and black streaks running horizontally through the picture. To eliminate corona, the point of discharge should be located by looking for a bluish glow surrounding a part associated with the dc high-voltage circuit. The high voltage should be measured, and if it is found excessive, it should be corrected before any other repair is made.

If the corona is being emitted from an area that is obviously dirty or shows signs of cracked insulation or other mechanical defects, these should first be corrected. In difficult cases, an anticorona dope or high-voltage insulating putty may be spread over the region of corona discharge in order to suppress it.

Arcing. Arcing is an electrical breakdown between two points in which a spark jumps between the points. This condition announces itself by loud snapping noises and flashes of light. Each time the high-voltage arc occurs the raster will also be lost. The location of the arcing is usually self-evident, as is the repair. In general, arcing is eliminated by moving the points of arcing apart, or by placing high-voltage putty between the two points.

15-10 high-voltage focus supplies

Good raster focus depends upon correct picture tube electrode voltages. Focus will deteriorate if the voltage on the focus electrode and the picture tube second anode depart significantly from normal.

Many color picture tubes that are designed for high-voltage electrostatic focus are designed to have the focus electrode voltage at approximately 20% of the second anode voltage. If 25,000 V were applied to the second anode, approximately 5000 V would have to be provided for the focus electrode. This voltage is usually adjustable through a range of 4 to 6 kV.

In picture tubes that use the hi-bi focus lens system the ratio is about 40% of the second anode voltage, making the required focus voltage 7 kV.

In the tripotential focus lens system (see Figure 15-15), two high-voltage electrodes are required for proper focus. One of them (pin 3) requires a focus voltage of 40% of the second anode voltage (12 kV), and the other (pin 1) must be 23% (7 kV). In this circuit the focus voltages are tapped off from R_{555}, the focus block adjustable resistor. The high-voltage source is obtained from a tap on the split diode high-voltage transformer, T_{551}.

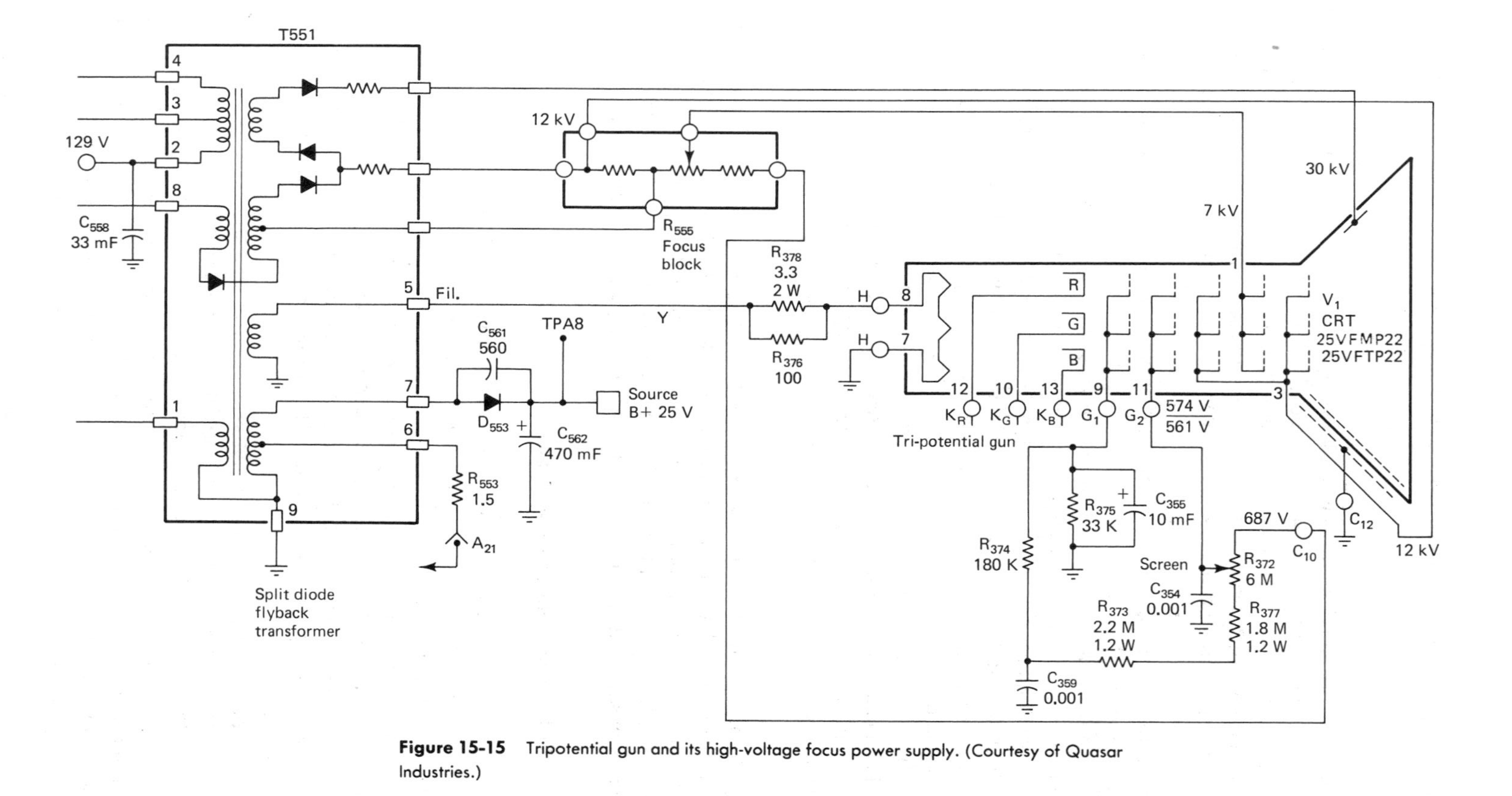

Figure 15-15 Tripotential gun and its high-voltage focus power supply. (Courtesy of Quasar Industries.)

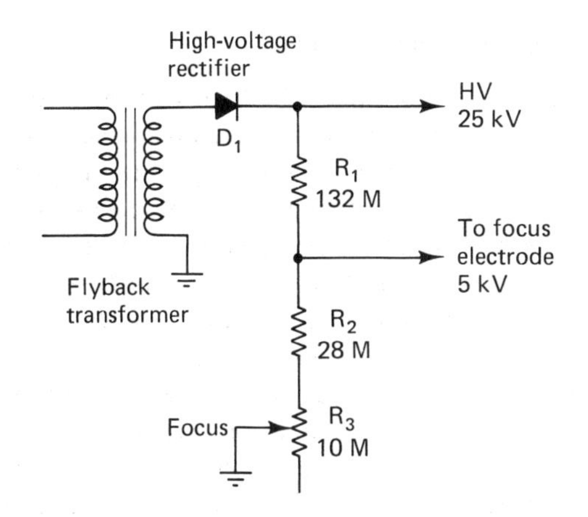

Figure 15-16 Typical voltage divider high-voltage focus supply.

Typical focus circuits. Unfortunately, there is no standard method for providing and controlling the focus electrode voltage. The schematic diagrams for four commonly used arrangements are shown in Figures 15-7, 15-13, 15-15, and 15-16.

The circuits shown in Figures 15-7 and 15-12 use a solid-state, high-voltage tripler to develop the second anode voltage. A tap is placed at a point inside the voltage multiplier that provides approximately 30% of the total voltage for focus. An adjustable voltage divider is used to set the exact value of dc delivered to the focus electrode.

In Figure 15-16 the focus voltage is obtained from an adjustable voltage divider placed across the second anode high-voltage supply. This takes care of high-voltage focus voltage tracking in a very simple way; the focus voltage is automatically a fixed percentage of the high voltage.

Focus circuit troubleshooting. Defects in the focus rectifier circuit can cause an out of focus raster or no raster at all. Poor focus is indicated by horizontal scanning lines that are fuzzy, broad, and indistinct.

Since the focus supply and the high-voltage power supply depend upon one another, the service technician must decide which circuit is the source of the trouble. If the high voltage is within 1000 V of its rated value and the raster is still out of focus, the trouble is in the focus supply. If the high voltage is found to be 10% low and the focus voltage is 50% low, chances are that the trouble is in the focus circuits. If the high voltage and the focus voltage are both low by the same percentage, the trouble is common to both and such things as the flyback transformer B+ boost circuit and the horizontal output amplifier must be checked.

A shorted selenium focus rectifier diode may cause a 50% reduction of high voltage, overheating of the horizontal output amplifier, and loss of raster.

The selenium focus diode should only be checked by direct substitution. It cannot be properly checked by an ordinary ohmmeter. Therefore, direct part substitution should be employed.

X-ray hazard. X-radiation may be produced by high-voltage tubes that operate at potential exceeding 10,000 V. In color and black & white television receivers the picture tube is the chief source of such radiation. In color receivers the picture tube x-radiation may be reduced by the use of external metal x-radiation shielding and by the use of picture tubes constructed of leaded glass. If the receiver is operated at abnormally high voltages, some excess x-radiation may penetrate the walls of these tubes. As little as a 1000-V increase in high voltage can cause the x-radiation produced to double. Increasing the high voltage by 5000 V may cause radiation levels to increase by 20 times. Therefore, it is extremely important to be sure that the manufacturer's specified high voltage is adhered to.

The recommended annual x-radiation dose from TV sets should not exceed 5% of the average dose from natural background radiation that is part of our everyday environment. As a result, the federal government has required that television receivers manufactured after January 15, 1970, be designed to keep x-radiation from exceeding 0.5 mR/h (milliroentgen per hour) measured at a point 2 in. (5 cm) from the surface of the receiver.

The Department of Health, Education and Welfare (HEW) has issued a number of regulations designed to protect the public from x-radiation produced by single defects that may occur in a color television receiver. One of these regulations (effective June 1, 1971) requires that the manufacturer provide circuits that produce an unusable picture in the event that a malfunction causes excessive high voltage. Such circuits are called high-voltage detection or *hold-down* circuits.

Unfortunately, there are no universally accepted circuits and there are no universally accepted ways of making an unusable picture. In some circuits the hold-down circuit merely lowers the high voltage, causing a dark narrow raster with poor focus. In other circuits the horizontal or vertical hold is electronically altered so that the picture loses sync. In other receivers excessive high voltage causes a blank raster to be produced. These trouble symptoms remain in the circuit until the defect producing the excessive high voltage is repaired.

Transistor high-voltage hold-down circuit. An example of a high-voltage hold-down circuit used in a transistor color receiver is shown in Figure 15-17. This circuit is designed so that an increase in high voltage above 33,000 V causes an SCR to conduct and ground the horizontal oscillator supply voltage. This causes the horizontal oscillator (Q_2) to cease functioning, resulting in a loss of high voltage and the raster.

The gate of the SCR is fed by a voltage divider that is placed directly across the 30,000-V second anode voltage. The zener diode (D_1) that is placed in series with the gate ensures that the SCR will turn on at precisely the desired voltage. If the high voltage increases and this voltage is reached, the positive gate voltage turns the SCR on. The SCR will remain turned on until either the receiver is turned off or the cause of the excessive high voltage is eliminated. The circuit action described above is summarized in Figure 15-17 by an arrow diagram.

Horizontal frequency hold-down circuit. In the circuit shown in Figure 15-18 any excessive increase in high voltage or CRT beam current will increase the frequency of the horizontal oscillator, causing a loss of picture horizontal sync, disabling the receiver.

Under normal conditions Q_1 and Q_2 are both cut off. This is accomplished in the NPN transistor Q_1 by returning the base to ground through R_3 and connecting the emitter

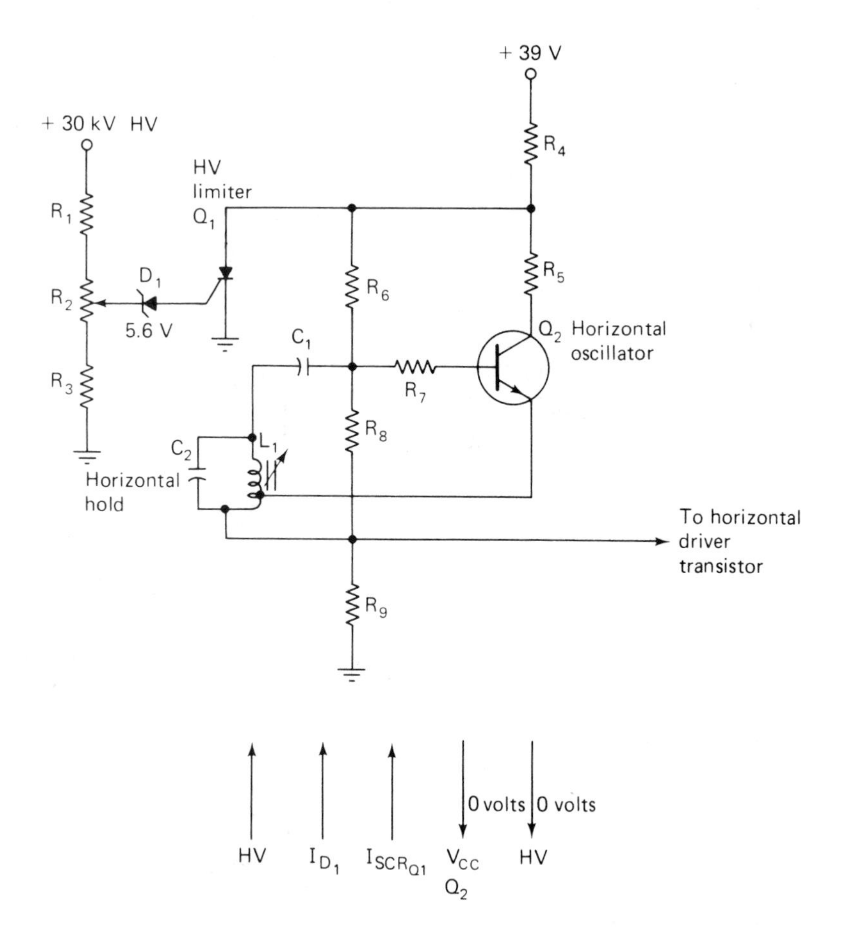

Figure 15-17 High-voltage hold-down circuit used in a transistor color receiver.

to the positive source through R_4, reverse biasing Q_1. D_2 is a zener diode that clamps the emitter voltage of Q_1 to about 7.6 V.

Q_2 is normally at cutoff because both the emitter and base of the PNP transistor return to a common B supply; therefore, its bias voltage is zero.

If the high voltage should increase, diode D_1 will rectify the increased amplitude horizontal retrace pulse, charging C_1 to a larger voltage. This, in turn, will make the base of Q_1 more positive, causing it to turn on. Collector current flowing through R_5 will lower the base voltage of Q_2, allowing it to conduct. Q_2 may be thought of as a voltage (bias)-controlled resistor that is in parallel with the hold control R_9. When Q_2 conducts, the effective resistance of the hold control decreases, forcing the horizontal oscillator frequency to increase, disabling the receiver.

If CRT beam current should become excessive, the beam current flowing through R_4, R_{10}, R_{11}, and R_{12} will cause an increased voltage drop across these resistors, which will shift the bias of Q_1 toward conduction. If Q_1 is forced into conduction, the picture will lose horizontal sync, disabling the receiver.

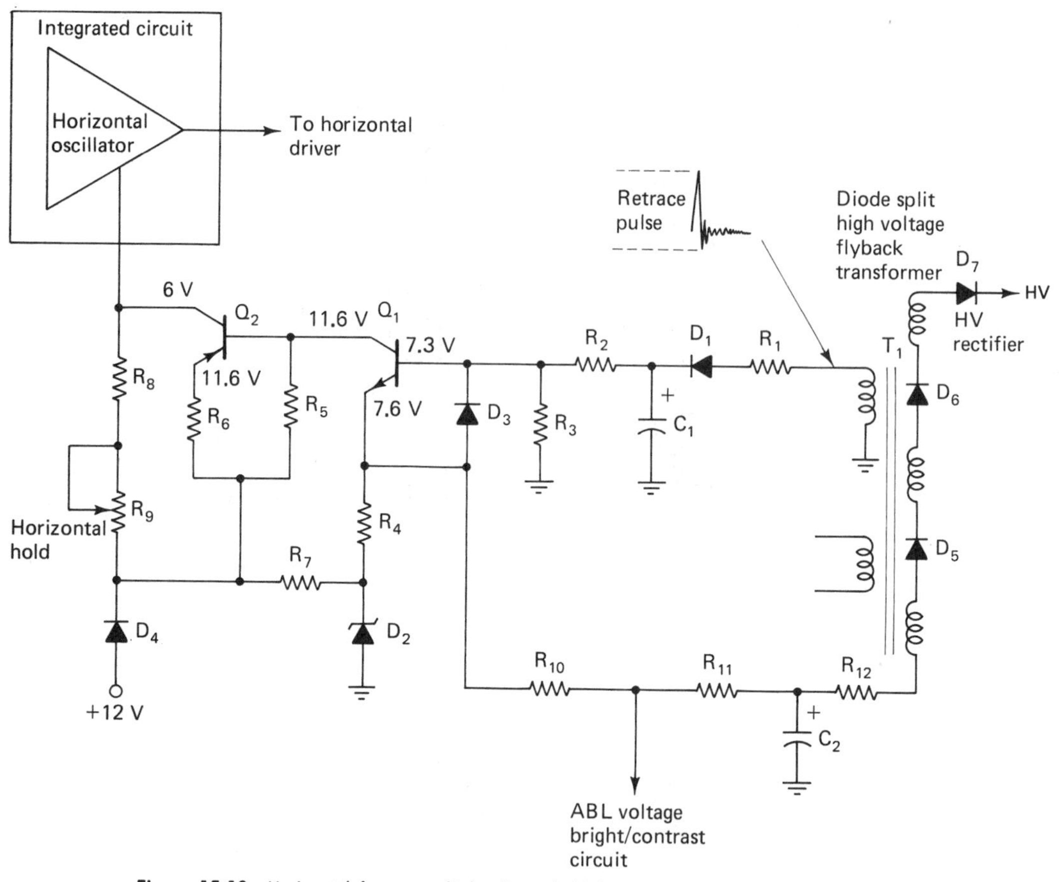

Figure 15-18 Horizontal frequency high-voltage hold-down circuit. (Courtesy of Quasar Industries.)

Troubleshooting. High-voltage hold-down circuits can produce trouble symptoms that are in addition to the actual circuit defect. This complicates an already complex and often difficult troubleshooting problem. The first step in troubleshooting color receivers that include hold-down circuitry is to lower the ac input line voltage by means of a variac or a light dimmer switch. This is to prevent excessive high voltage from being produced if the high-voltage circuit is defective. Then disconnect or disable the hold-down circuit. For example, disconnecting the SCR (Q_1) of Figure 15-17, or disconnecting the collector of Q_2 in Figure 15-18 will disable these hold-down circuits. When the hold-down circuit has been deactivated, the high voltage should be measured; excessive voltage indicates that the trouble must be in the high-voltage circuit. Under these conditions, the reason for the increase in high voltage will have to be investigated and then corrected. If the high voltage is normal, the defect is likely to be associated with the hold-down circuit.

15-12 high-voltage regulation

The electron beam current of a black & white picture tube generally does not exceed 500 μA. The beam current of a color picture tube may be two or three times the value. At maximum brightness (a second anode voltage of 25,000 V) the color picture tube may dissipate as much as 30 W.

In both types of picture tubes the beam current flow path is a closed loop flowing from the cathode of the picture tube to the second anode, through the high-voltage source and its internal resistance, and then back to the cathode. This current develops a voltage drop across the internal resistance of the power supply, which tends to reduce the second anode potential. A large reduction of this accelerating potential may result in an increase in deflection, loss of brightness, and poor focus. This condition is called *blooming* and may occur in both black & white and color receivers.

Picture tube electron beam current level is controlled by the brightness control and the peak-to-peak swing of the video signal. Thus, increasing the brightness control tends to lower the second anode voltage. If the internal resistance of the high-voltage power supply is too large, or if the beam current is too large, blooming will result. The higher beam current of the color picture tube means that these receivers are more subject to blooming than are black & white receivers.

Regulator circuits. In an attempt to keep the second anode voltage constant under varying brightness conditions, color receivers incorporate special high-voltage regulator circuits.

Many transistor color receivers do not have direct high-voltage regulation because compared to vacuum tubes, transistor horizontal output circuits are essentially low-impedance devices. As such, changes in the load current passing through them causes small internal voltage drops. The net result is that the high voltage remains relatively stable. Nevertheless, many transistor color receivers do incorporate some form of high-voltage regulation.

Regulation of the sweep and high voltage in many color receivers is done by regulating the B supply voltages that feed the horizontal output stage. In other circuits the duration of retrace or the conduction time of the horizontal output transistor is varied.

The saturable reactor high-voltage regulator. An example of a high-voltage regulator that controls the duration of retrace time is shown in Figure 15-19. In this circuit a saturable reactor transformer T_1 is used to regulate the 30-kV second anode voltage. This transformer has three windings: a control winding, a shunt winding, and a series winding. The collector current of the output transistor (Q_1) flows through the primary of the output transformer (T_2) and the control winding. An increase in the current flowing through the control winding of the saturable reactor results in a decrease in the inductance of all three windings. A decrease in current causes the inductances to increase. The shunt winding of T_1 is connected in parallel with the primary of the flyback transistor T_2. The series winding of T_1 is connected in series with the yoke and pincushioning coils.

An increase in picture tube brightness represents an additional load that tends to reduce the high voltage and narrow picture width. These changes are accompanied by an increase in output transistor (Q_1) collector current. This in turn causes an increase in current flow through the control winding of the saturable reactor (T_1), forcing a decrease in the inductance of the shunt winding. Since this winding is effectively in parallel with the yoke and its resonating capacitor C_1, the ringing frequency of the circuit increases. As a result, the flyback pulse becomes narrower and increases in amplitude, thus maintaining the high voltage.

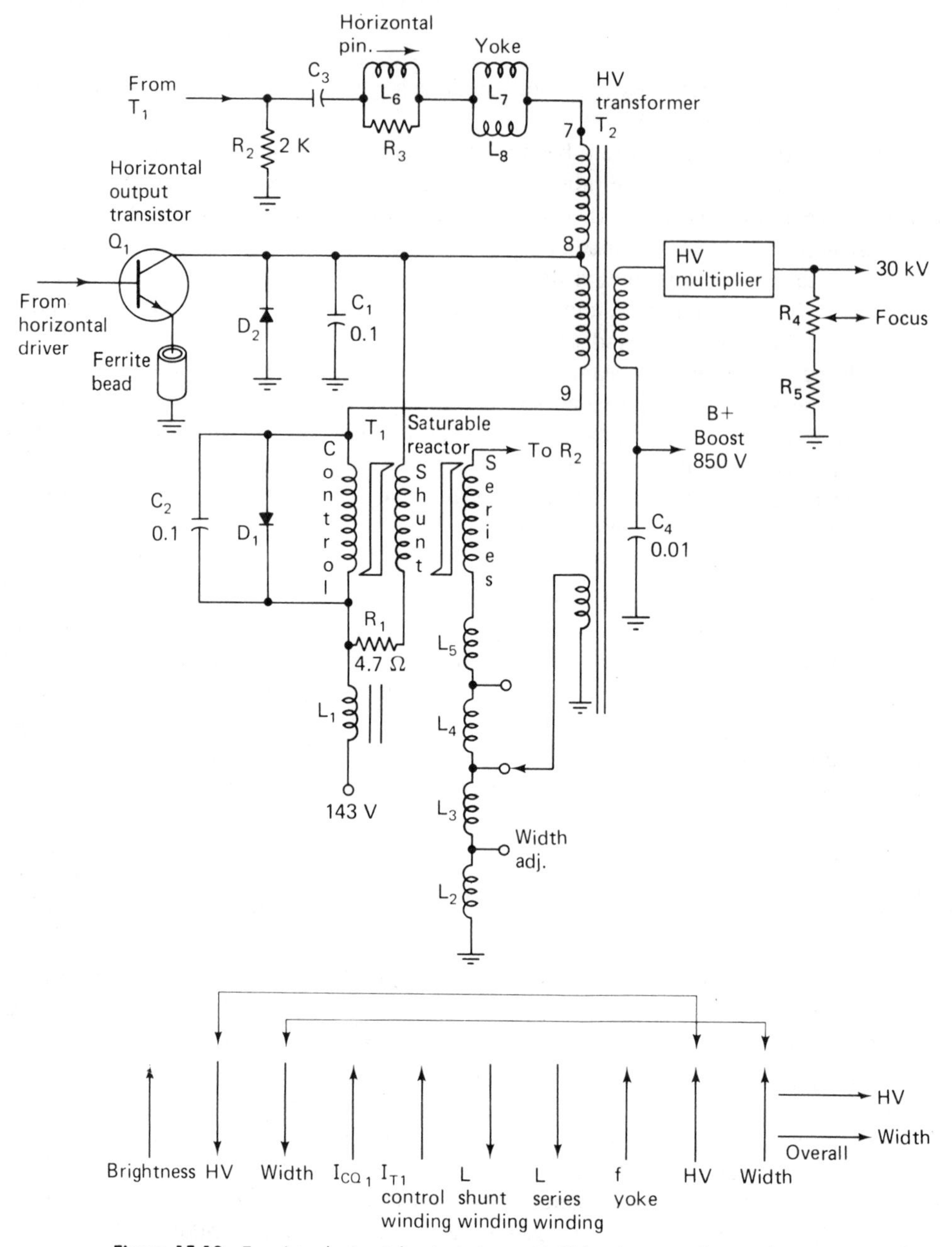

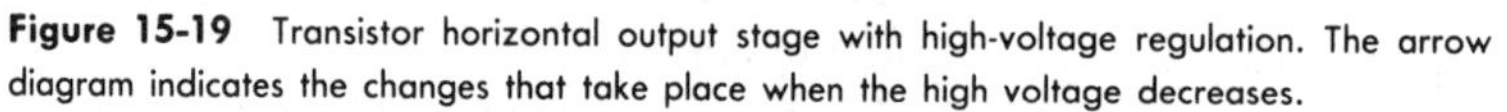

Figure 15-19 Transistor horizontal output stage with high-voltage regulation. The arrow diagram indicates the changes that take place when the high voltage decreases.

Raster width is maintained because of magnetic saturation of T_1, which reduces the inductance of the series winding. Consequently, the total impedance of the yoke circuit is reduced. This allows a greater yoke current and maintains picture width. Reduced picture brightness decreases the load on the horizontal system and reverses the preceding action. The above circuit behavior is summarized in an arrow diagram associated with Figure 15-19.

Capacitor C_2 and diode D_1 are used to prevent the saturable reactor from ringing. R_1 limits the current flowing through the shunt winding. The ferrite bead on the emitter lead of Q_1 suppresses a form of Barkhausen interference that may appear on the left side of the screen.

The Zenith high-voltage regulation system. Zenith has introduced a system of horizontal deflection self-regulation that provides high-voltage, ac line and load regulation. In addition, this circuit also maintains a constant raster width. This is done by a complex circuit that varies the horizontal output transistor's collector current pulse width in response to power supply loading.

A block diagram of this circuit is shown in Figure 15-20. An examination of this diagram indicates that there are three circuits which have been changed or added in comparison to the basic horizontal output circuit. First, the horizontal output stage has been modified by the addition of a diode switching circuit. Second, the single driver stage has been replaced by two drivers, called the forward and reverse horizontal driver stages. Third, the horizontal drive signal, obtained from a countdown circuit, feeds a pulse width modulator circuit, which in turn drives the forward horizontal driver. In addition, a high-voltage shutdown is indicated. The schematic diagram of the Zenith System 3 self-regulating circuit is shown in Figure 15-21.

The pulse width modulator. The pulse-width modulator IC chip has two basic functions:

1. It amplifies the horizontal square wave driving the chip.
2. It develops an output whose pulse width is determined by the system voltage regulation.

The input signal to the IC is applied to pin 7 and the output signal is at pin 6. The output is then coupled via R_{41}, C_{23}, and R_{39} to pin 4, where its wave shape is seen to have changed into a triangular waveform. Pin 4 also has a highly regulated 12-V dc voltage applied to it, via a factory-adjusted resistor R_{32}. This combined voltage is compared in the IC to a dc voltage on pin 3 that varies in direct proportion to the ac line and the 19-V scan derived dc voltage. This comparison determines the pulse width (duty cycle) of the

Figure 15-20 Block diagram of the Zenith horizontal deflection self-regulation system.

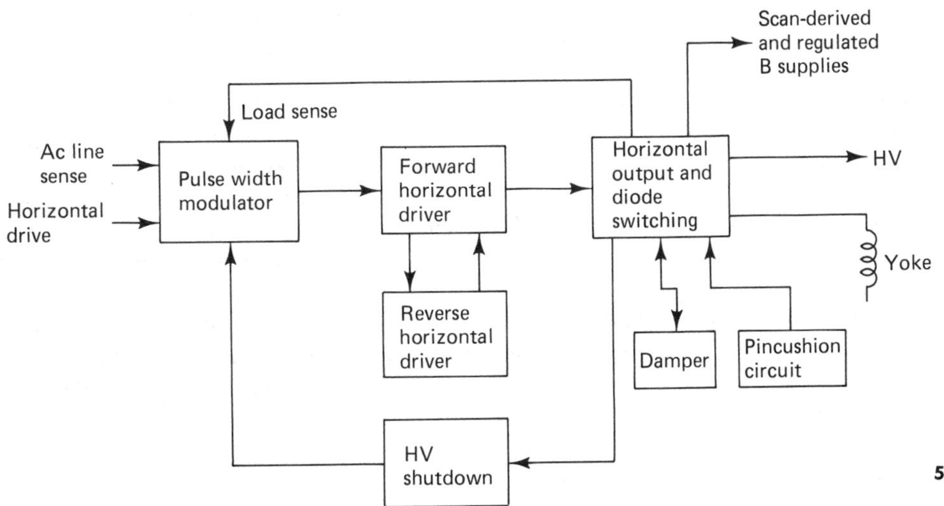

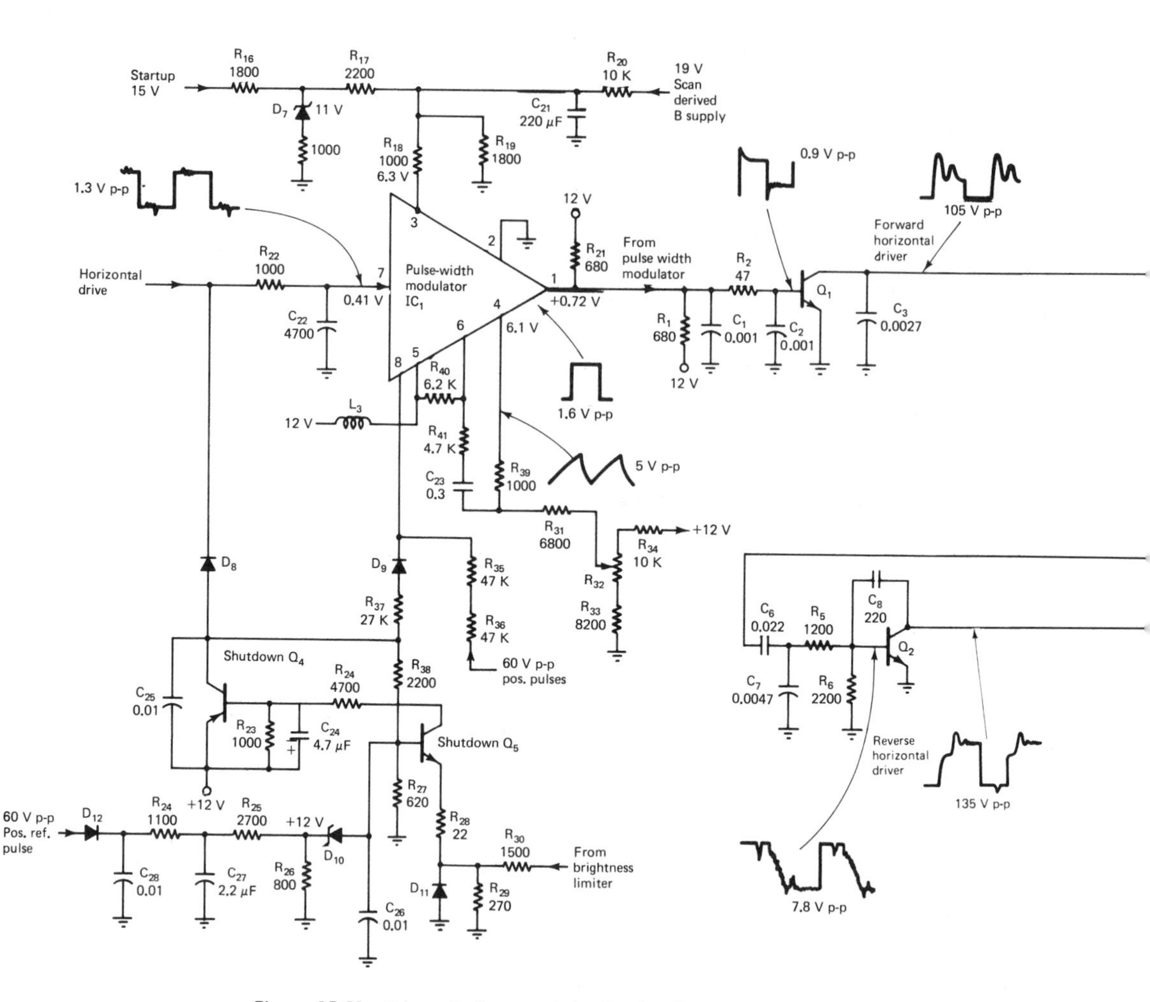

Figure 15-21 Schematic diagram of the Zenith self-regulation system.

output signal appearing on pin 1. The 19-V scan-derived voltage is actually a small sample of the high voltage since both are developed by the same flyback transformer. The pulse width output of the modulator is inversely related to the amplitude of the high voltage.

Horizontal driver. In this direct-drive system the horizontal driver (Q_1) is on when the horizontal output stage (Q_3) is conducting. Transformer T_1 has a low-impedance primary that is driven by Q_1 with a current ramp. The direct drive produces a constant drive level independent of duty cycle. A disadvantage is the negative drive current, which prevents a short fall time in the output transistor (Q_3). To overcome this problem a reverse driver transistor (Q_2) has been added to the circuit. Q_2 obtains its input from the collector of the forward driver transistor Q_1.

Diode D_1 and its associated circuit form a clamp circuit which clips the peak of Q_1's collector voltage and provides the necessary wave shaping before coupling it to the base of Q_2. Q_2 turns on after Q_1 turns off. The current flowing through the primary of T_1 drives the base of Q_3 negative, providing a "clean" output pulse that switches the horizontal output amplifier Q_3 in about 0.5 μs.

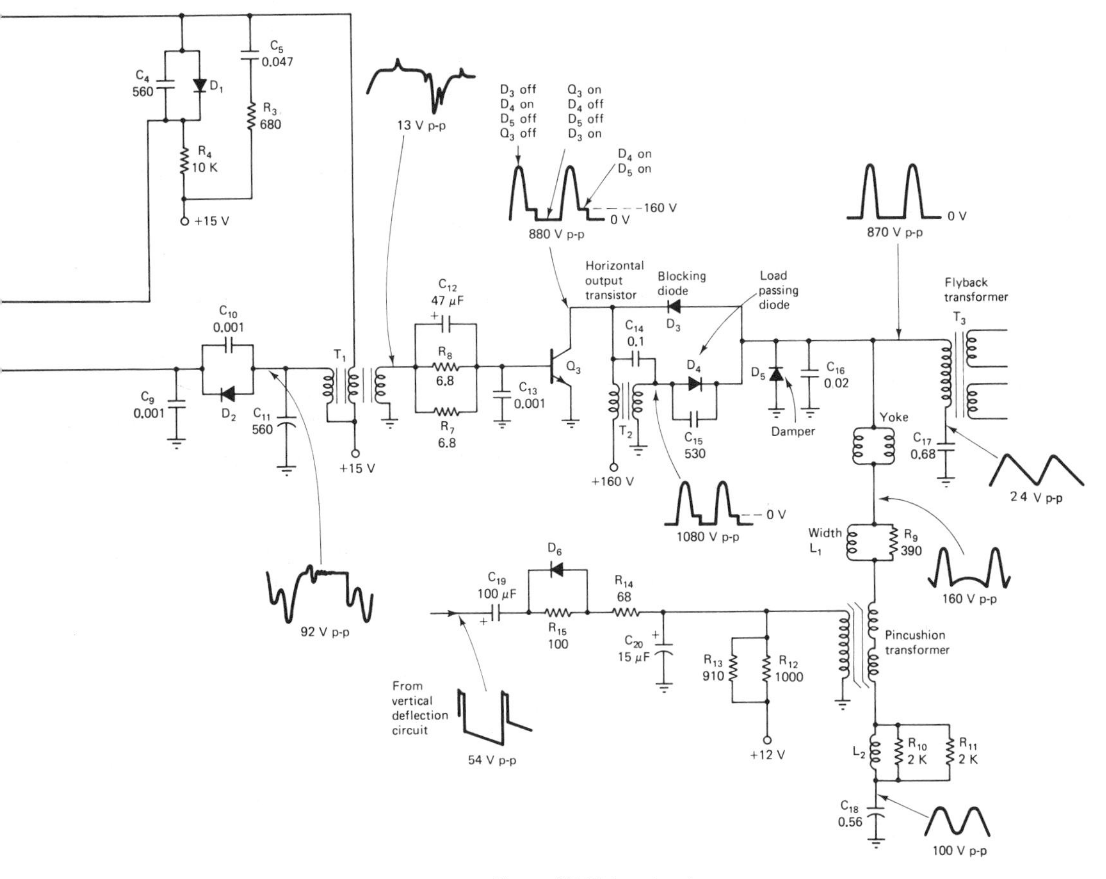

Figure 15-21 *(continued)*

Horizontal output amplifier. Examining the schematic diagram of the horizontal output amplifier found in Figure 15-21 will reveal that it is essentially the same circuit as that discussed in the conventional horizontal output circuit shown in Figure 15-7. Both use NPN transistors as horizontal output amplifiers (Q_3 and Q_1), and both have damper diodes (D_5 and D_1) that eliminate ringing immediately after retrace. Both circuits utilize resonating capacitors (C_{16} and C_4). In both circuits the yoke is driven directly and has an "S"-shaping capacitor in series with its (C_{18} and C_6). In addition, both circuits utilize the flyback transformer (T_3 and T_2) for purposes of developing ac high voltage, ac scan-derived low voltages, and pulses.

In the Zenith circuit a number of important changes have taken place. First, the input signal is obtained from the pulse width modulator and its pulse width is variable, depending upon high voltage and ac line voltage fluctuations. Second, the B supply for Q_3 is obtained via the primary of the regulation transformer (T_2). The flyback transformer (T_3) supplies the regulated high voltage and scan-derived B supply voltages, but in this system acts like a yoke with a large "S" capacitor (C_{17}) in series. Third, the Zenith circuit incorporates two diodes: "the load passing diode" (D_4) and "the blocking diode" (D_3). These diodes are not found in conventional horizontal output circuits.

Under normal conditions, the pulse width driving the horizontal output stage is such that it turns the transistor (Q_3) on approximately 18 μs after retrace time and remains on until the end of trace. Input pulse-width variations can cause "turn-on" time to vary from 5 to 30 μs. This turn-on-time variation provides system regulation.

During the time that the horizontal output transistor (Q_3) and the blocking diode D_3 are both on, the yoke current is increasing and moving the picture tube electron beam from the center of the screen to the right side. The source of this current is the discharge of the yoke S capacitor. The addition of diode D_3 between the horizontal output transistor collector and the damper cathode separates the collector current from the damper current. This minimizes horizontal linearity problems. When retrace begins, Q_3 is driven into cutoff by its input signal. The sudden stop of current through the primary of T_2 produces a large positive pulse on the collector of Q_3, in addition to the retrace pulse appearing across the flyback and yoke. During retrace time the blocking diode (D_3) is off and the load-passing diode (D_4) is turned on. The conduction of D_4 aids in charging the retrace capacitor (C_{16}) to a constant peak voltage level. In conventional horizontal circuits, during retrace time the sweep system restores lost power from the B+ power source. In this circuit, the load-passing diode (D_4) injects the required power needs during retrace time. D_4 continues to conduct until the pulse width modulator again turns on the horizontal output amplifier.

At the end of retrace, the picture tube electron beam is on the left side of the screen and damper D_5 is turned on. During this time D_4 continues to conduct. Damper conduction forces the beam back toward the center of the screen. At this time Q_3 is turned on and the cycle repeats itself.

Examination of Q_3 collector voltage waveform indicates that it has a "step" which does not appear in conventional horizontal output circuits. This step represents the amplitude of the B supply voltage that appears on the collector of Q_3 when it stops conducting. Variations in ac line voltage or horizontal output loading cause the duration of the step to vary. Under normal conditions, the step duration is approximately one-third of the duration of the waveform. When the circuit is "loaded" or a reduction in ac line voltage has occurred, the step duration is shortened. This means that the horizontal output transistor conducts longer, providing the energy needed to maintain a constant peak voltage across the damper diode (D_5).

From a troubleshooting point of view, this circuit will not operate at all if either D_3 or D_4 open or short. Defective diodes can cause the shut-down circuit to be activated, can cause motorboating, and can result in other components being damaged.

15-13 the SCR horizontal deflection circuit

In recent years RCA and a number of other television manufacturers have been using SCRs (silicon-controlled rectifiers) in the horizontal deflection circuits of their color television receivers. In those receivers a pair of SCRs is used in conjunction with two special diodes and a number of resonant circuits to form a system of switches that generates the sawtooth deflection current.

The circuit also generates the high-voltage pulses used to obtain the dc voltages necessary for picture tube operation. One advantage of the SCR deflection system is that it can be operated directly from the rectified ac line voltage. This is in contrast to transistor horizontal deflection circuits, which must be supplied by a transformer-operated low-voltage dc power supply.

In the SCR deflection circuit the yoke does not require a matching transformer. Consequently, the purpose of the flyback transformer is to develop the high-voltage pulses required by other circuits. SCR deflection circuits are similar to the older deflection circuits in that they provide such functions as pincushion correction, convergence, high-voltage regulation, and high-voltage hold-down.

Circuit operation. A simplified schematic diagram of an SCR horizontal deflection circuit and its associated waveforms are shown in Figure 15-22. The circuit consists of two silicon-controlled rectifiers, each of which is shunted by a special fast-recovery diode. In both cases, the anode of the SCR is tied to the cathode of the diode. This indicates that the diode and SCR conduct at different times. SCR_T and diode D_T form the *trace* switch and shunt the yoke and its series-connected capacitor C_Y. The trace switch is responsible for almost all the horizontal trace portion of the raster. Retrace is controlled by the conduction of SCR_C and D_C, which is called the *commutating switch.*

The gate voltage for SCR_T is obtained during the action of the circuit from a transformer T_1. This voltage is coupled to the gate by C_G, L_G, and R_G, which provide the required time delay and wave shaping. The gate voltage for SCR_C is a pulse obtained from the horizontal deflection oscillator and it corresponds to a time just before retrace.

Resonant circuits. The coils and capacitors that make up the SCR deflection circuit form resonant circuits that are alternately allowed to oscillate and then are turned off by the switching action of the SCRs and the diodes. The resonant circuit responsible for trace is formed by the yoke and C_Y. This circuit resonates at approximately 10 kHz. Another resonant circuit consisting of the yoke, C_Y, C_R, C_A, and L_R determines retrace time. This circuit is resonant at approximately 35.5 kHz. In addition, C_A and L_R form a third resonant circuit that rings at approximately 100 kHz during retrace. Furthermore, the flyback transformer (T_2) and the primary winding L_f and C_f form a resonant circuit that oscillates during retrace at a frequency which is the third harmonic of the retrace frequency of oscillation. This is done to lower the peak voltages across the SCRs and to increase the amplitude of the high-voltage pulse delivered to the rectifiers. A fifth circuit formed by the linearity circuit L_L, C_L, and C_Y forms another resonant circuit that rings at the second harmonic of the scan frequency. This current is added to the trace portion of the deflection sawtooth flowing through the yoke. The coil (L_L) is variable so that the timing of this oscillation may be changed, thus modifying the linearity of the deflection sawtooth.

High-voltage generation and regulation. High voltage is obtained in a manner similar to that used in other horizontal deflection systems. The horizontal retrace pulse developed across D_T (Figure 15-22) is stepped up by means of a high-voltage transformer T_2. This voltage is then rectified by a solid-state multiplier to provide the 26 kV necessary for picture tube

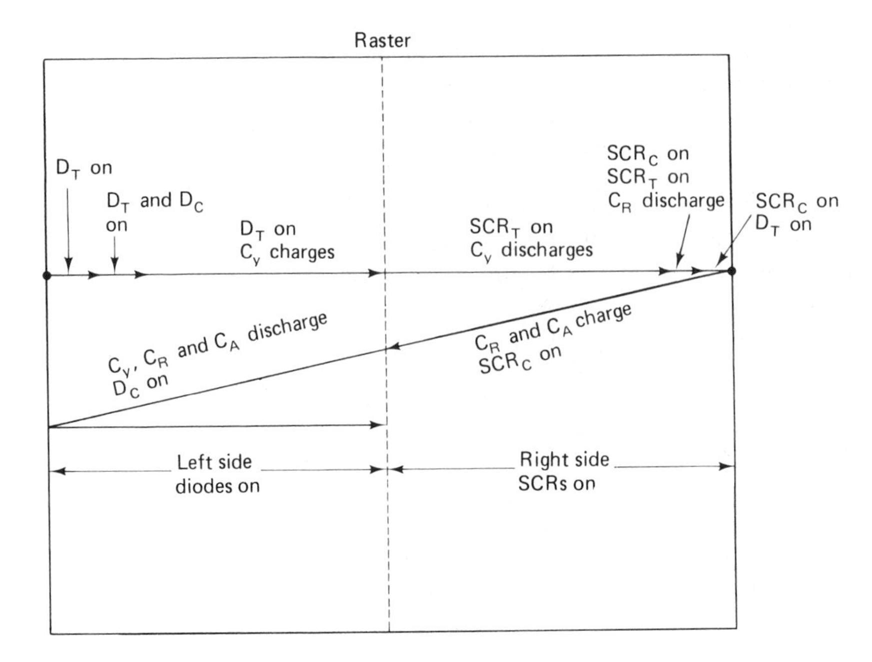

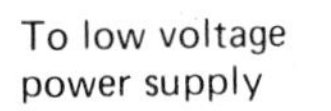

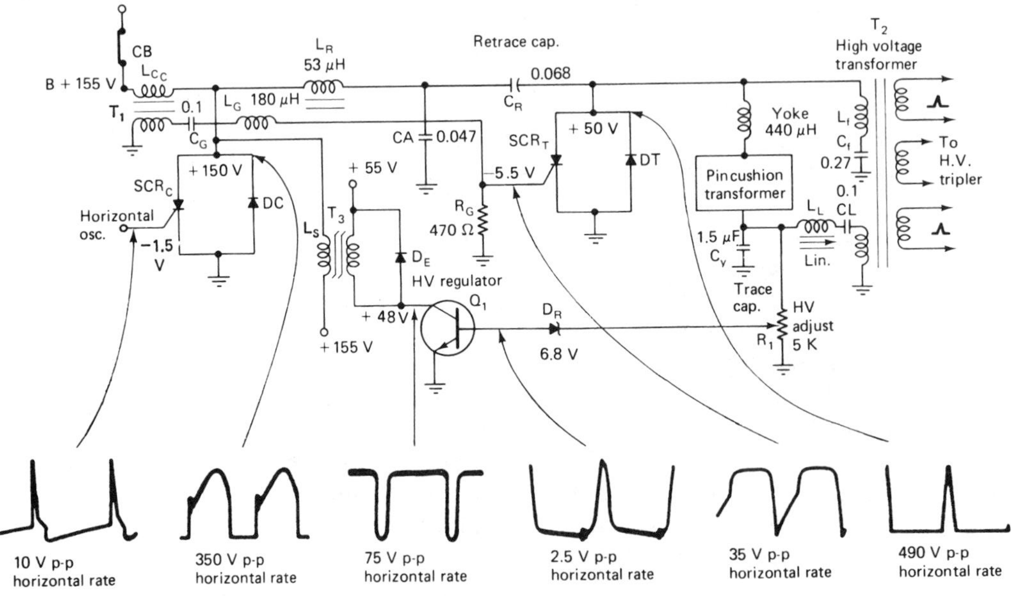

Figure 15-22 Schematic diagram of an SCR deflection circuit and its associated waveforms. The diagram of the raster shows one horizontal line of trace and retrace indicating those SCRs and diodes that are conducting at the designated position.

operation. Other windings on the transformer provide pulses for the focus and screen voltages as well as other functions of the receiver. The inductance (L_f) of the high-voltage transformer is in series with a capacitor C_f, forming a resonant circuit that is tuned to the third harmonic of the retrace frequency. When properly adjusted, third harmonic tuning will increase the peak of the high voltage delivered to the rectifier and will reduce the peak voltage that appears across the trace switch to zero when it opens and again when it closes.

The high voltage is regulated by controlling the energy that is made available to the horizontal output trace circuit. This energy is in the form of the charge stored in C_R and C_A during trace. These capacitors are charged through L_{CC} and L_S and form a resonant circuit whose period is twice the horizontal scan interval. Control of high voltage is accomplished by varying the resonant frequency of this circuit.

The regulator circuit shown in Figure 15-22 consists of Q_1, T_3, and zener diode D_R. This circuit is essentially a scan regulator that keeps picture size constant regardless of power supply voltage variations or beam current loading. In this system the commutating circuit is tuned so that a change in high voltage results in a proportional change in yoke current. The change in current causes a voltage drop to appear across R_1, the high-voltage control, which in turn affects the conduction of Q_1, and saturable reactor T_3. This, in turn, controls the amount of charge on C_R and C_A at time T_0. The resonant frequency of the tuned circuit formed by C_R, C_A, and the parellel combination of L_{CC} and L_S is made variable by changing the dc current flow through the primary of the saturable reactor T_3.

If there is an increase in high voltage resulting from a decrease in beam current or some other cause, the following circuit action takes place. The increase in high voltage causes the voltage across C_Y to increase, which in turn forces an increase in the voltage applied to the reference zener diode (D_R). As a result, the forward bias of the regulator transistor (Q_1) increases, causing an increase in dc collector current flowing through the primary of the saturable reactor transformer T_3. The inductance of the parallel combination of L_{CC} and L_S is then decreased, causing the resonant frequency to increase. This causes the voltage on C_R to decrease, which reduces the energy made available to the output circuit. Therefore, the high voltage is reduced to its former level. The reverse action would occur if the high voltage tended to decrease. In general, the high-voltage regulator is designed to maintain the high voltage constant for line voltage variations ranging from 105 to 130 V ac. In addition, an increase in picture tube beam current from 0 mA to 1.5 mA will restrict the high-voltage change to a range extending from 26.5 to 25 kV.

Troubleshooting the SCR deflection circuit. Defects in the SCR deflection circuit can be classified into four basic categories: (1) B+ overload circuit breaker trips and the set goes dead; (2) circuit breaker does not trip and there is no high voltage; (3) linearity problems; and (4) interference problems.

When the B+ overload circuit breaker trips and the set goes dead, this trouble symptom can be caused by defects in the horizontal deflection circuit or shorts elsewhere in the receiver. The circuit breaker is located in series with one leg of the ac power line and may be tripped by any cause of excessive line current. To isolate whether the trouble is in the horizontal deflection circuit or in some other circuit, the horizontal output stage may be removed from the rest of the receiver. This may be done by disconnecting the yoke and convergence plugs which contain interconnecting jumpers that supply power to T_1 and T_3. If the circuit breaker continues to trip, the short is not in the horizontal circuit.

The key to troubleshooting the SCR deflection circuit is the fact that capacitor C_R divides the circuit into two parts: the trace switch (D_T and SCR_T) and the commutating switch (D_C and SCR_C). If a dc short occurs in the trace switch or associated components, the circuit breaker will not trip because the capacitor (C_R) isolates the circuit from B+. Shorts in the commutating switch and its associated components can cause the circuit breaker to trip because these components return directly to the B supply. These shorts can easily be located by checking the circuits with an ohmmeter.

Severe ac shorts can also cause repeated power supply overload. To determine whether or not the short circuit is caused by an ac overload, place a jumper across SCR_T. This prevents any ac current from flowing through the yoke or flyback transformer. If the B supply is still overloaded, the defect must be caused by a dc short. If the short removes the overload, look for shorts in such components as the flyback transformer and its loads, focus rectifier diode, shorted screen rectifier diodes, and so on. In addition to circuit defects, improper SCR_C trigger wave shape or frequency can cause the same defects as a severe ac short. This can be easily checked with an oscilloscope and should be one of the first checks made.

If the symptoms are no high voltage and if the circuit breaker does not trip, suspect the following components: the trace diode D_T, the trace SCR, the flyback transformer, C_Y the yoke return capacitor, the hold-down diode, the high-voltage rectifier, and the horizontal oscillator module that feeds the gate of SCR_C.

Linearity problems are generally associated with defects in the high-voltage regulator circuit. Defects in the saturable reactor (T_3) or its associated components can cause raster piecrusting.

Interferences similar to Barkhausen interference and various types of foldovers may have a number of different causes. A summary of these interferences, their locations on the raster, and their causes are shown in Figure 15-23. Foldover on the left side of the screen may be caused by an open retrace diode (D_C). Foldover in the center may be caused by an open trace diode.

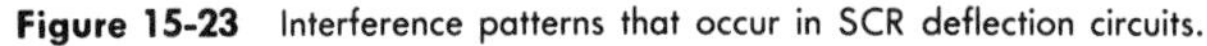

Figure 15-23 Interference patterns that occur in SCR deflection circuits.

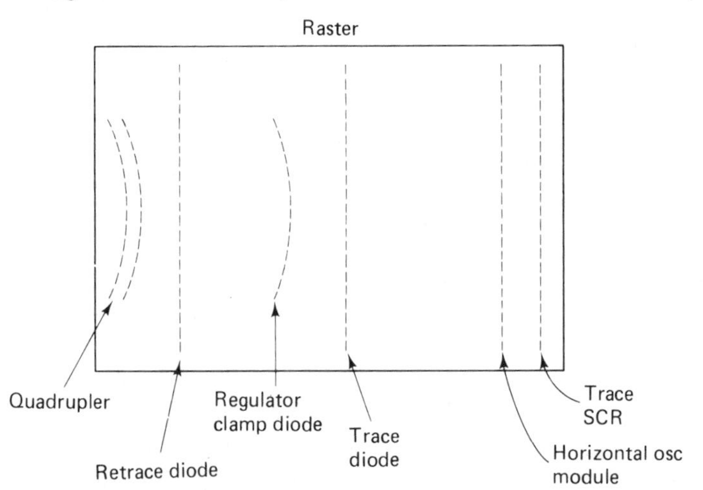

An open yoke or flyback transformer can cause excessively high voltages to appear across the trace SCR or diode and damage them. It is good practice when replacing SCR_T or D_T to check the yoke and the flyback for opens.

The yoke and high-voltage sections of the receiver require approximately 100 W of power to be properly activated. This power level requires a peak current of approximately 14 A to flow through C_R. If this capacitor has any appreciable power loss, it will overheat. Ordinary capacitors cannot be used to replace C_R or the other capacitors in the circuit because they are not designed to have small enough losses. Exact replacement capacitors must be obtained from the manufacturer if the circuit is to have a reasonable life expectancy after the capacitor has been replaced.

15-14 the vacuum tube horizontal output amplifier

The schematic diagram of a horizontal output amplifier and its associated circuits commonly found in color receivers is shown in Figure 15-24. An examination of the circuits will show that basically this circuit consists of five major components: the horizontal output amplifier V_1, the damper diode V_2, the high-voltage rectifier V_3, the horizontal output transformer T_1 (often called the *flyback transformer*), and the yoke Y_1. The horizontal output amplifier operating Class B is designed to act as a switch that supplies power to the flyback transformer and drives a sawtooth current through the yoke.

The flyback transformer is an autotransformer wound on a rectangular ferrite core and has three essential functions: (1) it acts as a step-down matching transformer between the horizontal output amplifier and the yoke; (2) it acts as a step-up high-voltage transformer that drives the high-voltage rectifier; and (3) in receivers that use vacuum tube rectifiers the flyback transformer will have two or three turns of high-voltage insulated wire wrapped around the transformer core. These turns act as a step-down winding that provides the 1 to 3 V required by the heater of the high-voltage rectifier.

The damper V_2 acts like a switch that suppresses shock-excited oscillations that occur immediately after retrace. It is also responsible for horizontal scanning on the left side of the picture tube screen. The horizontal output amplifier grid is driven by a trapezoidal voltage of approximately 250 V p-p that is obtained from the horizontal deflection oscillator. The input signal is then rectified by grid to cathode conduction and C_1 and R_1 develop a negative grid leak bias of approximately -80 V. This bias sets the operating point of the horizontal output amplifier so that it operates Class B. As a result, the tube acts like a switch, conducting for approximately one-half of the input signal and turned off for the other half. The path of plate current flow may be traced from the cathode of the horizontal output stage through the windings of the flyback transformer to the damper and then to the positive terminal of the power supply.

If for any reason the horizontal oscillator were to cease functioning, horizontal drive would be lost and the horizontal output amplifier's bias would drop to zero. This in turn would cause excessive plate current to flow and would cause the plate of the horizontal output tube and sometimes the damper to glow red. Such an increase in plate current may cause damage to both of the tubes and the flyback transformer. To protect the circuit, many receivers include a fuse ($^1/_4$ or $^3/_8$ A, 250 V) in series with the plate of the damper.

Parasitic suppressor resistors (R_5) of 100 Ω are often inserted in series with the control and screen grids of the horizontal output amplifier. These resistors prevent unde-

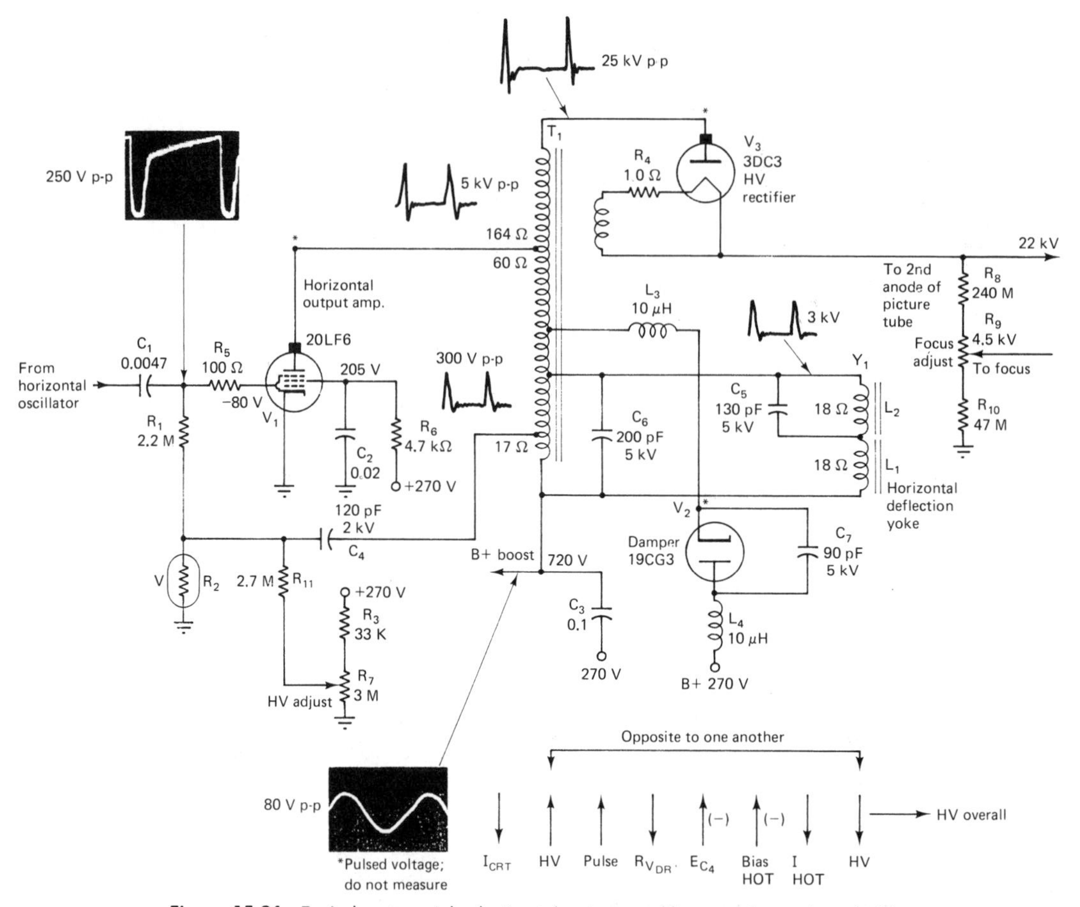

Figure 15-24 Typical vacuum tube horizontal output amplifier used in a color television receiver. This circuit uses VDR high-voltage regulation.The arrow diagram shows the operation of the VDR regulator.

sirable oscillations from being generated by counteracting any negative resistance introduced by the tube and by damping any tuned circuits that may be formed by distributed *LC* circuit elements.

The screen grid of the horizontal output amplifier is made 205 V positive by returning it to the B supply.

The screen bypass capacitor C_2 acts to prevent any signal voltage from being developed across R_6. If C_2 were to open, negative feedback would be introduced and tend to reduce picture width.

A shorted screen capacitor will cause R_6 to burn up and disable the horizontal output stage. This will produce a no-raster and sound normal condition.

B+ boost. By means of a cleverly arranged circuit it is possible to convert some of the ringing energy into a high-voltage B supply called B+ *boost.* Converting wasted energy in this way makes the circuit more efficient. For this reason, B+ boost arrangements are used in both color and black & white television receivers.

The B+ boost voltage developed in the horizontal output circuit is used as a power source for such circuits as the following:

1. First anode voltage or screen voltages in the picture tube
2. Power supply for the vertical deflection output amplifier stage
3. Sound detector plate supply voltage
4. Horizontal oscillator plate supply voltage
5. Plate supply voltage of the horizontal output amplifier

High-voltage focus supplies. Good raster focus depends upon correct picture tube electrode voltages. Focus will deteriorate if the voltage on the focus electrode and the picture tube second anode depart significantly from normal.

In Figure 15-24 the focus voltage is obtained from an adjustable voltage divider placed across the second anode high-voltage supply. This takes care of high-voltage focus voltage tracking in a very simple way; the focus voltage is automatically a fixed percentage of the high voltage.

High-voltage rectifiers. Both color and black & white picture tube operation requires that the second anode of the tube be energized with dc voltages that may be as high as 30,000 V. To obtain this voltage, the ac high-voltage retrace pulse developed by the flyback transformer is converted into high-voltage dc by a half-wave rectifier diode, as shown in Figure 15-24.

The VDR high-voltage regulator. A system of high-voltage regulation that operates at low voltages and therefore is not a source of x-rays is the voltage-dependent resistor (VDR) high-voltage regulator. A schematic diagram of this circuit is shown in Figure 15-24. In this system the grid bias of the horizontal output stage is the "handle" that controls the high voltage. An increase in bias lowers the plate current flow. This reduces the energy available for the development of the high voltage and forces a reduction in high voltage. Similarly, a decrease in bias causes an increase in high voltage.

In this circuit the grid bias of the horizontal output tube (HOT) is obtained from two sources. First, the input signal itself develops grid leak bias. The coupling capacitor C_1 and grid resistors R_1 and R_2 are used for this purpose. Second, a tap on the flyback transformer develops a positive pulse, which is coupled by C_4 to VDR R_2. The *VDR* rectifies the pulse, charging C_2 to the peak value of the pulse. The amplitude of the pulse and the dc voltage developed across C_2 are directly proportional to the high voltage. When C_4 discharges, it develops an additional voltage across R_2 that adds to the original grid leak bias of the horizontal output tube.

A VDR is a nonlinear resistor whose resistance decreases as the voltage across the resistor increases, and therefore acts like a diode in the presence of a pulse. This can be explained as follows. If a small sine-wave voltage is impressed across the VDR, a distorted current sine wave of symmetrical wave shape will flow. In this case, no rectification takes place. If a pulse is applied to the VDR, the negative portion of the pulse will appear in the high-resistance region of the VDR's characteristic and the positive pulse will appear in the relatively low resistance region. Thus, rectification takes place because current flows more easily in one direction than in the other direction.

The regulator action of the VDR grid bias regulator is summarized by the arrow diagram in Figure 15-24. In this diagram it is assumed that brightness has decreased and that the picture beam current has decreased. This tends to cause an increase in dc high voltage as well as an increase in the peak pulse amplitude fed to the VDR. As a result, the resistance of the VDR decreases, allowing the coupling capacitor C_4 to charge to a greater voltage. The dc voltage developed across C_4 is added to the HOT grid bias, increasing the bias. This in turn causes the HOT plate current to decrease, forcing the high voltage to decrease, resulting in an overall constant high voltage.

QUESTIONS

1. In what ways does the horizontal output stage of a color receiver differ from that of a black & white receiver?
2. Explain the function and operation of the damper.
3. What class of operation is the horizontal output transistor operated in and what kind of bias is used to set the operating point?
4. What are two causes of a red-hot horizontal output tube plate?
5. Discuss the different width controls used in transistor receivers.
6. What kind of capacitor is the resonating capacitor, (C_3 in Figure 15-7)? What would occur if it opened? What would occur if it is replaced with an ordinary capacitor?
7. How can yokes and flyback transformers be tested?
8. What is the purpose of a yoke balancing capacitor?
9. What circuit defects might cause the loss of high voltage?
10. What is blooming? What causes it?
11. What effect does the loss of horizontal drive have in a vacuum tube and in a transistor horizontal output amplifier?
12. Why is high-voltage regulation important in color receivers?
13. What is the purpose of high-voltage hold-down circuits?
14. How does a PNP horizontal deflection amplifier differ from one that uses an NPN transistor?
15. In what ways does the horizontal output transformer in a transistor horizontal output stage differ from those used in a vacuum tube circuit?
16. What is the purpose of the third harmonic tuning of the horizontal output transformer?
17. What is a saturable reactor, and how does it work?
18. How many tuned circuits are important to the operation of the SCR deflection circuit? Identify each one.
19. In Figure 15-21 what would be the result of D_T shorting? Of D_C shorting? Of C_R opening? C_Y opening? Of the gate voltage of SCR_C being lost? Of SCR_T opening and SCR_C opening?
20. Explain the proper approach to servicing an SCR deflection circuit in which the low-voltage circuit breaker trips each time the set is turned on.
21. Referring to Figure 15-3, assume that C_4 became shorted. What would be the trouble symptoms? How would you go about troubleshooting this circuit?
22. What would be the effect on circuit operation (Figure 15-3) if R_4 opened?
23. In terms of circuit operation, what does the horizontal retrace pulse represent?
24. What is the purpose of the ferrite beads shown in Figure 15-3?
25. What are the characteristics of a gate-controlled switch (GCS)?
26. What will occur to a GCS if the horizontal drive is lost?
27. What is a split high-voltage transformer, and where is it used?

28. What kind of high-voltage regulation do most black & white solid-state receivers use?
29. List the important characteristics of the Zenith high-voltage regulator.
30. What is the purpose of the pulse width modulator, the blocking diode, the load passing diode, and the forward and reverse drivers?
31. Under what conditions does a VDR act like a diode?
32. Referring to Figure 15-24, assume that L_3 opened. Explain why the plate voltage of V_1 would become slightly negative (-1 V).
33. What is the effect on transistor horizontal output circuit operation if the damper diode (D_1 in Figure 15-7) opened? Explain. What if it shorted?

chapter 16

The Bandpass Amplifier and Associated Circuits

16-1 introduction

Almost all of the circuits discussed in previous sections of this book have been those which can be used in either black & white or color television receivers. This chapter and the remaining chapters are devoted to those circuits, found only in color receivers.

Figure 16-1 is the block diagram of a typical color receiver. The shaded areas indicate those sections of the receiver that will be discussed in this chapter. Four blocks are shaded, two stages of chroma amplification, the color killer, and the automatic color control (ACC) circuit. Chroma amplifiers are often called "bandpass amplifiers" (BPAs).

A more detailed block diagram of this portion of the receiver indicating all of the circuits and their associated controls is shown in Figure 16-2. Examination of this diagram and its waveshapes indicates that the input to the bandpass amplifier is a composite video signal obtained from the video amplifier stage. The composite video signal consists of as many as five important components. These are (1) the luminance or *Y* signal corresponding to the black & white picture content; (2) the chrominance signal representing the color content of the picture; (3) the color burst; (4) the horizontal sync, vertical sync, and blanking pulses; and (5) a dc component corresponding to the average scene brightness of the picture. When the composite video signal passes through the bandpass amplifier only the chrominance signal appears at the output of the bandpass amplifier. In some receivers the color burst is prevented from appearing at the output by a horizontal blanking pulse that disables the bandpass amplifier during the horizontal blanking interval. The output waveform shown here is produced by a color bar generator.

On the basis of the input and output signals of the bandpass amplifier it is clear that the purpose of this stage is to act as a filter to separate the chrominance signal from the rest of the composite video signal. The bandpass amplifier is, in fact, a tuned amplifier

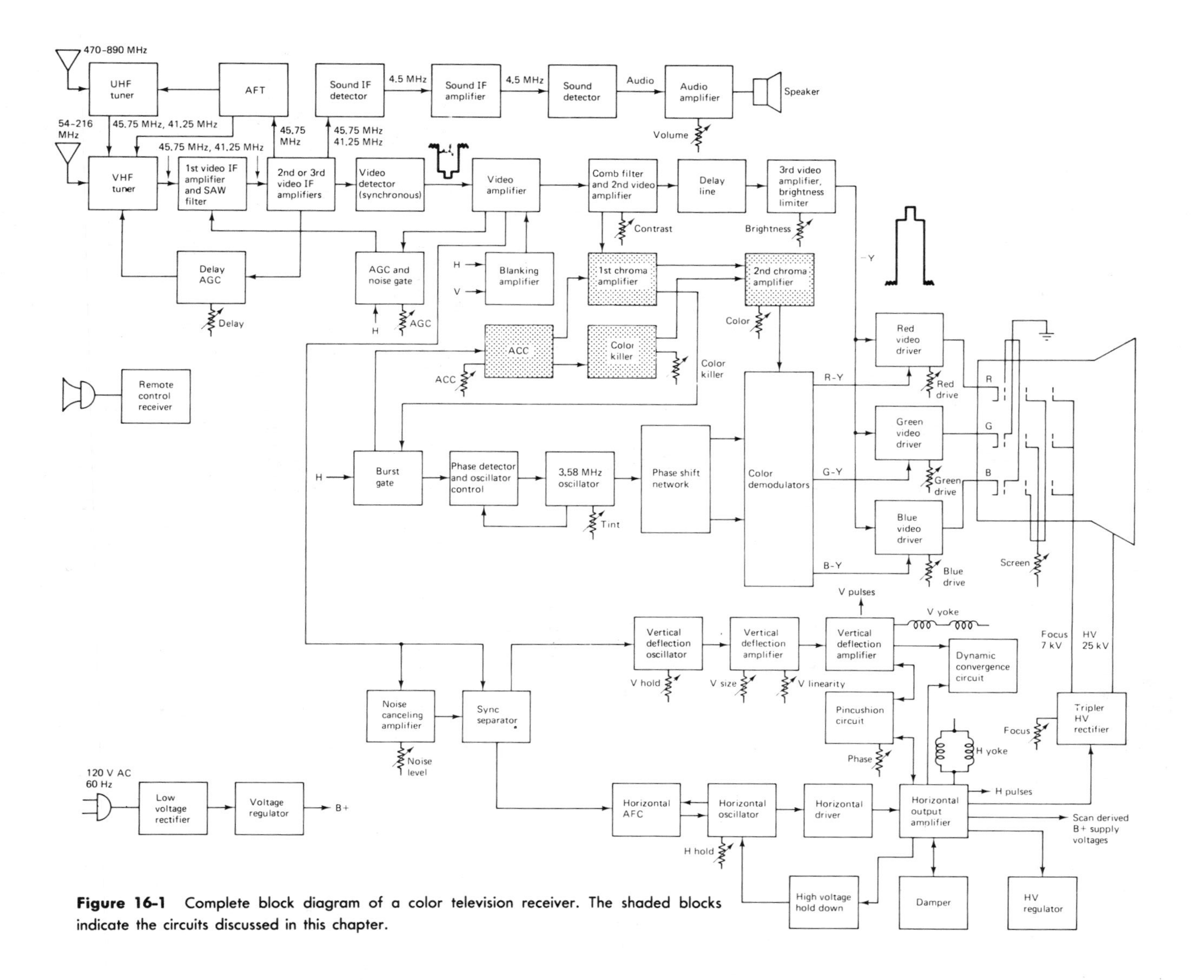

Figure 16-1 Complete block diagram of a color television receiver. The shaded blocks indicate the circuits discussed in this chapter.

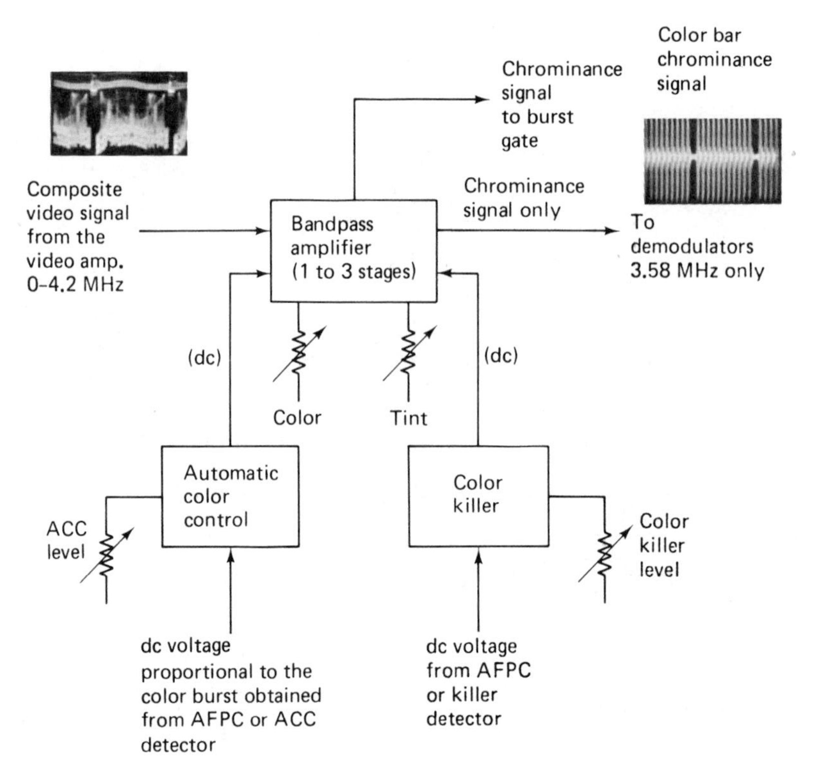

Figure 16-2 Block diagram of the bandpass amplifier and its associated circuits.

that is very similar to a video IF amplifier and is often called the *color IF amplifier.* The only real difference between the two is that the video IF amplifier operates at a frequency centered at approximately 44 MHz and the bandpass amplifier operates at a frequency of 3.58 MHz.

Controls. The amplitude of the chrominance signal corresponds to *color saturation.* Therefore, it is not suprising to find that the *color control* is associated with the bandpass amplifier. This control determines the amplitude of the chrominance signal and therefore the color intensity of the picture.

The phase of the chrominance signal corresponds to picture *color.* In some receivers a *tint control* that adjusts picture color is located in the bandpass amplifier. This control changes the phase of the chrominance signal.

The color killer. The color killer is a circuit that is used to disable the color sections of the receiver during black & white reception. Disabling the color sections of the receiver is necessary to prevent spurious chrominance signals that can pass through the bandpass amplifier from reaching the picture tube and causing undesired coloration of black & white pictures. The block diagram of Figure 16-2 shows that a *color killer level control* is associated with the circuit. This control determines the input level that activates the color killer. In most cases, both the input and output voltages of the color killer are dc levels; no ac signals are required for its operation.

ACC circuits. Automatic color control circuits are identical in purpose and design to AGC circuits used in IF amplifier circuits. The basic function of the ACC circuit is to keep the output level of the bandpass amplifier constant under varying input signal conditions. The effectiveness of this circuit can be adjusted by the *ACC level control.*

As shown in Figure 16-2, the bandpass amplifier has two outputs: One output feeds the demodulators and the other output feeds the burst gate. In some receivers, however, the burst gate is fed by the video amplifier stage.

16-2 the color bar generator

The composite video signal appears at the output of the video detector after demodulation. This signal consists of the luminance (*Y*) signal, the chrominance signal, the color burst, sync pulses, and blanking pulses. In practice, the amplitude of the composite video signal is constantly varying due to picture content. From a troubleshooting point of view, this waveform is not too useful because it cannot be used as a standard of comparison. The purpose of a color bar generator is to act as a substitute transmitter. As such, it supplies to the receiver a known, nonvarying color pattern that can be used for adjustment and troubleshooting purposes.

Types of color bar generators. Currently, two types of color bar generators are in use: the NTSC (National Television Systems Committee) generator and the gated rainbow color bar generator. Figure 16-3(a) and (b) shows the color bar pattern that these two generators produce on the picture tube screen. Accompanying each pattern is the composite video signal for one horizontal line that is necessary to produce these patterns.

NTSC generator. An examination of the seven-bar NTSC color bar pattern and its composite video signal shows that this signal contains a *Y*-signal component. The *Y* signal corresponds to the shades of gray that would be produced in a black & white receiver by the seven bars of color that constitute the pattern. The presence of the *Y* signal is indicated in the composite video signal by the fact that the color bursts are shifted vertically by different amounts.

Each of the seven color bars corresponds to a burst of 3.58 MHz. The bursts differ from one another in phase angle and amplitude. The amplitude of the burst represents the saturation of the color bar and the phase corresponds to the color. The NTSC color bar pattern consists of the three primary colors (red, green, and blue), their complementary colors (cyan, magenta, and yellow) at 100% saturation, and a white bar.

The reference color phase is established by the color burst located on the back porch of the horizontal blanking pulse.

The gated rainbow color bar generator. The gated rainbow color bar generator is the most commonly used color bar generator for color TV service. It produces a color bar pattern consisting of 10 color bars ranging in color from shades of red on the left side, through blue in the center, to green on the far right side. In Figure 16-3(b) each color in the color bar pattern is identified and is lined up with its associated modulating waveform. Notice

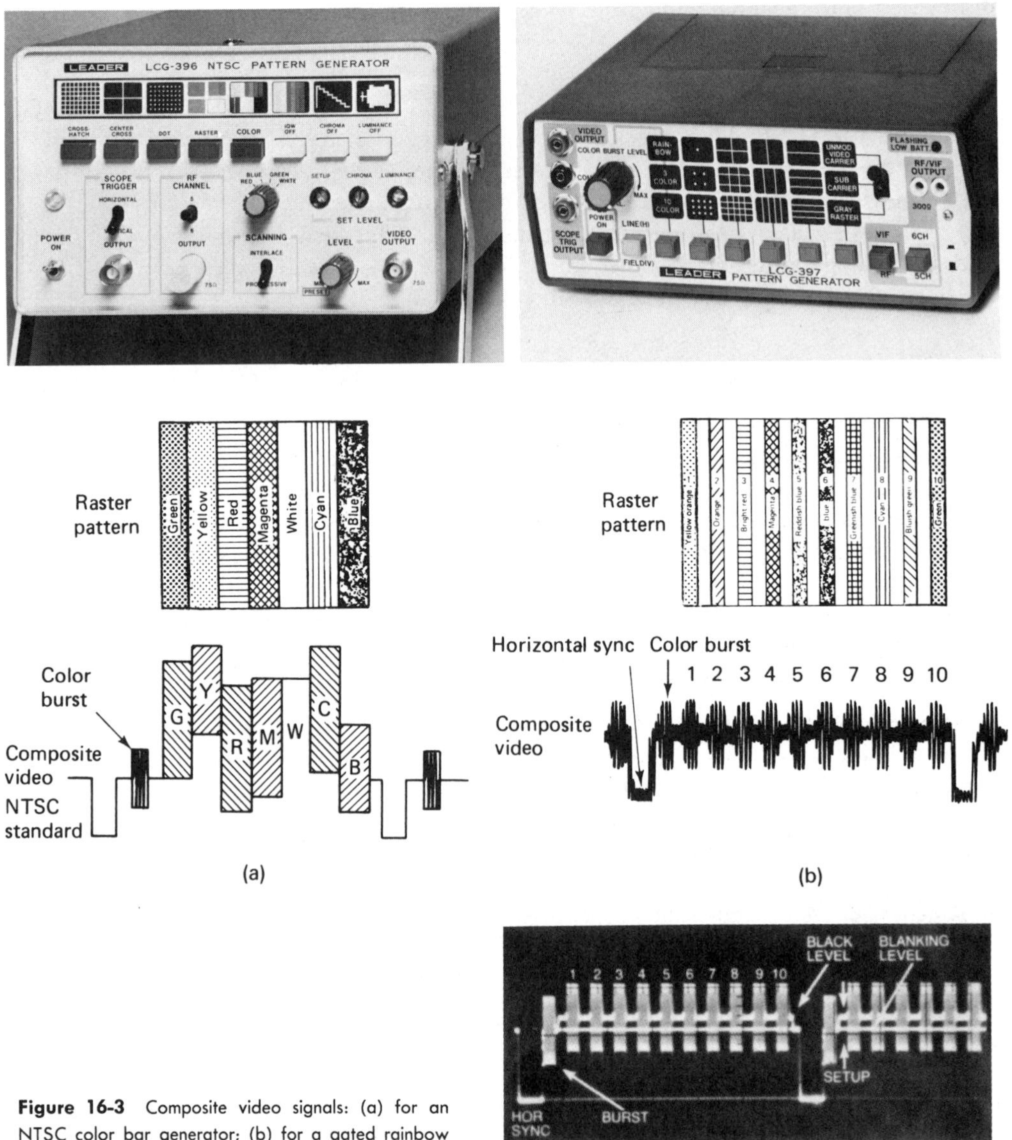

Figure 16-3 Composite video signals: (a) for an NTSC color bar generator; (b) for a gated rainbow color bar generator; (c) at the output of the video detector. (Photos courtesy of Leader Instrument Corp.)

that the composite video signal responsible for this pattern consists of a horizontal sync pulse and 11 equal amplitude bursts of 3.58 MHz. The burst to the right of the horizontal sync pulse is the color burst. The other bursts differ in phase from one another and correspond to different colors.

The modulating signal used in the rainbow color bar generator does not contain a *Y* signal. This means that the color bar pattern will exhibit a full range of colors of equal saturation and brightness. The brightness of the display depends upon the gray scale settings (bias) of the picture tube in the receiver.

Color bar generators often have *chroma controls* that can vary the amplitude of all eleven 3.58-MHz bursts. Increasing their amplitude increases the color saturation and makes the bars appear more intense. Increasing the color control output also increases the reference color burst amplitude and may be used to check the effectiveness of the color sync section of the receiver.

The color bar pattern. The basic principle of color bar generator operation is based on two facts. First, two signals of different frequency have a phase difference that is constantly changing. For example, if two sine waves differ in frequency by the horizontal scanning rate (15,750 Hz), their relative phase will change by 360° per horizontal line. Second, it is the phase of the chrominance signal that determines the color seen on the picture tube screen.

To see how these principles are put into operation, refer to the simplified block diagram of a color bar generator in Figure 16-4. Notice that the color bar generator produces an amplitude modulated RF signal whose frequency of operation may either be channel 3 or 4. This option is given so that the service technician can select a channel that does not have a local transmission that might cause interference.

The modulating signal consists of horizontal and vertical sync and blanking pulses combined with a 3.563811-MHz sine-wave subcarrier. A *Y* signal is not included in the modulating signal. The frequency of the color bar generator's modulating subcarrier

Figure 16-4 Simplified block diagram of a rainbow color bar generator.

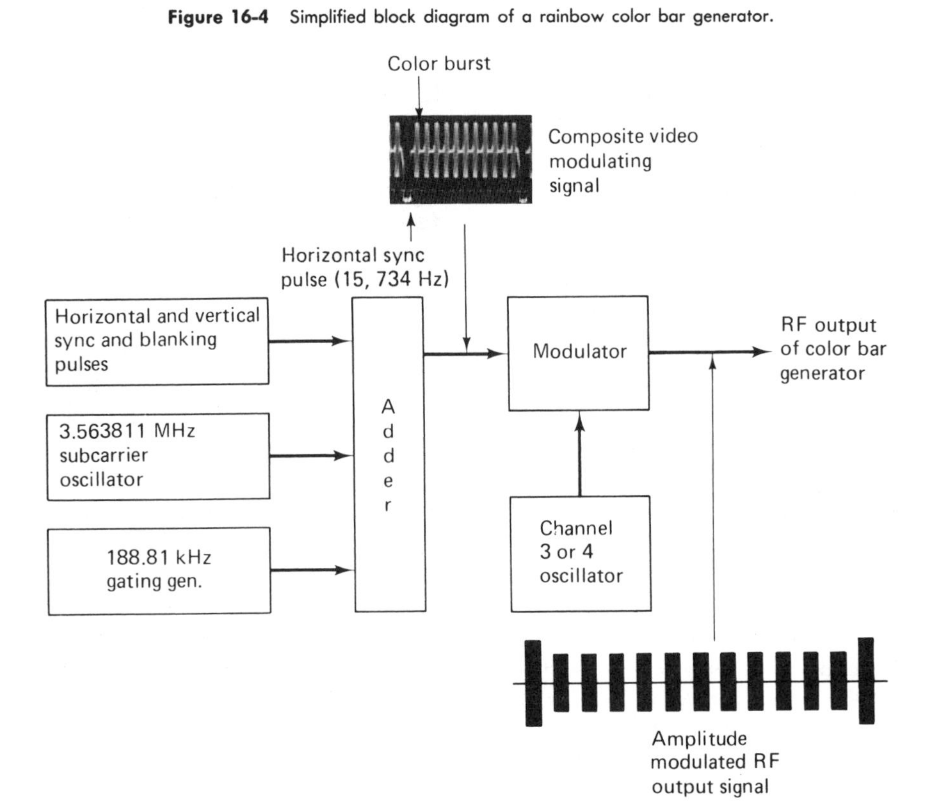

Figure 16-5 Normal color bar pattern. See color plate.

(3.563811 MHz) is exactly 15,734 Hz lower than the 3.579545-MHz subcarrier frequency generated in the receiver.

The modulated signal is gated on and off by a 188.81-kHz square wave. Since 188.81 kHz is the twelfth harmonic of the horizontal scanning frequency (15,734.26 Hz), the color bar generator modulating signal will be interrupted 12 times during one horizontal line. One of the pulses corresponds to the horizontal sync pulse. This pulse, plus the remaining 11 bursts of 3.563811 MHz, divides the 360° of chrominance phase change into 11 clearly defined 30° intervals. Each burst is 15° wide, as is each space between bursts; thus a 30° difference exists between the center of two adjacent bursts.

The first burst following the sync pulse is the color burst corresponding to 0°. The sync pulse and the color burst occur during horizontal retrace and are not seen on the picture tube display. The remaining bursts produce the gated color bar pattern shown in Figure 16-5. Each bar may be identified by phase angle and color. The phase angle of the color bars also corresponds to color signals, such as the $R - Y$, $B - Y$, $G - Y$, I, and Q. Identification of each bar is shown in Figure 16-6. Notice that the color bar pattern begins on the left edge at 30° and terminates on the right edge at 315°. Phase angles from 0° to 15° corresponding to the color burst and colors such as yellow (13°) are blanked out during horizontal retrace and do not appear on the picture tube screen. The horizontal sync pulse occupies the phase interval corresponding to 330 to 345°.

Receiver operation with the color bar generator. The modulated RF output of the rainbow color bar generator is fed into the antenna terminals of the receiver. The signal is then processed by the tuner, video IF amplifiers, video detector, video amplifier, and bandpass amplifier of the color receiver before it is fed into color demodulators. See Figure 16-7. If the alignment of all of these stages is proper, the waveform feeding the color demodulators will be essentially the same as that which modulated the RF carrier in the color bar generator. The main difference is that the horizontal sync pulse (15,734 Hz) has been removed by the 3.58-MHz tuned circuits in the bandpass amplifier.

The input to each demodulator consists of the gated rainbow chrominance signal

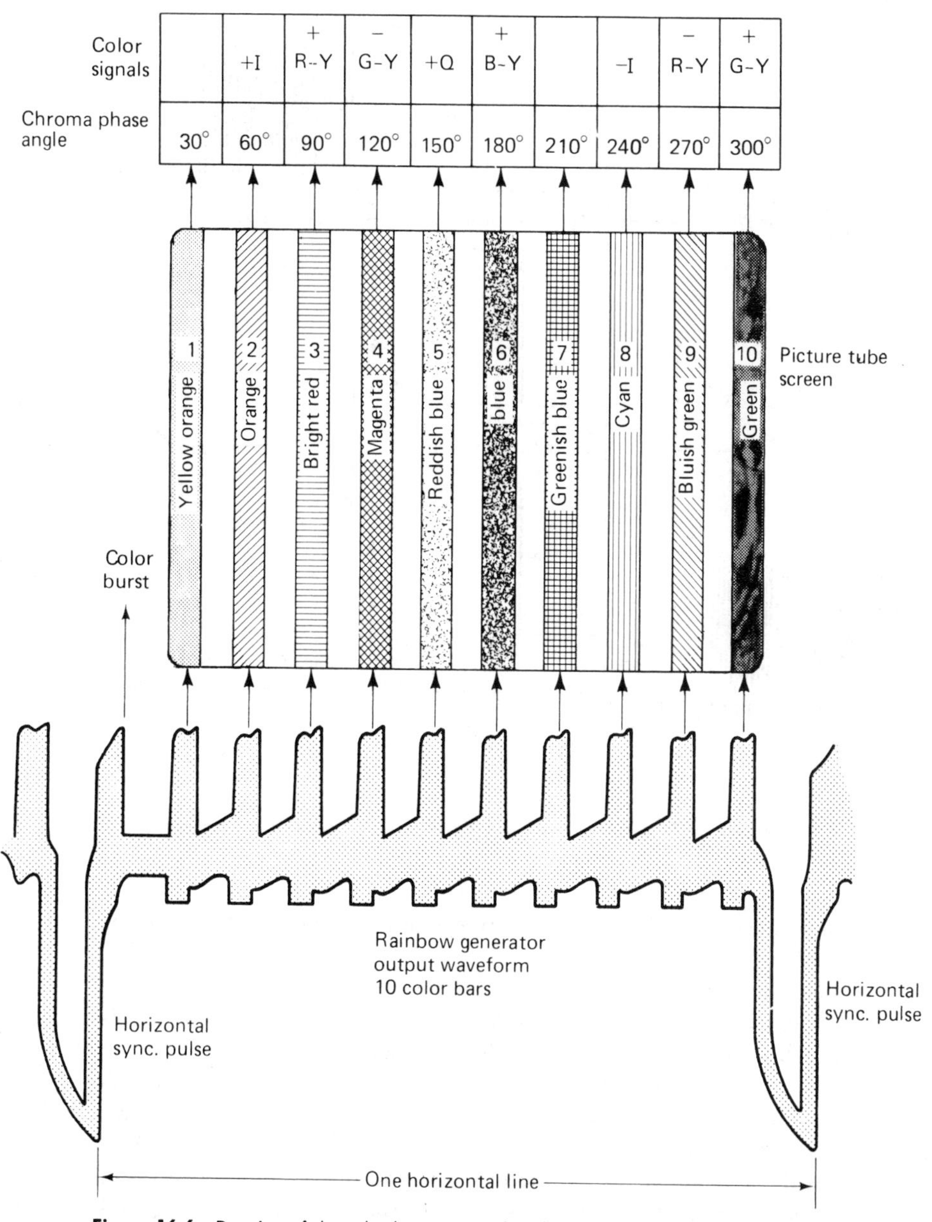

Figure 16-6 Drawing of the color bar pattern plus phase, color, and signal identification of each bar.

(3.563811 MHz) and a locally generated constant amplitude subcarrier oscillator signal (3.579545 MHz). The subcarrier oscillator signals fed to each demodulator differ from one another by 90°. Consequently, the outputs of the mixer-like demodulators will be two sine waves corresponding to $R - Y$ and $B - Y$ that differ in phase by 90°. These sine waves

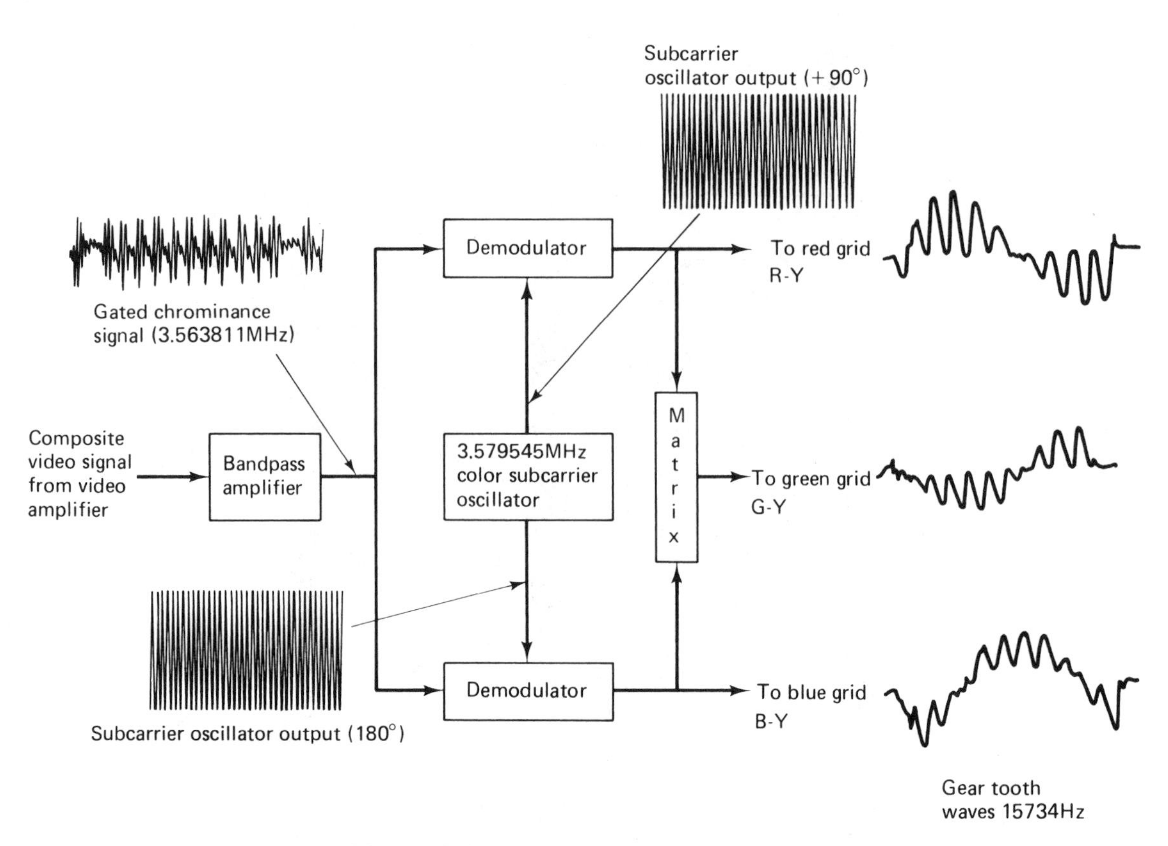

Figure 16-7 Simplified block diagram of the color section of the receiver.

will have a frequency of 15,734 Hz, corresponding to one cycle per horizontal line of deflection.

Each sine-wave output is serrated because of the interruption in the gated rainbow chrominance signal feeding the inputs of the demodulators. The two output signals are then passed through a matrix that derives a third signal, the $G - Y$ signal. The three out of phase signals are then amplified and may be fed to the grids or the cathodes of the picture tube. These voltages cause the bias of their respective electron guns to vary, thereby intensity modulating the electron beams. The result of this is shown in Figure 16-8. Notice that at the time that the voltage on the red picture tube grid is maximum positive (bar 3) the blue grid is zero and the green grid is negative. Making the grid of a tube more positive increases both beam current and a raster brightness. Making it more negative decreases both to cutoff. Consequently, under normal conditions, the third bar of the color bar pattern will be red, the sixth bar will be blue, and the tenth bar will be green. The other bars will vary in color, depending upon the relative amplitudes of the pulses fed to each grid of the picture tube.

The color bar generator is extremely useful. It can be used for RF, IF, and video signal tracing and as a standard color bar pattern for determining such things as color fit, color sync adjustment, range of tint and color controls, and loss of one or more colors.

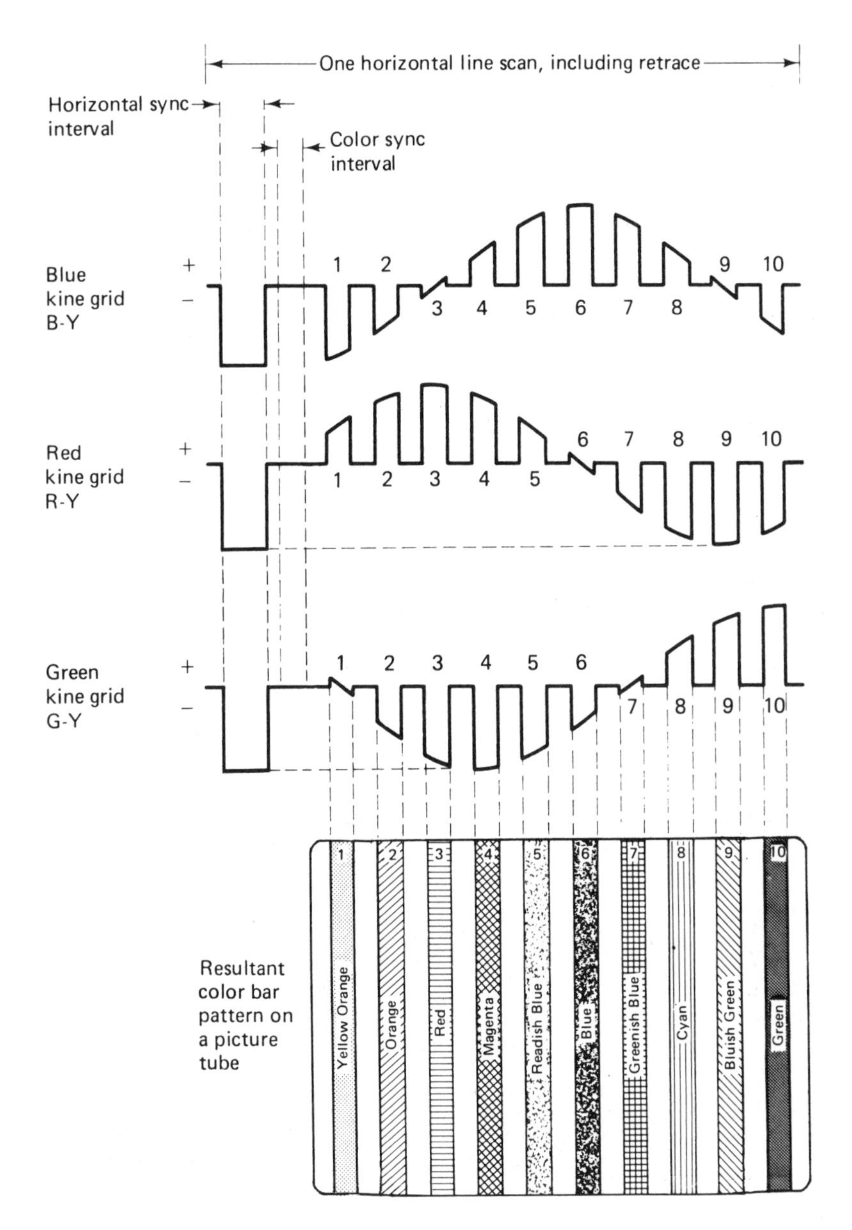

Figure 16-8 Picture tube signals necessary to produce the color bar pattern.

16-3 the transistor bandpass amplifier

The purpose of the bandpass amplifier is to separate the chrominance signal from the composite video signal, amplify it, and then pass the chrominance signal on to the demodulators. This is done by means of one or more low-distortion Class A tuned amplifiers that are called *bandpass amplifiers.* The name of the amplifier indicates that these amplifiers act as very special filters that pass a narrow range of signals centered around 3.58 MHz.

In general, the overall bandwidth of the bandpass amplifier is designed to be 1 MHz. An example of a five-stage transistor bandpass amplifier circuit is shown in Figure 16-9.

Examination of this circuit shows that the composite video input to the first chroma amplifier (Q_1) is fed through a tuned circuit consisting of C_1 and L_1. The bandwidth of the tuned circuit is adjusted by R_1 and R_2. This tuned circuit has two functions. First, it eliminates the low-frequency components of the input composite video signal and passes only to the chrominance signal. Second, the shape of its response curve is adjusted to compensate for the *roll-off* of the chrominance sidebands caused by the frequency response characteristics of the video IF and video amplifiers. This action provides the input to Q_1 with a reasonably flat band of frequencies ranging from 3.08 to 4.08 MHz. The alignment and frequency response characteristics of the bandpass amplifier will be discussed in detail in Section 16-5.

The base bias of the first chroma amplifier (Q_1) is determined by the dc voltage fed to it by the ACC circuit. The output of the first chroma amplifier is *RC* coupled to the base of the second chroma amplifier (Q_2). Fixed bias for this circuit is obtained from a voltage divider circuit consisting of R_5 and R_6. The collector circuit is impedance coupled to a broadly tuned circuit consisting of L_2 and C_8 whose bandwidth is sufficient to pass the entire range of chrominance signals.

Because of the automatic chroma control (ACC) circuit, the signal output of the first chroma amplifier is free of unwanted chrominance amplitude variations. The signal is then amplified by the second chroma amplifier to a level sufficient to drive the third chroma amplifier Q_3. This stage is arranged as a common base amplifier that has a relatively low-input and a high-output impedance.

The color control diode. The chrominance signal output of the second chroma amplifier is coupled through R_{10} to the base of Q_3 and to the cathode of D_1, the color control diode. This diode is forward biased by a voltage divider formed by R_7, R_8, and R_9, the color control potentiometer. In this circuit the color control varies a dc voltage that in turn controls the amplitude of the chrominance signal.

Varying R_9 changes the degree of D_1's forward bias and thereby changes its effective resistance. As a matter of fact, the color control diode can be thought of as a voltage-controlled variable resistor. Increasing its forward bias decreases the effective resistance of the diode.

To see how R_9 can control the amplitude of the chroma signal by changing a dc voltage, consider what would happen if D_1 were to be excessively forward biased (R_9 at +30 V position) so that it would be effectively a short circuit. Under these conditions, all of the output signal current flowing through R_{10} would be shorted to ground via C_9. As a result, the base chroma signal current of Q_3 would be zero and the picture tube would produce a black & white picture. If R_9 were to be adjusted so that the forward bias of D_1 would be reduced to zero, the diode would become an open circuit. Consequently, all of the signal current would be injected into the emitter of Q_3, resulting in a color picture of maximum saturation. At other resistance settings of R_9 the conduction of D_1 would cause the signal current to divide between the emitter of Q_3 and C_9, resulting in intermediate levels of picture color saturation.

The output of Q_3 is then directly coupled to the base of Q_4, the bandpass driver. R_{14} and R_{15} act as the load resistances for Q_3 and the bias resistors for Q_4. The bandpass

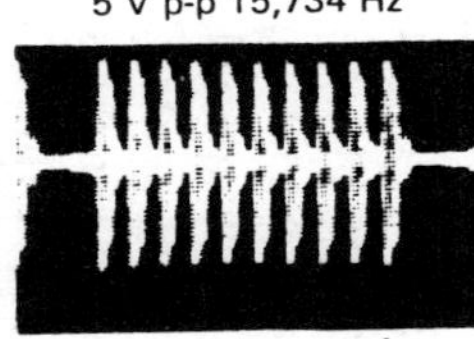

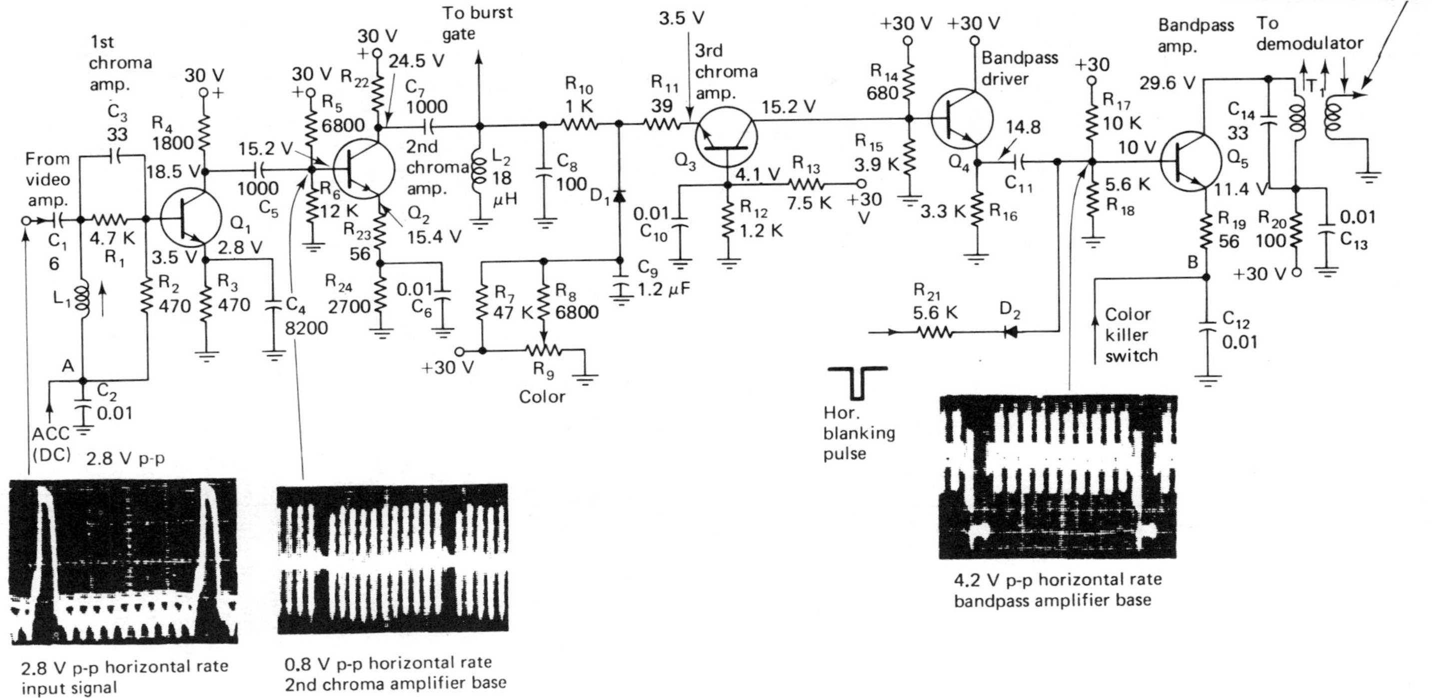

Figure 16-9 Five-stage transitor bandpass amplifer.

driver (Q_4) is an emitter follower whose high-input and low-output impedance characteristics are used to match the high output impedance of the third chroma amplifier to the low input impedance of the bandpass amplifier (Q_5). C_{11} is a coupling capacitor and R_{17} and R_{18} establish the bias for Q_5. R_{21} and D_2 couple a negative-going horizontal blanking pulse into the base of Q_5 This pulse drives Q_5 into cutoff during the color burst interval and prevents it from reaching the picture tube. T_1 is a tuned transformer that controls the bandwidth of chroma signal and provides impedance matching between the bandpass amplifier and the demodulator stages. The emitter of Q_5 is returned to ground through the color killer stage. In this circuit the color killer acts like a switch that disables the bandpass amplifier when it is open (black & white transmission) and returns the bandpass amplifier to normal operation when it is closed (color transmission).

BPA wave shapes. An examination of the rainbow color bar generator waveforms that pass through the amplifier (see Figure 16-9) discloses a number of important facts. Notice that the input base (Q_2) voltage has 11 bursts and that the horizontal sync pulse at the input of the amplifier (Q_1) has been eliminated from it. The elimination of the horizontal sync pulse is the result of frequency discrimination of the input tuned circuit (L_1) which is tuned to 3.58 MHz and rejects the 15,734-Hz horizontal sync pulse.

The output waveform (Q_5) is larger in amplitude than the input signal and has only 10 bursts. This is the result of the horizontal blanking pulse fed to the base of the bandpass amplifier (Q_5). This pulse drives the amplifier into cutoff during the horizontal blanking interval. As a result, the color burst is eliminated from the bandpass amplifier output. The difference in wave shape between input and output waveforms is due to the alignment and bandpass characteristics of the amplifier's tuned circuits.

The tint control. The amplitude of the chrominance signal corresponds to color saturation and is adjusted by the color control. Chrominance signal phase represents the picture color and may be adjusted by the tint control. Varying the tint control causes the picture color to change by changing the phase relationship between the chrominance signal and the local color subcarrier. Thus, the tint control may be located in a number of different stages of the receiver. Locating the tint control in the burst gate, phase detector, reactance control circuit, or in the color subcarrier oscillator stage allows a direct control on the phase of the reference subcarrier injected into the demodulators. Another way of doing the same thing is to locate the phase control in the bandpass amplifier. In this arrangement the picture color is adjusted by shifting the phase of the chrominance signal before it reaches the color demodulators and is compared to the reference phase of the subcarrier.

The phase splitter tint control circuit. An example of a tint control that uses a transistor phase splitter as a means of changing the chrominance phase is shown in Figure 16-10. In this circuit the chrominance signal is fed to the base of a phase splitter (Q_1) by C_1. The transistor is operated Class A, and its bias is set by the combined effect of R_1, R_2, R_4, and R_7. Since emitter resistor R_4 is not bypassed, an ac signal voltage (v_e) will be developed across it which will be 180° out of phase with the ac collector voltage (v_c). Connected between the collector and the emitter is a variable phase shift network formed by R_5, R_6, and C_2. Varying R_5 causes the output chrominance signal to shift its phase with respect to the input signal at the base of Q_1. This, in turn, causes the picture color to vary.

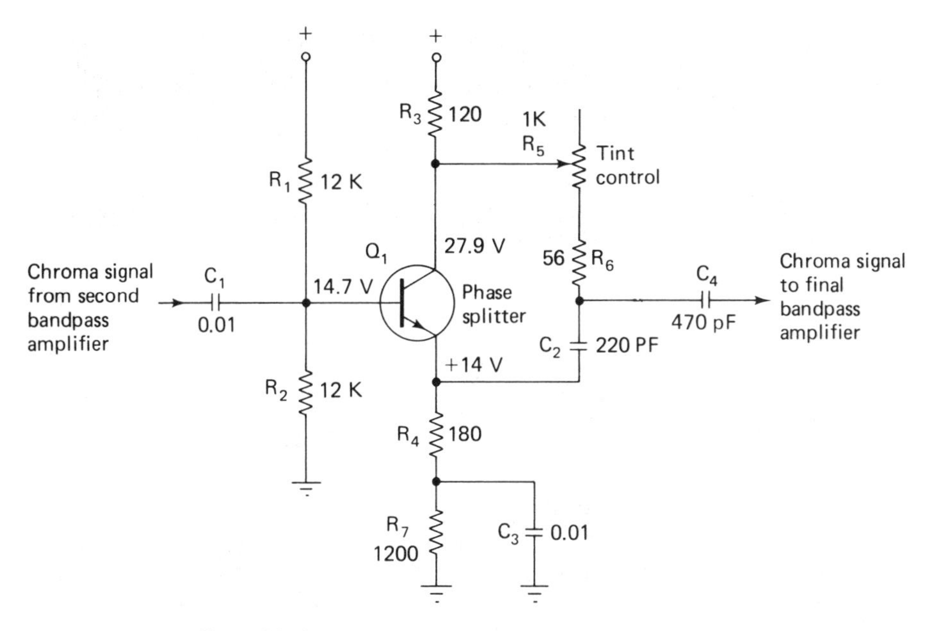

Figure 16-10 Tint control used in a bandpass amplifier circuit.

16-4 the IC bandpass amplifier

Many television receivers use modular construction. In this type of receiver entire sections of the receiver are constructed on plug-in boards called *modules.* As few as 3 or as many as 10 or 12 modules may be used in a color receiver.

A typical module may be seen in Figure 16-11(a). Each module may consist of two or more complete stages of the receiver. These stages may be constructed of discrete components or may be combined in integrated circuits (IC). Most modules are a mixture of one or more ICs and a number of discrete components such as transistors, coils, capacitors, and resistors. Figure 16-11(b) shows the schematic diagram of the module shown in Figure 16-11(a).

Servicing modular receivers often consists of simply locating the defective module and replacing it with a good one. The defective module is then either repaired or returned to the manufacturer in exchange for a good one.

Typical IC BPA. Figure 16-12 illustrates the block and schematic diagram of an integrated circuit commonly used as a bandpass (chroma) amplifier in a chroma processor module. This IC is part of a module (Figure 16-11) that consists of three ICs, three transistors, and associated resistors, capacitors, coils, and controls.

The block diagram of the chroma amplifier IC [Figure 16-12(a)] discloses that the integrated circuit consists of a power distributor circuit (bias circuit) and a two-stage bandpass amplifier that is associated with a color killer circuit. External to the integrated circuit chip are the color control, a preset color master (CM) control, color master switch, and the color killer level adjustment. The CM control and switch allow the operator to

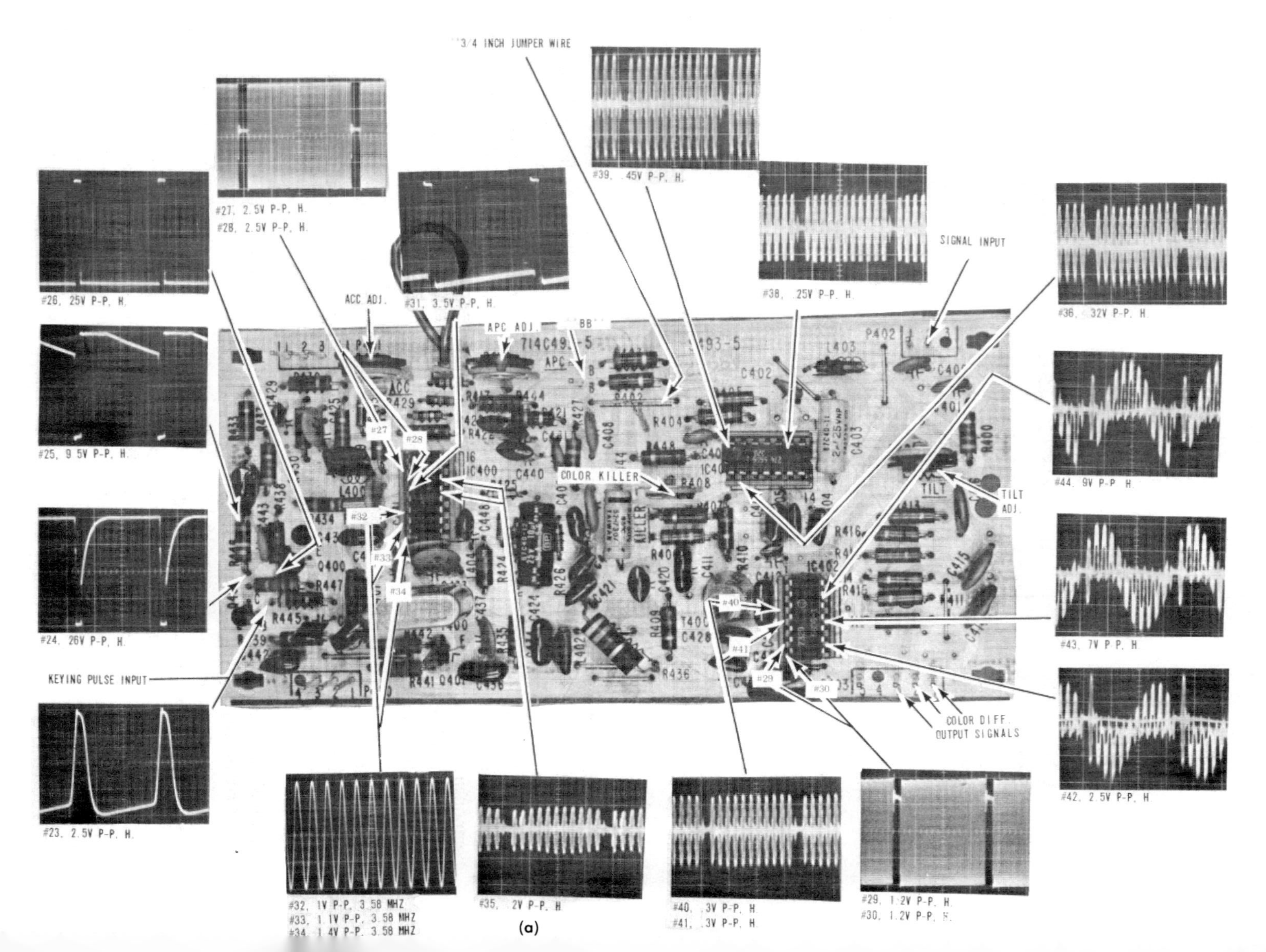

3/4 INCH JUMPER WIRE
#39, .45V P-P, H.
#38, .25V P-P, H.
SIGNAL INPUT
#36, .32V P-P, H
#27, 2.5V P-P, H.
#28, 2.5V P-P, H.
#26, 25V P-P, H.
#31, 3.5V P-P, H
ACC ADJ.
APC ADJ.
COLOR KILLER
TILT ADJ.
#44, 9V P-P, H.
#25, 9.5V P-P, H.
#24, 26V P-P, H.
#43, 7V P-P, H.
KEYING PULSE INPUT
COLOR DIFF. OUTPUT SIGNALS
#42, 2.5V P-P, H.
#23, 2.5V P-P, H.
#32, 1V P-P, 3.58 MHZ
#33, 1.1V P-P, 3.58 MHZ
#34, 1.4V P-P, 3.58 MHZ
#35, .2V P-P, H
#40, .3V P-P, H.
#41, .3V P-P, H.
#29, 1.2V P-P, H.
#30, 1.2V P-P, H.

(a)

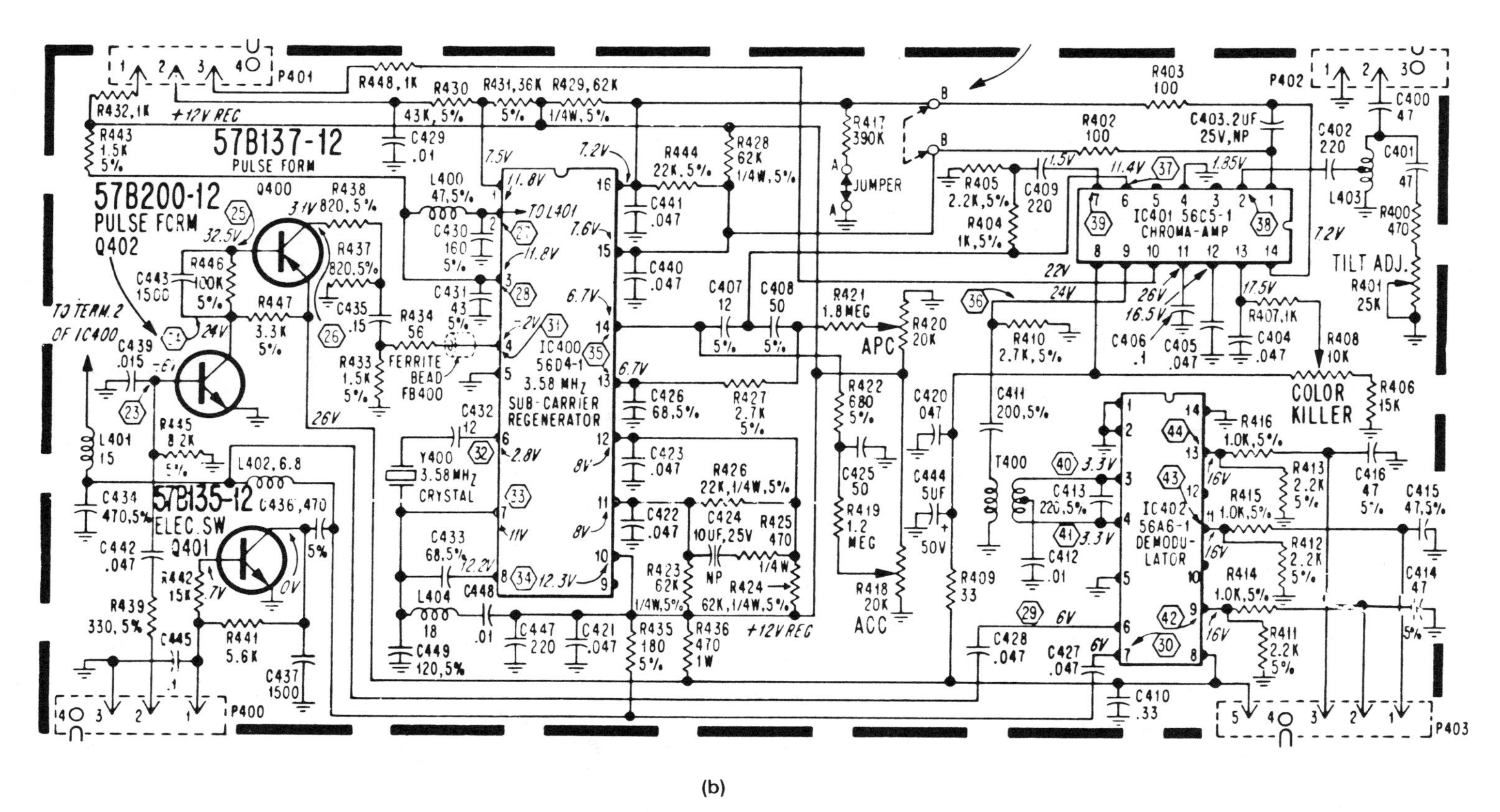

(b)

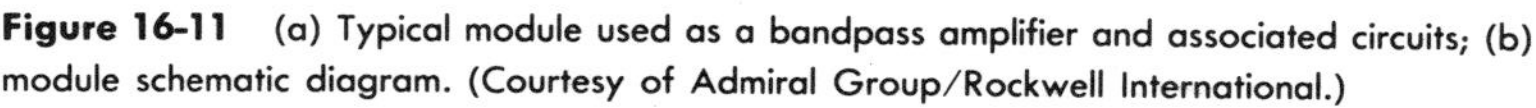

Figure 16-11 (a) Typical module used as a bandpass amplifier and associated circuits; (b) module schematic diagram. (Courtesy of Admiral Group/Rockwell International.)

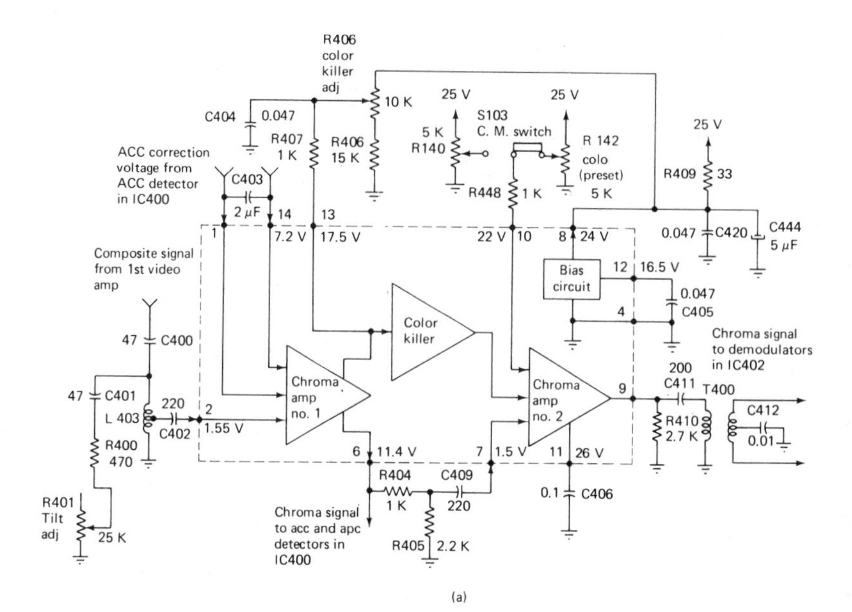

(a)

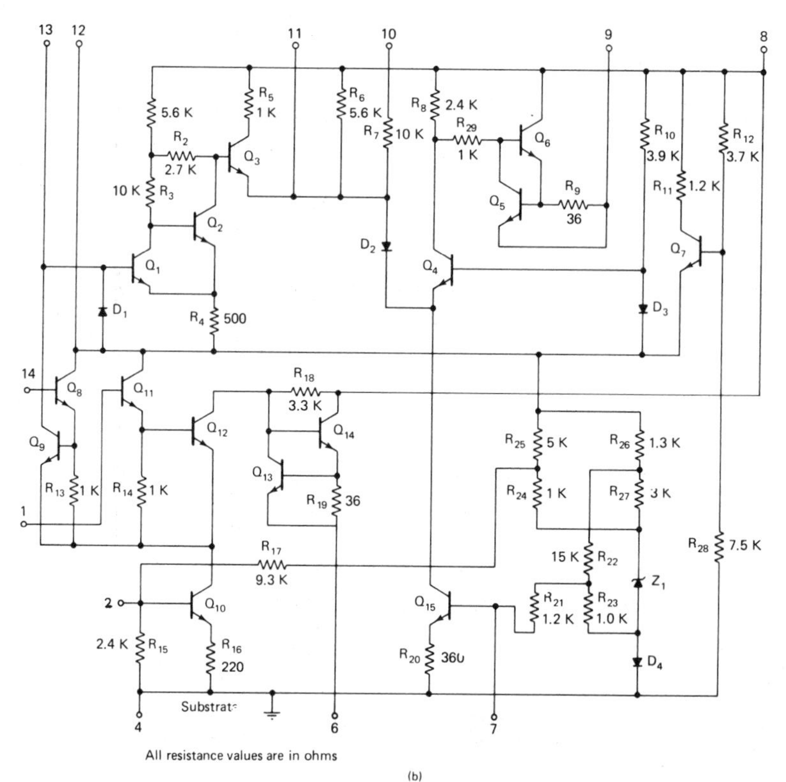

(b)

Figure 16-12 Block (a) and schematic (b) diagrams of a chroma bandpass amplifier. (Courtesy of Admiral Group/Rockwell International.)

switch from manual control of color saturation to a preset saturation control and avoids the need for saturation adjustment.

Power is supplied to pin 8 of the IC by R_{409} and decoupling capacitors C_{420} and C_{444}. The composite video signal is fed through a bandpass filter network C_{400}, L_{403}, C_{401}, R_{400}, and tilt adjust R_{401}. This tuned circuit rejects the low-frequency components of the input signal and only allows the chrominance signal to reach the input of the first chroma amplifier, terminal (2) of the IC. The tuned circuit is not slug tuned. The only adjustment provided to match the bandpass amplifier tuned circuit to that of the video amplifier is the tilt adjustment. This control is able to shift the position and amplitude of the tuned circuit peak response. Proper adjustment is obtained by observing the color bar pattern and adjusting the tilt control for proper registration of the color bars so that color smear is eliminated.

The output of the first chroma amplifier (pin 6) is then *RC* coupled (R_{404}, C_{409}, and R_{405}) to the input of the second chroma amplifier. This output is also used to provide chroma signal input to the ACC and APC (automatic phase control) circuits located in another IC. The first chroma amplifier is used as an automatic gain control amplifier and is therefore fed (pins 1 and 14) by a dc ACC correction voltage.

The first chroma amplifier has a second internal output that drives the color killer circuit. During color broadcasts the color killer is disabled so that the second chroma amplifier may operate normally. However, during black & white transmissions the loss of the color burst allows the color killer to operate, and it disables the second chroma amplifier.

The output of the second chroma amplifier (pin 9) is fed to a double-tuned circuit consisting of R_{410}, C_{411}, T_{400}, and C_{412}. This fixed tuned circuit is designed to provide a broadband, double-humped response centered at 3.58 MHz and extending from 3.08 to 4.08 MHz. The center-tapped secondary provides two outputs equal in amplitude and 180° out of phase that are fed to the demodulators. Since the output tuned circuit cannot be adjusted, alignment consists entirely of the tilt adjustment discussed above.

An examination of the schematic diagram [Figure 16-12(b)] of the IC chroma amplifier reveals that it consists of 15 transistors, 4 diodes, 1 zener diode, and a great many resistors. No capacitors and no inductors are included in the chip. The transistors, resistors, and diodes are used to form differential amplifiers, voltage regulators, current regulators, and a Schmitt trigger is required for proper circuit action.

16-5 bandpass amplifier alignment

At the transmitter, *I* and *Q* signals are used to modulate the RF carrier. The *I* signal upper sidebands extend above the color subcarrier by 0.5 MHz and below the subcarrier by 1.5 MHz. The *Q* signal has a sideband spectrum that is symmetrical and extends 0.5 MHz above and below the color subcarrier.

In an attempt to reduce costs, modern receivers do not use the full *I* and *Q* signal information that is transmitted. The bandwidth of the bandpass amplifier used in those receivers is limited to 0.5 MHz on either side of the color subcarrier for a total bandwidth of 1 MHz. Thus, the highest chrominance signal frequency appearing at the output of the BPA is 4.08 MHz and the lowest is 3.08 MHz. *I* signal frequencies higher than 0.5 MHz are simply discarded by the video IF and BPA tuned circuits. To provide proper color reproduction, the tuned circuits must be aligned in order to have the proper bandpass and shape. Improper alignment can cause such defects as no color, weak color, color smear, poor color fit, and oscillation. The BPA alignment procedure is different from that of video IF alignment.

The chrominance signal on its way to the color demodulators must pass through the RF tuner, the video IF amplifier, the video amplifier, and the bandpass amplifier. Each of these circuits has its own frequency response that affects the operation of the other circuits. Thus, the RF tuner is designed to have a relatively flat response for each channel. In the video IF amplifier the bandwidth requirements are simplified by attenuating the chrominance subcarrier to the 50% level of the video IF response. This is shown in Figure 16-13. The roll-off characteristics of the video amplifier also affect the amplitude and phase of the chrominance signal feeding the bandpass amplifier. Because of these frequency response characteristics, the chrominance sidebands that are separated from the IF picture carrier by 3.08 MHz are considerably stronger than those at 4.08 MHz.

Figure 16-13 Frequency response of video IF amplifier, video amplifier, bandpass amplifier, and the overall response at the output of the bandpass amplifier.

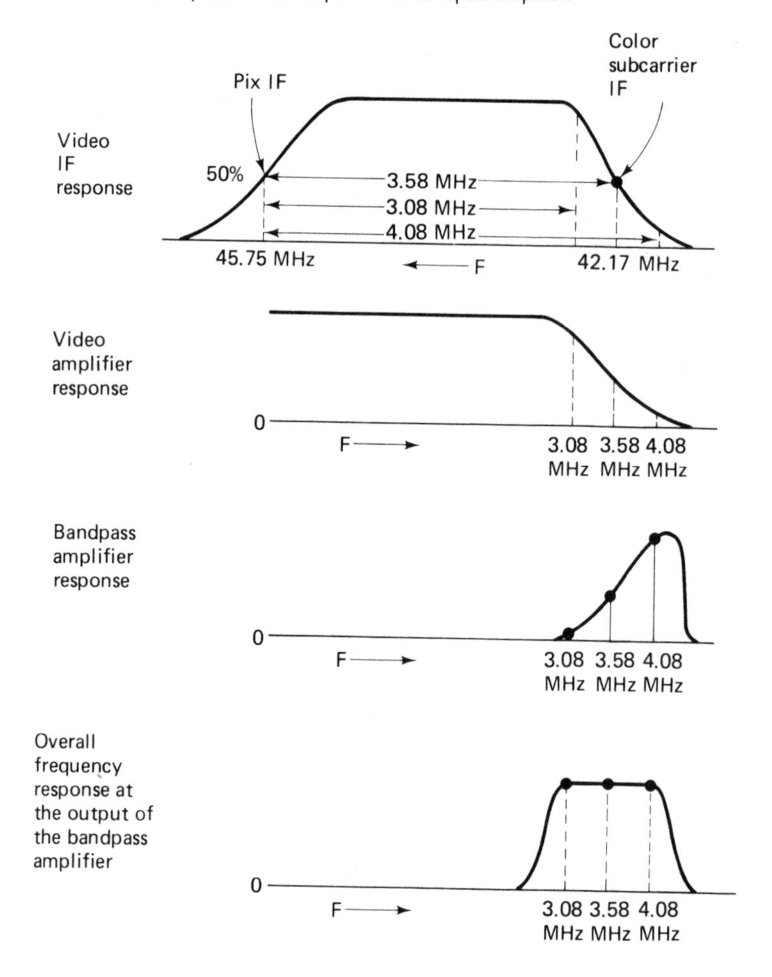

If these conditions are not compensated for, severe phase distortion will be introduced, seriously affecting picture color and color registration. The method of compensation used to achieve a flat overall frequency response is to adjust the input tuned circuit of the BPA so that its frequency response characteristic is opposite in slope to that of the IF and video amplifier. See Figure 16-13. This means that the input tuned circuit provides maximum output at 4.08 MHz and minimum output at 3.08 MHz. If the combination of all of these frequency responses is correct, the desired overall response at the output of the BPA will be flat. Under these conditions, the phase and amplitude characteristics of the chrominance signal will be restored to normal.

The obvious conclusion to the foregoing is that bandpass alignment is critical because it must match the BPA response to the existing video IF and video amplifier responses. Obtaining the desired overall response requires a special alignment technique.

VSM. Bandpass amplifier alignment should always be performed according to the manufacturer's specifications. These specifications are usually broken down into two steps. First, the BPA is sweep aligned in the same way that a video IF amplifier is sweep aligned. This is done to ensure that the interstage and output tuned circuits are correctly adjusted for a flat topped response 1 MHz wide (3.08 to 4.08 MHz). Second, video sweep modulation (VSM) is used to adjust the BPA input tuned circuit for a flat topped *overall* receiver response.

VSM is a technique in which an IF picture carrier (45.75 MHz), obtained from a marker generator, is amplitude modulated by a frequency modulated sweep that is varying from 0 to 5 MHz at a 60-Hz rate. A block diagram indicating the basic circuit hookup is shown in Figure 16-14. Notice that the sweep generator is adjusted for an FM output that is varying from 0 to 5 MHz. The constant amplitude output of the generator is then fed through a calibrated absorption marker to an RF modulator. The absorption marker reduces the amplitude of the sweep generator at 3.08, 3.58, and 4.08 MHz. Also feeding this modulator is an unmodulated marker generator set at 45.75 MHz. As a result, the output of the RF modulator is a frequency modulated AM signal centered at the picture IF (45.75 MHz). This signal is then injected at the input to the mixer stage of the receiver.

If the video IF amplifiers have been aligned properly, the output of the video detector will be a constant amplitude FM signal that is sweeping from 0 to 5 MHz. This signal then passes through part of the video amplifier on its way to the BPA. The combined roll-off characteristics of both the video IF amplifier and the video amplifier cause the high-frequency portion of the sweep envelope to become attenuated. The BPA tuned circuit is tuned to 3.58 MHz. As such, it eliminates all of the low-frequency (lower than 3 MHz) signals feeding the input of the BPA. When the input and output circuits are properly adjusted, they compensate for the roll-off introduced by the video IF amplifier and the video amplifier. At this point the output of the BPA is an AM envelope that is centered at 3.58 MHz and is relatively flat topped. Since the vertical deflection amplifiers of most modern oscilloscopes respond to frequencies in excess of 4 MHz, this sweep modulated envelope may be observed directly. The usual procedure is to demodulate the output of the BPA before observing it with the oscilloscope. The resultant desired response curve shown in Figure 16-14(a) will have a number of marker dips that are due to the absorption markers introduced at the output of the sweep generator.

Chroma bar sweep. Alignment of the BPA may also be obtained by the use of a special generator (VA48) produced by Sencore. See Sections 1-8 and 6-11. This generator when operated in its Chroma Bar Sweep mode produces an RF output that is modulated by three equal-amplitude chroma signals. The signals are in the form of bursts and are the color carrier, a 3.08-MHz signal that is used to indicate the lower limit of the BPA response, and another signal at 4.08 MHz used to indicate the upper limit of the BPA response.

To use the generator for BPA alignment the RF output leads of the generator are attached to the antenna terminals (or video IF input) of the receiver and then using an oscilloscope the output of the bandpass amplifier is observed. If the system is properly aligned, the three bursts should be of equal amplitude, as indicated in Figure 16-14(b). The

Figure 16-14 (a) Video sweep modulation (VSM) method of BPA alignment.

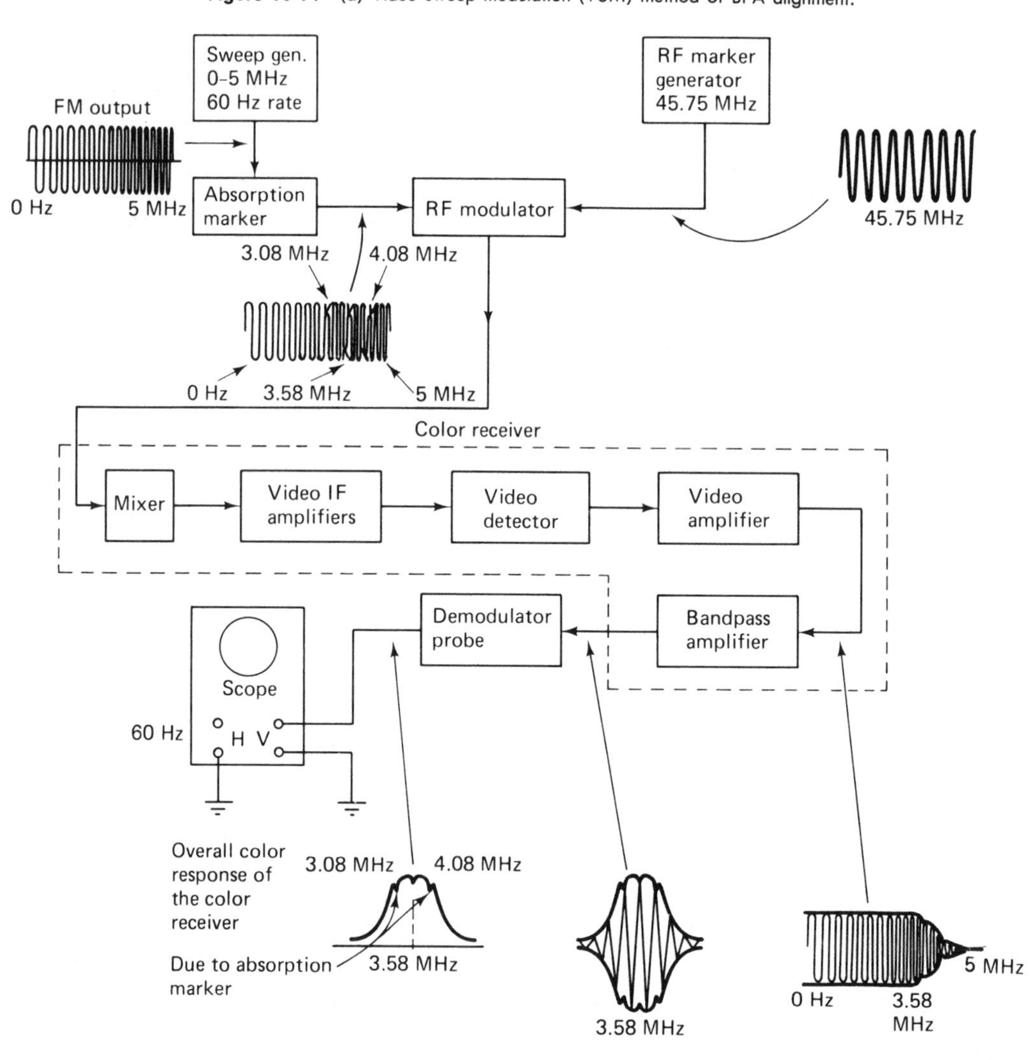

(a)

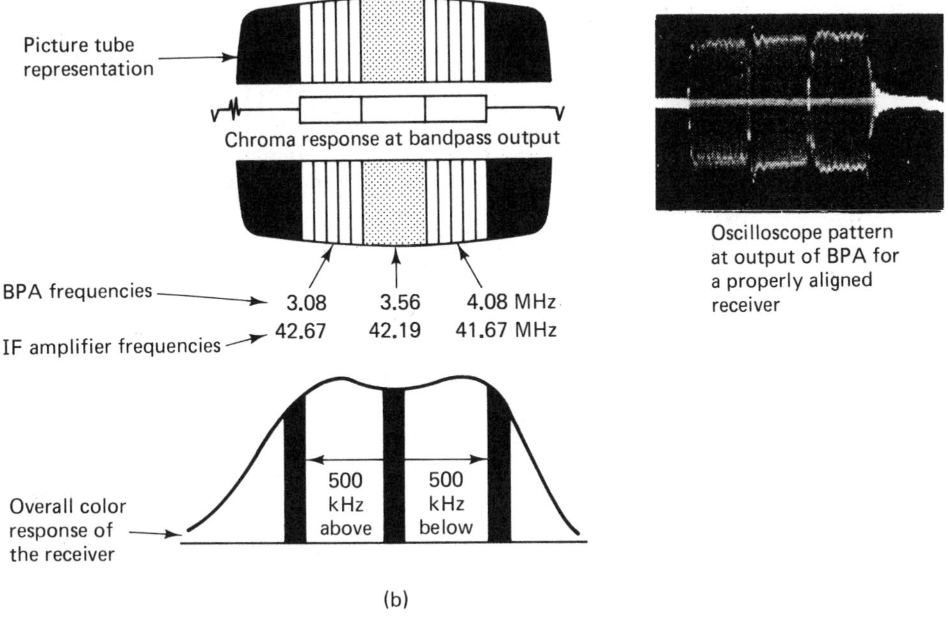

Figure 16–14 (*Cont.*) (b) Chroma bar sweep patterns.

visual display on the picture tube should appear as three thick vertical bars of equal brightness. See Figure 16-14(b). If the BPA alignment is defective (video IF alignment is proper), the bars will not be of equal amplitude. BPA alignment consists of adjusting the tuned circuits until the three bars at the BPA output are of equal amplitude.

16-6 bandpass amplifier—vacuum tube type

Typical BPA circuit. An example of a typical vacuum tube bandpass amplifier is shown in Figure 16-15. An examination of the circuit discloses that both the input and output circuits contain adjustable tuned circuits tuned to 3.58 MHz. To minimize distortion, the bandpass amplifiers are operated Class A. The tube's operating bias is determined by the combination of self-bias produced by cathode resistor R_1 and the negative dc voltages that are fed to the grid from the color killer and ACC circuits. In addition, a positive horizontal blanking pulse of approximately 30 V p-p is coupled to the cathode of the tube. This pulse disables the bandpass amplifier by forcing the amplifier into cutoff during the horizontal blanking and retrace period. As a result, the color burst is prevented from reaching the grids of the picture tube. The color burst corresponds to a color of greenish-yellow, and if it were allowed to excite the grids of the picture tube, it might cause horizontal retrace to produce a greenish-yellow cast to the raster. The 820-pF capacitor C_1 is a cathode bypass capacitor that prevents degeneration of the chrominance signal.

During black & white broadcasts the color killer develops a large negative voltage (−8 V) which is coupled to the grid of the tube via R_2, R_3, and R_5. This voltage is sufficient to disable the bandpass amplifier by driving it far into cutoff.

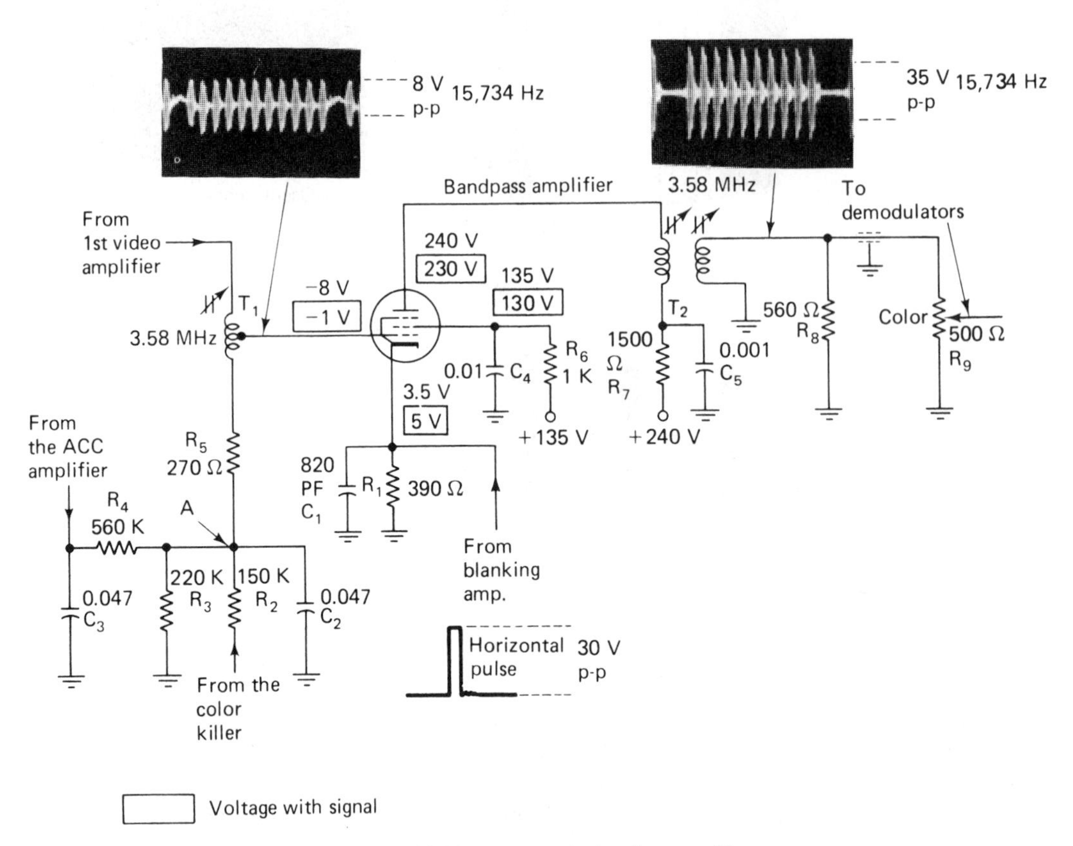

Figure 16-15 Vacuum tube bandpass amplifier.

During color reception the color killer voltage drops (−1 V) and the bandpass amplifier is allowed to function properly. The dc grid voltage, however, is now varied by the automatic color control (ACC) circuit in the same manner as AGC varies the bias in the video IF amplifier stage of the receiver. Resistor R_4 is an isolating resistor that minimizes any interaction between the ACC and color killer circuits. Capacitors C_2 and C_3 are bypass capacitors that prevent ac feedback between the color killer, bandpass amplifier, and ACC circuits.

A shielded cable is used to connect the output of T_2 to the color control R_9. This control is a voltage divider similar to the volume control found in the audio amplifier circuit. Varying the amplitude of the chrominance signal changes the saturation of the color. Increasing the color saturation causes the picture colors to become more intense.

16-7 troubleshooting the bandpass amplifier

Defects in the bandpass amplifier may cause such trouble symptoms as no color, weak color, excessive color, and color smear.

No color is a symptom that may be caused by defects in many stages of the color receiver. Listed below are some of the more common ones:

1. Improper fine tuning.
2. Defective video IF or bandpass amplifier alignment.
3. Defective bandpass amplifier.
4. Color control turned down.
5. Defective or improperly adjusted color killer.
6. Defective or improperly adjusted ACC.
7. Inoperative 3.58-MHz oscillator.
8. A defective burst gate of 3.58-MHz phase detector indirectly causing loss of color. This is because these circuits control the color killer and ACC sections of the receiver.
9. A defective burst gate feeding a ringing amplifier.
10. A defective ringing amplifier.

The first four causes of no color refer to circuits that have already been discussed in this text. The remaining causes and circuits are to be discussed in the following sections of this book.

Since there are so many causes for a no-color condition, it is important to be able to isolate the specific stage that is at fault. This is done in two steps. First, all "obvious defects" are eliminated; for example, bad tubes or transistors, broken wires, improper fine tuning, improper color control, adjustment, color killer or ACC adjustments, and so on. Second, the defective stage is eliminated by cause-and-effect reasoning. If this is not possible, such techniques as signal injection, signal tracing, or a combination of the two are used. That is, a gated rainbow color bar generator is applied at the receiver input and then wave shapes are observed at the output of the video detector, the input to the BPA, and the output of the BPA, and all the other circuits listed as possible causes of no color. The defective stage is isolated when a wave shape that is normally found at a given point is missing. The normal wave shape and its amplitude are found in the receiver service manual or must be known from experience.

Once the stage is isolated, the defective component must be found. This is done by voltage and resistance checks.

Weak or excessive color are both problems that may be caused by defects in the same stages that caused no color. The main difference is the degree by which the gain of the BPA changes. If the BPA's gain is reduced, weak color is produced. If the gain increases, the color will be highly saturated and excessive.

Excessive color can be caused by defects in the ACC which can cause the BPA gain to increase greatly. A bias clamp can quickly isolate whether the ACC circuit or the BPA is at fault.

Color smear refers to the condition in which the color does not register or fit in the area where it belongs. For example, the red lips of a young lady smearing onto her cheek or other similar conditions can be the result of defects that cause a change in the BPA alignment. The troubleshooting procedures that should be most helpful are signal tracing and alignment. Changes in alignment may cause changes in the amplitude and wave shape of the gated rainbow waveforms. Signal tracing can be used to locate these changes and pinpoint the area of the defect. Alignment may be useful in locating smear causing defects. In this method adjustments that have improper or little effect on the response curve are looked for. Components associated with the alignment adjustment are then checked for opens, leakage, and shorts.

16-8 the color killer

The purpose of the color killer is to automatically disable the chrominance chain during black & white broadcasting. This is done so that noise generated in the BPA will not appear on the picture tube screen and intermix with the black & white picture information. If unchecked, the noise generated in the BPA may appear on the screen as either colored snow, called *confetti,* or as annoying scintillations of red, green, and blue that accompany vertical lines in the picture. These scintillations are sometimes referred to as *worms.* This type of interference is due to black & white picture information whose frequency falls in the chrominance bandpass, thereby being processed by the color circuits.

The color killer is designed to act as a switch that disables the chrominance signal path during black & white transmission. In Figure 16-16 a block diagram shows the input and output connections of the color killer. The input is a dc control voltage that may be obtained from the automatic frequency and phase control (AFPC) circuit or from a separate phase detector circuit that is used as a color killer detector. Some color killer circuits are also pulse operated. They obtain horizontal retrace pulses from the flyback transformer.

The dc output of the color killer acts as a bias voltage and may feed the *bandpass amplifier,* the *color demodulators,* or the *color difference amplifiers.* When it is activated, the color killer produces a bias voltage that is sufficiently large and of correct polarity to drive these stages far into cutoff. Unfortunately, there is no universally accepted color killer circuit. Color killers may be divided into two groups: those that are not conducting when the BPA is off and those that are conducting when the BPA is off.

Figure 16-16 Color killer and its input and output connections. The dashed lines indicate alternate output connections.

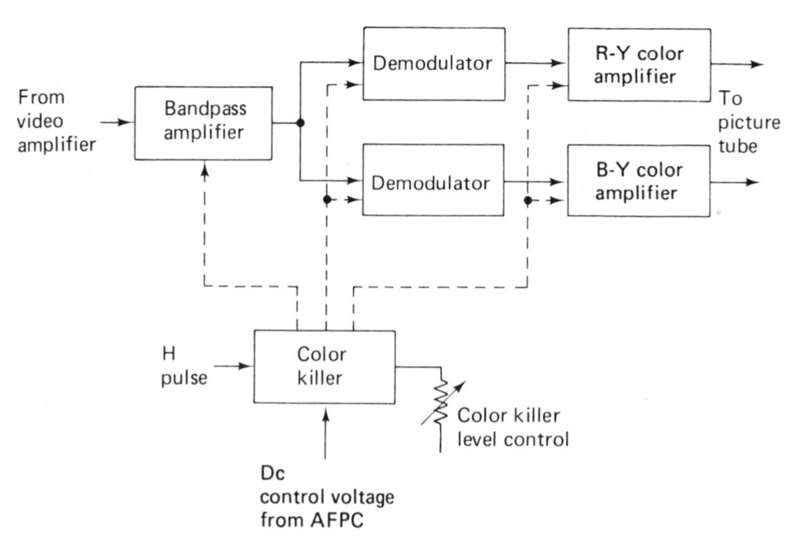

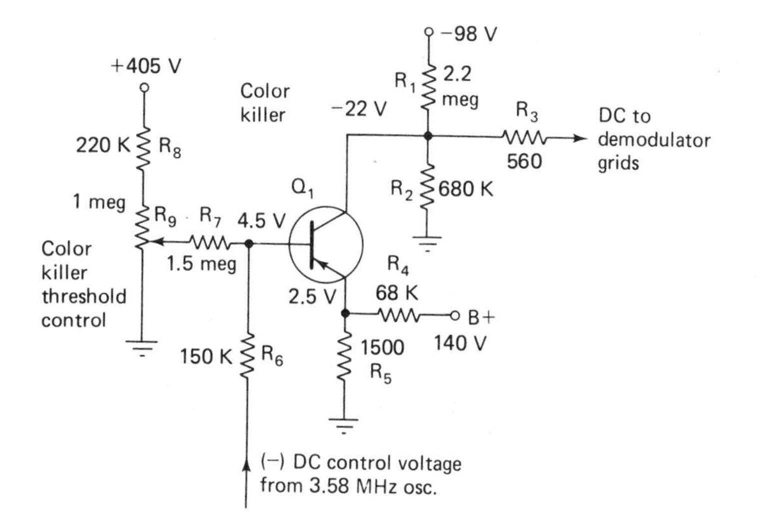

Figure 16-17 Transistor color killer.

The transistor color killer. An example of a transistor color killer is shown in Figure 16-17. An examination of the circuit shows that the transistor is a PNP and that its resultant base voltage is the sum of the dc control voltage obtained from the grid of the 3.58-MHz oscillator and the dc voltage obtained from the color killer threshold control (R_9). The collector is fed from a negative 98-V source through a voltage divider made of R_1 and R_2. The positive emitter voltage is developed by another voltage divider formed by R_4 and R_5.

Black & white reception. During black & white transmission the dc control voltage developed by the 3.58-MHz oscillator is relatively small. As a result, the base voltage of Q_1 is more positive than its emitter, forcing the transistor (Q_1) into cutoff. Since the transistor is effectively out of the circuit, its collector voltage is −22 V, the same as the voltage drop across R_2. This large negative voltage is coupled to the color demodulators, cutting them off and preventing any color from reaching the picture tube.

Color reception. When a color program is received, the negative dc keying voltage increases and the killer transistor is driven into saturation. A saturated transistor may be thought of as a short circuit. Therefore, the collector is now tied to the emitter, increasing the collector voltage to a positive 2 V. This voltage is sufficient to activate the demodulators and allow normal color operation. Note that in this circuit when the killer transistor is on, the demodulators are also on, and vice versa.

Color killer adjustment. To adjust the color killer control, the receiver is tuned to an unused channel to produce a snow-filled raster. The color killer control then adjusts the bias of the color killer to the point where the color snow is just turned off. This adjustment means that black & white signals that range from the very weak to the very strong will be able to activate the color killer.

The IC color killer. Color processing integrated circuits also include color killer circuits. A circuit of this type is shown in the integrated circuit diagrams of Figure 16-12. The block diagram of the chroma amplifier discloses that the color killer is connected between chroma amplifier 1 and chroma amplifier 2. In the schematic diagram of this circuit the color killer is a Schmitt trigger and is formed by Q_1 and Q_2.

A Schmitt trigger is a circuit that has two stable output stages: a high-voltage state and a low-voltage state. These output states are controlled by two input trigger levels of different amplitude that are applied sequentially. The first level is larger than the second level and produces a "high" output. The second level is smaller and produces a "low" output. The difference in input trigger level is often referred to as *hysteresis.*

Figure 16-18 Internal circuit of an IC color killer.

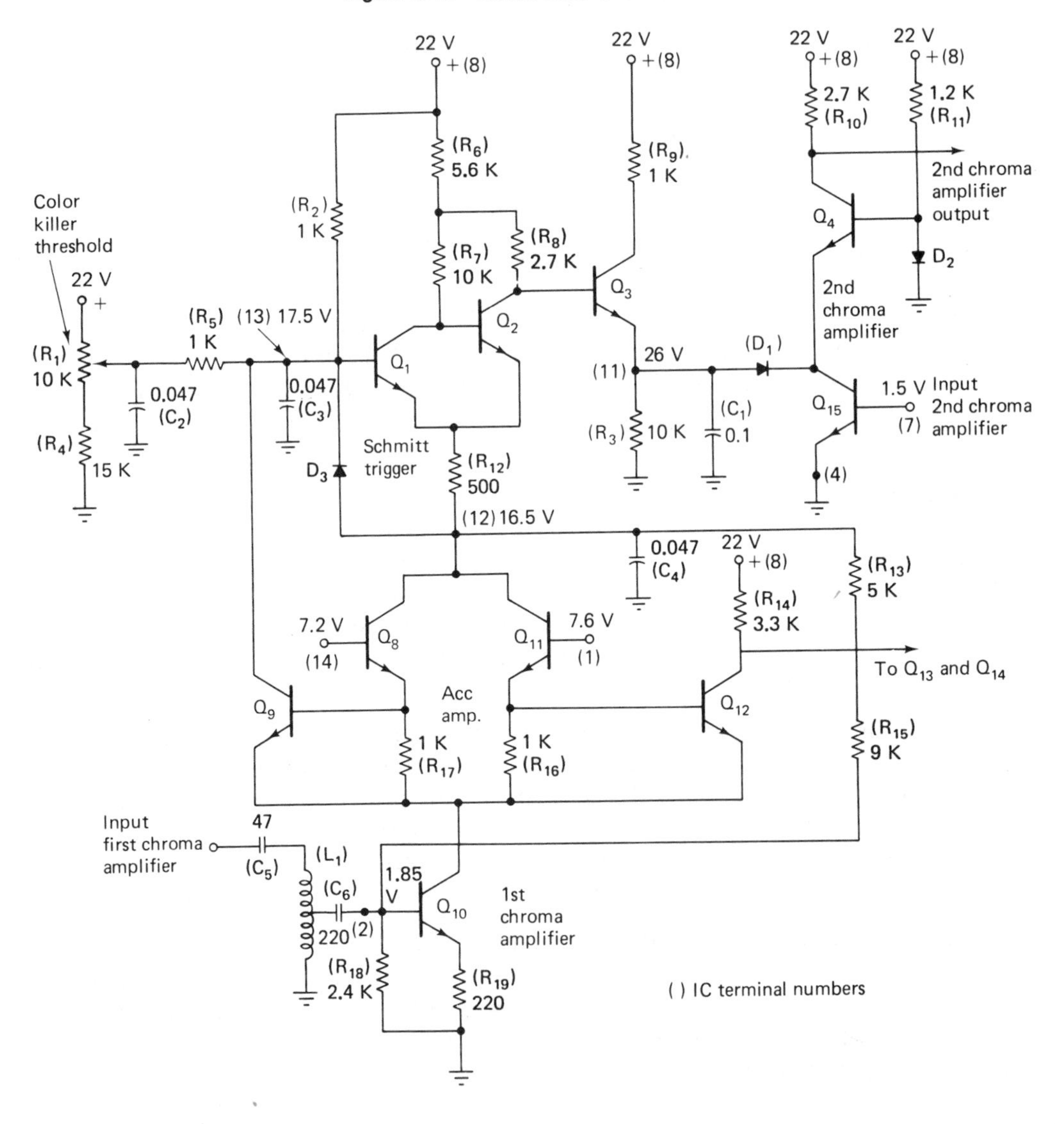

The operation of the color killer used in this IC (redrawn in Figure 16-18) may be divided into two modes of operation: black & white and color. During black & white transmission color signals are not transmitted. Therefore, there will be no input to the base of Q_{10}, the first chroma bandpass amplifier (Figure 16-18, terminal 2). Since this circuit is forward biased, a quiescent level of collector current flows. This current divides through the differential pair of transistors Q_9 and Q_{12}. The exact division depends upon the ACC voltage at the bases of Q_8 and Q_{11} (terminals 1 and 14). The ACC voltage produced by a black & white signal is such that Q_9 is cut off. The dc voltage at the base of Q_1 is due to the setting of the color killer threshold control R_1 and the conduction state of Q_9. Non-conduction of Q_9 reduces the voltage drop across the external resistor R_2, which increases the forward bias on the base of Q_1 and forces the transistor into saturation. The 0.2-V drop between collector and emitter of Q_1 is also the base-to-emitter voltage of Q_2, forcing it into cutoff. As a result, the collector voltage of Q_2 increases (high) and forces Q_3 into conduction. The current flowing through R_3 (terminal 11) develops a voltage drop that forward biases D_1. This ties the collector of the second chroma amplifier Q_{15} to C_1, a large externally connected bypass capacitor (terminal 11). The capacitor effectively shorts out any chroma signal that might have reached this point by reducing the gain of the second chroma amplifier stage (Q_{15}) to zero.

The reception of a color signal reverses the conditions just described. The input chroma signal changes the ACC input to Q_8 and Q_{11} so that Q_9 is allowed to conduct. This lowers the voltage at the base of Q_1 cutting it off. Q_2 is forced into conduction and the Schmitt trigger now reverses its output. The low voltage at the base of Q_3 forces it into cutoff, which reduces the voltage drop across R_3 to zero. This reverse biases D_1, effectively taking it and C_1 out of the amplifier circuit. The gain of the second chroma amplifier now returns to normal and a color picture is now allowed to appear on the screen of the picture tube.

16-9 troubleshooting the color killer

A defective color killer can cause the following trouble symptoms:

1. No color
2. Color snow in the black & white picture
3. Weak color
4. Excessive color; extremely saturated color

No color. In the transistor color killer circuit of Figure 16-17 a no-color symptom means that the transistor must be open or cut off. If the color killer is at fault, bias clamping the collector with +2 or +3 V will restore the BPA to normal operation. If shifting the bias clamp to the base and adjusting it for a voltage of approximately 2 V causes the color to return, the trouble has been localized to the bias circuit of the killer. Any defect that tends to shift the transistor further into cutoff is to be suspected. For example, no color could be caused by R_5 decreasing, R_4 increasing, R_9 open, and R_6 open.

Color snow. Color snow in the black & white picture means that the color killer is not functioning.

The transistor color killer of Figure 16-17 may become inoperative and produce color snow if it is always conducting. This can be caused by such defects as R_5 open, Q_1 shorted, R_9 open, or R_1 open.

Since this circuit is dc operated, the best method for troubleshooting is to use a bias clamp to verify and to establish the location of the defect. Voltage and resistance checks are then made to locate the defective component.

Weak and excessive color. Weak or excessive color may be the result of color killer faults that cause the BPA bias to change or ac circuit changes that affect the gain or alignment of the BPA. In the case of weak color, bias changes in one direction can reduce the gain of the BPA and shifts in bias in the opposite direction can increase the gain and cause excessive color.

16-10 automatic color control

The purpose of the automatic color control (ACC) circuit is to keep the output chrominance signal of the bandpass amplifier constant under varying input signal conditions. For example, the input signal to the BPA tends to vary over a wide range when changing channels because there are differences in channel signal strength at the antenna and differences in tuner RF alignment and gain from one channel to another. ACC allows the receiver to be switched from one channel to another without having to readjust the color control.

Automatic color control circuits are similar to AGC circuits used in the RF and video IF stages of the receiver. The AGC circuit produces a dc output that is controlled by the amplitude of the horizontal sync pulses. Automatic color control circuits can be considered auxiliary AGC circuits that provide a dc output that is proportional to the amplitude of the *color burst.*

Notice in the block diagram of Figure 16-2 that the input to the ACC circuit is a dc voltage control that is obtained from either the AFPC circuit or a separate ACC detector similar to the phase detector used for the color killer circuit. The output of the ACC circuit is a dc voltage that is used to control the bias of the bandpass amplifier. As is the case in AGC circuits, changing the bias of a vacuum tube or a transistor will cause a change in amplifier gain. For example, under normal operating conditions, an increase in chroma signal strength will cause the bias of the BPA to change in such a direction as to decrease the gain of the amplifier, thereby compensating for the increase in signal strength. The opposite will occur if the chroma signal level decreases.

The all-transistor ACC. Figure 16-19 is the schematic diagram of an ACC circuit that automatically controls the gain of a transistor bandpass amplifier (Q_2). In this circuit the tuned bandpass amplifier is called a *chroma amplifier.* The gain of the amplifier (Q_2) is controlled by varying its base-to-emitter bias. Analysis of the circuit will show that an increase in chroma signal strength will cause the ACC circuit to move the bias of the BPA toward cutoff, identifying it as a form of reverse AGC. In some transistor receivers a form of forward AGC may be used to control the gain of the BPA.

The input to the ACC circuit of Figure 16-19 is obtained from the 3.58-MHz subcarrier oscillator whose output level is directly related to the chroma signal level. This RF signal is rectified by the ACC rectifier D_1 and develops a positive dc voltage at the

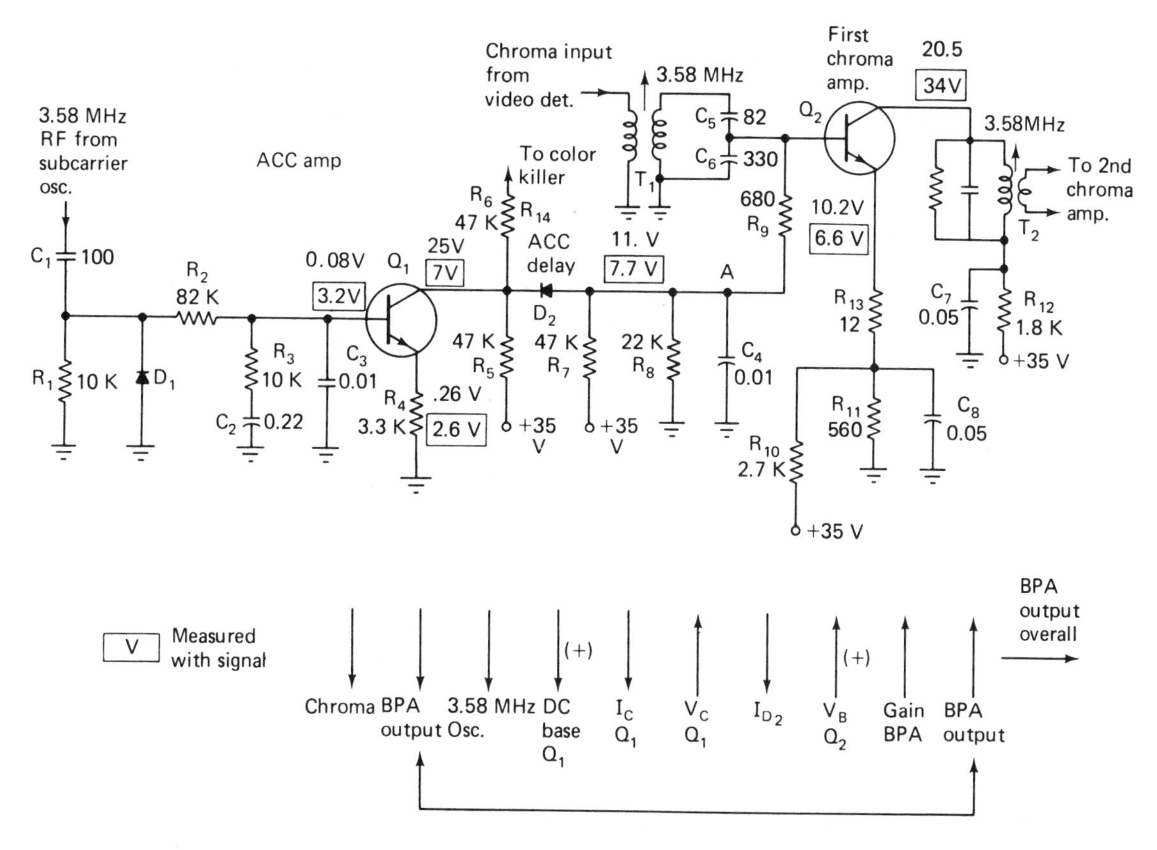

Figure 16-19 All-transistor ACC circuit.

base of Q_1, the ACC amplifier. If no color is present, this dc voltage is only 0.08 V and Q_1 is cut off. When a chroma signal of normal amplitude is received, the base voltage rises to 3.2 V, forcing Q_1 to conduct. The flow of collector current forces the collector voltage to drop from 25 V, (the collector voltage when the transistor was cut off), to 7 V. The voltages found at the base, collector, and emitter of Q_1 depend upon the strength of the chroma signal. In practice, the voltages may be different from the voltages indicated on the schematic diagram. For the voltages indicated, the ACC delay diode will be forward biased, coupling the dc collector voltage to the base of Q_2, the bandpass amplifier. This decrease in base voltage moves the operating point of the bandpass amplifier toward cutoff and lowers the gain of the stage.

The ACC delay diode D_2 is designed to prevent the ACC bias from reaching the BPA until the chroma signal level increases to the point where it overcomes the reverse bias of the delay diode. This is done so that the BPA will operate at maximum gain for weak color signals and have automatic color control from normal or strong signals. Thus, for weak signals, the RF subcarrier oscillator voltage will be small, perhaps only 1.5 V. This will cause the collector voltage of Q_1 to drop from 25 V to approximately 14 V. Since the cathode of the delay diode is more positive than the 11 V on its anode, the diode remains reverse biased and the BPA is not affected by the ACC circuit.

The arrow diagram accompanying the schematic diagram of Figure 16-19 shows the effect on circuit operation if the input chroma signal amplitude decreases, thereby causing a decrease in chroma output. This undesired condition is compensated for in the following manner. The reduced chroma signal causes the RF subcarrier level to decrease, which after rectification decreases the base voltage of the ACC amplifier Q_1. This in turn forces the collector current to decrease, thus increasing the collector voltage. The forward bias of the delay diode is decreased, thus reducing the diode forward current and increasing the base voltage of the BPA, Q_2. Since this amplifier is using reverse AGC, moving the bias away from cutoff tends to increase the gain and signal output of the stage. Notice that the decrease in output signal level caused by a decreased chroma signal has been compensated for by an increase in BPA gain. Thus, the net overall chroma signal output of the BPA tends to remain constant.

16-11 troubleshooting the ACC circuit

Defects in an ACC circuit can cause the following trouble symptoms.

1. No color
2. Weak color
3. Excessive color
4. No ACC action

Since the first three trouble symptoms can be caused by many circuits, it is important to isolate the trouble to the stage causing the trouble. This can be done by a bias clamp.

For example, placing a correctly adjusted bias clamp at the proper point in the ACC circuit will restore the receiver to normal operation if the defect is in the ACC circuit. The proper position to place the bias clamp in the circuit shown in Figure 16-19 is where the ACC circuit feeds the bandpass amplifier (identified by the letter *A*). In this circuit, the correct bias clamp voltage should be approximately 7 V.

A *no-color* condition caused by the ACC circuit can be caused by the BPA becoming reverse biased. In Figure 16-19 a no-color trouble symptom can be caused by such component failures as C_4 short, and R_7 open.

Weak color may be caused by the same kind of circuit defects that caused no color. In this case, however, the components have probably changed value instead of opening or shorting, and instead of shifting the bias of the transistor into cutoff, it is moved closer to zero, reducing the gain.

Excessive color with the color control in its normal position may be caused by an increase of the BPA (Q_2) bias. In the circuit of Figure 16-19 this can be caused by such component failures as D_2 open, R_8 open, Q_1 open, C_3 shorted, and R_4 open.

Loss of ACC action forcing the manual color control to be readjusted whenever channels are changed, is caused by circuit defects associated with the ACC amplifier. When this trouble symptom is associated with the circuit of Figure 16-19, it may be caused by Q_1 open, D_1 shorted or open, D_2 open, or C_1 open.

QUESTIONS

1. What is the primary function of the bandpass amplifier?
2. What characteristic of the color picture does the color control adjust?
3. Explain how it is possible for the tint control to be found in the bandpass amplifier.
4. What is the function of the color killer circuit?
5. What is the function of the color bar generator?
6. What are the differences between the NTSC color bar generator and the rainbow color bar generator?
7. Why is the color bar generator's subcarrier oscillator at a frequency of 3.563811 MHz while the receiver's subcarrier is at a frequency of 3.579545 MHz?
8. Explain why the input wave shape in Figure 16-9 has 11 bursts and the output wave shape has only 10 bursts.
9. List the component failures that might cause the BPA in Figure 16-15 to oscillate.
10. Explain how D_1 in Figure 16-9 is used to control the amplitude of the chrominance signal feeding the emitter of the third chroma amplifier.
11. Explain why the tint control might be found in the bandpass amplifier, burst gate, phase detector, reactance control circuit, and the color subcarrier oscillator circuit.
12. Explain how the tint control of Figure 16-10 operates.
13. What is a module?
14. Explain how and why the bandpass amplifier must be aligned by the video sweep modulation (VSM) method of alignment.
15. The color killer may be used to disable any one of the three stages of the color receiver. What are these stages?
16. Why can a defective color killer or ACC circuit cause a trouble symptom such as no color?
17. Explain how to align a bandpass amplifier using a Chroma Bar Sweep generator.
18. List four trouble symptoms and their possible causes due to a defective bandpass amplifier.
19. List three trouble symptoms that might be caused by (1) a defective ACC and (2) a color killer stage.

chapter 17

Color Sync Circuits

17-1 introduction

The phase of the chrominance signal represents the color of the picture being transmitted. To accurately reproduce this color, the receiver must regenerate and reinsert into the chrominance signal the 3.58-MHz subcarrier that was suppressed at the transmitter. This reinserted 3.58-MHz subcarrier must have the exact phase (0°) of the subcarrier at the transmitter. Any slight deviation of the carrier from the reference phase (0°) will result in a picture of changed color. For example, if a phase error occurs, normal flesh tones may be shifted so that green or magenta people result.

The color burst riding on the back porch of the horizontal blanking pulse is a sample of 8 to 10 cycles of the suppressed 3.58-MHz reference subcarrier. This reference color burst has a phase of 0° and is used to synchronize the receiver's 3.58-MHz oscillator.

It is the function of the color sync circuits to separate the color burst from the composite video signal, compare it to the phase of the local 3.58-MHz oscillator, and correct the oscillator's phase if the two signals are not phase synchronized.

Block diagram. Figure 17-1 is the block diagram of a complete color receiver. The shaded blocks are representative of those color circuits that are discussed in this chapter. These three circuits are designed to generate the 3.58-MHz subcarrier and to ensure its phase and frequency synchronization with the color burst. Other systems of subcarrier regeneration are also being used and are discussed in this chapter.

The burst gate is designed to separate the color burst from the composite video signal feeding its input. The isolated color burst is then fed to an automatic frequency and phase control (AFPC) circuit consisting of a *phase detector,* an *oscillator control circuit,* and the *3.58-MHz oscillator.*

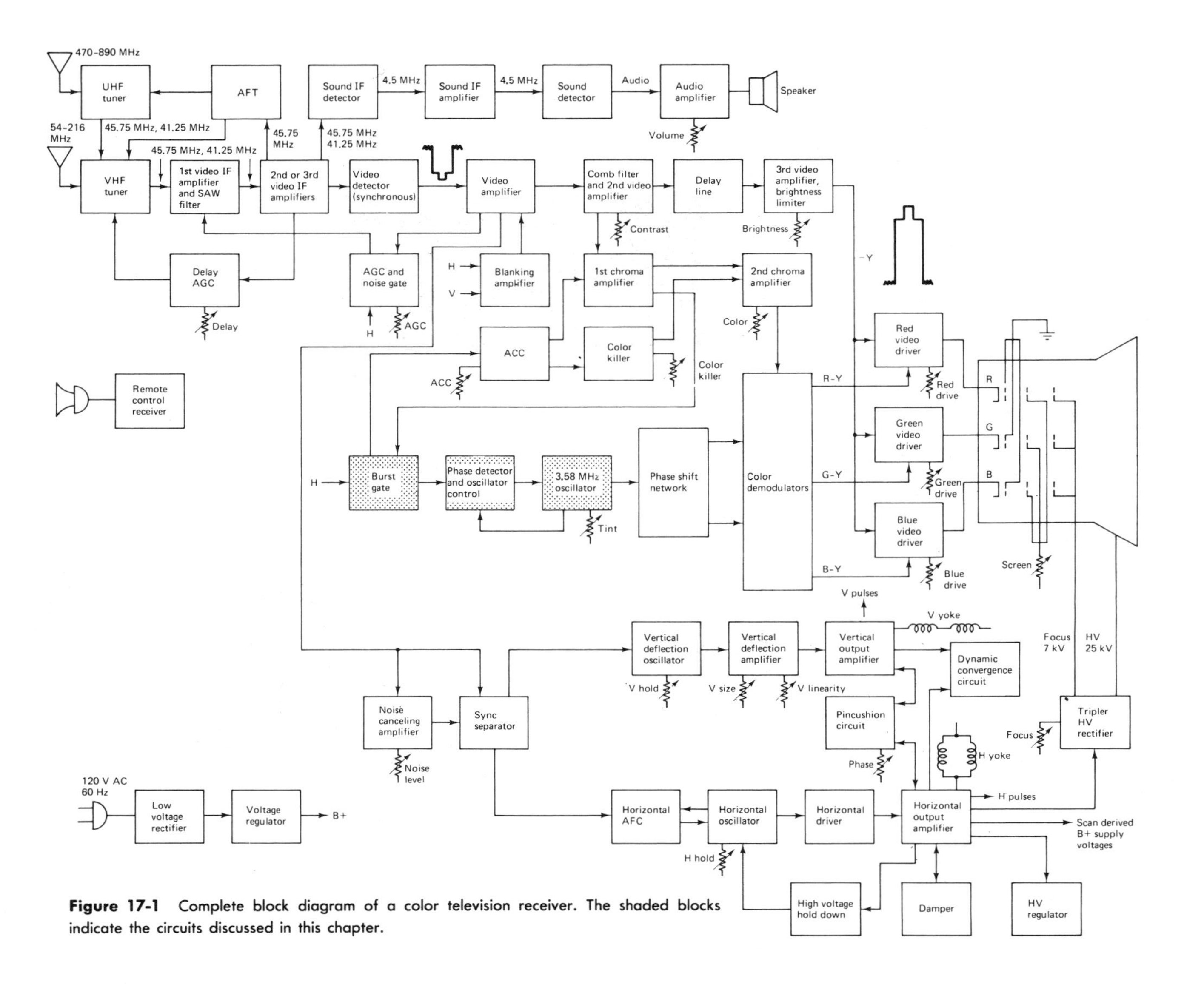

Figure 17-1 Complete block diagram of a color television receiver. The shaded blocks indicate the circuits discussed in this chapter.

The phase detector compares the color burst to the locally generated 3.58-MHz subcarrier signal and develops a dc error voltage that is proportional to any phase difference. Any dc error voltage that is developed is then fed to the oscillator control circuit that is able to shift the phase of the 3.58-MHz oscillator and return it to its proper zero phase position.

17-2 the burst gate

The *burst gate*, or *burst amplifier* as it is often called, separates the color burst from the chrominance signal and is essentially a gated Class B or C amplifier that is tuned to 3.58 MHz. Typically, the bandpass of the tuned circuit is approximately 0.5 MHz.

The detailed block diagram of a burst gate and its associated waveforms is shown in Figure 17-2 and indicates that the circuit has two inputs. One of the inputs is a large-amplitude horizontal retrace pulse obtained from the flyback transformer. This pulse momentarily drives the amplifier out of cutoff and allows the amplifier to function normally.

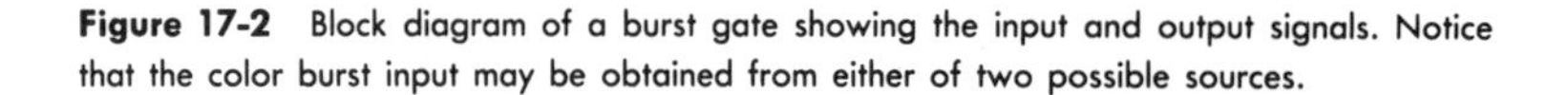

Figure 17-2 Block diagram of a burst gate showing the input and output signals. Notice that the color burst input may be obtained from either of two possible sources.

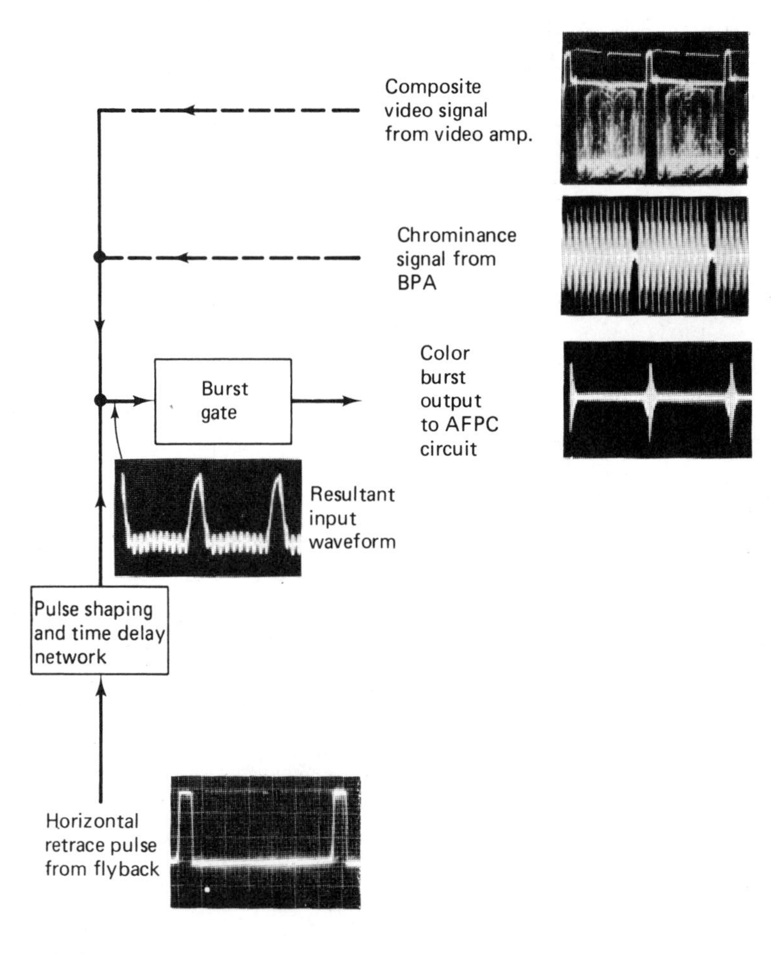

The other input is the chrominance signal, which may be obtained from either the video amplifier or the bandpass amplifier.

Burst gate timing. For proper operation, the timing of the horizontal pulse and the color burst must be such that the amplifier is turned on and operative during the time that the color burst is at the input of the burst gate. Under these conditions, the output of the burst gate will consist solely of the color burst. The horizontal retrace pulse formed at the flyback transformer is generated during the horizontal sync pulse interval. Since this occurs *before* the color burst, a pulse-shaping and time delay network is placed between the flyback transformer and the burst gate. The location of this network is shown in the block diagram of Figure 17-2. If this network delays the flyback pulse by the correct amount (see Figure 17-3), the pulse will occur at the same time as the color burst, and the output color burst will have maximum amplitude and duration. If some pulse-shaping circuit defect causes the horizontal pulse to be in step with the horizontal sync pulse as shown in Figure 17-3, the

Figure 17-3 Effect on burst gate operation caused by changes in horizontal pulse timing.

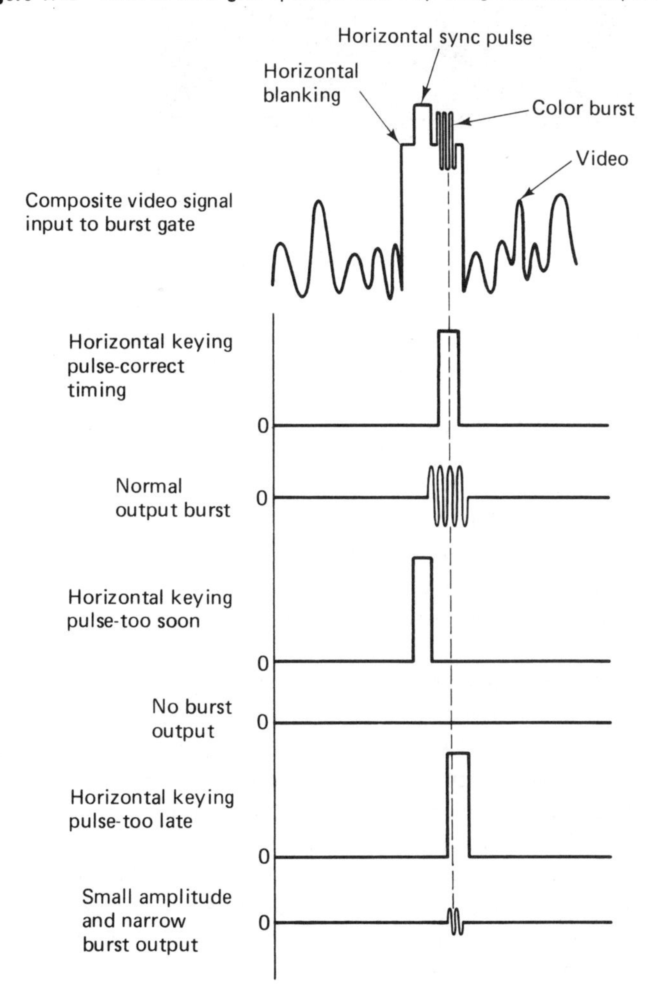

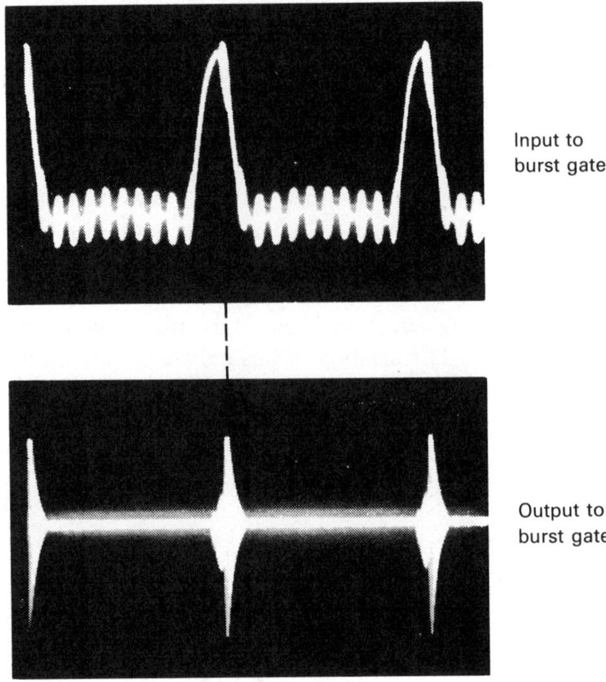

Figure 17-4 Input and output waveforms of a burst gate. The input signal is a color bar generator.

burst gate will conduct before the color burst appears and there will be no output. This will result in the loss of color synchronization. See Figure 17-6. That is, the black & white picture will be normal, but the color portion of the picture will not be in its proper place at the proper time and will appear to drift across the picture.

If a time delay circuit defect introduces excessive time delay and shifts the position of the horizontal pulse to that shown in Figure 17-3, the burst gate output will be smaller in amplitude and shorter in duration than is normal. This in turn may result in erratic or critical color synchronization.

Proper timing between the horizontal pulse and the color burst may be checked by observing the wave shape at the input to the burst gate. (See Figures 17-2, 17-4, and 17-5.) Notice that the 11-pulse bar pattern is superimposed on the horizontal pulse and that the color burst is positioned on the peak of the horizontal pulse. See Figure 17-4.

Transistor burst gate. The transistor burst gate shown in Figure 17-5 uses a PNP transistor arranged in the common emitter configuration. The operating bias of the transistor is determined by resistors R_2, R_3, and R_5. The base and emitter voltages shown in the schematic indicate that the transistor is normally reverse biased. This means that the transistor is operated Class C and is pulsed into conduction during horizontal retrace time. Also indicated in the schematic is the fact that the base of the burst gate receives two input signals: the horizontal keying pulse and the chrominance signal including the color burst.

The negative-going horizontal keying pulse is obtained from the collector circuit of the horizontal retrace blanker and is coupled to the base of the burst gate via C_1 and R_1. Pulse delay is controlled by R_1 and C_2 which returns to ground through the bandpass

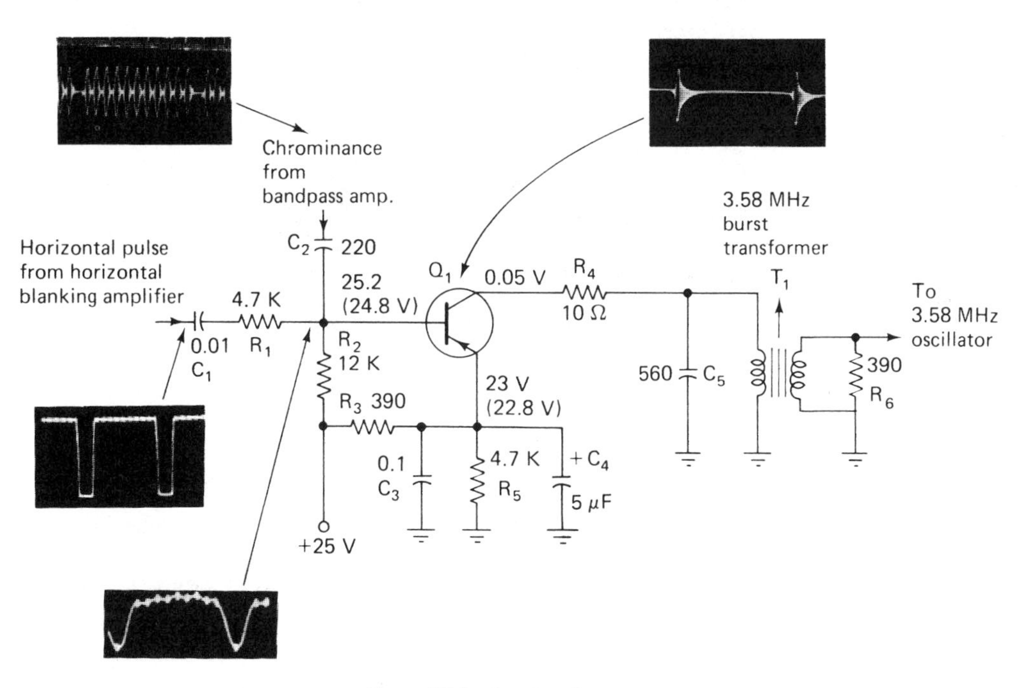

Figure 17-5 Transistor burst gate.

amplifier. The chrominance signal is obtained from the bandpass amplifier and is coupled to the base of the burst gate by C_2.

The function of the horizontal keying pulse is to force the burst gate into conduction during the time that the color burst is present at the base of the transistor. Thus, the output collector voltage developed across tuned transformer T_1 will be an amplified color burst with all other chrominance information missing. The burst signal is then coupled to the 3.58-MHz oscillator via the secondary winding of tuned burst transformer T_1.

Alignment of the tuned transformer (T_1) is important because it affects the amplitude and phase of the color burst output feeding the 3.58-MHz color oscillator circuit. When the tuned transformer is at resonance, it is resistive and produces no undesired phase shift and a maximum secondary output voltage. If the transformer tuning is not at resonance, the phase of the output color burst will be changed. This, in turn, will cause a change in picture color.

17-3 troubleshooting the burst gate

A defective burst gate can cause the following trouble symptoms:

1. Loss of color sync
2. No color
3. Wrong color

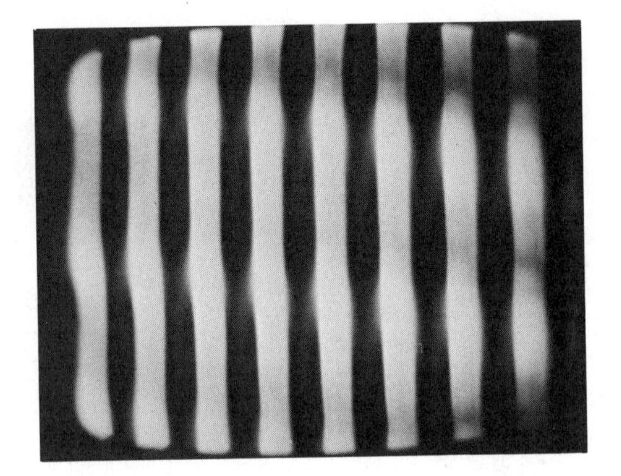

Figure 17-6 Loss of color sync. See color plate.

Loss of color sync. The exact phase and frequency of the local color oscillator depend upon the color burst. If the burst is not able to reach the color oscillator, it will become free-running and cause the color to run erratically through the picture. This is called *loss of color sync.* Figure 17-6 is an example of how a color bar pattern appears when there is a loss of color sync. The number of diagonal color bars depend upon how far out of color sync the local color oscillator has drifted. If the frequency of the local 3.58-MHz oscillator is exactly right, but it has only shifted in phase, the picture will have wrong color. If the frequency of the 3.58-MHz oscillator has shifted slightly from its correct value of 3.579545 MHz, the color information will no longer be in sync with the black & white picture information and will appear as moving diagonal stripes running through the picture.

The fastest way to isolate a defective burst gate is by signal tracing with an oscilloscope. For example, if the input to the base has a wave shape similar to that shown in Figure 17-4 or Figure 17-5, but no output burst at the collector, the trouble is definitely localized to the burst gate stage.

Loss of color sync in a receiver using a transistor burst gate such as that of Figure 17-5 may be caused by such circuit defects as a defective burst gate transistor Q_1, open coupling capacitors C_1 and C_2, open delay network resistor R_1, open bias resistor R_3, open damping resistor R_4, defective or misaligned output transformer T_1, or shorted tuning capacitor C_5.

No color. The burst gate can be indirectly responsible for the loss of color. This can be better understood by referring to the block diagram of Figure 17-1. Notice that the burst gate feeds the ACC circuit. The ACC in turn feeds the color killer. Hence, if the burst gate becomes defective, the color burst will not reach the ACC and the ACC will not deactivate the color killer. This allows the color killer to disable the bandpass amplifier causing a no-color condition. If the color killer level control is adjusted to deactivate the color killer, the bandpass amplifier will be allowed to function normally and produce a color picture exhibiting a no-color sync condition.

The burst gate can also be directly responsible for loss of color in those subcarrier regeneration systems that do not use a local 3.58-MHz oscillator. In these systems a defective burst gate removes the 3.58-MHz input to the demodulators and produces a no-color symptom.

Wrong color. Defects in the burst gate can also produce symptoms of color change. In some cases, the changes may be erratic. Color shift may be caused by a change in color burst phase. Such phase shifts can be the result of misalignment of the output tuned circuit or of some defect that might cause such a misalignment. For example, if C_5 of Figure 17-5 were open, the phase of the color burst would shift and flesh tones would be purple. The tint control would also be inoperative.

Erratic color. To understand how erratic color may be produced, refer to Figure 17-5 and notice that if R_2, R_3, or R_5 were to change value so that Q_1 would be forward biased, the burst gate will no longer be able to separate the color burst from the chrominance signal and the full chrominance signal will appear at the output of the burst gate. This in turn forces the phase of the local 3.58-MHz oscillator to shift with the changing chroma signal and causes the colors in the picture to change erratically.

17-4 automatic frequency phase control

The section of the color receiver that forces the color subcarrier oscillator to assume the same phase as the color burst is called the *automatic frequency and phase control* (AFPC). In practice, for a color to be reproduced as transmitted, the phase error in demodulation should not exceed $\pm 5°$. To allow the operator to adjust the phase and to compensate for any phase error, a tint control is used and may have a phase control of 70° or more. Three such systems are being used. The block diagram of each system is shown in Figure 17-7.

The AFPC system shown in Figure 17-7(a) uses a phase detector to compare the 3.58-MHz local oscillator frequency to that of the color burst and it generates a dc error voltage. The dc error voltage then activates an oscillator control circuit that forces the local 3.58-MHz oscillator into phase synchronization with the color burst. This circuit arrangement is called a phase-locked loop (PLL).

Included in the block diagram are the waveforms found in the circuit and their relative phase relationships. Notice that the burst gate has dual burst outputs, 0° and 180°. The color oscillator's output also feeds the phase detector and is shifted 90° with respect to the color burst. In addition, the $R-Y$ (90°) and $B-Y$ (180°) outputs of the oscillator feed the receiver's color demodulators.

By comparison, the block diagrams [Figure 17-7(b) and (c)] of the ringing oscillator systems of AFPC shows that these systems are considerably simpler than the phase detector system. In these arrangements the output of the burst gate directly drives the 3.58-MHz reference crystal. Thus, in the *crystal injection-locked oscillator* circuit [Figure 17-7(b)] the color burst forces the crystal to ring at the desired frequency and it phase locks the oscillator to the color burst.

The *crystal ringing* amplifier circuit [Figure 17-7(c)] is not an oscillator. The color burst forces the crystal to ring at the proper frequency and phase. This signal is then amplified by buffer amplifiers and is fed to the demodulators.

17-5 the phase detector

The schematic diagram of a typical phase detector is shown in Figure 17-8(a). In this circuit the burst gate transistor (Q_1) drives the phase detector through a tuned transformer (T_1) whose secondary winding has a grounded center tap. As a result, the transformer

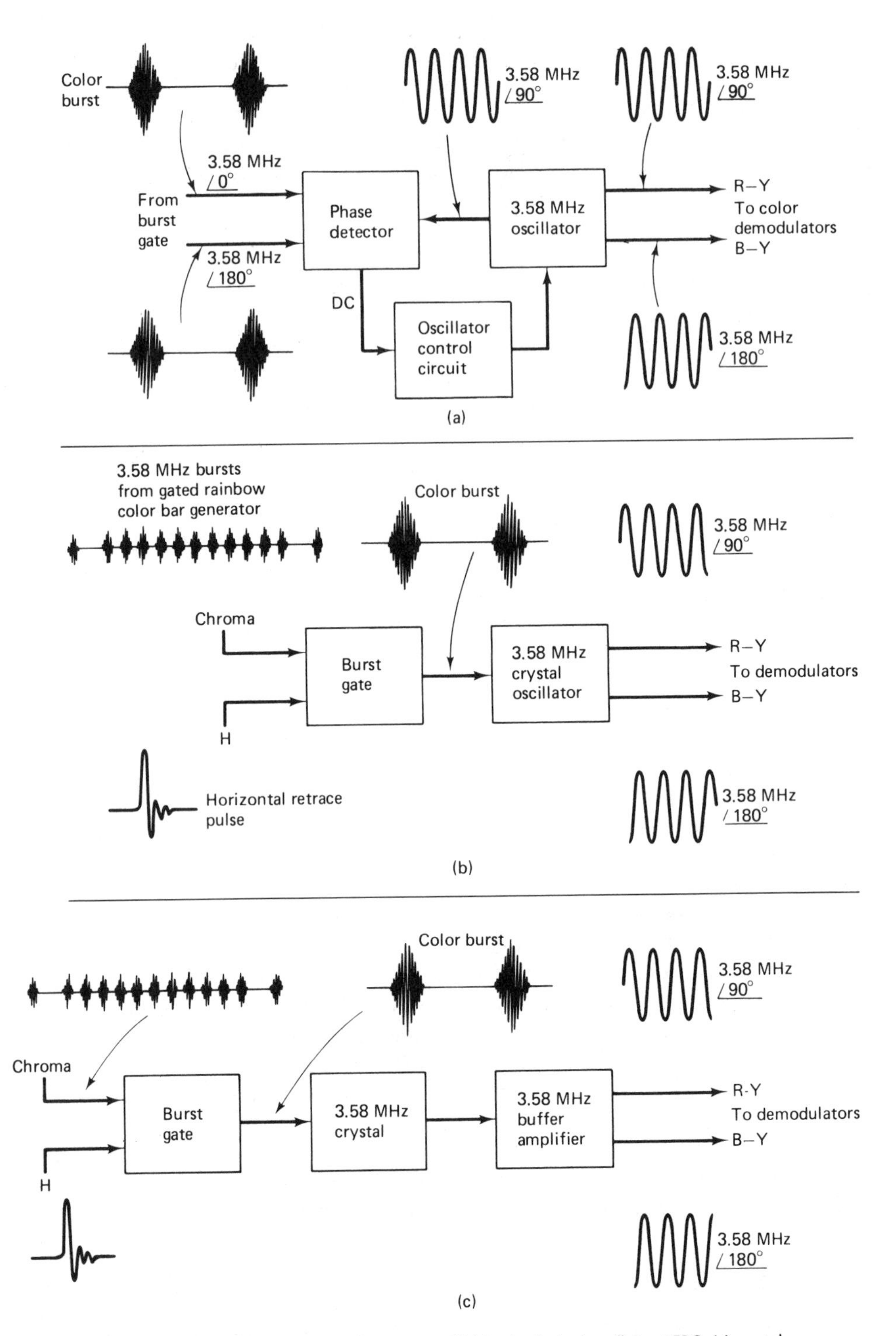

Figure 17-7 (a) Phase detector AFPC (PLL); (b) injection-locked oscillator AFPC; (c) crystal ringing AFPC.

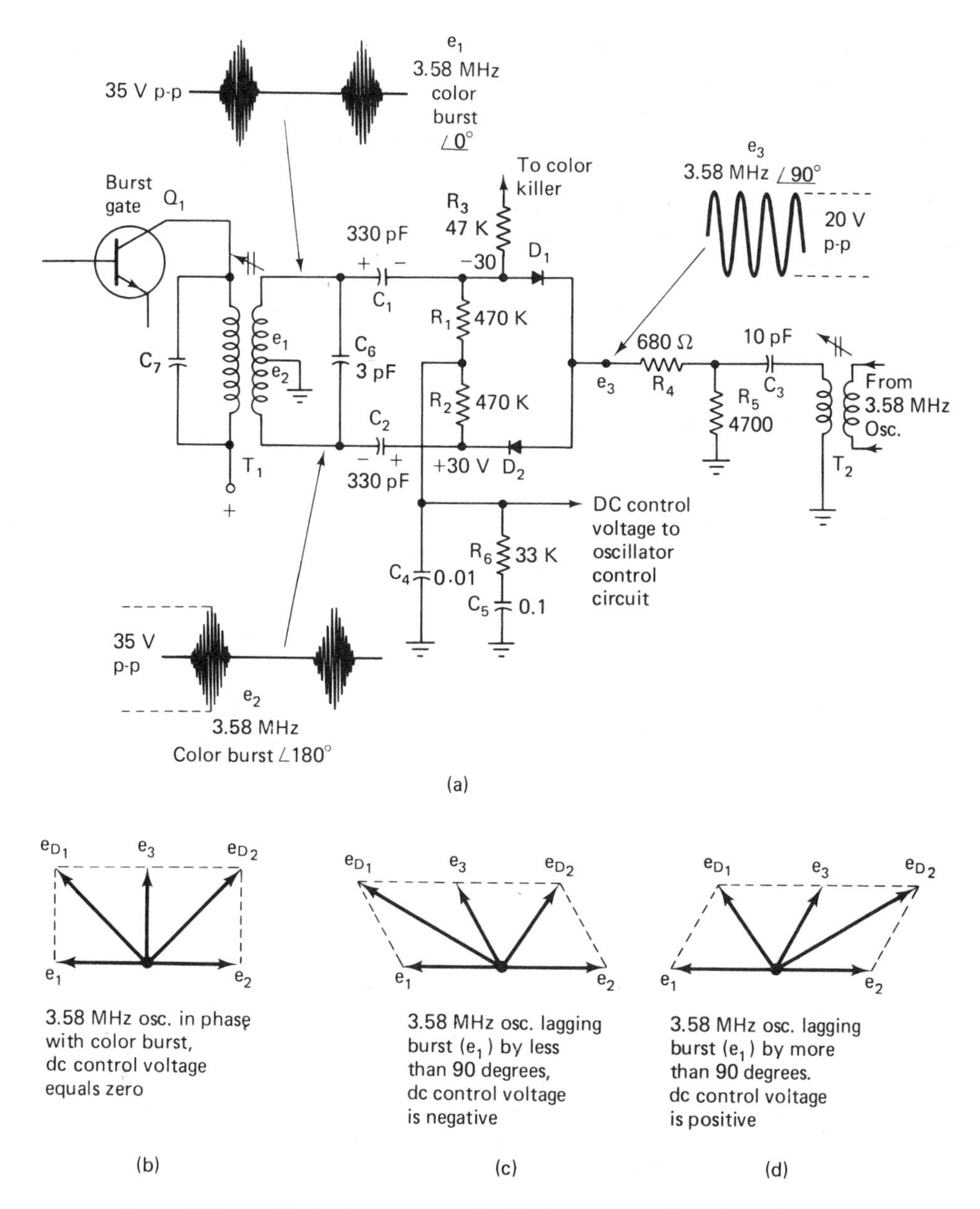

Figure 17-8 (a) Diode phase detector; (b)–(d) phase relations that exist in the phase detector circuit under different conditions.

provides two burst outputs (e_1 and e_2) that are equal in amplitude but opposite in phase. These bursts are coupled to the two phase detector diodes (D_1 and D_2) through coupling capacitors C_1 and C_2. For proper phase detector operation, it is important that these capacitors and their associated load resistors R_1 and R_2 are matched to one another.

The 3.58-MHz oscillator transformer (T_2) develops a voltage (e_3) that feeds the phase detector diodes through a phase shift network consisting of T_2, C_3, R_4, and R_5. These components are responsible for placing the color oscillator signal 90° out of phase with the two burst signals feeding the other side of diodes D_1 and D_2. Since all of the signal voltages acting in the phase detector are sinusoids, they may be represented by a vector diagram. Thus, during a color broadcast the normal phase relationships that exist in the phase detector when the color burst and the local 3.58-MHz oscillator are in phase are shown in the vector diagram of Figure 17-8(b). Notice that the ac voltages across the diodes (e_{D1} and e_{D2}) are the vector sums of the oscillator voltage e_3 and the color burst signals (e_1 and e_2) feeding the diodes. Under normal conditions e_3 is 90° out of phase with both e_1 and e_2. As a result, the ac voltages e_{D1} and e_{D2} across the diodes are equal to one another and the diodes conduct equally.

When diode D_1 conducts, conventional current flows in the direction of the diode arrow and C_1 charges with the polarity indicated. Similarly, when D_2 conducts, C_2 charges in the opposite direction. Between conduction periods the capacitors discharge. C_1 discharges through T_1, *up* through C_4, and back to C_1 through R_1. C_2 discharges through R_2, *down* through C_4, and then back to C_2 through T_1. If the circuit is balanced, that is, if all of the components are matched, the net voltage developed across C_4 will be zero. The voltage across C_4 is the dc error voltage used to activate the oscillator control circuit. Since the voltage developed across C_4 is zero, no correction of the 3.58-MHz reference oscillator phase will occur.

It should also be noticed that as a result of the rectification of e_{D1} and e_{D2} fairly large voltages (between 5 V and 20 V) are developed at the junctions of C_1, R_1 and C_2, R_2. The magnitude of these voltages is determined by the strength of the color burst. Strong color signals will produce larger voltages than weak signals will. During black & white transmission the color burst is not transmitted; this causes the diode voltages e_{D1} and e_{D2} to become equal to the oscillator voltage e_3. As a result, the dc voltage developed at the diodes will drop to a range of 3 to 10 V. In many receivers this dc voltage is used to key the color killer and automatic color control circuits. These facts may also be used to check the phase detector for proper operation.

Phase error. If the color oscillator shifts in phase so that it lags the color burst by less than 90°, the circuit phase relationship may change to that shown in Figure 17-8(c). The net result of this is to make the diode voltage e_{D1} larger than e_{D2}. This, in turn, causes D_1 to conduct more heavily than D_2, charging C_1 to a larger voltage than that developed across C_2. During the discharge of these capacitors a greater current (conventional) flows up through C_4 than down. This leaves a net negative control voltage across C_4. This voltage then causes the oscillator control circuit to shift the phase of the 3.58-MHz oscillator back to that of Figure 17-8(b) so that a zero control voltage appears at the output of the phase detector. R_6 and C_5 at the output of the phase detector form an antihunt circuit that prevents the oscillator control system from overcorrecting or undercorrecting phase errors.

Figure 17-8(d) illustrates the phase relationship that exists if the 3.58-MHz oscillator were to lag the color burst (e_1) by more than 90°. In this case, diode voltage e_{D2} becomes larger than e_{D1} and results in a positive voltage across C_4. This positive voltage then forces the oscillator control circuit to shift the phase of the 3.58-MHz oscillator so that the oscillator (e_3) and burst voltages (e_1 and e_2) are 90° out of phase and the dc error voltage produced by the phase detector is zero.

Color killer and ACC detectors. Many receivers may use essentially the same phase detector circuit discussed in the preceding section as a *color killer detector,* an *automatic color control (ACC) detector,* or a *color demodulator.* In the case of the color killer and ACC detectors, the object of these detector circuits is to convert the color burst into a dc voltage that will control the color killer or the gain of the BPA. An example of a color killer phase detector used in conjunction with an AFPC phase detector is shown in Figure 17-9. The application of the phase detector as a color demodulator will be discussed in Chapter 18.

In this circuit the center tapped secondary of the burst transformer T_1 feeds the color burst to both detectors. The color burst is then coupled to the detector diodes through *RC* networks (C_1, C_2, C_3, C_4, R_1, R_2, R_3, and R_4) that are identical to one another. The dc output of each detector is taken from the junction of its matched pair of resistors and is developed across the respective filter capacitors C_5 and C_6. The dc error voltage developed across C_6, the AFPC *phase detector* output capacitor, is then fed to the oscillator control

Figure 17-9 Color killer detector and AFPC diode phase detector.

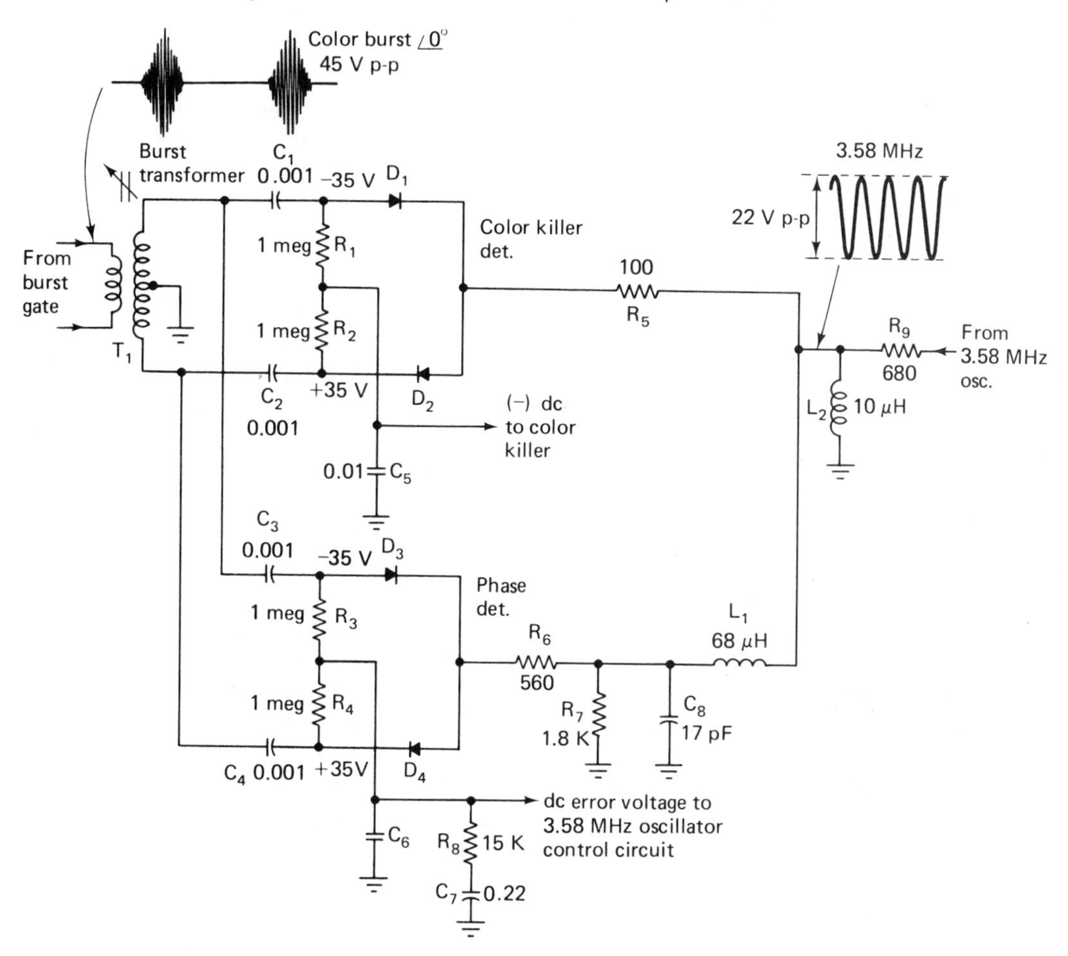

circuits. Similarly, the dc voltage developed across C_5, the *color killer detector* output capacitor, is fed to the color killer circuit.

The only real difference between these two circuits is the phase shift network between the color oscillator and the detectors. Notice that a complex phase shift network consisting of L_1, L_2, C_8, R_9, and R_7 is placed between the 3.58-MHz oscillator and the AFPC phase detector, but only a small resistor (R_5) isolates the 3.58-MHz oscillator from the color killer detector. This arrangement is used so that the oscillator voltage feeding the *AFPC phase detector* will be 90° out of phase with the color burst and so that the oscillator voltage feeding the *color killer detector* will be in phase (0°) with the color oscillator. As a result, during color reception the dc output of the AFPC phase detector is normal and will depend upon the relative phase between the color burst and the local 3.58-MHz oscillator. Thus it may vary from a positive to a negative value. During black & white transmission the color burst will be missing and the output of the detector will be zero.

During black & white reception the output of the *color killer detector* will be zero, for the same reason as above, and during color broadcasts it will become highly negative. This is because e_3 is in phase with e_1. See Figure 17-8(b). This negative keying voltage is then used to disable the color killer, which in turn allows the bandpass amplifier to operate normally.

Some color receivers use the output of the color killer detector as an ACC source. This is possible because this detector produces a dc output voltage that is directly proportional to the signal level and may be used to control the gain of the bandpass amplifier.

17-6 the oscillator control circuit

The AFPC block diagram of Figure 17-7(a) indicates that the oscillator control circuit is situated between the output of the color phase detector and the 3.58-MHz crystal oscillator. The function of this circuit is to convert the dc error voltage generated by the phase's detector into a variable reactance that will control the phase of the 3.58-MHz oscillator.

Three circuits that are commonly used for this purpose are (1) varactor oscillator control circuits, (2) reactance FET control circuits, and (3) transistor reactance circuits.

The varactor oscillator control. The oscillator control circuit shown in Figure 17-10 is based on the operation of a varactor. This circuit element is a reverse-biased silicon diode that is arranged to operate as a voltage-controlled capacitor.

In this circuit the dc error voltage output of the phase detector is filtered by C_1, C_2, and R_1 before it passes through the isolating resistor R_2 and acts as the control voltage for the varactor D_1. The voltage divider resistors R_3, R_4, and R_5 are used to set the reverse operating bias of the varactor. This in turn determines the capacitance of the varactor when the dc error voltage is zero. Changing the bias of the varactor causes a change in its effective capacitance. In some receivers the variable resistor R_4 is a front panel tint control. In other receiver circuits R_4 may only be used as an AFPC alignment adjustment.

From an ac point of view, the varactor is effectively in parallel with the crystal (X_1) because the anode of the diode is tied directly to one side of X_1 and the cathode of D_1 returns to the other side of the crystal through capacitors C_3, C_4, and C_5. Thus, if the error were to become positive, the varactor reverse bias would become smaller, and the varactor's

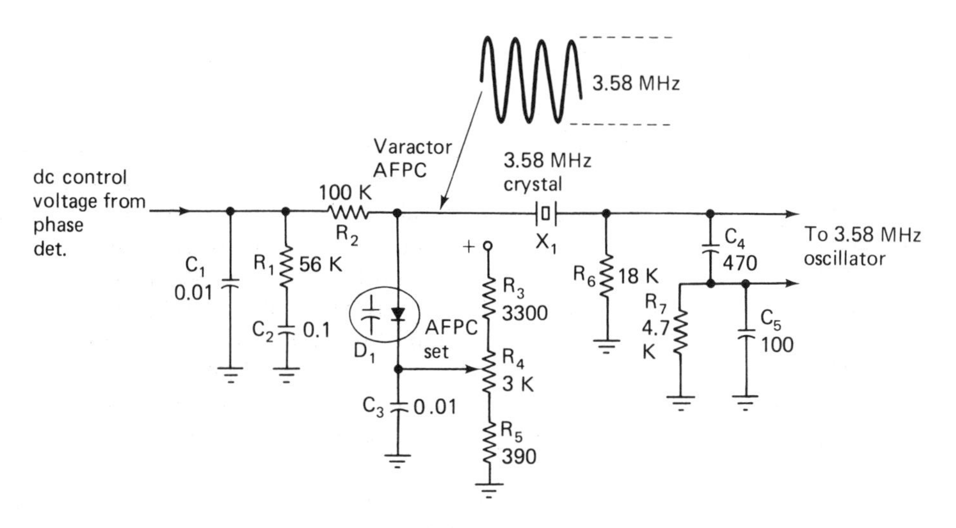

Figure 17-10 Varactor control circuit.

effective capacity would increase. This in turn would have the effect of shifting the phase of the crystal oscillator in the direction necessary to reduce the positive error voltage to zero. A negative error voltage would produce a similar result by making the varactor capacitance smaller.

The reactance FET. Another circuit in which a reactance can be controlled by a dc voltage is the reactance FET. In this circuit arrangement, shown in Figure 17-11, the ac drain current of the Class A FET amplifier (Q_1) is made to lead its ac drain voltage by 90°. This is done by the phase shift network C_7 and R_5 that is connected across the oscillator tuned

Figure 17-11 Reactance FET circuit.

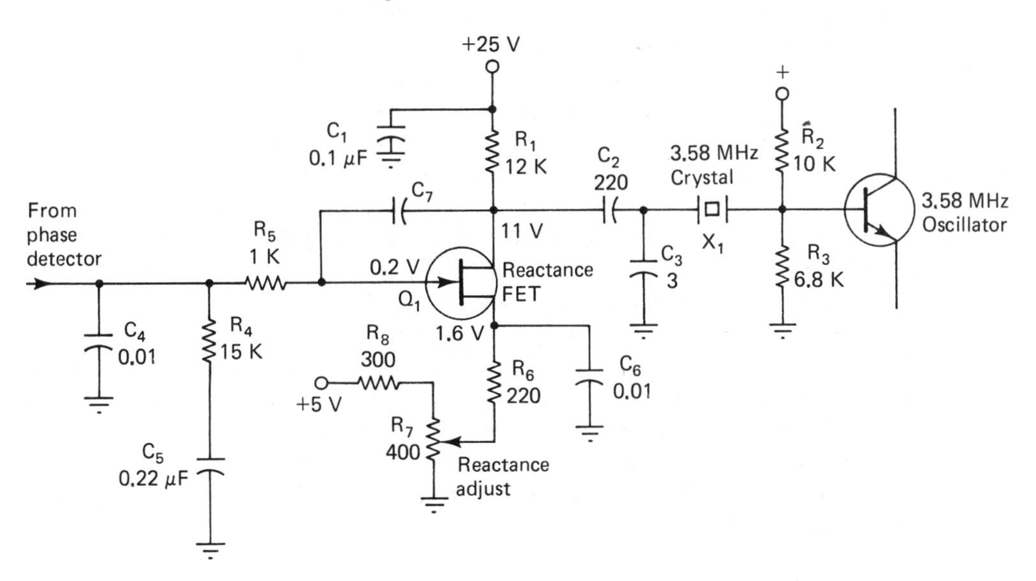

circuit formed by the crystal X_1. As a result, the oscillator "sees" the reactance FET as though it were a capacitive reactance.

The effective capacitive reactance of the FET may be changed by changing the dc bias of the FET. Shifting the FET's bias causes the ac drain current to change amplitude, which by Ohm's law is equivalent to a change in the FET's effective reactance. Since capacitive reactance and capacitance are inversely related, an increase in the FET's reactance is equivalent to a decrease in the tuning capacitance of the oscillator tuned circuit. If the capacitance change is large enough the oscillator frequency will increase, but small changes in reactance FET effective capacitance will only cause the oscillator to shift phase.

The reactance FET operating bias is established by R_6 and R_7 the source resistor. Capacitor C_6 is a short circuit at the operating frequency of 3.58 MHz and acts to prevent degenerative feedback. The dc error voltage obtained from the phase detector is filtered by R_4, C_5, and C_4. R_4 and C_5 also form an antihunt circuit similar to those used in the horizontal AFC.

Transistor reactance circuit theory, circuit configuration, and operation are for all purposes identical to that of the reactance FET, and will not be discussed here.

17-7 the color oscillator

The color oscillator is a precision RF generator operating at the exact frequency of 3.579545 MHz. The output of this generator is fed to the color demodulators where it is reinserted into the chrominance sidebands. In effect, it restores the subcarrier that was suppressed at the transmitter. The frequency and phase of the color oscillator must be identical to that of the subcarrier if high-fidelity color reproduction is to be obtained.

Crystals. To help achieve this high degree of frequency and phase accuracy, all color oscillators are crystal controlled. These crystals are constructed of a quartz wafer that is mounted between two metal plates.

Quartz and other crystal materials are piezoelectric; that is, they generate a voltage when mechanically stressed, and they become mechanically stressed when a voltage is applied to them. This means that if a crystal is suddenly shocked and made to vibrate, it will produce a damped oscillatory output voltage similar to that produced by a shock-excited tuned circuit. In fact, the crystal used in a color oscillator may be considered to be equivalent to an LC tuned circuit. If an LC tuned circuit is shock excited, the damped oscillation will decay at a rate that depends upon the resistive losses in the circuit. This is conveniently expressed by the Q of the circuit. In an ordinary LC circuit a Q of 500 is considered very high. A crystal like that used in color oscillators is equivalent to a tuned circuit having a Q of 300,000 or more. This means that the bandpass of the crystal at 3.58 MHz is only 12 cycles, making it essentially a single-frequency device.

The frequency of crystal oscillation depends upon the thickness of the quartz crystal. During manufacture this thickness can be controlled so precisely that it is possible to obtain crystals that oscillate at the exact color subcarrier frequency of 3.579545 MHz. Because the crystal is the equivalent of an LC tuned circuit, it is possible to shift the frequency or phase of its oscillation slightly by shunting it with a small variable capacitor such as a varactor or reactance FET.

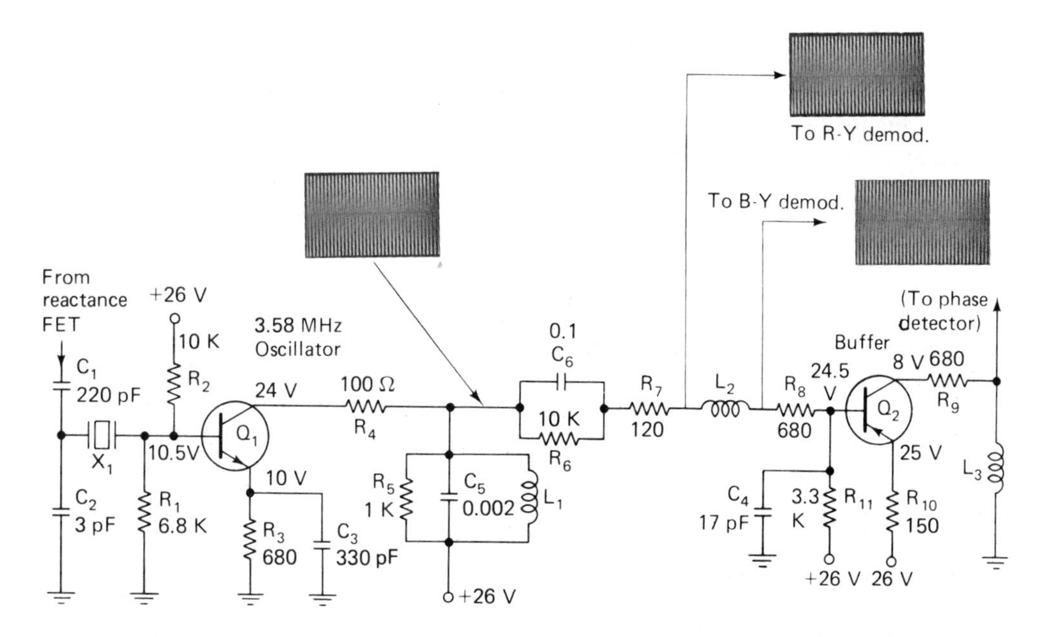

Figure 17-12 Transistor 3.58-MHz color oscillator used in a phase detector AFPC system.

The transistor oscillator. The schematic diagram of a transistor 3.58-MHz oscillator used in a phase detector AFPC circuit is shown in Figure 17-12. This oscillator is arranged as the transistor equivalent of a TPTG (tuned plate tuned grid) oscillator. In this circuit the input and output circuits are both tuned. The crystal X_1, the transistors input capacitance and the capacitors C_2 and C_3 form the input tuned circuit, and the collector tuned circuit is formed by L_1 and C_5. Positive feedback is provided by the internal collector to base capacitance of Q_1. The phase of the oscillator is controlled by a reactance FET that is coupled into the oscillator by C_1.

Resistors R_1, R_2, and R_3 establish the forward starting bias of the oscillator. This bias is necessary if the oscillator is to oscillate when the receiver power supply is turned on.

The sine-wave output of the oscillator is direct coupled through an R/L phase shift network made up of R_6, R_7, R_8, L_2, and C_6 to the base of the buffer amplifier Q_2. As a result, the dc collector voltage of the 3.58-MHz oscillator (Q_1) establishes the base voltage of the buffer amplifier Q_2. The phase shift network adjusts the output of the oscillator so that the proper phase and amplitude signals are delivered to the $R - Y$ and $B - Y$ demodulators.

The buffer amplifier (Q_2) amplifies and inverts the oscillator signal before passing it through the phase shift network R_9 and L_3 and delivering the signal to the AFPC phase detector. The phase of the oscillator voltage at this point should be 90° out of phase with the color burst. In addition, the buffer amplifier acts to isolate the oscillator and demodulators outputs of the oscillator from the phase detector.

17-8 alignment of the phase detector AFPC

Examination of the complete circuit of the typical phase detector AFPC circuit shown in Figure 17-13 will show that there are three tuned circuits (T_1, T_2, and L_1) and one potentiometer (R_{19}) that must be adjusted if proper AFPC action is to be obtained.

The circuit consists of a burst gate (Q_1) that couples the color burst through T_1 to the phase detector diodes D_1 and D_2. The dc error voltage produced by the phase detector is then used to activate the varactor D_3, which in turn controls the phase of the 3.58-MHz crystal oscillator (Q_2). The sine-wave output of the crystal oscillator is amplified by the CW amplifier (Q_3). At this point the 3.58-MHz oscillator output is divided and a portion of it is coupled by C_{12} back to the phase detector. The remaining 3.58-MHz signal is then coupled via C_{13}, C_{14}, and C_{15} to the bases of the first and second tint amplifiers (Q_4 and Q_5). The ac collector current of each tint amplifier transistor differs in phase because the impedance in the emitter of Q_4 is capacitive and that in the emitter of Q_5 is inductive. Since the total ac collector current flowing through the primary of T_2 is the vector sum of the individual ac collector currents, changing the amplitude of either one will cause the total current to shift in phase. This in turn results in a change in picture color. The amplitude of Q_5's collector current is adjusted by the tint control R_{34}, which is part of a voltage divider circuit consisting of R_{32}, R_{33}, R_{34}, and R_{35}. This circuit determines the base bias voltage of Q_5 and is therefore able to control the transistor's gain and its ac collector current.

The three objectives of phase detector AFPC alignment are as follows: (1) The burst transformer T_1 must be tuned to resonance (3.58 MHz). (2) The oscillator voltage feeding the phase detector must be 90° out of phase with the color burst. (3) The oscillator phase must be correct when the dc error voltage generated by the phase detector is zero.

The only equipment needed to perform the AFPC alignment are a color bar generator, a VTVM, and some jumper leads. It should be noted that the exact procedure can only be obtained from the manufacturer's service literature and may vary somewhat from one receiver to another.

The alignment procedure for this AFPC circuit is as follows: The tint control is set to midrange, the color bar generator is connected to the antenna terminals of the receiver, and the receiver fine tuning, color control, AGC, and color killer adjustments are made to produce the best color picture possible.

The first step in the alignment procedure is to tune the oscillator CW amplifier tuned circuit L_1, C_{12}, C_{13}, and C_{14} to resonance. The usual method for doing this is to rectify the 3.58-MHz signal developed across the tuned circuit and then measure it with a dc VTVM. Resonance is indicated when the voltage is maximum. Since the phase detector diodes D_1 and D_2 are connected to L_1, they will rectify the 3.58-MHz oscillator voltage and develop dc voltages across R_{10} and R_{11} that are directly proportional to the oscillator voltage. To avoid confusion with the color burst that contributes to the dc voltage developed by the diodes, the burst gate is disabled by shorting the base of Q_1 to ground. The VTVM is connected to point A through a 10-kΩ isolating resistor, and L_1 is tuned for maximum voltage reading.

In this circuit tuned transformer T_2 couples the 3.58-MHz oscillator signal into the demodulators. For proper operation, this tuned circuit must also be tuned to resonance. To indicate when resonance is obtained, an oscilloscope using an LCP and a series isolating resistor of approximately 470 kΩ may be connected to the $R - Y$ output, point B, T_2 is then adjusted for a maximum peak-to-peak 3.58-MHz signal.

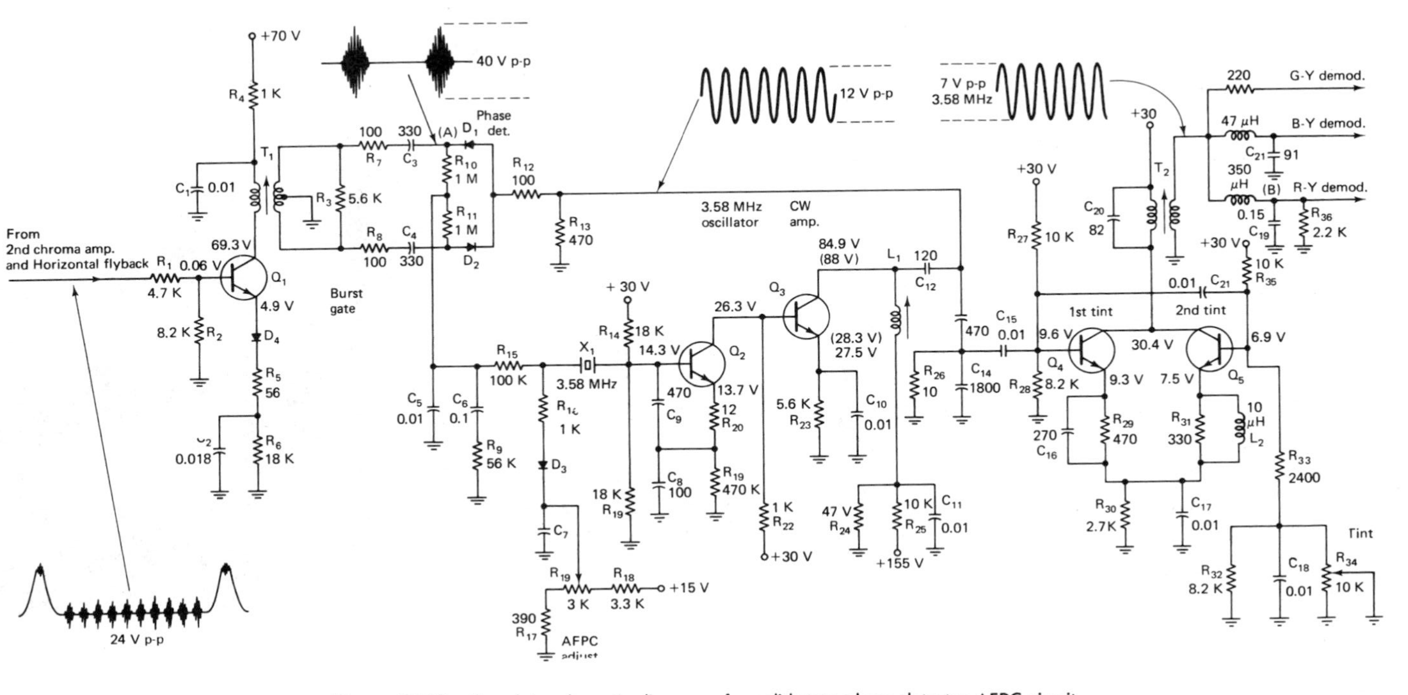

Figure 17-13 Complete schematic diagram of a solid-state phase detector AFPC circuit.

At this point, the color killer should be disabled and the color saturation should be increased to maximum. Since the burst gate is disabled, the resultant color bar picture will be out of color sync. To adjust the free-running frequency and phase of the 3.58-MHz oscillator so that it is approximately correct, the varactor AFPC adjust potentiometer R_{19} is adjusted until the color bars stand still or slowly drift across the screen. This condition is called *zero beat.*

To obtain maximum color burst, the burst transformer T_1 now must be adjusted to resonance. First, the burst gate is restored to normal by removing the base to ground short. Then an output indicator dc VTVM is connected to point A, and the transformer (T_1) is adjusted for maximum voltage reading.

At this point the major adjustments have been made and all that remains is to check the tint control range and a touch-up if necessary. In a normal gated rainbow color bar pattern such as that shown in Figure 16-5 notice that under proper conditions (the tint control is midrange), the third bar is red, the sixth bar is blue, and the tenth bar is green. Rotating the tint control should move the red, blue, and green bars approximately two bars to the left and two bars to the right. This represents a total phase shift of as much as 60° each way, or a total 120°. If this cannot be done, T_1 should be readjusted. If this does not help, the AFPC circuit should be realigned. Continued difficulty indicates a defect in the circuit that must be repaired.

17-9 troubleshooting the phase detector AFPC

Defects in the phase detector AFPC can cause the following trouble symptoms:

1. Loss of color sync (see Figure 17-6 on the color plate)
2. Wrong color (see Figure 17-14 on the color plate)
3. No color [see Figure 5-2(c) on the color plate]
4. Weak color

These trouble symptoms do not affect the quality of the black & white picture.

Figure 17-14 AFPC trouble symptoms, wrong color. See the color plate.

Loss of color sync. This trouble symptom indicates that the AFPC system uses a color oscillator and that it is functioning but drifting in phase. For this condition to occur, one of the following four things must have occurred:

1. The burst gate is defective and not delivering the color burst to the phase detector.
2. The phase detector is unbalanced or inoperative and is not producing the proper error voltage.
3. The oscillator control circuit is defective and is not able to translate the dc error voltage into proper phase control of the 3.58-MHz oscillator.
4. The 3.58-MHz oscillator has changed frequency and is too far off frequency to respond to the reactance control device.

The burst gate. To check for proper burst gate operation, an oscilloscope is used to determine whether or not burst signals of equal amplitude are reaching the phase detector diodes. If normal amplitude color bursts are present, the burst gate must be functioning normally. If the color burst at the output of the burst gate is weak or missing, the trouble has been localized to the burst gate.

The phase detector. To check the phase detector, first the burst and oscillator input wave shapes are checked with an oscilloscope. These wave shapes should appear similar to that shown in Figure 17-8. If any of these wave shapes is missing, loss of color sync will result. A sensitive test of phase detector operation is to disable the burst gate (short input of burst gate to ground) and measure dc voltage developed at the junction of the diodes and their associated matched resistors. If these voltages are unequal, there is a defective component in the phase detector circuit.

The oscillator control circuit. To check the oscillator control circuit, a bias clamp (dc power supply) whose output can be varied from +3 to −3 V is connected at the output of the phase detector. The bias clamp is a substitute error voltage. If varying the bias clamp voltage is able to restore normal color operation, the oscillator control and the color oscillator must be functioning properly and the phase detector is at fault. If varying the bias clamp causes little or no variation in color sync, it is an indication that the oscillator control circuit is not functioning. If the color sync varies but is not restored to normal, the color oscillator may be far off its proper frequency.

Wrong color. Wrong color may be the result of an oscillator phase error that has been introduced by misalignment or a malfunction of the AFPC. Such a defect will cause an error voltage to be produced when no error exists. A powerful method for isolating the defective stage is to go through an AFPC alignment. If one of the adjustments seems to be unresponsive, it is a good probability that the trouble is related to that part of the circuit that includes the adjustment. If the AFPC alignment is proper, the next step is to isolate the defective stage. The troubleshooting procedure essentially follows the same pattern as that for loss of color sync. First, check the phase detector by waveform and dc voltage measurements. Second, use a bias clamp as a substitute error voltage to check the range of the oscillator control circuit. Any departure from normal requires that all the components associated with the suspected stage be checked for changed value resistors, leaky capacitors, a defective crystal, or diodes and transistors that have had a change in their characteristics.

No color. A no-color trouble symptom caused by the AFPC circuit may be caused by either a nonoperative oscillator or a color killer circuit that has not been disabled because of some defect in the AFPC circuit.

An inoperative 3.58-MHz oscillator does not allow the color demodulator to function because the color demodulators are frequency changers similar to the mixer stage of the tuner. Demodulators have two inputs: the chroma signal (3.58 ± 0.5 MHz) and the 3.58-MHz oscillator signal. The output of the demodulator is the difference signal (0 to 0.5 MHz). Thus, the loss of the oscillator signal will cause the demodulators to cease functioning which in turn causes loss of color. An oscilloscope check of the oscillator will quickly determine whether or not the oscillator is generating a 3.58-MHz sine wave of correct amplitude.

In Figure 17-8 it can be seen that in some circuits the phase detector is used to disable the color killer. This means that if the burst gate or phase detector becomes defective, the dc voltage used to disable the color killer may not be developed, and the BPA will be cut off. This situation is easily remedied by adjusting the color killer control so that the color killer is disabled. This may cause a picture that exhibits a loss of color sync.

Restoring an out-of-color-sync picture in this manner is important because it indicates that the trouble must be in the phase detector or burst gate.

Weak color. An oscillator that has a weak output will produce a weak color picture because the output of the demodulator is directly related to the oscillator level driving it. The fastest way to troubleshoot this problem is to signal trace the output of the oscillator with an oscilloscope. If the oscillator levels driving the demodulator are less than that indicated in the service literature, a complete check of the 3.58-MHz oscillator is in order. In the transistor circuit of Figure 17-12 Q_1, R_1, R_2, R_3, R_4, C_1, and C_6 might be likely causes.

17-10 injection-locked oscillator AFPC

The schematic diagram of a transistor injection-locked oscillator AFPC system is shown in Figure 17-15. Comparing this circuit to that of the multistaged phase detector AFPC shown in Figure 17-13 reveals its relative simplicity. Notice that this circuit is a simple Clapp-type oscillator driven by the burst gate. In the absence of a color burst (during black & white transmission) the oscillator is free-running and its phase is allowed to drift randomly.

In this circuit the transistor oscillator (Q_1) obtains positive feedback by returning some of the emitter signal to the base via C_1. Proper feedback phasing is obtained by making the emitter impedance capacitive. The necessary emitter impedance is provided by the parallel combination of R_1 and C_2.

The oscillator tuned circuit is formed by the circuit consisting of C_1, C_2, C_3, C_4, C_7, the 3.58-MHz crystal X_1, and the tuned transformer T_1. Injected into this tuned circuit via T_1 and C_7 is the color burst. The color burst is an ac voltage that drives a current through the circuit. For the purposes of understanding this circuit, the color burst may be considered as though it were an impedance ($Z = E/I$), which changes the phase and frequency of the oscillator so that it locks with the color burst.

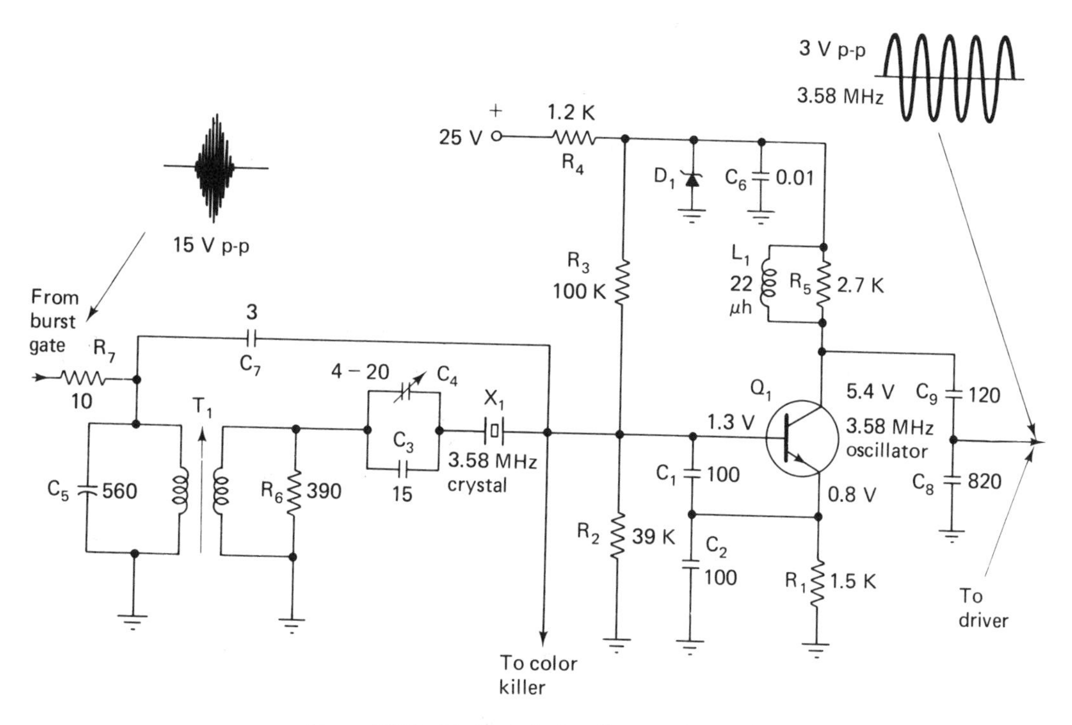

Figure 17-15 Injection-locked oscillator AFPC system.

This circuit has the advantage that it is relatively simple in both construction and alignment. It is, however, sensitive to noise, and its lock-in range is proportional to the amplitude of the input burst.

Oscillator starting bias is provided by R_1, R_2, and R_3. The collector circuit of the oscillator is tuned to 3.58 MHz by L_1, C_8, and C_9. In addition, C_8 and C_9 form an impedance matching and coupling circuit that is used to feed the driver stage. Oscillator stability is provided by the zener diode voltage regulator (D_1 and R_4) connected between the B supply and the collector circuit of the oscillator.

17-11 crystal ringing AFPC

The crystal ringing system of AFPC (see Figure 17-7(c)) may be compared to chimes. The 3.58-MHz crystal is shock excited by as little as one cycle of color burst, forcing it to ring in frequency and phase with the color burst. The crystal has a high enough effective Q so that it will continue to ring at approximately the same level between bursts. The output of the 3.58-MHz crystal is then amplified and fed to the demodulators. Since this system does not have a free-running oscillator, the absence of the color burst (black & white transmission) will cause the demodulators to become inoperative, removing the need for a color killer circuit. If, however, the receiver uses ACC, a color killer circuit is required to compensate for the increased BPA gain that occurs during the black & white reception.

A typical crystal ringing circuit is shown in Figure 17-16. The crystal (X_1) is driven by a 4-V peak-to-peak color burst obtained from the burst gate (Q_1). The sine-wave output

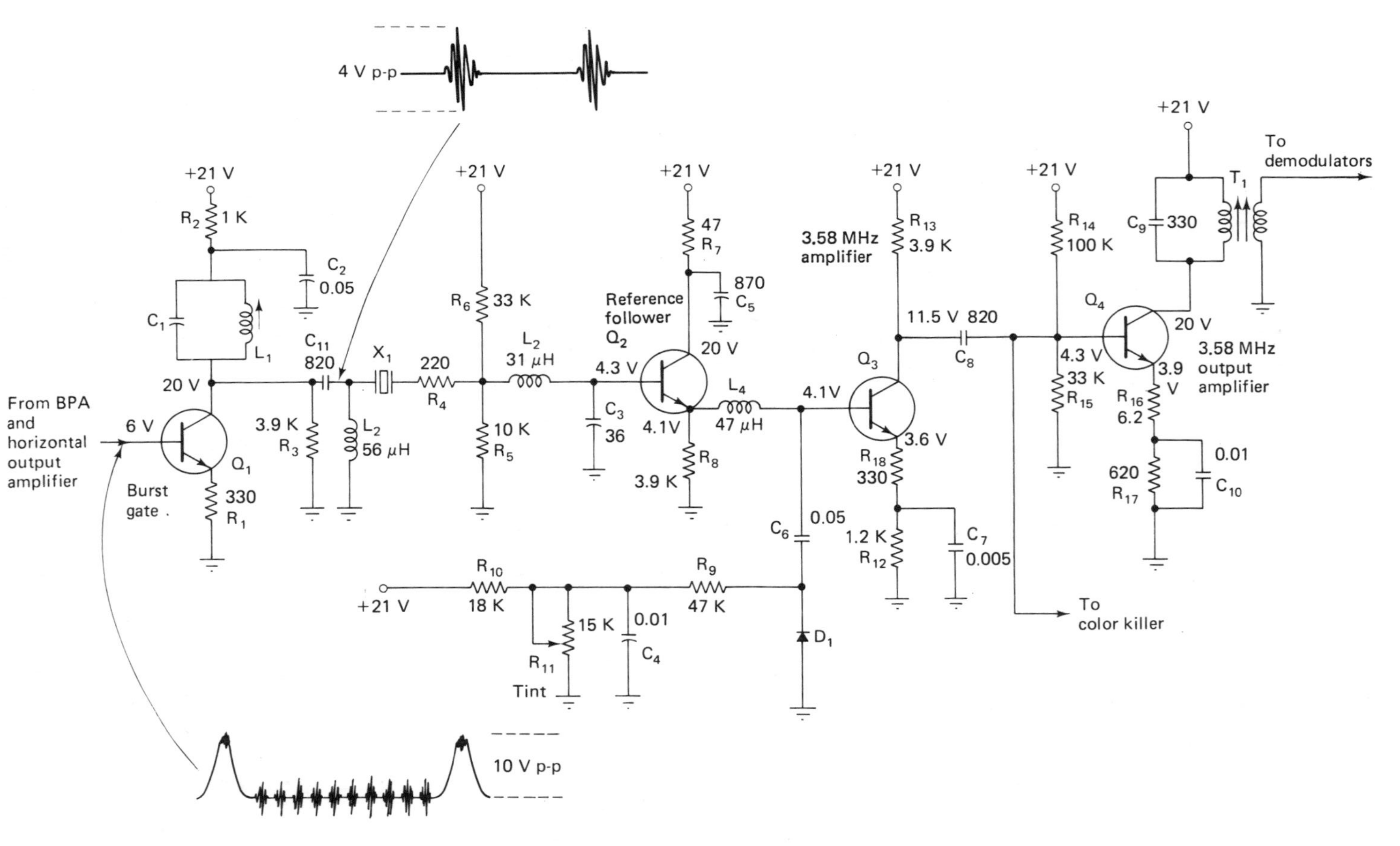

Figure 17-16 Crystal ringing AFPC circuit.

of the crystal is then dc coupled to the base of an emitter follower (Q_2). An emitter follower is a power amplifier whose input resistance is very high and whose output resistance is low. As such, the stage acts as an impedance matching stage between the crystal and Q_3, the 3.58-MHz amplifier stage. It prevents loading of the crystal, thereby allowing the crystal to ring effectively.

Capacitor C_6 connects a varactor diode (D_1) between the base of the 3.58-MHz amplifier (Q_3) and ground. The tint control R_{11} in conjunction with R_{10} form a variable dc voltage divider that adjusts the varactor bias. Changing this bias changes the effective capacitance of the varactor and shifts the phase of the 3.58-MHz signal. This, in turn, causes a change in picture color.

The 3.58-MHz sine-wave output of the 3.58-MHz amplifier is RC coupled via C_8 to the color killer and the base of the 3.58-MHz output amplifier (Q_4). A tuned transformer (T_1) is used to couple the output of this stage to the color demodulators. Loss of color sync is not possible in this circuit because there is no free-running oscillator that can drift. In this circuit a defective burst gate or ringing amplifier will cause a loss of color.

Alignment. The alignment of the crystal ringing AFPC circuit consists of adjusting the burst gate tuned circuit (L_1) and the 3.58-MHz output transformer (T_1) for a maximum output as measured at the secondary of the demodulator driver transformer T_1. This is done under signal conditions with the tint control in the center of its range. The output indicator is a dc VTVM that is connected to the output of one of the diode demodulators (not shown in this diagram).

17-12 the integrated circuit subcarrier processor

Figure 17-17 shows the block diagram of an integrated circuit (IC) often found in modular television receivers that is used as the 3.58-MHz subcarrier regenerator. The block diagram of the IC discloses that the integrated circuit combines the functions of automatic color control (ACC), automatic phase control (APC), 3.58-MHz oscillator, and 3.58-MHz output amplifier in one package. In addition, a horizontal keyer circuit and a shunt regulator bias power supply are also included in this IC chip.

The ACC and APC functions are performed by sampling the color burst during the horizontal blanking interval. This gating action is provided by a gate transistor circuit found inside the IC. Normally, this transistor is off and is only turned on by a positive pulse obtained from the horizontal output transformer. This pulse is delayed so that it keys the gate into an on condition when the color burst is present. As a result of the gating action, the ACC and APC circuits develop control voltages that depend upon the relatively constant amplitude of the color burst and are independent of the changes in color content caused by program material.

The phase detector APC circuit has three inputs: a chroma signal (pin 13), a 3.58-MHz oscillator input, and a horizontal keying pulse obtained from the horizontal keyer circuit. The horizontal pulse allows the color burst to be compared to the 3.58-MHz oscillator voltage and to develop a dc error voltage. This voltage is then filtered by the RC network connected between pins 11 and 12 before being applied to the oscillator circuit.

The design of the 3.58-MHz crystal oscillator is such that the dc error voltage is able to control its phase and frequency. The 3.58-MHz crystal is operated at series resonance

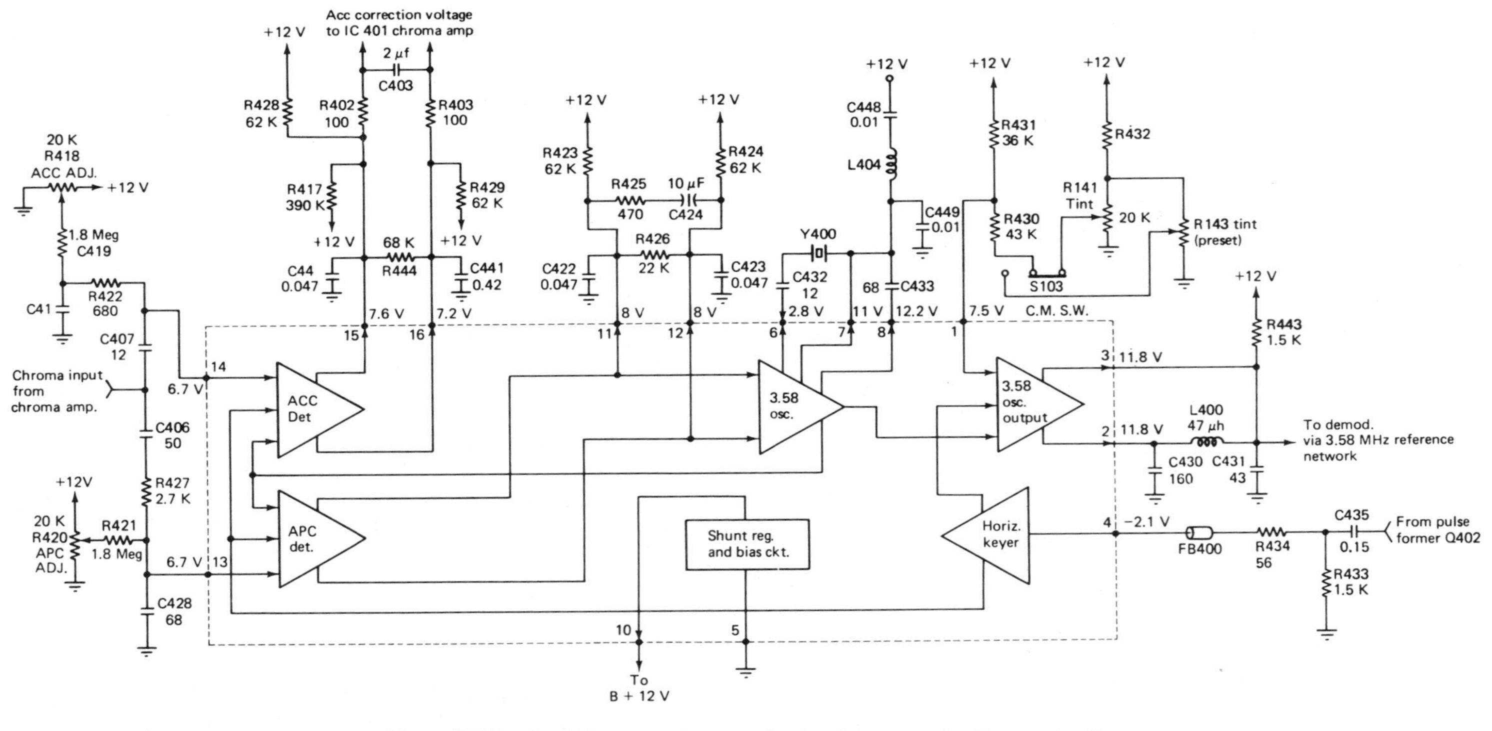

Figure 17-17 IC AFPC system. (Courtesy of Admiral Group/Rockwell International.)

and is connected between pins 6 and 7 of the IC. A feedback capacitor C_{433} (68 pF) is connected between pins 7 and 8. This combination of the phase detector and dc controlled oscillator forms a phase locked loop system that permits the oscillator to track the phase and frequency of the color burst.

The phase of the oscillator output is differentially controlled at terminals 2 and 3 by the dc tint control input to terminal 1. The *RLC* network connected to pins 2 and 3 is designed for a 90° phase delay, so that the range of the tint control is $\pm 45°$. The oscillator is gated so that during the color burst interval the oscillator output is disabled. This has the same effect as blanking the burst from the chroma channel and removes the need for 3.58-MHz traps after demodulation.

The gated ACC detector is similar to that of the APC detector. It is also fed by three inputs: the chroma signal (pin 14), the 3.58-MHz oscillator, and the horizontal pulse from the horizontal keyer. The dc output of this circuit is obtained from pins 15 and 16. This output in turn drives another IC containing the BPA and an ACC amplifier. When this circuit is keyed on by the horizontal pulse, the amplitude of the color burst determines the conduction of the ACC detector transistor and determines the dc output voltage.

Also built into this integrated circuit is a shunt regulator that holds the B supply (pin 10) at approximately 12 V.

Alignment. Alignment of this IC subcarrier regenerator is simple and straightforward. The entire system only contains three controls: an ACC adjust control, an APC adjust control, and a tint control. These controls are variable resistors; no tuned circuits are involved in the alignment procedure. It is assumed that the VHF tuner and the video IF alignment have been completed prior to the alignment of this circuit.

The ACC adjustment is done under signal conditions with a dc VTVM. A voltmeter is connected to pin 6 of the IC and a short is connected between pins 1 and 14 (ACC input). The voltmeter reading is noted and the short is removed. The ACC control (R_{418}) is then adjusted to bring the meter back to the previously noted reading. When the ACC control is properly adjusted, reconnecting the short from pins 1 to 14 will cause no change in the voltmeter reading.

Before the APC control (R_{420}) is adjusted the tint control (R_{141}) is placed in its mechanical center of its range. A strong local station is tuned in and then the color killer is disabled. The color control is adjusted for maximum color. As a means of weakening the color burst and forcing the receiver to lose color sync (*barber poling*), the fine tuning is misadjusted. The final step is to adjust the APC control (R_{420}) for minimum barber poling.

17-13 the vacuum tube burst gate

The schematic diagram of a typical vacuum tube burst gate is shown in Figure 17-18. Included with the schematic diagram are the wave forms associated with the circuit. The grid input signal is the composite of the chrominance signal and the horizontal keying pulse. If the receiver antenna input signal is obtained from a color bar generator, the input signal to the burst gate will consist of (1) a horizontal sync pulse and eleven 3.58-MHz bursts that are coupled by C_3 into the grid of the burst gate and (2) a horizontal pulse that is introduced into the grid of V_1 by the time delay network R_2, C_4, and R_3. The grid voltage waveform shown is the resultant sum of these two signals.

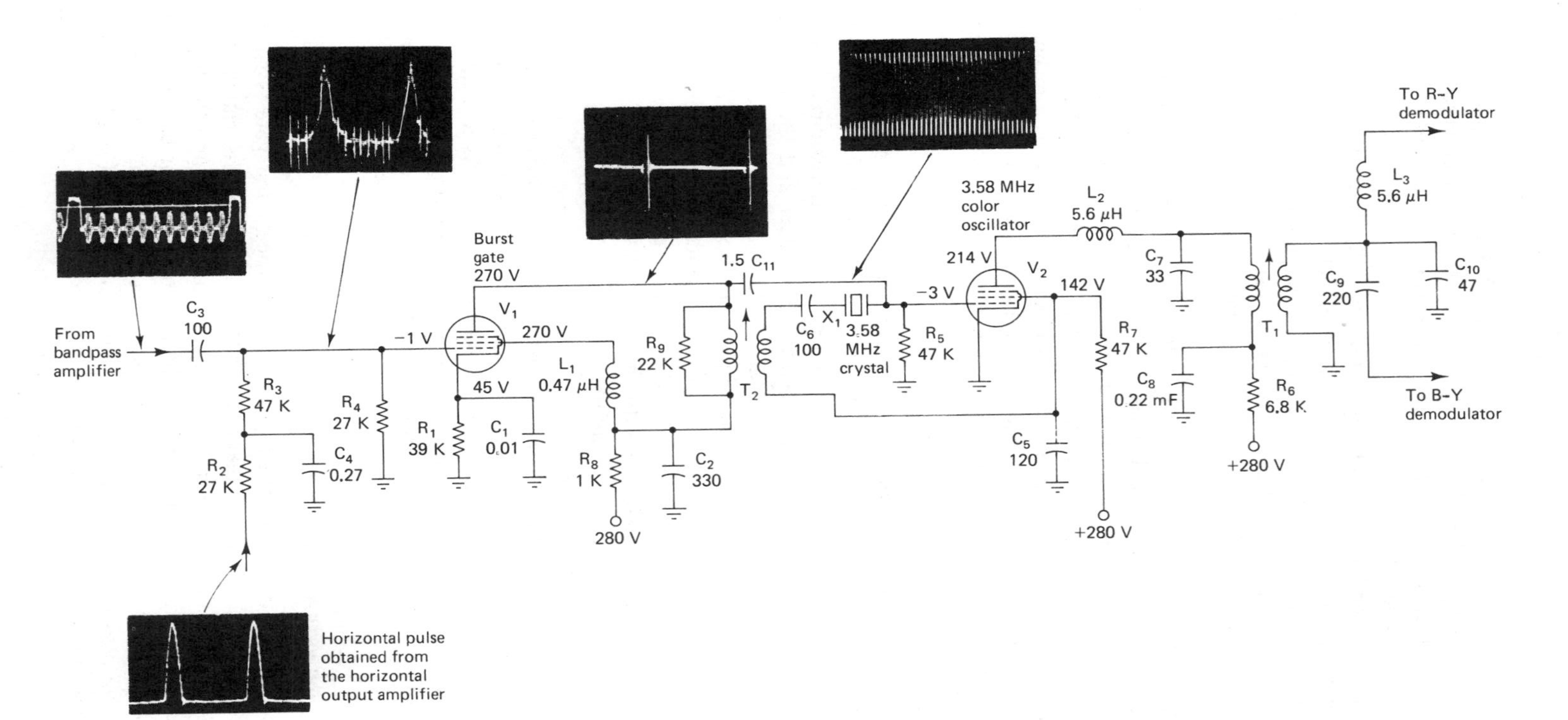

Figure 17-18 Vacuum tube burst gate and injection-locked color oscillator.

In this circuit a form of pulsed cathode bias is used. The horizontal input pulse drives the pentode burst gate into conduction developing a large voltage drop across the 39,000-Ω cathode resistor R_1. The capacitor C_1 charges to the peak of the voltage drop across R_1 when the tube conducts, and then it slowly discharges through R_1 between pulses. The average value of voltage developed during these periods of charge and discharge is the bias voltage that sets the operating point of the tube. Since this voltage is 45 V, and is much larger than the grid cutoff voltage, the tube may be considered to be operated as a Class C amplifier. As such, it conducts only for a short interval during the peak of the applied voltage.

Because of the large grid to cathode bias the resultant ac components of grid voltage are positioned on the tube's dynamic grid transfer curve so that only the most extreme positive peak of this voltage causes the tube to conduct. When the timing between the horizontal keying pulse and the color burst is correct, the color burst will be added to the peak of the horizontal keying pulse and it will then activate the burst gate during its conduction time. As a result, the output plate current of the burst gate will consist of the color burst and the peak of the horizontal keying pulse. Since the plate circuit of the burst gate is tuned (T_1) to 3.58-MHz, the horizontal sync pulse component of the plate current is filtered out, leaving only the color burst. See the waveform in Figure 17-18. The plate voltage of this pulse may be as large as 180 V p-p.

Oscillator circuits. The crystal oscillator circuits used in color receivers may be vacuum tube or transistor varieties of the TPTG, Pierce, or other similar type. Vacuum tube crystal oscillators are usually of the electron coupled oscillator (ECO) type similar to that shown in Figure 17-18. In this arrangement the screen grid is used as the plate of a triode oscillator, and the cathode to plate circuit of the tube is used as a pentode amplifier. The main advantage of this system is the relative isolation of the triode crystal oscillator from the loads fed by the plate circuit.

In the circuit of Figure 17-18 the screen grid is bypassed by a very small capacitor C_5. Thus, a sizable ac voltage may be developed across the screen load resistor R_7. As a result, the cathode, control grid, and screen grid may be considered to form a triode amplifier capable of considerable amplification. The ac voltage developed on the screen grid is returned to the control grid through the crystal X_1. This serves two functions. The crystal provides the feedback path and it acts as one of the frequency selective circuit elements needed to produce a sine-wave output from the oscillator. This crystal oscillator circuit arrangement is called a *Pierce oscillator* and is the equivalent of the ordinary L/C ultra-audion Colpitts oscillator.

Grid leak bias is developed by the capacitance of X_1 and R_5. The phase and frequency of the injection oscillator are controlled by the phase and frequency of the injected color burst.

The plate current of the oscillator will vary at a 3.58-MHz rate because of the fluctuation imparted to the electron stream by the ac voltages on the control and screen grids. This, in turn, develops an ac voltage drop across the output tuned transformer T_1. The secondary of T_1 feeds the demodulators with 3.58-MHz oscillator voltages of the proper phase and amplitude.

QUESTIONS

1. What does the phase of the chrominance signal represent?
2. What is the phase of the color burst? What is its purpose?
3. What are the functions of the color sync circuits?
4. What is the purpose of the burst gate?
5. What is the function of the AFPC circuit?
6. Why are burst gates usually operated as Class C amplifiers?
7. Explain why the horizontal keying pulse is delayed. What effect does improper timing have on burst gate operation?
8. What will be the effect on the picture if the alignment of T_1 in Figure 17-5 is changed?
9. Explain how a defective burst gate can cause the following trouble symtoms: (a) loss of color sync, (b) no color, (c) wrong color.
10. Explain how shorting C_2 in Figure 17-13 will cause erratic changes in color.
11. What is the function of the oscillator control circuit?
12. Describe the three types of AFPC circuits in current use.
13. Why must the components in the phase detector AFPC be matched?
14. In what way does the AFPC phase detector differ from the ACC and color killer detectors?
15. What trouble symptoms can be caused by a defective oscillator control circuit?
16. Why may a quartz crystal be compared to an *LC* tuned circuit?
17. Why is the oscillator shown in Figure 17-18 called an electron-coupled oscillator (ECO)?
18. What is the purpose of the buffer (Q_2) in Figure 17-12?
19. List three objectives of phase detector AFPC alignment.
20. List four trouble symptoms caused by a defective phase detector AFPC.
21. What are the differences between phase detector, injection-locked oscillator, and the ringing amplifier types of AFPC?
22. Why doesn't a receiver using the ringing amplifier AFPC require a color killer circuit?
23. Can a defective ringing amplifier AFPC cause loss of color sync? Explain.
24. List those components which might be responsible for an approximately zero voltage reading at the base of Q_2 in Figure 17-16.
25. Referring to Figure 17-8, what would be two probable effects of D_1 opening.

chapter 18

Color Demodulators and Output Amplifier Circuits

18-1 introduction

The operation of the color demodulator and its associated matrix are closely related to the basic principles of color television presented in Chapter 3. In order to provide a better understanding of the color demodulator and the matrix requirements, we will now give a simplified review of these principles.

Color transmission. At the color transmitter the color video signal generated by the color camera is divided by a matrix into three low-frequency signals: the Y signal, the I signal, and the Q signal. The Y signal represents the brightness information of the picture and has a bandwidth of 4.2 MHz. Color information is represented by the I and Q video signals that have bandwidths of 0 to 1.5 MHz and 0 to 0.5 MHz, respectively. The frequency of the I and Q signals is then increased by a pair of balanced modulators. The sideband outputs of the balanced modulators are centered around the suppressed subcarrier frequency of 3.579545 MHz. Sidebands produced at the output of the I balanced modulator have a frequency range extending from 2.08 to 4.08 MHz, and those produced by the Q balanced modulator range from 3.08 to 4.08 MHz. The outputs of the I and Q balanced modulators are then combined to form the chrominance signal. The amplitude of this signal represents saturation and its phase corresponds to the color of the scene being televised. Finally, the chrominance signal is added to the Y signal, the color burst, and the required sync and blanking pulses to form the composite video signal. This signal is then used to amplitude modulate the television transmitter.

Color reception. At the receiver the process is reversed. The high-frequency chrominance signal is lowered in frequency by the demodulator. Figure 18-1 shows the location of the

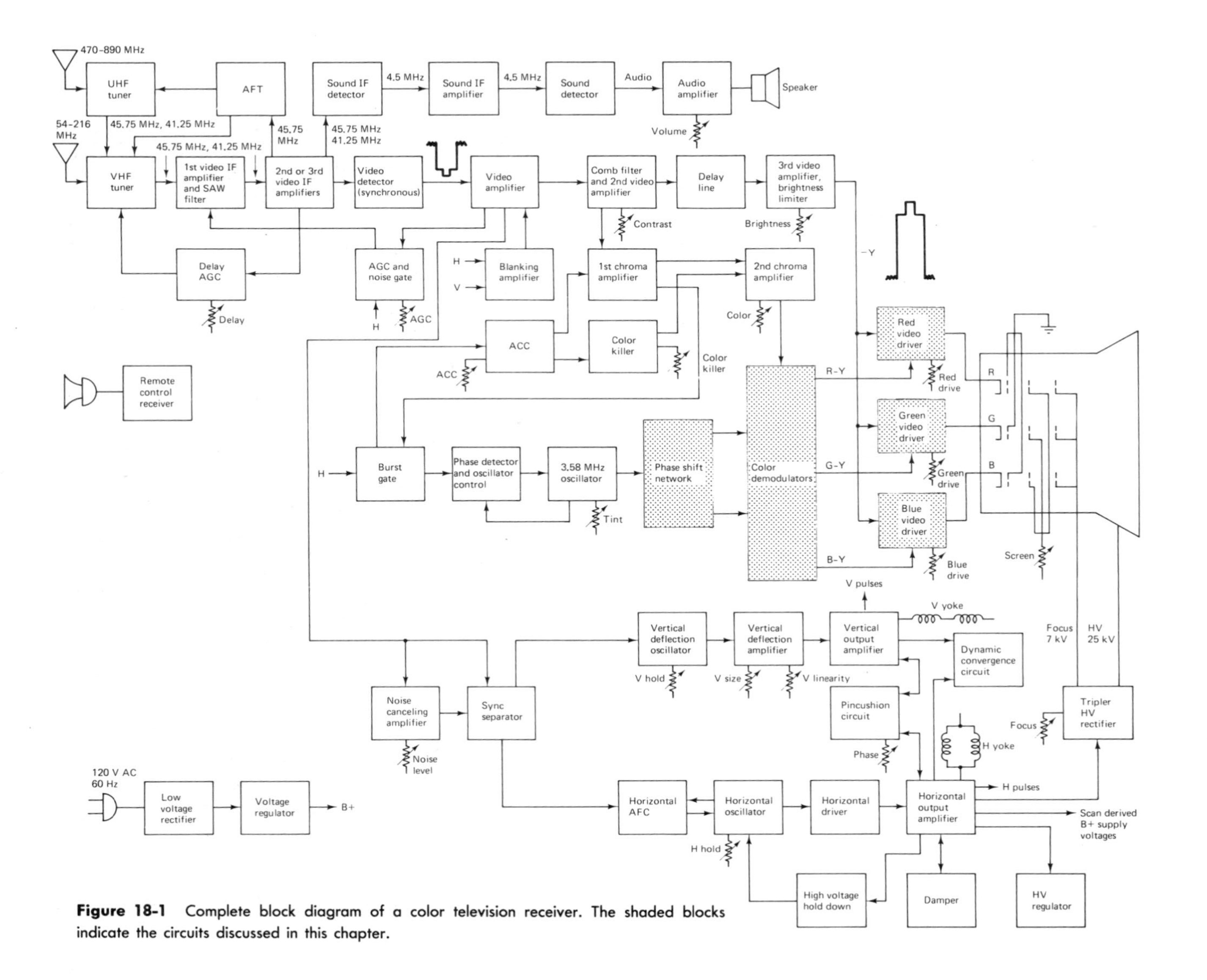

Figure 18-1 Complete block diagram of a color television receiver. The shaded blocks indicate the circuits discussed in this chapter.

color demodulators in relationship to the other circuits of a complete color receiver. From this diagram it can be seen that the demodulators are fed the high-frequency chrominance signal by the 2nd chroma amplifier. The demodulators are also fed by the 3.58-MHz oscillator. The low-frequency outputs of the mixer-like demodulators are then fed into three color video drivers (red, blue, and green) that increase the level of the color signals to that necessary to drive the picture tube. These amplifier stages act as a matrix that converts the three outputs ($R - Y$, $B - Y$, and $G - Y$) of the demodulators and the $-Y$ signal into their original form (R, G, B). In actual practice, there are many other demodulator and amplifier arrangements that are possible. These systems will be discussed in a later section of this chapter.

The shaded blocks in Figure 18-1 correspond to those stages discussed in this chapter.

18-2 color demodulators

Demodulator circuits are designed to produce an output that is dependent upon both phase and amplitude of the chrominance signal. Color demodulators may be thought of as a combination phase modulation and amplitude modulation detector. Such a detector is called a *synchronous detector.* In a typical demodulator arrangement the chrominance signal and the 3.58-MHz oscillator signal are fed into a circuit that is very similar to the mixer stage in a super-heterodyne receiver. The main difference between these circuits is that the synchronous detector uses a locally generated reference signal that is the same frequency as the color subcarrier, whereas the local oscillator of the receiver's mixer operates at a frequency that is much higher than that of the input RF signal.

High-level and low-level demodulators. Color demodulators may be classified according to the amount of chrominance signal amplification done before demodulation. High-level demodulators are circuits that directly drive the picture tube. In these circuits almost all of the color signal amplification needed to drive the picture tube precedes demodulation. This form of demodulation is primarily done in vacuum tube type of receivers. Low-level demodulators require that the demodulator output signals receive additional amplification in order to drive the picture tube.

Demodulator signal flow. A more detailed block diagram of the demodulator section of a color receiver is shown in Figure 18-2. Accompanying each block are the typical gated rainbow waveforms normally found at that point in the circuit. These waveforms are those that would be obtained if a gated rainbow color bar generator were connected to the input of the receiver. In this block diagram $R - Y$ and $B - Y$ demodulators are used.

The *Y* signal. Notice in Figure 18-2 that the composite video signal obtained from the video detector drives the video amplifier. At this point the signal is divided. Part of the signal goes to the picture tube cathode and part of the signal goes to the bandpass amplifier. The video signal going to the cathodes of the picture tube is called the Y signal and corresponds to brightness information.

Actually, the signal fed to the cathodes of the picture tube is a $-Y$ signal. This means that the composite video signal feeding the cathodes has a positive going sync pulse.Under these conditions, white levels have a magnitude close to zero, whereas at the

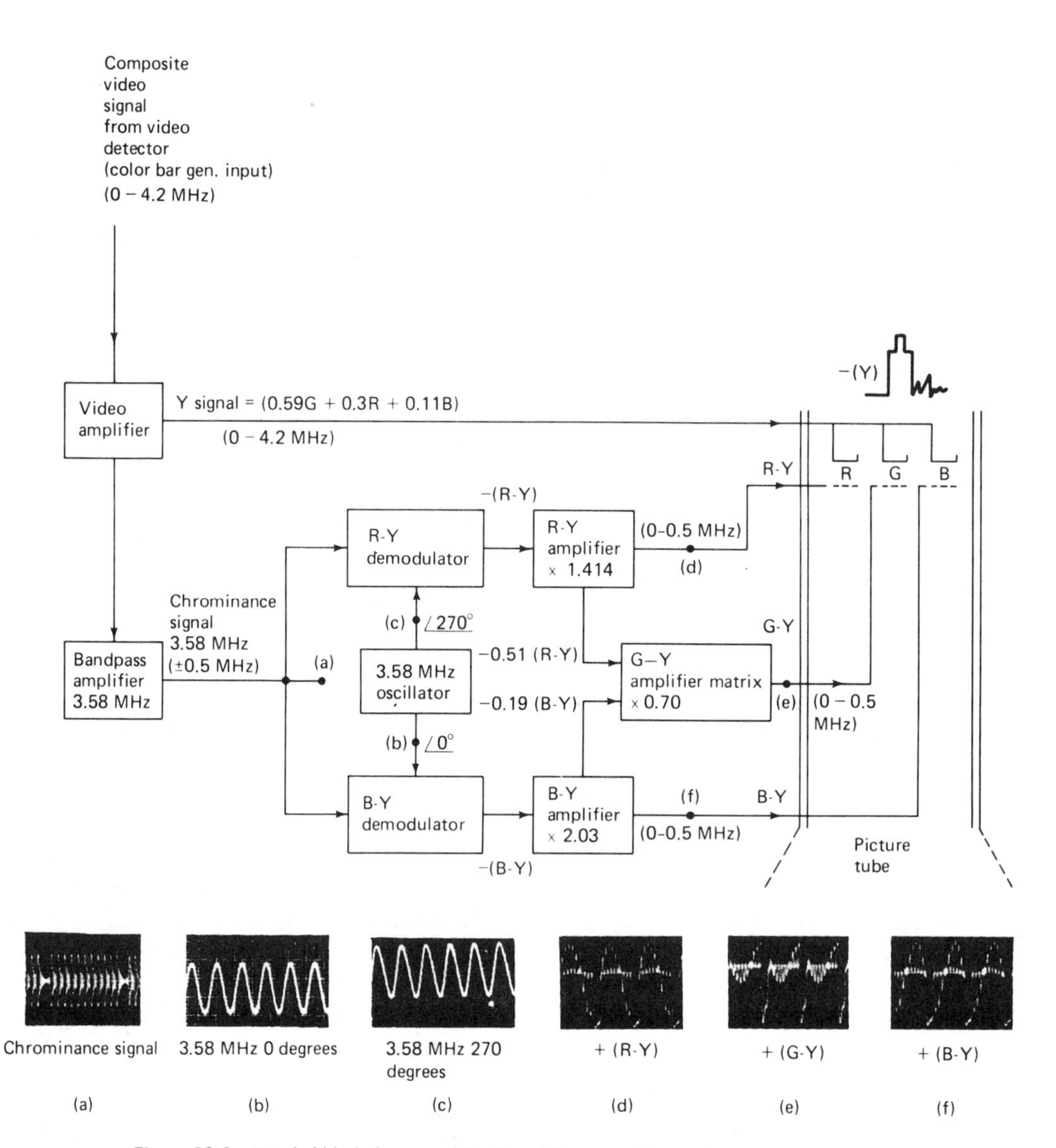

Figure 18-2 Detailed block diagram of the demodulator amplifier matrix portion of a color receiver. Important waveforms found at various parts of the circuit are also shown.

transmitter the white level of the $+Y$ signal is maximum positive. Minus Y at the cathode of a picture tube may be considered to be a $+Y$ at the grid. If the three grids of the picture tube are fed color difference signals such as $R - Y$, $B - Y$, and $G - Y$, the $-Y$ signal on the cathode will cancel (matrix with) the Y-signal components $[(R - Y) - (-Y) = R]$, leaving the desired red (R), green (G), and blue (B) signal voltages between the grids and cathodes of the picture tube.

Demodulator input signals. At the bandpass amplifier the color signal information is separated from the composite video signal. The chrominance output of the bandpass amplifier is then fed equally to both color demodulators. At this point the input signals to the demodulators consist of the chrominance signal and a constant amplitude 3.58-MHz oscillator voltage. The oscillator voltage is coupled to the demodulators by phase shift networks.

The chrominance signal found at the input to the demodulator is centered at 3.58 MHz. Associated with it are sidebands ranging in frequency from 3.08 MHz to 4.08 MHz, for a bandwidth of 1.0 MHz. The input waveform is a train of bursts produced by the gated rainbow color bar generator.

Demodulator output signals. Each demodulator has a 3.58-MHz subcarrier oscillator signal input. In this circuit arrangement the $R - Y$ demodulator oscillator voltage has a phase of 270° with respect to the color subcarrier (0°) and phase of the $B - Y$ demodulator oscillator voltage is 0°. This produces demodulator outputs that correspond to $-(R - Y)$ and $-(B - Y)$ color difference signals. In actual practice, the output of the demodulator may be either positive $(R - Y$ or $B - Y)$ or negative $[-(R - Y)$ or $-(B - Y)]$, depending upon the phase of the oscillator. Deciding which demodulator output to use depends upon the type and number of amplifier stages between the demodulator and the grids of the picture tube. For the circuit described in the block diagram of Figure 18-2, the color difference signals at the grids of the picture tube must all be positive $[+(R - Y), +(B - Y)$, and $+(G - Y)]$. This signal phase will then combine with the $-Y$ signal on the cathode to produce the required red (R), green (G), and blue (B) color signals that determine the instantaneous picture tube bias and beam current of each gun. Obtaining positive color difference output signals from the inverting driver amplifiers requires that their input signals be negative $[-(R - Y)$ and $-(B - Y)]$.

The reason that two demodulators are commonly used is that a demodulator may also be considered to be a kind of trigonometric computer because it is able to convert a vector quantity (chroma signal) into its right-angle components (polar into rectangular conversion). To do this two demodulators must be used, one for each right-angle component. Figure 18-3 shows the phasor diagram for a chrominance signal corresponding to the color red. It is the function of the $R - Y$ demodulator to produce an output that corresponds to the $R - Y$ component of the red chrominance signal appearing at the input of the demodulators. Similarly, it is the function of the $B - Y$ demodulator to develop an output that corresponds to the $B - Y$ component of the red chrominance signal.

The color matrix. The color difference signals appearing at the output of the demodulator must be combined properly in order to reconstruct the original (R, G, and B) color signals generated by the color cameras at the transmitter. The circuit used for this purpose is called a *matrix.* Because a matrix can solve an equation, it may be thought of as a kind of computer that performs the operations of addition and multiplication. Color television receivers may have two matrixes. This is indicated in Figure 18-2. The *first matrix* is located between the $R - Y$ and $B - Y$ amplifiers and generates a $G - Y$ signal by combining -0.19 $(B - Y)$ with $-0.51(R - Y)$. The *second matrix* is the grid and cathode of the picture tube. The grids are fed positive color difference signals and the cathode is fed a $-Y$ signal. The resultant voltage between the grid and cathode is the signal difference between grid

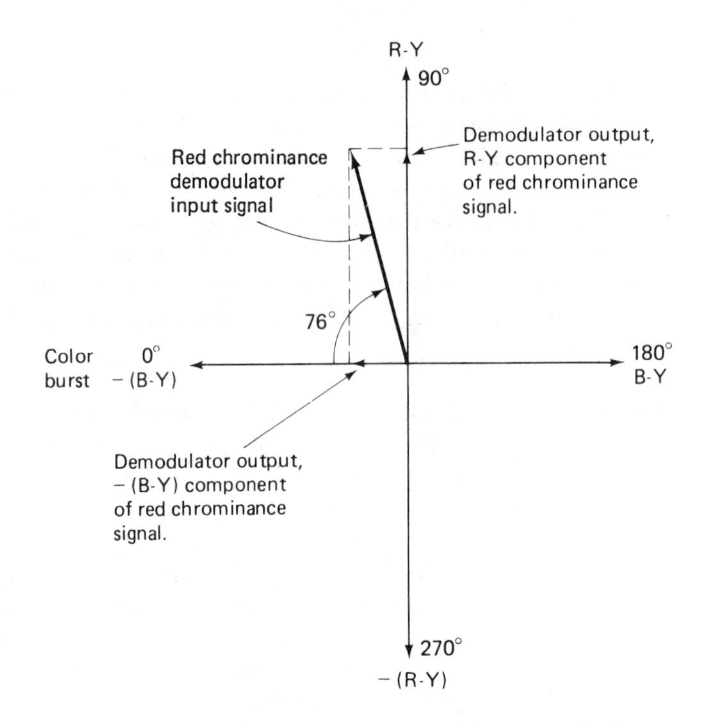

Figure 18-3 Vector components of a red chrominance signal.

and cathode corresponding to the original red (R), green (G), and blue (B) signals generated at the transmitter.

In addition, the matrix amplifier also compensates for the chroma signal compression that was introduced at the transmitter as a means of preventing overmodulation. This is indicated in Figure 18-2 by the multiplication factors associated with each color difference amplifier. Thus, the $R - Y$ amplifier provides a signal boost of 1.414 times, and that of the $B - Y$ amplifier 2.03 times. The $G - Y$ amplifier reduces the level by 0.70 times.

Demodulator waveforms. The waveforms shown in the block diagram of Figure 18-2 assume that a gated color bar generator is connected to the input of the receiver. As a result, the chrominance input to the demodulators will be a string of 11 bursts. The oscillator input to each demodulator will be a constant amplitude sine wave of 3.58 MHz. These waveforms will differ in phase. Finally, the output of the demodulators will be a *geartooth* waveform.

The geartooth waveform seen at the output of the demodulator is the result of the interrupted nature of the chroma signal feeding the demodulator. During each color burst the demodulator produces an output. Between bursts the output drops to zero. Since the chroma bursts go through a 360° phase shift in one horizontal line, the demodulator output will be an interrupted sine wave (geartooth) whose frequency will be at the horizontal rate.

18-3 the transistor demodulator

Color demodulators may use diodes, triodes, pentodes, special tubes, or ICs as their active circuit element. No matter which circuit element is used, the basic principles are the same. That is, color demodulators must produce an output that is sensitive to both the phase and the amplitude of the input chrominance signal.

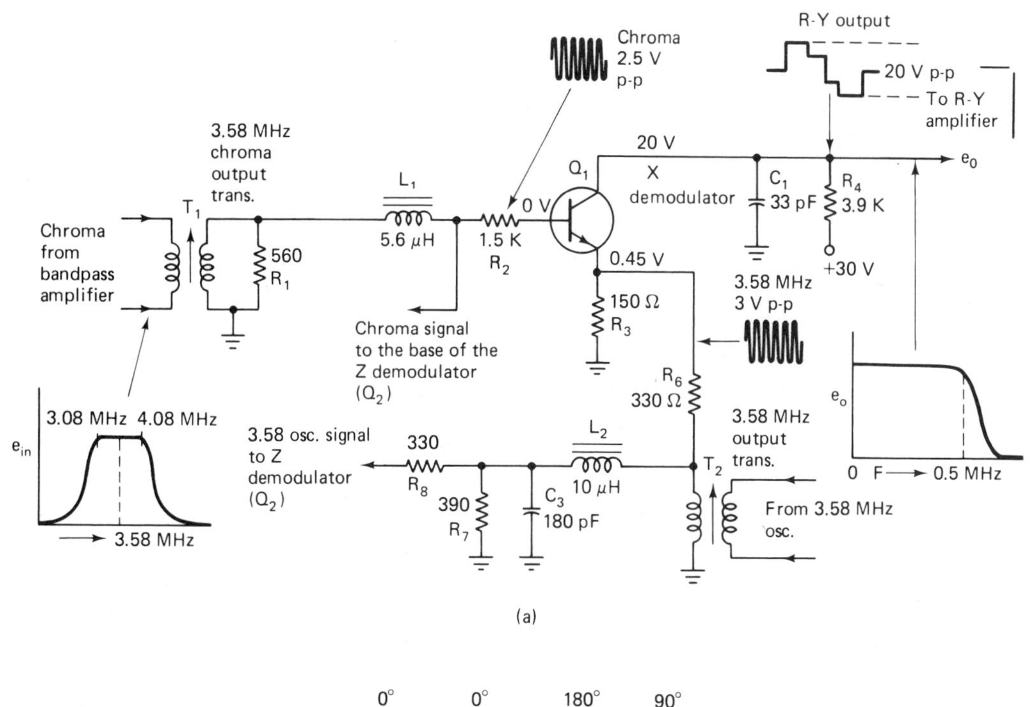

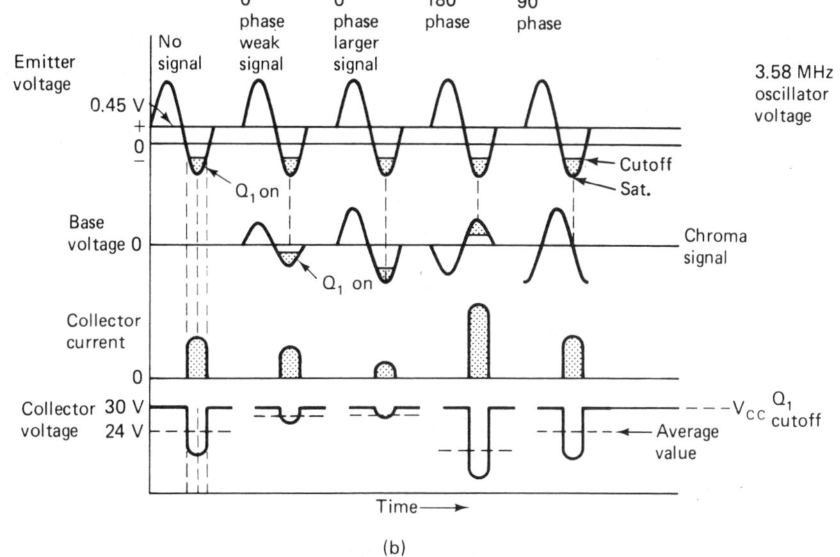

Figure 18-4 (a) Transistor color demodulator; (b) ladder diagram of the input and output waveforms in the circuit for different chroma phase angles.

The schematic diagram of a transistor demodulator is shown in Figure 18-4(a). The chrominance signal output of the bandpass amplifier is fed without any phase shift to the base of Q_1, the X demodulator, and to the base of Q_2, the Z demodulator. This is done by the tuned transformer T_1. The overall frequency response of the BPA indicates that the center frequency of the circuit is 3.58 MHz and that the bandpass is 1.0 MHz (3.08 to 4.08 MHz).

The 3.58-MHz color oscillator drives the emitter of Q_1 by means of the 3.58-MHz oscillator output transformer (T_2) and the voltage divider resistors R_3 and R_6. The oscillator output transformer T_2 also drives the emitter of a Z demodulator not shown in this diagram. The phase of the oscillator voltage feeding the Z demodulator is shifted 64° with respect to that of the X signal by a phase shift network consisting of L_2, C_3, and R_7.

The collector of Q_1 returns to the 30-V B supply by means of the load resistor R_4. Bypass capacitor C_1 in conjunction with the transistor and R_4 form a low-pass filter that effectively eliminates any residual 3.58-MHz components from the demodulator output circuit. As shown in Figure 18-4(a), the frequency range normally appearing at the output of modern demodulator circuits falls between 0 and 0.5 MHz.

Examination of the demodulator circuit [Figure 18-4(a)] shows that the base and emitter circuits do not return to any B supply voltage. Therefore, the no-signal bias of the transistor is zero and the transistor is operating Class B.

The ladder diagram. Figure 18-4(b) is a ladder diagram that compares the voltages and currents that exist at various points in the circuit. The topmost voltage is the emitter voltage. The ac component (3 V p-p) in this diagram is the 3.58-MHz oscillator voltage that is injected into the emitter circuit. The dc voltage (0.45 V) developed across R_3 is the result of the transistor's being forced into conduction during the negative half-cycle of the oscillator signal. Shown in the next lower voltage on the ladder diagram is the chrominance signal that is applied to the base of the demodulator transistor. Below that is the resultant collector current. It should be noticed that this current is in the form of a short-duration pulse that occurs at a 3.58-MHz rate. This is because the oscillator voltage is switching the transistor on and off. When the transistor is in cutoff, the collector current is zero. On the most negative swing of the oscillator voltage the transistor is driven toward saturation and the collector current will be some value determined by the level of the base input signal.

The collector voltage is shown in the bottom-most waveform. When the transistor is cut off, the collector voltage will be maximum. During the period of transistor conduction the collector current flows through the load resistor R_4. This produces a voltage drop that subtracts from the supply voltage and lowers the collector-to-ground voltage. The average value (dc component) will increase or decrease, depending upon the degree of transistor conduction.

Chroma amplitude changes. The effect of changing the amplitude and phase of the chrominance signal (base voltage) on demodulator operation may now be compared. In the first case, a *no-signal* condition is assumed. The resultant collector voltage is the reference for all the other conditions indicated. In this case, there is no base voltage at all. Therefore, the collector current flow is completely determined by the oscillator voltage injected at the emitter of the transistor. For this level of collector current flow, the average value of collector voltage is assumed to be 24 V.

The next case assumes that a small chrominance signal having the same phase (0° phase) as the 3.58-MHz oscillator voltage is applied to the base of the demodulator. Since the transistor is an NPN, the positive half-cycle of signal will tend to increase collector current and the negative half-cycle will tend to decrease it. However, the oscillator voltage on the emitter effectively keeps the transistor in cutoff during the positive portion of the base voltage (chrominance signal) so that the collector current remains unchanged. During the time that the oscillator voltage turns the transistor on, the negative half-cycle of the base voltage reduces the level of collector current flow. This, in turn, causes the peak-to-peak change of collector voltage to become smaller, and results in a larger average (dc) collector voltage.

The result of maintaining the same phase (0°) but increasing the level of the chrominance signal is seen in the next case. The net result of this is that the collector current decreases still further and forces the average (dc) collector voltage to increase to a higher value.

It can be seen from these two examples that the ac output of the color demodulator is dependent upon the *amplitude* of the chrominance signal. *The amplitude of the chrominance signal corresponds to the saturation of the color being transmitted.*

Chroma phase changes. In the fourth case the chrominance signal is assumed to be 180° out of phase with the color oscillator signal. As a result, during the time that the color oscillator drives the demodulator transistor on, the base voltage increases the forward bias of the transistor, forcing the collector current to increase. This in turn increases the amplitude of the peak-to-peak change in collector voltage and lowers its average (dc) value to a level below the no-signal value. Notice that in this case the output level of the demodulator was affected by the *phase* of the input chrominance signal. *It is the phase of the chrominance signal which represents the color being telecast.*

From the above it can be seen that changing the phase of the chrominance signal from 0° to 180° will cause the dc output of the demodulator to increase and decrease above and below the no-signal value. Consequently, when the chrominance signal is midway between the extremes, or at 90°, the output of the demodulator will be the same as the no-signal output. This is illustrated in the last column of the diagram in Figure 18-4(b). Notice that when the emitter voltage (oscillator signal) is maximum negative, the transistor is turned on and the base voltage (chrominance input signal) is passing through zero—the no-signal condition. As a result, the average collector current flow will be the same as that produced when there is no signal applied to the base of the demodulator. In other words, the demodulator is blind to any signal that is 90° out of phase with the oscillator voltage.

Summary. The following conclusions may be drawn from the discussion above:

1. A color demodulator produces an output that is directly related to the *amplitude* of the chrominance signal.
2. Color demodulators produce outputs that depend upon the *phase* of the chrominance signal. Greatest output occurs when the chrominance signal is either in phase (0°) or 180° out of phase with the oscillator voltage. This fact is very important because it means that if it is desired for a demodulator to produce a given output such as $R - Y$, $B - Y$, $G - Y$, and so on, *the phase of the oscillator must be the same as that of the desired output.*
3. The color demodulator is blind (does not respond) to any chrominance signal that is 90° out of phase with the oscillator voltage driving the demodulator.

The demodulator phase. Figure 18-5 is the phase diagram of a gated color bar chrominance signal for one horizontal line. Notice that in this diagram 0° corresponds to the phase of

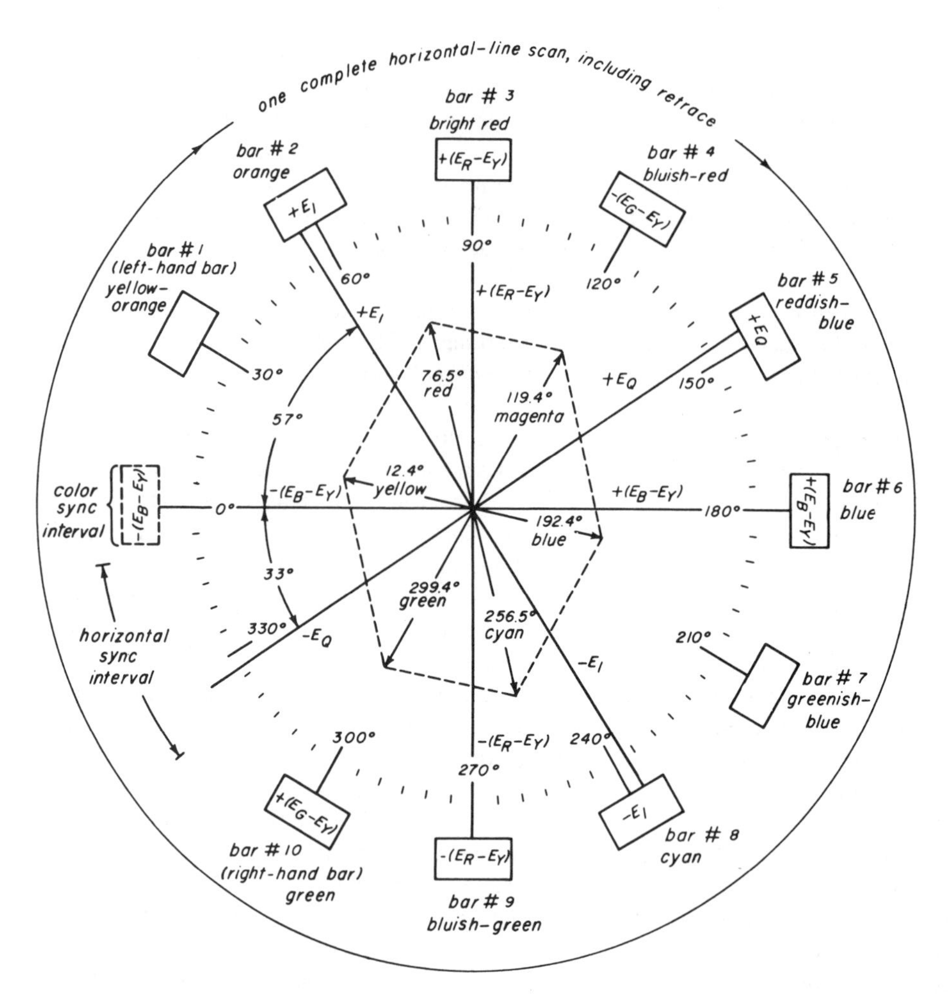

Figure 18-5 Phase relationships of a gated rainbow color bar generator.

the color burst and that the various colors and color difference signals are represented by different phase angles. Table 18-1 summarizes these phase relationships.

Thus, if it is desired to have a demodulator whose output corresponds to the $R - Y$ signal, it would be necessary for the color oscillator signal that is injected into the demodulator to have the same phase as the $R - Y$ component of the chrominance signal. That is, the oscillator voltage would be placed at a phase of 90° with respect to the color burst. Similarly the $B - Y$ demodulator would use an oscillator voltage whose phase is 180°. It can be seen that a color demodulator can produce any output signal simply by setting the phase of the oscillator to the phase of the desired signal.

It should be understood that a demodulator produces a maximum *change* in output level for signals that are in phase, or 180° out of phase with the injected oscillator voltage. This means that an $R - Y$ demodulator will produce a maximum output of opposite phase for both $R - Y$ and $-(R - Y)$ signals.

The phase difference between the oscillator voltages injected into the $R - Y$ and

TABLE 18-1

Signal	Phase Angle	Signal	Phase Angle
Red	76.5°	$+X$	282.0°
Blue	192.4°	$+Z$	346.0°
Green	299.4°	I	57.0°
$(R - Y)$	90.0°	Q	147.0°
$(B - Y)$	180.0°	$-(B - Y)$	0.0°
$(G - Y)$	300.0°	$-(G - Y)$	120.0°
$-X$	102.0°	$-(R - Y)$	270.0°
$-Z$	166.0°	Magenta	119.4°
Cyan	256.5°	Yellow	12.4°

$B - Y$ demodulators was stated as being 90°, but in many modern receivers the phase angle difference between demodulators can be as much as 105° or as little as 64°. The 105° demodulator phase difference was selected as the best angle to complement the picture tube phosphors and in particular to take advantage of the high-efficiency yttrium red phosphor.

Some older color receivers used X and Z demodulators. In this scheme the oscillator signal injected into the X demodulator has a phase of 102° and the oscillator voltage injected into the Z demodulator has a phase of 166°. The difference in phase between oscillator voltages is then only 64°.

18-4 demodulator types

In practice, various demodulators have been used. Figure 18-6 shows the block diagrams of some of the more commonly used arrangements.

All of the demodulator block diagrams shown in Figure 18-6 are assumed to be low-level demodulators. Demodulator systems such as that in Figure 18-6(a) are often used as high-level demodulators. In this case, the demodulators would not use amplifiers and would drive the grids of the picture tube directly.

In the demodulator systems of Figure 18-6(a), (b), and (c) the outputs of the demodulators feed color difference amplifiers that, in turn, drive the *grids* of the color picture tube. However, in the RGB demodulator system shown in Figure 18-6(d) the amplified outputs are used to drive the *cathodes* of the color picture tube.

R—Y/B—Y demodulators. In the demodulator system shown in Figure 18-6(a) an $R - Y$ demodulator is fed by a 3.58-MHz oscillator whose phase is 270° with respect to the color burst. As a result, the demodulator produces a $-(R - Y)$ signal output. The second demodulator is designed to produce a $-(B - Y)$ output. This is done by injecting into the demodulator an oscillator voltage that is in phase with the color burst (0°). The outputs of both demodulators are then passed through an amplifier matrix to produce a $+(R - Y)$ and a $+(B - Y)$ output as well as the $+(G - Y)$ signal. The matrix usually takes the form of three amplifiers. The $R - Y$ and $B - Y$ amplifier outputs are combined in the proper ratios to drive an amplifier that produces a $G - Y$ signal output.

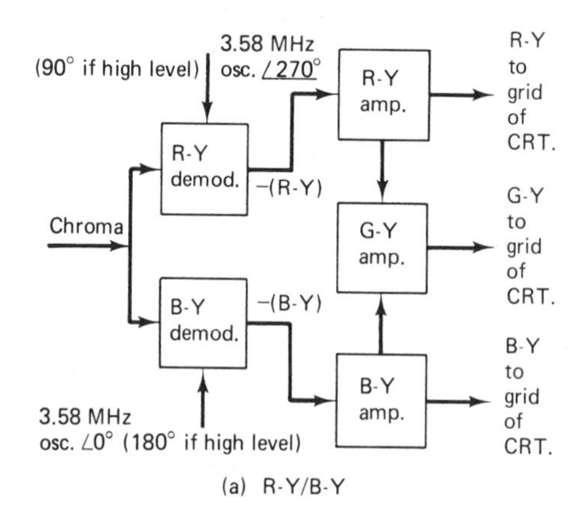

(a) R-Y/B-Y

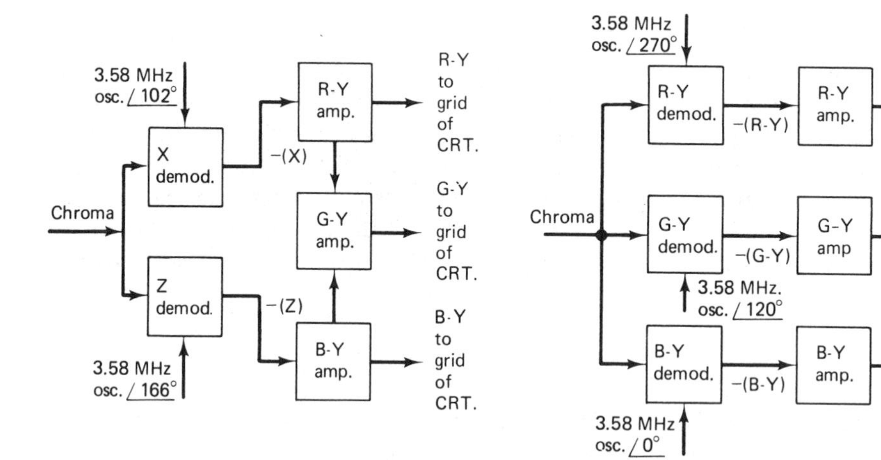

(b) X/Z

(c) R-Y/G-Y/B-Y

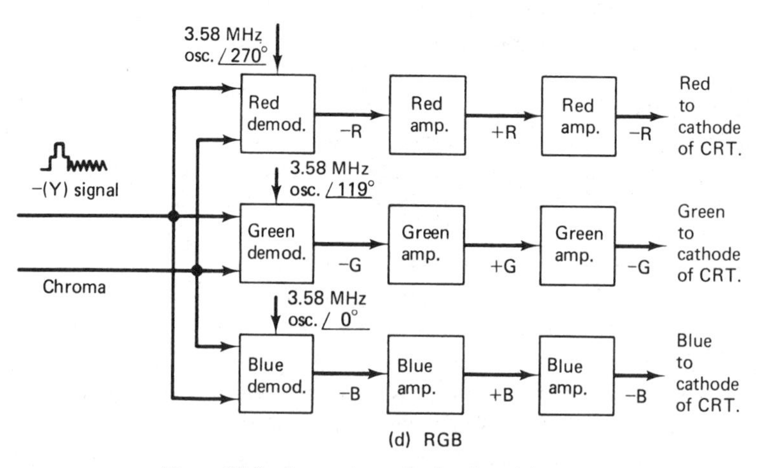

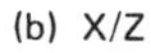

(d) RGB

Figure 18-6 Commonly used color demodulator systems.

In high-level demodulators, the outputs of the demodulator must feed the picture tube directly and be $+(R - Y)$ and $+(B - Y)$. Therefore, the oscillator voltage injected into the $R - Y$ demodulator is set at 90° and the $B - Y$ demodulator has an oscillator phase of 180°. Matrixing to obtain the $G - Y$ signal is done by resistive voltage dividers connected between the demodulator outputs.

X and Z demodulator. In Figure 18-6(b) is the block diagram of the X and Z type of demodulator. In this system the 3.58-MHz oscillator voltages feeding the demodulators are 102° and 166° respectively, that is, there is a phase difference of 64°. As a result, the $-X$ and $-Z$ output of the demodulators must be converted into $R - Y$, $B - Y$, and $G - Y$ chroma signals before they can drive the picture tube grids. This is done by a matrix formed by the $R - Y$, $B - Y$, and $G - Y$ color difference amplifiers.

R—Y/B—Y/G—Y demodulators. In all of the previous demodulator systems only two demodulators were used, one to develop the $R - Y$ signal and the other to regenerate the $B - Y$ (or $G - Y$) signal. As a result, it was necessary to use a matrix to extract the third signal, $(G - Y)$, required for proper picture tube operation.

An alternative approach not requiring any matrix is shown in Figure 18-6(c). In this system three demodulators are used, one for each desired color difference output. The $R - Y$ demodulator is fed by an oscillator voltage whose phase is set to 270° and produces a $-(R - Y)$ output. This signal is then inverted by the $R - Y$ amplifier that drives the grid of the picture tube. The phase of the oscillator voltage feeding the $B - Y$ demodulator is 0° and that of the $G - Y$ demodulator is 120° and produces $-(B - Y)$ and $-(G - Y)$ outputs, respectively. Color amplifiers then invert these signals and drive the picture tube grids.

RGB demodulators. All modern solid-state receivers use demodulator schemes similar to that shown in Figure 18-6(d). In this RGB system three demodulators are used to extract from their input 3.58-MHz chrominance signal, low-frequency color signals corresponding to (−R) red, (−G) green, and (−B) blue. Negative color signals are necessary since the cathode of the picture tube is being driven. These voltages are obtained by injecting properly phased oscillator voltages into the demodulators and then performing the necessary matrixing with a $-Y$ signal that is also injected into the demodulators.

The oscillator voltage injected into the red demodulator has a phase angle of 270° with respect to the color burst [producing a $-(R - Y)$ demodulator output]. The oscillator injection voltage for the green demodulator is set to 119° (producing a $G - Y$ output) and that of the blue demodulator to 0° [producing a $-(B - Y)$ output]. Since each demodulator is followed by a two-stage amplifier, there is no overall phase reversal and the cathodes of the picture tube are driven with the same negative phase color signal produced at the output of the demodulators. The control grids of the picture tube are tied together and return to a dc bias supply.

The output of each demodulator consists of a voltage that is the sum of the $-Y$ signal level and the voltage $[-(R - Y), -(G - Y)$, and $-(B - Y)]$ corresponding to the color being transmitted. For example, $-Y$ added to $-(R - Y)$ results in a $-R$ output. Similarly, $-Y$ added to $-(G - Y)$ and $-(B - Y)$ produces $-G$ and $-B$ outputs, respectively.

During black & white picture transmission only the $-Y$ signal passes through the RGB demodulators and their associated color amplifiers to reach the cathodes of the picture tube. Under these conditions, the color demodulators are inoperative.

During color transmission, however, the color demodulators are activated and both color and Y signal information is passed through the color amplifiers on its way to the picture tube. On this basis it can be seen that the major advantage of the RGB system of demodulation is that it eliminates the need for an expensive separate video output amplifier transistor.

18-5 demodulator circuits

The RGB integrated circuit demodulator. The block diagram and a schematic diagram of a low-level integrated circuit demodulator are shown in Figure 18-7. An examination of the block diagram shows that there are four input signals applied to the IC "chip." The chroma signal is obtained from a transformer that produces a balanced, oppositely phased output. These two signals are applied to the IC pins 3 and 4. Two 3.58-MHz oscillator signals of 74° and 177° are applied to pins 6 and 7, respectively.

Inside the IC a multitude of transistors, diodes, and resistors performs the functions indicated in the block diagram. Thus, the chroma signal is amplified by two input chroma amplifiers. The output of each amplifier and a reference oscillator signal are fed to a doubly balanced synchronous detector. These circuits are balanced for both reference oscillator and chroma signal. The doubly balanced arrangement provides cancellation of the reference signal, thus eliminating the need for elaborate filtering at the output of the IC. The outputs of the demodulators feed a matrix that produces the required $R - Y$, $G - Y$, and $B - Y$ signals. Output amplifiers then deliver these signals to the "outside world" at IC pins 9, 11, and 13.

Matrix amplifiers. Since this circuit is a low-level demodulator, the output must be amplified to a level sufficient to drive the picture tube.

In addition, matrixing of the $R - Y$, $G - Y$, and $B - Y$ signals into red, green, and blue signals must also be provided.

These two requirements are met by the dc coupled color amplifier circuit shown in Figure 18-8. In this circuit both the chroma and the black & white signals (Y) are fed to the cathode of the picture tube. The control grids are at ac ground (C_{15}) and are all tied together and return to B+ through R_{25}.

Proper picture tube operation requires that the Y signal at the cathode of the picture tube have a minus phase. The color signals must also have a minus phase. This means that the demodulator output signals feeding the bases of the inverting color amplifiers must be positively phased color difference signals. At the same time, a $-Y$ signal feeds the emitters of the color amplifiers via a noninverting emitter follower amplifier Q_4. Driving the emitter of an amplifier produces a noninverted output at the collector. Therefore, the $-Y$ signal at the emitters of all three amplifiers appears at the collectors of each of the amplifiers amplified but noninverted.

To illustrate the operation of this amplifier matrix, assume that a picture of a single vertical green bar against a black background is being televised. The $R - Y$ and $B - Y$ outputs of the demodulator for one horizontal line will consist of negative-going pulses (see Figure 3-13). The output of the $G - Y$ demodulator will be somewhat smaller and positive

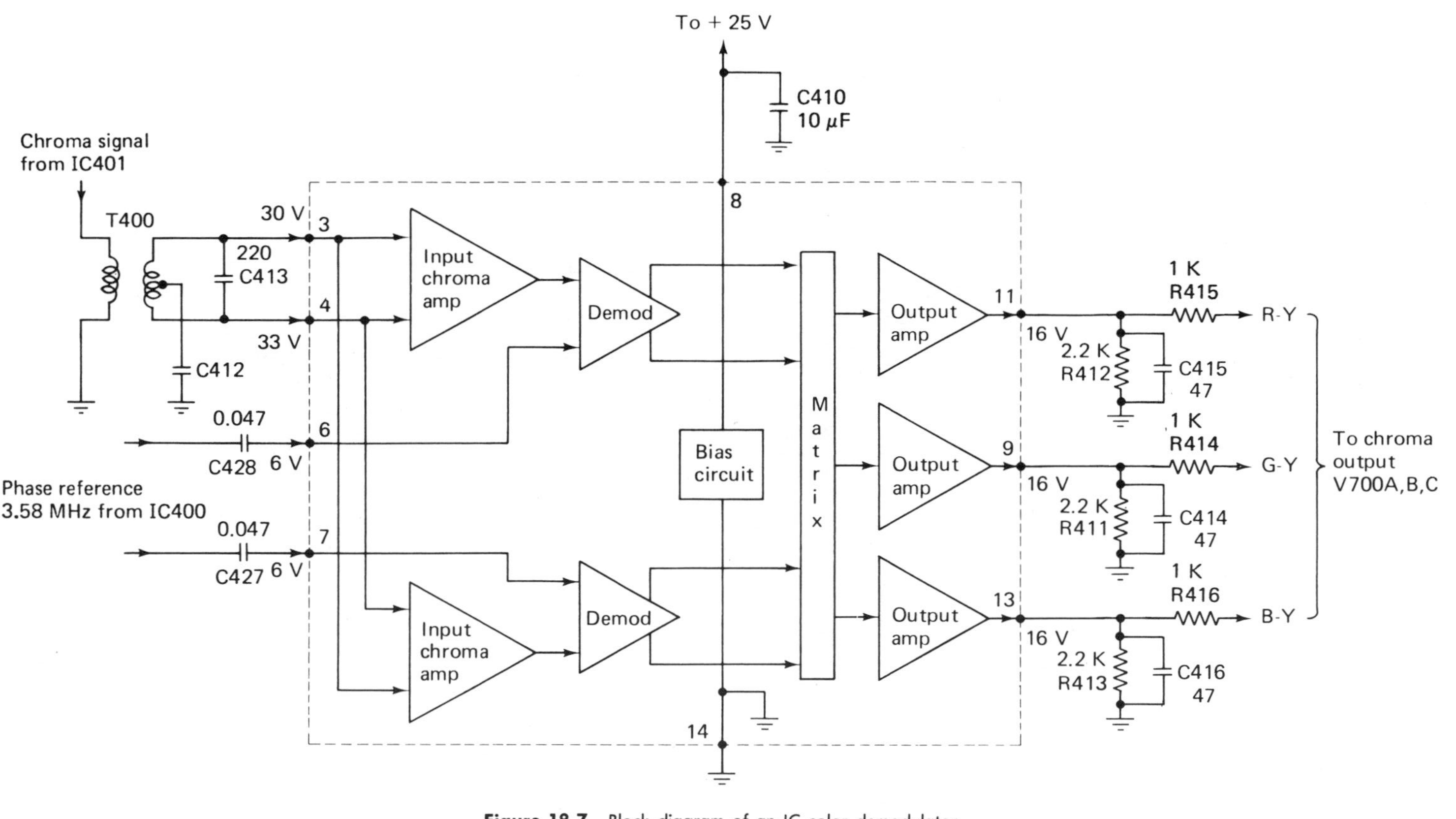

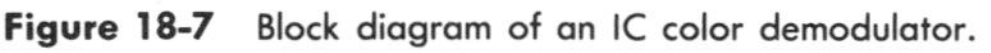

Figure 18-7 Block diagram of an IC color demodulator.

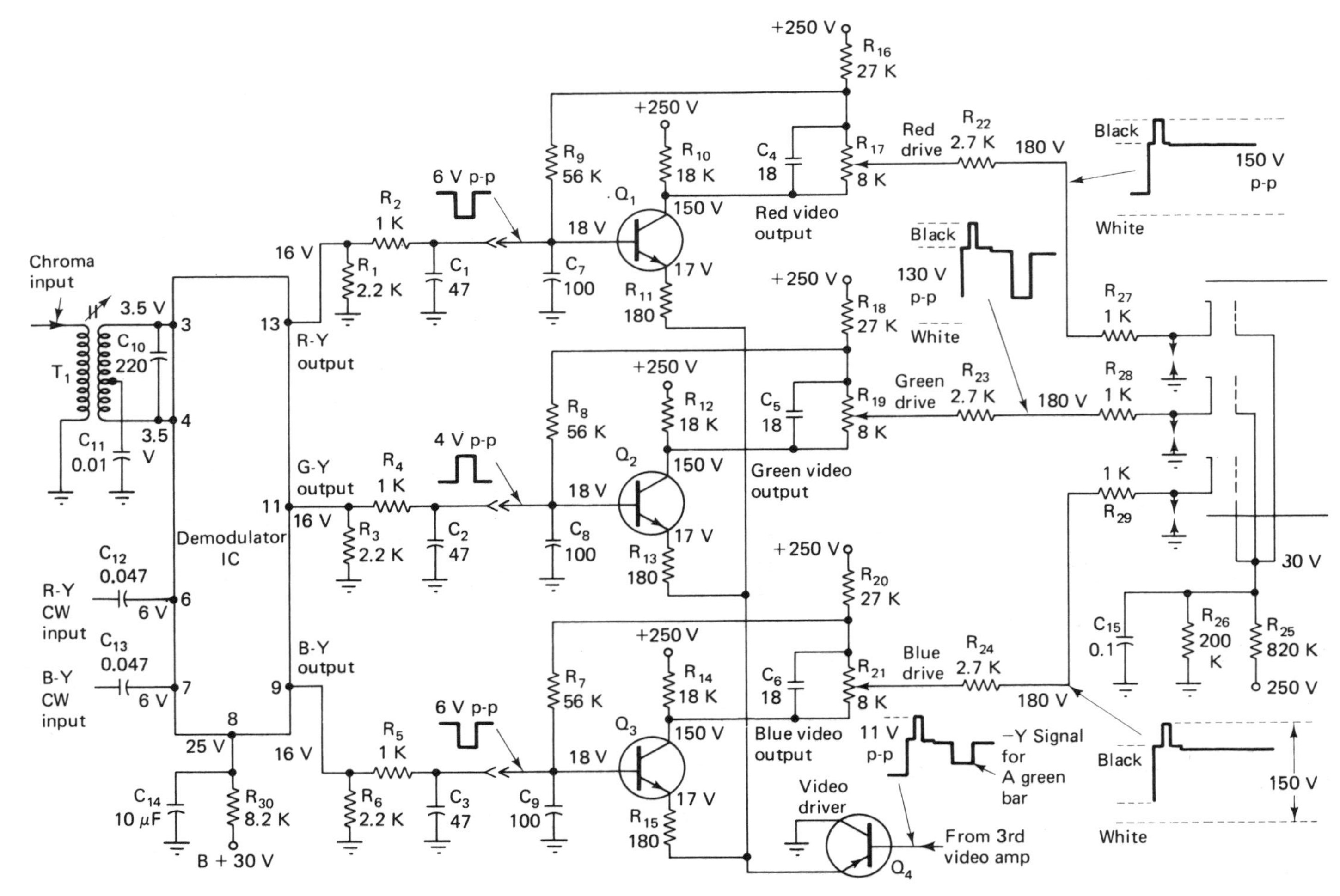

Figure 18-8 RGB IC color demodulator system.

going. At the same time that the color difference signals are applied to the base of the color amplifiers, the composite video signal ($-Y$) is applied to the base of the video driver Q_4. The $-Y$ waveform for one horizontal line shows that the horizontal blanking level corresponds to black (picture tube cutoff) and that the green bar is approximately one-half this level and represents a shade of gray.

The net result of applying the color difference signals to the bases of the video output transistor and the $-Y$ signal to their emitters may be seen by observing the waveforms at the cathodes of the picture tube. Notice that the red output and blue output amplifiers are cut off and have produced output signal levels that are of the same amplitude as black. This means that the red and blue CRT guns are driven into cutoff. The green signal output amplifier (Q_2) is conducting and produces an output waveform that is negative going. This waveform causes the grid to cathode bias of the CRT to reduce, thus increasing the green light output of the picture tube.

Compared to other methods, the RGB matrixing system has several advantages: (1) By eliminating the need for a high-level video amplifier, the number of high-voltage transistors required to drive the picture tube is reduced from four to three. (2) The RGB system provides better CRT arc protection since the CRT control grids are at ac ground potential. (3) Wider bandwidths are available for the color and luminance information, which makes for crisper pictures. This is because each amplifier drives one cathode of the picture tube instead of all three, as in the case in circuits such as that of Figure 10-24. As a result, the total shunt distributed capacitance is smaller, which allows better high-frequency response.

The diode demodulator. Essentially, a color demodulator is an amplitude and phase-sensitive detector. As such, it may be compared to the phase detector circuit used in the automatic phase and frequency control (AFPC) circuit discussed in Chapter 17 with reference to Figure 17-8.

In summary, the output voltage of the diode phase detector is zero when the input signals are 90° out of phase, and it is either positive or negative when the phase between them is altered. Maximum output occurs when the oscillator and the burst are in phase or 180° out of phase.

The voltage and phase characteristics of this detector are exactly what are required of a color demodulator. Thus, a diode color demodulator is identical to the circuit used in the phase detector AFPC circuit. When this circuit is used as a demodulator, the color burst is replaced by the chrominance signal, and the phase of the 3.58-MHz oscillator is adjusted to that needed to produce the desired output. For example, if the demodulator is designed to produce a $-(R - Y)$ output, the oscillator is placed at a phase of 270°. This means that the demodulator will produce maximum output for chrominance signals whose phase is either 270° or 90°, but it will be "blind" (zero output) to any chrominance signal whose phase is 0° or 180°. Intermediate outputs will be produced for chrominance signals whose phase falls between these two extremes.

A chrominance signal of any one color is the resultant of both $R - Y$ and $B - Y$ components. To demodulate a chrominance signal means that both $R - Y$ and $B - Y$ components must be resolved. This is done by using two demodulators, one for $R - Y$ and the other for $B - Y$. It should be noted that it is common practice to designate demodulators as $R - Y$ or $B - Y$ even though their actual outputs may be $-(R - Y)$ or $-(B - Y)$.

Practical diode demodulator circuit. Practical examples of low-level diode demodulators and their associated matrix amplifiers are shown in Figure 18-9. The chrominance signal is fed via L_1 to the junctions of diodes D_1, D_2 and D_3, D_4. The oscillator signal is fed to the diodes via the 3.58-MHz oscillator output transformer T_1. Alignment of each of the secondary windings of T_1 determines the phase of the oscillator signal driving each demodulator. As explained earlier, the grids of the picture tube require signals that are positive ($R - Y$, $B - Y$, and $G - Y$). Since these signals are obtained from inverting amplifiers, the inputs

Figure 18-9 Diode color demodulators feeding three direct-coupled color amplifiers.

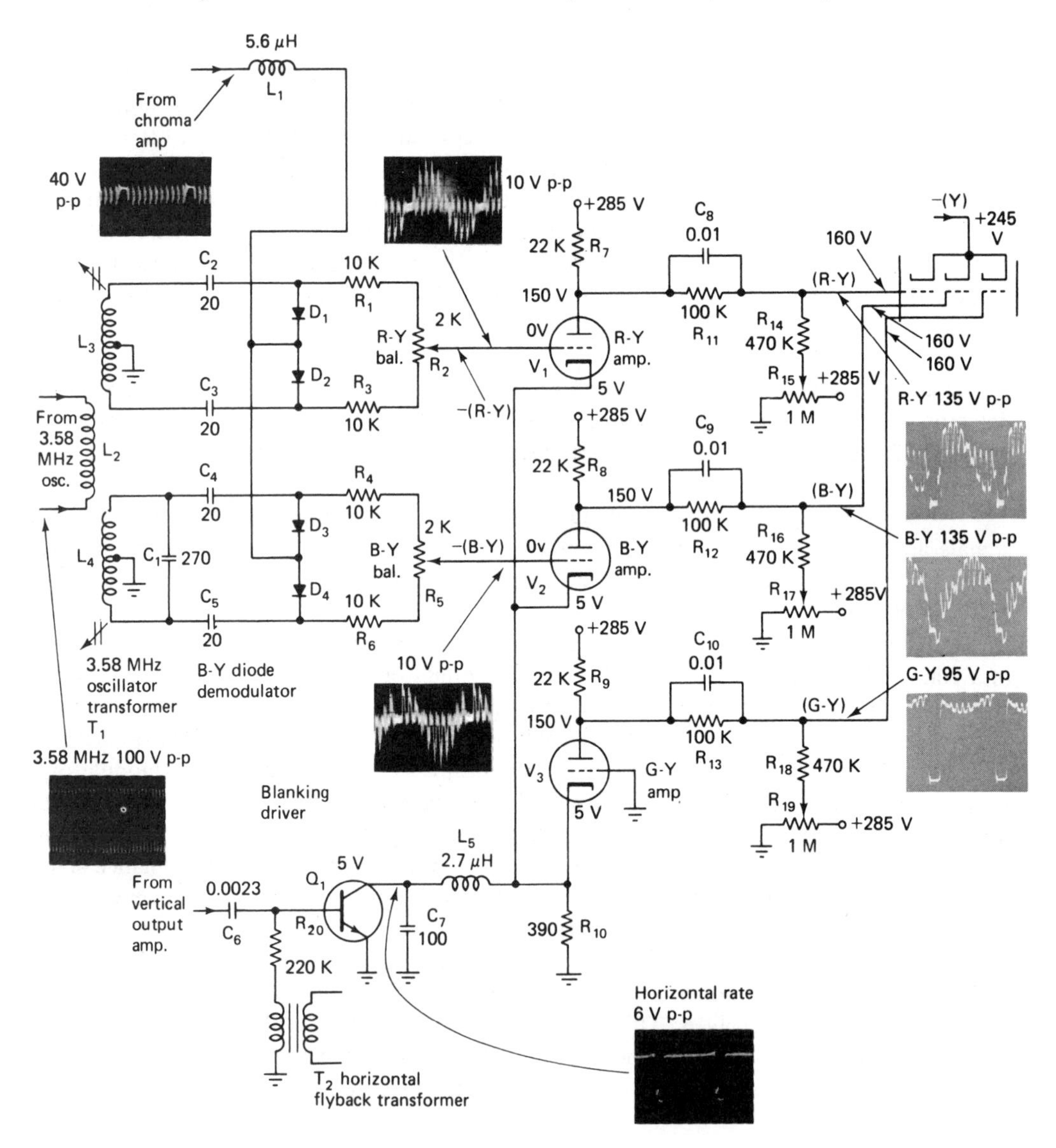

to the amplifiers must be negative. Thus, the output of the $R - Y$ demodulator is $-(R - Y)$, and its oscillator must have a phase of 270°. The output of the $B - Y$ demodulator is $-(B - Y)$ and its oscillator has a phase of 0°.

Color difference amplifiers. Examination of the color amplifier circuits reveals a number of important characteristics. Notice that the three tubes are identical and have identical load resistors (R_7, R_8, and R_9). Since the grid of the $G - Y$ amplifier is grounded, the $G - Y$ amplifier is operated as a grounded grid amplifier, the other two amplifiers are connected as grounded cathode amplifiers. From this it may be concluded that the gain of the $R - Y$ and $B - Y$ amplifiers are equal, but that of the $G - Y$ amplifier is somewhat less. This is verified by the output waveforms of each amplifier. The $G - Y$ amplifier output is designed to be less than that of the other amplifiers for modulation compensation and because the green phosphor of the picture tube is more efficient than that of red and blue. Therefore, it requires less drive for the same level of light output.

Matrixing of the $G - Y$ signal is accomplished by adding the correct proportions of $R - Y$ and $B - Y$ signals,

$$G - Y = -0.51(R - Y) - 0.19(B - Y).$$

This is done in this circuit by the signal developed across the common cathode resistor R_{10}. The $-(R - Y)$ and $-(B - Y)$ outputs of the demodulators are applied to the grids of the difference amplifiers. This in turn produces plate current variations in the amplifiers that generate voltage drops of the same input phase across R_{10}. The $G - Y$ voltage formed at the cathode is then amplified. The $G - Y$ grounded grid amplifier (V_3), however, does not invert the signal and drives the grid of the green gun of the picture tube with a $G - Y$ signal.

Blanking. If the color burst is allowed to reach the picture tube grids, it will produce an undesirable greenish-yellow smear across the picture tube face. The purpose of the horizontal blanking stage is to prevent this from happening. Since the color burst occurs during the horizontal blanking interval, retrace blanking of the color burst is accomplished by pulses obtained from the horizontal output stage.

In Figure 18-9 a transistor Q_1 is used as the blanking driver. The base of this stage is driven by two inputs. A vertical blanking pulse is obtained via C_6, and a horizontal pulse is obtained from a winding on the flyback transformer. Both of these pulses are positive going. Since the transistor is an NPN type, these pulses drive the transistor into saturation. The base current causes C_6 to charge and develop signal bias, which keeps the transistor cut off between pulses. During retrace pulse time transistor saturation causes the collector voltage to drop, thus developing a relatively large negative-going pulse across the common cathode resistor R_{10}. This negative-going signal acts to increase the conduction of all three color difference amplifiers. The net result is to cause their plate voltages to drop, producing very large (100 V p-p) negative-going pulses. These pulses are dc coupled to the grids of the picture tube, turning all three guns off during color burst intervals. Since these pulses are present with or without color program material, they also help to establish the dc bias voltages of the picture tube.

Dc coupling. The color diode demodulators and color difference amplifiers are dc coupled. That is, no coupling capacitors are placed between the demodulators and the amplifiers or between the amplifiers and the picture tube grids. The parallel combination of resistors and capacitors ($R_{11}C_8$, $R_{12}C_9$, and $R_{13}C_{10}$) found between the plates of the amplifiers and the grids of the picture tube are used as frequency compensation networks. They are used to improve the frequency response of the amplifier by compensating for the input impedance of the picture tube. The resistor shunting each capacitor couples the dc voltage across the capacitor and provides dc coupling.

18-6 troubleshooting

Dc coupling means that any changes in the dc operating conditions of either the demodulator or the amplifier stages will be communicated to the picture tube. Since the dc voltages on the electrodes of the picture tube establish its operating bias, they also establish the picture tube gray scale conditions. From a troubleshooting point of view, this means that troubles in a dc coupled demodulator amplifier system can result in changes in gray scale. Since a change in gray scale affects both color and black & white reception, it will be referred to as a *raster color* defect. Circuit defects that cause no change in gray scale but do affect the quality of the color reproduction will be termed *picture color* defects.

Raster color. A number of examples will help illustrate the difference between raster color and picture color defects. A common raster color defect is the *loss of one color.* This may be caused by a defective picture tube gun or some change in picture tube operating voltages. For example, let us assume that in Figure 18-9, R_9, the plate load of the $G - Y$ amplifier, opens. This will cause the plate voltage of this tube to drop toward zero. As a result, the picture tube grid voltage will also drop, cutting off the green gun. This will cause a raster color defect in which the raster is magenta for black & white and the color bar pattern does not display any green bars. See Figure 18-10 on the color plate. Of course, if the load resistance of the $R - Y$ or $B - Y$ amplifiers opened, raster color loss of red or blue would occur. See Figures 18-11 and 18-12 on the color plate.

Isolation of the defective stage that is causing a raster color defect is most easily accomplished by voltage checks of the picture tube electrodes. A voltage check of the picture tube control grid indicating a voltage much lower than normal would establish the cause of a color loss. Moving the voltmeter back to the color difference amplifier and checking its electrode voltages will indicate whether the trouble is to be found at the input or output of the stage. Isolation of the defective component is then done with an ohmmeter.

An alternative method is to signal trace the geartooth waveforms normally found on the grids of the picture tube. This is done with a gated rainbow color bar generator applied to the antenna input of the receiver. Starting at the grids of the picture tube, the geartooth waveforms are observed for a missing waveform. The missing waveform identifies the defective amplifier/demodulator chain. By observing the plate and grid of each stage back toward the bandpass amplifier, the defective area is identified by the point where the geartooth is found. Voltage and resistance checks in the suspected area will then identify the defective component.

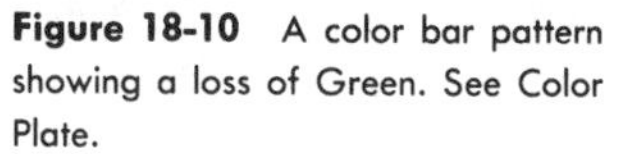

Figure 18-10 A color bar pattern showing a loss of Green. See Color Plate.

Figure 18-11 A color bar pattern showing a loss of Red. See Color Plate.

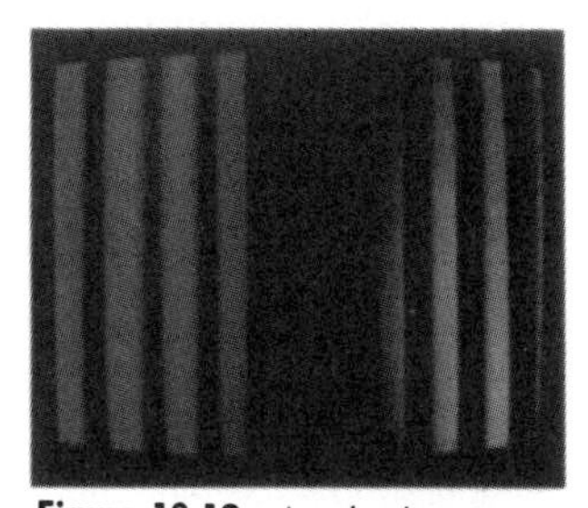

Figure 18-12 A color bar pattern showing a loss of Blue. See Color Plate.

Figure 18-13 An excessive blue raster, coloring the picture blue. See Color Plate.

Excessive color. Another raster color defect resulting in *excessive color* can be caused by a short between grid and cathode of the picture tube or some other defect that decreases picture tube bias for one gun. In this circuit decreased bias can only be caused by a more positive picture tube grid. This, in turn, can be caused by any defect that makes the plate current of any of the amplifiers decrease. For example, assume that the cathode of the $B - Y$ amplifier (V_2) opens. Plate current will drop to zero, forcing the plate voltage to rise. The rise in plate voltage is dc coupled to the grid of the picture tube, forcing the blue gun bias to decrease and turn up the brightness of the blue field. As a result, the picture tube will display a blue raster on both black & white and color programs. See Figure 18-13 on the color plate. In addition, the excessive blue beam current may cause blooming.

Raster color resulting in either excessive color or loss of color can also be caused by defects in the demodulator. Since the demodulator is two stages removed from the picture tube, defects in the demodulator make themselves felt at the picture tube because of dc coupling between stages. For example, if one of the demodulator diodes, such as D_2, opened, the color demodulator would become unbalanced. In this case, the dc output voltage of the $R - Y$ demodulator would change from zero to some negative value. Since this voltage is also the grid voltage of the $R - Y$ amplifier, the plate current of the amplifier would be reduced. This, in turn, would increase the plate and picture tube control grid voltages and would cause an increased red raster. This defect would affect both black & white and color reception because the 3.58-MHz color oscillator feeding the demodulator is responsible for developing the dc output voltage of the color demodulators under both conditions.

If D_1 opened, the demodulator would become unbalanced in the reverse direction. This, in turn, would cause a raster color decrease in red. Similarly, defects in the $B - Y$ demodulator would also cause changes in blue raster color.

Picture color. Picture color defects that do not affect raster color are also possible in a dc coupled color demodulator/amplifier chain. One possible cause of such a trouble symptom in Figure 18-9 is a shorted tuning capacitor C_1. In this case, the oscillator voltage is prevented from reaching the diodes and the $B - Y$ demodulator becomes inoperative, but the balance of the circuit is not affected and no dc voltage shifts occur. As a result, the gray scale will be normal, but during color reception blue will be missing. This defect will have the same appearance as that of Figure 18-12 on the color plate.

Blooming. Defects in the color amplifiers can cause blooming. For example, assume that a picture expands, goes out of focus, and becomes darker as the brightness control is increased. These are the classic symptoms of blooming. Blooming may be caused by three general circuit defects: (1) a defective picture tube, (2) poor high-voltage regulation, and (3) excessive beam current resulting from changes in picture tube electrode voltages. The easiest thing to do is to first check the picture tube electrode voltages (grids, cathodes, screens, and focus electrode). Any change in bias causing a large increase in beam current would be immediately suspect. In this case, all three picture tube grid voltages as well as the plate voltages of the amplifiers are found to be much higher than normal. This indicates that the defect must be common to all three amplifiers. The only component common to all three stages is R_{10}, the cathode resistor. Checking it with an ohmmeter will show that it is open.

Ac coupling. Color amplifiers that are ac coupled to the grids of the picture tube avoid some of the raster color problems that are common in dc-coupled amplifiers because changes in the dc levels of demodulators or amplifiers are blocked by the coupling capacitors and are not coupled to the grids of the picture tube. As a result, raster color symptoms similar to those of Figures 18-10 to 18-13 on the color plate can only be caused by defects in the coupling circuit between the coupling capacitors and the grids of the picture tube. Picture color defects in ac coupled circuits may be divided into two general categories: (1) loss of one color and (2) an increase in intensity of one color. In both cases, there will be little or no change in raster color (gray scale). See Figures 18-12 and 18-13 on the color plate. A loss of red in a color program or color bar pattern may be caused by a defective $R - Y$ or red demodulator or a red color amplifier. Similarly, the loss of blue would localize the trouble to the $B - Y$ or blue demodulator or a blue color amplifier. Similarly, the loss of green would indicate a defective $G - Y$ demodulator or a green color amplifier.

Oscilloscope signal tracing of the demodulator and amplifier stages will quickly identify whether the defect is at the input or output of the color amplifier stage. The specific component causing the trouble is then identified by substitution or resistance checks.

Signal substitution. An alternative method of isolating the defective stage is signal substitution. In this case, an audio generator operating at approximately 360 Hz or any other convenient frequency may be used to inject a signal into the base (grid) or collector (plate) circuit of the color difference amplifier. For example, if the signal is injected into the input of a normal $R - Y$ amplifier, six red and six black horizontal bars will be displayed. Green bars will be produced if the signal is injected at the input of the $G - Y$ amplifier, and blue bars will be produced if the signal is applied to the input of the $B - Y$ amplifier. Inability to produce color bars indicates a defective stage.

Signal injection at the demodulator requires a 3.58-MHz signal that is unmodulated. This signal may be obtained from an RF signal generator and may be used to substitute for either the chrominance signal or the subcarrier oscillator.

The visual effect of injecting an unmodulated RF signal into the demodulator of a normal receiver is to produce color bar patterns on the picture tube screen. The pattern screen will depend upon the exact phase and frequency of the injected signal. When the *frequency difference* between the injected demodulator signals is equal to the horizontal scanning frequency, a rainbow of red, blue, and green vertical bands will appear on the screen. The number of repeated rainbows will increase as the difference frequency between the signal generator and the color oscillator increase in multiples of the horizontal scanning rate. If the frequency difference between the signal generator and the color oscillator is made one-half the horizontal scanning rate, the picture will become a single band of color covering the entire raster. The specific color produced depends upon the phase of the signal generator. Thus, it is possible to have a completely red, green, or blue raster by simply varying the frequency of the signal generator.

Loss of one color. Loss of one picture color may be the result of a reduction or loss of oscillator voltage feeding the demodulators. For example, if L_4 (Figure 18-9) were to open, the 3.58-MHz oscillator voltage would not reach the $B - Y$ demodulator. As a result, the demodulator would cease functioning and its $-(B - Y)$ output would drop to zero. This in turn would cause a loss of blue in the picture. A picture color loss like this would not be accompanied by any raster color changes since the picture tube bias voltages would be unaffected.

RGB troubleshooting. Since the RGB system (Figure 18-8) is a dc-coupled system, defects in the circuits between the demodulator and the cathode of the picture tube may cause both raster color defects and picture color defects. Any circuit fault that shifts the operating point of any of the output amplifiers will cause changes in the picture tube operating conditions. This, in turn, will result in changes in gray scale and possibly blooming.

Defects that tend to cause any of the output transistors to conduct more heavily will lower the picture tube cathode voltage and cause an increase in beam current of the affected electron gun. Thus, an excessive red screen on both black & white and color could be caused by such defects as a defective demodulator IC, R_1 or R_2 open, R_9 decreased in value, Q_1 shorted, or a shorted spark gap. Similar defects in the blue and green amplifiers will result in excessive blue and green rasters.

Some circuit defects may cause a reduction in amplifier conduction increasing the picture tube cathode voltage, forcing a reduction in the light output of the corresponding picture tube electron gun. Such a loss of one raster color will affect both color and black & white reception. For example, assume that there is a yellow raster during black & white reception and that there is a lack of blue during color reception. These symptoms indicate a raster color defect caused by a dc operating voltage change in the blue color amplifier.

Defective components such as R_{15} open, Q_3 open, R_7 open, C_9 shorted, R_{24} open, R_{29} open, a defective demodulator IC, or a defective picture tube gun could be the cause of this trouble symptom.

Picture color defects. Picture color defects are characterized by normal black & white reception but poor color reproduction. The symptoms may appear as (1) loss of color, (2) loss of one color, (3) wrong color, or (4) weak color.

Each of the symptoms above may be caused by a defective demodulator IC. If signal tracing of the input signals discloses normal chroma and oscillator wave shapes but loss of one or more output wave shapes, the IC may be suspected as being faulty. Before replacing it, all of the associated components such as dropping resistors and bypass capacitors should be checked. If these components are good, the IC should be changed.

Loss of color can also be caused by component failures such as C_{10} or C_{11} shorted or a defective chroma input transformer T_1.

Loss of one color may be caused by a defective IC or such component defects as R_1, R_3, or R_6 reduced in value or defective oscillator capacitors C_{12} and C_{13}.

Wrong colors may be caused by such things as misalignment or defective chroma input transformer T_1, open C_{10}, leaky C_{11}, or defective oscillator coupling capacitors C_{12} and C_{13}.

Weak color may be caused by a defective IC or defects associated with the chroma input circuit. Such things as a defective transformer T_1 and a leaky C_{10} may be responsible.

No raster. In addition to the defects noted above, a *no-raster* condition can be produced by defects in either the demodulator or the color amplifier circuits. For example, if Q_4 were to open, the emitter currents for all three amplifiers would drop to zero. This would cause the collector voltages of all three amplifiers (Q_1, Q_2, and Q_3) to increase, cutting off the three guns of the picture tube. Similarly, if the power supply dropping resistor R_{30} feeding the demodulator IC were to open, a no-raster condition may result because the dc voltages at the outputs of the demodulators would decrease, tending to turn off the color amplifier. This, in turn, would increase the collector voltage and force the picture tube toward cutoff.

18-7 the vectorscope

The amplitude of the chrominance signal represents the color saturation of a scene. The phase of the chrominance signal corresponds to the color. This information can be conveniently displayed on the oscilloscope in the form of a Lissajous pattern. The resultant display is called a *vectorgram,* and the oscilloscope that produces this display is called a *vectorscope.*

To obtain a vectorgram, a gated rainbow generator must be used as the receiver signal source. The geartooth $R - Y$ and $B - Y$ outputs appearing at the grids of the picture tube are connected to the vertical and horizontal inputs of the oscilloscope. This is shown in Figure 18-14. Since both the $R - Y$ and $B - Y$ inputs to the oscilloscope are interrupted sine waves that differ from one another by 90°, the resultant Lissajous pattern will be in the basic form of a circle that collapses toward the center whenever the sine wave is interrupted. This daisylike pattern of 10 petals is shown in Figure 18-15. In constructing this diagram the $R - Y$ signal is plotted so that its amplitude changes are positioned

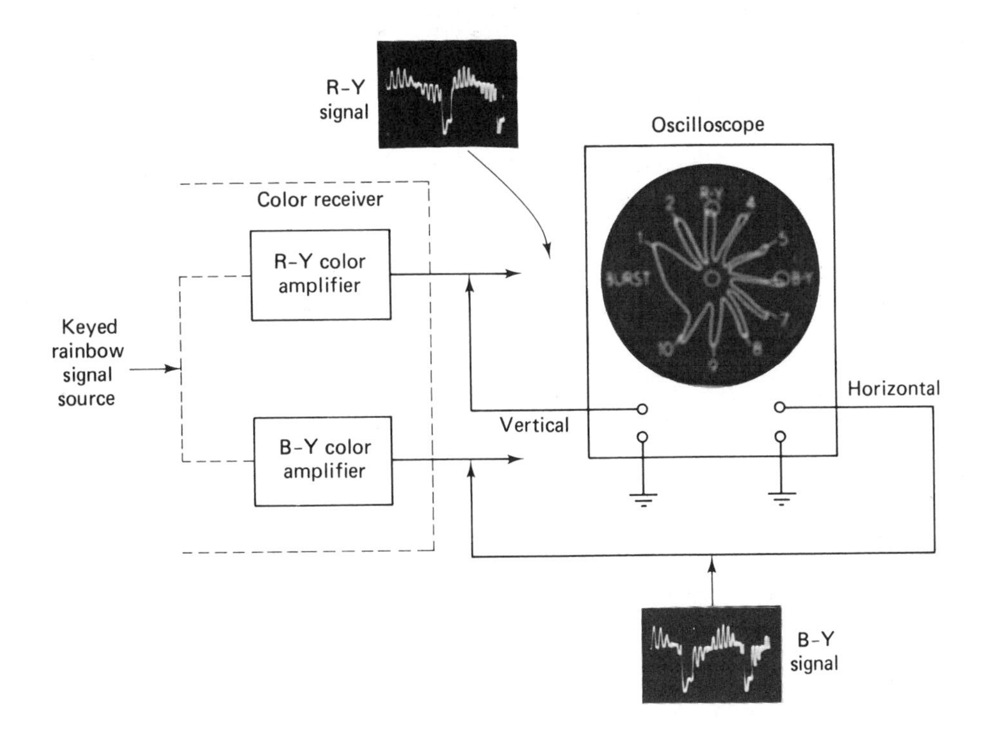

Figure 18-14 Oscilloscope connections required to produce a vectorgram.

Figure 18-15 Lissajous pattern produced by the $R - Y$ and $B - Y$ color signals.

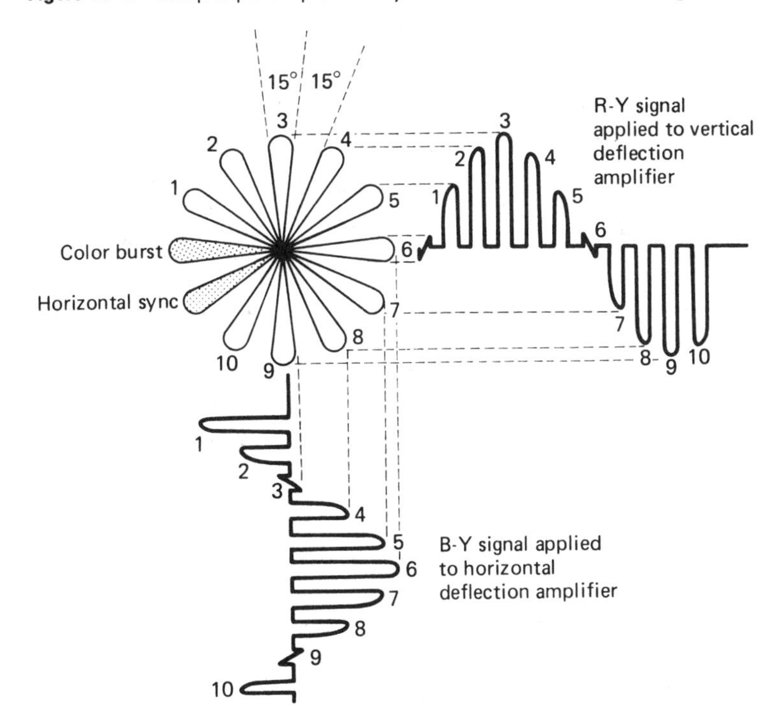

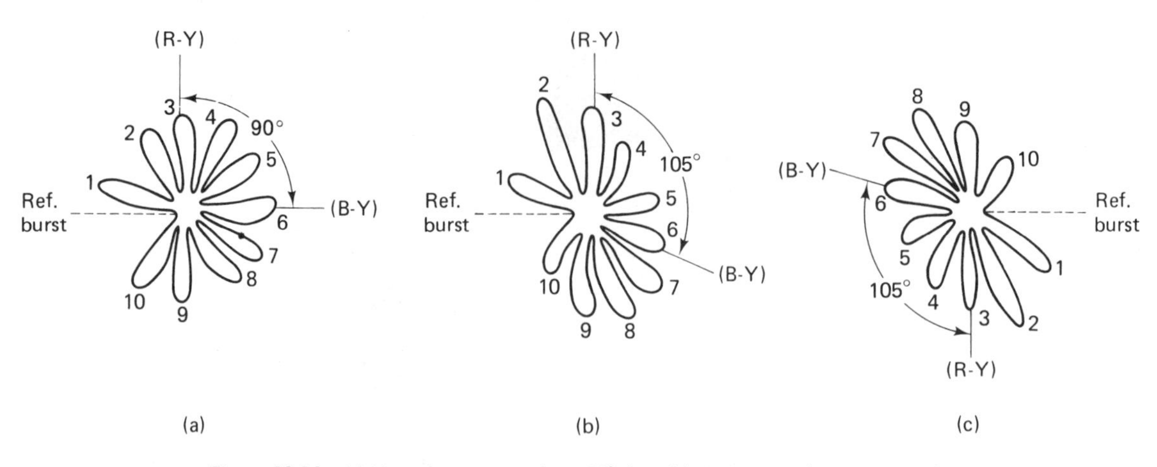

Figure 18-16 (a) Normal vectorgram for a 90° demodulation system; (b) vectorgram for a 105° demodulation system; (c) vectorgram for a 105° demodulation system using cathode drive.

vertically. The $B - Y$ signal is arranged so that its amplitude variations are positioned horizontally. The resultant Lissajous pattern is produced by plotting the points of intersection at each instant of time. Since there are 10 color bursts, the vectorgram will consist of 10 petals. The horizontal sync and color burst are not present because they have been blanked out. The position of each petal represents the phase angle of each color in the color bar pattern. Petal 1 corresponds to 30°, petal 3 ($R - Y$) corresponds to 90°, petal 6 ($B - Y$) corresponds to 180°, and petal 10 ($G - Y$) corresponds to 300°.

From the above a number of conclusions may be drawn: (1) Loss of $R - Y$ will produce a vectorgram consisting of one horizontal line. A loss of the $B - Y$ signal will cause the vectorgram to form a single vertical line. (2) Changing the receiver *color control* will cause both $R - Y$ and $B - Y$ signals to vary in amplitude and cause the diameter of the vector pattern to change. Similarly, the receiver's fine tuning will have a large effect on the size of the vectorgram. Proper fine tuning generally produces the best shaped and largest vectorgram. Defects such as a weak $R - Y$ amplifier will show themselves as a flattened oval. (3) A change in phase (representing a change in color) produced by adjusting the *tint control* will cause the petals to rotate.

If the demodulators are operated with a 90° phase difference, the vectorgram will be such that petal 3 will be vertical and petal 6 will be horizontal and pointing to the right. See Figure 18-16(a). If the demodulators are 105° apart, the vectorgram will become somewhat elliptical. Thus, when the $R - Y$ petal is vertical, the $B - Y$ petal will be 105° away from it [Figure 18-16(b)]. Because the vectorscope is a sensitive phase indicator, it may be used for AFPC alignment.

Modern receivers feed the color signals to the picture tube cathodes instead of to the grids. Vectorgrams produced from these receivers will produce an inverted pattern similar to that of Figure 18-16(c).

18-8 automatic tint control and VIR

Automatic tint control circuits are designed to compensate for undesirable changes in tint, or hue, caused by small phase changes in the broadcast signal. These phase changes result from differences between cameras and other studio equipment. The visual effect of these changes is most evident in the color ranges associated with flesh tones. It was not uncommon to see green people turn into purple people and then return to normal color as commercials are changed or when cameras are switched or when one switches from one channel to another. Much work has been done at the transmitter to minimize these variations.

In 1969 receiver manufacturers began introducing circuits into the receiver that compensated for this problem. These circuits take advantage of the fact that the colors associated with flesh tones are found at a chrominance signal phase angle of 57°. This angle corresponds to the I signal that the transmitter generates. Phase errors resulting from camera changes or other variations in transmission characteristics that move the $+I(57°)$ vector toward the burst cause greenish people. Phase errors away from the burst cause red-faced people.

Off-color flesh tones can be made less objectionable by widening the demodulation phase angle. This means that the $R - Y$ demodulator would demodulate at an angle of less than 90° and the $B - Y$ demodulator would demodulate at an angle greater than 180°. The net result of this is to produce an $R - Y$ output that has a greater output for orange colors. The $B - Y$ output will become weaker in the red area and stronger in the blue. Figure 18-17(a) shows a vectorgram of a normal $R - Y/B - Y$ receiver. Notice that the third petal is at 90° ($R - Y$) and the sixth petal is at 180° ($B - Y$). When the demodulation phase angle is widened, the vectorgram, as indicated in Figure 18-17(b), becomes a distorted ellipse whose major axis lies on 57°.

Widening the demodulation angle results in less flesh-tone color changes, but it causes considerable color errors elsewhere. Fortunately, this color distortion is generally overlooked as long as the people look natural to the viewer.

Unfortunately, there is no universal circuit used for ATC; each manufacturer has its own scheme. In general, two techniques have been used. Some manufacturers simply widen the demodulator phase angle. Others accomplish the same thing by modifying the chroma signal phase angle. In some circuits not only is the phase angle widened when ATC is switched on, but the gain of the $B - Y$ amplifiers is also reduced, and the gray scale is shifted from white to sepia (brown-white). These methods were generally discontinued as studio control of phase became tighter.

Figure 18-17 Vectorgram before (a) and after (b) ATC is introduced into the receiver.

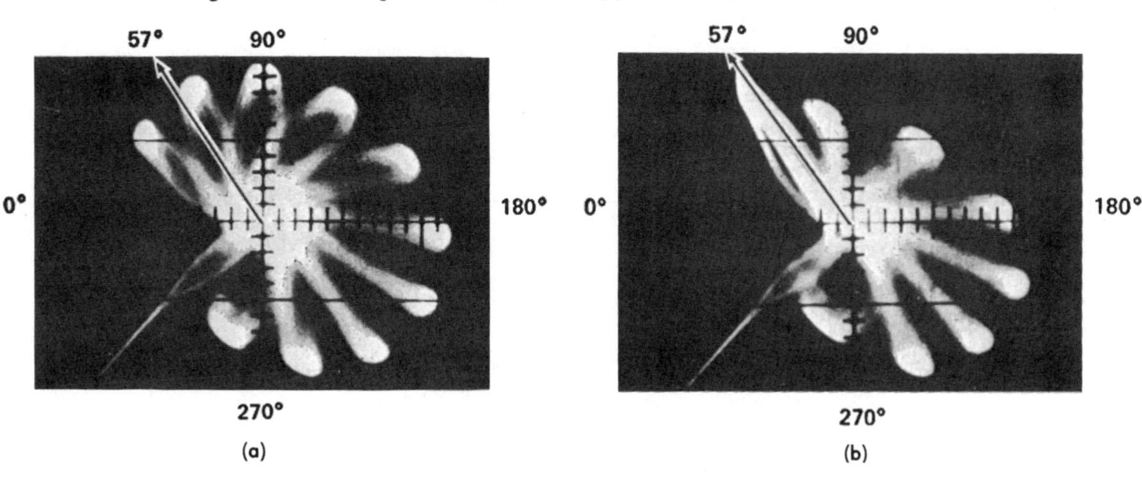

VIR. A method of providing the television receiver with the capability of truly automatic tint and intensity control was introduced by General Electric (GE) in 1977. This is done by detecting and processing the vertical interval reference (VIR) signal transmitted on the line 19 of each vertical deflection field of the composite video signal. See Section 3-3.

The original GE circuitry needed to detect and process the VIR signal was found on a separate module consisting of five integrated circuits and 30 transistors. Other manufacturers have since introduced single IC units that can perform the same functions.

In Figure 18-18 is the simplified diagram of the GE VIR "Broadcast-Controlled" color system. The circuit consists of five major stages. Each of these stages may, in turn, be broken up into many additional stages.

The operation of this system is initiated by the *line 19 recognizer stage.* This circuit uses a digital counter in conjunction with enabling and decoder circuits to identify line 19 of each field. During line 19 of each field the output of this stage is a 63-μs pulse. This pulse feeds the *pulse slicer* and *tint controller stages.* The pulse slicer stage makes use of two monostable multivibrators to generate two pulses. One is called the chroma reference pulse

Figure 18-18 Simplified block diagram of the GE VIR ATC module.

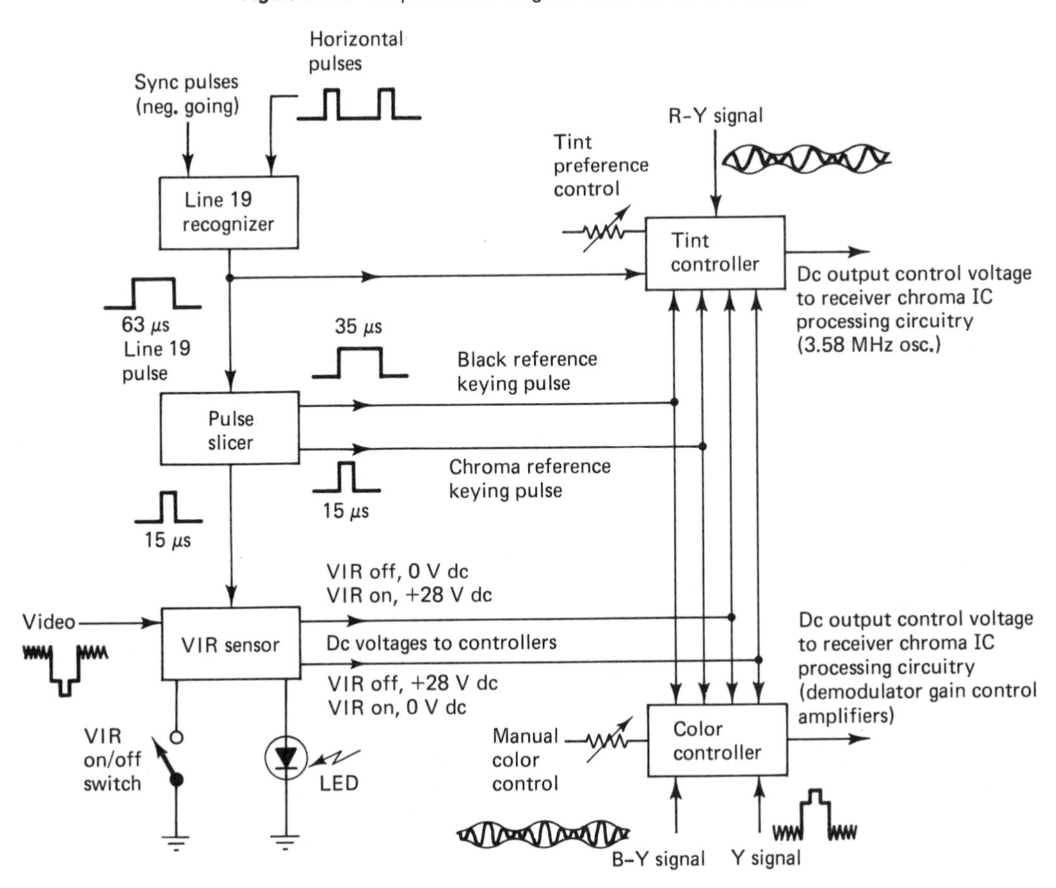

and has a 15-μs duration. The other pulse is of 35-μs duration and is called the black reference pulse. The narrower pulse occurs at the beginning of line 19 immediately following horizontal blanking. It therefore occurs during the color reference interval of the VIR. The 35-μs pulse occurs shortly after the 15-μs pulse and during the black & white reference interval of the VIR. These pulses are both fed to the *tint* and *color controller* stages. In addition, the chroma reference pulse is also used to feed the *VIR sensor stage.* This stage compares the chroma reference pulse (15 μs) and the composite video signal by feeding these signals into a monostable multivibrator and thereby developing a pair of dc control voltages.

The video signal is first passed through a low-pass filter amplifier circuit which removes the color reference signal. In digital circuit terminology this means that line 19 will present a "low" for the 24 μs that the VIR color reference is present. If the VIR color reference is not being transmitted, line 19 will be "high" during the color reference interval. During VIR transmission the combination of the 15-μs chroma reference pulse "high" and the 24-μs "low" will cause the monostable multivibrator to produce two outputs (+28 V and 0 V) that will activate the tint and color controller stages of the system. If the VIR signal is not transmitted, these voltages will interchange, turning off the tint and color controller stages. In addition, the VIR sensor stage also drives an LED that indicates whether or not a VIR signal is being transmitted.

The tint controller stage has six inputs. Two inputs are dc voltage obtained from the VIR sensor that enable or disable the stage depending upon whether or not the VIR signal is being transmitted. In addition, the chroma reference pulse (15 μs), the black reference pulse (35 μs), the line 19 pulse (63 μs) and the $R - Y$ signal are also fed into this stage.

The line 19 pulse (63 μs) is used to activate or deactivate a tint preference amplifier during VIR transmission. This circuit and its associated control allows the operator to adjust the color to his or her own preference.

The transmitted VIR color reference signal corresponds to a $-(B - Y)$ signal; therefore, during the VIR interval the $R - Y$ signal feeding the tint controller should normally be zero. Any other $R - Y$ signal level indicates a tint error. The $R - Y$ signal feeding the tint controller is compared to the 15-μs chroma reference pulse and the 35-μs black reference pulse; a dc error voltage is then generated that is proportional to the tint error. This dc voltage is fed to an oscillator control circuit that forces the 3.58-MHz color oscillator to assume its proper phase, returning the $R - Y$ signal level to zero.

The color controller circuit and the tint controller circuits have similar inputs. Both are fed by the black and chroma reference pulses, and both are controlled by the dc error voltages developed by the VIR sensor. However, the tint controller is fed an $R - Y$ signal, whereas the color controller is fed a $B - Y$ and a Y signal. It should be recalled that the color reference is a $-(B - Y)$ signal and that during the VIR chroma interval the $B - Y$ color demodulator will produce an appreciable positive signal output for a blue color signal and a negative output for a green yellow $-(B - Y)$ signal. In the color controller the $B - Y$ signal and the Y signal are matrixed producing a blue signal output. The amplitude of this output signal depends upon the color saturation. To understand this, consider that the reference phase (0°) of the VIR signal feeding the $B - Y$ color demodulator will produce an output amplitude equal to its corresponding Y-signal amplitude. (Note that the peak of the color reference and the Y signal are both 20 IRE units in amplitude.) After matrixing in the color controller, the blue output signal will be zero. Any saturation error will make the $B - Y$ signal larger or smaller than the Y-signal level, forcing the blue matrix output to shift above or below its zero level. This shift in level is, in turn, detected by the comparitor circuit in the color controller. A dc error voltage is then developed which feeds the demodulator gain control amplifiers, forcing the output level of the $B - Y$ demodulator to return to normal.

18-9 other color television systems

The color television system in use in the United States today is based on the NTSC (National Television Systems Committee) standards for the generation of color signals. These standards were adopted more than 30 years ago (1953) and have remained virtually unchanged since then. The NTSC system, adopted by about 31 countries, has certain inherent limitations that have given rise to alternative systems such as PAL (Phase Alternate Lines), adopted in about 45 countries, and SECAM (Sequential Color and Memory), adopted in about 29 countries. See Table 18-2 and its accompanying map for a listing of worldwide television broadcasting systems.

The NTSC system has a number of important shortcomings at both the receiver and the transmitter. At the receiver, crosstalk between demodulator output signals causes color distortion. This means that color reproduction of the transmitted scene will never be exact. At the transmitter, phase changes in the chrominance signal that take place whenever video tape recorders are switched or whenever changes between local and network television systems take place, cause changes in the receiver color that can require the use of VIR networks. In addition, the NTSC system is sensitive to transmission path differences that introduce phase errors that result in color changes in the picture.

The PAL system. In the PAL system the disadvantages of the NTSC system are overcome by reversing the phase of the color information on adjacent horizontal lines. In this way, phase errors may be averaged out by the eye so that color hues are accurately reproduced. This system is not perfect. It is, however, the least expensive method. A more expensive method called Deluxe PAL uses a delay line to electrically accumulate adjacent line color signals and display the information after integration on the picture tube screen.

At the transmitter end of the PAL system a color burst is sent out at the start of each line. Its function is to synchronize the receiver color oscillator for reinsertion of the exact carrier into the $R - Y$ and $B - Y$ demodulators. The color burst (10 cycles, 4.43 MHz) used in the PAL system is not exactly in phase with the color subcarrier. On alternate lines it shifts $\pm 45°$ about the zero reference phase. As such, it is often called the *swing burst.*

The PAL receiver. The PAL color receiver is essentially the same as the NTSC color receiver, but there are some important differences in the demodulator and AFPC circuits.

The block diagram of the color demodulator section of the PAL receiver is shown in Figure 18-19. Notice that the output of the bandpass amplifier divides into three outputs that feed an "adder," a "subtractor," and a PAL delay line of one horizontal line period.

The output of the delay line also feeds the adder and the subtractor. As a result, the chrominance signal for any line is added to and subtracted from the corresponding signals for the previous line which have been held up by the delay line. Since the $R - Y$ signal at the transmitter is reversed on each line, the adder cancels out the $R - Y$ signals and leaves only the $B - Y$ chrominance information feeding the $B - Y$ demodulator.

By similar reasoning, the $B - Y$ signals are canceled in the subtractor, and the $R - Y$ chrominance signals that are phase reversing on alternate lines add, producing an $R - Y$ chrominance signal output that feeds the $R - Y$ demodulator. It is in this way that the transmitted chrominance phase errors are prepared to be canceled out in the demodulators.

Two synchronous demodulators are used to produce the $R - Y$ and $B - Y$ color difference signals. The oscillator input to each demodulator comes from a 4.43-MHz oscillator. The $B - Y$ demodulator is fed by the $B - Y$ chrominance signal and an oscillator voltage that is in phase with the color subcarrier. The oscillator voltage feeding the $R - Y$ demodulator is phase shifted 90° and then passed through a PAL switch which on alternate lines reverse the phase of the 4.43-MHz oscillator voltage. As a result, the output of the $R - Y$ demodulator is a reproduction of the $R - Y$ transmitter modulating signal.

The outputs of the $R - Y$ and $B - Y$ demodulators then feed the amplifier-matrix in the same manner as in the NTSC color receivers.

Color sync circuits. The PAL receiver requires an AFPC circuit that is similar to the phase detector AFPC circuits used in NTSC receivers but two additional circuits are used in the PAL receiver. These circuits are shown in Figure 18-20.

Since the color burst is reversing its phase between +45° and −45°, the output of the phase detector will be a dc error voltage and a ripple voltage whose frequency will be one-half the scanning rate (approximately 7.8 kHz). If there is no phase error, the average value of this ripple will be zero. The dc error voltage is used to activate a reactance control circuit. The ripple is fed to an "identification amplifier" that drives a phase switch. This arrangement identifies the line being received and switches in the right phase of oscillator drive to the $R - Y$ demodulator for that line.

SECAM. SECAM is a color television system developed in France in the early 1960s. Its name, *Sequential Color and Memory,* suggests the basic principles of its encoding system.

In the SECAM system, color transmission is performed in a line sequential manner where only one of the two color difference signals is transmitted at a time. First, the red color difference signal is transmitted on one line, then the blue color difference signal is transmitted on the following line. This process is repeated for the remaining lines of the raster.

At the transmitter two frequency modulated subcarriers are used to represent the $R - Y$ and $B - Y$ color difference signals. The frequency of the subcarrier is placed above 4 MHz, thereby reducing interference and improving resolution. Hue and saturation of the chromaticity information are represented by frequency modulation of the subcarrier. Color bars of 100% saturation and 75% amplitude produce a deviation of ±350 kHz. However, due to preemphasis, peak deviations of ±750 kHz are possible. It is the FM characteristic that makes the SECAM system immune to differential gain, differential phase, and time base error. Such immunity is important to proper studio video tape recorder operation.

Table 18.2 Worldwide Television Broadcasting Systems

Country	Broadcasting Standard		Color System
	VHF	UHF	
Algeria	B		PAL
Argentina	B		PAL
Australia	N		PAL
Austria	B	G	PAL
Barbados	M		NTSC
Belgium	C	H	PAL
Bermuda	M		NTSC
Bolivia	N		NTSC
Brazil	M	M	PAL
Bulgaria	D		SECAM
Cambodia	M		
Canada	M	M	NTSC
Canary Islands	B		PAL
Chile	M		NTSC
China	D		PAL
Colombia	M		NTSC
Congo	D		SECAM
Costa Rica	M		NTSC
Cuba	M		NTSC
Cyprus	B	G	
Czechoslovakia	D	K	SECAM
Denmark	B	G	PAL
Dominican Rep.	M	M	NTSC
East Germany	B	G	SECAM
Ecuador	M		NTSC
El Salvador	M		NTSC
England	A	I	PAL
Ethiopia	B		
Finland	B	G	PAL
France	E	L	SECAM
Gabon	K		SECAM
Gambia	I		
Ghana	B		PAL
Gibraltar	B		
Greece	B		SECAM
Greenland	M		NTSC
Guadeloupe	K_1		SECAM
Guam	M		NTSC
Guatemala	M		NTSC
Guyana	M		NTSC
Haiti	M		SECAM
Honduras	M		
Hong Kong		I	PAL
Hungary	D	K	SECAM
Iceland	B		PAL
India	B		
Indonesia	B		PAL
Iran	B		SECAM
Iraq	B		SECAM
Ireland	I / A		PAL
Israel	B	G	PAL
Italy	B	G	PAL
Ivory Coast	K_1		SECAM
Jamaica	M		
Japan	M	M	NTSC
Jordan	B		PAL
Kenya	B		PAL
Korea	M	M	NTSC
Kuwait	B		PAL
Lebanon	B		SECAM
Liberia	B		PAL
Libya	B		SECAM
Luxembourg	C	L	SECAM
Malagasy Rep.	K_1		
Malawi	B		
Malaysia	B		PAL
Malta	B		
Mexico	M		NTSC
Micronesia	M		NTSC
Monaco	E	L	SECAM
Mongolia	D		
Morocco	D		SECAM
Netherlands	B	G	PAL
New Caledonia	K_1		SECAM
New Zealand	B		PAL
Nicaragua	M		NTSC
Niger	K		SECAM
Nigeria	B		PAL
North Korea	D		
Norway	B	G	PAL
Oman	B		PAL
Pakistan	B		PAL
Panama	M		NTSC
Paraguay	M		
Peru	M		
Philippines	M	M	NTSC
Poland	D	K	SECAM
Portugal	B	G	PAL
Puerto Rico	M	M	NTSC
Qatar	B		PAL
Romania	D		
Rwanda	D		
Saudi Arabia	B		SECAM
Senegal	K_1		SECAM
Sierra Leone	B		PAL
Singapore	B		PAL
South Africa	I	I	PAL
Spain	B	G	PAL
Sudan	B		PAL
Surinam	M		NTSC
Sweden	B	G	PAL
Switzerland	B	G	PAL
Syria	B		SECAM
Tahiti	K_1		SECAM
Taiwan	M		NTSC
Tanzania	B		PAL
Thailand	B		PAL
Trinidad/Tobago	M		NTSC
Tunisia	B		SECAM
Turkey	B		PAL
U.S.A.	M	M	NTSC
U.S.S.R.	D	K	SECAM
Uganda	B		PAL
United Arab Emirates	B		PAL
Upper Volta	D		
Uruguay	N		NTSC
Venezuela	M		NTSC
Virgin Islands	M		NTSC
West Germany	B	G	PAL
Western Samoa	M		NTSC
Yemen	B		
Yugoslavia	C	L	SECAM
Zambia	B		PAL
Zaire	D		SECAM
Zimbabwe	B		

Key to Broadcasting Standards

Broad-casting System	No. of Lines	Channel Band-width	Sound Modula-tion	F_S-F_P
A	405	5MHz	A_3	−3.5MHz
B	625	7MHz	F_3kHz ± 50	+5.5MHz
C	625	7MHz	A_3	+5.5MHz
D	625	8MHz	F_3kHz ± 50	+6.5MHz
E	819	14MHz	A_3	+11.15MHz
F	819	7MHz	A_3	+5.5MHz
G, H	625	8MHz	F_3kHz ± 50	5.5MHz
I	625	8MHz	F_2kHz ± 50	+6MHz
K_1	625	8MHz	F_3kHz ± 50	+6.5MHz
L	625	8MHz	A_3	+6.5MHz
M	525	6MHz	F_3kHz ± 25	+4.5MHz
N	625	6MHz	F_3khz ± 25	+4.5MHz

Courtesy of Leader Instrument Corp.

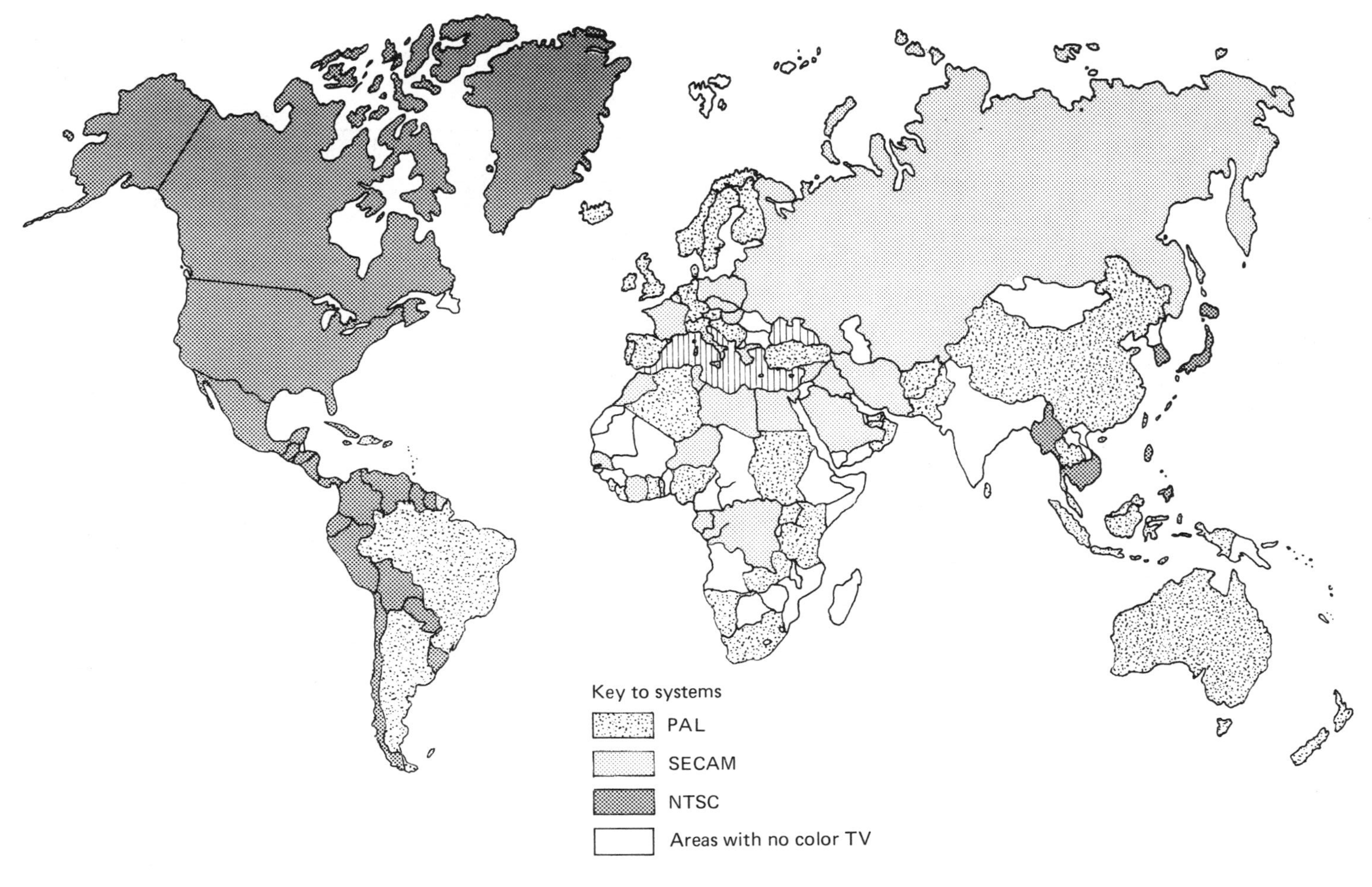

Table 18.2 (*Cont.*)

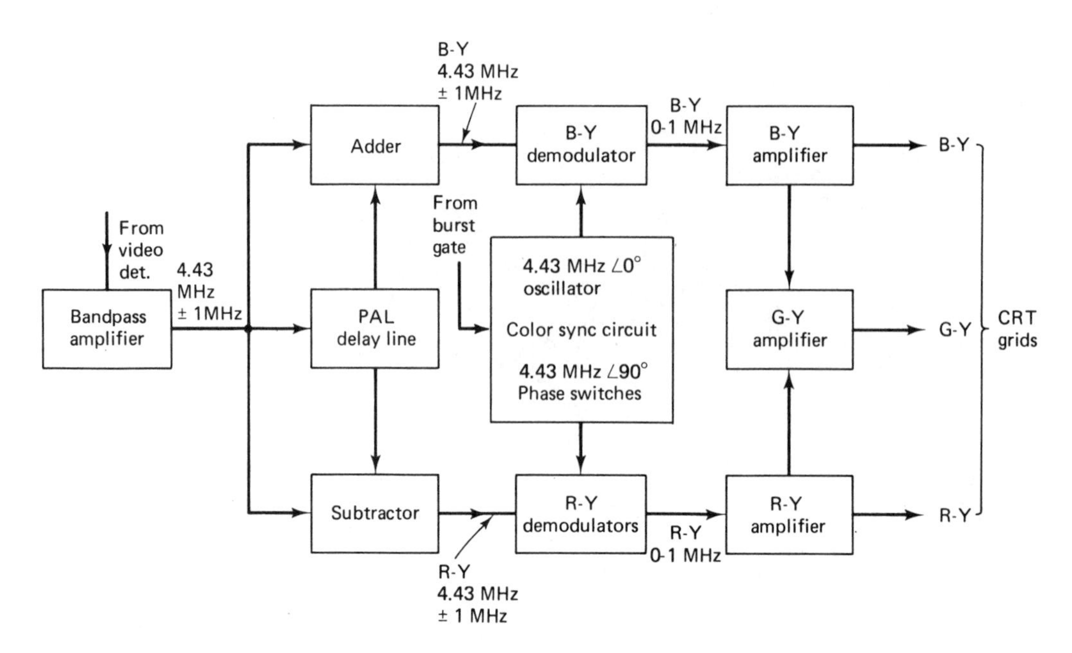

Figure 18-19 Demodulator circuit of a PAL receiver.

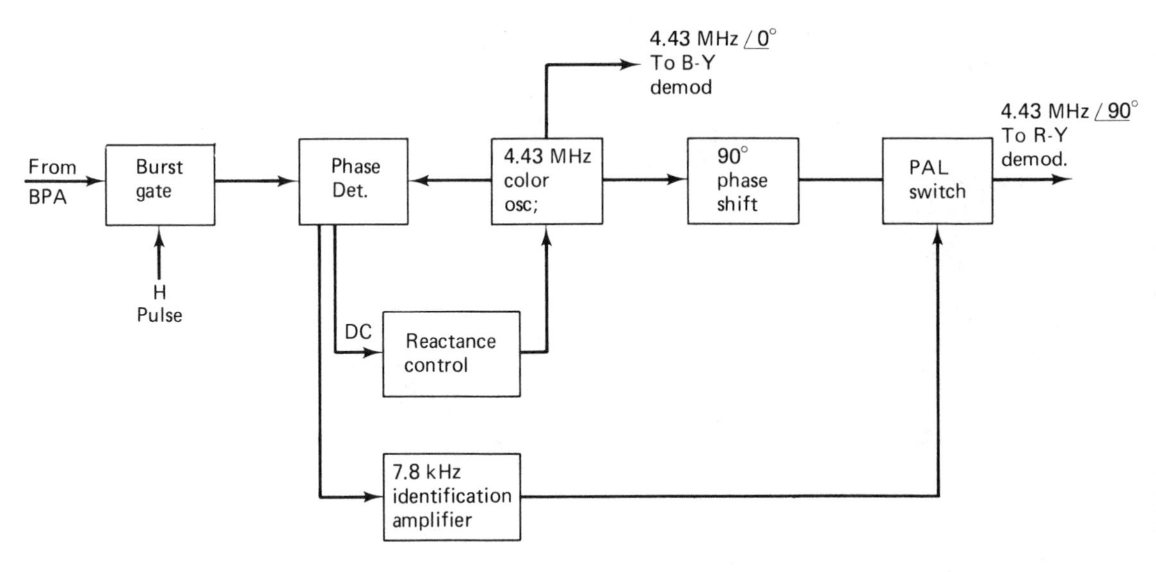

Figure 18-20 PAL color sync circuits.

The color signals are limited to a high frequency of 1.5 MHz and undergo a high-frequency preemphasis. To further improve the system's noise immunity and to achieve compatability for saturated colors, a special circuit called *anti-cloche* or *mise en forme* increases subcarrier amplitude as the deviation increases.

The SECAM standards call for the subcarrier amplitude to be approximately 16% of the overall signal, and allow about 2 kHz tolerance for defining a specific color. These standards result in a system where unstable video tape recorders are able to render color signals of high fidelity.

Identification of the proper sequence of color lines in each field is accomplished by identification pulses that are generated during vertical blanking. This signal consists of a sawtooth modulated subcarrier. The sawtooth is negative going when the red difference signal is normally transmitted. At the receiver, the identification pulses generate positive and negative switching pulses in the receiver decoder to produce the correct color sequence.

At the receiver, an ultrasonic delay line is used as a one-line memory device to produce a simultaneous decoder output of both color difference signals. One of the components of the color difference signal appearing at the output of the chrominance amplifier is the currently arrived signal. The other component is the signal that was picked up during the preceding line period and held in the receiver delay line for one horizontal line period.

Routing these color difference signals to their correct FM demodulators is accomplished by an electronic switch operating at the horizontal line frequency. The switch is driven by a bistable circuit that is triggered by pulses from the receiver's horizontal deflection generators.

One of the major advantages of the SECAM system is the reduced number of receiver controls necessary for proper operation. Figure 18-21 compares the NTSC, PAL, and SECAM receivers. Notice that the NTSC receiver requires four controls—brightness, contrast, color, and tint. The PAL receiver uses three controls because it does not require a tint control. In this system phase errors tend to be translated into small shifts in saturation that are more acceptable to the observer.

The SECAM receiver only requires a brightness and contrast control; color controls are not needed. This is because the color signals are constant amplitude frequency modulated signals. Assuming picture tube gray scale has been adjusted properly, the chrominance information is automatically as accurate as the originating source.

Because the SECAM system uses normal sync and blanking pulses, the SECAM signal is fully compatible with black & white receivers.

Figure 18-21 Comparison of (a) an NTSC receiver; (b) a PAL receiver, and (c) a SECAM receiver.

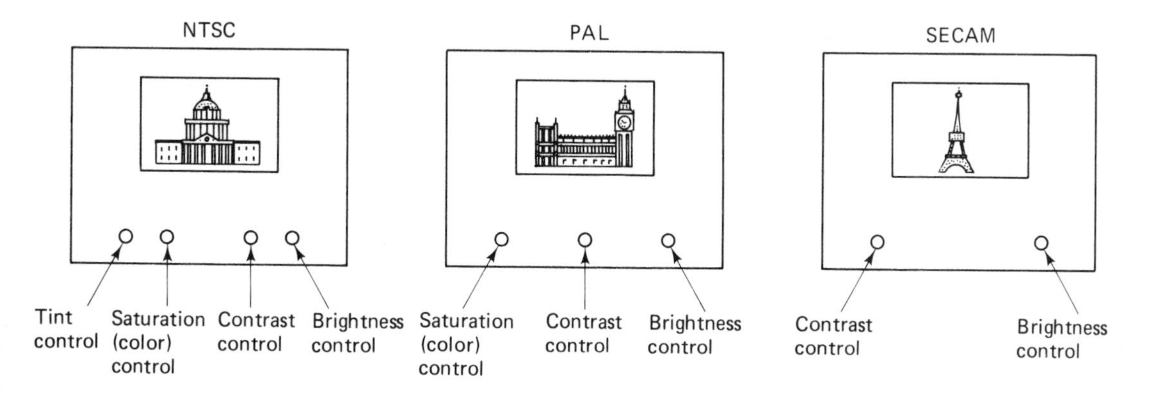

QUESTIONS

1. What is the function of the color demodulator stage?
2. What is the difference between a high-level demodulator and a low-level demodulator?
3. What is the Y signal? Why is a $-Y$ signal used to drive the cathode of the picture tube?
4. Explain the purpose and indicate the location of the two matrixes found in the output circuit of a color receiver, such as that shown in Figure 18-2.
5. What factors determine the output amplitude of a color demodulator?
6. Explain why a demodulator is "blind" to any chroma signal that is 90° out of phase with the 3.58-MHz oscillator voltage.
7. What is the phase of the 3.58-MHz oscillator if the color demodulator is to produce a $-(G - Y)$ output?
8. If the phase of the oscillator feeding a color demodulator is set to 256.5°, what is its output signal?
9. Why do some demodulators operate with a demodulation phase angle of 64° and others with a demodulation phase angle of 105°?
10. Explain why the diode color demodulator is similar to the AFPC phase detector.
11. Why is the circuit of Figure 18-8 considered a low-level demodulator?
12. What is the difference between raster color and picture color trouble symptoms? List the typical symptoms produced by each defect.
13. Explain how to identify and locate the cause of raster color defects.
14. Explain how to identify and locate the cause of picture color defects.
15. Explain how defects in the color amplifiers can cause blooming.
16. What would be the symptoms of a defective (cutoff) $R - Y$ amplifier in a dc-coupled amplifier system and in an ac-coupled amplifier system?
17. What would be the symptoms of a defective $G - Y$ amplifier (such as base shorted to emitter in Figure 18-8) in an ac-coupled amplifier system and in a dc-coupled system?
18. What is the effect on a vectorgram when the tint control and the color controls are varied?
19. What are the two methods by which automatic tint control (ATC) is introduced into a color receiver?
20. How does the VIR system of automatic tint control differ from the older systems of ATC?
21. What is the VIR signal?
22. Explain the purpose of the adder, subtractor, delay, identification, amplifier, and phase switch in the PAL receiver.
23. Explain the characteristics and advantages of the SECAM color TV system.
24. Referring to Figure 18–8, what would be the probable effect if C_1 shorted?
25. Referring to Figure 18–9 what would be the probable effect if Q_1 shorted?

chapter 19

Video Tape Recorders

19-1 introduction

Video tape recorders have been in commercial use since mid-1950. However, widespread home video tape recorder use became important in 1976 when Sony introduced the Betamax video cassette recorder. The term "video tape recorder" (VTR) usually refers to those machines that feed their tape from a separate supply reel to a separate take-up reel. A video cassette recorder (VCR) is a type of VTR where the tape supply and take-up reels are contained in a housing similar to that used in audio tape recording. Due to rapid public acceptance of VCRs, these devices have become an important part of the TV servicing market.

VCR formats. Two important VCR formats are the Video Home Systems (VHS) and Beta systems. The VHS format, developed by JVC, is used by such manufacturers as JVC, RCA, GE, and Hitachi. The Beta system has been used by such manufacturers as Sony, Zenith, Sanyo, and Sears. Photos of a VHS and a Beta video tape recorder are shown in Figure 19-1. Both systems are alike in that they both use a plug-in tape cassette. As seen in Figures 9-2(a) and (b), the construction of the video cassette is similar to that used in audio recording except that the video cassettes are much larger and use 1/2-in.-wide tape. In addition, the video cassette cannot be flipped over since the tape moves only in one direction.

VHS cassettes are larger ($7.4 \times 4.1 \times 1$ in.) than the Beta cassette ($6.1 \times 3.8 \times 1$ in.) and therefore can hold more tape. Prerecorded video tapes may be protected from erasure by knocking out a tab in the rear of the tape housing as is done in audio cassettes.

Recording time is also an important difference between the two systems. The Beta format can record up to 5 hours at a tape speed of 0.53 in. per second. The VHS format allows up to 8 hours of recording time at a tape speed of 0.44 in. per second. Thinner tapes

Figure 19-1 A top load VHS video tape recorder, and a front load Beta video tape recorder. Courtesy Panasonic and Zenith.

may eventually allow both systems to extend recording time. VCRs have two or three recording speeds:standard play (SP or Beta I), long play (LP or Beta II), and super long play (SLP or Beta III).

Cassette loading. A major difference between the two systems is the method of threading the tape so that it makes contact with the recording heads, the rollers, and the capstan. Tape is extracted from the cassette and threaded into the VHS recorder only for the play and record modes of operation. The mechanism unthreads the tape and returns it to its cassette whenever the stop button is activated. In the Beta machines threading occurs immediately after inserting the cartridge into its holder and pressing it down into its latching position. The tape remains threaded for all operations and will unthread only when the cassette is ejected.

Additional features. Both VHS and Beta formats may also offer such features as still frame, frame-by-frame advance, slow motion, high-speed search (as much as 15 times normal speed)

Figure 19-2 a) A video cassette for a Beta type VCR. b) A video cassette for a VHS type VCR.

in both forward and reverse directions. Older VHS VCRs could not perform the reverse search function.

VCRs are usually equipped with tape counters that can be used during fast forward or reverse to locate a desired segment of the recorded tape. More sophisticated methods include an automatic cue signal that is recorded at the beginning of a recording, forcing the recorder to come to a halt whenever a cue mark is reached.

With the exception of portable machines, all consumer-type VCRs have built-in tuners and timers that allow them to record off-the-air TV programs. Portable machines can have tuner/timers as accessories. Most of the modern tuners use electronic tuning. Older machines used mechanical tuners. Built-in tuners may be used to record one channel while watching another. Those machines equipped with timers can automatically turn on and record many programs over a 2-week period. Between programmed recordings the VCR turns itself off.

19-2 principles of magnetic recording

Magnetic recording is based on the principle that a current passing through a coil of wire will produce a magnetic field. If pieces of high-retentivity magnetic material are placed near the coil, they will become permanently magnetized.

In audio and video magnetic recording practice the coil of wire is an iron-core electromagnet called the "recording head." The recording head is a U-shaped piece of low-retentivity magnetic material that has a coil of wire wound around it. The open end of the U is bent to an almost closed position, forming a very narrow space called the "head gap." See Figure 19-3. This space is actually filled by a nonmagnetic insulator, usually a special glass, which maintains a constant head gap width. The head gap construction concentrates the magnetic field produced by the electromagnet in the space in front of the gap. Moving through this space and in contact with the head is a plastic tape (Mylar or acetate) coated with a layer of high-magnetic-retentivity iron oxide.

In the record mode of operation, variations in the signal current passing through the recording head will cause variations in gap magnetic strength. This, in turn, will cause variations in the intensity of the magnetization of the iron oxide particles on the plastic tape. These magnetic intensity variations stored on the tape represent audio, digital, or video information that may be stored indefinitely.

In the playback mode of operation the magnetized tape is pulled past the playback head. The same head may be used for both recording and for playback. The variations in magnetic intensity stored on the tape, induce a voltage into the coil wound on the head. This voltage is very weak and must be adequately amplified to drive the output speakers in an audio system, or the picture tube in a video system.

Magnetic recording limitations. The simple recording and playback system just described is called the *direct-drive system.* The direct-drive recording system has both low-frequency and high-frequency limits that are determined by the *width of the head gap* and the relative

Figure 19-3 Basic recording head and its associated video tape.

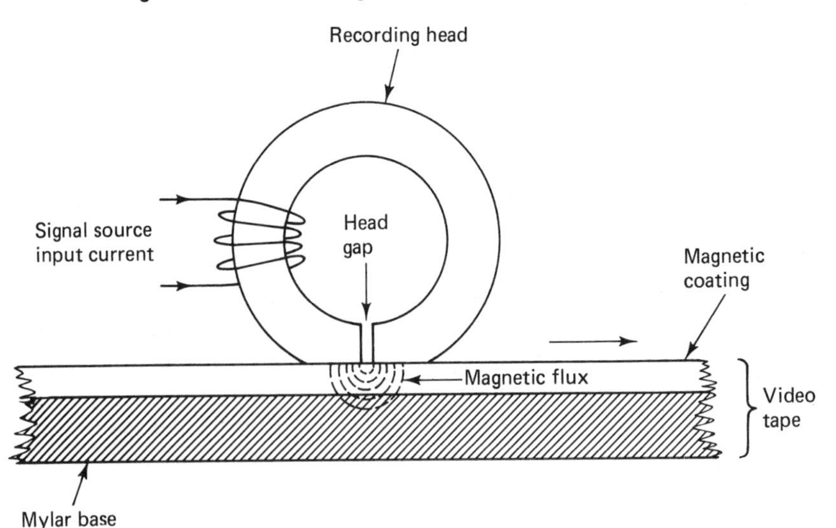

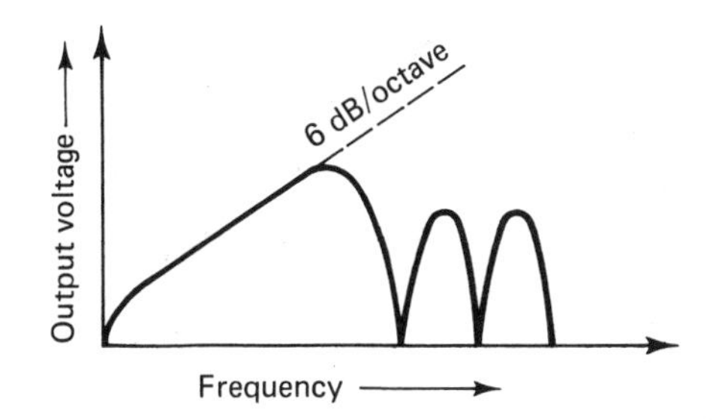

Figure 19-4 Frequency response of a reproducing head.

speed between the tape and the recording head. The output voltage of the playback head is determined by the rate of change of the magnetic field produced by the moving tape. For a constant-tape speed, assuming that the recording level is constant, increasing the frequency of the recorded material will cause the output of the *playback* head to increase. As shown in Figure 19-4, the increase in output grows at a rate of 6 dB per octave. An octave is a doubling or halving of frequency. This means that if the frequency is doubled, the output will also double. Maximum output occurs when the wavelength of the recorded signal on the tape is twice as long as the head gap. As the frequency is increased further, the output drops and a point is reached where the wavelength is equal to the gap width and the output becomes zero. This is shown in Figure 19-5. As the frequency is increased further, the wavelength becomes smaller than the head gap width and the signal dropoff is much more rapid. These shorter wavelengths attenuate from full output to zero in one octave or less.

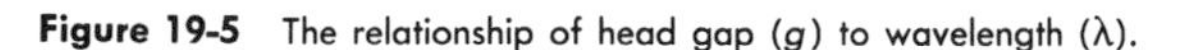

Figure 19-5 The relationship of head gap (g) to wavelength (λ).

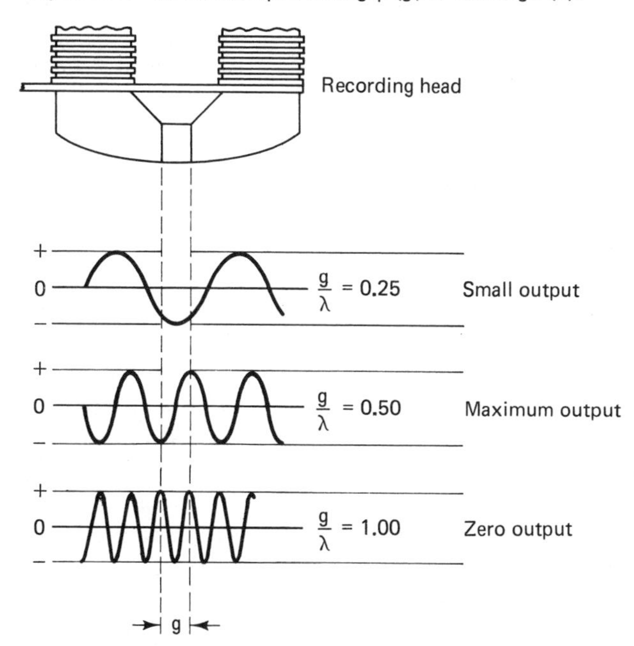

As the frequency is lowered, the wavelength increases and the output decreases at a 6-dB rate until the output gets so small that it cannot be distinguished from the system noise.

To increase the upper cutoff frequency, the gap width may be reduced and/or the tape-to-head speed may be increased. Both these methods individually used will increase the high-frequency limit, but the total bandwidth that can be recorded is not increased due to the 6-dB per octave rolloff that takes place at lower frequencies.

The direct recording system allows nine octaves of range before low-frequency noise or high-frequency cancellation renders the output unusable. A nine-octave frequency range would, for example, range from 30 to 15,000 Hz.

Video signals range in frequency from about 20 Hz to 4 MHz. This corresponds to an 18-octave span. A range of 4 to 8 MHz also covers a 4-MHz bandwidth, but occupies only a one-octave frequency change.

Frequency compensation techniques. To record video information, three things must be done to increase the high-frequency limit and to compress the 18-octave video frequency spread into less than one octave:

1. To increase the high-frequency response, the head gap is made smaller, 0.35 μm in VHS and 0.6 μm in the Beta system.
2. The head-to-tape speed must be increased without increasing the amount of tape that is used for a given recording. This may be done by moving *both* the *head* and the *tape*. This technique again increases the high-frequency limit of recording. In the VHS system a 2-hour tape moves at 3.34 cm/s (4 cm/s in Beta I) and a 6-hour recording at 1.11 cm/s (1.33 cm/s for Beta III, a 5-hour recording). The recording head rotates at 1800 r/min for both VHS and Beta systems.
3. To change the 18-octave bandwidth of the video signal to a more manageable range, the video signal is increased in frequency and is made to *frequency modulate* a 3.4-MHz carrier in the VHS system and a 3.5-MHz carrier in the Beta system.

Frequency modulation of the luminance video signal has the advantage of producing a recording system of high noise immunity. Head wear, dirty playback heads, and worn tapes can introduce all kinds of signal amplitude fluctuations which are removed by *limiter* circuits, leaving only the original FM signal.

Recording bias is not necessary when recording an FM signal since any distortion introduced by recording without bias will also be removed by the limiter circuit. Another advantage of FM recording is that the dc level of the video signal is represented by a constant frequency corresponding to the scene's average brightness. Direct-drive recording produces an output which is dependent upon the *change* in tape magnetization levels. A dc level would produce no output. In contrast, the FM carrier will provide an accurate output for any dc level.

The main advantage of using an FM system is that it allows the 18-octave video signal information to be compressed into fewer than three octaves of bandwidth. This is shown in Figure 19-6.

It should be recalled that in FM the amplitude of the modulating signal determines

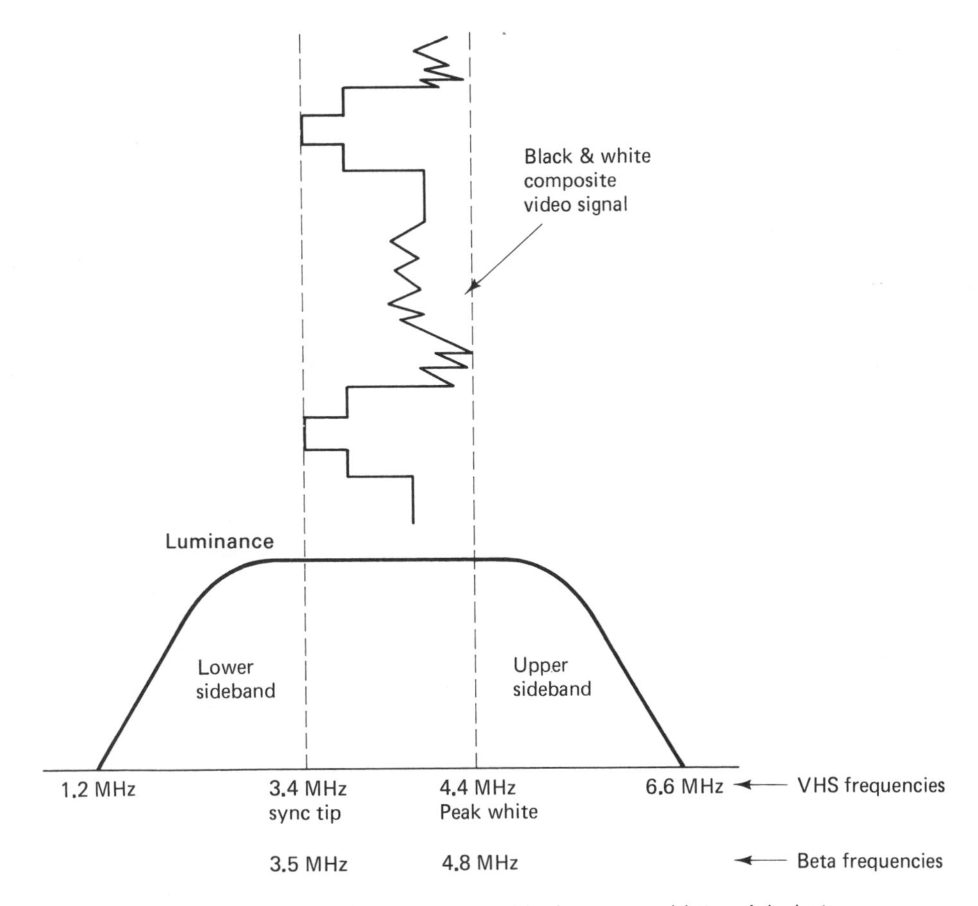

Figure 19-6 Video signal spectrum produced by frequency modulation of the luminance signal.

the frequency deviation, and that the frequency of the modulating signal determines the rate of deviation. Therefore, the maximum amplitude sync pulse tip forces the FM oscillator to a low-frequency limit of 3.4 MHz in the VHS system and 3.5 MHz in Beta. When the video signal is maximum white, the FM oscillator is forced to shift to maximum frequency, 4.4 MHz in VHS and 4.8 MHz in Beta.

Notice that in the VHS system the video signal deviation is 1 MHz (4.4 MHz − 3.4 MHz = 1 MHz) and in the Beta system it is 1.3 MHz. However, FM introduces sidebands which extend beyond these limits. The recorded FM video spectrum must accommodate signals ranging from about 1 to 6 MHz. In both VHS and Beta systems the frequency range of the FM signal is less than three octaves, which is well within the limits of a moving recording head.

In addition to head gap reduction, moving head recording, and recording the video signal in FM, video cassette recorders compensate for playback rolloff by a process called head equalization. Recorders may compensate for rolloff in the recording circuitry. Most systems provide the needed compensation in the playback circuits.

Color recording. In the color television system it is the phase of the chrominance signal that represents picture color. Any change in chrominance phase results in a change in color.

Phase errors may be introduced by several mechanical factors. The *writing speed and time base stability* of home VCRs is not good enough for color signal information to be recorded together with the luminance signal. To minimize this problem, the rotating head speed must be kept constant. This is done during playback by servo control systems that make small head speed adjustments, compensating for any deviations from the recording mode of operation.

Another factor that is a cause of phase error is *tape tension.* Tapes tend to stretch by different amounts, as the amount of tape on the take-up and supply reels change. Tension differences are also present when a tape recorded on one machine is played back on another. If tape tension is increased, horizontal sync pulse width will increase and the color subcarrier frequency will decrease slightly. A decrease in tension will produce an opposite result. This type of color subcarrier frequency or phase error is known as *time base* error and can result in poor color reproduction.

Writing speed and time base errors can be minimized by heterodyning the color signal to a lower frequency. See Figure 19-7. This "color under" recording system lowers the color signal to 629 kHz in the VHS system, and 688 kHz in Beta. Frequency changing is produced by mixing the chroma signal with a stable oscillator signal in an appropriate converter. The lower-frequency chroma signal then uses the FM luminance signal as a recording bias for direct recording.

Figure 19-7 Frequency spectrum of the luminance and chrominance (color under) signals.

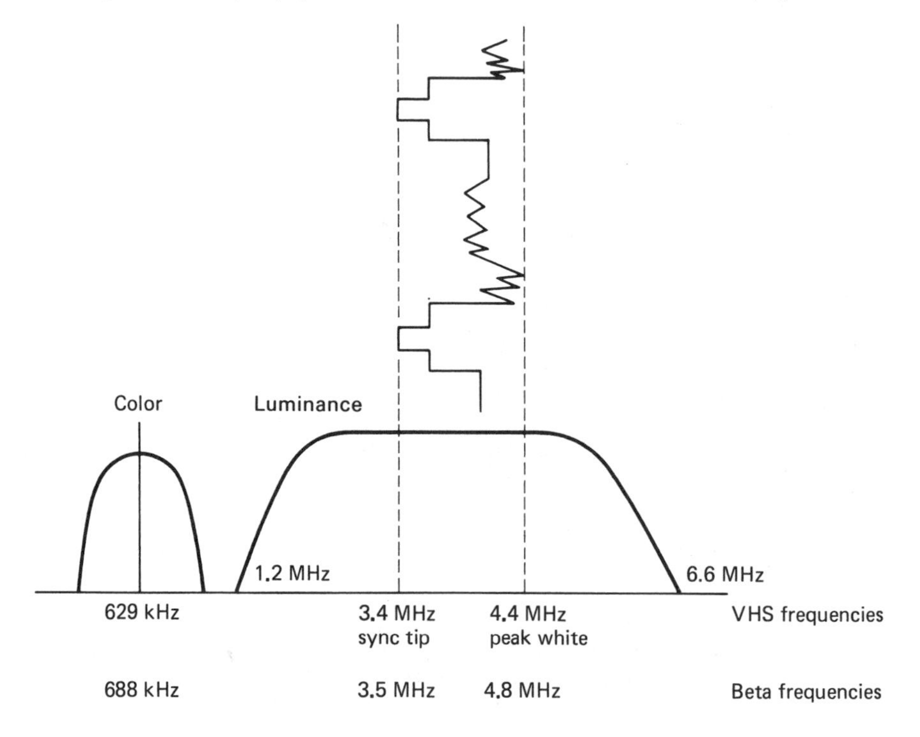

During playback, the low-frequency color information is heterodyned back into the NTSC color signal (3.58 MHz). The heterodyning oscillator is phase controlled to compensate for any phase changes that may affect color reproduction. The amount of time base and other phase error correction needed is controlled by reference to the recovered horizontal sync pulses.

19-3 recording and playback heads

The combination of a rotating recording head and slow moving tape has allowed VCRs to record as much as 8 hours of programming on one VHS cassette. High relative tape-to-head speed is important to be able to record the high video frequencies. However, equally important is economical tape use.

Quadraplex recording. Commercial broadcast-type video tape decks use four heads mounted on a 2-in. disk that spins at 240 revolutions per second. This system is known as the Quadraplex system. The heads are mounted perpendicular to the tape that is moving past the heads. This is shown in Figure 19-8. The tape is 2 in. wide and travels at 15 in. per second. The relative tape-to-head speed is in excess of 1500 in. per second. A full hour of recording can be stored in a reel about 10 in. in diameter. The video information is recorded in tracks that run at right angles to the direction of tape travel. There are 16.4 horizontal lines in each track, and 16 tracks are needed for a complete TV field. The Quadraplex system is too expensive in both equipment and tape to allow home use.

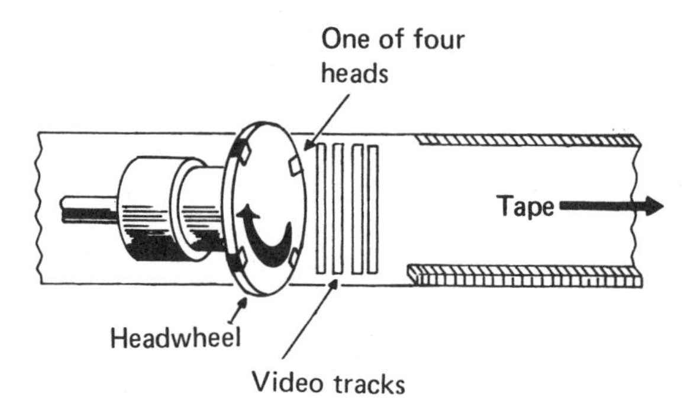

Figure 19-8 Quadraplex four-head recording system.

Helical scanning. The "helical scan" or "slant track" recording system was invented by N. Sawazaki in 1954 and allows recording one complete field (262.5) Hz on one track. This system reduced the cost of video tape recorders and reduced the amount of tape required for recording any given length of time.

In the helical scanning system the moving recording headwheel has two heads mounted 180° apart, as shown in Figure 19-9. The headwheel rotates at a speed of 30 r/s, so that each head records a complete field. One complete rotation of the headwheel records a full frame (two fields) or 525 horizontal lines. The tape is wrapped around the headwheel at an angle to the slightly protruding heads. See Figure 19-10. It is possible to use one head and wrap the tape completely around the head in a *full-wrap* configuration. VHS and

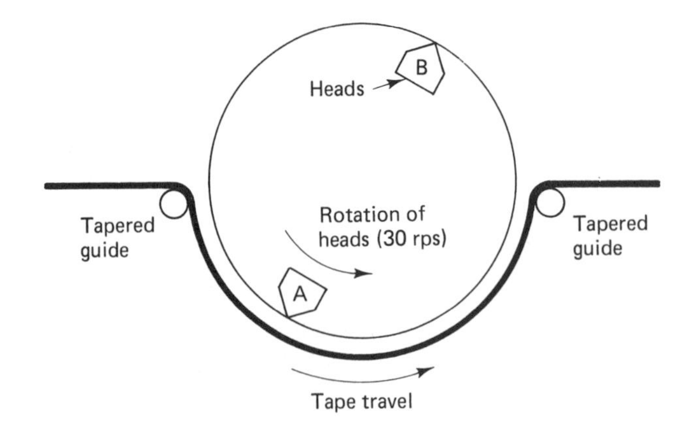

Figure 19-9 Helical two-head drumwheel.

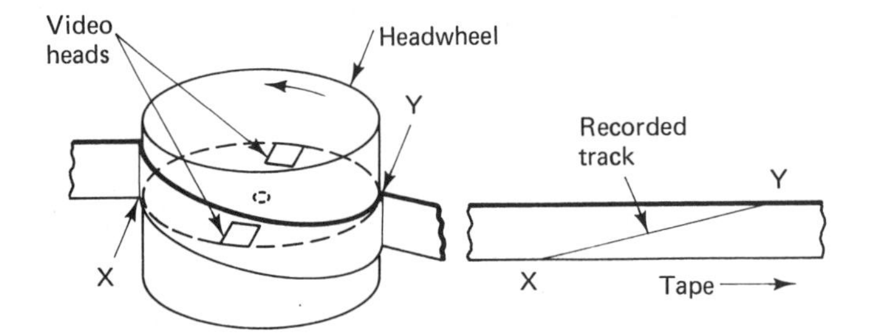

Figure 19-10 Relationship between the tape and the headwheel. The headwheel axis is tilted so that the video heads write diagonal tracks across the tape.

Beta systems wrap the tape in a *half-wrap* arrangement, as shown. The usual method for forming the head-to-drum path is to slant (angle mount) the headwheel drum while leaving the supply and take-up reels parallel to the deck they are mounted on. When the tape is not moving, the track is at an angle of 5.302° with respect to the tape edge. When the tape is moving, this angle changes to 5.33°. This fact becomes important during stop-action operation.

The various heads, tape guides, and capstan used in a typical VCR are shown in Figure 19-11. Notice that the tape first passes over a full erase head that when activated demagnetizes the tape. The tape is then positioned by guides and rollers to pass over the rotating video heads where the picture information is recorded. The tape is pulled by the capstan past the audio and control heads. These heads can record, playback, and erase.

Four-head VHS system. Some VHS 6-hour recording systems use four heads. One pair is used for the 2-hour speed and the other pair is used for the 4- and 6-hour recording speeds. Each pair of heads makes different track widths. In addition, they have four dissimilar widths. Slow-speed recordings require thinner diagonal tracks and associated heads than do standard-speed recordings. Using a slow-speed head for 2-hour (fast) playback, sacrifices signal-to-noise ratio. In addition, narrow-track 2-hour tapes play back very noisily on wide-track 2-hour heads, which pick up noise on either side of the recorded track. Using a wide-

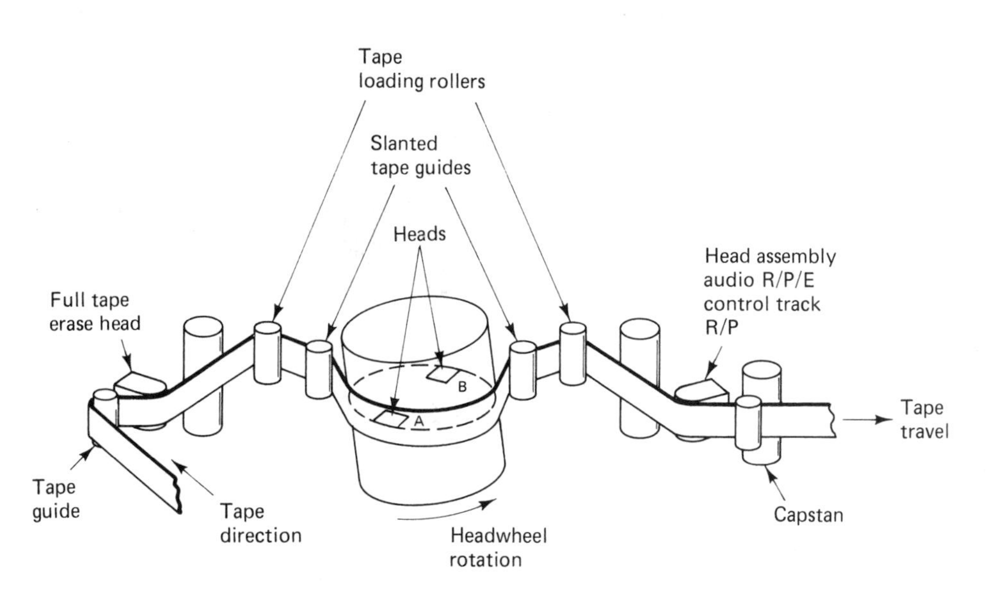

Figure 19-11 Arrangement of heads, tape guides, and rollers found in a typical VCR.

width head for standard 2-hour recording and playback, and a narrow width head for long-play recordings, improves the signal-to-noise ratio of the system.

In some recorders that have a still-frame feature, the head widths of each pair of heads are asymmetrical. In one VHS machine, the 2-hour head widths were 70 and 90 μm wide: the long-play head widths were 26 and 31 μm wide. Using asymmetrical head widths produces tracks that offset the noise bars inherent in still-frame operation. The noise bars are made to occur during blanking and are outside the visible portion of the frame.

Some recent VCRs have five heads. The fifth head is used to provide jitter-free stop-action performance. The fifth head is located near the A head. During stop-action mode, the A head is disabled and the fifth head becomes operative as an additional B head. The two heads only playback and interleave the B tracks making this a *still-field* system. Since the heads only pick up one signal, noise is minimized.

Head switching. During the *record mode* of operation the video heads are connected in parallel, and the video signal information is fed to them continuously. Since the tape is wrapped around the head drum for slightly more than 180°, each head records slightly more than 262.5 horizontal lines (one field) in one diagonal swipe of about 3.829 inches across the tape. As a result, there is a small time interval when both heads are in contact with the tape. This time interval is called *overlap.* The results of this are shown in Figure 19-12. To ensure a proper head starting point for each swipe, the vertical sync pulse is used to adjust the phase of the head so that each head swipe begins with the vertical sync pulse interval.

During the *playback mode* of operation, it is important that the information recorded by the A head be played back by the A head and its associated amplifier chain. Similarly, the B head must be synchronized with the B track on the tape. This is accomplished by a head-switching pulse that is generated in the head servo system. An electronic switch is activated by this pulse and it turns on the proper head preamplifier when it is "reading" the tape, and turns off the other head preamplifier. This eliminates the possibility of noise being introduced into the video display by this head.

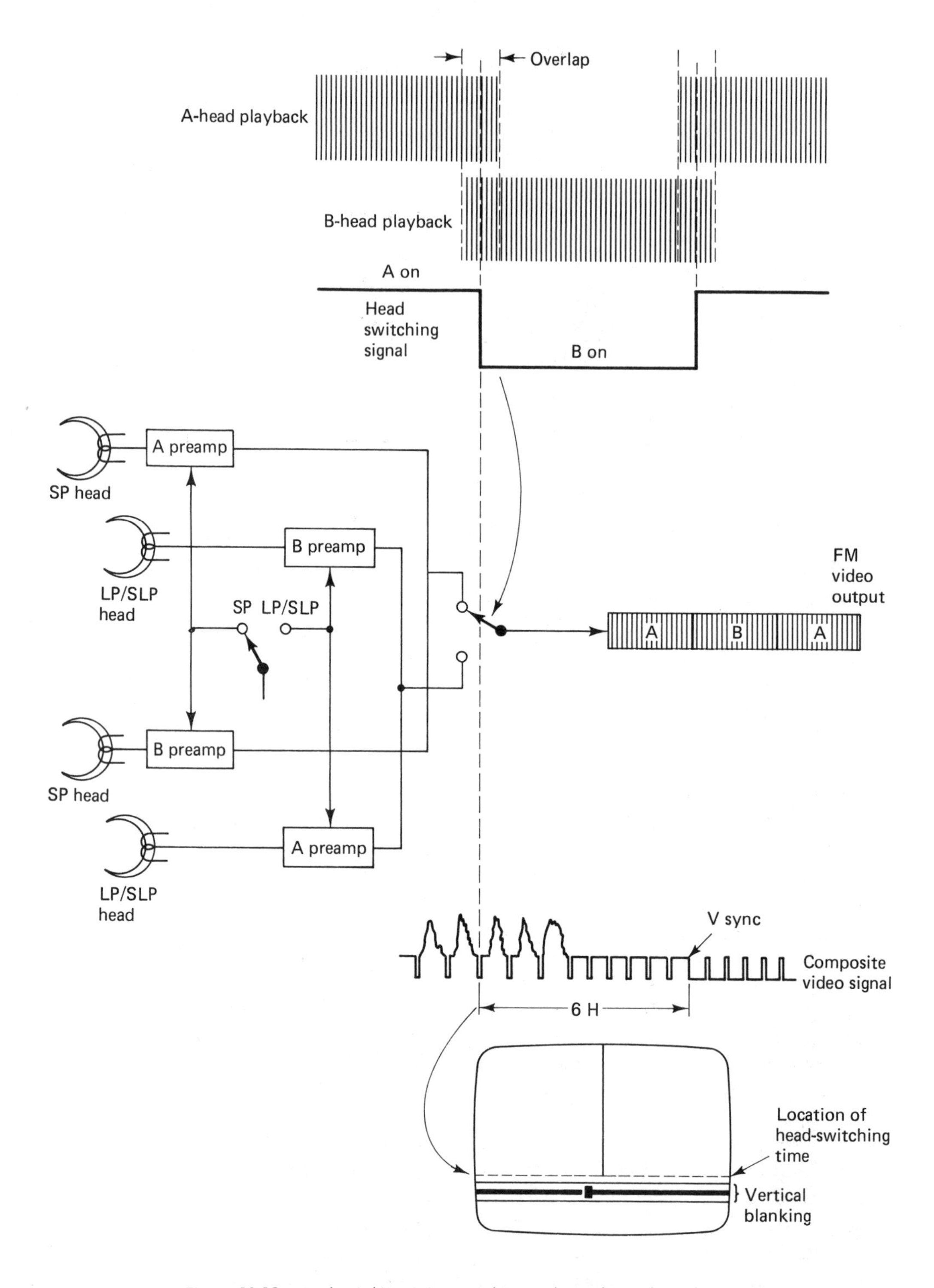

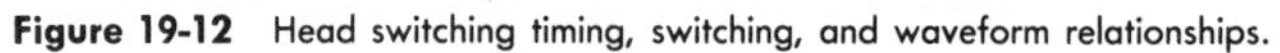
Figure 19-12 Head switching timing, switching, and waveform relationships.

Proper adjustment of head timing occurs when head switching takes place six horizontal lines before the start of the vertical sync pulse interval. This adjustment is found in the head servo circuit.

Machines that have four heads must also have four preamplifiers. They have an A and B preamplifier for SP (2-hour) play, and an A and B preamplifier for the LP and SLP (4-hour and 6-hour) operation. A two-position SP,LP/SLP switch is used for both record and playback operation to select the desired heads and their associated preamplifiers. The switch activates a relay or some other circuit that enables the desired head's preamplifiers, and disables the other heads and their preamplifiers.

19-4 tape format

The information being recorded is positioned on the tape as shown in Figure 19-13. Each invisible magnetic track corresponds to one TV field containing all of the video, color, sync, and blanking information. One head referred to as A will record all the odd fields (1, 3, 5, 7, etc.) and head B records all the even fields (2, 4, 6, 8, etc.).

The sound information is recorded along one edge of the tape. Along the other edge is the control track, which consists of pulses generated and recorded by a separate and fixed head of the VCR. This signal is used during playback as a reference to properly position the rotating heads. The signal that is recorded on the control track is a 30-Hz pulse that is formed by recording every other vertical sync pulse. Assuming that the control track pulse is recorded whenever head A records its field, the electromechanical servo system uses the pulse during playback to force head A to precisely retrace its path. Head B is located 180° apart from head A and if correctly positioned it will automatically be in the proper position to retrace the track laid down by head B during recording. The control track acts like the sprocket holes found in movie film. In both cases, the control track and

Figure 19-13 VHS and Beta tape format for 1/2-in. video tape.

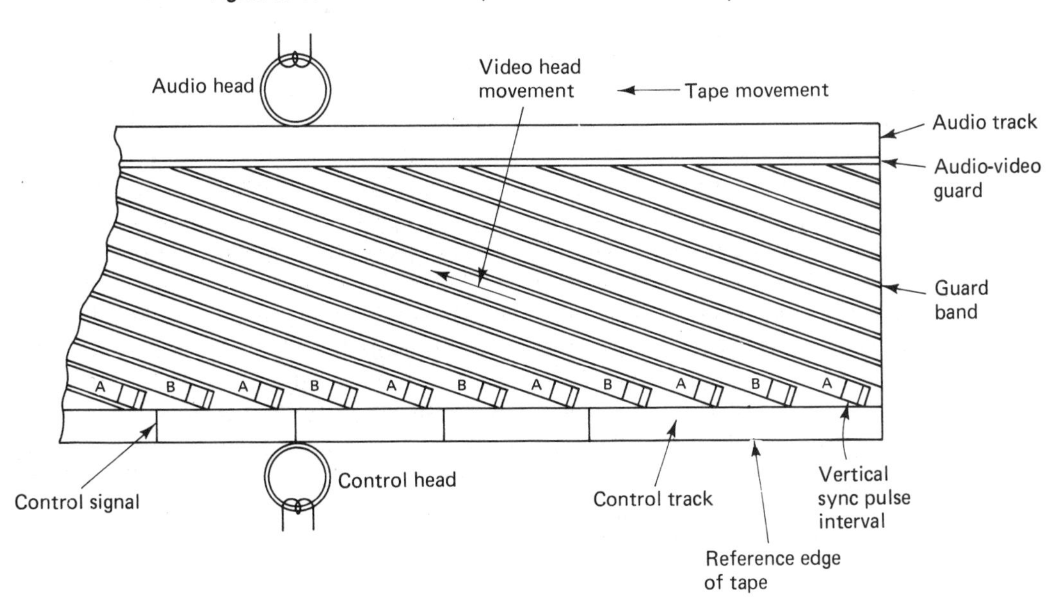

the sprocket holes are being used as a reference that is used for proper picture synchronization.

Guardbands and crosstalk. During playback each of the rotating heads must respond to one track at a time. If a head should overlap two tracks as it does during LP and SLP operation, it would produce beats and herringbone patterns in the reproduced picture. This condition is called "crosstalk."

There are three systems for avoiding crosstalk. First, in the standard 2-hour play system a "guardband" or space is placed between each track of the video information. This is shown in Figure 19-14. The main disadvantage of this technique is that no information is recorded in the guardbands; therefore, more tape per minute must be used.

Azimuth recording. A second method used to minimize crosstalk is called azimuth recording. Azimuth refers to the angle that the vertical gap of the recording head presents to the moving tape. For maximum playback output the vertical head gap should be at right angles to the tape movement. Tilting the gap at some other angle effectively increases the gap width, reducing the high frequency response of the head.

In the VHS recording and playback system, the gap of one head is tilted 6° from the vertical in one direction, and the other head is tilted 6° in the opposite direction. See Figure 19-14(b). In Beta systems the tilt is $\pm 7°$. This means that the azimuth angle between adjacent tracks is 12° for VHS and 14° for Beta. During playback each head will be in perfect alignment with its recorded track, but will be 12° or 14° out of alignment for any crosstalk from an adjacent track. Crosstalk from adjacent tracks is thereby minimized, and the need for a guardband with its unused tape is eliminated.

19-5 color crosstalk minimization

Azimuth alignment used as a means of crosstalk minimization is primarily effective at high frequencies. Therefore, it is the FM black & white portions of the video signal which benefit from this technique. An additional method of crosstalk rejection is required for the relatively low frequency color signals.

Beta crosstalk elimination. Both the VHS and the Beta systems of video recording use a scheme of color signal phase shifting in conjunction with a comb filter (see Section 10-8) to reduce color crosstalk. In the Beta system, the color signal is inverted on alternate lines of one field (A). This counteracts the normal transmitted sequence in which the color phase of each horizontal line is reversed. The second field (B) is not inverted. A block diagram of how this is done as well as the results on the video tape is shown in Figure 19-15.

When the tape is played back, the 688-kHz chroma signal obtained from the tape head is frequency converted into 3.58-MHz color information. This is shown in Figure 19-16. Also, the 4.27-MHz source feeding the converter is switched at both a vertical and a horizontal rate. The vertical rate is synchronized with field B such that its color phase is not alternated and remains at zero phase. During the next 1/60 of a second the electronic switch is keyed at a horizontal rate such that on alternate horizontal lines the phase of the 4.27-MHz signal is reversed. This phase reversal is carried over into the 3.58-MHz output of the frequency converter feeding the comb filter.

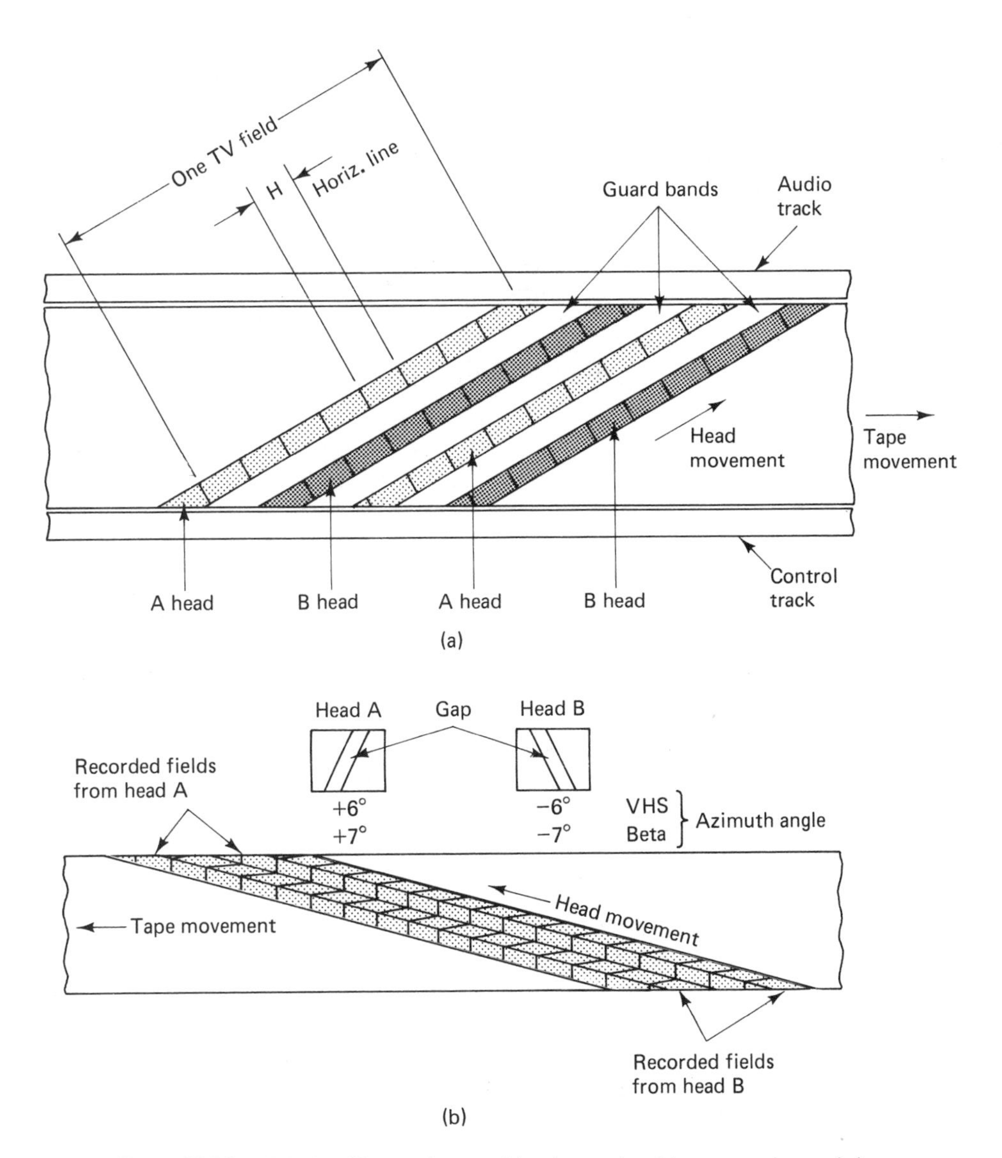

Figure 19-14 (a) During SP operation guard bands are placed between each recorded track; (b) Azimuth recording eliminates the guard bands and minimizes crosstalk.

The combination of a 1-H delay line and a bridge are used to form a comb filter. Assuming that the video signal has been transmitting for some time, notice that when horizontal line 2 is appearing at the output of the frequency converter, line 1 is being outputted from the delay line. Both of these outputs feed the bridge. The bridge acts as a subtractor, taking the difference between lines 2 and 1.

Shown in the waveform ladder diagram are the color signals for three horizontal lines developed during the A and B field recordings. Above these waveforms is the horizontal switching waveform feeding the electronic switch. Only one cycle of the color signal sine

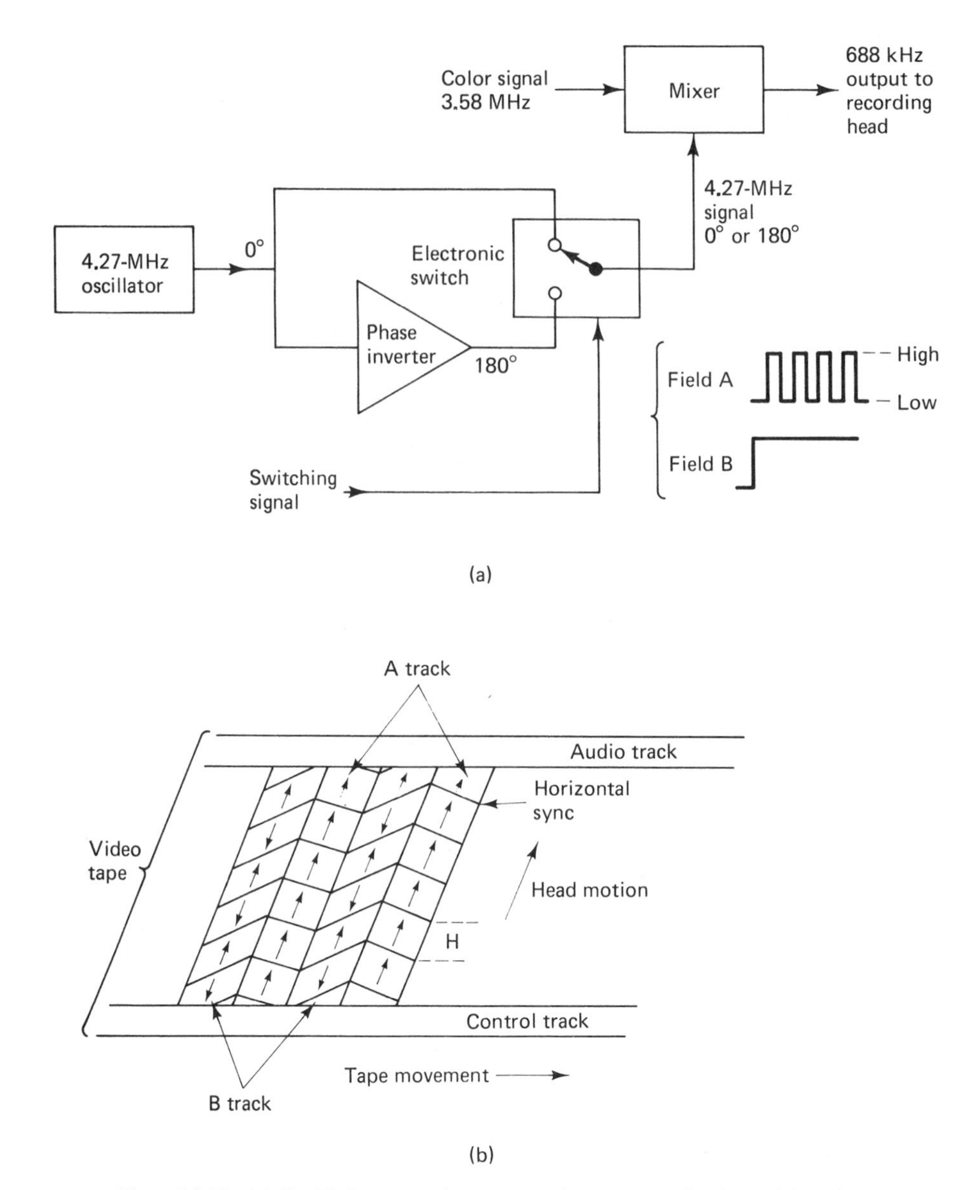

Figure 19-15 (a) Simplified circuit used in Beta recorders to reverse the phase of the color signal on B-track fields; (b) relative phase of the A tracks for each horizontal line indicating that they are all of the same phase. The B track is not changed, so that each horizontal line reverses its phase.

wave is shown for ease of presentation. The phase of the A field is the same from one horizontal line to the next, because alternate lines have been reversed during the recording. The B field was not altered and therefore the phase of the color signal reverses for each alternate line. The dashed sine waves indicate the crosstalk from A to B and from B to A. The frequency of the color signals in the first two rows is 688 kHz, and the signals shown in the bottom three rows are at 3.58 MHz.

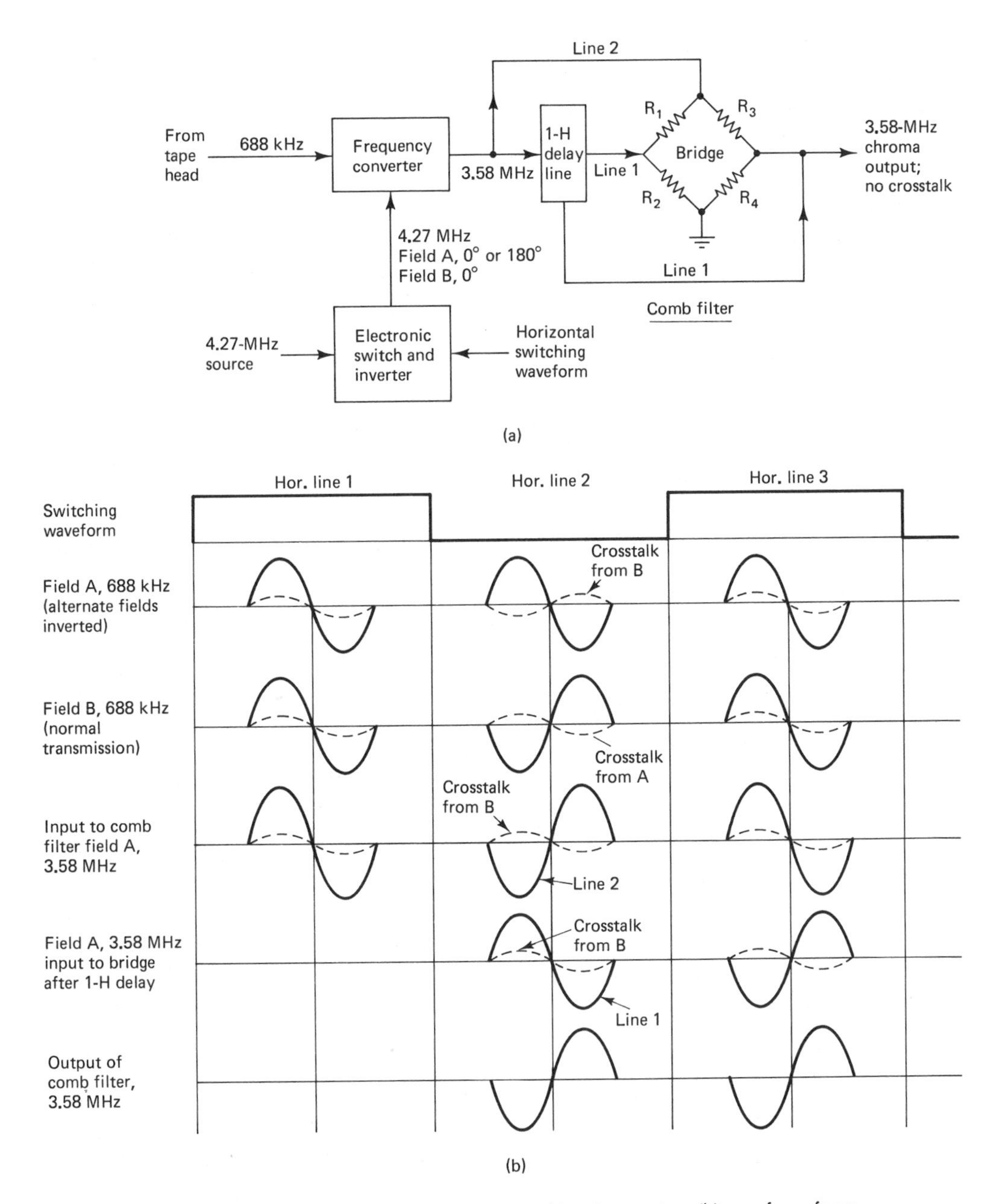

Figure 19-16 (a) Simplified playback circuit used in a Beta system; (b) waveforms founa in the playback circuit.

Notice that field A has been altered so that the phase of line 2 has been reversed. In addition, line 1 has passed through the 1-H delay line and now presents itself at the bridge at the same time as line 2. The bridge acts as a subtractor. Since lines 1 and 2 are 180° out of phase, they add $[(-) - (+) = -2]$, and the crosstalk which is in phase cancels $[(+) - (+) = 0]$.

During the B field normal color transmission takes place, and field A crosstalk into field B occurs. Notice that horizontal line 1 is the same for both fields A and B, and that line 2 for the B field has the same phase relationship as line 2 for the A field input to the comb filter. Thus, cancellation of crosstalk will again occur in the same manner as before.

VHS color crosstalk elimination. As shown in Figure 19-17(a), In the VHS system of color crosstalk minimization the phase of the 629-kHz color signal is advanced in phase 90° for each successive horizontal line. Thus, the first horizontal line of track A has a phase of +90°, the second +180°, the third +270°, and the fourth would be 0°. This phase rotation continues for all the remaining lines in the A field. The B field also introduces color phase rotation for each horizontal line, but the phase delay is in the reverse direction. The result of this is that the even-numbered lines of each field are 180° out of phase and the odd-numbered lines are in phase. Shown in Figure 19-17(b) is a vector representation of the playback phase relationships existing between the desired signal of one track and the crosstalk from the other track.

During playback the phase of the color signal for each line is restored to its original phase. This is done by means of the circuit shown in block diagram form in Figure 19-17(c). Associated with the block diagram are vectors indicating the phase relationships of the desired signal and its crosstalk signal components after phase restoration. Phase restoration causes the desired color signal to align into the same phase for each successive line. The crosstalk component continues rotating at 90° increments per horizontal rate. After passing through a signal splitter that provides two identical signals, the two signals are fed into a comb filter. The combination of the delayed signal [Figure 19-17(c)] and the nondelayed signal in the comb filter results in the addition of the desired color signal and the cancellation of the crosstalk component [Figure 19-17(d)].

Sync beat interference. As a consequence of both VHS and Beta systems' attempt to increase record and playback time, adjacent tracks are made to overlap. This is shown in Figure 19-18. The result of this is that the horizontal sync pulse information (3.4 MHz FM) will overlap into neighboring tracks, causing noticeable stationary beats with other video information. Azimuth recording minimizes crosstalk production, but the beats produced by the sync pulses have no true azimuth and will not be canceled by the heads.

To minimize sync beat interference a method called FM interleaving recording is used. This method does not directly eliminate the beat. Instead, the beat is arranged so that it is 180° out of phase on two adjacent lines of the picture, optically canceling one another.

FM interleaving recording is used when recording in VHS LP/SLP and is accomplished by changing the sync tip frequency of track A to 3.407867 MHz, which is H/2 or 7867 Hz higher than the sync tip frequency of 3.4 MHz for track B. Peak white becomes 4.407867 MHz for track A and 4.4 MHz for track B. This one-sided shift in frequency causes the beat to interleave, solving the problem.

Playback is essentially the same as before except for "flicker" compensation of the brightness when the VCR is in the LP/SLP mode of operation.

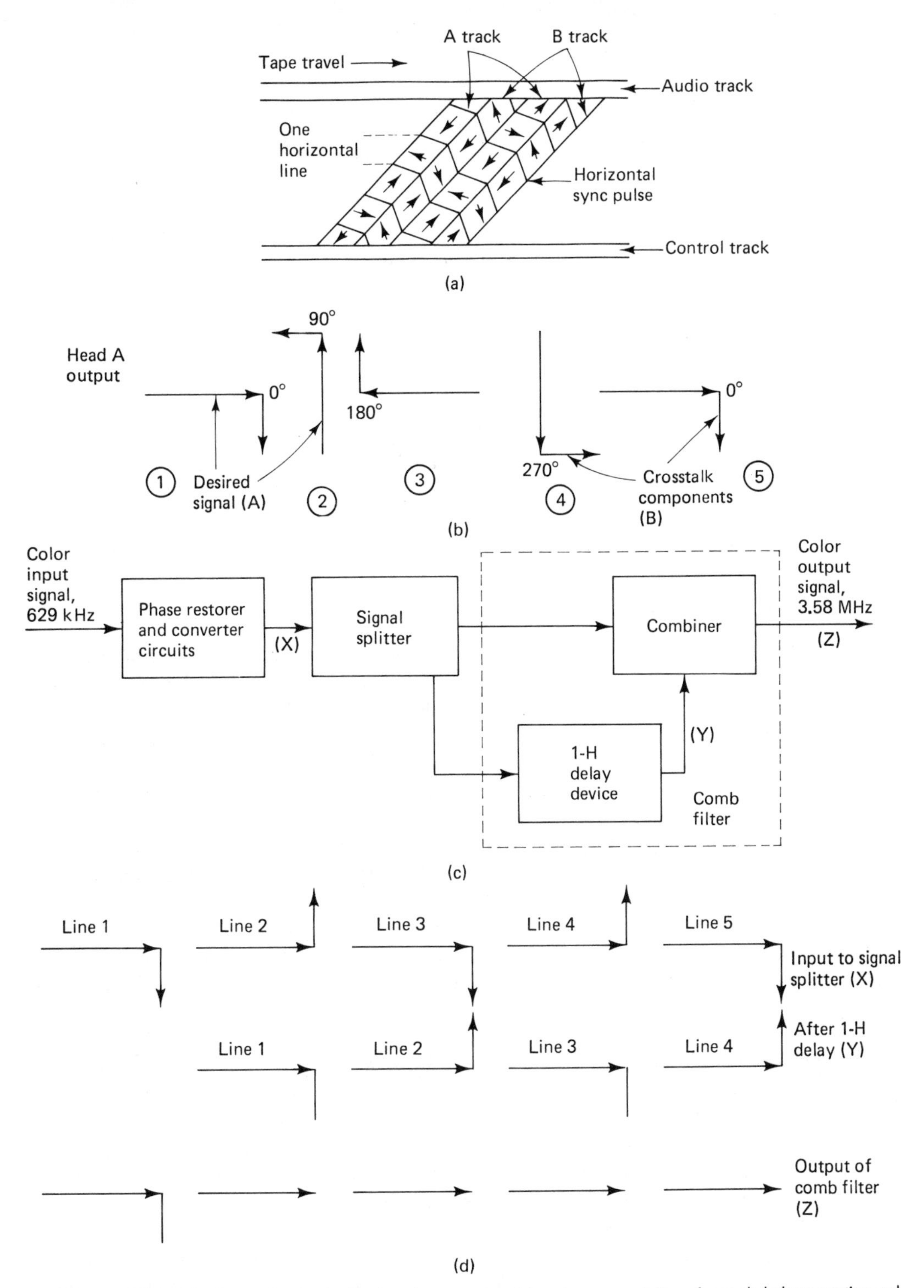

Figure 19-17 (a) VHS recording tracks showing phase rotation; (b) vector representation of recorded phase rotation and crosstalk; (c) simplified block diagram of the VHS playback circuit and the phase relationships of the desired signal and its associated crosstalk after phase restoration; (d) result of 1-H delay and the output of the comb filter.

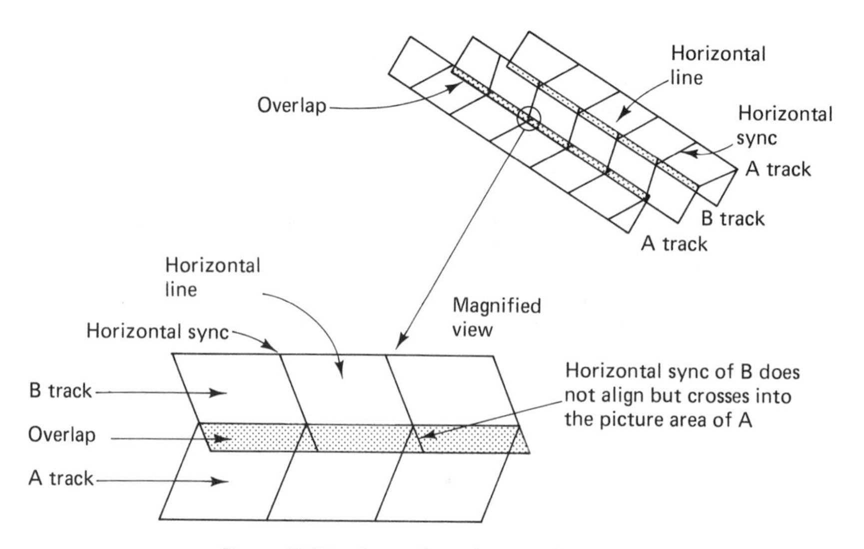

Figure 19-18 Cause of sync beat interference.

19-6 VCR systems

The complexity of the VCR system is evident from an examination of the parts list of a modern unit. A recent RCA VHS video cassette recorder having all the latest features had 27 circuit boards, 349 transistors, 305 diodes, 59 ICs, and 4 microprocessors. Beta systems have a similar parts count. It should be expected that with greater use of ICs the number of transistors and circuit boards will probably decrease in the near future.

To determine the number of stages used in a VCR system, we may consider that one to several transistors constitutes a stage that performs a specific function in a VCR. In addition, ICs and microprocessors usually contain many stages. On this basis it can be estimated that the complete VCR system must have in excess of 300 stages. This system complexity is made evident in Figure 19-19(a).

Besides the purely electronic sections of the systems, there is also a rather complex mechanical section of the VCR which is necessary for the system to operate. Thus, the VCR may be broken into two major categories: the electronic system and the mechanical system. See Figure 19-19(b).

Electronic systems. The electronic systems of the VCR may be divided into four major areas. First, there are the circuits necessary for processing and recording black & white video, color, and sound information. This information is obtained from such sources as TV receivers, cameras, and other VCRs. Second, there must be playback circuits available to extract the recorded information from the tape and restore it into a composite video signal plus audio such that it may be reproduced on a television receiver or monitor. The third major area of VCR electronics must provide electronic servo circuits. These circuits interface with the mechanical tape transport components and regulate tape speed, preventing picture distortion. In addition, the VCR incorporates various electronically controlled safety features that are able to stop the recorder in the event of such things as excessive humidity, end of

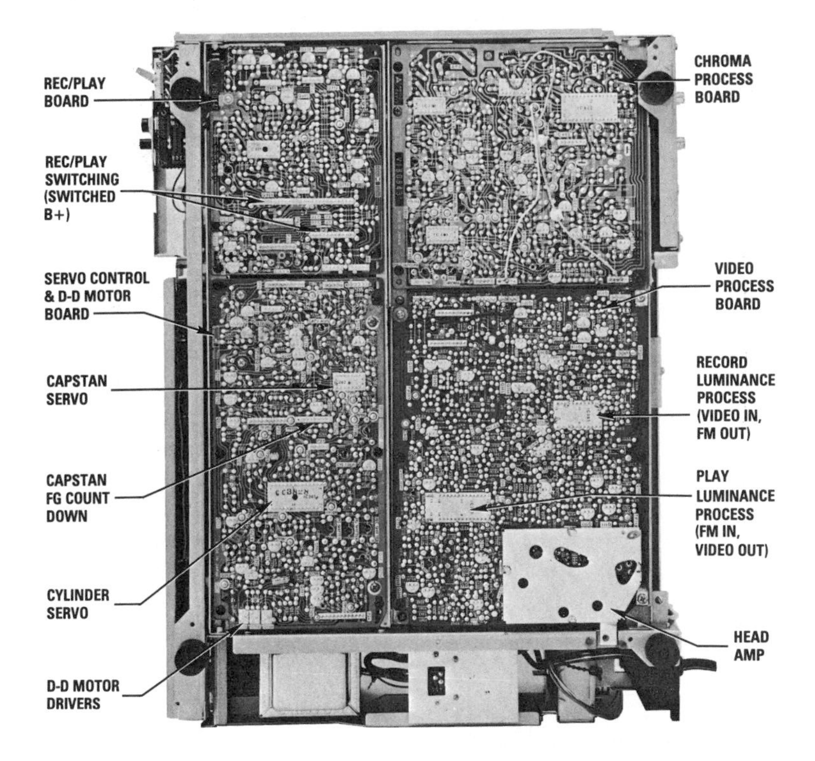

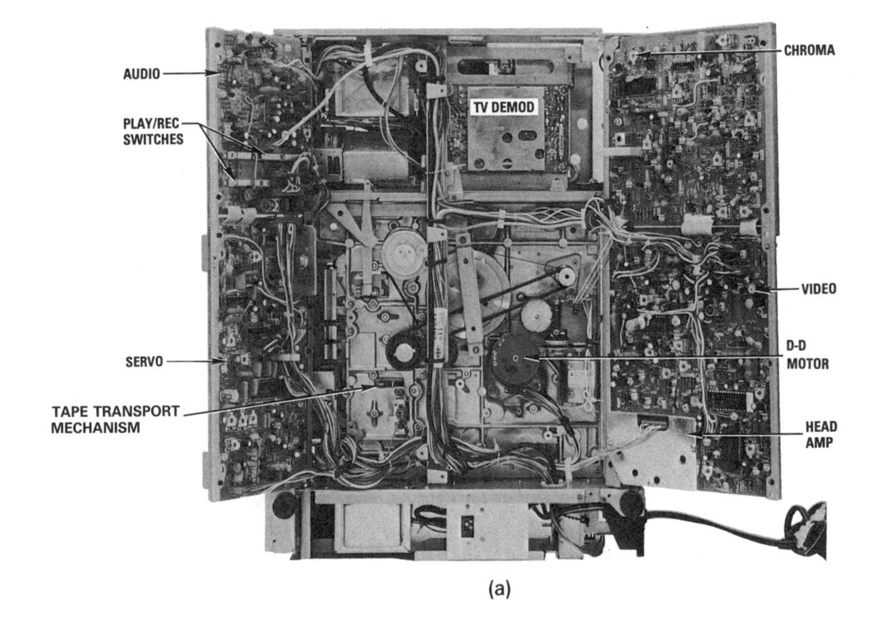

(a)

Figure 19-19 (a) Typical VCR with its cover removed; (b) basic electronic and mechanical systems that constitute a video cassette recorder. [(a) Courtesy of RCA, Consumer Electronics.]

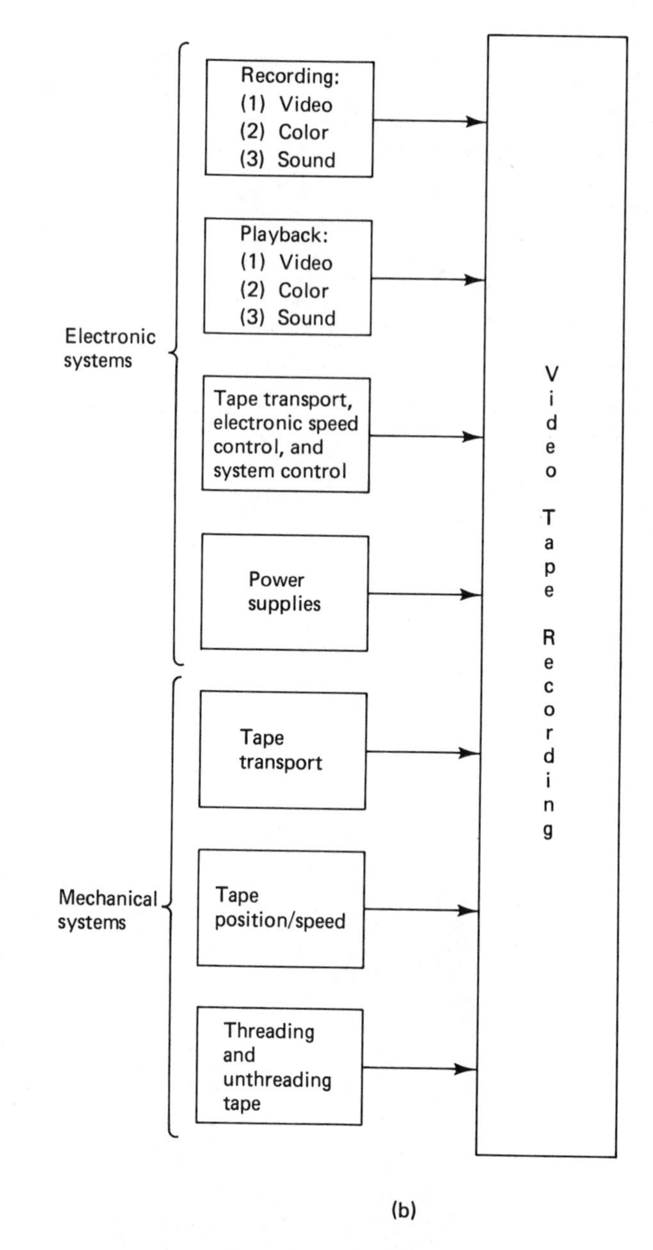

Figure 19-19 *(continued)*

tape, head motor protection, power interruption, and programmable timer circuit operation. The fourth major area of VCR electronics is the power supply. This circuit provides the energy required by all the other circuits in the VCR.

Mechanical systems. The mechanical VCR systems may be broken up into three major categories:

1. A tape transport system, which is responsible for the tape take-up and supply reels, tape tension, tape torque, stop brakes, and the function selector button assembly.
2. A tape position and speed section, which involves the servo-controlled rotating head; the fixed control, audio, and erase heads; the servo-controlled capstan; and the tape guides and rollers that orient the tape for its passage over the heads.
3. The threading mechanism, which extracts the tape from its cassette and wraps it around the headwheel as well as properly positioning the tape on the audio, control, and erase heads.

19-7 VHS recording system

Figure 19-20 is the simplified block diagram of those circuits used by a VHS system during the record mode of operation. Notice that of the 18 blocks shown there are 7 shaded blocks that are devoted to processing the luminance signal, and all of the remaining 11 blocks are concerned with color processing.

The composite video signal (1 V p-p with negative sync) is obtained from a TV camera or from the video output of a TV receiver and is fed into a filter network where the luminance and chroma signals are split. The luminance signal passes through a low-pass filter that is roughly flat to about 4 MHz. This attenuates both the high-frequency video signal components and the 4.5-MHz sound carrier. The luminance signal then passes through a keyed AGC-controlled amplifier that keeps the signal at a constant level.

The 3.58-MHz color subcarrier is then removed by a low-pass filter. The output of this stage is the luminance (Y) signal, which has no chroma components.

The Y signal is then fed into a nonlinear preemphasis network, which provides greater preemphasis for low-level signals than for higher-level signals. As a result, greater preemphasis is introduced during LP operation, thereby improving the system's signal-to-noise ratio.

Due to the action of the preemphasis circuit, the Y output signal contains excessive overshoots and undershoots at the leading and trailing edges of step-like waveforms. To avoid FM overmodulation, these peaks are clamped. Overshoots produce peak voltages that exceed the sync pulse amplitude and are clipped by an adjustable "dark" clip circuit. Undershoots have pulse peaks that exceed the white level of modulation and these are clipped by the "white" clip circuit.

The video output of the clipping circuit, frequency modulates an astable multivibrator. The output of the modulator is a pulse whose frequency depends upon the amplitude of the video signal, and whose rate of frequency change depends upon the frequency of the video signal.

As a means of avoiding beats produced during LP operation by the overlapping video tape tracks, a system of frequency interleaving is used. This is done by increasing the frequency of the FM modulator for one head each time it sweeps the tape. The other head

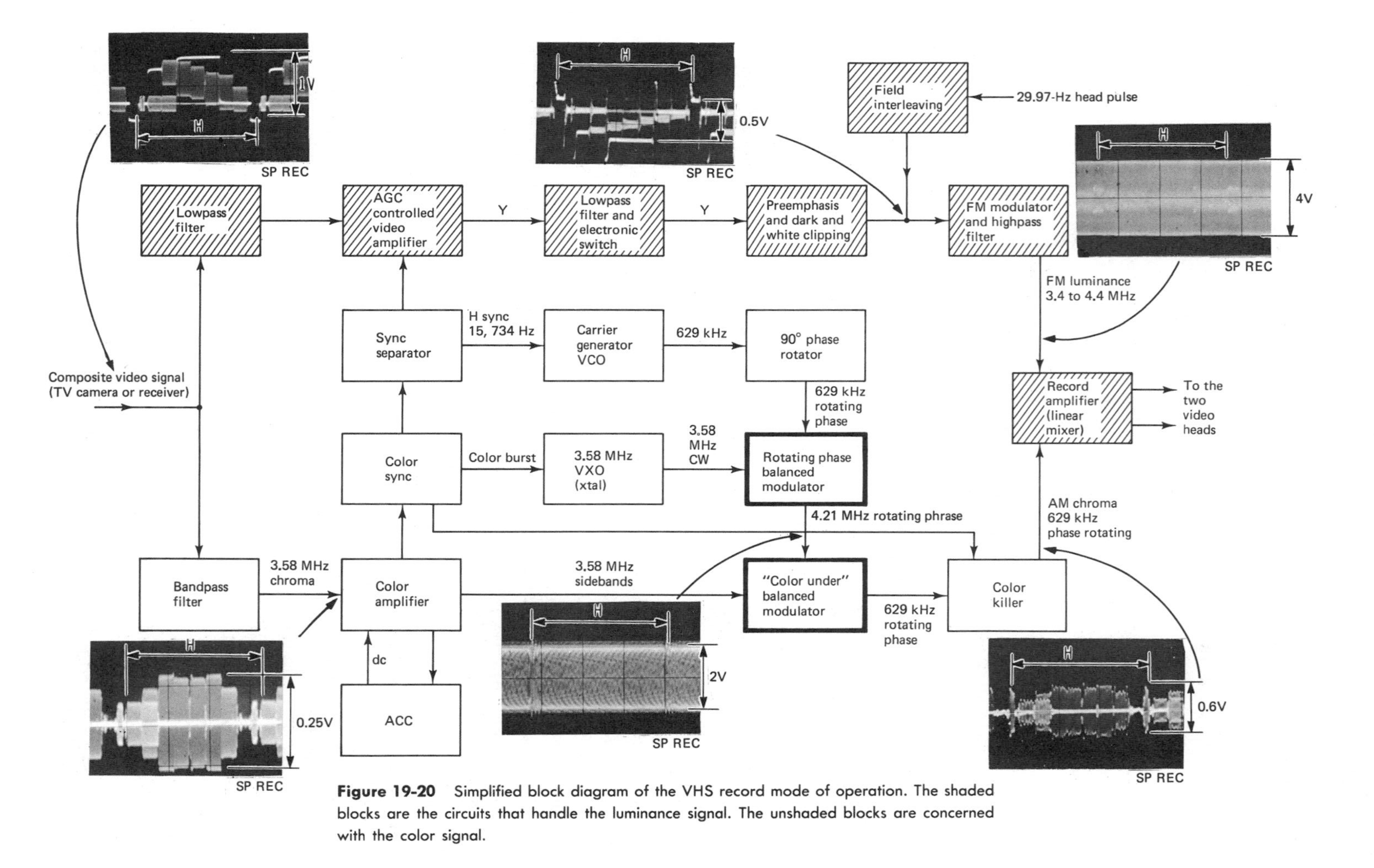

Figure 19-20 Simplified block diagram of the VHS record mode of operation. The shaded blocks are the circuits that handle the luminance signal. The unshaded blocks are concerned with the color signal.

is operated normally. The frequency of the FM modulator is increased by adding a dc voltage to the modulator input signal each time the active head contacts the tape.

The frequency-modulated video signal is then passed into an electronic switch high-pass filter combination, which inserts a high-pass filter (HPF) during color operation and removes it during black & white recordings.

As a consequence of video signal frequency modulation, sidebands are produced that extend into the band of frequencies that are used for color (629 kHz $\pm$ 0.5 MHz). To avoid interference when the *Y* signal is combined with the color signal, these sidebands are removed during color recording and are passed without attenuation during black & white recording.

The luminance and color signals are combined in the record mixing amplifier, which drives the recording heads.

VHS chroma recording. Chroma recording must accomplish at least three major objectives. First, it must take the 3.58 MHz color signal and lower its frequency to 629 kHz. Second, it must compensate for color crosstalk by rotating the phase of the 629 kHz "color under" signal. Third, it must combine with the luminance signal to drive the recording heads.

Examination of the simplified block diagram (Figure 19-20) of the VHS record mode of operation will reveal that the chroma record section has two balanced modulators. All of the circuitry in the chroma section is used to process the chroma signal into a suitable form for application to one of the balanced modulators.

Referring to the block diagram, the video signal is passed through a bandpass filter and burst amplifier. This circuit separates the 3.58 MHz chroma signal from the composite video signal and boosts the burst by 6 dB when the tape recorder is in the SP mode of operation. No boost is provided in the LP/SLP modes of operation. The chroma signal is then amplified and the signal level is automatically controlled by the associated ACC system. At this point the output signal consists of constant-amplitude 3.58 MHz sidebands, which are fed into one of the inputs of the "color under" balanced modulator. It is the function of this circuit to "down convert" the 3.58 MHz color signal to a 629 kHz rotating phase signal.

To generate the 629 kHz signal, a 4.21 MHz signal is mixed with the 3.58 MHz color signal in the balanced modulator. The difference frequency output is the 629 kHz "color under" signal. The 4.21 MHz input to the "color under" balanced modulator is itself obtained from the "rotating phase" balanced modulator. One of the inputs to this balanced modulator is a 3.58 MHz CW signal obtained from a 3.58 MHz crystal-operated phase detector-controlled oscillator (VXO). Control of this oscillator is obtained by comparing the phase of the color burst to that of a 3.58 MHz crystal oscillator, and generating a dc error voltage that forces the phase of the VXO into the correct color phase.

The other input to the "rotating phase" balanced modulator is a 629 kHz signal obtained by using a sync separator to obtain the horizontal sync pulse and then phase locking it to a VCO that is operating at a frequency of 160 times the horizontal rate (2.517 MHz). This signal is then fed into a system of counters and phase shift circuits which produces 90° rotation at 629 kHz.

The 629 kHz phase rotating signal is then *added* to the 3.58 MHz signal in the "rotating phase" balanced modulator to produce the 4.21 MHz rotating phase signal which feeds the "color under" balanced modulator.

The output of the "color under" balanced modulator is at 629 kHz and is passed through a color killer stage that blocks or passes the color signal. It blocks signals when the color is absent and passes the signal when color is present.

At the recording amplifier the color 629-kHz phase rotating signal is added to the frequency-modulated luminance signal. This composite signal is then used to drive the recording head. The luminance signal acts as the recording bias for the color signal in the manner of audio tape recording.

VHS playback. The VHS system of playback can be divided into four important circuit areas:

1. The luminance section
2. ACC color amplifier control
3. Crosstalk and jitter compensation
4. Color and black & white matrixing

A simplified block diagram of the circuits used during VHS playback is shown in Figure 19-21. To allow rapid recognition, the luminance as well as shared sections of the block diagram are shaded.

During playback the two rotating heads develop video signals that are fed into separate preamplifers that are switched on when their respective heads are "reading" the tape. Summing these outputs produces a single output. During color playback this output is then filtered, leaving an FM luminance signal.

Dropout compensation. If, during record or playback mode of operation, tape defects caused by dust or missing oxide coating cause the tape-head combination to momentarily lose contact, the result will be a momentary loss of picture information with a possible loss of sync.

Dropouts are annoying and cannot be completely avoided. Therefore, VCRs are equipped with special dropout compensating circuits. These circuits are used during the playback mode of operation and are placed in the luminance chain of circuits between the head amplifiers and the limiter/demodulator circuit. Dropout compensation depends upon the fact that the FM carrier disappears during these dropouts.

The dropout compensator uses a delay line that delays the luminance signal for the duration of one horizontal line. If a dropout occurs, the loss of the FM carrier activates an electric switch that disconnects the luminance amplifier and limiter circuit and inserts the output of the delay line. This substitutes the contents of the previous line which is still coming through the delay line and replaces the dropped out information. The defective horizontal line is not seen, but the previous line is seen twice. Dropouts of a few lines will also be compensated for, since the same line will continue to be recirculated as long as the FM carrier is missing.

FM demodulator. The output of the dropout compensator feeds a limiter FM demodulator circuit. This circuit removes all amplitude variations and converts the FM luminance signal (3.4 to 4.4 MHz) into a black & white video signal (0 to 4 MHz). In order to compensate for the record preemphasis and to improve the signal-to-noise character of the system, the signal is then passed through an appropriate deemphasis network. The output of this circuit

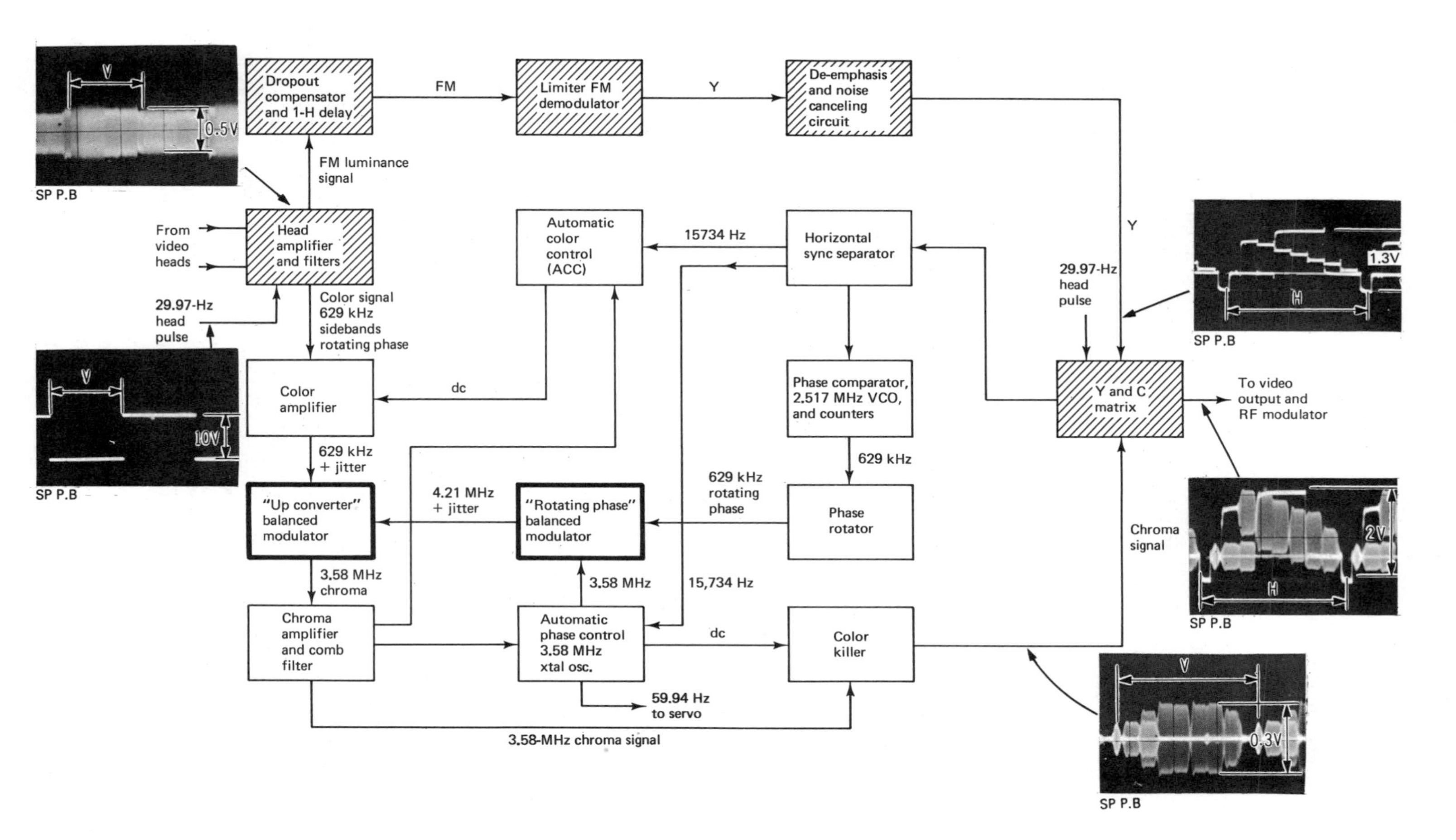

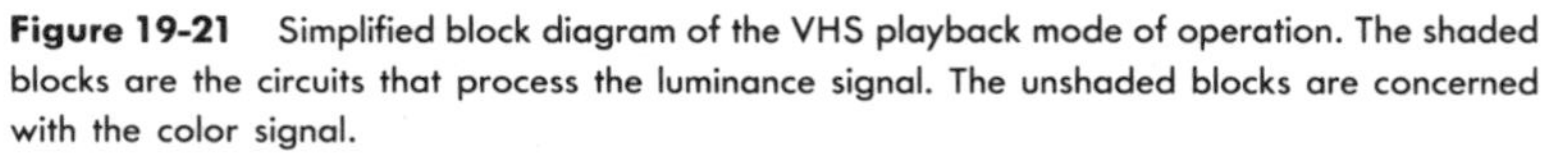

Figure 19-21 Simplified block diagram of the VHS playback mode of operation. The shaded blocks are the circuits that process the luminance signal. The unshaded blocks are concerned with the color signal.

feeds a mixer amplifier that combines the luminance signal (*Y*) and the color signal (*C*), producing an output composite video signal of 1 V p-p having a negative-going sync pulse. This signal drives an RF modulator which provides a channel 3 or 4 output to a TV receiver.

Color playback. The color playback circuits must accomplish at least four important objectives. First, they must up-convert the 629-kHz rotating phase signal into the original 3.58-MHz chroma signal. Second, these circuits must use the rotating phase characteristics of the color signal to compensate for color crosstalk. Third, the chroma playback circuits must minimize "jitter." The fourth objective of the chroma circuits is to compensate for "color flicker."

Color flicker appears as a 30-Hz flicker in the saturated colors in the picture. This problem is due to differences in video head alignment and sensitivity. These differences result in variations in the playback of low-frequency signals. Thus, head A can produce more output than head B, and vice versa. Head output variations can also be produced when tapes recorded on other machines are played back. These head output differences can be partially compensated for by adjustment of a balancing potentiometer.

To compensate for "color flicker" the playback system incorporates an ACC circuit that automatically adjusts the gain of a chroma amplifier to produce a constant color output.

The unshaded blocks of Figure 19-21 are those that are associated with color playback circuits. The color signal (629 kHz) is separated from the composite video signal in the head amplifier. This signal is then processed by an ACC-controlled color amplifier and fed into the "up converter" balanced modulator (frequency changer) that up converts the "color under" signal (629 kHz) into its original 3.58 MHz video form. In order for the up converter balanced modulator to perform its function, a second input signal at 4.21 MHz and of rotating phase is required. The difference signal between 629 kHz and 4.21 MHz is the desired 3.58 MHz color signal.

The 4.21 MHz signal is obtained from the "rotating phase" balanced modulator. To generate the 4.21 MHz rotating phase signal, the "rotating phase" balanced modulator has two inputs; one is at 3.58 MHz and is obtained from an automatic phase-controlled crystal oscillator operating at 3.58 MHz. The other input is a rotating phase 629 kHz signal that is developed by a rotating phase generator circuit. This circuit separates the horizontal sync pulse from the composite video signal and phase locks a 2.517 MHz oscillator rotating phase switch and frequency divider circuit to generate the 629 kHz signal.

Jitter. Time base errors are often called *jitter* and are the result of slight video head speed and tape transport speed variations. These velocity changes have no noticeable effect on black & white pictures, but will produce severe chroma phase errors, making color pictures unwatchable. To minimize the effects of jitter, the 629 kHz color signal containing the jitter signal components is mixed with the 4.21 MHz signal which also contains the same jitter information. The jitter is the same because the 4.21 MHz signal is phase locked to the picture's horizontal sync pulse. The difference frequency of 3.58 MHz produced in the "up converter" balanced modulator, will be the color signal free of the jitter components.

This jitter-free signal is then passed through a chroma amplifier and comb filter circuit which removes the color crosstalk produced by the overlapping tracks. (See the discussion in Section 19-5.) The crosstalk-free color signal is then passed through a color killer circuit to the luminance and color matrix, where these two signals are combined. During black & white playback the color killer disconnects the color circuits from the matrix.

19-8 the Beta recording system

The simplified block diagram of the Beta system record mode of operation is shown in Figure 19-22. A comparison of this block diagram with that of the VHS record mode, will reveal that the luminance sections of the two systems are almost identical. In both cases the luminance signal is separated from the composite video signal by a low-pass filter (LPF) and processed by a video amplifier and Beta II/Beta III equalization circuit. Black & white clipping circuits and preemphasis prepare the Y signal for the necessary frequency modulation. In the Beta system the frequency deviation ranges from 3.5 to 4.8 MHz. The Y output of the FM modulator then feeds the record amplifier, which matrixes the color signal to produce the necessary signal to drive the video heads.

Beta color record. The two main objectives of the Beta system color record mode of operation are: first, to lower the 3.58 MHz color signal to a frequency of 688 kHz, and second, to arrange the color signal so that color crosstalk can be canceled during playback.

Examining the Beta record block diagram shows that a bandpass filter is used to extract the chroma signal from the composite video signal and to feed it to an ACC-controlled color amplifier. The output (3.58 MHz) of this amplifier is fed into one of the inputs of the color frequency converter. The ACC portion of this circuit develops its dc control voltage via a burst gate and 3.58 MHz ringer circuit. This circuit keeps the chroma signal at a constant level despite signal variations due to tuner alignment, multipath reception, and antenna or cable system variables.

The second input to the color frequency converter is a 4.7 MHz phase-alternated signal. The phase alternation is used to minimize color crosstalk. The output of the frequency converter is the difference between the 3.58 MHz color signal and the 4.27 MHz phase alternated signal, which is the "color under" 688 kHz chroma signal.

The 4.27 MHz phase-alternated signal is developed by a combination of a horizontal sync-controlled phase-locked loop (PLL) operating at a frequency of 692 kHz and a mixer stage that "up converts" a 3.57 MHz crystal oscillator output to the desired 4.27 MHz. Phase alternation is provided by a phase inverter that inverts every horizontal line recorded by the A head, and passes unchanged the information recorded by the B head.

Beta playback. As was done in the VHS system, the Beta playback system may be broken up into four major sections:

1. The luminance strip
2. ACC control of the chroma signal
3. Color crosstalk and jitter compensation circuits
4. The luminance and color matrix circuit

A simplified block diagram of the Beta playback circuit is shown in Figure 19-23.

The two outputs of the video heads are coupled through rotary matching transformers to the head amplifiers. The rotating transformers are part of the scanning head.

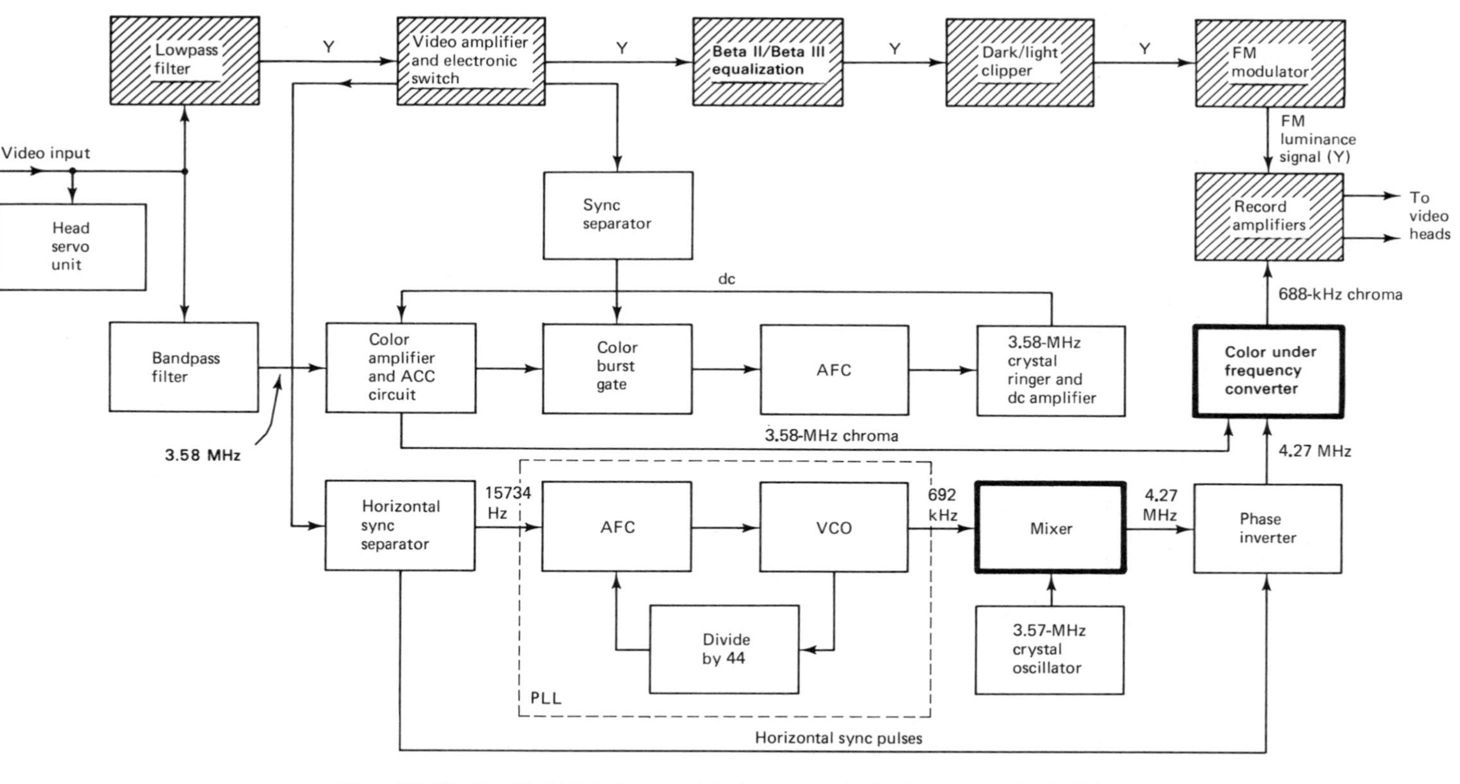

Figure 19-22 Simplified block diagram of the Beta record mode of operation. The shaded blocks are circuits that process the luminance signal. The unshaded blocks are the circuits concerned with the color signal.

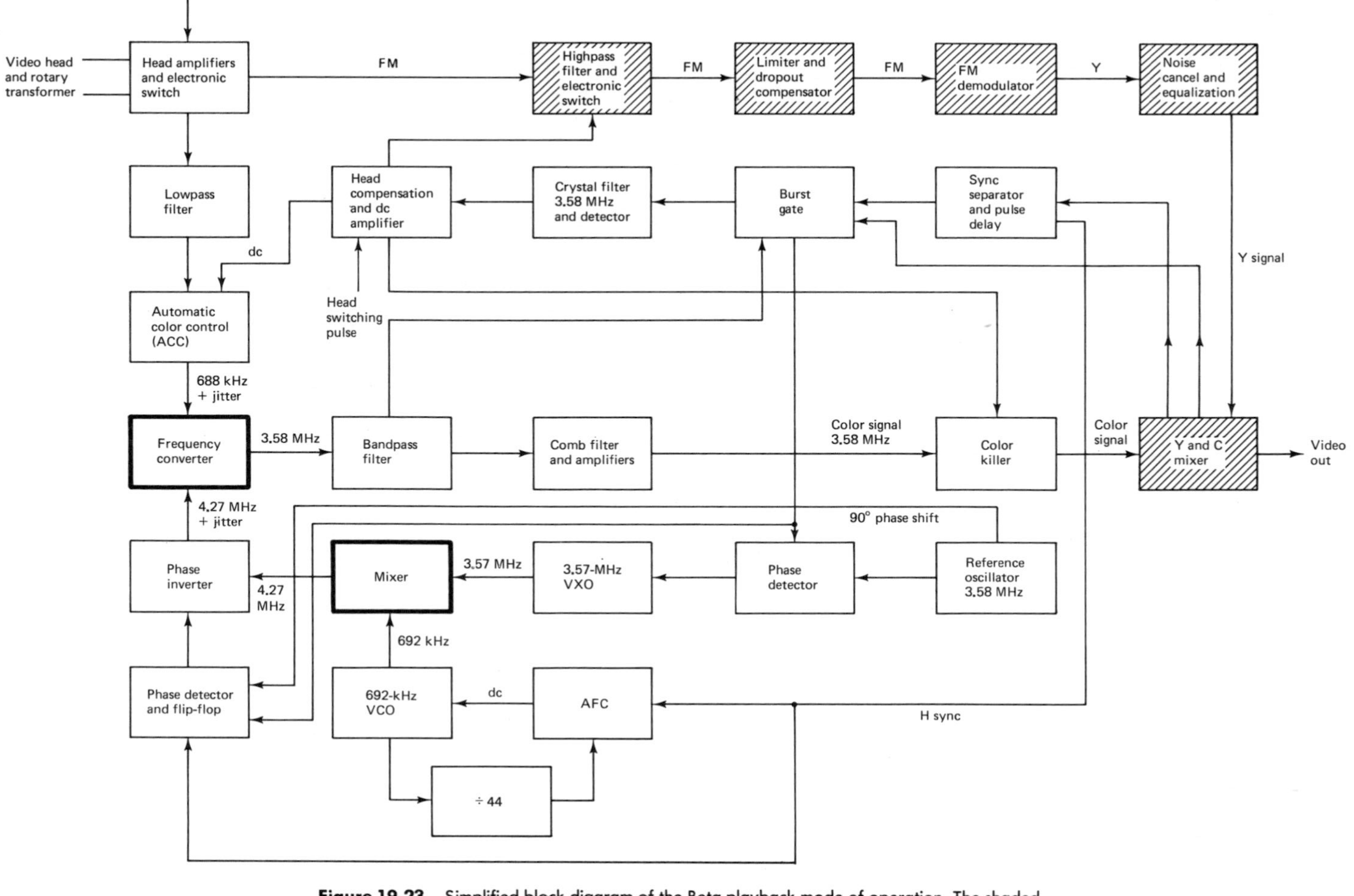

Figure 19-23 Simplified block digaram of the Beta playback mode of operation. The shaded blocks are circuits that process the luminance signal. The unshaded blocks are the circuits concerned with the color signal.

One winding of the ferrite core transformer is stationary, while the other winding is part of the rotating drum. The composite video signal is then divided by low- and high-pass filters into the FM luminance signal and the 688 kHz chroma signal.

The luminance signal passes through a limiter-dropout compensator circuit and an FM demodulator circuit, as was done in the VHS system of playback. The video output of the demodulator is then provided with the necessary deemphasis, equalization, and noise cancellation to provide a good signal-to-noise ratio to the system. The recovered luminance signal is then fed into a *Y/C* mixer, where it is combined with the recovered chroma signal, forming the original composite video signal.

Beta color playback. Color processing begins with the color signal being separated from the video head output by a low-pass filter. This signal is then amplified by an automatic color control amplifier. The ACC dc control voltage is developed by sampling the composite video signal color burst amplitude. This is done with a circuit consisting of a burst gate, a 3.58 MHz ringing filter, a detector, and a dc amplifier.

The 688 kHz color signal is then fed into a frequency converter, where its frequency is up converted to 3.58 MHz by mixing it with a 4.27 MHz signal. Both signals have time base errors (jitter) signal components which are canceled in the up frequency converter, leaving a stable 3.58 MHz color signal output.

The 4.27 MHz signal is developed in a mixer that is also an up converter. It combines a 692 kHz signal that is phase locked by an AFC system to the horizontal sync pulse, with a 3.57 MHz signal which is developed by a 3.58 MHz reference oscillator and APC circuit.

Crosstalk cancellation occurs by phase alternation of the 4.27 MHz signal so that every other 688 kHz horizontal line recorded by the A head is returned to its original phase. The 688 kHz output video signal of the B head is not inverted. After down conversion, the 3.58 MHz video signal is passed through the comb filter to remove the crosstalk.

The output color signal of the comb filter is passed through a color killer circuit that passes it on to the *Y/C* mixer if color is being processed, and blocks it during black & white playback.

19-9 the servo system

Most VCRs have three motors: one for head drive, one for capstan drive, and another for threading and unthreading of the cassette video tape. Some modern VCRs use five motors, a head motor, a capstan motor, a loading motor, a supply reel motor, and a take up reel motor. In those systems that use two motors, belts are used between the capstan motor and the threading mechanism.

The dc head drive motor and the dc capstan motor must have their speed and phase tightly controlled to prevent such trouble symptoms as picture tearing, Donald Duck voices, and rapid moving figures, exceptionally noisy picture playback, and noise bars through the picture.

The separate head and capstan servo systems in the VCR are electronic control devices that permit automatic control over the speed and position (phase) of the head and capstan motors. Servo systems are feedback systems that require information as to the present speed and position of the heads and of the capstan.

In most VCRs the capstan is used to maintain a constant linear tape speed and a constant tape position. The video head speed and position must complement those of the capstan. This must be done in order to place the video tracks on the tape in their required pattern. Servo control during record function assumes that the width of the guardbands in SP, and the degree of overlap in LP or SLP will be constant. In addition, servo control forces the vertical sync to be placed at the same point on each recorded track. This point is approximately 10 horizontal lines after the beginning of a track. To achieve these two requirements, the speed and phase of the video head must be controlled. The speed of the video head while recording a color program is based on the fact that each head records one field during its time of tape contact. Since the color TV field rate is 59.94 Hz, the speed of head rotation must be 59.94/2 (two heads) or 29.97 r/s, which is the same as 1798.2 r/min, usually stated as 1800 r/min for simplicity. Control of speed alone is not sufficient. The position of the heads must be such as to record the vertical sync pulse at the same point on each vertical field. To provide this, the video head is also phase locked.

The capstan must also be controlled in both speed and position, and as such it has its own speed and servo system. In all modern VCRs, the speed of the capstan is changed to obtain SP, LP, or SLP operation. Similarly, it is the phase of the capstan that is varied when the operator tracking control is adjusted.

The head servo system. During the record mode of operation the servo control system has three functions: (1) it controls the speed of the video heads, (2) it controls the position (phase) of the video heads, and (3) it records the control pulses on the control track, that are used as the reference signal during playback. The head servo system contains two important control systems: a speed control circuit and a phase control circuit. The speed control circuit keeps the head at a constant rotational speed so that one field is scanned during the pass of one head over the tape. The phase control circuit ensures that the heads start at the same point (vertical sync) on each pass across the tape.

Video head phase control. Referring to Figure 19-24, it can be seen that the rotation of the video head motor also generates a train of pulses. This pulse generator is constructed as part of the head drum assembly. Embedded in the assembly and 180° apart are two permanent magnets of opposite polarity. The magnets rotate at the same speed as the head drum (29.97 r/s for color), and when passing the pulse generator coil they induce a pulse into the coil. Due to the polarity of the magnets, the polarity of the pulses will be alternately positive and negative pulses.

The A magnet passes the pulse generator (PG) coil at the same time that the video head A starts to scan the tape. Therefore, a negative pulse is generated by the PG coil at the start of each field scan. Similarly, a positive-going pulse generated by the PG coil, indicates that the B head is starting to scan the tape.

The pulses produced by the PG coil are responsible for the phase control (fine tuning) of the head drum motor. These pulses are first amplified and then passed through polarity-identifying diode circuits, which route the positive and negative pulses to their respective monostable multivibrators (MMs). One MM stage triggers on the positive pulses and the other triggers on the negative pulses. The delays of the MMs are adjustable by means of two variable resistors, R_1 and R_2. These resistors are adjusted to compensate for any errors in the physical positioning of the PG head magnets with respect to the video

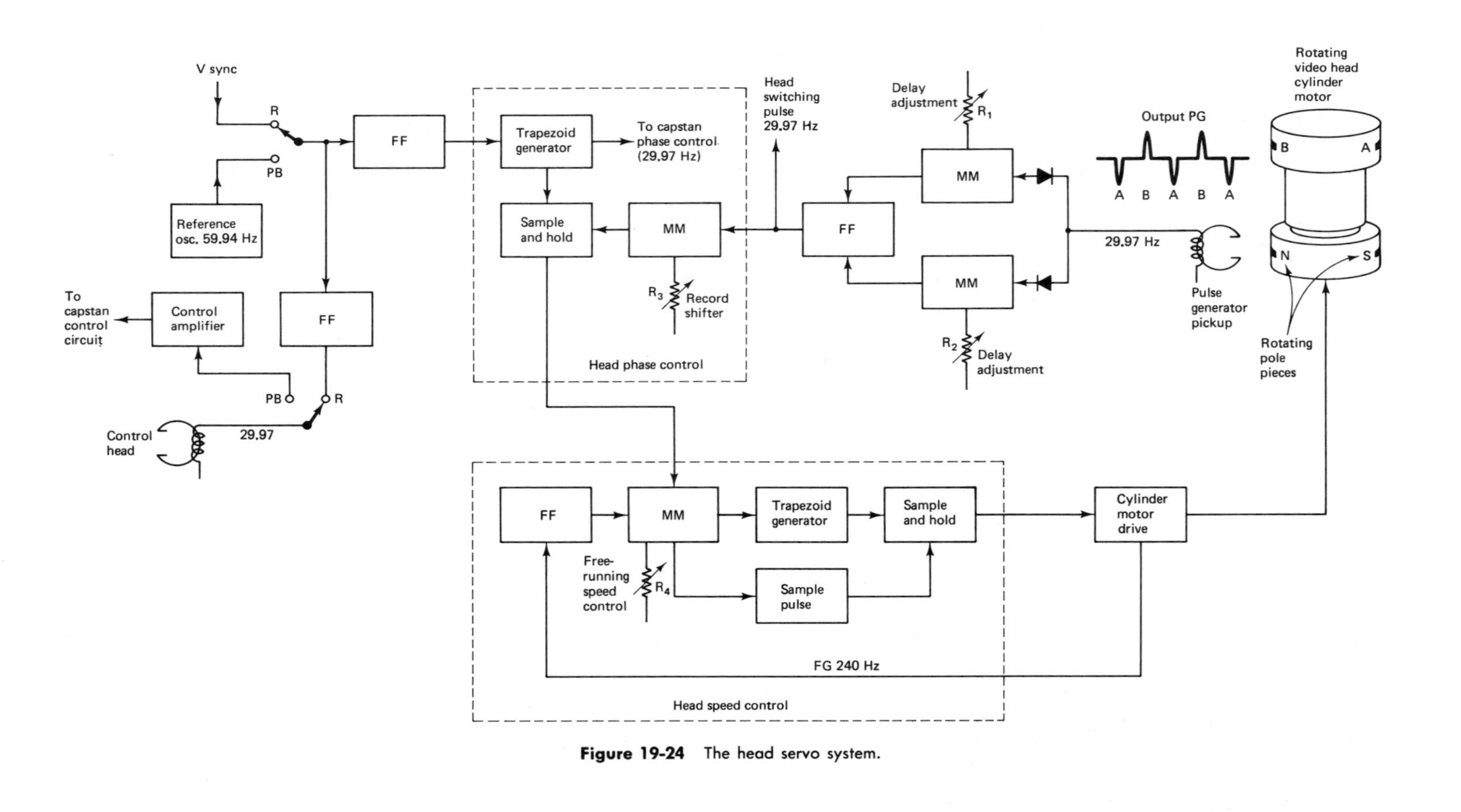

Figure 19-24 The head servo system.

heads. This adjustment becomes important during playback, where any timing error can cause head-switching noise and/or horizontal jitter in the picture.

The output of the MMs set and reset a flip-flop (FF) stage, which under normal conditions produces a 29.97 Hz (color) square-wave output. The actual output frequency depends upon the speed of the head rotation. This pulse is then fed to the signal processing circuit and is used for head switching during playback. During the record mode of operation the video heads are connected in parallel and are active all the time. During playback the head-switching pulse turns on the appropriate head amplifier when that head is scanning the tape, and turns off the other head amplifier.

In addition to driving the head-switching circuits the FF circuit also feeds the "head phase control" circuits. This block of circuits consists of an MM, a sample-and-hold circuit, and a trapezoid generator. During the record mode, the trapezoid generator circuit is driven by a 59.94 Hz (color) vertical sync pulse obtained from the video signal being recorded. During playback a 3.579545 MHz crystal reference oscillator and a counter circuit are used to generate the vertical sync pulse.

Sample and hold. As shown in Figure 19-25, the trapezoid generator is essentially an interrupted sawtooth generator that is driven by pulses obtained from the outputs of FF or

Figure 19-25 Timing comparator circuit used as a sample-and-hold circuit.

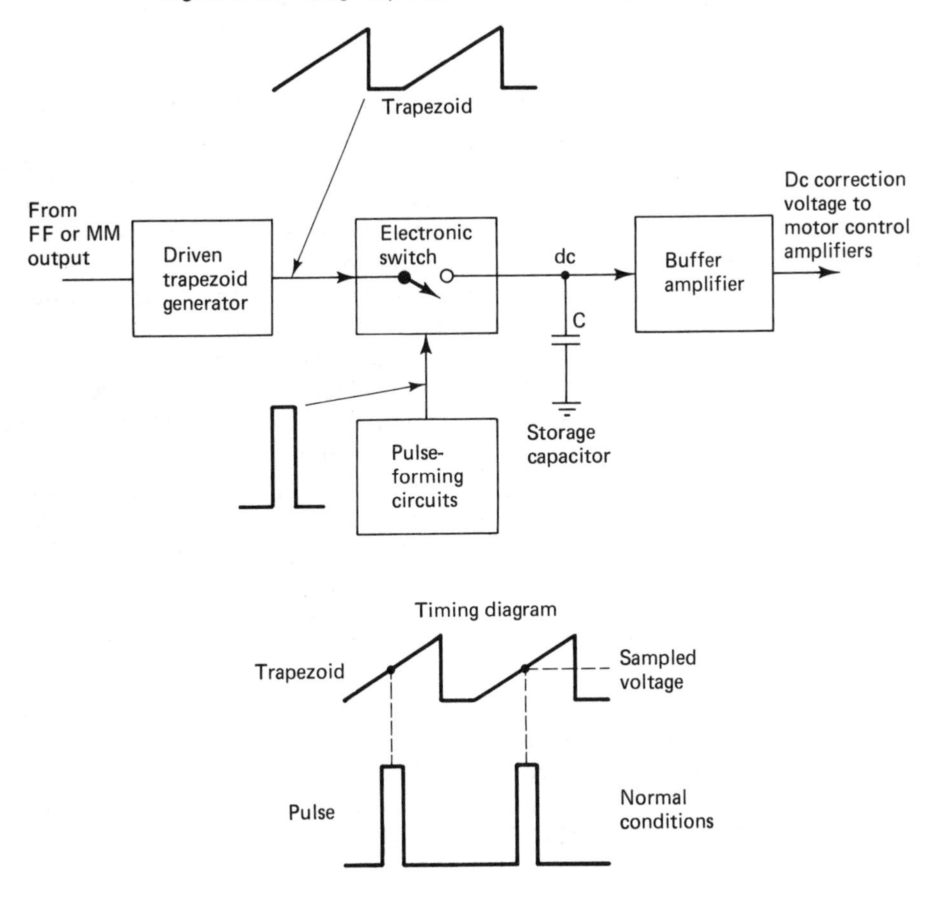

MM circuits. The trapezoid generator develops a ramp which rises to a peak, then drops to zero, stays at that level for a short time, and then repeats itself.

The sample-and-hold circuit acts like a series switch that opens and closes in response to the narrow MM pulse driving it. During the time the switch is closed, the ramp voltage charges a capacitor to the level of the ramp at the time the switch is closed.

If the head motor runs slightly slower than normal, the timing between the ramp and the pulse will be such that the sampling pulse occurs nearer to the ramp peak, and the capacitor will charge to a higher voltage. This, in turn, will result in the motor being driven harder, increasing its speed and compensating for the error. A lower-capacitor charge voltage will be developed if the pulse timing is such that it occurs near the beginning of the ramp. The lower dc voltage across the capacitor will then cause the motor speed to decrease, again providing compensation for the speed error. To prevent the capacitor from discharging, a high-impedance amplifier buffer is placed between it and the following stages.

Referring to Figure 19-24, it can be seen that timing between the pulse and the trapezoid is adjusted by R_3, which controls the delay time of the monostable multivibrator. When the adjustment is correct, head switching occurs six horizontal lines before the start of the vertical sync pulses, (three horizontal lines before vertical blanking). If this adjustment is incorrect, forcing the head switching to be late, noise can be introduced during the vertical sync interval causing possible vertical sync problems. If head switching occurs too soon, a small noise band may appear at the bottom of the picture just before the vertical blanking bar.

Video head speed control. The video head motor is called a direct-drive dc brushless motor. It consists of an eight-pole ring magnet and a position detection vane that is mounted on the same shaft. Inside the eight-pole ring magnet are three drive coils. When properly driven by two of the three coils, a rotating magnetic field is created, forcing the ring magnets to follow. The position indicator senses the present situation and acts to turn on the next sequentially arranged main drive coils to continue the rotation.

Also placed inside the ring magnets and driven by a 65 kHz signal source are three position-detecting transformers whose coupling is changed when the eight-position indicator vanes, mounted in the rotating housing, pass the primary and secondary windings of each transformer. When a vane is near the transformer, coupling is maximum and the secondary output becomes maximum. When the vane position is between the transformers, coupling is reduced and the secondary output is reduced. The net result of this is that the three secondary outputs of the position indicator transformers develop three amplitude-modulated 65 kHz signals. These signals are used to drive the main drive windings of the motor. Since there are eight position vanes that rotate at a 30 Hz rate, the amplitude modulation rate is $8 \times 30 = 240$ Hz. After detection and processing, the 240 Hz frequency generator (FG) sine-wave signal is then used to generate the signal feedback needed to provide head-speed control.

A comparison of the head speed control and head phase control block diagrams (Figure 19-24) will show that they are similar. Both use trapezoid and sample-and-hold circuits to develop the correction voltages needed to control the head motor.

However, in the head speed control system, the pulse position (MM output) on the trapezoid is controlled by the "fine tuning" voltage generated by the head phase control system as well as the adjustment of the "free-running speed control" R_4. Proper adjustment of R_4 maintains a constant video head speed of 1800 r/min. Misadjustment of R_4 affects picture horizontal stability.

Control track pulses. During the record mode of operation the vertical sync pulse input is processed so that a 29.97 Hz (color) square wave is applied to the windings of the control head. This results in the control head "writing" a 29.97 Hz control pulse onto the control track of the tape.

During playback operation, the head servo system is controlled by a reference oscillator operating at 3.58 MHz and then divided down to 29.97 Hz. This square-wave signal is then fed into the head speed and phase control servo system.

The control head produces a 29.97 Hz output during playback, which is used to control the speed and phase of the capstan servo system.

Capstan servo system. The purpose of the capstan is to transport the tape at a constant speed. This is done, as shown in Figure 19-26, by passing the tape between the capstan and a pressure roller. The capstan is the shaft of a flywheel, and the pressure roller presses the tape up against the shaft. In the VHS SP mode, the capstan moves the tape at 1.3 in. per second (ips), in LP operation it moves at 0.66 ips, and in SLP operation it moves at 0.44 ips. In Beta VCRs the speeds are 1.57 ips, 0.787 ips, and 0.52 ips, respectively. In playback the tape speeds must remain the same as they were during record operation. This is important when using tapes recorded on another machine.

Figure 19-26 Capstan motor and associated components.

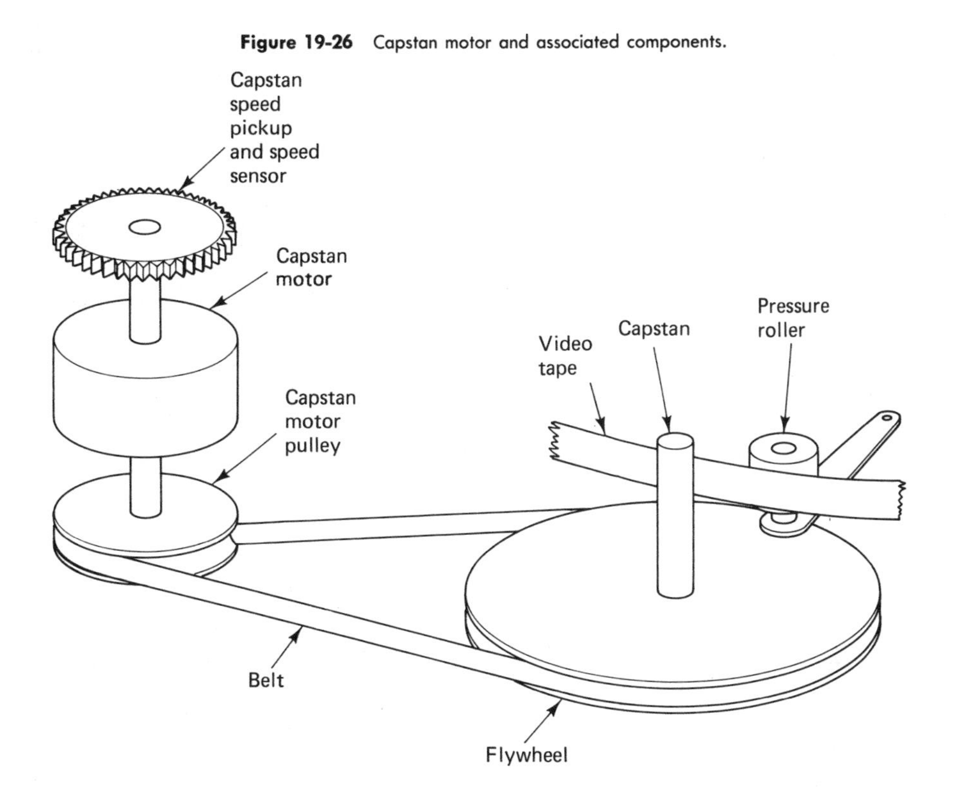

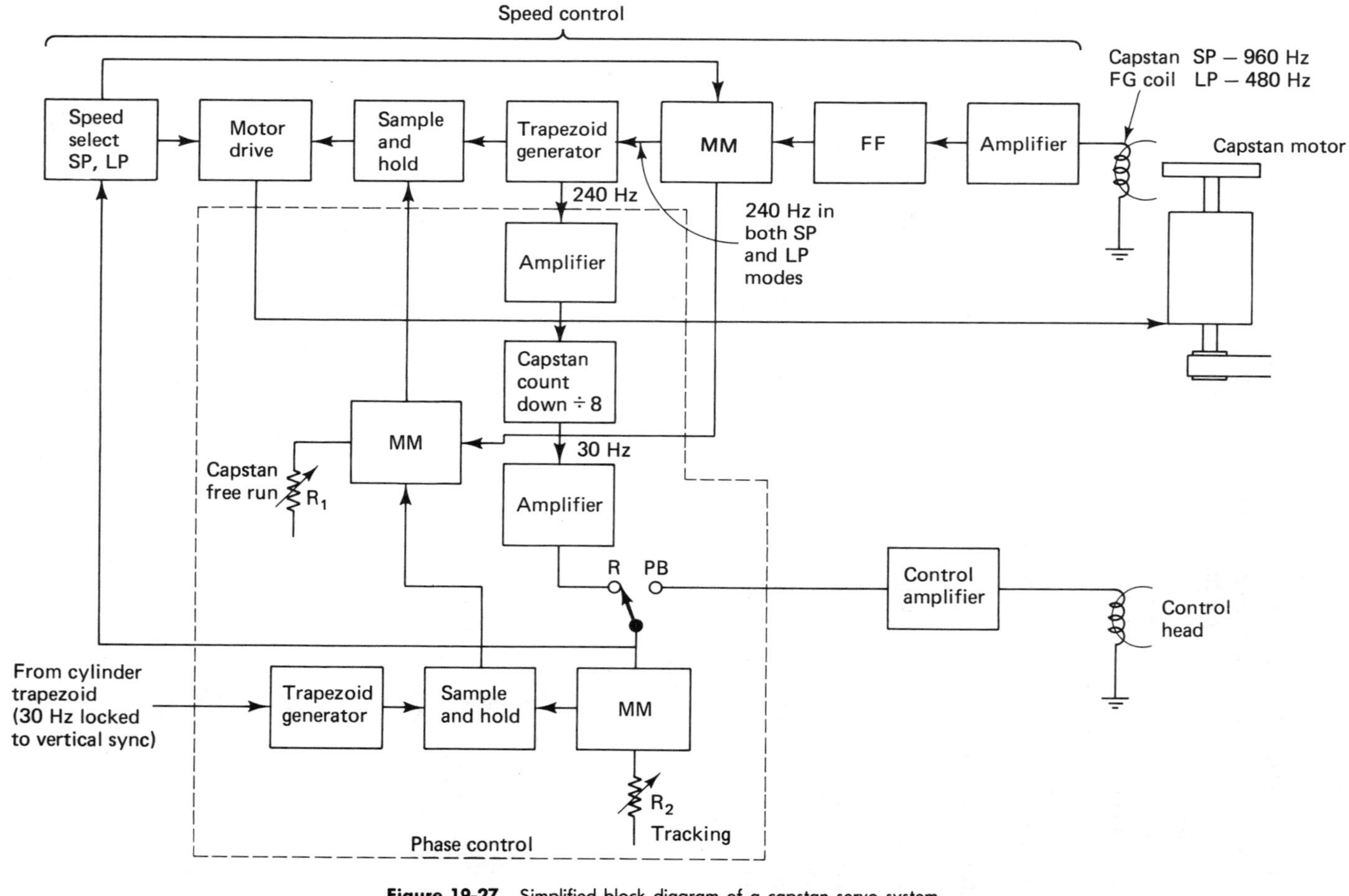

Figure 19-27 Simplified block diagram of a capstan servo system.

The capstan flywheel is belt driven by the dc capstan motor, which also has mounted its shaft a magnetized geartooth plate. Mounted near the teeth is a speed sensor, which generates a sine-wave output. The output frequency of the frequency generator (FG) depends upon the speed of capstan rotation. When the unit is in SP mode, its frequency output is 960 Hz, and 480 Hz in LP mode of operation. Three speed machines may have somewhat different frequencies: SP, 1440 Hz; LP, 720 Hz; and SLP, 480 Hz.

As shown in Figure 19-27, the output of the capstan FG coil is processed through an amplifier and flip-flop (FF) circuit that produces an output that is always 240 Hz for both SP and LP operation. In three-speed machines the output frequency may also be 240 Hz. The speed selection circuit determines the frequency division necessary to produce the required 240 Hz output.

The speed control portion of the circuit is similar to that used in the video head servo system. The trapezoid generator and a monostable multivibrator (MM) feed the sample-and-hold circuit, developing the dc correction voltage necessary to drive the motor. Fine tuning of the motor speed is provided by an MM input from the phase control sample-and-hold circuit, and the capstan free-run adjustment resistor R_1.

Proper adjustment of R_1 locks the capstan servo to the system signals and maintains proper capstan motor speed. If R_1 is misadjusted, the picture may suffer from loss of vertical sync as well as being very noisy.

The phase control portion of the capstan servo is similar to that of the video head servo system. The trapezoid generator 240 Hz pulse is amplified and divided by 8 to produce a 30 Hz pulse. This pulse is fed into the phase control MM, which together with the phase control trapezoid generator feeds a sample-and-hold circuit. The dc output of this circuit is then fed to the capstan speed control MM, modifying its time delay characteristics. Fine tuning of the phase control MM is provided by a tracking control R_2. This control adjusts tracking errors when playing back tapes made on other machines. Misadjustment of this control will cause noise bars that move vertically through the picture.

19-10 mechanical considerations

In the VHS system, one swipe of the head (one field) during SLP operation (0.44 in./s or 11.176 mm/s) corresponds to a tape movement of 0.00734 in. (0.18645 mm) in 1/59.94 of a second. This, in turn, represents a tape movement of .0144 in. (0.366 mm) per horizontal line. Similar results are obtained for the Beta system. Hence, any microscopically small displacement of the video tape from its normal position will result in a loss of system performance, and poor signal-to-noise ratio.

Very small mechanical faults can result in changes in track length that cause microsecond timing errors which have a noticeably large impact on picture formation. To minimize tape distance errors, the manufacturer uses precision production jigs and highly specialized gauges to set the positions of tape guides, the location of the stationary heads, and the assembly of the video head.

From a servicing point of view the foregoing indicates that a great deal of care must be taken when doing any repairs, maintenance, or adjustments of the mechanical aspects of the VCR. When performing service on the mechanical portions of the VCR, the technician must have the proper service literature, as well as the required test plates, jigs, gauges, special tools, and alignment tapes necessary for any adjustments. Examples of these

Back Tension Test Tape

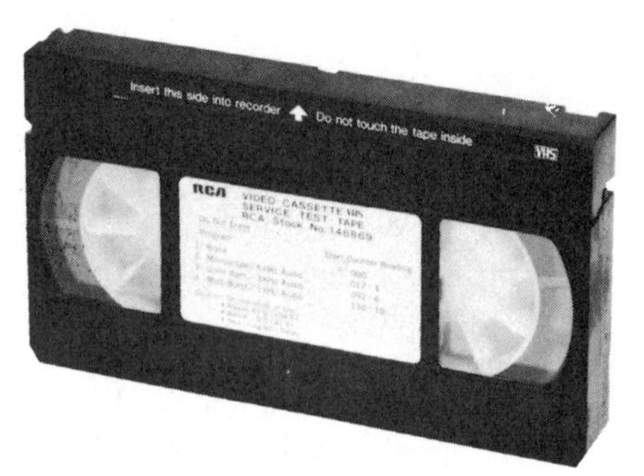

Service Test Tape (6 Hr.)

Height Reference Plate

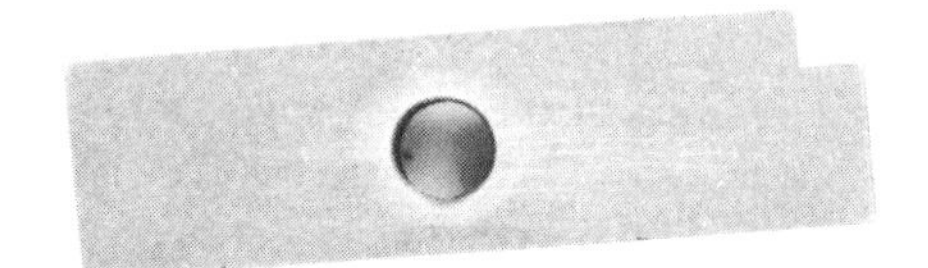

Reel Table Height Jig

Torque Gauge

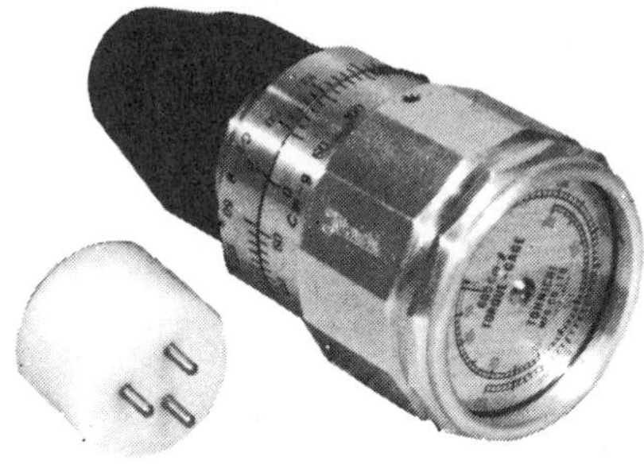

Torque Gauge

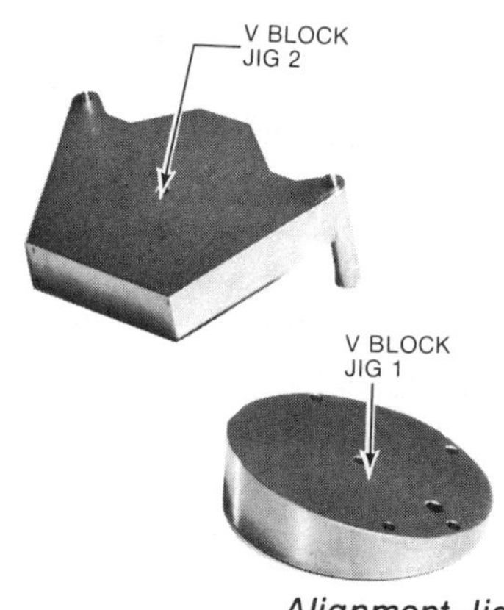

Alignment Jigs

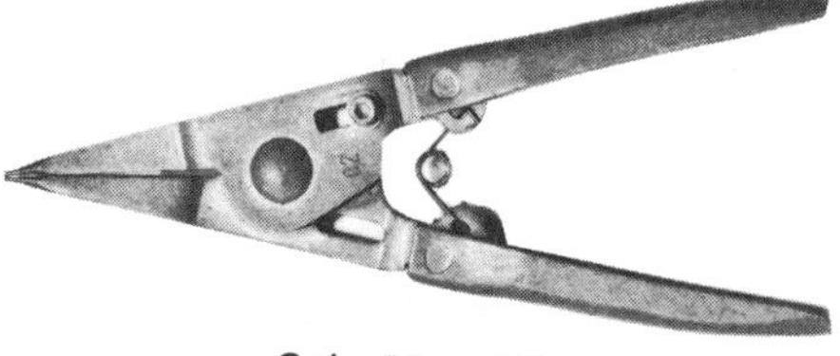

Grip Ring Pliers

(a)

Figure 19-28 (a) Typical alignment tools and test fixtures for use with VHS video tape recorders; (b) The waveform and picture tube display of the color bar signal found on an alignment tape. [(a) Courtesy of RCA, Consumer Electronics.]

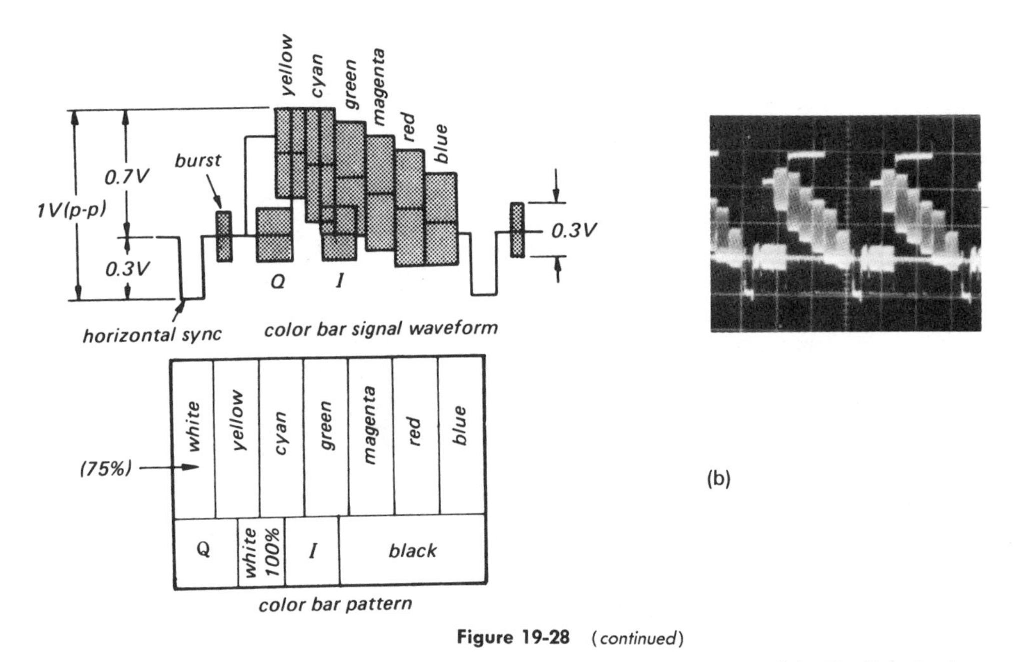

Figure 19-28 (*continued*)

devices for the VHS recording system are shown in Figure 19-28(a). Similar devices are used for Beta VCR adjustments. Also shown in Figure 19-28(b) is the waveform and picture tube display produced by the color bar alignment tape.

Beta tape threading. For the VCR to operate, the tape inside the cassette must be extracted from the cassette and properly wrapped around the head drum. The threading mechanism which performs this operation is extremely complex, and is an area of major difference between VHS and Beta systems.

Figure 19-29 shows the Beta threading mechanism. In the Beta format inserting the tape into its holder and activating the play or record mode of operation [see parts (a) and (b)] activates the threading motor and causes a single swinging arm attached to a rotating threading ring to draw the tape out of the cassette. The rotation of the threading ring [parts (c) and (d)] ensures that the tape is then wrapped around the head wheel, the stationary audio and control track head, and the erase head in a complex path resembling the letter U. The tape remains wrapped around the video head wheel and transport path during all modes of operation, including play, record, rewind, search, and fast forward. Unthreading is the reverse operation of threading.

VHS tape threading. As shown in Figure 19-30, VHS tape decks use two movable arms and their associated guide pins and rollers to pull the tape out of the cassette. During threading operation, the arms move [see arrows in part (a)] the tape so that it is in contact with the headwheel and the other heads mounted on the deck. Arm travel is limited by precisely positioned guide anchors into which the guide pins seat. The guide pins and guide rollers direct the tape to and from the video head wheel. At the end of threading [see part (b)] the pinch wheel and the tape tension arm are released, providing the necessary tape tension and speed due to the capstan pinch wheel pressure. The final shape of the tape path roughly forms the letter M.

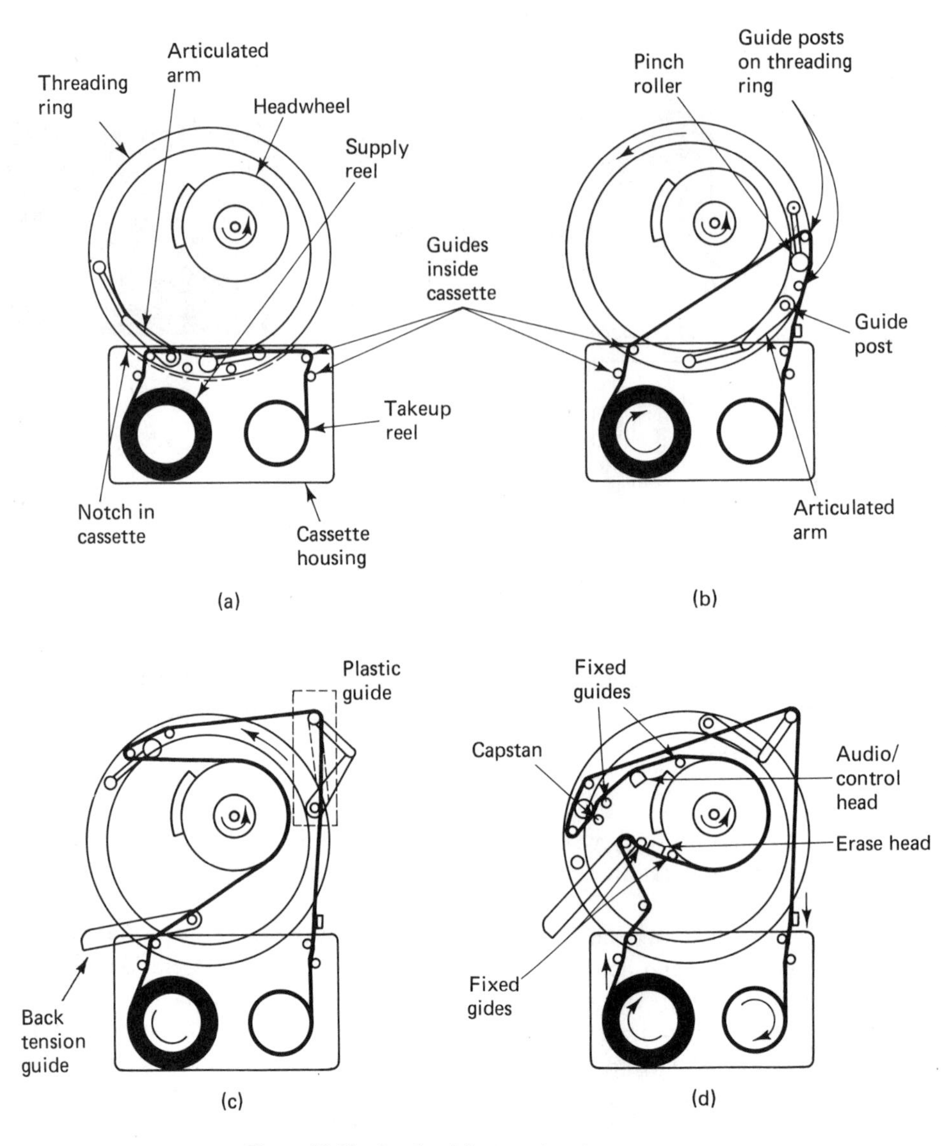

Figure 19-29 Details of the Beta threading process.

Loading time of the VHS cassette is slightly faster than that of Beta. However, older VHS machines must unthread during fast forward and rewind modes of operation. Modern machines do not.

The Beta machines can go directly from play to rewind on fast forward without unthreading. A VHS deck requires a few seconds for the tape to retract from the heads. After fast forward or rewind, a small delay is also experienced to allow the tape to reload.

In modern VHS systems, due to very tight servo control of the capstan, the video head, and the take-up and feed reel motors, it is possible to have reverse search without unthreading. This tight servo control is made possible by the use of microprocessors.

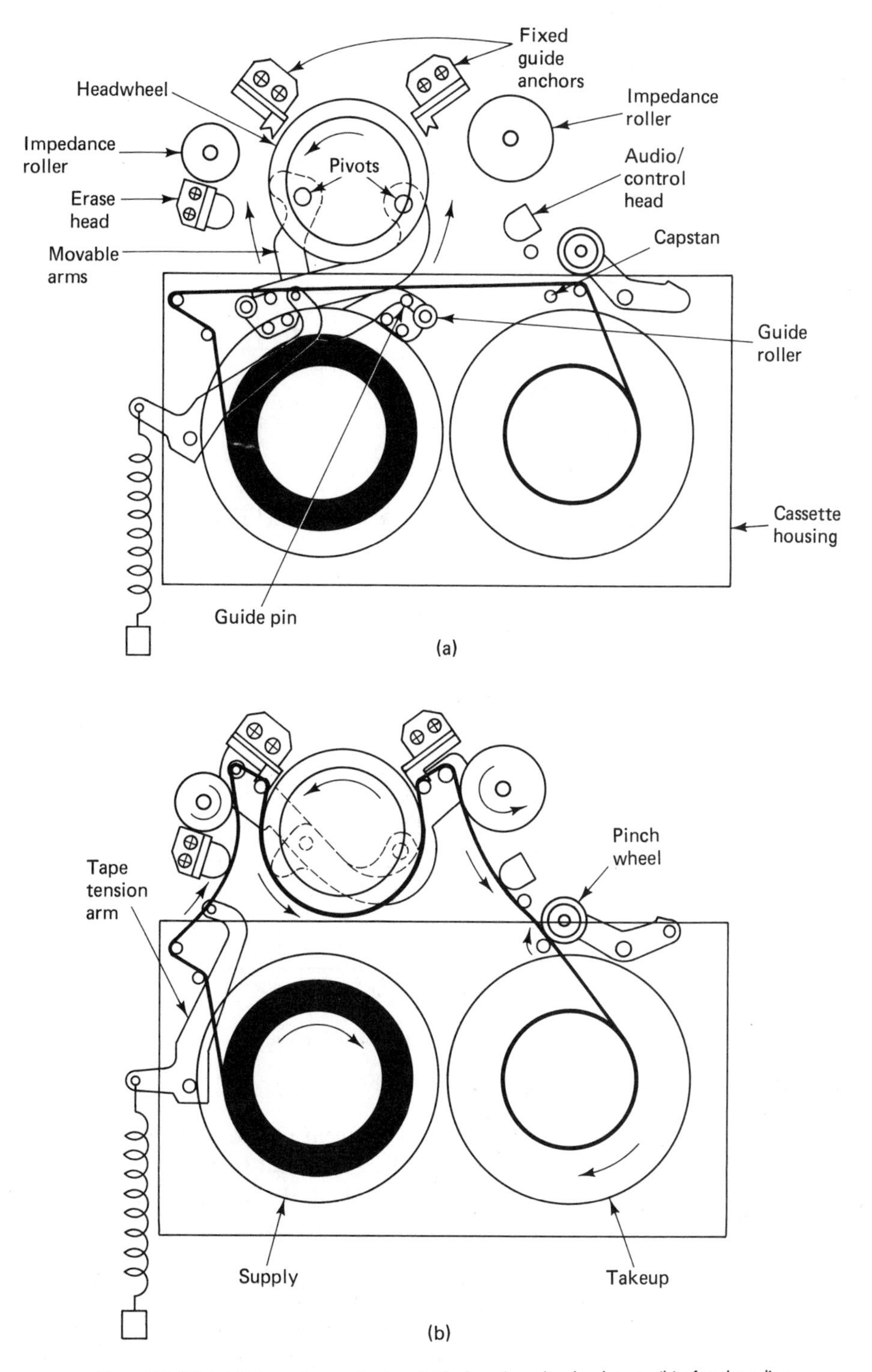

Figure 19-30 VHS threading mechanism: (a) before threading has begun; (b) after threading has been completed.

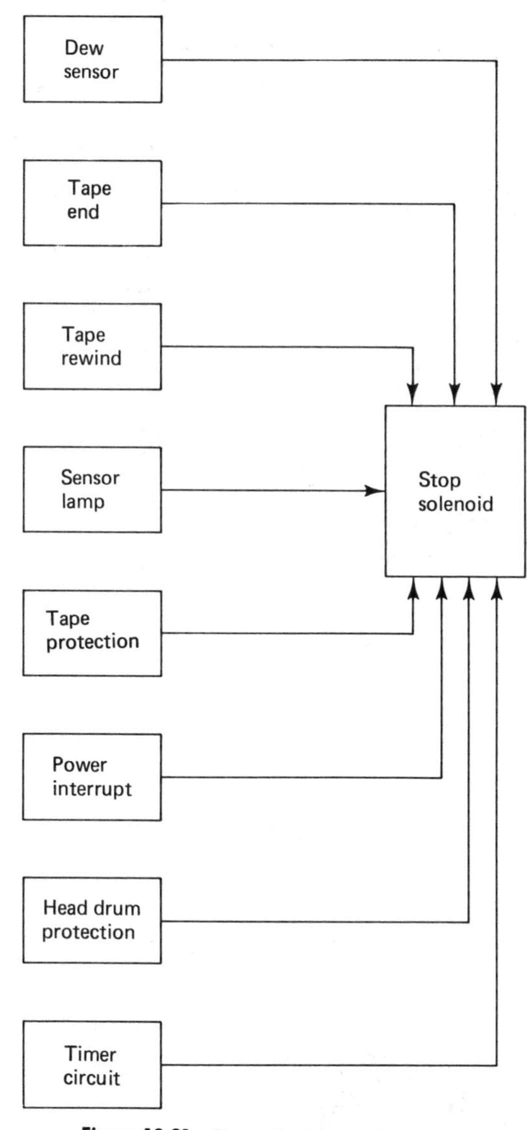

Figure 19-31 Stop solenoid control circuits.

19-11 VCR protection circuits

Various conditions of operation and environment can cause faulty operation and damage to the mechanical components of the VCR. To prevent such problems, VCRs have incorporated a number of safety circuits. When activated, the safety circuits trigger a stop solenoid that disables the threading/loading motor, which is the same as depressing the operator-controlled "stop" button.

Figure 19-31 is a block diagram that shows the various conditions that trigger the shutdown solenoid. The solenoid is an electromechanical device similar to a relay, except that it has an iron rod that is pulled into the electromagnet when it is energized. This mechanical motion is then used to release all latches on the function keys bringing the machine to a halt. The following paragraphs discuss each of the shutdown conditions.

Dew sensor. If the head drum base becomes wet due to moisture or condensation (dew), damage to the moving head/tape combination may result. A special sensor whose resistance increases when it is wet is used to activate a circuit that causes the shutdown solenoid to be activated, stopping the unit. When activated, an LED will light, indicating the cause of the shutdown.

Tape end. During normal playback, record, or fast forward operation, it is undesirable for the tape to end suddenly. This might cause the tape to break. The cassette is designed so that the VHS tape has a translucent leader at its beginning and end. A sensor light or infrared LED is placed so that its light passes through the translucent tape exciting a photo transistor or infrared photo diode just before the tape reaches its end. This energizes the stop solenoid, protecting the system from tape breakage. In Beta systems the tape leader is a metallic film that activates an inductive sensor which in turn initiates shutdown.

Tape rewind. During rewind operation a sudden stop due to tape runout can cause the tape to break. To prevent this from happening, a light source or infrared LED shines through the translucent VHS tape leader near the end of the tape. This excites a phototransistor, whose output drives the stop solenoid. The Beta system uses a metallic film leader as was explained before.

Sensor lamp. If the light source used to shine through the translucent VHS tape during rewind or fast forward operation burns out, tape damage may occur. The sensor lamp is made part of a circuit, which in the event of lamp burnout, generates a stop solenoid command voltage. Failure of the sensor lamp is one of the more common causes of an inoperative VCR.

Tape protection. Inability of the VCR to load in a few seconds, or the inability of the take-up reel to accept its tape properly, will cause activation of the stop solenoid. This circuit may contain as many as five switches and a Hall effect generator. This device generates a voltage in the presence of a magnetic field. The Hall effect generator is associated with the take-up reel and generates pulses as long as the reel is turning. If the reel should stop turning, the Hall device would no longer generate its pulses, activating the stop solenoid. The switches involved in this circuit are the play switch, the tape slack switch, the pause switch, the load completion switch, and the unload completion switch. If any of these switches is improperly activated, the stop solenoid would be activated.

Power interrupt. If for any reason the ac power is interrupted, or if the VCR is turned off, this circuit momentarily activates the stop solenoid and returns all of the function buttons to their off positions.

Head drum protection. If for any reason the head drum fails to rotate, the VCR will activate the stop solenoid preventing damage to the unit.

Timer circuit. When the VCR is used in a programmed timer application, the play/record buttons are depressed. However, it is not desired for the unit to operate until the timer has reached its set time. Then it is desired that the VCR should operate. The function of the timer is to activate the stop solenoid until the set time, after which it is deactivated, allowing the VCR to operate normally. When the program has been recorded for the set period of time, the timer circuit then will activate the stop solenoid, turning off the machine and unlatching all depressed function buttons.

Other solenoids. In addition to the stop solenoid, VCRs using "soft touch" function buttons may contain as many as six more solenoids to perform all of the desired functions. In addition to the function button solenoid, a "pinch solenoid" and a "brake solenoid" are also used. During playback and record operation, the pinch solenoid is engaged, forcing a pressure roller against the tape, which in turn is pressed against the capstan, ensuring smooth tape travel. The pinch solenoid must also be disengaged during the "stop" and "pause" modes.

Brake solenoids are used in Beta systems to free the brakes on the take-up and supply reel turntables during loading or threading operations. Solenoids are also used for such functions as select, fast forward, play, rewind, and reverse modes of operation. These solenoids require adjustment in order for the VCR to operate properly. Adjustment procedures are listed in the manufacturer's service literature.

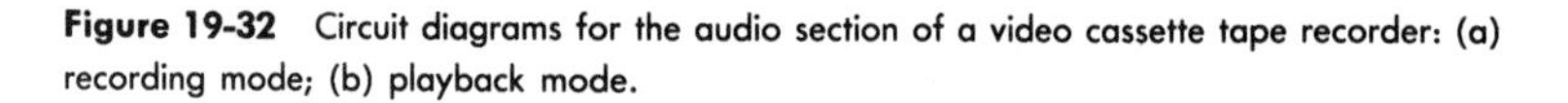

Figure 19-32 Circuit diagrams for the audio section of a video cassette tape recorder: (a) recording mode; (b) playback mode.

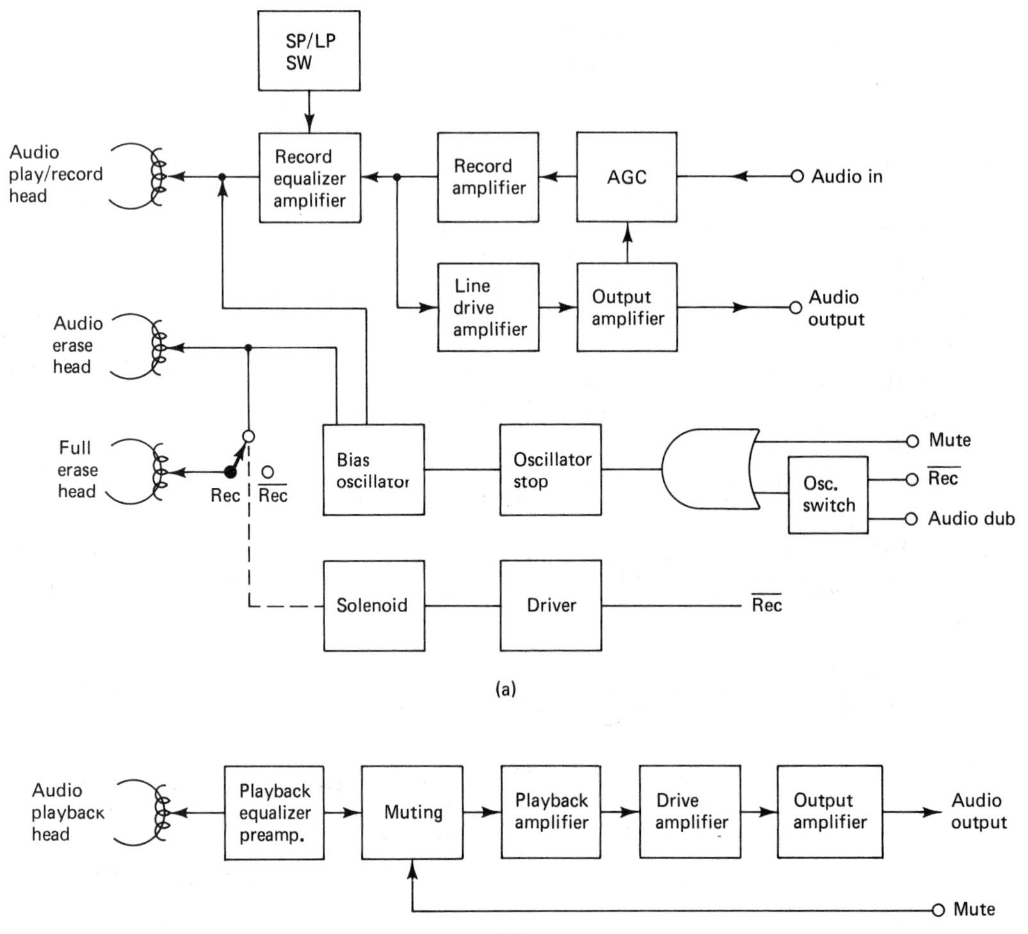

19-12 audio selection

The basic construction of the audio selection of the VCR is the same as in an audio tape recorder circuit. As shown in Figure 19-32, it consists of control and muting circuits, a playback equalizer amplifier, line amplifier, record equalizer amplifier, AGC, and bias oscillator. The bias oscillator generates a signal of 65 kHz that is used by the audio record head, audio erase head, and full erase head.

Beta stereo system. Until recently VCRs used monaural recording and playback. Newer units have incorporated stereo recording and playback. These units use the video helical scanner to record the stereo audio information. The standard audio track is also used for monaural recording and playback, thereby preventing obsolescence of existing VCR units. In the Beta system, stereo sound is frequency modulated and recorded together with the video chrominance and the video luminance signals by the video heads. To accommodate the FM audio signal, the FM Y signal bandwidth is slightly reduced, so that the audio carriers falls between the "color under" signal and the Y signal.

To reduce interference a number of techniques are used: (1) the luminance carrier is moved 400 kHz higher in frequency so that it ranges between 3.9 and 5.2 MHz; (2) by selecting the proper audio carrier frequencies, frequency interleaving slips the audio sidebands between the video sidebands; (3) Audio crosstalk due to overlapping audio tracks is minimized by using two alternating pairs of FM carriers for successive tracks. During video field "A", the left audio channel uses a carrier of 1.38 MHz, and the right channel uses 1.68 MHz. During video track "B", the FM carriers are at 1.53 and 1.83 MHz respectively.

VHS stereo system. VHS stereo recording is similar to the Beta system in that it uses two FM carriers, (1.3 MHz for the left channel and 1.7 MHz for the right channel) which are placed between the color and the luminance RF signals. To minimize crosstalk, a separate pair of heads is used for the audio FM carriers that are set at an azimuth angle $\pm$ 30°. The audio is recorded first and the video is recorded over it without erasing the audio, because the audio is recorded in a deeper portion of the tape coating than the video information. This is called "depth multiplex".

QUESTIONS

1. What are three important differences between VHS and Beta VCRs?
2. How are prerecorded video tapes protected against erasure?
3. What are the major differences between VHS and Beta systems of tape threading and unthreading?
4. Describe the operation of the direct-drive recording system.
5. What do the abreviations SP, LP, and SLP mean?
6. What limitations does the width of the recording head gap have on magnetic recording?
7. What is an octave?
8. Under what conditions will the output of a playback head be maximum?
9. What is the maximum recording frequency range possible with the direct recording system?
10. What three things must be done to compress the 18-octave video signal and to increase the high-frequency limit?
11. What are the advantages of frequency modulation for video recording?
12. How do VHS and Beta systems differ in video recording techniques?
13. What is meant by "color under" recording?

14. Why is "color under" recording used?
15. What is the cause of time base errors?
16. What are the differences between the Quadraplex and helical scanning methods of recording?
17. Why are four- and five-head helical scanning methods used?
18. Describe the operation of head switching during record and playback modes of operation.
19. Describe the standard video tape format used by both VHS and Beta systems.
20. Under what conditions are guardbands used, and when are they not used?
21. What is azimuth recording, and why is it used?
22. How is color crosstalk minimized in the Beta system?
23. How is color crosstalk minimized in the VHS system?
24. What is sync beat interference, and how is it minimized?
25. What are the four major electronic areas in the video cassette recorder?
26. What are the three major mechanical areas in the video cassette recorder?
27. Describe the operation of the VHS recording system.
28. What are three major objectives of VHS chroma recording?
29. What is the purpose of phase rotation?
30. What are the four important circuit areas found in the VHS playback system?
31. What is dropout compensation, and how does it work?
32. What are the four important objectives of VHS color playback?
33. What is jitter, and how is it minimized?
34. Explain the differences between the Beta and VHS systems of recording.
35. Explain how jitter and color crosstalk are minimized in the Beta system.
36. How many motors do VCRs usually have?
37. How many servo control systems are used in the VCR, and what is their function?
38. What is a sample-and-hold circuit, and how does it operate?
39. What is the difference between speed control and phase control?
40. What are the adjustments found on the head servo system, and how are they adjusted?
41. What are the controls found on the capstan servo system, and how are they adjusted?
42. What is the effect of improper timing of head switching?
43. Where are the input pulses for the head servo and capstan servo systems obtained?
44. What kind of motors are used for the head and capstan?
45. What are the input and output pulses used in the capstan servo system?
46. What is the speed of rotation of the recording head?
47. What is the purpose of the capstan?
48. Why are special jigs, gauges, and test tapes required for VCR service?
49. Describe the VHS and Beta tape threading systems.
50. List the various conditions that can cause the stop solenoid to be activated.

45. What are the input and output pulses used in the capstan servo system?
46. What is the speed of rotation of the recording head?
47. What is the purpose of the capstan?
48. Why are special jigs, gauges, and test tapes required for VCR service?
49. Describe the VHS and Beta tape threading systems.
50. List the various conditions that can cause the stop solenoid to be activated.

Index